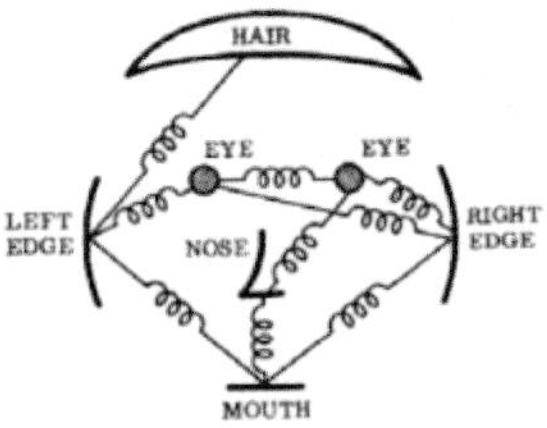

第1章 概述 1

什么是计算机视觉？·简史·本书概述·课程大纲样例
·标记法说明

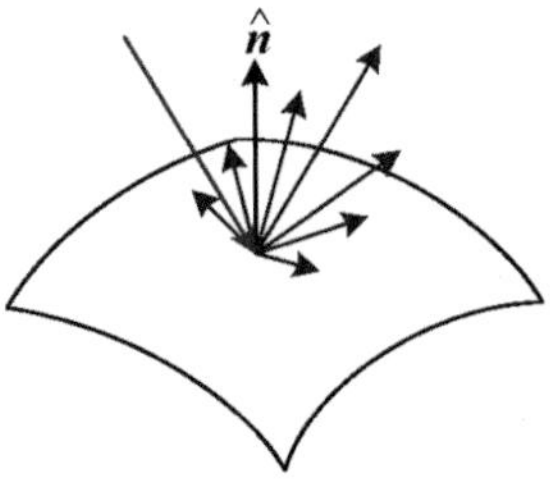

第2章 图像形成 25

几何基元和变换·光度测定学的图像形成·数字摄像机

第3章 图像处理 77

点算子·线性滤波·更多的邻域算子·傅里叶变换
·金字塔与小波·几何变换·全局优化

第4章 特征检测与匹配 157

点和块·边缘·线条

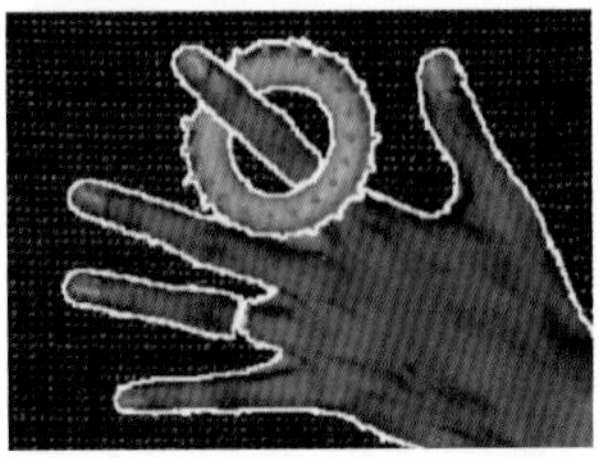

第5章 分割 205

活动轮廓·分裂与归并·均值移位和模态发现
·规范图割·图割和基于能量的方法

第6章 基于特征的配准 237

基于2D和3D特征的配准·姿态估计·几何内参数标定

第7章 由运动到结构 263

三角测量·二视图由运动到结构·因子分解
·光束平差法·限定结构和运动

第8章　稠密运动估计　293

平移配准・参数化运动・基于样条的运动・光流・层次运动

第9章　图像拼接　327

运动模型・全局配准・合成

第10章　计算摄影学　359

光度学标定・高动态范围成像・超分辨率和模糊去除・图像抠图和合成・纹理分析与合成

第11章　立体视觉对应　409

极线几何学・稀疏对应・稠密对应・局部方法・全局优化・多视图立体视觉

第12章　3D重建　443

由X到形状・主动距离获取・表面表达・基于点的表达・体积表达・基于模型的重建・恢复纹理映射与反照率

第13章　基于图像的绘制　477

视图插值・层次深度图像・光场与发光图・环境影像形板・基于视频的绘制

第14章　识别　507

物体检测・人脸识别・实例识别・类别识别・上下文与场景理解・识别数据库和测试集

第1章 概述

图1.1 人类视觉系统可以毫不费力地解释这张照片中存在的由半透明材质和阴影形成的微妙变化，并从背景中将物体正确地分割出来

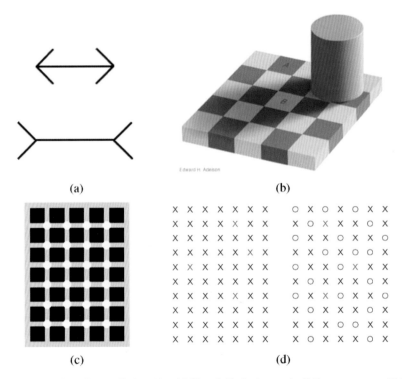

图1.3 一些常见的光学错觉和其可以给我们关于视觉系统的启示。(a)经典的Muller-Lyer错觉，两条水平线条的长度显得不同，可能是由于想象的透视效果。(b)阴影中的"白"块B和亮处的"黑"块A实际上具有相同的绝对亮度值。该感觉是由于亮度恒常性，即视觉系统当解释颜色时试图减少亮度的特性。图像承蒙Ted Adelson的允许，*http://web.mit.edu/persci/people/adelson/checkershadow_illusion.html*。(c)Hermann网格错觉的一个变型，承蒙Hany Farid许可使用，*http://www.cs.dartmouth.edu/_farid/illusions/hermann.html*。当你移动眼睛扫视该图时，灰色的斑点会出现在交叉处。(d)在图的左半部分数一数红色的X。现在，在右半部分数。是不是明显困难了？其解释与弹出(pop-out)效应有关(Treisman 1985)，它告诉我们大脑中有关并行感知和集成通道的操作

第2章 图像形成

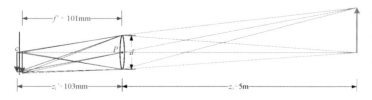

图2.21 在受色像差影响的镜头中,不同波长的光线(例如,红色和蓝色箭头)以不同的焦距f聚焦,因此有不同的深度z_i',最后产生了几何(平面内的)平移和失焦

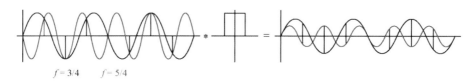

图2.24 一维信号的混叠:蓝色的正弦波在$f = 3/4$且红色正弦波在$f = 5/4$,当用$f = 2$采样时,有相同的数字采样。即便是在用100%填充率的方框型滤光器卷积这两个信号之后,尽管幅度不再相同,但在采样过的红色信号看起来像蓝色信号的较低幅度的翻转版本的意义下,它们仍然是走样的。(右侧的图像为了便于观察而放大了。实际的正弦幅度是其原来数值的30%和-18%)

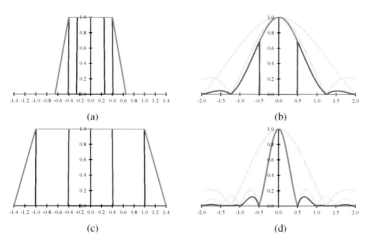

图2.26 点扩散函数(PSF)样例:(a)中的模糊圆盘(蓝)的直径等于像素间距的一半,而(c)中的直径是像素间距的两倍。传感芯片的横向填充率是80%,显示为棕色。这两个核的卷积给出点扩散函数,显示为绿色。PSF (MTF)的傅里叶响应绘制在(b)和(d)中。混叠出现在奈奎斯特频率之外的区域,显示为红色

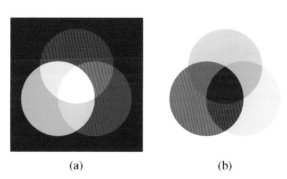

图2.27 基色和合成色:(a)加性色红、绿、蓝可以混合产生青、洋红、黄和白;(b)减性色青、洋红、黄可以混合产生红、绿、蓝和黑

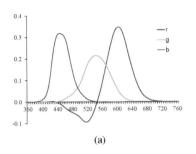 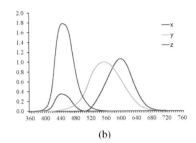

图2.28 标准CIE彩色匹配函数：(a)通过与纯色基R = 700.0 nm，G = 546.1 nm和B = 435.8 nm匹配获得的 $\bar{r}(\lambda), \bar{g}(\lambda), \bar{b}(\lambda)$ 色谱；(b) $\bar{x}(\lambda), \bar{y}(\lambda), \bar{z}(\lambda)$ 彩色匹配函数，它们是 $\bar{r}(\lambda), \bar{g}(\lambda), \bar{b}(\lambda)$ 谱的线性组合

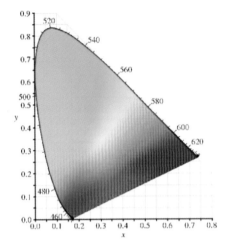

图2.29 CIE色度图，显示了彩色和其对应的(x,y)值。纯谱色绕着曲线的外端布置

图2.30 Bayer RGB样式：(a)色彩滤波器阵列布置；(b)插值后的像素值，未知量(估计的量)显示为小写字母

第3章 图像处理

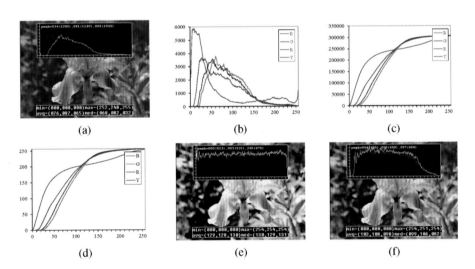

图3.7 直方图分析和均衡化：(a)原始图像；(b)各个彩色通道的直方图和亮度直方图；(c)累积分布函数；(d)均衡化(转换)函数；(e)全局的直方图均衡化；(f)局部的直方图均衡化

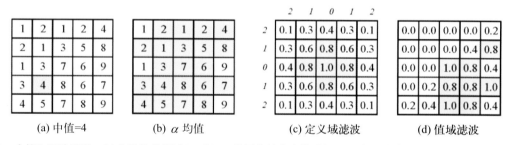

图3.10 邻域滤波(卷积)：左边图像与中间图像的卷积产生右边图像。目标图像中绿色标记的像素是利用原图像中蓝色标记的像素计算得到的

图3.17 区域求和表：(a)原始图像；(b)区域求和表；(c)区域和的计算。区域求和表中的每个值$s(i,j)$(红色)是利用它三个相邻(蓝色)的像素递归计算出来的(3.31)。区域和s(绿色)是利用矩形的四个角点(紫色)的值计算出来的(3.32)，粗体表示计算时取正值，斜体表示计算时取负值

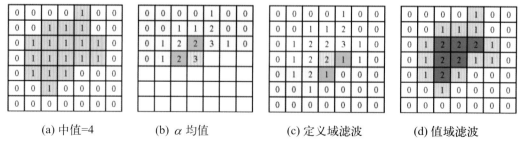

图3.19 中值和双边滤波：(a)中值像素(绿色)；(b)α-截尾均值像素的选择；(c)定义域滤波(沿着边界的数字是像素的距离)；(d)值域滤波

图3.22 城街距离变换：(a)原始的二值图像；(b)自顶向底(正向)光栅扫描：橙色像素的值是利用绿色像素的值的计算出来的；(c)自底向上(逆向)光栅扫描：绿色像素的值将与原始的橙色像素的值合并；(d)距离变换的最终结果

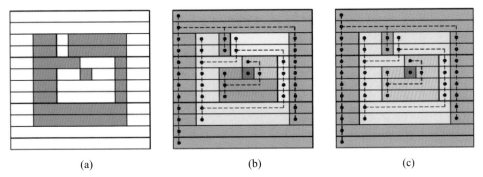

图3.23 连通量的计算：(a)原始的灰度图像；(b)水平行程(结点)通过垂直(图)边缘(蓝色的虚线)连接；对行程用从父节点继承下来的唯一的彩色作伪彩色处理；(c)合并邻接段后重新着色

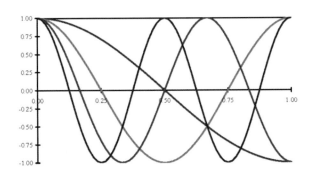

图3.26 离散余弦变换(DCT)的基函数：第一个DC(即常数)基是蓝色的水平线，第二个是棕色的半周期波形，等等。这些基广泛应用于图像和视频压缩标准，例如JPEG

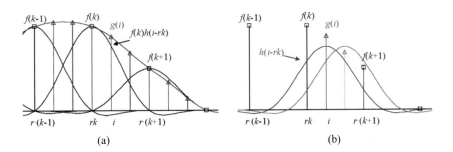

图3.27 信号插值，$g(i) = \sum_k f(k)h(i-rk)$：(a)输入值的带权求和；(b)多相滤波器描述

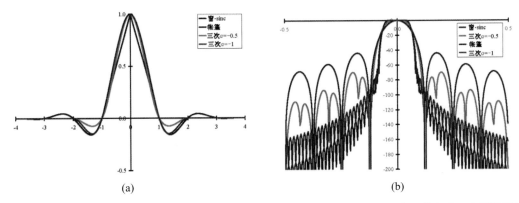

图3.29 (a)一些窗sinc函数和(b)它们的log傅里叶变换：蓝色为提升余弦窗sinc，绿色和紫色为三次插值器(a=-1和a=-0.5)，棕色为tent函数。它们都常用于进行高精度低通滤波操作

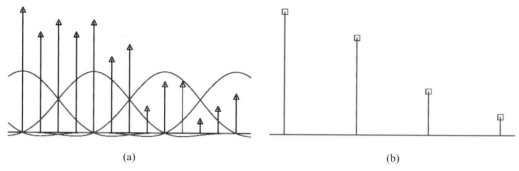

图3.30 信号降采样：(a)原始信号；(b)进行下采样前与低通滤波器卷积过的信号

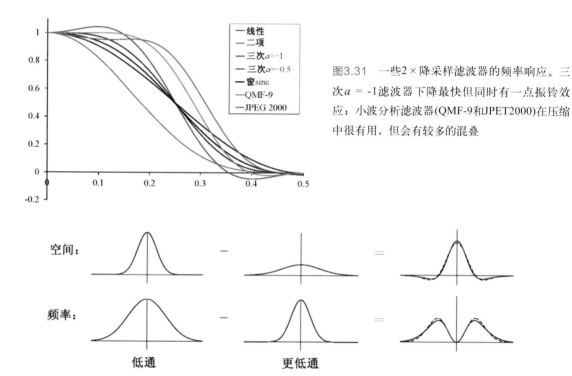

图3.31 一些2×降采样滤波器的频率响应。三次$a = -1$滤波器下降最快但同时有一点振铃效应；小波分析滤波器(QMF-9和JPET2000)在压缩中很有用，但会有较多的混叠

图3.35 两个低通滤波器的差是一个带通滤波器。蓝色虚线显示了对一个高斯的半倍频程拉普拉斯密切拟合

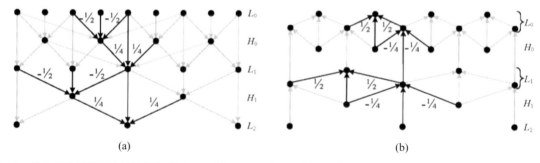

图3.39 提升变换表示为信号处理流程图：(a)分析阶段首先从偶数邻居中预测奇数值，存储差分小波，然后通过加入小波的片段来补偿较粗糙的偶数值；(b)合成阶段简单地反转计算流和一些滤波器和操作的符号。浅蓝色线条展示了用四阶代替二阶进行预测和纠正的状况

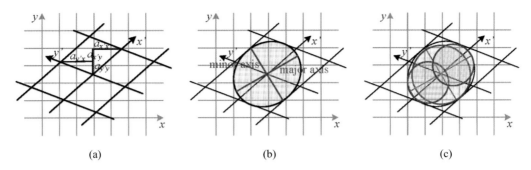

图3.48 各向异性的纹理滤波：(a)变换矩阵A的雅克比行列式和引入的水平和竖直重采样率$\{a_{x'x}, a_{x'y}, a_{y'x}, a_{y'y}\}$；(b)一个EWA平滑核的椭圆足迹；(c)使用沿主轴的多样例的各向异性滤波。位于直线交叉处的图像像素

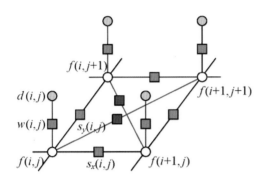

图3.56 $\mathcal{N}_4$邻域马尔科夫随机场的图模型。蓝色的边是为$\mathcal{N}_8$邻域添加的。白色圆圈是未知的$f(i,j)$而深色圆圈是输入数据$d(i,j)$。黑色方块$s_x(i,j), s_y(i,j)$表示随机场中任意的相邻的节点间交互作用势，$w(i,j)$表示数据惩罚函数。同样的图模型可以用于描述一个离散版本的一阶正则化问题(图3.55)

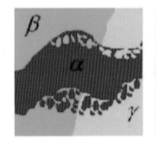 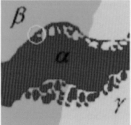 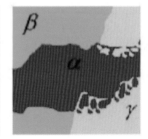 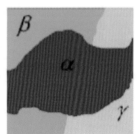

(a) 初始标注　　　　　(b) 标准移动　　　　　(c) $\alpha - \beta$ 交换　　　　　(d) α 扩展

图3.58 多级别图优化(Boykov, Veksler和Zabih 2001)© 2001 IEEE：(a)初始问题构造；(b)标准移动仅改变一个像素；(c) $\alpha - \beta$ 交换最优地互换所有具有α和β标注的像素；(d) α扩展移动最优地在当前像素值和α标签中进行选择

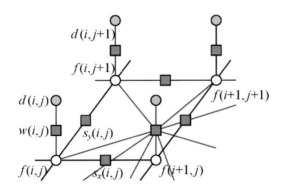

图3.59 一个有更复杂测量模型的马尔科夫随机场的图模型。额外的彩色边展示了未知值的结合(比如，在一幅锐化图像中)是如何产生测量值的(一幅有噪声的模糊图像)。得到的图模型仍旧是一个经典的MRF，而且同样容易进行采样，但是一些推理算法(例如，基于图割的)可能不再可用，因为网络复杂性增加了，由于推理过程中的状态改变变得更纠结，后验MRF有更大的团块

图3.61 图像分割(Boykov and Funka-Lea 2006)© 2006 Springer：用户在前景物体上用红色画几笔，在背景上用蓝色画几笔。系统计算前景和背景的颜色分布，解一个二值MRF。光滑权重由亮度梯度(边缘)来调整，这使其成为条件随机场(CRF)

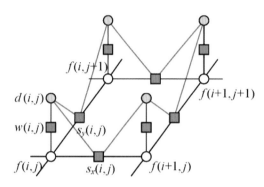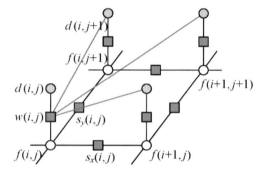

图 3.62 条件随机场 (CRF) 的图模型。额外的绿色边显示了传感数据的结合如何影响其 MRF 先验模型的光滑性，即式 (3.113) 中的 $s_x(i, j)$ 和 $s_y(i, j)$ 依赖于邻近的值。这些额外的连接(因子)使得光滑性依赖于输入数据。但是，它们使从这个 MRF 中采样更复杂

图3.63 一个区分性随机场(DRF)的图模型。额外的绿色边显示了传感数据的结合，例如，$d(i,j+1)$，影响$f(i,j)$的数据项。因此产生式模型变得更复杂，即，我们不能将一个简单的函数应用于未知变量和添加噪声

第4章 特征检测与匹配

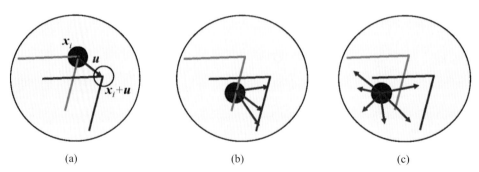

图4.4 不同图像块的孔径问题：(a)稳定的(类似于角点的)流；(b)经典的孔径问题；(c)无纹理的区域。两个图像I_0(黄)和I_1(红)重叠了。红色矢量u指出了图块中心之间的移位，$W(x_i)$加权函数(图块窗口)如图中的黑色圆圈所示

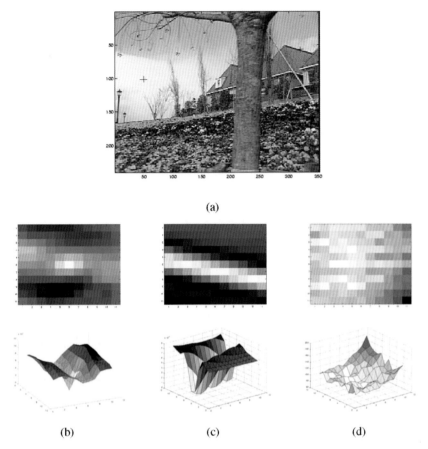

图4.5 显示成灰度值图像和表面图的三个自相关表面$E_{AC}(\Delta u)$：(a)自相关表面计算的位置在原始图像上用三个红十字表示；(b)这个图像块取自于花坛(有良好的唯一最小值)；(c)这个图像块取自于屋顶(一维的孔径问题)；(d)这个图像块取自于云朵(没有好的峰值)。图像(b)~(d)中的每一个网格点都是Δu的一个值

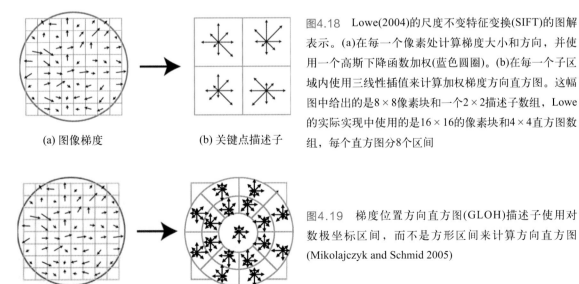

图4.18 Lowe(2004)的尺度不变特征变换(SIFT)的图解表示。(a)在每一个像素处计算梯度大小和方向，并使用一个高斯下降函数加权(蓝色圆圈)。(b)在每一个子区域内使用三线性插值来计算加权梯度方向直方图。这幅图中给出的是8×8像素块和一个2×2描述子数组，Lowe的实际实现中使用的是16×16的像素块和4×4直方图数组，每个直方图分8个区间

图4.19 梯度位置方向直方图(GLOH)描述子使用对数极坐标区间，而不是方形区间来计算方向直方图(Mikolajczyk and Schmid 2005)

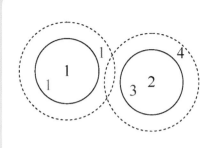
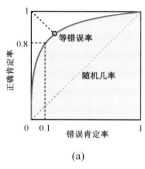
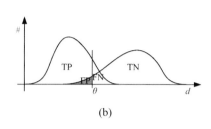

图4.22 误报和漏报：黑色数字1和2是特征，将与其他图像中的一个特征的数据库进行匹配。在目前阈值设置下(实线圈)，绿色1是一个正确肯定(正确的匹配)，蓝色1是一个漏报(未能匹配)，红色3是一个误报(不正确的匹配)。如果我们将阈值设置设高一点(虚线圆圈)，蓝色1变成一个正确肯定但棕色4变成了一个额外的误报

图4.23 ROC曲线和它相关的比率。(a)对于一特定的特征提取和匹配算法的组合ROC曲线绘制了正确肯定率相对于错误肯定率的曲线。理想情况下，正确肯定率应该接近1，而错误肯定率应该接近0。ROC曲线下面的面积(AUC)通常用作算法性能的一个单一(标量)度量。有时也用等错误率作为另一个替换的度量。(b)肯定数目(匹配的数目)和否定数目(非匹配的数目)的分布作为特征间距离d的一个函数。随着阈值 θ 的增加，正确肯定(TP)和误报(FP)的数目也随着增加

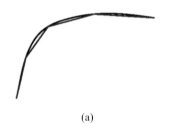
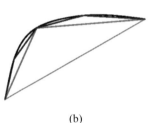
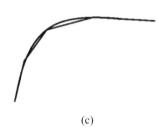

图4.40 用折线或者B样条曲线来近似一个曲线(用黑色表示)：(a)原始曲线用一个红色的折线来近似；(b)通过迭代地寻找与目前的近似离得最远的点做逐次近似；(c)拟合折线顶点的光滑的内插样条用深蓝色显示

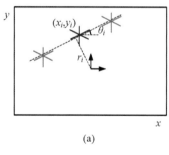
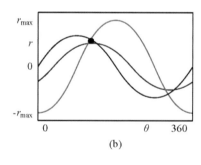

图4.41 原始哈夫变换：(a)每一个点为所有可能的直线 $r_i\theta = x_i\cos\theta + y_i\sin\theta$ 进行投票进行；(b)每束线给出 (r, θ) 处的一个正弦函数；它们的交点确定了所要求的直线方程

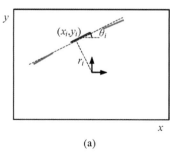
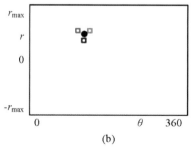

图4.42 带方向的哈夫变换：(a)一个位于极坐标 (r, θ) 处的边界基元用 $\hat{n}_i = (\cos\theta_i, \sin\theta_i)$ 和 $r_i = \hat{n}_i \cdot x_i$ 重新参数化；(b) (r, θ) 显示了分别标记为红绿蓝三个边界基元投票的累加器阵列

第5章 分割

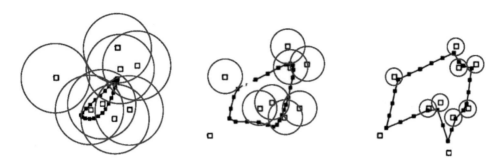

图5.3 弹性网:空心正方形表示城市而通过直线线段连接起来的实心正方形是旅行点。蓝圈表示每个城市吸引力的近似范围,它随着时间递减。在该弹性网的贝叶斯解释下,蓝圈对应于圆形高斯的一个标准差,该高斯从某未知旅行点生成每座城市

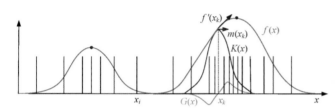

图5.17 核密度估计,它的微分以及均值位移的一维可视化。核密度估计$f(x)$通过将稀疏的输入样本x_i与和函数$K(x)$卷积得到。该函数的微分$f'(x)$可以通过将输入与微分核$G(x)$卷积得到。在当前估计x_k附近估计局部替换向量会产生均值移位向量$f(x_k)$,在多维设定中,它指向与函数梯度$\nabla f(x_k)$相同的方向。红色点表示$f(x)$的局部极大值,均值移位会收敛到该点

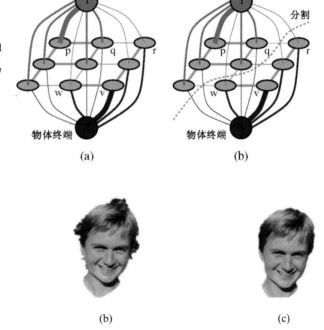

图5.23 用于区域分割的图割(Boykov and Jolly 2001)©2001 IEEE:(a)能量函数编码为最大流问题;(b)最小割确定区域边界

图5.24 GrabCut图像分割(Rother, Kolmogorov, and Blake 2004)©2004 ACM:(a)用户画一个红色的边界框;(b)算法猜测物体与背景的颜色分布并执行二值分割;(c)该过程采用更好的区域统计量重复进行

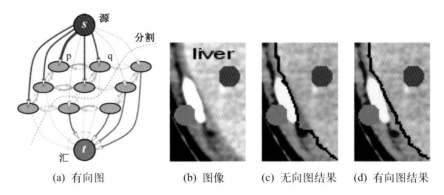

(a) 有向图　　　　　(b) 图像　　　(c) 无向图结果　　(d) 有向图结果

图5.25　采用有向图割的分割(Boykov and Funka-Lea 2006)©2006 Springer：(a)有向图；(b)带种子点的图像；(c)无向图错误地沿明亮物体延续边界；(d)有向图正确地将明亮灰色区域从它的更暗的背景中分割出来

第6章　基于特征的配准

图6.3　由三幅图像组成的用平移模型自动配准后平均在一起的全景图

第7章　由运动到结构

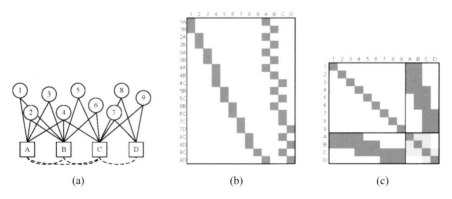

图7.9　(a)由运动到结构玩具问题的二分图；(b)关联的雅可比矩阵J；(c)Hessian矩阵A。数字表示3D点，字母表示摄像机。虚线弧和浅蓝色正方形指示结构(点)变量减少时出现的替代项

第8章 稠密运动估计

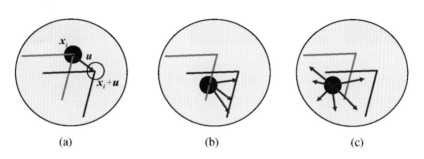

图 8.3 不同图像区域的孔径问题，由橙色和红色的 L 型结构指示，画在同一幅图像中以方便解释光流：(a) 以 x_i 为中心的窗口 $w(x_i)$（黑色）可以被唯一的匹配到第二幅图像（红色）中 x_i+u 为中心的对应结构上；(b) 中心在边界上的窗口体现了传统的孔径问题，它可以匹配到一个 1D 可能位置的族上；(c) 在完全无纹理的区域内，匹配变得毫无约束

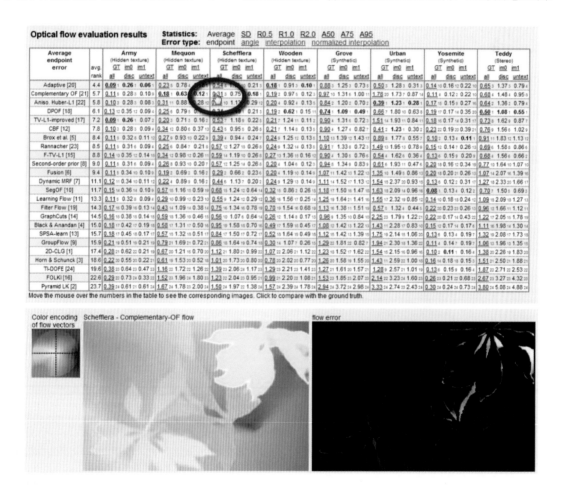

图8.12 24种光流算法评估结果，October 2009, *http://vision.middlebury.edu/flow/*, (Baker, Scharstein, Lewis *et al.* 2009)。移动鼠标到有下划线的分数之下，用户可以交互查看对应光流和误差图的关系。点击分数可以看计算结果和真实流。每个分数旁边，蓝色数字表示当前列对应的排名。粗体表示每列最小(优)值。表用平均排名排序(计算全部 24列，八个序列，每个三个区域掩模)。平均排名近似在选定度量/统计方法下的性能量度

图8.13　时空体中的切片(Szeliski 1999)©1999 IEEE: (a和b)花园序列的两帧，(c)完全时空体中的一个水平切片，箭头表示估计流的潜在关键帧位置。注意花园序列的颜色不正确，正确的颜色(黄色花朵)参见图8.15

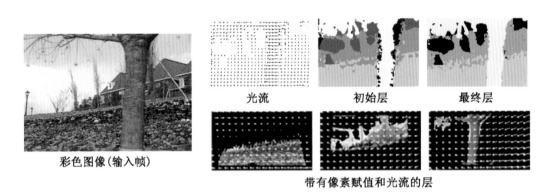

图8.15　层次运动估计结果(Wang and Adelson 1994)©1994 IEEE

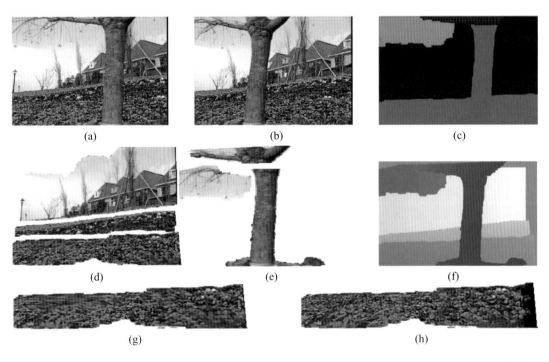

图8.16　分层立体重建(Baker, Szeliski, and Anandan 1998)©1998 IEEE：(a)初始和(b)最终输入图像；(c)初始分为六层的分割；(d)和(e)六层的子画面；(f)平面形子画面的深度图(深色表示近处)；最前面一层在残差深度估计之前(g)和之后(h)。注意花园序列的色彩不正确；正确的颜色(黄花)见图8.15

第9章 图像拼接

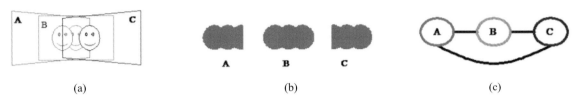

图9.15 计算差分区域(RODs)(Uyttendaele, Eden, and Szeliski 2001)©2001 IEEE：(a)包含一个运动人脸的三幅互相重叠的图像；(b)对应的差分区域；(c)一致的差分区域的图

图9.16 照片蒙太奇(Agarwala, Dontcheva, Agrawala et al. 2004)©2004 ACM。从一个包含五幅图像(其中四幅图像显示在左侧)的集合中，照片蒙太奇方法可以很快地构造出一幅合成的全家福，其中每个人都微笑着看着镜头(右侧)。用户可以只是浏览照片堆叠，粗略根据源图像中他们想加入最终合成图像中的人来描绘线条。这种用户施加的线条以及计算得到的区域(中间)是用左侧源图像的边界来标记颜色的

图9.18 泊松图像编辑(Perez, Gangnet, and Blake 2003)©2003 ACM；(a)选择一只狗和两个儿童的图像作为源图像，然后粘贴到目标游泳池里面；(b)简单的粘贴算法不能在边界处匹配相应的颜色；(c)泊松图像融合可以掩盖这些差异

第10章 计算机摄影学

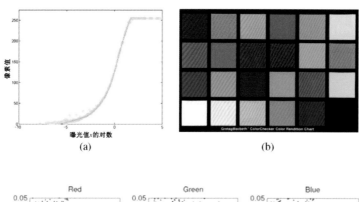

图10.3 辐射响应标定:(a)典型摄像机响应函数,显示了单彩色通道下,入射辐照度的对数(曝光值)和输出的8位像素值间的映射关系(Devevec and Malik 1997)©1997 ACM;(b)选色图板

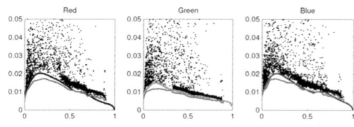

图10.4 由单一彩色照片获得的噪声水平函数(NLF)的估计(Liu, Szeliski, Kang et al. 2008)©2008 IEEE。彩色曲线是NLF的估计,它被拟合为在带噪声的图像与分段光滑的图像之间测量得到的偏差的概率下包络。通过求取29幅图像均值而得到的NLF的真值在图中用灰色标出

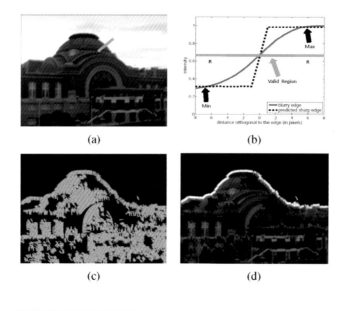

图10.9 未采用标定图案的PSF估计(Joshi, Szeliski, and Kriegman 2008)©2008 IEEE:(a)输入图像和蓝色截面(轮廓)的位置;(b)拍摄的和预测的阶梯边缘的轮廓;(c和d)边缘附近所预测彩色的位置和值

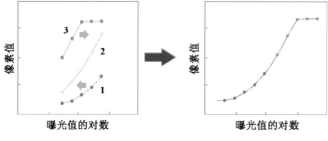

图10.13 采用多种曝光度进行辐射度标定(Debevec and Malik 1997)。对应像素值被绘制为曝光度对数(辐照度)的函数。移动左边的曲线来调节每个像素未知的辐射亮度,直到它们连接成一条平滑的曲线

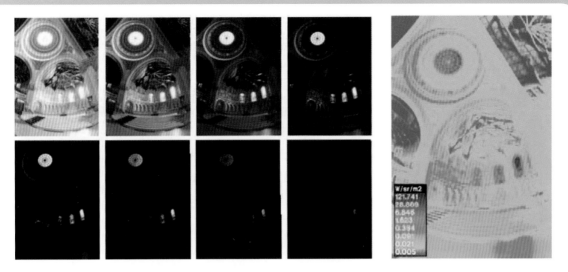

图10.15 一组由胶片摄影机拍摄的包围曝光图像以及用伪彩色显示的结果辐照度图像(Debevec and Malik 1997)© 1997 ACM

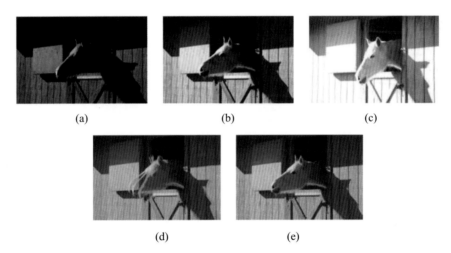

图10.16 融合多种曝光图像来制作高动态范围合成图像(Kang, Uyttendaele, Winder et al. 2003): (a~c)三种曝光程度不同的图像; (d)采用传统算法进行合成(注意由于马头运动而导致的重影); (e)带有运动补偿的合成

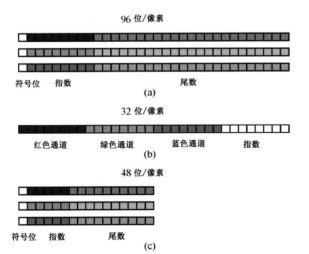

图10.19 HDR图像编码格式: (a)Portable PixMap格式(.ppm); (b)辐射亮度格式(.pic, .hdr); (c)OpenEXR格式(.exr)

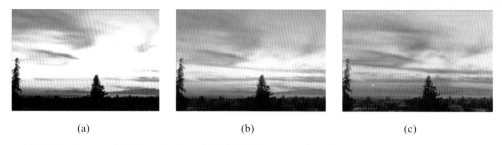

图10.20 全局色调映射：(a)输入的HDR图像，线性映射后的；(b)每个彩色通道分别独立作用γ；(c)γ只作用于亮度(色彩较少冲洗)。HDR原图承蒙Paul Debevec许可使用，*http://ict.debevec.org/~debevec/Research/HDR/*。处理后的效果图承蒙Frédo Durand许可使用，MIT 6.815/6.865计算摄影学课程

图10.22 采用双边滤波的局部色调映射(Durand and Dorsey 2002)：(a)低通和高通经双边滤波后的亮度对数图像和色彩(色度)图像；(b)色调映射后的效果图像(经衰减低通亮度对数图像后)没有光晕。处理后的图像承蒙Frédo Durand许可使用，MIT 6.815/6.865计算摄影学课程

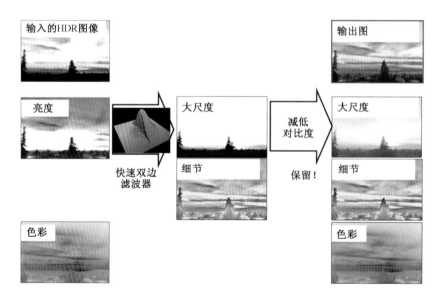

图10.24 采用双边滤波器的色局部调映射(Durand and Dorsey 2002)：算法流程总结。图像承蒙Frédo Durand许可使用，MIT 6.815/6.865计算摄影学课程

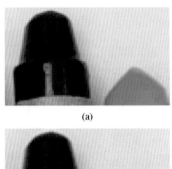
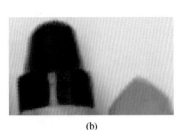

图10.34 Bayer RGB样式：(a)色彩滤波器矩阵布局；(b)插值后的像素彩色值，其中未知(猜测)的值用小写字母表示

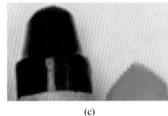
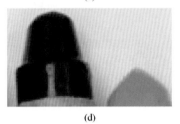

图10.35 CFA去马赛克效果(Bennett, Uyttendaele, Zitnick et al. 2006)©2006 Springer：(a)完整分辨率原图像(经彩色重采样后的版本作为算法的输入)；(b)双线性插值的结果，在蓝色蜡笔的笔尖有彩色边缘现象并且左侧(垂直)边缘有拉链效应；(c)Malvar, He, and Cutler(2004)的高质量线性插值结果(注意黄色蜡笔上明显的光晕/格子状瑕疵)；(d)Bennett, Uyttendaele, Zitnick et al. (2006)中采用局部二彩色先验的效果

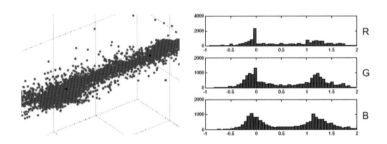

图10.36 由5×5的局部邻域内计算得到的二彩色模型(Bennett, Uyttendaele, Zitnick et al. 2006)©2006 Springer。经过2-均值聚类并重投影到连接两个主导彩色(图中红点)的直线上，可见大部分像素的彩色值落在设定的直线附近。沿这条直线的分布投影到RGB坐标轴上后可以看到峰值在0和1处，分别为两个主导彩色处

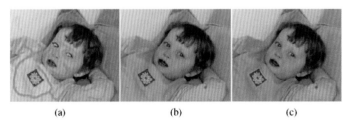

图10.37 基于优化的彩色化方法(Levin, Lischinski, and Weiss 2004)©2004 ACM：(a)灰度图像及覆盖的彩色笔画；(b)彩色化后的效果图；(c)导出灰度图及笔画颜色值的原图。原图由Rotem Weiss提供

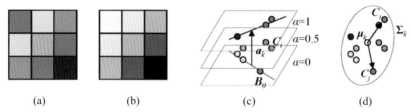

图10.43 色彩线抠图(Levin, Lischinski, Weiss 2008)：(a)3×3的局部小色块；(b)可能的 α 值；(c)前景和背景色彩线、连接它们间最近两点的向量 a_k 以及一组 α 为常量的平行平面，$\alpha_i = a_k \cdot (C_i - B_0)$。(d)颜色样本的散布图和两个颜色 C_i 和 C_j 到平均值 μ_k 的偏移

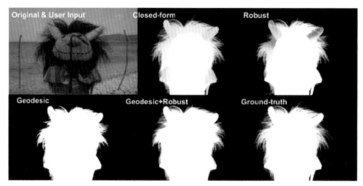

图10.45 以笔画为输入的抠图效果比较。Wang and Cohen(2007a)逐个介绍了这些方法

图10.46 基于笔画的分割结果(Themann, Rother, Rav-Acha et al. 2008)©2008 IEEE

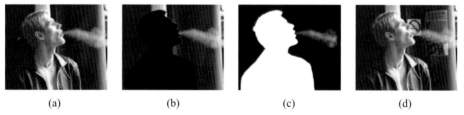

图10.47 烟抠图(Chuang, Agarwala, Curless et al. 2002)©2002 ACM：(a)输入的视频帧；(b)移去前景物体后；(c)估计的透明度遮罩；(d)将新物体插入到背景中

(a) 前景图　　(b) 背景图　　(c) 蓝屏合成图　　(d) 我们的方法　　(e) 参考图

图10.48 阴影抠图(Chuang, Goldman, Curless et al. 2003)©2003 ACM。阴影抠图能够正确使用阴影减弱场景中的光亮，并在3D几何体上披上阴影(d)，而不是简单地使新场景中的阴影部分变暗(c)

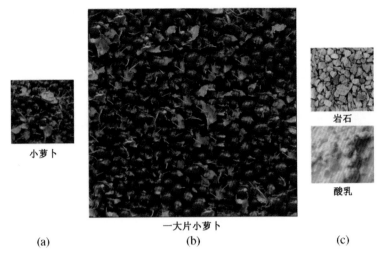

图10.49 纹理分析:(a)给定一小块纹理,目标是合成大片纹理;(b)外观相似的大片纹理;(c)一些难以合成的半结构化纹理(图像承蒙Alyosha Efros许可使用)

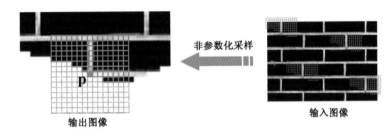

图10.50 采用非参数化采样的纹理合成(Efros and Leung 1999)。最新的像素p的值从原纹理(输入图像)中有相似局部(部分)块的像素中随机选取(图像承蒙Alyosha Efros许可使用)

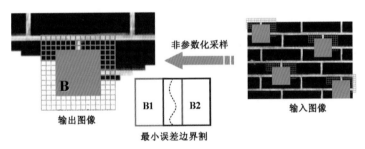

图10.51 用图像衍缝算法进行纹理合成(Efros and Freeman 2001)。算法从原纹理中一次复制一个图像块,而不是每次只生成单一像素。选定图像块间的重叠区域的过渡将由动态规划进行优化(图像承蒙Alyosha Efros许可使用)

图10.52 图像修复(空洞填充):(a和b)沿等照度线方向的传播(Bertalmio, Sapiro, Caselles et al. 2000)©2000 ACM;(c和d)基于样例的修图,填充顺序以置信度为基础(Criminisi, Pérez, and Toyama 2004)

第11章 立体视觉对应

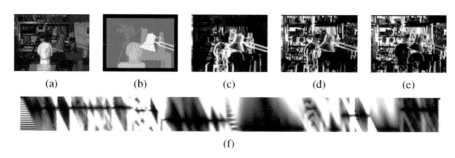

图11.5 一个典型视差空间图像(DSI)的切片(Scharstein and Szeliski 2002)©2002 Springer: (a)原始彩色图像; (b)视差图真值; (c~e)对于$d=10,16,21$的三个(x,y)切片; (f)对于$y=151$的一个(x,d)切片((b)中的虚线)。在(c~e)中各种深色(匹配)区域都可见,比如书架、桌子和罐以及头像雕塑,在图(f)中可以明显看到表现为水平直线的三个不同视差水平。这些DIS中深色的条纹表示在这个视差上匹配的区域。(小的深色区域通常是由于无明显纹理结构区域所造成的) Bobick and Intille(1999)讨论了更多DSI的例子

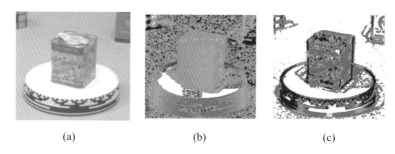

图11.10 立体视觉深度估计中的不确定性(Szeliski 1991b): (a)输入图像; (b)估计的深度图像(蓝色表示深度更近); (c)估计的置信度(红色表示置信度更高)。从中可以看出,纹理更明显的区域拥有更高的置信度

图11.12 基于分割的立体视觉匹配(Zitnick, Kang, Uyttendaele et al. 2004)©2004 ACM: (a)输入彩色图像; (b)基于颜色的分割; (c)初始视差估计; (d)最终分段平滑视差; (e)在视差空间分布中的片段上的MRF邻居定义(Zitnick and Kang 2007)© 2007 Springer

图11.13 使用自适应过分割和抠图的立体视觉匹配(Taguchi, Wilburn, and Zitnick 2008)© 2008 IEEE: (a) 分割边界在优化过程中被求精,这样能够得到更精确的结果(例如,底部行中的薄的绿叶); (b) 在片段边界抽取的透明度遮罩,得到视觉效果更好的合成结果(中间列)

第12章　3D重建

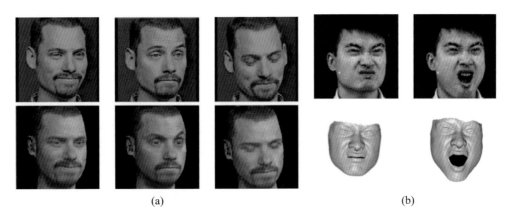

(a)　　　　　　　　　　　　　　　　　　(b)

图12.18　利用3D形变模型的头部和表情跟踪与动画再现：(a)在五段输入视频流中直接拟合的模型(Pighin, Szeliski, and Salesin 2002)©2002 Springer，下面一行显示动画再现的合成的纹理映射的3D模型，其姿态和表情参数与上面一行的输入图像相符；(b)帧速率时空立体视觉表面模型拟合的模型(Zhang, Snavely, Curless et al. 2004)©2004 ACM，上面一行显示覆盖了合成绿色标记点的输入图像，下一行显示拟合的3D表面模型

第13章　基于图像的绘制

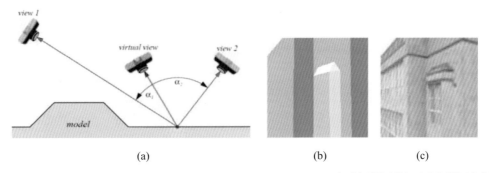

(a)　　　　　　　　　(b)　　　　　　(c)

图13.3　视图相关纹理映射(Debevec, Taylor, and Malik 1996)©1996 ACM：(a)每个源视图的权重和源视图与新(虚拟)视图的相对角度有关；(b)简化的3D几何模型；(c)使用视图相关纹理映射后，几何模型看起来更加细致(如内嵌的窗户)

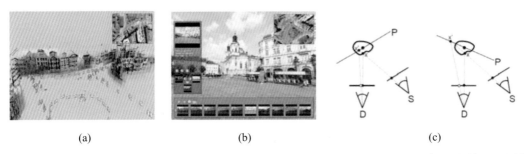

(a)　　　　　　　　　(b)　　　　　　　　(c)

图13.4　照片游览系统(Snavely, Seitz, and Szeliski 2006)©2006 ACM：(a)场景的3D概览图，半透明刷色和线段被绘制在平面代理上；(b)用户选择感兴趣区域时，系统底部会显示一些对应的缩略图；(c)最佳稳定的平面代理(替代)选择(Snavely, Garg, Seitz et al. 2008)© 2008 ACM

图13.6 带深度的子画面(Shade, Gortler, He *et al.* 1988)©1998 ACM：(a)Alpha遮罩的彩色子画面；(b)对应的相对深度或视差；(c)不使用相对深度的绘制结果；(d)使用相对深度的绘制结果(注意物体边界的弧线部分)

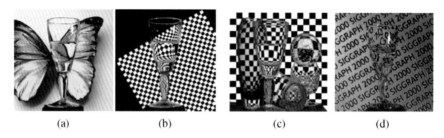

图13.10 环境影像形板：(a和b)将一个折射物体放在不同的背景前，这些背景的光照模式会被正确地折射(Zongker, Werner, Curless *et al.* 1999)；(c)多次折射可以用混合高斯模型处理；(d)实时影像形板可以利用单个分级彩色背景抽取得到(Chuang, Zongker, Hindorff *et al.* 2000)© 2000 ACM

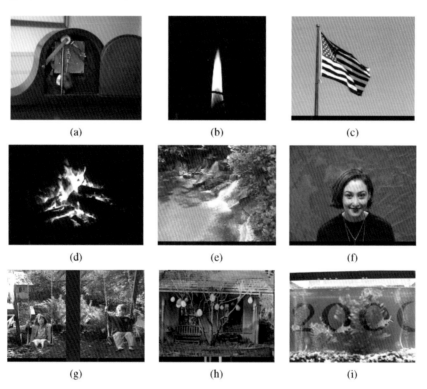

图13.13 视频纹理(Schodl, Szeliski, Salesin *et al.*2000)©2000 ACM：(a)一个运动方向正确匹配的钟摆；(b)蜡烛火焰，其中弧线显示了时间转移；(c)画面中旗子是通过在跳转处进行变形操作得到的；(d)在生成篝火的视频纹理时，使用了更长的淡入淡出；(e)生成瀑布的视频纹理时，多个序列被淡入淡出在一起；(f)动画笑脸；(g)对两个荡秋千的小孩分别进行动画绘制；(h)多个气球被自动分割成独立的运动区域；(i)一个合成的包括水泡、植物和鱼的鱼缸。与这些图片对应的视频可以在*http://www.cc.gatech.edu/gvu/perception/projects/videotexture/*找到

第14章 识别

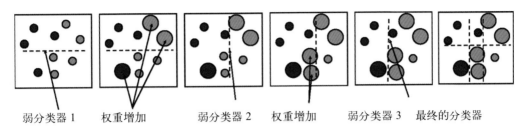

图14.7 boosting的图解说明，经Svetlana Lazebnik授权使用，源自Paul Viola和David Lowe的解释。在选择每个弱分类器(决策树桩或者超平面)后，提高了被分错的数据点的权重。最终的检测器是简单弱分类器的线性组合

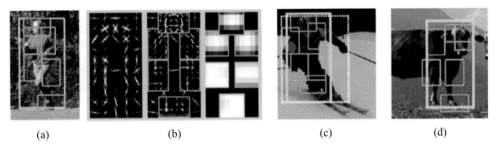

图14.9 基于部件的物体检测(Felzenszwalb, McAllester, and Ramanan 2008)©2008 IEEE：(a)输入照片及其关联的人(蓝色)和部件(黄色)检测结果；(b)检测模型用一个粗糙的模板、几个较高分辨率的部件模板及关于每个部件位置的空间模型定义；(c)滑雪者正确的检测结果；(d)作为错误检测结果的奶牛(当作人标注出来)

图14.10 针对人、自行车和马的基于部件的物体检测结果(Felzenszwalb, McAllester, and Ramanan 2008)©2008 IEEE。前三列给出的是正确的检测结果，而最右列给出的是错误的检测结果

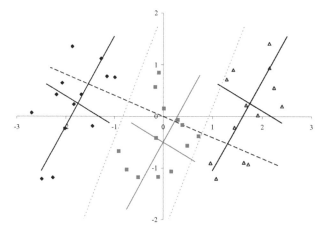

图14.16 Fisher线性鉴别分析的简单示例。样本来自三个不同的类别,在图中沿着其放缩到$2\sigma_i$的主轴用不同的颜色表示。(倾斜坐标轴的交点是类别均值m_k)短划线是(主要)Fisher线性鉴别方向而虚线表示类间的线性区分。注意区分性方向如何融合类间和类内散布矩阵的主方向

图14.35 Xerox 10类数据集的样例图像(Csurka, Dance, Perronnin et al. 2006)©2007 Springer。想象一下尝试写一个程序,从其他的图像中区分这样的图像

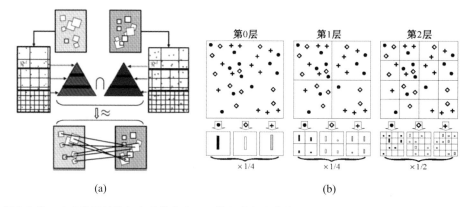

图14.37 用金字塔匹配方法比较特征向量的集合:(a)特征空间金字塔匹配核(Grauman and Darrell 2007b)构建高维特征空间的一个金字塔,并用它计算特征向量集合之间的距离(和隐含的对应);(b)空间金字塔匹配(Lazebnik, Schmid, and Ponce 2006)Ó 2006 IEEE把图像分成区域池的金字塔,然后计算每个空间箱内单独的视觉词的直方图(分布)

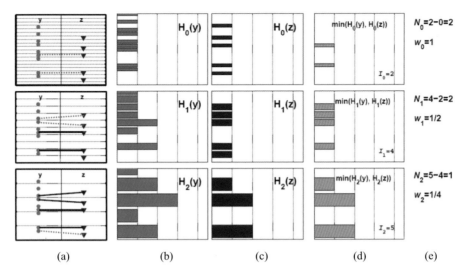

图 14.38 使用金字塔匹配核比较特征向量的汇集的一维图解 (Grauman and Darrell 2007b): (a) 特征向量 (点集) 在金字塔箱中的分布; (b 和 c) 对于两幅图像, 在 B_{il} 和 B_{il}' 箱中点数的直方图; (d) 直方图的交 (最小值); (e) 每一层相似性的分值, 加权相加得到最终的距离 / 相似性度量

图14.39 "图像-图像"对比"图像-类别"距离比较(Boiman, Shechtman, and Irani 2008)©2008 IEEE。在左上角的查询图像, 可能不与下面一行中任何一个数据库图像里的的特征分布相匹配。但是, 如果将查询中的每个特征匹配到所有类别图像中最类似的那一个, 就可以找到一个好的匹配

图14.46 用几百万幅照片进行场景补全(Hays and Efros 2007)©2007 ACM: (a)原图; (b)消除不需要的前景后; (c)比较真实的场景匹配, 用户选择的匹配用红色高亮显示出来了; (d)经过替换和融合的输出图像

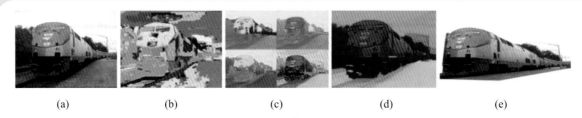

图14.47 自动照片弹出(Hoiem, Efros, and Hebert 2005a)Ó 2005 ACM：(a)输入图像；(b)超像素(superpixels)分为；(c)多个区域；(d)标签代表的地面(绿色)、直立物(红色)和天空(蓝色)；(e)产生的分段线性3D模型的新视图

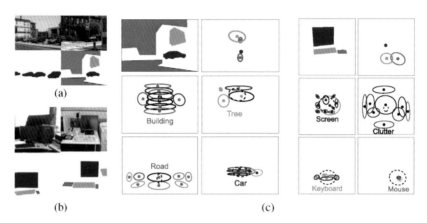

图14.50 物体识别的上下文场景模型(Sudderth, Torralba, Freeman et al. 2008)©2008 Springer：(a)一些街景和它们对应的标签(紫 = 建筑, 红 = 汽车, 绿 = 树, 蓝 = 路)；(b)一些办公场景(红 = 计算机屏幕, 绿 = 键盘, 蓝 = 鼠标)；(c)从这些标记的场景中学习到的上下文模型。上面一行显示了样例标记图像和其他物体相对于中心红色(汽车或者屏幕)物体的分布情况。下面一行显示了组成每个物体的部件分布情况

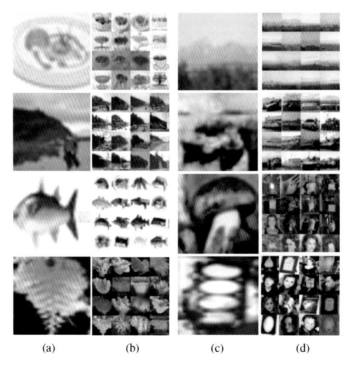

图14.52 使用小图像的识别(Torralba, Freeman, and Fergus 2008)©2008 IEEE：(a)和(c)列显示了输入图像样例，(b)和(d)列显示了在有8千万个小图像的数据库中其对应的16个最近邻

图 14.53 ImageNet(Deng, Dong, Socher et al. 2009)©2009 IEEE。这个数据库包含对每个名词有 500 个仔细核实的图像，其中每个名词是 WordNet 层次结构的 14 841 个名词（截止 2010 年 4 月）中的一个

图 14.54 PASCAL Visual Object Classes Challenge 2008 (VOC2008) 数据库中的样例图像 (Everingham, Van Gool, Williams et al. 2008)。原始图像来自 flickr(http://www.flickr.com/)。数据库的权利在 http://pascallin.ecs.soton.ac.uk/challenges/VOC/voc2008/ 中解释了说明

计算机视觉
——算法与应用

(美) Richard Szeliski 著

艾海舟 兴军亮 等译

清华大学出版社
北京

重印说明：为方便广大读者阅读、参考和延伸阅读，重印时附加了一个小手册，其中包含彩图、参考文献和公式，供大家进一步探索计算机视觉领域的奥秘。

参考文献

Abdel-Hakim, A. E. and Farag, A. A. (2006). CSIFT: A SIFT descriptor with color invariant characterstics. In *IEEE Computer Society Conference on Computer Vision and Pattern Recognition (CVPR'2006)*, pp. 1978–1983, New York City, NY.

Adelson, E. H. and Bergen, J. (1991). The plenoptic function and the elements of early vision. In *Computational Models of Visual Processing*, pp. 3–20.

Adelson, E. H., Simoncelli, E., and Hingorani, R. (1987). Orthogonal pyramid transforms for image coding. In *SPIE Vol. 845, Visual Communications and Image Processing II*, pp. 50–58, Cambridge, Massachusetts.

Adiv, G. (1989). Inherent ambiguities in recovering 3-D motion and structure from a noisy flow field. *IEEE Transactions on Pattern Analysis and Machine Intelligence*, 11(5):477–490.

Agarwal, A. and Triggs, B. (2006). Recovering 3D human pose from monocular images. *IEEE Transactions on Pattern Analysis and Machine Intelligence*, 28(1):44–58.

Agarwal, S. and Roth, D. (2002). Learning a sparse representation for object detection. In *Seventh European Conference on Computer Vision (ECCV 2002)*, pp. 113–127, Copenhagen.

Agarwal, S., Snavely, N., Seitz, S. M., and Szeliski, R. (2010). Bundle adjustment in the large. In *Eleventh European Conference on Computer Vision (ECCV 2010)*, Heraklion, Crete.

Agarwal, S., Snavely, N., Simon, I., Seitz, S. M., and Szeliski, R. (2009). Building Rome in a day. In *Twelfth IEEE International Conference on Computer Vision (ICCV 2009)*, Kyoto, Japan.

Agarwal, S., Furukawa, Y., Snavely, N., Curless, B., Seitz, S. M., and Szeliski, R. (2010). Reconstructing Rome. *Computer*, 43(6):40–47.

Agarwala, A. (2007). Efficient gradient-domain compositing using quadtrees. *ACM Transactions on Graphics*, 26(3).

Agarwala, A., Hertzmann, A., Seitz, S., and Salesin, D. (2004). Keyframe-based tracking for rotoscoping and animation. *ACM Transactions on Graphics (Proc. SIGGRAPH 2004)*, 23(3):584–591.

Agarwala, A., Agrawala, M., Cohen, M., Salesin, D., and Szeliski, R. (2006). Photographing long scenes with multi-viewpoint panoramas. *ACM Transactions on Graphics (Proc. SIGGRAPH 2006)*, 25(3):853–861.

Agarwala, A., Dontcheva, M., Agrawala, M., Drucker, S., Colburn, A., Curless, B., Salesin, D. H., and Cohen, M. F. (2004). Interactive digital photomontage. *ACM Transactions on Graphics (Proc. SIGGRAPH 2004)*, 23(3):292–300.

Agarwala, A., Zheng, K. C., Pal, C., Agrawala, M., Cohen, M., Curless, B., Salesin, D., and Szeliski, R. (2005). Panoramic video textures. *ACM Transactions on Graphics (Proc. SIGGRAPH 2005)*, 24(3):821–827.

Aggarwal, J. K. and Nandhakumar, N. (1988). On the computation of motion from sequences of images—a review. *Proceedings of the IEEE*, 76(8):917–935.

Agin, G. J. and Binford, T. O. (1976). Computer description of curved objects. *IEEE Transactions on Computers*, C-25(4):439–449.

Ahonen, T., Hadid, A., and Pietikäinen, M. (2006). Face description with local binary patterns: Application to face recognition. *IEEE Transactions on Pattern Analysis and Machine Intelligence*, 28(12):2037–2041.

Akenine-Möller, T. and Haines, E. (2002). *Real-Time Rendering*. A K Peters, Wellesley, Massachusetts, second edition.

Al-Baali, M. and Fletcher., R. (1986). An efficient line search for nonlinear least squares. *Journal Journal of Optimization Theory and Applications*, 48(3):359–377.

Alahari, K., Kohli, P., and Torr, P. (2011). Dynamic hybrid algorithms for discrete MAP MRF inference. *IEEE Transactions on Pattern Analysis and Machine Intelligence*.

Alexa, M., Behr, J., Cohen-Or, D., Fleishman, S., Levin, D., and Silva, C. T. (2003). Computing and rendering point set surfaces. *IEEE Transactions on Visualization and Computer Graphics*, 9(1):3–15.

Aliaga, D. G., Funkhouser, T., Yanovsky, D., and Carlbom, I. (2003). Sea of images. *IEEE Computer Graphics and Applications*, 23(6):22–30.

Allen, B., Curless, B., and Popović, Z. (2003). The space of human body shapes: reconstruction and parameterization from range scans. *ACM Transactions on Graphics (Proc. SIGGRAPH 2003)*, 22(3):587–594.

Allgower, E. L. and Georg, K. (2003). *Introduction to Numerical Continuation Methods*. Society for Industrial and Applied Mathematics.

Aloimonos, J. (1990). Perspective approximations. *Image and Vision Computing*, 8:177–192.

Alpert, S., Galun, M., Basri, R., and Brandt, A. (2007). Image segmentation by probabilistic bottom-up aggregation and cue integration. In *IEEE Computer Society Conference on Computer Vision and Pattern Recognition (CVPR 2007)*, Minneapolis, MN.

Amini, A. A., Weymouth, T. E., and Jain, R. C. (1990). Using dynamic programming for solving variational problems in vision. *IEEE Transactions on Pattern Analysis and Machine Intelligence*, 12(9):855–867.

Anandan, P. (1984). Computing dense displacement fields with confidence measures in scenes containing occlusion. In *Image Understanding Workshop*, pp. 236–246, New Orleans.

Anandan, P. (1989). A computational framework and an algorithm for the measurement of visual motion. *International Journal of Computer Vision*, 2(3):283–310.

Anandan, P. and Irani, M. (2002). Factorization with uncertainty. *International Journal of Computer Vision*, 49(2-3):101–116.

Anderson, E., Bai, Z., Bischof, C., Blackford, S., Demmel, J. W. *et al.* (1999). *LAPACK Users' Guide*. Society for Industrial and Applied Mathematics, 3rd edition.

Andrieu, C., de Freitas, N., Doucet, A., and Jordan, M. I. (2003). An introduction to MCMC for machine learning. *Machine Learning*, 50(1-2):5–43.

Andriluka, M., Roth, S., and Schiele, B. (2008). People-tracking-by-detection and people-detection-by-tracking. In *IEEE Computer Society Conference on Computer Vision and Pattern Recognition (CVPR 2008)*, Anchorage, AK.

Andriluka, M., Roth, S., and Schiele, B. (2009). Pictorial structures revisited: People detection and articulated pose estimation. In *IEEE Computer Society Conference on Computer Vision and Pattern Recognition (CVPR 2009)*, Miami Beach, FL.

Andriluka, M., Roth, S., and Schiele, B. (2010). ocular 3d pose estimation and tracking by detection. In *IEEE Computer Society Conference on Computer Vision and Pattern Recognition (CVPR 2010)*, San Francisco, CA.

Anguelov, D., Srinivasan, P., Koller, D., Thrun, S., Rodgers, J., and Davis, J. (2005). SCAPE: Shape completion and animation of people. *ACM Transactions on Graphics (Proc. SIGGRAPH 2005)*, 24(3):408–416.

Ansar, A., Castano, A., and Matthies, L. (2004). Enhanced real-time stereo using bilateral filtering. In *International Symposium on 3D Data Processing, Visualization, and Transmission (3DPVT)*.

Antone, M. and Teller, S. (2002). Scalable extrinsic calibration of omni-directional image networks. *International Journal of Computer Vision*, 49(2-3):143–174.

Arbeláez, P., Maire, M., Fowlkes, C., and Malik, J. (2010). *Contour Detection and Hierarchical Image Segmentation*. Technical Report UCB/EECS-2010-17, EECS Department, University of California, Berkeley. Submitted to PAMI.

Argyriou, V. and Vlachos, T. (2003). Estimation of sub-pixel motion using gradient cross-correlation. *Electronic Letters*, 39(13):980–982.

Arikan, O. and Forsyth, D. A. (2002). Interactive motion generation from examples. *ACM Transactions on Graphics*, 21(3):483–490.

Arnold, R. D. (1983). *Automated Stereo Perception*. Technical Report AIM-351, Artificial Intelligence Laboratory, Stanford University.

Arya, S., Mount, D. M., Netanyahu, N. S., Silverman, R., and Wu, A. Y. (1998). An optimal algorithm for approximate nearest neighbor searching in fixed dimensions. *Journal of the ACM*, 45(6):891–923.

Ashdown, I. (1993). Near-field photometry: A new approach. *Journal of the Illuminating Engineering Society*, 22(1):163–180.

Atkinson, K. B. (1996). *Close Range Photogrammetry and Machine Vision*. Whittles Publishing, Scotland, UK.

Aurich, V. and Weule, J. (1995). Non-linear Gaussian filters performing edge preserving diffusion. In *17th DAGM-Symposium*, pp. 538–545, Bielefeld.

Avidan, S. (2001). Support vector tracking. In *IEEE Computer Society Conference on Computer Vision and Pattern Recognition (CVPR'2001)*, pp. 283–290, Kauai, Hawaii.

Avidan, S., Baker, S., and Shan, Y. (2010). Special issue on Internet Vision. *Proceedings of the IEEE*, 98(8):1367–1369.

Axelsson, O. (1996). *Iterative Solution Methods*. Cambridge University Press, Cambridge.

Ayache, N. (1989). *Vision Stéréoscopique et Perception Multisensorielle*. InterEditions, Paris.

Azarbayejani, A. and Pentland, A. P. (1995). Recursive estimation of motion, structure, and focal length. *IEEE Transactions on Pattern Analysis and Machine Intelligence*, 17(6):562–575.

Azuma, R. T., Baillot, Y., Behringer, R., Feiner, S. K., Julier, S., and MacIntyre, B. (2001). Recent advances in augmented reality. *IEEE Computer Graphics and Applications*, 21(6):34–47.

Bab-Hadiashar, A. and Suter, D. (1998a). Robust optic flow computation. *International Journal of Computer Vision*, 29(1):59–77.

Bab-Hadiashar, A. and Suter, D. (1998b). Robust total least squares based optic flow computation. In *Asian Conference on Computer Vision (ACCV'98)*, pp. 566–573, Hong Kong.

Badra, F., Qumsieh, A., and Dudek, G. (1998). Rotation and zooming in image mosaicing. In *IEEE Workshop on Applications of Computer Vision (WACV'98)*, pp. 50–55, Princeton.

Bae, S., Paris, S., and Durand, F. (2006). Two-scale tone management for photographic look. *ACM Transactions on Graphics*, 25(3):637–645.

Baeza-Yates, R. and Ribeiro-Neto, B. (1999). *Modern Information Retrieval*. Addison Wesley.

Bai, X. and Sapiro, G. (2009). Geodesic matting: A framework for fast interactive image and video segmentation and matting. *International Journal of Computer Vision*, 82(2):113–132.

Bajcsy, R. and Kovacic, S. (1989). Multiresolution elastic matching. *Computer Vision, Graphics, and Image Processing*, 46(1):1–21.

Baker, H. H. (1977). Three-dimensional modeling. In *Fifth International Joint Conference on Artificial Intelligence (IJCAI-77)*, pp. 649–655.

Baker, H. H. (1982). *Depth from Edge and Intensity Based Stereo*. Technical Report AIM-347, Artificial Intelligence Laboratory, Stanford University.

Baker, H. H. (1989). Building surfaces of evolution: The weaving wall. *International Journal of Computer Vision*, 3(1):50–71.

Baker, H. H. and Binford, T. O. (1981). Depth from edge and intensity based stereo. In *IJCAI81*, pp. 631–636.

Baker, H. H. and Bolles, R. C. (1989). Generalizing epipolar-plane image analysis on the spatiotemporal surface. *International Journal of Computer Vision*, 3(1):33–49.

Baker, S. and Kanade, T. (2002). Limits on super-resolution and how to break them. *IEEE Transactions on Pattern Analysis and Machine Intelligence*, 24(9):1167–1183.

Baker, S. and Matthews, I. (2004). Lucas-Kanade 20 years on: A unifying framework: Part 1: The quantity approximated, the warp update rule, and the gradient descent approximation. *International Journal of Computer Vision*, 56(3):221–255.

Baker, S. and Nayar, S. (1999). A theory of single-viewpoint catadioptric image formation. *International Journal of Computer Vision*, 5(2):175–196.

Baker, S. and Nayar, S. K. (2001). Single viewpoint catadioptric cameras. In Benosman, R. and Kang, S. B. (eds), *Panoramic Vision: Sensors, Theory, and Applications*, pp. 39–71, Springer, New York.

Baker, S., Gross, R., and Matthews, I. (2003). *Lucas-Kanade 20 Years On: A Unifying Framework: Part 3*. Technical Report CMU-RI-TR-03-35, The Robotics Institute, Carnegie Mellon University.

Baker, S., Gross, R., and Matthews, I. (2004). *Lucas-Kanade 20 Years On: A Unifying Framework: Part 4*. Technical Report CMU-RI-TR-04-14, The Robotics Institute, Carnegie Mellon University.

Baker, S., Szeliski, R., and Anandan, P. (1998). A layered approach to stereo reconstruction. In *IEEE Computer Society Conference on Computer Vision and Pattern Recognition (CVPR'98)*, pp. 434–441, Santa Barbara.

Baker, S., Gross, R., Ishikawa, T., and Matthews, I. (2003). *Lucas-Kanade 20 Years On: A Unifying Framework: Part 2*. Technical Report CMU-RI-TR-03-01, The Robotics Institute, Carnegie Mellon University.

Baker, S., Black, M., Lewis, J. P., Roth, S., Scharstein, D., and Szeliski, R. (2007). A database and evaluation methodology for optical flow. In *Eleventh International Conference on Computer Vision (ICCV 2007)*, Rio de Janeiro, Brazil.

Baker, S., Scharstein, D., Lewis, J., Roth, S., Black, M. J., and Szeliski, R. (2009). *A Database and Evaluation Methodology for Optical Flow*. Technical Report MSR-TR-2009-179, Microsoft Research.

Ballard, D. H. (1981). Generalizing the Hough transform to detect arbitrary patterns. *Pattern Recognition*, 13(2):111–122.

Ballard, D. H. and Brown, C. M. (1982). *Computer Vision*. Prentice-Hall, Englewood Cliffs, New Jersey.

Banno, A., Masuda, T., Oishi, T., and Ikeuchi, K. (2008). Flying laser range sensor for large-scale site-modeling and its applications in Bayon digital archival project. *International Journal of Computer Vision*, 78(2-3):207–222.

Bar-Hillel, A., Hertz, T., and Weinshall, D. (2005). Object class recognition by boosting a part based model. In *IEEE Computer Society Conference on Computer Vision and Pattern Recognition (CVPR'2005)*, pp. 701–708, San Diego, CA.

Bar-Joseph, Z., El-Yaniv, R., Lischinski, D., and Werman, M. (2001). Texture mixing and texture movie synthesis using statistical learning. *IEEE Transactions on Visualization and Computer Graphics*, 7(2):120–135.

Bar-Shalom, Y. and Fortmann, T. E. (1988). *Tracking and data association*. Academic Press, Boston.

Barash, D. (2002). A fundamental relationship between bilateral filtering, adaptive smoothing, and the nonlinear diffusion equation. *IEEE Transactions on Pattern Analysis and Machine Intelligence*, 24(6):844–847.

Barash, D. and Comaniciu, D. (2004). A common framework for nonlinear diffusion, adaptive smoothing, bilateral filtering and mean shift. *Image and Vision Computing*, 22(1):73–81.

Barbu, A. and Zhu, S.-C. (2003). Graph partition by Swendsen–Wang cuts. In *Ninth International Conference on Computer Vision (ICCV 2003)*, pp. 320–327, Nice, France.

Barbu, A. and Zhu, S.-C. (2005). Generalizing Swendsen–Wang to sampling arbitrary posterior probabilities. *IEEE Transactions on Pattern Analysis and Machine Intelligence*, 27(9):1239–1253.

Barkans, A. C. (1997). High quality rendering using the Talisman architecture. In *Proceedings of the Eurographics Workshop on Graphics Hardware*.

Barnard, S. T. (1989). Stochastic stereo matching over scale. *International Journal of Computer Vision*, 3(1):17–32.

Barnard, S. T. and Fischler, M. A. (1982). Computational stereo. *Computing Surveys*, 14(4):553–572.

Barnes, C., Jacobs, D. E., Sanders, J., Goldman, D. B., Rusinkiewicz, S., Finkelstein, A., and Agrawala, M. (2008). Video puppetry: A performative interface for cutout animation. *ACM Transactions on Graphics*, 27(5).

Barreto, J. P. and Daniilidis, K. (2005). Fundamental matrix for cameras with radial distortion. In *Tenth International Conference on Computer Vision (ICCV 2005)*, pp. 625–632, Beijing, China.

Barrett, R., Berry, M., Chan, T. F., Demmel, J., Donato, J. et al. (1994). *Templates for the Solution of Linear Systems: Building Blocks for Iterative Methods, 2nd Edition*. SIAM, Philadelphia, PA.

Barron, J. L., Fleet, D. J., and Beauchemin, S. S. (1994). Performance of optical flow techniques. *International Journal of Computer Vision*, 12(1):43–77.

Barrow, H. G. and Tenenbaum, J. M. (1981). Computational vision. *Proceedings of the IEEE*, 69(5):572–595.

Bartels, R. H., Beatty, J. C., and Barsky, B. A. (1987). *An Introduction to Splines for use in Computer Graphics and Geeometric Modeling*. Morgan Kaufmann Publishers, Los Altos.

Bartoli, A. (2003). Towards gauge invariant bundle adjustment: A solution based on gauge dependent damping. In *Ninth International Conference on Computer Vision (ICCV 2003)*, pp. 760–765, Nice, France.

Bartoli, A. and Sturm, P. (2003). Multiple-view structure and motion from line correspondences. In *Ninth International Conference on Computer Vision (ICCV 2003)*, pp. 207–212, Nice, France.

Bartoli, A., Coquerelle, M., and Sturm, P. (2004). A framework for pencil-of-points structure-from-motion. In *Eighth European Conference on Computer Vision (ECCV 2004)*, pp. 28–40, Prague.

Bascle, B., Blake, A., and Zisserman, A. (1996). Motion deblurring and super-resolution from an image sequence. In *Fourth European Conference on Computer Vision (ECCV'96)*, pp. 573–582, Cambridge, England.

Bathe, K.-J. (2007). *Finite Element Procedures*. Prentice-Hall, Inc., Englewood Cliffs, New Jersey.

Batra, D., Sukthankar, R., and Chen, T. (2008). Learning class-specific affinities for image labelling. In *IEEE Computer Society Conference on Computer Vision and Pattern Recognition (CVPR 2008)*, Anchorage, AK.

Baudisch, P., Tan, D., Steedly, D., Rudolph, E., Uyttendaele, M., Pal, C., and Szeliski, R. (2006). An exploration of user interface designs for real-time panoramic photography. *Australian Journal of Information Systems*, 13(2).

Baumberg, A. (2000). Reliable feature matching across widely separated views. In *IEEE Computer Society Conference on Computer Vision and Pattern Recognition (CVPR'2000)*, pp. 774–781, Hilton Head Island.

Baumberg, A. M. and Hogg, D. C. (1996). Generating spatiotemporal models from examples. *Image and Vision Computing*, 14(8):525–532.

Baumgart, B. G. (1974). *Geometric Modeling for Computer Vision*. Technical Report AIM-249, Artificial Intelligence Laboratory, Stanford University.

Bay, H., Ferrari, V., and Van Gool, L. (2005). Wide-baseline stereo matching with line segments. In *IEEE Computer Society Conference on Computer Vision and Pattern Recognition (CVPR'2005)*, pp. 329–336, San Diego, CA.

Bay, H., Tuytelaars, T., and Van Gool, L. (2006). SURF: Speeded up robust features. In *Ninth European Conference on Computer Vision (ECCV 2006)*, pp. 404–417.

Bayer, B. E. (1976). Color imaging array. US Patent No. 3,971,065.

Beardsley, P., Torr, P., and Zisserman, A. (1996). 3D model acquisition from extended image sequences. In *Fourth European Conference on Computer Vision (ECCV'96)*, pp. 683–695, Cambridge, England.

Beare, R. (2006). A locally constrained watershed transform. *IEEE Transactions on Pattern Analysis and Machine Intelligence*, 28(7):1063–1074.

Becker, S. and Bove, V. M. (1995). Semiautomatic 3-D model extraction from uncalibrated 2-D camera views. In *SPIE Vol. 2410, Visual Data Exploration and Analysis II*, pp. 447–461, San Jose.

Beier, T. and Neely, S. (1992). Feature-based image metamorphosis. *Computer Graphics (SIGGRAPH '92)*, 26(2):35–42.

Beis, J. S. and Lowe, D. G. (1999). Indexing without invariants in 3D object recognition. *IEEE Transactions on Pattern Analysis and Machine Intelligence*, 21(10):1000–1015.

Belhumeur, P. N. (1996). A Bayesian approach to binocular stereopsis. *International Journal of Computer Vision*, 19(3):237–260.

Belhumeur, P. N., Hespanha, J. P., and Kriegman, D. J. (1997). Eigenfaces vs. Fisherfaces: Recognition using class specific linear projection. *IEEE Transactions on Pattern Analysis and Machine Intelligence*, 19(7):711–720.

Belongie, S. and Malik, J. (1998). Finding boundaries in natural images: a new method using point descriptors and area completion. In *Fifth European Conference on Computer Vision (ECCV'98)*, pp. 751–766, Freiburg, Germany.

Belongie, S., Malik, J., and Puzicha, J. (2002). Shape matching and object recognition using shape contexts. *IEEE Transactions on Pattern Analysis and Machine Intelligence*, 24(4):509–522.

Belongie, S., Fowlkes, C., Chung, F., and Malik, J. (2002). Spectral partitioning with indefinite kernels using the Nyström extension. In *Seventh European Conference on Computer Vision (ECCV 2002)*, pp. 531–543, Copenhagen.

Bennett, E., Uyttendaele, M., Zitnick, L., Szeliski, R., and Kang, S. B. (2006). Video and image Bayesian demosaicing with a two color image prior. In *Ninth European Conference on Computer Vision (ECCV 2006)*, pp. 508–521, Graz.

Benosman, R. and Kang, S. B. (eds). (2001). *Panoramic Vision: Sensors, Theory, and Applications*, Springer, New York.

Berg, T. (2008). Internet vision. SUNY Stony Brook Course CSE 690, http://www.tamaraberg.com/teaching/Fall_08/.

Berg, T. and Forsyth, D. (2006). Animals on the web. In *IEEE Computer Society Conference on Computer Vision and Pattern Recognition (CVPR'2006)*, pp. 1463–1470, New York City, NY.

Bergen, J. R., Anandan, P., Hanna, K. J., and Hingorani, R. (1992). Hierarchical model-based motion estimation. In *Second European Conference on Computer Vision (ECCV'92)*, pp. 237–252, Santa Margherita Liguere, Italy.

Bergen, J. R., Burt, P. J., Hingorani, R., and Peleg, S. (1992). A three-frame algorithm for estimating two-component image motion. *IEEE Transactions on Pattern Analysis and Machine Intelligence*, 14(9):886–896.

Berger, J. O. (1993). *Statistical Decision Theory and Bayesian Analysis*. Springer, New York, second edition.

Bertalmio, M., Sapiro, G., Caselles, V., and Ballester, C. (2000). Image inpainting. In *ACM SIGGRAPH 2000 Conference Proceedings*, pp. 417–424.

Bertalmio, M., Vese, L., Sapiro, G., and Osher, S. (2003). Simultaneous structure and texture image inpainting. *IEEE Transactions on Image Processing*, 12(8):882–889.

Bertero, M., Poggio, T. A., and Torre, V. (1988). Ill-posed problems in early vision. *Proceedings of the IEEE*, 76(8):869–889.

Besag, J. (1986). On the statistical analysis of dirty pictures. *Journal of the Royal Statistical Society B*, 48(3):259–302.

Besl, P. (1989). Active optical range imaging sensors. In Sanz, J. L. (ed.), *Advances in Machine Vision*, chapter 1, pp. 1–63, Springer-Verlag.

Besl, P. J. and Jain, R. C. (1985). Three-dimensional object recognition. *Computing Surveys*, 17(1):75–145.

Besl, P. J. and McKay, N. D. (1992). A method for registration of 3-D shapes. *IEEE Transactions on Pattern Analysis and Machine Intelligence*, 14(2):239–256.

Betrisey, C., Blinn, J. F., Dresevic, B., Hill, B., Hitchcock, G. *et al.* (2000). Displaced filtering for patterned displays. In *Society for Information Display Symposium,*, pp. 296–299.

Beymer, D. (1996). Feature correspondence by interleaving shape and texture computations. In *IEEE Computer Society Conference on Computer Vision and Pattern Recognition (CVPR'96)*, pp. 921–928, San Francisco.

Bhat, D. N. and Nayar, S. K. (1998). Ordinal measures for image correspondence. *IEEE Transactions on Pattern Analysis and Machine Intelligence*, 20(4):415–423.

Bickel, B., Botsch, M., Angst, R., Matusik, W., Otaduy, M., Pfister, H., and Gross, M. (2007). Multi-scale capture of facial geometry and motion. *ACM Transactions on Graphics*, 26(3).

Billinghurst, M., Kato, H., and Poupyrev, I. (2001). The MagicBook: a transitional AR interface. *Computers & Graphics*, 25:745–753.

Bimber, O. (2006). Computational photography—the next big step. *Computer*, 39(8):28–29.

Birchfield, S. and Tomasi, C. (1998). A pixel dissimilarity measure that is insensitive to image sampling. *IEEE Transactions on Pattern Analysis and Machine Intelligence*, 20(4):401–406.

Birchfield, S. and Tomasi, C. (1999). Depth discontinuities by pixel-to-pixel stereo. *International Journal of Computer Vision*, 35(3):269–293.

Birchfield, S. T., Natarajan, B., and Tomasi, C. (2007). Correspondence as energy-based segmentation. *Image and Vision Computing*, 25(8):1329–1340.

Bishop, C. M. (2006). *Pattern Recognition and Machine Learning*. Springer, New York, NY.

Bitouk, D., Kumar, N., Dhillon, S., Belhumeur, P., and Nayar, S. K. (2008). Face swapping: Automatically replacing faces in photographs. *ACM Transactions on Graphics*, 27(3).

Björck, A. (1996). *Numerical Methods for Least Squares Problems*. Society for Industrial and Applied Mathematics.

Björck, A. and Dahlquist, G. (2010). *Numerical Methods in Scientific Computing*. Volume II, Society for Industrial and Applied Mathematics.

Black, M., Yacoob, Y., Jepson, A. D., and Fleet, D. J. (1997). Learning parameterized models of image motion. In *IEEE Computer Society Conference on Computer Vision and Pattern Recognition (CVPR'97)*, pp. 561–567, San Juan, Puerto Rico.

Black, M. J. and Anandan, P. (1996). The robust estimation of multiple motions: Parametric and piecewise-smooth flow fields. *Computer Vision and Image Understanding*, 63(1):75–104.

Black, M. J. and Jepson, A. D. (1996). Estimating optical flow in segmented images using variable-order parametric models with local deformations. *IEEE Transactions on Pattern Analysis and Machine Intelligence*, 18(10):972–986.

Black, M. J. and Jepson, A. D. (1998). EigenTracking: robust matching and tracking of articulated objects using a view-based representation. *International Journal of Computer Vision*, 26(1):63–84.

Black, M. J. and Rangarajan, A. (1996). On the unification of line processes, outlier rejection, and robust statistics with applications in early vision. *International Journal of Computer Vision*, 19(1):57–91.

Black, M. J., Sapiro, G., Marimont, D. H., and Heeger, D. (1998). Robust anisotropic diffusion. *IEEE Transactions on Image Processing*, 7(3):421–432.

Blackford, L. S., Demmel, J., Dongarra, J., Duff, I., Hammarling, S. *et al.* (2002). An updated set of basic linear algebra subprograms (BLAS). *ACM Transactions on Mathematical Software*, 28(2):135–151.

Blake, A. and Isard, M. (1998). *Active Contours: The Application of Techniques from Graphics, Vision, Control Theory and Statistics to Visual Tracking of Shapes in Motion.* Springer Verlag, London.

Blake, A. and Zisserman, A. (1987). *Visual Reconstruction.* MIT Press, Cambridge, Massachusetts.

Blake, A., Curwen, R., and Zisserman, A. (1993). A framework for spatio-temporal control in the tracking of visual contour. *International Journal of Computer Vision*, 11(2):127–145.

Blake, A., Kohli, P., and Rother, C. (eds). (2010). *Advances in Markov Random Fields*, MIT Press.

Blake, A., Zimmerman, A., and Knowles, G. (1985). Surface descriptions from stereo and shading. *Image and Vision Computing*, 3(4):183–191.

Blake, A., Rother, C., Brown, M., Perez, P., and Torr, P. (2004). Interactive image segmentation using an adaptive GMMRF model. In *Eighth European Conference on Computer Vision (ECCV 2004)*, pp. 428–441, Prague.

Blanz, V. and Vetter, T. (1999). A morphable model for the synthesis of 3D faces. In *ACM SIGGRAPH 1999 Conference Proceedings*, pp. 187–194.

Blanz, V. and Vetter, T. (2003). Face recognition based on fitting a 3D morphable model. *IEEE Transactions on Pattern Analysis and Machine Intelligence*, 25():1063–1074.

Bleyer, M., Gelautz, M., Rother, C., and Rhemann, C. (2009). A stereo approach that handles the matting problem via image warping. In *IEEE Computer Society Conference on Computer Vision and Pattern Recognition (CVPR 2009)*, Miami Beach, FL.

Blinn, J. (1998). *Dirty Pixels.* Morgan Kaufmann Publishers, San Francisco.

Blinn, J. F. (1994a). Jim Blinn's corner: Compositing, part 1: Theory. *IEEE Computer Graphics and Applications*, 14(5):83–87.

Blinn, J. F. (1994b). Jim Blinn's corner: Compositing, part 2: Practice. *IEEE Computer Graphics and Applications*, 14(6):78–82.

Blinn, J. F. and Newell, M. E. (1976). Texture and reflection in computer generated images. *Communications of the ACM*, 19(10):542–547.

Blostein, D. and Ahuja, N. (1987). Shape from texture: Integrating texture-element extraction and surface estimation. *IEEE Transactions on Pattern Analysis and Machine Intelligence*, 11(12):1233–1251.

Bobick, A. F. (1997). Movement, activity and action: the role of knowledge in the perception of motion. *Proceedings of the Royal Society of London*, B 352:1257–1265.

Bobick, A. F. and Intille, S. S. (1999). Large occlusion stereo. *International Journal of Computer Vision*, 33(3):181–200.

Boden, M. A. (2006). *Mind As Machine: A History of Cognitive Science*. Oxford University Press, Oxford, England.

Bogart, R. G. (1991). View correlation. In Arvo, J. (ed.), *Graphics Gems II*, pp. 181–190, Academic Press, Boston.

Boiman, O., Shechtman, E., and Irani, M. (2008). In defense of nearest-neighbor based image classification. In *IEEE Computer Society Conference on Computer Vision and Pattern Recognition (CVPR 2008)*, Anchorage, AK.

Boissonat, J.-D. (1984). Representing 2D and 3D shapes with the Delaunay triangulation. In *Seventh International Conference on Pattern Recognition (ICPR'84)*, pp. 745–748, Montreal, Canada.

Bolles, R. C., Baker, H. H., and Hannah, M. J. (1993). The JISCT stereo evaluation. In *Image Understanding Workshop*, pp. 263–274.

Bolles, R. C., Baker, H. H., and Marimont, D. H. (1987). Epipolar-plane image analysis: An approach to determining structure from motion. *International Journal of Computer Vision*, 1:7–55.

Bookstein, F. L. (1989). Principal warps: Thin-plate splines and the decomposition of deformations. *IEEE Transactions on Pattern Analysis and Machine Intelligence*, 11(6):567–585.

Borenstein, E. and Ullman, S. (2008). Combined top-down/bottom-up segmentation. *IEEE Transactions on Pattern Analysis and Machine Intelligence*, 30(12):2109–2125.

Borgefors, G. (1986). Distance transformations in digital images. *Computer Vision, Graphics and Image Processing*, 34(3):227–248.

Bouchard, G. and Triggs, B. (2005). Hierarchical part-based visual object categorization. In *IEEE Computer Society Conference on Computer Vision and Pattern Recognition (CVPR'2005)*, pp. 709–714, San Diego, CA.

Bougnoux, S. (1998). From projective to Euclidean space under any practical situation, a criticism of self-calibration. In *Sixth International Conference on Computer Vision (ICCV'98)*, pp. 790–798, Bombay.

Bouguet, J.-Y. and Perona, P. (1999). 3D photography using shadows in dual-space geometry. *International Journal of Computer Vision*, 35(2):129–149.

Boult, T. E. and Kender, J. R. (1986). Visual surface reconstruction using sparse depth data. In *IEEE Computer Society Conference on Computer Vision and Pattern Recognition (CVPR'86)*, pp. 68–76, Miami Beach.

Bourdev, L. and Malik, J. (2009). Poselets: Body part detectors trained using 3D human pose annotations. In *Twelfth International Conference on Computer Vision (ICCV 2009)*, Kyoto, Japan.

Bovik, A. (ed.). (2000). *Handbook of Image and Video Processing*, Academic Press, San Diego.

Bowyer, K. W., Kranenburg, C., and Dougherty, S. (2001). Edge detector evaluation using empirical ROC curves. *Computer Vision and Image Understanding*, 84(1):77–103.

Box, G. E. P. and Muller, M. E. (1958). A note on the generation of random normal deviates. *Annals of Mathematical Statistics*, 29(2).

Boyer, E. and Berger, M. O. (1997). 3D surface reconstruction using occluding contours. *International Journal of Computer Vision*, 22(3):219–233.

Boykov, Y. and Funka-Lea, G. (2006). Graph cuts and efficient N-D image segmentation. *International Journal of Computer Vision*, 70(2):109–131.

Boykov, Y. and Jolly, M.-P. (2001). Interactive graph cuts for optimal boundary and region segmentation of objects in N-D images. In *Eighth International Conference on Computer Vision (ICCV 2001)*, pp. 105–112, Vancouver, Canada.

Boykov, Y. and Kolmogorov, V. (2003). Computing geodesics and minimal surfaces via graph cuts. In *Ninth International Conference on Computer Vision (ICCV 2003)*, pp. 26–33, Nice, France.

Boykov, Y. and Kolmogorov, V. (2004). An experimental comparison of min-cut/max-flow algorithms for energy minimization in vision. *IEEE Transactions on Pattern Analysis and Machine Intelligence*, 26(9):1124–1137.

Boykov, Y. and Kolmogorov, V. (2010). Basic graph cut algorithms. In Blake, A., Kohli, P., and Rother, C. (eds), *Advances in Markov Random Fields*, MIT Press.

Boykov, Y., Veksler, O., and Zabih, R. (1998). A variable window approach to early vision. *IEEE Transactions on Pattern Analysis and Machine Intelligence*, 20(12):1283–1294.

Boykov, Y., Veksler, O., and Zabih, R. (2001). Fast approximate energy minimization via graph cuts. *IEEE Transactions on Pattern Analysis and Machine Intelligence*, 23(11):1222–1239.

Boykov, Y., Veksler, O., and Zabih, R. (2010). Optimizing multi-label MRFs by move making algorithms. In Blake, A., Kohli, P., and Rother, C. (eds), *Advances in Markov Random Fields*, MIT Press.

Boykov, Y., Kolmogorov, V., Cremers, D., and Delong, A. (2006). An integral solution to surface evolution PDEs via Geo-cuts. In *Ninth European Conference on Computer Vision (ECCV 2006)*, pp. 409–422.

Bracewell, R. N. (1986). *The Fourier Transform and its Applications*. McGraw-Hill, New York, 2nd edition.

Bradley, D., Boubekeur, T., and Heidrich, W. (2008). Accurate multi-view reconstruction using robust binocular stereo and surface meshing. In *IEEE Computer Society Conference on Computer Vision and Pattern Recognition (CVPR 2008)*, Anchorage, AK.

Bradsky, G. and Kaehler, A. (2008). *Learning OpenCV: Computer Vision with the OpenCV Library*. O'Reilly, Sebastopol, CA.

Brandt, A. (1986). Algebraic multigrid theory: The symmetric case. *Applied Mathematics and Computation*, 19(1-4):23–56.

Bregler, C. and Malik, J. (1998). Tracking people with twists and exponential maps. In *IEEE Computer Society Conference on Computer Vision and Pattern Recognition (CVPR'98)*, pp. 8–15, Santa Barbara.

Bregler, C., Covell, M., and Slaney, M. (1997). Video rewrite: Driving visual speech with audio. In *ACM SIGGRAPH 1997 Conference Proceedings*, pp. 353–360.

Bregler, C., Malik, J., and Pullen, K. (2004). Twist based acquisition and tracking of animal and human kinematics. *International Journal of Computer Vision*, 56(3):179–194.

Breu, H., Gil, J., Kirkpatrick, D., and Werman, M. (1995). Linear time Euclidean distance transform algorithms. *IEEE Transactions on Pattern Analysis and Machine Intelligence*, 17(5):529–533.

Brice, C. R. and Fennema, C. L. (1970). Scene analysis using regions. *Artificial Intelligence*, 1(3-4):205–226.

Briggs, W. L., Henson, V. E., and McCormick, S. F. (2000). *A Multigrid Tutorial*. Society for Industrial and Applied Mathematics, Philadelphia, second edition.

Brillaut-O'Mahoney, B. (1991). New method for vanishing point detection. *Computer Vision, Graphics, and Image Processing*, 54(2):289–300.

Brinkmann, R. (2008). *The Art and Science of Digital Compositing*. Morgan Kaufmann Publishers, San Francisco, 2nd edition.

Brooks, R. A. (1981). Symbolic reasoning among 3-D models and 2-D images. *Artificial Intelligence*, 17:285–348.

Brown, D. C. (1971). Close-range camera calibration. *Photogrammetric Engineering*, 37(8):855–866.

Brown, L. G. (1992). A survey of image registration techniques. *Computing Surveys*, 24(4):325–376.

Brown, M. and Lowe, D. (2002). Invariant features from interest point groups. In *British Machine Vision Conference*, pp. 656–665, Cardiff, Wales.

Brown, M. and Lowe, D. (2003). Unsupervised 3D object recognition and reconstruction in unordered datasets. In *International Conference on 3D Imaging and Modelling*, pp. 1218–1225, Nice, France.

Brown, M. and Lowe, D. (2007). Automatic panoramic image stitching using invariant features. *International Journal of Computer Vision*, 74(1):59–73.

Brown, M., Hartley, R., and Nistér, D. (2007). Minimal solutions for panoramic stitching. In *IEEE Computer Society Conference on Computer Vision and Pattern Recognition (CVPR 2007)*, Minneapolis, MN.

Brown, M., Szeliski, R., and Winder, S. (2004). *Multi-Image Matching Using Multi-Scale Oriented Patches*. Technical Report MSR-TR-2004-133, Microsoft Research.

Brown, M., Szeliski, R., and Winder, S. (2005). Multi-image matching using multi-scale oriented patches. In *IEEE Computer Society Conference on Computer Vision and Pattern Recognition (CVPR'2005)*, pp. 510–517, San Diego, CA.

Brown, M. Z., Burschka, D., and Hager, G. D. (2003). Advances in computational stereo. *IEEE Transactions on Pattern Analysis and Machine Intelligence*, 25(8):993–1008.

Brox, T., Bregler, C., and Malik, J. (2009). Large displacement optical flow. In *IEEE Computer Society Conference on Computer Vision and Pattern Recognition (CVPR 2009)*, Miami Beach, FL.

Brox, T., Bruhn, A., Papenberg, N., and Weickert, J. (2004). High accuracy optical flow estimation based on a theory for warping. In *Eighth European Conference on Computer Vision (ECCV 2004)*, pp. 25–36, Prague.

Brubaker, S. C., Wu, J., Sun, J., Mullin, M. D., and Rehg, J. M. (2008). On the design of cascades of boosted ensembles for face detection. *International Journal of Computer Vision*, 77(1-3):65–86.

Bruhn, A., Weickert, J., and Schnörr, C. (2005). Lucas/Kanade meets Horn/Schunck: Combining local and global optic flow methods. *International Journal of Computer Vision*, 61(3):211–231.

Bruhn, A., Weickert, J., Kohlberger, T., and Schnörr, C. (2006). A multigrid platform for real-time motion computation with discontinuity-preserving variational methods. *International Journal of Computer Vision*, 70(3):257–277.

Buades, A., Coll, B., and Morel, J.-M. (2008). Nonlocal image and movie denoising. *International Journal of Computer Vision*, 76(2):123–139.

Bălan, A. O. and Black, M. J. (2008). The naked truth: Estimating body shape under clothing. In *Tenth European Conference on Computer Vision (ECCV 2008)*, pp. 15–29, Marseilles.

Buchanan, A. and Fitzgibbon, A. (2005). Damped Newton algorithms for matrix factorization with missing data. In *IEEE Computer Society Conference on Computer Vision and Pattern Recognition (CVPR'2005)*, pp. 316–322, San Diego, CA.

Buck, I., Finkelstein, A., Jacobs, C., Klein, A., Salesin, D. H., Seims, J., Szeliski, R., and Toyama, K. (2000). Performance-driven hand-drawn animation. In *Symposium on Non Photorealistic Animation and Rendering*, pp. 101–108, Annecy.

Buehler, C., Bosse, M., McMillan, L., Gortler, S. J., and Cohen, M. F. (2001). Unstructured Lumigraph rendering. In *ACM SIGGRAPH 2001 Conference Proceedings*, pp. 425–432.

Bugayevskiy, L. M. and Snyder, J. P. (1995). *Map Projections: A Reference Manual*. CRC Press.

Burger, W. and Burge, M. J. (2008). *Digital Image Processing: An Algorithmic Introduction Using Java*. Springer, New York, NY.

Burl, M. C., Weber, M., and Perona, P. (1998). A probabilistic approach to object recognition using local photometry and global geometry. In *Fifth European Conference on Computer Vision (ECCV'98)*, pp. 628–641, Freiburg, Germany.

Burns, J. B., Hanson, A. R., and Riseman, E. M. (1986). Extracting straight lines. *IEEE Transactions on Pattern Analysis and Machine Intelligence*, PAMI-8(4):425–455.

Burns, P. D. and Williams, D. (1999). Using slanted edge analysis for color registration measurement. In *IS&T PICS Conference*, pp. 51–53.

Burt, P. J. and Adelson, E. H. (1983a). The Laplacian pyramid as a compact image code. *IEEE Transactions on Communications*, COM-31(4):532–540.

Burt, P. J. and Adelson, E. H. (1983b). A multiresolution spline with applications to image mosaics. *ACM Transactions on Graphics*, 2(4):217–236.

Burt, P. J. and Kolczynski, R. J. (1993). Enhanced image capture through fusion. In *Fourth International Conference on Computer Vision (ICCV'93)*, pp. 173–182, Berlin, Germany.

Byröd, M. and øAström, K. (2009). Bundle adjustment using conjugate gradients with multiscale preconditioning. In *British Machine Vision Conference (BMVC 2009)*.

Cai, D., He, X., Hu, Y., Han, J., and Huang, T. (2007). Learning a spatially smooth subspace for face recognition. In *IEEE Computer Society Conference on Computer Vision and Pattern Recognition (CVPR 2007)*, Minneapolis, MN.

Campbell, N. D. F., Vogiatzis, G., Hernández, C., and Cipolla, R. (2008). Using multiple hypotheses to improve depth-maps for multi-view stereo. In *Tenth European Conference on Computer Vision (ECCV 2008)*, pp. 766–779, Marseilles.

Can, A., Stewart, C., Roysam, B., and Tanenbaum, H. (2002). A feature-based, robust, hierarchical algorithm for registering pairs of images of the curved human retina. *IEEE Transactions on Pattern Analysis and Machine Intelligence*, 24(3):347–364.

Canny, J. (1986). A computational approach to edge detection. *IEEE Transactions on Pattern Analysis and Machine Intelligence*, PAMI-8(6):679–698.

Cao, Z., Yin, Q., Tang, X., and Sun, J. (2010). Face recognition with learning-based descriptor. In *IEEE Computer Society Conference on Computer Vision and Pattern Recognition (CVPR 2010)*, San Francisco, CA.

Capel, D. (2004). *Image Mosaicing and Super-resolution. Distinguished Dissertation Series, British Computer Society*, Springer-Verlag.

Capel, D. and Zisserman, A. (1998). Automated mosaicing with super-resolution zoom. In *IEEE Computer Society Conference on Computer Vision and Pattern Recognition (CVPR'98)*, pp. 885–891, Santa Barbara.

Capel, D. and Zisserman, A. (2000). Super-resolution enhancement of text image sequences. In *Fifteenth International Conference on Pattern Recognition (ICPR'2000)*, pp. 600–605, Barcelona, Spain.

Capel, D. and Zisserman, A. (2003). Computer vision applied to super resolution. *IEEE Signal Processing Magazine*, 20(3):75–86.

Capel, D. P. (2001). *Super-resolution and Image Mosaicing*. Ph.D. thesis, University of Oxford.

Caprile, B. and Torre, V. (1990). Using vanishing points for camera calibration. *International Journal of Computer Vision*, 4(2):127–139.

Carneiro, G. and Jepson, A. (2005). The distinctiveness, detectability, and robustness of local image features. In *IEEE Computer Society Conference on Computer Vision and Pattern Recognition (CVPR'2005)*, pp. 296–301, San Diego, CA.

Carneiro, G. and Lowe, D. (2006). Sparse flexible models of local features. In *Ninth European Conference on Computer Vision (ECCV 2006)*, pp. 29–43.

Carnevali, P., Coletti, L., and Patarnello, S. (1985). Image processing by simulated annealing. *IBM Journal of Research and Development*, 29(6):569–579.

Carranza, J., Theobalt, C., Magnor, M. A., and Seidel, H.-P. (2003). Free-viewpoint video of human actors. *ACM Transactions on Graphics (Proc. SIGGRAPH 2003)*, 22(3):569–577.

Carroll, R., Agrawala, M., and Agarwala, A. (2009). Optimizing content-preserving projections for wide-angle images. *ACM Transactions on Graphics*, 28(3).

Caselles, V., Kimmel, R., and Sapiro, G. (1997). Geodesic active contours. *International Journal of Computer Vision*, 21(1):61–79.

Catmull, E. and Smith, A. R. (1980). 3-D transformations of images in scanline order. *Computer Graphics (SIGGRAPH '80)*, 14(3):279–285.

Celniker, G. and Gossard, D. (1991). Deformable curve and surface finite-elements for free-form shape design. *Computer Graphics (SIGGRAPH '91)*, 25(4):257–266.

Chakrabarti, A., Scharstein, D., and Zickler, T. (2009). An empirical camera model for internet color vision. In *British Machine Vision Conference (BMVC 2009)*, London, UK.

Cham, T. J. and Cipolla, R. (1998). A statistical framework for long-range feature matching in uncalibrated image mosaicing. In *IEEE Computer Society Conference on Computer Vision and Pattern Recognition (CVPR'98)*, pp. 442–447, Santa Barbara.

Cham, T.-J. and Rehg, J. M. (1999). A multiple hypothesis approach to figure tracking. In *IEEE Computer Society Conference on Computer Vision and Pattern Recognition (CVPR'99)*, pp. 239–245, Fort Collins.

Champleboux, G., Lavallée, S., Sautot, P., and Cinquin, P. (1992). Accurate calibration of cameras and range imaging sensors, the NPBS method. In *IEEE International Conference on Robotics and Automation*, pp. 1552–1558, Nice, France.

Champleboux, G., Lavallée, S., Szeliski, R., and Brunie, L. (1992). From accurate range imaging sensor calibration to accurate model-based 3-D object localization. In *IEEE Computer Society Conference on Computer Vision and Pattern Recognition (CVPR'92)*, pp. 83–89, Champaign, Illinois.

Chan, A. B. and Vasconcelos, N. (2009). Layered dynamic textures. *IEEE Transactions on Pattern Analysis and Machine Intelligence*, 31(10):1862–1879.

Chan, T. F. and Vese, L. A. (1992). Active contours without edges. *IEEE Transactions on Image Processing*, 10(2):266–277.

Chan, T. F., Osher, S., and Shen, J. (2001). The digital TV filter and nonlinear denoising. *IEEE Transactions on Image Processing*, 10(2):231–241.

Chang, M. M., Tekalp, A. M., and Sezan, M. I. (1997). Simultaneous motion estimation and segmentation. *IEEE Transactions on Image Processing*, 6(9):1326–1333.

Chaudhuri, S. (2001). *Super-Resolution Imaging*. Springer.

Chaudhuri, S. and Rajagopalan, A. N. (1999). *Depth from Defocus: A Real Aperture Imaging Approach*. Springer.

Cheeseman, P., Kanefsky, B., Hanson, R., and Stutz, J. (1993). *Super-Resolved Surface Reconstruction From Multiple Images*. Technical Report FIA-93-02, NASA Ames Research Center, Artificial Intelligence Branch.

Chellappa, R., Wilson, C., and Sirohey, S. (1995). Human and machine recognition of faces: A survey. *Proceedings of the IEEE*, 83(5):705–740.

Chen, B., Neubert, B., Ofek, E., Deussen, O., and Cohen, M. F. (2009). Integrated videos and maps for driving directions. In *UIST '09: Proceedings of the 22nd annual ACM symposium on User interface software and technology*, pp. 223–232, Victoria, BC, Canada, New York, NY, USA.

Chen, C.-Y. and Klette, R. (1999). Image stitching - comparisons and new techniques. In *Computer Analysis of Images and Patterns (CAIP'99)*, pp. 615–622, Ljubljana.

Chen, J. and Chen, B. (2008). Architectural modeling from sparsely scanned range data. *International Journal of Computer Vision*, 78(2-3):223–236.

Chen, J., Paris, S., and Durand, F. (2007). Real-time edge-aware image processing with the bilateral grid. *ACM Transactions on Graphics*, 26(3).

Chen, S. and Williams, L. (1993). View interpolation for image synthesis. In *ACM SIGGRAPH 1993 Conference Proceedings*, pp. 279–288.

Chen, S. E. (1995). QuickTime VR – an image-based approach to virtual environment navigation. In *ACM SIGGRAPH 1995 Conference Proceedings*, pp. 29–38, Los Angeles.

Chen, Y. and Medioni, G. (1992). Object modeling by registration of multiple range images. *Image and Vision Computing*, 10(3):145–155.

Cheng, L., Vishwanathan, S. V. N., and Zhang, X. (2008). Consistent image analogies using semi-supervised learning. In *IEEE Computer Society Conference on Computer Vision and Pattern Recognition (CVPR 2008)*, Anchorage, AK.

Cheng, Y. (1995). Mean shift, mode seeking, and clustering. *IEEE Transactions on Pattern Analysis and Machine Intelligence*, 17(8):790–799.

Chiang, M.-C. and Boult, T. E. (1996). Efficient image warping and super-resolution. In *IEEE Workshop on Applications of Computer Vision (WACV'96)*, pp. 56–61, Sarasota.

Chiu, K. and Raskar, R. (2009). Computer vision on tap. In *Second IEEE Workshop on Internet Vision*, Miami Beach, Florida.

Chou, P. B. and Brown, C. M. (1990). The theory and practice of Bayesian image labeling. *International Journal of Computer Vision*, 4(3):185–210.

Christensen, G., Joshi, S., and Miller, M. (1997). Volumetric transformation of brain anatomy. *IEEE Transactions on Medical Imaging*, 16(6):864–877.

Christy, S. and Horaud, R. (1996). Euclidean shape and motion from multiple perspective views by affine iterations. *IEEE Transactions on Pattern Analysis and Machine Intelligence*, 18(11):1098–1104.

Chuang, Y.-Y., Curless, B., Salesin, D. H., and Szeliski, R. (2001). A Bayesian approach to digital matting. In *IEEE Computer Society Conference on Computer Vision and Pattern Recognition (CVPR'2001)*, pp. 264–271, Kauai, Hawaii.

Chuang, Y.-Y., Agarwala, A., Curless, B., Salesin, D. H., and Szeliski, R. (2002). Video matting of complex scenes. *ACM Transactions on Graphics (Proc. SIGGRAPH 2002)*, 21(3):243–248.

Chuang, Y.-Y., Goldman, D. B., Curless, B., Salesin, D. H., and Szeliski, R. (2003). Shadow matting. *ACM Transactions on Graphics (Proc. SIGGRAPH 2003)*, 22(3):494–500.

Chuang, Y.-Y., Goldman, D. B., Zheng, K. C., Curless, B., Salesin, D. H., and Szeliski, R. (2005). Animating pictures with stochastic motion textures. *ACM Transactions on Graphics (Proc. SIGGRAPH 2005)*, 24(3):853–860.

Chuang, Y.-Y., Zongker, D., Hindorff, J., Curless, B., Salesin, D. H., and Szeliski, R. (2000). Environment matting extensions: Towards higher accuracy and real-time capture. In *ACM SIGGRAPH 2000 Conference Proceedings*, pp. 121–130, New Orleans.

Chui, C. K. (1992). *Wavelet Analysis and Its Applications*. Academic Press, New York.

Chum, O. and Matas, J. (2005). Matching with PROSAC—progressive sample consensus. In *IEEE Computer Society Conference on Computer Vision and Pattern Recognition (CVPR'2005)*, pp. 220–226, San Diego, CA.

Chum, O. and Matas, J. (2010). Large-scale discovery of spatially related images. *IEEE Transactions on Pattern Analysis and Machine Intelligence*, 32(2):371–377.

Chum, O., Philbin, J., and Zisserman, A. (2008). Near duplicate image detection: min-hash and tf-idf weighting. In *British Machine Vision Conference (BMVC 2008)*, Leeds, England.

Chum, O., Philbin, J., Sivic, J., Isard, M., and Zisserman, A. (2007). Total recall: Automatic query expansion with a generative feature model for object retrieval. In *Eleventh International Conference on Computer Vision (ICCV 2007)*, Rio de Janeiro, Brazil.

Cipolla, R. and Blake, A. (1990). The dynamic analysis of apparent contours. In *Third International Conference on Computer Vision (ICCV'90)*, pp. 616–623, Osaka, Japan.

Cipolla, R. and Blake, A. (1992). Surface shape from the deformation of apparent contours. *International Journal of Computer Vision*, 9(2):83–112.

Cipolla, R. and Giblin, P. (2000). *Visual Motion of Curves and Surfaces*. Cambridge University Press, Cambridge.

Cipolla, R., Drummond, T., and Robertson, D. P. (1999). Camera calibration from vanishing points in images of architectural scenes. In *British Machine Vision Conference (BMVC99)*.

Claus, D. and Fitzgibbon, A. (2005). A rational function lens distortion model for general cameras. In *IEEE Computer Society Conference on Computer Vision and Pattern Recognition (CVPR'2005)*, pp. 213–219, San Diego, CA.

Clowes, M. B. (1971). On seeing things. *Artificial Intelligence*, 2:79–116.

Cohen, L. D. and Cohen, I. (1993). Finite-element methods for active contour models and balloons for 2-D and 3-D images. *IEEE Transactions on Pattern Analysis and Machine Intelligence*, 15(11):1131–1147.

Cohen, M. and Wallace, J. (1993). *Radiosity and Realistic Image Synthesis*. Morgan Kaufmann.

Cohen, M. F. and Szeliski, R. (2006). The Moment Camera. *Computer*, 39(8):40–45.

Collins, R. T. (1996). A space-sweep approach to true multi-image matching. In *IEEE Computer Society Conference on Computer Vision and Pattern Recognition (CVPR'96)*, pp. 358–363, San Francisco.

Collins, R. T. and Liu, Y. (2003). On-line selection of discriminative tracking features. In *Ninth International Conference on Computer Vision (ICCV 2003)*, pp. 346–352, Nice, France.

Collins, R. T. and Weiss, R. S. (1990). Vanishing point calculation as a statistical inference on the unit sphere. In *Third International Conference on Computer Vision (ICCV'90)*, pp. 400–403, Osaka, Japan.

Comaniciu, D. and Meer, P. (2002). Mean shift: A robust approach toward feature space analysis. *IEEE Transactions on Pattern Analysis and Machine Intelligence*, 24(5):603–619.

Comaniciu, D. and Meer, P. (2003). An algorithm for data-driven bandwidth selection. *IEEE Transactions on Pattern Analysis and Machine Intelligence*, 25(2):281–288.

Conn, A. R., Gould, N. I. M., and Toint, P. L. (2000). *Trust-Region Methods*. Society for Industrial and Applied Mathematics, Philadephia.

Cook, R. L. and Torrance, K. E. (1982). A reflectance model for computer graphics. *ACM Transactions on Graphics*, 1(1):7–24.

Coorg, S. and Teller, S. (2000). Spherical mosaics with quaternions and dense correlation. *International Journal of Computer Vision*, 37(3):259–273.

Cootes, T., Edwards, G. J., and Taylor, C. J. (2001). Active appearance models. *IEEE Transactions on Pattern Analysis and Machine Intelligence*, 23(6):681–685.

Cootes, T., Cooper, D., Taylor, C., and Graham, J. (1995). Active shape models—their training and application. *Computer Vision and Image Understanding*, 61(1):38–59.

Cootes, T., Taylor, C., Lanitis, A., Cooper, D., and Graham, J. (1993). Building and using flexible models incorporating grey-level information. In *Fourth International Conference on Computer Vision (ICCV'93)*, pp. 242–246, Berlin, Germany.

Cootes, T. F. and Taylor, C. J. (2001). Statistical models of appearance for medical image analysis and computer vision. In *Medical Imaging*.

Coquillart, S. (1990). Extended free-form deformations: A sculpturing tool for 3D geometric modeling. *Computer Graphics (SIGGRAPH '90)*, 24(4):187–196.

Cormen, T. H. (2001). *Introduction to Algorithms*. MIT Press, Cambridge, Massachusetts.

Cornelis, N., Leibe, B., Cornelis, K., and Van Gool, L. (2008). 3D urban scene modeling integrating recognition and reconstruction. *International Journal of Computer Vision*, 78(2-3):121–141.

Corso, J. and Hager, G. (2005). Coherent regions for concise and stable image description. In *IEEE Computer Society Conference on Computer Vision and Pattern Recognition (CVPR'2005)*, pp. 184–190, San Diego, CA.

Costeira, J. and Kanade, T. (1995). A multi-body factorization method for motion analysis. In *Fifth International Conference on Computer Vision (ICCV'95)*, pp. 1071–1076, Cambridge, Massachusetts.

Costen, N., Cootes, T. F., Edwards, G. J., and Taylor, C. J. (1999). Simultaneous extraction of functional face subspaces. In *IEEE Computer Society Conference on Computer Vision and Pattern Recognition (CVPR'99)*, pp. 492–497, Fort Collins.

Couprie, C., Grady, L., Najman, L., and Talbot, H. (2009). Power watersheds: A new image segmentation framework extending graph cuts, random walker and optimal spanning forest. In *Twelfth International Conference on Computer Vision (ICCV 2009)*, Kyoto, Japan.

Cour, T., Bénézit, F., and Shi, J. (2005). Spectral segmentation with multiscale graph decomposition. In *IEEE Computer Society Conference on Computer Vision and Pattern Recognition (CVPR'2005)*, pp. 1123–1130, San Diego, CA.

Cox, D., Little, J., and O'Shea, D. (2007). *Ideals, Varieties, and Algorithms: An Introduction to Computational Algebraic Geometry and Commutative Algebra*. Springer.

Cox, I. J. (1994). A maximum likelihood N-camera stereo algorithm. In *IEEE Computer Society Conference on Computer Vision and Pattern Recognition (CVPR'94)*, pp. 733–739, Seattle.

Cox, I. J., Roy, S., and Hingorani, S. L. (1995). Dynamic histogram warping of image pairs for constant image brightness. In *IEEE International Conference on Image Processing (ICIP'95)*, pp. 366–369.

Cox, I. J., Hingorani, S. L., Rao, S. B., and Maggs, B. M. (1996). A maximum likelihood stereo algorithm. *Computer Vision and Image Understanding*, 63(3):542–567.

Crandall, D. and Huttenlocher, D. (2007). Composite models of objects and scenes for category recognition. In *IEEE Computer Society Conference on Computer Vision and Pattern Recognition (CVPR 2007)*, Minneapolis, MN.

Crandall, D., Felzenszwalb, P., and Huttenlocher, D. (2005). Spatial priors for part-based recognition using statistical models. In *IEEE Computer Society Conference on Computer Vision and Pattern Recognition (CVPR'2005)*, pp. 10–17, San Diego, CA.

Crandall, D., Backstrom, L., Huttenlocher, D., and Kleinberg, J. (2009). Mapping the world's photos. In *18th Int. World Wide Web Conference*, pp. 761–770, Madrid.

Crandall, D. J. and Huttenlocher, D. P. (2006). Weakly supervised learning of part-based spatial models for visual object recognition. In *Ninth European Conference on Computer Vision (ECCV 2006)*, pp. 16–29.

Crane, R. (1997). *A Simplified Approach to Image Processing*. Prentice Hall, Upper Saddle River, NJ.

Craswell, N. and Szummer, M. (2007). Random walks on the click graph. In *ACM SIGIR Conference on Research and Development in Informaion Retrieval*, pp. 239–246, New York, NY.

Cremers, D. and Soatto, S. (2005). Motion competition: A variational framework for piecewise parametric motion segmentation. *International Journal of Computer Vision*, 62(3):249–265.

Cremers, D., Rousson, M., and Deriche, R. (2007). A review of statistical approaches to level set segmentation: integrating color, texture, motion and shape. *International Journal of Computer Vision*, 72(2):195–215.

Crevier, D. (1993). *AI: The Tumultuous Search for Artificial Intelligence*. BasicBooks, New York, NY.

Criminisi, A., Pérez, P., and Toyama, K. (2004). Region filling and object removal by exemplar-based inpainting. *IEEE Transactions on Image Processing*, 13(9):1200–1212.

Criminisi, A., Reid, I., and Zisserman, A. (2000). Single view metrology. *International Journal of Computer Vision*, 40(2):123–148.

Criminisi, A., Sharp, T., and Blake, A. (2008). Geos: Geodesic image segmentation. In *Tenth European Conference on Computer Vision (ECCV 2008)*, pp. 99–112, Marseilles.

Criminisi, A., Cross, G., Blake, A., and Kolmogorov, V. (2006). Bilayer segmentation of live video. In *IEEE Computer Society Conference on Computer Vision and Pattern Recognition (CVPR'2006)*, pp. 53–60, New York City, NY.

Criminisi, A., Shotton, J., Blake, A., and Torr, P. (2003). Gaze manipulation for one-to-one teleconferencing. In *Ninth International Conference on Computer Vision (ICCV 2003)*, pp. 191–198, Nice, France.

Criminisi, A., Kang, S. B., Swaminathan, R., Szeliski, R., and Anandan, P. (2005). Extracting layers and analyzing their specular properties using epipolar-plane-image analysis. *Computer Vision and Image Understanding*, 97(1):51–85.

Criminisi, A., Shotton, J., Blake, A., Rother, C., and Torr, P. H. S. (2007). Efficient dense stereo with occlusion by four-state dynamic programming. *International Journal of Computer Vision*, 71(1):89–110.

Crow, F. C. (1984). Summed-area table for texture mapping. *Computer Graphics (SIGGRAPH '84)*, 18(3):207–212.

Crowley, J. L. and Stern, R. M. (1984). Fast computation of the difference of low-pass transform. *IEEE Transactions on Pattern Analysis and Machine Intelligence*, 6(2):212–222.

Csurka, G. and Perronnin, F. (2008). A simple high performance approach to semantic segmentation. In *British Machine Vision Conference (BMVC 2008)*, Leeds.

Csurka, G., Dance, C. R., Perronnin, F., and Willamowski, J. (2006). Generic visual categorization using weak geometry. In Ponce, J., Hebert, M., Schmid, C., and Zisserman, A. (eds), *Toward Category-Level Object Recognition*, pp. 207–224, Springer, New York.

Csurka, G., Dance, C. R., Fan, L., Willamowski, J., and Bray, C. (2004). Visual categorization with bags of keypoints. In *ECCV International Workshop on Statistical Learning in Computer Vision*, Prague.

Cui, J., Yang, Q., Wen, F., Wu, Q., Zhang, C., Van Gool, L., and Tang, X. (2008). Transductive object cutout. In *IEEE Computer Society Conference on Computer Vision and Pattern Recognition (CVPR 2008)*, Anchorage, AK.

Curless, B. (1999). From range scans to 3D models. *Computer Graphics*, 33(4):38–41.

Curless, B. and Levoy, M. (1995). Better optical triangulation through spacetime analysis. In *Fifth International Conference on Computer Vision (ICCV'95)*, pp. 987–994, Cambridge, Massachusetts.

Curless, B. and Levoy, M. (1996). A volumetric method for building complex models from range images. In *ACM SIGGRAPH 1996 Conference Proceedings*, pp. 303–312, New Orleans.

Cutler, R. and Davis, L. S. (2000). Robust real-time periodic motion detection, analysis, and applications. *IEEE Transactions on Pattern Analysis and Machine Intelligence*, 22(8):781–796.

Cutler, R. and Turk, M. (1998). View-based interpretation of real-time optical flow for gesture recognition. In *IEEE International Conference on Automatic Face and Gesture Recognition*, pp. 416–421, Nara, Japan.

Dai, S., Baker, S., and Kang, S. B. (2009). An MRF-based deinterlacing algorithm with exemplar-based refinement. *IEEE Transactions on Image Processing*, 18(5):956–968.

Dalal, N. and Triggs, B. (2005). Histograms of oriented gradients for human detection. In *IEEE Computer Society Conference on Computer Vision and Pattern Recognition (CVPR'2005)*, pp. 886–893, San Diego, CA.

Dalal, N., Triggs, B., and Schmid, C. (2006). Human detection using oriented histograms of flow and appearance. In *Ninth European Conference on Computer Vision (ECCV 2006)*, pp. 428–441.

Dana, K. J., van Ginneken, B., Nayar, S. K., and Koenderink, J. J. (1999). Reflectance and texture of real world surfaces. *ACM Transactions on Graphics*, 18(1):1–34.

Danielsson, P. E. (1980). Euclidean distance mapping. *Computer Graphics and Image Processing*, 14(3):227–248.

Darrell, T. and Pentland, A. (1991). Robust estimation of a multi-layered motion representation. In *IEEE Workshop on Visual Motion*, pp. 173–178, Princeton, New Jersey.

Darrell, T. and Pentland, A. (1995). Cooperative robust estimation using layers of support. *IEEE Transactions on Pattern Analysis and Machine Intelligence*, 17(5):474–487.

Darrell, T. and Simoncelli, E. (1993). "Nulling" filters and the separation of transparent motion. In *IEEE Computer Society Conference on Computer Vision and Pattern Recognition (CVPR'93)*, pp. 738–739, New York.

Darrell, T., Gordon, G., Harville, M., and Woodfill, J. (2000). Integrated person tracking using stereo, color, and pattern detection. *International Journal of Computer Vision*, 37(2):175–185.

Darrell, T., Baker, H., Crow, F., Gordon, G., and Woodfill, J. (1997). Magic morphin mirror: face-sensitive distortion and exaggeration. In *ACM SIGGRAPH 1997 Visual Proceedings*, Los Angeles.

Datta, R., Joshi, D., Li, J., and Wang, J. Z. (2008). Image retrieval: Ideas, influences, and trends of the new age. *ACM Computing Surveys*, 40(2).

Daugman, J. (2004). How iris recognition works. *IEEE Transactions on Circuits and Systems for Video Technology*, 14(1):21–30.

David, P., DeMenthon, D., Duraiswami, R., and Samet, H. (2004). SoftPOSIT: Simultaneous pose and correspondence determination. *International Journal of Computer Vision*, 59(3):259–284.

Davies, R., Twining, C., and Taylor, C. (2008). *Statistical Models of Shape*. Springer-Verlag, London.

Davis, J. (1998). Mosaics of scenes with moving objects. In *IEEE Computer Society Conference on Computer Vision and Pattern Recognition (CVPR'98)*, pp. 354–360, Santa Barbara.

Davis, J., Ramamoorthi, R., and Rusinkiewicz, S. (2003). Spacetime stereo: A unifying framework for depth from triangulation. In *IEEE Computer Society Conference on Computer Vision and Pattern Recognition (CVPR'2003)*, pp. 359–366, Madison, WI.

Davis, J., Nahab, D., Ramamoorthi, R., and Rusinkiewicz, S. (2005). Spacetime stereo: A unifying framework for depth from triangulation. *IEEE Transactions on Pattern Analysis and Machine Intelligence*, 27(2):296–302.

Davis, L. (1975). A survey of edge detection techniques. *Computer Graphics and Image Processing*, 4(3):248–270.

Davis, T. A. (2006). *Direct Methods for Sparse Linear Systems*. SIAM.

Davis, T. A. (2008). Multifrontal multithreaded rank-revealing sparse QR factorization. *ACM Trans. on Mathematical Software*, (submitted).

Davison, A., Reid, I., Molton, N. D., and Stasse, O. (2007). MonoSLAM: Real-time single camera SLAM. *IEEE Transactions on Pattern Analysis and Machine Intelligence*, 29(6):1052–1067.

de Agapito, L., Hayman, E., and Reid, I. (2001). Self-calibration of rotating and zooming cameras. *International Journal of Computer Vision*, 45(2):107–127.

de Berg, M., Cheong, O., van Kreveld, M., and Overmars, M. (2006). *Computational Geometry: Algorithms and Applications*. Springer, New York, NY, third edition.

De Bonet, J. (1997). Multiresolution sampling procedure for analysis and synthesis of texture images. In *ACM SIGGRAPH 1997 Conference Proceedings*, pp. 361–368, Los Angeles.

De Bonet, J. S. and Viola, P. (1999). Poxels: Probabilistic voxelized volume reconstruction. In *Seventh International Conference on Computer Vision (ICCV'99)*, pp. 418–425, Kerkyra, Greece.

De Castro, E. and Morandi, C. (1987). Registration of translated and rotated images using finite Fourier transforms. *IEEE Transactions on Pattern Analysis and Machine Intelligence*, PAMI-9(5):700–703.

de Haan, G. and Bellers, E. B. (1998). Deinterlacing—an overview. *Proceedings of the IEEE*, 86:1839–1857.

De la Torre, F. and Black, M. J. (2003). A framework for robust subspace learning. *International Journal of Computer Vision*, 54(1/2/3):117–142.

Debevec, P. (1998). Rendering synthetic objects into real scenes: Bridging traditional and image-based graphics with global illumination and high dynamic range photography. In *ACM SIGGRAPH 1998 Conference Proceedings*, pp. 189–198.

Debevec, P. (2006). Virtual cinematography: Relighting through computation. *Computer*, 39(8):57–65.

Debevec, P., Hawkins, T., Tchou, C., Duiker, H.-P., Sarokin, W., and Sagar, M. (2000). Acquiring the reflectance field of a human face. In *ACM SIGGRAPH 2000 Conference Proceedings*, pp. 145–156.

Debevec, P., Wenger, A., Tchou, C., Gardner, A., Waese, J., and Hawkins, T. (2002). A lighting reproduction approach to live-action compositing. *ACM Transactions on Graphics (Proc. SIGGRAPH 2002)*, 21(3):547–556.

Debevec, P. E. (1999). Image-based modeling and lighting. *Computer Graphics*, 33(4):46–50.

Debevec, P. E. and Malik, J. (1997). Recovering high dynamic range radiance maps from photographs. In *ACM SIGGRAPH 1997 Conference Proceedings*, pp. 369–378.

Debevec, P. E., Taylor, C. J., and Malik, J. (1996). Modeling and rendering architecture from photographs: A hybrid geometry- and image-based approach. In *ACM SIGGRAPH 1996 Conference Proceedings*, pp. 11–20, New Orleans.

Debevec, P. E., Yu, Y., and Borshukov, G. D. (1998). Efficient view-dependent image-based rendering with projective texture-mapping. In *Eurographics Rendering Workshop 1998*, pp. 105–116.

DeCarlo, D. and Santella, A. (2002). Stylization and abstraction of photographs. *ACM Transactions on Graphics (Proc. SIGGRAPH 2002)*, 21(3):769–776.

DeCarlo, D., Metaxas, D., and Stone, M. (1998). An anthropometric face model using variational techniques. In *ACM SIGGRAPH 1998 Conference Proceedings*, pp. 67–74.

Delingette, H., Hebert, M., and Ikeuichi, K. (1992). Shape representation and image segmentation using deformable surfaces. *Image and Vision Computing*, 10(3):132–144.

Dellaert, F. and Collins, R. (1999). Fast image-based tracking by selective pixel integration. In *ICCV Workshop on Frame-Rate Vision*, pp. 1–22.

Delong, A., Osokin, A., Isack, H. N., and Boykov, Y. (2010). Fast approximate energy minimization with label costs. In *IEEE Computer Society Conference on Computer Vision and Pattern Recognition (CVPR 2010)*, San Francisco, CA.

DeMenthon, D. I. and Davis, L. S. (1995). Model-based object pose in 25 lines of code. *International Journal of Computer Vision*, 15(1-2):123–141.

Demmel, J., Dongarra, J., Eijkhout, V., Fuentes, E., Petitet, A. *et al.* (2005). Self-adapting linear algebra algorithms and software. *Proceedings of the IEEE*, 93(2):293–312.

Dempster, A., Laird, N. M., and Rubin, D. B. (1977). Maximum likelihood from incomplete data via the EM algorithm. *Journal of the Royal Statistical Society B*, 39(1):1–38.

Deng, J., Dong, W., Socher, R., Li, L.-J., Li, K., and Fei-Fei, L. (2009). ImageNet: A large-scale hierarchical image database. In *IEEE Computer Society Conference on Computer Vision and Pattern Recognition (CVPR 2009)*, Miami Beach, FL.

Deriche, R. (1987). Using Canny's criteria to derive a recursively implemented optimal edge detector. *International Journal of Computer Vision*, 1(2):167–187.

Deriche, R. (1990). Fast algorithms for low-level vision. *IEEE Transactions on Pattern Analysis and Machine Intelligence*, 12(1):78–87.

Deutscher, J. and Reid, I. (2005). Articulated body motion capture by stochastic search. *International Journal of Computer Vision*, 61(2):185–205.

Deutscher, J., Blake, A., and Reid, I. (2000). Articulated body motion capture by annealed particle filtering. In *IEEE Computer Society Conference on Computer Vision and Pattern Recognition (CVPR'2000)*, pp. 126–133, Hilton Head Island.

Dev, P. (1974). *Segmentation Processes in Visual Perception: A Cooperative Neural Model*. COINS Technical Report 74C-5, University of Massachusetts at Amherst.

Dhond, U. R. and Aggarwal, J. K. (1989). Structure from stereo—a review. *IEEE Transactions on Systems, Man, and Cybernetics*, 19(6):1489–1510.

Dick, A., Torr, P. H. S., and Cipolla, R. (2004). Modelling and interpretation of architecture from several images. *International Journal of Computer Vision*, 60(2):111–134.

Dickinson, S., Leonardis, A., Schiele, B., and Tarr, M. J. (eds). (2007). *Object Categorization: Computer and Human Vision Perspectives*, Cambridge University Press, New York.

Dickmanns, E. D. and Graefe, V. (1988). Dynamic monocular machine vision. *Machine Vision and Applications*, 1:223–240.

Diebel, J. (2006). *Representing Attitude: Euler Angles, Quaternions, and Rotation Vectors*. Technical Report, Stanford University. http://ai.stanford.edu/~diebel/attitude.html.

Diebel, J. R., Thrun, S., and Brünig, M. (2006). A Bayesian method for probable surface reconstruction and decimation. *ACM Transactions on Graphics*, 25(1).

Dimitrijevic, M., Lepetit, V., and Fua, P. (2006). Human body pose detection using Bayesian spatio-temporal templates. *Computer Vision and Image Understanding*, 104(2-3):127–139.

Dinh, H. Q., Turk, G., and Slabaugh, G. (2002). Reconstructing surfaces by volumetric regularization using radial basis functions. *IEEE Transactions on Pattern Analysis and Machine Intelligence*, 24(10):1358–1371.

Divvala, S., Hoiem, D., Hays, J., Efros, A. A., and Hebert, M. (2009). An empirical study of context in object detection. In *IEEE Computer Society Conference on Computer Vision and Pattern Recognition (CVPR 2009)*, Miami, FL.

Dodgson, N. A. (1992). *Image Resampling*. Technical Report TR261, Wolfson College and Computer Laboratory, University of Cambridge.

Dollàr, P., Belongie, S., and Perona, P. (2010). The fastest pedestrian detector in the west. In *British Machine Vision Conference (BMVC 2010)*, Aberystwyth, Wales, UK.

Dollàr, P., Wojek, C., Schiele, B., and Perona, P. (2009). Pedestrian detection: A benchmark. In *IEEE Computer Society Conference on Computer Vision and Pattern Recognition (CVPR 2009)*, Miami Beach, FL.

Doretto, G. and Soatto, S. (2006). Dynamic shape and appearance models. *IEEE Transactions on Pattern Analysis and Machine Intelligence*, 28(12):2006–2019.

Doretto, G., Chiuso, A., Wu, Y. N., and Soatto, S. (2003). Dynamic textures. *International Journal of Computer Vision*, 51(2):91–109.

Dorkó, G. and Schmid, C. (2003). Selection of scale-invariant parts for object class recognition. In *Ninth International Conference on Computer Vision (ICCV 2003)*, pp. 634–640, Nice, France.

Dorsey, J., Rushmeier, H., and Sillion, F. (2007). *Digital Modeling of Material Appearance*. Morgan Kaufmann, San Francisco.

Douglas, D. H. and Peucker, T. K. (1973). Algorithms for the reduction of the number of points required to represent a digitized line or its caricature. *The Canadian Cartographer*, 10(2):112–122.

Drori, I., Cohen-Or, D., and Yeshurun, H. (2003). Fragment-based image completion. *ACM Transactions on Graphics (Proc. SIGGRAPH 2003)*, 22(3):303–312.

Duda, R. O. and Hart, P. E. (1972). Use of the Hough transform to detect lines and curves in pictures. *Communications of the ACM*, 15(1):11–15.

Duda, R. O., Hart, P. E., and Stork, D. G. (2001). *Pattern Classification*. John Wiley & Sons, New York, 2nd edition.

Dupuis, P. and Oliensis, J. (1994). An optimal control formulation and related numerical methods for a problem in shape reconstruction. *Annals of Applied Probability*, 4(2):287–346.

Durand, F. and Dorsey, J. (2002). Fast bilateral filtering for the display of high-dynamic-range images. *ACM Transactions on Graphics (Proc. SIGGRAPH 2002)*, 21(3):257–266.

Durand, F. and Szeliski, R. (2007). Computational photography. *IEEE Computer Graphics and Applications*, 27(2):21–22. Guest Editors' Introduction to Special Issue.

Durbin, R. and Willshaw, D. (1987). An analogue approach to the traveling salesman problem using an elastic net method. *Nature*, 326:689–691.

Durbin, R., Szeliski, R., and Yuille, A. (1989). An analysis of the elastic net approach to the travelling salesman problem. *Neural Computation*, 1(3):348–358.

Eck, M., DeRose, T., Duchamp, T., Hoppe, H., Lounsbery, M., and Stuetzle, W. (1995). Multiresolution analysis of arbitrary meshes. In *ACM SIGGRAPH 1995 Conference Proceedings*, pp. 173–182, Los Angeles.

Eden, A., Uyttendaele, M., and Szeliski, R. (2006). Seamless image stitching of scenes with large motions and exposure differences. In *IEEE Computer Society Conference on Computer Vision and Pattern Recognition (CVPR'2006)*, pp. 2498–2505, New York, NY.

Efros, A. A. and Freeman, W. T. (2001). Image quilting for texture synthesis and transfer. In *ACM SIGGRAPH 2001 Conference Proceedings*, pp. 341–346.

Efros, A. A. and Leung, T. K. (1999). Texture synthesis by non-parametric sampling. In *Seventh International Conference on Computer Vision (ICCV'99)*, pp. 1033–1038, Kerkyra, Greece.

Efros, A. A., Berg, A. C., Mori, G., and Malik, J. (2003). Recognizing action at a distance. In *Ninth International Conference on Computer Vision (ICCV 2003)*, pp. 726–733, Nice, France.

Eichner, M. and Ferrari, V. (2009). Better appearance models for pictorial structures. In *British Machine Vision Conference (BMVC 2009)*.

Eisemann, E. and Durand, F. (2004). Flash photography enhancement via intrinsic relighting. *ACM Transactions on Graphics*, 23(3):673–678.

Eisert, P., Steinbach, E., and Girod, B. (2000). Automatic reconstruction of stationary 3-D objects from multiple uncalibrated camera views. *IEEE Transactions on Circuits and Systems for Video Technology*, 10(2):261–277.

Eisert, P., Wiegand, T., and Girod, B. (2000). Model-aided coding: a new approach to incorporate facial animation into motion-compensated video coding. *IEEE Transactions on Circuits and Systems for Video Technology*, 10(3):344–358.

Ekman, P. and Friesen, W. V. (1978). *Facial Action Coding System: A Technique for the Measurement of Facial Movement*. Consulting Psychologists press, Palo Alto, CA.

El-Melegy, M. and Farag, A. (2003). Nonmetric lens distortion calibration: Closed-form solutions, robust estimation and model selection. In *Ninth International Conference on Computer Vision (ICCV 2003)*, pp. 554–559, Nice, France.

Elder, J. H. (1999). Are edges incomplete? *International Journal of Computer Vision*, 34(2/3):97–122.

Elder, J. H. and Goldberg, R. M. (2001). Image editing in the contour domain. *IEEE Transactions on Pattern Analysis and Machine Intelligence*, 23(3):291–296.

Elder, J. H. and Zucker, S. W. (1998). Local scale control for edge detection and blur estimation. *IEEE Transactions on Pattern Analysis and Machine Intelligence*, 20(7):699–716.

Engels, C., Stewénius, H., and Nistér, D. (2006). Bundle adjustment rules. In *Photogrammetric Computer Vision (PCV'06)*, Bonn, Germany.

Engl, H. W., Hanke, M., and Neubauer, A. (1996). *Regularization of Inverse Problems*. Kluwer Academic Publishers, Dordrecht.

Enqvist, O., Josephson, K., and Kahl, F. (2009). Optimal correspondences from pairwise constraints. In *Twelfth International Conference on Computer Vision (ICCV 2009)*, Kyoto, Japan.

Estrada, F. J. and Jepson, A. D. (2009). Benchmarking image segmentation algorithms. *International Journal of Computer Vision*, 85(2):167–181.

Estrada, F. J., Jepson, A. D., and Chennubhotla, C. (2004). Spectral embedding and min-cut for image segmentation. In *British Machine Vision Conference (BMVC 2004)*, pp. 317–326, London.

Evangelidis, G. D. and Psarakis, E. Z. (2008). Parametric image alignment using enhanced correlation coefficient maximization. *IEEE Transactions on Pattern Analysis and Machine Intelligence*, 30(10):1858–1865.

Everingham, M., Van Gool, L., Williams, C. K. I., Winn, J., and Zisserman, A. (2008). The PASCAL Visual Object Classes Challenge 2008 (VOC2008) Results. http://www.pascal-network.org/challenges/VOC/voc2008/workshop/index.html.

Everingham, M., Van Gool, L., Williams, C. K. I., Winn, J., and Zisserman, A. (2010). The PASCAL visual object classes (VOC) challenge. *International Journal of Computer Vision*, 88(2):147–168.

Ezzat, T., Geiger, G., and Poggio, T. (2002). Trainable videorealistic speech animation. *ACM Transactions on Graphics (Proc. SIGGRAPH 2002)*, 21(3):388–398.

Fabbri, R., Costa, L. D. F., Torelli, J. C., and Bruno, O. M. (2008). 2D Euclidean distance transform algorithms: A comparative survey. *ACM Computing Surveys*, 40(1).

Fairchild, M. D. (2005). *Color Appearance Models*. Wiley, 2nd edition.

Fan, R.-E., Chen, P.-H., and Lin, C.-J. (2005). Working set selection using second order information for training support vector machines. *Journal of Machine Learning Research*, 6:1889–1918.

Fan, R.-E., Chang, K.-W., Hsieh, C.-J., Wang, X.-R., and Lin, C.-J. (2008). LIBLINEAR: A library for large linear classification. *Journal of Machine Learning Research*, 9:1871–1874.

Farbman, Z., Fattal, R., Lischinski, D., and Szeliski, R. (2008). Edge-preserving decompositions for multi-scale tone and detail manipulation. *ACM Transactions on Graphics (Proc. SIGGRAPH 2008)*, 27(3).

Farenzena, M., Fusiello, A., and Gherardi, R. (2009). Structure-and-motion pipeline on a hierarchical cluster tree. In *IEEE International Workshop on 3D Digital Imaging and Modeling (3DIM 2009)*, Kyoto, Japan.

Farin, G. (1992). From conics to NURBS: A tutorial and survey. *IEEE Computer Graphics and Applications*, 12(5):78–86.

Farin, G. E. (1996). *Curves and Surfaces for Computer Aided Geometric Design: A Practical Guide*. Academic Press, Boston, Massachusetts, 4th edition.

Fattal, R. (2007). Image upsampling via imposed edge statistics. *ACM Transactions on Graphics*, 26(3).

Fattal, R. (2009). Edge-avoiding wavelets and their applications. *ACM Transactions on Graphics*, 28(3).

Fattal, R., Lischinski, D., and Werman, M. (2002). Gradient domain high dynamic range compression. *ACM Transactions on Graphics (Proc. SIGGRAPH 2002)*, 21(3):249–256.

Faugeras, O. (1993). *Three-dimensional computer vision: A geometric viewpoint*. MIT Press, Cambridge, Massachusetts.

Faugeras, O. and Keriven, R. (1998). Variational principles, surface evolution, PDEs, level set methods, and the stereo problem. *IEEE Transactions on Image Processing*, 7(3):336–344.

Faugeras, O. and Luong, Q.-T. (2001). *The Geometry of Multiple Images*. MIT Press, Cambridge, MA.

Faugeras, O. D. (1992). What can be seen in three dimensions with an uncalibrated stereo rig? In *Second European Conference on Computer Vision (ECCV'92)*, pp. 563–578, Santa Margherita Liguere, Italy.

Faugeras, O. D. and Hebert, M. (1987). The representation, recognition and positioning of 3-D shapes from range data. In Kanade, T. (ed.), *Three-Dimensional Machine Vision*, pp. 301–353, Kluwer Academic Publishers, Boston.

Faugeras, O. D., Luong, Q.-T., and Maybank, S. J. (1992). Camera self-calibration: Theory and experiments. In *Second European Conference on Computer Vision (ECCV'92)*, pp. 321–334, Santa Margherita Liguere, Italy.

Favaro, P. and Soatto, S. (2006). *3-D Shape Estimation and Image Restoration: Exploiting Defocus and Motion-Blur*. Springer.

Fawcett, T. (2006). An introduction to ROC analysis. *Pattern Recognition Letters*, 27(8):861–874.

Fei-Fei, L. and Perona, P. (2005). A Bayesian hierarchical model for learning natural scene categories. In *IEEE Computer Society Conference on Computer Vision and Pattern Recognition (CVPR'2005)*, pp. 524–531, San Diego, CA.

Fei-Fei, L., Fergus, R., and Perona, P. (2006). One-shot learning of object categories. *IEEE Transactions on Pattern Analysis and Machine Intelligence*, 28(4):594–611.

Fei-Fei, L., Fergus, R., and Torralba, A. (2009). ICCV 2009 short course on recognizing and learning object categories. In *Twelfth International Conference on Computer Vision (ICCV 2009)*, Kyoto, Japan. http://people.csail.mit.edu/torralba/shortCourseRLOC/.

Feilner, M., Van De Ville, D., and Unser, M. (2005). An orthogonal family of quincunx wavelets with continuously adjustable order. *IEEE Transactions on Image Processing*, 14(4):499–520.

Feldmar, J. and Ayache, N. (1996). Rigid, affine, and locally affine registration of free-form surfaces. *International Journal of Computer Vision*, 18(2):99–119.

Fellbaum, C. (ed.). (1998). *WordNet: An Electronic Lexical Database*, Bradford Books.

Felzenszwalb, P., McAllester, D., and Ramanan, D. (2008). A discriminatively trained, multiscale, deformable part model. In *IEEE Computer Society Conference on Computer Vision and Pattern Recognition (CVPR 2008)*, Anchorage, AK.

Felzenszwalb, P. F. and Huttenlocher, D. P. (2004a). *Distance Transforms of Sampled Functions*. Technical Report TR2004-1963, Cornell University Computing and Information Science.

Felzenszwalb, P. F. and Huttenlocher, D. P. (2004b). Efficient graph-based image segmentation. *International Journal of Computer Vision*, 59(2):167–181.

Felzenszwalb, P. F. and Huttenlocher, D. P. (2005). Pictorial structures for object recognition. *International Journal of Computer Vision*, 61(1):55–79.

Felzenszwalb, P. F. and Huttenlocher, D. P. (2006). Efficient belief propagation for early vision. *International Journal of Computer Vision*, 70(1):41–54.

Felzenszwalb, P. F., Girshick, R. B., McAllester, D., and Ramanan, D. (2010). Object detection with discriminatively trained part-based models. *IEEE Transactions on Pattern Analysis and Machine Intelligence*, 32(9):1627–1645.

Ferencz, A., Learned-Miller, E. G., and Malik, J. (2008). Learning to locate informative features for visual identification. *International Journal of Computer Vision*, 77(1-3):3–24.

Fergus, R. (2007a). Combined segmentation and recognition. In *CVPR 2007 Short Course on Recognizing and Learning Object Categories*. http://people.csail.mit.edu/torralba/shortCourseRLOC/.

Fergus, R. (2007b). Part-based models. In *CVPR 2007 Short Course on Recognizing and Learning Object Categories*. http://people.csail.mit.edu/torralba/shortCourseRLOC/.

Fergus, R. (2009). Classical methods for object recognition. In *ICCV 2009 Short Course on Recognizing and Learning Object Categories*, Kyoto, Japan. http://people.csail.mit.edu/torralba/shortCourseRLOC/.

Fergus, R., Perona, P., and Zisserman, A. (2004). A visual category filter for Google images. In *Eighth European Conference on Computer Vision (ECCV 2004)*, pp. 242–256, Prague.

Fergus, R., Perona, P., and Zisserman, A. (2005). A sparse object category model for efficient learning and exhaustive recognition. In *IEEE Computer Society Conference on Computer Vision and Pattern Recognition (CVPR'2005)*, pp. 380–387, San Diego, CA.

Fergus, R., Perona, P., and Zisserman, A. (2007). Weakly supervised scale-invariant learning of models for visual recognition. *International Journal of Computer Vision*, 71(3):273–303.

Fergus, R., Fei-Fei, L., Perona, P., and Zisserman, A. (2005). Learning object categories from Google's image search. In *Tenth International Conference on Computer Vision (ICCV 2005)*, pp. 1816–1823, Beijing, China.

Fergus, R., Singh, B., Hertzmann, A., Roweis, S. T., and Freeman, W. T. (2006). Removing camera shake from a single photograph. *ACM Transactions on Graphics*, 25(3):787–794.

Ferrari, V., Marin-Jimenez, M., and Zisserman, A. (2009). Pose search: retrieving people using their pose. In *IEEE Computer Society Conference on Computer Vision and Pattern Recognition (CVPR 2009)*, Miami Beach, FL.

Ferrari, V., Marin-Jimenez, M. J., and Zisserman, A. (2008). Progressive search space reduction for human pose estimation. In *IEEE Computer Society Conference on Computer Vision and Pattern Recognition (CVPR 2008)*, Anchorage, AK.

Ferrari, V., Tuytelaars, T., and Van Gool, L. (2006a). Object detection by contour segment networks. In *Ninth European Conference on Computer Vision (ECCV 2006)*, pp. 14–28.

Ferrari, V., Tuytelaars, T., and Van Gool, L. (2006b). Simultaneous object recognition and segmentation from single or multiple model views. *International Journal of Computer Vision*, 67(2):159–188.

Field, D. J. (1987). Relations between the statistics of natural images and the response properties of cortical cells. *Journal of the Optical Society of America A*, 4(12):2379–2394.

Finkelstein, A. and Salesin, D. H. (1994). Multiresolution curves. In *ACM SIGGRAPH 1994 Conference Proceedings*, pp. 261–268.

Fischler, M. A. and Bolles, R. C. (1981). Random sample consensus: A paradigm for model fitting with applications to image analysis and automated cartography. *Communications of the ACM*, 24(6):381–395.

Fischler, M. A. and Elschlager, R. A. (1973). The representation and matching of pictorial structures. *IEEE Transactions on Computers*, 22(1):67–92.

Fischler, M. A. and Firschein, O. (1987). *Readings in Computer Vision*. Morgan Kaufmann Publishers, Inc., Los Altos.

Fischler, M. A., Firschein, O., Barnard, S. T., Fua, P. V., and Leclerc, Y. (1989). *The Vision Problem: Exploiting Parallel Computation*. Technical Note 458, SRI International, Menlo Park.

Fitzgibbon, A. W. and Zisserman, A. (1998). Automatic camera recovery for closed and open image sequences. In *Fifth European Conference on Computer Vision (ECCV'98)*, pp. 311–326, Freiburg, Germany.

Fitzgibbon, A. W., Cross, G., and Zisserman, A. (1998). Automatic 3D model construction for turn-table sequences. In *European Workshop on 3D Structure from Multiple Images of Large-Scale Environments (SMILE)*, pp. 155–170, Freiburg.

Fleet, D. and Jepson, A. (1990). Computation of component image velocity from local phase information. *International Journal of Computer Vision*, 5(1):77–104.

Fleuret, F. and Geman, D. (2001). Coarse-to-fine face detection. *International Journal of Computer Vision*, 41(1/2):85–107.

Flickner, M., Sawhney, H., Niblack, W., Ashley, J., Huang, Q. *et al.* (1995). Query by image and video content: The QBIC system. *Computer*, 28(9):23–32.

Foley, J. D., van Dam, A., Feiner, S. K., and Hughes, J. F. (1995). *Computer Graphics: Principles and Practice*. Addison-Wesley, Reading, MA, 2 edition.

Förstner, W. (1986). A feature-based correspondence algorithm for image matching. *Intl. Arch. Photogrammetry & Remote Sensing*, 26(3):150–166.

Förstner, W. (2005). Uncertainty and projective geometry. In Bayro-Corrochano, E. (ed.), *Handbook of Geometric Computing*, pp. 493–534, Springer, New York.

Forsyth, D. and Ponce, J. (2003). *Computer Vision: A Modern Approach*. Prentice Hall, Upper Saddle River, NJ.

Forsyth, D. A. and Fleck, M. M. (1999). Automatic detection of human nudes. *International Journal of Computer Vision*, 32(1):63–77.

Forsyth, D. A., Arikan, O., Ikemoto, L., O'Brien, J., and Ramanan, D. (2006). Computational studies of human motion: Part 1, tracking and motion synthesis. *Foundations and Trends in Computer Graphics and Computer Vision*, 1(2/3):77–254.

Fossati, A., Dimitrijevic, M., Lepetit, V., and Fua, P. (2007). Bridging the gap between detection and tracking for 3D monocular video-based motion capture. In *IEEE Computer Society Conference on Computer Vision and Pattern Recognition (CVPR 2007)*, Minneapolis, MN.

Frahm, J.-M. and Koch, R. (2003). Camera calibration with known rotation. In *Ninth International Conference on Computer Vision (ICCV 2003)*, pp. 1418–1425, Nice, France.

Freeman, M. (2008). *Mastering HDR Photography*. Amphoto Books, New York.

Freeman, W., Perona, P., and Schölkopf, B. (2008). Guest editorial: Special issue on machine learning for vision. *International Journal of Computer Vision*, 77(1-3):1.

Freeman, W. T. (1992). *Steerable Filters and Local Analysis of Image Structure*. Ph.D. thesis, Massachusetts Institute of Technology.

Freeman, W. T. and Adelson, E. H. (1991). The design and use of steerable filters. *IEEE Transactions on Pattern Analysis and Machine Intelligence*, 13(9):891–906.

Freeman, W. T., Jones, T. R., and Pasztor, E. C. (2002). Example-based super-resolution. *IEEE Computer Graphics and Applications*, 22(2):56–65.

Freeman, W. T., Pasztor, E. C., and Carmichael, O. T. (2000). Learning low-level vision. *International Journal of Computer Vision*, 40(1):25–47.

Frey, B. J. and MacKay, D. J. C. (1997). A revolution: Belief propagation in graphs with cycles. In *Advances in Neural Information Processing Systems*.

Friedman, J., Hastie, T., and Tibshirani, R. (2000). Additive logistic regression: a statistical view of boosting. *Annals of Statistics*, 38(2):337–374.

Frisken, S. F., Perry, R. N., Rockwood, A. P., and Jones, T. R. (2000). Adaptively sampled distance fields: A general representation of shape for computer graphics. In *ACM SIGGRAPH 2000 Conference Proceedings*, pp. 249–254.

Fritz, M. and Schiele, B. (2008). Decomposition, discovery and detection of visual categories using topic models. In *IEEE Computer Society Conference on Computer Vision and Pattern Recognition (CVPR 2008)*, Anchorage, AK.

Frome, A., Singer, Y., Sha, F., and Malik, J. (2007). Learning globally-consistent local distance functions for shape-based image retrieval and classification. In *Eleventh International Conference on Computer Vision (ICCV 2007)*, Rio de Janeiro, Brazil.

Fua, P. (1993). A parallel stereo algorithm that produces dense depth maps and preserves image features. *Machine Vision and Applications*, 6(1):35–49.

Fua, P. and Leclerc, Y. G. (1995). Object-centered surface reconstruction: Combining multi-image stereo and shading. *International Journal of Computer Vision*, 16(1):35–56.

Fua, P. and Sander, P. (1992). Segmenting unstructured 3D points into surfaces. In *Second European Conference on Computer Vision (ECCV'92)*, pp. 676–680, Santa Margherita Liguere, Italy.

Fuh, C.-S. and Maragos, P. (1991). Motion displacement estimation using an affine model for image matching. *Optical Engineering*, 30(7):881–887.

Fukunaga, K. and Hostetler, L. D. (1975). The estimation of the gradient of a density function, with applications in pattern recognition. *IEEE Transactions on Information Theory*, 21:32–40.

Furukawa, Y. and Ponce, J. (2007). Accurate, dense, and robust multi-view stereopsis. In *IEEE Computer Society Conference on Computer Vision and Pattern Recognition (CVPR 2007)*, Minneapolis, MN.

Furukawa, Y. and Ponce, J. (2008). Accurate calibration from multi-view stereo and bundle adjustment. In *IEEE Computer Society Conference on Computer Vision and Pattern Recognition (CVPR 2008)*, Anchorage, AK.

Furukawa, Y. and Ponce, J. (2009). Carved visual hulls for image-based modeling. *International Journal of Computer Vision*, 81(1):53–67.

Furukawa, Y. and Ponce, J. (2011). Accurate, dense, and robust multi-view stereopsis. *IEEE Transactions on Pattern Analysis and Machine Intelligence*.

Furukawa, Y., Curless, B., Seitz, S. M., and Szeliski, R. (2009a). Manhattan-world stereo. In *IEEE Computer Society Conference on Computer Vision and Pattern Recognition (CVPR 2009)*, Miami, FL.

Furukawa, Y., Curless, B., Seitz, S. M., and Szeliski, R. (2009b). Reconstructing building interiors from images. In *Twelfth IEEE International Conference on Computer Vision (ICCV 2009)*, Kyoto, Japan.

Furukawa, Y., Curless, B., Seitz, S. M., and Szeliski, R. (2010). Towards internet-scale multi-view stereo. In *IEEE Computer Society Conference on Computer Vision and Pattern Recognition (CVPR 2010)*, San Francisco, CA.

Fusiello, A., Roberto, V., and Trucco, E. (1997). Efficient stereo with multiple windowing. In *IEEE Computer Society Conference on Computer Vision and Pattern Recognition (CVPR'97)*, pp. 858–863, San Juan, Puerto Rico.

Fusiello, A., Trucco, E., and Verri, A. (2000). A compact algorithm for rectification of stereo pairs. *Machine Vision and Applications*, 12(1):16–22.

Gai, J. and Kang, S. B. (2009). Matte-based restoration of vintage video. *IEEE Transactions on Image Processing*, 18:2185–2197.

Gal, R., Wexler, Y., Ofek, E., Hoppe, H., and Cohen-Or, D. (2010). Seamless montage for texturing models. In *Proceedings of Eurographics 2010*.

Gallagher, A. C. and Chen, T. (2008). Multi-image graph cut clothing segmentation for recognizing people. In *IEEE Computer Society Conference on Computer Vision and Pattern Recognition (CVPR 2008)*, Anchorage, AK.

Gallup, D., Frahm, J.-M., Mordohai, P., and Pollefeys, M. (2008). Variable baseline/resolution stereo. In *IEEE Computer Society Conference on Computer Vision and Pattern Recognition (CVPR 2008)*, Anchorage, AK.

Gamble, E. and Poggio, T. (1987). *Visual integration and detection of discontinuities: the key role of intensity edges.* A. I. Memo 970, Artificial Intelligence Laboratory, Massachusetts Institute of Technology.

Gammeter, S., Bossard, L., Quack, T., and Van Gool, L. (2009). I know what you did last summer: Object-level auto-annotation of holiday snaps. In *Twelfth International Conference on Computer Vision (ICCV 2009)*, Kyoto, Japan.

Gao, W., Chen, Y., Wang, R., Shan, S., and Jiang, D. (2003). Learning and synthesizing MPEG-4 compatible 3-D face animation from video sequence. *IEEE Transactions on Circuits and Systems for Video Technology*, 13(11):1119–1128.

Garding, J. (1992). Shape from texture for smooth curved surfaces in perspective projection. *Journal of Mathematical Imaging and Vision*, 2:329–352.

Gargallo, P., Prados, E., and Sturm, P. (2007). Minimizing the reprojection error in surface reconstruction from images. In *Eleventh International Conference on Computer Vision (ICCV 2007)*, Rio de Janeiro, Brazil.

Gavrila, D. M. (1999). The visual analysis of human movement: A survey. *Computer Vision and Image Understanding*, 73(1):82–98.

Gavrila, D. M. and Davis, L. S. (1996). 3D model-based tracking of humans in action: A multi-view approach. In *IEEE Computer Society Conference on Computer Vision and Pattern Recognition (CVPR'96)*, pp. 73–80, San Francisco.

Gavrila, D. M. and Philomin, V. (1999). Real-time object detection for smart vehicles. In *Seventh International Conference on Computer Vision (ICCV'99)*, pp. 87–93, Kerkyra, Greece.

Geiger, D. and Girosi, F. (1991). Parallel and deterministic algorithms for MRFs: Surface reconstruction. *IEEE Transactions on Pattern Analysis and Machine Intelligence*, 13(5):401–412.

Geiger, D., Ladendorf, B., and Yuille, A. (1992). Occlusions and binocular stereo. In *Second European Conference on Computer Vision (ECCV'92)*, pp. 425–433, Santa Margherita Liguere, Italy.

Gelb, A. (ed.). (1974). *Applied Optimal Estimation.* MIT Press, Cambridge, Massachusetts.

Geller, T. (2008). Overcoming the uncanny valley. *IEEE Computer Graphics and Applications*, 28(4):11–17.

Geman, S. and Geman, D. (1984). Stochastic relaxation, Gibbs distribution, and the Bayesian restoration of images. *IEEE Transactions on Pattern Analysis and Machine Intelligence*, PAMI-6(6):721–741.

Gennert, M. A. (1988). Brightness-based stereo matching. In *Second International Conference on Computer Vision (ICCV'88)*, pp. 139–143, Tampa.

Gersho, A. and Gray, R. M. (1991). *Vector Quantization and Signal Compression*. Springer.

Gershun, A. (1939). The light field. *Journal of Mathematics and Physics*, XVIII:51–151.

Gevers, T., van de Weijer, J., and Stokman, H. (2006). Color feature detection. In Lukac, R. and Plataniotis, K. N. (eds), *Color Image Processing: Methods and Applications*, CRC Press.

Giblin, P. and Weiss, R. (1987). Reconstruction of surfaces from profiles. In *First International Conference on Computer Vision (ICCV'87)*, pp. 136–144, London, England.

Gionis, A., Indyk, P., and Motwani, R. (1999). Similarity search in high dimensions via hashing. In *25th International Conference on Very Large Data Bases (VLDB'99)*, pp. 518–529.

Girod, B., Greiner, G., and Niemann, H. (eds). (2000). *Principles of 3D Image Analysis and Synthesis*, Kluwer, Boston.

Glassner, A. S. (1995). *Principles of Digital Image Synthesis*. Morgan Kaufmann Publishers, San Francisco.

Gleicher, M. (1995). Image snapping. In *ACM SIGGRAPH 1995 Conference Proceedings*, pp. 183–190.

Gleicher, M. (1997). Projective registration with difference decomposition. In *IEEE Computer Society Conference on Computer Vision and Pattern Recognition (CVPR'97)*, pp. 331–337, San Juan, Puerto Rico.

Gleicher, M. and Witkin, A. (1992). Through-the-lens camera control. *Computer Graphics (SIGGRAPH '92)*, 26(2):331–340.

Glocker, B., Komodakis, N., Tziritas, G., Navab, N., and Paragios, N. (2008). Dense image registration through MRFs and efficient linear programming. *Medical Image Analysis*, 12(6):731–741.

Glocker, B., Paragios, N., Komodakis, N., Tziritas, G., and Navab, N. (2008). Optical flow estimation with uncertainties through dynamic MRFs. In *IEEE Computer Society Conference on Computer Vision and Pattern Recognition (CVPR 2008)*, Anchorage, AK.

Gluckman, J. (2006a). Higher order image pyramids. In *Ninth European Conference on Computer Vision (ECCV 2006)*, pp. 308–320.

Gluckman, J. (2006b). Scale variant image pyramids. In *IEEE Computer Society Conference on Computer Vision and Pattern Recognition (CVPR'2006)*, pp. 1069–1075, New York City, NY.

Goesele, M., Curless, B., and Seitz, S. (2006). Multi-view stereo revisited. In *IEEE Computer Society Conference on Computer Vision and Pattern Recognition (CVPR'2006)*, pp. 2402–2409, New York City, NY.

Goesele, M., Fuchs, C., and Seidel, H.-P. (2003). Accuracy of 3D range scanners by measurement of the slanted edge modulation transfer function. In *Fourth International Conference on 3-D Digital Imaging and Modeling*, Banff.

Goesele, M., Snavely, N., Curless, B., Hoppe, H., and Seitz, S. M. (2007). Multi-view stereo for community photo collections. In *Eleventh International Conference on Computer Vision (ICCV 2007)*, Rio de Janeiro, Brazil.

Gold, S., Rangarajan, A., Lu, C., Pappu, S., and Mjolsness, E. (1998). New algorithms for 2D and 3D point matching: Pose estimation and correspondence. *Pattern Recognition*, 31(8):1019–1031.

Goldberg, A. V. and Tarjan, R. E. (1988). A new approach to the maximum-flow problem. *Journal of the ACM*, 35(4):921–940.

Goldluecke, B. and Cremers, D. (2009). Superresolution texture maps for multiview reconstruction. In *Twelfth International Conference on Computer Vision (ICCV 2009)*, Kyoto, Japan.

Goldman, D. B. (2011). Vignette and exposure calibration and compensation. *IEEE Transactions on Pattern Analysis and Machine Intelligence*.

Golovinskiy, A., Matusik, W., ster, H. P., Rusinkiewicz, S., and Funkhouser, T. (2006). A statistical model for synthesis of detailed facial geometry. *ACM Transactions on Graphics*, 25(3):1025–1034.

Golub, G. and Van Loan, C. F. (1996). *Matrix Computation, third edition*. The John Hopkins University Press, Baltimore and London.

Gomes, J. and Velho, L. (1997). *Image Processing for Computer Graphics*. Springer-Verlag, New York.

Gomes, J., Darsa, L., Costa, B., and Velho, L. (1999). *Warping and Morphing of Graphical Objects*. Morgan Kaufmann Publishers, San Francisco.

Gong, M., Yang, R., Wang, L., and Gong, M. (2007). A performance study on different cost aggregation approaches used in realtime stereo matching. *International Journal of Computer Vision*, 75(2):283–296.

Gonzales, R. C. and Woods, R. E. (2008). *Digital Image Processing*. Prentice-Hall, Upper Saddle River, NJ, 3rd edition.

Gooch, B. and Gooch, A. (2001). *Non-Photorealistic Rendering*. A K Peters, Ltd, Natick, Massachusetts.

Gordon, I. and Lowe, D. G. (2006). What and where: 3D object recognition with accurate pose. In Ponce, J., Hebert, M., Schmid, C., and Zisserman, A. (eds), *Toward Category-Level Object Recognition*, pp. 67–82, Springer, New York.

Gorelick, L., Blank, M., Shechtman, E., Irani, M., and Basri, R. (2007). Actions as space-time shapes. *IEEE Transactions on Pattern Analysis and Machine Intelligence*, 29(12):2247–2253.

Gortler, S. J. and Cohen, M. F. (1995). Hierarchical and variational geometric modeling with wavelets. In *Symposium on Interactive 3D Graphics*, pp. 35–43, Monterey, CA.

Gortler, S. J., Grzeszczuk, R., Szeliski, R., and Cohen, M. F. (1996). The Lumigraph. In *ACM SIGGRAPH 1996 Conference Proceedings*, pp. 43–54, New Orleans.

Goshtasby, A. (1989). Correction of image deformation from lens distortion using Bézier patches. *Computer Vision, Graphics, and Image Processing*, 47(4):385–394.

Goshtasby, A. (2005). *2-D and 3-D Image Registration*. Wiley, New York.

Gotchev, A. and Rosenhahn, B. (eds). (2009). *Proceedings of the 3DTV Conference: The True Vision—Capture, Transmission and Display of 3D Video*, IEEE Computer Society Press.

Govindu, V. M. (2006). Revisiting the brightness constraint: Probabilistic formulation and algorithms. In *Ninth European Conference on Computer Vision (ECCV 2006)*, pp. 177–188.

Grady, L. (2006). Random walks for image segmentation. *IEEE Transactions on Pattern Analysis and Machine Intelligence*, 28(11):1768–1783.

Grady, L. (2008). A lattice-preserving multigrid method for solving the inhomogeneous Poisson equations used in image analysis. In *Tenth European Conference on Computer Vision (ECCV 2008)*, pp. 252–264, Marseilles.

Grady, L. and Ali, S. (2008). Fast approximate random walker segmentation using eigenvector precomputation. In *IEEE Computer Society Conference on Computer Vision and Pattern Recognition (CVPR 2008)*, Anchorage, AK.

Grady, L. and Alvino, C. (2008). Reformulating and optimizing the Mumford–Shah functional on a graph — a faster, lower energy solution. In *Tenth European Conference on Computer Vision (ECCV 2008)*, pp. 248–261, Marseilles.

Grauman, K. and Darrell, T. (2005). Efficient image matching with distributions of local invariant features. In *IEEE Computer Society Conference on Computer Vision and Pattern Recognition (CVPR'2005)*, pp. 627–634, San Diego, CA.

Grauman, K. and Darrell, T. (2007a). Pyramid match hashing: Sub-linear time indexing over partial correspondences. In *IEEE Computer Society Conference on Computer Vision and Pattern Recognition (CVPR 2007)*, Minneapolis, MN.

Grauman, K. and Darrell, T. (2007b). The pyramid match kernel: Efficient learning with sets of features. *Journal of Machine Learning Research*, 8:725–760.

Grauman, K., Shakhnarovich, G., and Darrell, T. (2003). Inferring 3D structure with a statistical image-based shape model. In *Ninth International Conference on Computer Vision (ICCV 2003)*, pp. 641–648, Nice, France.

Greene, N. (1986). Environment mapping and other applications of world projections. *IEEE Computer Graphics and Applications*, 6(11):21–29.

Greene, N. and Heckbert, P. (1986). Creating raster Omnimax images from multiple perspective views using the elliptical weighted average filter. *IEEE Computer Graphics and Applications*, 6(6):21–27.

Greig, D., Porteous, B., and Seheult, A. (1989). Exact maximum a posteriori estimation for binary images. *Journal of the Royal Statistical Society, Series B*, 51(2):271–279.

Gremban, K. D., Thorpe, C. E., and Kanade, T. (1988). Geometric camera calibration using systems of linear equations. In *IEEE International Conference on Robotics and Automation*, pp. 562–567, Philadelphia.

Griffin, G., Holub, A., and Perona, P. (2007). *Caltech-256 Object Category Dataset*. Technical Report 7694, California Institute of Technology.

Grimson, W. E. L. (1983). An implementation of a computational theory of visual surface interpolation. *Computer Vision, Graphics, and Image Processing*, 22:39–69.

Grimson, W. E. L. (1985). Computational experiments with a feature based stereo algorithm. *IEEE Transactions on Pattern Analysis and Machine Intelligence*, PAMI-7(1):17–34.

Gross, R., Matthews, I., and Baker, S. (2006). Active appearance models with occlusion. *Image and Vision Computing*, 24(6):593–604.

Gross, R., Shi, J., and Cohn, J. F. (2005). Quo vadis face recognition? In *IEEE Workshop on Empirical Evaluation Methods in Computer Vision*, San Diego.

Gross, R., Baker, S., Matthews, I., and Kanade, T. (2005). Face recognition across pose and illumination. In Li, S. Z. and Jain, A. K. (eds), *Handbook of Face Recognition*, Springer.

Gross, R., Sweeney, L., De la Torre, F., and Baker, S. (2008). Semi-supervised learning of multi-factor models for face de-identification. In *IEEE Computer Society Conference on Computer Vision and Pattern Recognition (CVPR 2008)*, Anchorage, AK.

Gross, R., Matthews, I., Cohn, J., Kanade, T., and Baker, S. (2010). Multi-PIE. *Image and Vision Computing*, 28(5):807–813.

Grossberg, M. D. and Nayar, S. K. (2001). A general imaging model and a method for finding its parameters. In *Eighth International Conference on Computer Vision (ICCV 2001)*, pp. 108–115, Vancouver, Canada.

Grossberg, M. D. and Nayar, S. K. (2004). Modeling the space of camera response functions. *IEEE Transactions on Pattern Analysis and Machine Intelligence*, 26(10):1272–1282.

Gu, C., Lim, J., Arbelaez, P., and Malik, J. (2009). Recognition using regions. In *IEEE Computer Society Conference on Computer Vision and Pattern Recognition (CVPR 2009)*, Miami Beach, FL.

Gu, X., Gortler, S. J., and Hoppe, H. (2002). Geometry images. *ACM Transactions on Graphics*, 21(3):355–361.

Guan, P., Weiss, A., Bǎlan, A. O., and Black, M. J. (2009). Estimating human shape and pose from a single image. In *Twelfth International Conference on Computer Vision (ICCV 2009)*, Kyoto, Japan.

Guennebaud, G. and Gross, M. (2007). Algebraic point set surfaces. *ACM Transactions on Graphics*, 26(3).

Guennebaud, G., Germann, M., and Gross, M. (2008). Dynamic sampling and rendering of algebraic point set surfaces. *Computer Graphics Forum*, 27(2):653–662.

Guenter, B., Grimm, C., Wood, D., Malvar, H., and Pighin, F. (1998). Making faces. In *ACM SIGGRAPH 1998 Conference Proceedings*, pp. 55–66.

Guillaumin, M., Verbeek, J., and Schmid, C. (2009). Is that you? Metric learning approaches for face identification. In *Twelfth International Conference on Computer Vision (ICCV 2009)*, Kyoto, Japan.

Gulbins, J. and Gulbins, R. (2009). *Photographic Multishot Techniques: High Dynamic Range, Super-Resolution, Extended Depth of Field, Stitching*. Rocky Nook.

Habbecke, M. and Kobbelt, L. (2007). A surface-growing approach to multi-view stereo reconstruction. In *IEEE Computer Society Conference on Computer Vision and Pattern Recognition (CVPR 2007)*, Minneapolis, MN.

Hager, G. D. and Belhumeur, P. N. (1998). Efficient region tracking with parametric models of geometry and illumination. *IEEE Transactions on Pattern Analysis and Machine Intelligence*, 20(10):1025–1039.

Hall, R. (1989). *Illumination and Color in Computer Generated Imagery*. Springer-Verlag, New York.

Haller, M., Billinghurst, M., and Thomas, B. (2007). *Emerging Technologies of Augmented Reality: Interfaces and Design*. IGI Publishing.

Hampel, F. R., Ronchetti, E. M., Rousseeuw, P. J., and Stahel, W. A. (1986). *Robust Statistics : The Approach Based on Influence Functions*. Wiley, New York.

Han, F. and Zhu, S.-C. (2005). Bottom-up/top-down image parsing by attribute graph grammar. In *Tenth International Conference on Computer Vision (ICCV 2005)*, pp. 1778–1785, Beijing, China.

Hanna, K. J. (1991). Direct multi-resolution estimation of ego-motion and structure from motion. In *IEEE Workshop on Visual Motion*, pp. 156–162, Princeton, New Jersey.

Hannah, M. J. (1974). *Computer Matching of Areas in Stereo Images*. Ph.D. thesis, Stanford University.

Hannah, M. J. (1988). Test results from SRI's stereo system. In *Image Understanding Workshop*, pp. 740–744, Cambridge, Massachusetts.

Hansen, M., Anandan, P., Dana, K., van der Wal, G., and Burt, P. (1994). Real-time scene stabilization and mosaic construction. In *IEEE Workshop on Applications of Computer Vision (WACV'94)*, pp. 54–62, Sarasota.

Hanson, A. R. and Riseman, E. M. (eds). (1978). *Computer Vision Systems*, Academic Press, New York.

Haralick, R. M. and Shapiro, L. G. (1985). Image segmentation techniques. *Computer Vision, Graphics, and Image Processing*, 29(1):100–132.

Haralick, R. M. and Shapiro, L. G. (1992). *Computer and Robot Vision*. Addison-Wesley, Reading, MA.

Haralick, R. M., Lee, C.-N., Ottenberg, K., and Nölle, M. (1994). Review and analysis of solutions of the three point perspective pose estimation problem. *International Journal of Computer Vision*, 13(3):331–356.

Hardie, R. C., Barnard, K. J., and Armstrong, E. E. (1997). Joint MAP registration and high-resolution image estimation using a sequence of undersampled images. *IEEE Transactions on Image Processing*, 6(12):1621–1633.

Haritaoglu, I., Harwood, D., and Davis, L. S. (2000). W^4: Real-time surveillance of people and their activities. *IEEE Transactions on Pattern Analysis and Machine Intelligence*, 22(8):809–830.

Harker, M. and O'Leary, P. (2008). Least squares surface reconstruction from measured gradient fields. In *IEEE Computer Society Conference on Computer Vision and Pattern Recognition (CVPR 2008)*, Anchorage, AK.

Harris, C. and Stephens, M. J. (1988). A combined corner and edge detector. In *Alvey Vision Conference*, pp. 147–152.

Hartley, R. and Kang, S. B. (2007). Parameter-free radial distortion correction with center of distortion estimation. *IEEE Transactions on Pattern Analysis and Machine Intelligence*, 31(8):1309–1321.

Hartley, R., Gupta, R., and Chang, T. (1992). Estimation of relative camera positions for uncalibrated cameras. In *Second European Conference on Computer Vision (ECCV'92)*, pp. 579–587, Santa Margherita Liguere, Italy.

Hartley, R. I. (1994a). Projective reconstruction and invariants from multiple images. *IEEE Transactions on Pattern Analysis and Machine Intelligence*, 16(10):1036–1041.

Hartley, R. I. (1994b). Self-calibration from multiple views of a rotating camera. In *Third European Conference on Computer Vision (ECCV'94)*, pp. 471–478, Stockholm, Sweden.

Hartley, R. I. (1997a). In defense of the 8-point algorithm. *IEEE Transactions on Pattern Analysis and Machine Intelligence*, 19(6):580–593.

Hartley, R. I. (1997b). Self-calibration of stationary cameras. *International Journal of Computer Vision*, 22(1):5–23.

Hartley, R. I. (1998). Chirality. *International Journal of Computer Vision*, 26(1):41–61.

Hartley, R. I. and Kang, S. B. (2005). Parameter-free radial distortion correction with centre of distortion estimation. In *Tenth International Conference on Computer Vision (ICCV 2005)*, pp. 1834–1841, Beijing, China.

Hartley, R. I. and Sturm, P. (1997). Triangulation. *Computer Vision and Image Understanding*, 68(2):146–157.

Hartley, R. I. and Zisserman, A. (2004). *Multiple View Geometry*. Cambridge University Press, Cambridge, UK.

Hartley, R. I., Hayman, E., de Agapito, L., and Reid, I. (2000). Camera calibration and the search for infinity. In *IEEE Computer Society Conference on Computer Vision and Pattern Recognition (CVPR'2000)*, pp. 510–517, Hilton Head Island.

Hasinoff, S. W. and Kutulakos, K. N. (2008). Light-efficient photography. In *Tenth European Conference on Computer Vision (ECCV 2008)*, pp. 45–59, Marseilles.

Hasinoff, S. W., Durand, F., and Freeman, W. T. (2010). Noise-optimal capture for high dynamic range photography. In *IEEE Computer Society Conference on Computer Vision and Pattern Recognition (CVPR 2010)*, San Francisco, CA.

Hasinoff, S. W., Kang, S. B., and Szeliski, R. (2006). Boundary matting for view synthesis. *Computer Vision and Image Understanding*, 103(1):22–32.

Hasinoff, S. W., Kutulakos, K. N., Durand, F., and Freeman, W. T. (2009). Time-constrained photography. In *Twelfth International Conference on Computer Vision (ICCV 2009)*, Kyoto, Japan.

Hastie, T., Tibshirani, R., and Friedman, J. (2001). *The Elements of Statistical Learning: Data Mining, Inference, and Prediction*. Springer-Verlag, New York.

Hayes, B. (2008). Computational photography. *American Scientist*, 96:94–99.

Hays, J. and Efros, A. A. (2007). Scene completion using millions of photographs. *ACM Transactions on Graphics*, 26(3).

Hays, J., Leordeanu, M., Efros, A. A., and Liu, Y. (2006). Discovering texture regularity as a higher-order correspondence problem. In *Ninth European Conference on Computer Vision (ECCV 2006)*, pp. 522–535.

He, L.-W. and Zhang, Z. (2005). Real-time whiteboard capture and processing using a video camera for teleconferencing. In *IEEE International Conference on Acoustics, Speech, and Signal Processing (ICASSP 2005)*, pp. 1113–1116, Philadelphia.

He, X. and Zemel, R. S. (2008). Learning hybrid models for image annotation with partially labeled data. In *Advances in Neural Information Processing Systems*.

He, X., Zemel, R. S., and Carreira-Perpiñán, M. A. (2004). Multiscale conditional random fields for image labeling. In *IEEE Computer Society Conference on Computer Vision and Pattern Recognition (CVPR'2004)*, pp. 695–702, Washington, DC.

He, X., Zemel, R. S., and Ray, D. (2006). Learning and incorporating top-down cues in image segmentation. In *Ninth European Conference on Computer Vision (ECCV 2006)*, pp. 338–351.

Healey, G. E. and Kondepudy, R. (1994). Radiometric CCD camera calibration and noise estimation. *IEEE Transactions on Pattern Analysis and Machine Intelligence*, 16(3):267–276.

Healey, G. E. and Shafer, S. A. (1992). *Color. Physics-Based Vision: Principles and Practice*, Jones & Bartlett, Cambridge, MA.

Heath, M. D., Sarkar, S., Sanocki, T., and Bowyer, K. W. (1998). Comparison of edge detectors. *Computer Vision and Image Understanding*, 69(1):38–54.

Hebert, M. (2000). Active and passive range sensing for robotics. In *IEEE International Conference on Robotics and Automation*, pp. 102–110, San Francisco.

Hecht, E. (2001). *Optics*. Pearson Addison Wesley, Reading, MA, 4th edition.

Heckbert, P. (1986). Survey of texture mapping. *IEEE Computer Graphics and Applications*, 6(11):56–67.

Heckbert, P. (1989). *Fundamentals of Texture Mapping and Image Warping*. Master's thesis, The University of California at Berkeley.

Heeger, D. J. (1988). Optical flow using spatiotemporal filters. *International Journal of Computer Vision*, 1(1):279–302.

Heeger, D. J. and Bergen, J. R. (1995). Pyramid-based texture analysis/synthesis. In *ACM SIGGRAPH 1995 Conference Proceedings*, pp. 229–238.

Heisele, B., Serre, T., and Poggio, T. (2007). A component-based framework for face detection and identification. *International Journal of Computer Vision*, 74(2):167–181.

Heisele, B., Ho, P., Wu, J., and Poggio, T. (2003). Face recognition: component-based versus global approaches. *Computer Vision and Image Understanding*, 91(1-2):6–21.

Herley, C. (2005). Automatic occlusion removal from minimum number of images. In *International Conference on Image Processing (ICIP 2005)*, pp. 1046–1049–16, Genova.

Hernandez, C. and Schmitt, F. (2004). Silhouette and stereo fusion for 3D object modeling. *Computer Vision and Image Understanding*, 96(3):367–392.

Hernandez, C. and Vogiatzis, G. (2010). Self-calibrating a real-time monocular 3d facial capture system. In *Fifth International Symposium on 3D Data Processing, Visualization and Transmission (3DPVT'10)*, Paris.

Hernandez, C., Vogiatzis, G., and Cipolla, R. (2007). Probabilistic visibility for multi-view stereo. In *IEEE Computer Society Conference on Computer Vision and Pattern Recognition (CVPR 2007)*, Minneapolis, MN.

Hernandez, C., Vogiatzis, G., Brostow, G. J., Stenger, B., and Cipolla, R. (2007). Non-rigid photometric stereo with colored lights. In *Eleventh International Conference on Computer Vision (ICCV 2007)*, Rio de Janeiro, Brazil.

Hershberger, J. and Snoeyink, J. (1992). *Speeding Up the Douglas-Peucker Line-Simplification Algorithm*. Technical Report TR-92-07, Computer Science Department, The University of British Columbia.

Hertzmann, A., Jacobs, C. E., Oliver, N., Curless, B., and Salesin, D. H. (2001). Image analogies. In *ACM SIGGRAPH 2001 Conference Proceedings*, pp. 327–340.

Hiep, V. H., Keriven, R., Pons, J.-P., and Labatut, P. (2009). Towards high-resolution large-scale multi-view stereo. In *IEEE Computer Society Conference on Computer Vision and Pattern Recognition (CVPR 2009)*, Miami Beach, FL.

Hillman, P., Hannah, J., and Renshaw, D. (2001). Alpha channel estimation in high resolution images and image sequences. In *IEEE Computer Society Conference on Computer Vision and Pattern Recognition (CVPR'2001)*, pp. 1063–1068, Kauai, Hawaii.

Hilton, A., Fua, P., and Ronfard, R. (2006). Modeling people: Vision-based understanding of a person's shape, appearance, movement, and behaviour. *Computer Vision and Image Understanding*, 104(2-3):87–89.

Hilton, A., Stoddart, A. J., Illingworth, J., and Windeatt, T. (1996). Reliable surface reconstruction from multiple range images. In *Fourth European Conference on Computer Vision (ECCV'96)*, pp. 117–126, Cambridge, England.

Hinckley, K., Sinclair, M., Hanson, E., Szeliski, R., and Conway, M. (1999). The Video-Mouse: a camera-based multi-degree-of-freedom input device. In *12th annual ACM symposium on User interface software and technology*, pp. 103–112.

Hinterstoisser, S., Benhimane, S., Navab, N., Fua, P., and Lepetit, V. (2008). Online learning of patch perspective rectification for efficient object detection. In *IEEE Computer Society Conference on Computer Vision and Pattern Recognition (CVPR 2008)*, Anchorage, AK.

Hinton, G. E. (1977). *Relaxation and its Role in Vision*. Ph.D. thesis, University of Edinburgh.

Hirschmüller, H. (2008). Stereo processing by semiglobal matching and mutual information. *IEEE Transactions on Pattern Analysis and Machine Intelligence*, 30(2):328–341.

Hirschmüller, H. and Scharstein, D. (2009). Evaluation of stereo matching costs on images with radiometric differences. *IEEE Transactions on Pattern Analysis and Machine Intelligence*, 31(9):1582–1599.

Hjaltason, G. R. and Samet, H. (2003). Index-driven similarity search in metric spaces. *ACM Transactions on Database Systems*, 28(4):517–580.

Hofmann, T. (1999). Probabilistic latent semantic indexing. In *ACM SIGIR Conference on Research and Development in Informaion Retrieval*, pp. 50–57, Berkeley, CA.

Hogg, D. (1983). Model-based vision: A program to see a walking person. *Image and Vision Computing*, 1(1):5–20.

Hoiem, D., Efros, A. A., and Hebert, M. (2005a). Automatic photo pop-up. *ACM Transactions on Graphics (Proc. SIGGRAPH 2005)*, 24(3):577–584.

Hoiem, D., Efros, A. A., and Hebert, M. (2005b). Geometric context from a single image. In *Tenth International Conference on Computer Vision (ICCV 2005)*, pp. 654–661, Beijing, China.

Hoiem, D., Efros, A. A., and Hebert, M. (2008a). Closing the loop in scene interpretation. In *IEEE Computer Society Conference on Computer Vision and Pattern Recognition (CVPR 2008)*, Anchorage, AK.

Hoiem, D., Efros, A. A., and Hebert, M. (2008b). Putting objects in perspective. *International Journal of Computer Vision*, 80(1):3–15.

Hoiem, D., Rother, C., and Winn, J. (2007). 3D LayoutCRF for multi-view object class recognition and segmentation. In *IEEE Computer Society Conference on Computer Vision and Pattern Recognition (CVPR 2007)*, Minneapolis, MN.

Hoover, A., Jean-Baptiste, G., Jiang, X., Flynn, P. J., Bunke, H. *et al.* (1996). An experimental comparison of range image segmentation algorithms. *IEEE Transactions on Pattern Analysis and Machine Intelligence*, 18(7):673–689.

Hoppe, H. (1996). Progressive meshes. In *ACM SIGGRAPH 1996 Conference Proceedings*, pp. 99–108, New Orleans.

Hoppe, H., DeRose, T., Duchamp, T., McDonald, J., and Stuetzle, W. (1992). Surface reconstruction from unorganized points. *Computer Graphics (SIGGRAPH '92)*, 26(2):71–78.

Horn, B. K. P. (1974). Determining lightness from an image. *Computer Graphics and Image Processing*, 3(1):277–299.

Horn, B. K. P. (1975). Obtaining shape from shading information. In Winston, P. H. (ed.), *The Psychology of Computer Vision*, pp. 115–155, McGraw-Hill, New York.

Horn, B. K. P. (1977). Understanding image intensities. *Artificial Intelligence*, 8(2):201–231.

Horn, B. K. P. (1986). *Robot Vision*. MIT Press, Cambridge, Massachusetts.

Horn, B. K. P. (1987). Closed-form solution of absolute orientation using unit quaternions. *Journal of the Optical Society of America A*, 4(4):629–642.

Horn, B. K. P. (1990). Height and gradient from shading. *International Journal of Computer Vision*, 5(1):37–75.

Horn, B. K. P. and Brooks, M. J. (1986). The variational approach to shape from shading. *Computer Vision, Graphics, and Image Processing*, 33:174–208.

Horn, B. K. P. and Brooks, M. J. (eds). (1989). *Shape from Shading*, MIT Press, Cambridge, Massachusetts.

Horn, B. K. P. and Schunck, B. G. (1981). Determining optical flow. *Artificial Intelligence*, 17:185–203.

Horn, B. K. P. and Weldon Jr., E. J. (1988). Direct methods for recovering motion. *International Journal of Computer Vision*, 2(1):51–76.

Hornung, A., Zeng, B., and Kobbelt, L. (2008). Image selection for improved multi-view stereo. In *IEEE Computer Society Conference on Computer Vision and Pattern Recognition (CVPR 2008)*, Anchorage, AK.

Horowitz, S. L. and Pavlidis, T. (1976). Picture segmentation by a tree traversal algorithm. *Journal of the ACM*, 23(2):368–388.

Horry, Y., Anjyo, K.-I., and Arai, K. (1997). Tour into the picture: Using a spidery mesh interface to make animation from a single image. In *ACM SIGGRAPH 1997 Conference Proceedings*, pp. 225–232.

Hough, P. V. C. (1962). Method and means for recognizing complex patterns. *U. S. Patent*, 3,069,654.

Houhou, N., Thiran, J.-P., and Bresson, X. (2008). Fast texture segmentation using the shape operator and active contour. In *IEEE Computer Society Conference on Computer Vision and Pattern Recognition (CVPR 2008)*, Anchorage, AK.

Howe, N. R., Leventon, M. E., and Freeman, W. T. (2000). Bayesian reconstruction of 3D human motion from single-camera video. In *Advances in Neural Information Processing Systems*.

Hsieh, Y. C., McKeown, D., and Perlant, F. P. (1992). Performance evaluation of scene registration and stereo matching for cartographic feature extraction. *IEEE Transactions on Pattern Analysis and Machine Intelligence*, 14(2):214–238.

Hu, W., Tan, T., Wang, L., and Maybank, S. (2004). A survey on visual surveillance of object motion and behaviors. *IEEE Transactions on Systems, Man, and Cybernetics, Part C: Applications and Reviews*, 34(3):334–352.

Hua, G., Brown, M., and Winder, S. (2007). Discriminant embedding for local image descriptors. In *Eleventh International Conference on Computer Vision (ICCV 2007)*, Rio de Janeiro, Brazil.

Huang, G. B., Ramesh, M., Berg, T., and Learned-Miller, E. (2007). *Labeled Faces in the Wild: A Database for Studying Face Recognition in Unconstrained Environments.* Technical Report 07-49, University of Massachusetts, Amherst.

Huang, T. S. (1981). *Image Sequence Analysis.* Springer-Verlag, Berlin, Heidelberg.

Huber, P. J. (1981). *Robust Statistics.* John Wiley & Sons, New York.

Huffman, D. A. (1971). Impossible objects and nonsense sentences. *Machine Intelligence*, 8:295–323.

Huguet, F. and Devernay, F. (2007). A variational method for scene flow estimation from stereo sequences. In *Eleventh International Conference on Computer Vision (ICCV 2007)*, Rio de Janeiro, Brazil.

Huttenlocher, D. P., Klanderman, G., and Rucklidge, W. (1993). Comparing images using the Hausdorff distance. *IEEE Transactions on Pattern Analysis and Machine Intelligence*, 15(9):850–863.

Huynh, D. Q., Hartley, R., and Heyden, A. (2003). Outlier correcton in image sequences for the affine camera. In *Ninth International Conference on Computer Vision (ICCV 2003)*, pp. 585–590, Nice, France.

Iddan, G. J. and Yahav, G. (2001). 3D imaging in the studio (and elsewhere...). In *Three-Dimensional Image Capture and Applications IV*, pp. 48–55.

Igarashi, T., Nishino, K., and Nayar, S. (2007). The appearance of human skin: A survey. *Foundations and Trends in Computer Graphics and Computer Vision*, 3(1):1–95.

Ikeuchi, K. (1981). Shape from regular patterns. *Artificial Intelligence*, 22(1):49–75.

Ikeuchi, K. and Horn, B. K. P. (1981). Numerical shape from shading and occluding boundaries. *Artificial Intelligence*, 17:141–184.

Ikeuchi, K. and Miyazaki, D. (eds). (2007). *Digitally Archiving Cultural Objects*, Springer, Boston, MA.

Ikeuchi, K. and Sato, Y. (eds). (2001). *Modeling From Reality*, Kluwer Academic Publishers, Boston.

Illingworth, J. and Kittler, J. (1988). A survey of the Hough transform. *Computer Vision, Graphics, and Image Processing*, 44:87–116.

Intille, S. S. and Bobick, A. F. (1994). Disparity-space images and large occlusion stereo. In *Third European Conference on Computer Vision (ECCV'94)*, Stockholm, Sweden.

Irani, M. and Anandan, P. (1998). Video indexing based on mosaic representations. *Proceedings of the IEEE*, 86(5):905–921.

Irani, M. and Peleg, S. (1991). Improving resolution by image registration. *Graphical Models and Image Processing*, 53(3):231–239.

Irani, M., Hsu, S., and Anandan, P. (1995). Video compression using mosaic representations. *Signal Processing: Image Communication*, 7:529–552.

Irani, M., Rousso, B., and Peleg, S. (1994). Computing occluding and transparent motions. *International Journal of Computer Vision*, 12(1):5–16.

Irani, M., Rousso, B., and Peleg, S. (1997). Recovery of ego-motion using image stabilization. *IEEE Transactions on Pattern Analysis and Machine Intelligence*, 19(3):268–272.

Isaksen, A., McMillan, L., and Gortler, S. J. (2000). Dynamically reparameterized light fields. In *ACM SIGGRAPH 2000 Conference Proceedings*, pp. 297–306.

Isard, M. and Blake, A. (1998). CONDENSATION—conditional density propagation for visual tracking. *International Journal of Computer Vision*, 29(1):5–28.

Ishiguro, H., Yamamoto, M., and Tsuji, S. (1992). Omni-directional stereo. *IEEE Transactions on Pattern Analysis and Machine Intelligence*, 14(2):257–262.

Ishikawa, H. (2003). Exact optimization for Markov random fields with convex priors. *IEEE Transactions on Pattern Analysis and Machine Intelligence*, 25(10):1333–1336.

Ishikawa, H. and Veksler, O. (2010). Convex and truncated convex priors for multi-label MRFs. In Blake, A., Kohli, P., and Rother, C. (eds), *Advances in Markov Random Fields*, MIT Press.

Isidoro, J. and Sclaroff, S. (2003). Stochastic refinement of the visual hull to satisfy photometric and silhouette consistency constraints. In *Ninth International Conference on Computer Vision (ICCV 2003)*, pp. 1335–1342, Nice, France.

Ivanchenko, V., Shen, H., and Coughlan, J. (2009). Elevation-based stereo implemented in real-time on a GPU. In *IEEE Workshop on Applications of Computer Vision (WACV 2009)*, Snowbird, Utah.

Jacobs, C. E., Finkelstein, A., and Salesin, D. H. (1995). Fast multiresolution image querying. In *ACM SIGGRAPH 1995 Conference Proceedings*, pp. 277–286.

Jähne, B. (1997). *Digital Image Processing*. Springer-Verlag, Berlin.

Jain, A. K. and Dubes, R. C. (1988). *Algorithms for Clustering Data*. Prentice Hall, Englewood Cliffs, New Jersey.

Jain, A. K., Bolle, R. M., and Pankanti, S. (eds). (1999). *Biometrics: Personal Identification in Networked Society*, Kluwer.

Jain, A. K., Duin, R. P. W., and Mao, J. (2000). Statistical pattern recognition: A review. *IEEE Transactions on Pattern Analysis and Machine Intelligence*, 22(1):4–37.

Jain, A. K., Topchy, A., Law, M. H. C., and Buhmann, J. M. (2004). Landscape of clustering algorithms. In *International Conference on Pattern Recognition (ICPR 2004)*, pp. 260–263.

Jenkin, M. R. M., Jepson, A. D., and Tsotsos, J. K. (1991). Techniques for disparity measurement. *CVGIP: Image Understanding*, 53(1):14–30.

Jensen, H. W., Marschner, S. R., Levoy, M., and Hanrahan, P. (2001). A practical model for subsurface light transport. In *ACM SIGGRAPH 2001 Conference Proceedings*, pp. 511–518.

Jeong, Y., Nistér, D., Steedly, D., Szeliski, R., and Kweon, I.-S. (2010). Pushing the envelope of modern methods for bundle adjustment. In *IEEE Computer Society Conference on Computer Vision and Pattern Recognition (CVPR 2010)*, San Francisco, CA.

Jia, J. and Tang, C.-K. (2003). Image registration with global and local luminance alignment. In *Ninth International Conference on Computer Vision (ICCV 2003)*, pp. 156–163, Nice, France.

Jia, J., Sun, J., Tang, C.-K., and Shum, H.-Y. (2006). Drag-and-drop pasting. *ACM Transactions on Graphics*, 25(3):631–636.

Jiang, Z., Wong, T.-T., and Bao, H. (2003). Practical super-resolution from dynamic video sequences. In *IEEE Computer Society Conference on Computer Vision and Pattern Recognition (CVPR'2003)*, pp. 549–554, Madison, WI.

Johnson, A. E. and Hebert, M. (1999). Using spin images for efficient object recognition in cluttered 3D scenes. *IEEE Transactions on Pattern Analysis and Machine Intelligence*, 21(5):433–448.

Johnson, A. E. and Kang, S. B. (1997). Registration and integration of textured 3-D data. In *International Conference on Recent Advances in 3-D Digital Imaging and Modeling*, pp. 234–241, Ottawa.

Jojic, N. and Frey, B. J. (2001). Learning flexible sprites in video layers. In *IEEE Computer Society Conference on Computer Vision and Pattern Recognition (CVPR'2001)*, pp. 199–206, Kauai, Hawaii.

Jones, D. G. and Malik, J. (1992). A computational framework for determining stereo correspondence from a set of linear spatial filters. In *Second European Conference on Computer Vision (ECCV'92)*, pp. 397–410, Santa Margherita Liguere, Italy.

Jones, M. J. and Rehg, J. M. (2001). Statistical color models with application to skin detection. *International Journal of Computer Vision*, 46(1):81–96.

Joshi, N., Matusik, W., and Avidan, S. (2006). Natural video matting using camera arrays. *ACM Transactions on Graphics*, 25(3):779–786.

Joshi, N., Szeliski, R., and Kriegman, D. J. (2008). PSF estimation using sharp edge prediction. In *IEEE Computer Society Conference on Computer Vision and Pattern Recognition (CVPR 2008)*, Anchorage, AK.

Joshi, N., Zitnick, C. L., Szeliski, R., and Kriegman, D. J. (2009). Image deblurring and denoising using color priors. In *IEEE Computer Society Conference on Computer Vision and Pattern Recognition (CVPR 2009)*, Miami, FL.

Ju, S. X., Black, M. J., and Jepson, A. D. (1996). Skin and bones: Multi-layer, locally affine, optical flow and regularization with transparency. In *IEEE Computer Society Conference on Computer Vision and Pattern Recognition (CVPR'96)*, pp. 307–314, San Francisco.

Ju, S. X., Black, M. J., and Yacoob, Y. (1996). Cardboard people: a parameterized model of articulated image motion. In *2nd International Conference on Automatic Face and Gesture Recognition*, pp. 38–44, Killington, VT.

Juan, O. and Boykov, Y. (2006). Active graph cuts. In *IEEE Computer Society Conference on Computer Vision and Pattern Recognition (CVPR'2006)*, pp. 1023–1029, New York City, NY.

Jurie, F. and Dhome, M. (2002). Hyperplane approximation for template matching. *IEEE Transactions on Pattern Analysis and Machine Intelligence*, 24(7):996–1000.

Jurie, F. and Schmid, C. (2004). Scale-invariant shape features for recognition of object categories. In *IEEE Computer Society Conference on Computer Vision and Pattern Recognition (CVPR'2004)*, pp. 90–96, Washington, DC.

Kadir, T., Zisserman, A., and Brady, M. (2004). An affine invariant salient region detector. In *Eighth European Conference on Computer Vision (ECCV 2004)*, pp. 228–241, Prague.

Kaftory, R., Schechner, Y., and Zeevi, Y. (2007). Variational distance-dependent image restoration. In *IEEE Computer Society Conference on Computer Vision and Pattern Recognition (CVPR 2007)*, Minneapolis, MN.

Kahl, F. and Hartley, R. (2008). Multiple-view geometry under the l_∞-norm. *IEEE Transactions on Pattern Analysis and Machine Intelligence*, 30(11):1603–1617.

Kakadiaris, I. and Metaxas, D. (2000). Model-based estimation of 3D human motion. *IEEE Transactions on Pattern Analysis and Machine Intelligence*, 22(12):1453–1459.

Kakumanu, P., Makrogiannis, S., and Bourbakis, N. (2007). A survey of skin-color modeling and detection methods. *Pattern Recognition*, 40(3):1106–1122.

Kamvar, S. D., Klein, D., and Manning, C. D. (2002). Interpreting and extending classical agglomerative clustering algorithms using a model-based approach. In *International*

Conference on Machine Learning, pp. 283–290.

Kanade, T. (1977). *Computer Recognition of Human Faces*. Birkhauser, Basel.

Kanade, T. (1980). A theory of the origami world. *Artificial Intelligence*, 13:279–311.

Kanade, T. (ed.). (1987). *Three-Dimensional Machine Vision*, Kluwer Academic Publishers, Boston.

Kanade, T. (1994). Development of a video-rate stereo machine. In *Image Understanding Workshop*, pp. 549–557, Monterey.

Kanade, T. and Okutomi, M. (1994). A stereo matching algorithm with an adaptive window: Theory and experiment. *IEEE Transactions on Pattern Analysis and Machine Intelligence*, 16(9):920–932.

Kanade, T., Rander, P. W., and Narayanan, P. J. (1997). Virtualized reality: constructing virtual worlds from real scenes. *IEEE MultiMedia Magazine*, 4(1):34–47.

Kanade, T., Yoshida, A., Oda, K., Kano, H., and Tanaka, M. (1996). A stereo machine for video-rate dense depth mapping and its new applications. In *IEEE Computer Society Conference on Computer Vision and Pattern Recognition (CVPR'96)*, pp. 196–202, San Francisco.

Kanatani, K. and Morris, D. D. (2001). Gauges and gauge transformations for uncertainty description of geometric structure with indeterminacy. *IEEE Transactions on Information Theory*, 47(5):2017–2028.

Kang, S. B. (1998). *Depth Painting for Image-based Rendering Applications*. Technical Report, Compaq Computer Corporation, Cambridge Research Lab.

Kang, S. B. (1999). A survey of image-based rendering techniques. In *Videometrics VI*, pp. 2–16, San Jose.

Kang, S. B. (2001). Radial distortion snakes. *IEICE Trans. Inf. & Syst.*, E84-D(12):1603–1611.

Kang, S. B. and Jones, M. (2002). Appearance-based structure from motion using linear classes of 3-D models. *International Journal of Computer Vision*, 49(1):5–22.

Kang, S. B. and Szeliski, R. (1997). 3-D scene data recovery using omnidirectional multi-baseline stereo. *International Journal of Computer Vision*, 25(2):167–183.

Kang, S. B. and Szeliski, R. (2004). Extracting view-dependent depth maps from a collection of images. *International Journal of Computer Vision*, 58(2):139–163.

Kang, S. B. and Weiss, R. (1997). Characterization of errors in compositing panoramic images. In *IEEE Computer Society Conference on Computer Vision and Pattern Recognition (CVPR'97)*, pp. 103–109, San Juan, Puerto Rico.

Kang, S. B. and Weiss, R. (1999). Characterization of errors in compositing panoramic images. *Computer Vision and Image Understanding*, 73(2):269–280.

Kang, S. B. and Weiss, R. (2000). Can we calibrate a camera using an image of a flat, textureless Lambertian surface? In *Sixth European Conference on Computer Vision (ECCV 2000)*, pp. 640–653, Dublin, Ireland.

Kang, S. B., Szeliski, R., and Anandan, P. (2000). The geometry-image representation tradeoff for rendering. In *International Conference on Image Processing (ICIP-2000)*, pp. 13–16, Vancouver.

Kang, S. B., Szeliski, R., and Chai, J. (2001). Handling occlusions in dense multi-view stereo. In *IEEE Computer Society Conference on Computer Vision and Pattern Recognition (CVPR'2001)*, pp. 103–110, Kauai, Hawaii.

Kang, S. B., Szeliski, R., and Shum, H.-Y. (1997). A parallel feature tracker for extended image sequences. *Computer Vision and Image Understanding*, 67(3):296–310.

Kang, S. B., Szeliski, R., and Uyttendaele, M. (2004). *Seamless Stitching using Multi-Perspective Plane Sweep*. Technical Report MSR-TR-2004-48, Microsoft Research.

Kang, S. B., Li, Y., Tong, X., and Shum, H.-Y. (2006). Image-based rendering. *Foundations and Trends in Computer Graphics and Computer Vision*, 2(3):173–258.

Kang, S. B., Uyttendaele, M., Winder, S., and Szeliski, R. (2003). High dynamic range video. *ACM Transactions on Graphics (Proc. SIGGRAPH 2003)*, 22(3):319–325.

Kang, S. B., Webb, J., Zitnick, L., and Kanade, T. (1995). A multibaseline stereo system with active illumination and real-time image acquisition. In *Fifth International Conference on Computer Vision (ICCV'95)*, pp. 88–93, Cambridge, Massachusetts.

Kannala, J., Rahtu, E., Brandt, S. S., and Heikkila, J. (2008). Object recognition and segmentation by non-rigid quasi-dense matching. In *IEEE Computer Society Conference on Computer Vision and Pattern Recognition (CVPR 2008)*, Anchorage, AK.

Kass, M. (1988). Linear image features in stereopsis. *International Journal of Computer Vision*, 1(4):357–368.

Kass, M., Witkin, A., and Terzopoulos, D. (1988). Snakes: Active contour models. *International Journal of Computer Vision*, 1(4):321–331.

Kato, H., Billinghurst, M., Poupyrev, I., Imamoto, K., and Tachibana, K. (2000). Virtual object manipulation on a table-top AR environment. In *International Symposium on Augmented Reality (ISAR 2000)*.

Kaufman, L. and Rousseeuw, P. J. (1990). *Finding Groups in Data: An Introduction to Cluster Analysis*. John Wiley & Sons, Hoboken.

Kazhdan, M., Bolitho, M., and Hoppe, H. (2006). Poisson surface reconstruction. In *Eurographics Symposium on Geometry Processing*, pp. 61–70.

Ke, Y. and Sukthankar, R. (2004). PCA-SIFT: a more distinctive representation for local image descriptors. In *IEEE Computer Society Conference on Computer Vision and Pattern Recognition (CVPR'2004)*, pp. 506–513, Washington, DC.

Kehl, R. and Van Gool, L. (2006). Markerless tracking of complex human motions from multiple views. *Computer Vision and Image Understanding*, 104(2-3):190–209.

Kenney, C., Zuliani, M., and Manjunath, B. (2005). An axiomatic approach to corner detection. In *IEEE Computer Society Conference on Computer Vision and Pattern Recognition (CVPR'2005)*, pp. 191–197, San Diego, CA.

Keren, D., Peleg, S., and Brada, R. (1988). Image sequence enhancement using sub-pixel displacements. In *IEEE Computer Society Conference on Computer Vision and Pattern Recognition (CVPR'88)*, pp. 742–746, Ann Arbor, Michigan.

Kim, J., Kolmogorov, V., and Zabih, R. (2003). Visual correspondence using energy minimization and mutual information. In *Ninth International Conference on Computer Vision (ICCV 2003)*, pp. 1033–1040, Nice, France.

Kimmel, R. (1999). Demosaicing: image reconstruction from color CCD samples. *IEEE Transactions on Image Processing*, 8(9):1221–1228.

Kimura, S., Shinbo, T., Yamaguchi, H., Kawamura, E., and Nakano, K. (1999). A convolver-based real-time stereo machine (SAZAN). In *IEEE Computer Society Conference on Computer Vision and Pattern Recognition (CVPR'99)*, pp. 457–463, Fort Collins.

Kindermann, R. and Snell, J. L. (1980). *Markov Random Fields and Their Applications*. American Mathematical Society.

King, D. (1997). *The Commissar Vanishes*. Henry Holt and Company.

Kirby, M. and Sirovich, L. (1990). Application of the Karhunen–Lòeve procedure for the characterization of human faces. *IEEE Transactions on Pattern Analysis and Machine Intelligence*, 12(1):103–108.

Kirkpatrick, S., Gelatt, C. D. J., and Vecchi, M. P. (1983). Optimization by simulated annealing. *Science*, 220:671–680.

Kirovski, D., Jojic, N., and Jancke, G. (2004). Tamper-resistant biometric IDs. In *ISSE 2004 - Securing Electronic Business Processes: Highlights of the Information Security Solutions Europe 2004 Conference*, pp. 160–175.

Kittler, J. and Föglein, J. (1984). Contextual classification of multispectral pixel data. *Image and Vision Computing*, 2(1):13–29.

Klaus, A., Sormann, M., and Karner, K. (2006). Segment-based stereo matching using belief propagation and a self-adapting dissimilarity measure. In *International Conference on Pattern Recognition (ICPR 2006)*, pp. 15–18.

Klein, G. and Murray, D. (2007). Parallel tracking and mapping for small AR workspaces. In *International Symposium on Mixed and Augmented Reality (ISMAR 2007)*, Nara.

Klein, G. and Murray, D. (2008). Improving the agility of keyframe-based slam. In *Tenth European Conference on Computer Vision (ECCV 2008)*, pp. 802–815, Marseilles.

Klein, S., Staring, M., and Pluim, J. P. W. (2007). Evaluation of optimization methods for nonrigid medical image registration using mutual information and B-splines. *IEEE Transactions on Image Processing*, 16(12):2879–2890.

Klinker, G. J. (1993). *A Physical Approach to Color Image Understanding*. A K Peters, Wellesley, Massachusetts.

Klinker, G. J., Shafer, S. A., and Kanade, T. (1990). A physical approach to color image understanding. *International Journal of Computer Vision*, 4(1):7–38.

Koch, R., Pollefeys, M., and Van Gool, L. J. (2000). Realistic surface reconstruction of 3D scenes from uncalibrated image sequences. *Journal Visualization and Computer Animation*, 11:115–127.

Koenderink, J. J. (1990). *Solid Shape*. MIT Press, Cambridge, Massachusetts.

Koethe, U. (2003). Integrated edge and junction detection with the boundary tensor. In *Ninth International Conference on Computer Vision (ICCV 2003)*, pp. 424–431, Nice, France.

Kohli, P. (2007). *Minimizing Dynamic and Higher Order Energy Functions using Graph Cuts*. Ph.D. thesis, Oxford Brookes University.

Kohli, P. and Torr, P. H. S. (2005). Effciently solving dynamic Markov random fields using graph cuts. In *Tenth International Conference on Computer Vision (ICCV 2005)*, pp. 922–929, Beijing, China.

Kohli, P. and Torr, P. H. S. (2007). Dynamic graph cuts for efficient inference in markov random fields. *IEEE Transactions on Pattern Analysis and Machine Intelligence*, 29(12):2079–2088.

Kohli, P. and Torr, P. H. S. (2008). Measuring uncertainty in graph cut solutions. *Computer Vision and Image Understanding*, 112(1):30–38.

Kohli, P., Kumar, M. P., and Torr, P. H. S. (2009). $\mathcal{P}^3$ & beyond: Move making algorithms for solving higher order functions. *IEEE Transactions on Pattern Analysis and Machine Intelligence*, 31(9):1645–1656.

Kohli, P., Ladický, L., and Torr, P. H. S. (2009). Robust higher order potentials for enforcing label consistency. *International Journal of Computer Vision*, 82(3):302–324.

Kokaram, A. (2004). On missing data treatment for degraded video and film archives: a survey and a new Bayesian approach. *IEEE Transactions on Image Processing*, 13(3):397–415.

Kolev, K. and Cremers, D. (2008). Integration of multiview stereo and silhouettes via convex functionals on convex domains. In *Tenth European Conference on Computer Vision (ECCV 2008)*, pp. 752–765, Marseilles.

Kolev, K. and Cremers, D. (2009). Continuous ratio optimization via convex relaxation with applications to multiview 3D reconstruction. In *IEEE Computer Society Conference on Computer Vision and Pattern Recognition (CVPR 2009)*, Miami Beach, FL.

Kolev, K., Klodt, M., Brox, T., and Cremers, D. (2009). Continuous global optimization in multiview 3D reconstruction. *International Journal of Computer Vision*, 84(1):80–96.

Koller, D. and Friedman, N. (2009). *Probabilistic Graphical Models: Principles and Techniques*. MIT Press, Cambridge, Massachusetts.

Kolmogorov, V. (2006). Convergent tree-reweighted message passing for energy minimization. *IEEE Transactions on Pattern Analysis and Machine Intelligence*, 28(10):1568–1583.

Kolmogorov, V. and Boykov, Y. (2005). What metrics can be approximated by geo-cuts, or global optimization of length/area and flux. In *Tenth International Conference on Computer Vision (ICCV 2005)*, pp. 564–571, Beijing, China.

Kolmogorov, V. and Zabih, R. (2002). Multi-camera scene reconstruction via graph cuts. In *Seventh European Conference on Computer Vision (ECCV 2002)*, pp. 82–96, Copenhagen.

Kolmogorov, V. and Zabih, R. (2004). What energy functions can be minimized via graph cuts? *IEEE Transactions on Pattern Analysis and Machine Intelligence*, 26(2):147–159.

Kolmogorov, V., Criminisi, A., Blake, A., Cross, G., and Rother, C. (2006). Probabilistic fusion of stereo with color and contrast for bi-layer segmentation. *IEEE Transactions on Pattern Analysis and Machine Intelligence*, 28(9):1480–1492.

Komodakis, N. and Paragios, N. (2008). Beyond loose LP-relaxations: Optimizing MRFs by repairing cycles. In *Tenth European Conference on Computer Vision (ECCV 2008)*, pp. 806–820, Marseilles.

Komodakis, N. and Tziritas, G. (2007a). Approximate labeling via graph cuts based on linear programming. *IEEE Transactions on Pattern Analysis and Machine Intelligence*, 29(8):1436–1453.

Komodakis, N. and Tziritas, G. (2007b). Image completion using efficient belief propagation via priority scheduling and dynamic pruning. *IEEE Transactions on Image Processing*, 29(11):2649–2661.

Komodakis, N., Paragios, N., and Tziritas, G. (2007). MRF optimization via dual decomposition: Message-passing revisited. In *Eleventh International Conference on Computer Vision (ICCV 2007)*, Rio de Janeiro, Brazil.

Komodakis, N., Tziritas, G., and Paragios, N. (2007). Fast, approximately optimal solutions for single and dynamic MRFs. In *IEEE Computer Society Conference on Computer Vision and Pattern Recognition (CVPR 2007)*, Minneapolis, MN.

Komodakis, N., Tziritas, G., and Paragios, N. (2008). Performance vs computational efficiency for optimizing single and dynamic MRFs: Setting the state of the art with primal dual strategies. *Computer Vision and Image Understanding*, 112(1):14–29.

Konolige, K. (1997). Small vision systems: Hardware and implementation. In *Eighth International Symposium on Robotics Research*, pp. 203–212, Hayama, Japan.

Kopf, J., Cohen, M. F., Lischinski, D., and Uyttendaele, M. (2007). Joint bilateral upsampling. *ACM Transactions on Graphics*, 26(3).

Kopf, J., Uyttendaele, M., Deussen, O., and Cohen, M. F. (2007). Capturing and viewing gigapixel images. *ACM Transactions on Graphics*, 26(3).

Kopf, J., Lischinski, D., Deussen, O., Cohen-Or, D., and Cohen, M. (2009). Locally adapted projections to reduce panorama distortions. *Computer Graphics Forum (Proceedings of EGSR 2009)*, 28(4).

Koutis, I. (2007). *Combinatorial and algebraic tools for optimal multilevel algorithms*. Ph.D. thesis, Carnegie Mellon University. Technical Report CMU-CS-07-131.

Koutis, I. and Miller, G. L. (2008). Graph partitioning into isolated, high conductance clusters: theory, computation and applications to preconditioning. In *Symposium on Parallel Algorithms and Architectures*, pp. 137–145, Munich.

Koutis, I., Miller, G. L., and Tolliver, D. (2009). Combinatorial preconditioners and multilevel solvers for problems in computer vision and image processing. In *5th International Symposium on Visual Computing (ISVC09)*, Las Vegas.

Kovar, L., Gleicher, M., and Pighin, F. (2002). Motion graphs. *ACM Transactions on Graphics*, 21(3):473–482.

Košecká, J. and Zhang, W. (2005). Extraction, matching and pose recovery based on dominant rectangular structures. *Computer Vision and Image Understanding*, 100(3):174–293.

Kraus, K. (1997). *Photogrammetry*. Dümmler, Bonn.

Krishnan, D. and Fergus, R. (2009). Fast image deconvolution using hyper-Laplacian priors. In *Advances in Neural Information Processing Systems*.

Kuglin, C. D. and Hines, D. C. (1975). The phase correlation image alignment method. In *IEEE 1975 Conference on Cybernetics and Society*, pp. 163–165, New York.

Kulis, B. and Grauman, K. (2009). Kernelized locality-sensitive hashing for scalable image search. In *Twelfth International Conference on Computer Vision (ICCV 2009)*, Kyoto, Japan.

Kumar, M. P. (2008). *Combinatorial and Convex Optimization for Probabilistic Models in Computer Vision*. Ph.D. thesis, Oxford Brookes University.

Kumar, M. P. and Torr, P. H. S. (2006). Fast memory-efficient generalized belief propagation. In *Ninth European Conference on Computer Vision (ECCV 2006)*, pp. 451–463.

Kumar, M. P. and Torr, P. H. S. (2008). Improved moves for truncated convex models. In *Advances in Neural Information Processing Systems*.

Kumar, M. P., Torr, P. H. S., and Zisserman, A. (2008). Learning layered motion segmentations of video. *International Journal of Computer Vision*, 76(3):301–319.

Kumar, M. P., Torr, P. H. S., and Zisserman, A. (2010). OBJCUT: Efficient segmentation using top-down and bottom-up cues. *IEEE Transactions on Pattern Analysis and Machine Intelligence*, 32(3).

Kumar, M. P., Veksler, O., and Torr, P. H. S. (2010). Improved moves for truncated convex models. *Journal of Machine Learning Research*, (sumbitted).

Kumar, M. P., Zisserman, A., and H.S.Torr, P. (2009). Efficient discriminative learning of parts-based models. In *Twelfth International Conference on Computer Vision (ICCV 2009)*, Kyoto, Japan.

Kumar, R., Anandan, P., and Hanna, K. (1994). Direct recovery of shape from multiple views: A parallax based approach. In *Twelfth International Conference on Pattern Recognition (ICPR'94)*, pp. 685–688, Jerusalem, Israel.

Kumar, R., Anandan, P., Irani, M., Bergen, J., and Hanna, K. (1995). Representation of scenes from collections of images. In *IEEE Workshop on Representations of Visual Scenes*, pp. 10–17, Cambridge, Massachusetts.

Kumar, S. and Hebert, M. (2003). Discriminative random fields: A discriminative framework for contextual interaction in classification. In *Ninth International Conference on Computer Vision (ICCV 2003)*, pp. 1150–1157, Nice, France.

Kumar, S. and Hebert, M. (2006). Discriminative random fields. *International Journal of Computer Vision*, 68(2):179–202.

Kundur, D. and Hatzinakos, D. (1996). Blind image deconvolution. *IEEE Signal Processing Magazine*, 13(3):43–64.

Kutulakos, K. N. (2000). Approximate N-view stereo. In *Sixth European Conference on Computer Vision (ECCV 2000)*, pp. 67–83, Dublin, Ireland.

Kutulakos, K. N. and Seitz, S. M. (2000). A theory of shape by space carving. *International Journal of Computer Vision*, 38(3):199–218.

Kwatra, V., Essa, I., Bobick, A., and Kwatra, N. (2005). Graphcut textures: Image and video synthesis using graph cuts. *ACM Transactions on Graphics (Proc. SIGGRAPH 2005)*, 24(5):795–802.

Kwatra, V., Schödl, A., Essa, I., Turk, G., and Bobick, A. (2003). Graphcut textures: Image and video synthesis using graph cuts. *ACM Transactions on Graphics (Proc. SIGGRAPH 2003)*, 22(3):277–286.

Kybic, J. and Unser, M. (2003). Fast parametric elastic image registration. *IEEE Transactions on Image Processing*, 12(11):1427–1442.

Lafferty, J., McCallum, A., and Pereira, F. (2001). Conditional random fields: Probabilistic models for segmenting and labeling sequence data. In *International Conference on Machine Learning*.

Lafortune, E. P. F., Foo, S.-C., Torrance, K. E., and Greenberg, D. P. (1997). Non-linear approximation of reflectance functions. In *ACM SIGGRAPH 1997 Conference Proceedings*, pp. 117–126, Los Angeles.

Lai, S.-H. and Vemuri, B. C. (1997). Physically based adaptive preconditioning for early vision. *IEEE Transactions on Pattern Analysis and Machine Intelligence*, 19(6):594–607.

Lalonde, J.-F., Hoiem, D., Efros, A. A., Rother, C., Winn, J., and Criminisi, A. (2007). Photo clip art. *ACM Transactions on Graphics*, 26(3).

Lampert, C. H. (2008). Kernel methods in computer vision. *Foundations and Trends in Computer Graphics and Computer Vision*, 4(3):193–285.

Langer, M. S. and Zucker, S. W. (1994). Shape from shading on a cloudy day. *Journal Optical Society America, A*, 11(2):467–478.

Lanitis, A., Taylor, C. J., and Cootes, T. F. (1997). Automatic interpretation and coding of face images using flexible models. *IEEE Transactions on Pattern Analysis and Machine Intelligence*, 19(7):742–756.

Larlus, D. and Jurie, F. (2008). Combining appearance models and Markov random fields for category level object segmentation. In *IEEE Computer Society Conference on Computer Vision and Pattern Recognition (CVPR 2008)*, Anchorage, AK.

Larson, G. W. (1998). LogLuv encoding for full-gamut, high-dynamic range images. *Journal of Graphics Tools*, 3(1):15–31.

Larson, G. W., Rushmeier, H., and Piatko, C. (1997). A visibility matching tone reproduction operator for high dynamic range scenes. *IEEE Transactions on Visualization and Computer Graphics*, 3(4):291–306.

Laurentini, A. (1994). The visual hull concept for silhouette-based image understanding. *IEEE Transactions on Pattern Analysis and Machine Intelligence*, 16(2):150–162.

Lavallée, S. and Szeliski, R. (1995). Recovering the position and orientation of free-form objects from image contours using 3-D distance maps. *IEEE Transactions on Pattern Analysis and Machine Intelligence*, 17(4):378–390.

Laveau, S. and Faugeras, O. D. (1994). 3-D scene representation as a collection of images. In *Twelfth International Conference on Pattern Recognition (ICPR'94)*, pp. 689–691, Jerusalem, Israel.

Lazebnik, S., Schmid, C., and Ponce, J. (2005). A sparse texture representation using local affine regions. *IEEE Transactions on Pattern Analysis and Machine Intelligence*, 27(8):1265–1278.

Lazebnik, S., Schmid, C., and Ponce, J. (2006). Beyond bags of features: Spatial pyramid matching for recognizing natural scene categories. In *IEEE Computer Society Conference on Computer Vision and Pattern Recognition (CVPR'2006)*, pp. 2169–2176, New York City, NY.

Le Gall, D. (1991). MPEG: A video compression standard for multimedia applications. *Communications of the ACM*, 34(4):46–58.

Leclerc, Y. G. (1989). Constructing simple stable descriptions for image partitioning. *International Journal of Computer Vision*, 3(1):73–102.

LeCun, Y., Huang, F. J., and Bottou, L. (2004). Learning methods for generic object recognition with invariance to pose and lighting. In *IEEE Computer Society Conference on Computer Vision and Pattern Recognition (CVPR'2004)*, pp. 97–104, Washington, DC.

Lee, J., Chai, J., Reitsma, P. S. A., Hodgins, J. K., and Pollard, N. S. (2002). Interactive control of avatars animated with human motion data. *ACM Transactions on Graphics*, 21(3):491–500.

Lee, M.-C., ge Chen, W., lung Bruce Lin, C., Gu, C., Markoc, T., Zabinsky, S. I., and Szeliski, R. (1997). A layered video object coding system using sprite and affine motion model. *IEEE Transactions on Circuits and Systems for Video Technology*, 7(1):130–145.

Lee, M. E. and Redner, R. A. (1990). A note on the use of nonlinear filtering in computer graphics. *IEEE Computer Graphics and Applications*, 10(3):23–29.

Lee, M. W. and Cohen, I. (2006). A model-based approach for estimating human 3D poses in static images. *IEEE Transactions on Pattern Analysis and Machine Intelligence*, 28(6):905–916.

Lee, S., Wolberg, G., and Shin, S. Y. (1996). Data interpolation using multilevel b-splines. *IEEE Transactions on Visualization and Computer Graphics*, 3(3):228–244.

Lee, S., Wolberg, G., Chwa, K.-Y., and Shin, S. Y. (1996). Image metamorphosis with scattered feature constraints. *IEEE Transactions on Visualization and Computer Graphics*, 2(4):337–354.

Lee, Y. D., Terzopoulos, D., and Waters, K. (1995). Realistic facial modeling for animation. In *ACM SIGGRAPH 1995 Conference Proceedings*, pp. 55–62.

Lei, C. and Yang, Y.-H. (2009). Optical flow estimation on coarse-to-fine region-trees using discrete optimization. In *Twelfth International Conference on Computer Vision (ICCV 2009)*, Kyoto, Japan.

Leibe, B. and Schiele, B. (2003). Analyzing appearance and contour based methods for object categorization. In *IEEE Computer Society Conference on Computer Vision and Pattern Recognition (CVPR'2003)*, pp. 409–415, Madison, WI.

Leibe, B., Leonardis, A., and Schiele, B. (2008). Robust object detection with interleaved categorization and segmentation. *International Journal of Computer Vision*, 77(1-3):259–289.

Leibe, B., Seemann, E., and Schiele, B. (2005). Pedestrian detection in crowded scenes. In *IEEE Computer Society Conference on Computer Vision and Pattern Recognition (CVPR'2005)*, pp. 878–885, San Diego, CA.

Leibe, B., Cornelis, N., Cornelis, K., and Van Gool, L. (2007). Dynamic 3D scene analysis from a moving vehicle. In *IEEE Computer Society Conference on Computer Vision and Pattern Recognition (CVPR 2007)*, Minneapolis, MN.

Leibowitz, D. (2001). *Camera Calibration and Reconstruction of Geometry from Images*. Ph.D. thesis, University of Oxford.

Lempitsky, V. and Boykov, Y. (2007). Global optimization for shape fitting. In *IEEE Computer Society Conference on Computer Vision and Pattern Recognition (CVPR 2007)*,

Minneapolis, MN.

Lempitsky, V. and Ivanov, D. (2007). Seamless mosaicing of image-based texture maps. In *IEEE Computer Society Conference on Computer Vision and Pattern Recognition (CVPR 2007)*, Minneapolis, MN.

Lempitsky, V., Blake, A., and Rother, C. (2008). Image segmentation by branch-and-mincut. In *Tenth European Conference on Computer Vision (ECCV 2008)*, pp. 15–29, Marseilles.

Lempitsky, V., Roth, S., and Rother., C. (2008). FlowFusion: Discrete-continuous optimization for optical flow estimation. In *IEEE Computer Society Conference on Computer Vision and Pattern Recognition (CVPR 2008)*, Anchorage, AK.

Lempitsky, V., Rother, C., and Blake, A. (2007). Logcut - efficient graph cut optimization for Markov random fields. In *Eleventh International Conference on Computer Vision (ICCV 2007)*, Rio de Janeiro, Brazil.

Lengyel, J. and Snyder, J. (1997). Rendering with coherent layers. In *ACM SIGGRAPH 1997 Conference Proceedings*, pp. 233–242, Los Angeles.

Lensch, H. P. A., Kautz, J., Goesele, M., Heidrich, W., and Seidel, H.-P. (2003). Image-based reconstruction of spatial appearance and geometric detail. *ACM Transactions on Graphics*, 22(2):234–257.

Leonardis, A. and Bischof, H. (2000). Robust recognition using eigenimages. *Computer Vision and Image Understanding*, 78(1):99–118.

Leonardis, A., Jaklič, A., and Solina, F. (1997). Superquadrics for segmenting and modeling range data. *IEEE Transactions on Pattern Analysis and Machine Intelligence*, 19(11):1289–1295.

Lepetit, V. and Fua, P. (2005). Monocular model-based 3D tracking of rigid objects. *Foundations and Trends in Computer Graphics and Computer Vision*, 1(1).

Lepetit, V., Pilet, J., and Fua, P. (2004). Point matching as a classification problem for fast and robust object pose estimation. In *IEEE Computer Society Conference on Computer Vision and Pattern Recognition (CVPR'2004)*, pp. 244–250, Washington, DC.

Lepetit, V., Pilet, J., and Fua, P. (2006). Keypoint recognition using randomized trees. *IEEE Transactions on Pattern Analysis and Machine Intelligence*, 28(9):1465–1479.

Leung, T. K., Burl, M. C., and Perona, P. (1995). Finding faces in cluttered scenes using random labeled graph matching. In *Fifth International Conference on Computer Vision (ICCV'95)*, pp. 637–644, Cambridge, Massachusetts.

Levenberg, K. (1944). A method for the solution of certain problems in least squares. *Quarterly of Applied Mathematics*, 2:164–168.

Levin, A. (2006). Blind motion deblurring using image statistics. In *Advances in Neural Information Processing Systems*.

Levin, A. and Szeliski, R. (2004). Visual odometry and map correlation. In *IEEE Computer Society Conference on Computer Vision and Pattern Recognition (CVPR'2004)*, pp. 611–618, Washington, DC.

Levin, A. and Szeliski, R. (2006). *Motion Uncertainty and Field of View*. Technical Report MSR-TR-2006-37, Microsoft Research.

Levin, A. and Weiss, Y. (2006). Learning to combine bottom-up and top-down segmentation. In *Ninth European Conference on Computer Vision (ECCV 2006)*, pp. 581–594.

Levin, A. and Weiss, Y. (2007). User assisted separation of reflections from a single image using a sparsity prior. *IEEE Transactions on Pattern Analysis and Machine Intelligence*, 29(9):1647–1654.

Levin, A., Acha, A. R., and Lischinski, D. (2008). Spectral matting. *IEEE Transactions on Pattern Analysis and Machine Intelligence*, 30(10):1699–1712.

Levin, A., Lischinski, D., and Weiss, Y. (2004). Colorization using optimization. *ACM Transactions on Graphics*, 23(3):689–694.

Levin, A., Lischinski, D., and Weiss, Y. (2008). A closed form solution to natural image matting. *IEEE Transactions on Pattern Analysis and Machine Intelligence*, 30(2):228–242.

Levin, A., Zomet, A., and Weiss, Y. (2004). Separating reflections from a single image using local features. In *IEEE Computer Society Conference on Computer Vision and Pattern Recognition (CVPR'2004)*, pp. 306–313, Washington, DC.

Levin, A., Fergus, R., Durand, F., and Freeman, W. T. (2007). Image and depth from a conventional camera with a coded aperture. *ACM Transactions on Graphics*, 26(3).

Levin, A., Weiss, Y., Durand, F., and Freeman, B. (2009). Understanding and evaluating blind deconvolution algorithms. In *IEEE Computer Society Conference on Computer Vision and Pattern Recognition (CVPR 2009)*, Miami Beach, FL.

Levin, A., Zomet, A., Peleg, S., and Weiss, Y. (2004). Seamless image stitching in the gradient domain. In *Eighth European Conference on Computer Vision (ECCV 2004)*, pp. 377–389, Prague.

Levoy, M. (1988). Display of surfaces from volume data. *IEEE Computer Graphics and Applications*, 8(3):29–37.

Levoy, M. (2006). Light fields and computational imaging. *Computer*, 39(8):46–55.

Levoy, M. (2008). Technical perspective: Computational photography on large collections of images. *Communications of the ACM*, 51(10):86.

Levoy, M. and Hanrahan, P. (1996). Light field rendering. In *ACM SIGGRAPH 1996 Conference Proceedings*, pp. 31–42, New Orleans.

Levoy, M. and Whitted, T. (1985). *The Use of Points as a Display Primitive*. Technical Report 85-022, University of North Carolina at Chapel Hill.

Levoy, M., Ng, R., Adams, A., Footer, M., and Horowitz, M. (2006). Light field microscopy. *ACM Transactions on Graphics*, 25(3):924–934.

Levoy, M., Pulli, K., Curless, B., Rusinkiewicz, S., Koller, D. *et al.* (2000). The digital Michelangelo project: 3D scanning of large statues. In *ACM SIGGRAPH 2000 Conference Proceedings*, pp. 131–144.

Lew, M. S., Sebe, N., Djeraba, C., and Jain, R. (2006). Content-based multimedia information retrieval: State of the art and challenges. *ACM Transactions on Multimedia Computing, Communications and Applications*, 2(1):1–19.

Leyvand, T., Cohen-Or, D., Dror, G., and Lischinski, D. (2008). Data-driven enhancement of facial attractiveness. *ACM Transactions on Graphics*, 27(3).

Lhuillier, M. and Quan, L. (2002). Match propagation for image-based modeling and rendering. *IEEE Transactions on Pattern Analysis and Machine Intelligence*, 24(8):1140–1146.

Lhuillier, M. and Quan, L. (2005). A quasi-dense approach to surface reconstruction from uncalibrated images. *IEEE Transactions on Pattern Analysis and Machine Intelligence*, 27(3):418–433.

Li, H. and Hartley, R. (2007). The 3D–3D registration problem revisited. In *Eleventh International Conference on Computer Vision (ICCV 2007)*, Rio de Janeiro, Brazil.

Li, L.-J. and Fei-Fei, L. (2010). Optimol: Automatic object picture collection via incremental model learning. *International Journal of Computer Vision*, 88(2):147–168.

Li, S. (1995). *Markov Random Field Modeling in Computer Vision*. Springer-Verlag.

Li, S. Z. and Jain, A. K. (eds). (2005). *Handbook of Face Recognition*, Springer.

Li, X., Wu, C., Zach, C., Lazebnik, S., and Frahm, J.-M. (2008). Modeling and recognition of landmark image collections using iconic scene graphs. In *Tenth European Conference on Computer Vision (ECCV 2008)*, pp. 427–440, Marseilles.

Li, Y. and Huttenlocher, D. P. (2008). Learning for optical flow using stochastic optimization. In *Tenth European Conference on Computer Vision (ECCV 2008)*, pp. 379–391, Marseilles.

Li, Y., Crandall, D. J., and Huttenlocher, D. P. (2009). Landmark classification in large-scale image collections. In *Twelfth International Conference on Computer Vision (ICCV 2009)*, Kyoto, Japan.

Li, Y., Wang, T., and Shum, H.-Y. (2002). Motion texture: a two-level statistical model for character motion synthesis. *ACM Transactions on Graphics*, 21(3):465–472.

Li, Y., Shum, H.-Y., Tang, C.-K., and Szeliski, R. (2004). Stereo reconstruction from multiperspective panoramas. *IEEE Transactions on Pattern Analysis and Machine Intelligence*, 26(1):44–62.

Li, Y., Sun, J., Tang, C.-K., and Shum, H.-Y. (2004). Lazy snapping. *ACM Transactions on Graphics (Proc. SIGGRAPH 2004)*, 23(3):303–308.

Liang, L., Xiao, R., Wen, F., and Sun, J. (2008). Face alignment via component-based discriminative search. In *Tenth European Conference on Computer Vision (ECCV 2008)*, pp. 72–85, Marseilles.

Liang, L., Liu, C., Xu, Y.-Q., Guo, B., and Shum, H.-Y. (2001). Real-time texture synthesis by patch-based sampling. *ACM Transactions on Graphics*, 20(3):127–150.

Liebowitz, D. and Zisserman, A. (1998). Metric rectification for perspective images of planes. In *IEEE Computer Society Conference on Computer Vision and Pattern Recognition (CVPR'98)*, pp. 482–488, Santa Barbara.

Lim, J. (1990). *Two-Dimensional Signal and Image Processing*. Prentice-Hall, Englewood, NJ.

Lim, J. J., Arbeláez, P., Gu, C., and Malik, J. (2009). Context by region ancestry. In *Twelfth International Conference on Computer Vision (ICCV 2009)*, Kyoto, Japan.

Lin, D., Kapoor, A., Hua, G., and Baker, S. (2010). Joint people, event, and location recognition in personal photo collections using cross-domain context. In *Eleventh European Conference on Computer Vision (ECCV 2010)*, Heraklion, Crete.

Lin, W.-C., Hays, J., Wu, C., Kwatra, V., and Liu, Y. (2006). Quantitative evaluation of near regular texture synthesis algorithms. In *IEEE Computer Society Conference on Computer Vision and Pattern Recognition (CVPR'2006)*, pp. 427–434, New York City, NY.

Lindeberg, T. (1990). Scale-space for discrete signals. *IEEE Transactions on Pattern Analysis and Machine Intelligence*, 12(3):234–254.

Lindeberg, T. (1993). Detecting salient blob-like image structures and their scales with a scale-space primal sketch: a method for focus-of-attention. *International Journal of Computer Vision*, 11(3):283–318.

Lindeberg, T. (1994). Scale-space theory: A basic tool for analysing structures at different scales. *Journal of Applied Statistics*, 21(2):224–270.

Lindeberg, T. (1998a). Edge detection and ridge detection with automatic scale selection. *International Journal of Computer Vision*, 30(2):116–154.

Lindeberg, T. (1998b). Feature detection with automatic scale selection. *International Journal of Computer Vision*, 30(2):79–116.

Lindeberg, T. and Garding, J. (1997). Shape-adapted smoothing in estimation of 3-D shape cues from affine deformations of local 2-D brightness structure. *Image and Vision Computing*, 15(6):415–434.

Lippman, A. (1980). Movie maps: An application of the optical videodisc to computer graphics. *Computer Graphics (SIGGRAPH '80)*, 14(3):32–43.

Lischinski, D., Farbman, Z., Uyttendaele, M., and Szeliski, R. (2006a). Interactive local adjustment of tonal values. *ACM Transactions on Graphics (Proc. SIGGRAPH 2006)*, 25(3):646–653.

Lischinski, D., Farbman, Z., Uyttendaele, M., and Szeliski, R. (2006b). Interactive local adjustment of tonal values. *ACM Transactions on Graphics*, 25(3):646–653.

Litvinov, A. and Schechner, Y. Y. (2005). Radiometric framework for image mosaicking. *Journal of the Optical Society of America A*, 22(5):839–848.

Litwinowicz, P. (1997). Processing images and video for an impressionist effect. In *ACM SIGGRAPH 1997 Conference Proceedings*, pp. 407–414.

Litwinowicz, P. and Williams, L. (1994). Animating images with drawings. In *ACM SIGGRAPH 1994 Conference Proceedings*, pp. 409–412.

Liu, C. (2009). *Beyond Pixels: Exploring New Representations and Applications for Motion Analysis*. Ph.D. thesis, Massachusetts Institute of Technology.

Liu, C., Yuen, J., and Torralba, A. (2009). Nonparametric scene parsing: Label transfer via dense scene alignment. In *IEEE Computer Society Conference on Computer Vision and Pattern Recognition (CVPR 2009)*, Miami Beach, FL.

Liu, C., Freeman, W. T., Adelson, E., and Weiss, Y. (2008). Human-assisted motion annotation. In *IEEE Computer Society Conference on Computer Vision and Pattern Recognition (CVPR 2008)*, Anchorage, AK.

Liu, C., Szeliski, R., Kang, S. B., Zitnick, C. L., and Freeman, W. T. (2008). Automatic estimation and removal of noise from a single image. *IEEE Transactions on Pattern Analysis and Machine Intelligence*, 30(2):299–314.

Liu, F., Gleicher, M., Jin, H., and Agarwala, A. (2009). Content-preserving warps for 3d video stabilization. *ACM Transactions on Graphics*, 28(3).

Liu, X., Chen, T., and Kumar, B. V. (2003). Face authentication for multiple subjects using eigenflow. *Pattern Recognition*, 36(2):313–328.

Liu, Y., Collins, R. T., and Tsin, Y. (2004). A computational model for periodic pattern perception based on frieze and wallpaper groups. *IEEE Transactions on Pattern Analysis and Machine Intelligence*, 26(3):354–371.

Liu, Y., Lin, W.-C., and Hays, J. (2004). Near-regular texture analysis and manipulation. *ACM Transactions on Graphics*, 23(3):368–376.

Livingstone, M. (2008). *Vision and Art: The Biology of Seeing*. Abrams, New York.

Lobay, A. and Forsyth, D. A. (2006). Shape from texture without boundaries. *International Journal of Computer Vision*, 67(1):71–91.

Lombaert, H., Sun, Y., Grady, L., and Xu, C. (2005). A multilevel banded graph cuts method for fast image segmentation. In *Tenth International Conference on Computer Vision (ICCV 2005)*, pp. 259–265, Beijing, China.

Longere, P., Delahunt, P. B., Zhang, X., and Brainard, D. H. (2002). Perceptual assessment of demosaicing algorithm performance. *Proceedings of the IEEE*, 90(1):123–132.

Longuet-Higgins, H. C. (1981). A computer algorithm for reconstructing a scene from two projections. *Nature*, 293:133–135.

Loop, C. and Zhang, Z. (1999). Computing rectifying homographies for stereo vision. In *IEEE Computer Society Conference on Computer Vision and Pattern Recognition (CVPR'99)*, pp. 125–131, Fort Collins.

Lorensen, W. E. and Cline, H. E. (1987). Marching cubes: A high resolution 3D surface construction algorithm. *Computer Graphics (SIGGRAPH '87)*, 21(4):163–169.

Lorusso, A., Eggert, D., and Fisher, R. B. (1995). A comparison of four algorithms for estimating 3-D rigid transformations. In *British Machine Vision Conference (BMVC95)*, pp. 237–246, Birmingham, England.

Lourakis, M. I. A. and Argyros, A. A. (2009). SBA: A software package for generic sparse bundle adjustment. *ACM Transactions on Mathematical Software*, 36(1).

Lowe, D. G. (1988). Organization of smooth image curves at multiple scales. In *Second International Conference on Computer Vision (ICCV'88)*, pp. 558–567, Tampa.

Lowe, D. G. (1989). Organization of smooth image curves at multiple scales. *International Journal of Computer Vision*, 3(2):119–130.

Lowe, D. G. (1999). Object recognition from local scale-invariant features. In *Seventh International Conference on Computer Vision (ICCV'99)*, pp. 1150–1157, Kerkyra, Greece.

Lowe, D. G. (2004). Distinctive image features from scale-invariant keypoints. *International Journal of Computer Vision*, 60(2):91–110.

Lucas, B. D. and Kanade, T. (1981). An iterative image registration technique with an application in stereo vision. In *Seventh International Joint Conference on Artificial Intelligence (IJCAI-81)*, pp. 674–679, Vancouver.

Luong, Q.-T. and Faugeras, O. D. (1996). The fundamental matrix: Theory, algorithms, and stability analysis. *International Journal of Computer Vision*, 17(1):43–75.

Luong, Q.-T. and Viéville, T. (1996). Canonical representations for the geometries of multiple projective views. *Computer Vision and Image Understanding*, 64(2):193–229.

Lyu, S. and Simoncelli, E. (2008). Nonlinear image representation using divisive normalization. In *IEEE Computer Society Conference on Computer Vision and Pattern Recognition (CVPR 2008)*, Anchorage, AK.

Lyu, S. and Simoncelli, E. (2009). Modeling multiscale subbands of photographic images with fields of Gaussian scale mixtures. *IEEE Transactions on Pattern Analysis and Machine Intelligence*, 31(4):693–706.

Ma, W.-C., Jones, A., Chiang, J.-Y., Hawkins, T., Frederiksen, S., Peers, P., Vukovic, M., Ouhyoung, M., and Debevec, P. (2008). Facial performance synthesis using deformation-driven polynomial displacement maps. *ACM Transactions on Graphics*, 27(5).

Ma, Y., Derksen, H., Hong, W., and Wright, J. (2007). Segmentation of multivariate mixed data via lossy data coding and compression. *IEEE Transactions on Pattern Analysis and Machine Intelligence*, 29(9):1546–1562.

MacDonald, L. (ed.). (2006). *Digital Heritage: Applying Digital Imaging to Cultural Heritage*, Butterworth-Heinemann.

Madsen, K., Nielsen, H. B., and Tingleff, O. (2004). Methods for non-linear least squares problems. Informatics and Mathematical Modelling, Technical University of Denmark (DTU).

Maes, F., Collignon, A., Vandermeulen, D., Marchal, G., and Suetens, P. (1997). Multi-modality image registration by maximization of mutual information. *IEEE Transactions on Medical Imaging*, 16(2):187–198.

Magnor, M. (2005). *Video-Based Rendering*. A. K. Peters, Wellesley, MA.

Magnor, M. and Girod, B. (2000). Data compression for light-field rendering. *IEEE Transactions on Circuits and Systems for Video Technology*, 10(3):338–343.

Magnor, M., Ramanathan, P., and Girod, B. (2003). Multi-view coding for image-based rendering using 3-D scene geometry. *IEEE Transactions on Circuits and Systems for Video Technology*, 13(11):1092–1106.

Mahajan, D., Huang, F.-C., Matusik, W., Ramamoorthi, R., and Belhumeur, P. (2009). Moving gradients: A path-based method for plausible image interpolation. *ACM Transactions on Graphics*, 28(3).

Maimone, M., Cheng, Y., and Matthies, L. (2007). Two years of visual odometry on the Mars exploration rovers. *Journal of Field Robotics*, 24(3).

Maire, M., Arbelaez, P., Fowlkes, C., and Malik, J. (2008). Using contours to detect and localize junctions in natural images. In *IEEE Computer Society Conference on Computer Vision and Pattern Recognition (CVPR 2008)*, Anchorage, AK.

Maitin-Shepard, J., Cusumano-Towner, M., Lei, J., and Abbeel, P. (2010). Cloth grasp point detection based on multiple-view geometric cues with application to robotic towel folding. In *IEEE International Conference on Robotics and Automation*, Anchorage, AK.

Maitre, M., Shinagawa, Y., and Do, M. N. (2008). Symmetric multi-view stereo reconstruction from planar camera arrays. In *IEEE Computer Society Conference on Computer Vision and Pattern Recognition (CVPR 2008)*, Anchorage, AK.

Maji, S., Berg, A., and Malik, J. (2008). Classification using intersection kernel support vector machines is efficient. In *IEEE Computer Society Conference on Computer Vision and Pattern Recognition (CVPR 2008)*, Anchorage, AK.

Malik, J. and Rosenholtz, R. (1997). Computing local surface orientation and shape from texture for curved surfaces. *International Journal of Computer Vision*, 23(2):149–168.

Malik, J., Belongie, S., Leung, T., and Shi, J. (2001). Contour and texture analysis for image segmentation. *International Journal of Computer Vision*, 43(1):7–27.

Malisiewicz, T. and Efros, A. A. (2008). Recognition by association via learning per-exemplar distances. In *IEEE Computer Society Conference on Computer Vision and Pattern Recognition (CVPR 2008)*, Anchorage, AK.

Malladi, R., Sethian, J. A., and Vemuri, B. C. (1995). Shape modeling with front propagation. *IEEE Transactions on Pattern Analysis and Machine Intelligence*, 17(2):158–176.

Mallat, S. G. (1989). A theory for multiresolution signal decomposition: the wavelet representation. *IEEE Transactions on Pattern Analysis and Machine Intelligence*, PAMI-11(7):674–693.

Malvar, H. S. (1990). Lapped transforms for efficient transform/subband coding. *IEEE Transactions on Acoustics, Speech, and Signal Processing*, 38(6):969–978.

Malvar, H. S. (1998). Biorthogonal and nonuniform lapped transforms for transform coding with reduced blocking and ringing artifacts. *IEEE Transactions on Signal Processing*, 46(4):1043–1053.

Malvar, H. S. (2000). Fast progressive image coding without wavelets. In *IEEE Data Compressions Conference*, pp. 243–252, Snowbird, UT.

Malvar, H. S., He, L.-W., and Cutler, R. (2004). High-quality linear interpolation for demosaicing of Bayer-patterned color images. In *IEEE International Conference on Acoustics, Speech, and Signal Processing (ICASSP'04)*, pp. 485–488, Montreal.

Mancini, T. A. and Wolff, L. B. (1992). 3D shape and light source location from depth and reflectance. In *IEEE Computer Society Conference on Computer Vision and Pattern Recognition (CVPR'92)*, pp. 707–709, Champaign, Illinois.

Manjunathi, B. S. and Ma, W. Y. (1996). Texture features for browsing and retrieval of image data. *IEEE Transactions on Pattern Analysis and Machine Intelligence*, 18(8):837–842.

Mann, S. and Picard, R. W. (1994). Virtual bellows: Constructing high-quality images from video. In *First IEEE International Conference on Image Processing (ICIP-94)*, pp. 363–367, Austin.

Mann, S. and Picard, R. W. (1995). On being 'undigital' with digital cameras: Extending dynamic range by combining differently exposed pictures. In *IS&T's 48th Annual Conference*, pp. 422–428, Washington, D. C.

Manning, C. D., Raghavan, P., and Schütze, H. (2008). *Introduction to Information Retrieval*. Cambridge University Press.

Marquardt, D. W. (1963). An algorithm for least-squares estimation of nonlinear parameters. *Journal of the Society for Industrial and Applied Mathematics*, 11(2):431–441.

Marr, D. (1982). *Vision: A Computational Investigation into the Human Representation and Processing of Visual Information*. W. H. Freeman, San Francisco.

Marr, D. and Hildreth, E. (1980). Theory of edge detection. *Proceedings of the Royal Society of London*, B 207:187–217.

Marr, D. and Nishihara, H. K. (1978). Representation and recognition of the spatial organization of three-dimensional shapes. *Proc. Roy. Soc. London, B*, 200:269–294.

Marr, D. and Poggio, T. (1976). Cooperative computation of stereo disparity. *Science*, 194:283–287.

Marr, D. C. and Poggio, T. (1979). A computational theory of human stereo vision. *Proceedings of the Royal Society of London*, B 204:301–328.

Marroquin, J., Mitter, S., and Poggio, T. (1985). Probabilistic solution of ill-posed problems in computational vision. In *Image Understanding Workshop*, pp. 293–309, Miami Beach.

Marroquin, J., Mitter, S., and Poggio, T. (1987). Probabilistic solution of ill-posed problems in computational vision. *Journal of the American Statistical Association*, 82(397):76–89.

Marroquin, J. L. (1983). *Design of Cooperative Networks*. Working Paper 253, Artificial Intelligence Laboratory, Massachusetts Institute of Technology.

Martin, D., Fowlkes, C., and Malik, J. (2004). Learning to detect natural image boundaries using local brightness, color, and texture cues. *IEEE Transactions on Pattern Analysis and Machine Intelligence*, 26(5):530–549.

Martin, D., Fowlkes, C., Tal, D., and Malik, J. (2001). A database of human segmented natural images and its application to evaluating segmentation algorithms and measuring ecological statistics. In *Eighth International Conference on Computer Vision (ICCV 2001)*, pp. 416–423, Vancouver, Canada.

Martin, W. N. and Aggarwal, J. K. (1983). Volumetric description of objects from multiple views. *IEEE Transactions on Pattern Analysis and Machine Intelligence*, PAMI-5(2):150–158.

Martinec, D. and Pajdla, T. (2007). Robust rotation and translation estimation in multiview reconstruction. In *IEEE Computer Society Conference on Computer Vision and Pattern Recognition (CVPR 2007)*, Minneapolis, MN.

Massey, M. and Bender, W. (1996). Salient stills: Process and practice. *IBM Systems Journal*, 35(3&4):557–573.

Matas, J., Chum, O., Urban, M., and Pajdla, T. (2004). Robust wide baseline stereo from maximally stable extremal regions. *Image and Vision Computing*, 22(10):761–767.

Matei, B. C. and Meer, P. (2006). Estimation of nonlinear errors-in-variables models for computer vision applications. *IEEE Transactions on Pattern Analysis and Machine Intelligence*, 28(10):1537–1552.

Matsushita, Y. and Lin, S. (2007). Radiometric calibration from noise distributions. In *IEEE Computer Society Conference on Computer Vision and Pattern Recognition (CVPR 2007)*, Minneapolis, MN.

Matsushita, Y., Ofek, E., Ge, W., Tang, X., and Shum, H.-Y. (2006). Full-frame video stabilization with motion inpainting. *IEEE Transactions on Pattern Analysis and Machine*

Intelligence, 28(7):1150–1163.

Matthews, I. and Baker, S. (2004). Active appearance models revisited. *International Journal of Computer Vision*, 60(2):135–164.

Matthews, I., Xiao, J., and Baker, S. (2007). 2D vs. 3D deformable face models: Representational power, construction, and real-time fitting. *International Journal of Computer Vision*, 75(1):93–113.

Matthies, L., Kanade, T., and Szeliski, R. (1989). Kalman filter-based algorithms for estimating depth from image sequences. *International Journal of Computer Vision*, 3(3):209–236.

Matusik, W., Buehler, C., and McMillan, L. (2001). Polyhedral visual hulls for real-time rendering. In *12th Eurographics Workshop on Rendering Techniques*, pp. 115–126, London.

Matusik, W., Buehler, C., Raskar, R., Gortler, S. J., and McMillan, L. (2000). Image-based visual hulls. In *ACM SIGGRAPH 2000 Conference Proceedings*, pp. 369–374.

Mayhew, J. E. W. and Frisby, J. P. (1980). The computation of binocular edges. *Perception*, 9:69–87.

Mayhew, J. E. W. and Frisby, J. P. (1981). Psychophysical and computational studies towards a theory of human stereopsis. *Artificial Intelligence*, 17(1-3):349–408.

McCamy, C. S., Marcus, H., and Davidson, J. G. (1976). A color-rendition chart. *Journal of Applied Photogrammetric Engineering*, 2(3):95–99.

McCane, B., Novins, K., Crannitch, D., and Galvin, B. (2001). On benchmarking optical flow. *Computer Vision and Image Understanding*, 84(1):126–143.

McGuire, M., Matusik, W., Pfister, H., Hughes, J. F., and Durand, F. (2005). Defocus video matting. *ACM Transactions on Graphics (Proc. SIGGRAPH 2005)*, 24(3):567–576.

McInerney, T. and Terzopoulos, D. (1993). A finite element model for 3D shape reconstruction and nonrigid motion tracking. In *Fourth International Conference on Computer Vision (ICCV'93)*, pp. 518–523, Berlin, Germany.

McInerney, T. and Terzopoulos, D. (1996). Deformable models in medical image analysis: A survey. *Medical Image Analysis*, 1(2):91–108.

McInerney, T. and Terzopoulos, D. (1999). Topology adaptive deformable surfaces for medical image volume segmentation. *IEEE Transactions on Medical Imaging*, 18(10):840–850.

McInerney, T. and Terzopoulos, D. (2000). T-snakes: Topology adaptive snakes. *Medical Image Analysis*, 4:73–91.

McLauchlan, P. F. (2000). A batch/recursive algorithm for 3D scene reconstruction. In *IEEE Computer Society Conference on Computer Vision and Pattern Recognition (CVPR'2000)*, pp. 738–743, Hilton Head Island.

McLauchlan, P. F. and Jaenicke, A. (2002). Image mosaicing using sequential bundle adjustment. *Image and Vision Computing*, 20(9-10):751–759.

McLean, G. F. and Kotturi, D. (1995). Vanishing point detection by line clustering. *IEEE Transactions on Pattern Analysis and Machine Intelligence*, 17(11):1090–1095.

McMillan, L. and Bishop, G. (1995). Plenoptic modeling: An image-based rendering system. In *ACM SIGGRAPH 1995 Conference Proceedings*, pp. 39–46.

McMillan, L. and Gortler, S. (1999). Image-based rendering: A new interface between computer vision and computer graphics. *Computer Graphics*, 33(4):61–64.

Meehan, J. (1990). *Panoramic Photography*. Watson-Guptill.

Meer, P. and Georgescu, B. (2001). Edge detection with embedded confidence. *IEEE Transactions on Pattern Analysis and Machine Intelligence*, 23(12):1351–1365.

Meilă, M. and Shi, J. (2000). Learning segmentation by random walks. In *Advances in Neural Information Processing Systems*.

Meilă, M. and Shi, J. (2001). A random walks view of spectral segmentation. In *Workshop on Artificial Intelligence and Statistics*, pp. 177–182, Key West, FL.

Meltzer, J. and Soatto, S. (2008). Edge descriptors for robust wide-baseline correspondence. In *IEEE Computer Society Conference on Computer Vision and Pattern Recognition (CVPR 2008)*, Anchorage, AK.

Meltzer, T., Yanover, C., and Weiss, Y. (2005). Globally optimal solutions for energy minimization in stereo vision using reweighted belief propagation. In *Tenth International Conference on Computer Vision (ICCV 2005)*, pp. 428–435, Beijing, China.

Mémin, E. and Pérez, P. (2002). Hierarchical estimation and segmentation of dense motion fields. *International Journal of Computer Vision*, 44(2):129–155.

Menet, S., Saint-Marc, P., and Medioni, G. (1990a). Active contour models: overview, implementation and applications. In *IEEE International Conference on Systems, Man and Cybernetics*, pp. 194–199, Los Angeles.

Menet, S., Saint-Marc, P., and Medioni, G. (1990b). B-snakes: implementation and applications to stereo. In *Image Understanding Workshop*, pp. 720–726, Pittsburgh.

Merrell, P., Akbarzadeh, A., Wang, L., Mordohai, P., Frahm, J.-M., Yang, R., Nister, D., and Pollefeys, M. (2007). Real-time visibility-based fusion of depth maps. In *Eleventh International Conference on Computer Vision (ICCV 2007)*, Rio de Janeiro, Brazil.

Mertens, T., Kautz, J., and Reeth, F. V. (2007). Exposure fusion. In *Proceedings of Pacific Graphics 2007*, pp. 382–390.

Metaxas, D. and Terzopoulos, D. (2002). Dynamic deformation of solid primitives with constraints. *ACM Transactions on Graphics (Proc. SIGGRAPH 2002)*, 21(3):309–312.

Metropolis, N., Rosenbluth, A. W., Rosenbluth, M. N., Teller, A. H., and Teller, E. (1953). Equations of state calculations by fast computing machines. *Journal of Chemical Physics*, 21:1087–1091.

Meyer, C. D. (2000). *Matrix Analysis and Applied Linear Algebra*. Society for Industrial and Applied Mathematics, Philadephia.

Meyer, Y. (1993). *Wavelets: Algorithms and Applications*. Society for Industrial and Applied Mathematics, Philadephia.

Mikolajczyk, K. and Schmid, C. (2004). Scale & affine invariant interest point detectors. *International Journal of Computer Vision*, 60(1):63–86.

Mikolajczyk, K. and Schmid, C. (2005). A performance evaluation of local descriptors. *IEEE Transactions on Pattern Analysis and Machine Intelligence*, 27(10):1615–1630.

Mikolajczyk, K., Schmid, C., and Zisserman, A. (2004). Human detection based on a probabilistic assembly of robust part detectors. In *Eighth European Conference on Computer Vision (ECCV 2004)*, pp. 69–82, Prague.

Mikolajczyk, K., Tuytelaars, T., Schmid, C., Zisserman, A., Matas, J., Schaffalitzky, F., Kadir, T., and Van Gool, L. J. (2005). A comparison of affine region detectors. *International Journal of Computer Vision*, 65(1-2):43–72.

Milgram, D. L. (1975). Computer methods for creating photomosaics. *IEEE Transactions on Computers*, C-24(11):1113–1119.

Milgram, D. L. (1977). Adaptive techniques for photomosaicking. *IEEE Transactions on Computers*, C-26(11):1175–1180.

Miller, I., Campbell, M., Huttenlocher, D., Kline, F.-R., Nathan, A. *et al.* (2008). Team Cornell's Skynet: Robust perception and planning in an urban environment. *Journal of Field Robotics*, 25(8):493–527.

Mitiche, A. and Bouthemy, P. (1996). Computation and analysis of image motion: A synopsis of current problems and methods. *International Journal of Computer Vision*, 19(1):29–55.

Mitsunaga, T. and Nayar, S. K. (1999). Radiometric self calibration. In *IEEE Computer Society Conference on Computer Vision and Pattern Recognition (CVPR'99)*, pp. 374–380, Fort Collins.

Mittal, A. and Davis, L. S. (2003). M$_2$ tracker: A multi-view approach to segmenting and tracking people in a cluttered scene. *International Journal of Computer Vision*, 51(3):189–203.

Mičušík, B. and Košecká, J. (2009). Piecewise planar city 3D modeling from street view panoramic sequences. In *IEEE Computer Society Conference on Computer Vision and Pattern Recognition (CVPR 2009)*, Miami Beach, FL.

Mičušìk, B., Wildenauer, H., and Košecká, J. (2008). Detection and matching of rectilinear structures. In *IEEE Computer Society Conference on Computer Vision and Pattern Recognition (CVPR 2008)*, Anchorage, AK.

Moeslund, T. B. and Granum, E. (2001). A survey of computer vision-based human motion capture. *Computer Vision and Image Understanding*, 81(3):231–268.

Moeslund, T. B., Hilton, A., and Krüger, V. (2006). A survey of advances in vision-based human motion capture and analysis. *Computer Vision and Image Understanding*, 104(2-3):90–126.

Moezzi, S., Katkere, A., Kuramura, D., and Jain, R. (1996). Reality modeling and visualization from multiple video sequences. *IEEE Computer Graphics and Applications*, 16(6):58–63.

Moghaddam, B. and Pentland, A. (1997). Probabilistic visual learning for object representation. *IEEE Transactions on Pattern Analysis and Machine Intelligence*, 19(7):696–710.

Moghaddam, B., Jebara, T., and Pentland, A. (2000). Bayesian face recognition. *Pattern Recognition*, 33(11):1771–1782.

Mohan, A., Papageorgiou, C., and Poggio, T. (2001). Example-based object detection in images by components. *IEEE Transactions on Pattern Analysis and Machine Intelligence*, 23(4):349–361.

Möller, K. D. (1988). *Optics*. University Science Books, Mill Valley, CA.

Montemerlo, M., Becker, J., Bhat, S., Dahlkamp, H., Dolgov, D. *et al.* (2008). Junior: The Stanford entry in the Urban Challenge. *Journal of Field Robotics*, 25(9):569–597.

Moon, P. and Spencer, D. E. (1981). *The Photic Field*. MIT Press, Cambridge, Massachusetts.

Moons, T., Van Gool, L., and Vergauwen, M. (2010). 3D reconstruction from multiple images. *Foundations and Trends in Computer Graphics and Computer Vision*, 4(4).

Moosmann, F., Nowak, E., and Jurie, F. (2008). Randomized clustering forests for image classification. *IEEE Transactions on Pattern Analysis and Machine Intelligence*, 30(9):1632–1646.

Moravec, H. (1977). Towards automatic visual obstacle avoidance. In *Fifth International Joint Conference on Artificial Intelligence (IJCAI'77)*, p. 584, Cambridge, Massachusetts.

Moravec, H. (1983). The Stanford cart and the CMU rover. *Proceedings of the IEEE*, 71(7):872–884.

Moreno-Noguer, F., Lepetit, V., and Fua, P. (2007). Accurate non-iterative $O(n)$ solution to the PnP problem. In *Eleventh International Conference on Computer Vision (ICCV 2007)*, Rio de Janeiro, Brazil.

Mori, G. (2005). Guiding model search using segmentation. In *Tenth International Conference on Computer Vision (ICCV 2005)*, pp. 1417–1423, Beijing, China.

Mori, G., Ren, X., Efros, A., and Malik, J. (2004). Recovering human body configurations: Combining segmentation and recognition. In *IEEE Computer Society Conference on Computer Vision and Pattern Recognition (CVPR'2004)*, pp. 326–333, Washington, DC.

Mori, M. (1970). The uncanny valley. *Energy*, 7(4):33–35. http://www.androidscience.com/theuncannyvalley/proceedings2005/uncannyvalley.html.

Morimoto, C. and Chellappa, R. (1997). Fast 3D stabilization and mosaic construction. In *IEEE Computer Society Conference on Computer Vision and Pattern Recognition (CVPR'97)*, pp. 660–665, San Juan, Puerto Rico.

Morita, T. and Kanade, T. (1997). A sequential factorization method for recovering shape and motion from image streams. *IEEE Transactions on Pattern Analysis and Machine Intelligence*, 19(8):858–867.

Morris, D. D. and Kanade, T. (1998). A unified factorization algorithm for points, line segments and planes with uncertainty models. In *Sixth International Conference on Computer Vision (ICCV'98)*, pp. 696–702, Bombay.

Morrone, M. and Burr, D. (1988). Feature detection in human vision: A phase dependent energy model. *Proceedings of the Royal Society of London B*, 235:221–245.

Mortensen, E. N. (1999). Vision-assisted image editing. *Computer Graphics*, 33(4):55–57.

Mortensen, E. N. and Barrett, W. A. (1995). Intelligent scissors for image composition. In *ACM SIGGRAPH 1995 Conference Proceedings*, pp. 191–198.

Mortensen, E. N. and Barrett, W. A. (1998). Interactive segmentation with intelligent scissors. *Graphical Models and Image Processing*, 60(5):349–384.

Mortensen, E. N. and Barrett, W. A. (1999). Toboggan-based intelligent scissors with a four parameter edge model. In *IEEE Computer Society Conference on Computer Vision and Pattern Recognition (CVPR'99)*, pp. 452–458, Fort Collins.

Mueller, P., Zeng, G., Wonka, P., and Van Gool, L. (2007). Image-based procedural modeling of facades. *ACM Transactions on Graphics*, 26(3).

Mühlich, M. and Mester, R. (1998). The role of total least squares in motion analysis. In *Fifth European Conference on Computer Vision (ECCV'98)*, pp. 305–321, Freiburg, Germany.

Muja, M. and Lowe, D. G. (2009). Fast approximate nearest neighbors with automatic algorithm configuration. In *International Conference on Computer Vision Theory and Applications (VISAPP)*, Lisbon, Portugal.

Mumford, D. and Shah, J. (1989). Optimal approximations by piecewise smooth functions and variational problems. *Comm. Pure Appl. Math.*, XLII(5):577–685.

Munder, S. and Gavrila, D. M. (2006). An experimental study on pedestrian classification. *IEEE Transactions on Pattern Analysis and Machine Intelligence*, 28(11):1863–1868.

Mundy, J. L. (2006). Object recognition in the geometric era: A retrospective. In Ponce, J., Hebert, M., Schmid, C., and Zisserman, A. (eds), *Toward Category-Level Object Recognition*, pp. 3–28, Springer, New York.

Mundy, J. L. and Zisserman, A. (eds). (1992). *Geometric Invariance in Computer Vision*. MIT Press, Cambridge, Massachusetts.

Murase, H. and Nayar, S. K. (1995). Visual learning and recognition of 3-D objects from appearance. *International Journal of Computer Vision*, 14(1):5–24.

Murphy, E. P. (2005). *A Testing Procedure to Characterize Color and Spatial Quality of Digital Cameras Used to Image Cultural Heritage*. Master's thesis, Rochester Institute of Technology.

Murphy, K., Torralba, A., and Freeman, W. T. (2003). Using the forest to see the trees: A graphical model relating features, objects, and scenes. In *Advances in Neural Information Processing Systems*.

Murphy-Chutorian, E. and Trivedi, M. M. (2009). Head pose estimation in computer vision: A survey. *IEEE Transactions on Pattern Analysis and Machine Intelligence*, 31(4):607–626.

Murray, R. M., Li, Z. X., and Sastry, S. S. (1994). *A Mathematical Introduction to Robotic Manipulation*. CRC Press.

Mutch, J. and Lowe, D. G. (2008). Object class recognition and localization using sparse features with limited receptive fields. *International Journal of Computer Vision*, 80(1):45–57.

Nagel, H. H. (1986). Image sequences—ten (octal) years—from phenomenology towards a theoretical foundation. In *Eighth International Conference on Pattern Recognition (ICPR'86)*, pp. 1174–1185, Paris.

Nagel, H.-H. and Enkelmann, W. (1986). An investigation of smoothness constraints for the estimation of displacement vector fields from image sequences. *IEEE Transactions on Pattern Analysis and Machine Intelligence*, PAMI-8(5):565–593.

Nakamura, Y., Matsuura, T., Satoh, K., and Ohta, Y. (1996). Occlusion detectable stereo—occlusion patterns in camera matrix. In *IEEE Computer Society Conference on Computer Vision and Pattern Recognition (CVPR'96)*, pp. 371–378, San Francisco.

Nakao, T., Kashitani, A., and Kaneyoshi, A. (1998). Scanning a document with a small camera attached to a mouse. In *IEEE Workshop on Applications of Computer Vision (WACV'98)*, pp. 63–68, Princeton.

Nalwa, V. S. (1987). Edge-detector resolution improvement by image interpolation. *IEEE Transactions on Pattern Analysis and Machine Intelligence*, PAMI-9(3):446–451.

Nalwa, V. S. (1993). *A Guided Tour of Computer Vision*. Addison-Wesley, Reading, MA.

Nalwa, V. S. and Binford, T. O. (1986). On detecting edges. *IEEE Transactions on Pattern Analysis and Machine Intelligence*, PAMI-8(6):699–714.

Narasimhan, S. G. and Nayar, S. K. (2005). Enhancing resolution along multiple imaging dimensions using assorted pixels. *IEEE Transactions on Pattern Analysis and Machine Intelligence*, 27(4):518–530.

Narayanan, P., Rander, P., and Kanade, T. (1998). Constructing virtual worlds using dense stereo. In *Sixth International Conference on Computer Vision (ICCV'98)*, pp. 3–10, Bombay.

Nayar, S., Watanabe, M., and Noguchi, M. (1995). Real-time focus range sensor. In *Fifth International Conference on Computer Vision (ICCV'95)*, pp. 995–1001, Cambridge, Massachusetts.

Nayar, S. K. (2006). Computational cameras: Redefining the image. *Computer*, 39(8):30–38.

Nayar, S. K. and Branzoi, V. (2003). Adaptive dynamic range imaging: Optical control of pixel exposures over space and time. In *Ninth International Conference on Computer Vision (ICCV 2003)*, pp. 1168–1175, Nice, France.

Nayar, S. K. and Mitsunaga, T. (2000). High dynamic range imaging: Spatially varying pixel exposures. In *IEEE Computer Society Conference on Computer Vision and Pattern Recognition (CVPR'2000)*, pp. 472–479, Hilton Head Island.

Nayar, S. K. and Nakagawa, Y. (1994). Shape from focus. *IEEE Transactions on Pattern Analysis and Machine Intelligence*, 16(8):824–831.

Nayar, S. K., Ikeuchi, K., and Kanade, T. (1991). Shape from interreflections. *International Journal of Computer Vision*, 6(3):173–195.

Nayar, S. K., Watanabe, M., and Noguchi, M. (1996). Real-time focus range sensor. *IEEE Transactions on Pattern Analysis and Machine Intelligence*, 18(12):1186–1198.

Negahdaripour, S. (1998). Revised definition of optical flow: Integration of radiometric and geometric cues for dynamic scene analysis. *IEEE Transactions on Pattern Analysis and Machine Intelligence*, 20(9):961–979.

Nehab, D., Rusinkiewicz, S., Davis, J., and Ramamoorthi, R. (2005). Efficiently combining positions and normals for precise 3d geometry. *ACM Transactions on Graphics (Proc. SIGGRAPH 2005)*, 24(3):536–543.

Nene, S. and Nayar, S. K. (1997). A simple algorithm for nearest neighbor search in high dimensions. *IEEE Transactions on Pattern Analysis and Machine Intelligence*, 19(9):989–1003.

Nene, S. A., Nayar, S. K., and Murase, H. (1996). *Columbia Object Image Library (COIL-100)*. Technical Report CUCS-006-96, Department of Computer Science, Columbia University.

Netravali, A. and Robbins, J. (1979). Motion-compensated television coding: Part 1. *Bell System Tech.*, 58(3):631–670.

Nevatia, R. (1977). A color edge detector and its use in scene segmentation. *IEEE Transactions on Systems, Man, and Cybernetics*, SMC-7(11):820–826.

Nevatia, R. and Binford, T. (1977). Description and recognition of curved objects. *Artificial Intelligence*, 8:77–98.

Ng, A. Y., Jordan, M. I., and Weiss, Y. (2001). On spectral clustering: Analysis and an algorithm. In *Advances in Neural Information Processing Systems*, pp. 849–854.

Ng, R. (2005). Fourier slice photography. *ACM Transactions on Graphics (Proc. SIGGRAPH 2005)*, 24(3):735–744.

Ng, R., Levoy, M., Bréedif, M., Duval, G., Horowitz, M., and Hanrahan, P. (2005). *Light Field Photography with a Hand-held Plenoptic Camera*. Technical Report CSTR 2005-02, Stanford University.

Nielsen, M., Florack, L. M. J., and Deriche, R. (1997). Regularization, scale-space, and edge-detection filters. *Journal of Mathematical Imaging and Vision*, 7(4):291–307.

Nielson, G. M. (1993). Scattered data modeling. *IEEE Computer Graphics and Applications*, 13(1):60–70.

Nir, T., Bruckstein, A. M., and Kimmel, R. (2008). Over-parameterized variational optical flow. *International Journal of Computer Vision*, 76(2):205–216.

Nishihara, H. K. (1984). Practical real-time imaging stereo matcher. *OptEng*, 23(5):536–545.

Nistér, D. (2003). Preemptive RANSAC for live structure and motion estimation. In *Ninth International Conference on Computer Vision (ICCV 2003)*, pp. 199–206, Nice, France.

Nistér, D. (2004). An efficient solution to the five-point relative pose problem. *IEEE Transactions on Pattern Analysis and Machine Intelligence*, 26(6):756–777.

Nistér, D. and Stewénius, H. (2006). Scalable recognition with a vocabulary tree. In *IEEE Computer Society Conference on Computer Vision and Pattern Recognition (CVPR'2006)*, pp. 2161–2168, New York City, NY.

Nistér, D. and Stewénius, H. (2008). Linear time maximally stable extremal regions. In *Tenth European Conference on Computer Vision (ECCV 2008)*, pp. 183–196, Marseilles.

Nistér, D., Naroditsky, O., and Bergen, J. (2006). Visual odometry for ground vehicle applications. *Journal of Field Robotics*, 23(1):3–20.

Noborio, H., Fukada, S., and Arimoto, S. (1988). Construction of the octree approximating three-dimensional objects by using multiple views. *IEEE Transactions on Pattern Analysis and Machine Intelligence*, PAMI-10(6):769–782.

Nocedal, J. and Wright, S. J. (2006). *Numerical Optimization*. Springer, New York, second edition.

Nomura, Y., Zhang, L., and Nayar, S. K. (2007). Scene collages and flexible camera arrays. In *Eurographics Symposium on Rendering*.

Nordström, N. (1990). Biased anisotropic diffusion: A unified regularization and diffusion approach to edge detection. *Image and Vision Computing*, 8(4):318–327.

Nowak, E., Jurie, F., and Triggs, B. (2006). Sampling strategies for bag-of-features image classification. In *Ninth European Conference on Computer Vision (ECCV 2006)*, pp. 490–503.

Obdržálek, S. and Matas, J. (2006). Object recognition using local affine frames on maximally stable extremal regions. In Ponce, J., Hebert, M., Schmid, C., and Zisserman, A. (eds), *Toward Category-Level Object Recognition*, pp. 83–104, Springer, New York.

Oh, B. M., Chen, M., Dorsey, J., and Durand, F. (2001). Image-based modeling and photo editing. In *ACM SIGGRAPH 2001 Conference Proceedings*, pp. 433–442.

Ohlander, R., Price, K., and Reddy, D. R. (1978). Picture segmentation using a recursive region splitting method. *Computer Graphics and Image Processing*, 8(3):313–333.

Ohta, Y. and Kanade, T. (1985). Stereo by intra- and inter-scanline search using dynamic programming. *IEEE Transactions on Pattern Analysis and Machine Intelligence*, PAMI-7(2):139–154.

Ohtake, Y., Belyaev, A., Alexa, M., Turk, G., and Seidel, H.-P. (2003). Multi-level partition of unity implicits. *ACM Transactions on Graphics (Proc. SIGGRAPH 2003)*, 22(3):463–470.

Okutomi, M. and Kanade, T. (1992). A locally adaptive window for signal matching. *International Journal of Computer Vision*, 7(2):143–162.

Okutomi, M. and Kanade, T. (1993). A multiple baseline stereo. *IEEE Transactions on Pattern Analysis and Machine Intelligence*, 15(4):353–363.

Okutomi, M. and Kanade, T. (1994). A stereo matching algorithm with an adaptive window: Theory and experiment. *IEEE Transactions on Pattern Analysis and Machine Intelligence*, 16(9):920–932.

Oliensis, J. (2005). The least-squares error for structure from infinitesimal motion. *International Journal of Computer Vision*, 61(3):259–299.

Oliensis, J. and Hartley, R. (2007). Iterative extensions of the Sturm/Triggs algorithm: Convergence and nonconvergence. *IEEE Transactions on Pattern Analysis and Machine Intelligence*, 29(12):2217–2233.

Oliva, A. and Torralba, A. (2001). Modeling the shape of the scene: a holistic representation of the spatial envelope. *International Journal of Computer Vision*, 42(3):145–175.

Oliva, A. and Torralba, A. (2007). The role of context in object recognition. *Trends in Cognitive Sciences*, 11(12):520–527.

Olsson, C., Eriksson, A. P., and Kahl, F. (2008). Improved spectral relaxation methods for binary quadratic optimization problems. *Computer Vision and Image Understanding*, 112(1):3–13.

Omer, I. and Werman, M. (2004). Color lines: Image specific color representation. In *IEEE Computer Society Conference on Computer Vision and Pattern Recognition (CVPR'2004)*, pp. 946–953, Washington, DC.

Ong, E.-J., Micilotta, A. S., Bowden, R., and Hilton, A. (2006). Viewpoint invariant exemplar-based 3D human tracking. *Computer Vision and Image Understanding*, 104(2-3):178–189.

Opelt, A., Pinz, A., and Zisserman, A. (2006). A boundary-fragment-model for object detection. In *Ninth European Conference on Computer Vision (ECCV 2006)*, pp. 575–588.

Opelt, A., Pinz, A., Fussenegger, M., and Auer, P. (2006). Generic object recognition with boosting. *IEEE Transactions on Pattern Analysis and Machine Intelligence*, 28(3):614–641.

OpenGL-ARB. (1997). *OpenGL Reference Manual: The Official Reference Document to OpenGL, Version 1.1*. Addison-Wesley, Reading, MA, 2nd edition.

Oppenheim, A. V. and Schafer, A. S. (1996). *Signals and Systems*. Prentice Hall, Englewood Cliffs, New Jersey, 2nd edition.

Oppenheim, A. V., Schafer, R. W., and Buck, J. R. (1999). *Discrete-Time Signal Processing*. Prentice Hall, Englewood Cliffs, New Jersey, 2nd edition.

Oren, M. and Nayar, S. (1997). A theory of specular surface geometry. *International Journal of Computer Vision*, 24(2):105–124.

O'Rourke, J. and Badler, N. I. (1980). Model-based image analysis of human motion using constraint propagation. *IEEE Transactions on Pattern Analysis and Machine Intelligence*, 2(6):522–536.

Osher, S. and Paragios, N. (eds). (2003). *Geometric Level Set Methods in Imaging, Vision, and Graphics*, Springer.

Osuna, E., Freund, R., and Girosi, F. (1997). Training support vector machines: An application to face detection. In *IEEE Computer Society Conference on Computer Vision and Pattern Recognition (CVPR'97)*, pp. 130–136, San Juan, Puerto Rico.

O'Toole, A. J., Jiang, F., Roark, D., and Abdi, H. (2006). Predicting human face recognition. In Zhao, W.-Y. and Chellappa, R. (eds), *Face Processing: Advanced Methods and Models*, Elsevier.

O'Toole, A. J., Phillips, P. J., Jiang, F., Ayyad, J., Pénard, N., and Abdi, H. (2009). Face recognition algorithms surpass humans matching faces over changes in illumination. *IEEE Transactions on Pattern Analysis and Machine Intelligence*, 29(9):1642–1646.

Ott, M., Lewis, J. P., and Cox, I. J. (1993). Teleconferencing eye contact using a virtual camera. In *INTERACT'93 and CHI'93 conference companion on Human factors in computing systems*, pp. 109–110, Amsterdam.

Otte, M. and Nagel, H.-H. (1994). Optical flow estimation: advances and comparisons. In *Third European Conference on Computer Vision (ECCV'94)*, pp. 51–60, Stockholm, Sweden.

Oztireli, C., Guennebaud, G., and Gross, M. (2008). Feature preserving point set surfaces. *Computer Graphics Forum*, 28(2):493–501.

Özuysal, M., Calonder, M., Lepetit, V., and Fua, P. (2010). Fast keypoint recognition using random ferns. *IEEE Transactions on Pattern Analysis and Machine Intelligence*, 32(3).

Paglieroni, D. W. (1992). Distance transforms: Properties and machine vision applications. *Graphical Models and Image Processing*, 54(1):56–74.

Pal, C., Szeliski, R., Uyttendaele, M., and Jojic, N. (2004). Probability models for high dynamic range imaging. In *IEEE Computer Society Conference on Computer Vision and Pattern Recognition (CVPR'2004)*, pp. 173–180, Washington, DC.

Palmer, S. E. (1999). *Vision Science: Photons to Phenomenology*. The MIT Press, Cambridge, Massachusetts.

Pankanti, S., Bolle, R. M., and Jain, A. K. (2000). Biometrics: The future of identification. *Computer*, 21(2):46–49.

Papageorgiou, C. and Poggio, T. (2000). A trainable system for object detection. *International Journal of Computer Vision*, 38(1):15–33.

Papandreou, G. and Maragos, P. (2008). Adaptive and constrained algorithms for inverse compositional active appearance model fitting. In *IEEE Computer Society Conference on Computer Vision and Pattern Recognition (CVPR 2008)*, Anchorage, AK.

Papenberg, N., Bruhn, A., Brox, T., Didas, S., and Weickert, J. (2006). Highly accurate optic flow computation with theoretically justified warping. *International Journal of Computer Vision*, 67(2):141–158.

Papert, S. (1966). *The Summer Vision Project*. Technical Report AIM-100, Artificial Intelligence Group, Massachusetts Institute of Technology. http://hdl.handle.net/1721.1/6125.

Paragios, N. and Deriche, R. (2000). Geodesic active contours and level sets for the detection and tracking of moving objects. *IEEE Transactions on Pattern Analysis and Machine Intelligence*, 22(3):266–280.

Paragios, N. and Sgallari, F. (2009). Special issue on scale space and variational methods in computer vision. *International Journal of Computer Vision*, 84(2).

Paragios, N., Faugeras, O. D., Chan, T., and Schnörr, C. (eds). (2005). *Third International Workshop on Variational, Geometric, and Level Set Methods in Computer Vision (VLSM 2005)*, Springer.

Paris, S. and Durand, F. (2006). A fast approximation of the bilateral filter using a signal processing approach. In *Ninth European Conference on Computer Vision (ECCV 2006)*, pp. 568–580.

Paris, S. and Durand, F. (2007). A topological approach to hierarchical segmentation using mean shift. In *IEEE Computer Society Conference on Computer Vision and Pattern Recognition (CVPR 2007)*, Minneapolis, MN.

Paris, S., Kornprobst, P., Tumblin, J., and Durand, F. (2008). Bilateral filtering: Theory and applications. *Foundations and Trends in Computer Graphics and Computer Vision*, 4(1):1–73.

Park, M., Brocklehurst, K., Collins, R. T., and Liu, Y. (2009). Deformed lattice detection in real-world images using mean-shift belief propagation. *IEEE Transactions on Pattern Analysis and Machine Intelligence*, 31(10):1804–1816.

Park, S. C., Park, M. K., and Kang, M. G. (2003). Super-resolution image reconstruction: A technical overview. *IEEE Signal Processing Magazine*, 20:21–36.

Parke, F. I. and Waters, K. (1996). *Computer Facial Animation*. A K Peters, Wellesley, Massachusetts.

Parker, J. A., Kenyon, R. V., and Troxel, D. E. (1983). Comparison of interpolating methods for image resampling. *IEEE Transactions on Medical Imaging*, MI-2(1):31–39.

Pattanaik, S. N., Ferwerda, J. A., Fairchild, M. D., and Greenberg, D. P. (1998). A multiscale model of adaptation and spatial vision for realistic image display. In *ACM SIGGRAPH 1998 Conference Proceedings*, pp. 287–298, Orlando.

Pauly, M., Keiser, R., Kobbelt, L. P., and Gross, M. (2003). Shape modeling with point-sampled geometry. *ACM Transactions on Graphics (Proc. SIGGRAPH 2003)*, 21(3):641–650.

Pavlidis, T. (1977). *Structural Pattern Recognition*. Springer-Verlag, Berlin; New York.

Pavlidis, T. and Liow, Y.-T. (1990). Integrating region growing and edge detection. *IEEE Transactions on Pattern Analysis and Machine Intelligence*, 12(3):225–233.

Pavlović, V., Sharma, R., and Huang, T. S. (1997). Visual interpretation of hand gestures for human-computer interaction: A review. *IEEE Transactions on Pattern Analysis and Machine Intelligence*, 19(7):677–695.

Pearl, J. (1988). *Probabilistic reasoning in intelligent systems: networks of plausible inference*. Morgan Kaufmann Publishers, Los Altos.

Peleg, R., Ben-Ezra, M., and Pritch, Y. (2001). Omnistereo: Panoramic stereo imaging. *IEEE Transactions on Pattern Analysis and Machine Intelligence*, 23(3):279–290.

Peleg, S. (1981). Elimination of seams from photomosaics. *Computer Vision, Graphics, and Image Processing*, 16(1):1206–1210.

Peleg, S. and Herman, J. (1997). Panoramic mosaics by manifold projection. In *IEEE Computer Society Conference on Computer Vision and Pattern Recognition (CVPR'97)*, pp. 338–343, San Juan, Puerto Rico.

Peleg, S. and Rav-Acha, A. (2006). Lucas-Kanade without iterative warping. In *International Conference on Image Processing (ICIP-2006)*, pp. 1097–1100, Atlanta.

Peleg, S., Rousso, B., Rav-Acha, A., and Zomet, A. (2000). Mosaicing on adaptive manifolds. *IEEE Transactions on Pattern Analysis and Machine Intelligence*, 22(10):1144–1154.

Penev, P. and Atick, J. (1996). Local feature analysis: A general statistical theory for object representation. *Network Computation and Neural Systems*, 7:477–500.

Pentland, A. P. (1984). Local shading analysis. *IEEE Transactions on Pattern Analysis and Machine Intelligence*, PAMI-6(2):170–179.

Pentland, A. P. (1986). Perceptual organization and the representation of natural form. *Artificial Intelligence*, 28(3):293–331.

Pentland, A. P. (1987). A new sense for depth of field. *IEEE Transactions on Pattern Analysis and Machine Intelligence*, PAMI-9(4):523–531.

Pentland, A. P. (1994). Interpolation using wavelet bases. *IEEE Transactions on Pattern Analysis and Machine Intelligence*, 16(4):410–414.

Pérez, P., Blake, A., and Gangnet, M. (2001). JetStream: Probabilistic contour extraction with particles. In *Eighth International Conference on Computer Vision (ICCV 2001)*, pp. 524–531, Vancouver, Canada.

Pérez, P., Gangnet, M., and Blake, A. (2003). Poisson image editing. *ACM Transactions on Graphics (Proc. SIGGRAPH 2003)*, 22(3):313–318.

Perona, P. (1995). Deformable kernels for early vision. *IEEE Transactions on Pattern Analysis and Machine Intelligence*, 17(5):488–499.

Perona, P. and Malik, J. (1990a). Detecting and localizing edges composed of steps, peaks and roofs. In *Third International Conference on Computer Vision (ICCV'90)*, pp. 52–57, Osaka, Japan.

Perona, P. and Malik, J. (1990b). Scale space and edge detection using anisotropic diffusion. *IEEE Transactions on Pattern Analysis and Machine Intelligence*, 12(7):629–639.

Peters, J. and Reif, U. (2008). *Subdivision Surfaces*. Springer.

Petschnigg, G., Agrawala, M., Hoppe, H., Szeliski, R., Cohen, M., and Toyama, K. (2004). Digital photography with flash and no-flash image pairs. *ACM Transactions on Graphics (Proc. SIGGRAPH 2004)*, 23(3):664–672.

Pfister, H., Zwicker, M., van Baar, J., and Gross, M. (2000). Surfels: Surface elements as rendering primitives. In *ACM SIGGRAPH 2000 Conference Proceedings*, pp. 335–342.

Pflugfelder, R. (2008). *Self-calibrating Cameras in Video Surveillance*. Ph.D. thesis, Graz University of Technology.

Philbin, J. and Zisserman, A. (2008). Object mining using a matching graph on very large image collections. In *Indian Conference on Computer Vision, Graphics and Image Processing*, Bhubaneswar, India.

Philbin, J., Chum, O., Isard, M., Sivic, J., and Zisserman, A. (2007). Object retrieval with large vocabularies and fast spatial matching. In *IEEE Computer Society Conference on Computer Vision and Pattern Recognition (CVPR 2007)*, Minneapolis, MN.

Philbin, J., Chum, O., Sivic, J., Isard, M., and Zisserman, A. (2008). Lost in quantization: Improving particular object retrieval in large scale image databases. In *IEEE Computer Society Conference on Computer Vision and Pattern Recognition (CVPR 2008)*, Anchorage, AK.

Phillips, P. J., Moon, H., Rizvi, S. A., and Rauss, P. J. (2000). The FERET evaluation methodology for face recognition algorithms. *IEEE Transactions on Pattern Analysis and Machine Intelligence*, 22(10):1090–1104.

Phillips, P. J., Scruggs, W. T., O'Toole, A. J., Flynn, P. J., Bowyer, K. W. *et al.* (2010). FRVT 2006 and ICE 2006 large-scale experimental results. *IEEE Transactions on Pattern Analysis and Machine Intelligence*, 32(5):831–846.

Phong, B. T. (1975). Illumination for computer generated pictures. *Communications of the ACM*, 18(6):311–317.

Pickup, L. C. (2007). *Machine Learning in Multi-frame Image Super-resolution*. Ph.D. thesis, University of Oxford.

Pickup, L. C. and Zisserman, A. (2009). Automatic retrieval of visual continuity errors in movies. In *ACM International Conference on Image and Video Retrieval*, Santorini, Greece.

Pickup, L. C., Capel, D. P., Roberts, S. J., and Zisserman, A. (2007). Overcoming registration uncertainty in image super-resolution: Maximize or marginalize? *EURASIP Journal on Advances in Signal Processing*, 2010(Article ID 23565).

Pickup, L. C., Capel, D. P., Roberts, S. J., and Zisserman, A. (2009). Bayesian methods for image super-resolution. *The Computer Journal*, 52.

Pighin, F., Szeliski, R., and Salesin, D. H. (2002). Modeling and animating realistic faces from images. *International Journal of Computer Vision*, 50(2):143–169.

Pighin, F., Hecker, J., Lischinski, D., Salesin, D. H., and Szeliski, R. (1998). Synthesizing realistic facial expressions from photographs. In *ACM SIGGRAPH 1998 Conference Proceedings*, pp. 75–84, Orlando.

Pilet, J., Lepetit, V., and Fua, P. (2008). Fast non-rigid surface detection, registration, and realistic augmentation. *International Journal of Computer Vision*, 76(2).

Pinz, A. (2005). Object categorization. *Foundations and Trends in Computer Graphics and Computer Vision*, 1(4):255–353.

Pizer, S. M., Amburn, E. P., Austin, J. D., Cromartie, R., Geselowitz, A. *et al.* (1987). Adaptive histogram equalization and its variations. *Computer Vision, Graphics, and Image Processing*, 39(3):355–368.

Platel, B., Balmachnova, E., Florack, L., and ter Haar Romeny, B. (2006). Top-points as interest points for image matching. In *Ninth European Conference on Computer Vision (ECCV 2006)*, pp. 418–429.

Platt, J. C. (2000). Optimal filtering for patterned displays. *IEEE Signal Processing Letters*, 7(7):179–180.

Pock, T., Unger, M., Cremers, D., and Bischof, H. (2008). Fast and exact solution of total variation models on the GPU. In *CVPR 2008 Workshop on Visual Computer Vision on GPUs (CVGPU)*, Anchorage, AK.

Poelman, C. J. and Kanade, T. (1997). A paraperspective factorization method for shape and motion recovery. *IEEE Transactions on Pattern Analysis and Machine Intelligence*, 19(3):206–218.

Poggio, T. and Koch, C. (1985). Ill-posed problems in early vision: from computational theory to analogue networks. *Proceedings of the Royal Society of London*, B 226:303–323.

Poggio, T., Gamble, E., and Little, J. (1988). Parallel integration of vision modules. *Science*, 242(4877):436–440.

Poggio, T., Torre, V., and Koch, C. (1985). Computational vision and regularization theory. *Nature*, 317(6035):314–319.

Poggio, T., Little, J., Gamble, E., Gillet, W., Geiger, D. *et al.* (1988). The MIT vision machine. In *Image Understanding Workshop*, pp. 177–198, Boston.

Polana, R. and Nelson, R. C. (1997). Detection and recognition of periodic, nonrigid motion. *International Journal of Computer Vision*, 23(3):261–282.

Pollard, S. B., Mayhew, J. E. W., and Frisby, J. P. (1985). PMF: A stereo correspondence algorithm using a disparity gradient limit. *Perception*, 14:449–470.

Pollefeys, M. and Van Gool, L. (2002). From images to 3D models. *Communications of the ACM*, 45(7):50–55.

Pollefeys, M., Nistér, D., Frahm, J.-M., Akbarzadeh, A., Mordohai, P. *et al.* (2008). Detailed real-time urban 3D reconstruction from video. *International Journal of Computer Vision*, 78(2-3):143–167.

Ponce, J., Hebert, M., Schmid, C., and Zisserman, A. (eds). (2006). *Toward Category-Level Object Recognition*, Springer, New York.

Ponce, J., Berg, T., Everingham, M., Forsyth, D., Hebert, M. *et al.* (2006). Dataset issues in object recognition. In Ponce, J., Hebert, M., Schmid, C., and Zisserman, A. (eds), *Toward Category-Level Object Recognition*, pp. 29–48, Springer, New York.

Pons, J.-P., Keriven, R., and Faugeras, O. (2005). Modelling dynamic scenes by registering multi-view image sequences. In *IEEE Computer Society Conference on Computer Vision and Pattern Recognition (CVPR'2005)*, pp. 822–827, San Diego, CA.

Pons, J.-P., Keriven, R., and Faugeras, O. (2007). Multi-view stereo reconstruction and scene flow estimation with a global image-based matching score. *International Journal of Computer Vision*, 72(2):179–193.

Porter, T. and Duff, T. (1984). Compositing digital images. *Computer Graphics (SIGGRAPH '84)*, 18(3):253–259.

Portilla, J. and Simoncelli, E. P. (2000). A parametric texture model based on joint statistics of complex wavelet coefficients. *International Journal of Computer Vision*, 40(1):49–71.

Portilla, J., Strela, V., Wainwright, M., and Simoncelli, E. P. (2003). Image denoising using scale mixtures of Gaussians in the wavelet domain. *IEEE Transactions on Image Processing*, 12(11):1338–1351.

Potetz, B. and Lee, T. S. (2008). Efficient belief propagation for higher-order cliques using linear constraint nodes. *Computer Vision and Image Understanding*, 112(1):39–54.

Potmesil, M. (1987). Generating octree models of 3D objects from their silhouettes in a sequence of images. *Computer Vision, Graphics, and Image Processing*, 40:1–29.

Pratt, W. K. (2007). *Digital Image Processing*. Wiley-Interscience, Hoboken, NJ, 4th edition.

Prazdny, K. (1985). Detection of binocular disparities. *Biological Cybernetics*, 52:93–99.

Pritchett, P. and Zisserman, A. (1998). Wide baseline stereo matching. In *Sixth International Conference on Computer Vision (ICCV'98)*, pp. 754–760, Bombay.

Proesmans, M., Van Gool, L., and Defoort, F. (1998). Reading between the lines – a method for extracting dynamic 3D with texture. In *Sixth International Conference on Computer Vision (ICCV'98)*, pp. 1081–1086, Bombay.

Protter, M. and Elad, M. (2009). Super resolution with probabilistic motion estimation. *IEEE Transactions on Image Processing*, 18(8):1899–1904.

Pullen, K. and Bregler, C. (2002). Motion capture assisted animation: texturing and synthesis. *ACM Transactions on Graphics*, 21(3):501–508.

Pulli, K. (1999). Multiview registration for large data sets. In *Second International Conference on 3D Digital Imaging and Modeling (3DIM'99)*, pp. 160–168, Ottawa, Canada.

Pulli, K., Abi-Rached, H., Duchamp, T., Shapiro, L., and Stuetzle, W. (1998). Acquisition and visualization of colored 3D objects. In *International Conference on Pattern Recognition (ICPR'98)*, pp. 11–15.

Quack, T., Leibe, B., and Van Gool, L. (2008). World-scale mining of objects and events from community photo collections. In *Conference on Image and Video Retrieval*, pp. 47–56, Niagara Falls.

Quam, L. H. (1984). Hierarchical warp stereo. In *Image Understanding Workshop*, pp. 149–155, New Orleans.

Quan, L. and Lan, Z. (1999). Linear N-point camera pose determination. *IEEE Transactions on Pattern Analysis and Machine Intelligence*, 21(8):774–780.

Quan, L. and Mohr, R. (1989). Determining perspective structures using hierarchical Hough transform. *Pattern Recognition Letters*, 9(4):279–286.

Rabinovich, A., Vedaldi, A., Galleguillos, C., Wiewiora, E., and Belongie, S. (2007). Objects in context. In *Eleventh International Conference on Computer Vision (ICCV 2007)*, Rio de Janeiro, Brazil.

Rademacher, P. and Bishop, G. (1998). Multiple-center-of-projection images. In *ACM SIGGRAPH 1998 Conference Proceedings*, pp. 199–206, Orlando.

Raginsky, M. and Lazebnik, S. (2009). Locality-sensitive binary codes from shift-invariant kernels. In *Advances in Neural Information Processing Systems*.

Raman, S. and Chaudhuri, S. (2007). A matte-less, variational approach to automatic scene compositing. In *Eleventh International Conference on Computer Vision (ICCV 2007)*, Rio de Janeiro, Brazil.

Raman, S. and Chaudhuri, S. (2009). Bilateral filter based compositing for variable exposure photography. In *Proceedings of Eurographics 2009*.

Ramanan, D. and Baker, S. (2009). Local distance functions: A taxonomy, new algorithms, and an evaluation. In *Twelfth International Conference on Computer Vision (ICCV 2009)*, Kyoto, Japan.

Ramanan, D., Forsyth, D., and Zisserman, A. (2005). Strike a pose: Tracking people by finding stylized poses. In *IEEE Computer Society Conference on Computer Vision and Pattern Recognition (CVPR'2005)*, pp. 271–278, San Diego, CA.

Ramanarayanan, G. and Bala, K. (2007). Constrained texture synthesis via energy minimization. *IEEE Transactions on Visualization and Computer Graphics*, 13(1):167–178.

Ramer, U. (1972). An iterative procedure for the polygonal approximation of plane curves. *Computer Graphics and Image Processing*, 1(3):244–256.

Ramnath, K., Koterba, S., Xiao, J., Hu, C., Matthews, I., Baker, S., Cohn, J., and Kanade, T. (2008). Multi-view AAM fitting and construction. *International Journal of Computer Vision*, 76(2):183–204.

Raskar, R. and Tumblin, J. (2010). *Computational Photography: Mastering New Techniques for Lenses, Lighting, and Sensors*. A K Peters, Wellesley, Massachusetts.

Raskar, R., Tan, K.-H., Feris, R., Yu, J., and Turk, M. (2004). Non-photorealistic camera: Depth edge detection and stylized rendering using multi-flash imaging. *ACM Transactions on Graphics*, 23(3):679–688.

Rav-Acha, A., Kohli, P., Fitzgibbon, A., and Rother, C. (2008). Unwrap mosaics: A new representation for video editing. *ACM Transactions on Graphics*, 27(3).

Rav-Acha, A., Pritch, Y., Lischinski, D., and Peleg, S. (2005). Dynamosaics: Video mosaics with non-chronological time. In *IEEE Computer Society Conference on Computer Vision and Pattern Recognition (CVPR'2005)*, pp. 58–65, San Diego, CA.

Ravikumar, P., Agarwal, A., and Wainwright, M. J. (2008). Message-passing for graph-structured linear programs: Proximal projections, convergence and rounding schemes. In *International Conference on Machine Learning*, pp. 800–807.

Ray, S. F. (2002). *Applied Photographic Optics*. Focal Press, Oxford, 3rd edition.

Rehg, J. and Kanade, T. (1994). Visual tracking of high DOF articulated structures: an application to human hand tracking. In *Third European Conference on Computer Vision (ECCV'94)*, pp. 35–46, Stockholm, Sweden.

Rehg, J. and Witkin, A. (1991). Visual tracking with deformation models. In *IEEE International Conference on Robotics and Automation*, pp. 844–850, Sacramento.

Rehg, J., Morris, D. D., and Kanade, T. (2003). Ambiguities in visual tracking of articulated objects using two- and three-dimensional models. *International Journal of Robotics Research*, 22(6):393–418.

Reichenbach, S. E., Park, S. K., and Narayanswamy, R. (1991). Characterizing digital image acquisition devices. *Optical Engineering*, 30(2):170–177.

Reinhard, E., Stark, M., Shirley, P., and Ferwerda, J. (2002). Photographic tone reproduction for digital images. *ACM Transactions on Graphics (Proc. SIGGRAPH 2002)*, 21(3):267–276.

Reinhard, E., Ward, G., Pattanaik, S., and Debevec, P. (2005). *High Dynamic Range Imaging: Acquisition, Display, and Image-Based Lighting*. Morgan Kaufmann.

Rhemann, C., Rother, C., and Gelautz, M. (2008). Improving color modeling for alpha matting. In *British Machine Vision Conference (BMVC 2008)*, Leeds.

Rhemann, C., Rother, C., Rav-Acha, A., and Sharp, T. (2008). High resolution matting via interactive trimap segmentation. In *IEEE Computer Society Conference on Computer Vision and Pattern Recognition (CVPR 2008)*, Anchorage, AK.

Rhemann, C., Rother, C., Wang, J., Gelautz, M., Kohli, P., and Rott, P. (2009). A perceptually motivated online benchmark for image matting. In *IEEE Computer Society Conference on Computer Vision and Pattern Recognition (CVPR 2009)*, Miami Beach, FL.

Richardson, I. E. G. (2003). *H.264 and MPEG-4 Video Compression: Video Coding for Next Generation Multimedia*. Wiley.

Rioul, O. and Vetterli, M. (1991). Wavelets and signal processing. *IEEE Signal Processing Magazine*, 8(4):14–38.

Rioux, M. and Bird, T. (1993). White laser, synced scan. *IEEE Computer Graphics and Applications*, 13(3):15–17.

Rioux, M., Bechthold, G., Taylor, D., and Duggan, M. (1987). Design of a large depth of view three-dimensional camera for robot vision. *Optical Engineering*, 26(12):1245–1250.

Riseman, E. M. and Arbib, M. A. (1977). Computational techniques in the visual segmentation of static scenes. *Computer Graphics and Image Processing*, 6(3):221–276.

Ritter, G. X. and Wilson, J. N. (2000). *Handbook of Computer Vision Algorithms in Image Algebra*. CRC Press, Boca Raton, 2nd edition.

Robert, C. P. (2007). *The Bayesian Choice: From Decision-Theoretic Foundations to Computational Implementation*. Springer-Verlag, New York.

Roberts, L. G. (1965). Machine perception of three-dimensional solids. In Tippett, J. T., Borkowitz, D. A., Clapp, L. C., Koester, C. J., and Vanderburgh Jr., A. (eds), *Optical and Electro-Optical Information Processing*, pp. 159–197, MIT Press, Cambridge, Massachusetts.

Robertson, D. and Cipolla, R. (2004). An image-based system for urban navigation. In *British Machine Vision Conference*, pp. 656–665, Kingston.

Robertson, D. P. and Cipolla, R. (2002). Building architectural models from many views using map constraints. In *Seventh European Conference on Computer Vision (ECCV 2002)*, pp. 155–169, Copenhagen.

Robertson, D. P. and Cipolla, R. (2009). Architectural modelling. In Varga, M. (ed.), *Practical Image Processing and Computer Vision*, John Wiley.

Robertson, N. and Reid, I. (2006). A general method for human activity recognition in video. *Computer Vision and Image Understanding*, 104(2-3):232–248.

Roble, D. (1999). Vision in film and special effects. *Computer Graphics*, 33(4):58–60.

Roble, D. and Zafar, N. B. (2009). Don't trust your eyes: cutting-edge visual effects. *Computer*, 42(7):35–41.

Rogez, G., Rihan, J., Ramalingam, S., Orrite, C., and Torr, P. H. S. (2008). Randomized trees for human pose detection. In *IEEE Computer Society Conference on Computer Vision and Pattern Recognition (CVPR 2008)*, Anchorage, AK.

Rogmans, S., Lu, J., Bekaert, P., and Lafruit, G. (2009). Real-time stereo-based views synthesis algorithms: A unified framework and evaluation on commodity GPUs. *Signal Processing: Image Communication*, 24:49–64.

Rohr, K. (1994). Towards model-based recognition of human movements in image sequences. *Computer Vision, Graphics, and Image Processing*, 59(1):94–115.

Román, A. and Lensch, H. P. A. (2006). Automatic multiperspective images. In *Eurographics Symposium on Rendering*, pp. 83–92.

Román, A., Garg, G., and Levoy, M. (2004). Interactive design of multi-perspective images for visualizing urban landscapes. In *IEEE Visualization 2004*, pp. 537–544, Minneapolis.

Romdhani, S. and Vetter, T. (2003). Efficient, robust and accurate fitting of a 3D morphable model. In *Ninth International Conference on Computer Vision (ICCV 2003)*, pp. 59–66, Nice, France.

Romdhani, S., Torr, P. H. S., Schölkopf, B., and Blake, A. (2001). Computationally efficient face detection. In *Eighth International Conference on Computer Vision (ICCV 2001)*, pp. 695–700, Vancouver, Canada.

Rosales, R. and Sclaroff, S. (2000). Inferring body pose without tracking body parts. In *IEEE Computer Society Conference on Computer Vision and Pattern Recognition (CVPR'2000)*, pp. 721–727, Hilton Head Island.

Rosenfeld, A. (1980). Quadtrees and pyramids for pattern recognition and image processing. In *Fifth International Conference on Pattern Recognition (ICPR'80)*, pp. 802–809, Miami Beach.

Rosenfeld, A. (ed.). (1984). *Multiresolution Image Processing and Analysis*, Springer-Verlag, New York.

Rosenfeld, A. and Davis, L. S. (1979). Image segmentation and image models. *Proceedings of the IEEE*, 67(5):764–772.

Rosenfeld, A. and Kak, A. C. (1976). *Digital Picture Processing*. Academic Press, New York.

Rosenfeld, A. and Pfaltz, J. L. (1966). Sequential operations in digital picture processing. *Journal of the ACM*, 13(4):471–494.

Rosenfeld, A., Hummel, R. A., and Zucker, S. W. (1976). Scene labeling by relaxation operations. *IEEE Transactions on Systems, Man, and Cybernetics*, SMC-6:420–433.

Rosten, E. and Drummond, T. (2005). Fusing points and lines for high performance tracking. In *Tenth International Conference on Computer Vision (ICCV 2005)*, pp. 1508–1515, Beijing, China.

Rosten, E. and Drummond, T. (2006). Machine learning for high-speed corner detection. In *Ninth European Conference on Computer Vision (ECCV 2006)*, pp. 430–443.

Roth, S. and Black, M. J. (2007a). On the spatial statistics of optical flow. *International Journal of Computer Vision*, 74(1):33–50.

Roth, S. and Black, M. J. (2007b). Steerable random fields. In *Eleventh International Conference on Computer Vision (ICCV 2007)*, Rio de Janeiro, Brazil.

Roth, S. and Black, M. J. (2009). Fields of experts. *International Journal of Computer Vision*, 82(2):205–229.

Rother, C. (2002). A new approach for vanishing point detection in architectural environments. *Image and Vision Computing*, 20(9-10):647–656.

Rother, C. (2003). Linear multi-view reconstruction of points, lines, planes and cameras using a reference plane. In *Ninth International Conference on Computer Vision (ICCV 2003)*, pp. 1210–1217, Nice, France.

Rother, C. and Carlsson, S. (2002). Linear multi view reconstruction and camera recovery using a reference plane. *International Journal of Computer Vision*, 49(2/3):117–141.

Rother, C., Kolmogorov, V., and Blake, A. (2004). "GrabCut"—interactive foreground extraction using iterated graph cuts. *ACM Transactions on Graphics (Proc. SIGGRAPH 2004)*, 23(3):309–314.

Rother, C., Bordeaux, L., Hamadi, Y., and Blake, A. (2006). Autocollage. *ACM Transactions on Graphics*, 25(3):847–852.

Rother, C., Kohli, P., Feng, W., and Jia, J. (2009). Minimizing sparse higher order energy functions of discrete variables. In *IEEE Computer Society Conference on Computer Vision and Pattern Recognition (CVPR 2009)*, Miami Beach, FL.

Rother, C., Kolmogorov, V., Lempitsky, V., and Szummer, M. (2007). Optimizing binary MRFs via extended roof duality. In *IEEE Computer Society Conference on Computer Vision and Pattern Recognition (CVPR 2007)*, Minneapolis, MN.

Rother, C., Kumar, S., Kolmogorov, V., and Blake, A. (2005). Digital tapestry. In *IEEE Computer Society Conference on Computer Vision and Pattern Recognition (CVPR'2005)*, pp. 589–596, San Diego, CA.

Rothganger, F., Lazebnik, S., Schmid, C., and Ponce, J. (2006). 3D object modeling and recognition using local affine-invariant image descriptors and multi-view spatial constraints. *International Journal of Computer Vision*, 66(3):231–259.

Rousseeuw, P. J. (1984). Least median of squares regresssion. *Journal of the American Statistical Association*, 79:871–880.

Rousseeuw, P. J. and Leroy, A. M. (1987). *Robust Regression and Outlier Detection*. Wiley, New York.

Rousson, M. and Paragios, N. (2008). Prior knowledge, level set representations, and visual grouping. *International Journal of Computer Vision*, 76(3):231–243.

Roweis, S. (1998). EM algorithms for PCA and SPCA. In *Advances in Neural Information Processing Systems*, pp. 626–632.

Rowland, D. A. and Perrett, D. I. (1995). Manipulating facial appearance through shape and color. *IEEE Computer Graphics and Applications*, 15(5):70–76.

Rowley, H. A., Baluja, S., and Kanade, T. (1998a). Neural network-based face detection. *IEEE Transactions on Pattern Analysis and Machine Intelligence*, 20(1):23–38.

Rowley, H. A., Baluja, S., and Kanade, T. (1998b). Rotation invariant neural network-based face detection. In *IEEE Computer Society Conference on Computer Vision and Pattern Recognition (CVPR'98)*, pp. 38–44, Santa Barbara.

Roy, S. and Cox, I. J. (1998). A maximum-flow formulation of the N-camera stereo correspondence problem. In *Sixth International Conference on Computer Vision (ICCV'98)*, pp. 492–499, Bombay.

Rozenfeld, S., Shimshoni, I., and Lindenbaum, M. (2007). Dense mirroring surface recovery from 1d homographies and sparse correspondences. In *IEEE Computer Society*

Conference on Computer Vision and Pattern Recognition (CVPR 2007), Minneapolis, MN.

Rubner, Y., Tomasi, C., and Guibas, L. J. (2000). The earth mover's distance as a metric for image retrieval. *International Journal of Computer Vision*, 40(2):99–121.

Rumelhart, D. E., Hinton, G. E., and Williams, R. J. (1986). Learning internal representations by error propagation. In Rumelhart, D. E., McClelland, J. L., and the PDP research group (eds), *Parallel distributed processing: Explorations in the microstructure of cognition*, pp. 318–362, Bradford Books, Cambridge, Massachusetts.

Rusinkiewicz, S. and Levoy, M. (2000). Qsplat: A multiresolution point rendering system for large meshes. In *ACM SIGGRAPH 2000 Conference Proceedings*, pp. 343–352.

Russ, J. C. (2007). *The Image Processing Handbook*. CRC Press, Boca Raton, 5th edition.

Russell, B., Efros, A., Sivic, J., Freeman, W., and Zisserman, A. (2006). Using multiple segmentations to discover objects and their extent in image collections. In *IEEE Computer Society Conference on Computer Vision and Pattern Recognition (CVPR'2006)*, pp. 1605–1612, New York City, NY.

Russell, B. C., Torralba, A., Murphy, K. P., and Freeman, W. T. (2008). LabelMe: A database and web-based tool for image annotation. *International Journal of Computer Vision*, 77(1-3):157–173.

Russell, B. C., Torralba, A., Liu, C., Fergus, R., and Freeman, W. T. (2007). Object recognition by scene alignment. In *Advances in Neural Information Processing Systems*.

Ruzon, M. A. and Tomasi, C. (2000). Alpha estimation in natural images. In *IEEE Computer Society Conference on Computer Vision and Pattern Recognition (CVPR'2000)*, pp. 18–25, Hilton Head Island.

Ruzon, M. A. and Tomasi, C. (2001). Edge, junction, and corner detection using color distributions. *IEEE Transactions on Pattern Analysis and Machine Intelligence*, 23(11):1281–1295.

Ryan, T. W., Gray, R. T., and Hunt, B. R. (1980). Prediction of correlation errors in stereo-pair images. *Optical Engineering*, 19(3):312–322.

Saad, Y. (2003). *Iterative Methods for Sparse Linear Systems*. Society for Industrial and Applied Mathematics, second edition.

Saint-Marc, P., Chen, J. S., and Medioni, G. (1991). Adaptive smoothing: A general tool for early vision. *IEEE Transactions on Pattern Analysis and Machine Intelligence*, 13(6):514–529.

Saito, H. and Kanade, T. (1999). Shape reconstruction in projective grid space from large number of images. In *IEEE Computer Society Conference on Computer Vision and Pattern Recognition (CVPR'99)*, pp. 49–54, Fort Collins.

Samet, H. (1989). *The Design and Analysis of Spatial Data Structures*. Addison-Wesley, Reading, Massachusetts.

Sander, P. T. and Zucker, S. W. (1990). Inferring surface trace and differential structure from 3-D images. *IEEE Transactions on Pattern Analysis and Machine Intelligence*, 12(9):833–854.

Sapiro, G. (2001). *Geometric Partial Differential Equations and Image Analysis*. Cambridge University Press.

Sato, Y. and Ikeuchi, K. (1996). Reflectance analysis for 3D computer graphics model generation. *Graphical Models and Image Processing*, 58(5):437–451.

Sato, Y., Wheeler, M., and Ikeuchi, K. (1997). Object shape and reflectance modeling from observation. In *ACM SIGGRAPH 1997 Conference Proceedings*, pp. 379–387, Los Angeles.

Savarese, S. and Fei-Fei, L. (2007). 3D generic object categorization, localization and pose estimation. In *Eleventh International Conference on Computer Vision (ICCV 2007)*, Rio de Janeiro, Brazil.

Savarese, S. and Fei-Fei, L. (2008). View synthesis for recognizing unseen poses of object classes. In *Tenth European Conference on Computer Vision (ECCV 2008)*, pp. 602–615, Marseilles.

Savarese, S., Chen, M., and Perona, P. (2005). Local shape from mirror reflections. *International Journal of Computer Vision*, 64(1):31–67.

Savarese, S., Andreetto, M., Rushmeier, H. E., Bernardini, F., and Perona, P. (2007). 3D reconstruction by shadow carving: Theory and practical evaluation. *International Journal of Computer Vision*, 71(3):305–336.

Sawhney, H. S. (1994). Simplifying motion and structure analysis using planar parallax and image warping. In *Twelfth International Conference on Pattern Recognition (ICPR'94)*, pp. 403–408, Jerusalem, Israel.

Sawhney, H. S. and Ayer, S. (1996). Compact representation of videos through dominant multiple motion estimation. *IEEE Transactions on Pattern Analysis and Machine Intelligence*, 18(8):814–830.

Sawhney, H. S. and Hanson, A. R. (1991). Identification and 3D description of 'shallow' environmental structure over a sequence of images. In *IEEE Computer Society Con-*

ference on Computer Vision and Pattern Recognition (CVPR'91), pp. 179–185, Maui, Hawaii.

Sawhney, H. S. and Kumar, R. (1999). True multi-image alignment and its application to mosaicing and lens distortion correction. *IEEE Transactions on Pattern Analysis and Machine Intelligence*, 21(3):235–243.

Sawhney, H. S., Kumar, R., Gendel, G., Bergen, J., Dixon, D., and Paragano, V. (1998). VideoBrush: Experiences with consumer video mosaicing. In *IEEE Workshop on Applications of Computer Vision (WACV'98)*, pp. 56–62, Princeton.

Sawhney, H. S., Arpa, A., Kumar, R., Samarasekera, S., Aggarwal, M., Hsu, S., Nister, D., and Hanna, K. (2002). Video flashlights: real time rendering of multiple videos for immersive model visualization. In *Proceedings of the 13th Eurographics Workshop on Rendering*, pp. 157–168, Pisa, Italy.

Saxena, A., Sun, M., and Ng, A. Y. (2009). Make3D: Learning 3D scene structure from a single still image. *IEEE Transactions on Pattern Analysis and Machine Intelligence*, 31(5):824–840.

Schaffalitzky, F. and Zisserman, A. (2000). Planar grouping for automatic detection of vanishing lines and points. *Image and Vision Computing*, 18:647–658.

Schaffalitzky, F. and Zisserman, A. (2002). Multi-view matching for unordered image sets, or "How do I organize my holiday snaps?". In *Seventh European Conference on Computer Vision (ECCV 2002)*, pp. 414–431, Copenhagen.

Scharr, H., Black, M. J., and Haussecker, H. W. (2003). Image statistics and anisotropic diffusion. In *Ninth International Conference on Computer Vision (ICCV 2003)*, pp. 840–847, Nice, France.

Scharstein, D. (1994). Matching images by comparing their gradient fields. In *Twelfth International Conference on Pattern Recognition (ICPR'94)*, pp. 572–575, Jerusalem, Israel.

Scharstein, D. (1999). *View Synthesis Using Stereo Vision*. Volume 1583, Springer-Verlag.

Scharstein, D. and Pal, C. (2007). Learning conditional random fields for stereo. In *IEEE Computer Society Conference on Computer Vision and Pattern Recognition (CVPR 2007)*, Minneapolis, MN.

Scharstein, D. and Szeliski, R. (1998). Stereo matching with nonlinear diffusion. *International Journal of Computer Vision*, 28(2):155–174.

Scharstein, D. and Szeliski, R. (2002). A taxonomy and evaluation of dense two-frame stereo correspondence algorithms. *International Journal of Computer Vision*, 47(1):7–42.

Scharstein, D. and Szeliski, R. (2003). High-accuracy stereo depth maps using structured light. In *IEEE Computer Society Conference on Computer Vision and Pattern Recognition (CVPR'2003)*, pp. 195–202, Madison, WI.

Schechner, Y. Y., Nayar, S. K., and Belhumeur, P. N. (2009). Multiplexing for optimal lighting. *IEEE Transactions on Pattern Analysis and Machine Intelligence*, 29(8):1339–1354.

Schindler, G., Brown, M., and Szeliski, R. (2007). City-scale location recognition. In *IEEE Computer Society Conference on Computer Vision and Pattern Recognition (CVPR 2007)*, Minneapolis, MN.

Schindler, G., Krishnamurthy, P., Lublinerman, R., Liu, Y., and Dellaert, F. (2008). Detecting and matching repeated patterns for automatic geo-tagging in urban environments. In *IEEE Computer Society Conference on Computer Vision and Pattern Recognition (CVPR 2008)*, Anchorage, AK.

Schlesinger, D. and Flach, B. (2006). *Transforming an arbitrary minsum problem into a binary one*. Technical Report TUD-FI06-01, Dresden University of Technology.

Schlesinger, M. I. (1976). Syntactic analysis of two-dimensional visual signals in noisy conditions. *Kibernetika*, 4:113–130.

Schlesinger, M. I. and Giginyak, V. V. (2007a). Solution to structural recognition (max,+)-problems by their equivalent transformations – part 1. *Control Systems and Computers*, 2007(1):3–15.

Schlesinger, M. I. and Giginyak, V. V. (2007b). Solution to structural recognition (max,+)-problems by their equivalent transformations – part 2. *Control Systems and Computers*, 2007(2):3–18.

Schmid, C. and Mohr, R. (1997). Local grayvalue invariants for image retrieval. *IEEE Transactions on Pattern Analysis and Machine Intelligence*, 19(5):530–534.

Schmid, C. and Zisserman, A. (1997). Automatic line matching across views. In *IEEE Computer Society Conference on Computer Vision and Pattern Recognition (CVPR'97)*, pp. 666–671, San Juan, Puerto Rico.

Schmid, C., Mohr, R., and Bauckhage, C. (2000). Evaluation of interest point detectors. *International Journal of Computer Vision*, 37(2):151–172.

Schneiderman, H. and Kanade, T. (2004). Object detection using the statistics of parts. *International Journal of Computer Vision*, 56(3):151–177.

Schödl, A. and Essa, I. (2002). Controlled animation of video sprites. In *ACM Symposium on Computater Animation*, San Antonio.

Schödl, A., Szeliski, R., Salesin, D. H., and Essa, I. (2000). Video textures. In *ACM SIGGRAPH 2000 Conference Proceedings*, pp. 489–498, New Orleans.

Schoenemann, T. and Cremers, D. (2008). High resolution motion layer decomposition using dual-space graph cuts. In *IEEE Computer Society Conference on Computer Vision and Pattern Recognition (CVPR 2008)*, Anchorage, AK.

Schölkopf, B. and Smola, A. (eds). (2002). *Learning with Kernels: Support Vector Machines, Regularization, Optimization and Beyond*. MIT Press, Cambridge, Massachusetts.

Schraudolph, N. N. (2010). Polynomial-time exact inference in NP-hard binary MRFs via reweighted perfect matching. In *13th International Conference on Artificial Intelligence and Statistics (AISTATS)*, pp. 717–724.

Schröder, P. and Sweldens, W. (1995). Spherical wavelets: Efficiently representing functions on the sphere. In *ACM SIGGRAPH 1995 Conference Proceedings*, pp. 161–172.

Schultz, R. R. and Stevenson, R. L. (1996). Extraction of high-resolution frames from video sequences. *IEEE Transactions on Image Processing*, 5(6):996–1011.

Sclaroff, S. and Isidoro, J. (2003). Active blobs: region-based, deformable appearance models. *Computer Vision and Image Understanding*, 89(2-3):197–225.

Scott, G. L. and Longuet-Higgins, H. C. (1990). Feature grouping by relocalization of eigenvectors of the proximity matrix. In *British Machine Vision Conference*, pp. 103–108.

Sebastian, T. B. and Kimia, B. B. (2005). Curves vs. skeletons in object recognition. *Signal Processing*, 85(2):246–263.

Sederberg, T. W. and Parry, S. R. (1986). Free-form deformations of solid geometric models. *Computer Graphics (SIGGRAPH '86)*, 20(4):151–160.

Sederberg, T. W., Gao, P., Wang, G., and Mu, H. (1993). 2D shape blending: An intrinsic solution to the vertex path problem. In *ACM SIGGRAPH 1993 Conference Proceedings*, pp. 15–18.

Seitz, P. (1989). Using local orientation information as image primitive for robust object recognition. In *SPIE Vol. 1199, Visual Communications and Image Processing IV*, pp. 1630–1639.

Seitz, S. (2001). The space of all stereo images. In *Eighth International Conference on Computer Vision (ICCV 2001)*, pp. 26–33, Vancouver, Canada.

Seitz, S. and Szeliski, R. (1999). Applications of computer vision to computer graphics. *Computer Graphics*, 33(4):35–37. Guest Editors' introduction to the Special Issue.

Seitz, S., Curless, B., Diebel, J., Scharstein, D., and Szeliski, R. (2006). A comparison and evaluation of multi-view stereo reconstruction algorithms. In *IEEE Computer Society Conference on Computer Vision and Pattern Recognition (CVPR'2006)*, pp. 519–526, New York, NY.

Seitz, S. M. and Baker, S. (2009). Filter flow. In *Twelfth International Conference on Computer Vision (ICCV 2009)*, Kyoto, Japan.

Seitz, S. M. and Dyer, C. M. (1996). View morphing. In *ACM SIGGRAPH 1996 Conference Proceedings*, pp. 21–30, New Orleans.

Seitz, S. M. and Dyer, C. M. (1997). Photorealistic scene reconstruction by voxel coloring. In *IEEE Computer Society Conference on Computer Vision and Pattern Recognition (CVPR'97)*, pp. 1067–1073, San Juan, Puerto Rico.

Seitz, S. M. and Dyer, C. M. (1999). Photorealistic scene reconstruction by voxel coloring. *International Journal of Computer Vision*, 35(2):151–173.

Seitz, S. M. and Dyer, C. R. (1997). View invariant analysis of cyclic motion. *International Journal of Computer Vision*, 25(3):231–251.

Serra, J. (1982). *Image Analysis and Mathematical Morphology*. Academic Press, New York.

Serra, J. and Vincent, L. (1992). An overview of morphological filtering. *Circuits, Systems and Signal Processing*, 11(1):47–108.

Serre, T., Wolf, L., and Poggio, T. (2005). Object recognition with features inspired by visual cortex. In *IEEE Computer Society Conference on Computer Vision and Pattern Recognition (CVPR'2005)*, pp. 994–1000, San Diego, CA.

Sethian, J. (1999). *Level Set Methods and Fast Marching Methods*. Cambridge University Press, Cambridge, 2nd edition.

Shade, J., Gortler, S., He, L., and Szeliski, R. (1998). Layered depth images. In *ACM SIGGRAPH 1998 Conference Proceedings*, pp. 231–242, Orlando.

Shade, J., Lischinski, D., Salesin, D., DeRose, T., and Snyder, J. (1996). Hierarchical images caching for accelerated walkthroughs of complex environments. In *ACM SIGGRAPH 1996 Conference Proceedings*, pp. 75–82, New Orleans.

Shafer, S. A. (1985). Using color to separate reflection components. *COLOR Research and Applications*, 10(4):210–218.

Shafer, S. A., Healey, G., and Wolff, L. (1992). *Physics-Based Vision: Principles and Practice*. Jones & Bartlett, Cambridge, MA.

Shafique, K. and Shah, M. (2005). A noniterative greedy algorithm for multiframe point correspondence. *IEEE Transactions on Pattern Analysis and Machine Intelligence*, 27(1):51–65.

Shah, J. (1993). A nonlinear diffusion model for discontinuous disparity and half-occlusion in stereo. In *IEEE Computer Society Conference on Computer Vision and Pattern Recognition (CVPR'93)*, pp. 34–40, New York.

Shakhnarovich, G., Darrell, T., and Indyk, P. (eds). (2006). *Nearest-Neighbor Methods in Learning and Vision: Theory and Practice*, MIT Press.

Shakhnarovich, G., Viola, P., and Darrell, T. (2003). Fast pose estimation with parameter-sensitive hashing. In *Ninth International Conference on Computer Vision (ICCV 2003)*, pp. 750–757, Nice, France.

Shan, Y., Liu, Z., and Zhang, Z. (2001). Model-based bundle adjustment with application to face modeling. In *Eighth International Conference on Computer Vision (ICCV 2001)*, pp. 644–641, Vancouver, Canada.

Sharon, E., Galun, M., Sharon, D., Basri, R., and Brandt, A. (2006). Hierarchy and adaptivity in segmenting visual scenes. *Nature*, 442(7104):810–813.

Shashua, A. and Toelg, S. (1997). The quadric reference surface: Theory and applications. *International Journal of Computer Vision*, 23(2):185–198.

Shashua, A. and Wexler, Y. (2001). Q-warping: Direct computation of quadratic reference surfaces. *IEEE Transactions on Pattern Analysis and Machine Intelligence*, 23(8):920–925.

Shaw, D. and Barnes, N. (2006). Perspective rectangle detection. In *Workshop on Applications of Computer Vision at ECCV'2006*.

Shewchuk, J. R. (1994). An introduction to the conjugate gradient method without the agonizing pain. Unpublished manuscript, available on author's homepage (http://www.cs.berkeley.edu/~jrs/). An earlier version appeared as a Carnegie Mellon University Technical Report, CMU-CS-94-125.

Shi, J. and Malik, J. (2000). Normalized cuts and image segmentation. *IEEE Transactions on Pattern Analysis and Machine Intelligence*, 8(22):888–905.

Shi, J. and Tomasi, C. (1994). Good features to track. In *IEEE Computer Society Conference on Computer Vision and Pattern Recognition (CVPR'94)*, pp. 593–600, Seattle.

Shimizu, M. and Okutomi, M. (2001). Precise sub-pixel estimation on area-based matching. In *Eighth International Conference on Computer Vision (ICCV 2001)*, pp. 90–97, Vancouver, Canada.

Shirley, P. (2005). *Fundamentals of Computer Graphics*. A K Peters, Wellesley, Massachusetts, second edition.

Shizawa, M. and Mase, K. (1991). A unified computational theory of motion transparency and motion boundaries based on eigenenergy analysis. In *IEEE Computer Society Conference on Computer Vision and Pattern Recognition (CVPR'91)*, pp. 289–295, Maui, Hawaii.

Shoemake, K. (1985). Animating rotation with quaternion curves. *Computer Graphics (SIGGRAPH '85)*, 19(3):245–254.

Shotton, J., Blake, A., and Cipolla, R. (2005). Contour-based learning for object detection. In *Tenth International Conference on Computer Vision (ICCV 2005)*, pp. 503–510, Beijing, China.

Shotton, J., Johnson, M., and Cipolla, R. (2008). Semantic texton forests for image categorization and segmentation. In *IEEE Computer Society Conference on Computer Vision and Pattern Recognition (CVPR 2008)*, Anchorage, AK.

Shotton, J., Winn, J., Rother, C., and Criminisi, A. (2009). Textonboost for image understanding: Multi-class object recognition and segmentation by jointly modeling appearance, shape and context. *International Journal of Computer Vision*, 81(1):2–23.

Shufelt, J. (1999). Performance evaluation and analysis of vanishing point detection techniques. *IEEE Transactions on Pattern Analysis and Machine Intelligence*, 21(3):282–288.

Shum, H.-Y. and He, L.-W. (1999). Rendering with concentric mosaics. In *ACM SIGGRAPH 1999 Conference Proceedings*, pp. 299–306, Los Angeles.

Shum, H.-Y. and Szeliski, R. (1999). Stereo reconstruction from multiperspective panoramas. In *Seventh International Conference on Computer Vision (ICCV'99)*, pp. 14–21, Kerkyra, Greece.

Shum, H.-Y. and Szeliski, R. (2000). Construction of panoramic mosaics with global and local alignment. *International Journal of Computer Vision*, 36(2):101–130. Erratum published July 2002, 48(2):151–152.

Shum, H.-Y., Chan, S.-C., and Kang, S. B. (2007). *Image-Based Rendering*. Springer, New York, NY.

Shum, H.-Y., Han, M., and Szeliski, R. (1998). Interactive construction of 3D models from panoramic mosaics. In *IEEE Computer Society Conference on Computer Vision and Pattern Recognition (CVPR'98)*, pp. 427–433, Santa Barbara.

Shum, H.-Y., Ikeuchi, K., and Reddy, R. (1995). Principal component analysis with missing data and its application to polyhedral modeling. *IEEE Transactions on Pattern*

Analysis and Machine Intelligence, 17(9):854–867.

Shum, H.-Y., Kang, S. B., and Chan, S.-C. (2003). Survey of image-based representations and compression techniques. *IEEE Transactions on Circuits and Systems for Video Technology*, 13(11):1020–1037.

Shum, H.-Y., Wang, L., Chai, J.-X., and Tong, X. (2002). Rendering by manifold hopping. *International Journal of Computer Vision*, 50(2):185–201.

Shum, H.-Y., Sun, J., Yamazaki, S., Li, Y., and Tang, C.-K. (2004). Pop-up light field: An interactive image-based modeling and rendering system. *ACM Transactions on Graphics*, 23(2):143–162.

Sidenbladh, H. and Black, M. J. (2003). Learning the statistics of people in images and video. *International Journal of Computer Vision*, 54(1):189–209.

Sidenbladh, H., Black, M. J., and Fleet, D. J. (2000). Stochastic tracking of 3D human figures using 2D image motion. In *Sixth European Conference on Computer Vision (ECCV 2000)*, pp. 702–718, Dublin, Ireland.

Sigal, L. and Black, M. J. (2006). Predicting 3D people from 2D pictures. In *AMDO 2006 - IV Conference on Articulated Motion and Deformable Objects*, pp. 185–195, Mallorca, Spain.

Sigal, L., Balan, A., and Black, M. J. (2010). Humaneva: Synchronized video and motion capture dataset and baseline algorithm for evaluation of articulated human motion. *International Journal of Computer Vision*, 87(1-2):4–27.

Sigal, L., Bhatia, S., Roth, S., Black, M. J., and Isard, M. (2004). Tracking loose-limbed people. In *IEEE Computer Society Conference on Computer Vision and Pattern Recognition (CVPR'2004)*, pp. 421–428, Washington, DC.

Sillion, F. and Puech, C. (1994). *Radiosity and Global Illumination*. Morgan Kaufmann.

Sim, T., Baker, S., and Bsat, M. (2003). The CMU pose, illumination, and expression database. *IEEE Transactions on Pattern Analysis and Machine Intelligence*, 25(12):1615–1618.

Simard, P. Y., Bottou, L., Haffner, P., and Cun, Y. L. (1998). Boxlets: a fast convolution algorithm for signal processing and neural networks. In *Advances in Neural Information Processing Systems 13*, pp. 571–577.

Simon, I. and Seitz, S. M. (2008). Scene segmentation using the wisdom of crowds. In *Tenth European Conference on Computer Vision (ECCV 2008)*, pp. 541–553, Marseilles.

Simon, I., Snavely, N., and Seitz, S. M. (2007). Scene summarization for online image collections. In *Eleventh International Conference on Computer Vision (ICCV 2007)*, Rio de Janeiro, Brazil.

Simoncelli, E. P. (1999). Bayesian denoising of visual images in the wavelet domain. In Müller, P. and Vidakovic, B. (eds), *Bayesian Inference in Wavelet Based Models*, pp. 291–308, Springer-Verlag, New York.

Simoncelli, E. P. and Adelson, E. H. (1990a). Non-separable extensions of quadrature mirror filters to multiple dimensions. *Proceedings of the IEEE*, 78(4):652–664.

Simoncelli, E. P. and Adelson, E. H. (1990b). Subband transforms. In Woods, J. (ed.), *Subband Coding*, pp. 143–191, Kluwer Academic Press, Norwell, MA.

Simoncelli, E. P., Adelson, E. H., and Heeger, D. J. (1991). Probability distributions of optic flow. In *IEEE Computer Society Conference on Computer Vision and Pattern Recognition (CVPR'91)*, pp. 310–315, Maui, Hawaii.

Simoncelli, E. P., Freeman, W. T., Adelson, E. H., and Heeger, D. J. (1992). Shiftable multiscale transforms. *IEEE Transactions on Information Theory*, 38(3):587–607.

Singaraju, D., Grady, L., and Vidal, R. (2008). Interactive image segmentation via minimization of quadratic energies on directed graphs. In *IEEE Computer Society Conference on Computer Vision and Pattern Recognition (CVPR 2008)*, Anchorage, AK.

Singaraju, D., Rother, C., and Rhemann, C. (2009). New appearance models for natural image matting. In *IEEE Computer Society Conference on Computer Vision and Pattern Recognition (CVPR 2009)*, Miami Beach, FL.

Singaraju, D., Grady, L., Sinop, A. K., and Vidal, R. (2010). A continuous valued MRF for image segmentation. In Blake, A., Kohli, P., and Rother, C. (eds), *Advances in Markov Random Fields*, MIT Press.

Sinha, P., Balas, B., Ostrovsky, Y., and Russell, R. (2006). Face recognition by humans: Nineteen results all computer vision researchers should know about. *Proceedings of the IEEE*, 94(11):1948–1962.

Sinha, S., Mordohai, P., and Pollefeys, M. (2007). Multi-view stereo via graph cuts on the dual of an adaptive tetrahedral mesh. In *Eleventh International Conference on Computer Vision (ICCV 2007)*, Rio de Janeiro, Brazil.

Sinha, S. N. and Pollefeys, M. (2005). Multi-view reconstruction using photo-consistency and exact silhouette constraints: A maximum-flow formulation. In *Tenth International Conference on Computer Vision (ICCV 2005)*, pp. 349–356, Beijing, China.

Sinha, S. N., Steedly, D., and Szeliski, R. (2009). Piecewise planar stereo for image-based rendering. In *Twelfth IEEE International Conference on Computer Vision (ICCV 2009)*,

Kyoto, Japan.

Sinha, S. N., Steedly, D., Szeliski, R., Agrawala, M., and Pollefeys, M. (2008). Interactive 3D architectural modeling from unordered photo collections. *ACM Transactions on Graphics (Proc. SIGGRAPH Asia 2008)*, 27(5).

Sinop, A. K. and Grady, L. (2007). A seeded image segmentation framework unifying graph cuts and random walker which yields a new algorithm. In *Eleventh International Conference on Computer Vision (ICCV 2007)*, Rio de Janeiro, Brazil.

Sivic, J. and Zisserman, A. (2003). Video Google: A text retrieval approach to object matching in videos. In *Ninth International Conference on Computer Vision (ICCV 2003)*, pp. 1470–1477, Nice, France.

Sivic, J. and Zisserman, A. (2009). Efficient visual search of videos cast as text retrieval. *IEEE Transactions on Pattern Analysis and Machine Intelligence*, 31(4):591–606.

Sivic, J., Everingham, M., and Zisserman, A. (2009). "Who are you?"—Learning person specific classifiers from video. In *IEEE Computer Society Conference on Computer Vision and Pattern Recognition (CVPR 2009)*, Miami Beach, FL.

Sivic, J., Zitnick, C. L., and Szeliski, R. (2006). Finding people in repeated shots of the same scene. In *British Machine Vision Conference (BMVC 2006)*, pp. 909–918, Edinburgh.

Sivic, J., Russell, B., Zisserman, A., Freeman, W. T., and Efros, A. A. (2008). Unsupervised discovery of visual object class hierarchies. In *IEEE Computer Society Conference on Computer Vision and Pattern Recognition (CVPR 2008)*, Anchorage, AK.

Sivic, J., Russell, B. C., Efros, A. A., Zisserman, A., and Freeman, W. T. (2005). Discovering objects and their localization in images. In *Tenth International Conference on Computer Vision (ICCV 2005)*, pp. 370–377, Beijing, China.

Slabaugh, G. G., Culbertson, W. B., Slabaugh, T. G., Culbertson, B., Malzbender, T., and Stevens, M. (2004). Methods for volumetric reconstruction of visual scenes. *International Journal of Computer Vision*, 57(3):179–199.

Slama, C. C. (ed.). (1980). *Manual of Photogrammetry*. American Society of Photogrammetry, Falls Church, Virginia, fourth edition.

Smelyanskiy, V. N., Cheeseman, P., Maluf, D. A., and Morris, R. D. (2000). Bayesian super-resolved surface reconstruction from images. In *IEEE Computer Society Conference on Computer Vision and Pattern Recognition (CVPR'2000)*, pp. 375–382, Hilton Head Island.

Smeulders, A. W. M., Worring, M., Santini, S., Gupta, A., and Jain, R. C. (2000). Content-based image retrieval at the end of the early years. *IEEE Transactions on Pattern Analysis and Machine Intelligence*, 22(12):477–490.

Sminchisescu, C. and Triggs, B. (2001). Covariance scaled sampling for monocular 3D body tracking. In *IEEE Computer Society Conference on Computer Vision and Pattern Recognition (CVPR'2001)*, pp. 447–454, Kauai, Hawaii.

Sminchisescu, C., Kanaujia, A., and Metaxas, D. (2006). Conditional models for contextual human motion recognition. *Computer Vision and Image Understanding*, 104(2-3):210–220.

Sminchisescu, C., Kanaujia, A., Li, Z., and Metaxas, D. (2005). Discriminative density propagation for 3D human motion estimation. In *IEEE Computer Society Conference on Computer Vision and Pattern Recognition (CVPR'2005)*, pp. 390–397, San Diego, CA.

Smith, A. R. and Blinn, J. F. (1996). Blue screen matting. In *ACM SIGGRAPH 1996 Conference Proceedings*, pp. 259–268, New Orleans.

Smith, B. M., Zhang, L., Jin, H., and Agarwala, A. (2009). Light field video stabilization. In *Twelfth International Conference on Computer Vision (ICCV 2009)*, Kyoto, Japan.

Smith, S. M. and Brady, J. M. (1997). SUSAN—a new approach to low level image processing. *International Journal of Computer Vision*, 23(1):45–78.

Smolic, A. and Kauff, P. (2005). Interactive 3-D video representation and coding technologies. *Proceedings of the IEEE*, 93(1):98–110.

Snavely, N., Seitz, S. M., and Szeliski, R. (2006). Photo tourism: Exploring photo collections in 3D. *ACM Transactions on Graphics (Proc. SIGGRAPH 2006)*, 25(3):835–846.

Snavely, N., Seitz, S. M., and Szeliski, R. (2008a). Modeling the world from Internet photo collections. *International Journal of Computer Vision*, 80(2):189–210.

Snavely, N., Seitz, S. M., and Szeliski, R. (2008b). Skeletal graphs for efficient structure from motion. In *IEEE Computer Society Conference on Computer Vision and Pattern Recognition (CVPR 2008)*, Anchorage, AK.

Snavely, N., Garg, R., Seitz, S. M., and Szeliski, R. (2008). Finding paths through the world's photos. *ACM Transactions on Graphics (Proc. SIGGRAPH 2008)*, 27(3).

Snavely, N., Simon, I., Goesele, M., Szeliski, R., and Seitz, S. M. (2010). Scene reconstruction and visualization from community photo collections. *Proceedings of the IEEE*, 98(8):1370–1390.

Soatto, S., Yezzi, A. J., and Jin, H. (2003). Tales of shape and radiance in multiview stereo. In *Ninth International Conference on Computer Vision (ICCV 2003)*, pp. 974–981, Nice, France.

Soille, P. (2006). Morphological image compositing. *IEEE Transactions on Pattern Analysis and Machine Intelligence*, 28(5):673–683.

Solina, F. and Bajcsy, R. (1990). Recovery of parametric models from range images: The case for superquadrics with global deformations. *IEEE Transactions on Pattern Analysis and Machine Intelligence*, 12(2):131–147.

Sontag, D. and Jaakkola, T. (2007). New outer bounds on the marginal polytope. In *Advances in Neural Information Processing Systems*.

Sontag, D., Meltzer, T., Globerson, A., Jaakkola, T., and Weiss, Y. (2008). Tightening LP relaxations for MAP using message passing. In *Uncertainty in Artificial Intelligence (UAI)*.

Soucy, M. and Laurendeau, D. (1992). Multi-resolution surface modeling from multiple range views. In *IEEE Computer Society Conference on Computer Vision and Pattern Recognition (CVPR'92)*, pp. 348–353, Champaign, Illinois.

Srinivasan, S., Chellappa, R., Veeraraghavan, A., and Aggarwal, G. (2005). Electronic image stabilization and mosaicking algorithms. In Bovik, A. (ed.), *Handbook of Image and Video Processing*, Academic Press.

Srivasan, P., Liang, P., and Hackwood, S. (1990). Computational geometric methods in volumetric intersections for 3D reconstruction. *Pattern Recognition*, 23(8):843–857.

Stamos, I., Liu, L., Chen, C., Wolberg, G., Yu, G., and Zokai, S. (2008). Integrating automated range registration with multiview geometry for the photorealistic modeling of large-scale scenes. *International Journal of Computer Vision*, 78(2-3):237–260.

Stark, J. A. (2000). Adaptive image contrast enhancement using generalizations of histogram equalization. *IEEE Transactions on Image Processing*, 9(5):889–896.

Stauffer, C. and Grimson, W. (1999). Adaptive background mixture models for real-time tracking. In *IEEE Computer Society Conference on Computer Vision and Pattern Recognition (CVPR'99)*, pp. 246–252, Fort Collins.

Steedly, D. and Essa, I. (2001). Propagation of innovative information in non-linear least-squares structure from motion. In *Eighth International Conference on Computer Vision (ICCV 2001)*, pp. 223–229, Vancouver, Canada.

Steedly, D., Essa, I., and Dellaert, F. (2003). Spectral partitioning for structure from motion. In *Ninth International Conference on Computer Vision (ICCV 2003)*, pp. 996–1003, Nice, France.

Steedly, D., Pal, C., and Szeliski, R. (2005). Efficiently registering video into panoramic mosaics. In *Tenth International Conference on Computer Vision (ICCV 2005)*, pp. 1300–1307, Beijing, China.

Steele, R. and Jaynes, C. (2005). Feature uncertainty arising from covariant image noise. In *IEEE Computer Society Conference on Computer Vision and Pattern Recognition (CVPR'2005)*, pp. 1063–1070, San Diego, CA.

Steele, R. M. and Jaynes, C. (2006). Overconstrained linear estimation of radial distortion and multi-view geometry. In *Ninth European Conference on Computer Vision (ECCV 2006)*, pp. 253–264.

Stein, A., Hoiem, D., and Hebert, M. (2007). Learning to extract object boundaries using motion cues. In *Eleventh International Conference on Computer Vision (ICCV 2007)*, Rio de Janeiro, Brazil.

Stein, F. and Medioni, G. (1992). Structural indexing: Efficient 3-D object recognition. *IEEE Transactions on Pattern Analysis and Machine Intelligence*, 14(2):125–145.

Stein, G. (1995). Accurate internal camera calibration using rotation, with analysis of sources of error. In *Fifth International Conference on Computer Vision (ICCV'95)*, pp. 230–236, Cambridge, Massachusetts.

Stein, G. (1997). Lens distortion calibration using point correspondences. In *IEEE Computer Society Conference on Computer Vision and Pattern Recognition (CVPR'97)*, pp. 602–608, San Juan, Puerto Rico.

Stenger, B., Thayananthan, A., Torr, P. H. S., and Cipolla, R. (2006). Model-based hand tracking using a hierarchical bayesian filter. *IEEE Transactions on Pattern Analysis and Machine Intelligence*, 28(9):1372–1384.

Stewart, C. V. (1999). Robust parameter estimation in computer vision. *SIAM Reviews*, 41(3):513–537.

Stiller, C. and Konrad, J. (1999). Estimating motion in image sequences: A tutorial on modeling and computation of 2D motion. *IEEE Signal Processing Magazine*, 16(4):70–91.

Stollnitz, E. J., DeRose, T. D., and Salesin, D. H. (1996). *Wavelets for Computer Graphics: Theory and Applications*. Morgan Kaufmann, San Francisco.

Strang, G. (1988). *Linear Algebra and its Applications*. Harcourt, Brace, Jovanovich, Publishers, San Diego, 3rd edition.

Strang, G. (1989). Wavelets and dilation equations: A brief introduction. *SIAM Reviews*, 31(4):614–627.

Strecha, C., Fransens, R., and Van Gool, L. (2006). Combined depth and outlier estimation in multi-view stereo. In *IEEE Computer Society Conference on Computer Vision and Pattern Recognition (CVPR'2006)*, pp. 2394–2401, New York City, NY.

Strecha, C., Tuytelaars, T., and Van Gool, L. (2003). Dense matching of multiple wide-baseline views. In *Ninth International Conference on Computer Vision (ICCV 2003)*, pp. 1194–1201, Nice, France.

Strecha, C., von Hansen, W., Van Gool, L., Fua, P., and Thoennessen, U. (2008). On benchmarking camera calibration and multi-view stereo. In *IEEE Computer Society Conference on Computer Vision and Pattern Recognition (CVPR 2008)*, Anchorage, AK.

Sturm, P. (2005). Multi-view geometry for general camera models. In *IEEE Computer Society Conference on Computer Vision and Pattern Recognition (CVPR'2005)*, pp. 206–212, San Diego, CA.

Sturm, P. and Ramalingam, S. (2004). A generic concept for camera calibration. In *Eighth European Conference on Computer Vision (ECCV 2004)*, pp. 1–13, Prague.

Sturm, P. and Triggs, W. (1996). A factorization based algorithm for multi-image projective structure and motion. In *Fourth European Conference on Computer Vision (ECCV'96)*, pp. 709–720, Cambridge, England.

Su, H., Sun, M., Fei-Fei, L., and Savarese, S. (2009). Learning a dense multi-view representation for detection, viewpoint classification and synthesis of object categories. In *Twelfth International Conference on Computer Vision (ICCV 2009)*, Kyoto, Japan.

Sudderth, E. B., Torralba, A., Freeman, W. T., and Willsky, A. S. (2008). Describing visual scenes using transformed objects and parts. *International Journal of Computer Vision*, 77(1-3):291–330.

Sullivan, S. and Ponce, J. (1998). Automatic model construction and pose estimation from photographs using triangular splines. *IEEE Transactions on Pattern Analysis and Machine Intelligence*, 20(10):1091–1096.

Sun, D., Roth, S., Lewis, J. P., and Black, M. J. (2008). Learning optical flow. In *Tenth European Conference on Computer Vision (ECCV 2008)*, pp. 83–97, Marseilles.

Sun, J., Zheng, N., and Shum, H. (2003). Stereo matching using belief propagation. *IEEE Transactions on Pattern Analysis and Machine Intelligence*, 25(7):787–800.

Sun, J., Jia, J., Tang, C.-K., and Shum, H.-Y. (2004). Poisson matting. *ACM Transactions on Graphics (Proc. SIGGRAPH 2004)*, 23(3):315–321.

Sun, J., Li, Y., Kang, S. B., and Shum, H.-Y. (2006). Flash matting. *ACM Transactions on Graphics*, 25(3):772–778.

Sun, J., Yuan, L., Jia, J., and Shum, H.-Y. (2004). Image completion with structure propagation. *ACM Transactions on Graphics (Proc. SIGGRAPH 2004)*, 24(3):861–868.

Sun, M., Su, H., Savarese, S., and Fei-Fei, L. (2009). A multi-view probabilistic model for 3D object classes. In *IEEE Computer Society Conference on Computer Vision and Pattern Recognition (CVPR 2009)*, Miami Beach, FL.

Sung, K.-K. and Poggio, T. (1998). Example-based learning for view-based human face detection. *IEEE Transactions on Pattern Analysis and Machine Intelligence*, 20(1):39–51.

Sutherland, I. E. (1974). Three-dimensional data input by tablet. *Proceedings of the IEEE*, 62(4):453–461.

Swain, M. J. and Ballard, D. H. (1991). Color indexing. *International Journal of Computer Vision*, 7(1):11–32.

Swaminathan, R., Kang, S. B., Szeliski, R., Criminisi, A., and Nayar, S. K. (2002). On the motion and appearance of specularities in image sequences. In *Seventh European Conference on Computer Vision (ECCV 2002)*, pp. 508–523, Copenhagen.

Sweldens, W. (1996). Wavelets and the lifting scheme: A 5 minute tour. *Z. Angew. Math. Mech.*, 76 (Suppl. 2):41–44.

Sweldens, W. (1997). The lifting scheme: A construction of second generation wavelets. *SIAM J. Math. Anal.*, 29(2):511–546.

Swendsen, R. H. and Wang, J.-S. (1987). Nonuniversal critical dynamics in Monte Carlo simulations. *Physical Review Letters*, 58(2):86–88.

Szeliski, R. (1986). *Cooperative Algorithms for Solving Random-Dot Stereograms*. Technical Report CMU-CS-86-133, Computer Science Department, Carnegie Mellon University.

Szeliski, R. (1989). *Bayesian Modeling of Uncertainty in Low-Level Vision*. Kluwer Academic Publishers, Boston.

Szeliski, R. (1990a). Bayesian modeling of uncertainty in low-level vision. *International Journal of Computer Vision*, 5(3):271–301.

Szeliski, R. (1990b). Fast surface interpolation using hierarchical basis functions. *IEEE Transactions on Pattern Analysis and Machine Intelligence*, 12(6):513–528.

Szeliski, R. (1991a). Fast shape from shading. *CVGIP: Image Understanding*, 53(2):129–153.

Szeliski, R. (1991b). Shape from rotation. In *IEEE Computer Society Conference on Computer Vision and Pattern Recognition (CVPR'91)*, pp. 625–630, Maui, Hawaii.

Szeliski, R. (1993). Rapid octree construction from image sequences. *CVGIP: Image Understanding*, 58(1):23–32.

Szeliski, R. (1994). Image mosaicing for tele-reality applications. In *IEEE Workshop on Applications of Computer Vision (WACV'94)*, pp. 44–53, Sarasota.

Szeliski, R. (1996). Video mosaics for virtual environments. *IEEE Computer Graphics and Applications*, 16(2):22–30.

Szeliski, R. (1999). A multi-view approach to motion and stereo. In *IEEE Computer Society Conference on Computer Vision and Pattern Recognition (CVPR'99)*, pp. 157–163, Fort Collins.

Szeliski, R. (2006a). Image alignment and stitching: A tutorial. *Foundations and Trends in Computer Graphics and Computer Vision*, 2(1):1–104.

Szeliski, R. (2006b). Locally adapted hierarchical basis preconditioning. *ACM Transactions on Graphics (Proc. SIGGRAPH 2006)*, 25(3):1135–1143.

Szeliski, R. and Coughlan, J. (1997). Spline-based image registration. *International Journal of Computer Vision*, 22(3):199–218.

Szeliski, R. and Golland, P. (1999). Stereo matching with transparency and matting. *International Journal of Computer Vision*, 32(1):45–61. Special Issue for Marr Prize papers.

Szeliski, R. and Hinton, G. (1985). Solving random-dot stereograms using the heat equation. In *IEEE Computer Society Conference on Computer Vision and Pattern Recognition (CVPR'85)*, pp. 284–288, San Francisco.

Szeliski, R. and Ito, M. R. (1986). New Hermite cubic interpolator for two-dimensional curve generation. *IEE Proceedings E*, 133(6):341–347.

Szeliski, R. and Kang, S. B. (1994). Recovering 3D shape and motion from image streams using nonlinear least squares. *Journal of Visual Communication and Image Representation*, 5(1):10–28.

Szeliski, R. and Kang, S. B. (1995). Direct methods for visual scene reconstruction. In *IEEE Workshop on Representations of Visual Scenes*, pp. 26–33, Cambridge, Massachusetts.

Szeliski, R. and Kang, S. B. (1997). Shape ambiguities in structure from motion. *IEEE Transactions on Pattern Analysis and Machine Intelligence*, 19(5):506–512.

Szeliski, R. and Lavallée, S. (1996). Matching 3-D anatomical surfaces with non-rigid deformations using octree-splines. *International Journal of Computer Vision*, 18(2):171–186.

Szeliski, R. and Scharstein, D. (2004). Sampling the disparity space image. *IEEE Transactions on Pattern Analysis and Machine Intelligence*, 26(3):419–425.

Szeliski, R. and Shum, H.-Y. (1996). Motion estimation with quadtree splines. *IEEE Transactions on Pattern Analysis and Machine Intelligence*, 18(12):1199–1210.

Szeliski, R. and Shum, H.-Y. (1997). Creating full view panoramic image mosaics and texture-mapped models. In *ACM SIGGRAPH 1997 Conference Proceedings*, pp. 251–258, Los Angeles.

Szeliski, R. and Tonnesen, D. (1992). Surface modeling with oriented particle systems. *Computer Graphics (SIGGRAPH '92)*, 26(2):185–194.

Szeliski, R. and Torr, P. (1998). Geometrically constrained structure from motion: Points on planes. In *European Workshop on 3D Structure from Multiple Images of Large-Scale Environments (SMILE)*, pp. 171–186, Freiburg, Germany.

Szeliski, R. and Weiss, R. (1998). Robust shape recovery from occluding contours using a linear smoother. *International Journal of Computer Vision*, 28(1):27–44.

Szeliski, R., Avidan, S., and Anandan, P. (2000). Layer extraction from multiple images containing reflections and transparency. In *IEEE Computer Society Conference on Computer Vision and Pattern Recognition (CVPR'2000)*, pp. 246–253, Hilton Head Island.

Szeliski, R., Tonnesen, D., and Terzopoulos, D. (1993a). Curvature and continuity control in particle-based surface models. In *SPIE Vol. 2031, Geometric Methods in Computer Vision II*, pp. 172–181, San Diego.

Szeliski, R., Tonnesen, D., and Terzopoulos, D. (1993b). Modeling surfaces of arbitrary topology with dynamic particles. In *IEEE Computer Society Conference on Computer Vision and Pattern Recognition (CVPR'93)*, pp. 82–87, New York.

Szeliski, R., Uyttendaele, M., and Steedly, D. (2008). *Fast Poisson Blending using Multi-Splines*. Technical Report MSR-TR-2008-58, Microsoft Research.

Szeliski, R., Winder, S., and Uyttendaele, M. (2010). *High-quality multi-pass image resampling*. Technical Report MSR-TR-2010-10, Microsoft Research.

Szeliski, R., Zabih, R., Scharstein, D., Veksler, O., Kolmogorov, V., Agarwala, A., Tappen, M., and Rother, C. (2008). A comparative study of energy minimization methods for Markov random fields with smoothness-based priors. *IEEE Transactions on Pattern Analysis and Machine Intelligence*, 30(6):1068–1080.

Szummer, M. and Picard, R. W. (1996). Temporal texture modeling. In *IEEE International Conference on Image Processing (ICIP-96)*, pp. 823–826, Lausanne.

Tabb, M. and Ahuja, N. (1997). Multiscale image segmentation by integrated edge and region detection. *IEEE Transactions on Image Processing*, 6(5):642–655.

Taguchi, Y., Wilburn, B., and Zitnick, C. L. (2008). Stereo reconstruction with mixed pixels using adaptive over-segmentation. In *IEEE Computer Society Conference on Computer Vision and Pattern Recognition (CVPR 2008)*, Anchorage, AK.

Tanaka, M. and Okutomi, M. (2008). Locally adaptive learning for translation-variant MRF image priors. In *IEEE Computer Society Conference on Computer Vision and Pattern Recognition (CVPR 2008)*, Anchorage, AK.

Tao, H., Sawhney, H. S., and Kumar, R. (2001). A global matching framework for stereo computation. In *Eighth International Conference on Computer Vision (ICCV 2001)*, pp. 532–539, Vancouver, Canada.

Tappen, M. F. (2007). Utilizing variational optimization to learn Markov random fields. In *IEEE Computer Society Conference on Computer Vision and Pattern Recognition (CVPR 2007)*, Minneapolis, MN.

Tappen, M. F. and Freeman, W. T. (2003). Comparison of graph cuts with belief propagation for stereo, using identical MRF parameters. In *Ninth International Conference on Computer Vision (ICCV 2003)*, pp. 900–907, Nice, France.

Tappen, M. F., Freeman, W. T., and Adelson, E. H. (2005). Recovering intrinsic images from a single image. *IEEE Transactions on Pattern Analysis and Machine Intelligence*, 27(9):1459–1472.

Tappen, M. F., Russell, B. C., and Freeman, W. T. (2003). Exploiting the sparse derivative prior for super-resolution and image demosaicing. In *Third International Workshop on Statistical and Computational Theories of Vision*, Nice, France.

Tappen, M. F., Liu, C., Freeman, W., and Adelson, E. (2007). Learning Gaussian conditional random fields for low-level vision. In *IEEE Computer Society Conference on Computer Vision and Pattern Recognition (CVPR 2007)*, Minneapolis, MN.

Tardif, J.-P. (2009). Non-iterative approach for fast and accurate vanishing point detection. In *Twelfth International Conference on Computer Vision (ICCV 2009)*, Kyoto, Japan.

Tardif, J.-P., Sturm, P., and Roy, S. (2007). Plane-based self-calibration of radial distortion. In *Eleventh International Conference on Computer Vision (ICCV 2007)*, Rio de Janeiro, Brazil.

Tardif, J.-P., Sturm, P., Trudeau, M., and Roy, S. (2009). Calibration of cameras with radially symmetric distortion. *IEEE Transactions on Pattern Analysis and Machine Intelligence*, 31(9):1552–1566.

Taubin, G. (1995). Curve and surface smoothing without shrinkage. In *Fifth International Conference on Computer Vision (ICCV'95)*, pp. 852–857, Cambridge, Massachusetts.

Taubman, D. S. and Marcellin, M. W. (2002). Jpeg2000: standard for interactive imaging. *Proceedings of the IEEE*, 90(8):1336–1357.

Taylor, C. J. (2003). Surface reconstruction from feature based stereo. In *Ninth International Conference on Computer Vision (ICCV 2003)*, pp. 184–190, Nice, France.

Taylor, C. J., Debevec, P. E., and Malik, J. (1996). Reconstructing polyhedral models of architectural scenes from photographs. In *Fourth European Conference on Computer Vision (ECCV'96)*, pp. 659–668, Cambridge, England.

Taylor, C. J., Kriegman, D. J., and Anandan, P. (1991). Structure and motion in two dimensions from multiple images: A least squares approach. In *IEEE Workshop on Visual Motion*, pp. 242–248, Princeton, New Jersey.

Taylor, P. (2009). *Text-to-Speech Synthesis*. Cambridge University Press, Cambridge.

Tek, K. and Kimia, B. B. (2003). Symmetry maps of free-form curve segments via wave propagation. *International Journal of Computer Vision*, 54(1-3):35–81.

Tekalp, M. (1995). *Digital Video Processing*. Prentice Hall, Upper Saddle River, NJ.

Telea, A. (2004). An image inpainting technique based on fast marching method. *Journal of Graphics Tools*, 9(1):23–34.

Teller, S., Antone, M., Bodnar, Z., Bosse, M., Coorg, S., Jethwa, M., and Master, N. (2003). Calibrated, registered images of an extended urban area. *International Journal of Computer Vision*, 53(1):93–107.

Teodosio, L. and Bender, W. (1993). Salient video stills: Content and context preserved. In *ACM Multimedia 93*, pp. 39–46, Anaheim, California.

Terzopoulos, D. (1983). Multilevel computational processes for visual surface reconstruction. *Computer Vision, Graphics, and Image Processing*, 24:52–96.

Terzopoulos, D. (1986a). Image analysis using multigrid relaxation methods. *IEEE Transactions on Pattern Analysis and Machine Intelligence*, PAMI-8(2):129–139.

Terzopoulos, D. (1986b). Regularization of inverse visual problems involving discontinuities. *IEEE Transactions on Pattern Analysis and Machine Intelligence*, PAMI-8(4):413–424.

Terzopoulos, D. (1988). The computation of visible-surface representations. *IEEE Transactions on Pattern Analysis and Machine Intelligence*, PAMI-10(4):417–438.

Terzopoulos, D. (1999). Visual modeling for computer animation: Graphics with a vision. *Computer Graphics*, 33(4):42–45.

Terzopoulos, D. and Fleischer, K. (1988). Deformable models. *The Visual Computer*, 4(6):306–331.

Terzopoulos, D. and Metaxas, D. (1991). Dynamic 3D models with local and global deformations: Deformable superquadrics. *IEEE Transactions on Pattern Analysis and Machine Intelligence*, 13(7):703–714.

Terzopoulos, D. and Szeliski, R. (1992). Tracking with Kalman snakes. In Blake, A. and Yuille, A. L. (eds), *Active Vision*, pp. 3–20, MIT Press, Cambridge, Massachusetts.

Terzopoulos, D. and Waters, K. (1990). Analysis of facial images using physical and anatomical models. In *Third International Conference on Computer Vision (ICCV'90)*, pp. 727–732, Osaka, Japan.

Terzopoulos, D. and Witkin, A. (1988). Physically-based models with rigid and deformable components. *IEEE Computer Graphics and Applications*, 8(6):41–51.

Terzopoulos, D., Witkin, A., and Kass, M. (1987). Symmetry-seeking models and 3D object reconstruction. *International Journal of Computer Vision*, 1(3):211–221.

Terzopoulos, D., Witkin, A., and Kass, M. (1988). Constraints on deformable models: Recovering 3D shape and nonrigid motion. *Artificial Intelligence*, 36(1):91–123.

Thayananthan, A., Iwasaki, M., and Cipolla, R. (2008). Principled fusion of high-level model and low-level cues for motion segmentation. In *IEEE Computer Society Conference on Computer Vision and Pattern Recognition (CVPR 2008)*, Anchorage, AK.

Thirthala, S. and Pollefeys, M. (2005). The radial trifocal tensor: A tool for calibrating the radial distortion of wide-angle cameras. In *IEEE Computer Society Conference on Computer Vision and Pattern Recognition (CVPR'2005)*, pp. 321–328, San Diego, CA.

Thomas, D. B., Luk, W., Leong, P. H., and Villasenor, J. D. (2007). Gaussian random number generators. *ACM Computing Surveys*, 39(4).

Thrun, S., Burgard, W., and Fox, D. (2005). *Probabilistic Robotics*. The MIT Press, Cambridge, Massachusetts.

Thrun, S., Montemerlo, M., Dahlkamp, H., Stavens, D., Aron, A. *et al.* (2006). Stanley, the robot that won the DARPA Grand Challenge. *Journal of Field Robotics*, 23(9):661–692.

Tian, Q. and Huhns, M. N. (1986). Algorithms for subpixel registration. *Computer Vision, Graphics, and Image Processing*, 35:220–233.

Tikhonov, A. N. and Arsenin, V. Y. (1977). *Solutions of Ill-Posed Problems*. V. H. Winston, Washington, D. C.

Tipping, M. E. and Bishop, C. M. (1999). Probabilistic principal components analysis. *Journal of the Royal Statistical Society, Series B*, 61(3):611–622.

Toint, P. L. (1987). On large scale nonlinear least squares calculations. *SIAM J. Sci. Stat. Comput.*, 8(3):416–435.

Tola, E., Lepetit, V., and Fua, P. (2010). DAISY: An efficient dense descriptor applied to wide-baseline stereo. *IEEE Transactions on Pattern Analysis and Machine Intelligence*, 32(5):815–830.

Tolliver, D. and Miller, G. (2006). Graph partitioning by spectral rounding: Applications in image segmentation and clustering. In *IEEE Computer Society Conference on Computer Vision and Pattern Recognition (CVPR'2006)*, pp. 1053–1060, New York City, NY.

Tomasi, C. and Kanade, T. (1992). Shape and motion from image streams under orthography: A factorization method. *International Journal of Computer Vision*, 9(2):137–154.

Tomasi, C. and Manduchi, R. (1998). Bilateral filtering for gray and color images. In *Sixth International Conference on Computer Vision (ICCV'98)*, pp. 839–846, Bombay.

Tombari, F., Mattoccia, S., and Di Stefano, L. (2007). Segmentation-based adaptive support for accurate stereo correspondence. In *Pacific-Rim Symposium on Image and Video Technology*.

Tombari, F., Mattoccia, S., Di Stefano, L., and Addimanda, E. (2008). Classification and evaluation of cost aggregation methods for stereo correspondence. In *IEEE Computer Society Conference on Computer Vision and Pattern Recognition (CVPR 2008)*, Anchorage, AK.

Tommasini, T., Fusiello, A., Trucco, E., and Roberto, V. (1998). Making good features track better. In *IEEE Computer Society Conference on Computer Vision and Pattern Recognition (CVPR'98)*, pp. 178–183, Santa Barbara.

Torborg, J. and Kajiya, J. T. (1996). Talisman: Commodity realtime 3D graphics for the PC. In *ACM SIGGRAPH 1996 Conference Proceedings*, pp. 353–363, New Orleans.

Torr, P. H. S. (2002). Bayesian model estimation and selection for epipolar geometry and generic manifold fitting. *International Journal of Computer Vision*, 50(1):35–61.

Torr, P. H. S. and Fitzgibbon, A. W. (2004). Invariant fitting of two view geometry. *IEEE Transactions on Pattern Analysis and Machine Intelligence*, 26(5):648–650.

Torr, P. H. S. and Murray, D. (1997). The development and comparison of robust methods for estimating the fundamental matrix. *International Journal of Computer Vision*, 24(3):271–300.

Torr, P. H. S., Szeliski, R., and Anandan, P. (1999). An integrated Bayesian approach to layer extraction from image sequences. In *Seventh International Conference on Computer Vision (ICCV'99)*, pp. 983–990, Kerkyra, Greece.

Torr, P. H. S., Szeliski, R., and Anandan, P. (2001). An integrated Bayesian approach to layer extraction from image sequences. *IEEE Transactions on Pattern Analysis and Machine Intelligence*, 23(3):297–303.

Torralba, A. (2003). Contextual priming for object detection. *International Journal of Computer Vision*, 53(2):169–191.

Torralba, A. (2007). Classifier-based methods. In *CVPR 2007 Short Course on Recognizing and Learning Object Categories.* http://people.csail.mit.edu/torralba/shortCourseRLOC/.

Torralba, A. (2008). Object recognition and scene understanding. MIT Course 6.870, http://people.csail.mit.edu/torralba/courses/6.870/6.870.recognition.htm.

Torralba, A., Freeman, W. T., and Fergus, R. (2008). 80 million tiny images: a large dataset for non-parametric object and scene recognition. *IEEE Transactions on Pattern Analysis and Machine Intelligence*, 30(11):1958–1970.

Torralba, A., Murphy, K. P., and Freeman, W. T. (2004). Contextual models for object detection using boosted random fields. In *Advances in Neural Information Processing Systems*.

Torralba, A., Murphy, K. P., and Freeman, W. T. (2007). Sharing visual features for multiclass and multiview object detection. *IEEE Transactions on Pattern Analysis and Machine Intelligence*, 29(5):854–869.

Torralba, A., Weiss, Y., and Fergus, R. (2008). Small codes and large databases of images for object recognition. In *IEEE Computer Society Conference on Computer Vision and Pattern Recognition (CVPR 2008)*, Anchorage, AK.

Torralba, A., Murphy, K. P., Freeman, W. T., and Rubin, M. A. (2003). Context-based vision system for place and object recognition. In *Ninth International Conference on Computer Vision (ICCV 2003)*, pp. 273–280, Nice, France.

Torrance, K. E. and Sparrow, E. M. (1967). Theory for off-specular reflection from roughened surfaces. *Journal of the Optical Society of America A*, 57(9):1105–1114.

Torresani, L., Hertzmann, A., and Bregler, C. (2008). Non-rigid structure-from-motion: Estimating shape and motion with hierarchical priors. *IEEE Transactions on Pattern Analysis and Machine Intelligence*, 30(5):878–892.

Toyama, K. (1998). *Prolegomena for Robust Face Tracking*. Technical Report MSR-TR-98-65, Microsoft Research.

Toyama, K., Krumm, J., Brumitt, B., and Meyers, B. (1999). Wallflower: Principles and practice of background maintenance. In *Seventh International Conference on Computer Vision (ICCV'99)*, pp. 255–261, Kerkyra, Greece.

Tran, S. and Davis, L. (2002). 3D surface reconstruction using graph cuts with surface constraints. In *Seventh European Conference on Computer Vision (ECCV 2002)*, pp. 219–231, Copenhagen.

Trefethen, L. N. and Bau, D. (1997). *Numerical Linear Algebra*. SIAM.

Treisman, A. (1985). Preattentive processing in vision. *Computer Vision, Graphics, and Image Processing*, 31(2):156–177.

Triggs, B. (1996). Factorization methods for projective structure and motion. In *IEEE Computer Society Conference on Computer Vision and Pattern Recognition (CVPR'96)*, pp. 845–851, San Francisco.

Triggs, B. (2004). Detecting keypoints with stable position, orientation, and scale under illumination changes. In *Eighth European Conference on Computer Vision (ECCV 2004)*, pp. 100–113, Prague.

Triggs, B., McLauchlan, P. F., Hartley, R. I., and Fitzgibbon, A. W. (1999). Bundle adjustment — a modern synthesis. In *International Workshop on Vision Algorithms*, pp. 298–372, Kerkyra, Greece.

Trobin, W., Pock, T., Cremers, D., and Bischof, H. (2008). Continuous energy minimization via repeated binary fusion. In *Tenth European Conference on Computer Vision (ECCV 2008)*, pp. 677–690, Marseilles.

Troccoli, A. and Allen, P. (2008). Building illumination coherent 3D models of large-scale outdoor scenes. *International Journal of Computer Vision*, 78(2-3):261–280.

Trottenberg, U., Oosterlee, C. W., and Schuller, A. (2000). *Multigrid*. Academic Press.

Trucco, E. and Verri, A. (1998). *Introductory Techniques for 3-D Computer Vision*. Prentice Hall, Upper Saddle River, NJ.

Tsai, P. S. and Shah, M. (1994). Shape from shading using linear approximation. *Image and Vision Computing*, 12:487–498.

Tsai, R. Y. (1987). A versatile camera calibration technique for high-accuracy 3D machine vision metrology using off-the-shelf TV cameras and lenses. *IEEE Journal of Robotics and Automation*, RA-3(4):323–344.

Tschumperlé, D. (2006). Curvature-preserving regularization of multi-valued images using PDEs. In *Ninth European Conference on Computer Vision (ECCV 2006)*, pp. 295–307.

Tschumperlé, D. and Deriche, R. (2005). Vector-valued image regularization with PDEs: A common framework for different applications. *IEEE Transactions on Pattern Analysis and Machine Intelligence*, 27:506–517.

Tsin, Y., Kang, S. B., and Szeliski, R. (2006). Stereo matching with linear superposition of layers. *IEEE Transactions on Pattern Analysis and Machine Intelligence*, 28(2):290–301.

Tsin, Y., Ramesh, V., and Kanade, T. (2001). Statistical calibration of CCD imaging process. In *Eighth International Conference on Computer Vision (ICCV 2001)*, pp. 480–487, Vancouver, Canada.

Tu, Z., Chen, X., Yuille, A. L., and Zhu, S.-C. (2005). Image parsing: Unifying segmentation, detection, and recognition. *International Journal of Computer Vision*, 63(2):113–140.

Tumblin, J. and Rushmeier, H. E. (1993). Tone reproduction for realistic images. *IEEE Computer Graphics and Applications*, 13(6):42–48.

Tumblin, J. and Turk, G. (1999). LCIS: A boundary hierarchy for detail-preserving contrast reduction. In *ACM SIGGRAPH 1999 Conference Proceedings*, pp. 83–90, Los Angeles.

Tumblin, J., Agrawal, A., and Raskar, R. (2005). Why I want a gradient camera. In *IEEE Computer Society Conference on Computer Vision and Pattern Recognition (CVPR'2005)*, pp. 103–110, San Diego, CA.

Turcot, P. and Lowe, D. G. (2009). Better matching with fewer features: The selection of useful features in large database recognition problems. In *ICCV Workshop on Emergent Issues in Large Amounts of Visual Data (WS-LAVD)*, Kyoto, Japan.

Turk, G. and Levoy, M. (1994). Zippered polygonal meshes from range images. In *ACM SIGGRAPH 1994 Conference Proceedings*, pp. 311–318.

Turk, G. and O'Brien, J. (2002). Modelling with implicit surfaces that interpolate. *ACM Transactions on Graphics*, 21(4):855–873.

Turk, M. and Pentland, A. (1991a). Eigenfaces for recognition. *Journal of Cognitive Neuroscience*, 3(1):71–86.

Turk, M. and Pentland, A. (1991b). Face recognition using eigenfaces. In *IEEE Computer Society Conference on Computer Vision and Pattern Recognition (CVPR'91)*, pp. 586–591, Maui, Hawaii.

Tuytelaars, T. and Mikolajczyk, K. (2007). Local invariant feature detectors. *Foundations and Trends in Computer Graphics and Computer Vision*, 3(1).

Tuytelaars, T. and Van Gool, L. (2004). Matching widely separated views based on affine invariant regions. *International Journal of Computer Vision*, 59(1):61–85.

Tuytelaars, T., Van Gool, L., and Proesmans, M. (1997). The cascaded Hough transform. In *International Conference on Image Processing (ICIP'97)*, pp. 736–739.

Ullman, S. (1979). The interpretation of structure from motion. *Proceedings of the Royal Society of London*, B-203:405–426.

Unnikrishnan, R., Pantofaru, C., and Hebert, M. (2007). Toward objective evaluation of image segmentation algorithms. *IEEE Transactions on Pattern Analysis and Machine Intelligence*, 29(6):828–944.

Unser, M. (1999). Splines: A perfect fit for signal and image processing. *IEEE Signal Processing Magazine*, 16(6):22–38.

Urmson, C., Anhalt, J., Bagnell, D., Baker, C., Bittner, R. *et al.* (2008). Autonomous driving in urban environments: Boss and the urban challenge. *Journal of Field Robotics*, 25(8):425–466.

Urtasun, R., Fleet, D. J., and Fua, P. (2006). Temporal motion models for monocular and multiview 3D human body tracking. *Computer Vision and Image Understanding*, 104(2-3):157–177.

Uyttendaele, M., Eden, A., and Szeliski, R. (2001). Eliminating ghosting and exposure artifacts in image mosaics. In *IEEE Computer Society Conference on Computer Vision and Pattern Recognition (CVPR'2001)*, pp. 509–516, Kauai, Hawaii.

Uyttendaele, M., Criminisi, A., Kang, S. B., Winder, S., Hartley, R., and Szeliski, R. (2004). Image-based interactive exploration of real-world environments. *IEEE Computer Graphics and Applications*, 24(3):52–63.

Vaillant, R. and Faugeras, O. D. (1992). Using extremal boundaries for 3-D object modeling. *IEEE Transactions on Pattern Analysis and Machine Intelligence*, 14(2):157–173.

Vaish, V., Szeliski, R., Zitnick, C. L., Kang, S. B., and Levoy, M. (2006). Reconstructing occluded surfaces using synthetic apertures: Shape from focus vs. shape from stereo. In *IEEE Computer Society Conference on Computer Vision and Pattern Recognition (CVPR'2006)*, pp. 2331–2338, New York, NY.

van de Weijer, J. and Schmid, C. (2006). Coloring local feature extraction. In *Ninth European Conference on Computer Vision (ECCV 2006)*, pp. 334–348.

van den Hengel, A., Dick, A., Thormhlen, T., Ward, B., and Torr, P. H. S. (2007). Videotrace: Rapid interactive scene modeling from video. *ACM Transactions on Graphics*, 26(3).

Van Huffel, S. and Lemmerling, P. (eds). (2002). *Total Least Squares and Errors-in-Variables Modeling*, Springer.

Van Huffel, S. and Vandewalle, J. (1991). *The Total Least Squares Problem: Computational Aspects and Analysis*. Society for Industrial and Applied Mathematics, Philadephia.

van Ouwerkerk, J. D. (2006). Image super-resolution survey. *Image and Vision Computing*, 24(10):1039–1052.

Varma, M. and Ray, D. (2007). Learning the discriminative power-invariance trade-off. In *Eleventh International Conference on Computer Vision (ICCV 2007)*, Rio de Janeiro, Brazil.

Vasconcelos, N. (2007). From pixels to semantic spaces: Advances in content-based image retrieval. *Computer*, 40(7):20–26.

Vasilescu, M. A. O. and Terzopoulos, D. (2007). Multilinear (tensor) image synthesis, analysis, and recognition. *IEEE Signal Processing Magazine*, 24(6):118–123.

Vedaldi, A. and Fulkerson, B. (2008). VLFeat: An open and portable library of computer vision algorithms. http://www.vlfeat.org/.

Vedaldi, A., Gulshan, V., Varma, M., and Zisserman, A. (2009). Multiple kernels for object detection. In *Twelfth International Conference on Computer Vision (ICCV 2009)*, Kyoto, Japan.

Vedula, S., Baker, S., and Kanade, T. (2005). Image-based spatio-temporal modeling and view interpolation of dynamic events. *ACM Transactions on Graphics*, 24(2):240–261.

Vedula, S., Baker, S., Rander, P., Collins, R., and Kanade, T. (2005). Three-dimensional scene flow. *IEEE Transactions on Pattern Analysis and Machine Intelligence*, 27(3):475–480.

Veeraraghavan, A., Raskar, R., Agrawal, A., Mohan, A., and Tumblin, J. (2007). Dappled photography: Mask enhanced cameras for heterodyned light fields and coded aperture refocusing. *ACM Transactions on Graphics*, 26(3).

Veksler, O. (1999). *Efficient Graph-based Energy Minimization Methods in Computer Vision*. Ph.D. thesis, Cornell University.

Veksler, O. (2001). Stereo matching by compact windows via minimum ratio cycle. In *Eighth International Conference on Computer Vision (ICCV 2001)*, pp. 540–547, Vancouver, Canada.

Veksler, O. (2003). Fast variable window for stereo correspondence using integral images. In *IEEE Computer Society Conference on Computer Vision and Pattern Recognition (CVPR'2003)*, pp. 556–561, Madison, WI.

Veksler, O. (2007). Graph cut based optimization for MRFs with truncated convex priors. In *IEEE Computer Society Conference on Computer Vision and Pattern Recognition (CVPR 2007)*, Minneapolis, MN.

Verbeek, J. and Triggs, B. (2007). Region classification with Markov field aspect models. In *IEEE Computer Society Conference on Computer Vision and Pattern Recognition (CVPR 2007)*, Minneapolis, MN.

Vergauwen, M. and Van Gool, L. (2006). Web-based 3D reconstruction service. *Machine Vision and Applications*, 17(2):321–329.

Vetter, T. and Poggio, T. (1997). Linear object classes and image synthesis from a single example image. *IEEE Transactions on Pattern Analysis and Machine Intelligence*, 19(7):733–742.

Vezhnevets, V., Sazonov, V., and Andreeva, A. (2003). A survey on pixel-based skin color detection techniques. In *GRAPHICON03*, pp. 85–92.

Vicente, S., Kolmogorov, V., and Rother, C. (2008). Graph cut based image segmentation with connectivity priors. In *IEEE Computer Society Conference on Computer Vision and Pattern Recognition (CVPR 2008)*, Anchorage, AK.

Vidal, R., Ma, Y., and Sastry, S. S. (2010). *Generalized Principal Component Analysis*. Springer.

Viéville, T. and Faugeras, O. D. (1990). Feedforward recovery of motion and structure from a sequence of 2D-lines matches. In *Third International Conference on Computer Vision (ICCV'90)*, pp. 517–520, Osaka, Japan.

Vincent, L. and Soille, P. (1991). Watersheds in digital spaces: An efficient algorithm based on immersion simulations. *IEEE Transactions on Pattern Analysis and Machine Intelligence*, 13(6):583–596.

Vineet, V. and Narayanan, P. J. (2008). CUDA cuts: Fast graph cuts on the GPU. In *CVPR 2008 Workshop on Visual Computer Vision on GPUs (CVGPU)*, Anchorage, AK.

Viola, P. and Wells III, W. (1997). Alignment by maximization of mutual information. *International Journal of Computer Vision*, 24(2):137–154.

Viola, P., Jones, M. J., and Snow, D. (2003). Detecting pedestrians using patterns of motion and appearance. In *Ninth International Conference on Computer Vision (ICCV 2003)*, pp. 734–741, Nice, France.

Viola, P. A. and Jones, M. J. (2004). Robust real-time face detection. *International Journal of Computer Vision*, 57(2):137–154.

Vlasic, D., Baran, I., Matusik, W., and Popović, J. (2008). Articulated mesh animation from multi-view silhouettes. *ACM Transactions on Graphics*, 27(3).

Vlasic, D., Brand, M., Pfister, H., and Popović, J. (2005). Face transfer with multilinear models. *ACM Transactions on Graphics (Proc. SIGGRAPH 2005)*, 24(3):426–433.

Vogiatzis, G., Torr, P., and Cipolla, R. (2005). Multi-view stereo via volumetric graph-cuts. In *IEEE Computer Society Conference on Computer Vision and Pattern Recognition (CVPR'2005)*, pp. 391–398, San Diego, CA.

Vogiatzis, G., Hernandez, C., Torr, P., and Cipolla, R. (2007). Multi-view stereo via volumetric graph-cuts and occlusion robust photo-consistency. *IEEE Transactions on Pattern Analysis and Machine Intelligence*, 29(12):2241–2246.

von Ahn, L. and Dabbish, L. (2004). Labeling images with a computer game. In *CHI'04: SIGCHI Conference on Human Factors in Computing Systems*, pp. 319–326, Vienna, Austria.

von Ahn, L., Liu, R., and Blum, M. (2006). Peekaboom: A game for locating objects in images. In *CHI'06: SIGCHI Conference on Human Factors in Computing Systems*, pp. 55–64, Montréal, Québec, Canada.

Wainwright, M. J. and Jordan, M. I. (2008). Graphical models, exponential families, and variational inference. *Foundations and Trends in Machine Learning*, 1(1-2):1–305.

Wainwright, M. J., Jaakkola, T. S., and Willsky, A. S. (2005). MAP estimation via agreement on trees: message-passing and linear programming. *IEEE Transactions on Information Theory*, 51(11):3697–3717.

Waithe, P. and Ferrie, F. (1991). From uncertainty to visual exploration. *IEEE Transactions on Pattern Analysis and Machine Intelligence*, 13(10):1038–1049.

Walker, E. L. and Herman, M. (1988). Geometric reasoning for constructing 3D scene descriptions from images. *Artificial Intelligence*, 37:275–290.

Wallace, G. K. (1991). The JPEG still picture compression standard. *Communications of the ACM*, 34(4):30–44.

Wallace, J. R., Cohen, M. F., and Greenberg, D. P. (1987). A two-pass solution to the rendering equation: A synthesis of ray tracing and radiosity methods. *Computer Graphics (SIGGRAPH '87)*, 21(4):311–320.

Waltz, D. L. (1975). Understanding line drawings of scenes with shadows. In Winston, P. H. (ed.), *The Psychology of Computer Vision*, McGraw-Hill, New York.

Wang, H. and Oliensis, J. (2010). Shape matching by segmentation averaging. *IEEE Transactions on Pattern Analysis and Machine Intelligence*, 32(4):619–635.

Wang, J. and Cohen, M. F. (2005). An iterative optimization approach for unified image segmentation and matting. In *Tenth International Conference on Computer Vision (ICCV 2005)*, Beijing, China.

Wang, J. and Cohen, M. F. (2007a). Image and video matting: A survey. *Foundations and Trends in Computer Graphics and Computer Vision*, 3(2).

Wang, J. and Cohen, M. F. (2007b). Optimized color sampling for robust matting. In *IEEE Computer Society Conference on Computer Vision and Pattern Recognition (CVPR 2007)*, Minneapolis, MN.

Wang, J. and Cohen, M. F. (2007c). Simultaneous matting and compositing. In *IEEE Computer Society Conference on Computer Vision and Pattern Recognition (CVPR 2007)*, Minneapolis, MN.

Wang, J., Agrawala, M., and Cohen, M. F. (2007). Soft scissors: An interactive tool for realtime high quality matting. *ACM Transactions on Graphics*, 26(3).

Wang, J., Thiesson, B., Xu, Y., and Cohen, M. (2004). Image and video segmentation by anisotropic kernel mean shift. In *Eighth European Conference on Computer Vision (ECCV 2004)*, pp. 238–249, Prague.

Wang, J., Bhat, P., Colburn, R. A., Agrawala, M., and Cohen, M. F. (2005). Video cutout. *ACM Transactions on Graphics (Proc. SIGGRAPH 2005)*, 24(3):585–594.

Wang, J. Y. A. and Adelson, E. H. (1994). Representing moving images with layers. *IEEE Transactions on Image Processing*, 3(5):625–638.

Wang, L., Kang, S. B., Szeliski, R., and Shum, H.-Y. (2001). Optimal texture map reconstruction from multiple views. In *IEEE Computer Society Conference on Computer Vision and Pattern Recognition (CVPR'2001)*, pp. 347–354, Kauai, Hawaii.

Wang, Y. and Zhu, S.-C. (2003). Modeling textured motion: Particle, wave and sketch. In *Ninth International Conference on Computer Vision (ICCV 2003)*, pp. 213–220, Nice, France.

Wang, Z., Bovik, A. C., and Simoncelli, E. P. (2005). Structural approaches to image quality assessment. In Bovik, A. C. (ed.), *Handbook of Image and Video Processing*, pp. 961–974, Elsevier Academic Press.

Wang, Z., Bovik, A. C., Sheikh, H. R., and Simoncelli, E. P. (2004). Image quality assessment: From error visibility to structural similarity. *IEEE Transactions on Image Processing*, 13(4):600–612.

Wang, Z.-F. and Zheng, Z.-G. (2008). A region based stereo matching algorithm using cooperative optimization. In *IEEE Computer Society Conference on Computer Vision and Pattern Recognition (CVPR 2008)*, Anchorage, AK.

Ward, G. (1992). Measuring and modeling anisotropic reflection. *Computer Graphics (SIGGRAPH '92)*, 26(4):265–272.

Ward, G. (1994). The radiance lighting simulation and rendering system. In *ACM SIGGRAPH 1994 Conference Proceedings*, pp. 459–472.

Ward, G. (2003). Fast, robust image registration for compositing high dynamic range photographs from hand-held exposures. *Journal of Graphics Tools*, 8(2):17–30.

Ward, G. (2004). High dynamic range image encodings. http://www.anyhere.com/gward/hdrenc/hdr_encodings.html.

Ware, C., Arthur, K., and Booth, K. S. (1993). Fish tank virtual reality. In *INTERCHI'03*, pp. 37–42, Amsterdam.

Warren, J. and Weimer, H. (2001). *Subdivision Methods for Geometric Design: A Constructive Approach*. Morgan Kaufmann.

Watanabe, M. and Nayar, S. K. (1998). Rational filters for passive depth from defocus. *International Journal of Computer Vision*, 27(3):203–225.

Watt, A. (1995). *3D Computer Graphics*. Addison-Wesley, Harlow, England, third edition.

Weber, J. and Malik, J. (1995). Robust computation of optical flow in a multi-scale differential framework. *International Journal of Computer Vision*, 14(1):67–81.

Weber, M., Welling, M., and Perona, P. (2000). Unsupervised learning of models for recognition. In *Sixth European Conference on Computer Vision (ECCV 2000)*, pp. 18–32, Dublin, Ireland.

Wedel, A., Cremers, D., Pock, T., and Bischof, H. (2009). Structure- and motion-adaptive regularization for high accuracy optic flow. In *Twelfth International Conference on Computer Vision (ICCV 2009)*, Kyoto, Japan.

Wedel, A., Rabe, C., Vaudrey, T., Brox, T., Franke, U., and Cremers, D. (2008). Efficient dense scene flow from sparse or dense stereo data. In *Tenth European Conference on Computer Vision (ECCV 2008)*, pp. 739–751, Marseilles.

Wei, C. Y. and Quan, L. (2004). Region-based progressive stereo matching. In *IEEE Computer Society Conference on Computer Vision and Pattern Recognition (CVPR'2005)*, pp. 106–113, Washington, D. C.

Wei, L.-Y. and Levoy, M. (2000). Fast texture synthesis using tree-structured vector quantization. In *ACM SIGGRAPH 2000 Conference Proceedings*, pp. 479–488.

Weickert, J. (1998). *Anisotropic Diffusion in Image Processing*. Tuebner, Stuttgart.

Weickert, J., ter Haar Romeny, B. M., and Viergever, M. A. (1998). Efficient and reliable schemes for nonlinear diffusion filtering. *IEEE Transactions on Image Processing*, 7(3):398–410.

Weinland, D., Ronfard, R., and Boyer, E. (2006). Free viewpoint action recognition using motion history volumes. *Computer Vision and Image Understanding*, 104(2-3):249–257.

Weiss, Y. (1997). Smoothness in layers: Motion segmentation using nonparametric mixture estimation. In *IEEE Computer Society Conference on Computer Vision and Pattern Recognition (CVPR'97)*, pp. 520–526, San Juan, Puerto Rico.

Weiss, Y. (1999). Segmentation using eigenvectors: A unifying view. In *Seventh International Conference on Computer Vision (ICCV'99)*, pp. 975–982, Kerkyra, Greece.

Weiss, Y. (2001). Deriving intrinsic images from image sequences. In *Eighth International Conference on Computer Vision (ICCV 2001)*, pp. 7–14, Vancouver, Canada.

Weiss, Y. and Adelson, E. H. (1996). A unified mixture framework for motion segmentation: Incorporating spatial coherence and estimating the number of models. In *IEEE Computer Society Conference on Computer Vision and Pattern Recognition (CVPR'96)*, pp. 321–326, San Francisco.

Weiss, Y. and Freeman, B. (2007). What makes a good model of natural images? In *IEEE Computer Society Conference on Computer Vision and Pattern Recognition (CVPR 2007)*, Minneapolis, MN.

Weiss, Y. and Freeman, W. T. (2001a). Correctness of belief propagation in Gaussian graphical models of arbitrary topology. *Neural Computation*, 13(10):2173–2200.

Weiss, Y. and Freeman, W. T. (2001b). On the optimality of solutions of the max-product belief propagation algorithm in arbitrary graphs. *IEEE Transactions on Information Theory*, 47(2):736–744.

Weiss, Y., Torralba, A., and Fergus, R. (2008). Spectral hashing. In *Advances in Neural Information Processing Systems*.

Weiss, Y., Yanover, C., and Meltzer, T. (2010). Linear programming and variants of belief propagation. In Blake, A., Kohli, P., and Rother, C. (eds), *Advances in Markov Random Fields*, MIT Press.

Wells, III, W. M. (1986). Efficient synthesis of Gaussian filters by cascaded uniform filters. *IEEE Transactions on Pattern Analysis and Machine Intelligence*, 8(2):234–239.

Weng, J., Ahuja, N., and Huang, T. S. (1993). Optimal motion and structure estimation. *IEEE Transactions on Pattern Analysis and Machine Intelligence*, 15(9):864–884.

Wenger, A., Gardner, A., Tchou, C., Unger, J., Hawkins, T., and Debevec, P. (2005). Performance relighting and reflectance transformation with time-multiplexed illumination. *ACM Transactions on Graphics (Proc. SIGGRAPH 2005)*, 24(3):756–764.

Werlberger, M., Trobin, W., Pock, T., Bischof, H., Wedel, A., and Cremers, D. (2009). Anisotropic Huber-L1 optical flow. In *British Machine Vision Conference (BMVC 2009)*, London.

Werner, T. (2007). A linear programming approach to max-sum problem: A review. *IEEE Transactions on Pattern Analysis and Machine Intelligence*, 29(7):1165–1179.

Werner, T. and Zisserman, A. (2002). New techniques for automated architectural reconstruction from photographs. In *Seventh European Conference on Computer Vision (ECCV 2002)*, pp. 541–555, Copenhagen.

Westin, S. H., Arvo, J. R., and Torrance, K. E. (1992). Predicting reflectance functions from complex surfaces. *Computer Graphics (SIGGRAPH '92)*, 26(4):255–264.

Westover, L. (1989). Interactive volume rendering. In *Workshop on Volume Visualization*, pp. 9–16, Chapel Hill.

Wexler, Y., Fitzgibbon, A., and Zisserman, A. (2002). Bayesian estimation of layers from multiple images. In *Seventh European Conference on Computer Vision (ECCV 2002)*, pp. 487–501, Copenhagen.

Wexler, Y., Shechtman, E., and Irani, M. (2007). Space-time completion of video. *IEEE Transactions on Pattern Analysis and Machine Intelligence*, 29(3):463–476.

Weyrich, T., Lawrence, J., Lensch, H. P. A., Rusinkiewicz, S., and Zickler, T. (2008). Principles of appearance acquisition and representation. *Foundations and Trends in Computer Graphics and Computer Vision*, 4(2):75–191.

Weyrich, T., Matusik, W., Pfister, H., Bickel, B., Donner, C. *et al.* (2006). Analysis of human faces using a measurement-based skin reflectance model. *ACM Transactions on Graphics*, 25(3):1013–1024.

Wheeler, M. D., Sato, Y., and Ikeuchi, K. (1998). Consensus surfaces for modeling 3D objects from multiple range images. In *Sixth International Conference on Computer Vision (ICCV'98)*, pp. 917–924, Bombay.

White, R. and Forsyth, D. (2006). Combining cues: Shape from shading and texture. In *IEEE Computer Society Conference on Computer Vision and Pattern Recognition (CVPR'2006)*, pp. 1809–1816, New York City, NY.

White, R., Crane, K., and Forsyth, D. A. (2007). Capturing and animating occluded cloth. *ACM Transactions on Graphics*, 26(3).

Wiejak, J. S., Buxton, H., and Buxton, B. F. (1985). Convolution with separable masks for early image processing. *Computer Vision, Graphics, and Image Processing*, 32(3):279–290.

Wilburn, B., Joshi, N., Vaish, V., Talvala, E.-V., Antunez, E. *et al.* (2005). High performance imaging using large camera arrays. *ACM Transactions on Graphics (Proc. SIGGRAPH 2005)*, 24(3):765–776.

Wilczkowiak, M., Brostow, G. J., Tordoff, B., and Cipolla, R. (2005). Hole filling through photomontage. In *British Machine Vision Conference (BMVC 2005)*, pp. 492–501, Oxford Brookes.

Williams, D. and Burns, P. D. (2001). Diagnostics for digital capture using MTF. In *IS&T PICS Conference*, pp. 227–232.

Williams, D. J. and Shah, M. (1992). A fast algorithm for active contours and curvature estimation. *Computer Vision, Graphics, and Image Processing*, 55(1):14–26.

Williams, L. (1983). Pyramidal parametrics. *Computer Graphics (SIGGRAPH '83)*, 17(3):1–11.

Williams, L. (1990). Performace driven facial animation. *Computer Graphics (SIGGRAPH '90)*, 24(4):235–242.

Williams, O., Blake, A., and Cipolla, R. (2003). A sparse probabilistic learning algorithm for real-time tracking. In *Ninth International Conference on Computer Vision (ICCV 2003)*, pp. 353–360, Nice, France.

Williams, T. L. (1999). *The Optical Transfer Function of Imaging Systems*. Institute of Physics Publishing, London.

Winder, S. and Brown, M. (2007). Learning local image descriptors. In *IEEE Computer Society Conference on Computer Vision and Pattern Recognition (CVPR 2007)*, Minneapolis, MN.

Winkenbach, G. and Salesin, D. H. (1994). Computer-generated pen-and-ink illustration. In *ACM SIGGRAPH 1994 Conference Proceedings*, pp. 91–100, Orlando, Florida.

Winn, J. and Shotton, J. (2006). The layout consistent random field for recognizing and segmenting partially occluded objects. In *IEEE Computer Society Conference on Computer Vision and Pattern Recognition (CVPR'2006)*, pp. 37–44, New York City, NY.

Winnemöller, H., Olsen, S. C., and Gooch, B. (2006). Real-time video abstraction. *ACM Transactions on Graphics*, 25(3):1221–1226.

Winston, P. H. (ed.). (1975). *The Psychology of Computer Vision*, McGraw-Hill, New York.

Wiskott, L., Fellous, J.-M., Krüger, N., and von der Malsburg, C. (1997). Face recognition by elastic bunch graph matching. *IEEE Transactions on Pattern Analysis and Machine Intelligence*, 19(7):775–779.

Witkin, A. (1981). Recovering surface shape and orientation from texture. *Artificial Intelligence*, 17(1-3):17–45.

Witkin, A. (1983). Scale-space filtering. In *Eighth International Joint Conference on Artificial Intelligence (IJCAI-83)*, pp. 1019–1022.

Witkin, A., Terzopoulos, D., and Kass, M. (1986). Signal matching through scale space. In *Fifth National Conference on Artificial Intelligence (AAAI-86)*, pp. 714–719, Philadelphia.

Witkin, A., Terzopoulos, D., and Kass, M. (1987). Signal matching through scale space. *International Journal of Computer Vision*, 1:133–144.

Wolberg, G. (1990). *Digital Image Warping*. IEEE Computer Society Press, Los Alamitos.

Wolberg, G. and Pavlidis, T. (1985). Restoration of binary images using stochastic relaxation with annealing. *Pattern Recognition Letters*, 3:375–388.

Wolff, L. B., Shafer, S. A., and Healey, G. E. (eds). (1992a). *Radiometry. Physics-Based Vision: Principles and Practice*, Jones & Bartlett, Cambridge, MA.

Wolff, L. B., Shafer, S. A., and Healey, G. E. (eds). (1992b). *Shape Recovery. Physics-Based Vision: Principles and Practice*, Jones & Bartlett, Cambridge, MA.

Wood, D. N., Finkelstein, A., Hughes, J. F., Thayer, C. E., and Salesin, D. H. (1997). Multiperspective panoramas for cel animation. In *ACM SIGGRAPH 1997 Conference Proceedings*, pp. 243–250, Los Angeles.

Wood, D. N., Azuma, D. I., Aldinger, K., Curless, B., Duchamp, T., Salesin, D. H., and Stuetzle, W. (2000). Surface light fields for 3D photography. In *ACM SIGGRAPH 2000 Conference Proceedings*, pp. 287–296.

Woodford, O., Reid, I., Torr, P. H., and Fitzgibbon, A. (2008). Global stereo reconstruction under second order smoothness priors. In *IEEE Computer Society Conference on Computer Vision and Pattern Recognition (CVPR 2008)*, Anchorage, AK.

Woodham, R. J. (1981). Analysing images of curved surfaces. *Artificial Intelligence*, 17:117–140.

Woodham, R. J. (1994). Gradient and curvature from photometric stereo including local confidence estimation. *Journal of the Optical Society of America, A*, 11:3050–3068.

Wren, C. R., Azarbayejani, A., Darrell, T., and Pentland, A. P. (1997). Pfinder: Real-time tracking of the human body. *IEEE Transactions on Pattern Analysis and Machine Intelligence*, 19(7):780–785.

Wright, S. (2006). *Digital Compositing for Film and Video*. Focal Press, 2nd edition.

Wu, C. (2010). SiftGPU: A GPU implementation of scale invariant feature transform (SIFT). http://www.cs.unc.edu/~ccwu/siftgpu/.

Wyszecki, G. and Stiles, W. S. (2000). *Color Science: Concepts and Methods, Quantitative Data and Formulae*. John Wiley & Sons, New York, 2nd edition.

Xiao, J. and Shah, M. (2003). Two-frame wide baseline matching. In *Ninth International Conference on Computer Vision (ICCV 2003)*, pp. 603–609, Nice, France.

Xiao, J. and Shah, M. (2005). Motion layer extraction in the presence of occlusion using graph cuts. *IEEE Transactions on Pattern Analysis and Machine Intelligence*, 27(10):1644–1659.

Xiong, Y. and Turkowski, K. (1997). Creating image-based VR using a self-calibrating fisheye lens. In *IEEE Computer Society Conference on Computer Vision and Pattern Recognition (CVPR'97)*, pp. 237–243, San Juan, Puerto Rico.

Xiong, Y. and Turkowski, K. (1998). Registration, calibration and blending in creating high quality panoramas. In *IEEE Workshop on Applications of Computer Vision (WACV'98)*, pp. 69–74, Princeton.

Xu, L., Chen, J., and Jia, J. (2008). A segmentation based variational model for accurate optical flow estimation. In *Tenth European Conference on Computer Vision (ECCV 2008)*, pp. 671–684, Marseilles.

Yang, D., El Gamal, A., Fowler, B., and Tian, H. (1999). A 640x512 CMOS image sensor with ultra-wide dynamic range floating-point pixel level ADC. *IEEE Journal of Solid State Circuits*, 34(12):1821–1834.

Yang, L. and Albregtsen, F. (1996). Fast and exact computation of Cartesian geometric moments using discrete Green's theorem. *Pattern Recognition*, 29(7):1061–1073.

Yang, L., Meer, P., and Foran, D. (2007). Multiple class segmentation using a unified framework over mean-shift patches. In *IEEE Computer Society Conference on Computer Vision and Pattern Recognition (CVPR 2007)*, Minneapolis, MN.

Yang, L., Jin, R., Sukthankar, R., and Jurie, F. (2008). Unifying discriminative visual codebook generation with classifier training for object category recognition. In *IEEE Computer Society Conference on Computer Vision and Pattern Recognition (CVPR 2008)*, Anchorage, AK.

Yang, M.-H., Ahuja, N., and Tabb, M. (2002). Extraction of 2D motion trajectories and its application to hand gesture recognition. *IEEE Transactions on Pattern Analysis and Machine Intelligence*, 24(8):1061–1074.

Yang, M.-H., Kriegman, D. J., and Ahuja, N. (2002). Detecting faces in images: A survey. *IEEE Transactions on Pattern Analysis and Machine Intelligence*, 24(1):34–58.

Yang, Q., Wang, L., Yang, R., Stewénius, H., and Nistér, D. (2009). Stereo matching with color-weighted correlation, hierarchical belief propagation and occlusion handling.

IEEE Transactions on Pattern Analysis and Machine Intelligence, 31(3):492–504.

Yang, Y., Yuille, A., and Lu, J. (1993). Local, global, and multilevel stereo matching. In *IEEE Computer Society Conference on Computer Vision and Pattern Recognition (CVPR'93)*, pp. 274–279, New York.

Yanover, C., Meltzer, T., and Weiss, Y. (2006). Linear programming relaxations and belief propagation — an empirical study. *Journal of Machine Learning Research*, 7:1887–1907.

Yao, B. Z., Yang, X., Lin, L., Lee, M. W., and Zhu, S.-C. (2010). I2T: Image parsing to text description. *Proceedings of the IEEE*, 98(8):1485–1508.

Yaou, M.-H. and Chang, W.-T. (1994). Fast surface interpolation using multiresolution wavelets. *IEEE Transactions on Pattern Analysis and Machine Intelligence*, 16(7):673–689.

Yatziv, L. and Sapiro, G. (2006). Fast image and video colorization using chrominance blending. *IEEE Transactions on Image Processing*, 15(5):1120–1129.

Yedidia, J. S., Freeman, W. T., and Weiss, Y. (2001). Understanding belief propagation and its generalization. In *International Joint Conference on Artificial Intelligence (IJCAI 2001)*.

Yezzi, Jr., A. J., Kichenassamy, S., Kumar, A., Olver, P., and Tannenbaum, A. (1997). A geometric snake model for segmentation of medical imagery. *IEEE Transactions on Medical Imaging*, 16(2):199–209.

Yilmaz, A. and Shah, M. (2006). Matching actions in presence of camera motion. *Computer Vision and Image Understanding*, 104(2-3):221–231.

Yilmaz, A., Javed, O., and Shah, M. (2006). Object tracking: A survey. *ACM Computing Surveys*, 38(4).

Yin, P., Criminisi, A., Winn, J., and Essa, I. (2007). Tree-based classifiers for bilayer video segmentation. In *IEEE Computer Society Conference on Computer Vision and Pattern Recognition (CVPR 2007)*, Minneapolis, MN.

Yoon, K.-J. and Kweon, I.-S. (2006). Adaptive support-weight approach for correspondence search. *IEEE Transactions on Pattern Analysis and Machine Intelligence*, 28(4):650–656.

Yserentant, H. (1986). On the multi-level splitting of finite element spaces. *Numerische Mathematik*, 49:379–412.

Yu, S. X. and Shi, J. (2003). Multiclass spectral clustering. In *Ninth International Conference on Computer Vision (ICCV 2003)*, pp. 313–319, Nice, France.

Yu, Y. and Malik, J. (1998). Recovering photometric properties of architectural scenes from photographs. In *ACM SIGGRAPH 1996 Conference Proceedings*, pp. 207–218, Orlando.

Yu, Y., Debevec, P., Malik, J., and Hawkins, T. (1999). Inverse global illumination: Recovering reflectance models of real scenes from photographs. In *ACM SIGGRAPH 1999 Conference Proceedings*, pp. 215–224.

Yuan, L., Sun, J., Quan, L., and Shum, H.-Y. (2007). Image deblurring with blurred/noisy image pairs. *ACM Transactions on Graphics*, 26(3).

Yuan, L., Sun, J., Quan, L., and Shum, H.-Y. (2008). Progressive inter-scale and intra-scale non-blind image deconvolution. *ACM Transactions on Graphics*, 27(3).

Yuan, L., Wen, F., Liu, C., and Shum, H.-Y. (2004). Synthesizing dynamic texture with closed-loop linear dynamic system. In *Eighth European Conference on Computer Vision (ECCV 2004)*, pp. 603–616, Prague.

Yuille, A. (1991). Deformable templates for face recognition. *Journal of Cognitive Neuroscience*, 3(1):59–70.

Yuille, A. (2002). CCCP algorithms to minimize the Bethe and Kikuchi free energies: Convergent alternatives to belief propagation. *Neural Computation*, 14(7):1691–1722.

Yuille, A. (2010). Loopy belief propagation, mean-field and Bethe approximations. In Blake, A., Kohli, P., and Rother, C. (eds), *Advances in Markov Random Fields*, MIT Press.

Yuille, A. and Poggio, T. (1984). *A Generalized Ordering Constraint for Stereo Correspondence*. A. I. Memo 777, Artificial Intelligence Laboratory, Massachusetts Institute of Technology.

Yuille, A., Vincent, L., and Geiger, D. (1992). Statistical morphology and Bayesian reconstruction. *Journal of Mathematical Imaging and Vision*, 1(3):223–238.

Zabih, R. and Woodfill, J. (1994). Non-parametric local transforms for computing visual correspondence. In *Third European Conference on Computer Vision (ECCV'94)*, pp. 151–158, Stockholm, Sweden.

Zach, C. (2008). Fast and high quality fusion of depth maps. In *Fourth International Symposium on 3D Data Processing, Visualization and Transmission (3DPVT'08)*, Atlanta.

Zach, C., Gallup, D., and Frahm, J.-M. (2008). Fast gain-adaptive KLT tracking on the GPU. In *CVPR 2008 Workshop on Visual Computer Vision on GPUs (CVGPU)*, Anchorage, AK.

Zach, C., Klopschitz, M., and Pollefeys, M. (2010). Disambiguating visual relations using loop constraints. In *IEEE Computer Society Conference on Computer Vision and Pattern Recognition (CVPR 2010)*, San Francisco, CA.

Zach, C., Pock, T., and Bischof, H. (2007a). A duality based approach for realtime TV-L1 optical flow. In *Pattern Recognition (DAGM 2007)*.

Zach, C., Pock, T., and Bischof, H. (2007b). A globally optimal algorithm for robust TV-L^1 range image integration. In *Eleventh International Conference on Computer Vision (ICCV 2007)*, Rio de Janeiro, Brazil.

Zanella, V. and Fuentes, O. (2004). An approach to automatic morphing of face images in frontal view. In *Mexican International Conference on Artificial Intelligence (MICAI 2004)*, pp. 679–687, Mexico City.

Zebedin, L., Bauer, J., Karner, K., and Bischof, H. (2008). Fusion of feature- and area-based information for urban buildings modeling from aerial imagery. In *Tenth European Conference on Computer Vision (ECCV 2008)*, pp. 873–886, Marseilles.

Zelnik-Manor, L. and Perona, P. (2007). Automating joiners. In *Symposium on Non Photorealistic Animation and Rendering*, Annecy.

Zhang, G., Jia, J., Wong, T.-T., and Bao, H. (2008). Recovering consistent video depth maps via bundle optimization. In *IEEE Computer Society Conference on Computer Vision and Pattern Recognition (CVPR 2008)*, Anchorage, AK.

Zhang, J., McMillan, L., and Yu, J. (2006). Robust tracking and stereo matching under variable illumination. In *IEEE Computer Society Conference on Computer Vision and Pattern Recognition (CVPR'2006)*, pp. 871–878, New York City, NY.

Zhang, J., Marszalek, M., Lazebnik, S., and Schmid, C. (2007). Local features and kernels for classification of texture and object categories: a comprehensive study. *International Journal of Computer Vision*, 73(2):213–238.

Zhang, L., Curless, B., and Seitz, S. (2003). Spacetime stereo: Shape recovery for dynamic scenes. In *IEEE Computer Society Conference on Computer Vision and Pattern Recognition (CVPR'2003)*, pp. 367–374, Madison, WI.

Zhang, L., Dugas-Phocion, G., Samson, J.-S., and Seitz, S. M. (2002). Single view modeling of free-form scenes. *Journal of Visualization and Computer Animation*, 13(4):225–235.

Zhang, L., Snavely, N., Curless, B., and Seitz, S. M. (2004). Spacetime faces: High resolution capture for modeling and animation. *ACM Transactions on Graphics*, 23(3):548–558.

Zhang, R., Tsai, P.-S., Cryer, J. E., and Shah, M. (1999). Shape from shading: A survey. *IEEE Transactions on Pattern Analysis and Machine Intelligence*, 21(8):690–706.

Zhang, Y. and Kambhamettu, C. (2003). On 3D scene flow and structure recovery from multiview image sequences. *IEEE Transactions on Systems, Man, and Cybernetics*, 33(4):592–606.

Zhang, Z. (1994). Iterative point matching for registration of free-form curves and surfaces. *International Journal of Computer Vision*, 13(2):119–152.

Zhang, Z. (1998a). Determining the epipolar geometry and its uncertainty: A review. *International Journal of Computer Vision*, 27(2):161–195.

Zhang, Z. (1998b). On the optimization criteria used in two-view motion analysis. *IEEE Transactions on Pattern Analysis and Machine Intelligence*, 20(7):717–729.

Zhang, Z. (2000). A flexible new technique for camera calibration. *IEEE Transactions on Pattern Analysis and Machine Intelligence*, 22(11):1330–1334.

Zhang, Z. and He, L.-W. (2007). Whiteboard scanning and image enhancement. *Digital Signal Processing*, 17(2):414–432.

Zhang, Z. and Shan, Y. (2000). A progressive scheme for stereo matching. In *Second European Workshop on 3D Structure from Multiple Images of Large-Scale Environments (SMILE 2000)*, pp. 68–85, Dublin, Ireland.

Zhang, Z., Deriche, R., Faugeras, O., and Luong, Q. (1995). A robust technique for matching two uncalibrated images through the recovery of the unknown epipolar geometry. *Artificial Intelligence*, 78:87–119.

Zhao, G. and Pietikäinen, M. (2007). Dynamic texture recognition using local binary patterns with an application to facial expressions. *IEEE Transactions on Pattern Analysis and Machine Intelligence*, 29(6):915–928.

Zhao, W., Chellappa, R., Phillips, P. J., and Rosenfeld, A. (2003). Face recognition: A literature survey. *ACM Computing Surveys*, 35(4):399–358.

Zheng, J. Y. (1994). Acquiring 3-D models from sequences of contours. *IEEE Transactions on Pattern Analysis and Machine Intelligence*, 16(2):163–178.

Zheng, K. C., Kang, S. B., Cohen, M., and Szeliski, R. (2007). Layered depth panoramas. In *IEEE Computer Society Conference on Computer Vision and Pattern Recognition (CVPR 2007)*, Minneapolis, MN.

Zheng, Y., Lin, S., and Kang, S. B. (2006). Single-image vignetting correction. In *IEEE Computer Society Conference on Computer Vision and Pattern Recognition (CVPR'2006)*, pp. 461–468, New York City, NY.

Zheng, Y., Yu, J., Kang, S.-B., Lin, S., and Kambhamettu, C. (2008). Single-image vignetting correction using radial gradient symmetry. In *IEEE Computer Society Conference on Computer Vision and Pattern Recognition (CVPR 2008)*, Anchorage, AK.

Zheng, Y., Zhou, X. S., Georgescu, B., Zhou, S. K., and Comaniciu, D. (2006). Example based non-rigid shape detection. In *Ninth European Conference on Computer Vision (ECCV 2006)*, pp. 423–436.

Zheng, Y.-T., Zhao, M., Song, Y., Adam, H., Buddemeier, U., Bissacco, A., Brucher, F., Chua, T.-S., and Neven, H. (2009). Tour the world: building a web-scale landmark recognition engine. In *IEEE Computer Society Conference on Computer Vision and Pattern Recognition (CVPR 2009)*, Miami Beach, FL.

Zhong, J. and Sclaroff, S. (2003). Segmenting foreground objects from a dynamic, textured background via a robust Kalman filter. In *Ninth International Conference on Computer Vision (ICCV 2003)*, pp. 44–50, Nice, France.

Zhou, C., Lin, S., and Nayar, S. (2009). Coded aperture pairs for depth from defocus. In *Twelfth International Conference on Computer Vision (ICCV 2009)*, Kyoto, Japan.

Zhou, Y., Gu, L., and Zhang, H.-J. (2003). Bayesian tangent shape model: Estimating shape and pose parameters via Bayesian inference. In *IEEE Computer Society Conference on Computer Vision and Pattern Recognition (CVPR'2003)*, pp. 109–116, Madison, WI.

Zhu, L., Chen, Y., Lin, Y., Lin, C., and Yuille, A. (2008). Recursive segmentation and recognition templates for 2D parsing. In *Advances in Neural Information Processing Systems*.

Zhu, S.-C. and Mumford, D. (2006). A stochastic grammar of images. *Foundations and Trends in Computer Graphics and Computer Vision*, 2(4).

Zhu, S. C. and Yuille, A. L. (1996). Region competition: Unifying snakes, region growing, and Bayes/MDL for multiband image segmentation. *IEEE Transactions on Pattern Analysis and Machine Intelligence*, 18(9):884–900.

Zhu, Z. and Kanade, T. (2008). Modeling and representations of large-scale 3D scenes. *International Journal of Computer Vision*, 78(2-3):119–120.

Zisserman, A., Giblin, P. J., and Blake, A. (1989). The information available to a moving observer from specularities. *Image and Vision Computing*, 7(1):38–42.

Zitnick, C. L. and Kanade, T. (2000). A cooperative algorithm for stereo matching and occlusion detection. *IEEE Transactions on Pattern Analysis and Machine Intelligence*, 22(7):675–684.

Zitnick, C. L. and Kang, S. B. (2007). Stereo for image-based rendering using image over-segmentation. *International Journal of Computer Vision*, 75(1):49–65.

Zitnick, C. L., Jojic, N., and Kang, S. B. (2005). Consistent segmentation for optical flow estimation. In *Tenth International Conference on Computer Vision (ICCV 2005)*, pp. 1308–1315, Beijing, China.

Zitnick, C. L., Kang, S. B., Uyttendaele, M., Winder, S., and Szeliski, R. (2004). High-quality video view interpolation using a layered representation. *ACM Transactions on Graphics (Proc. SIGGRAPH 2004)*, 23(3):600–608.

Zitov'aa, B. and Flusser, J. (2003). Image registration methods: A survey. *Image and Vision Computing*, 21:997–1000.

Zoghlami, I., Faugeras, O., and Deriche, R. (1997). Using geometric corners to build a 2D mosaic from a set of images. In *IEEE Computer Society Conference on Computer Vision and Pattern Recognition (CVPR'97)*, pp. 420–425, San Juan, Puerto Rico.

Zongker, D. E., Werner, D. M., Curless, B., and Salesin, D. H. (1999). Environment matting and compositing. In *ACM SIGGRAPH 1999 Conference Proceedings*, pp. 205–214.

Zorin, D., Schröder, P., and Sweldens, W. (1996). Interpolating subdivision for meshes with arbitrary topology. In *ACM SIGGRAPH 1997 Conference Proceedings*, pp. 189–192, New Orleans.

公式

第 1 章 概述

本章无。

第 2 章 图像形成

$$x = \left[\begin{array}{c} x \\ y \end{array}\right]. \tag{2.1}$$

$$\tilde{x} = (\tilde{x}, \tilde{y}, \tilde{w}) = \tilde{w}(x, y, 1) = \tilde{w}\bar{x}, \tag{2.2}$$

$$\bar{x} \cdot \tilde{l} = ax + by + c = 0. \tag{2.3}$$

$$\tilde{x} = \tilde{l}_1 \times \tilde{l}_2, \tag{2.4}$$

$$\tilde{l} = \tilde{x}_1 \times \tilde{x}_2. \tag{2.5}$$

$$\tilde{x}^T Q \tilde{x} = 0. \tag{2.6}$$

$$\bar{x} \cdot \tilde{m} = ax + by + cz + d = 0. \tag{2.7}$$

$$\hat{n} = (\cos\theta\cos\phi, \sin\theta\cos\phi, \sin\phi), \tag{2.8}$$

$$r = (1-\lambda)p + \lambda q, \tag{2.9}$$

$$\tilde{r} = \mu\tilde{p} + \lambda\tilde{q}. \tag{2.10}$$

$$r = p + \lambda\hat{d}. \tag{2.11}$$

$$L = \tilde{p}\tilde{q}^T - \tilde{q}\tilde{p}^T, \tag{2.12}$$

$$\bar{x}^T Q \bar{x} = 0 \tag{2.13}$$

$$\boldsymbol{x}' = \begin{bmatrix} \boldsymbol{I} & \boldsymbol{t} \end{bmatrix} \bar{\boldsymbol{x}} \tag{2.14}$$

$$\bar{\boldsymbol{x}}' = \begin{bmatrix} \boldsymbol{I} & \boldsymbol{t} \\ \boldsymbol{0}^T & 1 \end{bmatrix} \bar{\boldsymbol{x}} \tag{2.15}$$

$$\boldsymbol{x}' = \begin{bmatrix} \boldsymbol{R} & \boldsymbol{t} \end{bmatrix} \bar{\boldsymbol{x}} \tag{2.16}$$

$$\boldsymbol{R} = \begin{bmatrix} \cos\theta & -\sin\theta \\ \sin\theta & \cos\theta \end{bmatrix} \tag{2.17}$$

$$\boldsymbol{x}' = \begin{bmatrix} s\boldsymbol{R} & \boldsymbol{t} \end{bmatrix} \bar{\boldsymbol{x}} = \begin{bmatrix} a & -b & t_x \\ b & a & t_y \end{bmatrix} \bar{\boldsymbol{x}}, \tag{2.18}$$

$$\boldsymbol{x}' = \begin{bmatrix} a_{00} & a_{01} & a_{02} \\ a_{10} & a_{11} & a_{12} \end{bmatrix} \bar{\boldsymbol{x}}. \tag{2.19}$$

$$\tilde{\boldsymbol{x}}' = \tilde{\boldsymbol{H}} \tilde{\boldsymbol{x}}, \tag{2.20}$$

$$x' = \frac{h_{00}x + h_{01}y + h_{02}}{h_{20}x + h_{21}y + h_{22}} \text{ and } y' = \frac{h_{10}x + h_{11}y + h_{12}}{h_{20}x + h_{21}y + h_{22}}. \tag{2.21}$$

$$\tilde{\boldsymbol{l}}' \cdot \tilde{\boldsymbol{x}}' = \tilde{\boldsymbol{l}}'^T \tilde{\boldsymbol{H}} \tilde{\boldsymbol{x}} = (\tilde{\boldsymbol{H}}^T \tilde{\boldsymbol{l}}')^T \tilde{\boldsymbol{x}} = \tilde{\boldsymbol{l}} \cdot \tilde{\boldsymbol{x}} = 0, \tag{2.22}$$

$$\begin{aligned} x' &= s_x x + t_x \\ y' &= s_y y + t_y, \end{aligned}$$

$$\begin{aligned} x' &= a_0 + a_1 x + a_2 y + a_6 x^2 + a_7 xy \\ y' &= a_3 + a_4 x + a_5 y + a_7 x^2 + a_6 xy, \end{aligned}$$

$$\begin{aligned} x' &= a_0 + a_1 x + a_2 y + a_6 xy \\ y' &= a_3 + a_4 x + a_5 y + a_7 xy, \end{aligned}$$

$$\boldsymbol{x}' = \begin{bmatrix} \boldsymbol{I} & \boldsymbol{t} \end{bmatrix} \bar{\boldsymbol{x}} \tag{2.23}$$

$$\boldsymbol{x}' = \begin{bmatrix} \boldsymbol{R} & \boldsymbol{t} \end{bmatrix} \bar{\boldsymbol{x}} \tag{2.24}$$

$$\boldsymbol{x}' = \boldsymbol{R}(\boldsymbol{x} - \boldsymbol{c}) = \boldsymbol{R}\boldsymbol{x} - \boldsymbol{R}\boldsymbol{c}, \tag{2.25}$$

$$x' = \begin{bmatrix} sR & t \end{bmatrix} \bar{x}. \tag{2.26}$$

$$x' = \begin{bmatrix} a_{00} & a_{01} & a_{02} & a_{03} \\ a_{10} & a_{11} & a_{12} & a_{13} \\ a_{20} & a_{21} & a_{22} & a_{23} \end{bmatrix} \bar{x}. \tag{2.27}$$

$$\tilde{x}' = \tilde{H}\tilde{x}, \tag{2.28}$$

$$v_\parallel = \hat{n}(\hat{n} \cdot v) = (\hat{n}\hat{n}^T)v, \tag{2.29}$$

$$v_\perp = v - v_\parallel = (I - \hat{n}\hat{n}^T)v. \tag{2.30}$$

$$v_\times = \hat{n} \times v = [\hat{n}]_\times v, \tag{2.31}$$

$$[\hat{n}]_\times = \begin{bmatrix} 0 & -\hat{n}_z & \hat{n}_y \\ \hat{n}_z & 0 & -\hat{n}_x \\ -\hat{n}_y & \hat{n}_x & 0 \end{bmatrix}. \tag{2.32}$$

$$u = u_\perp + v_\parallel = (I + \sin\theta [\hat{n}]_\times + (1-\cos\theta)[\hat{n}]_\times^2)v. \tag{2.33}$$

$$R(\hat{n}, \theta) = I + \sin\theta [\hat{n}]_\times + (1-\cos\theta)[\hat{n}]_\times^2, \tag{2.34}$$

$$R(\omega) \approx I + \sin\theta [\hat{n}]_\times \approx I + [\theta\hat{n}]_\times = \begin{bmatrix} 1 & -\omega_z & \omega_y \\ \omega_z & 1 & -\omega_x \\ -\omega_y & \omega_x & 1 \end{bmatrix}, \tag{2.35}$$

$$\frac{\partial Rv}{\partial \omega^T} = -[v]_\times = \begin{bmatrix} 0 & z & -y \\ -z & 0 & x \\ y & -x & 0 \end{bmatrix}. \tag{2.36}$$

$$R(\hat{n}, \theta) = \lim_{k \to \infty} (I + \frac{1}{k}[\theta\hat{n}]_\times)^k = \exp[\omega]_\times. \tag{2.37}$$

$$\begin{aligned} \exp[\omega]_\times &= I + \theta[\hat{n}]_\times + \frac{\theta^2}{2}[\hat{n}]_\times^2 + \frac{\theta^3}{3!}[\hat{n}]_\times^3 + \cdots \\ &= I + (\theta - \frac{\theta^3}{3!} + \cdots)[\hat{n}]_\times + (\frac{\theta^2}{2} - \frac{\theta^3}{4!} + \cdots)[\hat{n}]_\times^2 \\ &= I + \sin\theta[\hat{n}]_\times + (1-\cos\theta)[\hat{n}]_\times^2, \end{aligned} \tag{2.38}$$

$$q = (v, w) = (\sin\frac{\theta}{2}\hat{n}, \cos\frac{\theta}{2}), \tag{2.39}$$

$$\begin{aligned}\boldsymbol{R}(\hat{\boldsymbol{n}},\theta) &= \boldsymbol{I} + \sin\theta[\hat{\boldsymbol{n}}]_\times + (1-\cos\theta)[\hat{\boldsymbol{n}}]_\times^2 \\ &= \boldsymbol{I} + 2w[\boldsymbol{v}]_\times + 2[\boldsymbol{v}]_\times^2.\end{aligned} \qquad (2.40)$$

$$\boldsymbol{R}(\boldsymbol{q}) = \begin{bmatrix} 1-2(y^2+z^2) & 2(xy-zw) & 2(xz+yw) \\ 2(xy+zw) & 1-2(x^2+z^2) & 2(yz-xw) \\ 2(xz-yw) & 2(yz+xw) & 1-2(x^2+y^2) \end{bmatrix}. \qquad (2.41)$$

$$\boldsymbol{q}_2 = \boldsymbol{q}_0 \boldsymbol{q}_1 = (\boldsymbol{v}_0 \times \boldsymbol{v}_1 + w_0 \boldsymbol{v}_1 + w_1 \boldsymbol{v}_0, w_0 w_1 - \boldsymbol{v}_0 \cdot \boldsymbol{v}_1), \qquad (2.42)$$

$$\boldsymbol{q}_2 = \boldsymbol{q}_0/\boldsymbol{q}_1 = \boldsymbol{q}_0 \boldsymbol{q}_1^{-1} = (\boldsymbol{v}_0 \times \boldsymbol{v}_1 + w_0 \boldsymbol{v}_1 - w_1 \boldsymbol{v}_0, -w_0 w_1 - \boldsymbol{v}_0 \cdot \boldsymbol{v}_1). \qquad (2.43)$$

$$\boldsymbol{q}_2 = \boldsymbol{q}_r^\alpha \boldsymbol{q}_0, \qquad (2.44)$$

$$\boldsymbol{q}_2 = \frac{\sin(1-\alpha)\theta}{\sin\theta} \boldsymbol{q}_0 + \frac{\sin\alpha\theta}{\sin\theta} \boldsymbol{q}_1, \qquad (2.45)$$

$$\boldsymbol{x} = [\boldsymbol{I}_{2\times 2}|\boldsymbol{0}]\,\boldsymbol{p}. \qquad (2.46)$$

$$\tilde{\boldsymbol{x}} = \begin{bmatrix} 1 & 0 & 0 & 0 \\ 0 & 1 & 0 & 0 \\ 0 & 0 & 0 & 1 \end{bmatrix} \tilde{\boldsymbol{p}}, \qquad (2.47)$$

$$\boldsymbol{x} = [s\boldsymbol{I}_{2\times 2}|\boldsymbol{0}]\,\boldsymbol{p}. \qquad (2.48)$$

$$\tilde{\boldsymbol{x}} = \begin{bmatrix} a_{00} & a_{01} & a_{02} & a_{03} \\ a_{10} & a_{11} & a_{12} & a_{13} \\ 0 & 0 & 0 & 1 \end{bmatrix} \tilde{\boldsymbol{p}}. \qquad (2.49)$$

$$\bar{\boldsymbol{x}} = \mathcal{P}_z(\boldsymbol{p}) = \begin{bmatrix} x/z \\ y/z \\ 1 \end{bmatrix}. \qquad (2.50)$$

$$\tilde{\boldsymbol{x}} = \begin{bmatrix} 1 & 0 & 0 & 0 \\ 0 & 1 & 0 & 0 \\ 0 & 0 & 1 & 0 \end{bmatrix} \tilde{\boldsymbol{p}}, \qquad (2.51)$$

$$\tilde{\boldsymbol{x}} = \begin{bmatrix} 1 & 0 & 0 & 0 \\ 0 & 1 & 0 & 0 \\ 0 & 0 & -z_{\text{far}}/z_{\text{range}} & z_{\text{near}} z_{\text{far}}/z_{\text{range}} \\ 0 & 0 & 1 & 0 \end{bmatrix} \tilde{\boldsymbol{p}}, \qquad (2.52)$$

$$p = \begin{bmatrix} R_s & | & c_s \end{bmatrix} \begin{bmatrix} s_x & 0 & 0 \\ 0 & s_y & 0 \\ 0 & 0 & 0 \\ 0 & 0 & 1 \end{bmatrix} \begin{bmatrix} x_s \\ y_s \\ 1 \end{bmatrix} = M_s \bar{x}_s. \tag{2.53}$$

$$\tilde{x}_s = \alpha M_s^{-1} p_c = K p_c. \tag{2.54}$$

$$\tilde{x}_s = K \begin{bmatrix} R & | & t \end{bmatrix} p_w = P p_w, \tag{2.55}$$

$$P = K[R|t] \tag{2.56}$$

$$K = \begin{bmatrix} f_x & s & c_x \\ 0 & f_y & c_y \\ 0 & 0 & 1 \end{bmatrix}, \tag{2.57}$$

$$K = \begin{bmatrix} f & s & c_x \\ 0 & af & c_y \\ 0 & 0 & 1 \end{bmatrix}, \tag{2.58}$$

$$K = \begin{bmatrix} f & 0 & c_x \\ 0 & f & c_y \\ 0 & 0 & 1 \end{bmatrix}. \tag{2.59}$$

$$\tan \frac{\theta}{2} = \frac{W}{2f} \quad \text{or} \quad f = \frac{W}{2} \left[\tan \frac{\theta}{2} \right]^{-1}. \tag{2.60}$$

$$x'_s = (2x_s - W)/S \text{ and } y'_s = (2y_s - H)/S, \quad \text{where} \quad S = \max(W, H). \tag{2.61}$$

$$f^{-1} = \tan \frac{\theta}{2}. \tag{2.62}$$

$$P = K \begin{bmatrix} R & | & t \end{bmatrix}. \tag{2.63}$$

$$\tilde{P} = \begin{bmatrix} K & 0 \\ 0^T & 1 \end{bmatrix} \begin{bmatrix} R & t \\ 0^T & 1 \end{bmatrix} = \tilde{K} E, \tag{2.64}$$

$$x_s \sim \tilde{P} \bar{p}_w, \tag{2.65}$$

$$d = \frac{s_3}{z} (\hat{n}_0 \cdot p_w + c_0), \tag{2.66}$$

$$\tilde{p}_w = \tilde{P}^{-1} x_s. \tag{2.67}$$

$$\tilde{\boldsymbol{x}}_0 \sim \tilde{\boldsymbol{K}}_0 \boldsymbol{E}_0 \boldsymbol{p} = \tilde{\boldsymbol{P}}_0 \boldsymbol{p}. \tag{2.68}$$

$$\boldsymbol{p} \sim \boldsymbol{E}_0^{-1} \tilde{\boldsymbol{K}}_0^{-1} \tilde{\boldsymbol{x}}_0 \tag{2.69}$$

$$\tilde{\boldsymbol{x}}_1 \sim \tilde{\boldsymbol{K}}_1 \boldsymbol{E}_1 \boldsymbol{p} = \tilde{\boldsymbol{K}}_1 \boldsymbol{E}_1 \boldsymbol{E}_0^{-1} \tilde{\boldsymbol{K}}_0^{-1} \tilde{\boldsymbol{x}}_0 = \tilde{\boldsymbol{P}}_1 \tilde{\boldsymbol{P}}_0^{-1} \tilde{\boldsymbol{x}}_0 = \boldsymbol{M}_{10} \tilde{\boldsymbol{x}}_0. \tag{2.70}$$

$$\tilde{\boldsymbol{x}}_1 \sim \tilde{\boldsymbol{H}}_{10} \tilde{\boldsymbol{x}}_0, \tag{2.71}$$

$$\tilde{\boldsymbol{x}}_1 \sim \boldsymbol{K}_1 \boldsymbol{R}_1 \boldsymbol{R}_0^{-1} \boldsymbol{K}_0^{-1} \tilde{\boldsymbol{x}}_0 = \boldsymbol{K}_1 \boldsymbol{R}_{10} \boldsymbol{K}_0^{-1} \tilde{\boldsymbol{x}}_0, \tag{2.72}$$

$$x_s = f \frac{\boldsymbol{r}_x \cdot \boldsymbol{p} + t_x}{\boldsymbol{r}_z \cdot \boldsymbol{p} + t_z} + c_x \tag{2.73}$$
$$y_s = f \frac{\boldsymbol{r}_y \cdot \boldsymbol{p} + t_y}{\boldsymbol{r}_z \cdot \boldsymbol{p} + t_z} + c_y, \tag{2.74}$$

$$x_s = s \frac{\boldsymbol{r}_x \cdot \boldsymbol{p} + t_x}{1 + \eta_z \boldsymbol{r}_z \cdot \boldsymbol{p}} + c_x \tag{2.75}$$
$$y_s = s \frac{\boldsymbol{r}_y \cdot \boldsymbol{p} + t_y}{1 + \eta_z \boldsymbol{r}_z \cdot \boldsymbol{p}} + c_y \tag{2.76}$$

$$\begin{aligned} x_c &= \frac{\boldsymbol{r}_x \cdot \boldsymbol{p} + t_x}{\boldsymbol{r}_z \cdot \boldsymbol{p} + t_z} \\ y_c &= \frac{\boldsymbol{r}_y \cdot \boldsymbol{p} + t_y}{\boldsymbol{r}_z \cdot \boldsymbol{p} + t_z}. \end{aligned} \tag{2.77}$$

$$\begin{aligned} \hat{x}_c &= x_c(1 + \kappa_1 r_c^2 + \kappa_2 r_c^4) \\ \hat{y}_c &= y_c(1 + \kappa_1 r_c^2 + \kappa_2 r_c^4), \end{aligned} \tag{2.78}$$

$$\begin{aligned} x_s &= f x'_c + c_x \\ y_s &= f y'_c + c_y. \end{aligned} \tag{2.79}$$

$$L(\hat{\boldsymbol{v}}; \lambda), \tag{2.80}$$

$$f_r(\theta_i, \phi_i, \theta_r, \phi_r; \lambda). \tag{2.81}$$

$$f_r(\theta_i, \theta_r, |\phi_r - \phi_i|; \lambda) \text{ or } f_r(\hat{\boldsymbol{v}}_i, \hat{\boldsymbol{v}}_r, \hat{\boldsymbol{n}}; \lambda), \tag{2.82}$$

$$L_r(\hat{\boldsymbol{v}}_r; \lambda) = \int L_i(\hat{\boldsymbol{v}}_i; \lambda) f_r(\hat{\boldsymbol{v}}_i, \hat{\boldsymbol{v}}_r, \hat{\boldsymbol{n}}; \lambda) \cos^+ \theta_i \, d\hat{\boldsymbol{v}}_i, \tag{2.83}$$

$$\cos^+ \theta_i = \max(0, \cos \theta_i). \tag{2.84}$$

$$L_r(\hat{\boldsymbol{v}}_r; \lambda) = \sum_i L_i(\lambda) f_r(\hat{\boldsymbol{v}}_i, \hat{\boldsymbol{v}}_r, \hat{\boldsymbol{n}}; \lambda) \cos^+ \theta_i. \tag{2.85}$$

$$f_d(\hat{\boldsymbol{v}}_i, \hat{\boldsymbol{v}}_r, \hat{\boldsymbol{n}}; \lambda) = f_d(\lambda), \tag{2.86}$$

$$L_d(\hat{\boldsymbol{v}}_r; \lambda) = \sum_i L_i(\lambda) f_d(\lambda) \cos^+ \theta_i = \sum_i L_i(\lambda) f_d(\lambda) [\hat{\boldsymbol{v}}_i \cdot \hat{\boldsymbol{n}}]^+, \tag{2.87}$$

$$[\hat{\boldsymbol{v}}_i \cdot \hat{\boldsymbol{n}}]^+ = \max(0, \hat{\boldsymbol{v}}_i \cdot \hat{\boldsymbol{n}}). \tag{2.88}$$

$$\hat{\boldsymbol{s}}_i = \boldsymbol{v}_\parallel - \boldsymbol{v}_\perp = (2\hat{\boldsymbol{n}}\hat{\boldsymbol{n}}^T - \boldsymbol{I})\boldsymbol{v}_i. \tag{2.89}$$

$$f_s(\theta_s; \lambda) = k_s(\lambda) \cos^{k_e} \theta_s, \tag{2.90}$$

$$f_s(\theta_s; \lambda) = k_s(\lambda) \exp(-c_s^2 \theta_s^2). \tag{2.91}$$

$$f_a(\lambda) = k_a(\lambda) L_a(\lambda). \tag{2.92}$$

$$L_r(\hat{\boldsymbol{v}}_r; \lambda) = k_a(\lambda) L_a(\lambda) + k_d(\lambda) \sum_i L_i(\lambda)[\hat{\boldsymbol{v}}_i \cdot \hat{\boldsymbol{n}}]^+ + k_s(\lambda) \sum_i L_i(\lambda)(\hat{\boldsymbol{v}}_r \cdot \hat{\boldsymbol{s}}_i)^{k_e}. \tag{2.93}$$

$$L_r(\hat{\boldsymbol{v}}_r; \lambda) = L_i(\hat{\boldsymbol{v}}_r, \hat{\boldsymbol{v}}_i, \hat{\boldsymbol{n}}; \lambda) + L_b(\hat{\boldsymbol{v}}_r, \hat{\boldsymbol{v}}_i, \hat{\boldsymbol{n}}; \lambda) \tag{2.94}$$
$$= c_i(\lambda) m_i(\hat{\boldsymbol{v}}_r, \hat{\boldsymbol{v}}_i, \hat{\boldsymbol{n}}) + c_b(\lambda) m_b(\hat{\boldsymbol{v}}_r, \hat{\boldsymbol{v}}_i, \hat{\boldsymbol{n}}), \tag{2.95}$$

$$\frac{1}{z_o} + \frac{1}{z_i} = \frac{1}{f}, \tag{2.96}$$

$$f/\# = N = \frac{f}{d}, \tag{2.97}$$

$$\frac{\delta o \cos \alpha}{r_o^2} \pi \left(\frac{d}{2}\right)^2 \cos \alpha = \delta o \frac{\pi}{4} \frac{d^2}{z_o^2} \cos^4 \alpha. \tag{2.98}$$

$$\frac{\delta o}{\delta i} = \frac{z_o^2}{z_i^2}. \tag{2.99}$$

$$\delta o \frac{\pi}{4} \frac{d^2}{z_o^2} \cos^4 \alpha = \delta i \frac{\pi}{4} \frac{d^2}{z_i^2} \cos^4 \alpha \approx \delta i \frac{\pi}{4} \left(\frac{d}{f}\right)^2 \cos^4 \alpha, \tag{2.100}$$

$$E = L \frac{\pi}{4} \left(\frac{d}{f}\right)^2 \cos^4 \alpha, \tag{2.101}$$

$$f_s \geq 2 f_{\max}. \tag{2.102}$$

$$\begin{bmatrix} X \\ Y \\ Z \end{bmatrix} = \frac{1}{0.17697} \begin{bmatrix} 0.49 & 0.31 & 0.20 \\ 0.17697 & 0.81240 & 0.01063 \\ 0.00 & 0.01 & 0.99 \end{bmatrix} \begin{bmatrix} R \\ G \\ B \end{bmatrix}. \tag{2.103}$$

$$x = \frac{X}{X+Y+Z}, \quad y = \frac{Y}{X+Y+Z}, \quad z = \frac{Z}{X+Y+Z}, \tag{2.104}$$

$$L^* = 116 f\left(\frac{Y}{Y_n}\right), \tag{2.105}$$

$$f(t) = \begin{cases} t^{1/3} & t > \delta^3 \\ t/(3\delta^2) + 2\delta/3 & \text{else}, \end{cases} \tag{2.106}$$

$$a^* = 500 \left[f\left(\frac{X}{X_n}\right) - f\left(\frac{Y}{Y_n}\right) \right] \text{ and } b^* = 200 \left[f\left(\frac{Y}{Y_n}\right) - f\left(\frac{Z}{Z_n}\right) \right], \tag{2.107}$$

$$\begin{bmatrix} X \\ Y \\ Z \end{bmatrix} = \begin{bmatrix} 0.412453 & 0.357580 & 0.180423 \\ 0.212671 & 0.715160 & 0.072169 \\ 0.019334 & 0.119193 & 0.950227 \end{bmatrix} \begin{bmatrix} R_{709} \\ G_{709} \\ B_{709} \end{bmatrix}. \tag{2.108}$$

$$\begin{aligned} R &= \int L(\lambda) S_R(\lambda) d\lambda, \\ G &= \int L(\lambda) S_G(\lambda) d\lambda, \\ B &= \int L(\lambda) S_B(\lambda) d\lambda, \end{aligned} \tag{2.109}$$

$$B = V^\gamma, \tag{2.110}$$

$$Y' = Y^{\frac{1}{\gamma}}, \tag{2.111}$$

$$Y'_{601} = 0.299R' + 0.587G' + 0.114B', \tag{2.112}$$

$$Y'_{709} = 0.2125R' + 0.7154G' + 0.0721B'. \tag{2.113}$$

$$U = 0.492111(B' - Y') \text{ and } V = 0.877283(R' - Y'), \tag{2.114}$$

$$\begin{bmatrix} Y' \\ C_b \\ C_r \end{bmatrix} = \begin{bmatrix} 0.299 & 0.587 & 0.114 \\ -0.168736 & -0.331264 & 0.5 \\ 0.5 & -0.418688 & -0.081312 \end{bmatrix} \begin{bmatrix} R' \\ G' \\ B' \end{bmatrix} + \begin{bmatrix} 0 \\ 128 \\ 128 \end{bmatrix}, \tag{2.115}$$

$$r = \frac{R}{R+G+B}, \ g = \frac{G}{R+G+B}, \ b = \frac{B}{R+G+B} \tag{2.116}$$

$$MSE = \frac{1}{n} \sum_{\boldsymbol{x}} \left[I(\boldsymbol{x}) - \hat{I}(\boldsymbol{x}) \right]^2, \tag{2.117}$$

$$RMS = \sqrt{MSE}. \tag{2.118}$$

$$PSNR = 10 \log_{10} \frac{I_{\max}^2}{MSE} = 20 \log_{10} \frac{I_{\max}}{RMS}, \tag{2.119}$$

$$D = \sum_i (\tilde{\boldsymbol{x}} \cdot \tilde{\boldsymbol{l}}_i)^2, \tag{2.120}$$

第 3 章　图像处理

$$g(\boldsymbol{x}) = h(f(\boldsymbol{x})) \text{ or } g(\boldsymbol{x}) = h(f_0(\boldsymbol{x}), \ldots, f_n(\boldsymbol{x})), \tag{3.1}$$

$$g(i,j) = h(f(i,j)). \tag{3.2}$$

$$g(\boldsymbol{x}) = af(\boldsymbol{x}) + b. \tag{3.3}$$

$$g(\boldsymbol{x}) = a(\boldsymbol{x})f(\boldsymbol{x}) + b(\boldsymbol{x}), \tag{3.4}$$

$$h(f_0 + f_1) = h(f_0) + h(f_1). \tag{3.5}$$

$$g(\boldsymbol{x}) = (1-\alpha)f_0(\boldsymbol{x}) + \alpha f_1(\boldsymbol{x}). \tag{3.6}$$

$$g(\boldsymbol{x}) = [f(\boldsymbol{x})]^{1/\gamma}, \tag{3.7}$$

$$C = (1-\alpha)B + \alpha F. \tag{3.8}$$

$$c(I) = \frac{1}{N}\sum_{i=0}^{I} h(i) = c(I-1) + \frac{1}{N}h(I), \tag{3.9}$$

$$f_{s,t}(I) = (1-s)(1-t)f_{00}(I) + s(1-t)f_{10}(I) + (1-s)tf_{01}(I) + stf_{11}(I) \tag{3.10}$$

$$h_{k,l}(I(i,j)) \mathrel{+}= w(i,j,k,l), \tag{3.11}$$

$$g(i,j) = \sum_{k,l} f(i+k, j+l)h(k,l). \tag{3.12}$$

$$g = f \otimes h. \tag{3.13}$$

$$g(i,j) = \sum_{k,l} f(i-k, j-l) h(k,l) = \sum_{k,l} f(k,l) h(i-k, j-l), \tag{3.14}$$

$$g = f * h, \tag{3.15}$$

$$h \circ (f_0 + f_1) = h \circ f_0 + h \circ f_1, \tag{3.16}$$

$$g(i,j) = f(i+k, j+l) \Leftrightarrow (h \circ g)(i,j) = (h \circ f)(i+k, j+l), \tag{3.17}$$

$$g(i,j) = \sum_{k,l} f(i-k, j-l) h(k,l; i,j), \tag{3.18}$$

$$\boldsymbol{g} = \boldsymbol{H} \boldsymbol{f}, \tag{3.19}$$

$$\boldsymbol{K} = \boldsymbol{v} \boldsymbol{h}^T \tag{3.20}$$

$$\boldsymbol{K} = \sum_i \sigma_i \boldsymbol{u}_i \boldsymbol{v}_i^T \tag{3.21}$$

$$g_{\text{sharp}} = f + \gamma (f - h_{\text{blur}} * f). \tag{3.22}$$

$$g_{\text{unsharp}} = f(1 - \gamma h_{\text{blur}} * f). \tag{3.23}$$

$$G(x, y; \sigma) = \frac{1}{2\pi\sigma^2} e^{-\frac{x^2 + y^2}{2\sigma^2}}, \tag{3.24}$$

$$\nabla^2 f = \frac{\partial^2 f}{\partial x^2} + \frac{\partial^2 y}{\partial y^2}, \tag{3.25}$$

$$\nabla^2 G(x, y; \sigma) = \left(\frac{x^2 + y^2}{\sigma^4} - \frac{2}{\sigma^2} \right) G(x, y; \sigma), \tag{3.26}$$

$$\hat{\boldsymbol{u}} \cdot \nabla (G * f) = \nabla_{\hat{\boldsymbol{u}}} (G * f) = (\nabla_{\hat{\boldsymbol{u}}} G) * f. \tag{3.27}$$

$$G_{\hat{\boldsymbol{u}}} = u G_x + v G_y = u \frac{\partial G}{\partial x} + v \frac{\partial G}{\partial y}, \tag{3.28}$$

$$G_{\hat{\boldsymbol{u}} \hat{\boldsymbol{u}}} = u^2 G_{xx} + 2uv G_{xy} + v^2 G_{yy}. \tag{3.29}$$

$$s(i,j) = \sum_{k=0}^{i}\sum_{l=0}^{j} f(k,l). \tag{3.30}$$

$$s(i,j) = s(i-1,j) + s(i,j-1) - s(i-1,j-1) + f(i,j). \tag{3.31}$$

$$S(i_0\ldots i_1, j_0\ldots j_1) = \sum_{i=i_0}^{i_1}\sum_{j=j_0}^{j_1} s(i_1,j_1) - s(i_1,j_0-1) - s(i_0-1,j_1) + s(i_0-1,j_0-1). \tag{3.32}$$

$$\sum_{k,l} w(k,l) |f(i+k, j+l) - g(i,j)|^p, \tag{3.33}$$

$$g(i,j) = \frac{\sum_{k,l} f(k,l) w(i,j,k,l)}{\sum_{k,l} w(i,j,k,l)}. \tag{3.34}$$

$$d(i,j,k,l) = \exp\left(-\frac{(i-k)^2 + (j-l)^2}{2\sigma_d^2}\right), \tag{3.35}$$

$$r(i,j,k,l) = \exp\left(-\frac{\|f(i,j) - f(k,l)\|^2}{2\sigma_r^2}\right). \tag{3.36}$$

$$w(i,j,k,l) = \exp\left(-\frac{(i-k)^2 + (j-l)^2}{2\sigma_d^2} - \frac{\|f(i,j) - f(k,l)\|^2}{2\sigma_r^2}\right). \tag{3.37}$$

$$d(i,j,k,l) = \exp\left(-\frac{(i-k)^2 + (j-l)^2}{2\sigma_d^2}\right) \tag{3.38}$$

$$= \begin{cases} 1, & |k-i| + |l-j| = 0, \\ \lambda = e^{-1/2\sigma_d^2}, & |k-i| + |l-j| = 1. \end{cases} \tag{3.39}$$

$$f^{(t+1)}(i,j) = \frac{f^{(t)}(i,j) + \eta \sum_{k,l} f^{(t)}(k,l) r(i,j,k,l)}{1 + \eta \sum_{k,l} r(i,j,k,l)} \tag{3.40}$$

$$= f^{(t)}(i,j) + \frac{\eta}{1+\eta R}\sum_{k,l} r(i,j,k,l)[f^{(t)}(k,l) - f^{(t)}(i,j)],$$

$$\theta(f,t) = \begin{cases} 1 & \text{if } f \geq t, \\ 0 & \text{else,} \end{cases} \tag{3.41}$$

$$c = f \otimes s \tag{3.42}$$

$$d_1(k,l) = |k| + |l| \tag{3.43}$$

$$d_2(k,l) = \sqrt{k^2+l^2}. \tag{3.44}$$

$$D(i,j) = \min_{k,l:b(k,l)=0} d(i-k,j-l), \tag{3.45}$$

$$M = \sum_{(x,y)\in\mathcal{R}} \begin{bmatrix} x-\overline{x} \\ y-\overline{y} \end{bmatrix} \begin{bmatrix} x-\overline{x} & y-\overline{y} \end{bmatrix}, \tag{3.46}$$

$$s(x) = \sin(2\pi f x + \phi_i) = \sin(\omega x + \phi_i) \tag{3.47}$$

$$o(x) = h(x) * s(x) = A\sin(\omega x + \phi_o), \tag{3.48}$$

$$s(x) = e^{j\omega x} = \cos\omega x + j\sin\omega x. \tag{3.49}$$

$$o(x) = h(x) * s(x) = Ae^{j\omega x + \phi}. \tag{3.50}$$

$$H(\omega) = \mathcal{F}\{h(x)\} = Ae^{j\phi}, \tag{3.51}$$

$$h(x) \stackrel{\mathcal{F}}{\leftrightarrow} H(\omega). \tag{3.52}$$

$$H(\omega) = \int_{-\infty}^{\infty} h(x)e^{-j\omega x}dx, \tag{3.53}$$

$$H(k) = \frac{1}{N}\sum_{x=0}^{N-1} h(x)e^{-j\frac{2\pi k x}{N}}, \tag{3.54}$$

$$\text{box}(x) = \begin{cases} 1 & \text{if } |x| \leq 1 \\ 0 & \text{else} \end{cases} \tag{3.55}$$

$$\text{sinc}(\omega) = \frac{\sin\omega}{\omega}, \tag{3.56}$$

$$\text{tent}(x) = \max(0, 1-|x|), \tag{3.57}$$

$$G(x;\sigma) = \frac{1}{\sqrt{2\pi}\sigma}e^{-\frac{x^2}{2\sigma^2}}, \tag{3.58}$$

$$LoG(x;\sigma) = (\frac{x^2}{\sigma^4} - \frac{1}{\sigma^2})G(x;\sigma) \tag{3.59}$$

$$-\frac{\sqrt{2\pi}}{\sigma}\omega^2 G(\omega;\sigma^{-1}) \tag{3.60}$$

$$\text{rcos}(x) = \frac{1}{2}(1 + \cos \pi x)\text{box}(x), \tag{3.61}$$

$$s(x, y) = \sin(\omega_x x + \omega_y y). \tag{3.62}$$

$$H(\omega_x, \omega_y) = \int_{-\infty}^{\infty} \int_{-\infty}^{\infty} h(x, y) e^{-j(\omega_x x + \omega_y y)} dx\, dy, \tag{3.63}$$

$$H(k_x, k_y) = \frac{1}{MN} \sum_{x=0}^{M-1} \sum_{y=0}^{N-1} h(x, y) e^{-j2\pi \frac{k_x x + k_y y}{MN}}, \tag{3.64}$$

$$\langle [S(\omega_x, \omega_y)]^2 \rangle = P_s(\omega_x, \omega_y), \tag{3.65}$$

$$o(x, y) = s(x, y) + n(x, y), \tag{3.66}$$

$$O(\omega_x, \omega_y) = S(\omega_x, \omega_y) + N(\omega_x, \omega_y), \tag{3.67}$$

$$p(S|O) = \frac{p(O|S)p(S)}{p(O)}, \tag{3.68}$$

$$p(S) = e^{-\frac{(S-\mu)^2}{2P_s}}, \tag{3.69}$$

$$p(S) = e^{-\frac{(S-O)^2}{2P_n}}. \tag{3.70}$$

$$\begin{aligned}
-\log p(S|O) &= -\log p(O|S) - \log p(S) + C & (3.71) \\
&= {}^{1}\!/\!_{2} P_n^{-1}(S-O)^2 + {}^{1}\!/\!_{2} P_s^{-1} S^2 + C, & (3.72)
\end{aligned}$$

$$S_{\text{opt}} = \frac{P_n^{-1}}{P_n^{-1} + P_s^{-1}} O = \frac{P_s}{P_s + P_n} O = \frac{1}{1 + P_n/P_s} O. \tag{3.73}$$

$$W(\omega_x, \omega_y) = \frac{1}{1 + \sigma_n^2 / P_s(\omega_x, \omega_y)} \tag{3.74}$$

$$o(x, y) = b(x, y) * s(x, y) + n(x, y), \tag{3.75}$$

$$F(k) = \sum_{i=0}^{N-1} \cos\left(\frac{\pi}{N}(i + \frac{1}{2})k\right) f(i), \tag{3.76}$$

$$F(k, l) = \sum_{i=0}^{N-1} \sum_{j=0}^{N-1} \cos\left(\frac{\pi}{N}(i + \frac{1}{2})k\right) \cos\left(\frac{\pi}{N}(j + \frac{1}{2})l\right) f(i, j). \tag{3.77}$$

$$g(i,j) = \sum_{k,l} f(k,l) h(i-rk, j-rl). \tag{3.78}$$

$$h(x) = \begin{cases} 1 - (a+3)x^2 + (a+2)|x|^3 & \text{if } |x| < 1 \\ a(|x|-1)(|x|-2)^2 & \text{if } 1 \leq |x| < 2 \\ 0 & \text{otherwise,} \end{cases} \tag{3.79}$$

$$g(i,j) = \sum_{k,l} f(k,l) h(ri-k, rj-l), \tag{3.80}$$

$$g(i,j) = \frac{1}{r} \sum_{k,l} f(k,l) h(i-k/r, j-l/r) \tag{3.81}$$

$$\boxed{c \mid b \mid a \mid b \mid c}, \tag{3.82}$$

$$\frac{1}{16} \boxed{1 \mid 4 \mid 6 \mid 4 \mid 1}, \tag{3.83}$$

$$\text{DoG}\{I; \sigma_1, \sigma_2\} = G_{\sigma_1} * I - G_{\sigma_2} * I = (G_{\sigma_1} - G_{\sigma_2}) * I. \tag{3.84}$$

$$\text{LoG}\{I; \sigma\} = \nabla^2 (G_\sigma * I) = (\nabla^2 G_\sigma) * I, \tag{3.85}$$

$$\nabla^2 = \frac{\partial^2}{\partial x^2} + \frac{\partial^2}{\partial y^2} \tag{3.86}$$

$$g(\boldsymbol{x}) = h(f(\boldsymbol{x})), \tag{3.87}$$

$$g(\boldsymbol{x}) = f(\boldsymbol{h}(\boldsymbol{x})) \tag{3.88}$$

$$g(x,y) = \sum_{k,l} f(k,l) h(x-k, y-l), \tag{3.89}$$

$$g(\boldsymbol{A}\boldsymbol{x}) \Leftrightarrow |\boldsymbol{A}|^{-1} G(\boldsymbol{A}^{-T} \boldsymbol{f}). \tag{3.90}$$

$$l = \log_2 r. \tag{3.91}$$

$$\mathcal{E}_1 = \int f_x^2(x)\, dx \tag{3.92}$$

$$\mathcal{E}_2 = \int f_{xx}^2(x)\, dx. \tag{3.93}$$

$$\mathcal{E}_1 = \int f_x^2(x,y) + f_y^2(x,y)\, dx\, dy = \int \|\nabla f(x,y)\|^2\, dx\, dy \tag{3.94}$$

$$\mathcal{E}_2 = \int f_{xx}^2(x,y) + 2f_{xy}^2(x,y) + f_{yy}^2(x,y)\, dx\, dy, \tag{3.95}$$

$$\mathcal{E}_d = \sum_i [f(x_i, y_i) - d_i]^2. \tag{3.96}$$

$$\mathcal{E}_d = \int [f(x,y) - d(x,y)]^2\, dx\, dy. \tag{3.97}$$

$$\mathcal{E} = \mathcal{E}_d + \lambda \mathcal{E}_s, \tag{3.98}$$

$$\begin{aligned} E_1 &= \sum_{i,j} s_x(i,j)[f(i+1,j) - f(i,j) - g_x(i,j)]^2 \\ &\quad + s_y(i,j)[f(i,j+1) - f(i,j) - g_y(i,j)]^2 \end{aligned} \tag{3.99}$$

$$\begin{aligned} E_2 &= h^{-2} \sum_{i,j} c_x(i,j)[f(i+1,j) - 2f(i,j) + f(i-1,j)]^2 \\ &\quad + 2c_m(i,j)[f(i+1,j+1) - f(i+1,j) - f(i,j+1) + f(i,j)]^2 \\ &\quad + c_y(i,j)[f(i,j+1) - 2f(i,j) + f(i,j-1)]^2, \end{aligned} \tag{3.100}$$

$$E_d = \sum_{i,j} w(i,j)[f(i,j) - d(i,j)]^2, \tag{3.101}$$

$$E = E_d + \lambda E_s = \boldsymbol{x}^T \boldsymbol{A} \boldsymbol{x} - 2\boldsymbol{x}^T \boldsymbol{b} + c, \tag{3.102}$$

$$\boldsymbol{A}\boldsymbol{x} = \boldsymbol{b}, \tag{3.103}$$

$$\begin{aligned} E_{1r} &= \sum_{i,j} s_x(i,j)\rho(f(i+1,j) - f(i,j)) \\ &\quad + s_y(i,j)\rho(f(i,j+1) - f(i,j)), \end{aligned} \tag{3.104}$$

$$\rho(x) = \min(x^2, V). \tag{3.105}$$

$$p(\boldsymbol{x}|\boldsymbol{y}) = \frac{p(\boldsymbol{y}|\boldsymbol{x})p(\boldsymbol{x})}{p(\boldsymbol{y})}, \tag{3.106}$$

$$-\log p(\boldsymbol{x}|\boldsymbol{y}) = -\log p(\boldsymbol{y}|\boldsymbol{x}) - \log p(\boldsymbol{x}) + C, \tag{3.107}$$

$$E(\boldsymbol{x},\boldsymbol{y}) = E_d(\boldsymbol{x},\boldsymbol{y}) + E_p(\boldsymbol{x}). \tag{3.108}$$

$$E_p(\boldsymbol{x}) = \sum_{\{(i,j),(k,l)\}\in\mathcal{N}} V_{i,j,k,l}(f(i,j),f(k,l)), \tag{3.109}$$

$$E_d(i,j) = w\delta(f(i,j),d(i,j)) \tag{3.110}$$

$$E_p(i,j) = E_x(i,j) + E_y(i,j) = s\delta(f(i,j),f(i+1,j)) + s\delta(f(i,j),f(i,j+1)). \tag{3.111}$$

$$E_d(i,j) = w(i,j)\rho_d(f(i,j)-d(i,j)) \tag{3.112}$$

$$E_p(i,j) = s_x(i,j)\rho_p(f(i,j)-f(i+1,j)) + s_y(i,j)\rho_p(f(i,j)-f(i,j+1)), \tag{3.113}$$

$$\rho_p(d) = |d|^p, \ p < 1, \tag{3.114}$$

$$E_p(i,j) = s_x(i,j)V(l(i,j),l(i+1,j)) + s_y(i,j)V(l(i,j),l(i,j+1)), \tag{3.115}$$

$$E_p = \sum_{(p,q)\in\mathcal{N}} V_{p,q}(l_p,l_q), \tag{3.116}$$

$$p(\boldsymbol{x}|\boldsymbol{y}) \propto p(\boldsymbol{y}|\boldsymbol{x})p(\boldsymbol{x}), \tag{3.117}$$

$$\begin{aligned} E(\boldsymbol{x}|\boldsymbol{y}) &= E_d(\boldsymbol{x},\boldsymbol{y}) + E_s(\boldsymbol{x},\boldsymbol{y}) \\ &= \sum_p V_p(x_p,\boldsymbol{y}) + \sum_{(p,q)\in\mathcal{N}} V_{p,q}(x_p,x_q,\boldsymbol{y}), \end{aligned} \tag{3.118}$$

第 4 章 特征检测与匹配

$$E_{\text{WSSD}}(\boldsymbol{u}) = \sum_i w(\boldsymbol{x}_i)[I_1(\boldsymbol{x}_i + \boldsymbol{u}) - I_0(\boldsymbol{x}_i)]^2, \tag{4.1}$$

$$E_{\text{AC}}(\Delta \boldsymbol{u}) = \sum_i w(\boldsymbol{x}_i)[I_0(\boldsymbol{x}_i + \Delta \boldsymbol{u}) - I_0(\boldsymbol{x}_i)]^2 \tag{4.2}$$

$$\begin{align}
E_{\text{AC}}(\Delta \boldsymbol{u}) &= \sum_i w(\boldsymbol{x}_i)[I_0(\boldsymbol{x}_i + \Delta \boldsymbol{u}) - I_0(\boldsymbol{x}_i)]^2 \tag{4.3} \\
&\approx \sum_i w(\boldsymbol{x}_i)[I_0(\boldsymbol{x}_i) + \nabla I_0(\boldsymbol{x}_i) \cdot \Delta \boldsymbol{u} - I_0(\boldsymbol{x}_i)]^2 \tag{4.4} \\
&= \sum_i w(\boldsymbol{x}_i)[\nabla I_0(\boldsymbol{x}_i) \cdot \Delta \boldsymbol{u}]^2 \tag{4.5} \\
&= \Delta \boldsymbol{u}^T A \Delta \boldsymbol{u}, \tag{4.6}
\end{align}$$

$$\nabla I_0(\boldsymbol{x}_i) = \left(\frac{\partial I_0}{\partial x}, \frac{\partial I_0}{\partial y}\right)(\boldsymbol{x}_i) \tag{4.7}$$

$$\boldsymbol{A} = w * \begin{bmatrix} I_x^2 & I_x I_y \\ I_x I_y & I_y^2 \end{bmatrix}, \tag{4.8}$$

$$\det(\boldsymbol{A}) - \alpha\, \text{trace}(\boldsymbol{A})^2 = \lambda_0 \lambda_1 - \alpha(\lambda_0 + \lambda_1)^2 \tag{4.9}$$

$$\lambda_0 - \alpha \lambda_1 \tag{4.10}$$

$$\frac{\det \boldsymbol{A}}{\text{tr } \boldsymbol{A}} = \frac{\lambda_0 \lambda_1}{\lambda_0 + \lambda_1}, \tag{4.11}$$

$$\boldsymbol{H} = \begin{bmatrix} D_{xx} & D_{xy} \\ D_{xy} & D_{yy} \end{bmatrix}, \tag{4.12}$$

$$\frac{\text{Tr}(\boldsymbol{H})^2}{\text{Det}(\boldsymbol{H})} > 10. \tag{4.13}$$

$$\text{TPR} = \frac{\text{TP}}{\text{TP+FN}} = \frac{\text{TP}}{\text{P}}; \tag{4.14}$$

$$\text{FPR} = \frac{\text{FP}}{\text{FP+TN}} = \frac{\text{FP}}{\text{N}}; \tag{4.15}$$

$$\text{PPV} = \frac{\text{TP}}{\text{TP+FP}} = \frac{\text{TP}}{\text{P'}}; \tag{4.16}$$

$$\text{ACC} = \frac{\text{TP+TN}}{\text{P+N}}. \tag{4.17}$$

$$\text{NNDR} = \frac{d_1}{d_2} = \frac{\|D_A - D_B\|}{\|D_A - D_C\|}, \tag{4.18}$$

$$\boldsymbol{J}(\boldsymbol{x}) = \nabla I(\boldsymbol{x}) = (\frac{\partial I}{\partial x}, \frac{\partial I}{\partial y})(\boldsymbol{x}). \tag{4.19}$$

$$\boldsymbol{J}_\sigma(\boldsymbol{x}) = \nabla[G_\sigma(\boldsymbol{x}) * I(\boldsymbol{x})] = [\nabla G_\sigma](\boldsymbol{x}) * I(\boldsymbol{x}), \tag{4.20}$$

$$\nabla G_\sigma(\boldsymbol{x}) = (\frac{\partial G_\sigma}{\partial x}, \frac{\partial G_\sigma}{\partial y})(\boldsymbol{x}) = [-x \ -y]\frac{1}{\sigma^3}\exp\left(-\frac{x^2+y^2}{2\sigma^2}\right) \tag{4.21}$$

$$S_\sigma(\boldsymbol{x}) = \nabla \cdot \boldsymbol{J}_\sigma(\boldsymbol{x}) = [\nabla^2 G_\sigma](\boldsymbol{x}) * I(\boldsymbol{x}). \tag{4.22}$$

$$\nabla^2 G_\sigma(\boldsymbol{x}) = \frac{1}{\sigma^3}\left(2 - \frac{x^2+y^2}{2\sigma^2}\right)\exp\left(-\frac{x^2+y^2}{2\sigma^2}\right) \tag{4.23}$$

$$\nabla^2 G_\sigma(\boldsymbol{x}) = \frac{1}{\sigma^3}\left(1 - \frac{x^2}{2\sigma^2}\right)G_\sigma(x)G_\sigma(y) + \frac{1}{\sigma^3}\left(1 - \frac{y^2}{2\sigma^2}\right)G_\sigma(y)G_\sigma(x) \tag{4.24}$$

$$\boldsymbol{x}_z = \frac{\boldsymbol{x}_i S(\boldsymbol{x}_j) - \boldsymbol{x}_j S(\boldsymbol{x}_i)}{S(\boldsymbol{x}_j) - S(\boldsymbol{x}_i)}. \tag{4.25}$$

$$\theta = \tan^{-1} n_y/n_x, \tag{4.26}$$

$$\hat{\boldsymbol{m}} = (\cos\theta\cos\phi, \sin\theta\cos\phi, \sin\phi), \tag{4.27}$$

$$\boldsymbol{v}_{ij} = \hat{\boldsymbol{m}}_i \times \hat{\boldsymbol{m}}_j \tag{4.28}$$

$$w_{ij} = \|\boldsymbol{v}_{ij}\| l_i l_j. \tag{4.29}$$

$$A_i = |(\boldsymbol{p}_{i0} \times \boldsymbol{p}_{i1}) \cdot \boldsymbol{v}|, \tag{4.30}$$

$$\mathcal{E} = \sum_i \rho(A_i) = \boldsymbol{v}^T \left(\sum_i w_i(A_i) \boldsymbol{m}_i \boldsymbol{m}_i^T \right) \boldsymbol{v} = \boldsymbol{v}^T \boldsymbol{M} \boldsymbol{v}, \tag{4.31}$$

$$s_{i+1} = s_i + \|\boldsymbol{x}_{i+1} - \boldsymbol{x}_i\|. \tag{4.32}$$

$$d = x n_x + y n_y \tag{4.33}$$

第 5 章 分割

$$\mathcal{E}_{\text{int}} = \int \alpha(s)\|\boldsymbol{f}_s(s)\|^2 + \beta(s)\|\boldsymbol{f}_{ss}(s)\|^2 \, ds, \tag{5.1}$$

$$\begin{aligned} E_{\text{int}} &= \sum_i \alpha(i)\|f(i+1)-f(i)\|^2/h^2 \\ &\quad + \beta(i)\|f(i+1)-2f(i)+f(i-1)\|^2/h^4, \end{aligned} \tag{5.2}$$

$$\mathcal{E}_{\text{image}} = w_{\text{line}}\mathcal{E}_{\text{line}} + w_{\text{edge}}\mathcal{E}_{\text{edge}} + w_{\text{term}}\mathcal{E}_{\text{term}}, \tag{5.3}$$

$$E_{\text{edge}} = \sum_i -\|\nabla I(\boldsymbol{f}(i))\|^2, \tag{5.4}$$

$$E_{\text{edge}} = \sum_i -|(G_\sigma * \nabla^2 I)(\boldsymbol{f}(i))|^2. \tag{5.5}$$

$$E_{\text{spring}} = k_i\|\boldsymbol{f}(i) - \boldsymbol{d}(i)\|^2, \tag{5.6}$$

$$p(\boldsymbol{d}(j)) = \sum_i p_{ij} \;\; \text{with} \;\; p_{ij} = e^{-d_{ij}^2/(2\sigma^2)} \tag{5.7}$$

$$d_{ij} = \|\boldsymbol{f}(i) - \boldsymbol{d}(j)\| \tag{5.8}$$

$$E_{\text{slippery}} = -\sum_j \log p(\boldsymbol{d}(j)) = -\sum_j \log\left[\sum e^{-\|\boldsymbol{f}(i)-\boldsymbol{d}(j)\|^2/2\sigma^2}\right]. \tag{5.9}$$

$$\boldsymbol{f}(s) = \sum_k B_k(s)\boldsymbol{x}_k \tag{5.10}$$

$$\boldsymbol{F} = \boldsymbol{B}\boldsymbol{X} \tag{5.11}$$

$$F = \begin{bmatrix} f^T(0) \\ \vdots \\ f^T(N) \end{bmatrix}, \quad B = \begin{bmatrix} B_0(s_0) & \ldots & B_K(s_0) \\ \vdots & \ddots & \vdots \\ B_0(s_N) & \ldots & B_K(s_N) \end{bmatrix}, \text{ and } X = \begin{bmatrix} x^T(0) \\ \vdots \\ x^T(K) \end{bmatrix}. \tag{5.12}$$

$$E_{\text{loc}}(x_k) = \frac{1}{2}(x_k - \bar{x}_k)^T C_k^{-1}(x_k - \bar{x}_k). \tag{5.13}$$

$$C = \frac{1}{P}\sum_p (x_p - \bar{x})(x_p - \bar{x})^T, \tag{5.14}$$

$$C = \Phi \, \text{diag}(\lambda_0 \ldots \lambda_{K-1}) \, \Phi^T. \tag{5.15}$$

$$x = \bar{x} + \hat{\Phi} \, b, \tag{5.16}$$

$$E_{\text{shape}} = \frac{1}{2} b^T \, \text{diag}(\lambda_0 \ldots \lambda_{M-1}) \, b = \sum_m b_m^2 / 2\lambda_m. \tag{5.17}$$

$$x_t = A x_{t-1} + w_t, \tag{5.18}$$

$$\begin{aligned} \frac{d\phi}{dt} &= |\nabla\phi|\text{div}\left(g(I)\frac{\nabla\phi}{|\nabla\phi|}\right) \\ &= g(I)|\nabla\phi|\text{div}\left(\frac{\nabla\phi}{|\nabla\phi|}\right) + \nabla g(I) \cdot \nabla\phi, \end{aligned} \tag{5.19}$$

$$Int(R) = \min_{e \in MST(R)} w(e). \tag{5.20}$$

$$Dif(R_1, R_2) = \min_{e=(v_1,v_2)|v_1 \in R_1, v_2 \in R_2} w(e). \tag{5.21}$$

$$MInt(R_1, R_2) = \min(Int(R_1) + \tau(R_1), Int(R_2) + \tau(R_2)), \tag{5.22}$$

$$\sigma_{local}^+ = \min(\Delta_i^+, \Delta_j^+), \tag{5.23}$$

$$\sigma_{local}^- = \frac{\Delta_i^- + \Delta_j^-}{2}, \tag{5.24}$$

$$\frac{\sum_{j \in C} p_{ij}}{\sum_{j \in V} p_{ij}} > \phi, \tag{5.25}$$

$$d(\boldsymbol{x}_i, \boldsymbol{\mu}_k; \boldsymbol{\Sigma}_k) = \|\boldsymbol{x}_i - \boldsymbol{\mu}_k\|_{\boldsymbol{\Sigma}_k^{-1}} = (\boldsymbol{x}_i - \boldsymbol{\mu}_k)^T \boldsymbol{\Sigma}_k^{-1}(\boldsymbol{x}_i - \boldsymbol{\mu}_k) \tag{5.26}$$

$$p(\boldsymbol{x}|\{\pi_k, \boldsymbol{\mu}_k, \boldsymbol{\Sigma}_k\}) = \sum_k \pi_k \mathcal{N}(\boldsymbol{x}|\boldsymbol{\mu}_k, \boldsymbol{\Sigma}_k), \tag{5.27}$$

$$\mathcal{N}(\boldsymbol{x}|\boldsymbol{\mu}_k, \boldsymbol{\Sigma}_k) = \frac{1}{|\boldsymbol{\Sigma}_k|} e^{-d(\boldsymbol{x}, \boldsymbol{\mu}_k; \boldsymbol{\Sigma}_k)} \tag{5.28}$$

$$z_{ik} = \frac{1}{Z_i} \pi_k \mathcal{N}(\boldsymbol{x}|\boldsymbol{\mu}_k, \boldsymbol{\Sigma}_k) \quad \text{with} \quad \sum_k z_{ik} = 1, \tag{5.29}$$

$$\boldsymbol{\mu}_k = \frac{1}{N_k} \sum_i z_{ik} \boldsymbol{x}_i, \tag{5.30}$$

$$\boldsymbol{\Sigma}_k = \frac{1}{N_k} \sum_i z_{ik} (\boldsymbol{x}_i - \boldsymbol{\mu}_k)(\boldsymbol{x}_i - \boldsymbol{\mu}_k)^T, \tag{5.31}$$

$$\pi_k = \frac{N_k}{N}, \tag{5.32}$$

$$N_k = \sum_i z_{ik}. \tag{5.33}$$

$$f(\boldsymbol{x}) = \sum_i K(\boldsymbol{x} - \boldsymbol{x}_i) = \sum_i k\left(\frac{\|\boldsymbol{x} - \boldsymbol{x}_i\|^2}{h^2}\right), \tag{5.34}$$

$$\nabla f(\boldsymbol{x}) = \sum_i (\boldsymbol{x}_i - \boldsymbol{x}) G(\boldsymbol{x} - \boldsymbol{x}_i) = \sum_i (\boldsymbol{x}_i - \boldsymbol{x}) g\left(\frac{\|\boldsymbol{x} - \boldsymbol{x}_i\|^2}{h^2}\right), \tag{5.35}$$

$$g(r) = -k'(r), \tag{5.36}$$

$$\nabla f(\boldsymbol{x}) = \left[\sum_i G(\boldsymbol{x} - \boldsymbol{x}_i)\right] \boldsymbol{m}(\boldsymbol{x}), \tag{5.37}$$

$$\boldsymbol{m}(\boldsymbol{x}) = \frac{\sum_i \boldsymbol{x}_i G(\boldsymbol{x} - \boldsymbol{x}_i)}{\sum_i G(\boldsymbol{x} - \boldsymbol{x}_i)} - \boldsymbol{x} \tag{5.38}$$

$$\boldsymbol{y}_{k+1} = \boldsymbol{y}_k + \boldsymbol{m}(\boldsymbol{y}_k) = \frac{\sum_i \boldsymbol{x}_i G(\boldsymbol{y}_k - \boldsymbol{x}_i)}{\sum_i G(\boldsymbol{y}_k - \boldsymbol{x}_i)}. \tag{5.39}$$

$$k_E(r) = \max(0, 1 - r), \tag{5.40}$$

$$k_N(r) = \exp\left(-\frac{1}{2}r\right). \tag{5.41}$$

$$K(\boldsymbol{x}_j) = k\left(\frac{\|\boldsymbol{x}_r\|^2}{h_r^2}\right) k\left(\frac{\|\boldsymbol{x}_s\|^2}{h_s^2}\right), \tag{5.42}$$

$$cut(A,B) = \sum_{i \in A, j \in B} w_{ij}, \tag{5.43}$$

$$Ncut(A,B) = \frac{cut(A,B)}{assoc(A,V)} + \frac{cut(A,B)}{assoc(B,V)}, \tag{5.44}$$

$$\min_{\boldsymbol{y}} \frac{\boldsymbol{y}^T(\boldsymbol{D}-\boldsymbol{W})\boldsymbol{y}}{\boldsymbol{y}^T \boldsymbol{D} \boldsymbol{y}}, \tag{5.45}$$

$$(\boldsymbol{D}-\boldsymbol{W})\boldsymbol{y} = \lambda \boldsymbol{D}\boldsymbol{y}, \tag{5.46}$$

$$(\boldsymbol{I}-\boldsymbol{N})\boldsymbol{z} = \lambda \boldsymbol{z}, \tag{5.47}$$

$$w_{ij} = \exp\left(-\frac{\|\boldsymbol{F}_i - \boldsymbol{F}_j\|^2}{\sigma_F^2} - \frac{\|\boldsymbol{x}_i - \boldsymbol{x}_j\|^2}{\sigma_s^2}\right), \tag{5.48}$$

$$w_{ij}^{IC} = 1 - \max_{\boldsymbol{x} \in l_{ij}} p_{con}(\boldsymbol{x}), \tag{5.49}$$

$$E(f) = \sum_{i,j} E_r(i,j) + E_b(i,j), \tag{5.50}$$

$$E_r(i,j) = E_S(I(i,j); R(f(i,j))) \tag{5.51}$$

$$E_b(i,j) = s_x(i,j)\delta(f(i,j) - f(i+1,j)) + s_y(i,j)\delta(f(i,j) - f(i,j+1)) \tag{5.52}$$

$$E_S(I; \mu_k) = \|I - \mu_k\|^2. \tag{5.53}$$

第 6 章 基于特征的配准

$$x' = f(x; p), \tag{6.1}$$

$$E_{\text{LS}} = \sum_i \|r_i\|^2 = \sum_i \|f(x_i; p) - x'_i\|^2, \tag{6.2}$$

$$r_i = f(x_i; p) - x'_i = \hat{x}'_i - \tilde{x}'_i \tag{6.3}$$

$$\Delta x = x' - x = J(x)p, \tag{6.4}$$

$$\begin{aligned}
E_{\text{LLS}} &= \sum_i \|J(x_i)p - \Delta x_i\|^2 & (6.5) \\
&= p^T \left[\sum_i J^T(x_i)J(x_i)\right] p - 2p^T \left[\sum_i J^T(x_i)\Delta x_i\right] + \sum_i \|\Delta x_i\|^2 & (6.6) \\
&= p^T A p - 2p^T b + c. & (6.7)
\end{aligned}$$

$$Ap = b, \tag{6.8}$$

$$A = \sum_i J^T(x_i)J(x_i) \tag{6.9}$$

$$E_{\text{WLS}} = \sum_i \sigma_i^{-2} \|r_i\|^2. \tag{6.10}$$

$$E_{\text{CWLS}} = \sum_i \|r_i\|^2_{\Sigma_i^{-1}} = \sum_i r_i^T \Sigma_i^{-1} r_i = \sum_i \sigma_n^{-2} r_i^T A_i r_i. \tag{6.11}$$

$$E_{\text{PLS}} = \sum_{ij} \|(t_j + x_{ij}) - x_i\|^2, \tag{6.12}$$

$$
\begin{aligned}
E_{\text{NLS}}(\Delta \bm{p}) &= \sum_i \|\bm{f}(\bm{x}_i; \bm{p} + \Delta \bm{p}) - \bm{x}'_i\|^2 & (6.13) \\
&\approx \sum_i \|\bm{J}(\bm{x}_i; \bm{p})\Delta \bm{p} - \bm{r}_i\|^2 & (6.14) \\
&= \Delta \bm{p}^T \left[\sum_i \bm{J}^T \bm{J}\right] \Delta \bm{p} - 2\Delta \bm{p}^T \left[\sum_i \bm{J}^T \bm{r}_i\right] + \sum_i \|\bm{r}_i\|^2 & (6.15) \\
&= \Delta \bm{p}^T \bm{A} \Delta \bm{p} - 2\Delta \bm{p}^T \bm{b} + c, & (6.16)
\end{aligned}
$$

$$\bm{b} = \sum_i \bm{J}^T(\bm{x}_i) \bm{r}_i \qquad (6.17)$$

$$(\bm{A} + \lambda \text{diag}(\bm{A}))\Delta \bm{p} = \bm{b}, \qquad (6.18)$$

$$x' = \frac{(1+h_{00})x + h_{01}y + h_{02}}{h_{20}x + h_{21}y + 1} \text{ and } y' = \frac{h_{10}x + (1+h_{11})y + h_{12}}{h_{20}x + h_{21}y + 1}. \qquad (6.19)$$

$$\bm{J} = \frac{\partial \bm{f}}{\partial \bm{p}} = \frac{1}{D} \begin{bmatrix} x & y & 1 & 0 & 0 & 0 & -x'x & -x'y \\ 0 & 0 & 0 & x & y & 1 & -y'x & -y'y \end{bmatrix}, \qquad (6.20)$$

$$\begin{bmatrix} \hat{x}' - x \\ \hat{y}' - y \end{bmatrix} = \begin{bmatrix} x & y & 1 & 0 & 0 & 0 & -\hat{x}'x & -\hat{x}'y \\ 0 & 0 & 0 & x & y & 1 & -\hat{y}'x & -\hat{y}'y \end{bmatrix} \begin{bmatrix} h_{00} \\ \vdots \\ h_{21} \end{bmatrix}. \qquad (6.21)$$

$$\frac{1}{D}\begin{bmatrix} \hat{x}' - x \\ \hat{y}' - y \end{bmatrix} = \frac{1}{D}\begin{bmatrix} x & y & 1 & 0 & 0 & 0 & -\hat{x}'x & -\hat{x}'y \\ 0 & 0 & 0 & x & y & 1 & -\hat{y}'x & -\hat{y}'y \end{bmatrix} \begin{bmatrix} h_{00} \\ \vdots \\ h_{21} \end{bmatrix}. \qquad (6.22)$$

$$\begin{bmatrix} \hat{x}' - \tilde{x}' \\ \hat{y}' - \tilde{y}' \end{bmatrix} = \frac{1}{D}\begin{bmatrix} x & y & 1 & 0 & 0 & 0 & -\tilde{x}'x & -\tilde{x}'y \\ 0 & 0 & 0 & x & y & 1 & -\tilde{y}'x & -\tilde{y}'y \end{bmatrix} \begin{bmatrix} \Delta h_{00} \\ \vdots \\ \Delta h_{21} \end{bmatrix}. \qquad (6.23)$$

$$\begin{bmatrix} \hat{x}' - x \\ \hat{y}' - y \end{bmatrix} = \begin{bmatrix} x & y & 1 & 0 & 0 & 0 & -x^2 & -xy \\ 0 & 0 & 0 & x & y & 1 & -xy & -y^2 \end{bmatrix} \begin{bmatrix} \Delta h_{00} \\ \vdots \\ \Delta h_{21} \end{bmatrix}, \qquad (6.24)$$

$$E_{\text{RLS}}(\Delta \bm{p}) = \sum_i \rho(\|\bm{r}_i\|) \qquad (6.25)$$

$$\sum_i \psi(\|\boldsymbol{r}_i\|)\frac{\partial \|\boldsymbol{r}_i\|}{\partial \boldsymbol{p}} = \sum_i \frac{\psi(\|\boldsymbol{r}_i\|)}{\|\boldsymbol{r}_i\|}\boldsymbol{r}_i^T\frac{\partial \boldsymbol{r}_i}{\partial \boldsymbol{p}} = 0, \tag{6.26}$$

$$E_{\text{IRLS}} = \sum_i w(\|\boldsymbol{r}_i\|)\|\boldsymbol{r}_i\|^2, \tag{6.27}$$

$$\boldsymbol{r}_i = \tilde{\boldsymbol{x}}_i'(\boldsymbol{x}_i;\boldsymbol{p}) - \hat{\boldsymbol{x}}_i', \tag{6.28}$$

$$1 - P = (1 - p^k)^S \tag{6.29}$$

$$S = \frac{\log(1-P)}{\log(1-p^k)}. \tag{6.30}$$

$$E_{\text{R3D}} = \sum_i \|\boldsymbol{x}_i' - \boldsymbol{R}\boldsymbol{x}_i - \boldsymbol{t}\|^2, \tag{6.31}$$

$$\boldsymbol{C} = \sum_i \hat{\boldsymbol{x}}'\hat{\boldsymbol{x}}^T = \boldsymbol{U}\boldsymbol{\Sigma}\boldsymbol{V}^T. \tag{6.32}$$

$$x_i = \frac{p_{00}X_i + p_{01}Y_i + p_{02}Z_i + p_{03}}{p_{20}X_i + p_{21}Y_i + p_{22}Z_i + p_{23}} \tag{6.33}$$

$$y_i = \frac{p_{10}X_i + p_{11}Y_i + p_{12}Z_i + p_{13}}{p_{20}X_i + p_{21}Y_i + p_{22}Z_i + p_{23}}, \tag{6.34}$$

$$\boldsymbol{P} = \boldsymbol{K}[\boldsymbol{R}|\boldsymbol{t}]. \tag{6.35}$$

$$\hat{\boldsymbol{x}}_i = \mathcal{N}(\boldsymbol{K}^{-1}\boldsymbol{x}_i) = \boldsymbol{K}^{-1}\boldsymbol{x}_i/\|\boldsymbol{K}^{-1}\boldsymbol{x}_i\|, \tag{6.36}$$

$$\boldsymbol{p}_i = d_i\hat{\boldsymbol{x}}_i + \boldsymbol{c} \tag{6.37}$$

$$f_{ij}(d_i, d_j) = d_i^2 + d_j^2 - 2d_id_jc_{ij} - d_{ij}^2 = 0, \tag{6.38}$$

$$c_{ij} = \cos\theta_{ij} = \hat{\boldsymbol{x}}_i \cdot \hat{\boldsymbol{x}}_j \tag{6.39}$$

$$d_{ij}^2 = \|\boldsymbol{p}_i - \boldsymbol{p}_j\|^2. \tag{6.40}$$

$$g_{ijk}(d_i^2) = a_4d_i^8 + a_3d_i^6 + a_2d_i^4 + a_1d_i^2 + a_0 = 0. \tag{6.41}$$

$$\boldsymbol{x}_i = \boldsymbol{f}(\boldsymbol{p}_i; \boldsymbol{R}, \boldsymbol{t}, \boldsymbol{K}) \tag{6.42}$$

$$E_{\text{NLP}} = \sum_i \rho \left(\frac{\partial \boldsymbol{f}}{\partial \boldsymbol{R}} \Delta \boldsymbol{R} + \frac{\partial \boldsymbol{f}}{\partial t} \Delta t + \frac{\partial \boldsymbol{f}}{\partial \boldsymbol{K}} \Delta \boldsymbol{K} - \boldsymbol{r}_i \right), \tag{6.43}$$

$$\begin{aligned}
\boldsymbol{y}^{(1)} &= \boldsymbol{f}_{\text{T}}(\boldsymbol{p}_i; \boldsymbol{c}_j) = \boldsymbol{p}_i - \boldsymbol{c}_j, & (6.44) \\
\boldsymbol{y}^{(2)} &= \boldsymbol{f}_{\text{R}}(\boldsymbol{y}^{(1)}; \boldsymbol{q}_j) = \boldsymbol{R}(\boldsymbol{q}_j)\, \boldsymbol{y}^{(1)}, & (6.45) \\
\boldsymbol{y}^{(3)} &= \boldsymbol{f}_{\text{P}}(\boldsymbol{y}^{(2)}) = \frac{\boldsymbol{y}^{(2)}}{z^{(2)}}, & (6.46) \\
\boldsymbol{x}_i &= \boldsymbol{f}_{\text{C}}(\boldsymbol{y}^{(3)}; \boldsymbol{k}) = \boldsymbol{K}(\boldsymbol{k})\, \boldsymbol{y}^{(3)}. & (6.47)
\end{aligned}$$

$$\frac{\partial \boldsymbol{r}_i}{\partial \boldsymbol{p}^{(k)}} = \frac{\partial \boldsymbol{r}_i}{\partial \boldsymbol{y}^{(k)}} \frac{\partial \boldsymbol{y}^{(k)}}{\partial \boldsymbol{p}^{(k)}}, \tag{6.48}$$

$$\boldsymbol{x}_i = \begin{bmatrix} x_i \\ y_i \\ 1 \end{bmatrix} \sim \boldsymbol{K} \begin{bmatrix} \boldsymbol{r}_0 & \boldsymbol{r}_1 & \boldsymbol{t} \end{bmatrix} \begin{bmatrix} X_i \\ Y_i \\ 1 \end{bmatrix} \sim \tilde{\boldsymbol{H}} \boldsymbol{p}_i, \tag{6.49}$$

$$\hat{\boldsymbol{x}}_i = \begin{bmatrix} x_i - c_x \\ y_i - c_y \\ f \end{bmatrix} \sim \boldsymbol{R} \boldsymbol{p}_i = \boldsymbol{r}_i, \tag{6.50}$$

$$\boldsymbol{r}_i \cdot \boldsymbol{r}_j \sim (x_i - c_x)(x_j - c_y) + (y_i - c_y)(y_j - c_y) + f^2 = 0 \tag{6.51}$$

$$\tilde{\boldsymbol{H}}_{ij} = \boldsymbol{K}_i \boldsymbol{R}_i \boldsymbol{R}_j^{-1} \boldsymbol{K}_j^{-1} = \boldsymbol{K}_i \boldsymbol{R}_{ij} \boldsymbol{K}_j^{-1}. \tag{6.52}$$

$$\boldsymbol{R}_{10} \sim \boldsymbol{K}_1^{-1} \tilde{\boldsymbol{H}}_{10} \boldsymbol{K}_0 \sim \begin{bmatrix} h_{00} & h_{01} & f_0^{-1} h_{02} \\ h_{10} & h_{11} & f_0^{-1} h_{12} \\ f_1 h_{20} & f_1 h_{21} & f_0^{-1} f_1 h_{22} \end{bmatrix}, \tag{6.53}$$

$$h_{00}^2 + h_{01}^2 + f_0^{-2} h_{02}^2 = h_{10}^2 + h_{11}^2 + f_0^{-2} h_{12}^2 \tag{6.54}$$

$$h_{00} h_{10} + h_{01} h_{11} + f_0^{-2} h_{02} h_{12} = 0. \tag{6.55}$$

$$f_0^2 = \frac{h_{12}^2 - h_{02}^2}{h_{00}^2 + h_{01}^2 - h_{10}^2 - h_{11}^2} \text{ if } h_{00}^2 + h_{01}^2 \neq h_{10}^2 + h_{11}^2 \tag{6.56}$$

$$f_0^2 = -\frac{h_{02} h_{12}}{h_{00} h_{10} + h_{01} h_{11}} \text{ if } h_{00} h_{10} \neq -h_{01} h_{11}. \tag{6.57}$$

$$\tilde{\boldsymbol{H}}_{ij}(\boldsymbol{K}\boldsymbol{K}^T) \sim (\boldsymbol{K}\boldsymbol{K}^T)\tilde{\boldsymbol{H}}_{ij}^{-T}. \tag{6.58}$$

$$\begin{aligned}
\hat{x} &= x(1 + \kappa_1 r^2 + \kappa_2 r^4) \\
\hat{y} &= y(1 + \kappa_1 r^2 + \kappa_2 r^4),
\end{aligned} \tag{6.59}$$

第 7 章　由运动到结构

$$\|\boldsymbol{c}_j + d_j \hat{\boldsymbol{v}}_j - \boldsymbol{p}\|^2, \tag{7.1}$$

$$\boldsymbol{q}_j = \boldsymbol{c}_j + (\hat{\boldsymbol{v}}_j \hat{\boldsymbol{v}}_j^T)(\boldsymbol{p} - \boldsymbol{c}_j) = \boldsymbol{c}_j + (\boldsymbol{p} - \boldsymbol{c}_j)_\|, \tag{7.2}$$

$$r_j^2 = \|(\boldsymbol{I} - \hat{\boldsymbol{v}}_j \hat{\boldsymbol{v}}_j^T)(\boldsymbol{p} - \boldsymbol{c}_j)\|^2 = \|(\boldsymbol{p} - \boldsymbol{c}_j)_\perp\|^2. \tag{7.3}$$

$$\boldsymbol{p} = \left[\sum_j (\boldsymbol{I} - \hat{\boldsymbol{v}}_j \hat{\boldsymbol{v}}_j^T)\right]^{-1} \left[\sum_j (\boldsymbol{I} - \hat{\boldsymbol{v}}_j \hat{\boldsymbol{v}}_j^T)\boldsymbol{c}_j\right]. \tag{7.4}$$

$$x_j = \frac{p_{00}^{(j)} X + p_{01}^{(j)} Y + p_{02}^{(j)} Z + p_{03}^{(j)} W}{p_{20}^{(j)} X + p_{21}^{(j)} Y + p_{22}^{(j)} Z + p_{23}^{(j)} W} \tag{7.5}$$

$$y_j = \frac{p_{10}^{(j)} X + p_{11}^{(j)} Y + p_{12}^{(j)} Z + p_{13}^{(j)} W}{p_{20}^{(j)} X + p_{21}^{(j)} Y + p_{22}^{(j)} Z + p_{23}^{(j)} W}, \tag{7.6}$$

$$d_1 \hat{\boldsymbol{x}}_1 = \boldsymbol{p}_1 = \boldsymbol{R}\boldsymbol{p}_0 + \boldsymbol{t} = \boldsymbol{R}(d_0 \hat{\boldsymbol{x}}_0) + \boldsymbol{t}, \tag{7.7}$$

$$d_1 [\boldsymbol{t}]_\times \hat{\boldsymbol{x}}_1 = d_0 [\boldsymbol{t}]_\times \boldsymbol{R} \hat{\boldsymbol{x}}_0. \tag{7.8}$$

$$d_0 \hat{\boldsymbol{x}}_1^T ([\boldsymbol{t}]_\times \boldsymbol{R}) \hat{\boldsymbol{x}}_0 = d_1 \hat{\boldsymbol{x}}_1^T [\boldsymbol{t}]_\times \hat{\boldsymbol{x}}_1 = 0, \tag{7.9}$$

$$\hat{\boldsymbol{x}}_1^T \boldsymbol{E} \hat{\boldsymbol{x}}_0 = 0, \tag{7.10}$$

$$\boldsymbol{E} = [\boldsymbol{t}]_\times \boldsymbol{R} \tag{7.11}$$

$$(\hat{\boldsymbol{x}}_0, \boldsymbol{R}^{-1}\hat{\boldsymbol{x}}_1, -\boldsymbol{R}^{-1}\boldsymbol{t}) = (\boldsymbol{R}\hat{\boldsymbol{x}}_0, \hat{\boldsymbol{x}}_1, -\boldsymbol{t}) = \hat{\boldsymbol{x}}_1 \cdot (\boldsymbol{t} \times \boldsymbol{R}\hat{\boldsymbol{x}}_0) = \hat{\boldsymbol{x}}_1^T([\boldsymbol{t}]_\times \boldsymbol{R})\hat{\boldsymbol{x}}_0 = 0. \tag{7.12}$$

$$\begin{array}{rrrr} x_{i0}x_{i1}e_{00} & + & y_{i0}x_{i1}e_{01} & + & x_{i1}e_{02} & + \\ x_{i0}y_{i1}e_{00} & + & y_{i0}y_{i1}e_{11} & + & y_{i1}e_{12} & + \\ x_{i0}e_{20} & + & y_{i0}e_{21} & + & e_{22} & = & 0 \end{array} \tag{7.13}$$

$$[\boldsymbol{x}_{i1}\,\boldsymbol{x}_{i0}^T] \otimes \boldsymbol{E} = \boldsymbol{Z}_i \otimes \boldsymbol{E} = \boldsymbol{z}_i \cdot \boldsymbol{f} = 0, \tag{7.14}$$

$$\tilde{x}_i = s(x_i - \mu_x) \tag{7.15}$$
$$\tilde{y}_i = s(x_i - \mu_y) \tag{7.16}$$

$$\boldsymbol{E} = \boldsymbol{T}_1 \tilde{\boldsymbol{E}} \boldsymbol{T}_0. \tag{7.17}$$

$$\boldsymbol{E} = [\hat{\boldsymbol{t}}]_\times \boldsymbol{R} = \boldsymbol{U}\boldsymbol{\Sigma}\boldsymbol{V}^T = \begin{bmatrix} \boldsymbol{u}_0 & \boldsymbol{u}_1 & \hat{\boldsymbol{t}} \end{bmatrix} \begin{bmatrix} 1 & & \\ & 1 & \\ & & 0 \end{bmatrix} \begin{bmatrix} \boldsymbol{v}_0^T \\ \boldsymbol{v}_1^T \\ \boldsymbol{v}_2^T \end{bmatrix} \tag{7.18}$$

$$\boldsymbol{E} = \alpha \boldsymbol{E}_0 + (1-\alpha)\boldsymbol{E}_1. \tag{7.19}$$

$$\det|\alpha \boldsymbol{E}_0 + (1-\alpha)\boldsymbol{E}_1| = 0. \tag{7.20}$$

$$[\hat{\boldsymbol{t}}]_\times = \boldsymbol{S}\boldsymbol{Z}\boldsymbol{R}_{90°}\boldsymbol{S}^T = \begin{bmatrix} \boldsymbol{s}_0 & \boldsymbol{s}_1 & \hat{\boldsymbol{t}} \end{bmatrix} \begin{bmatrix} 1 & & \\ & 1 & \\ & & 0 \end{bmatrix} \begin{bmatrix} 0 & -1 & \\ 1 & 0 & \\ & & 1 \end{bmatrix} \begin{bmatrix} \boldsymbol{s}_0^T \\ \boldsymbol{s}_1^T \\ \hat{\boldsymbol{t}}^T \end{bmatrix}, \tag{7.21}$$

$$\boldsymbol{E} = [\hat{\boldsymbol{t}}]_\times \boldsymbol{R} = \boldsymbol{S}\boldsymbol{Z}\boldsymbol{R}_{90°}\boldsymbol{S}^T\boldsymbol{R} = \boldsymbol{U}\boldsymbol{\Sigma}\boldsymbol{V}^T, \tag{7.22}$$

$$\boldsymbol{R}_{90°}\boldsymbol{U}^T\boldsymbol{R} = \boldsymbol{V}^T \tag{7.23}$$

$$\boldsymbol{R} = \boldsymbol{U}\boldsymbol{R}_{90°}^T\boldsymbol{V}^T. \tag{7.24}$$

$$\boldsymbol{R} = \pm\boldsymbol{U}\boldsymbol{R}_{\pm 90°}^T\boldsymbol{V}^T \tag{7.25}$$

$$\sum_i (\boldsymbol{x}_{i0}, \boldsymbol{x}_{i1}, \boldsymbol{e})^2 = \sum_i ((\boldsymbol{x}_{i0} \times \boldsymbol{x}_{i1}) \cdot \boldsymbol{e})^2, \tag{7.26}$$

$$(y_{i0} - y_{i1})e_0 + (x_{i1} - x_{i0})e_1 + (x_{i0}y_{i1} - y_{i0}x_{i1})e_2 = 0. \tag{7.27}$$

$$\hat{\boldsymbol{x}}_1^T \boldsymbol{E} \hat{\boldsymbol{x}}_1 = \boldsymbol{x}_1^T \boldsymbol{K}_1^{-T} \boldsymbol{E} \boldsymbol{K}_0^{-1} \boldsymbol{x}_0 = \boldsymbol{x}_1^T \boldsymbol{F} \boldsymbol{x}_0 = 0, \tag{7.28}$$

$$\boldsymbol{F} = \boldsymbol{K}_1^{-T} \boldsymbol{E} \boldsymbol{K}_0^{-1} = [\boldsymbol{e}]_\times \tilde{\boldsymbol{H}} \tag{7.29}$$

$$\boldsymbol{F} = [\boldsymbol{e}]_\times \tilde{\boldsymbol{H}} = \boldsymbol{U}\boldsymbol{\Sigma}\boldsymbol{V}^T = \begin{bmatrix} \boldsymbol{u}_0 & \boldsymbol{u}_1 & \boldsymbol{e}_1 \end{bmatrix} \begin{bmatrix} \sigma_0 & & \\ & \sigma_1 & \\ & & 0 \end{bmatrix} \begin{bmatrix} \boldsymbol{v}_0^T \\ \boldsymbol{v}_1^T \\ \boldsymbol{e}_0^T \end{bmatrix}. \tag{7.30}$$

$$\tilde{\boldsymbol{H}} = \boldsymbol{K}_1^{-T} \boldsymbol{R} \boldsymbol{K}_0^{-1}, \tag{7.31}$$

$$\boldsymbol{F} = [\boldsymbol{e}]_\times \tilde{\boldsymbol{H}} = \boldsymbol{S} \boldsymbol{Z} \boldsymbol{R}_{90°} \boldsymbol{S}^T \tilde{\boldsymbol{H}} = \boldsymbol{U}\boldsymbol{\Sigma}\boldsymbol{V}^T \tag{7.32}$$

$$\tilde{\boldsymbol{H}} = \boldsymbol{U} \boldsymbol{R}_{90°}^T \hat{\boldsymbol{\Sigma}} \boldsymbol{V}^T, \tag{7.33}$$

$$\boldsymbol{P}_0 = [\boldsymbol{I}|\boldsymbol{0}] \quad \text{and} \quad \boldsymbol{P}_0 = [\tilde{\boldsymbol{H}}|\boldsymbol{e}], \tag{7.34}$$

$$\frac{\boldsymbol{u}_1^T \boldsymbol{D}_0 \boldsymbol{u}_1}{\sigma_0^2 \boldsymbol{v}_0^T \boldsymbol{D}_1 \boldsymbol{v}_0} = -\frac{\boldsymbol{u}_0^T \boldsymbol{D}_0 \boldsymbol{u}_1}{\sigma_0 \sigma_1 \boldsymbol{v}_0^T \boldsymbol{D}_1 \boldsymbol{v}_1} = \frac{\boldsymbol{u}_0^T \boldsymbol{D}_0 \boldsymbol{u}_0}{\sigma_1^2 \boldsymbol{v}_1^T \boldsymbol{D}_1 \boldsymbol{v}_1}, \tag{7.35}$$

$$\boldsymbol{D}_j = \boldsymbol{K}_j \boldsymbol{K}_j^T = \text{diag}(f_j^2, f_j^2, 1) = \begin{bmatrix} f_j^2 & & \\ & f_j^2 & \\ & & 1 \end{bmatrix} \tag{7.36}$$

$$e_{ij0}(f_0^2) = \boldsymbol{u}_i^T \boldsymbol{D}_0 \boldsymbol{u}_j = a_{ij} + b_{ij} f_0^2, \tag{7.37}$$

$$e_{ij1}(f_1^2) = \sigma_i \sigma_j \boldsymbol{v}_i^T \boldsymbol{D}_1 \boldsymbol{v}_j = c_{ij} + d_{ij} f_1^2. \tag{7.38}$$

$$e_{ij0}(f_0^2) = \lambda e_{ij1}(f_1^2) \tag{7.39}$$

$$\boldsymbol{x}_{ji} = \tilde{\boldsymbol{P}}_j \bar{\boldsymbol{p}}_i, \tag{7.40}$$

$$\bar{\boldsymbol{x}}_j = \frac{1}{N} \sum_i \boldsymbol{x}_{ji} = \tilde{\boldsymbol{P}}_j \frac{1}{N} \sum_i \bar{\boldsymbol{p}}_i = \tilde{\boldsymbol{P}}_j \bar{\boldsymbol{c}}, \tag{7.41}$$

$$\tilde{\boldsymbol{x}}_{ji} = \boldsymbol{M}_j \boldsymbol{p}_i, \tag{7.42}$$

$$\hat{X} = \begin{bmatrix} \tilde{x}_{11} & \cdots & \tilde{x}_{1i} & \cdots & \tilde{x}_{1N} \\ \vdots & & \vdots & & \vdots \\ \tilde{x}_{j1} & \cdots & \tilde{x}_{ji} & \cdots & \tilde{x}_{jN} \\ \vdots & & \vdots & & \vdots \\ \tilde{x}_{M1} & \cdots & \tilde{x}_{Mi} & \cdots & \tilde{x}_{MN} \end{bmatrix} = \begin{bmatrix} M_1 \\ \vdots \\ M_j \\ \vdots \\ M_M \end{bmatrix} \begin{bmatrix} p_1 & \cdots & p_i & \cdots & p_N \end{bmatrix} = \hat{M}\hat{S}. \tag{7.43}$$

$$\hat{X} = U\Sigma V^T = [UQ][Q^{-1}\Sigma V^T] \tag{7.44}$$

$$\begin{aligned} m_{j0} \cdot m_{j0} &= u_{2j}QQ^T u_{2j}^T &= 1, \\ m_{j0} \cdot m_{j1} &= u_{2j}QQ^T u_{2j+1}^T &= 0, \\ m_{j1} \cdot m_{j1} &= u_{2j+1}QQ^T u_{2j+1}^T &= 1, \end{aligned} \tag{7.45}$$

$$x_{ji} = s_j \frac{r_{xj} \cdot p_i + t_{xj}}{1 + \eta_j r_{zj} \cdot p_i} \tag{7.46}$$

$$y_{ji} = s_j \frac{r_{yj} \cdot p_i + t_{yj}}{1 + \eta_j r_{zj} \cdot p_i} \tag{7.47}$$

$$\hat{X} = \begin{bmatrix} d_{11}\tilde{x}_{11} & \cdots & d_{1i}\tilde{x}_{1i} & \cdots & d_{1N}\tilde{x}_{1N} \\ \vdots & & \vdots & & \vdots \\ d_{j1}\tilde{x}_{j1} & \cdots & d_{ji}\tilde{x}_{ji} & \cdots & d_{jN}\tilde{x}_{jN} \\ \vdots & & \vdots & & \vdots \\ d_{M1}\tilde{x}_{M1} & \cdots & d_{Mi}\tilde{x}_{Mi} & \cdots & d_{MN}\tilde{x}_{MN} \end{bmatrix} = \hat{M}\hat{S}. \tag{7.48}$$

$$x_{ij} = f(p_i, R_j, c_j, K_j), \tag{7.49}$$

$$f_{\mathrm{RD}}(x) = (1 + \kappa_1 r^2 + \kappa_2 r^4)x, \tag{7.50}$$

$$\begin{aligned} r_{ij} &= \tilde{x}_{ij} - \hat{x}_{ij}, & (7.51) \\ s_{ij}^2 &= r_{ij}^T \Sigma_{ij}^{-1} r_{ij}, & (7.52) \\ e_{ij} &= \hat{\rho}(s_{ij}^2), & (7.53) \end{aligned}$$

$$\frac{\partial e_{ij}}{\partial s_{ij}^2} = \hat{\rho}'(s_{ij}^2), \tag{7.54}$$

$$\frac{\partial s_{ij}^2}{\partial \tilde{x}_{ij}} = \Sigma_{ij}^{-1} r_{ij}. \tag{7.55}$$

$$A'_{cc} = A_{cc} - A_{pc}^T A_{pp}^{-1} A_{pc}, \tag{7.56}$$

第 8 章 稠密运动估计

$$E_{\text{SSD}}(\boldsymbol{u}) = \sum_i [I_1(\boldsymbol{x}_i + \boldsymbol{u}) - I_0(\boldsymbol{x}_i)]^2 = \sum_i e_i^2, \tag{8.1}$$

$$E_{\text{SRD}}(\boldsymbol{u}) = \sum_i \rho(I_1(\boldsymbol{x}_i + \boldsymbol{u}) - I_0(\boldsymbol{x}_i)) = \sum_i \rho(e_i). \tag{8.2}$$

$$E_{\text{SAD}}(\boldsymbol{u}) = \sum_i |I_1(\boldsymbol{x}_i + \boldsymbol{u}) - I_0(\boldsymbol{x}_i)| = \sum_i |e_i|. \tag{8.3}$$

$$\rho_{\text{GM}}(x) = \frac{x^2}{1 + x^2/a^2}, \tag{8.4}$$

$$E_{\text{WSSD}}(\boldsymbol{u}) = \sum_i w_0(\boldsymbol{x}_i) w_1(\boldsymbol{x}_i + \boldsymbol{u})[I_1(\boldsymbol{x}_i + \boldsymbol{u}) - I_0(\boldsymbol{x}_i)]^2, \tag{8.5}$$

$$A = \sum_i w_0(\boldsymbol{x}_i) w_1(\boldsymbol{x}_i + \boldsymbol{u}) \tag{8.6}$$

$$RMS = \sqrt{E_{\text{WSSD}}/A} \tag{8.7}$$

$$I_1(\boldsymbol{x} + \boldsymbol{u}) = (1 + \alpha) I_0(\boldsymbol{x}) + \beta, \tag{8.8}$$

$$E_{\text{BG}}(\boldsymbol{u}) = \sum_i [I_1(\boldsymbol{x}_i + \boldsymbol{u}) - (1+\alpha)I_0(\boldsymbol{x}_i) - \beta]^2 = \sum_i [\alpha I_0(\boldsymbol{x}_i) + \beta - e_i]^2. \tag{8.9}$$

$$E_{\text{CC}}(\boldsymbol{u}) = \sum_i I_0(\boldsymbol{x}_i) I_1(\boldsymbol{x}_i + \boldsymbol{u}). \tag{8.10}$$

$$E_{\text{NCC}}(\boldsymbol{u}) = \frac{\sum_i [I_0(\boldsymbol{x}_i) - \overline{I_0}]\,[I_1(\boldsymbol{x}_i + \boldsymbol{u}) - \overline{I_1}]}{\sqrt{\sum_i [I_0(\boldsymbol{x}_i) - \overline{I_0}]^2} \sqrt{\sum_i [I_1(\boldsymbol{x}_i + \boldsymbol{u}) - \overline{I_1}]^2}}, \tag{8.11}$$

$$\overline{I_0} = \frac{1}{N}\sum_i I_0(\boldsymbol{x}_i) \quad \text{and} \tag{8.12}$$

$$\overline{I_1} = \frac{1}{N}\sum_i I_1(\boldsymbol{x}_i + \boldsymbol{u}) \tag{8.13}$$

$$E_{\text{NSSD}}(\boldsymbol{u}) = \frac{1}{2}\frac{\sum_i \left[[I_0(\boldsymbol{x}_i) - \overline{I_0}] - [I_1(\boldsymbol{x}_i + \boldsymbol{u}) - \overline{I_1}]\right]^2}{\sqrt{\sum_i [I_0(\boldsymbol{x}_i) - \overline{I_0}]^2 + [I_1(\boldsymbol{x}_i + \boldsymbol{u}) - \overline{I_1}]^2}} \tag{8.14}$$

$$I_k^{(l)}(\boldsymbol{x}_j) \leftarrow \tilde{I}_k^{(l-1)}(2\boldsymbol{x}_j) \tag{8.15}$$

$$\hat{\boldsymbol{u}}^{(l-1)} \leftarrow 2\boldsymbol{u}^{(l)} \tag{8.16}$$

$$\mathcal{F}\{I_1(\boldsymbol{x} + \boldsymbol{u})\} = \mathcal{F}\{I_1(\boldsymbol{x})\}\,e^{-j\boldsymbol{u}\cdot\boldsymbol{\omega}} = \mathcal{I}_1(\boldsymbol{\omega})e^{-j\boldsymbol{u}\cdot\boldsymbol{\omega}}, \tag{8.17}$$

$$\mathcal{F}\{E_{\text{CC}}(\boldsymbol{u})\} = \mathcal{F}\left\{\sum_i I_0(\boldsymbol{x}_i)I_1(\boldsymbol{x}_i + \boldsymbol{u})\right\} = \mathcal{F}\{I_0(\boldsymbol{u})\bar{\ast}I_1(\boldsymbol{u})\} = \mathcal{I}_0(\boldsymbol{\omega})\mathcal{I}_1^*(\boldsymbol{\omega}), \tag{8.18}$$

$$f(\boldsymbol{u})\bar{\ast}g(\boldsymbol{u}) = \sum_i f(\boldsymbol{x}_i)g(\boldsymbol{x}_i + \boldsymbol{u}) \tag{8.19}$$

$$\begin{aligned}\mathcal{F}\{E_{\text{SSD}}(\boldsymbol{u})\} &= \mathcal{F}\left\{\sum_i [I_1(\boldsymbol{x}_i + \boldsymbol{u}) - I_0(\boldsymbol{x}_i)]^2\right\} \\ &= \delta(\boldsymbol{\omega})\sum_i [I_0^2(\boldsymbol{x}_i) + I_1^2(\boldsymbol{x}_i)] - 2\mathcal{I}_0(\boldsymbol{\omega})\mathcal{I}_1^*(\boldsymbol{\omega}).\end{aligned} \tag{8.20}$$

$$\begin{aligned}E_{\text{WCC}}(\boldsymbol{u}) &= \sum_i w_0(\boldsymbol{x}_i)I_0(\boldsymbol{x}_i)\,w_1(\boldsymbol{x}_i + \boldsymbol{u})I_1(\boldsymbol{x}_i + \boldsymbol{u}), \tag{8.21}\\ &= [w_0(\boldsymbol{x})I_0(\boldsymbol{x})]\bar{\ast}[w_1(\boldsymbol{x})I_1(\boldsymbol{x})] \tag{8.22}\end{aligned}$$

$$E_{\text{WSSD}}(\boldsymbol{u}) = \sum_i w_0(\boldsymbol{x}_i)w_1(\boldsymbol{x}_i + \boldsymbol{u})[I_1(\boldsymbol{x}_i + \boldsymbol{u}) - I_0(\boldsymbol{x}_i)]^2. \tag{8.23}$$

$$\mathcal{F}\{E_{\text{PC}}(\boldsymbol{u})\} = \frac{\mathcal{I}_0(\boldsymbol{\omega})\mathcal{I}_1^*(\boldsymbol{\omega})}{\|\mathcal{I}_0(\boldsymbol{\omega})\|\|\mathcal{I}_1(\boldsymbol{\omega})\|} \tag{8.24}$$

$$\begin{aligned}\mathcal{F}\{I_1(\boldsymbol{x}+\boldsymbol{u})\} &= \mathcal{I}_1(\boldsymbol{\omega})e^{-2\pi j u\cdot\boldsymbol{\omega}} = \mathcal{I}_0(\boldsymbol{\omega}) \text{ and}\\ \mathcal{F}\{E_{\text{PC}}(\boldsymbol{u})\} &= e^{-2\pi j u\cdot\boldsymbol{\omega}}.\end{aligned} \qquad (8.25)$$

$$I_1(\hat{\boldsymbol{R}}\boldsymbol{x}) = I_0(\boldsymbol{x}). \qquad (8.26)$$

$$\tilde{I}_0(r,\theta) = I_0(r\cos\theta, r\sin\theta) \text{ and } \tilde{I}_1(r,\theta) = I_1(r\cos\theta, r\sin\theta), \qquad (8.27)$$

$$\tilde{I}_1(r,\theta+\hat{\theta}) = \tilde{I}_0(r,\theta). \qquad (8.28)$$

$$I_1(e^{\hat{s}}\hat{\boldsymbol{R}}\boldsymbol{x}) = I_0(\boldsymbol{x}), \qquad (8.29)$$

$$\tilde{I}_0(s,\theta) = I_0(e^s\cos\theta, e^s\sin\theta) \text{ and } \tilde{I}_1(s,\theta) = I_1(e^s\cos\theta, e^s\sin\theta), \qquad (8.30)$$

$$\tilde{I}_1(s+\hat{s}, \theta+\hat{\theta}) = I_0(s,\theta). \qquad (8.31)$$

$$I_1(e^{\hat{s}}\hat{\boldsymbol{R}}\boldsymbol{x}+\boldsymbol{t}) = I_0(\boldsymbol{x}), \qquad (8.32)$$

$$\begin{aligned}E_{\text{LK-SSD}}(\boldsymbol{u}+\Delta\boldsymbol{u}) &= \sum_i [I_1(\boldsymbol{x}_i+\boldsymbol{u}+\Delta\boldsymbol{u}) - I_0(\boldsymbol{x}_i)]^2 & (8.33)\\ &\approx \sum_i [I_1(\boldsymbol{x}_i+\boldsymbol{u}) + \boldsymbol{J}_1(\boldsymbol{x}_i+\boldsymbol{u})\Delta\boldsymbol{u} - I_0(\boldsymbol{x}_i)]^2 & (8.34)\\ &= \sum_i [\boldsymbol{J}_1(\boldsymbol{x}_i+\boldsymbol{u})\Delta\boldsymbol{u} + e_i]^2, & (8.35)\end{aligned}$$

$$\boldsymbol{J}_1(\boldsymbol{x}_i+\boldsymbol{u}) = \nabla I_1(\boldsymbol{x}_i+\boldsymbol{u}) = \left(\frac{\partial I_1}{\partial x}, \frac{\partial I_1}{\partial y}\right)(\boldsymbol{x}_i+\boldsymbol{u}) \qquad (8.36)$$

$$e_i = I_1(\boldsymbol{x}_i+\boldsymbol{u}) - I_0(\boldsymbol{x}_i), \qquad (8.37)$$

$$I_x u + I_y v + I_t = 0, \qquad (8.38)$$

$$\boldsymbol{A}\Delta\boldsymbol{u} = \boldsymbol{b} \qquad (8.39)$$

$$\boldsymbol{A} = \sum_i \boldsymbol{J}_1^T(\boldsymbol{x}_i+\boldsymbol{u})\boldsymbol{J}_1(\boldsymbol{x}_i+\boldsymbol{u}) \qquad (8.40)$$

$$\boldsymbol{b} = -\sum_i e_i \boldsymbol{J}_1^T(\boldsymbol{x}_i + \boldsymbol{u}) \tag{8.41}$$

$$\boldsymbol{A} = \begin{bmatrix} \sum I_x^2 & \sum I_x I_y \\ \sum I_x I_y & \sum I_y^2 \end{bmatrix} \text{ and } \boldsymbol{b} = -\begin{bmatrix} \sum I_x I_t \\ \sum I_y I_t \end{bmatrix}. \tag{8.42}$$

$$\boldsymbol{J}_1(\boldsymbol{x}_i + \boldsymbol{u}) \approx \boldsymbol{J}_0(\boldsymbol{x}_i), \tag{8.43}$$

$$\Sigma_{\boldsymbol{u}} = \sigma_n^2 \boldsymbol{A}^{-1}, \tag{8.44}$$

$$E_{\text{LK-BG}}(\boldsymbol{u} + \Delta \boldsymbol{u}) = \sum_i [\boldsymbol{J}_1(\boldsymbol{x}_i + \boldsymbol{u})\Delta \boldsymbol{u} + e_i - \alpha I_0(\boldsymbol{x}_i) - \beta]^2 \tag{8.45}$$

$$I_1(\boldsymbol{x} + \boldsymbol{u}) \approx I_0(\boldsymbol{x}) + \sum_j \lambda_j B_j(\boldsymbol{x}), \tag{8.46}$$

$$E_{\text{LK-WSSD}}(\boldsymbol{u} + \Delta \boldsymbol{u}) = \sum_i w_0(\boldsymbol{x}_i) w_1(\boldsymbol{x}_i + \boldsymbol{u})[\boldsymbol{J}_1(\boldsymbol{x}_i + \boldsymbol{u})\Delta \boldsymbol{u} + e_i]^2. \tag{8.47}$$

$$E_{\text{LK-SRD}}(\boldsymbol{u} + \Delta \boldsymbol{u}) = \sum_i \rho(\boldsymbol{J}_1(\boldsymbol{x}_i + \boldsymbol{u})\Delta \boldsymbol{u} + e_i), \tag{8.48}$$

$$E_{\text{LK-PM}}(\boldsymbol{p} + \Delta \boldsymbol{p}) = \sum_i [I_1(\boldsymbol{x}'(\boldsymbol{x}_i; \boldsymbol{p} + \Delta \boldsymbol{p})) - I_0(\boldsymbol{x}_i)]^2 \tag{8.49}$$

$$\approx \sum_i [I_1(\boldsymbol{x}'_i) + \boldsymbol{J}_1(\boldsymbol{x}'_i)\Delta \boldsymbol{p} - I_0(\boldsymbol{x}_i)]^2 \tag{8.50}$$

$$= \sum_i [\boldsymbol{J}_1(\boldsymbol{x}'_i)\Delta \boldsymbol{p} + e_i]^2, \tag{8.51}$$

$$\boldsymbol{J}_1(\boldsymbol{x}'_i) = \frac{\partial I_1}{\partial \boldsymbol{p}} = \nabla I_1(\boldsymbol{x}'_i) \frac{\partial \boldsymbol{x}'}{\partial \boldsymbol{p}}(\boldsymbol{x}_i), \tag{8.52}$$

$$\boldsymbol{A} = \sum_i \boldsymbol{J}_{\boldsymbol{x}'}^T(\boldsymbol{x}_i)[\nabla I_1^T(\boldsymbol{x}'_i)\nabla I_1(\boldsymbol{x}'_i)]\boldsymbol{J}_{\boldsymbol{x}'}(\boldsymbol{x}_i) \tag{8.53}$$

$$\boldsymbol{b} = -\sum_i \boldsymbol{J}_{\boldsymbol{x}'}^T(\boldsymbol{x}_i)[e_i \nabla I_1^T(\boldsymbol{x}'_i)]. \tag{8.54}$$

$$\boldsymbol{A}_j = \sum_{i \in P_j} \nabla I_1^T(\boldsymbol{x}'_i)\nabla I_1(\boldsymbol{x}'_i) \tag{8.55}$$

$$\boldsymbol{b}_j = \sum_{i \in P_j} e_i \nabla I_1^T(\boldsymbol{x}'_i). \tag{8.56}$$

$$A \approx \sum_j J_{x'}^T(\hat{x}_j)[\sum_{i \in P_j} \nabla I_1^T(x'_i) \nabla I_1(x'_i)] J_{x'}(\hat{x}_j) = \sum_j J_{x'}^T(\hat{x}_j) A_j J_{x'}(\hat{x}_j) \tag{8.57}$$

$$b \approx -\sum_j J_{x'}^T(\hat{x}_j)[\sum_{i \in P_j} e_i \nabla I_1^T(x'_i)] = -\sum_j J_{x'}^T(\hat{x}_j) b_j, \tag{8.58}$$

$$\tilde{I}_1(x) = I_1(x'(x;p)), \tag{8.59}$$

$$E_{\text{LK-SS}}(\Delta p) = \sum_i [\tilde{I}_1(\tilde{x}(x_i; \Delta p)) - I_0(x_i)]^2 \tag{8.60}$$
$$\approx \sum_i [\tilde{J}_1(x_i)\Delta p + e_i]^2 \tag{8.61}$$
$$= \sum_i [\nabla \tilde{I}_1(x_i) J_{\tilde{x}}(x_i)\Delta p + e_i]^2. \tag{8.62}$$

$$J_{\tilde{x}} = \frac{\partial \tilde{x}}{\partial p}\Big|_{p=0} = \begin{bmatrix} x & y & 1 & 0 & 0 & 0 & -x^2 & -xy \\ 0 & 0 & 0 & x & y & 1 & -xy & -y^2 \end{bmatrix}. \tag{8.63}$$

$$E_{\text{LK-BM}}(\Delta p) = \sum_i [\tilde{I}_1(x_i) - I_0(\tilde{x}(x_i; \Delta p))]^2 \tag{8.64}$$
$$\approx \sum_i [\nabla I_0(x_i) J_{\tilde{x}}(x_i)\Delta p - e_i]^2. \tag{8.65}$$

$$u(x) = \sum_k a_k v_k(x). \tag{8.66}$$

$$E_{\text{SSD-OF}}(\{u_i\}) = \sum_i [I_1(x_i + u_i) - I_0(x_i)]^2. \tag{8.67}$$

$$u_i = \sum_j \hat{u}_j B_j(x_i) = \sum_j \hat{u}_j w_{i,j}, \tag{8.68}$$

$$E_{\text{SSD-OF}}(\{u_i\}) = \sum_i [I_1(x_i + u_i) - I_0(x_i)]^2. \tag{8.69}$$

$$E_{\text{HS}} = \int (I_x u + I_y v + I_t)^2 \, dx \, dy, \tag{8.70}$$

$$E_{\text{HSD}} = \sum_i u_i^T [J_i J_i^T] u_i + 2e_i J_i^T u_i + e_i^2. \tag{8.71}$$

$$I_t(x_0 + tu_0) = (1-t)I_0(x_0) + tI_1(x_0 + u_0). \tag{8.72}$$

第 9 章 图像拼接

$$\tilde{x}_1 \sim \tilde{P}_1 \tilde{P}_0^{-1} \tilde{x}_0 = M_{10} \tilde{x}_0. \tag{9.1}$$

$$\tilde{x}_1 \sim \tilde{H}_{10} \tilde{x}_0. \tag{9.2}$$

$$x_1 = \frac{(1+h_{00})x_0 + h_{01}y_0 + h_{02}}{h_{20}x_0 + h_{21}y_0 + 1} \text{ and } y_1 = \frac{h_{10}x_0 + (1+h_{11})y_0 + h_{12}}{h_{20}x_0 + h_{21}y_0 + 1}, \tag{9.3}$$

$$x_{k+1} = R_k x_k + t_k, \tag{9.4}$$

$$\tilde{H}_{10} = K_1 R_1 R_0^{-1} K_0^{-1} = K_1 R_{10} K_0^{-1}, \tag{9.5}$$

$$\begin{bmatrix} x_1 \\ y_1 \\ 1 \end{bmatrix} \sim \begin{bmatrix} f_1 & & \\ & f_1 & \\ & & 1 \end{bmatrix} R_{10} \begin{bmatrix} f_0^{-1} & & \\ & f_0^{-1} & \\ & & 1 \end{bmatrix} \begin{bmatrix} x_0 \\ y_0 \\ 1 \end{bmatrix} \tag{9.6}$$

$$\begin{bmatrix} x_1 \\ y_1 \\ f_1 \end{bmatrix} \sim R_{10} \begin{bmatrix} x_0 \\ y_0 \\ f_0 \end{bmatrix}, \tag{9.7}$$

$$\tilde{H}(\omega) = K_1 R(\omega) R_{10} K_0^{-1} = [K_1 R(\omega) K_1^{-1}][K_1 R_{10} K_0^{-1}] = D\tilde{H}_{10}. \tag{9.8}$$

$$D = K_1 R(\omega) K_1^{-1} \approx K_1 (I + [\omega]_\times) K_1^{-1} = \begin{bmatrix} 1 & -\omega_z & f_1 \omega_y \\ \omega_z & 1 & -f_1 \omega_x \\ -\omega_y/f_1 & \omega_x/f_1 & 1 \end{bmatrix}. \tag{9.9}$$

$$(h_{00}, h_{01}, h_{02}, h_{00}, h_{11}, h_{12}, h_{20}, h_{21}) = (0, -\omega_z, f_1\omega_y, \omega_z, 0, -f_1\omega_x, -\omega_y/f_1, \omega_x/f_1). \tag{9.10}$$

$$\begin{bmatrix} \hat{x}' - x \\ \hat{y}' - y \end{bmatrix} = \begin{bmatrix} -xy/f_1 & f_1 + x^2/f_1 & -y \\ -(f_1 + y^2/f_1) & xy/f_1 & x \end{bmatrix} \begin{bmatrix} \omega_x \\ \omega_y \\ \omega_z \end{bmatrix}, \tag{9.11}$$

$$(\sin\theta, h, \cos\theta) \propto (x, y, f), \tag{9.12}$$

$$x' = s\theta = s\tan^{-1}\frac{x}{f}, \tag{9.13}$$
$$y' = sh = s\frac{y}{\sqrt{x^2+f^2}}, \tag{9.14}$$

$$x = f\tan\theta = f\tan\frac{x'}{s}, \tag{9.15}$$
$$y = h\sqrt{x^2+f^2} = \frac{y'}{s}f\sqrt{1+\tan^2 x'/s} = f\frac{y'}{s}\sec\frac{x'}{s}. \tag{9.16}$$

$$(\sin\theta\cos\phi, \sin\phi, \cos\theta\cos\phi) \propto (x, y, f), \tag{9.17}$$

$$x' = s\theta = s\tan^{-1}\frac{x}{f}, \tag{9.18}$$
$$y' = s\phi = s\tan^{-1}\frac{y}{\sqrt{x^2+f^2}}, \tag{9.19}$$

$$x = f\tan\theta = f\tan\frac{x'}{s}, \tag{9.20}$$
$$y = \sqrt{x^2+f^2}\tan\phi = \tan\frac{y'}{s}f\sqrt{1+\tan^2 x'/s} = f\tan\frac{y'}{s}\sec\frac{x'}{s}. \tag{9.21}$$

$$(\cos\theta\sin\phi, \sin\theta\sin\phi, \cos\phi) = s(x, y, z). \tag{9.22}$$

$$x' = s\phi\cos\theta = s\frac{x}{r}\tan^{-1}\frac{r}{z}, \tag{9.23}$$
$$y' = s\phi\sin\theta = s\frac{y}{r}\tan^{-1}\frac{r}{z}, \tag{9.24}$$

$$E_{\text{pairwise-LS}} = \sum_i \|\boldsymbol{r}_i\|^2 = \|\tilde{\boldsymbol{x}}'_i(\boldsymbol{x}_i; \boldsymbol{p}) - \hat{\boldsymbol{x}}'_i\|^2. \tag{9.25}$$

$$\tilde{\boldsymbol{x}}_{ij} \sim \boldsymbol{K}_j \boldsymbol{R}_j \boldsymbol{x}_i \text{ and } \boldsymbol{x}_i \sim \boldsymbol{R}_j^{-1} \boldsymbol{K}_j^{-1} \tilde{\boldsymbol{x}}_{ij}, \qquad (9.26)$$

$$\tilde{\boldsymbol{x}}_{ik} \sim \tilde{\boldsymbol{H}}_{kj} \tilde{\boldsymbol{x}}_{ij} = \boldsymbol{K}_k \boldsymbol{R}_k \boldsymbol{R}_j^{-1} \boldsymbol{K}_j^{-1} \tilde{\boldsymbol{x}}_{ij}. \qquad (9.27)$$

$$E_{\text{all-pairs-2D}} = \sum_i \sum_{jk} c_{ij} c_{ik} \| \tilde{\boldsymbol{x}}_{ik}(\hat{\boldsymbol{x}}_{ij}; \boldsymbol{R}_j, f_j, \boldsymbol{R}_k, f_k) - \hat{\boldsymbol{x}}_{ik} \|^2, \qquad (9.28)$$

$$E_{\text{BA-2D}} = \sum_i \sum_j c_{ij} \| \tilde{\boldsymbol{x}}_{ij}(\boldsymbol{x}_i; \boldsymbol{R}_j, f_j) - \hat{\boldsymbol{x}}_{ij} \|^2, \qquad (9.29)$$

$$E_{\text{BA-3D}} = \sum_i \sum_j c_{ij} \| \tilde{\boldsymbol{x}}_i(\hat{\boldsymbol{x}}_{ij}; \boldsymbol{R}_j, f_j) - \boldsymbol{x}_i \|^2, \qquad (9.30)$$

$$E_{\text{all-pairs-3D}} = \sum_i \sum_{jk} c_{ij} c_{ik} \| \tilde{\boldsymbol{x}}_i(\hat{\boldsymbol{x}}_{ij}; \boldsymbol{R}_j, f_j) - \tilde{\boldsymbol{x}}_i(\hat{\boldsymbol{x}}_{ik}; \boldsymbol{R}_k, f_k) \|^2. \qquad (9.31)$$

$$\tilde{\boldsymbol{x}}_{ik} \sim \boldsymbol{K}_k \boldsymbol{R}_k \boldsymbol{x}_i. \qquad (9.32)$$

$$\hat{\boldsymbol{\imath}}^T \boldsymbol{R}_k \boldsymbol{R}_{\text{g}} \hat{\boldsymbol{\jmath}} = 0 \qquad (9.33)$$

$$\boldsymbol{r}_{\text{g}1} = \arg\min_{\boldsymbol{r}} \sum_k (\boldsymbol{r}^T \boldsymbol{r}_{k0})^2 = \arg\min_{\boldsymbol{r}} \boldsymbol{r}^T \left[\sum_k \boldsymbol{r}_{k0} \boldsymbol{r}_{k0}^T \right] \boldsymbol{r}. \qquad (9.34)$$

$$\bar{\boldsymbol{x}}_i \sim \sum_j c_{ij} \tilde{\boldsymbol{x}}_i(\hat{\boldsymbol{x}}_{ij}; \boldsymbol{R}_j, f_j) \bigg/ \sum_j c_{ij}, \qquad (9.35)$$

$$\boldsymbol{u}_{ij} = \bar{\boldsymbol{x}}_{ij} - \boldsymbol{x}_{ij}, \qquad (9.36)$$

$$C(\boldsymbol{x}) = \sum_k w_k(\boldsymbol{x}) \tilde{I}_k(\boldsymbol{x}) \bigg/ \sum_k w_k(\boldsymbol{x}), \qquad (9.37)$$

$$w_k(\boldsymbol{x}) = \arg\min_{\boldsymbol{y}} \{ \|\boldsymbol{y}\| \mid \tilde{I}_k(\boldsymbol{x} + \boldsymbol{y}) \text{ is invalid} \}, \qquad (9.38)$$

$$C(\boldsymbol{x}) = \tilde{I}_{l(\boldsymbol{x})}(\boldsymbol{x}), \qquad (9.39)$$

$$l = \arg\max_k w_k(\boldsymbol{x}) \qquad (9.40)$$

$$\mathcal{C}_D = \sum_{\boldsymbol{x}} D(\boldsymbol{x}, l(\boldsymbol{x})), \qquad (9.41)$$

$$\mathcal{C}_S = \sum_{(\boldsymbol{x},\boldsymbol{y})\in\mathcal{N}} S(\boldsymbol{x},\boldsymbol{y},l(\boldsymbol{x}),l(\boldsymbol{y})), \tag{9.42}$$

$$S(\boldsymbol{x},\boldsymbol{y},l_x,l_y) = \|\tilde{I}_{l_x}(\boldsymbol{x}) - \tilde{I}_{l_y}(\boldsymbol{x})\| + \|\tilde{I}_{l_x}(\boldsymbol{y}) - \tilde{I}_{l_y}(\boldsymbol{y})\|. \tag{9.43}$$

$$\min_{C(\boldsymbol{x})} \|\nabla C(\boldsymbol{x}) - \nabla \tilde{I}_{l(\boldsymbol{x})}(\boldsymbol{x})\|^2. \tag{9.44}$$

$$\begin{align}
C(\boldsymbol{x}+\hat{\boldsymbol{\imath}}) - C(\boldsymbol{x}) &= \tilde{I}_{l(\boldsymbol{x})}(\boldsymbol{x}+\hat{\boldsymbol{\imath}}) - \tilde{I}_{l(\boldsymbol{x})}(\boldsymbol{x}) \text{ and} \tag{9.45}\\
C(\boldsymbol{x}+\hat{\boldsymbol{\jmath}}) - C(\boldsymbol{x}) &= \tilde{I}_{l(\boldsymbol{x})}(\boldsymbol{x}+\hat{\boldsymbol{\jmath}}) - \tilde{I}_{l(\boldsymbol{x})}(\boldsymbol{x}), \tag{9.46}
\end{align}$$

第 10 章 计算摄影学

$$K = \arg\min_{K} \|B - D(I * K)\|^2, \tag{10.1}$$

$$b = \arg\min_{b} \|o - D(s * b)\|^2, \tag{10.2}$$

$$z_{ij} = f(E_i\, t_j), \tag{10.3}$$

$$f^{-1}(z_{ij}) = E_i\, t_j. \tag{10.4}$$

$$g(z_{ij}) = \log f^{-1}(z_{ij}) = \log E_i + \log t_j, \tag{10.5}$$

$$\lambda \sum_k g''(k)^2 = \lambda \sum [g(k-1) - 2g(k) + g(k+1)]^2, \tag{10.6}$$

$$w(z) = \begin{cases} z - z_{\min} & z \le (z_{\min} + z_{\max})/2 \\ z_{\max} - z & z > (z_{\min} + z_{\max})/2. \end{cases} \tag{10.7}$$

$$E = \sum_i \sum_j w(z_{i,j})[g(z_{i,j}) - \log E_i - \log t_j]^2 + \lambda \sum_k w(k) g''(k)^2. \tag{10.8}$$

$$f(E) = \alpha + \beta E^{\gamma}, \tag{10.9}$$

$$w(z) = g(z)/g'(z), \tag{10.10}$$

$$\log E_i = \frac{\sum_j w(z_{ij})[g(z_{ij}) - \log t_j]}{\sum_j w(z_{ij})}. \tag{10.11}$$

$$H(x,y) = \log L(x,y) \tag{10.12}$$

$$H_{\text{L}}(x,y) = B(x,y) * H(x,y), \tag{10.13}$$

$$H_{\text{H}}(x,y) = H(x,y) - H_{\text{L}}(x,y). \tag{10.14}$$

$$H'_{\text{H}}(x,y) = s\,H_{\text{H}}(x,y) \tag{10.15}$$

$$I(x,y) = H'_{\text{H}}(x,y) + H_{\text{L}}(x,y), \tag{10.16}$$

$$H'(x,y) = \nabla H(x,y). \tag{10.17}$$

$$G(x,y) = H'(x,y)\,\Phi(x,y). \tag{10.18}$$

$$\min \int\int \|\nabla I(x,y) - G(x,y)\|^2 dx\,dy \tag{10.19}$$

$$C_{\text{out}} = \left(\frac{C_{\text{in}}}{L_{\text{in}}}\right)^s L_{\text{out}}, \tag{10.20}$$

$$\min \int\int w_{\text{d}}(x,y)\|f(x,y) - g(x,y)\|^2 dx\,dy + \lambda \int\int w_{\text{s}}(x,y)\|\nabla f(x,y)\|^2 dx\,dy, \tag{10.21}$$

$$w_{\text{s}} = \frac{1}{\|\nabla H\|^\alpha + \epsilon} \tag{10.22}$$

$$r(i,j,k,l) = \exp\left(-\frac{\|f(i,j) - f(k,l)\|^2}{2\sigma_r^2}\right) \tag{10.23}$$

$$F^{Detail} = \frac{F + \epsilon}{F^{\text{Base}} + \epsilon}, \tag{10.24}$$

$$A^{Final} = (1-M)A^{NR}F^{Detail} + MA^{Base} \tag{10.25}$$

$$o_k(\boldsymbol{x}) = D\{b(\boldsymbol{x}) * s(\hat{\boldsymbol{h}}_k(\boldsymbol{x}))\} + n_k(\boldsymbol{x}), \tag{10.26}$$

$$\sum_k \|o_k(\boldsymbol{x}) - D\{b_k(\boldsymbol{x}) * s(\hat{\boldsymbol{h}}_k(\boldsymbol{x}))\}\|^2. \tag{10.27}$$

$$\sum_k \|\boldsymbol{o}_k - \boldsymbol{D}\boldsymbol{B}_k\boldsymbol{W}_k\boldsymbol{s}\|^2. \tag{10.28}$$

$$\sum_{(i,j)} \rho_p(s(i,j) - s(i+1,j)) + \rho_p(s(i,j) - s(i,j+1)). \tag{10.29}$$

$$C = (1-\alpha)B + \alpha F. \tag{10.30}$$

$$C - \alpha F = (1-\alpha)B, \tag{10.31}$$

$$P(F, B, \alpha | C) = P(C|F, B, \alpha) P(F) P(B) P(\alpha) / P(C). \tag{10.32}$$

$$L(F, B, \alpha | C) = L(C|F, B, \alpha) + L(F) + L(B) + L(\alpha) \tag{10.33}$$

$$L(C) = \tfrac{1}{2} \|C - [\alpha F + (1-\alpha)B]\|^2 / \sigma_C^2, \tag{10.34}$$

$$L(F) = (F - \overline{F})^T \Sigma_F^{-1} (F - \overline{F}). \tag{10.35}$$

$$\nabla \alpha = \frac{F - B}{\|F - B\|^2} \cdot \nabla C, \tag{10.36}$$

$$\alpha_i = \boldsymbol{a}_k \cdot (\boldsymbol{C}_i - \boldsymbol{B}_0) = \boldsymbol{a}_k \cdot \boldsymbol{C} + b_k, \tag{10.37}$$

$$E_\alpha = \sum_k \left(\sum_{i \in W_k} (\alpha_i - \boldsymbol{a}_k \cdot \boldsymbol{C}_i - b_k)^2 + \epsilon \|\boldsymbol{a}_k\| \right), \tag{10.38}$$

$$E_\alpha = \boldsymbol{\alpha}^T \boldsymbol{L} \boldsymbol{\alpha}, \tag{10.39}$$

$$L_{ij} = \sum_{k: i \in W_k \wedge j \in W_k} \left(\delta_{ij} - \frac{1}{M} \left(1 + (\boldsymbol{C}_i - \boldsymbol{\mu}_k)^T \hat{\boldsymbol{\Sigma}}_k^{-1} (\boldsymbol{C}_j - \boldsymbol{\mu}_k) \right) \right), \tag{10.40}$$

$$E_{B,F} = \sum_i \|C_i - [\alpha + F_i + (1-\alpha_i)B_i]\|^2 + \lambda |\nabla \alpha_i| (\|\nabla F_i\|^2 + \|\nabla B_i\|^2), \tag{10.41}$$

$$f(r) = 1 + \alpha_1 r + \alpha_2 r^2 + \cdots \tag{10.42}$$

第 11 章 立体视觉对应

$$d = f\frac{B}{Z}, \tag{11.1}$$

$$x' = x + d(x,y), \ y' = y \tag{11.2}$$

$$\tilde{\boldsymbol{x}}_k \sim \tilde{\boldsymbol{P}}_k \tilde{\boldsymbol{P}}^{-1} \boldsymbol{x}_s = \tilde{\boldsymbol{H}}_k \tilde{\boldsymbol{x}} + \boldsymbol{t}_k d = (\tilde{\boldsymbol{H}}_k + \boldsymbol{t}_k [0\ 0\ d])\tilde{\boldsymbol{x}}, \tag{11.3}$$

$$C(x,y,d) = \sum_k \rho(\tilde{I}(x,y,d,k) - I_r(x,y)). \tag{11.4}$$

$$c_i x_c + s_i y_c + r = d_i, \tag{11.5}$$

$$C(x,y,d) = w(x,y,d) * C_0(x,y,d), \tag{11.6}$$

$$Var(d) = \frac{\sigma_I^2}{a}, \tag{11.7}$$

$$E(d) = E_d(d) + \lambda E_s(d). \tag{11.8}$$

$$E_d(d) = \sum_{(x,y)} C(x,y,d(x,y)), \tag{11.9}$$

$$E_s(d) = \sum_{(x,y)} \rho(d(x,y) - d(x+1,y)) + \rho(d(x,y) - d(x,y+1)), \tag{11.10}$$

$$\rho_d(d(x,y) - d(x+1,y)) \cdot \rho_I(\|I(x,y) - I(x+1,y)\|), \tag{11.11}$$

$$C'(m,n) = C(m+n, m-n, y) \tag{11.12}$$

$$\begin{aligned}
D(m,n,M) &= \min(D(m{-}1,n{-}1,M), D(m{-}1,n,L), D(m{-}1,n{-}1,R)) + C'(m,n) \\
D(m,n,L) &= \min(D(m{-}1,n{-}1,M), D(m{-}1,n,L)) + O \\
D(m,n,R) &= \min(D(m,n{-}1,M), D(m,n{-}1,R)) + O,
\end{aligned} \qquad (11.13)$$

$$D(x,y,d) = C(x,y,d) + \min_{d'}\left\{D(x-1,y,d') + \rho_d(d-d')\right\}. \qquad (11.14)$$

$$C(x,y,d) = \sum_k \rho(\tilde{I}(x,y,d,k) - I_r(x,y)). \qquad (11.15)$$

第 12 章 3D 重建

$$I(x,y) = R(p(x,y), q(x,y)), \tag{12.1}$$

$$R(p,q) = \max\left(0, \rho\frac{pv_x + qv_y + v_z}{\sqrt{1+p^2+q^2}}\right), \tag{12.2}$$

$$\mathcal{E}_s = \int p_x^2 + p_y^2 + q_x^2 + q_y^2 \, dx \, dy = \int \|\nabla p\|^2 + \|\nabla q\|^2 \, dx \, dy, \tag{12.3}$$

$$\mathcal{E}_i = \int (p_y - q_x)^2 \, dx \, dy, \tag{12.4}$$

$$I_k = \rho \hat{\boldsymbol{n}} \cdot \boldsymbol{v}_k, \tag{12.5}$$

$$\boldsymbol{f}(\boldsymbol{x}) = \frac{\sum_i w_i(\boldsymbol{x}) \boldsymbol{d}_i}{\sum_i w_i(\boldsymbol{x})}, \tag{12.6}$$

$$w_i(\boldsymbol{x}) = K(\|\boldsymbol{x} - \boldsymbol{x}_i\|), \tag{12.7}$$

$$F(x,y,z) = \left(\left(\frac{x}{a_1}\right)^{2/\epsilon_2} + \left(\frac{y}{a_2}\right)^{2/\epsilon_2}\right)^{\epsilon_2/\epsilon_1} + \left(\frac{x}{a_1}\right)^{2/\epsilon_1} - 1 = 0 \tag{12.8}$$

$$f_r(\theta_i, \phi_i, \theta_r, \phi_r; \lambda), \tag{12.9}$$

$$f_v(x, y, \theta_i, \phi_i, \theta_r, \phi_r; \lambda). \tag{12.10}$$

$$f_e(x_i, y_i, \theta_i, \phi_i, x_e, y_e, \theta_e, \phi_e; \lambda), \tag{12.11}$$

第 13 章 基于图像的绘制

$$C_i = \alpha_i F_i + (1 - \alpha_i) B_i, \tag{13.1}$$

$$I_i = R_i \int A_i(\boldsymbol{x}) B(\boldsymbol{x}) d\boldsymbol{x}, \tag{13.2}$$

$$I_i = \sum_j R_{ij} \int G_{ij}(\boldsymbol{x}) B(\boldsymbol{x}) d\boldsymbol{x}, \tag{13.3}$$

$$G_{ij}(\boldsymbol{x}) = G_{2\mathrm{D}}(\boldsymbol{x}; \boldsymbol{c}_{ij}, \boldsymbol{\sigma}_{ij}, \theta_{ij}) \tag{13.4}$$

第 14 章 识别

$$h(\boldsymbol{x}) = \text{sign}\left[\sum_{j=0}^{m-1} \alpha_j h_j(\boldsymbol{x})\right], \tag{14.1}$$

$$h_j(\boldsymbol{x}) = a_j[f_j < \theta_j] + b_j[f_j \geq \theta_j] = \begin{cases} a_j & \text{if } f_j < \theta_j \\ b_j & \text{otherwise,} \end{cases} \tag{14.2}$$

$$e_j = \sum_{i=0}^{N-1} w_{i,j} e_{i,j}, \tag{14.3}$$

$$e_{i,j} = 1 - \delta(y_i, h_j(\boldsymbol{x}_i; f_j, \theta_j, s_j)). \tag{14.4}$$

$$\beta_j = \frac{e_j}{1-e_j} \quad \text{and} \quad \alpha_j = -\log \beta_j. \tag{14.5}$$

$$w_{i,j+1} \leftarrow w_{i,j} \beta_j^{1-e_{i,j}}, \tag{14.6}$$

$$h(\boldsymbol{x}) = \text{sign}\left[\sum_{j=0}^{m-1} \alpha_j h_j(\boldsymbol{x})\right]. \tag{14.7}$$

$$\tilde{\boldsymbol{x}} = \boldsymbol{m} + \sum_{i=0}^{M-1} a_i \boldsymbol{u}_i, \tag{14.8}$$

$$\boldsymbol{C} = \frac{1}{N} \sum_{j=0}^{N-1} (\boldsymbol{x}_j - \boldsymbol{m})(\boldsymbol{x}_j - \boldsymbol{m})^T. \tag{14.9}$$

$$\boldsymbol{C} = \boldsymbol{U}\boldsymbol{\Lambda}\boldsymbol{U}^T = \sum_{i=0}^{N-1} \lambda_i \boldsymbol{u}_i \boldsymbol{u}_i^T, \tag{14.10}$$

$$a_i = (\boldsymbol{x} - \boldsymbol{m}) \cdot \boldsymbol{u}_i, \tag{14.11}$$

$$\text{DIFS} = \|\tilde{\boldsymbol{x}} - \boldsymbol{m}\| = \sqrt{\sum_{i=0}^{M-1} a_i^2}, \tag{14.12}$$

$$\text{DIFS}(\boldsymbol{x}, \boldsymbol{y}) = \|\tilde{\boldsymbol{x}} - \tilde{\boldsymbol{y}}\| = \sqrt{\sum_{i=0}^{M-1} (a_i - b_i)^2}, \tag{14.13}$$

$$\text{DIFS}' = \|\tilde{\boldsymbol{x}} - \boldsymbol{m}\|_{\boldsymbol{C}^{-1}} = \sqrt{\sum_{i=0}^{M-1} a_i^2/\lambda_i^2}. \tag{14.14}$$

$$\hat{\boldsymbol{U}} = \boldsymbol{U}\boldsymbol{\Lambda}^{-1/2}. \tag{14.15}$$

$$\|\boldsymbol{x} - \boldsymbol{x}_k\| = \|\boldsymbol{a} - \boldsymbol{a}_k\|, \tag{14.16}$$

$$\boldsymbol{S}_{\text{W}} = \sum_{k=0}^{K-1} \boldsymbol{S}_k = \sum_{k=0}^{K-1} \sum_{i \in C_k} (\boldsymbol{x}_i - \boldsymbol{m}_k)(\boldsymbol{x}_i - \boldsymbol{m}_k)^T, \tag{14.17}$$

$$\boldsymbol{S}_{\text{B}} = \sum_{k=0}^{K-1} N_k (\boldsymbol{m}_k - \boldsymbol{m})(\boldsymbol{m}_k - \boldsymbol{m})^T, \tag{14.18}$$

$$\boldsymbol{S}_{\text{W}} = 3N \begin{bmatrix} 0.246 & 0.183 \\ 0.183 & 0.457 \end{bmatrix} \quad \text{and} \quad \boldsymbol{S}_{\text{B}} = N \begin{bmatrix} 6.125 & 0 \\ 0 & 0.375 \end{bmatrix}, \tag{14.19}$$

$$\boldsymbol{u}^* = \arg\max_{\boldsymbol{u}} \frac{\boldsymbol{u}^T \boldsymbol{S}_{\text{B}} \boldsymbol{u}}{\boldsymbol{u}^T \boldsymbol{S}_{\text{W}} \boldsymbol{u}}, \tag{14.20}$$

$$\boldsymbol{S}_{\text{B}} \boldsymbol{u} = \lambda \boldsymbol{S}_{\text{W}} \boldsymbol{u} \quad \text{or} \quad \lambda \boldsymbol{u} = \boldsymbol{S}_{\text{W}}^{-1} \boldsymbol{S}_{\text{B}} \boldsymbol{u}. \tag{14.21}$$

$$\boldsymbol{S}_{\text{W}}^{-1} \boldsymbol{S}_{\text{B}} = \begin{bmatrix} 11.796 & -0.289 \\ -4.715 & 0.3889 \end{bmatrix} \quad \text{and} \quad \boldsymbol{u} = \begin{bmatrix} 0.926 \\ -0.379 \end{bmatrix} \tag{14.22}$$

$$\boldsymbol{\Delta}_{ij} = \boldsymbol{x}_i - \boldsymbol{x}_j \tag{14.23}$$

$$p_I(\boldsymbol{\Delta}_i) = p_{\mathcal{N}}(\boldsymbol{\Delta}_i; \boldsymbol{\Sigma}_I) = \frac{1}{|2\pi\boldsymbol{\Sigma}_I|^{1/2}} \exp -\|\boldsymbol{\Delta}_i\|^2_{\boldsymbol{\Sigma}_I^{-1}}, \tag{14.24}$$

$$|2\pi\boldsymbol{\Sigma}_I|^{1/2} = (2\pi)^{M/2} \prod_{j=1}^{M} \lambda_j^{1/2} \tag{14.25}$$

$$\|\mathbf{\Delta}_i\|_{\mathbf{\Sigma}_I^{-1}}^2 = \mathbf{\Delta}_i^T \mathbf{\Sigma}_I^{-1} \mathbf{\Delta}_i = \|\mathbf{a}^I - \mathbf{a}_i^I\|^2 \tag{14.26}$$

$$\mathbf{a}^I = \hat{\mathbf{U}}^I \mathbf{x} \tag{14.27}$$

$$p(\mathbf{\Delta}_i) = \frac{p_I(\mathbf{\Delta}_i) l_I}{p_I(\mathbf{\Delta}_i) l_I + p_E(\mathbf{\Delta}_i) l_E}, \tag{14.28}$$

$$\begin{aligned} s &= \bar{s} + \mathbf{U}_s \mathbf{a} & (14.29) \\ t &= \bar{t} + \mathbf{U}_t \mathbf{a}, & (14.30) \end{aligned}$$

$$\mathbf{W} = \left[\frac{\partial \mathbf{x}^T}{\partial \mathbf{a}} \frac{\partial \mathbf{x}}{\partial \mathbf{a}}\right]^{-1} \frac{\partial \mathbf{x}^T}{\partial \mathbf{a}}, \tag{14.31}$$

$$d(\mathbf{x}_0, \mathbf{x}_1) = \|\mathbf{x}_0 - \mathbf{x}_1\|_{\mathbf{\Sigma}^{-1}} = \sqrt{(\mathbf{x}_0 - \mathbf{x}_1)^T \mathbf{\Sigma}^{-1} (\mathbf{x}_0 - \mathbf{x}_1)}. \tag{14.32}$$

$$t_i = \frac{n_{id}}{n_d} \log \frac{N}{N_i}. \tag{14.33}$$

$$\mathbf{t} = (t_1, \ldots, t_i, \ldots t_m). \tag{14.34}$$

$$\|\mathbf{t}_q - \mathbf{t}_j\|_p^p = 2 + \sum_{i|t_{iq}>0 \wedge t_{ij}>0} (|t_{iq} - t_{ij}|^p - |t_{iq}|^p - |t_{ij}|^p). \tag{14.35}$$

$$S = ((t_1, \mathbf{m}_1), \ldots, (t_m, \mathbf{m}_m)), \tag{14.36}$$

$$EMD(S, S') = \frac{\sum_i \sum_j f_{ij} d(\mathbf{m}_i, \mathbf{m}_j')}{\sum_i \sum_j f_{ij}}, \tag{14.37}$$

$$\chi^2(S, S') = \frac{1}{2} \sum_i \frac{(t_i - t_i')^2}{t_i + t_i'}, \tag{14.38}$$

$$K(S, S') = \exp\left(-\frac{1}{A} D(S, S')\right), \tag{14.39}$$

$$C_l = \sum_i \min(B_{il}, B_{il}') \tag{14.40}$$

$$D_\Delta = \sum_l w_l N_l \quad \text{with} \quad N_l = C_l - C_{l-1} \quad \text{and} \quad w_l = \frac{1}{d2^l} \tag{14.41}$$

$$E = \sum_i V_i(\mathbf{l}_i) + \sum_{ij \in E} V_{ij}(\mathbf{l}_i, \mathbf{l}_j) \tag{14.42}$$

$$\sum_i w_i y_i h(sf_i, \theta), \tag{14.43}$$

第 15 章 结语

本章无。

附录 A 线性代数与数值方法

$$A_{M\times N} = U_{M\times P}\Sigma_{P\times P}V_{P\times N}^T \tag{A.1}$$

$$= \begin{bmatrix} u_0 & \cdots & u_{p-1} \end{bmatrix} \begin{bmatrix} \sigma_0 & & \\ & \ddots & \\ & & \sigma_{p-1} \end{bmatrix} \begin{bmatrix} v_0^T \\ \cdots \\ v_{p-1}^T \end{bmatrix},$$

$$u_i \cdot u_j = v_i \cdot v_j = \delta_{ij}. \tag{A.2}$$

$$\sigma_0 \geq \sigma_1 \geq \cdots \geq \sigma_{p-1} \geq 0. \tag{A.3}$$

$$AV = U\Sigma \quad \text{or} \quad Av_j = \sigma_j u_j. \tag{A.4}$$

$$A = \sum_{j=0}^{t} \sigma_j u_j v_j^T, \tag{A.5}$$

$$C = U\Lambda U^T = \begin{bmatrix} u_0 & \cdots & u_{n-1} \end{bmatrix} \begin{bmatrix} \lambda_0 & & \\ & \ddots & \\ & & \lambda_{n-1} \end{bmatrix} \begin{bmatrix} u_0^T \\ \cdots \\ u_{n-1}^T \end{bmatrix}$$

$$= \sum_{i=0}^{n-1} \lambda_i u_i u_i^T. \tag{A.6}$$

$$\lambda_0 \geq \lambda_1 \geq \cdots \geq \lambda_{n-1} \tag{A.7}$$

$$C = \sum_i a_i a_i^T = AA^T, \tag{A.8}$$

$$x^T C x \geq 0, \ \forall x. \tag{A.9}$$

$$C = \frac{1}{n} \sum_i (x_i - \bar{x})(x_i - \bar{x})^T. \tag{A.10}$$

$$A = U\Sigma V^T, \tag{A.11}$$

$$C = AA^T = U\Sigma V^T V \Sigma U^T = U\Lambda U^T. \tag{A.12}$$

$$\hat{C} = A^T A, \tag{A.13}$$

$$U = AV\Sigma^{-1}. \tag{A.14}$$

$$\lambda_i u_i = C u_i \quad \text{or} \quad (\lambda_i I - C) u_i = 0. \tag{A.15}$$

$$|(\lambda I - C)| = 0. \tag{A.16}$$

$$A = QR, \tag{A.17}$$

$$C = LL^T = R^T R, \tag{A.18}$$

$$C = \begin{bmatrix} \gamma & c^T \\ c & C_{11} \end{bmatrix} \tag{A.19}$$

$$= \begin{bmatrix} \gamma^{1/2} & 0^T \\ c\gamma^{-1/2} & I \end{bmatrix} \begin{bmatrix} 1 & 0^T \\ 0 & C_{11} - c\gamma^{-1}c^T \end{bmatrix} \begin{bmatrix} \gamma^{1/2} & \gamma^{-1/2}c^T \\ 0 & I \end{bmatrix} \tag{A.20}$$

$$= R_0^T C_1 R_0, \tag{A.21}$$

$$C = R_0^T \ldots R_{n-1}^T R_{n-1} \ldots R_0 = R^T R. \tag{A.22}$$

$$E_{\text{LS}} = \sum_i \|r_i\|^2 = \sum_i \|f(x_i; p) - x_i'\|^2, \tag{A.23}$$

$$r_i = f(x_i; p) - x_i' = \hat{x}_i' - \tilde{x}_i' \tag{A.24}$$

$$E_{\text{VLS}} = \sum_i |y_i - (mx_i + b)|^2. \tag{A.25}$$

$$E_{\text{PLS}} = \sum_i |y_i - (\sum_{j=0}^p a_j x_i^j)|^2, \tag{A.26}$$

$$E_{\text{SLS}} = \sum_i |y_i - A\sin(2\pi f x_i + \phi)|^2 = \sum_i |y_i - (B\sin 2\pi f x_i + C\cos 2\pi f x_i)|^2, \tag{A.27}$$

$$E_{\text{LLS}} = \sum_i |a_i x - b_i|^2 = \|Ax - b\|^2. \tag{A.28}$$

$$E_{\text{LLS}} = x^T(A^T A)x - 2x^T(A^T b) + \|b\|^2, \tag{A.29}$$

$$(A^T A)x = A^T b. \tag{A.30}$$

$$C = A^T A = R^T R, \tag{A.31}$$

$$d = A^T b. \tag{A.32}$$

$$R^T z = d, \quad Rx = z, \tag{A.33}$$

$$Ax = QRx = b \quad \longrightarrow \quad Rx = Q^T b. \tag{A.34}$$

$$E_{\text{TLS}} = \sum_i (a_i x)^2 = \|Ax\|^2, \tag{A.35}$$

$$x = \arg\min_x x^T(A^T A)x \quad \text{such that} \quad \|x\|^2 = 1. \tag{A.36}$$

$$A = U\Sigma V, \tag{A.37}$$
$$A^T A = V\Sigma^2 V, \tag{A.38}$$
$$A^T A v_k = \sigma_k^2, \tag{A.39}$$

$$E_{\text{TLS-2D}} = \sum_i (ax_i + by_i + c)^2, \tag{A.40}$$

$$C = A^T A = \sum_i \begin{bmatrix} x_i \\ y_i \\ 1 \end{bmatrix} \begin{bmatrix} x_i & y_i & 1 \end{bmatrix}. \tag{A.41}$$

$$\hat{x}_i = x_i - \bar{x} \tag{A.42}$$
$$\hat{y}_i = y_i - \bar{y} \tag{A.43}$$

$$E_{\text{TLS-2Dm}} = \sum_i (a\hat{x}_i + b\hat{y}_i)^2. \tag{A.44}$$

$$E_{\text{NLS}}(\Delta \boldsymbol{p}) = \sum_i \|\boldsymbol{f}(\boldsymbol{x}_i; \boldsymbol{p} + \Delta \boldsymbol{p}) - \boldsymbol{x}'_i\|^2 \tag{A.45}$$

$$\approx \sum_i \|\boldsymbol{J}(\boldsymbol{x}_i; \boldsymbol{p})\Delta \boldsymbol{p} - \boldsymbol{r}_i\|^2, \tag{A.46}$$

$$\boldsymbol{p} \leftarrow \boldsymbol{p} + \alpha \Delta \boldsymbol{p}, \quad 0 < \alpha \leq 1. \tag{A.47}$$

$$\boldsymbol{C} = \sum_i \boldsymbol{J}^T(\boldsymbol{x}_i)\boldsymbol{J}(\boldsymbol{x}_i), \tag{A.48}$$

$$[\boldsymbol{C} + \lambda \operatorname{diag}(\boldsymbol{C})]\Delta \boldsymbol{p} = \boldsymbol{d}, \tag{A.49}$$

$$\boldsymbol{d} = \sum_i \boldsymbol{J}^T(\boldsymbol{x}_i)\boldsymbol{r}_i, \tag{A.50}$$

$$\beta_{k+1} = \frac{\nabla E(\boldsymbol{x}_{k+1})[\nabla E(\boldsymbol{x}_{k+1}) - \nabla E(\boldsymbol{x}_k)]}{\|\nabla E(\boldsymbol{x}_k)\|^2} \tag{A.51}$$

$$\hat{x} = \boldsymbol{S}\boldsymbol{x}. \tag{A.52}$$

$$\boldsymbol{A}\boldsymbol{S}^{-1}\hat{x} = \boldsymbol{S}^{-1}\boldsymbol{b} \quad \text{or} \quad \hat{\boldsymbol{A}}\hat{x} = \hat{\boldsymbol{b}}, \tag{A.53}$$

$$E_{\text{PLS}} = \hat{x}^T(\boldsymbol{S}^{-T}\boldsymbol{C}\boldsymbol{S}^{-1})\hat{x} - 2\hat{x}^T(\boldsymbol{S}^{-T}\boldsymbol{d}) + \|\hat{\boldsymbol{b}}\|^2. \tag{A.54}$$

附录 B 贝叶斯建模与推断

$$x_i = \frac{p_{00}X_i + p_{01}Y_i + p_{02}Z_i + p_{03}}{p_{20}X_i + p_{21}Y_i + p_{22}Z_i + p_{23}} \tag{B.1}$$

$$y_i = \frac{p_{10}X_i + p_{11}Y_i + p_{12}Z_i + p_{13}}{p_{20}X_i + p_{21}Y_i + p_{22}Z_i + p_{23}}, \tag{B.2}$$

$$x_i(p_{20}X_i + p_{21}Y_i + p_{22}Z_i + p_{23}) = p_{00}X_i + p_{01}Y_i + p_{02}Z_i + p_{03}, \tag{B.3}$$

$$y_i(p_{20}X_i + p_{21}Y_i + p_{22}Z_i + p_{23}) = p_{10}X_i + p_{11}Y_i + p_{12}Z_i + p_{13}, \tag{B.4}$$

$$\boldsymbol{y}_i = \boldsymbol{f}_i(\boldsymbol{x}) + \boldsymbol{n}_i. \tag{B.5}$$

$$L = p(\boldsymbol{y}|\boldsymbol{x}) = \prod_i p(\boldsymbol{y}_i|\boldsymbol{x}) = \prod_i p(\boldsymbol{y}_i|\boldsymbol{f}_i(\boldsymbol{x})) = \prod_i p(\boldsymbol{n}_i). \tag{B.6}$$

$$\boldsymbol{n}_i \sim \mathcal{N}(0, \boldsymbol{\Sigma}_i), \tag{B.7}$$

$$L = \prod_i |2\pi\boldsymbol{\Sigma}_i|^{-1/2} \exp\left(-\frac{1}{2}(\boldsymbol{y}_i - \boldsymbol{f}_i(\boldsymbol{x}))^T \boldsymbol{\Sigma}_i^{-1}(\boldsymbol{y}_i - \boldsymbol{f}_i(\boldsymbol{x}))\right) \tag{B.8}$$

$$= \prod_i |2\pi\boldsymbol{\Sigma}_i|^{-1/2} \exp\left(-\frac{1}{2}\|\boldsymbol{y}_i - \boldsymbol{f}_i(\boldsymbol{x})\|^2_{\boldsymbol{\Sigma}_i^{-1}}\right),$$

$$L = \prod_i (2\pi\sigma_i^2)^{-N_i/2} \exp\left(-\frac{1}{2\sigma_i^2}\|\boldsymbol{y}_i - \boldsymbol{f}_i(\boldsymbol{x})\|^2\right), \tag{B.9}$$

$$\boldsymbol{\Sigma} = \boldsymbol{\Phi} \operatorname{diag}(\lambda_0 \ldots \lambda_{N-1}) \boldsymbol{\Phi}^T. \tag{B.10}$$

$$E = -\log L = \frac{1}{2}\sum_i (\boldsymbol{y}_i - \boldsymbol{f}_i(\boldsymbol{x}))^T \Sigma_i^{-1}(\boldsymbol{y}_i - \boldsymbol{f}_i(\boldsymbol{x})) + k \tag{B.11}$$

$$= \frac{1}{2}\sum_i \|\boldsymbol{y}_i - \boldsymbol{f}_i(\boldsymbol{x})\|^2_{\Sigma_i^{-1}} + k, \tag{B.12}$$

$$E = -\log L = \sum_i \|\tilde{\boldsymbol{y}}_i - \hat{\boldsymbol{y}}_i\|_{\Sigma_i^{-1}} + k. \tag{B.13}$$

$$E_{\text{loss}}(\boldsymbol{x}, \boldsymbol{y}) = \int l(\boldsymbol{x} - \boldsymbol{z}) p(\boldsymbol{z}|\boldsymbol{y}) d\boldsymbol{z}. \tag{B.14}$$

$$\boldsymbol{f}_i(\boldsymbol{x}) = \boldsymbol{H}_i \boldsymbol{x}, \tag{B.15}$$

$$E = \sum_i \|\tilde{\boldsymbol{y}}_i - \boldsymbol{H}_i \boldsymbol{x}\|_{\Sigma_i^{-1}} = \sum_i (\tilde{\boldsymbol{y}}_i - \boldsymbol{H}_i \boldsymbol{x})^T C_i (\tilde{\boldsymbol{y}}_i - \boldsymbol{H}_i \boldsymbol{x}), \tag{B.16}$$

$$E_{\text{RLS}}(\Delta \boldsymbol{p}) = \sum_i \rho(\|\boldsymbol{r}_i\|) \tag{B.17}$$

$$\sum_i \psi(\|\boldsymbol{r}_i\|)\frac{\partial \|\boldsymbol{r}_i\|}{\partial \boldsymbol{p}} = \sum_i \frac{\psi(\|\boldsymbol{r}_i\|)}{\|\boldsymbol{r}_i\|}\boldsymbol{r}_i^T \frac{\partial \boldsymbol{r}_i}{\partial \boldsymbol{p}} = 0, \tag{B.18}$$

$$E_{\text{IRLS}} = \sum_i w(\|\boldsymbol{r}_i\|)\|\boldsymbol{r}_i\|^2, \tag{B.19}$$

$$MAD = \text{med}_i \|\boldsymbol{r}_i\|, \tag{B.20}$$

$$p(\boldsymbol{x}|\boldsymbol{y}) = \frac{p(\boldsymbol{y}|\boldsymbol{x})p(\boldsymbol{x})}{p(\boldsymbol{y})}, \tag{B.21}$$

$$-\log p(\boldsymbol{x}|\boldsymbol{y}) = -\log p(\boldsymbol{y}|\boldsymbol{x}) - \log p(\boldsymbol{x}) + \log p(\boldsymbol{y}), \tag{B.22}$$

$$E(\boldsymbol{x}, \boldsymbol{y}) = E_d(\boldsymbol{x}, \boldsymbol{y}) + E_p(\boldsymbol{x}). \tag{B.23}$$

$$E_p(\boldsymbol{x}) = \sum_{\{(i,j),(k,l)\}\in\mathcal{N}} V_{i,j,k,l}(f(i,j), f(k,l)), \tag{B.24}$$

$$p(\boldsymbol{x}) \propto e^{-E(\boldsymbol{x})/T}, \tag{B.25}$$

$$E(\boldsymbol{x}) = \sum_{(i,j)\in\mathcal{N}} V_{i,j}(x_i, x_j) + \sum_i V_i(x_i), \tag{B.26}$$

$$\tilde{E}_k(\boldsymbol{x}) = \sum_{i<k,\ j\leq k} V_{i,j}(x_i, x_j) = \sum_{i\in\mathcal{C}_k} \left[V_{i,k}(x_i, x_k) + \tilde{E}_i(\boldsymbol{x}) \right], \qquad \text{(B.27)}$$

$$\hat{E}_k(x_k) = \min_{\{x_i,\ i<k\}} \tilde{E}_k(\boldsymbol{x}) = \sum_{i\in\mathcal{C}_k} \min_{x_i} \left[V_{i,k}(x_i, x_k) + \hat{E}_i(x_i) \right] \qquad \text{(B.28)}$$

$$p(\boldsymbol{x}) = \prod_{(i,j)\in\mathcal{N}} \phi_{i,j}(x_i, x_j), \qquad \text{(B.29)}$$

$$\phi_{i,j}(x_i, x_j) = e^{-V_{i,j}(x_i, x_j)} \qquad \text{(B.30)}$$

$$\tilde{p}_k(\boldsymbol{x}) = \prod_{i<k,\ j\leq k} \phi_{i,j}(x_i, x_j) = \prod_{i\in\mathcal{C}_k} \tilde{p}_{i,k}(\boldsymbol{x}), \qquad \text{(B.31)}$$

$$\tilde{p}_{i,k}(\boldsymbol{x}) = \phi_{i,k}(x_i, x_k)\tilde{p}_i(\boldsymbol{x}). \qquad \text{(B.32)}$$

$$\hat{p}_k(x_k) = \max_{\{x_i,\ i<k\}} \tilde{p}_k(\boldsymbol{x}) = \prod_{i\in\mathcal{C}_k} \hat{p}_{i,k}(\boldsymbol{x}), \qquad \text{(B.33)}$$

$$\hat{p}_{i,k}(\boldsymbol{x}) = \max_{x_i} \phi_{i,k}(x_i, x_k)\hat{p}_i(\boldsymbol{x}). \qquad \text{(B.34)}$$

$$E(\boldsymbol{x}) = \sum_{(i,j)\in\mathcal{N}} V_{i,j}(x_i, x_j) + \sum_i V_i(x_i) \qquad \text{(B.35)}$$

$$= \sum_{i,j,\alpha,\beta} V_{i,j}(\alpha,\beta) x_{i,j;\alpha,\beta} + \sum_{i,\alpha} V_i(\alpha) x_{i;\alpha} \qquad \text{(B.36)}$$

$$\text{subject to} \quad x_{i;\alpha} = \sum_\beta x_{i,j;\alpha,\beta} \quad \forall (i,j)\in\mathcal{N},\alpha, \qquad \text{(B.37)}$$

$$x_{j;\beta} = \sum_\alpha x_{i,j;\alpha,\beta} \quad \forall (i,j)\in\mathcal{N},\beta, \quad \text{and} \qquad \text{(B.38)}$$

$$x_{i,\alpha},\ x_{i,j;\alpha,\beta} \in \{0,1\}. \qquad \text{(B.39)}$$

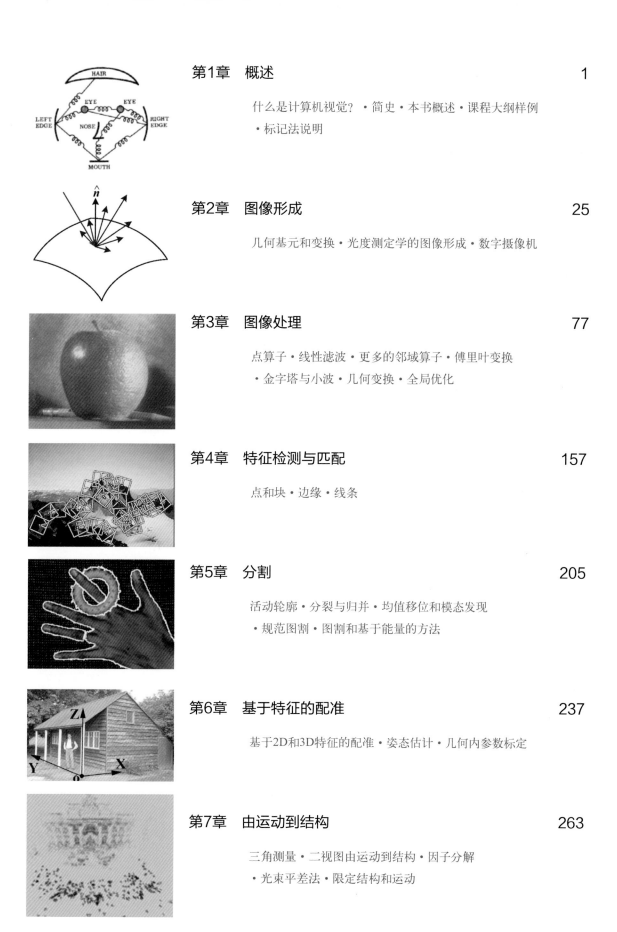

第1章 概述   1
什么是计算机视觉？·简史·本书概述·课程大纲样例·标记法说明

第2章 图像形成   25
几何基元和变换·光度测定学的图像形成·数字摄像机

第3章 图像处理   77
点算子·线性滤波·更多的邻域算子·傅里叶变换·金字塔与小波·几何变换·全局优化

第4章 特征检测与匹配   157
点和块·边缘·线条

第5章 分割   205
活动轮廓·分裂与归并·均值移位和模态发现·规范图割·图割和基于能量的方法

第6章 基于特征的配准   237
基于2D和3D特征的配准·姿态估计·几何内参数标定

第7章 由运动到结构   263
三角测量·二视图由运动到结构·因子分解·光束平差法·限定结构和运动

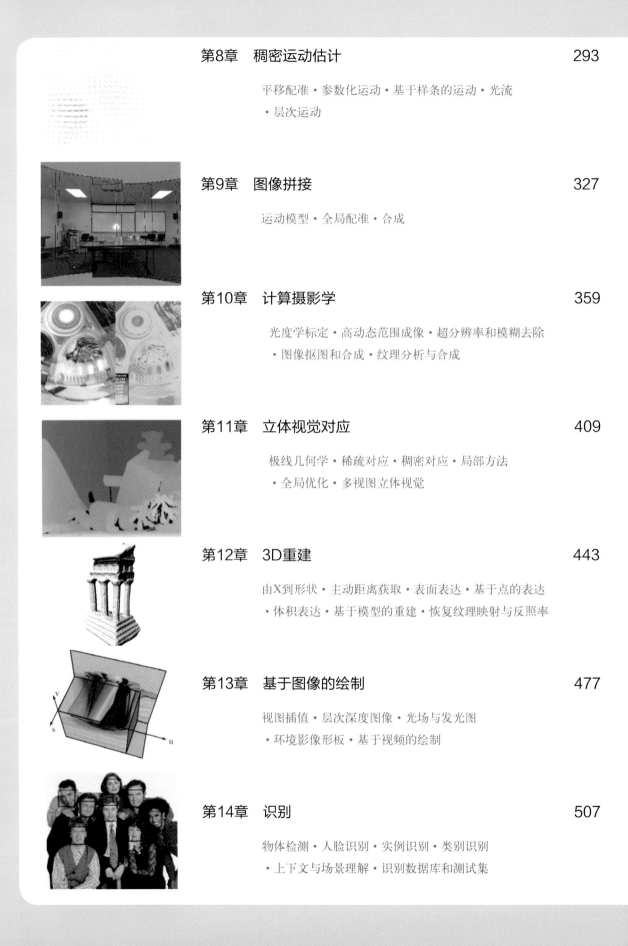

第8章　稠密运动估计　　293

平移配准・参数化运动・基于样条的运动・光流・层次运动

第9章　图像拼接　　327

运动模型・全局配准・合成

第10章　计算摄影学　　359

光度学标定・高动态范围成像・超分辨率和模糊去除・图像抠图和合成・纹理分析与合成

第11章　立体视觉对应　　409

极线几何学・稀疏对应・稠密对应・局部方法・全局优化・多视图立体视觉

第12章　3D重建　　443

由X到形状・主动距离获取・表面表达・基于点的表达・体积表达・基于模型的重建・恢复纹理映射与反照率

第13章　基于图像的绘制　　477

视图插值・层次深度图像・光场与发光图・环境影像形板・基于视频的绘制

第14章　识别　　507

物体检测・人脸识别・实例识别・类别识别・上下文与场景理解・识别数据库和测试集

第1章 概述

图1.1 人类视觉系统可以毫不费力地解释这张照片中存在的由半透明材质和阴影形成的微妙变化,并从背景中将物体正确地分割出来

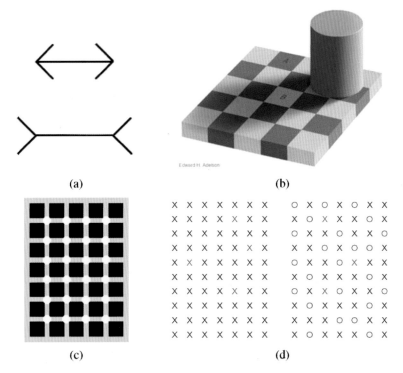

图1.3 一些常见的光学错觉和其可以给我们关于视觉系统的启示。(a)经典的Muller-Lyer错觉,两条水平线条的长度显得不同,可能是由于想象的透视效果。(b)阴影中的"白"块B和亮处的"黑"块A实际上具有相同的绝对亮度值。该感觉是由于亮度恒常性,即视觉系统当解释颜色时试图减少亮度的特性。图像承蒙Ted Adelson的允许,*http://web.mit.edu/persci/people/adelson/checkershadow illusion.html*。(c)Hermann网格错觉的一个变型,承蒙Hany Farid许可使用,*http://www.cs.dartmouth.edu/_farid/illusions/hermann.html*。当你移动眼睛扫视该图时,灰色的斑点会出现在交叉处。(d)在图的左半部分数一数红色的X。现在,在右半部分数。是不是明显困难了?其解释与弹出(pop-out)效应有关(Treisman 1985),它告诉我们大脑中有关并行感知和集成通道的操作

第2章　图像形成

图2.21　在受色像差影响的镜头中，不同波长的光线(例如，红色和蓝色箭头)以不同的焦距f聚焦，因此有不同的深度z_i'，最后产生了几何(平面内的)平移和失焦

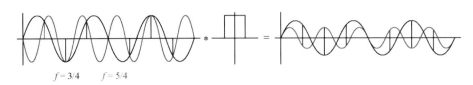

图2.24　一维信号的混叠：蓝色的正弦波在$f = 3/4$且红色正弦波在$f = 5/4$，当用$f = 2$采样时，有相同的数字采样。即便是在用100%填充率的方框型滤光器卷积这两个信号之后，尽管幅度不再相同，但在采样过的红色信号看起来像蓝色信号的较低幅度的翻转版本的意义下，它们仍然是走样的。(右侧的图像为了便于观察而放大了。实际的正弦幅度是其原来数值的30%和-18%)

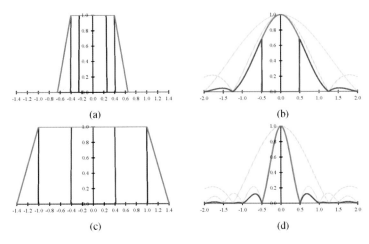

图2.26　点扩散函数(PSF)样例：(a)中的模糊圆盘(蓝)的直径等于像素间距的一半，而(c)中的直径是像素间距的两倍。传感芯片的横向填充率是80%，显示为棕色。这两个核的卷积给出点扩散函数，显示为绿色。PSF (MTF)的傅里叶响应绘制在(b)和(d)中。混叠出现在奈奎斯特频率之外的区域，显示为红色

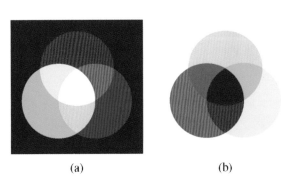

图2.27　基色和合成色：(a)加性色红、绿、蓝可以混合产生青、洋红、黄和白；(b)减性色青、洋红、黄可以混合产生红、绿、蓝和黑

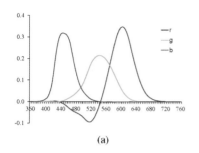
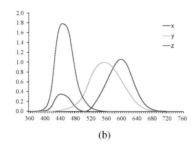

图2.28 标准CIE彩色匹配函数：(a)通过与纯色基 R = 700.0 nm，G = 546.1 nm和B = 435.8 nm匹配获得的 $\bar{r}(\lambda), \bar{g}(\lambda), \bar{b}(\lambda)$ 色谱；(b) $\bar{x}(\lambda), \bar{y}(\lambda), \bar{z}(\lambda)$ 彩色匹配函数，它们是 $\bar{r}(\lambda), \bar{g}(\lambda), \bar{b}(\lambda)$ 谱的线性组合

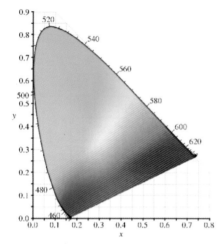

图2.29 CIE色度图，显示了彩色和其对应的(x,y)值。纯谱色绕着曲线的外端布置

图2.30 Bayer RGB样式：(a)色彩滤波器阵列布置；(b)插值后的像素值，未知量(估计的量)显示为小写字母

第3章　图像处理

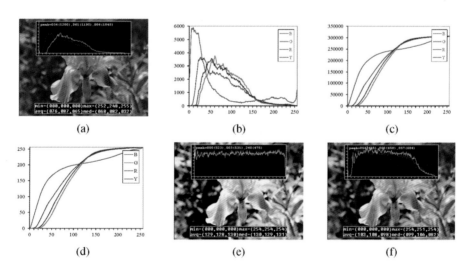

图3.7 直方图分析和均衡化：(a)原始图像；(b)各个彩色通道的直方图和亮度直方图；(c)累积分布函数；(d)均衡化(转换)函数；(e)全局的直方图均衡化；(f)局部的直方图均衡化

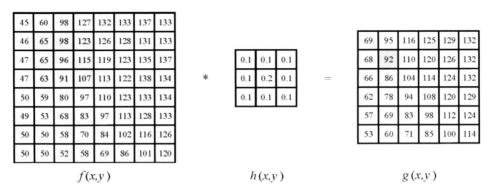

图3.10 邻域滤波(卷积)：左边图像与中间图像的卷积产生右边图像。目标图像中绿色标记的像素是利用原图像中蓝色标记的像素计算得到的

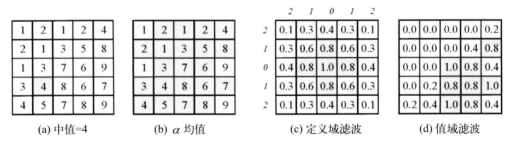

图3.17 区域求和表：(a)原始图像；(b)区域求和表；(c)区域和的计算。区域求和表中的每个值$s(i,j)$(红色)是利用它三个相邻(蓝色)的像素递归计算出来的(3.31)。区域和s(绿色)是利用矩形的四个角点(紫色)的值计算出来的(3.32)，粗体表示计算时取正值，斜体表示计算时取负值

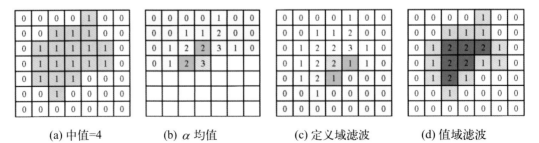

图3.19 中值和双边滤波：(a)中值像素(绿色)；(b)α-截尾均值像素的选择；(c)定义域滤波(沿着边界的数字是像素的距离)；(d)值域滤波

图3.22 城街距离变换：(a)原始的二值图像；(b)自顶向底(正向)光栅扫描：橙色像素的值是利用绿色像素的值的计算出来的；(c)自底向上(逆向)光栅扫描：绿色像素的值将与原始的橙色像素的值合并；(d)距离变换的最终结果

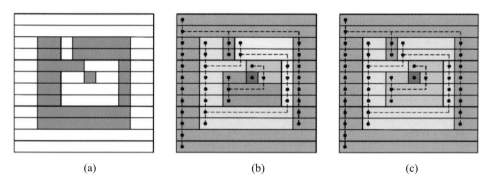

图3.23 连通量的计算：(a)原始的灰度图像；(b)水平行程(结点)通过垂直(图)边缘(蓝色的虚线)连接；对行程用从父节点继承下来的唯一的彩色作伪彩色处理；(c)合并邻接段后重新着色

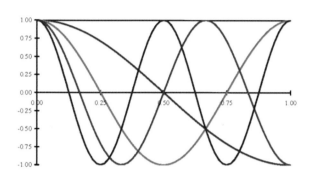

图3.26 离散余弦变换(DCT)的基函数：第一个DC(即常数)基是蓝色的水平线，第二个是棕色的半周期波形，等等。这些基广泛应用于图像和视频压缩标准，例如JPEG

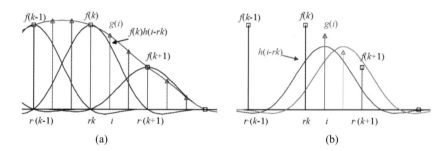

图3.27 信号插值，$g(i) = \sum_k f(k)h(i-rk)$：(a)输入值的带权求和；(b)多相滤波器描述

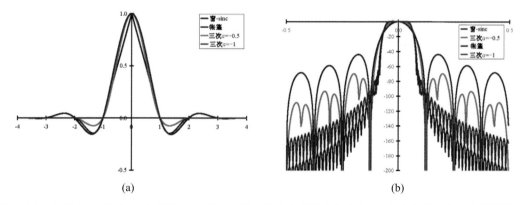

图3.29 (a)一些窗sinc函数和(b)它们的log傅里叶变换：蓝色为提升余弦窗sinc，绿色和紫色为三次插值器(a=-1和a=-0.5)，棕色为tent函数。它们都常用于进行高精度低通滤波操作

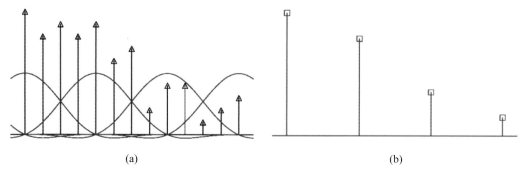

图3.30 信号降采样：(a)原始信号；(b)进行下采样前与低通滤波器卷积过的信号

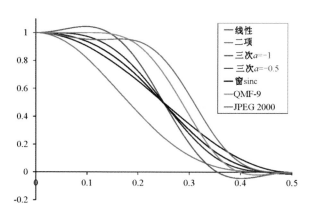

图3.31 一些2×降采样滤波器的频率响应。三次$a = -1$滤波器下降最快但同时有一点振铃效应；小波分析滤波器(QMF-9和JPET2000)在压缩中很有用，但会有较多的混叠

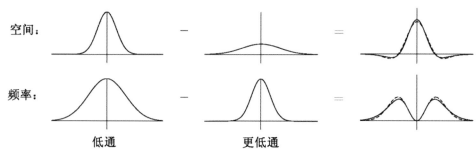

图3.35 两个低通滤波器的差是一个带通滤波器。蓝色虚线显示了对一个高斯的半倍频程拉普拉斯密切拟合

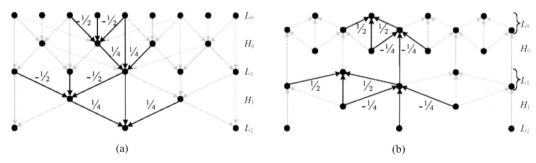

图3.39 提升变换表示为信号处理流程图：(a)分析阶段首先从偶数邻居中预测奇数值，存储差分小波，然后通过加入小波的片段来补偿较粗糙的偶数值；(b)合成阶段简单地反转计算流和一些滤波器和操作的符号。浅蓝色线条展示了用四阶代替二阶进行预测和纠正的状况

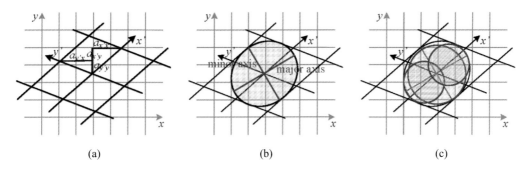

图3.48 各向异性的纹理滤波：(a)变换矩阵A的雅克比行列式和引入的水平和竖直重采样率 $\{a_{x'x}, a_{x'y}, a_{y'x}, a_{y'y}\}$；(b)一个EWA平滑核的椭圆足迹；(c)使用沿主轴的多样例的各向异性滤波。位于直线交叉处的图像像素

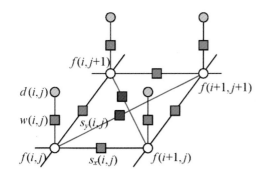

图3.56 $\mathcal{N}_4$ 邻域马尔科夫随机场的图模型。蓝色的边是为 $\mathcal{N}_8$ 邻域添加的。白色圆圈是未知的$f(i,j)$而深色圆圈是输入数据$d(i,j)$。黑色方块$s_x(i,j)$, $s_y(i,j)$表示随机场中任意的相邻的节点间交互作用势，$w(i,j)$表示数据惩罚函数。同样的图模型可以用于描述一个离散版本的一阶正则化问题(图3.55)

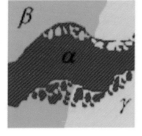 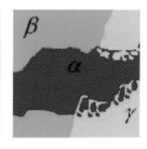 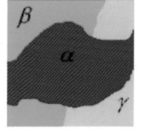

(a) 初始标注　　　(b) 标准移动　　　(c) $\alpha-\beta$ 交换　　　(d) α 扩展

图3.58 多级别图优化(Boykov, Veksler和Zabih 2001)© 2001 IEEE：(a)初始问题构造；(b)标准移动仅改变一个像素；(c)$\alpha-\beta$ 交换最优地互换所有具有 α 和 β 标注的像素；(d)α 扩展移动最优地在当前像素值和 α 标签中进行选择

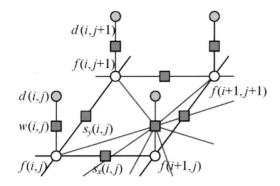

图3.59 一个有更复杂测量模型的马尔科夫随机场的图模型。额外的彩色边展示了未知值的结合(比如，在一幅锐化图像中)是如何产生测量值的(一幅有噪声的模糊图像)。得到的图模型仍旧是一个经典的MRF，而且同样容易进行采样，但是一些推理算法(例如，基于图割的)可能不再可用，因为网络复杂性增加了，由于推理过程中的状态改变变得更纠结，后验MRF有更大的团块

图3.61 图像分割(Boykov and Funka-Lea 2006)© 2006 Springer：用户在前景物体上用红色画几笔，在背景上用蓝色画几笔。系统计算前景和背景的颜色分布，解一个二值MRF。光滑权重由亮度梯度(边缘)来调整，这使其成为条件随机场(CRF)

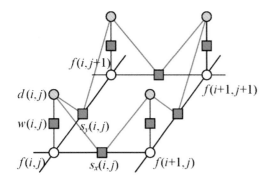
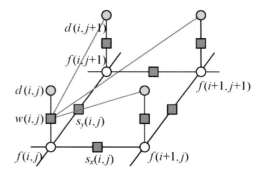

图 3.62 条件随机场 (CRF) 的图模型。额外的绿色边显示了传感数据的结合如何影响其 MRF 先验模型的光滑性，即式 (3.113) 中的 $s_x(i, j)$ 和 $s_y(i, j)$ 依赖于邻近的值。这些额外的连接(因子)使得光滑性依赖于输入数据。但是，它们使从这个 MRF 中采样更复杂

图3.63 一个区分性随机场(DRF)的图模型。额外的绿色边显示了传感数据的结合，例如，$d(i, j+1)$，影响$f(i, j)$的数据项。因此产生式模型变得更复杂，即，我们不能将一个简单的函数应用于未知变量和添加噪声

第4章 特征检测与匹配

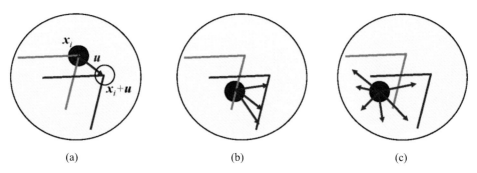

图4.4 不同图像块的孔径问题：(a)稳定的(类似于角点的)流；(b)经典的孔径问题；(c)无纹理的区域。两个图像I_0(黄)和I_1(红)重叠了。红色矢量u指出了图块中心之间的移位，$W(x_i)$加权函数(图块窗口)如图中的黑色圆圈所示

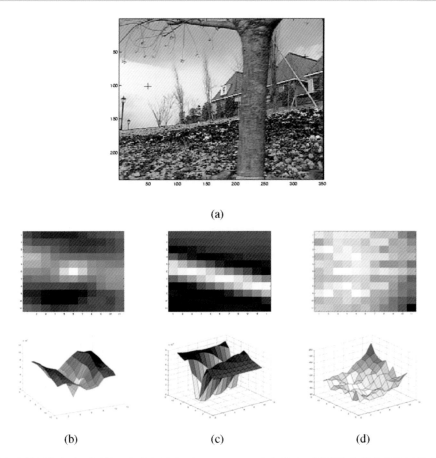

图4.5 显示成灰度值图像和表面图的三个自相关表面 $E_{AC}(\Delta u)$：(a)自相关表面计算的位置在原始图像上用三个红十字表示；(b)这个图像块取自于花坛(有良好的唯一最小值)；(c)这个图像块取自于屋顶(一维的孔径问题)；(d)这个图像块取自于云朵(没有好的峰值)。图像(b)~(d)中的每一个网格点都是 Δu 的一个值

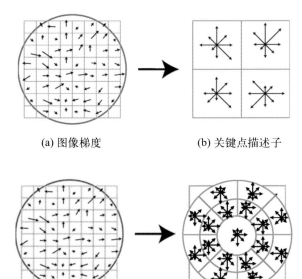

图4.18 Lowe(2004)的尺度不变特征变换(SIFT)的图解表示。(a)在每一个像素处计算梯度大小和方向，并使用一个高斯下降函数加权(蓝色圆圈)。(b)在每一个子区域内使用三线性插值来计算加权梯度方向直方图。这幅图中给出的是8×8像素块和一个2×2描述子数组，Lowe的实际实现中使用的是16×16的像素块和4×4直方图数组，每个直方图分8个区间

图4.19 梯度位置方向直方图(GLOH)描述子使用对数极坐标区间，而不是方形区间来计算方向直方图(Mikolajczyk and Schmid 2005)

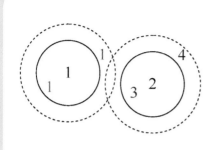 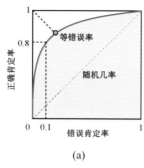 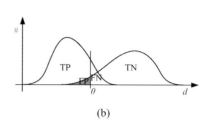

(a)   (b)

图4.22 误报和漏报：黑色数字1和2是特征，将与其他图像中的一个特征的数据库进行匹配。在目前阈值设置下(实线圈)，绿色1是一个正确肯定(正确的匹配)，蓝色1是一个漏报(未能匹配)，红色3是一个误报(不正确的匹配)。如果我们将阈值设置设高一点(虚线圆圈)，蓝色1变成一个正确肯定但棕色4变成了一个额外的误报

图4.23 ROC曲线和它相关的比率。(a)对于一特定的特征提取和匹配算法的组合ROC曲线绘制了正确肯定率相对于错误肯定率的曲线。理想情况下，正确肯定率应该接近1，而错误肯定率应该接近0。ROC曲线下面的面积(AUC)通常用作算法性能的一个单一(标量)度量。有时也用等错误率作为另一个替换的度量。(b)肯定数目(匹配的数目)和否定数目(非匹配的数目)的分布作为特征间距离d的一个函数。随着阈值 θ 的增加，正确肯定(TP)和误报(FP)的数目也随着增加

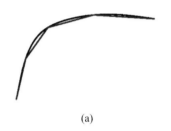 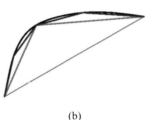 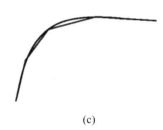

(a)   (b)   (c)

图4.40 用折线或者B样条曲线来近似一个曲线(用黑色表示)：(a)原始曲线用一个红色的折线来近似；(b)通过迭代地寻找与目前的近似离得最远的点做逐次近似；(c)拟合折线顶点的光滑的内插样条用深蓝色显示

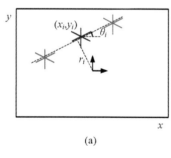 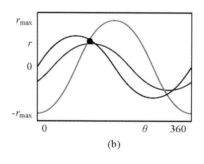

(a)   (b)

图4.41 原始哈夫变换：(a)每一个点为所有可能的直线 $r_i\theta = x_i\cos\theta + y_i\sin\theta$ 进行投票进行；(b)每束线给出 (r, θ) 处的一个正弦函数；它们的交点确定了所要求的直线方程

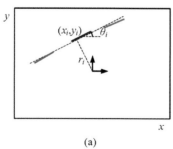 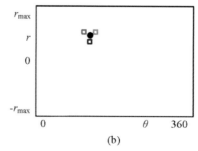

(a)   (b)

图4.42 带方向的哈夫变换：(a)一个位于极坐标 (r, θ) 处的边界基元用 $\hat{n}_i = (\cos\theta_i, \sin\theta_i)$ 和 $r_i = \hat{n}_i \cdot x_i$ 重新参数化；(b) (r, θ) 显示了分别标记为红绿蓝三个边界基元投票的累加器阵列

第5章 分割

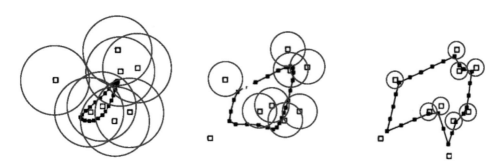

图5.3 弹性网：空心正方形表示城市而通过直线线段连接起来的实心正方形是旅行点。蓝圈表示每个城市吸引力的近似范围，它随着时间递减。在该弹性网的贝叶斯解释下，蓝圈对应于圆形高斯的一个标准差，该高斯从某未知旅行点生成每座城市

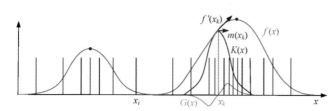

图5.17 核密度估计，它的微分以及均值位移的一维可视化。核密度估计$f(x)$通过将稀疏的输入样本x_i与和函数$K(x)$卷积得到。该函数的微分$f'(x)$可以通过将输入与微分核$G(x)$卷积得到。在当前估计x_k附近估计局部替换向量会产生均值移位向量$f(x_k)$，在多维设定中，它指向与函数梯度$\nabla f(x_k)$相同的方向。红色点表示$f(x)$的局部极大值，均值移位会收敛到该点

图5.23 用于区域分割的图割(Boykov and Jolly 2001)©2001 IEEE：(a)能量函数编码为最大流问题；(b)最小割确定区域边界

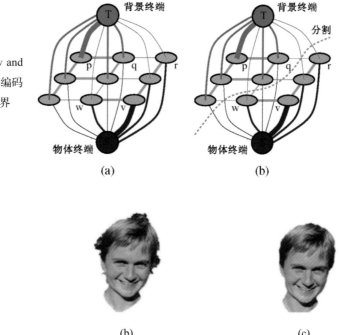

图5.24 GrabCut图像分割(Rother, Kolmogorov, and Blake 2004)©2004 ACM：(a)用户画一个红色的边界框；(b)算法猜测物体与背景的颜色分布并执行二值分割；(c)该过程采用更好的区域统计量重复进行

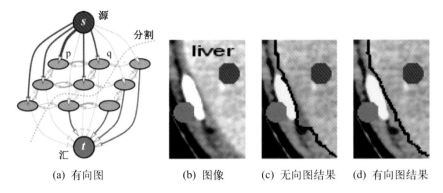

(a) 有向图　　(b) 图像　　(c) 无向图结果　　(d) 有向图结果

图5.25　采用有向图割的分割(Boykov and Funka-Lea 2006)©2006 Springer：(a)有向图；(b)带种子点的图像；(c)无向图错误地沿明亮物体延续边界；(d)有向图正确地将明亮灰色区域从它的更暗的背景中分割出来

第6章　基于特征的配准

图6.3　由三幅图像组成的用平移模型自动配准后平均在一起的全景图

第7章　由运动到结构

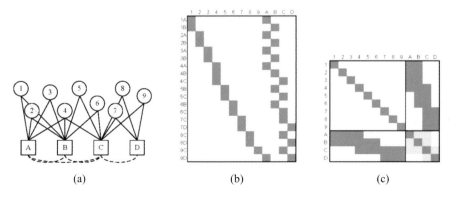

图7.9　(a)由运动到结构玩具问题的二分图；(b)关联的雅可比矩阵J；(c)Hessian矩阵A。数字表示3D点，字母表示摄像机。虚线弧和浅蓝色正方形指示结构(点)变量减少时出现的替代项

第8章 稠密运动估计

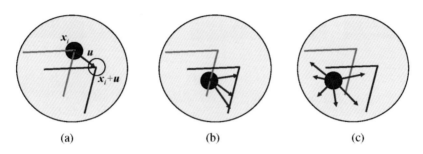

图 8.3 不同图像区域的孔径问题,由橙色和红色的 L 型结构指示,画在同一幅图像中以方便解释光流:(a) 以 x_i 为中心的窗口 $w(x_i)$(黑色)可以被唯一的匹配到第二幅图像(红色)中 x_i+u 为中心的对应结构上;(b) 中心在边界上的窗口体现了传统的孔径问题,它可以匹配到一个 1D 可能位置的族上;(c) 在完全无纹理的区域内,匹配变得毫无约束

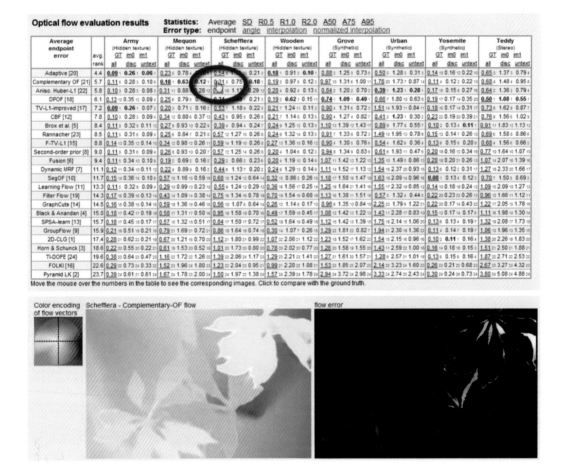

图8.12 24种光流算法评估结果,October 2009, http://vision.middlebury.edu/flow/, (Baker, Scharstein, Lewis et al. 2009)。移动鼠标到有下划线的分数之下,用户可以交互查看对应光流和误差图的关系。点击分数可以看计算结果和真实流。每个分数旁边,蓝色数字表示当前列对应的排名。粗体表示每列最小(优)值。表用平均排名排序(计算全部24列,八个序列,每个三个区域掩模)。平均排名近似在选定度量/统计方法下的性能量度

图8.13 时空体中的切片(Szeliski 1999)©1999 IEEE: (a和b)花园序列的两帧，(c)完全时空体中的一个水平切片，箭头表示估计流的潜在关键帧位置。注意花园序列的颜色不正确，正确的颜色(黄色花朵)参见图8.15

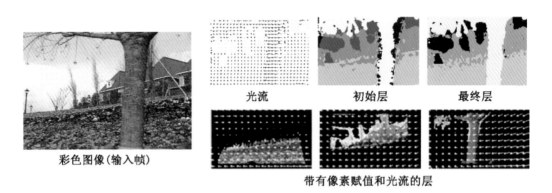

图8.15 层次运动估计结果(Wang and Adelson 1994)©1994 IEEE

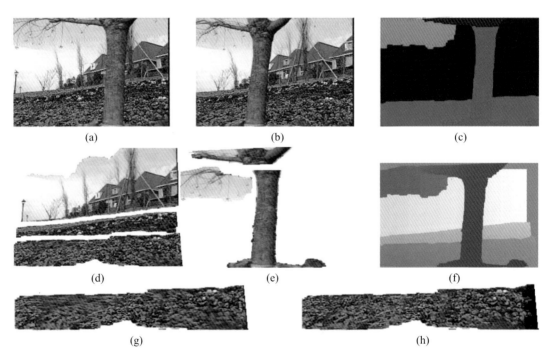

图8.16 分层立体重建(Baker, Szeliski, and Anandan 1998)©1998 IEEE：(a)初始和(b)最终输入图像；(c)初始分为六层的分割；(d)和(e)六层的子画面；(f)平面形子画面的深度图(深色表示近处)；最前面一层在残差深度估计之前(g)和之后(h)。注意花园序列的色彩不正确；正确的颜色(黄花)见图8.15

第9章 图像拼接

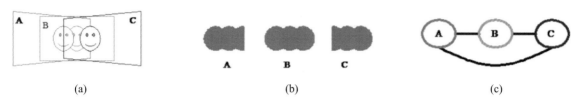

图9.15 计算差分区域(RODs)(Uyttendaele, Eden, and Szeliski 2001)©2001 IEEE：(a)包含一个运动人脸的三幅互相重叠的图像；(b)对应的差分区域；(c)一致的差分区域的图

图9.16 照片蒙太奇(Agarwala, Dontcheva, Agrawala *et al.* 2004)©2004 ACM。从一个包含五幅图像(其中四幅图像显示在左侧)的集合中，照片蒙太奇方法可以很快地构造出一幅合成的全家福，其中每个人都微笑着看着镜头(右侧)。用户可以只是浏览照片堆叠，粗略根据源图像中他们想加入最终合成图像中的人来描绘线条。这种用户施加的线条以及计算得到的区域(中间)是用左侧源图像的边界来标记颜色的

图9.18 泊松图像编辑(Perez, Gangnet, and Blake 2003)©2003 ACM：(a)选择一只狗和两个儿童的图像作为源图像，然后粘贴到目标游泳池里面；(b)简单的粘贴算法不能在边界处匹配相应的颜色；(c)泊松图像融合可以掩盖这些差异

第10章 计算机摄影学

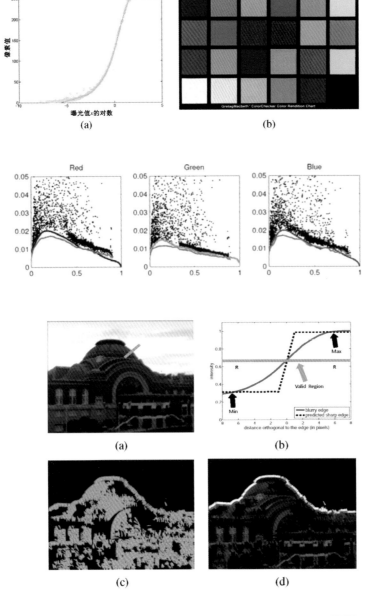

图10.3 辐射响应标定：(a)典型摄像机响应函数，显示了单彩色通道下，入射辐照度的对数(曝光值)和输出的8位像素值间的映射关系(Devevec and Malik 1997)©1997 ACM；(b)选色图板

图10.4 由单一彩色照片获得的噪声水平函数(NLF)的估计(Liu，Szeliski，Kang et al. 2008)©2008 IEEE。彩色曲线是NLF的估计，它被拟合为在带噪声的图像与分段光滑的图像之间测量得到的偏差的概率下包络。通过求取29幅图像均值而得到的NLF的真值在图中用灰色标出

图10.9 未采用标定图案的PSF估计(Joshi，Szeliski，and Kriegman 2008)©2008 IEEE：(a)输入图像和蓝色截面(轮廓)的位置；(b)拍摄的和预测的阶梯边缘的轮廓；(c和d)边缘附近所预测彩色的位置和值

图10.13 采用多种曝光度进行辐射度标定 (Debevec and Malik 1997)。对应像素值被绘制为曝光度对数 (辐照度) 的函数。移动左边的曲线来调节每个像素未知的辐射亮度，直到它们连接成一条平滑的曲线

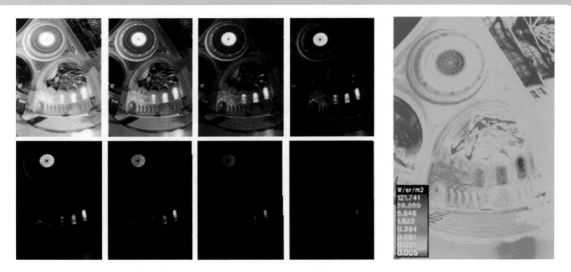

图10.15 一组由胶片摄影机拍摄的包围曝光图像以及用伪彩色显示的结果辐照度图像(Debevec and Malik 1997)© 1997 ACM

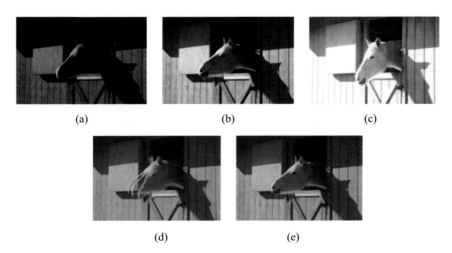

图10.16 融合多种曝光图像来制作高动态范围合成图像(Kang,Uyttendaele,Winder *et al*. 2003):(a~c)三种曝光程度不同的图像;(d)采用传统算法进行合成(注意由于马头运动而导致的重影);(e)带有运动补偿的合成

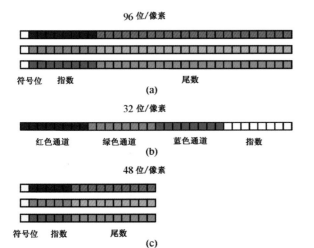

图10.19 HDR图像编码格式:(a)Portable PixMap格式(.ppm);(b)辐射亮度格式(.pic,.hdr);(c)OpenEXR格式(.exr)

图10.20 全局色调映射：(a)输入的HDR图像，线性映射后的；(b)每个彩色通道分别独立作用γ；(c)γ只作用于亮度(色彩较少冲洗)。HDR原图承蒙Paul Debevec许可使用，*http://ict.debevec.org/~debevec/Research/HDR/*。处理后的效果图承蒙Frédo Durand许可使用，MIT 6.815/6.865计算摄影学课程

图10.22 采用双边滤波的局部色调映射(Durand and Dorsey 2002)：(a)低通和高通经双边滤波后的亮度对数图像和色彩(色度)图像；(b)色调映射后的效果图像(经衰减低通亮度对数图像后)没有光晕。处理后的图像承蒙Frédo Durand许可使用，MIT 6.815/6.865计算摄影学课程

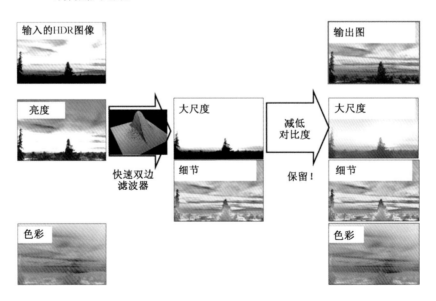

图10.24 采用双边滤波器的色局部调映射(Durand and Dorsey 2002)：算法流程总结。图像承蒙Frédo Durand许可使用，MIT 6.815/6.865计算摄影学课程

图10.34 Bayer RGB样式：(a)色彩滤波器矩阵布局；(b)插值后的像素彩色值，其中未知(猜测)的值用小写字母表示

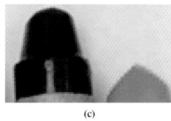

图10.35 CFA去马赛克效果(Bennett, Uyttendaele, Zitnick et al. 2006)©2006 Springer：(a)完整分辨率原图像(经彩色重采样后的版本作为算法的输入)；(b)双线性插值的结果，在蓝色蜡笔的笔尖有彩色边缘现象并且左侧(垂直)边缘有拉链效应；(c)Malvar, He, and Cutler(2004)的高质量线性插值结果(注意黄色蜡笔上明显的光晕/格子状瑕疵)；(d)Bennett, Uyttendaele, Zitnick et al. (2006)中采用局部二彩色先验的效果

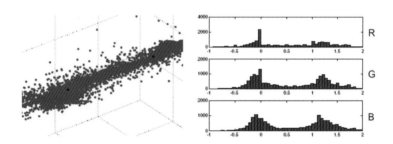

图10.36 由 5×5 的局部邻域内计算得到的二彩色模型 (Bennett, Uyttendaele, Zitnick et al. 2006)©2006 Springer。经过2-均值聚类并重投影到连接两个主导彩色(图中红点)的直线上，可见大部分像素的彩色值落在设定的直线附近。沿这条直线的分布投影到 RGB 坐标轴上后可以看到峰值在 0 和 1 处，分别为两个主导彩色处

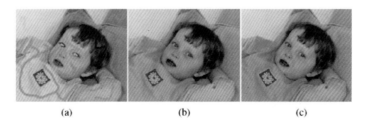

图10.37 基于优化的彩色化方法(Levin, Lischinski, and Weiss 2004)©2004 ACM：(a)灰度图像及覆盖的彩色笔画；(b)彩色化后的效果图；(c)导出灰度图及笔画颜色值的原图。原图由Rotem Weiss提供

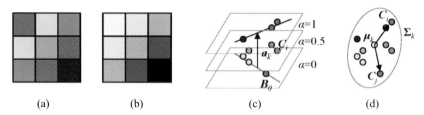

图10.43 色彩线抠图(Levin, Lischinski, Weiss 2008): (a)3×3的局部小色块; (b)可能的 α 值; (c)前景和背景色彩线、连接它们间最近两点的向量 a_k 以及一组 α 为常量的平行平面, $\alpha_i = a_k \cdot (C_i - B_0)$。(d)颜色样本的散布图和两个颜色 C_i 和 C_j 到平均值 μ_k 的偏移

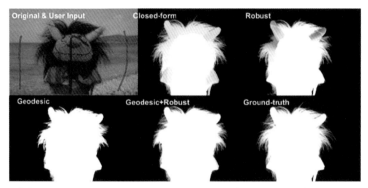

图10.45 以笔画为输入的抠图效果比较。Wang and Cohen(2007a)逐个介绍了这些方法

图10.46 基于笔画的分割结果(Themann, Rother, Rav-Acha et al. 2008)©2008 IEEE

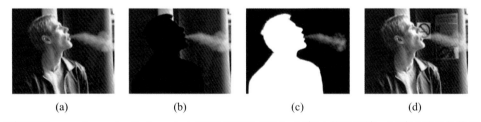

图10.47 烟抠图(Chuang, Agarwala, Curless et al. 2002)©2002 ACM: (a)输入的视频帧; (b)移去前景物体后; (c)估计的透明度遮罩; (d)将新物体插入到背景中

图10.48 阴影抠图(Chuang, Goldman, Curless et al. 2003)©2003 ACM。阴影抠图能够正确使用阴影减弱场景中的光亮, 并在3D几何体上披上阴影(d), 而不是简单地使新场景中的阴影部分变暗(c)

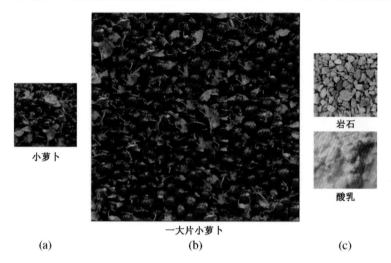

图10.49 纹理分析：(a)给定一小块纹理，目标是合成大片纹理；(b)外观相似的大片纹理；(c)一些难以合成的半结构化纹理(图像承蒙Alyosha Efros许可使用)

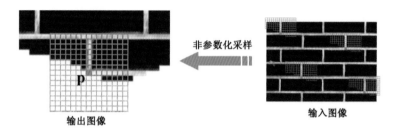

图10.50 采用非参数化采样的纹理合成(Efros and Leung 1999)。最新的像素p的值从原纹理(输入图像)中有相似局部(部分)块的像素中随机选取(图像承蒙Alyosha Efros许可使用)

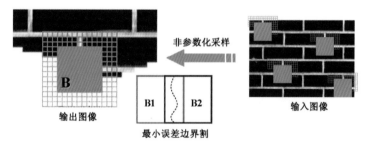

图10.51 用图像缉缝算法进行纹理合成(Efros and Freeman 2001)。算法从原纹理中一次复制一个图像块，而不是每次只生成单一像素。选定图像块间的重叠区域的过渡将由动态规划进行优化(图像承蒙Alyosha Efros许可使用)

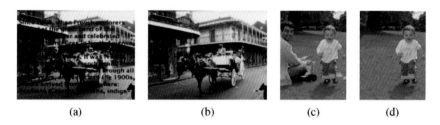

图10.52 图像修复(空洞填充)：(a和b)沿等照度线方向的传播(Bertalmio, Sapiro, Caselles et al. 2000)©2000 ACM；(c和d)基于样例的修图，填充顺序以置信度为基础(Criminisi, Pérez, and Toyama 2004)

第11章 立体视觉对应

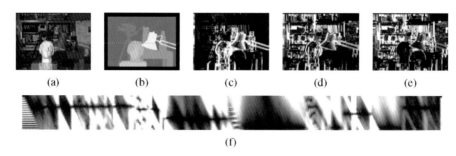

图11.5 一个典型视差空间图像(DSI)的切片(Scharstein and Szeliski 2002)©2002 Springer：(a)原始彩色图像；(b)视差图真值；(c~e)对于d=10,16,21的三个(x,y)切片；(f)对于y=151的一个(x,d)切片((b)中的虚线)。在(c~e)中各种深色(匹配)区域都可见，比如书架、桌子和罐以及头像雕塑，在图(f)中可以明显看到表现为水平直线的三个不同视差水平。这些DIS中深色的条纹表示在这个视差上匹配的区域。(小的深色区域通常是由于无明显纹理结构区域所造成的) Bobick and Intille(1999)讨论了更多DSI的例子

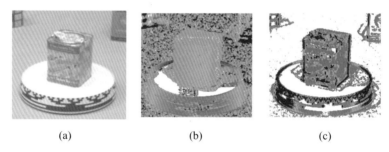

图11.10 立体视觉深度估计中的不确定性(Szeliski 1991b)：(a)输入图像；(b)估计的深度图像(蓝色表示深度更近)；(c)估计的置信度(红色表示置信度更高)。从中可以看出，纹理更明显的区域拥有更高的置信度

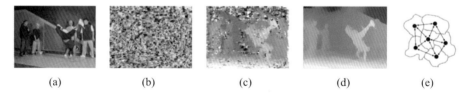

图11.12 基于分割的立体视觉匹配(Zitnick, Kang, Uyttendaele et al. 2004)©2004 ACM：(a)输入彩色图像；(b)基于颜色的分割；(c)初始视差估计；(d)最终分段平滑视差；(e)在视差空间分布中的片段上的MRF邻居定义(Zitnick and Kang 2007)© 2007 Springer

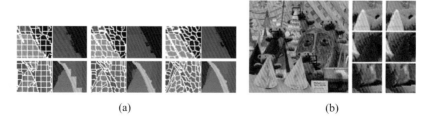

图 11.13 使用自适应过分割和抠图的立体视觉匹配 (Taguchi, Wilburn, and Zitnick 2008)© 2008 IEEE：(a) 分割边界在优化过程中被求精，这样能够得到更精确的结果 (例如，底部行中的薄的绿叶)；(b) 在片段边界抽取的透明度遮罩，得到视觉效果更好的合成结果 (中间列)

第12章 3D重建

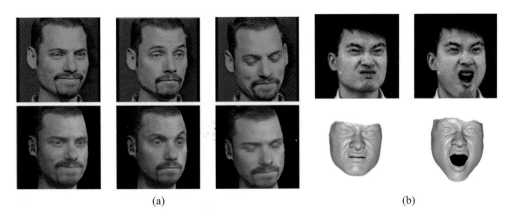

图12.18 利用3D形变模型的头部和表情跟踪与动画再现：(a)在五段输入视频流中直接拟合的模型(Pighin, Szeliski, and Salesin 2002)©2002 Springer，下面一行显示动画再现的合成的纹理映射的3D模型，其姿态和表情参数与上面一行的输入图像相符；(b)帧速率时空立体视觉表面模型拟合的模型(Zhang, Snavely, Curless et al. 2004)©2004 ACM，上面一行显示覆盖了合成绿色标记点的输入图像，下一行显示拟合的3D表面模型

第13章 基于图像的绘制

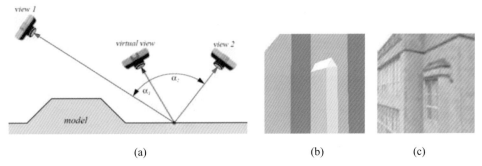

图13.3 视图相关纹理映射(Debevec, Taylor, and Malik 1996)©1996 ACM：(a)每个源视图的权重和源视图与新(虚拟)视图的相对角度有关；(b)简化的3D几何模型；(c)使用视图相关纹理映射后，几何模型看起来更加细致(如内嵌的窗户)

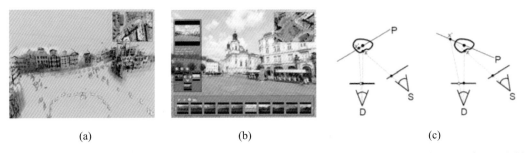

图13.4 照片游览系统(Snavely, Seitz, and Szeliski 2006)©2006 ACM：(a)场景的3D概览图，半透明刷色和线段被绘制在平面代理上；(b)用户选择感兴趣区域时，系统底部会显示一些对应的缩略图；(c)最佳稳定的平面代理(替代)选择(Snavely, Garg, Seitz et al. 2008)© 2008 ACM

图13.6 带深度的子画面(Shade, Gortler, He et al. 1988)©1998 ACM：(a)Alpha遮罩的彩色子画面；(b)对应的相对深度或视差；(c)不使用相对深度的绘制结果；(d)使用相对深度的绘制结果(注意物体边界的弧线部分)

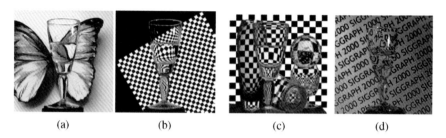

图13.10 环境影像形板：(a和b)将一个折射物体放在不同的背景前，这些背景的光照模式会被正确地折射(Zongker, Werner, Curless et al. 1999)；(c)多次折射可以用混合高斯模型处理；(d)实时影像形板可以利用单个分级彩色背景抽取得到(Chuang, Zongker, Hindorff et al. 2000)© 2000 ACM

图13.13 视频纹理(Schodl, Szeliski, Salesin et al.2000)©2000 ACM：(a)一个运动方向正确匹配的钟摆；(b)蜡烛火焰，其中弧线显示了时间转移；(c)画面中旗子是通过在跳转处进行变形操作得到的；(d)在生成篝火的视频纹理时，使用了更长的淡入淡出；(e)生成瀑布的视频纹理时，多个序列被淡入淡出在一起；(f)动画笑脸；(g)对两个荡秋千的小孩分别进行动画绘制；(h)多个气球被自动分割成独立的运动区域；(i)一个合成的包括水泡、植物和鱼的鱼缸。与这些图片对应的视频可以在*http://www.cc.gatech.edu/gvu/perception/projects/videotexture/*找到

第14章 识别

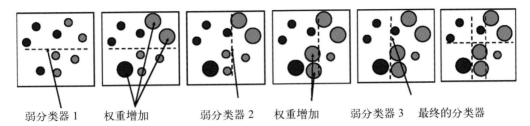

图14.7 boosting的图解说明,经Svetlana Lazebnik授权使用,源自Paul Viola和David Lowe的解释。在选择每个弱分类器(决策树桩或者超平面)后,提高了被分错的数据点的权重。最终的检测器是简单弱分类器的线性组合

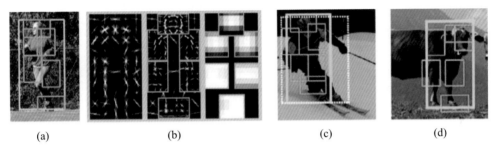

图14.9 基于部件的物体检测(Felzenszwalb, McAllester, and Ramanan 2008)©2008 IEEE:(a)输入照片及其关联的人(蓝色)和部件(黄色)检测结果;(b)检测模型用一个粗糙的模板、几个较高分辨率的部件模板及关于每个部件位置的空间模型定义;(c)滑雪者正确的检测结果;(d)作为错误检测结果的奶牛(当作人标注出来)

图14.10 针对人、自行车和马的基于部件的物体检测结果(Felzenszwalb, McAllester, and Ramanan 2008)©2008 IEEE。前三列给出的是正确的检测结果,而最右列给出的是错误的检测结果

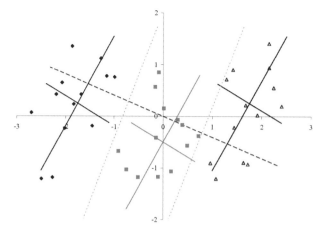

图14.16 Fisher线性鉴别分析的简单示例。样本来自三个不同的类别,在图中沿着其放缩到 $2\sigma_i$ 的主轴用不同的颜色表示。(倾斜坐标轴的交点是类别均值 m_k)短划线是(主要)Fisher线性鉴别方向而虚线表示类间的线性区分。注意区分性方向如何融合类间和类内散布矩阵的主方向

图14.35 Xerox 10类数据集的样例图像(Csurka, Dance, Perronnin et al. 2006)©2007 Springer。想象一下尝试写一个程序,从其他的图像中区分这样的图像

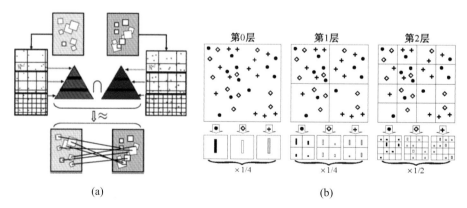

图14.37 用金字塔匹配方法比较特征向量的集合:(a)特征空间金字塔匹配核(Grauman and Darrell 2007b)构建高维特征空间的一个金字塔,并用它计算特征向量集合之间的距离(和隐含的对应);(b)空间金字塔匹配(Lazebnik, Schmid, and Ponce 2006)Ó 2006 IEEE把图像分成区域池的金字塔,然后计算每个空间箱内单独的视觉词的直方图(分布)

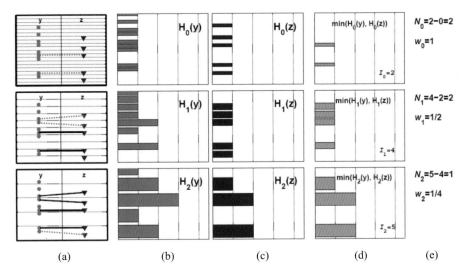

图 14.38 使用金字塔匹配核比较特征向量的汇集的一维图解 (Grauman and Darrell 2007b)：(a) 特征向量（点集）在金字塔箱中的分布；(b 和 c) 对于两幅图像，在 B_{il} 和 B_{il}' 箱中点数的直方图；(d) 直方图的交（最小值）；(e) 每一层相似性的分值，加权相加得到最终的距离/相似性度量

图14.39 "图像-图像"对比"图像-类别"距离比较(Boiman, Shechtman, and Irani 2008)©2008 IEEE。在左上角的查询图像，可能不与下面一行中任何一个数据库图像里的的特征分布相匹配。但是，如果将查询中的每个特征匹配到所有类别图像中最类似的那一个，就可以找到一个好的匹配

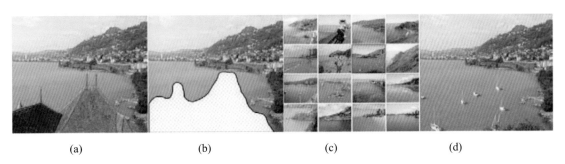

图14.46 用几百万幅照片进行场景补全(Hays and Efros 2007)©2007 ACM：(a)原图；(b)消除不需要的前景后；(c)比较真实的场景匹配，用户选择的匹配用红色高亮显示出来了；(d)经过替换和融合的输出图像

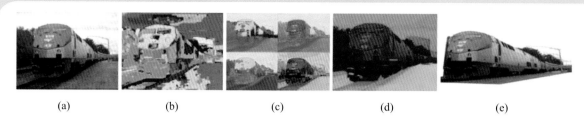

图14.47 自动照片弹出(Hoiem, Efros, and Hebert 2005a)Ó 2005 ACM：(a)输入图像；(b)超像素(superpixels)分为；(c)多个区域；(d)标签代表的地面(绿色)、直立物(红色)和天空(蓝色)；(e)产生的分段线性3D模型的新视图

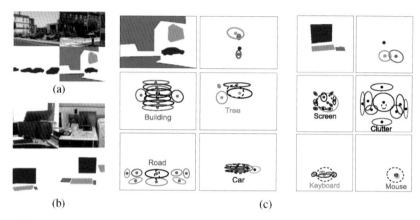

图14.50 物体识别的上下文场景模型(Sudderth, Torralba, Freeman et al. 2008)©2008 Springer：(a)一些街景和它们对应的标签(紫＝建筑，红＝汽车，绿＝树，蓝＝路)；(b)一些办公场景(红＝计算机屏幕，绿＝键盘，蓝＝鼠标)；(c)从这些标记的场景中学习到的上下文模型。上面一行显示了样例标记图像和其他物体相对于中心红色(汽车或者屏幕)物体的分布情况。下面一行显示了组成每个物体的部件分布情况

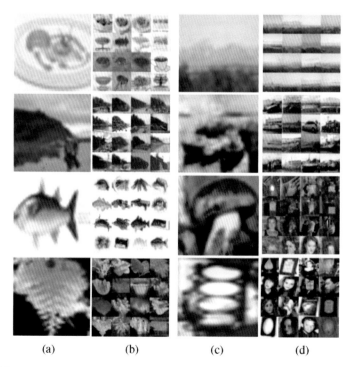

图14.52 使用小图像的识别(Torralba, Freeman, and Fergus 2008)©2008 IEEE：(a)和(c)列显示了输入图像样例，(b)和(d)列显示了在有8千万个小图像的数据库中其对应的16个最近邻

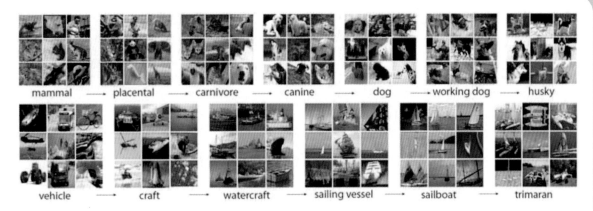

图 14.53 ImageNet(Deng, Dong, Socher et al. 2009)©2009 IEEE。这个数据库包含对每个名词有 500 个仔细核实的图像，其中每个名词是 WordNet 层次结构的 14 841 个名词（截止 2010 年 4 月）中的一个

图 14.54 PASCAL Visual Object Classes Challenge 2008 (VOC2008) 数据库中的样例图像 (Everingham, Van Gool, Williams et al. 2008)。原始图像来自 flickr(*http://www.flickr.com/*)。数据库的权利在 *http://pascallin.ecs.soton.ac.uk/challenges/VOC/voc2008/* 中解释了说明

计 算 机 视 觉
——算法与应用

(美) Richard Szeliski 著

艾海舟 兴军亮 等译

清华大学出版社

北京

内 容 简 介

《计算机视觉——算法与应用》探索了用于分析和解释图像的各种常用技术，描述了具有一定挑战性的视觉应用方面的成功实例，兼顾专业的医学成像和图像编辑与交织之类有趣的大众应用，以便学生能够将其应用于自己的照片和视频，从中获得成就感和乐趣。本书从科学的角度介绍基本的视觉问题，将成像过程的物理模型公式化，然后在此基础上生成对场景的逼真描述。作者还运用统计模型来分析和运用严格的工程方法来解决这些问题。

本书作为本科生和研究生"计算机视觉"课程的理想教材，适合计算机和电子工程专业学生使用，重点介绍现实中行之有效的基本技术，通过大量应用和练习来鼓励学生大胆创新。此外，本书的精心设计和编排，使其可以作为计算机视觉领域中一本独特的基础技术参考和最新研究成果文献。

Translation from the English language edition:
Computer Vision: Algorithms and Applications, 1st Edition by Richard Szeliski
Copyright © Richard Szeliski 2010
Springer is a part of Springer Science + Business Media
All Rights Reserved.

本书中文简体字翻译版由德国施普林格公司授权清华大学出版社在中华人民共和国境内(不包括中国香港、澳门特别行政区和中国台湾地区)独家出版发行。未经出版者预先书面许可，不得以任何方式复制或抄袭本书的任何部分。

北京市版权局著作权合同登记号　图字：01-2011-0579
本书封面贴有清华大学出版社防伪标签，无标签者不得销售。
版权所有，侵权必究。举报：010-62782989，beiqinquan@tup.tsinghua.edu.cn。

图书在版编目(CIP)数据

计算机视觉——算法与应用/(美)塞利斯基(Szeliski,R.)著；艾海舟，兴军亮等译. —北京：清华大学出版社，2012.1(2025.3 重印)
书名原文：Computer Vision: Algorithms and Applications
ISBN 978-7-302-26915-1

Ⅰ. ①计… Ⅱ. ①塞… ②艾… ③兴… Ⅲ. ①计算机视觉—应用 Ⅳ. ①TP302.7
中国版本图书馆 CIP 数据核字(2011)第 187910 号

责任编辑：文开琪
封面设计：杨玉兰
版式设计：北京东方人华科技有限公司
责任校对：周剑云
责任印制：沈　露

出版发行：清华大学出版社
　　　网　　址：https://www.tup.com.cn, https://www.wqxuetang.com
　　　地　　址：北京清华大学学研大厦 A 座　　邮　编：100084
　　　社 总 机：010-83470000　　　　　　　　邮　购：010-62786544
　　　投稿与读者服务：010-62776969, c-service@tup.tsinghua.edu.cn
　　　质量反馈：010-62772015, zhiliang@tup.tsinghua.edu.cn
印 装 者：三河市龙大印装有限公司
经　　销：全国新华书店
开　　本：185mm×260mm　　印　张：42.5　　插　页：16　　字　数：1020 千字
版　　次：2012 年 1 月第 1 版　　　　　　　　印　次：2025 年 3 月第 19 次印刷
定　　价：139.00 元

产品编号：036188-03

译者的话

在近两年"计算机视觉"课程的教学过程中,我向学生推荐这本当时尚未正式出版的教材(网上有不断更新的电子版草稿)作为参考书,我觉得这是一本难得的好教材。

为了帮助学生扫除阅读英文版时可能碰到的障碍,我承担了本书的翻译工作。有了中文版,广大读者——尤其是无法流畅阅读英文版的读者——学习起来无疑更轻松,从而使本书能够充分发挥作用。

在本书的翻译过程中,对于英文版中明显存在的少数纰漏(主要是排印或疏漏),我们逐一进行了订正。这些错误一般都很明显,因此译文中我们没有专门声明。有英文阅读能力的读者,在阅读本书的过程中,不妨参照英文版,这样做不仅可以加深对相关专业术语的理解,还能通过这一实践方式提升专业英语阅读能力。

参与本书翻译的人员如下(按照工作量的大小排序):

艾海舟(序、目录、第1章、第2章、第15章和术语)

兴军亮(第4章和第11章)

段根全(第6章和第14章,14.2节除外)

陈先捷(第10章)

曹翀(第3章的3.5~3.9节)

苏延超(第12章,14.2节)

忻海(第8章)

杜宇宁(第3章的3.1~3.4节)

王楠(第5章)

卜鹏洋(附录A~C)

张晨光(第9章)

刘力为(第7章)

刘之方(第13章)

全部译稿的最后审定由艾海舟负责。由于译者水平有限,书中难免存在纰漏,欢迎广大读者批评指正。在阅读过程中,如果发现问题,请发送电子邮件告知,以便今后重印时加以订正。

艾海舟

清华大学计算机系

电子邮件:ahz@mail.tsinghua.edu.cn

主页网址:http://media.cs.tsinghua.edu.cn/~ahz

序

本书萌芽于 2001 年，当时，华盛顿大学的 Steve Seitz 邀我和他一起讲一门课，课程名称是"面向计算机图形学的计算机视觉"。那个时候，计算机图形学领域正在越来越多地使用计算机视觉技术，用它来创建基于图像的真实物体的模型，用于产生视觉效果，用于通过计算摄影学技术来合并真实影像。我们决定聚焦于计算机视觉在若干有趣问题中的应用，例如使用个人照片的图像拼接和基于照片的 3D 建模等，这一想法引起了学生们的共鸣。

从那时起，华盛顿大学和斯坦福大学就一直使用类似的课程大纲和项目导向的课程结构来进行常规计算机视觉课程的教学(在斯坦福大学，在 2003 年这门课程由我和 David Fleet 共同讲授)。类似的课程大纲也被其他很多大学所采用，并被纳入计算摄影学相关的更专业的课程。(有关如何在课程中使用本书的建议，请参见 1.4 节的表 1.1。)

本书还反映了我在企业研究实验室(DEC 剑桥研究实验室和微软研究院)这二十年的计算机视觉研究经历。在从事研究的过程中，我主要关注在真实世界中具有实际应用的问题和在实践中行之有效的方法(算法)。因此，本书更强调在真实世界条件下有效的基本方法，而较少关注内在完美但难以实际应用的神秘的数学内容。

本书适用于计算机科学和电子工程专业高年级本科的计算机视觉课程。学生最好已经修过图像处理或计算机图形学课程，这样一来，便可以少花一些时间来学习一般性的数学背景知识，多花一些时间来学习计算机视觉技术。本书也适用于研究生的计算机视觉课程(通过专研更富有挑战性的应用和算法领域)，作为基本技术和近期研究文献的参考用书。为此，我尽量尝试引用每个子领域中最新的研究进展，即便其技术细节过于复杂而无法在本书中涉及。

在课程教学过程中，我们发现，要使学生从容应对真实图像及其带来的挑战，让他们尝试实现一些小的课程设计(通常一个建立在另一个基础之上)，是很有帮助的。随后，要求学生分成组选择各自的主题，完成最终的课程设计。(有时，这些课程设计甚至能转换为会议论文！)本书各章最后的习题包含有关小型中期课程设计题目的很多建议，也包含一些更开放的问题，这些问题的解决仍然是活跃的研究课题。只要有可能，我都会鼓励学生用他们自己的个人照片来测试他们的算法，因为这可以更好地激发他们的兴趣，往往会产生富有创造性的衍生问题，使他们更熟悉真实影像的多样性和复杂性。

在阐述和解决计算机视觉问题的过程中，我常常发现从三个高层途径获取灵感是有帮助的。

- **科学层面**：建立图像形成过程的详细模型，为了恢复感兴趣量而构建其逆过程的数学方法(必要时，做简化假设使其在数学上更容易处理)。

- **统计层面**：使用概率模型来量化产生输入图像的未知量先验似然率和噪声测量过程，然后推断所期望量的最可能的估计并分析其结果的不确定程度。使用的推断算法往往与用于逆转(科学的)图像形成过程的优化方法密切相关。
- **工程层面**：开发出易于描述和实现且已知在实践中行之有效的方法。测试这些方法，以便于了解其不足和失效模态，及其期望的计算代价(运行时的性能)。

以上这三个途径相互依存，并且贯穿本书始终。

我个人的研究和发展哲学(本书中的习题亦然)非常强调算法测试。在计算机视觉领域，提出一个算法在少数几幅图像上使某件事似乎可以做而不是把某件事做对，这太容易了。要想使算法有效，最理想的途径是使用一种"三部曲"策略。

首先，在干净的合成数据上测试算法，因为已知其精确结果。其次，在该数据上增加噪声，评测性能是怎样作为噪声水平的函数退化的。最后，在真实世界数据上测试算法，优先取自广泛输入源的数据，比如万维网上的照片。只有这样，我们才能确信该算法能够处理真实世界的复杂性，即不符合某种简化模型或假设的图像。

为了在这一过程中帮助学生，本书附带大量补充阅读材料，这些都可以在本书网站找到，网址为 *http://szeliski.org/Book*。具体资源类别(参见附录 C 的描述)如下：
- 指向万维网上可以找到的问题的常用数据集的链接；
- 指向软件库的链接，可帮助学生从基本任务入手，比如读/写图像或创建和操作图像；
- 与本书素材对应的幻灯片；
- 本书所引用的论文文献列表。

在本领域发表新论文的教师和研究人员可能对后两项资源更感兴趣，但即便是普通学生，迟早也会发现它们是很有用的。有些软件库包含广泛的计算机视觉算法的实现，能帮助你应对更难的项目(征得导师同意的情况下)。

致谢

我要感谢对本书写作有帮助的所有人，他们的研究热情、咨询和鼓励帮助我写就本书。

McGill 大学的 Steve Zucker 是第一个引导我涉足计算机视觉领域的人，他教导我们所有的学生要敢于质疑和辩论研究结果和研究方法，鼓励我攻读这个领域的研究生。

我的博士论文导师，卡内基·梅隆大学的 Takeo Kanade(金出武雄)和 Geoff Hinton，教给我良好的研究、写作和报告的基本方法。他们激发了我对视觉处理、3D 建模和统计方法的兴趣，与此同时，Larry Matthies 让我见识了卡尔曼滤波和立体匹配。

Demetri Terzopoulos 是我在涉足工业界后从事第一份研究工作的导师，他教给我成功发表论文的方法。Yvan Leclerc 和 Pascal Fua，我在斯坦福研究院(SRI International)短暂停留期间的同事，在可供选择的计算机视觉研究方法方面给予我新的观点。

在 DEC 剑桥研究实验室工作的六年里，我有幸与很多同事共同工作，包括 Ingrid Carlbom，Gudrun Klinker，Keith Waters，Richard Weiss，Stephane Lavallee 和 Sing Bing Kang(江胜明)，同时也指导了最初的一大批杰出的暑期实习生，包括 David Tonnesen，Sing Bing Kang(江胜明)，James Coughlan，Harry Shum(沈向洋)。正是在这里，我与 Daniel Scharstein 就此开始长期合作，他目前在 Middlebury 学院工作。

在微软研究院，我非常荣幸能和世界上最好的一些计算机视觉与计算机图形学领域的研究员一起工作，他们是：Michael Cohen，Hugues Hoppe，Stephen Gortler，Steve Shafer，Matthew Turk，Harry Shum(沈向洋)，Anandan，Phil Torr，Antonio Criminisi，Georg Petschnigg，Kentaro Toyama，Ramin Zabih，Shai Avidan，Sing Bing Kang(江胜明)，Matt Uyttendaele，Patrice Simard，Larry Zitnick，Richard Hartley，Simon Winder，Drew Steedly，Chris Pal，Nebojsa Jojic，Patrick Baudisch，Dani Lischinski，Matthew Brown，Simon Baker，Michael Goesele，Eric Stollnitz，David Nistér，Blaise Aguera y Arcas，Sudipta Sinha，Johannes Kopf，Neel Joshi，Krishnan Ramnath。我也非常幸运能有如此杰出的实习生，他们是 Polina Golland，Simon Baker，Mei Han(韩玫)，Arno Schödl，Ron Dror，Ashley Eden，Jinxiang Chai(柴金祥)，Rahul Swaminathan，Yanghai Tsin(秦漾海)，Sam Hasinoff，Anat Levin，Matthew Brown，Eric Bennett，Vaibhav Vaish，Jan-Michael Frahm，James Diebel，Ce Liu(刘策)，Josef Sivic，Grant Schindler，Colin Zheng，Neel Joshi，Sudipta Sinha，Zeev Farbman，Rahul Garg，Tim Cho，Yekeun Jeong，Richard Roberts，Varsha Hedau，Dilip Krishnan。

在微软工作时，我还有机会与在华盛顿大学的杰出的同事合作，我是该校的合聘教授。我要感谢 Tony DeRose 和 David Salesin，是他们最初鼓励我参与华盛顿大学正在进行的研究。我要感谢我的长期合作者 Brian Curless，Steve Seitz，Maneesh Agrawala，Sameer Agarwal，Yasu Furukawa。还要感谢我指导且和与我配合很好的学生，他们是：Frederic Pighin，Yung-Yu Chuang，Doug Zongker，Colin Zheng，Aseem Agarwala，Dan Goldman，Noah Snavely，Rahul Garg，Ryan Kaminsky。正如序开始时所提到的，本书发端于 Steve Seitz 邀请我一起讲授的视觉课程，源于 Steve 的鼓励、课程笔记和编辑输入。

我还要感谢许多其他的计算机视觉研究人员，他们给了我很多关于本书的建设性建议，包括：Sing Bing Kang(江胜明)，他是我的非正式版图书编辑；Vladimir Kolmogorov，他撰写了关于 MRF 推断的线性规划方法之附录 B.5.5；Daniel Scharstein，Richard Hartley，Simon Baker，Noah Snavely，Bill Freeman，Svetlana

Lazebnik，Matthew Turk，Jitendra Malik，Alyosha Efros，Michael Black，Brian Curless，Sameer Agarwal，Li Zhang(张力)，Deva Ramanan，Olga Veksler，Yuri Boykov，Carsten Rother，Phil Torr，Bill Triggs，Bruce Maxwell，Jana Kosecka，Eero Simoncelli，Aaron Hertzmann，Antonio Torralba，Tomaso Poggio，Theo Pavlidis，Baba Vemuri，Nando de Freitas，Chuck Dyer，Song Yi(宋毅)，Falk Schubert，Roman Pflugfelder，Marshall Tappen，James Coughlan，Sammy Rogmans，Klaus Strobel，Shanmuganathan，Andreas Siebert，Yongjun Wu(吴勇军)，Fred Pighin，Juan Cockburn，Ronald Mallet，Tim Soper，Georgios Evangelidis，Dwight Fowler，Itzik Bayaz，Daniel O'Connor，Srikrishna Bhat。Shena Deuchers 极为出色地完成了本书的排版编辑工作，提出了很多有价值的改进建议。Springer 出版社的 Wayne Wheeler 和 Simon Rees 在本书整个出版过程中对我帮助很大。Keith Price 的"Annotated Computer Vision Bibliography"(注解版计算机视觉参考文献)对追溯参考文献和查找相关工作起着非常重要的作用。

我期望这本书准确、信息可靠和及时，所以如果你有任何改进本书的建议，不妨发电子邮件告诉我。

最后，没有家人难以置信的支持和鼓励，本书不可能问世，或者说不值得费时费力。谨以此书献给我的父母，Zdzisław 和 Jadwiga，他们的爱、慷慨和成就总是激励着我；献给我的妹妹 Basia，因为她对我所付出的毕生的手足之情；特别献给 Lyn，Anne 和 Stephen，他们在所有事情(包括本书这个项目)上每天都给予我鼓励，使所有事情都有非凡的价值。

Wenatchee 湖畔
2010 年 8 月

目　录

第 1 章　概述 1
1.1　什么是计算机视觉？ 2
1.2　简史 8
1.3　本书概述 16
1.4　课程大纲样例 21
1.5　标记法说明 22
1.6　扩展阅读 22

第 2 章　图像形成 25
2.1　几何基元和变换 26
2.1.1　几何基元 26
2.1.2　2D 变换 29
2.1.3　3D 变换 32
2.1.4　3D 旋转 33
2.1.5　3D 到 2D 投影 37
2.1.6　镜头畸变 46
2.2　光度测定学的图像形成 47
2.2.1　照明 48
2.2.2　反射和阴影 49
2.2.3　光学 54
2.3　数字摄像机 57
2.3.1　采样与混叠 60
2.3.2　色彩 63
2.3.3　压缩 71
2.4　补充阅读 72
2.5　习题 73

第 3 章　图像处理 77
3.1　点算子 78
3.1.1　像素变换 79
3.1.2　彩色变换 81
3.1.3　合成与抠图 81
3.1.4　直方图均衡化 83
3.1.5　应用：色调调整 86
3.2　线性滤波 86
3.2.1　可分离的滤波 89
3.2.2　线性滤波示例 90
3.2.3　带通和导向滤波器ᅟ........... 91
3.3　更多的邻域算子 95
3.3.1　非线性滤波 95
3.3.2　形态学 99
3.3.3　距离变换 100
3.3.4　连通量 101
3.4　傅里叶变换 102
3.4.1　傅里叶变换对 105
3.4.2　二维傅里叶变换 107
3.4.3　维纳滤波 108
3.4.4　应用：锐化，模糊和去噪 ... 111
3.5　金字塔与小波 111
3.5.1　插值 112
3.5.2　降采样 114
3.5.3　多分辨率表达 116
3.5.4　小波 119
3.5.5　应用：图像融合 123
3.6　几何变换 125
3.6.1　参数化变换 125
3.6.2　基于网格的卷绕 131
3.6.3　应用：基于特征的变形 133
3.7　全局优化 133
3.7.1　正则化 134
3.7.2　马尔科夫随机场 138
3.7.3　应用：图像的恢复 147
3.8　补充阅读 147
3.9　习题 149

第4章 特征检测与匹配 157
4.1 点和块 159
4.1.1 特征检测器 160
4.1.2 特征描述子 169
4.1.3 特征匹配 172
4.1.4 特征跟踪 179
4.1.5 应用：表演驱动的动画 181
4.2 边缘 182
4.2.1 边缘检测 182
4.2.2 边缘连接 187
4.2.3 应用：边缘编辑和增强 189
4.3 线条 190
4.3.1 逐次近似 191
4.3.2 Hough 变换 191
4.3.3 消失点 194
4.3.4 应用：矩形检测 196
4.4 扩展阅读 197
4.5 习题 198

第5章 分割 205
5.1 活动轮廓 206
5.1.1 蛇行 207
5.1.2 动态蛇行和 CONDENSATION 211
5.1.3 剪刀 214
5.1.4 水平集 215
5.1.5 应用：轮廓跟踪和转描机 217
5.2 分裂与归并 218
5.2.1 分水岭 218
5.2.2 区域分裂(区分式聚类) 219
5.2.3 区域归并(凝聚式聚类) 219
5.2.4 基于图的分割 219
5.2.5 概率聚集 220
5.3 均值移位和模态发现 221
5.3.1 k-均值和高斯混合 222
5.3.2 均值移位 224
5.4 规范图割 227
5.5 图割和基于能量的方法 230
5.6 补充阅读 234
5.7 习题 235

第6章 基于特征的配准 237
6.1 基于 2D 和 3D 特征的配准 238
6.1.1 使用最小二乘的 2D 配准 238
6.1.2 应用：全景图 240
6.1.3 迭代算法 241
6.1.4 鲁棒最小二乘和 RANSAC 243
6.1.5 3D 配准 245
6.2 姿态估计 246
6.2.1 线性算法 246
6.2.2 迭代算法 248
6.2.3 应用：增强现实 249
6.3 几何内参数标定 250
6.3.1 标定模式 250
6.3.2 消失点 252
6.3.3 应用：单视图测量学 253
6.3.4 旋转运动 254
6.3.5 径向畸变 256
6.4 补充阅读 257
6.5 习题 258

第7章 由运动到结构 263
7.1 三角测量 264
7.2 二视图由运动到结构 266
7.2.1 投影(未标定的)重建 270
7.2.2 自标定 271
7.2.3 应用：视图变形 273
7.3 因子分解 274
7.3.1 透视与投影因子分解 276
7.3.2 应用：稀疏 3D 模型提取 277

	7.4	光束平差法 278
		7.4.1 挖掘稀疏性 280
		7.4.2 应用：匹配运动和增强现实 282
		7.4.3 不确定性和二义性 283
		7.4.4 应用：由因特网照片重建 284
	7.5	限定结构和运动 287
		7.5.1 基于线条的方法 287
		7.5.2 基于平面的方法 288
	7.6	补充阅读 289
	7.7	习题 290

第 8 章　稠密运动估计 293

	8.1	平移配准 294
		8.1.1 分层运动估计 297
		8.1.2 基于傅里叶的配准 298
		8.1.3 逐次求精 300
	8.2	参数化运动 305
		8.2.1 应用：视频稳定化 308
		8.2.2 学到的运动模型 308
	8.3	基于样条的运动 309
	8.4	光流 312
		8.4.1 多帧运动估计 315
		8.4.2 应用：视频去噪 316
		8.4.3 应用：去隔行扫描 316
	8.5	层次运动 317
		8.5.1 应用：帧插值 319
		8.5.2 透明层和反射 320
	8.6	补充阅读 321
	8.7	习题 322

第 9 章　图像拼接 327

	9.1	运动模型 329
		9.1.1 平面透视运动 329
		9.1.2 应用：白板和文档扫描 330
		9.1.3 旋转全景图 331

		9.1.4 缝隙消除 333
		9.1.5 应用：视频摘要和压缩 ... 334
		9.1.6 圆柱面和球面坐标 335
	9.2	全局配准 338
		9.2.1 光束平差法 338
		9.2.2 视差消除 341
		9.2.3 认出全景图 343
		9.2.4 直接配准和基于特征的配准 345
	9.3	合成 346
		9.3.1 合成表面的选择 346
		9.3.2 像素选择和加权（去虚影） 348
		9.3.3 应用：照片蒙太奇 352
		9.3.4 融合 353
	9.4	补充阅读 355
	9.5	习题 356

第 10 章　计算摄影学 359

	10.1	光度学标定 361
		10.1.1 辐射度响应函数 362
		10.1.2 噪声水平估计 363
		10.1.3 虚影 364
		10.1.4 光学模糊(空间响应)估计 365
	10.2	高动态范围成像 368
		10.2.1 色调映射 374
		10.2.2 应用：闪影术 380
	10.3	超分辨率和模糊去除 381
		10.3.1 彩色图像去马赛克 385
		10.3.2 应用：彩色化 387
	10.4	图像抠图和合成 388
		10.4.1 蓝屏抠图 389
		10.4.2 自然图像抠图 391
		10.4.3 基于优化的抠图 394
		10.4.4 烟、阴影和闪抠图 396
		10.4.5 视频抠图 397

10.5 纹理分析与合成 398
    10.5.1 应用：空洞填充
           与修图 400
    10.5.2 应用：非真实感绘制 401
10.6 补充阅读 403
10.7 习题 404

第 11 章 立体视觉对应 409

11.1 极线几何学 412
    11.1.1 矫正 412
    11.1.2 平面扫描 414
11.2 稀疏对应 416
11.3 稠密对应 418
11.4 局部方法 420
    11.4.1 亚像素估计
           与不确定性 422
    11.4.2 应用：基于立体视觉的
           头部跟踪 423
11.5 全局优化 424
    11.5.1 动态规划 425
    11.5.2 基于分割的方法 427
    11.5.3 应用：z-键控与背景
           替换 428
11.6 多视图立体视觉 429
    11.6.1 体积与 3D 表面重建 432
    11.6.2 由轮廓到形状 436
11.7 补充阅读 438
11.8 习题 439

第 12 章 3D 重建 443

12.1 由 X 到形状 444
    12.1.1 由阴影到形状与光度
           测量立体视觉 445
    12.1.2 由纹理到形状 447
    12.1.3 由聚焦到形状 448
12.2 主动距离获取 449
    12.2.1 距离数据归并 451

    12.2.2 应用：数字遗产 453
12.3 表面表达 454
    12.3.1 表面插值 454
    12.3.2 表面简化 455
    12.3.3 几何图像 456
12.4 基于点的表达 456
12.5 体积表达 457
12.6 基于模型的重建 459
    12.6.1 建筑结构 459
    12.6.2 头部和人脸 461
    12.6.3 应用：脸部动画 463
    12.6.4 完整人体建模与跟踪 465
12.7 恢复纹理映射与反照率 469
    12.7.1 估计 BRDF 470
    12.7.2 应用：3D 摄影学 471
12.8 补充阅读 472
12.9 习题 473

第 13 章 基于图像的绘制 477

13.1 视图插值 478
    13.1.1 视图相关的纹理映射 480
    13.1.2 应用：照片游览 481
13.2 层次深度图像 482
13.3 光场与发光图 484
    13.3.1 非结构化发光图 487
    13.3.2 表面光场 488
    13.3.3 应用：同心拼图 489
13.4 环境影像形板 490
    13.4.1 更高维光场 491
    13.4.2 从建模到绘制 492
13.5 基于视频的绘制 493
    13.5.1 基于视频的动画 493
    13.5.2 视频纹理 494
    13.5.3 应用：图片动画 497
    13.5.4 3D 视频 497

13.5.5 应用：基于视频的游览.................... 499
13.6 补充阅读 501
13.7 习题 503

第 14 章 识别 507

14.1 物体检测 509
    14.1.1 人脸检测 509
    14.1.2 行人检测 515
14.2 人脸识别 518
    14.2.1 特征脸 518
    14.2.2 活动表观与 3D 形状模型 525
    14.2.3 应用：个人照片收藏 528
14.3 实例识别 529
    14.3.1 几何配准 530
    14.3.2 大型数据库 531
    14.3.3 应用：位置识别 535
14.4 类别识别 537
    14.4.1 词袋 539
    14.4.2 基于部件的模型 542
    14.4.3 基于分割的识别 545
    14.4.4 应用：智能照片编辑 548
14.5 上下文与场景理解 550
    14.5.1 学习与大型图像收集 552
    14.5.2 应用：图像搜索 554
14.6 识别数据库和测试集 555
14.7 补充阅读 559
14.8 习题 562

第 15 章 结语 567

附录 A 线性代数与数值方法 569

A.1 矩阵分解 570
    A.1.1 奇异值分解 570
    A.1.2 特征值分解 571
    A.1.3 QR 因子分解 573
    A.1.4 乔里斯基分解 574
A.2 线性最小二乘 575
A.3 非线性最小二乘 578
A.4 直接稀疏矩阵方法 579
A.5 迭代方法 580
    A.5.1 共轭梯度 581
    A.5.2 预处理 582
    A.5.3 多重网格 583

附录 B 贝叶斯建模与推断 585

B.1 估计理论 586
B.2 最大似然估计与最小二乘 589
B.3 鲁棒统计学 590
B.4 先验模型与贝叶斯推断 591
B.5 马尔科夫随机场 592
    B.5.1 梯度下降与模拟退火 594
    B.5.2 动态规划 595
    B.5.3 置信传播 596
    B.5.4 图割 598
    B.5.5 线性规划 601
B.6 不确定性估计(误差分析) 602

附录 C 补充材料 604

C.1 数据集 605
C.2 软件 607
C.3 幻灯片与讲座 615
C.4 参考文献 615

词汇表 617

第1章
概 述

1.1 什么是计算机视觉？
1.2 简史
1.3 本书概述
1.4 课程大纲样例
1.5 标记法说明
1.6 扩展阅读

图 1.1 人类视觉系统可以毫不费力地解释这张照片中存在的由半透明材质和阴影形成的微妙变化，并从背景中将物体正确地分割出来

1.1 什么是计算机视觉？

作为人，我们显然很容易感知周围世界的三维结构。想一想，当我们观察身边桌上的一瓶花时，三维的感知是怎样的生动鲜明。通过光线和阴影在其表面显现出来的细微模式，可以分辨出每个花瓣的形状和其半透明性，毫不费力地从场景的背景中将每朵花分割出来(图 1.1)。观看一幅有边框的合影时，可以轻松地数出照片中所有的人(并说出名字)，甚至可以从他们的面部外观猜出其情感状态。感知心理学家已经花了几十年时间试图理解视觉系统是如何工作的，尽管他们能够想出光学错觉[①]来梳理其原理的某些部分(图 1.3)，但这个难题的完整解答仍然扑朔迷离(Marr 1982; Palmer 1999; Livingstone 2008)。

计算机视觉领域的研究人员同时也一直在研究恢复影像中物体的三维形状和外观的数学方法。目前我们已经有可靠的方法能够从几千幅部分重叠的照片精确地计算出环境的部分 3D 模型(图 1.2a)。如果有特定物体或建筑物正面足够多的一组视图，我们就可以使用立体匹配方法创建出稠密的 3D 表面模型(图 1.2b)。我们可以在复杂的背景中跟踪运动的人(图 1.2c)。使用人脸、衣服、头发的检测和识别相结合的方法，我们甚至可以试图找到照片中所有的人并说出他们的名字，此项任务可以获得中等的成功率(图 1.2d)。尽管有这些进展，但要让计算机解释图像的能力与两岁大的孩子(例如，数出照片中所有的动物)有一样的水平，这一梦想仍然是难以实现的。视觉为什么如此困难？部分原因为它是一个逆问题(inverse problem)，在信息不足的情况下，我们试图恢复一些未知量来给出完整的解答。因此，我们必须求助于基于物理的和基于概率的模型(model)来消除潜在解的歧义。然而，视觉世界的建模就其十分的复杂性远比产生话音的声道建模更困难。

在计算机视觉领域，我们使用的前向(forward)模型通常是在物理学(辐射测量学、光学和传感器设计)或计算机图形学中发展而来的。这两个领域都是在对如下过程进行建模：物体是如何运动和展示的，光线是怎样从表面反射、由空气散射、经由摄像机镜头(或人的眼睛)折射而最后投影到平的(或弯的)图像面上的。尽管计算机图形学还不完美(还没有完全由计算机生成的含有人物角色的电影可以成功地跨越这非比寻常的峡谷(uncanny valley)[②]，它将真正的人与机器人和计算机动画的人相区隔)，在有限的领域，比如渲染由日常物体构成的静态场景或像恐龙那样灭绝了的动物，真实感的幻觉是很完美的。

在计算机视觉领域，我们在试图做反过程，即描述我们从一幅或多幅图像中看到的世界，比如形状、照明和色彩分布。令人惊异的是，人和动物可以毫不费力地完成，而计算机视觉算法却很容易出错。没有在该领域工作过的人常低估该问题的

[①] http://www.michaelbach.de/ot/sze_muelue
[②] "非比寻常的峡谷"(uncanny valley)一词是机器人学家 Masahiro Mori 创造出来用于机器人学的(Mori 1970)。它也常用于计算机制作的动画电影中，比如《最终幻想》和《极地快车》(Geller 2008)。

困难程度。(工作中同事经常向我要软件,用来找到照片中所有的人并说出他们的名字,使他们可以继续做更"有趣"的工作。)"视觉应该是简单的"这一误解可以追溯到人工智能的早期岁月(参见 1.2 节),当时,人们最初认为智能的认知部分(逻辑证明和规划),本质上比感知部分更困难(Boden 2006)。

图 1.2 计算机视觉算法和应用的一些例子。(a)由运动到结构算法可以从几千幅部分重叠的照片重建起大规模复杂场景的稀疏 3D 点模型(Snavely, Seitz, and Szeliski 2006) 2006 ACM。(b)立体匹配算法可以从取自因特网的几百幅不同曝光的照片建立建筑物正面外观的详细 3D 模型(Goesele, Snavely, Curless *et al.* 2007)© 2007 IEEE。(c)人的跟踪算法可以在零乱的背景中跟踪行走的人(Sidenbladh, Black, and Fleet 2000)© 2000 Springer。(d)人脸检测算法,与基于颜色的衣服和头发检测算法相配合,可以定位并识别图像中的个人(Sivic, Zitnick, and Szeliski 2006)© 2006 Springer

好消息是计算机视觉如今正广泛应用于各种各样的实际应用之中,如下所示。

- **光学字符识别(OCR)**:阅读信上的手写邮政编码(图 1.4a) 和自动号码牌识别(ANPR)。
- **机器检验**:为保证质量而快速检验部件,使用立体视觉在专门的光照下测量飞机机翼或汽车车身配件的容差(图 1.4b),或使用 X 光视觉检查钢铸件的缺欠。
- **零售**:针对自动结账通道的物体识别(图 1.4c)。
- **3D 模型建立(摄影测量学)**:在诸如 Bing Maps 等系统中使用的从航空照片完全自动地构建 3D 模型。
- **医学成像**:注册手术前和手术中的成像(图 1.4d),或进行关于人老化过程中大脑形态的长期研究。

- **汽车安全**：在诸如雷达或激光雷达等主动视觉技术不好用时检测意外的障碍物，比如街道上的行人(图 1.4e，完全自动驾驶的例子请参见 Miller, Campbell, Huttenlocher *et al.* (2008); Montemerlo, Becker, Bhat *et al.* (2008); Urmson, Anhalt, Bagnell *et al.* (2008))。
- **匹配运动**：通过跟踪源视频中的特征点来估计摄像机的 3D 运动和环境的形状，将计算机生成的影像(CGI)与实景真人动作脚本相融合。这样的技术在好莱坞得到广泛使用(例如电影《侏罗纪公园》)(Roble 1999; Roble and Zafar 2009)；他们还需要使用精确的抠图来在前景和背景之间插入新的元素(Chuang, Agarwala, Curless *et al.* 2002)。
- **运动捕捉(mocap)**：使用从多台摄像机拍摄反光材料标记或其他视觉方法来捕捉演员的动作，以便用于计算机动画。
- **监视**：监控入侵者，分析高速公路的交通状况(图 1.4f)，监控游泳池以防溺水事件的发生。
- **指纹识别和生物测定学**：用于自动准入身份验证以及司法应用。

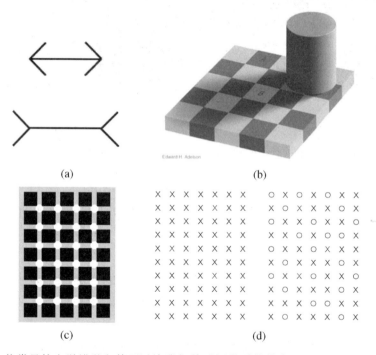

图 1.3 一些常见的光学错觉和其可以给我们关于视觉系统的启示。(a)经典的 Muller-Lyer 错觉，两条水平线条的长度显得不同，可能是由于想象的透视效果。(b)阴影中的"白"块 B 和亮处的"黑"块 A 实际上具有相同的绝对亮度值。该感觉是由于亮度恒常性，即视觉系统当解释颜色时试图减少亮度的特性。图像承蒙 Ted Adelson 的允许，*http://web.mit.edu/persci/people/adelson/checkershadow illusion.html*。(c)Hermann 网格错觉的一个变型，承蒙 Hany Farid 许可使用，*http://www.cs.dartmouth.edu/_farid/illusions/hermann.html*。当你移动眼睛扫视该图时，灰色的斑点会出现在交叉处。(d)在图的左半部分数一数红色的 X。现在，在右半部分数。是不是明显困难了？其解释与弹出(pop-out)效应有关(Treisman 1985)，它告诉我们大脑中有关并行感知和集成通道的操作

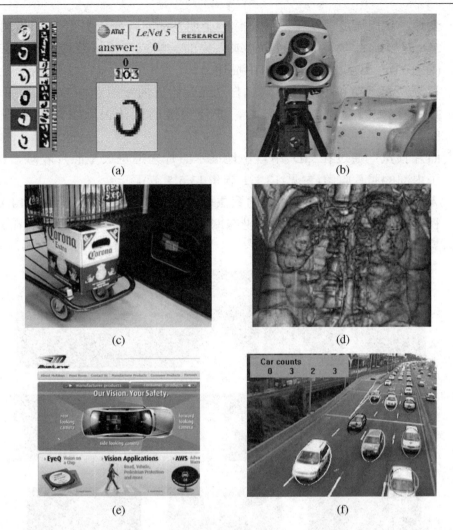

图 1.4 计算机视觉的一些工业应用：(a)光学字符识别(OCR)，http://yann.lecun.com/exdb/lenet/；(b)机械检验，http://www.cognitens.com/；(c)零售，http://www.evoretail.com/；(d)医学成像，http://www.clarontech.com/；(e)汽车安全，http://www.mobileye.com/；(f)监视和交通监控，http://www.honeywellvideo.com/，承蒙 Honeywell 国际公司许可使用

David Lowe 的工业视觉应用网页(*http://www.cs.ubc.ca/spider/lowe/vision.html*)列出了计算机视觉的其他很多有趣的工业应用。尽管前面提到的应用都非常重要，但是它们绝大多数属于种类相当特殊的成像和狭窄的领域。

在本书中，我们更多关注更广的消费类应用，比如使用你自己的个人照片和视频可以做的有趣的事情。具体包括下面这些应用。

- **拼图**：将有重叠的照片变成无缝拼接起来的单张全景画(图 1.5a)，该方法在第 9 章中描述。
- **曝光包围**：将在困难光照(强烈阳光和阴影)条件下拍摄的多个曝光照片融合成单张完美曝光的图像(图 1.5b)，该方法在 10.2 节中描述。

- **变形**：使用无缝的变形过渡将你一个朋友的照片变成另一个人的(图 1.5c)。
- **3D 建模**：将你所拍摄的物体或人物的一幅或多幅快照转变为其 3D 模型(图 1.5d)，该方法在 12.6 节中描述。
- **视频匹配运动和稳定化**：通过自动跟踪附近的参照点将 2D 图片或 3D 模型插入视频中(参见 7.4.2 节)[①]，或使用运动估计去除视频中的抖动(参见 8.2.1)。
- **照片预览**：通过在 3D 中自由地展示照片来巡览大型照片收藏，比如你房子内部的照片收藏(参见 13.1.2 节和 13.5.5 节)。
- **人脸检测**：用于改进照相机的聚焦以及更相关图像的搜索(参见 14.1.1 节)。
- **视觉身份认证**：当家庭成员坐在网络摄像头前时，自动给他们在家里的计算机上登录(参见 14.2 节)。

图 1.5 计算机视觉的一些消费层应用。(a)图像拼接：合并不同的视图(Szeliski and Shum 1997) ©1997 ACM。(b)曝光包围：合并不同的曝光。(c)在两幅照片间变形混合(Gomes, Darsa, Costa *et al*. 1999)©1999 Morgan Kaufmann。(d)将一组照片转化成 3D 模型 (Sinha, Steedly, Szeliski *et al*. 2008)©2008 ACM

---

① 关于这个主题的一个有趣的学生课程设计，请参见"PhotoBook"，*http://www.cc.gatech.edu/dvfx/videos/dvfx2005.html*。

这些应用的突出优点是它们已经为大多数学生所熟悉；它们至少是学生通过自己的个人媒体可以马上体会和使用的方法。由于计算机视觉是一个具有挑战性的题目，考虑到所覆盖数学的广度[①]和待解决问题固有的困难性，使工作问题相关而有趣可以大大地激励和鼓舞人们。

本书之所以重点关注于应用，主要还因为它们可以用于形成并约束视觉所特有的原本开放性的问题。例如，如果有人找我要好的边缘检测器，我的首个问题通常会是问为什么？他们试图解决什么类型的问题且为什么他们认为边缘检测是一个重要的成分？如果他们想定位人脸，我通常会指出最成功的人脸检测器使用肤色检测(习题 2.8)和 14.1.1 节的简单的团块(blob)特征；它们并不依赖边缘检测。如果他们是为了 3D 重建的目的而想匹配建筑物上的门窗边缘，我会告诉他们边缘是好主意，但是最好是调整其边缘检测器适合检测长的边缘(参见 3.2.3 节和 4.2 节)并将其连接起来形成具有共同消失点的直线，然后再匹配(参见 4.3 节)。

因此，与其抓住你可能听到的第一个方法，不如从你手头的问题回想所适用的方法。这种从问题到解答的反向工作方式是视觉研究的典型工程(engineering)方法，这反映了我自身在本领域的背景。首先，我提出详细的问题定义并决定问题的约束和技术参数。然后，我试图找到已知有效的那些方法，实现其中的一些，评测其性能，最终确定一个选择。为了使这一过程有效，有实际可行的测试数据(test data)是重要的，这既包括合成的数据，它可用于验证正确性并分析其噪声敏感性，也包括系统最终将要用到的那种类型的真实世界数据。

但是，本书并不只是一本工程教材(方法的出处)。对于基本的视觉问题，本书也采用了科学方法。我试图给手头的系统提出最可能的物理模型：场景是如何创建的，光线是如何与场景和大气影响相互作用的，以及传感器是如何工作的，包括噪声源和不确定性。随后的任务就是将获取过程逆转过来，提出场景的最可能的描述。

本书经常使用统计方法来阐述和解决计算机视觉问题。在适当之处，使用概率分布来对场景和带噪声的图像获取过程建模。将未知量与先验分布联合通常称作"贝叶斯建模"(附录 B)。可以将错误解答与风险或损失函数结合起来(B.2 节)，建立起你的推断算法来最小化期望风险(考虑机器人试图估计到障碍物的距离：通常低估比高估来得更安全)。对于统计方法，收集大量的训练数据来学习概率模型通常是有益的。最终，统计方法能够使你使用可证明的推断方法来估计最好的解答(或解答的分布)，并对估计结果的不确定性做量化。

因为这么多计算机视觉涉及逆问题的求解或未知量的估计，所以本书也很重视算法，特别是那些在实践中行之有效的算法。对于很多计算机视觉问题，提出问题的一个数学描述却不能与真实世界条件相匹配抑或不能给出未知量的稳定估计，这太容易了。我们需要的算法是，既要对噪声和对模型的偏离鲁棒(robust)，也要在运行时资源和空间效率高的算法。在本书中，我将详细叙述这些问题，在适用之处

[①] 这些方法包括物理学，欧氏和射影几何，统计学以及优化。它们使计算机视觉成为一个极有吸引力的研究领域和一个很好的学习在其他领域也有广泛用途的方法之途径。

使用贝叶斯方法,以确保鲁棒性、高效搜索、最小化以及确保效率高的线性系统解决算法。本书所描述的大多数算法都是高层次的,主要是步骤列表,必须由学生填补或阅读其他文献中更详细的描述。事实上,很多算法是在习题中勾勒出来的。

现在我已经描述了本书的宗旨和所使用的框架,本章的其余部分将用于另外两个主题。1.2 节是计算机视觉历史的简要梗概。想获得本书新素材之"精髓"而不太关心谁在何时发明什么的读者,可以简单地跳过这一节。

第二个主题是本书内容的概述,1.3 节,对于想要了解这一主题的所有人来说,阅读这个小节是有用的(或在中途跳入,因为它描述的是章节间的依赖关系)。这个大纲对于想要在本主题上安排一门或多门课程的教师也是有用的,因为它提供了基于本书内容的课程案例。

1.2 简　　史

在本节中,我提供一个简要的针对计算机视觉过去 30 年主要发展整理出来的梗概(图 1.6),至少包括我个人感兴趣的并且看起来经受住了时间检验的发展。对于各种概念的出处和本领域的演化不感兴趣的读者可以略过前往阅读 1.3 节的本书概述。

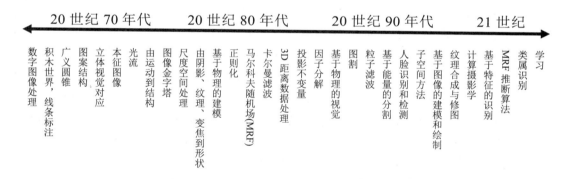

图 1.6　计算机视觉领域最活跃的一些主题大致的时间轴

20 世纪 70 年代　计算机视觉最初开始于 20 世纪 70 年代早期时,被视为模拟人类智能并赋予机器人智能行为的一项雄心勃勃议程的感知组成部分。在那时,人工智能和机器人学的一些早期先驱者(在 MIT,Stanford 和 CMU)认为,在解决诸如更高层次推理和规划的更困难问题的过程中解决"视觉输入"问题应该是一个简单的步骤。根据一个广为人知的故事,在 1966 年,MIT 的 Marvin Minsky 让他的本科生 Gerald Jay Sussman "在暑期将摄像机连接到计算机上,让计算机来描述它所看到的东西"(Boden 2006, p.781)①。今天,我们知道问题其实要比当时困难

---

① Boden (2006) 将 (Crevier 1993) 引用为最初的出处。实际的视觉备忘录是 Seymour Papert (1966) 写的,涉及整群学生。

一些[①]。

将计算机视觉与已经存在的数字图像处理领域(Rosenfeld and Pfaltz 1966; Rosenfeld and Kak 1976)相区别的是期望从图像恢复世界的三维结构并以此为跳板得出完整的场景理解。Winston(1975)和 Hanson and Riseman(1978)提供了这个早期时代的两本比较好的经典论文集。

场景理解的早期尝试涉及物体即"积木世界"的边缘抽取及随后的从 2D 线条的拓扑结构推断其 3D 结构(Roberts 1965)。那时提出了一些线条标注算法(Huffman 1971; Clowes 1971; Waltz 1975; Rosenfeld, Hummel, and Zucker 1976; Kanade 1980)。Nalwa (1993)对该领域进行了很好的回顾。那时边缘检测也是一个活跃的研究领域,对同时代工作的一个好的综述可见(Davis 1975)。

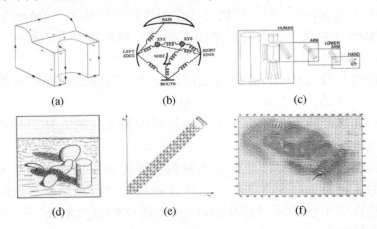

图 1.7 计算机视觉算法一些早期(20 世纪 70 年代)的例子:(a)线条标注(Nalwa 1993)©1993 Addison-Wesley;(b)图案结构(Fischler and Elschlager 1973)©1973 IEEE;(c)关节身体模型 (Marr 1982)©1982 David Marr;(d)本征图像 (Barrow and Tenenbaum 1981)©1973 IEEE;(e)立体视觉对应(Marr1982)©1982 David Marr;(f)光流(Nagel and Enkelmann 1986)©1986 IEEE

人们还对非多边形物体的三维建模进行了研究(Baumgart 1974; Baker 1977)。一种流行的做法是使用广义锥,即旋转体和封闭曲线扫描体(Agin and Binford 1976; Nevatia and Binford 1977),通常安排成部件关系[②](Hinton 1977; Marr 1982)(图 1.7c)。Fischler and Elschlager(1973)称这种部件的弹性安排为"图案结构"(pictorial structures)(图 1.7b)。这是目前物体识别领域受到推崇的方法(参见 14.4 节和 Felzenszwalb and Huttenlocher 2005)。Barrow and Tenenbaum(1981)在其关于本征图像(intrinsic image)(图 1.7d)的论文中,与 Marr(1982)的 $2_{1/2}$-D 简图(sketch)概念一起,提出了一种广受支持的理解亮度和阴影变化,并通过诸如表面朝向和阴影等图像形

[①] 要想知道过去四十年机器人视觉的长足发展,从毛巾折叠机器人就可见一斑,*http://rll.eecs.berkeley.edu/pr/icra10/* (Maitin-Shepard, Cusumano-Towner, Lei *et al*. 2010)。

[②] 在机器人和计算机动画领域,这些连接起来的部件图常常称为"运动链"(kinematic chain)。

成现象的效果对其进行解释的定性方法。这种方法在 Tappen, Freeman, and Adelson (2005)的工作中可以看到一点复兴的迹象。

那时也出现了一些更定量化的计算机视觉方法，包括众多基于特征的立体视觉对应(stereo correspondence)算法中最早出现的一些(图 1.7e)(Dev 1974; Marr and Poggio 1976; Moravec 1977; Marr and Poggio 1979; Mayhew and Frisby1981; Baker 1982; Barnard and Fischler 1982; Ohta and Kanade 1985; Grimson 1985; Pollard, Mayhew, and Frisby 1985; Prazdny 1985)和基于亮度的光流(optical flow)算法(图 1.7f)(Horn and Schunck 1981; Huang 1981; Lucas and Kanade 1981; Nagel 1986)。同时恢复 3D 结构和摄像机运动的早期工作(参见第 7 章)这时也开始出现(Ullman 1979; Longuet-Higgins 1981)。

David Marr(1982)总结了那个时代对视觉工作原理的认识[①]。Marr 特别介绍了其关于(视觉)信息处理系统的表达的三个层次概念。根据我个人的解释，对这三个层次做如下非常松散的释义。

- 计算理论：计算(任务)的目的是什么？该问题的已知或可以施加的约束是什么？
- 表达和算法：输入、输出和中间信息是如何表达的？使用哪些算法来计算所期望的结果？
- 硬件实现：表达和算法是如何映射到实际硬件即生物视觉系统或特殊的硅片上的？相反地，硬件的约束怎样才能用于指导表达和算法的选择？随着计算机视觉中使用图形芯片(GPU)和多核结构日益增长，这个问题再次变得相当重要。

正如早前提到的那样，我深信仔细分析问题的定义和已知的源于图像形成和先验的约束(科学的和统计的方法)必须与效率高的鲁棒的算法(工程的方法)相结合。因此在我们的领域中，Marr 的方法论在今天仍然是表达和解决问题的好向导，正如 25 年前一样。

20 世纪 80 年代　在 20 世纪 80 年代，很多研究关注于定量的图像和场景分析的更复杂的数学方法。

图像金字塔(参见 3.5 节)开始广泛用于完成诸如图像混合这样的任务(图 1.8a)和由粗到精的对应搜索(Rosenfeld 1980; Burt and Adelson 1983a,b; Rosenfeld 1984; Quam 1984; Anandan 1989)。使用尺度空间(scale-space)处理的概念也建立起了金字塔的连续版本(Witkin 1983; Witkin, Terzopoulos, and Kass 1986; Lindeberg 1990)。在 20 世纪 80 年代后期，小波(参见 3.5.4 节)在一些应用中开始取代或增强规范的图像金字塔(Adelson, Simoncelli, and Hingorani 1987; Mallat 1989; imoncelli and Adelson 1990a,b; Simoncelli, Freeman, Adelson *et al.* 1992)。

作为定量的形状线索使用的立体视觉扩展到由 X 到形状的各种各样的方法，包

---

① 视觉感知理论方面最新近的发展可参见(Palmer 1999; Livingstone 2008)。

括由阴影到形状(图 1.8b)(参见 12.1.1 节和 Horn 1975; Pentland 1984; Blake, Zimmerman, and Knowles 1985; Horn and Brooks 1986, 1989)、光度测定学立体视觉(photometric stereo)(参见 12.1.1 节和 Woodham 1981)、由纹理到形状(参见 12.1.2 节和 Witkin 1981; Pentland 1984; Malik and Rosenholtz 1997)以及由聚焦到形状(参见 12.1.3 节和 Nayar, Watanabe, and Noguchi 1995)。Horn(1986)给出了关于大多数这类方法的一个很好的论述。

这一时期，探寻更好的边缘和轮廓检测方法(图 1.8c)(参见 4.2 节)也是一个活跃的研究领域(Canny 1986; Nalwa and Binford 1986)，包括动态演化轮廓跟踪器的引入(5.1.1 节)，例如蛇行(Kass, Witkin, and Terzopoulos 1988)，还有基于三维物理量的模型的引入(图 1.8d)(Terzopoulos, Witkin, and Kass 1987; Kass, Witkin, and Terzopoulos 1988; Terzopoulos and Fleischer 1988; Terzopoulos, Witkin, and Kass 1988)。

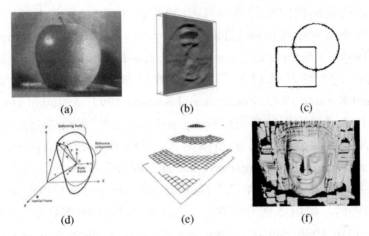

图 1.8 20 世纪 80 年代的计算机视觉算法的例子：(a)金字塔混合(Burt and Adelson 1983b)©1983 ACM；(b)由阴影到形状(Freeman and Adelson 1991)©1991 IEEE；(c)边缘检测(Freeman and Adelson 1991) ©1991 IEEE；(d)基于物理量的模型(Terzopoulos and Witkin 1988) ©1988 IEEE；(e)基于正则化的表面重建(Terzopoulos 1988)©1988 IEEE；(f)距离数据获取和归并(Banno, Masuda, Oishi et al. 2008)©2008 Springer

研究人员发现，很多立体视觉、流、由 X 到形状以及边缘检测算法，如果作为变分优化问题(参见 3.7 节)来处理，可以用相同的数学框架来统一或至少来描述，且可以使用正则化方法(图 1.8e)使其更鲁棒(适定的)(参见 3.7.1 节和 Terzopoulos 1983; Poggio, Torre, and Koch 1985; Terzopoulos 1986b; Blake and Zisserman 1987; Bertero, Poggio, and Torre 1988; Terzopoulos 1988)。差不多同时，Geman and Geman (1984)指出这类问题同样可以用离散马尔科夫随机场模型(Markov Random Field，MRF)来很好地表达(参见 3.7.2 节)，这样就能使用更好的(全局)搜索和优化算法，比如"模拟退火"(simulated annealing)。

稍后出现了使用卡尔曼滤波(Kalman filter)来对不确定性进行建模和更新的

MRF 算法的在线变形(Dickmanns and Graefe 1988; Matthies, Kanade, and Szeliski 1989; Szeliski 1989)。人们也尝试了将正则化及 MRF 算法映射到并行硬件(Poggio and Koch 1985; Poggio, Little, Gamble et al. 1988; Fischler, Firschein, Barnard et al. 1989)。Fischler and Firschein(1987)是一个很好的相关论文集,涵盖了所有的这些主题(立体视觉、流、正则化、MRF,甚至更高层视觉)。

三维距离数据(range data)处理(获取、归并、建模和识别,参见图 1.8f)继续成为这十年很活跃的研究领域(Agin and Binford 1976; Besl and Jain 1985; Faugeras and Hebert 1987; Curless and Levoy 1996)。Kanade(1987)编辑的书包含该领域中很多有趣的论文。

20 世纪 90 年代　虽然前面提到的很多主题都仍然在研究之中,但是其中一些问题明显更活跃。

在识别中使用投影不变量的研究呈现爆发性增长(Mundy and Zisserman1992),演变为解决从运动到结构问题的共同努力(参见第 7 章)。最初很多研究是针对投影重建问题的,它不需要摄像机标定的结果(Faugeras 1992; Hartley, Gupta, and Chang 1992; Hartley 1994a; Faugeras and Luong 2001; Hartley and Zisserman 2004)。与此同时,提出了因子分解方法(7.3 节)来高效地解决近似正交投影的问题(图 1.9a)(Tomasi and Kanade 1992; Poelman and Kanade 1997; Anandan and Irani 2002),接着后来扩展到了透视投影的情况(Christy and Horaud 1996; Triggs 1996)。最终,该领域开始使用完全的全局优化方法(参见 7.4 节和 Taylor, Kriegman, and Anandan 1991; Szeliski and Kang 1994; Azarbayejani and Pentland 1995),这在后来被认为与摄影测量学中常用的"光束平差法"(bundle adjustment)相同(Triggs, McLauchlan, Hartley et al. 1999)。使用这些方法建立了完全自动的(稀疏)3D 建模系统(Beardsley, Torr, and Zisserman 1996; Schaffalitzky and Zisserman 2002; Brown and Lowe 2003; Snavely, Seitz, and Szeliski 2006)。

使用颜色和亮度的精细测量,并与精确的辐射传输和形成彩色图像的物理模型相结合,这方面的工作始于 20 世纪 80 年代,构成一个称作"基于物理的视觉"(physics-based vision)的子领域。在本主题的一个三卷本汇集(Wolff, Shafer, and Healey 1992a; Healey and Shafer 1992; Shafer, Healey, and Wolff 1992)中可以找到该领域一个很好的综述。

光流方法(参见第 8 章)得到不断的改进(Nagel and Enkelmann 1986; Bolles, Baker, and Marimont 1987; Horn and Weldon Jr. 1988; Anandan 1989; Bergen, Anandan, Hanna et al. 1992; Black and Anandan 1996; Bruhn, Weickert, and Schnörr 2005; Papenberg, Bruhn, Brox et al. 2006),很好的综述参见(Nagel 1986; Barron, Fleet, and Beauchemin 1994; Baker, Black, Lewis et al. 2007)。类似地,在稠密立体视觉对应算法方面也取得了很多进展(参见第 11 章,Okutomi and Kanade(1993, 1994); Boykov, Veksler, and Zabih(1998); Birchfield and Tomasi(1999); Boykov, Veksler, and Zabih(2001),综述和比较参见 Scharstein and Szeliski(2002)),其中最大

的突破可能是使用"图割"(graph cut)方法的全局优化算法(图 1.9b)(Boykov, Veksler, and Zabih 2001)。

可以产生完整 3D 表面的多视角立体视觉算法(图 1.9c)(参见 11.6 节)也是一个重要的研究主题(Seitz and Dyer 1999; Kutulakos and Seitz 2000),至今仍然非常活跃(Seitz, Curless, Diebel et al. 2006)。从二值的轮廓产生 3D 体描述的方法(参见 11.6.2 节)仍在研究中(Potmesil 1987; Srivasan, Liang, and Hackwood 1990; Szeliski 1993; Laurentini 1994),还有基于跟踪的方法和重建光滑的遮挡轮廓的研究(参见 11.2.1 节和 Cipolla and Blake 1992; Vaillant and Faugeras 1992; Zheng 1994; Boyer and Berger 1997; Szeliski and Weiss 1998; Cipolla and Giblin 2000)。

跟踪算法也得到了很多改进,包括使用"活动轮廓"(active contours)方法(参见 5.1 节)的轮廓跟踪,例如蛇行(snake)(Kass,Witkin, and Terzopoulos 1988)、粒子滤波器(particle filter)(Blake and Isard 1998)和水平集(level set)方法(Malladi, Sethian, and Vemuri 1995),还有基于亮度的(直接)方法(Lucas and Kanade 1981; Shi and Tomasi 1994; Rehg and Kanade 1994),常用于跟踪人脸(图 1.9d)(Lanitis, Taylor, and Cootes 1997; Matthews and Baker 2004; Matthews, Xiao, and Baker 2007)和整个物体(Sidenbladh, Black, and Fleet 2000; Hilton, Fua, and Ronfard 2006; Moeslund, Hilton, and Krüger 2006)。

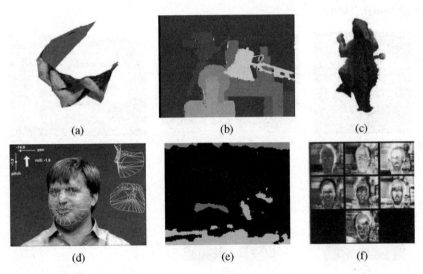

图 1.9　20 世纪 90 年代计算机视觉算法的例子:(a)基于因子分解的从运动到结构(Tomasi and Kanade 1992) © 1992 Springer;(b)稠密立体视觉匹配(Boykov, Veksler, and Zabih 2001);(c)多视角重建(Seitz and Dyer 1999)©1999 Springer;(d)人脸跟踪(Matthews, Xiao, and Baker 2007);(e)图像分割(Belongie, Fowlkes, Chung et al. 2002)©2002 Springer;(f)人脸识别(Turk and Pentland 1991a)

图像分割(参见第 5 章)(图 1.9e),从计算机视觉的早期开始就一直是一个重要的方向(Brice and Fennema 1970; Horowitz and Pavlidis 1976; Riseman and Arbib 1977;

Rosenfeld and Davis 1979; Haralick and Shapiro 1985; Pavlidis and Liow 1990)，也是一个活跃的研究主题，产生了基于最小能量的方法(Mumford and Shah 1989)和最小描述长度方法(Leclerc 1989)，规范化割(normalized cuts)方法(Shi and Malik 2000)以及均值移位(mean shift)方法(Comaniciu and Meer 2002)。

统计学习方法开始逐渐流行起来，最初应用于人脸识别的主分量本征脸分析(图 1.9f)(参见 14.2.1 节和 Turk and Pentland 1991a)和曲线跟踪的线性动态系统(参见 5.1.1 节和 Blake and Isard 1998)。

这十年以来，计算机视觉领域最显著的一个发展是与计算机图形学之间的交互增多了(Seitz and Szeliski 1999)，特别是在基于图像的建模和绘制这个交叉学科领域(参见第 13 章)。直接操作真实世界的影像来创建动画的想法最初是从图像变形方法开始变得显著起来的(图 1.5c)(参见 3.6.3 节和 Beier and Neely 1992)，后来用于视角插值(Chen and Williams 1993; Seitz and Dyer 1996)、全景图拼接(图 1.5a)(参见第 9 章和 Mann and Picard 1994; Chen 1995; Szeliski 1996; Szeliski and Shum 1997; Szeliski 2006a)和全光场绘制(图 1.10a)(参见 13.3 节和 Gortler, Grzeszczuk, Szeliski et al. 1996; Levoy and Hanrahan 1996; Shade, Gortler, He et al. 1998)。同时，也出现了从图像汇集自动创建具有真实感 3D 模型的基于图像的建模方法(图 1.10b)(Beardsley, Torr, and Zisserman 1996; Debevec, Taylor, and Malik 1996; Taylor, Debevec, and Malik 1996)。

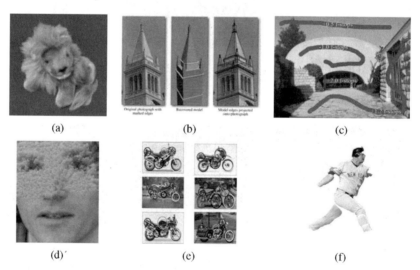

图 1.10 算机视觉算法的近期例子：(a)基于图像的绘制(Gortler, Grzeszczuk, Szeliski et al. 1996)；(b)基于图像的建模(Debevec, Taylor, and Malik 1996) © 1996 ACM；(c)交互色调映射(Lischinski, Farbman, Uyttendaele et al. 2006a)；(d)纹理合成(Efros and Freeman 2001)；(e)基于特征的识别(Fergus, Perona, and Zisserman 2007)；(f)基于区域的识别(Mori, Ren, Efros et al.2004) © 2004 IEEE

21 世纪 过去的十年继续见证了视觉与图形学相互影响的加深。特别明显的是，在基于图像的绘制这一标题下引入许多主题，例如图像拼接(参见第 9 章)、光场获取和绘制(参见13.3节)以及通过曝光包围(exposure bracketing)进行的高动态范围(high dynamic range, HDR))图像捕捉(图1.5b)(参见10.2节和Mann and Picard 1995; Debevec and Malik 1997)，这些主题被重新命名为计算摄影学(computational photography)(参见第10章)，作为对这些方法在日常数字摄影中日益增多应用的承认。例如，通过曝光包围创建高动态范围图像得到快速采纳，使得色调映射(tone mapping)(图1.10c) (参见10.2.1节)算法发展成为必须，以便将这样的图像转变为可显示的结果(Fattal, Lischinski, and Werman 2002; Durand and Dorsey 2002; Reinhard, Stark, Shirley et al. 2002; Lischinski, Farbman, Uyttendaele et al. 2006a)。除了归并多个曝光之外，也出现了将闪光图像和其对应的非闪光图像归并的方法(Eisemann and Durand 2004; Petschnigg, Agrawala, Hoppe et al. 2004)以及交互地或自动地从交叠的图像中选择不同区域的方法(Agarwala, Dontcheva, Agrawala et al. 2004)。

纹理合成(图1.10d)(参见10.5节)、绗缝(quilting)(Efros and Leung 1999; Efros and Freeman 2001; Kwatra, Schödl, Essa et al. 2003)和修图(inpainting)(Bertalmio, Sapiro, Caselles et al. 2000; Bertalmio, Vese, Sapiro et al. 2003; Criminisi, Pérez, and Toyama 2004)是可以划入计算摄影学的另外一些主题，因为它们将输入图像样例重新结合以产生新的照片。

过去十年的第二个显著趋势是物体识别中基于特征方法(与学习方法相结合)的显现(参见14.3节和Ponce, Hebert, Schmid et al. 2006)。该领域著名的论文有Fergus, Perona, and Zisserman(2007)的星群模型(constellation model)(图1.10e)和Felzenszwalb and Huttenlocher(2005)的图案结构(pictorial structures)。基于特征的方法也主导了其他识别任务，例如场景识别(Zhang, Marszalek, Lazebnik et al. 2007)和全景画以及位置识别(Brown and Lowe 2007; Schindler, Brown, and zeliski 2007)。虽然兴趣点(interest point)(基于块的)特征趋于主导着当前的研究，有些研究小组也在从事基于轮廓的识别研究(Belongie, Malik, and Puzicha 2002)和基于区域分割的识别研究(图1.10f)(Mori, Ren, Efros et al. 2004)。

过去十年的另外一个显著趋势是发展更高效求解复杂全局优化问题的算法(参见3.7节和B.5节以及Szeliski, Zabih, Scharstein et al. 2008; Blake, Kohli, and Rother 2010)。虽然这一趋势始于图割方面的工作(Boykov, Veksler, and Zabih 2001; Kohli and Torr 2007)，但在消息传递算法方面也取得了很多进展，例如，带环的置信传播(loopy belief propagation，LBP)(Yedidia, Freeman, and Weiss 2001; Kumar and Torr 2006)。

最后一个趋势是复杂的机器学习方法在计算机视觉问题中的应用，目前主导着我们领域中的很多视觉识别研究(参见14.5.1节和Freeman, Perona, and Schölkopf 2008)。这一趋势与因特网上日益增多的、可获得的、巨量的和部分标记数据相一致，这使学习物体类属(object category)更为可行，而不用人的精心监督。

1.3 本书概述

最后，我对本书的材料以及标记法注释和一些补充的一般性参考文献做一概览。由于计算机视觉是一个如此之宽的领域，所以我们完全有可能只研究其某些方面，(例如图像形成的几何和 3D 结构恢复)而不涉猎其他部分(例如反射和阴影的建模)。本书的一些章节只是松散地与其他部分关联，所以不要求读者严格按照顺序阅读所有的材料。

图 1.11 展示了本书内容的大致布局。由于计算机视觉涉及从场景的图像到结构描述(而计算机图形学则反过来)，因此我将章节安排除了在纵向根据其依赖性安排之外，在横向则按照其论述的主要成分展开。

从左到右，我们看到的主要列标题是图像(本质上是 2D 的)、几何(包含 3D 描述)、光度学(包含物体表观)(后两个的另一种标注也可以是形状和表观——参见第 13 章和 Kang, Szeliski, and Anandan(2000))。从上到下，我们看到建模和抽象的层次在增加，还有建立在前面算法基础上的方法。当然，这种分类法应该以有所保留的态度来看待，因为该图中的处理和依赖关系并不是严格有序的，还存在其他精细的依赖和联系(例如，有些识别方法使用 3D 信息)。横向排列的主题也应该松散地看待，因为多数视觉算法至少涉及两种不同的表达。[①]

全书中散布着应用样例，将各章中所论述的算法和数学材料与有用的真实世界应用关联起来。有很多应用出现在习题中，学生可以自己动手做。

每章的结尾，我提供了一组习题，借此学生可以用于实现、测试并改进每章所讲的算法和方法。有些习题可作为书面作业布置，有些可作为较短的一周课程设计，另外一些可作为开放的研究问题用作具有挑战性的最终课程设计。到本书结束时，积极性高的学生，他们实现了这些习题中的相当一部分，会建立起一个计算机视觉的软件库，可用于各种有趣的任务和项目。

作为一本参考书，我试图不仅提供指向本书所覆盖领域的最新研究结果的链接，而且尽可能讨论哪些方法和算法在实践中行之有效。习题可用于构建你自己的个人视觉算法库，它包含了你自己测试过并确认有效的算法。这从长期来看，比简单地从库中抽取不知其性能的算法更值得(假若你有时间的话)。

本书从第 2 章开始，概述创建我们所看到和捕捉的图像的图像形成过程。如果你想采用科学的(基于模型的)方法来研究计算机视觉，理解这个过程是必要的。渴望直接开始实现算法的学生(或者课程的时间很有限)可以向前跳过到第 3 章，晚些时候再来浏览该材料。在第 2 章，我们将图像形成分为三个主要部分。几何的图像形成(2.1 节)处理点、线、面，以及使用投影几何和其他模型(包括径向镜头畸变)是

[①] 关于与已知的人类视觉系统的有趣比较，例如在很大程度上是并行的是什么和在哪里的途径，参见某些关于人类感知的教材(Palmer 1999; Livingstone 2008)。

如何映射到图像上的。光度测定学的图像形成(2.2 节)涵盖辐射测量学和光学，前者描述光是如何与世界中的表面相互作用的，后者将光投影到传感器平面上。最后，2.3 节涵盖了传感器是如何工作的，包括诸如采样和混叠、色彩感知和摄像机内的压缩。

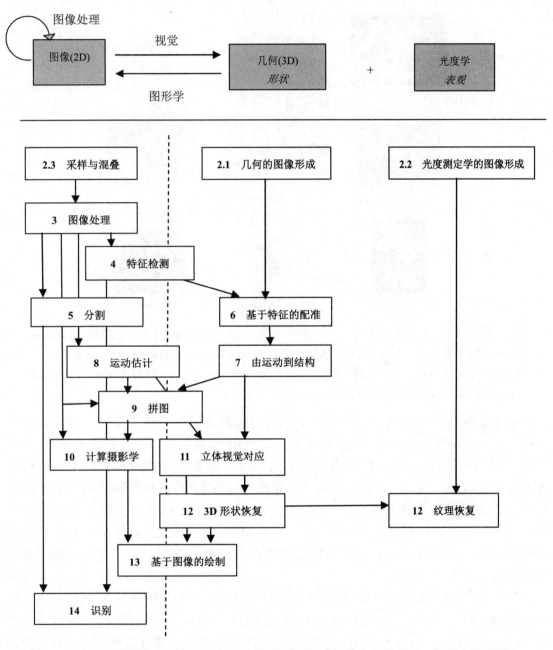

图 1.11 本书所覆盖的图像、几何、光度学以及主题分类之间的关系。主题大致根据它们是与基于图像的(左)、基于几何的(中)或基于表观的(右)的表达哪个更接近来沿着左右轴位置布置，而纵轴基于抽象层次的增加来安排。整个图应该以有所保留的态度来看待，因为有很多主题间另外的细微联系没有体现出来

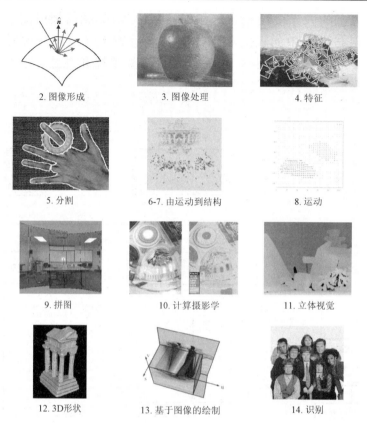

图 1.12 章节内容的图案提纲。源自 Brown, Szeliski, and Winder(2005); Comaniciu and Meer(2002); Snavely, Seitz, and Szeliski(2006); Nagel and Enkelmann(1986); Szeliski and Shum(1997); Debevec and Malik(1997); Gortler, Grzeszczuk, Szeliski *et al.* (1996); Viola and Jones(2004)——版权信息参见相应各章中的图

第 3 章覆盖图像处理,在几乎所有的计算机视觉应用中都需要它。这包括诸如线性和非线性滤波(3.3 节),傅里叶变换(3.4 节),图像金字塔和小波(3.5 节),诸如图像卷绕(image warping)的几何变换(3.6 节),诸如正则化和马尔科夫随机场(Markov Random Fields,MRFs))的全局优化方法(3.7 节)。虽然这些材料的大部分都包含在图像处理的课程和教材中,但优化方法的使用在计算机视觉中更为典型(尽管当今 MRF 也广泛用于图像处理)。关于 MRF 的一节也是首次引入使用贝叶斯推断的方法,这在附录 B 中有更抽象层次的介绍。第 3 章也介绍了包括无缝图像拼接和图像复原在内的应用。

在第 4 章中,我们介绍特征检测和匹配。很多当今的 3D 重建和识别方法是建立在抽取和匹配特征点(feature point)基础上的(4.1 节),因此,这是很多后续章节(第 6 章、第 7 章、第 9 章和第 14 章)所需要的重要方法。在 4.2 节和 4.3 节,我们还要介绍边缘和直线检测。

第 5 章介绍区域分割方法,包括活动轮廓检测和跟踪(5.1 节)。分割方法包括自顶向下(分裂)和自底向上(归并)方法,寻找聚类模态的均值移位(mean shift)方法,以及各种基于图的分割方法。所有这些方法都是广泛用于各种应用中的必要的构件，这些应用包括表演驱动的动画、交互图像编辑和识别。

第 6 章的内容是几何配准和摄像机标定。我们在 6.1 节介绍基本的基于特征的配准方法，说明这个问题如何用线性或非线性最小二乘来解决，具体的选择依赖于所牵涉的运动。我们也介绍了一些其他概念，比如不确定性加权和鲁棒回归，它们是使真实世界系统有效所必须的。基于特征的配准在 6.2 节用作 3D 姿态估计(外标定)的构成部分，而在 6.3 节用作摄像机(内)标定的构成部分。第 6 章也描述了这些方法，在活页动画、从手持摄像机做 3D 姿态估计和建筑物模型的单视图恢复这些任务中，做照片配准时的应用。

第 7 章覆盖了从运动到结构(structure from motion)的主题，涉及从一组跟踪出来的 2D 特征同时恢复 3D 摄像机运动和 3D 场景结构的工作。本章从 3D 点三角剖分(triangulation，详见 7.1 节)的较简单的问题开始，即当摄像机位置已知时从匹配的特征重建 3D 点。接着描述由运动到结构的两帧情形(7.2 节)，这种情形下存在代数方法，同时也介绍了鲁棒采样方法，例如可以容忍部分错误的特征匹配的 RANSAC 方法。第 7 章的另一半内容描述由运动到结构的多帧方法，包括因子分解(7.3 节)、光束平差法(7.4 节)及限定运动和结构模型(7.5 节)。此外也介绍了在视角变形、稀疏 3D 模型构建和匹配移动中的应用。

第 8 章我们返回来考虑直接处理图像亮度(相对于特征轨道上)的一个主题，即基于亮度的稠密运动估计(光流)。我们从最简单的运动模型即平移运动开始(8.1 节)，内容覆盖如下主题：分层(由粗到精)的运动估计、基于傅里叶变换的方法以及迭代求精。接着我们介绍参数化运动模型，可用于补偿摄像机的旋转和聚焦，还要介绍仿射即平面透视运动(8.2 节)。然后将其推广为基于样条的运动模型(8.3 节)，并最终推广到一般性的逐像素的光流(8.4 节)，包括分层的和学习到的运动模型(8.5 节)。这些方法的应用包括自动的变形，帧插值(缓慢运动)以及基于运动的用户界面。

第 9 章专注于图像拼接(image stitching)，即大的全景图和合成图。尽管拼图只是计算摄影学的一个例子(参见第 10 章)，其深度却足以单独成章。我们从介绍各种可能的运动模型开始(9.1 节)，包括平面运动和纯摄像机旋转。然后介绍全局配准(9.2 节)，这是普通的光束平差法的一个特殊(简化了的)情况，再介绍全景图识别(panorama recognition)，即自动发现哪些图像实际形成了交叠的全景图。最后，我们介绍图像合成(image composting)和混合(blending)主题(9.3 节)，涉及使用哪些图像、选择哪些像素并将其混合在一起以掩盖曝光差别。

图像拼接是一个奇妙的应用，将本书前面章节所覆盖的绝大部分材料联系在一起。它也构成了一个好的期中课程设计题目，可以在前面已经完成的包括图像卷绕和特征检测和匹配方法基础上建立起来。第 9 章也介绍了拼图的一些特殊的变型，包括白板和文本扫描、视频摘要、全景图法、全 360°球形全景图以及用于混合重复动作镜头在一起的交互式照片蒙太奇。

第 10 章介绍计算摄影学的其他例子，即从一幅或多幅输入照片创建新图像的过程，通常是基于对图像形成过程的精细建模和标定的(10.1 节)。计算摄影学方法包括归并多曝光照片创建高动态范围图像(high dynamic range image) (10.2 节)，通过模糊去除和超分辨率方法(super-resolution)来增加图像分辨率(10.3 节)，以及图像编辑和合成操作(10.4 节)。我们在 10.5 节中也介绍了纹理分析、合成和修图(空洞填充)主题，还介绍了非真实感绘制(10.5.2 节)。

在第 11 章中，我们关注于立体视觉对应，这可以看成是摄像机位置已知情况下的运动估计的一个特例(11.1 节)。这一附加的知识使得立体视觉算法可以在大大减小了的对应空间中搜索，且在很多情况下可以产生稠密的深度估计，从而能转化为可见表面的模型(11.3 节)。我们还介绍了多视角立体视觉算法，它可以建立真正的 3D 表面表达而不仅仅是单幅深度图(11.6 节)。立体视觉匹配的应用包括头部和注视跟踪，以及基于深度的背景替换(Z-键控，Z-keying)。

第 12 章涵盖了其他的 3D 形状和表观建模方法。这包括经典的由 X 到形状的方法，例如由阴影到形状，由纹理到形状，由聚焦到形状(12.1 节)，由光滑的遮挡轮廓(contour) (11.2.1 节)和剪影(silhouette) (12.5 节)到形状。与所有这些被动计算机视觉方法不同，另一种选择是使用主动的距离测定(12.2 节)，即将模式化的光投影到场景上，通过三角测量恢复 3D 几何。这些 3D 表达的处理通常都涉及插值或几何简化(12.3 节)，或使用诸如表面点集的其他表达(12.4 节)

从单幅或多幅图像到部分或全部 3D 的方法一起常称作基于图像的建模或 3D 摄影学。12.6 节探讨了三个特殊的应用领域(建筑物、人脸和人体)，它们使用基于模型的重建来对感知的数据进行参数化模型拟合。12.7 节探讨表观建模主题，即估计纹理图、反照度甚至有时是描述 3D 表面的表观的完整双向反射分布函数(BRDF)的方法。

在第 13 章中，我们讨论过去二十年来发展起来的众多基于图像的绘制方法，包括简单的方法，例如视角插值(13.1 节)、分层的深度图像(13.2 节)、子画面(sprite)和图层(13.2.1 节)，还有光场和照度图的更一般的框架(13.3 节)和诸如环境影像形板(environment matte)的更高阶场(13.4 节)。这些方法的应用包括使用照片导览来浏览照片的 3D 收藏和将 3D 模型作为物体制片来观察。

在第 13 章中，我们也讨论了基于视频的绘制，它是基于图像的绘制在时序上的扩展。我们所覆盖的主题包括基于视频的动画(13.5.1 节)，周期性视频转化为视频纹理(13.5.4 节)和从多个视频流重建 3D 视频(13.5.4 节)。这些方法的应用包括视频去噪、变形和基于 360°视频的导览。

第 14 章描述识别的不同方法。首先是检测和识别人脸的方法(14.1 节和 14.2 节)，然后是 14.3 节的寻找和识别特定物体(示例识别)。其次，我们介绍了识别的最困难的变形，即宽泛类属的识别，例如汽车、摩托车、马和其他动物(14.4 节)，并介绍了场景上下文在识别中的作用(14.5 节)。

为了支持本书作为教材来使用，附录和关联的网站含有更详细的数学主题和补充的材料。附录 A 涵盖了线性代数和数值方法，包括矩阵代数、最小二乘和迭代方法。附录 B 涵盖贝叶斯估计理论,包括最大似然估计、鲁棒统计学、马尔科夫随机场和不确定性建模。附录 C 给出可以获得的与本书互补的辅助材料,包括图像和数据集合、指向软件的链接、课程讲稿和在线的文献资料。

1.4 课程大纲样例

在单个学期的课程中讲授本书所覆盖的所有内容是非常困难的任务,恐怕不值得尝试。最好是简单地挑选一些与授课者优先强调的内容和为学生量身定做的小课程设计集合相关联的主题。

Steve Seitz 和我成功地使用过类似于表 1.1 的 10 周课程大纲(省略了带括号的周次),作为计算机视觉的本科和研究生层次的课程。本科的课程[1]趋向于轻数学而花更多的时间回顾基础,而研究生课程[2]则在方法上讲得更深入,并假设学生在视觉或相关的数学方法上已经有相当的基础。(参见 Stanford 的计算机视觉概论课程[3],它使用了类似的大纲)在 3D 摄影学[4]和计算摄影学[5]的主题上也有相关的课程。

表 1.1 10 周和 13 周课程的大纲样例。带括号的周次在较短的课程中不用。P1 和 P2 是两个前期的小课程设计,PP 是(学生选择的)最终课程设计提案截止的时候,而 FP 是最终课程设计汇报

周次	材料	课程设计
(1)	第 2 章 图像形成	
2.	第 3 章 图像处理	
3.	第 4 章 特征检测和匹配	P1
4.	第 6 章 基于特征的配准	
5.	第 9 章 图像拼接	P2
6.	第 8 章 稠密运动估计	
7.	第 7 章 由运动到结构	PP
8.	第 14 章 识别	
(9)	第 10 章 计算摄影学	
10.	第 11 章 立体视觉对应	
(11)	第 12 章 3D 重建	
12.	第 13 章 基于图像的绘制	
13.	最终课程设计汇报	FP

---

[1] http://www.cs.washington.edu/education/courses/455/
[2] http://www.cs.washington.edu/education/courses/576/
[3] http://vision.stanford.edu/teaching/cs223b/
[4] http://www.cs.washington.edu/education/courses/558/06sp/
[5] http://graphics.cs.cmu.edu/courses/15-463/

当 Steve Seitz 和我教这门课时，我们更喜欢在课程早期布置几个小的编程课程设计，而不是关注于书面作业或课堂小测验。在主题选择合适时，可以使这些课程设计一个建立在另一个基础之上。例如，在前期介绍特征匹配，就可用于第二个作业来做图像配准和拼图。另一种选择，直接(光流)方法可用于做配准，而更多关注图割接缝选择或多分辨率混合方法。

我们也要求学生期中提交最终课程设计提案(我们提供一组推荐主题供需要想法的人选择)，并预留最后一周的课用于汇报。幸运的话，有些最终的课程设计可以成为会议论文投稿！

无论你如何设计这门课抑或你怎样选择使用这本书，我鼓励你至少尝试几个小的程序设计任务，从而好好体验一下视觉方法是如何工作的，什么时候无效。更好的是，挑选一些你觉得有趣的且可用于你自己的照片中的主题，努力拓展你的创造疆域来拿出令人惊奇的结果。

1.5 标记法说明

不论好坏，在计算机视觉和多视角几何教材中所使用的标记法的差别随处可见(Faugeras 1993; Hartley and Zisserman 2004; Girod, Greiner, and Niemann 2000; Faugeras and Luong 2001; Forsyth and Ponce 2003)。在本书中，我使用惯例，这最初是在我高中物理课堂(后来在多元微积分和计算机图形学课程)上学到的，即向量 v 是粗体小写，矩阵 M 是粗体大写，标量$(T; s)$是大小写混用的斜体。除非特别说明，向量按照列向量来操作，即它们右乘矩阵，Mv，但它们有时写成分号分开的带括号的列表 $x = (x; y)$ 代替方括号的列向量 $x = (x\ y)^T$。一些常用的矩阵有旋转矩阵 R，标定矩阵 K，单位矩阵 I。齐次坐标(2.1 节)用向量上标注波浪号来表示，即在 p^2 中 $\tilde{x} = (\tilde{x}, \tilde{y}, \tilde{w}) = \tilde{w}(x, y, 1) = \tilde{w}\overline{x}$。矩阵形式的叉积运算记作 $[]_\times$。

1.6 扩展阅读

本书试图做到"包罗万象"，使学生可以实现这里所描述的基本的作业和算法而无需求助于其他参考资料。尽管如此，它的确假定读者熟悉一些基本概念，包括涉及线性代数和数值方法方面的(可通过附录 A 加以回顾)，还包括图像处理方面的(可通过第 3 章加以回顾)。想更深入学习这些主题的学生可以查阅(Golub and Van Loan 1996)了解矩阵代数，查阅(Strang 1988)了解线性代数。对于图像处理，有很多流行的教材，包括(Crane 1997; Gomes and Velho 1997; Jähne 1997; Pratt 2007; Russ 2007; Burger and Burge 2008; Gonzales and Woods 2008)。对于计算机图形学，受欢迎的教材有(Foley, van Dam, Feiner *et al.* 1995; Watt 1995)，而(Glassner 1995)提供了更深入的有关图像形成和绘制的论述。在统计学和机器学习方面，Chris

Bishop(2006)是一本非常好的含有大量习题的综合性入门教材。对于本书没有覆盖的材料，以及如果还想获得更多课程设计相关想法，学生还可以阅读计算机视觉的其他教材(Ballard and Brown 1982; Faugeras 1993; Nalwa 1993; Trucco and Verri 1998; Forsyth and Ponce 2003)。

　　但是，为了获得最新的想法和方法以及相关领域的最新的资料，除了阅读最新的文献外，没有可取代的途径[①]。在本书中，我力图引用每个领域的最新研究工作，以便学生可以直接阅读这些材料，在他们自己的工作中将其用于获取灵感。浏览最近几年的主要的计算机视觉和图形学会议论文集，例如 CVPR，ECCV，ICCV 和 SIGGRAPH，就可以获得丰富的新想法。这些会议中提供的教程，其讲稿或笔记通常可以在线获得，也是非常有价值的资源。

---

① 有关计算机视觉研究的综合性的文献和分类，Keith Price 的注解版计算机视觉文献(Annotated Computer Vision Bibliography) *http://www.visionbib.com/bibliography/contents.html* 是极有价值的资源。

第 2 章
图 像 形 成

2.1 几何基元和变换
2.2 光度测定学的图像形成
2.3 数字摄像机
2.4 补充阅读
2.5 习题

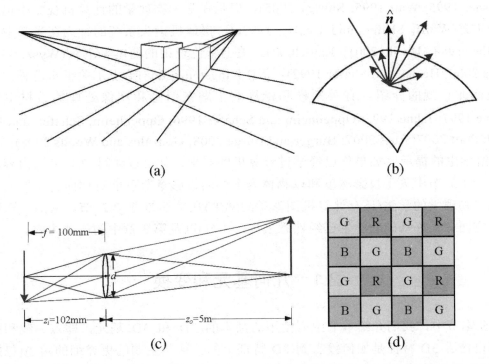

图 2.1 图像形成过程的几个组成部分：(a)透视投影；(b)光射到表面时的散射；(c)镜头光学；(d)Bayer 彩色滤波器阵列

在我们能智能地分析和处理图像之前，我们需要建立用以描述场景几何的词汇。我们也需要理解图像形成过程，即在给定一组光照条件、场景几何、表面特性和摄像机光学情况下产生一幅具体图像的过程。在本章中，我们阐述这一图像形成过程的简化模型。

2.1 节介绍在本书中所用到的基本几何基元(点、线和面)和将这些 3D 量投影到 2D 图像特征的几何变换(图 2.1a)。2.2 节描述照明、表面特性(图 2.1b)和摄像机光学(图 2.1c)是如何相互作用而产生落在图像传感器上的彩色数值的。2.3 节描述连续彩色图像在图像传感器内是如何转化为离散数字采样的(图 2.1d)以及如何避免(或至少刻画)采样缺欠的，例如走样。

本章所讲述的内容只是对一个非常丰富和精深主题集的简要概括，该主题集在传统上覆盖了几个不同的领域。关于点、线、面和投影的几何的更为完整的介绍可参见多视角几何方面的教材(Hartley and Zisserman 2004; Faugeras and Luong 2001)和计算机图形学方面的教材(Foley, van Dam, Feiner *et al.* 1995)。图像形成(合成)过程通常作为计算机图形学课程的一部分来讲授(Foley, van Dam, Feiner *et al.* 1995; Glassner 1995; Watt 1995; Shirley 2005)，但是在基于物理学的计算机视觉中也学习这个主题(Wolff, Shafer, and Healey 1992a)。摄像机镜头系统的行为在光学中研究(Möller 1988; Hecht 2001; Ray 2002)。关于彩色理论的两本好书是(Wyszecki and Stiles 2000; Healey and Shafer 1992)，而(Livingstone 2008)给出了关于彩色感知的更有趣而非正规的介绍。有关采样和走样的主题在信号和图像处理的教材中阐述(Crane 1997; Jähne 1997; Oppenheim and Schafer 1996; Oppenheim, Schafer, and Buck 1999; Pratt 2007; Russ 2007; Burger and Burge 2008; Gonzales and Woods 2008)。

给学生的提示 如果你已经学过计算机图形学，你可以跳过 2.1 节的内容，尽管在 2.1.5 节中关于投影深度和以物体为中心的投影章节可能对你而言是新的。类似地，物理学的学生(还有计算机图形学的学生)几乎都熟悉 2.2 节。最后，在图像处理方面有良好基础的学生已经熟悉采样(2.3 节)以及第 3 章中的一些内容。

2.1 几何基元和变换

在本节中，我们介绍本书中所使用的基本的 2D 和 3D 基元，即点、线和面。我们也描述 3D 特征是如何投影到 2D 特征上的。这些主题的更详细的阐述(包括入门和更直观的介绍)可见多视角几何方面的教材(Hartley and Zisserman 2004; Faugeras and Luong 2001)。

2.1.1 几何基元

几何基元构成了用以描述三维形状的基本构件。在本节中，我们介绍点、线和面。本书稍后的章节讨论曲线(5.1 节和 11.2 节)、表面(12.3 节)和体(12.5 节)。

2D 点　2D 点(图像中的像素坐标)可用一对数值来表示，$x = (x,y) \in \Re^2$，或者以另一种方式

$$x = \begin{bmatrix} x \\ y \end{bmatrix}. \tag{2.1}$$

(正如第 1 章中所陈述的，我们使用 $(x_1, x_2, \ldots)$ 记法来表示列向量。)

2D 点也可以用齐次坐标表示 $\tilde{x} = (\tilde{x}, \tilde{y}, \tilde{w}) \in \mathcal{P}^2$，其中仅在尺度上不同的矢量被视为等同的。$\mathcal{P}^2 = \Re^3 - (0,0,0)$ 称作 2D 投影空间。

齐次矢量 $\tilde{x}$ 可以通过除以最后一个元素 $\tilde{w}$ 转换回非齐次矢量 x，即

$$\tilde{x} = (\tilde{x}, \tilde{y}, \tilde{w}) = \tilde{w}(x, y, 1) = \tilde{w}\bar{x}, \tag{2.2}$$

其中 $\bar{x} = (x, y, 1)$ 是增广矢量。最后一个元素是 $\tilde{w} = 0$ 的齐次点称作理想点或无穷远点，它们没有等同的非齐次表达。

2D 直线　2D 直线也可以用齐次坐标表达 $\tilde{l} = (a, b, c)$。对应的直线方程是

$$\bar{x} \cdot \tilde{l} = ax + by + c = 0. \tag{2.3}$$

我们可以规范化直线方程，使得 $l = (\hat{n}_x, \hat{n}_y, d) = (\hat{n}, d)$ 而 $\|\hat{n}\| = 1$。这种情况下，$\hat{n}$ 是垂直于直线的法向量，d 是其到原点的距离(图 2.2a)。(该规范化的一个例外是无穷远处的直线 $\tilde{l} = (0, 0, 1)$，它包含所有无穷远处的(理想)点。)

我们可以将 $\hat{n}$ 表示为旋转角度 θ 的函数，$\hat{n} = (\hat{n}_x, \hat{n}_y) = (\cos\theta, \sin\theta)$(图 2.2a)。这种表达常用于霍夫变换检测直线的算法，这在 4.3.2 节讨论。(θ, d) 组合也称为极坐标。

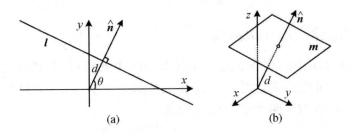

图 2.2　(a)2D 直线方程和(b)3D 平面方程，用法向量 $\hat{n}$ 和到原点的距离 d 表达

使用齐次坐标时，我们可以如下计算两条直线的交点

$$\tilde{x} = \tilde{l}_1 \times \tilde{l}_2, \tag{2.4}$$

其中×叉积算子。类似地，连接两点的直线可以写为

$$\tilde{l} = \tilde{x}_1 \times \tilde{x}_2. \tag{2.5}$$

当试图求取多条直线的交点，或反过来，通过多个点拟合一条直线时，可以使用最小二乘方法(6.1.1 节和附录 A.2)，这在习题 2.1 中讨论。

2D 圆锥曲线　还有其他的代数曲线可以用简单的多项式齐次方程来表示。例如，圆锥截面(取名于平面和 3D 圆锥的交)可以用二次方程写为

$$\tilde{x}^T Q \tilde{x} = 0. \tag{2.6}$$

二次方程在多视角几何和摄像机标定的研究中很有用(Hartley and Zisserman 2004;

Faugeras and Luong 2001)，但在本书中用的并不广泛。

3D 点 在三维中点的坐标可以写为非齐次坐标的 $\boldsymbol{x}=(x,y,z)\in\Re^3$ 或齐次坐标的 $\tilde{\boldsymbol{x}}=(\tilde{x},\tilde{y},\tilde{z},\tilde{w})\in\mathcal{P}^3$。正如前述，使用增广矢量 $\bar{\boldsymbol{x}}=(x,y,z,1)$ 而有 $\tilde{\boldsymbol{x}}=\tilde{w}\bar{\boldsymbol{x}}$ 来表示 3D 点有时是有用的。

3D 平面 3D 平面也可以表达为齐次坐标的 $\tilde{\boldsymbol{m}}=(a,b,c,d)$，对应的平面方程为

$$\bar{\boldsymbol{x}} \cdot \tilde{\boldsymbol{m}} = ax + by + cz + d = 0. \tag{2.7}$$

我们也可以规范化平面方程为 $\boldsymbol{m}=(\hat{n}_x,\hat{n}_y,\hat{n}_z,d)=(\hat{\boldsymbol{n}},d)$ 而 $\|\hat{\boldsymbol{n}}\|=1$。这种情况下，$\hat{\boldsymbol{n}}$ 是垂直于平面的法向量，d 是其到原点的距离(图 2.2b)。正如 2D 直线的情形，无穷远处的平面 $\tilde{\boldsymbol{m}}=(0,0,0,1)$，它包含了所有无穷远处的点，不能被规范化(即，它不具有唯一的法向抑或一个有限的距离)。

我们可以将 $\hat{\boldsymbol{n}}$ 表示为两个角度 (θ,ϕ) 的函数，

$$\hat{\boldsymbol{n}} = (\cos\theta\cos\phi, \sin\theta\cos\phi, \sin\phi), \tag{2.8}$$

即，使用球面坐标，但是这并不像极坐标那样广为使用，因为它们并没有在可能的法向量空间中均匀采样。

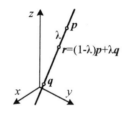

图 2.3 3D 直线方程，$\boldsymbol{r}=(1-\lambda)\boldsymbol{p}+\lambda\boldsymbol{q}$

3D 直线 在 3D 中的直线表达起来比起 2D 中的直线或 3D 中的平面来说就没有那么优美。一种可能的表达是使用直线上的两个点 $(\boldsymbol{p}, \boldsymbol{q})$。如图 2.3 所示，直线上的其他点都可以表示为这两个点的线性组合

$$\boldsymbol{r} = (1-\lambda)\boldsymbol{p} + \lambda\boldsymbol{q}, \tag{2.9}$$

如果我们限定 $0\leq\lambda\leq1$，我们得到连接 $\boldsymbol{p}$ 和 $\boldsymbol{q}$ 的线段。

如果我们使用齐次坐标，我们可以将直线写为

$$\tilde{\boldsymbol{r}} = \mu\tilde{\boldsymbol{p}} + \lambda\tilde{\boldsymbol{q}}. \tag{2.10}$$

一种特殊情况是当第二个点位于无穷远时，即 $\tilde{\boldsymbol{q}}=(\hat{d}_x,\hat{d}_y,\hat{d}_z,0)=(\hat{\boldsymbol{d}},0)$。这里 $\hat{\boldsymbol{d}}$ 即为直线的方向。我们可以将非齐次坐标的 3D 直线方程重写为

$$\boldsymbol{r} = \boldsymbol{p} + \lambda\hat{\boldsymbol{d}} \tag{2.11}$$

3D 直线的端点表达方式的缺点是它的自由度太多，即 6 个(每个端点 3 个)而不是实际上 3D 直线真正具有的 4 个。但是，如果我们将直线的两个点固定位于特定平面上，我们就获得了具有 4 个自由度的一个表达。例如，如果我们要表达几乎垂直的直线，则 $z=0$ 和 $z=1$ 就构成了两个合适的平面，即两个平面中的 (x, y) 坐标就提供了 4 个坐标来表达该直线。这种两平面参数化表达用于第 13 章中介绍的光场

和照度图(Lumigraph)基于图像的绘制系统中,表达当摄像机在物体前运动时它所看到的光束。这种两端点表达方式在表示线段时也是有用的,即便是在看不到两端点的确切位置(仅仅是猜测)时。

如果我们想表达所有可能的直线而不带有对任何特定方向的偏向,我们可以使用 Plücker 坐标(Hartley and Zisserman 2004,第 2 章;Faugeras and Luong 2001,第 3 章)。这些坐标是如下的 4×4 斜对称矩阵中的 6 个独立非 0 项:

$$L = \tilde{p}\tilde{q}^T - \tilde{q}\tilde{p}^T, \tag{2.12}$$

其中的 $\tilde{p}$ 和 $\tilde{q}$ 是直线上任意(非等同的)两点。这种表达只有 4 个自由度,因为 L 是齐次的且还满足 $\det(L) = 0$,这产生了对 Plücker 坐标的一个二次约束。

在实际中,这种最小表达在大多数应用中不是必要的。由于直线通常作为有限的直线片段为人所见,一个合适的 3D 直线模型可以通过估计其方向(这可能事先就知道,例如,对于建筑物)和其可见部分的某个点(参见 7.5.1 节)抑或使用两个端点来获得。但是,如果你对最小的直线参数化方法的更多细节感兴趣,Förstner(2005) 论述了在射影几何中 3D 直线的推断和建模的各种方法,也阐述了如何估计其所拟合模型的不确定性的方法。

3D 二次曲面 圆锥截面的 3D 类比物是二次曲面(Hartley and Zisserman 2004,第 2 章)。

$$\bar{x}^T Q \bar{x} = 0 \tag{2.13}$$

同样,尽管二次曲面在多视角几何的研究中很有用,而且还可以作为有用的建模基元(球、椭球、圆柱),但在本书中我们不作详细论述。

2.1.2 2D 变换

我们已经定义了基本基元,现在我们可以将注意力转到如何对其进行变换。最简单的变换发生在 2D 平面中,如图 2.4 所示。

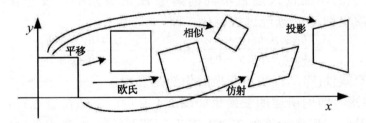

图 2.4 2D 平面变换的基本集

平移 2D 平移可以写为 $x' = x + t$,或者

$$x' = \begin{bmatrix} I & t \end{bmatrix} \bar{x} \tag{2.14}$$

其中 I 是 2×2 的单位矩阵,或者

$$\bar{x}' = \begin{bmatrix} I & t \\ 0^T & 1 \end{bmatrix} \bar{x} \qquad (2.15)$$

其中 **0** 是零矢量。使用 2×3 矩阵得到一个更紧凑的表达,而使用一个满秩 3×3 矩阵(这可以从 2×3 矩阵加上(0^T 1)行得到)可以使利用矩阵乘法形成变换链成为可能。注意,在两端都出现增广矢量 $\bar{x}$ 的任何公式中,它都可以用一个完整的齐次矢量 $\tilde{x}$ 代替。

旋转+平移 该变换也称 2D 刚体运动即 2D 欧氏变换(因其保持欧氏距离)。它可以写为 $x' = Rx + t$,或者

$$x' = \begin{bmatrix} R & t \end{bmatrix} \bar{x} \qquad (2.16)$$

其中

$$R = \begin{bmatrix} \cos\theta & -\sin\theta \\ \sin\theta & \cos\theta \end{bmatrix} \qquad (2.17)$$

是一个正交旋转矩阵,有 $RR^T = I$ 和 $|R| = 1$。

放缩旋转 也叫相似变换,该变换可以表示为 $x' = sRx + t$,其中 s 是一个任意的尺度因子。它也可以写作

$$x' = \begin{bmatrix} sR & t \end{bmatrix} \bar{x} = \begin{bmatrix} a & -b & t_x \\ b & a & t_y \end{bmatrix} \bar{x}, \qquad (2.18)$$

其中我们不再需要 $a^2 + b^2 = 1$。相似变换保持直线间的夹角。

仿射 仿射变换写作 $x' = A\bar{x}$,其中 A 是 2×3 矩阵,即

$$x' = \begin{bmatrix} a_{00} & a_{01} & a_{02} \\ a_{10} & a_{11} & a_{12} \end{bmatrix} \bar{x}. \qquad (2.19)$$

在仿射变换下平行线仍然保持平行。

投影 该变换也称透视变换或同态映射,作用在齐次坐标上,

$$\tilde{x}' = \tilde{H}\tilde{x}, \qquad (2.20)$$

其中 $\tilde{H}$ 是一个任意的 3×3 矩阵。注意 $\tilde{H}$ 是齐次的,即它只是在相差一个尺度量的情况下是定义了的,而仅尺度量不同的两个 $\tilde{H}$ 是等同的。要想获得非齐次结果 x,得到的齐次坐标 $\bar{x}'$ 必须经过规范化,即

$$x' = \frac{h_{00}x + h_{01}y + h_{02}}{h_{20}x + h_{21}y + h_{22}} \quad \text{和} \quad y' = \frac{h_{10}x + h_{11}y + h_{12}}{h_{20}x + h_{21}y + h_{22}} \qquad (2.21)$$

透视变换保持直线性(即,直线在变换后仍然是直线)。

2D 变换层次 前面所述的变换集如图 2.4 所示,总结在表 2.1 中。考虑它们的最简单的方式是将它们作为施加在 2D 齐次矢量上的(潜在受限的)3×3 矩阵集合看待。Hartley and Zisserman(2004)含有关于 2D 平面变换层次描述的更多细节。

上述变换构成一个嵌套的群集,即它们的复合是封闭的且有属于自身群的逆。(这在后面 3.6 节中在图像上施加这些变换时是重要的。)每个(较简单的)群是其下更复杂的群的一个子集。

表 2.1 2D 坐标变换的层次。每个变换还保持列在其行下的性质，即，相似变换不仅保持夹角而且还保持平行性和直线性。对于齐次坐标变换，将 2×3 矩阵以第三行 ($\mathbf{0}^T$ 1) 扩展形成一个完整的 3×3 矩阵

变换	矩阵	自由度数	保持	图标
平移	$[\mathbf{I}\|\mathbf{t}]_{2\times 3}$	2	方向	□
刚性(欧氏)	$[\mathbf{R}\|\mathbf{t}]_{2\times 3}$	3	长度	◇
相似	$[s\mathbf{R}\|\mathbf{t}]_{2\times 3}$	4	夹角	◇
仿射	$[\mathbf{A}]_{2\times 3}$	6	平行性	▱
投影	$[\tilde{\mathbf{H}}]_{3\times 3}$	8	直线性	⏢

辅助矢量 既然上述变换可以用于变换 2D 平面上的点，那么它们是否能直接地用于变换直线方程？我们来考虑齐次方程 $\tilde{\mathbf{l}}\cdot\tilde{\mathbf{x}}=0$。如果我们变换 $\mathbf{x}'=\tilde{\mathbf{H}}\mathbf{x}$，我们得到

$$\tilde{\mathbf{l}}'\cdot\tilde{\mathbf{x}}'=\tilde{\mathbf{l}}'^T\tilde{\mathbf{H}}\tilde{\mathbf{x}}=(\tilde{\mathbf{H}}^T\tilde{\mathbf{l}}')^T\tilde{\mathbf{x}}=\tilde{\mathbf{l}}\cdot\tilde{\mathbf{x}}=0, \tag{2.22}$$

即，$\tilde{\mathbf{l}}'=\tilde{\mathbf{H}}^{-T}\tilde{\mathbf{l}}$。因此，投影变换作用在诸如 2D 直线或 3D 法向的辅助矢量上可以用矩阵逆的转置来表示，它等同于 $\tilde{\mathbf{H}}$ 的伴随矩阵，这是由于投影变换矩阵是齐次的。Jim Blinn(1998)(在第 9 章和第 10 章)描述了辅助矢量的记法和操作的来龙去脉。

尽管上述变换是我们使用得最为广泛的，但有时我们还会用到其他一些变换。

拉伸/挤压 该变换改变图像的长宽比，它是仿射变换的一种受限形式。不幸的是，它不能整洁地嵌入表 2.1 所列出的群中。

$$\begin{aligned} x' &= s_x x + t_x \\ y' &= s_y y + t_y. \end{aligned}$$

平面状表面流 该 8 参数变换(Horn 1986; Bergen, Anandan, Hanna *et al.* 1992; Girod, Greiner, and Niemann 2000)。

$$\begin{aligned} x' &= a_0 + a_1 x + a_2 y + a_6 x^2 + a_7 xy \\ y' &= a_3 + a_4 x + a_5 y + a_7 x^2 + a_6 xy, \end{aligned}$$

出现在平面状表面经历小的 3D 运动时。因此，它可以被看作是对完整同态映射的小的运动估计。它的主要优点是其相关的运动参数 a_k 是线性的，而这常常是要估计的量。

双线性内插 该 8 参数变换(Wolberg 1990)

$$\begin{aligned} x' &= a_0 + a_1 x + a_2 y + a_6 xy \\ y' &= a_3 + a_4 x + a_5 y + a_7 xy, \end{aligned}$$

可用于内插由一个方块的四角点运动所引起的变形。(事实上，它可用于任意非共线的 4 点运动的内插。)尽管变形相关的运动参数是线性的，但是它一般并不

保持直线性(仅对平行于方块轴线的线条)。然而，它有时是相当有用的，例如，在用样条内插稀疏的网格中(8.3 节)。

2.1.3 3D 变换

三维坐标变换的集合与 2D 变换的情形非常相似，总结于表 2.2 中。正如在 2D 中，这些变换构成一个嵌套的群集。Hartley and Zisserman(2004，2.4 节)给出了该分层更详细的描述。

表 2.2 3D 坐标变换的层次。每个变换还保持列在其行下的性质，即，相似变换不仅保持夹角而且还保持平行性和直线性。对于齐次坐标变换，将 3×4 矩阵以第四行(0^T 1) 扩展形成一个完整的 4×4 矩阵。助记图标虽然画成 2D 的但它表达的是出现在完整 3D 立方体上的变换

变 换	矩 阵	自由度数	保 持	图 标
平移	$[I \mid t]_{3\times 4}$	3	方向	□
刚性(欧氏)	$[R \mid t]_{3\times 4}$	6	长度	◇
相似	$[sR \mid t]_{3\times 4}$	7	夹角	◇
仿射	$[A]_{3\times 4}$	12	平行性	▱
投影	$[\tilde{H}]_{4\times 4}$	15	直线性	▱

平移 3D 平移可以写为 $x' = x + t$，或者

$$x' = \begin{bmatrix} I & t \end{bmatrix} \bar{x} \tag{2.23}$$

其中 I 是 3×3 的单位矩阵。

旋转+平移 该变换也称作 3D 刚体运动或 3D 欧氏变换。它可以写为 $x' = Rx + t$，或者

$$x' = \begin{bmatrix} R & t \end{bmatrix} \bar{x} \tag{2.24}$$

其中 R 是一个 3×3 正交旋转矩阵，有 $RR^T = I$ 和 $|R| = 1$。注意，有时为了描述一个刚体运动，使用下式更为方便：

$$x' = R(x - c) = Rx - Rc, \tag{2.25}$$

其中 c 是旋转的中心(通常是摄像机中心)。

紧凑地参数化 3D 旋转并不简单，我们在下面还要做更详细的描述。

放缩旋转 3D 相似变换可以表示为 $x' = sRx + t$，其中 s 是一个任意的尺度因子。它也可以写作

$$x' = \begin{bmatrix} sR & t \end{bmatrix} \bar{x}. \tag{2.26}$$

该变换保持直线和平面间的夹角。

仿射 仿射变换写作 $x' = A\bar{x}$，其中 A 是 3×4 矩阵，即

$$x' = \begin{bmatrix} a_{00} & a_{01} & a_{02} & a_{03} \\ a_{10} & a_{11} & a_{12} & a_{13} \\ a_{20} & a_{21} & a_{22} & a_{23} \end{bmatrix} \bar{x}. \tag{2.27}$$

在仿射变换下平行的线和平面仍然保持平行。

投影 该变换也被冠以诸如 3D 透视变换(3D perspective transform)，同态映射(homography)或直射映射(collineation)之类的名称，作用在齐次坐标上，

$$\tilde{x}' = \tilde{H}\tilde{x}, \tag{2.28}$$

其中 $\tilde{H}$ 是一个任意的 4×4 齐次矩阵。正如在 2D 中，要想获得非齐次结果 x，得到的齐次坐标 $\tilde{x}'$ 必须经过规范化。透视变换保持直线性(即，直线在变换后仍然是直线)。

2.1.4 3D 旋转

2D 和 3D 坐标旋转变换最大的不同点是 3D 旋转矩阵 R 的参数化不那么简单明了，而是存在几种可能性。

欧拉角 旋转矩阵可以用围绕基轴的三个旋转的乘积形成，即 x, y 和 z，或 x, y 和 x。一般来说，这是一个糟糕的主意，因为结果依赖于所施加变换的顺序。更糟糕的是，在参数空间中并不是总能光滑地移动，即，有时小的旋转变化会引起一个或多个欧拉角的显著变化[1]。由于这些原因，在本书中我们甚至没有列出欧拉角的公式——有兴趣的读者可以参见其他教材或技术报告(Faugeras 1993; Diebel 2006)。注意，在有些应用中，如果知道旋转是单一轴变换的集合，它们就总是可以显示地表示为刚性变换集合。

轴/角(指数扭曲) 旋转可以用旋转轴 $\hat{n}$ 和角度 θ 来表示，或等同地用一个 3D 矢量 $\omega = \theta\hat{n}$ 来表示。图 2.5 展示了我们如何可以计算等同的旋转。首先，我们将矢量 v 投影到轴 $\hat{n}$ 上得到

$$v_\| = \hat{n}(\hat{n} \cdot v) = (\hat{n}\hat{n}^T)v, \tag{2.29}$$

这是 v 不受旋转影响的分量。其次，我们计算出 v 的相对于 $\hat{n}$ 的垂直余量，

$$v_\perp = v - v_\| = (I - \hat{n}\hat{n}^T)v. \tag{2.30}$$

我们可以用叉积将该矢量旋转 90°，

$$v_\times = \hat{n} \times v = [\hat{n}]_\times v, \tag{2.31}$$

其中 $[\hat{n}]_\times$ 是叉积算子的矩阵形式，矢量 $\hat{n} = (\hat{n}_x, \hat{n}_y, \hat{n}_z)$

$$[\hat{n}]_\times = \begin{bmatrix} 0 & -\hat{n}_z & \hat{n}_y \\ \hat{n}_z & 0 & -\hat{n}_x \\ -\hat{n}_y & \hat{n}_x & 0 \end{bmatrix}. \tag{2.32}$$

---

[1] 在机器人学中，这有时称为"万向节锁"(gimbal lock)。

注意，将该矢量再旋转 90°等同于再次做叉积，
$$v_{\times\times} = \hat{n} \times v_\times = [\hat{n}]_\times^2 v = -v_\perp,$$
因此，
$$v_\parallel = v - v_\perp = v + v_{\times\times} = (I + [\hat{n}]_\times^2)v.$$
现在我们可以计算旋转后的矢量 u 的平面内的分量为
$$u_\perp = \cos\theta v_\perp + \sin\theta v_\times = (\sin\theta[\hat{n}]_\times - \cos\theta[\hat{n}]_\times^2)v.$$
将所有这些项放在一起，我们得到最后的旋转矢量为
$$u = u_\perp + v_\parallel = (I + \sin\theta[\hat{n}]_\times + (1-\cos\theta)[\hat{n}]_\times^2)v. \tag{2.33}$$
所以，我们可以将以 $\hat{n}$ 为轴做 θ 角旋转的旋转矩阵写为
$$R(\hat{n},\theta) = I + \sin\theta[\hat{n}]_\times + (1-\cos\theta)[\hat{n}]_\times^2, \tag{2.34}$$
这称作 Rodriguez 公式(Ayache 1989)。

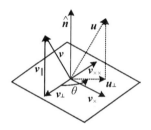

图 2.5　以 $\hat{n}$ 为轴做 θ 角的旋转

轴 $\hat{n}$ 和角度 θ 的乘积，$\omega = \theta\hat{n} = (\omega_x, \omega_y, \omega_z)$，是 3D 旋转的最小表示。如果 θ 是以角度存储的，通常角度的旋转，比如 90°的倍数，可以精确地表达出来(并转换为确切的矩阵)。不幸的是，这种表达不是唯一的，因为我们总是可以给 θ 乘以 360° (2π 弧度)的倍数而得到相同的旋转矩阵。同样，$(\hat{n}, \theta)$ 和 $(-\hat{n}, -\theta)$ 表示相同的旋转。

但是，对于小的旋转(例如，对旋转的修正)，这是一个很好的选择。特别地，对于小的(无限小或瞬时)旋转且 θ 用弧度表示时，Rodriguez 公式简化为
$$R(\omega) \approx I + \sin\theta[\hat{n}]_\times \approx I + [\theta\hat{n}]_\times = \begin{bmatrix} 1 & -\omega_z & \omega_y \\ \omega_z & 1 & -\omega_x \\ -\omega_y & \omega_x & 1 \end{bmatrix}, \tag{2.35}$$
这给出了在旋转参数 ω 和 R 之间的一个很好的线性化了的关系。我们还可以写出 $R(\omega)v \approx v + \omega \times v$，这在我们需要计算 Rv 相对于 ω 的导数时是很方便的，
$$\frac{\partial Rv}{\partial \omega^T} = -[v]_\times = \begin{bmatrix} 0 & z & -y \\ -z & 0 & x \\ y & -x & 0 \end{bmatrix}. \tag{2.36}$$

通过一个有限的角度导出旋转的另一种方法称作指数扭曲(exponential twist) (Murray, Li, and Sastry 1994)。旋转角度 θ 等同于 k 次旋转角度 θ/k。在 $k \to \infty$ 极限情况下，我们有
$$R(\hat{n}, \theta) = \lim_{k \to \infty}(I + \frac{1}{k}[\theta\hat{n}]_\times)^k = \exp[\omega]_\times. \tag{2.37}$$

如果将矩阵指数展开为泰勒级数(使用恒等式 $[\hat{n}]_\times^{k+2} = -[\hat{n}]_\times^k$, $k>0$),我们就得到

$$\begin{aligned}
\exp[\boldsymbol{\omega}]_\times &= \boldsymbol{I} + \theta[\hat{\boldsymbol{n}}]_\times + \frac{\theta^2}{2}[\hat{\boldsymbol{n}}]_\times^2 + \frac{\theta^3}{3!}[\hat{\boldsymbol{n}}]_\times^3 + \cdots \\
&= \boldsymbol{I} + (\theta - \frac{\theta^3}{3!} + \cdots)[\hat{\boldsymbol{n}}]_\times + (\frac{\theta^2}{2} - \frac{\theta^3}{4!} + \cdots)[\hat{\boldsymbol{n}}]_\times^2 \\
&= \boldsymbol{I} + \sin\theta[\hat{\boldsymbol{n}}]_\times + (1 - \cos\theta)[\hat{\boldsymbol{n}}]_\times^2,
\end{aligned} \quad (2.38)$$

这就产生了我们所熟悉的 Rodriguez 公式。

单位四元数 单位四元数表达与轴/角表达紧密相关。一个单位四元数是单位长度的 4 分量矢量,可写成 $\boldsymbol{q} = (q_x, q_y, q_z, q_\omega)$ 或简洁地写成 $\boldsymbol{q} = (x, y, z, \omega)$。单位四元数处于单位球 $\|\boldsymbol{q}\| = 1$ 上,且正相反的四元数(符号相反)$\boldsymbol{q}$ 和 $-\boldsymbol{q}$,表示相同的旋转(图 2.6)。除了该歧义(对偶覆盖)外,旋转的单位四元数表达是唯一的。而且,该表达是连续的,即,随着旋转矩阵连续地变化,我们可以看到连续的单位四元数表达,尽管在四元数球上的路径始终处于卷绕之中直至回到"原点"$\boldsymbol{q}_0 = (0, 0, 0, 1)$。由于下述的这样和那样的原因,四元数在计算机图形学中是姿态和姿态内插中常用的一种表达(Shoemake 1985)。

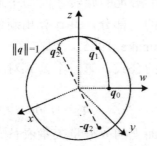

图 2.6 单位四元数处于单位球 $\|\boldsymbol{q}\| = 1$ 上。该图给出了穿过三个四元数 $\boldsymbol{q}_0$, $\boldsymbol{q}_1$ 和 $\boldsymbol{q}_2$ 的光滑轨迹。与 $\boldsymbol{q}_2$ 正相反的点即 $-\boldsymbol{q}_2$ 表示的旋转与 $\boldsymbol{q}_2$ 相同

四元数可以从轴/角表达通过以下公式导出

$$\boldsymbol{q} = (\boldsymbol{v}, w) = (\sin\frac{\theta}{2}\hat{\boldsymbol{n}}, \cos\frac{\theta}{2}), \quad (2.39)$$

其中 $\hat{\boldsymbol{n}}$ 和 θ 是旋转轴和角度。使用三角等式 $\sin\theta = 2\sin\frac{\theta}{2}\cos\frac{\theta}{2}$ 和 $(1 - \cos\theta) = 2\sin^2\frac{\theta}{2}$,Rodriguez 公式可以转化为

$$\begin{aligned}
\boldsymbol{R}(\hat{\boldsymbol{n}}, \theta) &= \boldsymbol{I} + \sin\theta[\hat{\boldsymbol{n}}]_\times + (1 - \cos\theta)[\hat{\boldsymbol{n}}]_\times^2 \\
&= \boldsymbol{I} + 2w[\boldsymbol{v}]_\times + 2[\boldsymbol{v}]_\times^2.
\end{aligned} \quad (2.40)$$

这对用四元数旋转矢量 $\boldsymbol{v}$ 提供了一种快速途径,即使用一系列叉积、伸缩和加法。为了获得作为 (x, y, z, ω) 的函数的关于 $\boldsymbol{R}(\boldsymbol{q})$ 的公式,回想一下

$$[\boldsymbol{v}]_\times = \begin{bmatrix} 0 & -z & y \\ z & 0 & -x \\ -y & x & 0 \end{bmatrix} \text{ 和 } [\boldsymbol{v}]_\times^2 = \begin{bmatrix} -y^2 - z^2 & xy & xz \\ xy & -x^2 - z^2 & yz \\ xz & yz & -x^2 - y^2 \end{bmatrix}$$

因此我们有

$$R(q) = \begin{bmatrix} 1-2(y^2+z^2) & 2(xy-zw) & 2(xz+yw) \\ 2(xy+zw) & 1-2(x^2+z^2) & 2(yz-xw) \\ 2(xz-yw) & 2(yz+xw) & 1-2(x^2+y^2) \end{bmatrix}. \tag{2.41}$$

这里的对角项通过将$1-2(y^2+z^2)$替换为$(x^2+w^2-y^2-z^2)$变得更为对称。

单位四元数最好的一面是存在简单的代数方法,用于复合用单位四元数表示的旋转。给定两个四元数$q_0 = (v_0, w_0)$和$q_1 = (v_1, w_1)$,四元数的乘法算子可以定义为

$$q_2 = q_0 q_1 = (v_0 \times v_1 + w_0 v_1 + w_1 v_0, w_0 w_1 - v_0 \cdot v_1), \tag{2.42}$$

具有性质$R(q_2) = R(q_0) R(q_1)$。注意,四元数乘法不是可交换的,正如 3D 旋转和矩阵乘法不可交换一样。

求四元数的逆很容易:只要反向v或w的符号(但不是两者都反向!)即可。(你可以验证一下,这具有将公式(2.41)中的 R 转置的期望效果。)因此,我们还可以定义四元数除法为

$$q_2 = q_0/q_1 = q_0 q_1^{-1} = (v_0 \times v_1 + w_0 v_1 - w_1 v_0, -w_0 w_1 - v_0 \cdot v_1). \tag{2.43}$$

这在需要在两个旋转之间做增量式旋转时有用。

特别地,如果我们想在两个给定的旋转之间的中途处确定一个旋转,就可以计算这个增量旋转,取出旋转角的一部分,并计算出新的旋转。这一过程称作球面线性内插或简称四元数插值(Shoemake 1985),列在算法 2.1 中。注意,Shoemake 给出的两个公式不同于这里给出的公式。第一个公式对 q_r 求 α 指数,然后乘以原来的四元数,

$$q_2 = q_r^\alpha q_0, \tag{2.44}$$

而第二个公式将四元数作为球面上的 4 分量矢量看待,使用

$$q_2 = \frac{\sin(1-\alpha)\theta}{\sin\theta} q_0 + \frac{\sin\alpha\theta}{\sin\theta} q_1, \tag{2.45}$$

其中$\theta = \cos^{-1}(q_0 \cdot q_1)$且点积直接作用于四元 4 分量矢量之间。所有这些公式都给出可比较的结果,但是需要关注 q_0 和 q_1 离得很近的情况,这正是为什么我推荐使用反正切来建立旋转角的原因。

哪种旋转表达更好呢? 选择 3D 旋转的表达部分取决于应用。轴/角表达是最小的,因此不需要对参数作任何额外的约束(每次更新后不需要重新规范化)。如果角度用度数表达,理解姿态(比如说,绕 x 轴做 90° 旋转)则比较容易,且也容易表达精确的旋转。角度以弧度表示时,计算 R 相对于 ω 的导数是很方便的(2.36)。

而另一方面,如果你想跟上光滑移动的摄像机,四元数表达则更好,因为该表达没有不连续性。在这种表达下,在旋转之间做内插和形成刚性变换链也都比较容易(Murray, Li, and Sastry 1994; Bregler and Malik 1998)。

我通常优先选择使用四元数,但是使用增量旋转来更新其估计,详见 6.2.2 节的描述。

> **procedure** $slerp(\boldsymbol{q}_0, \boldsymbol{q}_1, \alpha)$:
> 1. $\boldsymbol{q}_r = \boldsymbol{q}_1/\boldsymbol{q}_0 = (\boldsymbol{v}_r, w_r)$
> 2. if $w_r < 0$ then $\boldsymbol{q}_r \leftarrow -\boldsymbol{q}_r$
> 3. $\theta_r = 2\tan^{-1}(\|\boldsymbol{v}_r\|/w_r)$
> 4. $\hat{\boldsymbol{n}}_r = \mathcal{N}(\boldsymbol{v}_r) = \boldsymbol{v}_r/\|\boldsymbol{v}_r\|$
> 5. $\theta_\alpha = \alpha\,\theta_r$
> 6. $\boldsymbol{q}_\alpha = (\sin\frac{\theta_\alpha}{2}\hat{\boldsymbol{n}}_r, \cos\frac{\theta_\alpha}{2})$
> 7. **return** $\boldsymbol{q}_2 = \boldsymbol{q}_\alpha \boldsymbol{q}_0$

算法 2.1 球面线性内插(四元数插值)。首先，轴和整个角度是从四元数比值计算出来的。(该计算可以放在为动画而产生一组内插位置的内循环外。)然后计算一个增量四元数并乘以起始的旋转四元数

2.1.5 3D 到 2D 投影

现在我们已知如何表达 2D 和 3D 几何基元以及如何在空间上对其进行变换，我们需要规定 3D 基元是如何投影到图像平面上的。我们可以用 3D 到 2D 投影矩阵来做这件事。最简单的模型是正交投影，不需要除法就可以得到最终的(非齐次)结果。更常用的模型是透视投影，因为该模型更精确地反映了真实摄像机的性能。

正交投影和类透视投影

正交投影简单舍弃三维坐标 $\boldsymbol{p}$ 的 z 分量来得到 2D 点 $\boldsymbol{x}$。(在本节中，我们用 $\boldsymbol{p}$ 表示 3D 点而用 $\boldsymbol{x}$ 表示 2D 点。)这可以写作

$$\boldsymbol{x} = [\boldsymbol{I}_{2\times 2}|\boldsymbol{0}]\,\boldsymbol{p}. \tag{2.46}$$

如果使用齐次(投影)坐标，我们可以写作

$$\tilde{\boldsymbol{x}} = \begin{bmatrix} 1 & 0 & 0 & 0 \\ 0 & 1 & 0 & 0 \\ 0 & 0 & 0 & 1 \end{bmatrix} \tilde{\boldsymbol{p}}, \tag{2.47}$$

即，我们舍弃 z 分量但保留 ω 分量。正交投影是对长焦距(望远)镜头和物体当其自身的深度与其到摄像机的距离相比很浅时的近似模型(Sawhney and Hanson 1991)。它只对远心(telecentric)镜头才是精确的(Baker and Nayar 1999, 2001)。

在实践中，世界坐标(它可能用米来量度)需要按比例缩放以适合图像传感器(物理上以毫米量度，但最终以像素量度)。因为这个原因，归一的正交投影(scaled orthography)实际上更常用，

$$\boldsymbol{x} = [s\boldsymbol{I}_{2\times 2}|\boldsymbol{0}]\,\boldsymbol{p}. \tag{2.48}$$

该模型等同于首先将世界点投影在一个局部的正面平行(fronto-parallel)图像平

面上,然后将该图像用常规的透视投影缩放。这种缩放可以对场景所有部分都相同(图 2.7b),或对独立建模的不同物体各有不同(图 2.7c)。更重要的是缩放在估计由运动到结构时可以逐帧改变,这可以更好地对物体接近摄像机时出现的缩放变化进行建模。

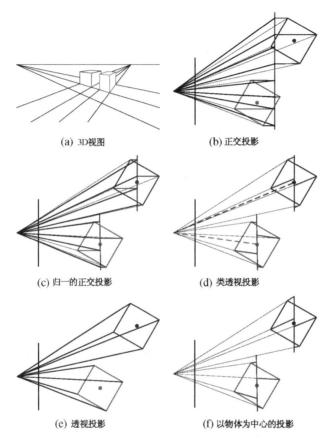

图 2.7 常用的投影模型:(a)3D 视图;(b)正交投影;(c)归一的正交投影;(d)类透视投影;(e)透视投影;(f)以物体为中心的投影。每个图给出的是投影的俯视图。注意地面上和盒子侧面的平行线在非透视投影中仍保持平行

在重建远离摄像机的物体的 3D 形状时,因为可以大大简化某些计算,归一的正交投影是常用的模型。例如,姿态(摄像机的方向)可以用简单的最小二乘来估计(6.2.1 节)。在正交投影下,正如 7.3 节的论述(Tomasi and Kanade 1992),结构和运动可以用因子分解(奇异值分解)同时估计出来。

一个紧密相关的投影模型是类透视投影(para-perspective)(Aloimonos 1990; Poelman and Kanade 1997)。在该模型中,物体点也是首先投影到一个平行于图像平面的局部参考面上。但是,不是正交地投影到该平面上,而是按照平行于到物体中心的视线来投影(图 2.7d)。然后按照通常的投影方式投影到最终的图像平面上,这也就对应于缩放。这两个投影的组合也就是仿射,可以写作

$$\tilde{x} = \begin{bmatrix} a_{00} & a_{01} & a_{02} & a_{03} \\ a_{10} & a_{11} & a_{12} & a_{13} \\ 0 & 0 & 0 & 1 \end{bmatrix} \tilde{p}. \tag{2.49}$$

注意，在如图 2.7b～d 所示的投影之后 3D 中的平行线仍然是平行的。类透视投影提供了比归一的正交投影更精确的投影模型，而不会出现每点透视的除法所增加的复杂性，每点透视的除法使得传统的因子分解方法失效(Poelman and Kanade 1997)。

透视投影

在计算机图形学和计算机视觉中最常用的投影是 3D 透视投影(图 2.7e)。这里，点映射到图像平面上是通过除以其 z 分量来实现的。使用非齐次坐标，这可以写为

$$\bar{x} = \mathcal{P}_z(p) = \begin{bmatrix} x/z \\ y/z \\ 1 \end{bmatrix}. \tag{2.50}$$

在齐次坐标下，投影具有简单的线性形式，

$$\tilde{x} = \begin{bmatrix} 1 & 0 & 0 & 0 \\ 0 & 1 & 0 & 0 \\ 0 & 0 & 1 & 0 \end{bmatrix} \tilde{p}, \tag{2.51}$$

即我们舍弃 p 点的 ω 分量。这样，在投影后，就不可能恢复该 3D 点到图像的距离了，对于 2D 的图像传感器，这是可以理解的。

在计算机图形学系统中常见的一种形式是一个两步投影，首先将 3D 坐标投影到范围在 $(x,y,z) \in [-1,1] \times [-1,1] \times [0,1]$ 的规范化设备坐标上，然后用视口变换(Watt 1995; OpenGL-ARB 1997)将这些坐标缩放成整数的像素坐标。该(原来的)透视投影就可以用 4×4 矩阵表示

$$\tilde{x} = \begin{bmatrix} 1 & 0 & 0 & 0 \\ 0 & 1 & 0 & 0 \\ 0 & 0 & -z_{\text{far}}/z_{\text{range}} & z_{\text{near}} z_{\text{far}}/z_{\text{range}} \\ 0 & 0 & 1 & 0 \end{bmatrix} \tilde{p} \tag{2.52}$$

其中 z_{near} 和 z_{far} 是近和远 z 裁剪面，$z_{\text{range}} = z_{\text{far}} - z_{\text{near}}$。注意，前两行实际上是以焦距和横纵比缩放的，以便可见光线映射到 $(x,y,z) \in [-1,1]^2$。保留而不是舍弃第三行的原因是可见性运算，例如 z-缓冲，需要对所绘制的每个图形单元有一个深度。

如果我们设 $z_{\text{near}} = 1$，$z_{\text{far}} \to \infty$，并将第三行反号，规范化的显示矢量的第 3 个元素就变为深度的倒数，即视差(Okutomi and Kanade 1993)。这在很多情况下相当方便，因为对于在室外移动的摄像机而言，到摄像机的深度的倒数往往为比直接 3D 距离更适定的(well-conditioned)参数化对象。

尽管常规的 2D 图像传感器没有办法测量到表面点的距离，但距离传感器(12.2 节)和立体视觉匹配算法(第 11 章)可以计算这样的数值。使用 4×4 矩阵的逆(2.1.5 节)就能将基于传感器的深度或视差数值 d 直接映射回 3D 位置，这是很方便的。如

果我们像公式(2.64)那样用一个满秩 4×4 矩阵来表示透视投影，就可以做到。

摄像机内参数 一旦我们将 3D 点使用一个投影矩阵穿过一个理想的针孔投影之后，我们还必须对结果坐标根据像素传感器的间距和传感器平面到原点的相对位置进行变换。图 2.8 展示了所涉及的几何知识。在本节中，我们首先用传感器同态映射 M_s 给出从 2D 像素坐标到 3D 射线的映射，因为用物理上可测量的量来解释它比较容易。然后，我们将这些量与更常用的摄像机内参数矩阵 K 联系起来，这个矩阵被用于将以摄像机为中心的 3D 点 p_c 映射到 2D 像素坐标 $\tilde{x}_s$。

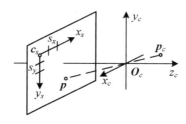

图 2.8 以摄像机为中心的 3D 点 p_c 在传感器平面位置 p 上的投影。O_c 是摄像机中心(节点)，c_s 是传感器平面坐标系的 3D 原点，s_x 和 s_y 是像素间隔

图像传感器返回以整数像素坐标(x_s，y_s)为索引的像素值，通常以图像的左上角为坐标起点而沿着向下和向右方向移动。(这个规范在所有的图像处理库中未得到遵守，但是调整到其他坐标系是直观易懂的。)将像素中心映射到 3D 坐标，我们首先用像素间距(s_x，s_y)缩放(x_s，y_s)的数值(对于固态传感器有时用微米表示)，然后相对于摄像机投影中心 O_c 以原点 c_s 和一个 3D 旋转 R_s 来描述传感器矩阵的方向(图 2.8)。

结合起来的 2D 到 3D 投影可以写作

$$p = \begin{bmatrix} R_s & | & c_s \end{bmatrix} \begin{bmatrix} s_x & 0 & 0 \\ 0 & s_y & 0 \\ 0 & 0 & 0 \\ 0 & 0 & 1 \end{bmatrix} \begin{bmatrix} x_s \\ y_s \\ 1 \end{bmatrix} = M_s \bar{x}_s. \tag{2.53}$$

3×3 矩阵 M_s 的前两列是对应于图像像素阵列的沿 x_s 和 y_s 方向单位步长的 3D 矢量，而第三列是 3D 图像阵列原点 c_s。

矩阵 M_s 用 8 个未知量来做参数化：三个参数描述旋转 R_s；三个参数描述平移 c_s；两个缩放因子(s_x，s_y)。注意，我们这里忽略了图像平面两个轴之间倾斜的可能性，因为固态制造技术事实上是可以忽略的。在实践中，除非我们有关于传感器间距或传感器方向的精确的外部信息，否则就只有 7 个自由度，因为从原点到传感器的距离仅根据外部图像测量本身是不能从传感器间距中梳理出来的。

但是，估计具有 7 个自由度的摄像机模型 M_s(即，其中前两列在适当的缩放后是正交的)并不现实，因此绝大多数实践者会假定更一般的 3×3 同态映射矩阵形式。

3D 像素中心 p 和以摄像机为中心的 3D 点 p_c 之间的关系由一个未知的缩放 s 给定，$p = s\,p_c$。因此我们可以将 p_c 和像素地址的齐次形式 $\tilde{x}_s$ 之间的完整投影写为

$$\tilde{x}_s = \alpha M_s^{-1} p_c = K p_c. \tag{2.54}$$

这个 3×3 矩阵 K 称作校准(标定)矩阵,描述摄像机的内参数(相对于摄像机在空间中的方向,称作外参数)。

由上所述,我们知道 K 在理论上有 7 个自由度,而在实践中有 8 个自由度(3×3 同态矩阵的全部自由度)。那么,为什么有关 3D 视觉和多视角几何的绝大多数教材(Faugeras 1993; Hartley and Zisserman 2004; Faugeras and Luong 2001)都将 K 作为具有 5 个自由度的上三角矩阵来处理呢?

这些教材通常并没有明确地说明,这是因为单靠外部测量本身我们不能恢复完整的 K 矩阵。基于外部 3D 点或者其他测量标定摄像机时(第 6 章)(Tsai 1987),我们使用一系列量测最终估计出摄像机的内参数 K 和外参数(R, t)

$$\tilde{x}_s = K \begin{bmatrix} R \mid t \end{bmatrix} p_w = P p_w, \tag{2.55}$$

其中 p_ω 是已知的 3D 世界坐标而

$$P = K[R|t] \tag{2.56}$$

是摄像机矩阵。审视该公式,我们发现可以用 R_1 后乘 K 且用 R_1^T 前乘($R|t$),仍然是一个有效的标定。因此,仅靠图像测量本身是不可能知道传感器真实的方向和真实的摄像机内参数的。

为 K 选择上三角形式似乎符合惯例。给定一个完整的 3×4 摄像机矩阵 $P = K(R|t)$,我们可以用 QR 因子分解计算出一个上三角 K 矩阵(Golub and Van Loan 1996)。(注意不幸的术语冲突:在矩阵代数教材中,R 表示上三角(对角线右侧)矩阵;在计算机视觉中,R 表示的是正交旋转)

有几种方式来写上三角矩阵 K。一种可能是

$$K = \begin{bmatrix} f_x & s & c_x \\ 0 & f_y & c_y \\ 0 & 0 & 1 \end{bmatrix}, \tag{2.57}$$

它使用相互独立的 x 和 y 维度的焦距 f_x 和 f_y。s 项刻画任何可能的传感器轴间的倾斜,这由传感器的安装没有与光轴垂直所引起,而(c_x, c_y)是以像素坐标表达的光心。另一种可能是

$$K = \begin{bmatrix} f & s & c_x \\ 0 & af & c_y \\ 0 & 0 & 1 \end{bmatrix}, \tag{2.58}$$

其中横纵比 a 被明确地表达出来且使用了一个常用的焦距 f。

在实践中,通过设置 $a = 1$ 和 $s = 0$,在很多应用中会使用如下更简单的形式,

$$K = \begin{bmatrix} f & 0 & c_x \\ 0 & f & c_y \\ 0 & 0 & 1 \end{bmatrix}. \tag{2.59}$$

通常情况下,通过将原点大致设置在图像的中心,即(c_x, c_y) = ($W/2$, $H/2$),其中 W 和 H 是图像的高和宽,就可以得到仅含一个未知量(即焦距 f)的完全可用的摄像机模型。

图 2.9 给出了这些量是如何构成一个简化成像模型的。注意，现在我们将图像平面置于节点(镜头的投影中心)之前了。y 轴的含义被翻转过来以便与多数成像库在处理纵(行)坐标时的方式相兼容。有些图形库(比如 Direct3D)使用左手坐标系，这可能导致某种混淆。

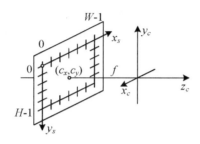

图 2.9 简化的摄像机内参数展示了焦距 f 和光心 (C_x, C_y)。图像宽度和高度分别为 W 和 H

关于焦距的注解

如何来表达焦距这个问题在实现计算机视觉算法和讨论其结果时常常引起困惑。这是因为焦距依赖于量测像素所使用的单位。

如果我们用整数值计数像素坐标，比如 $(0, W) \times (0, H)$，公式(2.59)中的焦距 f 和摄像机中心 (c_x, c_y) 可以表达为像素值。这些量是如何同我们更熟悉的摄影师所使用的焦距相联系的呢？

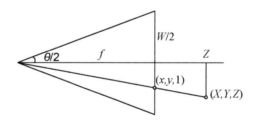

图 2.10 中心投影，展示了 3D 和 2D 坐标 p 和 x 之间的关系，也给出了焦距 f、图像宽度 W 和视场 θ 之间的关系

图 2.10 展示了焦距 f、图像宽度 W 和视场 θ 之间的关系，它们服从如下公式：

$$\tan \frac{\theta}{2} = \frac{W}{2f} \quad \text{or} \quad f = \frac{W}{2} \left[\tan \frac{\theta}{2} \right]^{-1}. \quad (2.60)$$

对于常规的胶片相机，$W = 35\text{mm}$，因此 f 也用毫米表达。由于我们使用数字图像，所以更方便的是用像素表示 W，使焦距 f 如公式(2.59)中那样能直接用于标定矩阵 K 之中。

另一种可能是缩放像素坐标使其沿着长轴落在 $[-1; 1)$ 而沿着短轴落在 $(-a^{-1}, a^{-1})$，其中 $a \geq 1$ 是图像横纵比(相比于早前介绍过的传感器单元的横纵比)。这可以用修改的规范化设备坐标来实现，

$$x'_s = (2x_s - W)/S \quad \text{and} \quad y'_s = (2y_s - H)/S, \quad \text{where} \quad S = \max(W, H). \quad (2.61)$$

这样做的好处是焦距 f 和光心 (c_x, c_y) 变得与图像分辨率无关，这在使用诸如图像

金字塔(3.5 节)等多分率图像处理算法时是很有用的[①]。使用 S 而不是 W 还使得对于景观(水平的)和画像(垂直的)照片的焦距相同，正如在 35mm 摄影中那样。(在一些计算机图形学教材和系统中，规范化设备坐标落在 $[-1,1]\times[-1,1]$ 上，这需要使用两个不同的焦距来描述摄像机内参数(Watt 1995; OpenGL-ARB 1997)。)在公式(2.60)中设置 $S = W = 2$，我们得到更简单的(无单位的)关系

$$f^{-1} = \tan\frac{\theta}{2}. \tag{2.62}$$

在各种焦距表达间的转换是直截了当的，例如，从无单位的 f 到以像素表达它，乘以 $W/2$，而从以像素表达的 f 到其等同的 35mm 焦距，则乘以 35/W。

摄像机矩阵

现在我们已经给出了如何参数化标定矩阵 K，便可以将摄像机内参数和外参数合并起来得到一个 3×4 摄像机矩阵

$$P = K\begin{bmatrix} R \mid t \end{bmatrix}. \tag{2.63}$$

有时使用一个可逆的 4×4 矩阵更好，这可以通过保留 P 矩阵的最后一行获得，

$$\tilde{P} = \begin{bmatrix} K & 0 \\ 0^T & 1 \end{bmatrix}\begin{bmatrix} R & t \\ 0^T & 1 \end{bmatrix} = \tilde{K}E, \tag{2.64}$$

其中 E 是一个 3D 刚体(欧氏)变换，$\tilde{K}$ 是满秩标定矩阵。4×4 摄像机矩阵 $\tilde{P}$ 可用于从 3D 世界坐标 $\bar{p}_\omega = (x_\omega, y_\omega, z_\omega, 1)$ 到屏幕坐标(加上视差) $x_\omega = (x_s, y_s, 1, d)$ 的直接映射，

$$x_s \sim \tilde{P}\bar{p}_w, \tag{2.65}$$

其中~表示相差一个缩放因子的等同性。注意，在乘以 $\tilde{P}$ 之后，该矢量被其第 3 个分量相除得到规范化形式 $x_\omega = (x_s, y_s, 1, d)$。

平面加平行视差(投影深度)

一般而言，当使用 4×4 矩阵 $\tilde{P}$ 时，我们可以自由地将最后一行重新映射到任何适合我们目的形式(而不仅限于作为视差的"标准"解释的深度的倒数)。

让我们将 $\tilde{P}$ 的最后一行重写为 $p_3 = s_3[\hat{n}_0 \mid c_0]$，其中 $\|\hat{n}_0\| = 1$，则我们有如下公式

$$d = \frac{s_3}{z}(\hat{n}_0 \cdot p_w + c_0), \tag{2.66}$$

其中 $z = p_2 \cdot \bar{p}_\omega = r_z \cdot (p_\omega - c)$ 是从摄像机中心 C(2.25)沿着光轴 Z 到 p_ω 的距离(图 2.11)。因此，我们将 d 解释为投影视差或 3D 场景点 p_ω 从参考面 $\hat{n}_0 \cdot p_w + c_0 = 0$ 来的投影深度(Szeliski and Coughlan 1997; Szeliski and Golland 1999; Shade, Gortler, He *et al.* 1998; Baker, Szeliski, and Anandan 1998)。(投影深度在使用平面加平行视差项的重建算法(Kumar, Anandan, and Hanna 1994; Sawhney 1994)中有时也称作"平行视差"。)设置 $\hat{n}_0 = 0$ 和 $c_0 = 1$，即将参考面设在无穷远，产生更标准的视差形式 $d = 1/z$ (Okutomi and Kanade 1993)。

[①] 为使金字塔中经过下采样后的转换真正精确，需要保留 W 和 H 的浮点值，因为如果它们在金字塔中的较高分辨率上曾是奇数，就可变为非整数。

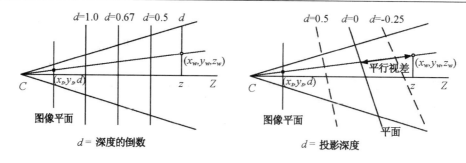

图 2.11 规则的视差(深度的倒数)和投影深度(到参考面的平行视差)

看待这一问题的另一种方式是，求 $\tilde{P}$ 的逆矩阵，使我们可以将像素加视差直接映射回 3D 点，

$$\tilde{p}_w = \tilde{P}^{-1} x_s. \tag{2.67}$$

一般而言，我们可以任意选择 $\tilde{P}$ 的形式，即使用任意的投影来采样空间，这是有好处的。当设置多视角立体重建算法时，这可以变得特别方便，因为它允许我们用一系列平面扫描(11.1.2 节)来采样空间，使这样的不同(投影)采样与所感知的图像运动有最好的匹配(Collins 1996; Szeliski and Golland 1999; Saito and Kanade 1999)。

从一个摄像机到另一个摄像机的映射

当我们从两个不同的摄像机位置或方向拍摄 3D 场景的两幅图像时会发生什么(图 2.12a)？使用公式(2.64)中的满秩的 4×4 摄像机矩阵 $\tilde{P} = \tilde{K}E$，我们可以将从世界坐标到屏幕坐标的投影写为

$$\tilde{x}_0 \sim \tilde{K}_0 E_0 p = \tilde{P}_0 p. \tag{2.68}$$

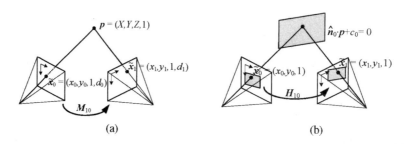

图 2.12 一点被投影到两个图像中：(a)3D 点坐标 $(X, Y, Z, 1)$ 和 2D 投影点 $(x, y, 1; d)$ 间的关系；(b)由所有的处在共同平面 $\hat{n}_0 \cdot p + c_0 = 0$ 上的点导出的平面同态映射

假设我们知道一幅图像中像素的 z-缓冲或视差值 d_0，就可以使用如下公式计算出 3D 点位置

$$p \sim E_0^{-1} \tilde{K}_0^{-1} \tilde{x}_0 \tag{2.69}$$

然后将其投影到另一幅图像中得到

$$\tilde{x}_1 \sim \tilde{K}_1 E_1 p = \tilde{K}_1 E_1 E_0^{-1} \tilde{K}_0^{-1} \tilde{x}_0 = \tilde{P}_1 \tilde{P}_0^{-1} \tilde{x}_0 = M_{10} \tilde{x}_0. \tag{2.70}$$

不幸的是，我们通常不能从普通摄影下的图像中获得像素的深度坐标。但是，

对于平面类场景，正如在上面公式(2.66)中所讨论的那样，我们可以用一个一般的平面方程 $\hat{\boldsymbol{n}}_0 \cdot \boldsymbol{p} + c_0 = 0$ 取代公式(2.64)中的最后一行 $\boldsymbol{P}_0$，将平面上的点映射到 $d_0 = 0$ 数值(图2.12b)。因此，如果将设置 $d_0 = 0$，我们可以忽略公式(2.70)中的 $\boldsymbol{M}_{10}$ 的最后一列和最后一行，因为我们不关心最后的 z-缓冲深度。因此，映射公式(2.70)简化为

$$\tilde{\boldsymbol{x}}_1 \sim \tilde{\boldsymbol{H}}_{10} \tilde{\boldsymbol{x}}_0, \tag{2.71}$$

其中 $\tilde{\boldsymbol{H}}_{10}$ 是一个一般的 3×3 摄像机矩阵 4×4 摄像机矩阵，$\tilde{\boldsymbol{x}}_1$ 和 $\tilde{\boldsymbol{x}}_0$ 现在是齐次坐标(即，3 分量向量)(Szeliski 1996)。这使将 8 参数同态映射用作平面状场景拼图的普通配准模型有了根据(Mann and Picard 1994; Szeliski 1996)。

我们不需要知道深度就可以进行摄像机间的映射的另一种特殊情况是摄像机在做纯旋转运动时(9.1.3 节)，即当 $\boldsymbol{t}_0 = \boldsymbol{t}_1$ 时。在这种情况下，我们可以写出

$$\tilde{\boldsymbol{x}}_1 \sim \boldsymbol{K}_1 \boldsymbol{R}_1 \boldsymbol{R}_0^{-1} \boldsymbol{K}_0^{-1} \tilde{\boldsymbol{x}}_0 = \boldsymbol{K}_1 \boldsymbol{R}_{10} \boldsymbol{K}_0^{-1} \tilde{\boldsymbol{x}}_0, \tag{2.72}$$

这也可以用 3×3 同态映射表达。如果我们假设标定矩阵具有已知的横纵比和投影中心(2.59)，该同态映射可以参数化为旋转量和两个未知的焦距。这种特殊的表达常用于图像拼接应用中(9.1.3 节)。

以物体为中心的投影

使用长焦距镜头时，单纯从图像量测本身来可靠地估计焦距往往会变得很困难。这是因为焦距和到物体的距离是紧密相关的，且很难将这两者的影响分离开。例如，从远焦镜头看到的物体尺度的变化或者可能是由变焦引起的，或者是由朝向用户的运动引起的。(这种现象在著名导演阿尔弗雷德·希区柯克(Alfred Hitchcock)的电影《迷魂记》中有生动的应用，其中变焦和摄像机运动的同时变化产生了一种令人焦虑的效果。)

如果我们写出对应于简单标定矩阵 $\boldsymbol{K}$ 的投影公式(2.59)，这种歧义就变得更清晰了，

$$x_s = f \frac{\boldsymbol{r}_x \cdot \boldsymbol{p} + t_x}{\boldsymbol{r}_z \cdot \boldsymbol{p} + t_z} + c_x \tag{2.73}$$

$$y_s = f \frac{\boldsymbol{r}_y \cdot \boldsymbol{p} + t_y}{\boldsymbol{r}_z \cdot \boldsymbol{p} + t_z} + c_y, \tag{2.74}$$

其中 $\boldsymbol{r}_x, \boldsymbol{r}_y, \boldsymbol{r}_z$ 是 $\boldsymbol{R}$ 的三行。如果到物体中心的距离 $t_z \gg \|\boldsymbol{p}\|$(物体的大小)，分母则近似为 t_z 而投影后物体的整个尺度取决于 f 对 t_z 的比率。因此很难解开这两个量。

为了更清楚地看这个问题，我们设 $\eta_z = t_z^{-1}$ 和 $s = \eta_z f$。然后，我们可以将以上公式重写为

$$x_s = s \frac{\boldsymbol{r}_x \cdot \boldsymbol{p} + t_x}{1 + \eta_z \boldsymbol{r}_z \cdot \boldsymbol{p}} + c_x \tag{2.75}$$

$$y_s = s \frac{\boldsymbol{r}_y \cdot \boldsymbol{p} + t_y}{1 + \eta_z \boldsymbol{r}_z \cdot \boldsymbol{p}} + c_y \tag{2.76}$$

(Szeliski and Kang 1994; Pighin, Hecker, Lischinski et al. 1998)。如果我们观察的是一个已知的物体(即 3D 坐标 $\boldsymbol{p}$ 是已知的)，投影的尺度 s 便可以可靠地估计出来。距离的倒数 η_z 现在从 s 的估计中基本解耦出来，且可以从随物体旋转时出现的透视

收缩量中估计出来。而且,随着镜头变得更长,即投影模型变成正交模型了,无需将透视成像模型换为正交模型,因为 $\eta_z \to 0$(相比于 f 和 t_z 同时趋向无穷)可使用同一个方程。这使我们可以在诸如因子分解等的正交重建方法和其投影/透视对等方法间构建起自然的联系(7.3 节)。

2.1.6 镜头畸变

上述成像模型都假定摄像机遵守线性投影模型,即世界中的直线产生图像中的直线。(这是作用在齐次坐标上的线性矩阵运算的自然结果。)不幸的是,很多广角镜头具有明显的径向畸变(失真),在直线投影时会彰显为可见的曲线。(2.2.3 节有关于镜头光学更详细的论述。)除非将这种畸变考虑进来,否则不可能创建出高度精确的具有真实感的重建。例如,没有考虑径向畸变的图像拼图常常会呈现出模糊效应,其原因在于像素混合前对应特征的误注册(第 9 章)。

幸运的是,补偿径向畸变在实践中并不那么困难。对于多数镜头,一个简单的二次畸变模型就可以产生很好的结果。设 (x_c, y_c) 是在透视除法后但在由焦距 f 缩放前并经过光心 (c_x, c_y) 偏移后所获得的像素坐标,即

$$\begin{aligned} x_c &= \frac{\boldsymbol{r}_x \cdot \boldsymbol{p} + t_x}{\boldsymbol{r}_z \cdot \boldsymbol{p} + t_z} \\ y_c &= \frac{\boldsymbol{r}_y \cdot \boldsymbol{p} + t_y}{\boldsymbol{r}_z \cdot \boldsymbol{p} + t_z}. \end{aligned} \tag{2.77}$$

径向畸变模型是说,所观察图像中的坐标按照其径向距离的一定比例偏离(桶形失真)或朝向(枕形失真)图像中心(图 2.13a 和 b)[①]。最简单的径向畸变模型使用低阶多项式,例如:

$$\begin{aligned} \hat{x}_c &= x_c(1 + \kappa_1 r_c^2 + \kappa_2 r_c^4) \\ \hat{y}_c &= y_c(1 + \kappa_1 r_c^2 + \kappa_2 r_c^4), \end{aligned} \tag{2.78}$$

其中 $r_c^2 = x_c^2 + y_c^2$ 和 κ_1 和 κ_2 被称为径向畸变参数[②]。在经过径向畸变这一步之后,最后的像素坐标可以用下式计算:

$$\begin{aligned} x_s &= f x_c' + c_x \\ y_s &= f y_c' + c_y. \end{aligned} \tag{2.79}$$

对于给定的镜头,有若干种方法可用于估计径向畸变参数,这在 6.3.5 节中介绍。

有时,上述简单的模型对由复杂镜头产生的真实畸变建模的精度不够(特别是在特别宽的视角处)。更完整的分析模型包含切向畸变和偏心畸变(Slama 1980),但是本书中没有覆盖这些畸变。

[①] 变形镜头,在故事片制作中广为使用,它并不遵从径向畸变模型。取而代之的是,作为最初的近似,我们可以认为它们导致不同的横向和纵向的缩放,即非方形的像素。

[②] 有时,x_c 和 $\hat{x}_c$ 之间的关系用其他方式来表达,即 $x_c = \hat{x}_c(1 + \kappa_1 \hat{r}_c^2 + \kappa_2 \hat{r}_c^4)$。如果我们通过除以 f 将图像像素映射到(卷绕的)射线,这就方便了。这样,我们便可以解除射线扭曲但仍有空间中的真实 3D 射线。

鱼眼镜头(图 2.13c)需要的模型不同于径向畸变的传统多项式模型。鱼眼镜头首先可以近似为偏离光轴的角度的等距投影(Xiong and Turkowski 1997)，这与公式(9.22～9.24)所描述的极坐标投影相同。Xiong and Turkowski(1997)描述了如何扩展该模型，即加上一个额外的对ϕ的二次矫正项，并介绍了如何从一组交叠的鱼眼图像使用直接(基于亮度的)非线性最小化算法来估计未知参数(投影中心、缩放因子s等)。

图 2.13 径向畸变: (a)桶形; (b)枕垫形; (c)鱼眼形。鱼眼图像从一端到另一端跨度几乎为180°

对于更大的，较不规范的畸变，使用样条函数的参数化畸变模型可能是必要的(Goshtasby 1989)。如果镜头不具有单投影中心，就可能有必要对对应于每个像素的3D线条(相对于方向)分别建模(Gremban, Thorpe, and Kanade 1988; Champleboux, Lavall´ee, Sautot et al. 1992; Grossberg and Nayar 2001; Sturm and Ramalingam 2004; Tardif, Sturm, Trudeau et al. 2009)。6.3.5 节有一些更详细的这样的方法的描述，它们阐述了如何矫正镜头畸变。

有一个与简单的径向畸变模型相关的微妙问题时常被掩盖。我们已经介绍了透视投影和最终传感器阵列投影步骤之间的非线性问题。因此，在一般情况下，我们不能后乘一个带旋转的任意的 3×3 矩阵 $\boldsymbol{K}$ 使其成为上三角形式而将其吸纳进整个旋转中。但是，这种情形并不像它最初看起来那么糟糕。对于很多应用，保持公式(2.59)的简单三角形式仍然是一个合适的模型。而且，如果我们将径向和其他畸变矫正到能够保持直线性的精度，就基本上将传感器转换回线性成像器，而且前述的分解仍然适用。

2.2 光度测定学的图像形成

在图像形成过程的建模中，我们描述了世界中 3D 几何特征是如何投影到图像中的 2D 特征的。但是，图像并不是由 2D 特征组成的。它们反而是由离散的色彩或亮度数值构成的。这些值从何而来？它们与环境中的光线、表面特性和几何、摄像机光学以及传感器特性的关系如何(图 2.14)？在本节中，我们建立一组模型来描述它们的相互关系并形式化图像形成的一般过程。这些主题更详细的阐述可在关于计算机图形学和图像合成的其他教材中找到(Glassner 1995; Weyrich, Lawrence, Lensch et al. 2008; Foley, van Dam, Feiner et al. 1995; Watt 1995; Cohen and Wallace 1993; Sillion and Puech 1994)。

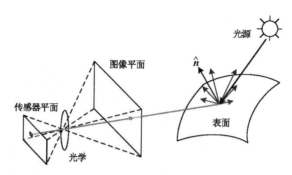

图 2.14 光度测定学的图像形成。光线从一个或多个光源发出，然后从物体表面反射。该光线的一部分朝向摄像机。该简化模型忽略了多次反射，这在现实世界场景中是常见的

2.2.1 照明

没有光就不存在图像。为了产生图像，场景必须用一个或多个光源照明。(荧光显微镜检查术和 X 射线断层摄影术等特定模态不符合这个模型，但是在本书中我们不处理这类情况。)光源通常可以分为点光源和面光源。

点光源起源于空间中的单个位置(例如，一个小灯泡)，可能在无穷远处(例如，太阳)。(注意对于某些应用，比如对软阴影(边缘区域)建模，太阳有可能必须当作面光源来处理。)除了其位置外，一个点光源具有强度和色谱，即在波长上的分布$L(\lambda)$。光源的强度以源点到其照亮的物体间的距离平方为比率衰减，因为同一个光在一个更大(球形)范围上传播。光源也可能具有带方向性(依赖)的衰减特性，但是在简化模型中我们忽略了这种情况。

面光源更复杂。简化的面光源，例如固定有散光罩的荧光蓬顶灯可以模型化为一个在所有方向上均匀发光的有限矩形区域(Cohen and Wallace 1993; Sillion and Puech 1994; Glassner 1995)。当分布强烈依赖于方向的时候，则要使用一个四维的光场模型(Ashdown 1993)。

环境贴图(environment map)(Greene 1986)最初叫"反射图"(reflection map)(Blinn and Newell 1976)常可以用来表达更为复杂的光分布，比如说，用于近似室外庭院中的物体上的入射照明。该表达将入射光方向$\hat{v}$映射到色彩值(即波长，λ)，

$$L(\hat{v}; \lambda), \tag{2.80}$$

同时，它也等同于假设所有光源是在无穷远处。环境贴图可以表达为立方体面的汇集(Greene 1986)，单个经度-纬度图(Blinn and Newell 1976)，或者反射球的像(Watt 1995)。获得真实世界环境贴图的粗略模型的一种方便的途径是，拍摄反射镜面球

的影像并将其解开到所需的环境贴图上(Debevec 1998)。Watt(1995)给出了关于环境贴图的很好的描述,包括对于三种最常用到的将方向映射到像素所需的公式的表达。

2.2.2 反射和阴影

当光线照射在物体表面上时,光线散开并反射(图 2.15a)。人们提出了多个不同的模型来描述这个交互作用过程。在本节中,我们首先介绍最一般的形式,双向反射分布函数,然后审视一些更特殊的模型,包括漫反射、镜面反射和补色渲染(Phong shading)模型。我们也要讨论如何用这些模型来计算对应于场景的全局光照。

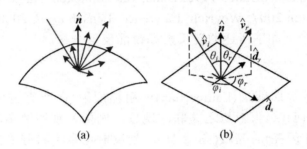

图 2.15 (a)光射到表面时,发生散射。(b)双向反射分布函数(BRDF) $f(\theta_i,\phi_i,\theta_r,\phi_r)$ 以入射光 $\hat{v}_i$ 和反射光 $\hat{v}_r$,光线方向与局部表面坐标架 $(\hat{d}_x,\hat{d}_y,\hat{n})$ 所构成的夹角为参数

双向反射分布函数(BRDF)

光线散射最一般的模型是双向反射分布函数(BRDF)[①]。相对于表面上某个局部坐标架构,BRDF 是一个四维函数,描述到达入射方向 $\hat{v}_i$ 的每个波长有多少在反射方向 $\hat{v}_r$ 上发出(图 2.15b)。函数可以用相对于表面坐标架构的入射和反射方向的角度项写成

$$f_r(\theta_i,\phi_i,\theta_r,\phi_r;\lambda). \tag{2.81}$$

BRDF 是互反的,即由于光线传播的物理特性,可以将 $\hat{v}_i$ 和 $\hat{v}_r$ 的作用调换但仍然会得到相同的答案(这有时称作 Helmholtz 互反性)。

大多数表面是各向同性的,即仅从光线传播的角度来考虑对表面方向没有偏向。(例外的是非各向同性表面,例如拉丝(有刮痕)铝,其上的反射取决于光线相对于刮痕的方向。)对于各向同性材料,我们可以将 BRDF 简化为

$$f_r(\theta_i,\theta_r,|\phi_r-\phi_i|;\lambda) \text{ 或 } f_r(\hat{v}_i,\hat{v}_r,\hat{n};\lambda), \tag{2.82}$$

因为量 θ_i,θ_r 和 ϕ_r-ϕ_i 可以从方向 $\hat{v}_i$,$\hat{v}_r$ 和 $\hat{n}$ 计算出来。

计算在给定光照条件下从表面上点 p 发出的方向为 $\hat{v}_r$ 的光的数量,我们对入射光 $L_i(\hat{v}_i;\lambda)$ 和 BRDF 的乘积求积分(有些作者称这一步为卷积)。将透视收缩因子 $\cos^+\theta_i$ 考虑进来,我们得到

[①] 实际上,还有光线传播更一般的模型,包括某些对沿着表面的空间变化、子表面散射以及大气效应等的建模,参见 12.7.1 节(Dorsey, Rushmeier, and Sillion 2007; Weyrich, Lawrence, Lensch *et al.* 2008)。

$$L_r(\hat{\boldsymbol{v}}_r;\lambda) = \int L_i(\hat{\boldsymbol{v}}_i;\lambda) f_r(\hat{\boldsymbol{v}}_i,\hat{\boldsymbol{v}}_r,\hat{\boldsymbol{n}};\lambda) \cos^+ \theta_i \, d\hat{\boldsymbol{v}}_i, \qquad (2.83)$$

其中

$$\cos^+ \theta_i = \max(0, \cos \theta_i). \qquad (2.84)$$

如果光源是离散的(有限个点光源)，我们可以用求和代替积分，

$$L_r(\hat{\boldsymbol{v}}_r;\lambda) = \sum_i L_i(\lambda) f_r(\hat{\boldsymbol{v}}_i,\hat{\boldsymbol{v}}_r,\hat{\boldsymbol{n}};\lambda) \cos^+ \theta_i. \qquad (2.85)$$

给定表面的 BRDF 可以通过物理建模(Torrance and Sparrow 1967; Cook and Torrance 1982; Glassner 1995)、启发式建模(Phong 1975)或通过经验观察(Ward 1992;Westin, Arvo, and Torrance 1992; Dana, van Ginneken, Nayar *et al.* 1999; Dorsey, Rushmeier, and Sillion 2007; Weyrich, Lawrence, Lensch *et al.* 2008)来获得[①]。典型的 BRDF 常常可以分解为如下所述的漫反射和镜面反射分量。

漫反射

漫反射分量(也称作朗伯(Lambertian)或粗糙反射)在所有方向上均匀地分散光线，是我们最常将(明暗)阴影与之关联的现象，例如，观察雕塑时所见到的随着表面法向光滑变化(非闪亮)的亮度(图 2.16)。漫反射也常常将强烈的物体颜色传递给光线，因为它是由物体内部材料对光的选择性吸收和发射特性所引起的效应(Shafer 1985; Glassner 1995)。

图 2.16 这个雕塑的特写镜头展示了漫反射(光滑的明暗变化)和镜面反射(光亮的高光)，以及在折皱和槽纹处由光可见性的减少和相互反射所引起的发暗的现象。照片承蒙 Caltech Vision Lab(*http://www.vision.caltech.edu/archive.html*)许可使用

尽管光在所有方向上均匀地散射，即 BRDF 是常量，

$$f_d(\hat{\boldsymbol{v}}_i,\hat{\boldsymbol{v}}_r,\hat{\boldsymbol{n}};\lambda) = f_d(\lambda), \qquad (2.86)$$

但光的数量依赖于入射光的方向和表面法向间的夹角 θ_i。这是因为暴露于给定量光下的表面区域在倾斜的角度下变得更大，当向外的表面法向指向背离光线的方向时，就会完全处在自身的阴影中(图 2.17a)。(想想看，你如何使自己朝向太阳或壁炉以获得最多热量，还有当闪光灯倾斜地投射在墙壁上时其亮度不如直射时来得亮。)因此漫反射下的阴影方程可以写作：

[①] *http://www1.cs.columbia.edu/CAVE/software/curet/*有关于某些经验性采样的 BRDF 的数据库。

$$L_d(\hat{v}_r;\lambda) = \sum_i L_i(\lambda) f_d(\lambda) \cos^+ \theta_i = \sum_i L_i(\lambda) f_d(\lambda) [\hat{v}_i \cdot \hat{n}]^+, \quad (2.87)$$

其中

$$[\hat{v}_i \cdot \hat{n}]^+ = \max(0, \hat{v}_i \cdot \hat{n}). \quad (2.88)$$

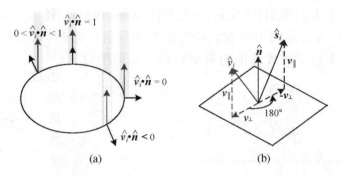

图 2.17 (a)由透视收缩引起的光量减少依赖于 $\hat{v}_i \cdot \hat{n}$，入射光的方向 $\hat{v}_i$ 和表面法向 $\hat{n}$ 间的夹角余弦。(b)镜面反射：入射光线方向 $\hat{v}_i$ 被反射到绕着表面法向的镜向方向 $\hat{s}_i$ 上

镜面反射

典型 BRDF 的第二个主要分量是镜面(光泽或高光)反射，强烈依赖于外射光的方向。考虑光从镜面表面的反射情况(图 2.17b)。入射光线以绕着表面法向 $\hat{n}$ 旋转 180°的方向被反射回来。用公式(2.29 和 2.30)中的相同记法，我们可以计算镜面反射方向 $\hat{s}_i$ 为

$$\hat{s}_i = v_\parallel - v_\perp = (2\hat{n}\hat{n}^T - I)v_i. \quad (2.89)$$

因此在给定方向 $\hat{v}_r$ 反射回的光量依赖于视线方向 $\hat{v}_r$ 和镜向 $\hat{s}_i$ 间的角度 $\theta_s = \cos^{-1}(\hat{v}_r \cdot \hat{s}_i)$。例如，Phong (1975)模型使用该角度的余弦的幂，

$$f_s(\theta_s;\lambda) = k_s(\lambda) \cos^{k_e} \theta_s, \quad (2.90)$$

而 Torrance and Sparrow(1967)微面元模型使用高斯函数，

$$f_s(\theta_s;\lambda) = k_s(\lambda) \exp(-c_s^2 \theta_s^2). \quad (2.91)$$

较大的指数 k_e(或高斯宽度的倒数 c_s)对应于镜面性更强的具有明显高光的表面，而较小的指数可以更好地对具有较软光泽的材料建模。

补色渲染

Phong(1975)将漫反射和镜面反射分量结合起来用他所称的环境照明(ambient illumination)项来表达。该项说明了这样的事实，物体通常不仅由点光源照明，而且还由对应于相互反射(例如，屋中的墙面间)或远光源(比如蓝色的天空)的一般的漫反射照明。在 Phong 模型中，环境项不依赖于表面法向，但是依赖于环境照明 $L_a(\lambda)$ 和物体 $k_a(\lambda)$ 两者的颜色，

$$f_a(\lambda) = k_a(\lambda) L_a(\lambda). \quad (2.92)$$

将所有这些项放在一起，我们就得到了 Phong 阴影(补色渲染，Phong shading)模型，

$$L_r(\hat{v}_r;\lambda) = k_a(\lambda)L_a(\lambda) + k_d(\lambda)\sum_i L_i(\lambda)[\hat{v}_i \cdot \hat{n}]^+ + k_s(\lambda)\sum_i L_i(\lambda)(\hat{v}_r \cdot \hat{s}_i)^{k_e}. \quad (2.93)$$

图 2.18 展示了 Phong 阴影模型分量的典型集合，它是偏离表面法向的角度的函数(在含有照明方向和观察者的平面中)。

典型情况下，环境和漫反射颜色分布 $k_a(\lambda)$ 和 $k_d(\lambda)$ 是相同的，因为它们都是由表面材料内部的子表面散射(体反射)引起的(Shafer 1985)。镜面反射分布 $k_s(\lambda)$ 经常是均匀(白)的，因为它是由不改变光颜色的界面反射引起的。(它的例外情况是金属材料，比如铜，不同于更普通的电解质材料，比如塑料。)

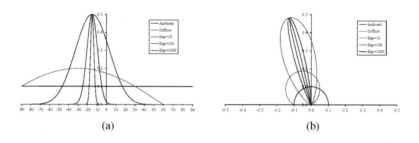

图 2.18 对于固定入射照明方向的补色渲染模型 BRDF 的横切面：(a)作为偏离表面法向的角度的函数的分量值；(b)极坐标图，Phong 指数 k_e 由 "Exp" 标签指示，光源位于偏离法向 30° 的方向

环境照明 $k_a(\lambda)$ 时常有来自于直接光源 $L_i(\lambda)$ 的不同的偏色，例如，在阳光充足的室外场景下它可能是蓝色的，而在室内烛光或白炽灯下它是黄色的。(在阴影区域中天空照明环境的出现是造成阴影部分看起来比光照到的场景部分更蓝的原因)。还要注意，Phong 模型(或任何其他阴影模型)的漫反射分量依赖于入射光源 $\hat{v}_i$ 的角度，而反射分量依赖于观察者 v_r 和镜面反射方向 $\hat{s}_i$(它本身依赖于入射光方向 $\hat{v}_i$ 和表面法向 $\hat{n}$)间的相对角度。

在计算机图形学中，Phong 阴影模型在物理精确度方面已经被近期发展的模型所取代，包括 Cook and Torrance(1982)提出的模型，该模型是建立在 Torrance and Sparrow(1967)原创的微面元模型基础上的。直到最近，多数计算机图形学硬件实现了 Phong 模型，但是最近出现的可编程像素着色器使得使用更复杂的模型成为可能。

色散反射模型

Torrance and Sparrow(1967)反射模型也构成了 Shafer's(1985)色散反射模型的基础，它指出在点光源照明下均匀材料的表观颜色依赖于两项的和，

$$\begin{aligned} L_r(\hat{v}_r;\lambda) &= L_i(\hat{v}_r,\hat{v}_i,\hat{n};\lambda) + L_b(\hat{v}_r,\hat{v}_i,\hat{n};\lambda) & (2.94) \\ &= c_i(\lambda)m_i(\hat{v}_r,\hat{v}_i,\hat{n}) + c_b(\lambda)m_b(\hat{v}_r,\hat{v}_i,\hat{n}), & (2.95) \end{aligned}$$

即，在界面 L_i 处反射的光的辐射和在表面体处反射的辐射 L_b。它们中的每一个又都是一个仅依赖于波长的相对功率谱 $c(\lambda)$ 和一个仅取决于几何的量 $m(\hat{v}_r,\hat{v}_i,\hat{n})$ 的简单乘积。(该模型可以很容易地从 Phong 模型的一般化版本中通过假定单光源和无

环境照明并且重新排列各项而导出。)在计算机视觉中，色散反射模型已经成功地用于分割阴影有很大变化的镜面色彩的物体(Klinker 1993)，近来也为局部双色模型提供了灵感，其应用包括 Bayer 模式去马赛克(Bayer pattern demosaicing)(Bennett, Uyttendaele, Zitnick et al. 2006)。

全局光照(光线追踪和辐射度)

迄今为止所介绍的简单阴影模型假设光线离开光源，从摄像机可见的表面反弹回来，因此在亮度或颜色上发生了变化，再达到摄像机。在现实中，光源可能被遮挡物所遮挡且光线可能在从光源到摄像机的路途中经过场景的多次反弹。

传统上有两种方法对这些效应建模。如果场景主体上是镜面的(典型的例子是由玻璃物体构成的场景和镜面或高度抛光过的球)，那么首选的方法是光线追踪或路径追踪(Glassner 1995; Akenine-Möller and Haines 2002; Shirley 2005)，它跟踪从摄像机经过多次弹射指向光源(或反过来)的各个光线。如果场景主要是由均匀反照率的简单几何照明体和表面构成，辐射度(全局光照)方法则是首选(Cohen and Wallace 1993; Sillion and Puech 1994; Glassner 1995)。也有将两种方法结合起来的方法(Wallace, Cohen, and Greenberg 1987)，还有更一般的光传播方法用于模拟诸如水涟漪造成的散焦效果。

基本的光线追踪算法将光线与摄像机图像中的每个像素相关联，寻找与它最近的表面的交。用前面讲述的简单的阴影方程(例如，公式(2.93))就可以计算出对于该面元可见的所有光源的基本贡献量。(另一种可选择的计算一个光源所照明表面的方法是计算一个阴影图(Williams 1983; Akenine-Möller and Haines 2002)。)附加的二次光线可以沿着镜面方向向场景中的其他物体投射，记录跟踪镜面反射所导致的任何衰减或颜色变化。

辐射度通过将光亮数值与场景中的矩形表面区域(包括面光源)相关联而起作用。场景中任意两块(彼此可见的)区域间交换的光量可以用形状因子(form factor)捕获，它依赖于其相对的方向和表面反射特性以及因随着光的远去它在更大有效球中分配所造成的 $1/r^2$ 下降(Cohen and Wallace 1993; Sillion and Puech 1994; Glassner 1995)。这样便可以建立起一个大型的线性方程组，将光源作为强制函数(右侧)来解决每块区域的最终亮度问题。一旦该方程组求解出来，场景就可以从任意期望的视点渲染出来。在一些情况下，使用计算机视觉方法有可能从照片恢复出场景的全局光照(Yu, Debevec, Malik et al. 1999)。

基本的辐射度算法不考虑某些近场效应，例如角和划痕内部的发暗，或由来自于其他表面的部分阴影所引起的有限环境光照(Nayar, Ikeuchi, and Kanade 1991; Langer and Zucker 1994)。

尽管所有这些全局光照效应都可以对场景的外观进而对其 3D 解释有很强的影响，但本书不会对它们做更详细的介绍。(12.7.1 节有针对从真实场景和物体恢复 BRDF 的讨论。)

2.2.3 光学

一旦从场景来的光线到达摄像机,它在到达传感器(模拟的胶卷或数字的硅片)之前肯定还会通过镜头。对于很多应用,将镜头当做理想的针孔就足够了,即通过共同的投影中心投影所有的射线(图 2.8 和图 2.9)。

但是,如果我们要处理诸如焦距、曝光、虚影和像差等问题,就需要提出更复杂的模型,这正是光学所研究的问题(Möller 1988; Hecht 2001; Ray 2002)。

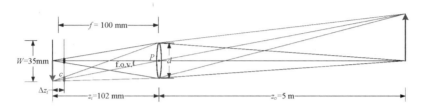

图 2.19 焦距为 f 的薄透镜,聚焦来自于镜头前距离 z_o 处的平面的光线于镜头后距离 z_i 处,有 $\frac{1}{z_o}+\frac{1}{z_i}=\frac{1}{f}$。如果焦平面(邻近 c 的灰线)前移,图像就不再是准确对焦的了,而且混淆圆 c(短的粗线段)依赖于相对于镜头光圈孔径 d 的图像平面运动的距离 Δz_i。视场(field of view,f.o.v)依赖于传感器宽度 W 和焦距 f(或者更确切地说,聚焦距离 z_i,它通常与 f 非常靠近)的比值

图 2.19 展示了最基本的镜头模型,即,由单片玻璃构成的具有很低曲率且两侧相同的薄透镜。根据透镜定律(这可以使用关于光线折射的简单的几何表达式推导出来),到物体的距离 z_o 和镜头后到形成聚焦图像的位置处的距离 z_i,这两者之间的关系可以表达为

$$\frac{1}{z_o}+\frac{1}{z_i}=\frac{1}{f}, \tag{2.96}$$

其中 f 称作镜头的焦距。如果我们让 $z_o \to \infty$,即我们调整镜头(移动图像平面)使无穷远处的物体对上焦,我们得到 $z_i = f$,这就是为什么我们能将焦距 f 的镜头当作等同于(作为初步的近似)距焦平面为 f 距离的针孔的原因(图 2.10),其视场由公式(2.60)给出。

如果焦平面从其合适对焦设置 z_i 上移开(例如,通过旋转镜头上的对焦环),z_o 处的物体就不再是对上焦的了,如图 2.19 中灰色平面所示。误对焦量由混淆圆 c(显示为灰色面上的短的粗线段)量度[①]。混淆圆的公式可以用相似三角形推导出,它依赖于相对于原聚焦距离 z_i 和光圈孔径(直径)d 在焦平面中运动的距离 Δz_i(见习题 2.4)。

场景中所允许的深度变化,它限定了混淆圆在可接受的数字下,通常称作"景

---

① 如果光圈不完全是圆,例如,如果它是由六角形的光阑引起的,有时在实际的模糊函数中或在朝向太阳拍摄时所看到的闪光效果中能看到这种效应(Levin, Fergus, Durand *et al.* 2007; Joshi, Szeliski, and Kriegman 2008)。

深"(depth of field)，是聚焦距离和光圈两者的函数，由很多镜头标志线图解式地展示出来(图 2.20)。由于该景深依赖于光圈孔径 d，我们还需要知道它是如何随着通常显示的 f-数(f-number)改变的，它通常标注为 $f/\#$ 或 N，且定义为

$$f/\# = N = \frac{f}{d}, \tag{2.97}$$

其中焦距 f 和光圈孔径 d 用同样的单位(比如，毫米)量测。

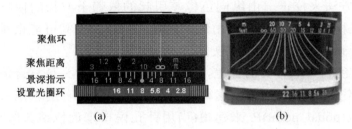

图 2.20 常规和变焦镜头的景深标注

f-数的通常写法是将 $f/\#$ 中的#用实际的数替换，即，$f/1.4$，$f/2$，$f/2.8$，……，$f/22$。(换一种方式，我们可以说 $N = 1.4$，如此等等)。解释这些数据的简单方式是注意这个公式：用 f-数去除焦距我们就得到光圈孔径 d。[①]

注意，通常的 f-数的级数是以全止进阶的，是 $\sqrt{2}$ 的倍数，因为逐次选择较小的 f-数就对应于两倍前透光孔的面积。(这种翻倍也称作以一次曝光值(exposure value)或 EV 来改变曝光。它与加倍曝光时间(例如从 1/125 到 1/250，参见习题 2.5)对到达传感器的光量的影响相同。)

现在你已经知道了如何在 f-数和光圈孔径间进行转换，便可以构建你自己的景深标绘图，将其作为焦距 f、混淆圆 c 和聚焦距离 z_o 的函数，正如在习题 2.4 中所解释的那样，看看它们与你看到的实际镜头有多匹配，就像图 2.20 所展示的那样。

当然，真实的镜头并不是无限薄的，因此受几何像差影响，除非使用了混合元件来对其校正。经典的五种赛德尔像差(Seidel aberration)，在使用三阶光学时出现，包括球形像差、彗差、像散、像场弯曲和畸变(Möller 1988; Hecht 2001; Ray 2002)。

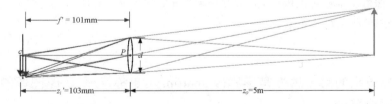

图 2.21 在受色像差影响的镜头中，不同波长的光线(例如，红色和蓝色箭头)以不同的焦距 f' 聚焦，因此有不同的深度 z_i'，最后产生了几何(平面内的)平移和失焦

[①] 这也解释了为什么对于聚焦镜头 f-数会随当前聚焦(焦距)设置而变化。

色像差

因为玻璃的折射作为波长的函数有少许的变化,所以简单的镜头会受色像差(chromatic aberration)的影响,这是指"不同颜色的光聚焦于略有不同的距离上"这一趋势(而且因此也具有略微不同的放大因子),如图2.21所示。依赖于波长的放大因子,即横向色像差,可以用每个颜色的径向偏离(2.1.6节)来建模,因此可以用6.3.5节描述的方法来校正。由纵向色像差引起的依赖于波长的模糊可以用10.1.4节描述的方法来校正。不幸的是,由纵向色像差引起的模糊可能很难去除,因为更高的频率会有较强的衰减因而很难恢复。

为了减小颜色和其他种类的像差,今日的多数照相镜头是复合镜头,由不同的玻璃元件(带有不同的涂层)构成。这样的镜头不再能用所有进入镜头的光线都必须穿过的单一节点(nodal point)P来建模(当用针孔模型来近似镜头时)。取而代之的是,这些镜头既有一个前节点,光线经过它进入镜头,还有一个后节点,光线经过它离开而到达传感器。在实践中,进行摄像机标定时,例如要确定一个旋转围绕的点以捕获无视差的全景图时(参见9.1.3节),我们只对前节点感兴趣。

但是,并不是所有的镜头都可以用单节点来建模。特别地,很宽的广角镜头,例如鱼眼镜头(2.1.6节)和某些由镜头和曲面镜组成的兼有反射和折射的成像系统(Baker and Nayar 1999)并没有单一的节点可以让所有获得的光线穿过。在这些情况下,如2.1.6节所述,应该优先选择在像素坐标和空间中的3D光线之间明确构建一个映射函数(查找表)(Gremban, Thorpe, and Kanade 1988; Champleboux, Lavallée, Sautot *et al*. 1992; Grossberg and Nayar 2001; Sturm and Ramalingam 2004; Tardif, Sturm, Trudeau *et al*. 2009)。

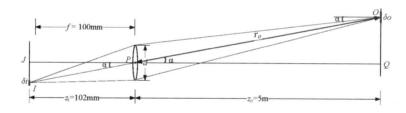

图 2.22 照射在面元 δ_i 的像素上的光量依赖于光圈空间 d 与焦距 f 的比值平方和离轴角度 α 的余弦的四次方,$\cos^4\alpha$

虚影

真实世界镜头的另一属性是虚影(vignetting,也称渐晕),即图像的亮度向图像边缘处过渡时会降低。

通常有两种现象对此效应有贡献(Ray 2002)。第一种称作"自然虚影",源于物体表面、投影的像素和镜头光圈的透视收缩效应,如图2.22所示。考虑离开物体表面大小为 δ_o 的面元的位于离轴角 α 的光线。由于该面元相对于摄像机镜头而透视收缩,到达镜头的光量以比率 $\cos\alpha$ 减少。到达镜头的光量还受通常的 $1/r^2$ 比率衰减；在这种情况下，距离 $r_o = z_o/\cos\alpha$。光线借以通过的实际的光圈面积以一额外的因子 $\cos\alpha$ 而透视收缩，即从 O 点看到的光圈是 $d \times d \cos\alpha$ 大小的椭圆。将所有这些因素合在一起，我们看到离开 O 穿过光圈到达位于 I 处的图像像素的光量与下式成比例：

$$\frac{\delta o \cos\alpha}{r_o^2}\pi\left(\frac{d}{2}\right)^2 \cos\alpha = \delta o \frac{\pi}{4}\frac{d^2}{z_o^2}\cos^4\alpha. \tag{2.98}$$

由于三角形 $\triangle OPQ$ 和 $\triangle IPJ$ 是相似的，所以物体表面 δ_o 的投影面积和图像像素 δ_i 与 $z_o : z_i$(的平方)具有相同的比率，

$$\frac{\delta o}{\delta i} = \frac{z_o^2}{z_i^2}. \tag{2.99}$$

将这些结合起来，我们得到最终的到达像素 i 的光量与光圈孔径 d，聚焦距离 $z_i \approx f$，离轴角 α 之间的关系，

$$\delta o \frac{\pi}{4}\frac{d^2}{z_o^2}\cos^4\alpha = \delta i \frac{\pi}{4}\frac{d^2}{z_i^2}\cos^4\alpha \approx \delta i \frac{\pi}{4}\left(\frac{d}{f}\right)^2 \cos^4\alpha, \tag{2.100}$$

这就是场景辐射 L 和到达像素传感器的光(辐照度)E 之间的基本辐射学关系。

$$E = L\frac{\pi}{4}\left(\frac{d}{f}\right)^2 \cos^4\alpha, \tag{2.101}$$

(Horn 1986; Nalwa 1993; Hecht 2001; Ray 2002)。注意在本公式中光量如何依赖于像素表面面积(这正是为什么傻瓜相机与数字单反相机相比噪声如此大的原因)、f 制光圈 $N=f/d$ (2.97)的倒数平方和余弦四次方 $\cos^4\alpha$ 的离轴衰减，这最后一项是自然虚影项。

另一种主要的虚影称作"机械虚影"，是由复合镜头中邻近镜头组件边缘处的光线的内部遮挡引起的，如果不进行真实镜头设计中的完整光线追踪，很难用数学方法来表达[1]。但是，与自然虚影不同的是，机械虚影可以通过减小光圈(增加 f-数)来减弱。它也可以用特殊的仪器来标定(与自然虚影一起)，例如 10.1.3 节讨论的积分球、均匀照明的目标或摄像机旋转。

2.3 数字摄像机

光最初从一个或多个光源出发，经世界中的单个或多个表面反射，再通过摄像机光学器件(镜头)最终到达成像传感器。到达该传感器的光子是如何转换成当我们看数字图像时所观察到的数字(R, G, B)数值的呢？在本节中，我们介绍一个简单的模型，来解释最重要的一些效应，包括曝光(增益和快门速度)、非线性映射、采样与走样以及噪声。基于 Healey and Kondepudy(1994); Tsin, Ramesh, and Kanade (2001); Liu, Szeliski, Kang et al. (2008)提出的摄像机模型，图 2.23 给出了现代数码

---

[1] 有一些经验模型在实践中很有效(Kang and Weiss 2000; Zheng, Lin, and Kang 2006)。

相机中出现的处理步骤的简单版本。Chakrabarti, Scharstein, and Zickler(2009)提出了24参数的复杂模型，可以更好地解释当今摄像机的处理步骤。

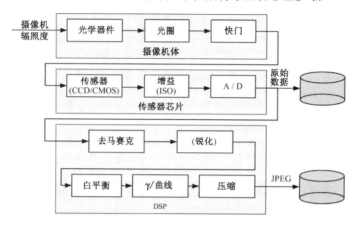

图2.23　图像成像流水线，展示了各种噪声源以及典型的数字后处理步骤

落在成像传感器上的光通常由一块活性传感区域拾取，在曝光持续时间(通常用快门速度来表达，单位是几分之一秒，例如 1/125，1/60，1/30)内累积，然后通过一组传感放大器。当今用在静态或视频摄像机中的两种主要的传感器是电荷耦合器件(CCD)和硅片上的互补型金属氧化物器件(CMOS)。

在 CCD 中，在曝光期间光子在每个活性势井中累积。然后在传送阶段，电荷以某种"桶组"方式从势井到势井传送直至存放在传感放大器处，由其放大信号并将其送至模数转换器(ADC)[①]。老式的 CCD 传感器易受电荷从一个过曝光像素溢出到邻近的像素时所引起的起霜(blooming)现象影响，但是多数更新的 CCD 具有抗起霜技术("排水槽"接收溢出的过量的电荷)。

在 CMOS 中，照射在传感器上的光子直接影响光检测器的电导率(或增益)，它可以被选择性地作为闸门来控制曝光时间，并在使用多路复用方案读出之前做局部放大。在传统上，在性能敏感的应用中，例如数字单反相机(SLR)，CCD 传感器超过 CMOS，而 CMOS 对于低功率应用更好，但今天，CMOS 用在大多数数码相机中。

影响数字图像传感器的主要因素是快门速度、采样间距、填充率、芯片尺寸、模拟增益、传感器噪声和模数转换器的分辨率(和品质)。这些参数的实际数值很多可以从数码图像所嵌入的 EXIF 标注信息中读出。而其他部分可以从摄像机制造商的说明书或摄像机评论或标定网站获得[②]。

快门速度　快门速度(曝光时间)直接控制到达传感器的光量，因此，决定着图像是欠曝光还是过曝光。(对于亮的场景，大的光圈或慢的快门速度会造成浅景深或运动模糊效果，摄影师有时会使用中性密度滤光片。)对于动态场景，快门速度

---

① 在数码静态相机中，每次捕获一幅完整的帧并顺序地读出。但是，如果要捕获视频，通常使用卷帘式快门，每次分别曝光和传送每一线。在老式的视频摄像机中，首先扫描偶数场(线)，然后是奇数场，这个过程称为"交错(隔行)扫描"。

② http://www.clarkvision.com/imagedetail/digital.sensor.performance.summary/

也决定了所产生的照片中的运动模糊量。通常，较高的快门速度(较小的运动模糊)使后续的分析更容易(去除这类模糊的方法参见 10.3 节)。但是，当拍摄的视频是用于显示的时候，为了避免频闪效应(stroboscopic effect)，可能还期望有一些运动模糊。

采样间距 采样间距是成像芯片上相邻传感器单元的物理间隔。具有较小采样间距的传感器具有更高的采样密度，因此对给定的活性芯片区域提供了更高的(根据像素的)分辨率。然而，较小的间距也意味着每个传感器具有较小的面积而不能累积更多的光子，使其光敏性降低而更易受噪声影响。

填充率 填充率是活性传感区域大小占理论上可获得的传感区域(横向和纵向采样间距的乘积)的比率。通常，更高的填充率较好，因为结果会捕获更多的光和较小的走样(参见 2.3.1 节)。但是，这必须与在活性传感区域添置额外电路的需要相平衡。摄像机的填充率可以使用光度测定学摄像机标定过程来凭经验确定(参见 10.1.4 节)。

芯片尺寸 视频和傻瓜相机通常使用小尺寸芯片(1/4 到 1/2 英寸的传感器[①])，而数码单反(DSLR)相机使用更接近传统 35mm 胶片帧的大小[②]。当整体器件的尺寸并不重要时，优先选择更大的芯片尺寸，因为这样每个传感器单元可以具有更好的光敏性。(对于紧凑的摄像机，更小的芯片意味着所有的光线器件可以成比例地缩小。)但是，较大的芯片在生产上成本更高，不仅因为每块晶片上可承载的芯片数量减少，而且还因为芯片瑕疵出现的概率随着芯片面积呈线性增长。

模拟增益 在模数转换前，感知到的信号通常由传感放大器来提升。在视频摄像机中，这些放大器的增益传统上由自动增益控制(AGC)逻辑所控制，它可以调整这些数值来获得好的整体曝光。在更新的数码静态相机中，使用者现在通过 ISO 设定对增益有一些的额外控制，这一般用 ISO 标准单位来表达，比如 100, 200 或 400。由于多数摄像机的自动曝光控制也调整光圈和快门速度，所以手工设置 ISO 从摄像机的控制中去除了一个自由度，就像手工设置光圈和快门速度一样。在理论上，更高的增益使摄像机在低光条件下可以更好地工作(减小了光圈已经最大后由曝光时间长而产生的运动模糊)。但是在实践中，更高的 ISO 设定通常会放大传感器噪声。

传感器噪声 在整个传感过程中，噪声从各种源叠加进来，可能包括固定样式的噪声、暗电流噪声、散粒噪声、放大器噪声和量化噪声(Healey and Kondepudy 1994; Tsin, Ramesh, and Kanade 2001)。呈现在采样图像中的最终的噪声量依赖于所有这些量，还依赖于入射光(由场景辐射和光圈所控制)、曝光时间和传感器增益。而且，对于低光条件(噪声起因于低光子数量)，泊松模型可能比高斯模型更合适。

正如在 10.1.1 节中所详细讨论的，Liu, Szeliski, Kang et al.(2008)使用该模型，

---

① 这些数指的是曾用在视频摄像机中老视像管的"管的直径"(http://www.dpreview.com/learn/?/Glossary/Camera System/sensor_sizes_01.htm)。佳能 SD800 相机中的 1/2.5″传感器实际上是 5.76mm×4.29mm，即 35mm 完整帧(36mm×24mm) DSLR 传感器(一边)大小的六分之一。

② 当 DSLR 芯片并不充满 35mm 完整帧时，会对镜头的焦距产生乘数效应。例如，仅有 35mm 帧的 0.6 倍大小的芯片将使 50mm 镜头的图像具有与 50/0.6=50×1.6=80mm 镜头一样的角度范围，正如公式(2.60)所表明的那样。

与 Grossberg and Nayar (2004) 所获得的摄像机响应函数(CRF)的经验数据库一起，来估计给定图像的噪声水平函数(NLF)，它预测每个像素处的作为亮度函数的整体噪声方差(为每个彩色通道分别估计 NLF)。当你拍摄图片之前可以拿到摄像机时，另一种方法是通过重复拍摄含有各种彩色和亮度的场景，例如图 10.3b 中所展示的 Macbeth 彩色图表(McCamy, Marcus, and Davidson 1976)，来预先标定 NLF。(当估计方差时，要保证扔掉具有大梯度的像素或降低其权重，因为曝光的微小偏移会影响这些像素所感知到的数值。)不幸的是，对于不同的曝光时间和增益设定，预先标定过程可能需要重复进行，其原因在于传感器系统内部所发生的复杂的相互作用。

在实践中，多数计算机视觉算法，比如图像去噪、边缘检测和立体视觉匹配，都会受益于噪声水平的估计，至少是初步的估计。在排除预先标定摄像机或重复拍摄相同场景的能力之后，最简单的方法是寻找接近常量值的区域并在这样的区域中估计噪声的方差(Liu, Szeliski, Kang et al. 2008)。

模数转换(ADC)分辨率 成像传感器内所发生的模拟处理链中的最后一步是模数转换(ADC)。尽管有各种方法可以用于实现这个处理，但有两个量是我们感兴趣的，即该处理的分辨率(它产生多少位)和其噪声水平(在实践中这些位有多少是有用的)。对于多数摄像机，注明的位数(对于压缩的 JPEG 图像是 8 位的，一些数码单反相机(DLSR)提供的原始格式数据名义上是 16 位的)超过实际可用的位数。最好的说明方法是简单地标定给定传感器的噪声，例如，通过重复拍摄同一个场景并绘制出所估计的作为亮度函数的噪声(习题 2.6)。

数字后处理 一旦到达传感器的辐照度数值被转换成数位，多数摄像机就会进行各种数字信号处理(DSP)运算来增强图像，然后再对像素值做压缩和存储。这包括彩色滤波阵列(CFA)去马赛克、白点设定以及通过γ函数对亮度值做映射以增加信号感知到的动态范围。我们在 2.3.2 节覆盖了这些主题，但是在我们这样做之前，先回到走样(混叠)这个主题，这是与传感器阵列填充率相关联的主题。

2.3.1 采样与混叠

当一个光场撞击传感器落在成像芯片的活性传感区域上时，会发生什么？到达每个活性感光元件上的光子被累积起来，然后被数字化。但是，如果芯片上的填充率低，要不然信号不是带宽有限的，就可能出现视觉上令人不快的混叠现象。

为了探究混叠现象，我们首先来观察一维信号(图 2.24)，其中我们有两个正弦波，一个的频率在 $f = 3/4$ 而另一个在 $f = 5/4$。如果我们用频率 $f = 2$ 采样这两个信号，就会看到它们产生了相同的采样(用黑色显示)，因此我们说它们混叠(走样)了。[①]为什么这是糟糕的效应？本质上，由于我们不知道究竟是这两种原始频率中的哪一个，所以无法重建原始信号。

[①] 别名是指某人的另一个名字，因此采样过的信号对应于两个不同的别名。(译者注：信号处理中使用的混叠(走样)的英文词 alias 原意是别名)。

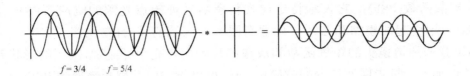

图 2.24 一维信号的混叠：蓝色的正弦波在 $f = 3/4$ 且红色正弦波在 $f = 5/4$，当用 $f = 2$ 采样时，有相同的数字采样。即便是在用 100%填充率的方框型滤光器卷积这两个信号之后，尽管幅度不再相同，但在采样过的红色信号看起来像蓝色信号的较低幅度的翻转版本的意义下，它们仍然是走样的。(右侧的图像为了便于观察而放大了。实际的正弦幅度是其原来数值的 30% 和 −18%)

事实上，香农(Shannon)的采样定理表明，从其瞬时样本重建信号的最小采样(Oppenheim and Schafer 1996; Oppenheim, Schafer, and Buck 1999)率必须至少是其最高频率的两倍[①]，

$$f_s \geq 2f_{\max}. \tag{2.102}$$

信号中的最大频率称作"奈奎斯特(Nyquist)频率"，最小采样频率的倒数 $r_s = 1/f_s$ 称作"奈奎斯特率"。

但是，你可能要问，既然成像芯片实际上平均了一块有限区域上的光场，那么点采样的结果仍然适用吗？传感器区域上的平均的确趋向于减弱某些较高的频率。但是，即使如图 2.24 的右侧图像那样填充率为 100%，在奈奎斯特限(采样频率的一半)上的频率仍然会产生走样的信号，尽管其幅值比其对应的带宽有限信号小。

之所以说走样是糟糕的，可以通过用一个劣质的滤波器(比如方框型滤波器)下采样一个信号来充分佐证。图 2.25 展示了高频线性调频脉冲(chirp)图像(因其频率随时间增长而得名)，以及分别用 25%填充率的区域传感器、100%填充率的传感器、高品质 9 阶滤波器(9-tap filter)来采样的结果。更多的下采样(降采样)滤波器的例子可以在 3.5.2 节和图 3.30 中找到。

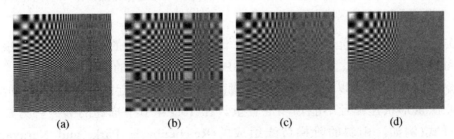

(a)　　　　　　(b)　　　　　　(c)　　　　　　(d)

图 2.25 二维信号的混叠：(a)全分辨率的原始图像；(b)用 25%填充率的方框型滤波器 4×下采样；(c)用 100%填充率的方框型滤波器 4×下采样；(d)用高品质 9 阶滤波器 4×下采样。注意，对于较低品质的滤波器，更高的频率是如何混叠成可见的频率的，而 9 阶滤波器完全抹掉了这些更高的频率

预测一个成像系统(或甚至是一个图像处理算法)产生的混叠量，最好的途径是

[①] 实际的定理陈述 f_s 必须至少两倍于信号带宽。但是，由于我们处理的不是广播之类的调制信号，因此最大频率足够了。

估计点扩散函数(PSF)，它表示特定像素传感器对理想点光源的响应。PSF 是由光学系统(镜头)引起的模糊和芯片传感器有限累积区域的结合(卷积)[①]。

如果我们知道镜头的模糊函数和成像芯片的的填充率(传感器区域的形状和间距)(选择性地加上，抗走样滤波器的响应)，我们可以将它们卷积起来得到 PSF。图 2.26a 展示了一个镜头的 PSF 的一维横截面，该镜头的模糊函数假定是半径等于像素间距的圆盘加上水平填充率是 80% 的传感芯片。对该 PSF 做傅里叶变换(3.4 节)，我们得到调制传递函数(MTF)，由它我们可以估计作为在 $f \leqslant f_s$ 奈奎斯特频率之外的傅里叶幅值面积的混叠量[②]。如果我们将镜头散焦使模糊函数的半径为 $2s$(图 2.26c)，会看到混叠量显著减少，而图像的细节量(靠近 $f = f_s$ 的频率)也减少。

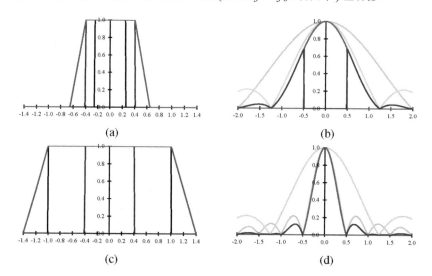

图 2.26 点扩散函数(PSF)样例：(a)中的模糊圆盘(蓝)的直径等于像素间距的一半，而(c)中的直径是像素间距的两倍。传感芯片的横向填充率是 80%，显示为棕色。这两个核的卷积给出点扩散函数，显示为绿色。PSF (MTF) 的傅里叶响应绘制在(b)和(d)中。混叠出现在奈奎斯特频率之外的区域，显示为红色

在实验室条件下，通过查看点光源，比如从后方照明的一块黑色硬纸板中的针孔看，可以估计出 PSF。但是，这个 PSF(实际上是针孔的像)的精度仅限于一个像素分辨率，且尽管它可以对比较大的模糊(比如由散焦引起的模糊)建模，但不能对 PSF 的亚像素形状建模且不能预测混叠量。在 10.1.4 节中描述的另一种方法是看一个标定样式(例如，由斜的阶梯边缘组成的(Reichenbach, Park, and Narayanswamy 1991;Williams and Burns 2001; Joshi, Szeliski, and Kriegman 2008))，其理想的表观可以重新合成为亚像素精度。

除了出现在图像获取期间，各种图像处理操作也有可能引入混叠，比如重采

[①] 成像芯片通常在成像芯元之前插入光学抗走样滤波器来减小或控制混叠量。
[②] 作为复数的 PSF 的傅里叶变换实际上称为"光学传递函数(OTF)" (Williams 1999)。其幅值称为"调制传递函数"(MTF)，而其相位称作"相位传递函数"(PTF)。

样、上采样和下采样。3.4 节和 3.5.2 节将讨论这些问题,并说明为什么仔细选择滤波器可以减少操作所注入的混叠量。

2.3.2 色彩

在 2.2 节,我们看到光和表面反射是如何为波长的函数的。当入射光照射到成像传感器时,来自光谱不同部分的光以某种方式整合为我们在数字图像中看到的离散的红、绿和蓝(RGB)色数值。这一过程是如何工作的?我们如何分析和处理彩色数值?

你可能会回想起童年时代时混合颜色后得到新颜色的魔幻过程。你可能记得,蓝+黄得到绿,红+蓝得到紫,红+绿得到棕。如果你在年龄更大的时候重温这一主题,你可能已经学过恰当的减色基实际上是青(亮的蓝绿色)、洋红(粉红色)和黄(图 2.27b),尽管在四色印刷(CMYK)中常常会用到黑色。(如果后来你曾上过任何绘画课程,你就知道颜色还有更多奇妙的名字,比如茜红色、天蓝色、黄绿色。)减色称作"减"是因为绘画中的颜料吸收光谱中的某些波长的原因。

后来,你可能已经学过加性基色(红、绿和蓝)以及它们如何相加(用幻灯片投影或在计算机监视器上)来产生青、洋红、黄和白,以及所有其他的我们在电视机和监视器上看到的典型色彩(图 2.27a)。

对于两种不同的颜色,比如红和绿,是通过什么过程相互作用而产生第三种颜色(如黄色的)呢?是波长以某种形式混合起来产生了一个新的波长?

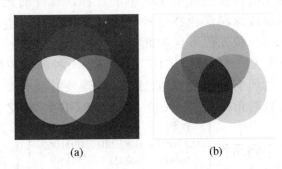

图 2.27 基色和合成色:(a)加性色红、绿、蓝可以混合产生青、洋红、黄和白;(b)减性色青、洋红、黄可以混合产生红、绿、蓝和黑

你可能知道,正确答案与物理上混合波长没有任何关系。而且你还可能知道,三基色的存在是人类视觉系统的三激励(或三原色)属性的结果,由于我们有三种不同椎体细胞,各自选择性地对彩色光谱的不同部分有响应(Glassner 1995; Wyszecki and Stiles 2000; Fairchild 2005; Reinhard, Ward, Pattanaik *et al.* 2005; Livingstone 2008)[①]。注意,对于机器视觉应用,比如遥感和地形分类,优先选择使用更多的波长。类似地,视觉监控应用也常常可以从近红外(NIR)范围的传感中受益。

---

① 参见 Mark Fairchild 的网页,*http://www.cis.rit.edu/fairchild/WhyIsColor/books links.html*。

CIE RGB 和 XYZ

为了测试和量化感知三原色理论，我们可以尝试用适当选择的三种基色的混合来再现所有单色的(单个波长)颜色。(纯波长的光可以用棱镜或特殊制造的彩色滤镜获得。)在 20 世纪 30 年代，国际照明协会(CIE)通过使用红(700.0nm 波长)，绿(546.1nm)和蓝(435.8nm)三基色所进行的彩色匹配实验标准化了 RGB 表示。

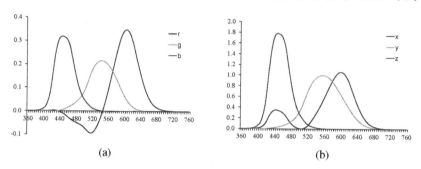

图 2.28 标准 CIE 彩色匹配函数：(a)通过与纯色基 R = 700.0nm, G = 546.1nm 和 B = 435.8nm 匹配获得的 $\bar{r}(\lambda), \bar{g}(\lambda), \bar{b}(\lambda)$ 色谱；(b) $\bar{x}(\lambda), \bar{y}(\lambda), \bar{z}(\lambda)$ 彩色匹配函数，它们是 $\bar{r}(\lambda), \bar{g}(\lambda), \bar{b}(\lambda)$ 谱的线性组合

图 2.28 展示了对标准观察者(即对大量受试者的感知结果做平均)所做的实验结果。你会注意到，对于蓝-绿值域中的某些纯谱，需要加上红光的负的量，即一定量的红必须加到正在匹配的彩色中以便获得一个彩色匹配。这些结果也为具有不同光谱能量分布的同色光的存在性提供了简单的解释，这样的光在感觉上是无法区分的。注意，两个纺织品或涂料颜色在一种照明下表现为具有不同光谱能量分布的同色光，并不意味着在另一种光照下也如此。

因为与负光混合所关联的问题，CIE 还发展了一种新的彩色空间，称作 XYZ，它含有在其正象限中的所有纯谱色。(它还将 Y 轴映射到亮度上，即感觉到的相对亮度，且将纯白映射到对角(等值的)矢量上。)从 RGB 到 XYZ 的转换为

$$\begin{bmatrix} X \\ Y \\ Z \end{bmatrix} = \frac{1}{0.17697} \begin{bmatrix} 0.49 & 0.31 & 0.20 \\ 0.17697 & 0.81240 & 0.01063 \\ 0.00 & 0.01 & 0.99 \end{bmatrix} \begin{bmatrix} R \\ G \\ B \end{bmatrix}. \tag{2.103}$$

而 CIE XYZ 标准的官方定义将这个矩阵规范化了，使对应于纯红的 Y 值为 1，一个更常用的形式是省略前面的分数，使第二行合起来为 1，即 RGB 三元组(1, 1, 1)映射到 Y 值 1。根据公式(2.103)线性地混合图 2.28a 中的 ($\bar{r}(\lambda), \bar{g}(\lambda), \bar{b}(\lambda)$) 曲线，我们得到显示在图 2.28b 中的结果曲线 ($\bar{x}(\lambda), \bar{y}(\lambda), \bar{z}(\lambda)$)。注意现在所有三个谱(彩色匹配函数)是如何只有正值且 $\bar{y}(\lambda)$ 曲线是如何匹配人所感知的亮度的。

如果我们以 X+Y+Z 之和除 XYZ 数值，我们得到色度坐标

$$x = \frac{X}{X+Y+Z}, \quad y = \frac{Y}{X+Y+Z}, \quad z = \frac{Z}{X+Y+Z}, \tag{2.104}$$

它们的和为 1。色度坐标舍弃了给定色彩样本的绝对亮度而只表示其纯色。如果我们扫描图 2.28b 中的单色的 λ 参数从 λ = 380nm 到 λ = 800nm，我们就得到图 2.29

所示的熟悉的色度图。该图显示了多数人可以感知到的每个彩色值的(x,y)值。(当然，本书中的 CMYK 重现过程并不能实际展开到可感觉到的彩色的整个色域。)外部的边缘表示所有纯单色数值在(x,y)空间上映射的位置，而下端的直线，它将两个端点相连，称作"紫线"。

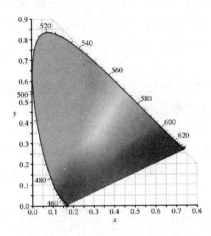

图 2.29 CIE 色度图，显示了彩色和其对应的(x,y)值。纯谱色绕着曲线的外端布置

因此，当我们想将亮度和色度分开来时，Yxy(亮度加上两个最具区分性的色度分量)是彩色数值的一种方便的表达形式。

L*a*b*色彩空间

尽管 XYZ 色彩空间有很多方便的性质，包括分离亮度和色度的能力，但是它实际上并不预测人对色彩或亮度差别的感知程度。

由于人类视觉系统的响应是粗略地对数关系(我们可以感知约 1%的相对亮度变化)，CIE 定义了 XYZ 空间的一个非线性重映射称作 L*a*b*(有时也称作 CIELAB)，其中亮度或色度上的差别在感觉更一致[①]。

亮度的 L*分量定义为

$$L^* = 116 f\left(\frac{Y}{Y_n}\right), \quad (2.105)$$

其中的 Y_n 是标称白的亮度值(Fairchild 2005)且

$$f(t) = \begin{cases} t^{1/3} & t > \delta^3 \\ t/(3\delta^2) + 2\delta/3 & \text{else,} \end{cases} \quad (2.106)$$

是三次根的有限斜率近似，$\delta = 6/29$。所产生的 0…100 标度粗略地度量等量的亮度可感知性。

以类似的方式，a* 和 b* 分量定义为

$$a^* = 500\left[f\left(\frac{X}{X_n}\right) - f\left(\frac{Y}{Y_n}\right)\right] \quad \text{and} \quad b^* = 200\left[f\left(\frac{Y}{Y_n}\right) - f\left(\frac{Z}{Z_n}\right)\right], \quad (2.107)$$

[①] 另一种称作"L*u*v*"的感知激发的色彩空间同时被提出来并被标准化 (Fairchild 2005)。

其中，(X_n, Y_n, Z_n)是测量的白点。图 2.32i~k 显示了一个样本彩色图像的 L*a*b*表达。

彩色摄像机

尽管前面的讨论告诉我们怎样唯一地描述任何彩色(谱分布)的被感知到的三激励刻画，但它并没有告诉我们 RGB 静态和视频摄像机实际上是如何工作的。它们只测量名义上的红(700.0nm 波长)，绿(546.1nm)和蓝(435.8nm)三个波长上的光量吗？彩色监视器是只发出精确的这些波长吗？如果是，它们是怎样发出负的红光来产生青色范围的彩色的？

事实上，RGB 视频摄像机的设计在历史上是基于可获得的在电视机中用的彩色荧光材料展开的。当标清的彩色电视被发明出来时，就在 RGB 数值和 XYZ 数值之间定义了一个映射，前者是在阴极射线管(CRT)中驱动三彩色枪的基准，后者明白地定义了所感知的彩色(这一标准称为"ITU-R BT.601")。随着高清电视(HDTV)和更新型的监视器的出现，形成了新的称作标准"ITU-R BT.709"，它规定了每个色基的 XYZ 数值，

$$\begin{bmatrix} X \\ Y \\ Z \end{bmatrix} = \begin{bmatrix} 0.412453 & 0.357580 & 0.180423 \\ 0.212671 & 0.715160 & 0.072169 \\ 0.019334 & 0.119193 & 0.950227 \end{bmatrix} \begin{bmatrix} R_{709} \\ G_{709} \\ B_{709} \end{bmatrix}. \quad (2.108)$$

在实践中，每个彩色摄像机根据其红、绿、蓝传感器的谱响应函数对光做积分，

$$\begin{aligned} R &= \int L(\lambda) S_R(\lambda) d\lambda, \\ G &= \int L(\lambda) S_G(\lambda) d\lambda, \\ B &= \int L(\lambda) S_B(\lambda) d\lambda, \end{aligned} \quad (2.109)$$

其中 $L(\lambda)$ 是给定像素处的入射光谱，$\{S_R(\lambda), S_G(\lambda), S_B(\lambda)\}$ 是对应传感器的红、绿、蓝光谱敏感性。

我们能知道摄像机实际具有的光谱敏感性吗？除非摄像机的制造商给我们提供该数据或者我们观察摄像机对单色光的整个谱的响应，这些光谱敏感性数据并没有在诸如 BT.709 的标准中规定。相反，更重要的是对于给定彩色的三激励数值要产生指定的 RGB 数值。制造商有使用其敏感性与标准 XYZ 定义不匹配的传感器的自由，只要它们后来能(通过线性变换)转换为标准的彩色。

类似地，尽管我们假定 TV 和计算机监视器产生由公式(2.108)规定的 RGB 数值，但我们并没有理由说他们不能使用数字逻辑将输入的 RGB 数值变换为不同的驱动每个彩色通道的信号。适当标定过的监视器为彩色管理软件应用提供了这种信息，使真实生活中的出现在屏幕上的和打印出来的彩色，都能尽可能地匹配。

色彩滤波器矩阵

尽管早期的彩色电视摄像机使用三个视像管来进行传感，再后来摄像机使用三块分开的 RGB 传感芯片，但今天的绝大多数数字静态和视频摄像机使用的是色彩

滤波器矩阵(CFA)，其中交替传感器用不同的色彩滤波器取代。[①]

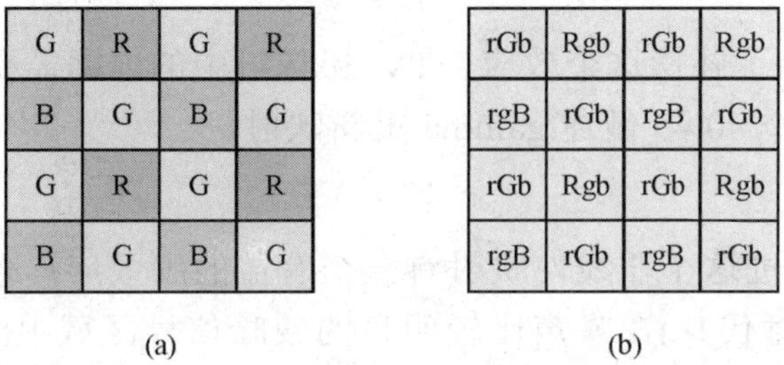

G	R	G	R
B	G	B	G
G	R	G	R
B	G	B	G

(a)

rGb	Rgb	rGb	Rgb
rgB	rGb	rgB	rGb
rGb	Rgb	rGb	Rgb
rgB	rGb	rgB	rGb

(b)

图 2.30　Bayer RGB 样式：(a)色彩滤波器阵列布置；(b)插值后的像素值，未知量(估计的量)显示为小写字母

今天的彩色摄像机中最普遍使用的样式是 Bayer 样式(Bayer 1976)，其中绿色传感器在传感器(棋盘样式)中占据了一半位置，红和蓝占据剩下的另一半(图 2.30)。绿色滤波器的数目是红和蓝的两倍，其原因是亮度信号主要是由绿色数值决定的，而且视觉系统对亮度中的高频信息远比色彩中的要敏感(参见 2.3.3 节，彩色图像压缩中利用的一个事实)。插值缺失的彩色数值以便所有像素都具有有效的 RGB 数值，这一过程称作"去马赛克"(demosaicing)，其详细论述可见 10.3.1 节。

类似地，彩色液晶(LCD)监视器通常使用位于每种液晶活性区域前端的交替红、绿、蓝滤波器条纹来模拟全彩色显示的体验。如前所述，因为视觉系统的亮度分辨率(敏锐度)比色彩高，所以能够做到通过数字化预滤波 RGB(和黑白)图像来增强感知清晰度(Betrisey, Blinn, Dresevic et al. 2000; Platt 2000)。

色彩平衡

在编码传感到的 RGB 数值之前，多数摄像机会进行某种形式的色彩平衡处理，以便将给定图像的白点移到靠近纯白(等同的 RGB 数值)的位置。如果色彩系统和照明是相同的(BT.709 系统使用日光照明 D_{65} 作为参考白)，这种变化可能微不足道。但是，如果照明是强有色的，比如室内白炽灯照明(这通常具有黄色或橘黄色色调)，这种补偿可能相当明显。

进行色彩矫正的简单方法是以不同的因子乘每个 RGB 数值(即对 RGB 彩色空间做对角型矩阵变换)。更复杂的变换，有时是映射到 XYZ 空间后再回来，实际上是做了彩色扭曲，即它们使用一般的 3×3 彩色变换矩阵[②]。习题 2.9 将探索这些问题中的一部分。

γ (Gamma)

在黑白电视的早期时代，CRT 中的用于显示 TV 信号的萤光材料对其输入电压的响应是非线性的。电压和其产生的亮度间的关系是由称作"gamma(γ)"的数刻画

[①] Foveon (*http://www.foveon.com*)设计的更新的芯片将红、绿、蓝传感器彼此上下堆放，但是它目前还没有被广泛采纳。
[②] 还记得彩色电视早期时代的老读者，自然会联想到电视机上的色调调整旋钮，它可能会产生非常古怪的结果。

的，因为其公式粗略为

$$B = V^\gamma, \quad (2.110)$$

其中的γ约为 2.2。为了补偿这个效应，TV 摄像机中的电路需要将感知到的亮度 Y 通过以典型的数值 1//γ=0.45 做逆 gamma 重新映射，

$$Y' = Y^{\frac{1}{\gamma}}, \quad (2.111)$$

在发射前信号通过这个非线性映射有一个有益的副效应：传输期间增加的噪声(记住，这是在模拟时代！)在噪声比较明显的较暗信号区域中(在接收器做 gamma 校正后)会被减少(图 2.31)[①]。(记住，我们的视觉系统对相对亮度差别是敏感的。)

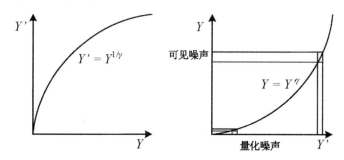

图 2.31 γ 压缩：(a)输入信号亮度 Y 和传送的信号 Y' 之间的关系是 $Y' = Y^{1/\gamma}$；(b)在接收器处，信号 Y' 以因子γ被指数化 $\hat{y} = Y'^\gamma$。传输期间引入的噪声压扁在黑色区域中，对应于视觉系统的噪声比较敏感的区域

在彩色电视发明后，红、绿、蓝信号被分别作 gamma 非线性校正后再合起来编码。今天，尽管在传输系统中我们不再有模拟噪声，但信号在压缩时仍然要量化(参见 2.3.3 节)，因此在传感数据上应用逆 gamma 仍然是有用的。

不幸的是，对于计算机视觉和计算机图形学这两个领域来说，图像中出现的 gamma 现象时常是有问题的。例如，诸如阴影这种辐射测量现象的合适的模拟(参见 2.2 节和公式(2.87))出现在线性辐射空间中。一旦所有计算完成之后，在显示之前需要施加适当的 gamma 校正。不幸的是，很多计算机图形学系统(比如阴影模型)直接作用于 RGB 数值且直接显示这些值。幸运的是，更新的彩色成像标准，比如 16 位 scRGB，使用线性空间，从而消除了这个问题(Glassner 1995)。

在计算机视觉中，这个情况更让人望而生畏。使用诸如光度测定学立体视觉的方法(12.1.1 节)或甚至使用精确的图像去模糊之类更简单的运算，来精确确定表面法向，需要在亮度的线性空间中测定。因此，在进行这类详细的定量计算时，首先要撤销 gamma 校正和在传感到的彩色数值中对每张图做彩色再平衡。Chakrabarti, Scharstein, and Zickler (2009)提出了一个复杂的 24-参数模型，与今天数码摄像机所进行的处理有很好的匹配，他们还提供了一个彩色图像的数据库供你自己的测试用。[②]

[①] 一个称作"补偿调整"的相关技术是录音带中使用的杜比(Dolby)降噪系统的基础。
[②] *http://vision.middlebury.edu/color/*

但是，对于其他视觉应用，比如特征检测或在立体视觉和运动估计中的信号匹配，这个线性化步骤通常不是必要的。事实上，确定是否有必要撤销 gamma 校正需要仔细斟酌，例如，在图像拼接中(参见习题 2.7)考虑对曝光变化做补偿的情况。

如果所有这些处理步骤听起来令人困惑，它们的确是，习题 2.10 将让你尝试使用经验研究来搞清楚这些现象中的一些，即拍摄彩色图版的照片并比较 RAW 和 JPEG 压缩后的彩色数值。

其他色彩空间

尽管 RGB 和 XYZ 是用于描述彩色信号光谱内容(因此而有三激励响应)的主要彩色空间，但在视频和静态图像编码以及计算机图形学中还发展了几种其他的表示方法。

用于视频传输的最早的彩色表示是 YIQ 标准，由北美为 NTSC 提出，与由欧洲为 PAL 提出的 YUV 标准紧密相关。在这两种情况下，都期望有一个亮度通道 Y(之所以这样称呼，是因为它只是粗略地模拟真实的亮度)可以与常规的黑白 TV 信号相兼容，还有两个低频的颜色通道。

在两种系统中，Y 信号(或更合适的说法，因其是经 gamma 压缩的而称为"Y 的亮度信号")由下式得到：

$$Y'_{601} = 0.299R' + 0.587G' + 0.114B', \tag{2.112}$$

其中 R′ G′ B′ 是 gamma 压缩过的彩色分量的三元组。使用为 HDTV 制定的更新的彩色定义 BT.709 时，公式是

$$Y'_{709} = 0.2125R' + 0.7154G' + 0.0721B'. \tag{2.113}$$

UV 分量是由(B′ - Y′)和(R′ - Y′)的缩放版本得出的，

$$U = 0.492111(B' - Y') \text{ and } V = 0.877283(R' - Y'), \tag{2.114}$$

然而 IQ 分量是 UV 分量经过 33°的旋转得到的。在复合(NTSC 和 PAL)视频中，色彩信号在调制并叠加在 Y′ 亮度信号的顶端之前经过横向的低通滤波。向后兼容是通过使老式黑白 TV 有效地忽略高频色彩信号(由于慢速的电子效应)或在最坏情况下将其作为高频模式叠加在主信号的顶端来获得的。

尽管在计算机视觉的早期时代这些转换是重要的，当时帧捕捉卡直接将复合 TV 信号数字化，今天所有的数字视频和静态图像压缩标准都基于更新的 YCbCr 转换。YCbCr 是与 YUV 紧密相关的(C_b 和 C_r 信号携带蓝和红的色差信号，具有比 UV 更有用的助记信息)，但使用不同的放缩因子来配合数字信号的 8 位范围。

对于视频，Y' 信号被重新缩放来配合数值的(16…235)范围，而 C_b 和 C_r 信号被缩放来配合(16…240)的范围(Gomes and Velho 1997; Fairchild 2005)。对于静态图像，JPEG 标准使用整个 8 位无保留数值的范围，

$$\begin{bmatrix} Y' \\ C_b \\ C_r \end{bmatrix} = \begin{bmatrix} 0.299 & 0.587 & 0.114 \\ -0.168736 & -0.331264 & 0.5 \\ 0.5 & -0.418688 & -0.081312 \end{bmatrix} \begin{bmatrix} R' \\ G' \\ B' \end{bmatrix} + \begin{bmatrix} 0 \\ 128 \\ 128 \end{bmatrix}, \tag{2.115}$$

其中 R′G′B′ 的数值是 8 位 gamma-压缩过的彩色分量(即我们打开或显示 JPEG 图像时所获得的实际的 RGB 数值)。对于多数应用，这个公式并不那么重要，因为你的图像读取软件将直接给你提供 8 位 gamma 压缩过的 R′G′B′ 数值。但是，如果你试图精细地去除图像块效应(习题 3.30)，该信息可能是有用的。

另一种你可能遇到的色彩模型是色调、饱和度、亮度(HSV)，它是 RGB 彩色立方体到色彩角、径向的饱和比例和亮度激励的数值上的非线性映射。更具体说来，亮度定义为平均或最大的彩色数值，饱和度定义为从对角线缩放的距离，而色调定义为绕着色盘的方向(Hall (1989); Foley, van Dam, Feiner *et al.* (1995)描述了确切的公式)。这种分解在图形学应用中很自然，例如在色彩拾取(它近似彩色描述的 Munsell 图版)应用中。图 2.32l～n 显示了样例彩色图像的 HSV 表达，其中饱和度是用灰度值编码的(饱和的=更暗)，而色调用彩色来描绘。

如果你想让你的计算机视觉算法只影响图像的亮度而不是其饱和度或色调，一个更简单的解决方法是使用定义在公式(2.104)中的 Yxy(亮度+色度)坐标或者甚至更简单的彩色比率，

$$r = \frac{R}{R+G+B}, \quad g = \frac{G}{R+G+B}, \quad b = \frac{B}{R+G+B} \tag{2.116}$$

(图 2.32e～h)。在处理亮度(2.112)之后，例如，经过直方图均衡化(3.1.4 节)，你可以用新老亮度的比值乘上每个彩色的比值来获得一个调整后的 RGB 三元组。

图 2.32　色彩空间变换：(a～d) RGB；(e～h) rgb；(i～k) L*a*b*；(l～n) HSV。注意，rgb，L*a*b*和 HSV 数值都重新缩放了以适合纸质印刷的动态范围

尽管所有这些色彩系统可能听起来都令人困惑，但最终用哪个常常并不见得那么重要。Poynton 在其彩色常见问题(*http://www.poynton.com/*)中，写到感知激发的 L*a*b* 系统与我们绝大多数情况下处理的 gamma-压缩过的 R′G′B′ 系统在性质上相似，由于两者在实际的亮度值和所处理的数字之间都有分数指数缩放(这近似对数响应)。正如在所有情况中一样，在决定使用哪种方法之前，仔细考虑你要解决的是什么问题[①]。

2.3.3 压缩

在摄像机处理流水线上的最后阶段，通常是某种形式的图像压缩(除非你使用无损压缩机制，比如 RAW 或 PNG)。

所有的视频和图像压缩算法都首先将信号转换为 YCbCr(或某种紧密相关的变形)，以便它们可以用更高的保真度(相较于色彩信号)来压缩亮度信号。(回想一下，人类视觉系统对色彩的频率响应不如对亮度变化的响应。)在视频中，通常对 Cb 和 Cr 在横向做因子为 2 的重采样，对于静态图像(JPEG)，重采样(平均化)会出现在横纵两个方向上。

图 2.33 在三种质量设定上的 JPEG 图像压缩。注意块状效应和高频走样("蚊子噪声"，mosquitonoise)从左到右是如何增加的

一旦亮度和色彩图像经过适当的重采样并分离成单独的图像，便会被送至块变换阶段。这里使用的最常见的方法是离散余弦变换(DCT)，它是离散傅里叶变换(DFT) (见 3.4.3 节)的实值变型。DCT 是对自然图像块的 Karhunen–Loève 即特征值分解的合理的近似，即同时将绝大多数能量压缩进前面的系数中且将像素间的联合协方差矩阵对角化了的分解(使变换系数在统计上相互独立)。MPEG 和 JPEG 都使用 8×8 DCT 变换(Wallace 1991; Le Gall 1991)，但是新的变形使用较小的 4×4 块或别的变换，比如小波变换(Taubman and Marcellin 2002)和重迭(lapped)变换(Malvar 1990, 1998, 2000)。

在变换编码之后，系数值被量化成一组小的整数值，可以用可变数位长度编码机制，比如 Huffman 编码或算术编码(Wallace 1991)，进行编码。(DC(最低的频率)系数也从前一个块的 DC 值中自适应地预测出来。"DC"项来自于"直流"，即

---

① 在会议中，如果你一下子找不到问题问，你总可以问演讲者为什么不使用感知色彩空间，比如 L*a*b*。反过来，如果他们的确用了 L*a*b*，你可以问他们是否有确切的证据说明这样比使用常规的色彩更有效。

信号的非正弦或非交替的部分。)量化中的步长是由 JPEG 文件的质量设定控制的主要变量(图 2.33)。

对于视频，通常也会执行基于块的运动补偿，即编码每个块与作为其预测的在前一帧通过块移位得到的一组像素值的差别做编码。(一个例外是老式 DV 摄录机使用的 motion-JPEG 机制，它只是一系列单独的 JPEG 格式压缩的图像帧。)尽管基本 MPEG 使用具有整数运动数值的 16×16 运动补偿块(Le Gall 1991)，但更新的标准使用自适应大小的块、亚像素运动并具有参考更老帧的块的能力。为了更好地从失败复原以及允许对视频流随机访问，预测得到的 P 帧交织在独立编码的 I 帧中。(双向预测的 B 帧有时也使用。)

压缩算法的品质通常使用峰值信噪比(PSNR)来报告，它是由均方误差导出的，

$$MSE = \frac{1}{n} \sum_{\boldsymbol{x}} \left[I(\boldsymbol{x}) - \hat{I}(\boldsymbol{x}) \right]^2, \tag{2.117}$$

其中 $I(x)$ 是原始的未压缩的图像，而 $\hat{I}(x)$ 是其压缩后的图像，或者等价地，用均方根误差报告，定义为

$$RMS = \sqrt{MSE}. \tag{2.118}$$

PSNR 定义为

$$PSNR = 10 \log_{10} \frac{I_{\max}^2}{MSE} = 20 \log_{10} \frac{I_{\max}}{RMS}, \tag{2.119}$$

其中 $I_{\max}$ 是最大的信号量程，例如对于 8 位图像是 255。

尽管这只是对图像压缩工作原理的高层次描绘，但为了在各种计算机视觉应用中对这些方法引入的人工效应进行补偿，理解这些是有用的。

2.4 补充阅读

正如在本章开始时所指出的，这只是有关一个非常丰富和精深的主题集的简要概括，该主题集在传统上覆盖了几个不同的领域。

关于点、线、面和投影的几何的更为完整的介绍，可见多视角几何方面的教材(Hartley and Zisserman 2004; Faugeras and Luong 2001)和计算机图形学方面的教材(Foley, van Dam, Feiner et al. 1995; Watt 1995; OpenGL-ARB 1997)。其中对于如下这些主题有更深入的阐述：诸如二次曲面、锥面和三次曲面的更高阶的基元，三视角和多视角几何。

图像形成(合成)过程通常作为计算机图形学课程的一部分来讲授(Foley, van Dam, Feiner et al. 1995; Glassner 1995; Watt 1995; Shirley 2005)，但在基于物理学的计算机视觉中也会学习这个主题(Wolff, Shafer, and Healey 1992a)。

摄像机镜头系统的行为在光学中研究(Möller 1988; Hecht 2001; Ray 2002)。

关于彩色理论的一些好书是由 Healey and Shafer(1992);Wyszecki and Stiles (2000); Fairchild(2005)撰写的，Livingstone(2008)针对彩色感知提供了一个更有趣而

非正规的介绍。Mark Fairchild 的关于彩色的书和链接[①]网页列出了很多其他的资料来源。

有关采样和走样的主题在信号和图像处理的教材中阐述(Crane 1997; Jähne 1997; Oppenheim and Schafer 1996; Oppenheim, Schafer, and Buck 1999; Pratt 2007; Russ 2007; Burger and Burge 2008; Gonzales and Woods 2008)。

2.5 习　　题

给学生的提示： 由于本章主要包含背景材料而没有很多可用的方法，因此习题比较轻松。如果你真想彻底搞明白多视角几何，我鼓励你阅读 Hartley and Zisserman(2004)并做其中的习题。类似地，如果你想做一些有关图像形成过程的习题，Glassner(1995)充满了具有挑战性的问题。

习题 2.1：最小二乘交点和线拟合——高级的

公式(2.4)给出了两条 2D 线的交点是怎样表达为它们的叉积的，假定这两条线是以齐次坐标表达的。

1. 如果你有超过两条的更多的线而想寻求一个点，使其到每条线的距离的平方和最小

$$D = \sum_i (\tilde{x} \cdot \tilde{l}_i)^2, \tag{2.120}$$

   如何计算这个量？(提示：将点积写成 $\tilde{x}^T \tilde{l}_i$，再将平方项改写成二次型 $\tilde{x}^T A \tilde{x}$。)

2. 为一串点拟合一条线，你可以计算这些点的质心(均值)及其关于均值的协方差矩阵。证明沿着协方差椭圆主轴穿过质心的线最小化了到点的距离平方和。

3. 这两种方法根本不同，尽管投影对偶性告诉我们点和线是可以互换的。为什么这两个算法如此明显地不同？是它们实际上最小化不同的目标吗？

习题 2.2：2D 变换编辑器

编写一个程序，使你能够交互地创建一组矩形并改变它们的"姿态"(2D 变换)。你应该实现下列步骤。

1. 打开一个空窗口("画布")。
2. 转动拖拉(橡皮筋)创建新矩形。
3. 选择形变模态(运动模型)：平移，刚性，相似，仿射，或透视。
4. 拖拉轮廓的任何一角来改变其变换。

该习题应该在一组像素坐标和变换类的基础上构建起来，或者自己实现，或者用一个软件库实现。还需要支持对所创建表达的保持(保存和载入)(对每个矩形，保

[①] http://www.cis.rit.edu/fairchild/WhyIsColor/books_links.html

存其变换)。

习题 2.3：3D 观察器

为 3D 点、线和多边形写一个简单的观察器。导入一组点和线命令(基元)以及观察变换。以交互方式改变物体或摄像机变换。该观察器可以是你在(习题 2.2)中创建的 2D 编辑器的扩展。只要将其观察变换用其 3D 对等形式替换即可。

(可选)加一个 z-缓冲，实现对多边形隐藏面的去除。

(可选)使用一个 3D 绘图库，只写观察器的控制。

习题 2.4：聚焦距离和景深

搞清楚镜头上的聚焦距离和景深指标是如何确定的。

1. 对于焦距为 f(比如说，100mm)的镜头，计算并绘制聚焦距离 z_o，将其作为从焦距走过的距离 $\Delta z_i = f - z_i$ 的函数。它能解释你从常规镜头上所见到的聚焦距离双曲变化吗(图 2.20)？

2. 计算给定焦距设置的景深 z_o(最小和最大的聚焦距离)，将其作为混淆圆直径 c(使其是传感器宽度的一个比率)、焦距 f、f-数 N 的函数(它与光圈空间 d 相关)。如图 2.20a 所示，它能解释镜头上界定在焦范围的通常的景深标志吗？

3. 现在来考虑一个具有可变焦距 f 的镜头。假设你变焦时镜头保持在焦上，即对于给定的聚焦距离 z_o，从后节点到传感器平面的距离 z_i 自动地调整。景深标志作为焦距的函数是如何变化的？你能够复制从镜头看到的如图 2.20b 所示的模拟景深曲线的二维绘图吗？

习题 2.5：f-数和快门速度

列出你的照相机提供的 f-数和快门速度。在较老型号的单反相机上，它们显示在镜头和快门速度拨盘上。在更新型的相机上，只能在手动调整曝光时查看电子取景器(或 LCD 屏幕/显示器)。

1. 这些构成几何级数吗；如果是，比率是多少？它们与曝光值(EV)有何关系？

2. 如果相机具有快门速度 1/60 和 1/125，你认为这两个速度是只差一个精确的两倍因子还是差一个因子 125/60=2.083？

3. 你认为这些数的精度如何？你能设计某种方法来测量光圈对到达传感器光量的确切影响以及实际的曝光时间的精确值吗？

习题 2.6：噪声水平标定

将相机固定在三脚架上，通过重复拍摄场景来估计噪声量。(如果你有单反相机的话，买一个遥控快门装备将是很好的投资。)另一种方式是，拍摄具有常量彩色的区域(比如棋盘格彩色图版)，通过给每个彩色区域拟合一个光滑函数来估计变化，然后再从预测函数上取得偏差。

1. 针对每个彩色通道分别绘制你所估计出来的作为噪声水平函数的变化量。

2. 改变相机的 ISO 设定。如果你不能这样做，就调低场景中的整体亮度(关掉灯，拉上帘，等待到黄昏)。噪声随着 ISO 增益变化大吗？

3. 将你的相机与另一种价码或不同制造年份的相机比较。是否有证据表明"你的相机物有所值"？数码相机的质量随着时间得到了改善吗？

习题 2.7：在图像拼接中的 Gamma 校正

这是一个相对简单的难题。假设你有两幅图像，它们是你想拼接的全景图中的部分(参见第 9 章)。这两幅图像是以不同的曝光拍摄的，因此你想调整其 RGB 值使其在接缝处匹配。为此，是否有必要对彩色值做 gamma 校正呢？

习题 2.8：肤色检测

基于色度或其他彩色属性设计一个简单的肤色检测器(Forsyth and Fleck 1999; Jones and Rehg 2001; Vezhnevets, Sazonov, and Andreeva 2003; Kakumanu, Makrogiannis, and Bourbakis 2007)。

1. 取各种各样的人物照片并计算其每个像素的 xy 色度值。
2. 剪切照片或另外用画笔工具拾取那些可能是肤色的像素(例如人脸和手臂)。
3. 对这些像素计算彩色(色度)分布。你可以使用某种像均值和方差度量这样简单的或像均值移位分割算法(参见 5.3.2 节)那样复杂的方法。你可以选择使用非肤色像素对背景分布建模。
4. 用你计算出来的分布寻找图像中的肤色区域。一种简单的可视化这一过程的方法是将所有非肤色像素涂成指定的一种颜色，比如白色或黑色。
5. 你的算法对色彩平衡(场景照明)的敏感度如何？
6. 更简单的色度量测，比如彩色比率(2.116)，是否同样有效？

习题 2.9：白点平衡——棘手的

进行白点平衡调整的常用(在摄像机内或后处理)方法是拍摄一张白纸的照片，调整一幅图像的 RGB 值使其成为中性色。

1. 给定"白色"的一个样本(R_w, G_w, B_w)，描述你如何调整一幅图像中的 RGB 值使其成为中性色(并没有很大程度地改变曝光)。
2. 你的变换是涉及一个简单的(每个通道)RGB 值的缩放还是需要一个完整的 3×3 彩色扭曲矩阵(或其他)？
3. 将 RGB 值转换为 XYZ。现在合适的校正是仅依赖 XY(或 xy)值吗？如果是，在转换回 RGB 空间时，你是否需要一个完整的 3×3 彩色扭曲矩阵来取得同样的效果？
4. 在直接的 RGB 模态下使用纯的对角缩放，而你在 XYZ 空间上工作却以一个扭曲而结束，你如何解释这种显然的二分歧义？哪种方法是正确的？(或者是否有可能两种方法实际上都不正确？)

如果你想知道你的相机究竟在做什么，继续做下一个习题。

习题 2.10：摄像机内彩色处理——挑战性的

如果你的相机支持 RAW 像素模式，就拍摄一对 RAW 和 JPEG 的图像，看看你能否推断出它在将 RAW 像素值转换为最终经色彩校正和 gamma 压缩后的 8 位 JPEG 像素值时做了什么。

1. 根据并存的 RAW 和映射了彩色的像素值推断你的色彩滤波器阵列的样式。如果能使你的工作更简单，这个阶段可以使用彩色棋盘图版。你可能会发现，将 RAW 图像分解成 4 个分开来的图像(对偶数和奇数列和行重采样)并将其每个新图像作为"虚拟"的传感器是有帮助的。
2. 通过拍摄含有强烈彩色边缘(比如在 10.3.1 节中所显示的那些)而有挑战性的场景来评价去马赛克算法的性能。
3. 如果你能在改变相机中彩色平衡值之后拍摄完全相同的照片，请比较一下这些设定是如何影响这个处理过程的。
4. 将你的结果与 Chakrabarti, Scharstein, and Zickler(2009)给出的或使用其彩色图像数据库[①]中的数据进行比较。

---

① *http://vision.middlebury.edu/color/*

第3章
图像处理

3.1 点算子
3.2 线性滤波
3.3 更多的邻域算子
3.4 傅里叶变换
3.5 金字塔与小波
3.6 几何变换
3.7 全局优化
3.8 补充阅读
3.9 习题

图 3.1 一些常用的图像处理算子：(a)原始图像；(b)对比度增强；(c)改变色调；(d)"多色调分色"(彩色量化)；(e)模糊；(f)旋转

目前我们已经介绍了如何通过 3D 场景元素、光照、摄像机光学和传感器的交互作用来形成图像，本章将介绍大多数计算机视觉应用中的第一步，使用图像处理方法对图像进行预处理，将其转化为便于进一步分析的形式。此类操作的一些例子有曝光校正、彩色平衡、图像噪声减少、图像锐化、通过旋转矫正图像(图 3.1)。有些人可能认为计算机视觉研究范围不应该包括图像预处理，但多数计算机视觉应用(例如计算摄影学，甚至识别)，为了获得满意的结果，需要考虑图像预处理的设计。

本章将回顾标准图像处理算子，即从一幅图像到另一幅图像的像素的值的映射。图像处理作为信号处理的后续课程，电机工程系常开设此课程(Oppenheim and Schafer 1996; Oppenheim, Schafer, and Buck 1999)。流行的图像处理教材有 Crane 1997, Gomes and Velho 1997, Jähne 1997, Pratt 2001, Gonzales and Woods 2002, Russ 2007, Burger and Burge 2008。

本章，我们从最简单的图像变换开始，即对每个像素的操作不依赖它的邻域像素(3.1 节)。这种变换也常称为点运算或点处理。接下来我们将介绍基于区域的邻域算子，即对每个像素的输出值依赖于其邻域像素(3.2 节和 3.3 节)。傅里叶变换是分析(有时为了加速)此类算子的一个便利工具，在 3.4 节我们将会介绍。邻域算子可以级联形成图像金字塔和小波，进而可以在不同的分辨率(尺度)下对图像进行分析和对特定的算子加速(3.5 节)。另一类重要的全局算子是几何变换，它的一些例子有旋转、切变和透视形变(3.6 节)。最后，我们将介绍图像处理中的全局优化方法，其中涉及能量函数的最小化，也即使用贝叶斯马尔科夫随机场模型进行最优估计(3.7 节)。

3.1 点算子

点运算是最简单的一类图像处理变换。图像中每个像素的输出值只取决于其输入值(可能会引入一些收集得到的全局信息或参数)。这类算子的一些例子有亮度和对比度的校正(图 3.2)，彩色的校正和变换。在一些图像处理文献中，这类算子也被为点运算(Crane 1997)。

本节，我们首先很快地回顾一些简单的点算子，例如亮度缩放和图像加法。接下来，我们将讨论怎样使用图像的彩色信息。然后，我们介绍图像的合成和抠图算子，它们在计算摄影学(第 10 章)和计算机图形学中有着重要应用。最后，我们详细介绍直方图均衡化的处理过程，并通过一个调整色调值(曝光和对比度)来改善图像显示效果的应用实例来结束本节。

图 3.2 一些局部的图像处理操作：(a)原始图像和它的三个彩色分量(每个通道)的直方图；(b)亮度增强(加上偏移量，$b=16$)；(c)对比度增强(乘上增益，$a=1.1$)；(d)伽马(局部)线性化($\gamma=1.2$)；(e)全局的直方图均衡化；(f)局部的直方图均衡化

3.1.1 像素变换

一般的图像处理算子是指一个或多个输入图像到一个输出图像的函数。在连续域中，可将其表示为

$$g(\boldsymbol{x}) = h(f(\boldsymbol{x})) \text{ or } g(\boldsymbol{x}) = h(f_0(\boldsymbol{x}),\ldots,f_n(\boldsymbol{x})), \tag{3.1}$$

其中，x 属于函数的 D 维定义域(对于图像通常 $D=2$)，函数 f 和 g 在某个值域上操作，该值域既可以是标量，也可以是向量，例如在彩色图像或二维运动中。对于离散(采样)图像，定义域由有限个像素位置组成，$x=(i,j)$，此时

$$g(i,j) = h(f(i,j)). \tag{3.2}$$

图 3.3 显示了一幅图像的不同表示，其中包括用彩色(表现)表示，数字网格表示，还有二维函数(绘制表面)表示。

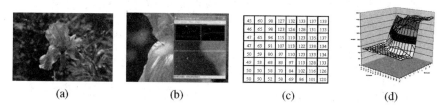

(a)   (b)   (c)   (d)

图 3.3 图像数据的可视化：(a)原始图像；(b)用图像查看工具裁剪部分图像并绘制扫描线；(c)数字栅格；(d)表面绘制。对于图(c)和图(d)，首先要将原始图像转化为灰度图

常用的两个点算子是乘以和加上一个常数，

$$g(\boldsymbol{x}) = af(\boldsymbol{x}) + b. \tag{3.3}$$

参数 a ($a>0$)和 b 常称作增益参数和偏差参数；这些参数有时被认为是分别用来控制图像的对比度和亮度的(图 3.2b-c)[①]。偏差和增益参数也可以随着空间位置的不同而变化，

$$g(\boldsymbol{x}) = a(\boldsymbol{x})f(\boldsymbol{x}) + b(\boldsymbol{x}), \tag{3.4}$$

例如，当摄影师模拟分级密度滤波器(graded density filter)来选择性地使天空变暗或者在光学系统中对虚影进行建模时。

乘法增益(包括全局和空间的变化)是一个线性算子，因为它遵从叠加原理，

$$h(f_0 + f_1) = h(f_0) + h(f_1). \tag{3.5}$$

(在 3.2 节中我们将讨论更多的线性平移不变算子)。而有些算子——例如对图像进行平方运算(此算子常用于经过带通滤波器的信号的能量的局部估计，见 3.5 节)——不是线性算子。

另一个常用的二元(两个输入)算子是线性混合算子，

$$g(\boldsymbol{x}) = (1-\alpha)f_0(\boldsymbol{x}) + \alpha f_1(\boldsymbol{x}). \tag{3.6}$$

其中 α 的变化范围是 $0 \to 1$，此算子可以实现两幅图像或视频间的时间上的淡入淡出，在幻灯片和电影制作中经常会看到。同时它也是图像变形(morphing)算法的一部分(3.6.3 节)。

伽马校正是图像预处理阶段经常使用的一个非线性算子，它可以去除输入辐射量和量化的像素值之间的非线性映射(2.3.2 节)。为了对传感器施加的伽马映射求逆，我们使用

$$g(\boldsymbol{x}) = [f(\boldsymbol{x})]^{1/\gamma}, \tag{3.7}$$

其中，对于大多数数字摄像机来说，γ 的合适取值是 2.2。

[①] 关键亮度值(平均亮度)和亮度值范围也反映了一幅图像的亮度特征(Kopf, Uyttendaele, Deussen et al. 2007)。

3.1.2 彩色变换

虽然彩色图像可以当作任意的向量值函数或者多个独立通道的汇集来处理，但是将其看作是与图像的形成过程(2.2 节)、传感器设计(2.3 节)和人的感知(2.3.2 节)强关联的高度相关的信号会更有意义。例如，通过在三个通道上分别加上同一个常数来增加图像的亮度，见图 3.2b。通过此操作，你看能达到预期的效果(即使图像亮一些)吗？你能看到图像中出现的一些未曾料想到的副作用或人为产物吗？

事实上，对图像的每个彩色通道加上同一个值，不但增加了每个像素的亮度，还影响了像素的色调和饱和度。那么我们应该怎样定义和利用这些量从而达到感觉上的预期效果呢？

正如 2.3.2 节的讨论，首先我们可以计算色度坐标(2, 104)或者简单的彩色比例(2, 116)，然后对亮度 Y 进行操作(例如亮度增强)，最后用相同色调和饱和度重新合成有效的 RGB 图像。图 2.32g～i 显示了一些不同彩色比例的图像，图中为了更好地显示结果，图像都乘以其灰度的中值。

类似地，色彩平衡(例如白炽光光照的补偿)可以通过对每个通道乘以不同的尺度因子来实现，也可以采用更复杂的处理过程，即将 RGB 映射到 XYZ 彩色空间，改变标称白色点，再重新映射到 RGB 空间(此过程可以记为一个线性3×3的彩色扭曲变换矩阵)。习题 2.9 和习题 3.1 是有关这方面问题的探讨。

另一个有趣的课程设计是增强一张带彩虹的图片的彩虹强度(习题 3.29)，建议你在掌握本章剩余内容后去尝试一下。

3.1.3 合成与抠图

在很多图像编辑和视觉效果应用中，经常会从一个场景中裁剪出前景物体，然后将其置于另一个背景之上(图 3.4)。从原始图像中抽取物体的过程常称为"抠图"(Smith and Blinn 1996)，而将物体插入另一幅图像的过程(没有明显的人为产物)常称为"合成"(Porter and Duff 1984, Blinn 1994a)。

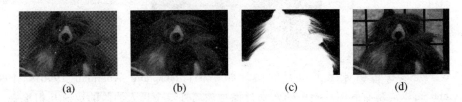

图 3.4　图像的抠图和合成(Chuang, Curless, Salesin *et al.* 2001)@2001 IEEE：(a)原图像；(b)抽取前景物体 F；(c)在灰度图中显示透明度遮罩 α；(d)新的合成图像 C

两个阶段之间的前景物体将用中间图像来表示，此图像称为透明度遮罩彩色图像(图 3.4b 和 c)。透明度遮罩图像除了图像本身的 RGB 三个彩色通道外，还包括第四个通道 α (或 A)。此通道用来描述每个像素的不透明度或者覆盖的程度(图 3.4c

和图 3.5b)。在物体中的像素是完全不透明的($\alpha=1$),而在物体外的像素是完全透明的($\alpha=0$)。在物体边界中的像素是在两个极端间平稳变化的,这样就可以避免只使用二值的不透明度引起的视觉感知上的锯齿边缘。

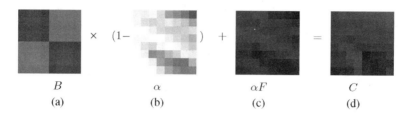

$$\begin{array}{cccc} B & \alpha & \alpha F & C \\ (a) & (b) & (c) & (d) \end{array}$$

图 3.5 图像合成公式 $C=(1-\alpha)B+\alpha F$。图像来源于图 3.4 中狮子右上角处靠近头发的区域

为了在旧的图像(背景图像)上合成新的图像(或前景图像),Porter and Duff (1984)首先提出了覆盖算子,Blinn(1994a)(1994b)对其进行了研究扩展,他们使用的算子如下:

$$C = (1-\alpha)B + \alpha F. \tag{3.8}$$

这个算子通过$(1-\alpha)$因子减弱了背景图像 B 的影响,加入了对应于前景层图像 F 的彩色值(和不透明度),如图 3.5 所示。

在很多情形下,前景图像的彩色用乘之前的形式表示会很方便,即直接存储(和使用)αF值。正如 Blinn(1994b)所述,用乘之前的 RGBA 的表示有很多优点,其中的一个优点是可以对透明度遮罩图像作模糊或重新采样(例如旋转)而不会带来额外的复杂度(仅仅是独立地处理每个 RGBA 通道)。但是,当利用局部彩色的一致性进行抠图时(Ruzon and Tomasi 2000; Chuang, Curless, Salesin *et al*. 2001),由于物体边缘附近彩色是一致的(或者说变化不大),此时则使用没有进行乘法操作的前景彩色 F。

图 3.6 光线在透明玻璃上反射的图片帧举例(Black and Anandan 1996)© 1996 Elsevier。可以清楚看到图片帧中的妇女肖像与玻璃外一个男人的脸的反射互相重叠

用于图像合成的算子并不仅仅只有覆盖算子一种。Porter and Duff(1984)描述了在图片编辑和视觉效果应用中其他一些有用的算子。在本书中,我们只关心另外一个常用的算子(参见习题 3.2)。

当光线在光滑透明的玻璃上反射时,就将穿过玻璃的光线和反射光线简单地相加(图 3.6)。这个模型在透明运动分析中很有用(Black and Anandan 1996; Szeliski,

Avidan, and Anandan 2000)。用运动的照相机观察场景时，透明运动常常会出现(8.5.2 节)。

抠图的实际过程，即从一幅或多幅图像中恢复前景、背景和透明度遮罩，有一段丰富的历史，10.4 节我们将对其进行介绍。Smith and Blinn(1996)写了一篇有关传统的蓝屏抠图方法的精彩综述，Toyama, Krumm, Brumitt *et al.*(1999)回顾了不同的抠图方法。近年来，计算摄影学中的许多工作与自然图像的抠图有关(Ruzon and Tomasi 2000; Chuang, Curless, Salesin *et al.* 2001; Wang and Cohen 2007a)，他们试图从一个单一的自然图像(图 3.4a)或扩展的视频序列(Chuang, Agarwala, Curless *et al.* 2002)中来抽取遮罩。所有的这些方法细节将在 10.4 节中描述。

3.1.4 直方图均衡化

在 3.1.1 节中，通过对亮度和增益的控制可以改善图像的显示，那么我们怎样来自动选择它们的最佳取值呢？一种方法是寻找图像中最亮的和最暗的像素值，将它们映射到纯白和纯黑。另一种方法是寻找图像中像素值的平均值作为中间灰度值，然后扩展范围以达到尽量充满可显示的值(Kopf, Uyttendaele, Deussen *et al.* 2007)。

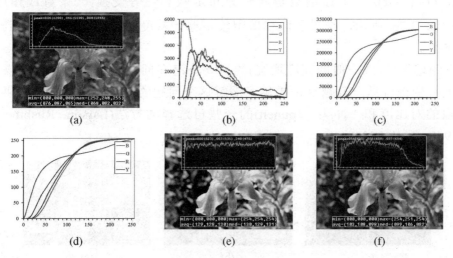

图 3.7 直方图分析和均衡化：(a)原始图像；(b)各个彩色通道的直方图和亮度直方图；(c)累积分布函数；(d)均衡化(转换)函数；(e)全局的直方图均衡化；(f)局部的直方图均衡化

我们应该怎样可视化图像的亮度值集合来测试这些启发式中的一些呢？答案是绘制各个彩色通道的直方图和亮度值直方图，如图 3.7b 所示[①]。通过这些分布，我们可以得到一些有意义的统计量(例如最大值、最小值和平均亮度值)。注意，图 3.7a 亮度直方图中黑色值和白色值过多，而大部分中间范围的值很少。如果我们将一些暗的值变亮，一些亮的值变暗，仍然使用整个可用的动态范围，岂不更好？你能想

---

① 直方图是对每个灰度级的值在图像像素中出现次数的简单统计。一个 8 位图像，需要 256 表项的累计表。对于更高的位深，需要使用合适表项的累计表(可能少于整个灰度级的数目)。

出一个可以实现这个想法的映射吗？

对于这个问题，一个常用的方法是对图像进行直方图均衡化，即寻找一个映射函数 $f(I)$，经过映射后直方图是平坦的。寻找此映射的方法与从概率密度分布函数产生随机样本的方法类似，其中首先要计算累积分布函数，如图 3.7c 所示。

可以把原始的直方图 $h(I)$ 看成一个班级在某次考试后的成绩分布。在一个特定的成绩和其学生所占百分比之间，怎样建立映射才可以使分数为总分 75% 的学生得分优于班里 3/4 的同学？答案是通过 $h(I)$ 的分布得到累积分布函数 $c(I)$，

$$c(I) = \frac{1}{N}\sum_{i=0}^{I} h(i) = c(I-1) + \frac{1}{N}h(I), \tag{3.9}$$

其中 N 是图像中像素的总个数或班级中学生总人数。对于任意给定的成绩或亮度，我们可以查出它对应的百分比 $c(I)$，此时可以决定此像素所对应的最终的值。对 8 位的像素的值进行操作时，坐标轴 I 和 c 要缩放到 $(0, 255)$。

图 3.7d 显示了将 $f(I) = c(I)$ 应用到原始图像的结果。我们可以看到，转换后的直方图变平坦了。此处"平坦"只是感觉上缺乏对比度，但看起来并不平滑。此时可以对不平坦的直方图采用局部补偿的方法改善其结果，例如使用映射函数 $f(I) = \alpha c(I) + (1-\alpha)I$。上述函数是累积分布函数和恒等变换(一条直线)的线性混合。正如图 3.7e 所示，结果图像在保留较多原始灰度图像分布的同时达到更有吸引力的平衡。

直方图均衡化(或一般的亮度增强)的另一个潜在问题是位于暗区域的噪声可能会被放大，变得可见。习题 3.6 提供了缓解这个问题的一些方法，同时提供了保持原始图像的对比度和"冲头"(punch)的一些可选择的方法(Larson, Rushmeier, and Piatko 1997; Stark 2000)。

图 3.8 局部自适应直方图均衡化：(a)原始图像；(b)基于块的直方图均衡化；(c)整个图像的局部自适应均衡化

局部自适应直方图均衡化

虽然全局的直方图均衡化方法很有效，但对于一些图像来说，不同区域采取不同均衡化方法，产生的效果可能会更好。例如图 3.8a，图像的亮度值变化范围很大。如果我们不再计算单一的曲线，而是将图像分成 $M \times M$ 像素块，分别对每个子块进行直方图均衡化，效果会怎样呢？正如图 3.8b 所示，结果图像中出现许多人为的区块效应，即在块的边界处亮度不连续。

消除人为区块效应的一个方法是使用移动窗口，即对所有的以每个像素为中心大小为 $M \times M$ 的块重新计算直方图。虽然此方法很慢(每个像素需要 M^2 次运算)，但是此方法可以加速，只需对进入块和离开块(在图像的光栅扫描中)的像素所在直方图中相应的项进行更新。注意，这种操作是一种非线性邻域操作，在 3.3.1 节中我们将对其进行详细研究。

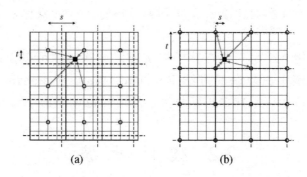

图 3.9 利用相对的 (s, t) 坐标进行局部直方图插值：(a)基于块的直方图，圆圈表示块的中心；(b)基于角点的"样条"直方图。像素落在网格交点处。计算 (s, t) 的值形成查找表(亮度数组)，黑色方块像素转换后的值是通过对其在查找表中四个邻接的值进行插值得到。块的边界用虚线表示

一个更有效的方法是和前面一样先将图像划分成不重叠块，对每块进行均衡化，然后在块与块之间对转换函数进行平滑插值。这种方法称为"自适应直方图均衡化"(adaptive histogram equalization，AHE)。CLAHE(Pizer，Amburn，Austin et al. 1987)只是在它基础上限制了对比度(限制增益)。[①]对于给定的像素 (i, j)，根据块的水平和垂直位置 (s, t) 计算其权重值，如图 3.9a 所示。为了混合四个查找函数 $\{f_{00}, \cdots, f_{11}\}$，将采用双线性混合函数，

$$f_{s,t}(I) = (1-s)(1-t)f_{00}(I) + s(1-t)f_{10}(I) + (1-s)tf_{01}(I) + stf_{11}(I) \tag{3.10}$$

(类似更高阶次的样条函数的产生可以参考 3.5.2 节)。注意，我们采用对给定的像素通过其在查找表中四个邻接的值混合出映射结果，而不采用查找表中四个邻接的值混合出一个输出像素(这样可能会非常慢)。

上述算法的一个变种是在每个 $M \times M$ 块的角点处设置查找表(如图 3.9b 所示，见习题 3.7)。除了混合四个查找表的值计算最终结果，我们也可以通过相位累积直方图将每个像素分配到查找表中四个邻接的值(注意在图 3.9b 中灰色箭头是双向的)，即

$$h_{k,l}(I(i,j)) \mathrel{+}= w(i,j,k,l), \tag{3.11}$$

其中 $w(i, j, k, l)$ 是像素 (i, j) 和查找表 (k, l) 之间的双线性权重函数。这是软直方图(soft histogramming)的一个例子，经常出现在一些其他的应用中，例如 SIFT 特征点描述子的构造(4.1.3 节)和词汇树(14.3.2 节)。

---

① 在 MATLAB 的 adapthist 函数中实现了此算法。

3.1.5 应用：色调调整

图像处理中，点运算一个最普遍的应用是对照片的对比度或色调进行操作，以使图片看起来更有吸引力或易于理解。你可以借用任何一个图片处理工具，像图 3.2 到图 3.7 那样，对图片的对比度、亮度和彩色进行不同的操作，这样可加深你对不同的点算子的认识。

习题 3.1，3.5 和 3.6 是一些点算子的实现，这些习题可以使你熟悉基本的图像处理操作。色调调整中更复杂的方法(Reinhard, Ward, Pattanaik *et al.* 2005; Bae, Paris, and Durand 2006)将在 10.2.1 节大动态范围的色调映射中进行介绍。

3.2 线 性 滤 波

局部自适应直方图均衡化是邻域算子或局部算子的一个例子。邻域算子(局部算子)是利用给定像素周围的像素的值决定此像素的最终的输出值(图 3.10)。邻域算子除了用于局部色调调整，还可以用于图像滤波，实现图像的平滑和锐化，图像边缘的增强或者图像噪声的去除(图 3.11b～d)。本节，我们将介绍线性滤波算子，它主要指用不同的权重结合一个小的邻域内的像素。在 3.3 节，我们将介绍非线性算子，例如形态学运算、距离变换。

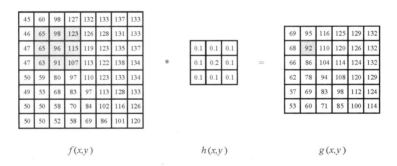

图 3.10 邻域滤波(卷积)：左边图像与中间图像的卷积产生右边图像。目标图像中绿色标记的像素是利用原图像中蓝色标记的像素计算得到的

线性滤波是一种常用的邻域算子，像素的输出值取决于输入像素的加权和(图 3.10)，

$$g(i,j) = \sum_{k,l} f(i+k, j+l)h(k,l) \tag{3.12}$$

其中权重核或掩模 $h(k,l)$ 常称为"滤波系数"。上面的相关算子可以简记为

$$g = f \otimes h \tag{3.13}$$

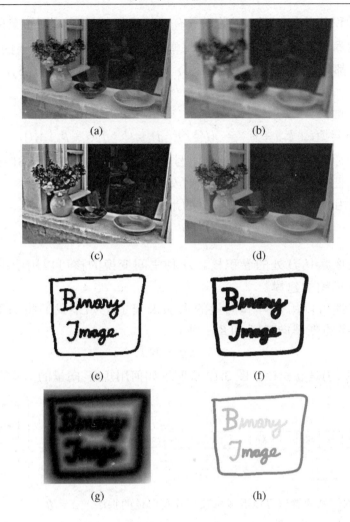

图 3.11 一些邻域算子：(a)原始图像；(b)平滑；(c)锐化；(d)保边平滑滤波；(e)图像的二值化；(f)膨胀；(g)距离变换；(h)连通量。对于膨胀和连通量，我们假设黑色像素是有效的，即在等式(3.41)~(3.45)中值为 1

式 3.12 常用的一个变形为

$$g(i,j) = \sum_{k,l} f(i-k, j-l) h(k,l) = \sum_{k,l} f(k,l) h(i-k, j-l) \tag{3.14}$$

其中 f 的偏移量的符号反向。此公式称为"卷积公式"，

$$g = f * h \tag{3.15}$$

其中 h 被称作脉冲响应函数。①之所以这么起名是因为核函数 h 与脉冲信号 $\delta(i,j)$（图像中除了位于原点的像素外其余像素的值都为 0）卷积的结果还是核函数，$h * \delta = h$，而相关运算产生翻转过来的信号。(你可以尝试证明这个结论。)

事实上，方程(3.14)可以被认为是输入像素的值 $f(k,l)$ 与平移的脉冲响应函数

---

① 对于连续函数的卷积可以记为 $g(x) = \int f(x-u) h(u) du$。

$h(i-k, j-l)$ 的乘积的叠加(求和)。卷积具有一些很好的性质，例如交换性和结合性。此外，两幅图像卷积的傅里叶变换等于每幅图像各自的傅里叶变换的乘积(3.4 节)。

相关和卷积都是线性移不变算子(LSI)。它们满足叠加原理(3.5)，

$$h \circ (f_0 + f_1) = h \circ f_0 + h \circ f_1, \tag{3.16}$$

和移位不变性原理，

$$g(i,j) = f(i+k, j+l) \Leftrightarrow (h \circ g)(i,j) = (h \circ f)(i+k, j+l), \tag{3.17}$$

移位不变性原理意味着平移一个信号与这个算子(○表示 LSI 算子)是可交换的。对移位不变性的另一种理解是算子"处处表现相同"。

偶尔，也会用到卷积或相关的移变的形式，例如

$$g(i,j) = \sum_{k,l} f(i-k, j-l) h(k,l;i,j), \tag{3.18}$$

其中 $h(k,l;i,j)$ 是像素 (i,j) 处的卷积核。这种空间变换的核可以用于由可变深度依赖散焦造成的图像模糊的建模。

相关和卷积都可以用矩阵和向量的乘法来表示。如果我们将二维图像 $f(i,j)$ 和 $g(i,j)$ 转化为有序的光栅向量 $\boldsymbol{f}$ 和 $\boldsymbol{g}$，则有

$$\boldsymbol{g} = \boldsymbol{H}\boldsymbol{f}, \tag{3.19}$$

其中矩阵 $\boldsymbol{H}$ (稀疏的)是卷积核。图 3.12 说明了如何用矩阵-向量的形式表示一维卷积。

$$\boxed{72\ |\ 88\ |\ 62\ |\ 52\ |\ 37} * \boxed{\tfrac{1}{4}\ |\ \tfrac{1}{2}\ |\ \tfrac{1}{4}} \Leftrightarrow \tfrac{1}{4} \begin{bmatrix} 2 & 1 & . & . & . \\ 1 & 2 & 1 & . & . \\ . & 1 & 2 & 1 & . \\ . & . & 1 & 2 & 1 \\ . & . & . & 1 & 2 \end{bmatrix} \begin{bmatrix} 72 \\ 88 \\ 62 \\ 52 \\ 37 \end{bmatrix}$$

图 3.12 一维信号的卷积可以看成稀疏矩阵和向量的乘法，$g = Hf$

填塞(边界效应)

细心的读者会发现图 3.12 中的矩阵乘法会产生边界效应，即采用这种形式的图像滤波会使角点处的像素变黑。这主要是因为当卷积核超出原始图像边界时，原始图像边界外的部分被认为是有效的，并用 0 值填塞。

为了抵消这种效应，可以采用一些对图像填塞或扩展的模式(图 3.13)。

- 0 填塞：将原图像之外的像素的值设置为 0(对于图像抠图中的透明度遮罩，此方式是很好的选择)。
- 常数填塞(边框彩色)：在原图像之外的像素的值设置为确定的边界值。
- 夹取填塞(复制或夹取边缘)：不限定地复制边缘像素的值。
- (圆形)重叠填塞(重复或平铺)：以"环状"形态环绕图像进行循环。
- 镜像填塞：像素围绕图像边界进行镜像反射。
- 延长：通过在边缘像素值中减去镜像信号的方式延长信号。

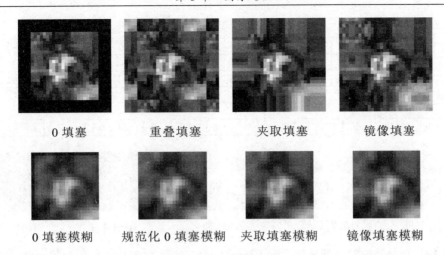

图 3.13 边界填塞(上面一行)和填塞图像的模糊结果(下面一行)。规范化 0 填塞模糊图像是以其对应的软遮罩的值除以(规范化)0 填塞的 RGBA 模糊图像的结果

在图形学文献(Akenine-Möller and Haines 2002, p. 124)中,这些方法称为"重叠模式"(OpenGL)或"纹理寻址模式"(Direct3D)。每种模式的公式这里将留给读者自己推导(习题 3.8)。

图 3.13 给出了使用以上方法填塞图像的效果和填塞图像的模糊结果。你可以观察到,0 填塞使边界变暗,夹取(复制)填塞将图像边界内部的值进行了延伸,镜像(反射)填塞保留了边界附近的彩色。扩展填塞(并没显示)使边界像素的值是常数(在模糊过程中)。

填塞的另一种模式是首先对 0 填塞的 RGBA 图像进行模糊,接着用图像的遮罩值分割结果图像,消除变暗效应。从图 3.13 中规范化 0 填塞模糊图像可以看出,此方法的模糊效果非常好。

3.2.1 可分离的滤波

对于卷积运算的实现,每个像素需要 K^2 (乘-加)操作, K 是卷积核的大小(宽或高),例如图 3.14a 的方框型滤波器。在许多情况下,这种运算可以采用如下计算方式来大幅度提高运算速度:先用一维行向量进行卷积,接着用一维列向量进行卷积(每个像素总共需要 $2K$ 操作)。如果一个卷积核可以采用上述方式来计算,则称其为可分离的。

显然,2 维核 K 的卷积和水平核 h 和垂直核 v 的连续卷积相对应,其对应关系可以用水平核和垂直核的乘积表示,如下:

$$K = vh^T \tag{3.20}$$

图 3.14 给出了一些例子。由于可分离性增加了计算效率,所以在计算机视觉应用中,卷积核的设计常常考虑其可分离性。

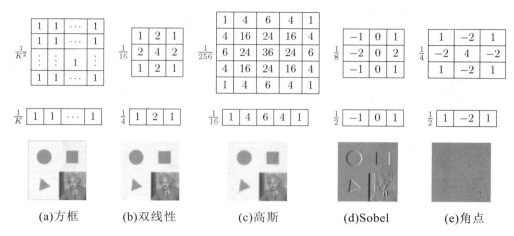

图 3.14 可分离的线性滤波器：对于图(a)~(e)，显示了 2D 滤波器核(第一层)，相对应的 1D 水平滤波器核(中间层)，滤波后的图像(底层)。Sobel 和角点滤波图像的像素的值是有符号的，在显示前分别放大了 2 倍和 4 倍，再加上一个灰度偏移量

如果给定一个核函数 K，那么我们该如何判断它是不是可分离的呢？我们可以利用核函数的解析式来进行判断(Freeman and Adelson 1991)。更直接的方法是将 2D 核函数看成一个 2D 矩阵 K 并且对此矩阵进行奇异值分解(SVD)，

$$K = \sum_i \sigma_i \boldsymbol{u}_i \boldsymbol{v}_i^T \tag{3.21}$$

(SVD 的定义可以参看附录 A.1.1)。当仅有第一个奇异值 σ_0 是非 0 值时，则核函数是可分离的，$\sqrt{\sigma_0}\boldsymbol{u}_0$ 和 $\sqrt{\sigma_0}\boldsymbol{v}_0^T$ 分别提供了垂直核函数和水平核函数(Perona 1995)。例如，高斯二阶导数核函数(式 3.26 和式 4.23)可以用两个可分离滤波器的和来实现(式 4.24)(Wiejak, Buxton, and Buxton 1985)。

如果核函数是不可分离的，但是你仍然想快速实现它，该怎么办呢？首先将核函数分解和 SVD 联系起来的 Perona(1995)建议使用序列(3.21)中的更多项，即将一些可分离的卷积累加。这种方式是否有效取决于 K 的大小、重要的奇异值的数量及其他一些考虑，例如高速缓存的连贯性和存储单元的局部性。

3.2.2 线性滤波示例

以上我们对图像的线性滤波处理进行了描述，下面我们将讨论一些常用的滤波器。

最简单的滤波器是移动平均或方框滤波器，它将 $K \times K$ 窗口中的像素值的平均值作为输出。这种滤波器等价于图像与全部元素值为 1 的核函数先进行卷积再进行尺度缩放(图 3.14a)。对于尺寸较大的核函数，一个有效的实现策略如下：在扫描行上用一个移动的窗口进行滑动(在可分离滤波器中)，新的窗口的和值等于上一个窗口的和值加上新的窗口中增加的像素的值并减去离开上一个窗口的像素的值。上述过程与区域求和表有关联，我们稍后介绍。

一个平滑的图像可以借用分段的线性"帐篷"(tent)函数对图像进行分步卷积得

到(也称"Bartlett 滤波器")。图 3.14b 显示一个 3×3 的这种滤波器,此滤波器也称"双线性核",因为它的输出是两个线性(一阶)样条的乘积(3.5.2 节)。

线性 tent 函数与自己卷积将产生三次近似样条。在 Burt and Adelson(1983a)的拉普拉斯金字塔的构造(3.5)中此样条称为"高斯核"(图 3.14c)。近似的高斯核也可以通过不断迭代卷积方框型滤波器(Wells 1986)得到。在应用中,要求滤波器具有旋转对称性时,需要仔细调整高斯模板,然后才可以使用(Freeman and Adelson 1991)(习题 3.10)。

我们刚刚讨论的核函数都是模糊(平滑)核或低通核(因为它们的作用是通过较低频信号衰减较高频信号)。那么如何度量这些核函数的效果呢?在 3.4 节中,我们将借用傅里叶分析查看这些滤波器确切的频率响应。同时,我们还会介绍 $sinc((\sin x)/x)$ 滤波器,此滤波器是理想低通滤波器的实现。

实际中,平滑核函数常用于减少高频噪声。接下来我们还会介绍更多不同的利用平滑去除噪声的方法(3.3.1 节,3.4 节和 3.7 节)。

令人惊奇的是,通过使用非锐化掩模处理,平滑函数可以对图像进行锐化。这是因为图像模糊操作将减少高频信息,从而在图像中增加一些原图像和模糊后图像的差别,图像就会更加锐利。

$$g_{\text{sharp}} = f + \gamma(f - h_{\text{blur}} * f) \tag{3.22}$$

事实上,在数字摄影方法发明之前,这种方法是在暗室条件下锐化图像的标准方法:通过散焦从原始底片上产生模糊("正面的")底片,然后按照下面的方式将两张底片重叠得到最后的图像,

$$g_{\text{unsharp}} = f(1 - \gamma h_{\text{blur}} * f). \tag{3.23}$$

虽然这种滤波器不是线性的,但是它的效果很好。

线性滤波算子也常用在边缘提取的预处理阶段(4.2 节)和兴趣点检测(4.1 节)的算法中。图 3.14d 显示了一个简单的 3×3 边缘提取算子,此算子称为"Sobel 算子",它是由水平的中心差分(如此命名是因为水平方向的导数以该像素为中心)和垂直的 tent 滤波器(为了平滑结果)构成的一个可分离的组合。正如你所看到的核函数下面的图像那样,这种滤波器有效地突出了水平边缘。

图 3.14e 中简单的角点检测器同时寻找水平和垂直方向的二阶导数。正如你看到的那样,这种算子不仅对正方形的角点有响应,而且对沿对角线方向的边缘也有响应。在 4.1 节将介绍具有旋转不变的更好的角点检测器。

3.2.3 带通和导向滤波器

Sobel 算子和角点算子是带通和带方向的滤波器的简单例子。可以按如下方式来构造更精细的核:首先用一个高斯滤波器(单位面积)平滑图像,

$$G(x, y; \sigma) = \frac{1}{2\pi\sigma^2} e^{-\frac{x^2+y^2}{2\sigma^2}}, \tag{3.24}$$

接下来采用一阶或二阶导数(Marr 1982, Witkin 1983, Freeman and Adelson1991)。这种滤波器被称为"带通滤波器",因为它们同时滤除了低频和高频。

二维图像的二阶导数(无方向)

$$\nabla^2 f = \frac{\partial^2 f}{\partial x^2} + \frac{\partial^2 f}{\partial y^2}, \tag{3.25}$$

称为"Laplacian 算子"。首先用高斯核平滑图像,然后用 Laplacian 算子作用于图像等价于直接用 LoG(Laplacian of Gaussian)滤波器与原图像卷积,

$$\nabla^2 G(x,y;\sigma) = \left(\frac{x^2+y^2}{\sigma^4} - \frac{2}{\sigma^2}\right) G(x,y;\sigma), \tag{3.26}$$

这种滤波器在一定程度上具有很好的尺度空间特性(Witkin 1983, Witkin, Terzopoulos, and Kass 1986)。5 点的 Laplacian 算子是对这种更为复杂的滤波器的一个紧凑的近似。

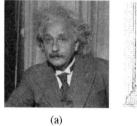

(a) (b) (c)

图 3.15 二阶导向滤波器(Freeman 1992)© 1992 IEEE:(a)爱因斯坦的原始图像;(b)通过二阶方向能量计算出的方向图;(c)方向结构增强了的原图

同样,Sobel 算子是方向或带方向的滤波器的一个简单近似。方向或带方向的滤波器获取方式如下:先用高斯核(或其他滤波器)平滑图像,再用方向导数 $\nabla \hat{u} = \frac{\partial}{\partial \hat{u}}$ 作用于图像,其中方向导数的计算借用了梯度场 ∇ 和单位方向 $\hat{u} = (\cos\theta, \sin\theta)$ 的点积,

$$\hat{u} \cdot \nabla(G * f) = \nabla_{\hat{u}}(G * f) = (\nabla_{\hat{u}} G) * f. \tag{3.27}$$

方向导数平滑滤波器,

$$G_{\hat{u}} = uG_x + vG_y = u\frac{\partial G}{\partial x} + v\frac{\partial G}{\partial y}, \tag{3.28}$$

其中 $\hat{u} = (u,v)$,是一个导向(steerable)滤波器,因为计算图像与 $G_{\hat{u}}$ 卷积的值可以先用一对滤波器 (G_x, G_y) 与原图像卷积,再用单位向量 $\hat{u}$ 与梯度场的乘积使滤波器具有导向(潜在的局部性)(Freeman and Adelson 1991)。这种方法的优点是可以用很少的代价评估整个滤波器族。

如果用二阶方向导数滤波器 $\nabla_{\hat{u}} \cdot \nabla_{\hat{u}} G$ 进行导向会怎样呢?二阶方向导数滤波器是指先求取一个(平滑的)方向导数,接着再次求取方向导数。例如,G_{xx} 是在 x 方向上的二阶方向导数。

乍一看,对于每个方向 $\hat{u}$,我们都需要计算不同的一阶方向导数,导向技巧没

法使用。令人惊奇的是，Freeman and Adelson (1991)指出，对于高斯方向导数，可以利用相对来说一组小数目的基函数计算任意阶次的导数的导向。例如，二阶方向导数只需要 3 个基函数，

$$G_{\hat{u}\hat{u}} = u^2 G_{xx} + 2uv G_{xy} + v^2 G_{yy}. \tag{3.29}$$

进一步，对于每个基滤波器，本身并没必要是可分离的，可以利用少许的可分离的滤波器的线性组合来计算(Freeman and Adelson 1991)。

这种出色的结果使构造具有越来越强的方向选择性的方向导数滤波器成为可能，也就是说，滤波器只在方向上具有很强的局部一致性的边缘有响应(图 3.15)。进一步，在给定的位置上，高阶的导向滤波器可以潜在地对超过单个边缘方向有响应，它们可以对条状边缘(细线)和典型的阶梯边缘(图 3.16)都有响应。但是为了做到这些，二阶和更高阶的滤波器需要使用完整的 Hilbert 变换对，相关描述可参考(Freeman and Adelson 1991)。

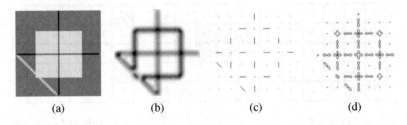

图 3.16　四阶导向滤波器(Freeman and Adelson 1991)© 1991 IEEE；(a)在不同的方向上包含条(线)和阶梯边缘的测试图像；(b)方向能量的平均值；(c)主方向；(d)将方向能量看成是角度的函数(极坐标)

导向滤波器常用来构造特征的描述子(4.1.3 节)和边缘检测器(4.2 节)。尽管 Freeman and Adelson(1991)提出的这种滤波器更适合检测线性(类似边缘)的结构，近期 Keothe(2003)做了进一步工作，他给出了如何利用 2×2 的边界张量对边缘和连接点("角点")特征进行编码。习题 3.12 是有关这种导向滤波器的实现以及其在寻找边缘和角点特征中的应用。

区域求和表(积分图像)

如果要对一幅图像用不同的方框滤波器(尤其在不同位置使用不同大小的滤波器)重复卷积时，你可以先计算区域求和表(Crow 1984)。区域求和表是指一定区域内所有像素的值的和，

$$s(i,j) = \sum_{k=0}^{i} \sum_{l=0}^{j} f(k,l). \tag{3.30}$$

它的一个有效计算方式是利用递归(光栅扫描)算法，

$$s(i,j) = s(i-1,j) + s(i,j-1) - s(i-1,j-1) + f(i,j). \tag{3.31}$$

图像 $s(i,j)$ 经常也称为"积分图像"(图 3.17)。实际中，如果将每行的求和分离，它的每个像素的值的计算只需要两次加法(Viola and Jones 2004)。为了计算矩

形$[i_0,i_1]\times[j_0,j_1]$内部区域的和(积分)，我们只需简单利用区域求和表中四个相关的值，

$$S(i_0\ldots i_1,j_0\ldots j_1)=\sum_{i=i_0}^{i_1}\sum_{j=j_0}^{j_1} s(i_1,j_1)-s(i_1,j_0-1)-s(i_0-1,j_1)+s(i_0-1,j_0-1). \quad (3.32)$$

区域求和表一个潜在的缺点是与原始图像相比，累计图像需要$\log M+\log N$额外的位，其中 M 和 N 是图像的宽度和高度。区域求和表也可以用于对其他卷积核的近似(Wolberg (1990, 6.5.2 节)对其进行了概述)。

在计算机视觉中，人脸检测(Viola and Jones 2004)利用区域求和表来计算简单的多尺度上的底层特征。这种特征是由相邻的带有正值和负值的矩形组成，常常也称作“方框型小波”(Simard, Bottou, Haffner *et al.* 1998)。从原理上讲，区域求和表也可以用于立体视觉和运动算法(11.4 节)中差分平方和(SSD)的求和计算。但在实际中，更常用的是可分离的移动平均滤波器(Kanade, Yoshida, Oda *et al.* 1996)，除非需要考虑许多形状和大小不同的窗口(Veksler 2003)。

3	2	7	2	3
1	5	1	3	4
5	1	**3**	5	1
4	3	2	1	6
2	4	1	4	8

(a)S=24

3	5	12	14	17
4	*11*	**19**	24	31
9	**17**	28	38	46
13	24	37	48	62
15	30	44	59	81

(b)S=28

3	5	12	*14*	17
4	11	19	24	31
9	17	28	38	46
13	24	37	**48**	62
15	30	44	59	81

(c)S=24

图 3.17 区域求和表：(a)原始图像；(b)区域求和表；(c)区域和的计算。区域求和表中的每个值 $s(i,j)$(红色)是利用它三个相邻(蓝色)的像素递归计算出来的(3.31)。区域和 s(绿色)是利用矩形的四个角点(紫色)的值计算出来的(3.32)，粗体表示计算时取正值，斜体表示计算时取负值

递归滤波器

区域求和的增量公式(3.31)是递归滤波器的一个例子。递归滤波器是指输出值取决于前一个滤波器的输出值。在信号处理的文献中，这种滤波器称为“无限脉冲响应(IIR)”，因为对于脉冲信号(只有一个非零值)，这种滤波器的输出是无限的。例如，对于区域求和表，这样的脉冲在脉冲的右下方产生无限大的以 1 为元素的矩形。而在本章前面我们研究的用一个有限区域核卷积图像的滤波器称作“有限脉冲响应(FIR)”。

二维的 IIR 滤波器和递归公式有时会用于计算一些涉及大的区域交互作用的量，例如二维距离函数(3.3.3 节)和连通量(3.3.4 节)。

然而，IIR 滤波器更常用于可分离的一维滤波阶段，计算大范围的平滑核，例如高斯和边缘滤波器的有效近似(Deriche 1990; Nielsen, Florack, and Deriche 1997)。基于金字塔的算法(3.5 节)也可用于实现大面积的平滑计算。

3.3 更多的邻域算子

正如我们前面看到的那样，线性滤波可以实现大量不同的图像变换。然而非线性滤波有时可以到达更好的效果，例如保边中值滤波器和双边滤波器。邻域算子的其他的一些例子还有对二值图像进行操作的形态学算子，用于计算距离变换和寻找连通量的半全局算子。

3.3.1 非线性滤波

到目前为止，我们所考虑的滤波器都是线性的，即两个信号的和的响应与它们各自响应的和相等。换句话说，每个像素的输出值是一些输入像素(3.19)的加权和。线性滤波器易于构造，并且易于从频率响应角度来进行分析(3.4 节)。

然而在很多情况下，使用邻域像素的非线性组合可能会得到更好的效果。考虑图 3.18e，噪声是散粒噪声，而不是高斯噪声，即图像偶尔会出现很大的值。这种情况下，用高斯滤波器对图像进行模糊，噪声像素是不会被去除的，它们只是转换为更柔和但仍然可见的散粒(图 3.18f)。

中值滤波

这种情况下，使用中值滤波器是一个较好的选择。中值滤波器选择每个像素的邻域像素的中值作为输出(图 3.19a)。通过使用随机选择算法(Cormen 2001)，中值可以在线性时间内计算出来。Tomasi and Manduchi(1998)和 Bovik(2000，Section 3.2)对中值滤波器进行了不断的改进。由于散粒噪声通常位于邻域内正确值的两端，中值滤波可以对这类异常像素进行过滤(图 3.18c)。

图 3.18 中值和双边滤波器：(a)带有高斯噪声的原始图像；(b)高斯滤波；(c)中值滤波；(d)双边滤波；(e)带有散粒噪声的原始图像；(f)高斯滤波；(g)中值滤波；(h)双边滤波。注意双边滤波不能去除散粒噪声，其原因是噪声像素与其邻域像素具有较大差异性

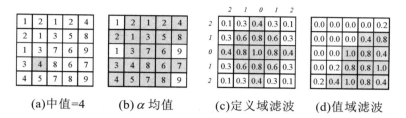

(a)中值=4　　(b)α均值　　(c)定义域滤波　　(d)值域滤波

图 3.19 中值和双边滤波：(a)中值像素(绿色)；(b)α-截尾均值像素的选择；(c)定义域滤波(沿着边界的数字是像素的距离)；(d)值域滤波

中值滤波除了中等计算速度代价外，它还有一个不足，即由于中值滤波只选择一个像素作为输出像素，所以一般很难有效去除规则的高斯噪声(Huber 1981; Hampel, Ronchetti, Rousseeuw et al. 1986; Stewart 1999)。这时采用α-截尾均值滤波(Lee and Redner 1990)(Crane 1997, p. 109)会得到更好的效果。α-截尾均值滤波是指去掉百分率为α的最小值和最大值后剩下的像素的均值(图 3.19b)。

另一种方法是计算加权中值，其中每个像素会被多次使用，次数依赖于其到中心的距离。这种方式等价于使以下的权重目标函数最小化，

$$\sum_{k,l} w(k,l)|f(i+k,j+l) - g(i,j)|^p, \tag{3.33}$$

其中 $g(i,j)$ 是理想输出值。当 $p=1$ 时，上式为权重中值。当 $p=2$ 时，上式为普通的加权均值，此时它等价于用权重的和进行归一化后的相关运算式(3.12)(Bovik 2000, Section 3.2)(Haralick and Shapiro 1992, Section 7.2.6)。此外，加权平均也与鲁棒统计学(附录 B.3)的一些方法(例如影响函数(Huber 1981; Hampel, Ronchetti, Rousseeuw et al. 1986))有着密切关系。

因为现在的照相机很少出现散粒噪声，所以非线性平滑的另一个特性可能会显得更重要。这种滤波器大的保边(edge preserving)效果更好，即当它们在滤除高频噪声时，边缘不容易被柔化。

考虑图 3.18a 的噪声图像。为了去除大部分噪声，高斯滤波器强制平滑掉高频的细节。而这些高频细节的大部分是明显的边缘。中值滤波效果会好一些，但是正如前面所述，对裂痕进行平滑时，它的效果并不是很好。更多的保边平滑方法参考(Tomasi and Manduchi 1998)。

尽管我们可以尝试α-截尾均值和加权中值方法，但是这些方法还是有一个使尖锐的转角圆滑的倾向，因为平滑区域中大部分像素来源于背景。

双边滤波器

如果我们将加权滤波器核和更好的抑制外点的方案结合起来会怎样呢？如果我们采取简单的抑制(以一种柔和的方式)策略，抑制与中心像素的值差别太大的像素，而不是抑制一个固定的百分比α会怎样呢？这就是双边滤波器思想的精髓，它在计算机视觉领域首次流行时是 Tomasi and Manduchi(1998)提出的。Chen，Pairs，and Durand(2007)和 Pairs，Kornprobst，Tumblin et al.(2008)引用了类似的早

期的工作(Aurich and Weule 1995, Smith and Brady 1997)及其在计算机视觉和计算摄影学中的大量应用。

在双边滤波器中，输出像素的值依赖于邻域像素的值的加权组合，

$$g(i,j) = \frac{\sum_{k,l} f(k,l) w(i,j,k,l)}{\sum_{k,l} w(i,j,k,l)}. \tag{3.34}$$

权重系数 $w(i,j,k,l)$ 取决于定义域核(图 3.19c)

$$d(i,j,k,l) = \exp\left(-\frac{(i-k)^2 + (j-l)^2}{2\sigma_d^2}\right), \tag{3.35}$$

和依赖于数据的值域核(图 3.19d)

$$r(i,j,k,l) = \exp\left(-\frac{\|f(i,j) - f(k,l)\|^2}{2\sigma_r^2}\right). \tag{3.36}$$

的乘积。它们相乘后，就会产生依赖于数据的双边权重函数

$$w(i,j,k,l) = \exp\left(-\frac{(i-k)^2 + (j-l)^2}{2\sigma_d^2} - \frac{\|f(i,j) - f(k,l)\|^2}{2\sigma_r^2}\right). \tag{3.37}$$

图 3.20 显示了对含有噪声的阶梯边缘的双边滤波的示例。注意，定义域核是普通的高斯核，值域核量测了与中心像素的表观(亮度)相似性，双边滤波器核是这两个核的乘积。

图 3.20 双边滤波器(Durand and Dorsey 2002)© 2002 ACM：(a)带有噪声的阶梯边缘输入；(b)定义域滤波(高斯)；(c)值域滤波(与中心像素的值的相似度)；(d)双边滤波；(e)阶梯边缘的滤波输出；(f)像素间的 3D 距离

值域滤波器(3.36)使用的是中心像素和邻域像素的矢量距离。对于彩色图像，这点很重要。因为，图像中任意一个彩色通道的边缘给出了材料变化的信号，都会导致此像素的影响权重的降低[①]。

---

① Tomasi and Manduchi (1998)表明使用矢量距离(而不是对每个彩色通道分别进行滤波)将会减少彩色边缘效应。他们同时建议在与感知一致的 CIELAB 彩色空间(2.3.2 节)来考虑彩色差别。

因为双边滤波器与常用的可分离的滤波器相比非常慢，所以出现了一些加速的方法(Durand and Dorsey 2002; Paris and Durand 2006; Chen, Paris, and Durand 2007; Paris, Kornprobst, Tumblin et al. 2008)。不幸的是，这些方法倾向于使用更多的内存(与常用的滤波器相比)，因此不能直接用于对全彩色图像进行滤波。

迭代自适应平滑和各向异性扩散

双边(或其他)滤波器也可以采用迭代的方式来使用，尤其是想获得更"卡通"的表观时(Tomasi and Manduchi 1998)。使用迭代滤波器时，常常使用一些很小的邻域。

作为一个例子，我们只使用四个最邻近的像素，即在公式 3.34 中限制 $|k-i|+|l-j|\leq 1$。可以观察到

$$d(i,j,k,l) = \exp\left(-\frac{(i-k)^2+(j-l)^2}{2\sigma_d^2}\right) \tag{3.38}$$

$$= \begin{cases} 1, & |k-i|+|l-j|=0, \\ \lambda=e^{-1/2\sigma_d^2}, & |k-i|+|l-j|=1. \end{cases} \tag{3.39}$$

因此我们可以将公式(3.34)重写为

$$\begin{aligned}f^{(t+1)}(i,j) &= \frac{f^{(t)}(i,j)+\eta\sum_{k,l}f^{(t)}(k,l)r(i,j,k,l)}{1+\eta\sum_{k,l}r(i,j,k,l)} \\ &= f^{(t)}(i,j)+\frac{\eta}{1+\eta R}\sum_{k,l}r(i,j,k,l)[f^{(t)}(k,l)-f^{(t)}(i,j)],\end{aligned} \tag{3.40}$$

其中 $R=\sum_{(k,l)}r(i,j,k,l)$，(k,l) 是 (i,j) 的 $\mathcal{N}_4$ 邻域。此时这种滤波器的迭代性质将十分明显。

正如 Barash(2002)所述，公式(3.40)与离散的各向异性扩散方程相同，是由 Perona and Malik(1990b)[①]首次提出的。各向异性扩散方程问世以来，已经得到扩展并应用在大量的问题中(Nielsen, Florack, and Deriche 1997; Black, Sapiro, Marimont et al. 1998; Weickert, ter Haar Romeny, and Viergever 1998; Weickert 1998)。有人也指出，公式(3.40)与其他一些自适应平滑方法(Saint-Marc, Chen, and Medioni 1991; Barash 2002; Barash and Comaniciu 2004)以及利用从图像统计数据中得到的非线性平滑条件进行贝叶斯正则化(Scharr, Black, and Haussecker 2003)的方法紧密相关。

值域核更一般的形式是 $r(i,j,k,l)=r(\|f(i,j)-f(k,l)\|)$，它常称为"增益"或"边缘停滞"函数或者称为"扩散系数"，可以是任何的当 $x\to\infty$ 时 $r'(x)\to 0$ 的单调增函数。Black, Sapiro, Marimont et al.(1998)指出各向异性扩散等价于图像梯度鲁棒惩罚函数的最小化(我们将在 3.7.1 节和 3.7.2 节讨论)。Scharr, Black, and Haussecker(2003)给出了怎样按一定的规则从图像的局部统计量推导出边缘停滞函数，同时他们将扩散邻域从 $\mathcal{N}_4$ 扩展到 $\mathcal{N}_8$，以便产生既包含旋转不变性，又有局部结构张量本征值信息的扩散算子。

值得注意的是，如果没有朝向原始图像的偏项，那么各向异性扩散和迭代自适

[①] 因子 $1/(1+\eta R)$ 在各向异性扩散中并没有出现，而是当 $\eta\to 0$ 时可以忽略。

应平滑将会趋于一个恒定的图像。除非迭代次数很少(例如为了加速)，用一个平滑项和一个数据保真项的联合最小化来形式化平滑问题会更合适，在 3.7.1 节和 3.7.2 节对其进行讨论。Scharr, Black, and Haussecker(2003)对此进行了介绍，其中以原理方式引入了此类偏项。

3.3.2 形态学

非线性滤波不但常用于灰度图像和彩色图像的增强，也广泛用于二值图像的处理。二值图像往往出现在图像阈值化操作之后，

$$\theta(f,t) = \begin{cases} 1 & \text{if } f \geq t, \\ 0 & \text{else,} \end{cases} \quad (3.41)$$

例如，在光学字符识别中将扫描的灰度图像首先转化为二值图像，然后做后续处理。

在二值图像中使用最普遍的算子是形态学算子，此类算子可以改变二值图像中物体的形状(Ritter and Wilson 2000, Chapter 7)。为了实现此类算子，我们首先用一个二值的结构元素与二值图像卷积，然后根据卷积结果的阈值选择二值的输出结果(通常，不会采取这种方式来描述形态学算子，但我发现此方式可以简单有效地统一描述这类算子。)结构元素可以是任何形状，既可以是简单的 3×3 方框滤波器，也可以很复杂的圆盘结构。它甚至可以是要在图像中搜寻的一个特殊的形状。

图 3.21 显示了用一个 3×3 的结构元素 s 卷积一个二值图像 f 后，采取下面所述操作所得到的结果。令整数

$$c = f \otimes s \quad (3.42)$$

表示当扫描整个图像时，每个像素运算结果(处于结构元素内)中 1 的个数，整数 S 表示结构元素的大小(像素的个数)。在二值形态学中，标准的操作包括：
- 膨胀：$dilate(f,s) = \theta(c,1)$；
- 腐蚀：$erode(f,s) = \theta(c,S)$；
- 过半：$maj(f,s) = \theta(c,S/2)$；
- 开运算：$open(f,s) = dilate(erode(f,s),s)$；
- 闭运算：$open(f,s) = erode(dilate(f,s),s)$。

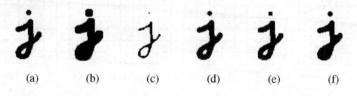

图 3.21 二值图像的形态学：(a)原始图像；(b)膨胀；(c)腐蚀；(d)过半；(e)开运算；(f)闭运算。所有操作使用的结构元素都是 5×5 的正方形。过半操作巧妙地使锐利的角落变圆滑。由于结构元素的宽度不够，开运算并没有将图中的点去除

从图 3.21 中我们可以看出，膨胀使物体扩张(变厚)，而腐蚀使其收缩(变细)。开运算和闭运算往往会保留大部分区域，去掉小的物体或洞，使边界平滑。

在本书中，我们不会过多地使用数学形态学，因为当你需要处理一些阈值化图像时，周围到处都有这种便利工具。关于形态学更加详细的介绍，可以参考其他的计算机视觉和图像处理的教材(Haralick and Shapiro 1992, Section 5.2)(Bovik 2000, Section 2.2)(Ritter and Wilson 2000, Section 7)，也可以参考有关这个主题的论文和书籍(Serra 1982; Serra and Vincent 1992; Yuille, Vincent, and Geiger 1992; Soille 2006)。

3.3.3 距离变换

距离变换通过使用两遍扫描的光栅算法(Rosenfeld and Pfaltz 1966; Danielsson 1980; Borgefors 1986; Paglieroni 1992; Breu, Gil, Kirkpatrick et al. 1995; Felzenszwalb and Huttenlocher 2004a; Fabbri, Costa, Torelli et al. 2008)，可以快速预计算到曲线或点集的距离。它应用广泛，包括水平集(5.1.4 节)，快速的斜切匹配(chamfer matching)(二值图像的配准)(Huttenlocher, Klanderman, and Rucklidge 1993)，图像拼接和图像混合的羽化(9.3.2 节)以及邻近点配准(12.2.1 节)。

二值图像 $b(i,j)$ 的距离变换 $D(i,j)$ 的定义如下：设 $d(k,l)$ 为像素之间的距离函数。常用的距离函数有城街距离或 Manhattan 距离

$$d_1(k,l) = |k| + |l| \tag{3.43}$$

和欧氏距离

$$d_2(k,l) = \sqrt{k^2 + l^2}. \tag{3.44}$$

此时距离变换的定义为

$$D(i,j) = \min_{k,l:b(k,l)=0} d(i-k, j-l), \tag{3.45}$$

即此像素与最邻近的背景像素(值为 0)的距离。

城街距离变换 D_1 的有效计算可以利用正向和逆向两遍简单的光栅扫描算法，如图 3.22 所示。在正向过程中，b 中的每个非 0 像素的值等于它北边或西边邻域像素的距离加上 1 的最小值。在逆向过程中，b 中的每个非 0 像素的值等于此像素的当前值 D 和它南边或东边邻域像素的距离加上 1 之后的三个值中的最小值(图 3.22)。

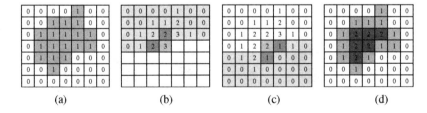

图 3.22 城街距离变换：(a)原始的二值图像；(b)自顶向下(正向)光栅扫描：橙色像素的值是利用绿色像素的值的计算出来的；(c)自底向上(逆向)光栅扫描：绿色像素的值将与原始的橙色像素的值合并；(d)距离变换的最终结果

欧氏距离变换的有效计算很复杂。此时，仅仅在两遍扫描时保持与边界距离的最小标量值是不够的。此时，应该使用正方形距离(斜边)规则，保留和比较由到边

界的距离 x 和 y 坐标组成的向量值。同时，搜索区域需要扩大以得到合理的结果。此处我们不再介绍算法的更多细节(Danielsson 1980, Borgefors 1986)，有兴趣的读者可以参看习题 3.13。

图 3.11g 显示了二值图像的距离变换。注意是如何从黑色(墨色)区域向外增长及原始图像的白色区域如何形成山脊。因为距离变换是从边界像素开始线性增长的，所以它有时也称为"烧草变换"，由于它描述了从黑色区域内部起火后烧到任何给定像素所需的时间，或者称为"斜切"(chamfer)，因为这与木工和工业设计中使用的类似形状相像。距离变换中的山脊形成了变换后区域的骨架(或中轴变换(MAT))，它由到两条(或更多)边界距离相等的像素构成(Tek and Kimia 2003; Sebastian and Kimia 2005)。

符号距离变换是基本距离变换的一个有用的扩展，它计算了所有像素到边界像素的距离(Lavallée and Szeliski 1995)。生成它的最简单方法是分别计算原始二值图像和它的补图的距离变换，然后在结合之前将它们中的一个求负。因为这种距离场一般是平滑的，所以可以使用定义在四叉树或八叉树数据结构上的样条来紧凑地存储它们(允许小的损失达到相对精确)(Lavallée and Szeliski 1995; Szeliski and Lavallée 1996; Frisken, Perry, Rockwood et al. 2000)。这种预先计算的符号距离变换在 2D 曲线和 3D 表面的有效配准和合并中非常有用(Huttenlocher, Klanderman, and Rucklidge 1993; Szeliski and Lavallée 1996; Curless and Levoy 1996)，尤其是如果存储和插值矢量型的距离变换，即从每个像素或体素到最邻近的边界或表面元素的指针。符号距离场也是水平集演化(5.1.4 节)的必要组成部分，此时它们称为"特征函数"。

3.3.4 连通量

另一个有用的半全局的图像操作是检测图像的连通量。连通量定义为具有相同输入值(标签)的邻接像素的区域。(在本节中，两个像素是邻接的是指它们是直接的 4 邻域邻居并且它们的输入值相同)。在许多不同的应用中都要用到连通量，例如在扫描文件中检测单个字母，在阈值化图像中检测物体(也可以称为单元)并计算它们的区域统计量。

考虑图 3.23a 中灰度图像。此图像有四个连通分量：最外层的白色像素集合、大的灰色像素环、白色闭合区域和单独的灰色像素。图 3.23c 是使用粉红色、绿色、蓝色和棕色对图 3.23a 进行伪彩色处理的结果。

为了计算图像的连通量，我们首先(概念上)将图像分裂成邻接像素的水平行程，此时用唯一的标签彩色对行程着色，当两条行程垂直邻近时，则尽可能的使用相同的标签。在第二阶段，将彩色不同但邻接的路径合并。

虽然以上描述很粗略，但是对于学生实现此算法应该足够了(习题 3.14)。Haralick and Shapiro(1992，Section 2.3)详细描述了不同的计算连通量的算法，其中包括一些不用建立大的着色(等值)表的算法。大多数图像处理库都有已经调试好的可以使用的连通量算法。

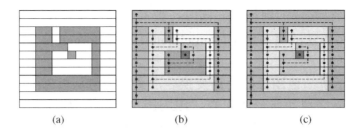

图 3.23 连通量的计算：(a)原始的灰度图像；(b)水平行程(结点)通过垂直(图)边缘(蓝色的虚线)连接；对行程用从父节点继承下来的唯一的彩色作伪彩色处理；(c)合并邻接段后重新着色

一旦一个二值或多值的图像被分割成连通量的形式，对于每个单独的区域 R 计算其统计量是很有用的。这类统计量包括：
- 面积(像素的个数)
- 周长(边界像素的个数)
- 质心(x 和 y 的平均值)
- 二阶矩

$$M = \sum_{(x,y) \in \mathcal{R}} \begin{bmatrix} x - \overline{x} \\ y - \overline{y} \end{bmatrix} \begin{bmatrix} x - \overline{x} & y - \overline{y} \end{bmatrix}, \tag{3.46}$$

通过对 M 的本征值分析可以计算长轴和短轴的方向和长度。[①]

这些统计量可以用于后续的处理，例如采取面积递减(为了优先考虑最大的区域)的方式对区域进行排序，或者对不同图像中的区域进行初步匹配。

3.4 傅里叶变换

在 3.2 节，我们提到傅里叶分析可以分析不同滤波器的频率特征。本节，我们将阐述如何通过傅里叶分析认识这些特征(换句话说，图像的频率信息)和怎样使用快速傅里叶变换(FFT)快速实现大尺寸核的卷积(不受核的尺寸限制)。有关傅里叶变换的全面介绍可参考 Bracewell (1986)；Glassner (1995)；Oppenheim and Schafer (1996)；Oppenheim, Schafer, and Buck (1999)。

对于一个给定的滤波器，我们应该如何分析它对高频、中频和低频的作用？一个可行的简单方案是用一个已知频率的正弦波通过滤波器，观察正弦波变弱程度。令

$$s(x) = \sin(2\pi f x + \phi_i) = \sin(\omega x + \phi_i) \tag{3.47}$$

是输入的正弦波，频率是 f，角频率是 $\omega = 2\pi f$，相位是 ϕ_i。注意，本节中我们将使用变量 x 和 y 表示图像的空间坐标，而不是前几节中的 i 和 j。这样做一方面是因为 i 和 j 用于表示虚数(是否使用取决于你是否在使用复变量或阅读电气工程文献)，另一方面是因为在频率空间中可以很清楚地区分水平(x)部分和垂直(y)分量。本节中，我们使用字母 j 表示虚数，因为在信号处理文献中普遍采用这种形式

[①] 对边界像素，使用格林定理也可以计算矩(Yang and Albregtsen 1996)。

(Bracewell 1986; Oppenheim and Schafer 1996; Oppenheim, Schafer, and Buck 1999)。

如果我们用正弦信号 $s(x)$ 和脉冲响应为 $h(x)$ 的滤波器进行卷积后，我们可以得到另一个频率相同但幅度 A 和相位 ϕ_o 不同的正弦波，

$$o(x) = h(x) * s(x) = A\sin(\omega x + \phi_o), \tag{3.48}$$

如图 3.24 所示。遇到这种情况，请记住卷积可以表示成多个由输入信号平移得到的信号(3.14)的加权和，同一频率的平移正弦波的和正好是这个频率所对应的正弦信号。①新的幅度值 A 称为增益或滤波器的幅度，相位差 $\Delta\phi = \phi_o - \phi_i$ 被称为偏移或相位。

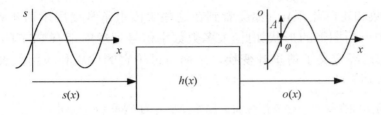

图 3.24 傅里叶变换可以认为是输入信号为正弦信号 $s(x) = e^{j\omega x}$，经过滤波器 $h(x)$ 后，产生的输出响应为正弦信号 $o(x) = h(x) * s(x) = Ae^{j\omega x + \phi}$

实际中，傅里叶变换使用的是更复杂的符号，即正弦波的复数值形式

$$s(x) = e^{j\omega x} = \cos\omega x + j\sin\omega x. \tag{3.49}$$

在这种情况下，它可以简写为，

$$o(x) = h(x) * s(x) = Ae^{j\omega x + \phi}. \tag{3.50}$$

傅里叶变换是对每个频率的幅度和相位响应的简单罗列，

$$H(\omega) = \mathcal{F}\{h(x)\} = Ae^{j\phi}, \tag{3.51}$$

即频率为 ω 的复数正弦波通过滤波器 $h(x)$ 的响应。傅里叶变换对也常常被记作

$$h(x) \overset{\mathcal{F}}{\leftrightarrow} H(\omega). \tag{3.52}$$

不幸的是，式(3.51)并没有给出实际计算傅里叶变换的公式。不过它给出了一个方法，即不断用正弦波和滤波器卷积，观察幅度和相位差。幸好，在连续域和离散域中都存在傅里叶变换式，在连续域中

$$H(\omega) = \int_{-\infty}^{\infty} h(x) e^{-j\omega x} dx, \tag{3.53}$$

在离散域中

$$H(k) = \frac{1}{N} \sum_{x=0}^{N-1} h(x) e^{-j\frac{2\pi k x}{N}}, \tag{3.54}$$

其中 N 是所分析的信号或区域的长度。这些公式既可以用于滤波器，例如 $h(x)$，也可以用于信号或图像，例如 $s(x)$ 或 $g(x)$。

---

① 如果 h 是一般(非线性)变换，此时会引入额外的谐波频率。对于 HiFi 发烧友，这曾经是无法忍受的，因为他们认为设备不应该存在谐波失真。现在，虽然数字音频做到了使声音绝对无失真，但是一些 HiFi 发烧友却在购买复古扬声器或能模拟这种失真的数字信号处理器，由于它们所具有的"更温暖的声音"。

傅里叶变换的离散形式(3.54)也称为"离散傅里叶变换(DFT)"。注意，虽然 k 取任何值，式(3.54)都是可以计算的，但是只有当 $k \in [-\frac{N}{2}, \frac{N}{2}]$ 时，它才有意义。因为当 k 取很大的值时，将与低频混叠，结果并没用提供附加信息，就像前面 2.3.1 节讨论混叠时描述的那样。

从表面上看，DFT 需要执行 $O(N^2)$ 操作(乘-加)。幸好有计算它的快速傅里叶变换 FFT(Fast Fourier Transform)算法，只需要 $O(N\log_2 N)$ 操作 (Bracewell 1986; Oppenheim, Schafer, and Buck 1999)。此处不再阐述算法的细节，只是简单说明一下。FFT 包括一连串的 $\log_2 N$ 的阶段，每个阶段根据半全局的排列实现 2×2 的小变换(用已知系数和矩阵相乘)。(你会看到在这些阶段中常常使用术语"蝶形"，因为在信号处理中所用图形的形状如此)大多数数字信号处理库都有 FFT 的实现。

现在我们已经定义了傅里叶变换，下面将讨论它的性质和应用。表 3.1 列出了一些有用的性质。

表3.1 傅里叶变换的一些有用性质。原始变换对为 $F(\omega) = F\{f(x)\}$

性 质	信 号	变 换
叠加	$f_1(x) + f_2(x)$	$F_1(\omega) + F_2(\omega)$
平移	$f(x - x_0)$	$F(\omega)e^{-j\omega x_0}$
反向	$f(-x)$	$F^*(\omega)$
卷积	$f(x) * h(x)$	$F(\omega)H(\omega)$
相关	$f(x) \otimes h(x)$	$F(\omega)H^*(\omega)$
乘	$f(x)h(x)$	$F(\omega) * H(\omega)$
微分	$f'(x)$	$j\omega F(\omega)$
定义域伸缩	$f(ax)$	$1/aF(\omega/a)$
实值图像	$f(x) = f^*(x) \Leftrightarrow$	$F(\omega) = F(-\omega)$
Parseval 定理	$\sum_x [f(x)]^2 =$	$\sum_\omega [F(\omega)]^2$

下面我们将详细描述这些性质。

- **叠加**：几个信号和的傅里叶变换等于每个信号傅里叶变换的和。因此，傅里叶变换是线性算子。
- **平移**：将一个信号平移后的傅里叶变换等于原始信号的傅里叶变换乘以线性移相(复数正弦波)。
- **反向**：反向信号的傅里叶变换等于原信号的傅里叶变换的共轭。
- **卷积**：两个信号的卷积的傅里叶变换等于它们的傅里叶变换的乘积。
- **相关**：两个信号的相关的傅里叶变换等于第一个信号的傅里叶变换与第二个信号的傅里叶变换的共轭的乘积。
- **乘**：两个信号乘积的傅里叶变换等于它们的傅里叶变换的卷积。
- **微分**：一个信号的微分的傅里叶变换等于此信号的傅里叶变换与频率的乘积。也就是说，微分使高频部分线性放大。

- **定义域缩放**：将一个信号拉伸后的傅里叶变换等于压扁原信号的傅里叶变换，反之亦然。
- **实值图像**：一个实数信号的傅里叶变换关于原点对称。这一点可以用于节省空间，也可以在变换过程中通过对信号的实部和虚部交替扫描使图像的 FFT 速度加倍。
- **Parseval 定理**：一个信号的能量(信号值的平方和)等于它的傅里叶变换的能量。

所有这些特性的证明相当简单(习题 3.15)。在本书的后面它们也将派上用场，例如设计最佳的维纳滤波器(3.4.3 节)，快速实现图像的相关运算(8.1.2 节)。

3.4.1 傅里叶变换对

现在我们放下这些性质，关注一些常用的滤波器和信号的傅里叶变换对，参见表 3.2。

表 3.2 一些有用的傅里叶变换对(连续的)。平移脉冲的傅里叶变换中的虚线指的是它的相位(线性)。其余的所有变换相位是 0(它们是实数值)。注意，表中的图没有按比例绘制，此处的目的只是为了说明滤波器或其响应的一般形状和特征。特别注意，高斯二阶导数是倒转绘制的，因为它很像"墨西哥帽"，所以有时也这样称呼它

名 称	信 号		变 换	
脉冲		$\delta(x)$	$\Leftrightarrow$	1
平移脉冲		$\delta(x-u)$	$\Leftrightarrow$	$e^{-j\omega u}$
方框型滤波器		$\text{box}(x/a)$	$\Leftrightarrow$	$a\,\text{sinc}(a\omega)$
tent		$\text{tent}(x/a)$	$\Leftrightarrow$	$a\,\text{sinc}^2(a\omega)$
高斯		$G(x;\sigma)$	$\Leftrightarrow$	$\frac{\sqrt{2\pi}}{\sigma}G(\omega;\sigma^{-1})$
高斯二阶导数		$(\frac{x^2}{\sigma^4}-\frac{1}{\sigma^2})G(x;\sigma)$	$\Leftrightarrow$	$-\frac{\sqrt{2\pi}}{\sigma}\omega^2 G(\omega;\sigma^{-1})$
Gabor		$\cos(\omega_0 x)G(x;\sigma)$	$\Leftrightarrow$	$\frac{\sqrt{2\pi}}{\sigma}G(\omega\pm\omega_0;\sigma^{-1})$
非锐化掩模		$(1+\gamma)\delta(x)-\gamma G(x;\sigma)$	$\Leftrightarrow$	$(1+\gamma)-\frac{\sqrt{2\pi}\gamma}{\sigma}G(\omega;\sigma^{-1})$
sinc窗函数		$\text{rcos}(x/(aW))$ $\text{sinc}(x/a)$	$\Leftrightarrow$	(参见图3.29)

有关这些变换对更详细的内容如下。
- 脉冲：脉冲的傅里叶变换是常数(所有的频率)。
- 平移脉冲：平移脉冲的傅里叶变换具有单位幅度和线性相位。
- 方框型滤波器：方框型滤波器(移动平均)

$$\text{box}(x) = \begin{cases} 1 & \text{if } |x| \leq 1 \\ 0 & \text{else} \end{cases} \tag{3.55}$$

的傅里叶变换是 sinc 函数

$$\text{sinc}(\omega) = \frac{\sin \omega}{\omega}, \tag{3.56}$$

sinc 函数有无限个旁瓣。sinc 滤波器是理想的低通滤波器。对于非单位的方框型滤波器，方框的宽度 α 和 sinc 函数的零点的间隔 $1/\alpha$ 成反比。

- tent：分段线性的 tent 函数

$$\text{tent}(x) = \max(0, 1 - |x|), \tag{3.57}$$

的傅里叶变换是 sinc^2。

- 高斯：宽度为 σ 的高斯函数(单位面积)，

$$G(x;\sigma) = \frac{1}{\sqrt{2\pi}\sigma} e^{-\frac{x^2}{2\sigma^2}}, \tag{3.58}$$

的傅里叶变换是宽度为 σ^{-1} 的高斯函数(单位高度)。

- 高斯二阶导数：宽度为 σ 的高斯二阶导数，

$$LoG(x;\sigma) = (\frac{x^2}{\sigma^4} - \frac{1}{\sigma^2})G(x;\sigma) \tag{3.59}$$

的傅里叶变换是带通响应

$$-\frac{\sqrt{2\pi}}{\sigma}\omega^2 G(\omega;\sigma^{-1}) \tag{3.60}$$

- Gabor：Gabor 偶函数是频率为 ω_0 的余弦和宽度为 σ 的高斯函数的乘积，它的傅里叶变换是中心位于 $\omega = \pm\omega_0$ 宽度为 σ^{-1} 的两个高斯函数的和。Gabor 奇函数使用的是正弦，它的傅里叶变换是与 Gabor 偶函数类似的两个高斯函数的差。Gabor 滤波器常用作带方向的滤波器和带通滤波器，因为与高斯函数的导数相比，它们可以选择的频率更多。

- 非锐化掩模：式(3.22)介绍了非锐化掩模。它的傅里叶变换是稍微增强了高频的单一响应。

- sinc 窗函数：表 3.2 中的 sinc 窗函数的响应函数与理想的低通滤波器近似，并随着两边的旁瓣增加(W 的增加)两者越接近。图 3.29 显示了此滤波器及其傅里叶变换的形状。在这些例子中，我们使用升余弦的一个旁瓣作为窗函数，

$$\text{rcos}(x) = \frac{1}{2}(1 + \cos \pi x)\text{box}(x), \tag{3.61}$$

这种窗函数也称 Hann 窗。Wolberg(1990)和 Oppenheim, Schafer, and Buck (1999)讨论了其他的窗函数，其中包括 Lanczos 窗(第一个旁瓣是正的正弦函数)。

此外，我们也计算了图 3.14 中小尺寸的离散核的傅里叶变换(表 3.3)。注意，由于移动平均滤波器并没有抑制所有的高频，所以结果中出现了振铃现象。二项式滤波器(Gomes and Velho 1997)在 Burt and Adelson(1983a)的拉普拉斯金字塔(3.5 节)中被用作"高斯函数"。它可以用于高频和低频的分离，但是结果仍然有大量的高频成分，所以在后面的降采样中导致混叠。Sobel 边缘检测器起最初线性地增强频率，但是它随后使高频衰减，结果对于精细边缘的检测会出现问题，例如黑白栏邻接时。在 3.5.2 节中我们将关注更多的小尺寸核的傅里叶变换，研究一些用于降采样(尺寸减小)前的预滤波中的较好的核。

表 3.3 图 3.14 中可分离核的傅里叶变换

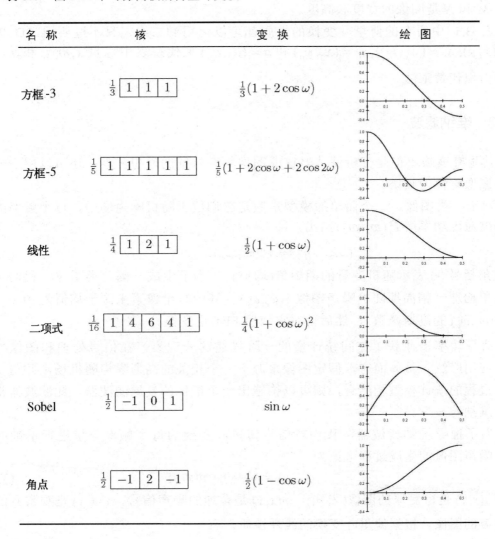

3.4.2 二维傅里叶变换

我们可以将一维信号及其变换中的公式和观点直接扩展到二维图像中。这里，我们不明确指出水平或垂直的频率是 ω_x 还是 ω_y, 只是构造频率 (ω_x, ω_y) 的有方向的正弦曲线,

$$s(x,y) = \sin(\omega_x x + \omega_y y). \tag{3.62}$$

这时, 二维的傅里叶变换的响应为

$$H(\omega_x, \omega_y) = \int_{-\infty}^{\infty} \int_{-\infty}^{\infty} h(x,y) e^{-j(\omega_x x + \omega_y y)} dx\, dy, \tag{3.63}$$

在离散域中为

$$H(k_x, k_y) = \frac{1}{MN} \sum_{x=0}^{M-1} \sum_{y=0}^{N-1} h(x,y) e^{-j 2\pi \frac{k_x x + k_y y}{MN}}, \tag{3.64}$$

其中 M 和 N 是图像的宽度和高度。

表 3.1 中所有的傅里叶变换的性质都可以运用到二维, 只不过要用 2D 向量 $x=(x,y)$, $w=(w_x, w_y)$, $x_0 = (x_0, y_0)$ 和 $a=(a_x, a_y)$ 来代替表中标量 x, ω, x_0 和 a, 用向量内积代替乘法。

3.4.3 维纳滤波

傅里叶变换不但是分析滤波器核或图像频率特征的理想工具, 还可以用于分析一类图像整体的频谱。

对于一类图像, 一个简单的模型是假定它们位于随机噪声场中, 每个频率的期望幅度通过功率谱 $P_s(\omega_x, \omega_y)$ 给出, 即

$$\langle [S(\omega_x, \omega_y)]^2 \rangle = P_s(\omega_x, \omega_y), \tag{3.65}$$

其中角括号 $\langle \cdot \rangle$ 表示随机变量的期望值(均值)。[1] 为了生成一幅此类图像, 我们可以先简单构造一幅高斯随机噪声图像 $S(\omega_x, \omega_y)$, 其中每个像素来源于均值为 0, 方差为 $P_s(\omega_x, \omega_y)$ 的高斯函数[2], 然后再对它使用 FFT 逆变换。

信号功率谱捕获了空间统计量的一阶描述这一观察, 在信号处理和图像处理中, 应用广泛。具体而言, 假定图像来源于一个相关的高斯噪声随机场, 并与一个测量过程的统计模型相结合, 则可以构造出一个最优的复原滤波器, 此滤波器常称为"维纳滤波"。[3]

为了推导维纳滤波器, 我们对信号傅里叶变换的每个频率分量进行单独的分析。噪声图像的生成过程可记为

$$o(x,y) = s(x,y) + n(x,y), \tag{3.66}$$

其中 $s(x,y)$ 是待复原的图像(未知), $n(x,y)$ 是叠加的噪声信号, $o(x,y)$ 是观察到的含有噪声的图像。根据傅里叶变换的线性性质, 有

---

[1] 经常也使用符号 $E[\cdot]$。
[2] 我们将直流分量(即常量)设置在 $S(0,0)$ 处, 其值为灰度值的平均值。高斯噪声的生成代码见附录 C.2 的算法 C.1。
[3] Wiener 的发音是 "veener", 因为在德国 "w" 发 "v" 的音。记得下次一定要点 "Wiener schnitzel"。

$$O(\omega_x, \omega_y) = S(\omega_x, \omega_y) + N(\omega_x, \omega_y), \tag{3.67}$$

其中上式中的每一项都是其相对应图像的傅里叶变换。

对于每个频率 (ω_x, ω_y)，从我们的图像谱中我们得知未知的变换分量 $S(\omega_x, \omega_y)$ 服从 0 均值，方差为 $P_s(\omega_x, \omega_y)$ 的高斯函数的先验分布。此外，我们得知噪声测量 $O(\omega_x, \omega_y)$ 的方差为 $P_n(\omega_x, \omega_y)$，即噪声的功率谱，通常假定它为常数(白噪声)，$P_n(\omega_x, \omega_y) = \sigma_n^2$。

根据贝叶斯法则(附录 B.4)，后验估计量 S 可以写成

$$p(S|O) = \frac{p(O|S)p(S)}{p(O)}, \tag{3.68}$$

其中 $p(O) = \int_S p(O|S)p(S)$ 是一个规范化常数，用于使 $p(S|O)$ 是一个合适的分布(积分为1)。先验分布 $p(S)$ 如下

$$p(S) = e^{-\frac{(S-\mu)^2}{2P_s}}, \tag{3.69}$$

其中 μ 为当前频率的期望的均值(除了原点，其余地方都为0)，测量分布 $P(O|S)$ 如下

$$p(O|S) = e^{-\frac{(S-O)^2}{2P_n}}. \tag{3.70}$$

对公式(3.68)两边分别取负对数，为了简便，令 $\mu = 0$，我们可以得到

$$\begin{aligned}-\log p(S|O) &= -\log p(O|S) - \log p(S) + C \tag{3.71}\\ &= 1/2 P_n^{-1}(S-O)^2 + 1/2 P_s^{-1} S^2 + C, \tag{3.72}\end{aligned}$$

此式称为"负的后验对数似然函数"。这个量的极小值很容易计算，

$$S_{\text{opt}} = \frac{P_n^{-1}}{P_n^{-1} + P_s^{-1}} O = \frac{P_s}{P_s + P_n} O = \frac{1}{1 + P_n/P_s} O. \tag{3.73}$$

则最优的维纳滤波器的傅里叶变换是

$$W(\omega_x, \omega_y) = \frac{1}{1 + \sigma_n^2/P_s(\omega_x, \omega_y)} \tag{3.74}$$

它就是去掉功率谱为 $P_s(\omega_x, \omega_y)$ 的图像噪声所需要的滤波器。

注意，此滤波器有一个定性的性质，即对于低频 $(P_s \gg \sigma_n^2)$，它具有单位增益，而对于高频，它使噪声减弱 P_s/σ_n^2 倍。图 3.25 显示了一维变换 $W(f)$ 和在通常的假定 $P(f) = f^{-2}$ (Field 1987)下滤波器核 $\omega(x)$ 的响应。习题 3.16 让你比较作为降噪算法的维纳滤波器与手动调节的高斯平滑。

以上推导维纳滤波器的方法可以很容易地扩展到观察图像是由原始图像经过噪声模糊后得到的情况，

$$o(x, y) = b(x, y) * s(x, y) + n(x, y), \tag{3.75}$$

其中 $b(x, y)$ 是已知的模糊核。此处我们不再推导此维纳滤波器的响应，而是将其作为一个习题(习题 3.17)。同时，我们也希望读者能将此方法与非锐化掩模和朴素逆滤波去模糊的方法进行比较。有关去模糊的更精致的方法，我们将在 3.7 节和 10.3 节讨论。

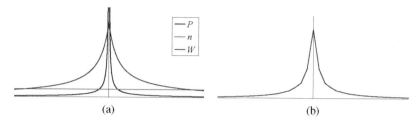

图 3.25 一维维纳滤波器：(a)信号功率谱 $P_s(f)$，噪声水平 σ^2 和维纳滤波器变换 $W(f)$；(b)维纳滤波器的空间核

离散余弦变换

离散余弦变换(DCT)是傅里叶变换的一个变种，它特别适合于目前流行的以块为单位的图像压缩。一维 DCT 的计算是通过宽度为 N 的块的每个像素与一系列不同频率的余弦值的点积来实现的，

$$F(k) = \sum_{i=0}^{N-1} \cos\left(\frac{\pi}{N}(i+\frac{1}{2})k\right) f(i), \tag{3.76}$$

其中 k 是系数(频率)索引，像素偏移量的 1/2 作为基对称系数(Wallace 1991)。图 3.26 显示了一些离散余弦基函数。你可以看到，第一个基函数(蓝色直线)是对块中像素的平均 DC 值进行编码，第二个是对稍微弯曲的倾斜线进行编码。

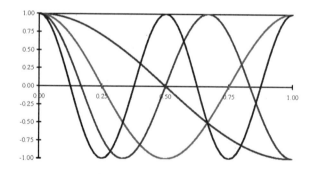

图 3.26 离散余弦变换(DCT)的基函数：第一个 DC(即常数)基是蓝色的水平线，第二个是棕色的半周期波形，等等。这些基广泛应用于图像和视频压缩标准，例如 JPEG

事实上，DCT 是自然图像中具有统计意义的一些小的区域的最优 Karhunen-Loeve 分解(可以通过对图像进行主分量分析 PCA 获得，14.2.1 节将介绍 PCA)的近似。KL 变换有效地对信号去相关(假定信号是用它的功率谱来描述的)，因此，从理论上它是达到最优压缩的。

类似的，二维 DCT 变换如下，

$$F(k,l) = \sum_{i=0}^{N-1} \sum_{j=0}^{N-1} \cos\left(\frac{\pi}{N}(i+\frac{1}{2})k\right) \cos\left(\frac{\pi}{N}(j+\frac{1}{2})l\right) f(i,j). \tag{3.77}$$

与 2D 的 FFT 一样，2D 的 DCT 是可分离实现的，即首先计算块中每行的

DCT，接下来对得到的结果计算每列的 DCT。与 FFT 一样，每个块的 DCT 的计算时间是 $O(N \log N)$。

正如我们在 2.3.3 节所述，DCT 广泛应用于目前图像和视频的压缩算法中，尽管它正逐渐被小波算法(Simoncelli and Adelson 1990b)(将在 3.5.4 节讨论)和 DCT 的有交叠的变种(Malvar 1990, Malvar 1998, Malvar 2000)(用于 JEPG XR 的新标准中[①])取代。JPEG 中由于每个块(常用的是 8×8)的变换和量化是独立的，因而会产生很多人为的区块效应(明显的边缘不连续)，而上述的这些新算法受此影响很小。习题 3.30 是针对去除 JPEG 压缩图像中人为区块效应的一些想法。

3.4.4 应用：锐化，模糊和去噪

图像处理的另一个普遍应用是通过使用锐化和去噪声的算子增强图像，此时需要特定的邻域处理。传统的方法，常采用线性滤波算子(参见 3.2 节和 3.4.3 节)。现在，普遍采用非线性滤波器(3.3.1 节)，例如加权中值或双边滤波器(3.34～3.37)、各向异性扩散(3.39～3.42)和非局部均值(Buades, Coll, and Morel 2008)。3.7.1 节介绍的变分方法，尤其是使用非二次(鲁棒)范式例如 L_1 范式(也称为"总变差")也经常会用到。图 3.19 显示了一些使用线性和非线性的滤波器去除噪声的例子。

度量图像去噪算法的效果时，一般会给出类似 PSNR(峰值信噪比)(2.119)的测量结果，其中 $I(x)$ 是原始图像(无噪声)，$\hat{I}(x)$ 是去噪后的图像；此方法主要针对噪声图像是由合成所产生的情况，这时无噪声的图像是已知的。一个更好的度量去噪质量的方法是使用基于感知相似性度量的策略，例如结构相似性(SSIM)索引(Wang, Bovik, Sheikh *et al.* 2004; Wang, Bovik, and Simoncelli 2005)。

习题 3.11，3.16，3.17，3.21 和 3.28 是一些这类算子的实现，你可以比较它们的效果。在 10.3 节，我们将讨论在模糊去除和与超分辨率相关任务中更复杂的方法。

3.5 金字塔与小波

迄今为止，本章中所研究的所有图像变换所产生的输出图像都与输入图像大小相同。但我们时常希望改变图像的分辨率以便进一步处理。例如，我们可能需要对一幅小图像进行插值使其分辨率与输出打印机或电脑屏幕的分辨率相符。或者，我们希望减小图像大小来加速算法执行或节省存储空间或传输时间。

有时，我们甚至不知道待处理图像应该有怎样的分辨率。例如，考虑在一幅图像中寻找人脸的任务(14.1.1 节)。由于我们并不知道人脸可能出现的尺寸，所以需要生成一个不同大小的图像组成的金字塔，扫描其中每一幅图像来寻找可能的人

[①] http://www.itu.int/rec/T-REC-T.832-200903-I/en

脸。生物视觉系统也会处理分层次的尺寸(Marr 1982)。这样一个金字塔也可以通过先在金字塔较粗级别找到一个较小实例,再仅在粗级别检测附近寻找全分辨率物体用来帮助加速(8.1.1 节)。最后,图像金字塔对于在保持细节的条件下进行图像融合等多尺度编辑操作非常有用。

在本节,我们首先讨论用于改变图像分辨率的好的滤波器,即上采样(插值,3.5.1 节)和下采样(降采样,3.5.2 节)。之后我们表述了多分辨率金字塔的概念,这可用于创造不同大小图像的一个完整的层次结构,并且使得一系列应用成为可能(3.5.3 节)。一个密切相关的概念就是小波,这个概念是一种有高频选择性和其他有用性质的特殊金字塔(3.5.4 节)。最后,我们给出了一个金字塔的有效应用,即以隐藏图像边界之间缝隙的方式将不同图像融合起来。

3.5.1 插值

要将一幅图像插值(也称"上采样")到较高分辨率,我们需要选择一些插值核来卷积图像,

$$g(i,j) = \sum_{k,l} f(k,l)h(i-rk, j-rl). \tag{3.78}$$

这个公式与离散卷积公式有关(3.14),我们只是将 $h()$ 中的 k 和 l 替换为 rk 和 rl,其中,r 是上采样率。图 3.27a 展现了如何将这个过程看作中心在每个输入样例 k 的采样带权插值核的叠加。另一种心智模型如图 3.27b 所示,核中心位于输出像素值 i(两种形式等价)。第二种形式有时也称为"多相滤波器"形式,由于核的值 $h(i)$ 可以存储为 r 个不同的核,每个都是为了与输入样例进行卷积而选择的,这些样例依赖于与上采样网格相关的相位 i。

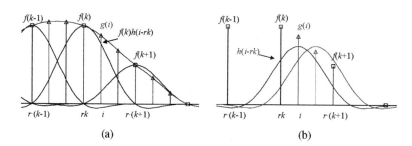

图 3.27 信号插值,$g(i) = \sum_k f(k)h(i-rk)$:(a)输入值的带权求和;(b)多相滤波器描述

什么样的核组成好的插值器?答案取决于具体的应用和所需的计算时间。表 3.2 和表 3.3 所示的任何一种平滑核都可以在合适的尺度变换后使用。[①]线性插值器(对应于帐篷(tent)核)产生分段线性插值曲线,这在应用于图像时会产生讨厌的褶皱(图 3.28a)。三次 B 样条的离散 1/2-像素采样在表 3.3 中看起来像是二项核,是一个

---

① 表 3.3 中的平滑核有一个单位面积。要将它们变成插值核,我们仅需要将它们按插值率 r 进行尺寸放大。

产生有较少高频细节软图像的近似核(插值图像不通过输入数据点)。三次 B 样条的方程最容易通过 tent 函数(线性 B 样条)与它本身进行卷积推导得到。

图 3.28 二维图像插值：(a)双线性；(b)双三次(a=-1)；(c)双三次(a=-0.5)；(d)窗 sinc(九阶)

大多数显卡使用双线性核(可选的与 MIP-映射结合——见 3.5.3 节)，而大多数图片编辑包使用双三次插值。三次插值器是一个 C_1(导数连续)分段-三次样条("样条"的概念和"分段多项式"相似)[①]，可用如下方程表示

$$h(x) = \begin{cases} 1 - (a+3)x^2 + (a+2)|x|^3 & \text{if } |x| < 1 \\ a(|x|-1)(|x|-2)^2 & \text{if } 1 \leq |x| < 2 \\ 0 & \text{otherwise,} \end{cases} \quad (3.79)$$

其中，a 指定 x = 1 处的导数(Parker，Kenyon and Troxel 1983)。a 的值通常设为-1，因为这样可以最好的匹配 sinc 函数的频率特性(图 3.29)。这也引入了少量的锐化，视觉上看起来更好一些。不幸的是，这个选择并不能线性插值直线(亮度斜坡带)，所以可能会出现一些可见的振铃效应。进行大量插值更好的一个选择是 a = -0.5，这会产生一个二次再生样条；它精确的插值线性和二次函数(Wolberg 1900，5.4.3 节)。图 3.29 展示了 a=-1 和 a=-0.5 的三次插值核及其傅里叶变换；图 3.28b 和 c 显示了它们应用于二维插值的情形。

样条可以精确确定控制点的导数并使用有效增量算法进行评估，因此长期以来被应用于函数和数据值插值(Bartels, Beatty and Barsky 1987；Farin 1992, 1996)。样条被广泛应用于几何建模和计算机辅助设计(CAD)应用，但是它们已经开始被子分表面替代(Zorin, Schroder, and Sweldens 1997；Peters and Reif 2008)。在计算机视觉

[①] "样条"的概念源自 draughtsman 的车间，它是一条用于绘制光滑曲线的柔性木头或金属。

中，样条常常可以用于弹性图像变形(3.6.2 节)、运动估计(8.3 节)和表面插值(12.3 节)。实际上，可以通过将图像表示为样条并用多分辨率框架处理它们来实现大部分的图像处理操作(Unser 1999)。

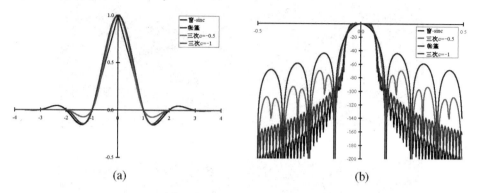

图 3.29 (a)一些窗 sinc 函数和(b)它们的 log 傅里叶变换：蓝色为提升余弦窗 sinc，绿色和紫色为三次插值器(a=-1 和 a=-0.5)，棕色为 tent 函数。它们都常用于进行高精度低通滤波操作

窗 sinc 函数通常被认为是品质最高的插值器，因为它既可以保留低分辨率图像的细节，又可以避免混叠。(也可以通过匹配过零点处的导数建立一个窗 sinc 的 C^1 分段-三次近似(Szeliski and Ito 1986))。但是，有些人不喜欢窗 sinc 引入的过度振铃效应和振铃频率的重复性(图 3.28d)。所以，有些摄影师比较喜欢多次少量重复插值(这往往会使原像素网格和最终图像不相关)。其他的可能包括使用双边滤波器作为插值器 (Kopf, Cohen, Lischinski *et al.* 2007)，用全局优化(3.6 节)或幻生(hallucinating)细节(10.3 节)。

3.5.2 降采样

插值可以用于增加图像的分辨率，而降采样则要求降低分辨率。[①]为了进行抽取，我们首先(概念上的)用一个低通滤波器来卷积图像(以避免混叠)，然后保持每第 r 个样例。在实际中，我们常常仅在每第 r 个样例处计算卷积，

$$g(i,j) = \sum_{k,l} f(k,l)h(ri-k, rj-l), \qquad (3.80)$$

如图 3.30 所示。注意，在这个情况下，平滑核 $h(k, l)$ 常常是一个插值核的拉伸和重缩放版本。或者，我们可写为

$$g(i,j) = \frac{1}{r}\sum_{k,l} f(k,l)h(i-k/r, j-l/r) \qquad (3.81)$$

插值和降采样中使用的核函数 $h(k, l)$ 是相同的。

一个常用的($r=2$)降采样滤波器是 Burt and Adelson(1983a)引入的二项滤波器。

---

① "降采样"的概念有一个可怕的词源，它与杀掉一个罗马单位中每第十个由于胆小而获罪的士兵有关。在信号处理中，它常用于表示任何一种下采样或降低比率操作。

如表 3.3 所示，这个核很好地分离了高频和低频，但仍旧留下了大量高频细节，这样会在下采样后产生混叠。但对于图像融合之类的应用(将在本节后面进行讨论)，这种混叠并不要紧。

但是，如果当下采样图像会被直接显示给用户，或者可能与其他分辨率图像进行融合(如在 MIP-映射中，3.5.3 节)，就需要更高质量的滤波器。对于高下采样率，窗 sinc 前置滤波器是一个很好的选择(图 3.29)。但是，对于低下采样率，例如 $r = 2$，就需要更仔细地设计滤波器。

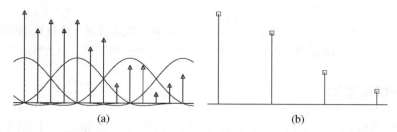

图 3.30 信号降采样：(a)原始信号；(b)进行下采样前与低通滤波器卷积过的信号

表 3.4 显示了一些常用的 $r = 2$ 下采样滤波器。

表 3.4 2×降采样滤波器的系数。这些滤波器都是长度为奇数、对称、归一化到有单位 DC 增益(和为 1)的。它们相关的频率响应如图 3.31 所示

$\|n\|$	线性	二项	三次 $a=-1$	三次 $a=-0.5$	窗sinc	QMF-9	JPEG 2000
0	0.50	0.3750	0.5000	0.50000	0.4939	0.5638	0.6029
1	0.25	0.2500	0.3125	0.28125	0.2684	0.2932	0.2669
2		0.0625	0.0000	0.00000	0.0000	-0.0519	-0.0782
3			-0.0625	-0.03125	-0.0153	-0.0431	-0.0169
4					0.0000	0.0198	0.0267

图 3.31 显示了它们相应的频率响应。这些滤波器如下所述。

- 线性(1, 2, 1)滤波器的响应相对较差。
- 二项(1, 4, 6, 4, 1)滤波器阻断了很多频率但是对于计算机视觉分析金字塔很有用。
- 式(3.79)中的三次滤波器；$a = -1$ 滤波器与 $a = -0.5$ 滤波器相比下降更快(图 3.31)。
- 一个余弦-窗 sinc 函数(表 3.2)。
- Simoncelli and Adelson(1990b)的 QMF-9 滤波器多用于小波去噪，有少许混叠(注意，原始滤波器的系数被归一化到增益的 $\sqrt{2}$ 倍，所以它们是"自反的"。
- JPEG 2000(Taubman and Marcellin 2002)中的 9/7 分析滤波器。

这些系数的一些全精度取值请参见原始论文。

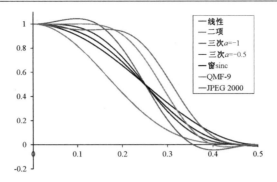

图 3.31 一些 2×降采样滤波器的频率响应。三次 $a=-1$ 滤波器下降最快但同时有一点振铃效应;小波分析滤波器(QMF-9 和 JPET2000)在压缩中很有用,但会有较多的混叠

3.5.3 多分辨率表达

既然我们已经描述了插值和降采样算法,我们就可以建立一个完全的图像金字塔(图 3.32)。正如我们之前所提到的,金字塔可用于加速由粗到精的搜索算法,以便在不同尺度上寻找物体或模式或者进行多分辨率融合操作。它们在计算机图形学硬件和软件中也广泛用于使用 MIP-映射来进行微小级别降采样,我们将在 3.6 节涉及这个问题。

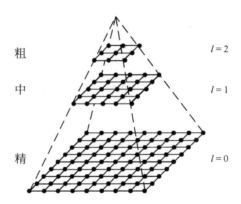

图 3.32 一个传统的图像金字塔:每级有其上一级一半的分辨率(宽度和高度),也就是四分之一的像素

计算机视觉中最有名(而且可能应用最广)的金字塔是 Burt and Adlson(1983a)的拉普拉斯(Laplacian)金字塔。为了建造这个金字塔,我们首先用大小为二的因子对原图像进行模糊和二次采样,并将它储存在金字塔的下一级(图 3.33)。由于金字塔内的邻接级是通过 $r=2$ 的采样率联系的,这种金字塔称为"八度金字塔"。Burt 和 Adelson 最初提出了一个形如下式的的五阶核

$$\boxed{c\ b\ a\ b\ c},\tag{3.82}$$

其中 $b=1/4$,$c=1/4-a/2$。在实际中,$a=3/8$,就会得到我们所熟悉的二项核,

$$\frac{1}{16}\boxed{1\ 4\ 6\ 4\ 1},\tag{3.83}$$

这个核用移位和相加非常容易实现。(这在乘法器很昂贵的时代很重要。)由于重复卷积二项核会收敛到高斯，所以他们称这个产生的金字塔为"高斯金字塔"。[①]

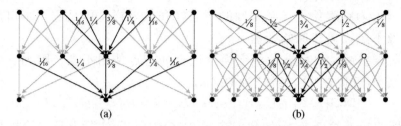

图 3.33 高斯金字塔的信号处理原理图表述：(a)分析和(b)重新合成阶段以使用相似计算的方式展示。白色圆圈表示通过↑2 上采样操作插入的零值。注意，重建滤波器系数是分析系数的两倍。计算过程如页面下行所示，不管我们是由粗到精或反过来进行这个过程

为了计算拉普拉斯金字塔，Burt 和 Adelson 首先对一幅分辨率较低的图像进行插值得到原图像的一个重建的低通版本(图 3.34b)。之后，从原图中减去这个低通版本得到带通拉普拉斯图像，存储这个结果以进行进一步处理。结果得到的金字塔有完美重建，即拉普拉斯图像加上基础级别的高斯(图 3.34b 中 L_2)足以精确重建原始图像。图 3.33 显示了同样的计算在一维情况下作为一个信号处理原理图的情形，这完全捕捉到了在分析和重新合成阶段中所进行的计算。

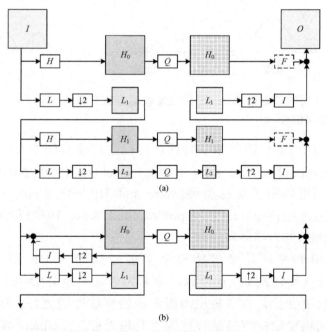

图 3.34 拉普拉斯金字塔：(a)处理阶段图像的概念流：图像通过高通和低通滤波，低通滤波过的图像在金字塔的下一阶段进行处理。在重建过程中，插值图像和(选择性滤波的)高通图像被一同加回去。Q 盒表示量化或一些其他金字塔处理，例如，通过核化降噪(将小的小波值设为 0)进行噪声去除。(b)高通滤波器的实际计算包括首先对下采样的低通图像进行插值，然后减去它。这在 Q 是恒等时会产生完美的重建。高通(或带通)图像一般称为拉普拉斯图像，而低通图像称为高斯图像

[①] 而且，这适用于任何平滑核(Wells 1986)。

Burt 和 Adelson 也描述了一个拉普拉斯金字塔的变形,用原始模糊图像替代重建金字塔来构造低通图像(将 L 盒的输出直接传送到图 3.34b 中的减法器)。这个变形走样更少,因为它避免了一个来回的下采样和上采样,但是它不是自反的,因为拉普拉斯图像不足以重建原始图像。

至于高斯金字塔,"拉普拉斯变换"这个词有点使用不当,因为它们的带通图像实际上是高斯(近似)差分,或记为 DoGs,

$$\text{DoG}\{I; \sigma_1, \sigma_2\} = G_{\sigma_1} * I - G_{\sigma_2} * I = (G_{\sigma_1} - G_{\sigma_2}) * I. \tag{3.84}$$

一个高斯的拉普拉斯变换(如我们在式(3.26)中所见)实际上是二阶导数,

$$\text{LoG}\{I; \sigma\} = \nabla^2(G_\sigma * I) = (\nabla^2 G_\sigma) * I, \tag{3.85}$$

其中

$$\nabla^2 = \frac{\partial^2}{\partial x^2} + \frac{\partial^2}{\partial y^2} \tag{3.86}$$

是一个函数的拉普拉斯(算子),图 3.35 展示了在空间和频率中高斯差分和高斯的拉普拉斯的样子。

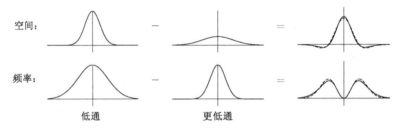

图 3.35 两个低通滤波器的差是一个带通滤波器。蓝色虚线显示了对一个高斯的半倍频程拉普拉斯密切拟合

高斯的拉普拉斯有着优雅的数学特性,这已在尺度空间中有广泛的研究(Witkin 1983; Witkin, Terzopoulos, and Kass 1986; Lindeberg 1990; Nielsen, Florack, and Deriche 1997),也可以用于边缘检测(Marr and Hildreth 1980; Perona and Malik 1990b)、立体视觉匹配(Witkin, Terzopoulos, and Kass 1987)和图像增强(Nielsen, Florack, and Deriche 1997)等多种应用。

一个不太常用的变种是半倍频程金字塔,如图 3.36a 所示。这些最初由 Crowley and Stern(1984)引入视觉领域,命名为"低通差分(DOLP)"变换。由于相邻级别中微小的尺度变化,作者指出由粗到精的算法性能更好。在图像处理领域,半倍频程金字塔同棋盘抽样网格结合称为"五角采样"(Feilner, Van De Ville, and Unser 2005)。在检测多尺度特征时(4.1.1 节),常使用半倍频程甚至四分之一频程的金字塔(Lowe 2004;Triggs 2004)。但是,在这种情况下,子抽样(下采样,subsampling)只在每半倍级别发生,即使用宽高斯重复模糊图像,直到得到一个完全半倍的分辨率改变(图 4.11)。

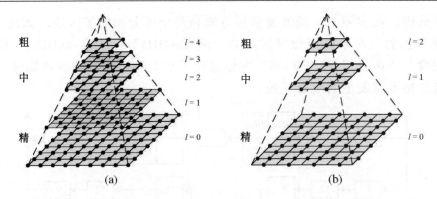

图 3.36 多分辨率金字塔：(a)半倍频程(五角)采样的金字塔(为了清楚表述，奇数级别被涂上灰色)；(b)小波金字塔-每个小波级别存储 3/4 的原始像素(通常是水平、垂直和混合梯度)，这样一来小波系数和原始像素的总数相同

3.5.4 小波

尽管金字塔在计算机视觉中有非常广泛的应用，但仍然有一些人选择小波分解作为替代。小波是在空间域和频域都定位一个信号的滤波器(类似表 3.2 中的 Gabor 滤波器)，而且它是在不同层次的尺度上定义的。小波提供了一种平滑的方式在不产生块效应的同时把信号分解为频率成分，并且它与金字塔密切相关。

小波最初产生于应用数学和信号处理领域，由 Mallat(1989)引入计算机视觉领域。Strang(1989)；Simoncelli and Adelson(1990b)；Rioul and Vetterli(1991)；Chui(1992)；Meyer(1993)都对这个问题提供了很好的介绍以及历史回顾，而 Chui(1992)提供了一个更好理解的回顾和应用综述。(Sweldens(1997))描述了比较新近的小波提升(lifting)方法，我们很快要对此进行讨论。

小波在计算机图形学领域广泛应用于进行多分辨率几何处理(Stollnitz, DeRose, and Salesin 1996)，在计算机视觉中也有相似的应用(Szeliski 1990b; Pentland 1994; Gortler and Cohen 1995; Yaou and Chang 1994; Lai and Vemuri 1997; Szeliski 2006b)，此外还用于多尺度有向滤波(Simoncelli, Freeman, Adelson *et al.* 1992)和去噪(Portilla, Strela, Wainwright *et al.* 2003)。

既然图像金字塔和小波都将一幅图像分解为空间和频率内的多分辨率描述，它们有什么不同呢？通常的答案是，传统金字塔过于完备，即它们比原图使用更多像素来描述图像分解，而小波提供了一个紧致框架，即它们保持分解图像与原图像大小相等(图 3.36b)。但是实际上，有一些小波族也是过完备的，以提供更好的移位能力或者方向导向(Simoncelli, Freeman, Adelson *et al.* 1992)。因此，更好的区别可能是，与常规的带通金字塔相比，小波的方向选择性更佳。

二维小波是如何构造的？图 3.37a 展示了一个沿着互补重建(合成)阶段(迭代的)由粗到精构造(分析)流水线的一个阶段的高层次流程图。在这个流程图里，降采样之前的高通滤波器保持 3/4 的原始像素，而 1/4 的低频系数被传递到下一个阶段以作进一步分析。在实际中，滤波通常被分解为两个可分离的子阶段，如图 3.37b 所示。这样得到的三个小波图像常常称为"高-高(HH)"，"高-低(HL)"和"低-高(LH)"图像。"高-低"和"低-高"图像强调了水平和垂直的边缘和梯度，而"高-高"图像包括不常发生的混合导数。

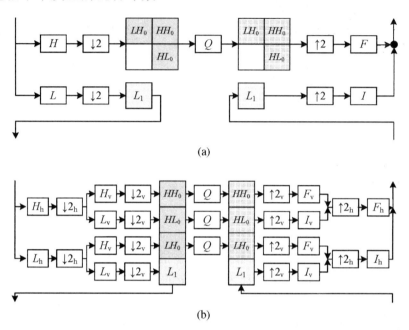

图 3.37　二维小波分解：(a)高层次流程图用单盒表示低通和高通变换；(b)可分离的实现，包括首先水平后垂直的进行小波变换。I 和 F 盒是从它的小波分量中重新合成图像所需要的插值和滤波盒

如何选择图 3.37b 中所示的高通 H 和低通 L 滤波器，如何计算相应的重建滤波器 I 和 F？滤波器可以设计为全部有有限脉冲响应吗？二十多年来，这一直是小波领域的研究主题。答案很大程度上依赖于目标应用，例如，小波是否被用于压缩、图像分析(特征查找)或者降噪。(Simoncelli and Adelson(1990b))展示了(表 4.1 中)一些实际应用中表现很好的奇数长度的正交镜像滤波器(QMF)系数。

由于小波滤波器的设计是一种微妙的艺术，是否可能有一种更好的方式？实际上，一个更简单的过程是将信号分为偶数和奇数分量，然后在每个序列上进行一般的可逆滤波操作来产生所谓的提升小波(图 3.38 和图 3.39)，Sweldens(1996)针对第二代小波提升方案提供了一个很好理解的介绍以及一个全面的回顾(Sweldens 1997)。

如图 3.38 所示，提升方案首先将序列分为奇数和偶数子分量，而不是先用高通滤波器和低通滤波器对整个输入序列(图像)滤波然后保持奇数子序列和偶数子序列。用一个低通滤波器 L 来滤波偶序列然后从偶序列中减去得到的结果，这一般是可逆的：简单进行同样的滤波，然后将结果加回去即可。进一步，这个操作可以本地进行，其结果明显地节省了空间。这同样适用于使用校正滤波器 C 对偶序列进行滤波,可以确保偶序列是低通的。一系列这种提升步骤可以用于在低计算消耗和保证可逆性的情况下产生更复杂的滤波响应。

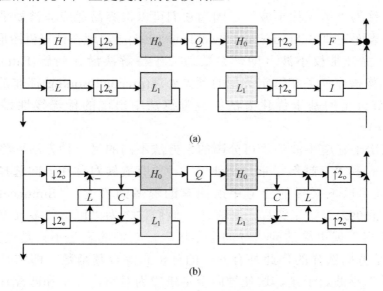

图 3.38 一维小波变换:(a)通常的高通+低通滤波器后接奇($\downarrow 2_o$)和偶($\downarrow 2_e$)下采样;(b)提升版本,首先选择奇数和偶数序列,然后以一种容易地可逆的方式应用一个低通预测阶段 L 和一个高通校正阶段 C

考虑图 3.39 中信号处理流程图有利于我们充分理解这个过程。在分析过程中,从奇数值中减去偶数值的平均值来得到一个高通小波系数。但是,偶数样例中仍旧包括低频信号的混淆样例。要对此进行补偿,一小部分高通小波被加回偶数序列,这样使它是适当地低通滤波过的。(很容易证明,有效的低通滤波是[-1/8,1/4,3/4,1/4,-1/8]),它的确是一个低通滤波器。)在综合中,同样的操作被颠倒过来并带有判断正确的符号改变。

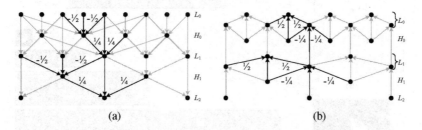

图 3.39 提升变换表示为信号处理流程图:(a)分析阶段首先从偶数邻居中预测奇数值,存储差分小波,然后通过加入小波的片段来补偿较粗糙的偶数值;(b)合成阶段简单地反转计算流和一些滤波器和操作的符号。浅蓝色线条展示了用四阶代替二阶进行预测和纠正的状况

当然,我们不需要将自己局限于二阶滤波。图 3.39 用浅蓝色箭头表示不影响其可逆转性情况下可以选择性加到提升方案中的额外的滤波系数。实际上,低通和高通滤波操作可以互换,例如,我们可以先在奇序列(加上中间值)上使用一个五阶

三次低通滤波器，后接一个四阶三次低通预测器来估计小波，不过我没见过有人记载这个方案。

提升小波称为"第二代小波"，因为它们可以很容易适应非常规采样拓扑，例如，多分辨率表面处理(Schröder and Sweldens 1995)等出现在计算机图形学应用中的问题。此外，提升带权小波，即，系数适应于待解决潜在问题(Fattal 2009)的小波，对低层图像操作任务和特定类型的预先处理(preconditioning)都非常有效，后者是指对 3.7 节讨论的基于优化方法的视觉算法中出现的稀疏线性系统的预处理(Szeliski 2006b)。

与广泛应用于建造小波的"可分离的"方法不同的另一种方法，将每个级别分解为水平、竖直和"穿越"子带，是使用一种更有旋转对称和方向选择性且也避免在 Nyquist 频率以下采样的信号中所固有的混淆的表达。[①]Simoncelli, Freeman, Adelson et al.(1992)介绍了这样一种表达，他们称之为"可移位多尺度变换的金字塔径向频率实现"或更简洁的"导向金字塔"。它们的表述不只是过完备(消除混淆问题)而且是方向选择的，此外有同一的分析和综合基函数，即它是自反转的，就像"常规的"小波。于是，这使导向金字塔成为对结构分析和匹配任务更有用的基础，而结构分析和匹配任务在计算机视觉中是很常用的。

图 3.40a 展示了这样一个在频域分解的例子。这里使用径向的弧来代替会产生棋盘格高频的递归地将频域分成 2×2 方块的方法。图 3.40b 展示了结果金字塔的子带。尽管表述是过完备的，即，这里有比输入像素更多的小波系数，额外的频率和方向选择性使这个表述特别适用于纹理分析和合成之类的任务(Portilla and Simoncelli 2000)和图像降噪(Portilla, Strela, Wainwright et al. 2003; Lyu and Simoncelli 2009)。

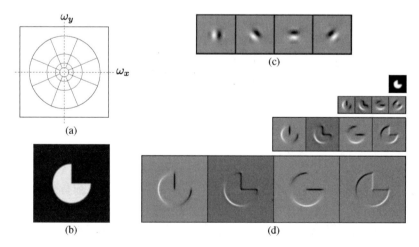

图 3.40　导向可移位多尺度变换(Simoncelli, Freeman, Adelson et al.1992)©1992 IEEE：(a) 径向多尺度频域分解；(b)原始图像；(c)四个导向滤波器的集合；(d)径向多尺度小波分解

[①] 这种混叠通常可以看作信号内容随着原始信号缓慢移位而在带之间移动。

3.5.5 应用：图像融合

3.5.3 节所展示的拉普拉斯金字塔有一个很迷人而有趣的应用，即生成混合合成图像，如图 3.41 所示(Burt and Adelson 1983b)。沿着中线将苹果和橘子图像直接拼在一起会产生一个明显的切痕，将它们夹在一起(正如(Burt and Adelson(1983b))称呼它们的过程)可以生成一个真实的杂交水果的漂亮的假象。其关键就是红苹果和橘子之间低频色彩变化被平滑地融合起来，而每个水果上高频纹理被更快地融合起来以避免两个纹理相互覆盖的时候产生"鬼影"现象。

图 3.41 拉普拉斯金字塔混合(Burt and Adelson 1983b)©1983 ACM：(a)苹果的原始图像；(b)橘子的原始图像；(c)常规拼接；(d)金字塔混合

要产生混合图像，每个源图像先被分解成它自己的拉普拉斯金字塔(图 3.42，左边列和中间列)。之后每个带被乘以一个大小正比于金字塔级别的平滑加权函数。最简单也最普通的产生这些权重的方法是采用一个二值掩模图像(图 3.43c)，从而根据这个掩模建立一个高斯金字塔。之后每个拉普拉斯金字塔图像被乘以相应的高斯掩模，这两个带权金字塔的和用于创建最终的图像(图 3.42，右列)。

如图 3.43 所示，这个过程可以应用于任意选定的掩模图像从而产生惊人的结果。将金字塔融合扩展到任意数量的图像上，其像素的出处用整数值标记图像来标示，也是直观易懂的(习题 3.20)。这在图像拼接和合成应用中相当常用，其中不同图像间的曝光可能有变化，详情参见 9.3.4 节。

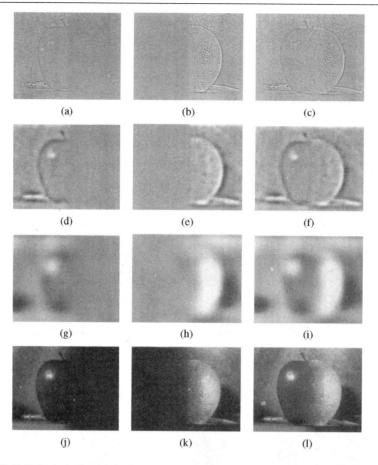

图 3.42 拉普拉斯金字塔混合细节(Burt and Adelson 1983b)©1983 ACM。前三行展示了拉普拉斯金字塔的高、中、低频部分(从 0, 2, 4 层提取)。左边一列和中间一列展示了原始苹果和橘子图像通过平滑插值函数加权,而右边一列展示了平均的贡献

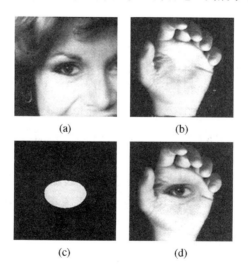

图 3.43 两幅任意形状的图像的拉普拉斯金字塔混合(Burt and Adelson 1983b)©1983 ACM:(a)第一幅输入图像;(b)第二幅输入图像;(c)区域掩模;(d)混合图像

3.6 几何变换

在前一小节中,我们讨论了如何用插值和下采样来改变一幅图像的分辨率。本节中,我们将讨论如何进行更普遍的变换,例如图像的旋转或常规的卷绕。相较于 3.1 节中的点操作,其中应用于一幅图像的函数变换了图像的值域范围,

$$g(\boldsymbol{x}) = h(f(\boldsymbol{x})), \tag{3.87}$$

这里我们关注变换定义域的函数(如图 3.44 所示),

$$g(\boldsymbol{x}) = f(h(\boldsymbol{x})) \tag{3.88}$$

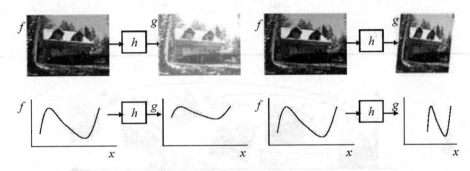

图 3.44 图像卷绕涉及改变一幅图像的定义域,而不是它的值域

我们首先研究最初在 2.1.2 节中介绍的全局参数化 2D 变换。(这样的变换称为"参数化变换",因其由很少的参数控制。)之后,我们将注意力转向诸如在网格上定义的变形等更多局部通用变形(3.6.2 节)。最后,在 3.6.3 节中,我们展示图像卷绕如何可以和渐隐渐现结合起来产生有趣的变形(其间的动画)。有志于进一步深入这些主题的读者,可以参考(Heckbert(1986))所做过的一个非常好的综述,还有 Wolberg(1990),Gomes, Darsa, Costa et al. (1999))和 Akenine-Möller and Haines(2002) 等非常易懂的教材。注意,Heckbert 所做的综述是针对纹理映射的,即计算机图形学领域所指的如何将图像卷绕贴到表面上这一问题。

3.6.1 参数化变换

参数化变换对一幅图像进行全局变形,其中变换的行为由少量参数控制。图 3.45 展示了这类变换的一些实例,它们以图 2.4 所示的 2D 几何变换为基础。这些变换的公式最初由表 2.1 给出,我们在表 3.5 中重新给出以便参考。

总体来说,给定一个通过公式 $x' = h(x)$ 确定的变换和一个源图像 $f(x)$,我们如何计算新图像中各像素点的值,如式(3.88)所示。在继续阅读之前,仔细考虑一下这个问题,看你是否可以解决它。

表 3.5 2D 坐标变换的层次。每个变换还保持列在其行下的性质，即，相似变换不仅保持夹角而且还保持平行性和直线性。对于齐次坐标变换，将 2×3 矩阵以第三行(0^T 1)扩展形成一个完整的 3×3 矩阵

变换	矩阵	自由度数	保持	图标
平移	$[I\|t]_{2\times3}$	2	方向	□
刚性(欧氏)	$[R\|t]_{2\times3}$	3	长度	◇
相似	$[sR\|t]_{2\times3}$	4	夹角	◇
仿射	$[A]_{2\times3}$	6	平行性	▱
投影	$[\tilde{H}]_{3\times3}$	8	直线性	⏢

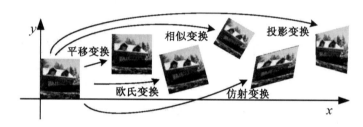

图 3.45 2D 几何图像变换的基本集合

如果你同大部分人一样，你会想到一种类似算法 3.1 的算法。这个过程称为"前向卷绕"或"前向映射"，如图 3.46a 所示。你可以想到这种方法的一些问题吗？

---

procedure forwardWarp(f, h, **out** g)
对于 f(x) 中的每个像素 x
    1. 计算目标位置 x′ = h(x)
    2. 将像素 f(x) 复制到 g(x′)

算法 3.1 将图像 f(x) 通过参数变换 x′ = h(x) 变换到图像 g(x′) 的前向卷绕算法

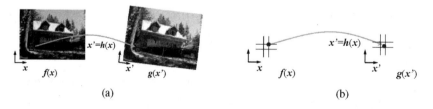

图 3.46 前向卷绕算法：(a)一个像素 f(x) 被复制到图像 g(x′) 中的相应位置 x′ = h(x)；(b)源像素位置和目标像素位置的细节

实际上，这种方法有很多限制。当 x′ 不是整数时，将一个像素 f(x) 复制 g 中

到 x' 位置的过程并没有很好的定义。在这种情况下我们该怎么办？你会怎么办？

你可以对 x' 取整到最近的整数坐标，并把像素复制到那里，但是进行变换时，得到的图像会有严重的混叠，像素会有很大偏移。你也可以使用加权(双线性)的方式将像素值"分配"给四个最近邻，记录每个像素点权重并在最后进行归一化。这个方法称为"体空间扫描(splatting)"，有时在图形学领域用于体绘制(Levoy and Whitted 1985; Levoy 1988; Westover 1989; Rusinkiewicz and Levoy 2000)。不幸的是，它既有一定的混叠又相当的模糊(丢失高分辨率细节)。

前向卷绕的第二个主要问题是裂缝和空洞的出现，特别是放大一幅图像时尤为明显。用邻近像素点填补这样的空洞会导致进一步的混叠和模糊。

那么我们可以做些什么来替代这个呢？一个比较好的解决方法是反向卷绕(算法 3.2)，目标图像 $g(x')$ 中的每个像素都是从原图像 $f(x)$ 中采样得到的(图 3.47)。

procedure inverseWarp(f, h, **out** g)
　对于 $g(x')$ 中的每个像素 x'
　1. 计算原图像位置 $x=h(x')$
　2. 对 $f(x)$ 在位置 x 处重采样并复制到 $g(x')$

算法 3.2　反向卷绕算法用于使用参数变换 $x'=h(x)$ 从一幅图像 $f(x)$ 生成一幅图像 $g(x')$

图 3.47　逆卷绕算法：(a)一个像素 $g(x')$ 被从 $f(x)$ 中对应位置 $x=h(x')$ 处重采样；(b)源像素位置和目标像素位置的细节

这和前向卷绕算法有什么不同？首先，由于 $\hat{h}=(x')$ (据推测)对 $g(x')$ 中所有像素都有定义，我们不再有空洞。更重要的是，这个方法很好地研究了在图像的非整数位置重采样的问题(常规的图像插值，见 3.5.2 节)，而且可以使用高质量滤波器来控制混叠。

函数 $\hat{h}(x')$ 从哪里来？它经常被计算为 $h(x)$ 的逆。实际上，表 3.5 所列所有参数变换的逆变换都有闭式解：简单地对指定变换的 3×3 矩阵求逆即可。

在其他情况下，将图像卷绕的问题形式化为给定一个从目标像素 x' 到原像素 x 的映射 $x=\hat{h}(x')$ 来重采样一幅源图像 $f(x)$ 会比较好。例如，在光流法(8.4 节)中，与计算将要到达的目标像素相反，我们将光流场估计为产生当前其光流正在被估计的像素的源像素的位置。类似地，当纠正径向畸变(2.1.6 节)时，我们通过计算最终(未畸变)图像中每个像素在原始(校正)图像中对应的像素位置来标定透镜。

什么样的插值适合重采样过程？我们在 3.5.2 节研究的滤波器，包括最近邻、双线性、双三次和窗 sinc 函数都可以使用。双线性常用于要求速度的地方(例如在一个块跟踪算法的内圈循环里面，见 8.1.3 节)，双三次和窗 sinc 函数更常用于视觉质量比较重要的地方。

要计算非整数位置 x 处 $f(x)$ 的值，我们简单地应用常用的 FIR 重采样滤波器，

$$g(x,y) = \sum_{k,l} f(k,l) h(x-k, y-l), \qquad (3.89)$$

这里 (x, y) 是亚像素坐标值，$h(x, y)$ 是一些插值或平滑核。回顾在 3.5.2 节中，进行降采样时，平滑核根据下采样率 r 被拉伸和重缩放。

不幸的是，对于一个一般的(无缩放)图像变换，重采样率 r 并没有被很好定义。考虑一个拉伸 x 维挤压 y 维的变换。重采样核应该沿着 x 维进行常规的插值，而在 y 方向进行平滑(来消除模糊图像中的锯齿)。这在处理一般的仿射或视角变换时会变得更复杂。

我们可以做些什么？幸运的是，傅里叶分析可以避免这些问题。表 3.1 所给的一维域缩放法则的二维推广为

$$g(\boldsymbol{Ax}) \Leftrightarrow |\boldsymbol{A}|^{-1} G(\boldsymbol{A}^{-T}\boldsymbol{f}). \qquad (3.90)$$

对于表 3.5 中除透视变换之外的所有变换，矩阵 $\boldsymbol{A}$ 均已定义。对于透视变换，矩阵 $\boldsymbol{A}$ 是透视变换线性化的导数(图 3.48a)，即对投影所引起的拉伸的局部仿射近似(Heckbert 1986; Wolberg 1990; Gomes, Darsa, Costa et al. 1999; Akenine-Möller and Haines 2002)。

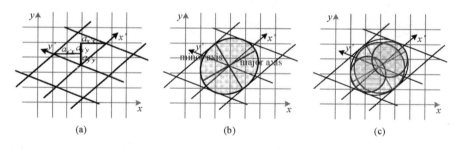

图 3.48 各向异性的纹理滤波：(a)变换矩阵 $\boldsymbol{A}$ 的雅克比行列式和引入的水平和竖直重采样率 $\{a_{xx}, a_{xy}, a_{yx}, a_{yy}\}$；(b)一个 EWA 平滑核的椭圆足迹；(c)使用沿主轴的多样例的各向异性滤波。位于直线交叉处的图像像素

为了防止混叠，我们需要用一个滤波器来预滤波图像 $f(x)$，该滤波器的频率响应恰好为最后想要的频谱通过 $\boldsymbol{A}^{-T}$ 变换后的投影(Szeliski, Winder, and Uyttendaele 2010)。总体来说(对于无缩放变换)，这个滤波器是不可分离的，所以计算速度很慢。因此，实际中会使用这个滤波器的一些近似，包括 MIP-映射、椭圆加权高斯平均和各向异性滤波(Akenine-Möller and Haines 2002)。

MIP-映射

Williams(1983)最初将 MIP-映射作为计算机图形学中一种用于纹理映射的快速

预滤波图像工具提出。一个 MIP 图[①]是一个标准图像金字塔(图 3.32)，每层用一个高质量滤波器进行预滤波而不是低质量的近似，例如(Burt and Adelson(1983b))的五阶二项式。从一个 MIP 图中重采样一幅图像，先要计算一个重采样率 r 的标量估计。例如，r 可以是矩阵 A 中绝对值的最大值(这可以抑制混淆)或者最小值(这可以减少模糊)。(Akenine-Möller and Haines(2002))更详细地讨论了这些问题。

一旦重采样率已经确定，可以用底数为 2 的对数表计算一个分数级的金字塔级别，

$$l = \log_2 r. \tag{3.91}$$

一个简单的解决办法是从下一个更高或更低的金字塔级别重采样纹理，这取决于是否要优先减少混淆或模糊。一个更好的解决方法是对两幅图像都进行重采样，并且用 l 的分数级成分将它们线性的混合在一起。因为大多 MIP 图的实现在每个级别内使用了双线性重采样，这个方法通常称为"三线性 MIP 映射"。计算机图形学渲染 API，例如 OpenGL 和 Direct3D，提供一些参数，基于想要的速度和质量的平衡，可以选择应该使用 MIP 映射(和采样率 r 的计算)的哪个变种。习题 3.22 让你更详细地检验这些权衡。

椭圆带权平均

Greene and Heckbert(1986)发明的椭圆带权平均(EWA)滤波器是建立在如下观察基础上的，即仿射映射 $x=Ax'$ 在每个源像素 x 附近定义了一个偏的二维坐标系统(图 3.48a)。对每个目标像素 x，计算 x' 中一个小像素格点到 x 的椭圆投影(图 3.48b)。之后，使用逆协方差矩阵为这个椭圆体的高斯对源图像 $g(x)$ 进行滤波。

尽管它名义上是一个高质量的滤波器(Akenine-Möller and Haines 2002)，但我们在工作中(Szeliski, Winder, and Uyttendaele 2010)发现，因为其中使用了一个高斯核，所以这种方法同高质量滤波器相比同时存在模糊和混叠问题。EWA 也非常慢，尽管已经有人提出了一些基于 MIP 映射的更快的变种((Szeliski, Winder, and Uyttendaele(2010))提供了一些额外的参考资料)。

各向异性滤波

滤波有向纹理的另一种替代方法，有时用图形学硬件(GPU)实现，是各向异性滤波器(Barkans 1997; Akenine-Möller and Haines 2002)。在这种方法中，一些不同分辨率(MIP 图中分数级的级别)的样例被联合在 EWA 的高斯主轴周围(图 3.48c)。

多通变换

要想卷绕图像同时不产生过度模糊或混叠，最优方法就是在每个像素点用一个理想低通滤波器自适应地预滤波源图像，例如，一个有向的、倾斜的 sinc 或低阶(例如，三次)近似(图 3.48a)。图 3.49 展示了这在一维情况下如何工作。信号首先

[①] "MIP" 表示 *multi in parvo*，意思是"多合一"。

(理论上)插值到一个连续的波形,再(理想地)低通滤波到在新的 Nyquist 率之下,然后重采样到最后想要的分辨率。实际上,插值和降采样步骤可以联系到一个多相(polyphase)数字滤波操作中(Szeliski, Winder, and Uyttendaele 2010)。

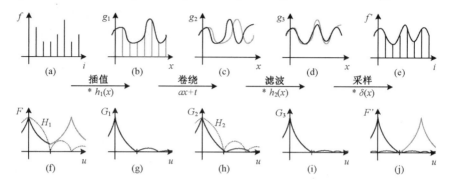

图 3.49 一维信号重采样(Szeliski, Winder,和 Uyttendaele 2010):(a)原始采样信号 $f(i)$;(b)插值信号 $g_1(x)$;(c)卷绕信号 $g_2(x)$;(d)滤波信号 $g_3(x)$;(e)采样信号 $f'(i)$。对应的频谱显示在信号下方,混叠部分用红色标出

对于参数变换,有向二维滤波和重采样操作可以用一系列一维重采样和剪切(shearing)变换(Catmull and Smith 1980; Heckbert 1989; Wolberg 1990; Gomes, Darsa, Costa *et al.* 1999; Szeliski, Winder, and Uyttendaele 2010)来近似。使用一系列一维变换的优点是它们(就基本运算操作而言)比大的、不可分离的二维滤波核更有效。

但是,为了避免混淆,可能有必要在应用剪切变换(Szeliski, Winder, and Uyttendaele 2010)之前在相反方向进行上采样。图 3.50 展现了针对旋转的这个过程,其中在水平剪切(和上采样)步骤之前添加了一个垂直上采样步骤。上面的图像展示了字母进行旋转时的外观,而下面的图像展示了相应的傅里叶变换。

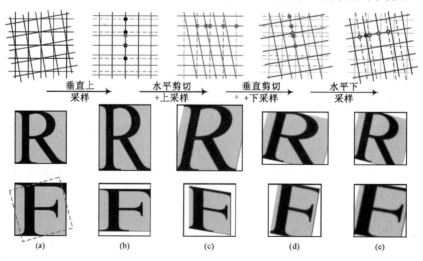

图 3.50 四通旋转(Szeliski, Winder,和 Uyttendaele 2010):(a)原始像素网格,图像及其傅里叶变换;(b)垂直上采样;(c)水平剪切和上采样;(d)垂直剪切和下采样;(e)水平下采样。除了最开始的两个阶段进行一般重采样之外,一般的仿射情况看起来很相似

3.6.2 基于网格的卷绕

尽管由少量全局参数确定的参数变换非常有用，但我们还是经常需要更自由的局部变形。

考虑，例如，人脸从皱眉到微笑的表观变化(图 3.51a)。怎样才能在这个情况下将嘴角向上卷翘而保持脸的其他部分不动呢？[①]要进行这样的变换，在图像的不同部分需要不同数量的运动。图 3.51 展示了一些常用的方法。

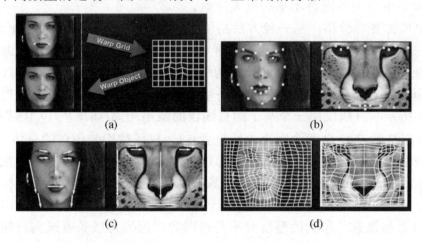

图 3.51 可供选择的图像卷绕方法(Gomes, Darsa, Costa et al. 1999)© 1999 Morgan Kaufmann：(a)稀疏控制点->变形网格；(b)控制点对应的稠密集；(c)有向直线对应；(d)一致的四边形网格

第一种方法，如图 3.51a 和图 3.51b 所示，要确定对应点的稀疏集。这些点的位移可以使用多种方法(Nielson 1993)插值到一个密集的位移场(第 8 章)。一个可能方法就是将一幅图像中的点集分成三角形(de Berg, Cheong, van Kreveld et al. 2006; Litwinowicz and Williams 1994; Buck, Finkelstein, Jacobs et al. 2000)，在每个三角形内使用由三个三角形顶点所确定的仿射运动模型(表 3.5)。如果目标图像根据新的顶点位置被分为三角形，则可以使用一个反卷绕算法(图 3.47)。如果源图像被分为三角形且被用作纹理图，则计算机图形学渲染算法可以用于绘制新图像(但一定要注意三角形边界以避免可能的混叠)。

另一种插值位移的稀疏集的方法包括移动邻近的四边网格节点，如图 3.51a 所示，用变分(能量最小化)内插式，例如正则化(Litwinowicz and Williams 1994)，见 3.7.1 节，或用位移的局部带权(径向基函数)组合(Nielson 1993)。(补充的散列数据插值方法可参见 12.3.1 节)如果使用四边形网格，可能会想到使用一个平滑内插式，例如一个四边形 B-样条(Farin 1996; Lee, Wolberg, Chwa et al. 1996)，插值移位

[①] Rowland and Perrett (1995); Pighin, Hecker, Lischinski et al. (1998); Blanz and Vetter (1999); Leyvand, Cohen-Or, Dror et al. (2008)展示了改变面部表情和外观的更复杂的示例。

到单独的像素值。[①]

在一些情况下，例如，如果一幅图像已经估计出一个密集深度图(Shade, Gortler, He et al. 1998)，我们就就只能知道每个像素的前向位移。如前面所提到的，在它们目标位置绘制源图像——即前向卷绕(图 3.46)——存在若干潜在的问题，包括混叠和小裂缝的出现。在这个情况下，一个替代方法是将位移场(或深度图)前向卷绕到新位置，在结果图中填补小的空洞，然后逆卷绕进行重采样(Shade, Gortler, He et al. 1998)。这通常比前向卷绕表现更好，因为位移场一般比图像更平滑，所以在位移场的前向卷绕中引入的混叠不那么引人注意。

要确定局部变形位移，另一种方法是使用相应的有向直线分割(Beier and Neely 1992)，如图 3.51c 和图 3.52 所示。沿着直线段的像素完全按照指定从源图像变换到目标图像，其他像素用一个这些位移的平滑内插式进行卷绕。每个直线段的对应确定了一个平移、旋转和放缩——即一个相似变换(表 3.5)——对于其附近的像素，如图 3.52a 所示。直线段用一个基于到直线段的最小距离(如果 $u \in [0,1]$ 则为图 3.52a 中的 v，否则是到 P 和 Q 两个距离中较短者)的加权函数影响图像的总体位移。

对于每个像素 X，每条直线对应目标位置 X' 和一个依赖于距离和直线段长度的权重一起计算(图 3.52b)。所有目标位置 X_i' 的带权平均成为最终目标位置。注意，(Beier 和 Neely)将这个算法描述为一个前向卷绕，一个等价的算法可以写为按照顺序穿过目标像素。得到的卷绕并不是相同的，因为直线长度或到直线的距离可能会不同。习题 3.23 让你实现 Beier–Neely(基于直线的)卷绕并且将它和一系列其他局部变形方法做比较。

还有另一种为了产生图像卷绕确定对应关系的方法是用蛇行(5.1.1 节)和 B 样条结合(Lee, Wolberg, Chwa et al. 1996)。这种方法用于苹果的 Shake 软件，而且在医学图像领域非常常用。

确定位移场的最后一个可能方案就是使用一个专门的适应于特定图像内容的网格，如图 3.51d 所示。手动确定这种网格会涉及大量的工作；(Gomes, Darsa, Costa et al.(1999))描述了一个相关的交互系统。一旦两个网格确定，中间的卷绕可以用线性插值产生，网格点的位移可以用样条插值。

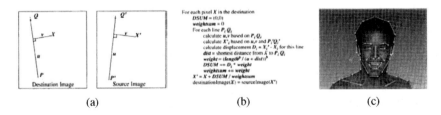

图 3.52　基于直线的图像卷绕(Beier and Neely 1992)© 1992 ACM：(a)距离计算和位置传递；(b)绘制算法；(c)两个用于变形的中间卷绕

[①] 注意，很多视频压缩标准使用的基于块的运动模型(Le Gall 1991)可以视为第 0 阶的(分段常数)的位移场。

3.6.3 应用：基于特征的变形

尽管卷绕可以用于改变单幅图像的外观或使其成为一个动画，但用一个通常称为"变形"的过程将两幅或更多图像卷绕并混合起来可以获得更强大的效果(Beier and Neely 1992; Lee, Wolberg, Chwa *et al.* 1996; Gomes, Darsa, Costa *et al.* 1999)。

图 3.53 展示了图像变形的本质。在两幅图像中进行简单地渐隐渐现(cross-dissolving)会导致上面一行所显示的鬼影，取而代之，每幅图像在融合之前经过向另一幅图像卷绕，如下面一行所示。如果建立了好的对应关系(用图 3.5 所展示的任何一种方法)，对应的特征便会对齐，因而不会有鬼影结果。

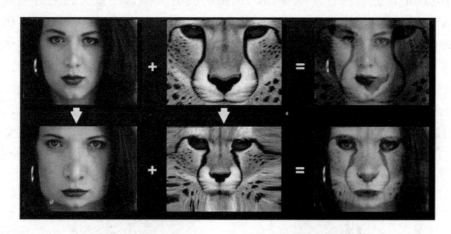

图 3.53 图像变形(Gomes, Darsa, Costa *et al.* 1999) © 1999 Morgan Kaufmann。上行：如果两幅图像直接混合，会导致可见的鬼影。下行：先将两幅图像都卷绕到同样的中间位置(例如，向另一幅图像的中间)，然后将得到的卷绕图像混合，产生一个无缝的变形

将以上过程重复应用于产生变形过程的每个中间帧，在每个间隔使用不同的融合(和不同数量的变形)。令 $t \in [0,1]$ 为描述插入帧序列的时间参数。融合中的两个卷绕图像的权重函数为$(1-t)$和 t。相反的，图像 0 在时间 t 所承受的运动量占由对应所确定的运动总量的 t。但是要注意，在定义将一幅图像部分卷绕到目标时，有何意义，特别是当预期的运动与线性相差甚远时(Sederberg, Gao, Wang *et al.* 1993)。习题 3.25 让你实现一个变形算法，并在这样有挑战的情况下进行测试。

3.7 全局优化

迄今为止，本章已经覆盖了很多图像处理算子，它们用一幅或多幅图像作为输入，产生一些滤波或变换版本。在很多应用中，更有用的是先用一些优化准则明确表达想要的变换的目标，再找到或推断出最符合这个准则的解决方法。

在本章最后一个小节中，我们介绍这种思路的两个不同但紧密相关的变形。第

一个方法，常称为"正则化"或"变分法"(3.7.1 节)，构造了一个描述解之特性的连续全局能量函数，然后用稀疏线性系统或相关的迭代方法找到一个最小能量解。第二种方法用贝叶斯统计学，对产生输入图像的有噪声的测量过程和关于解空间的先验假设进行建模，这些通常用一个马尔科夫随机场编码(3.7.2 节)。

这类问题的示例包括：从散列数据中进行表面插值(图 3.54)；图像去噪和缺失区域恢复(图 3.57)；将图像分为前景和背景区域(图 3.61)。

3.7.1 正则化

正则化理论首先由统计学家提出，试图用模型来拟合严重欠约束解空间的数据(Tikhonov and Arsenin 1977; Engl, Hanke, and Neubauer 1996)。考虑一下，例如找到一个平滑的表面穿过(或是靠近)一个测量数据点的集合的问题(图 3.54)。这样的问题被描述为病态的(ill-posed)，因为有很多可能的表面可以拟合这个数据。由于输入细微变化有时也会导致拟合的显著变化(例如，如果我们用多项式插值)，这样的问题也经常是不适定的(ill-conditioned)。由于我们试图从采样数据点 $d(x_i, y_i)$ 中恢复未知的函数 $f(x, y)$，这样的问题通常称为"逆问题"。很多计算机视觉任务可以视为逆问题，由于我们试图从一个有限的图像集合中恢复 3D 世界的完全描述。

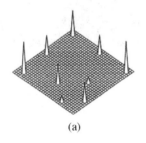 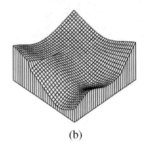

(a)　　　　　　　　(b)

图 3.54　一个简单的表面插值问题：(a)九个不同高度的数据点分布在一个网格上；(b)二阶，控制连续，薄板样条插值器，沿着其左侧边界有一处撕裂，沿着右侧有一个折痕(Szeliski 1989) © 1989 Springer

为了说明找平滑解的含义，我们在解空间上定义了一个范数。对于一维函数，我们可以对函数的一阶导数平方进行积分

$$\mathcal{E}_1 = \int f_x^2(x)\,dx \tag{3.92}$$

或者对二阶导数的平方进行积分

$$\mathcal{E}_2 = \int f_{xx}^2(x)\,dx. \tag{3.93}$$

(这里，我们用下标表示微分。)这种能量度量是泛函的样例，是将函数映射到标量值的算子。它们也通常称为"变分法"，因为它们度量一个函数的变化(非平滑性)。

在二维的情况(例如，对于图像、光流场或者表面)下，相应的平滑泛函为

$$\mathcal{E}_1 = \int f_x^2(x,y) + f_y^2(x,y)\, dx\, dy = \int \|\nabla f(x,y)\|^2\, dx\, dy \tag{3.94}$$

以及

$$\mathcal{E}_2 = \int f_{xx}^2(x,y) + 2f_{xy}^2(x,y) + f_{yy}^2(x,y)\, dx\, dy, \tag{3.95}$$

其中的混合项 $2f_{xy}^2$ 是用来使这个度量旋转不变的(Grimson 1983)。

一阶导数范数通常称为薄膜(memberance)，因为用这个度量插值一组数据点会得到一个类似帐篷的结构。(实际上，这个公式是对表面区域的小绕度近似，即肥皂泡最小化的目标。)二阶范数称为"薄板样条"，因为它近似了薄板(例如，柔性钢条)在小的变形下的行为。两者的混合称为"张力下的薄板样条"；这些公式的每个导数项乘以一个局部加权函数的版本称为"控制连续性样条"(Terzopoulos 1988)。图 3.54 展示了一个控制连续性插值算子拟合九个离散数据点的简单的样例。在实践中，更常见的是一阶平滑项用于图像和光流场(8.4 节)，而二阶平滑同表面相关联(12.3.1 节)。除了平滑项，正则化也需要一个数据项(或数据惩罚)。对于散列数据插值(Nielson 1993)，数据项度量函数 $f(x, y)$ 和一组数据点 $d_i = d(x_i, y_i)$ 的距离，

$$\mathcal{E}_d = \sum_i [f(x_i, y_i) - d_i]^2. \tag{3.96}$$

对于一个如噪声去除的问题，可以使用这个度量的一个连续版本，

$$\mathcal{E}_d = \int [f(x,y) - d(x,y)]^2\, dx\, dy. \tag{3.97}$$

要得到一个可以被最小化的全局能量，通常将两个能量项相加，

$$\mathcal{E} = \mathcal{E}_d + \lambda \mathcal{E}_s, \tag{3.98}$$

其中 $\mathcal{E}_s$ 是光滑惩罚($\mathcal{E}_1, \mathcal{E}_2$ 或一些带权混合)，λ 是正则化参数，控制解的光滑性。为了找到这个连续问题的最小值，通常先将函数 $f(x, y)$ 在一个常规网格上离散化。[①]进行这个离散化最有效的方法就是用有限元分析，即用一个分段连续样条来近似函数，然后进行分析集成(Bathe 2007)。

幸运的是，对于一阶和二阶光滑泛函，近似有限元的明智选择导致特别简单的离散形式(Terzopoulos 1983)。相应的离散平滑能量函数变成

$$\begin{aligned}E_1 &= \sum_{i,j} s_x(i,j)[f(i+1,j) - f(i,j) - g_x(i,j)]^2 \\ &\quad + s_y(i,j)[f(i,j+1) - f(i,j) - g_y(i,j)]^2\end{aligned} \tag{3.99}$$

和

$$\begin{aligned}E_2 &= h^{-2}\sum_{i,j} c_x(i,j)[f(i+1,j) - 2f(i,j) + f(i-1,j)]^2 \\ &\quad + 2c_m(i,j)[f(i+1,j+1) - f(i+1,j) - f(i,j+1) + f(i,j)]^2 \\ &\quad + c_y(i,j)[f(i,j+1) - 2f(i,j) + f(i,j-1)]^2,\end{aligned} \tag{3.100}$$

其中 h 是有限元网格的大小。H 因子的重要性仅仅体现在能量在多种分辨率下被离

[①] 另一种使用中心位于数据点的核基函数(Boult and Kender 1986; Nielson 1993)的方法在 12.3.1 小节有更详细的讨论。

散化时，例如在由粗到精或多网格方法中那样。

可选的光滑权重 $s_x(i,j)$ 和 $s_y(i,j)$ 在平面内控制水平和垂直泪点(或弱点)。对于其他问题，例如彩色化(Levin, Lischinski, and Weiss 2004)和交互式色调映射(Lischinski, Farbman, Uyttendaele *et al.* 2006a)，它们在插值色度或曝光区域时控制平滑度，并且常常被设为与局部光照梯度强度成反比。对于二阶问题，折皱变量 $c_x(i,j)$，$c_m(i,j)$ 和 $c_y(i,j)$ 控制表面内折皱的位置(Terzopoulos 1988; Szeliski 1990a)。

数据值 $g_x(i,j)$ 和 $g_y(i,j)$ 是算法使用的梯度数据项(约束)，例如光度测定学立体视觉(12.1.1 节)，HDR 色调映射(10.2.1 节)(Fattal, Lischinski, and Werman 2002)、泊松混合(9.3.4 节)(Pérez, Gangnet, and Blake 2003)和梯度域混合(9.3.4 节)(Levin, Zomet, Peleg *et al.* 2004)。它们在仅离散化常规一阶平滑泛函时都被设为零(3.94)。

二维的离散数据能量写为

$$E_d = \sum_{i,j} w(i,j)[f(i,j) - d(i,j)]^2, \tag{3.101}$$

其中局部权重 $w(i,j)$ 控制数据约束的强度。这些值在没有数据的地方被设为零，而在有数据的地方被设为数据量测的方差倒数(讨论参见(Szeliski(1989))和 3.7.2 节)。

现在，离散问题的总能量可以写为二次形式

$$E = E_d + \lambda E_s = \boldsymbol{x}^T \boldsymbol{A} \boldsymbol{x} - 2\boldsymbol{x}^T \boldsymbol{b} + c, \tag{3.102}$$

其中 $x = (f(0,0)\cdots f(m-1, n-1))$ 被称为"状态向量"。[①]

稀疏对称正定矩阵 $\boldsymbol{A}$ 称为 Hessian，因为它编码了能量函数的二阶导数。[②] 对于一维情况，一阶问题，$\boldsymbol{A}$ 是三对角矩阵；对于二维情况，一阶问题，它是每行有五个非零项的多频带的。我们称 b 为"带权数据向量"。最小化以上的二次形等价于解稀疏线性系统

$$\boldsymbol{A}\boldsymbol{x} = \boldsymbol{b}, \tag{3.103}$$

这可以通过使用附录 A.5 描述的各种稀疏矩阵方法来实现，例如多网格(Briggs, Henson, and McCormick 2000)和层次化预处理器(Szeliski 2006b)。

Poggio, Torre, and Koch(1985)和 Terzopoulos(1986b)为了解决表面插值等问题率先将正则化引入视觉领域，它迅速被其他视觉研究者采用来解决各类问题，例如边缘检测(4.2 节)、光流(8.4 节)、由阴影到形状(12.1 节)(Poggio, Torre, and Koch 1985; Horn and Brooks 1986; Terzopoulos 1986b; Bertero, Poggio, and Torre 1988; Brox, Bruhn, Papenberg *et al.* 2004)。Poggio, Torre, and Koch(1985)还展示了如何在一个电阻网格中实现一个方程(3.100 和 3.101)所定义的离散能量，如图 3.55 所示。在计算摄影学中(第 10 章)，正则化及其变种广泛应用于解决高动态变化色调映射(Fattal, Lischinski, and Werman 2002; Lischinski, Farbman, Uyttendaele *et al.* 2006a)、泊松和梯度域混合(Pérez, Gangnet, and Blake 2003; Levin, Zomet, Peleg *et al.* 2004; Agarwala, Dontcheva, Agrawala *et al.* 2004)、色彩化(Levin, Lischinski, and Weiss

---

[①] 我们使用 x 来代替 f，因为这种形式在数值分析领域更常见(Golub and Van Loan 1996)。
[②] 在数值分析中，$\boldsymbol{A}$ 称为"系数矩阵"(Saad 2003)；在有限元分析[Bathe 2007]中，它称为刚性矩阵。

2004)、自然图像抠图(Levin, Lischinski, and Weiss 2008)等问题。

鲁棒正则化

正则化最常用二次(L_2)范数(同方程(3.92~3.95)中的平方导数和(方程 3.100 和 3.101)中的平方差比较)形式化，它也可以用非二次鲁棒惩罚函数形式化(附录 B.3)。例如，方程(3.100)可以推广为

$$E_{1r} = \sum_{i,j} s_x(i,j)\rho(f(i+1,j) - f(i,j)) \\ + s_y(i,j)\rho(f(i,j+1) - f(i,j)), \quad (3.104)$$

其中 $\rho(x)$ 是一些单调增惩罚函数。例如，范式族 $\rho(x) = |x|^p$ 称为 p-范式。当 $p < 2$ 时，得到的光滑项变得更分段连续而不是整体光滑，这可以更好地对图像、光流场和 3D 表面的不连续性建模。

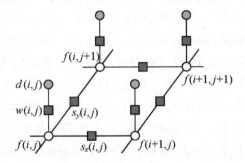

图 3.55 一阶正则化的图模型表述。白色圆圈是未知的 $f(i, j)$ 而深色圆圈是输入数据 $d(i, j)$。在电阻的网格表述中，d 和 f 的值编码输入输出电压，而黑色方块表示电导被设为 $s_x(i, j)$, $s_y(i, j)$ 和 $w(i, j)$ 的电阻。在弹簧-质量系统类比中，圆圈表示提升(elevation)而黑色方块表示弹簧。同样的图模型可以用于描绘一个一阶马尔科夫随机场(图 3.56)

一个鲁棒正则化的早期例子是 Blake and Zisserman(1987)引入的分度非凸性(GNC)算法。这里，数据和导数的范式被限定了一个最大值

$$\rho(x) = \min(x^2, V). \quad (3.105)$$

由于得到的问题是非凸的(它有很多局部最小值)，有人提出了一个延拓法，其中一个二次范数(凸的)被逐渐替换为非凸的鲁棒范数(Allgower and Georg 2003)。(在同一时期，Terzopoulos(1988)也在其表面插值问题中使用延拓来推断撕裂和折痕变量)。

现在，更常使用 L_1($p=1$)范式，也常称为"全变差"(Chan, Osher, and Shen 2001; Tschumperlé and Deriche 2005; Tschumperlé 2006; Kaftory, Schechner, and Zeevi 2007)。影响(导数)更快下降到 0 的其他范式，由(Black and Rangarajan (1996); Black, Sapiro, Marimont et al.(1998))提出，相关讨论参见附录 B.3。

更近一些，基于如下观测：图像导数的对数似然度分布遵从一个 $p \approx 0.5$–0.8 的斜率，因此是一个超拉普拉斯分布(Simoncelli 1999; Levin and Weiss 2007; Weiss and Freeman 2007; Krishnan and Fergus 2009)，使得 $p < 1$ 的超拉普拉斯范式非常受欢

迎。这类范式明显趋向于选择那些有很大不连续而不是小的不连续的解。相关讨论见 3.7.2 节(3.114)。

用 L_2 范式的最小二乘正则化问题，可以用线性系统解决，而其他 p-范式需要用不同的迭代方法，例如迭代重加权最小二乘(IRLS)，Levenberg–Marquardt，或在局部非线性子问题和全局二次正则化(Krishnan and Fergus 2009)之间交替。这些方法的讨论可参见 6.1.3 节、附录 A.3 和 B.3。

3.7.2 马尔科夫随机场

正如我们所见，正则化涉及定义在(分段)连续函数上的能量泛函的最小化，可以用来形式化和解决多种低层计算机视觉问题。一个替代方法是形式化一个贝叶斯模型，对噪声图像形成(测量)过程和在解空间上假设一个统计先验模型分别进行建模。在本节中，我们关注基于马尔科夫随机场的先验，它的对数似然度可以用局部邻域交互(或惩罚)项来描述(Kindermann and Snell 1980; Geman and Geman 1984; Marroquin, Mitter, and Poggio 1987; Li 1995; Szeliski, Zabih, Scharstein et al. 2008)。

贝叶斯建模的使用与正则化(参见附录 B)相比，有几个潜在的优点。可以统计地对测量过程建模，使我们能够从每个测量中尽可能地提取最多信息，而不是仅仅猜测赋予数据什么样的权重。类似地，先验分布的参数常可以通过观察我们建模的类中的样例学习得到(Roth and Black 2007a; Tappen 2007; Li and Huttenlocher 2008)。此外，由于我们的模型是概率的，所以(在原理上)估计针对待恢复的未知量的完全概率分布是可能的，特别是对解的不确定性建模，这在后续的处理阶段非常有用。最后，马尔科夫随机场模型可以在离散变量上定义，例如图像标注(其中变量没有合适的顺序)，而正则化对此并不适用。

回忆 3.4.3 节(或附录 B.4)中的式(3.68)，根据贝叶斯规则，对于一个给定的测量集合，y，$p(y|x)$，以及 x 的一个先验分布 $p(x)$，未知量 x 的后验分布由下式所得，

$$p(\boldsymbol{x}|\boldsymbol{y}) = \frac{p(\boldsymbol{y}|\boldsymbol{x})p(\boldsymbol{x})}{p(\boldsymbol{y})}, \tag{3.106}$$

其中，$p(y)$ 是一个归一化的常量，用于将 $p(x|y)$ 分布合理化(积分为 1)。对式(3.106)两边取负对数，我们得到

$$-\log p(\boldsymbol{x}|\boldsymbol{y}) = -\log p(\boldsymbol{y}|\boldsymbol{x}) - \log p(\boldsymbol{x}) + C, \tag{3.107}$$

称为"负后验对数似然度"。

给定一些测量 y，为了找到最可能(最大化后验概率，MAP)的解 x，我们简单地最小化这个负对数似然度，这也可以想象为一个能量，

$$E(\boldsymbol{x}, \boldsymbol{y}) = E_d(\boldsymbol{x}, \boldsymbol{y}) + E_p(\boldsymbol{x}). \tag{3.108}$$

(我们舍弃常数 C，因为它的值在能量最小化过程中不起作用)第一项 $E_d(x; y)$ 是数据能量或数据惩罚；它度量给定未知状态 x 下观察到的数据的负对数似然度。第二项 $E_p(x)$ 是先验能量，它在这里的作用类似于正则化中的光滑能量的作用。注意，MAP 估计有时并不是我们想要的，因为它选择后验分布中的"峰"而不是一些更

稳定的统计量——见附录 B.2 和(Levin, Weiss, Durand et al.(2009))中的讨论。

对于图像处理应用，未知的 x 是输出像素的集合
$$\boldsymbol{x} = [f(0,0) \ldots f(m-1, n-1)],$$
数据(最简单的情况)是输入像素
$$\boldsymbol{y} = [d(0,0) \ldots d(m-1, n-1)]$$
如图 3.56 所示。

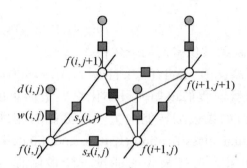

图 3.56 $\mathcal{N}_4$ 邻域马尔科夫随机场的图模型。(蓝色的边是为 $\mathcal{N}_8$ 邻域添加的。)白色圆圈是未知的 $f(i,j)$ 而深色圆圈是输入数据 $d(i,j)$。黑色方块 $s_x(i,j)$, $s_y(i,j)$ 表示随机场中任意的相邻的节点间交互作用势，$w(i,j)$ 表示数据惩罚函数。同样的图模型可以用于描述一个离散版本的一阶正则化问题(图 3.55)

对于一个马尔科夫随机场，概率 $p(x)$ 是一个 Gibbs 或 Boltzmann 分布，它们的负对数似然度(根据 Hammersley–Clifford 定理)可以写为成对交互作用势之和。
$$E_p(\boldsymbol{x}) = \sum_{\{(i,j),(k,l)\} \in \mathcal{N}} V_{i,j,k,l}(f(i,j), f(k,l)), \tag{3.109}$$
其中 $\mathcal{N}(i,j)$ 表示像素 (i,j) 的邻域。实际上，定理的一般版本是在一个更大的团(clique)的集上评价能量，这取决于马尔科夫随机场的阶次(Kindermann and Snell 1980; Geman and Geman 1984; Bishop 2006; Kohli, Ladický, and Torr 2009; Kohli, Kumar, and Torr 2009)。

在马尔科夫随机场建模中最常用的邻域是 $\mathcal{N}_4$ 邻域，其中每个像素 $f(i,j)$ 只与它最接近的邻居相互作用。图 3.56 中的模型，之前在图 3.55 中用于说明一阶正则化的离散版本，展示了 $\mathcal{N}_4$ MRF。$s_x(i,j)$ 和 $s_y(i,j)$ 黑方块表示随机场中相邻节点间任意的交互作用势，$w(i,j)$ 表示数据惩罚函数。这些方形节点也可以看作一个(无向)图模型(Bishop 2006)的因子图的因子。(严格的说，因子是(不适当的)概率函数，它们的乘积是(未归一化的)后验分布。)

如式(3.112)和式(3.113)所见，在这些交互作用势和正则化图像复原问题的离散版本之间有密切的联系。因此，作为初步近似，我们可以将在解决一个正则化问题时进行的能量最小化，等价于在一个 MRF 中进行的最大化后验推断。

尽管 $\mathcal{N}_4$ 邻域是最常用的，但在一些应用中 $\mathcal{N}_8$ (甚至更高阶)在诸如图像分割等问题中表现更好，因为它可以更好地对不同方向的不连续性建模(Boykov and Kolmogorov 2003; Rother, Kohli, Feng et al. 2009; Kohli, Ladický, and Torr 2009;

Kohli, Kumar, and Torr 2009)。

二值马尔科夫随机场

马尔科夫随机场最简单的例子就是一个二值场。这种场的例子包括 1 比特(黑和白)的扫描文件图像和分割为前景和背景区域的图像。

为了对一幅扫描图像降噪,我们设置数据惩罚项来反映扫描图像和最终图像的一致性,

$$E_d(i,j) = w\delta(f(i,j), d(i,j)) \tag{3.110}$$

而设置光滑惩罚项来反映邻域像素间的一致性,

$$E_p(i,j) = E_x(i,j) + E_y(i,j) = s\delta(f(i,j), f(i+1,j)) + s\delta(f(i,j), f(i,j+1)). \tag{3.111}$$

一旦用公式表示能量,我们该如何最小化它?最简单的方法就是进行梯度下降,每次翻转一个状态(如果该翻转能产生一个更低的能量的话)。这个方法称为"上下文分类"(contextual classification)(Kittler and Föglein 1984),迭代条件模态(ICM)(Besag 1986),如果先选择最大的能量下降最则称为"最高置信度优先"(HCF)(Chou and Brown 1990)。

不幸的是,这些下山法很容易在局部最小值处卡住。一个替代的方法就是向过程中加一些随机因素,通常称为"随机梯度下降"(Metropolis, Rosenbluth, Rosenbluth et al. 1953; Geman and Geman 1984)。当噪声量随时间下降时,这个方法称为"模拟退火"(Kirkpatrick, Gelatt, and Vecchi 1983; Carnevali, Coletti, and Patarnello 1985; Wolbergand Pavlidis 1985; Swendsen and Wang 1987),最初由 Geman and Geman(1984)提出而在计算机视觉中普及,之后被 Barnard(1989)用于立体视觉匹配和一些其他应用。

但是,即使是这个方法,也不能有那么好的结果(Boykov, Veksler, and Zabih 2001)。对于二值图像,一个更好的方法,由 Boykov, Veksler, and Zabih(2001)引入计算机视觉界,是将能量最小化重定义为一个最大流/最小割的图优化问题(Greig, Porteous, and Seheult 1989)。

这个方法在计算机视觉界的非正式称呼为"图割"(Boykov and Kolmogorov 2010)。对于简单的能量函数,例如,那些对于不相同邻域像素的惩罚是常数的,这个算法保证产生全局最小。Kolmogorov and Zabih(2004)形式化地刻画了这些结果成立时的二值能量势的类(正则化条件),而 Komodakis, Tziritas, and Paragios(2008)和 Rother, Kolmogorov, Lempitsky et al.(2007)所做的更新的工作提供了在它们不成立的条件下更好的算法。

除了以上所提到的方法外,人们还提出了一些其他优化方法来进行 MRF 能量最小化,例如(带环的)置信度传播和动态规划(对于一维问题)。更详细的讨论参见附录 B.5 以及 Szeliski, Zabih, Scharstein et al.(2008)的比较综述论文。

顺序值的马尔科夫随机场

除了二值图像,马尔科夫随机场可以应用于顺序值标签,例如灰度图像或深度

图。"顺序"表示标签有一个隐含的顺序，例如，较高的值是较亮的像素。在下一节中，我们会关注无序的标签，例如图像合成中的源图像标签。

在很多情况下，常常将二值数据和光滑先验项扩展为

$$E_d(i,j) = w(i,j)\rho_d(f(i,j) - d(i,j)) \tag{3.112}$$

和

$$E_p(i,j) = s_x(i,j)\rho_p(f(i,j) - f(i+1,j)) + s_y(i,j)\rho_p(f(i,j) - f(i,j+1)), \tag{3.113}$$

这是平方惩罚项(3.101)和(3.100)的鲁棒推广，最初在(3.105)中引入。如之前，权重$w(i,j)$，$s_x(i,j)$和$s_y(i,j)$可以用于局部控制数据加权和水平与竖直的光滑性。但是，这里没有使用平方惩罚，而是使用了一个几乎单调增的惩罚函数$\rho()$。(数据和光滑性项可以使用不同的函数。)例如，ρ_p可以是超拉普拉斯惩罚。

$$\rho_p(d) = |d|^p, \ p < 1, \tag{3.114}$$

这与一个二次或线性(总变差)惩罚相比，更好地编码了一幅图像中的梯度分布(主要是边缘)。[①]Levin and Weiss(2007)使用这样一个惩罚，通过鼓励梯度在一幅图像或另一幅图像中(而不是两者)来分离一个透射和反射的图像(图 8.17)。再后来，(Levin, Fergus, Durand et al.(2007))进一步使用超拉普拉斯作为先验用于图像去卷积(去模糊)，而(Krishnan and Fergus(2009))发展了一个更快的算法来解决这类问题。对于数据惩罚，ρ_d可以是二次的(对高斯噪声建模)或一个受污染的高斯(附录 B.3)的对数。

当ρ_p是一个二次函数时，得到的马尔科夫随机场称为"高斯马尔科夫随机场"(GMRF)，它的最小值可以通过稀疏线性系统求解来找到(3.103)。当加权函数是均匀的，GMRF 成为一个维纳滤波器(3.4.3 节)的特例。允许加权函数依赖于输入图像(如下描述，一种特殊的条件随机场)，使得可以运用相当复杂的图像处理算法，包括色彩化(Levin,Lischinski, and Weiss 2004)、交互色调映射(Lischinski, Farbman, Uyttendaele et al. 2006a)、自然图像抠图(Levin, Lischinski, andWeiss 2008)和图像复原(Tappen, Liu, Freeman et al. 2007)。

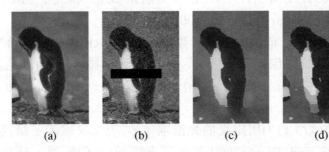

图 3.57 灰度图像去噪和修图：(a)原始图像；(b)受噪声污染和缺失数据(黑条)的图像；(c)用带环置信度传播恢复的图像；(d)用扩展移动图割修复的图像。图像来自 http://vision.middlebury.edu/ MRF/results/(Szeliski, Zabih, Scharstein et al. 200)

---

① 注意，与一个二次惩罚不同，水平和竖直导数的 p-范数的和不是旋转不变的。一个更好的方法可能是局部地估计梯度方向，在垂直和平行的分量上使用不同的范数，Roth and Black (2007b)称之为"导向随机场"。

当 ρ_d 或 ρ_p 是非二次函数，有时可以使用非线性最小二乘或迭代重加权最小二乘等梯度下降方法(附录 A.3)。但是，如果搜索空间有很多局部最小，如立体视觉匹配的情况(Barnard 1989; Boykov, Veksler, and Zabih 2001)，就需要更复杂的方法了。

图割方法到多值问题的扩展最初由 Boykov, Veksler, and Zabih (2001)提出。在他们的论文中，发展了两个不同的算法，称为"交换移动"(swap move)和"扩展移动"(expansion move)，在一系列二值标签子问题中迭代找到一个好的解(图 3.58)。注意，鉴于已证明这个问题对于一般的能量函数是 NP-难的，所以一个全局的解通常不可达。由于这两种算法都在内循环中都使用一个二值 MRF 优化，它们受二值标签情况中存在的关于能量函数的那种限制(Kolmogorovand Zabih 2004)。附录 B.5.4 更详细地讨论了这些算法以及一些最新发展的解决方法。

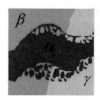   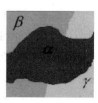

(a)初始标注　　(b)标准移动　　(c)$\alpha-\beta$ 交换　　(d)α 扩展

图 3.58　多级别图优化(Boykov, Veksler 和 Zabih 2001)© 2001 IEEE：(a)初始问题构造；(b)标准移动仅改变一个像素；(c)$\alpha-\beta$ 交换最优地互换所有具有 α 和 β 标注的像素；(d)α 扩展移动最优地在当前像素值和 α 标签中进行选择

另一个 MRF 推断方法是置信度传播(belief propagation，BP)。置信度传播最初用来对树进行推断，是精确的(Pearl 1988)，近期已被用于马尔科夫随机场等带环图(Freeman, Pasztor, and Carmichael 2000; Yedidia, Freeman, and Weiss 2001)。事实上，一些更好的立体视觉匹配办法使用带环的置信度传播(LBP)来进行推断(Sun, Zheng, and Shum 2003)。LBP 的详细讨论参见附录 B.5.3 和关于 MRF 优化的比较综述论文(Szeliski, Zabih, Scharstein *et al.* 2008)。

图 3.57 展示了用一个非二次能量函数(非高斯 MRF)进行图像降噪和修图(填补空洞)的一个例子。原始图像被噪声污染，一部分数据被移除了(黑色条)。在这种情况下，带环的置信度传播算法计算一个稍低的能量和一个比 alpha-扩展图割算法更光滑的图像。

当然，以上的公式(3.113)所定义的光滑项 $E_p(i, j)$ 仅仅展示了最简单的情况。更近期的工作中，Roth and Black(2009)提出了专家域(FoE)模型，计算了大量的指数局部滤波输出之和来得到光滑惩罚。Weiss and Freeman (2007)分析了这个方法并且将它同更简单的对于自然图像统计的超-拉普拉斯模型作比较。Lyu and Simoncelli (2009) 使用高斯缩放混合(GSMs)来构造一个不均匀的多尺度 MRF，有一个(正指数)GMRF 调整另一个高斯 MRF 的方差(幅值)。

还可以扩展观测模型来使采样的(噪声污染的)输入像素对应到未知的(隐)图像

像素的混合，如图 3.59 所示。这在试图对一幅图像去模糊时经常发生。尽管这类模型仍旧是传统的生成式马尔科夫随机场，但由于团块的大小变得更大，所以很难找到一个最优解。在这种情况下，可以使用梯度下降方法，例如迭代的重加权最小二乘(Joshi, Zitnick, Szeliski *et al.* 2009)。习题 3.31 会要求你探究这些问题。

无序的标签

另一个经常使用马尔科夫随机场多值标签的情况是无序标签，即，两个标签的数值不差别并没有语义含义。例如，如果我们从航空图像中对地貌分类，森林、田地、水域和路面的标签之间的数值差别是没有意义的。实际上，这些不同种类的地貌的相邻性有不同的似然度，于是用一个形如下式的先验项会更有意义

$$E_p(i,j) = s_x(i,j)V(l(i,j),l(i+1,j)) + s_y(i,j)V(l(i,j),l(i,j+1)), \quad (3.115)$$

其中$V(l_0, l_1)$是一个总体的兼容性或势函数。(注意我们也用$l(i, j)$替换$f(i, j)$以便更清楚地表示这些是标签而不是离散函数样例。)另一种写这个先验能量的方法(Boykov, Veksler, and Zabih 2001; Szeliski, Zabih, Scharstein *et al.* 2008)是

$$E_p = \sum_{(p,q)\in\mathcal{N}} V_{p,q}(l_p, l_q), \quad (3.116)$$

其中(p, q)是相邻像素，$V_{p,q}$是空间变化的势函数，用来度量每个相邻对。

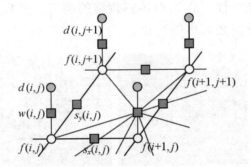

图 3.59 一个有更复杂测量模型的马尔科夫随机场的图模型。额外的彩色边展示了未知值的结合(比如，在一幅锐化图像中)是如何产生测量值的(一幅有噪声的模糊图像)。得到的图模型仍旧是一个经典的 MRF，而且同样容易进行采样，但是一些推理算法(例如，基于图割的)可能不再可用，因为网络复杂性增加了，由于推理过程中的状态改变变得更纠结，后验 MRF 有更大的团块

无序 MRF 标签的一个重要应用是图像合成中找缝隙(Davis 1998; Agarwala, Dontcheva, Agrawala *et al.* 2004)(见图 3.60，在 9.3.2 节中有更详细的解释)。这里，兼容性$V_{p,q}(l_p, l_q)$度量放置图像l_p中的像素p到图像l_q中的像素q所产生的视觉效果的质量。对于大多数 MRF，我们假设$V_{p,q}(1, 1)=0$，即从同一幅图像中选择相邻的像素会有很好的效果。但是，对于不同的标签，兼容性$V_{p,q}(l_p, l_q)$可能依赖于基底像素$I_{l_p}(p)$和$I_{l_q}(q)$。

考虑这样的情况：例如，一幅图像I_0全部是天蓝色，例如，$I_0(p) = I_0(q) = B$，而其他图像有一个从天蓝色$I_1(p) = B$的到森林绿色$I_1(q) = G$的转变，

在这种情况下，$V_{p,q}(1,0) = 0$ (色彩相符)，而 $V_{p,q}(0,1) > 0$ (色彩不符)。

图 3.60　一个无序标签 MRF(Agarwala, Dontcheva, Agrawala et al. 2004)©2004 ACM：左侧每幅源图像中的笔画被用作 MRF 优化中的限制，使用图割来进行求解。得到的多值标签场在中间图像中用彩色覆盖表示，最后的合成如右图所示

条件随机场

在一个经典的贝叶斯模型(3.106～3.108)中，
$$p(\boldsymbol{x}|\boldsymbol{y}) \propto p(\boldsymbol{y}|\boldsymbol{x})p(\boldsymbol{x}), \tag{3.117}$$
先验分布 $p(x)$ 与观测量 y 独立。但是，有时，修改我们的先验假设非常有用，比如说我们要估计的场的光滑性，作为对传感数据的响应。从概率观点来看这是否有道理，将在我们解释新模型之后再做讨论。

图 3.61　图像分割(Boykov and Funka-Lea 2006)© 2006 Springer：用户在前景物体上用红色画几笔，在背景上用蓝色画几笔。系统计算前景和背景的颜色分布，解一个二值 MRF。光滑权重由亮度梯度(边缘)来调整，这使其成为条件随机场(CRF)

考虑图 3.61 所示的交互图像分割问题(Boykov and Funka-Lea 2006)。在这个应用中，用户用笔画画出前景(红色)和背景(蓝色)，然后系统解决一个二值马尔科夫标签问题来估计前景物体的范围。除了最小化逐点度量像素色彩和所推断出的区域分布的相似度的一个数据项之外(5.5 节)，还修改了 MRF 使得图 3.56 和 (3.113) 中的光滑项 $s_x(x,y)$ 和 $s_y(x,y)$ 依赖于相邻像素间梯度的大小。[①]

由于现在光滑项依赖于数据，贝叶斯规则(3.117)不再适用。取而代之，我们用

---

① 另一个也使用检测到的边缘来调整一个深度或运动场的光滑性，并且因此集成了多个低层视觉模块的形式化由 Poggio, Gamble, and Little(1988)提出。

后验分布 $p(x\,|\,y)$ 的一个直接模型，该模型的负对数似然度可以写为

$$\begin{aligned}E(\boldsymbol{x}|\boldsymbol{y}) &= E_d(\boldsymbol{x},\boldsymbol{y})+E_s(\boldsymbol{x},\boldsymbol{y})\\ &= \sum_p V_p(x_p,\boldsymbol{y})+\sum_{(p,q)\in\mathcal{N}}V_{p,q}(x_p,x_q,\boldsymbol{y}),\end{aligned} \quad (3.118)$$

用式(3.116)中引入的概念。得到的概率分布称为"条件随机场"(CRF)，在 Lafferty, McCallum, and Pereira (2001)用于文本建模的早期工作的基础上，最初由 Kumarand Hebert(2003)引入计算机视觉领域。

图 3.62 展示了一个图模型，其中光滑项依赖于数据值。在这个特殊模型中，每个光滑项仅与它相邻的数据值对有关，即项是式(3.118)中为 $V_{p,\,q}(x_p, x_q, y_p, y_q)$ 的形式。

修改光滑项以响应输入数据的思想并不新鲜。例如，Boykov and Jolly (2001) 用这个想法来进行交互分割，如图 3.61 所示，现在它广泛应用于图像分割(5.5 节) (Blake, Rother, Brown et al. 2004; Rother, Kolmogorov, and Blake 2004)，降噪 (Tappen, Liu, Freeman et al. 2007)和物体识别(14.4.3 节)(Winn and Shotton 2006; Shotton, Winn, Rother et al. 2009)。

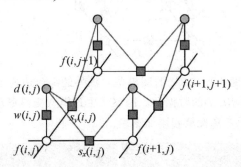

图 3.62 条件随机场(CRF)的图模型。额外的绿色边显示了传感数据的结合如何影响其 MRF 先验模型的光滑性，即式(3.113)中的 $s_x(i, j)$ 和 $s_y(i, j)$ 依赖于邻近的值。这些额外的连接(因子)使得光滑性依赖于输入数据。但是，它们使从这个 MRF 中采样更复杂

在立体视觉匹配中，鼓励视差不连续性与亮度边缘相符，甚至可追溯到优化和基于 MRF 的算法的早期时代(Poggio, Gamble, and Little 1988; Fua 1993; Bobick and Intille 1999; Boykov,Veksler, and Zabih 2001)，详细的讨论参见 11.5 节。

除了使用适合输入数据的光滑项，Kumar and Hebert(2003) 还在输入数据上对每个 $V_p(x_p; y)$ 项计算了一个邻域函数(如图 3.63 所示)来代替图 3.56 所示的经典的一元 MRF 数据项 $V_p(x_p; y_p)$。[1]因为这类邻域函数可以被认为是判别函数(一个常用于机器学习的概念(Bishop 2006))，所以他们称得到的图模型为"区分性随机场" (DRF)。在他们的论文中，Kumar andHebert(2006)展示了 DRF 在一些应用中胜于与之相似的 CRFs，例如结构检测(图 3.64)和二值图像降噪。

这里，有人又会争辩前面的立体视觉匹配算法也关注输入数据的邻域，或显式

---

[1] Kumar and Hebert (2006)称一元作用势 $V_p(x_p,y)$ 为"联合隐变量"，而逐对作用势 $V_{p,q}(x_p,y_p,y)$ 称为"交互作用势"。

地，因为它们计算相关性度量(Criminisi, Cross, Blake et al. 2006)作为数据项，或隐式地，因为即使是逐像素的视差代价也要关注左边或右边图像中的若干像素(Barnard 1989; Boykov, Veksler, and Zabih 2001)。

那么，用条件或区分性随机场来代替 MRF 各有哪些优缺点？

经典的贝叶斯推断(MRF)假设数据的先验分布与测量独立。这有很大的意义：如果你初次丢出一双骰子看到了一对六，认为它们总会出现这种情况是很不明智的。但是，如果在玩了很长一段时间后，你发现一个统计上明显的偏移，可能会想调整你的先验。CRF 所做的主要是根据观测数据选择或修改先验模型。但这可以看作对额外的隐变量或关于未知量(比如说，一个标签、深度或干净图像)和已知量(观测图像)之间的相关性进行部分推断。

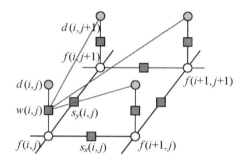

图 3.63　一个区分性随机场(DRF)的图模型。额外的绿色边显示了传感数据的结合，例如，$d(i, j+1)$，影响 $f(i, j)$ 的数据项。因此产生式模型变得更复杂，即，我们不能将一个简单的函数应用于未知变量和添加噪声

图 3.64　用一个 MRF(左)和 DRF(右)进行结构检测的结果(Kumar and Hebert 2006)©2006 Springer

在一些情况下，CRF 方法很有意义，而且实际上，是唯一的可以推进的有效方法。例如，在灰度图像色彩化(10.3.2 节)(Levin, Lischinski, and Weiss 2004)中，将输入灰度图像的连续性信息变换到未知的彩色图像的最好方法是修改局部光滑性限制。类似地，对于同时的分割和识别(Winn and Shotton 2006; Shotton, Winn, Rother et al. 2009)，允许强的彩色边缘来影响语义的图像标签连续性有很大意义。

在其他情况下，例如图像降噪，情况更微妙。用一个式(3.113)中非二次(鲁棒的)光滑项与在一个高斯 MRF(GMRF)(Tappen, Liu, Freeman et al. 2007)中根据局部梯度信息设置光滑性有相似的作用。(在更近期的工作中，Tanaka and Okutomi (2008)在一个相关的高斯 MRF 上使用一个更大的邻域和全协方差矩阵。)高斯 MRF

的优点是，如果可以正确推断平滑，得到的二次能量可以在一步中最小化。但是，当不连续性在输入数据中不能自明的情况下，例如逐片平滑稀疏数据插值(Blake and Zisserman 1987; Terzopoulos 1988)，经典的鲁棒平滑能量最小化可能更好。这样，如大部分计算机视觉算法一样，一个手上问题的细致分析和想要的鲁棒性和计算约束，是在选择最好的方法时所需要的。

如 Kumar and Hebert (2006),Tappen, Liu, Freeman et al. (2007) 和 Blake, Rother, Brown et al. (2004)所争辩，CRF 和 DRF 最大的优点，可能就是模型参数的学习有时可以更简单。然而在 MRF 和它们的变种中学习参数并不是本书中所讨论的问题，感兴趣的读者可以在最近发表的文章(Kumar and Hebert 2006; Roth and Black 2007a; Tappen, Liu, Freeman et al. 2007; Tappen 2007; Li and Huttenlocher 2008)中找到更多细节。

3.7.3 应用：图像的恢复

在 3.4.4 节中，我们看到了二维线性和非线性滤波器如何用于去除噪声和提高图像的锐度。但是，有时图像被更大的问题降级，例如划痕和斑点(Kokaram 2004)。在这种情况下，可以用贝叶斯方法(例如 MRF)来代替，这类方法可以对空间变化的测量噪声建模。一个替代的方法是用空洞填充或修图方法(Bertalmio, Sapiro,Caselles et al. 2000; Bertalmio, Vese, Sapiro et al. 2003; Criminisi, Pérez, and Toyama 2004)，5.1.4 节和 10.5.1 节对此进行了讨论。

图 3.57 展示了用马尔科夫随机场进行图像降噪和修图(空洞填充)。原始图像被噪声和一部分数据的移除破坏。在这个情况下，带环的置信度传播算法计算了一个稍低的能量和一个比 alpha-扩张图分割算法更光滑的图像。

3.8 补 充 阅 读

如果你对进一步探索图像处理主题很有兴趣，Lim(1990); Crane(1997); Gomes and Velho(1997); Jähne(1997); Pratt(2007); Russ(2007); Burger and Burge(2008); Gonzales and Woods(2008)这些教材都很受欢迎。这个领域内最优秀的会议和期刊是 ICIP 和 TIP。

对于图像合成算子，开创性的工作可见 Porter and Duff(1984)，而 Blinn (1994a,b)提供了一个更详细的教程。对于图像合成，首先让这个主题引起图形学领域注意的是 Smith and Blinn(1996)，而 Wang and Cohen(2007a)提供了一个详细的近期发展综述。

在线性滤波的领域中，Freeman and Adelson(1991)很好地介绍了可分离和导向有向带通滤波器，而 Perona(1995)展示了如何用可分离分量的和来近似任何滤波器。

非线性滤波方面的文献非常广泛，它包括双边滤波(Tomasi and Manduchi 1998;

Durand and Dorsey 2002; Paris and Durand 2006; Chen, Paris, and Durand 2007; Paris, Kornprobst, Tumblin et al. 2008)、相关的迭代算法(Saint-Marc, Chen, and Medioni 1991; Nielsen, Florack, and Deriche 1997; Black, Sapiro, Marimont et al. 1998; Weickert, ter Haar Romeny, and Viergever 1998; Weickert 1998; Barash 2002; Scharr, Black, and Haussecker 2003; Barash and Comaniciu 2004)和变分方法(Chan, Osher, and Shen 2001; Tschumperlé and Deriche 2005;Tschumperlé 2006; Kaftory, Schechner, and Zeevi 2007)之类的主题。

图像形态学很好的参考文献包括(Haralick and Shapiro 1992, Section 5.2; Bovik 2000, Section 2.2; Ritter and Wilson 2000, Section 7; Serra 1982; Serra and Vincent 1992; Yuille, Vincent, and Geiger 1992; Soille 2006)。

图像金字塔和金字塔混合的经典论文由 Burt and Adelson(1983a,b)所著。小波最初由 Mallat(1989)引入计算机视觉界，有很多好的教程、回顾论文和书(Strang 1989; Simoncelli and Adelson 1990b; Rioul and Vetterli 1991; Chui 1992; Meyer 1993; Sweldens 1997)。小波广泛应用于计算机图形学领域的多分辨率几何处理(Stollnitz, DeRose, and Salesin 1996)，并且已经被用于计算机视觉的类似应用(Szeliski 1990b; Pentland 1994; Gortler and Cohen 1995; Yaou and Chang 1994; Lai and Vemuri 1997; Szeliski 2006b)，也用于多尺度有向滤波(Simoncelli, Freeman, Adelson et al. 1992)和降噪(Portilla, Strela, Wainwright et al. 2003)。

图像金字塔常用线性滤波算子建造，而一些近期的工作已经开始研究非线性滤波器，因为这些可以更好地保护细节和其他显著特征。在计算机视觉领域中的一些有代表性的论文由 Gluckman(2006a,b); Lyu and Simoncelli(2008)所著，在计算形态学中由 Bae, Paris, and Durand(2006); Farbman, Fattal, Lischinski et al.(2008); Fattal (2009)所著。

图像卷绕和重采样的高质量算法在图像处理领域(Wolberg 1990; Dodgson 1992; Gomes, Darsa, Costa et al. 1999; Szeliski,Winder, and Uyttendaele 2010)和计算机图形学(Williams 1983; Heckbert 1986; Barkans 1997; Akenine-Möller and Haines 2002)中均有涉及，称为"纹理映射"。将图像卷绕和图像融合方法结合起来用于图像之间的变形，一系列有影响的论文和书(Beier and Neely 1992; Gomes, Darsa, Costa et al. 1999)都有涉及此内容。

计算机视觉问题的正则化方法是由 Poggio, Torre, and Koch (1985) 和 Terzopoulos(1986a,b, 1988)最初引入视觉领域的，并且持续成为一个受欢迎的用于形式化和解决低层视觉问题(Ju, Black, and Jepson 1996; Nielsen, Florack, and Deriche 1997; Nordström 1990; Brox, Bruhn, Papenberg et al. 2004; Levin, Lischinski, and Weiss 2008)的框架。更详细的数学解释和附加的应用可以参见应用数学和统计学文献(Tikhonov and Arsenin 1977; Engl, Hanke, and Neubauer 1996)。

马尔科夫随机场的文献真是巨多，还有视觉工作者很少注意到的优化和控制理论等相关领域的出版物。一个对于最新方法很好的导引是 Blake, Kohli, and Rother

(2010)所编辑的书。其他近期的包含很好的文献回顾或实验比较的文章有 Boykov and Funka-Lea 2006; Szeliski, Zabih, Scharstein et al. 2008; Kumar, Veksler, and Torr 2010。

马尔科夫随机场的开创性论文是 Geman and Geman(1984)，他们将这个形式介绍给计算机视觉研究者，也引入了直线处理的概念，以及用于控制是否实施光滑惩罚的额外的二值变量。Black and Rangarajan(1996)展示了独立直线处理如何替换为鲁棒的成对势；Boykov, Veksler, and Zabih(2001)发展了递归二值、图割算法来做多标签 MRFs 的优化；Kolmogorov and Zabih(2004)描述了这些方法的必要前提二值能量变量类；Freeman, Pasztor, and Carmichael(2000)推广了 MRF 推断的带环的置信度传播的使用。更多附加的参考文献可以参见 3.7.2 节、5.5 节和附录 B.5。

3.9 习　题

习题 3.1：色彩平衡

写一个简单的的应用，通过将每个色彩值乘以一个用户指定的不同常量的方法来改变一幅图像的色彩平衡。如果想把这个做得更好，你可以将这个做成有滑块的交互应用。

1. 如果在做乘法之前或之后进行伽玛变换，结果会有不同吗？为什么相同或者为什么不同？
2. 用你的数码相机以不同的色彩平衡设置拍摄同样的照片(大部分相机在菜单中有控制色彩平衡的选项)。你能恢复不同设置之间的色彩平衡比例吗？你可能需要将相机放到一个三角架上，手工或自动对齐图像来完成这项工作。或者，使用一个色彩棋盘格图表(图 10.3b)，参见 2.3 节和 10.1.1 节的讨论。
3. 如果可以得到相机的 RAW(原始数据)图像，自行去马赛克(10.3.1 节)或对图像分辨率下采样得到一个"真"RGB 图像。你的相机是在 RAW 值和一个 JPEG 中色彩平衡的值之间进行简单线性映射吗？一些高端相机有一种 RAW+JPEG 模式，可以简化这种比较。
4. 你能想到可能需要在图像上进行色彩扭曲(3.1.2 节)的理由吗？也可参考习题 2.9 获得一些启发。

习题 3.2：合成和反射

3.1.3 节中描述了在另一张图像上合成一张 alpha-抠图的过程。回答以下问题，可选择用实验验证它们：

1. 大部分拍摄到的图像都做过伽玛校正。这是否使基本合成方程 3.8 失效？如果是，该如何解决？
2. 加性模型(纯反射)可能受限。如果玻璃是有色的，特别是对于一个非黑白的色调将会怎样？如果玻璃脏了或者模糊了会怎样？你将如何建立有波纹的玻璃或者其他类型的折射物体的模型？

习题 3.3：蓝屏抠图

建立一个蓝色或绿色的背景，例如，通过买一大块彩色广告板。先拍摄一幅空背景的图像，再拍摄一幅在背景前有一个新物体的图像。用每个彩色像素点和它所对应的背景像素点的差得到形状，用 3.1.3 节描述的方法或 Smith and Blinn (1996) 所提出的方法进行抠图。

习题 3.4：差分键控特效

实现一个差分键控特效算法(见 3.1.3 节)(Toyama, Krumm, Brumitt *et al.* 1999)，包括以下步骤。

1. 计算一个"空"视频序列中每个像素点的均值和方差(或中值和鲁棒方差)。
2. 对每一帧，对每个像素为前景或背景做分类(将背景像素设为 RGBA=0)。
3. (可选)计算 alpha 通道，在一个新背景上合成。
4. (可选)用形态学方法对图像做后期处理(3.3.1 节)，标出连通分量(3.3.4 节)，计算它们的质心，在帧间跟踪它们。用这种方法建立"人的计数器"。

习题 3.5：图像效果

编写各种图像增强或效果的滤波器：对比度和曝光(量化)等。哪些有用(进行可感觉到的校正)，哪些比较有创造性(创造了不寻常的图像)？

习题 3.6：直方图均衡化

计算一幅图像的灰度级别(亮度)直方图，并将它均衡化使色调看起来更好(并且图像对于曝光设置不太敏感)。你可能想要使用以下步骤。

1. 将彩色图像转换为亮度(3.1.2 节)。
2. 计算直方图、累积分布以及补偿传递函数(3.1.4 节)。
3. (可选)试着通过保证一部分像素(例如，5%)被映射为纯黑色和白色来提高图像中的"力量(punch)"。
4. (可选)在传递函数中限制局部增益 $f(I)$。一个实现这个的方法是在进行积累式(3.9)时限制 $f'(I) < \gamma I$ 或 $f'(I) < \gamma$，同时保持任何未累积的值"备用"。(请自己搞清楚确切细节)
5. 通过查找表来补偿亮度通道，用色彩比式(2.116)重新生成彩色图像。
6. (可选)彩色值在原图像中被做截尾处理，即有一个或更多饱和的色彩通道，在被重映射到一个未截尾的值时可能看起来不自然。扩展算法以有用的方式来解决这个问题

习题 3.7：局部直方图均衡化

对每个图像块计算灰度级别(亮度)直方图，但是基于距离加到顶点上(一个样条)。

1. 以习题 3.6 为基础(亮度计算)。
2. 分配值(计数)给邻近节点(双线性)。
3. 转换到 CDF(查找函数)。
4. (可选)使用 CDF 的低通滤波。
5. 插值邻近 CDF 以得到最终查找表。

习题 3.8：邻域填充操作

对图 3.13 所示的每个填充模式，写一个将填充像素值 $f(I, j)$ 作为原始像素值 $f(k, l)$ 和图像的宽和高 (M, N) 之函数的计算公式。例如，对于图像复制(箝入)

$$\tilde{f}(i,j) = f(k,l), \quad \begin{array}{l} k = \max(0, \min(M-1, i)), \\ l = \max(0, \min(N-1, j)), \end{array}$$

提示：除了常规代数运算，你可能想用最大、最小、取余和绝对值操作。

- 更详细地描述各种模式的优缺点。
- (可选)通过绘制一个纹理坐标位于(0.0; 1.0)范围之外的纹理映射的矩形，检查你的显卡做了什么，使用不同的纹理箝入模式。

习题 3.9：可分离滤波器

用一个可分离的核实现卷积。输入应该是一个灰度或彩色图像和水平或垂直的核。确保你支持前面习题所建立的填充机制。后面的一些习题需要这个功能。如果你使用的图像处理包(例如 IPL)中已经包含可分离滤波器，可跳过这个习题。

(可选)用 Pietro Perona(1995)的方法来将卷积近似为一系列可分离核之和。让用户确定核的数目，给出近似程度的某种可感知的度量。

习题 3.10：离散高斯滤波器

实现一个高斯滤波器并讨论以下问题。

- 如果只在离散位置对一个连续高斯滤波器公式进行采样，你会得到想要的性质(例如系数之和会为 1)吗？类似地，如果对高斯函数的导数进行采样，采样值的和会为 0 或有消失更高阶矩吗？
- 将原始信号用 sinc 插值，用连续高斯模糊，然后在重采样前用 sinc 预滤波，这样做会不会更好？在频域中有没有一个更简单的方式来实现这个过程？
- 在傅里叶域产生一个高斯频率响应，然后做一个逆 FFT 变换得到一个离散滤波器会不会更合理呢？
- 滤波器的截断如何改变频率响应？这会引入某种附加失真吗？
- 得到的二维滤波器是否像它们的连续类比项一样旋转不变？是否有一些方法来改进这个？实际上，有任何的 2D 离散(可分离或不可分离的)滤波器是真的旋转不变的吗？

习题 3.11：锐化、模糊和噪声去除

实现一些软化、锐化和非线性扩散(选择性锐化或噪声去除)滤波器，例如高斯、中值和双边(3.3.1 节)，相关讨论参见 3.4.4 节。

找一些模糊或有噪声的图像(低光照条件下拍摄的图像是得到两者很好的方

法)，试着改善它们的外观和清晰度。

习题 3.12：导向滤波器

实现 Freeman and Adelson (1991) 的导向滤波算法。输入应该是一个灰度或彩色图像，输出应该是包括 $G_1^{0°}$ 和 $G_1^{90°}$ 的多带图像。滤波器系数可以在 Freeman and Adelson (1991) 的论文中找到。在你选择的一些图片上测试各阶滤波器，看看是否能够可靠地找到角点和交叉点特征。这些滤波器在后面检测直线等细长的结构时会非常有用(4.3 节)。

习题 3.13：距离变换

实现一些城市块和欧几里得距离变换的(光栅扫描)算法。你可以不看文献(Danielsson 1980; Borgefors 1986)完成它吗？如果可以，你遇到了什么问题，如何解决的？之后，你可以用所计算的距离函数在图像拼接过程中进行羽化(参见 9.3.2 节)。

习题 3.14：连通分量

实现 3.3.4 节或 Haralick and Shapiro 书中 2.3 节所介绍的一个连通分量算法，讨论它的计算复杂度。

- 阈值化或量化一幅图像以得到各种输入标签，然后计算你所找到的区域的面积统计量。
- 用你所找到的连通域在不同图像或视频帧中跟踪或匹配区域。

习题 3.15：傅里叶变换

证明表 3.1 中所列的傅里叶变换的性质，推导出表 3.2 和表 3.3 中所列出的傅里叶变换公式。这些习题对于你今后轻松使用傅里叶变换非常有帮助，在分析和设计很多计算机视觉算法的性能和效率时非常有用。

习题 3.16：维纳滤波

估计你个人照片集的频谱，用它在几张有不同程度噪声的图像上进行维纳滤波。

1. 收集几百张你的图片，将它们重缩放以适合 512×512 大小的窗口并进行裁减。
2. 对它们进行傅里叶变换，丢掉相位信息，对所有谱进行平均。
3. 挑选两张你最喜欢的图像，加上不同数量的高斯噪声，$\sigma_n \in$ {1; 2; 5; 10; 20}灰度级别。
4. 对每个图像和噪声的组合，肉眼确定什么宽度的高斯模糊滤波器可以得到最好的降噪结果。你需要在清晰度和噪声之间做出一个主观的决定。
5. 计算所有噪声图像经过维纳滤波的版本，将它们与你手工调整的高斯光滑图像进行对比。
6. (可选)你的图像谱有很多能量集中在水平和竖直轴($f_x = 0$ 和 $f_y = 0$)周围吗？对此，你能想到一个合理的解释吗？将你的图像例子旋转 45°会将这个能量移到对角线周围吗？如果不能，这个可能是由于傅里叶变换中的边界效应吗？你能建议一些减少这些效应的方法吗？

习题 3.17：用维纳滤波去模糊

用维纳滤波来去除一些图像中的模糊。

1. 修改维纳滤波的推导式(3.66)~(3.74)以加入模糊(3.75)。
2. 就噪声抑制和频率增强特性讨论所得到的维纳滤波器。
3. 假设模糊核是高斯的，图像频谱遵从一个逆频率法则，计算维纳滤波器的频率响应，将它和非锐化掩模进行比较。
4. 用不同半径的高斯模糊核模拟地模糊两幅样例图像，添加噪声，然后进行维纳滤波。
5. 用一个"抛物柱面"(圆盘)模糊核重复上述实验，是一个典型的有限光圈透镜的特性(2.2.3 节)。将这些结果与高斯模糊核作比较(确保检查你的频率图)。
6. 有人提出，常规的光圈不适合去模糊，因为它们在传感频谱(Veeraraghavan, Raskar, Agrawal *et al*. 2007)中引入零点。证明如果没有信号的先验模型，即 $P_s^{-1}I_1$，这确实是个问题。如果假设一个合理的功率谱，这还是个问题吗(我们仍旧会有分带或振铃缺陷吗)？

习题 3.18：高质量图像重采样

实现若干 3.5.2 节中提到的低通滤波器，还有关于表 3.2 和图 3.29 中所展示的窗 sinc 的讨论。可选实现其他一些滤波器(Wolberg 1990; Unser 1999)。

将你的滤波器应用于连续地改变一幅图像尺寸，既有放大(插值)又有缩小(降采样)；比较几种滤波器的动画结果。分别使用一张合成的线性调频脉冲图像(图 3.65a)和有很多高频细节的自然图像(图 3.65b 和 c)。①

你可能会发现编写一个简单的可视化程序来连续同时播放两个或更多滤波器的动画会非常有帮助，这可以让你在不同结果间快速切换。

讨论每个滤波器的优缺点，以及速度和品质之间的平衡。

(a)　　　　　　　　(b)　　　　　　　　(c)

图 3.65　用于测试重采样算法质量的样图：(a)一幅合成线性调频脉冲图像；(b)和(c)图像压缩领域的一些高频图像

习题 3.19：金字塔

建造一个图像金字塔。输入应该是一幅灰度或彩色图像，一个可分离的滤波器核和所希望的级数。至少实现以下的核：

---

① 这些特殊图像可以在本书网站上找到。

- 2×2 块滤波
- Burt and Adelson 的二项式核 1/16(1, 4, 6, 4, 1)(Burt and Adelson 1983a);
- 一个高质量的七或九阶滤波器
- 比较各种降采样滤波器的视觉质量。将你的输入图像移动 1 到 4 像素,比较得到的降采样(1/4 大小)图像序列

习题 3.20:金字塔混合

编写一个程序,输入两幅彩色图像和一个二维掩模图像,产生两幅图像混合的拉普拉斯金字塔。

1. 创建每个图像各自的拉普拉斯金字塔。
2. 创建两幅掩模图像的高斯金字塔(输入图像和它的补集)。
3. 将每幅图像乘以对应的掩模,对图像求和(如图 3.43 所示)。
4. 从混合的拉普拉斯金字塔中重建最终图像。

推广你的算法到输入 n 幅图像和一个有 $1\cdots n$ 标号(数值 0 可以代表"无输入")的情况。讨论带权相加阶段(步骤 3)是需要记录总的权重以便重新归一化,还是有效的数学方法可用。用你的算法混合两幅曝光不同的图像(来避免曝光不足和过曝光区域)或者产生两幅不同场景的创造性融合。

习题 3.21:小波建立和应用

实现 3.5.4 节或 Simoncelli and Adelson (1990b)所描述的小波族之一以及基本的拉普拉斯金字塔(习题 3.19)。将产生的表达应用于以下两个任务。

- 压缩:计算不同小波实现的每个带的熵,假定已经给定一个量化级别 (比如说, 1/4 灰度级,或保持舍入误差在可接受范围内)。量化小波系数,重建原始图像。哪种方法效果更好? (典型的结果可参见(Simoncelli and Adelson 1990b)或大量小波压缩论文的任何一个。)
- 降噪:在计算小波后,用核化(coring)降噪抑制小值,即用一个分段线性或其他 C_0 函数将小的数值设为零。用不同的小波和金字塔表达比较降噪结果。

习题 3.22:参数化图像卷绕

编写一段进行仿射和透视图像卷绕(也可以选择双线性)的代码。尝试多种插值子,汇报它们的视觉质量。具体讨论下面两个问题。

- 在一个 MIP-映射中,仅选择与所计算的分数级别相邻的较粗糙级别会产生一个更模糊的图像,而选择更细致级别会导致混叠。解释一下原因,讨论混合一个混淆和模糊的图像(三线性 MIP-映射)是否是一个好主意。
- 当水平和竖直重采样率的比例变得非常不同(异性),MIP-映射表现更糟糕。给出一些更好的方法来减少这类问题。

习题 3.23:局部图像卷绕

打开一幅图像,按照下述方法之一对它的外观进行变形。

1. 点击一定数量的像素,移动(拖拉)它们到新的位置。插值得到的稀疏位移场得到一个密集的运动场(3.6.2 节和 3.5.1 节)。

2. 在图像中画若干条线。移动这些线的终端来指定它们新的位置，用 3.6.2 节讨论的 Beier–Neely 插值算法(Beier and Neely 1992)得到一个密集的运动场。
3. 覆盖一个样条控制网格，每次移动一个网格点(或者选择变形的级别)。
4. 做一个稠密逐像素光流场，用一个软的画刷来设计一个水平和数值速度场。
5. (可选)：证明 Beier–Neely 卷绕是或不是随着线段长度变短(减少到点)简化为基于稀疏点的变形。

习题 3.24：前向卷绕

给定前面习题的一个位移场，写一个前向卷绕算法。

1. 写一个用体空间扫描(splatting)的前向卷绕，使用最近邻或软累积(3.6.1 节)。
2. 写一个二遍(two-pass)算法，前向卷绕位移场，填补小的空洞，然后用逆向卷绕(Shade, Gortler, He *et al.* 1998)。
3. 比较这两种算法的质量。

习题 3.25：基于特征的变形

扩展你在习题 3.23 中所写的卷绕的代码，引入两幅不同的图像，指定两幅图像之间的对应(基于点、线或网格)。

1. 创建一个将图像彼此部分卷绕的变形和渐隐渐现解(3.6.3 节)。
2. 试着用你的变形算法进行图像旋转，讨论它的表现是否如你所愿。

习题 3.26：2D 图像编辑器

扩展你在习题 2.2 中所写的程序，引入图像并允许你创建一个图像"库"。步骤如下。

1. 打开一幅新图像(在一个独立的窗口)。
2. 平移拉动(rubber-band)来剪裁一个子区域(或选择整幅图像)。
3. 将它粘贴到当前背景上。
4. 选择变形模式(运动模型)：平移、刚性、相似、仿射或透视。
5. 拉动轮廓的一角来改变它的变换。
6. (可选)改变图像的相对顺序，以及正在处理的当前图像。

用户应该看到各种图像片断在彼此上的合成。

这个习题应该建立在软件库支持的图像变换类的基础上。应该支持对所得外观的保持(保存和载入)(对每幅图像，保存它的变换)。

习题 3.27：3D 纹理映射观察器

扩展习题 2.3 中所创建的观察器，包含纹理映射多边形绘制。将每个多边形用坐标$(u; v; w)$嵌入到一幅图像中。

习题 3.28：图像降噪

实现本章所描述的各种图像降噪方法中的至少两种，并将它们在合成的噪声图像序列和真实世界(低光照)序列上作比较。算法的效果是否依赖于对噪声级别估计的正确选择？你能得出结论说明哪种方法更好吗？

习题 3.29：彩虹增强器-有挑战的

拍摄一幅包括一条彩虹的图像，如图 3.66 所示，增强彩虹的强度(饱和度)。

1. 在图像中画一道弧，标出彩虹的范围。
2. 对这条弧(最好使用线性化的像素值)拟合一个加性的彩虹函数(解释为什么是加性的)，用频谱作为截面，估计弧的宽带和增加色彩的数量。这是这个问题中最棘手的一部分，因为你需要分解(低频的)彩虹模式和隐藏在其后的自然图像。
3. 放大彩虹信号，将它加回到图像中，如果必要则重新应用伽玛函数产生最终图像。

图 3.66 这幅暗流的图中，有一道右侧可见的彩虹，你能否想到一种增强它的方法？(习题 3.29)

习题 3.30：图像去块效应——有挑战性的

现在你已经有一些很好的方法来区分信号和噪声，提出一个方法来去除 JPEG 在高压缩设置下产生的块状瑕疵(2.3.3 节)。你的方法可能简单到沿着块边界寻找意想不到的边缘，将量化阶段看作是一个变换系数空间的凸区域到对应量化值的投影。

1. JPEG 头信息中的压缩因子是否可帮助你得到更好的去块效果？
2. 由于在 DCT 变换 YCbCr 空间(2.115)中会发生量化，在这个空间里进行分析可能更好。另一方面，图像先验在一个 RGB 空间中更有意义(是吗？)。确定你怎样处理这两方面，讨论你的选择。
3. 由于 YCbCr 转换后跟随着一个色度子采样阶段(在 DCT 之前)，看看你是否可以用本章中所讨论的更好的恢复方法恢复一些丢失的高频色度信号。
4. 如果你的相机有一个 RAW+JPEG 模式，那么你距离无噪声真实像素值能有多近？

(这个建议可能不那么有用，因为相机通常在它们的 RAW+JPEG 模型上使用质量相当高的设置)。

习题 3.31：去模糊推断——有挑战性的

写一个图 3.59 对应的图模型来解决非盲图像去模糊问题，即其中事先知道模糊核。

你能想出哪些有效推断(优化)算法来解决这类问题？

第 4 章
特征检测与匹配

4.1 点与块
4.2 边缘
4.3 线条
4.4 补充阅读
4.5 习题

图 4.1 用于图像分析,图像描述和图像匹配的各种特征检测器和描述子:(a)点状兴趣算子(Brown, Szeliski, and Winder 2005)© 2005 IEEE;(b)区域状兴趣算子(Matas, Chum, Urban et al. 2004)© 2004 Elsevier;(c)边缘(Elder and Goldberg 2001)© 2001 IEEE;(d)直线(Sinha, Steedly, Szeliski et al. 2008)© 2008 ACM

特征检测和匹配是许多计算机视觉应用中的一个重要组成部分。比如图 4.2 中给出的两个图像对，对于第一对图像，我们可能希望配准(align)这两幅图像以使它们能够无缝地合成一幅复合的马赛克图(第 9 章)。对于第二对图像，我们可能希望建立一个密集的对应(correspondence)集合从而使得能够建立一个三维模型或者生成一个中间视图(第 11 章)。对于每一种情况，为了建立这样的一个配准或者对应集合，你将检测哪种特征(feature)来进行匹配？在继续阅读之前，请先用一些时间考虑一下这个问题。

图4.2 两对需要匹配的图像对。哪些种类的特征可以用来在这些图像之间建立对应呢？

你可能想到的第一类特征就是图像中的那些特殊位置，比如山的顶部、建筑物的角点处、门口或者特殊形状的雪块。这些局部的特征通常被叫"关键点特征"(keypoint feature)或者"兴趣点"(interest point)，或者还可以叫"角点"(corner)，它们通常用其位置点周围像素块的表观来描述(4.1 节)。另外一类特征是"边缘"(edge)，比如，对着天空的山峰的轮廓(4.2 节)。基于它们的方向和局部表观(边缘结构)，可以对这些不同类型的特征进行匹配，而且这些特征还可以作为图像序列中出现物体边界和遮挡(occlusion)情况的理想指示器。边缘可以聚成长的曲线和直线片段(straight line segment)，这些线段可以直接用于匹配或者用于分析寻找消失点，从而获取摄像机的内部参数和外部参数(4.3 节)。

在本章中，我们描述检测这些特征的一些实用方法，并且讨论如何在不同的图像之间建立特征对应。点特征现在已经被用于各种各样的应用之中，阅读和实现 4.1 节中给出的部分算法是一个非常不错的练习方法。边缘和线段提供的信息与基于关键点和基于区域的描述子构成互补，特别适合于描述物体边界和各种人造物体。尽管这些可供选择的特征非常有用，但在简短的概论性教程中可以跳过不作介绍。

4.1 点和块

点特征可以用来寻找一个不同图像中的对应位置的稀疏集合,这通常是计算摄像机姿态(第 7 章)的一个前期步骤,而摄像机姿态计算又是使用立体视觉匹配(第 11 章)计算稠密对应集合的一个前提。这样的对应也可以用于对齐不同的图像,例如,在拼接马赛克图像或者进行视频稳定化的时候(第 9 章)。在进行物体实例和类别识别时也广泛地用到它们(14.4 节)。关键点的一个重要优点是,它们甚至能够在出现拥挤(遮挡)、大的尺度和方向变化的情况下很好地做匹配。

从早期的立体视觉匹配开始(Hannah 1974; Moravec 1983; Hannah 1988),就开始使用基于点对应的方法了,近些年来,它们又被广泛用于图像拼接应用(Zoghlami, Faugeras, and Deriche 1997; Brown and Lowe 2007)和自动三维建模(Beardsley, Torr, and Zisserman 1996; Schaffalitzky and Zisserman 2002; Brown and Lowe 2003; Snavely, Seitz, and Szeliski 2006)。

为了获取特征点及其之间的对应关系,主要有两种方法。第一种方法是在第一幅图像中寻找那些可以使用局部搜索方法来精确跟踪的特征,比如相关或者最小二乘(4.1.4 节)。第二种方法是在所有考察的图像中独立地检测特征点然后再基于它们的局部表观进行匹配(4.1.3 节)。前一种方法在图像以相近的视角或被快速地连续(比如,视频帧序列)拍摄时更为合适,而后一种方法更适合于在图像中存在大量的运动或者表观变化的时候,比如在拼接全景图时(Brown and Lowe 2007)、在宽基线立体视觉(wide baseline stereo)中建立对应关系时(Schaffalitzky and Zisserman 2002)或者进行物体识别时(Fergus, Perona and Zisserman 2007)。

在本节中,我们将关键点检测和匹配流水线分为四个阶段。在"特征检测"(feature detection)(提取)阶段(4.1.1 节),从每一幅图像中寻找那些能在其他图像中较好匹配的位置。在"特征描述"(feature description)阶段(4.1.2 节),把检测到的关键点周围的每一个区域转化成一个更紧凑和稳定(不变的)的描述子(descriptor),描述子可以和其他描述子进行匹配。"特征匹配"(feature matching)阶段(4.1.3 节)则在其他图像中高效地搜索可能的匹配候选。"特征跟踪"(feature tracking)阶段(4.1.4 节)是第三个阶段的另一种替代方式,它只在检测到的特征点周围一个小的邻域内寻找匹配,因此更适合视频处理。

在 David Lowe(2004)的论文中,可以找到这些阶段的很好的例子,这篇论文描述了他的"尺度不变特征变换"(Scale Invariant Feature Transform,SIFT)的原理和改进。其他可选方法的详细描述可以在一系列的综述和评测论文中看到,这些论文还覆盖了特征检测(Schmid, Mohr and Bauckhage 2000; Mikolajczyk, Tuytelaars, Schmid *et al.* 2005; Tuytelaars and Mikolajczyk 2007)和特征描述(Mikolajczyk and Schmid 2005)。Shi and Tomasi(1994)和 Triggs(2004)也对特征检测方法做了一个很好的回顾。

4.1.1 特征检测器

我们怎样才能找到能够在其他图像中稳定匹配的图像位置,也就是说,什么是适合跟踪的特征(Shi and Tomasi 1994; Triggs 2004)?再看一下图 4.3 给出的图像对和三个样例图像块,看它们能够多好地被匹配和跟踪。或许你已经注意到,无明显纹理结构的图像块几乎不可能定位,而拥有较大对比度变化(梯度)的则比较容易定位,尽管一个单一方向的直线段存在着"孔径问题"(aperture problem)(Horn and Schunck 1981; Lucas and Kanade 1981; Anandan 1989),也就是说,仅可能沿边缘方向的法线方向进行对齐(图 4.4b)。拥有至少两个(明显)不同方向梯度的图像块最容易定位,如图 4.4a 所示。

图 4.3 图像对(上)和提取的块(下)。请注意,有些块可以比其他块以更高的精确度被定位或匹配

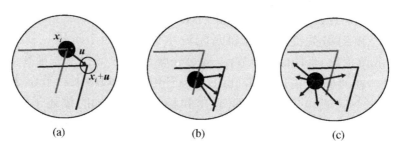

图 4.4 不同图像块的孔径问题:(a)稳定的(类似于角点的)流;(b)经典的孔径问题;(c)无纹理的区域。两个图像 I_0(黄)和 I_1(红)重叠了。红色矢量 u 指出了图块中心之间的移位,$W(x_i)$加权函数(图块窗口)如图中的黑色圆圈所示

这些直觉可以这样来形式化:用最简单的可能匹配策略来比较两个图像块,也就是,它们的(加权)差的平方和,

$$E_{\text{WSSD}}(u) = \sum_i w(x_i)[I_1(x_i + u) - I_0(x_i)]^2, \tag{4.1}$$

其中 I_0 和 I_1 是两幅需要比较的图像,$u = (u, v)$ 是平移向量,$\omega(x)$ 是在空间上变化的权重(或窗口)函数,求和变量 i 作用于块中的全体图像像素。值得一提的是,这个

形式化和我们后面用来估计整个图像间运动的形式化是相同的(8.1 节)。

在进行特征检测时，我们不知道该特征被匹配时会终止于哪些相对的其他图像位置的匹配。因此，我们只能在一个小的位置变化区域 $\Delta\boldsymbol{u}$ 内，通过与原图像块进行比较来计算这个匹配结果的稳定度，这就是通常所说的"自相关函数"(auto-correlation function)或自相关表面(图 4.5)[①]注意，对应于有纹理花坛的自相关表面(图 4.5b 和图 4.5a 右下象限中的红十字)存在一个很强的最小值，这表明它很容易定位。对应于房顶边缘的自相关表面在一个方向上存在很大的歧义性，而对应于云朵区域(图 4.5d)的自相关表面则没有稳定的最小值。

$$E_{\mathrm{AC}}(\Delta\boldsymbol{u}) = \sum_i w(\boldsymbol{x}_i)[I_0(\boldsymbol{x}_i + \Delta\boldsymbol{u}) - I_0(\boldsymbol{x}_i)]^2 \tag{4.2}$$

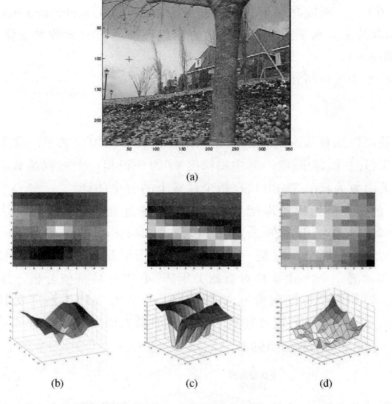

图 4.5 显示成灰度值图像和表面图的三个自相关表面 $E_{\mathrm{AC}}(\Delta\boldsymbol{u})$：(a)自相关表面计算的位置在原始图像上用三个红十字表示；(b)这个图像块取自于花坛(有良好的唯一最小值)；(c)这个图像块取自于屋顶(一维的孔径问题)；(d)这个图像块取自于云朵(没有好的峰值)。图像(b)~(d)中的每一个网格点都是 $\Delta\boldsymbol{u}$ 的一个值

使用图像函数 $I_0(x_i + \Delta\boldsymbol{u}) \approx I_0(x_i) + \nabla I_0(x_i) \cdot \Delta\boldsymbol{u}$ 的泰勒序列展开(Lucas and Kanade 1981; Shi and Tomasi 1994)，我们可以将自相关表面近似为

---

① 严格地说，一个自相关函数是两个块的乘积(3.12)；我这里以定性的方式使用这个术语。加权方差和通常称作 SSD 表面(8.1 节)。

$$E_{AC}(\Delta u) = \sum_i w(\boldsymbol{x}_i)[I_0(\boldsymbol{x}_i + \Delta \boldsymbol{u}) - I_0(\boldsymbol{x}_i)]^2 \quad (4.3)$$

$$\approx \sum_i w(\boldsymbol{x}_i)[I_0(\boldsymbol{x}_i) + \nabla I_0(\boldsymbol{x}_i) \cdot \Delta \boldsymbol{u} - I_0(\boldsymbol{x}_i)]^2 \quad (4.4)$$

$$= \sum_i w(\boldsymbol{x}_i)[\nabla I_0(\boldsymbol{x}_i) \cdot \Delta \boldsymbol{u}]^2 \quad (4.5)$$

$$= \Delta \boldsymbol{u}^T A \Delta \boldsymbol{u}, \quad (4.6)$$

其中

$$\nabla I_0(\boldsymbol{x}_i) = (\frac{\partial I_0}{\partial x}, \frac{\partial I_0}{\partial y})(\boldsymbol{x}_i) \quad (4.7)$$

是 x_i 处的"图像梯度"(image gradient)。梯度的计算可以采用多种方法(Schmid, Mohr, and Bauckhage 2000)。经典的 Harris 检测器(Harris and Stephens 1988)使用了滤波器(-2 -1 0 1 2)，而现在更为普遍的演变版本(Schmid, Mohr, and Bauckhage 2000; Triggs 2004)则是采用水平方向上和垂直方向的高斯函数的导数对图像进行卷积(通常情况下使用 $\sigma=1$)。

自相关矩阵 A 可以写作

$$A = w * \begin{bmatrix} I_x^2 & I_x I_y \\ I_x I_y & I_y^2 \end{bmatrix}, \quad (4.8)$$

其中，我们使用以加权核 ω 进行离散卷积来替换其中的加权求和。这个矩阵可以被解释为一个张量(多频带)图像，其中将梯度 ∇I 的外积和一个加权函数 ω 进行卷积，来获取自相关函数局部(二次)形状在每个像素上的一个估计。

矩阵 A 的逆矩阵给出了匹配块所在位置不确定度的一个下界。这一点最初由 Anandan(1984；1989)指出，在 8.1.3 节和公式(8.44)中会进一步讨论。因此对于哪一个图像块可以稳定匹配，它是一个非常有用的指示器。为了可视化和分析这个不确定度，最简单的一个方式就是对自相关矩阵 A 进行特征值分析，这就产生了两个特征值 (λ_0, λ_1) 和两个特征向量方向(图 4.6)。因为较大的不确定度取决于较小特征值，也就是 $a_0^{-1/2}$，所以通过寻找较小特征值的最大值寻找好的特征以便进行跟踪也就讲得通了(Shi and Tomasi 1994)。

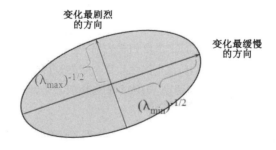

图 4.6 对应于自相关矩阵 A 的特征值分析的不确定性椭圆

Förstner–Harris

尽管 Anandan and Lucas and Kanade(1981)是第一个分析自相关矩阵不确定性结

构的，但是他们当时的背景是为光流测量确定性。Forstner(1986)和 Harris and Stephens(1988)是第一个提出使用从自相关矩阵导出的旋转不变标量测量的局部最大值来定位关键点以达到匹配稀疏特征之目的的。Schmid, Mohr, and Bauckhage (2000); Triggs(2004)等给出了更具体的关于特征检测算法的历史回顾。这两种方法都提出使用高斯权重窗口来替换之前使用的方形图像块，这使检测器的响应对于图像平面内旋转不敏感。

特征值的最小值 λ_0 (Shi and Tomasi 1994)不是唯一可以用来寻找关键点的量，另一个由 Harris and Stephens(1988)提出的更简单的量是

$$\det(\boldsymbol{A}) - \alpha\, \mathrm{trace}(\boldsymbol{A})^2 = \lambda_0\lambda_1 - \alpha(\lambda_0 + \lambda_1)^2 \tag{4.9}$$

其中 $\alpha = 0.05$。与特征值分析不同，这个量不要求使用平方根并且仍然保持旋转不变，同时也降低了那些 $\lambda_0 \gg \lambda_1$ 的边界类特征的权重。Triggs(2004)建议使用这个量

$$\lambda_0 - \alpha\lambda_1 \tag{4.10}$$

(比如，使用 $\alpha = 0.05$)，它也能够减小 1 维的边缘响应，但图像失真错误有时会使较小的特征值变大。他也指出了如何将基本的 2×2 Hessian 矩阵扩展到参数化运动中来，以检测那些能够在尺度和旋转变化下可准确定位的点。另一方面，Brown, Szeliski, and Winder(2005)使用调和均值

$$\frac{\det \boldsymbol{A}}{\mathrm{tr}\, \boldsymbol{A}} = \frac{\lambda_0\lambda_1}{\lambda_0 + \lambda_1}, \tag{4.11}$$

在 $\lambda_0 \approx \lambda_1$ 的区域，它是一个更平滑的函数。图 4.7 显示了不同兴趣点算子的等高线，从中我们可以看出这两个特征值是如何进行结合用于决定最终的兴趣值。

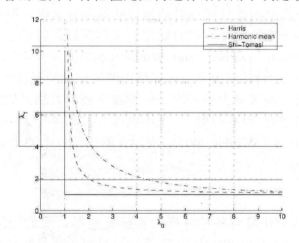

图 4.7 常见的关键点检测函数的等高线图(Brown, Szeliski, and Winder 2004)。每个检测器寻找矩阵 $\boldsymbol{A} = w * \nabla I \nabla I^T$ 的特征值 λ_0, λ_1 都大的点

Harmonic mean：调和均值

算法 4.1 总结了一个基本的基于自相关矩阵关键点检测器的步骤。图 4.8 给出了经典的 Harris 检测器和做高斯差分(DoG)检测器对应的检测响应结果。

1. 通过用高斯函数的导数对原始图像进行卷积来计算图像在水平方向和垂直方向上的导数 I_x 和 I_y(3.2.3 节)。
2. 计算对应于这些梯度外积的三个图像。(矩阵 A 是对称的，因此只需要计算三个量。)
3. 使用一个较大的高斯函数来对这些图像中的每一个进行卷积。
4. 使用前面公式中的任何一个来计算一个标量兴趣量。
5. 寻找一定阈值之上的局部最大值，并将它们作为检测到的特征点位置。

算法 4.1　一个基本特征检测算法的步骤

(a)　　　　　　　(b)　　　　　　　(c)

图 4.8　兴趣算子响应：(a)样本图像；(b)Harris 响应；(c)DoG 响应。圆圈的大小和颜色表明检测到的兴趣点的尺度。注意，这两个检测器趋向于在互补的位置产生响应

自适应非最大抑制(Adaptive non-maximal suppression，ANMS)

由于大多数特征点检测器只寻找兴趣函数的局部最大值，所以这通常会导致图像上特征点的非均匀分布，比如，在对比度较大的区域，特征点就会比较密集。为了缓解这个问题，Brown, Szeliski, and Winder(2005)就只检测那些同时是局部最大值且其响应明显大于(10%)其周围半径 r 区域内的响应的特征(图 4.9c 和 d)。他们设计了一种高效的为所有局部最大值关联一个抑制半径的方法，这种方法首先根据特征点的响应强度对其进行排序，然后通过不断减小抑制半径大小来建立第二个排序列表(Brown, Szeliski, and Winder 2005)。图 4.9 给出了一个使用 ANMS 选择最强的 n 个特征的定性比较。

衡量可重复性

计算机视觉领域中开发有各种各样的特征检测器，我们如何决定使用哪一个呢？Schmid, Mohr, and Bauckhage(2000)第一个提出了衡量特征检测器的可重复性思想，他们将这个可重复性定义为在一幅图像中检测到的关键点在另一幅变换过的图像中的对应位置的 ε(比如 $\varepsilon=1.5$)个像素范围内找到的频率。在他们的论文中，他们对平面图像进行各种变换，包括旋转、尺度变化，光照变化、视角变化以及增加噪声。他们同时也衡量了检测到的每一个特征点的"可用信息量"(Information Content)，这个信息量被定义为一个旋转不变的局部灰度描述子集合的熵。在他们所考查的特征点检测方法中，他们发现选用 $\sigma_d=1$(高斯导数函数的尺度)和

$\sigma_i = 2$ (积分高斯函数的尺度)改进(高斯导数)版本的 Harris 算子效果最好。

尺度不变

在很多情况下，在最精细的稳定尺度上检测特征点可能不是很合适。比如，在匹配那些缺乏高频细节的图像(比如，云朵)时，精细尺度的特征可能不存在。

这个问题的一种解决方法是在不同的尺度上提取特征点，比如，通过在图像金字塔的多个分辨率上都进行这样的操作，然后在同一个水平上进行特征匹配。这种方法对于待匹配的图像无较大尺度变化时比较适合，比如，在匹配从飞机上拍摄到的鸟瞰图序列时或者在做由固定焦距摄像机拍摄的全景图拼接时。图 4.10 给出了这种方法的一个输出结果，其中使用了 Brown, Szeliski, and Winder(2005)的多尺度、带方向的图像块检测器，结果给出了五个不同尺度的响应。

图 4.9 自适应非最大抑制(ANMS)(Brown, Szeliski, and Winder 2005)© 2005 IEEE：上面两个图给出了最强的 250 个和 500 个兴趣点，下面两幅图给出了使用自适应非最大抑制方法所选择的特征点以及相应的抑制半径 r。注意后面的特征如何在图像上拥有更均匀的空间分布

图 4.10 在五个金字塔级别上提取的多尺度带方向的图像块(Multi-scale oriented Patches，MOPS)(Brown, Szeliski, and Winder 2005)© 2005 IEEE。方框给出了特征的方向和描述子向量的采样区域

然而，对于大多数物体识别应用，图像中物体的尺度是不知道的。相比在多个不同尺度上提取特征后再全部匹配，提取所有位置和尺度都稳定的特征更高效(Lowe 2004; Mikolajczyk and Schmid 2004)。

早期针对尺度选择问题的研究是 Lindeberg(1993;1998b)进行的，他们首次提出使用高斯的拉普拉斯(LoG)函数的极值点作为兴趣点位置。在此基础上，Lowe(2004)提出了在一个子八度集合上计算高斯差分滤波器(图 4.11a)，寻找所得结构(图 4.11b)的三维(空间+尺度)最大值，然后使用一种二次拟合方法计算亚像素级别的空间+尺度的位置(Brown and Lowe 2002)。经过仔细的实验考查，子八度级别的数量被确定为三，对应一个四分八度的金字塔，这与 Triggs(2004)使用的金字塔是一样的。

和 Harris 算子一样，排除了指示函数(这里指 DoG)局部曲率非常不对称的像素。在实现上，需要首先计算差分图像 D 的局部 Hessian 矩阵，

$$\boldsymbol{H} = \left[\begin{array}{cc} D_{xx} & D_{xy} \\ D_{xy} & D_{yy} \end{array} \right], \tag{4.12}$$

然后除去满足下面条件的关键点

$$\frac{\mathrm{Tr}(\boldsymbol{H})^2}{\mathrm{Det}(\boldsymbol{H})} > 10. \tag{4.13}$$

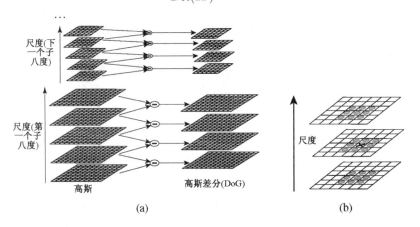

图 4.11　使用子八度差分高斯金字塔在尺度空间上检测特征(Lowe 2004)© 2004 Springer：(a)将子八度高斯金字塔的相邻级别相减来产生高斯差分图像；(b)通过一个像素与其 26 个相邻像素的比较得到所产生的三维体中的极值(最大值或者最小值)

虽然 Lowe 的尺度不变特征变换(SIFT)在实践中效果很好，但它的理论基础并不是基于最大化空间稳定性的，这与基于自相关性的检测器不同。(事实上，它的检测位置经常与那些方法所产生的检测位置互补，因此可以和那些方法一起使用。)为了给 Harris 角点检测提供一个尺度选择的机制，Mikolajczy and Schmid(2004)在每个检测到的 Harris 位置评估高斯拉普拉斯函数(在一个多尺度金字塔上)，然后只保留那些拉普拉斯函数取极值(比它高一级的尺度和低一级的尺度的值都大或者都小)的点。他们还提出和评估了一种对尺度和位置进行求精的可选择的迭代方法。Mikolajczyk, Tuytelaars, Schmid et al.(2005); Tuytelaars and Mikolajczyk(2007)讨论了

其他的一些尺度不变区域检测子的例子。

旋转不变和方向估计

除了处理尺度变化，大多数图像匹配和物体识别算法需要处理(至少)平面内图像旋转。处理这个问题的一种途径是设计出旋转不变的描述子(Schmid and Mohr 1997)，但是这些描述子区分性比较弱，也就是说，对于同一个描述子，它们映射出不同的查找块。

一个较好的方法是在检测到的每一个关键点估计一个"主导方向"(dominant orientation)。一旦估计出一个关键点的局部方向和尺度，就可以在检测出的关键点附近提取出一个特定尺度和方向的图像块(图 4.10 和图 4.17)。

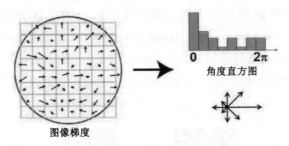

图 4.12 通过创建一个所有梯度方向(根据其梯度大小加权或阈值化小的梯度值之后)的直方图，可以估计出一个主导方向，然后可以找到这个分布上明显的峰值。(Lowe 2004)© 2004 Springer

关键点周围一个区域内的平均梯度是最简单的可选方向估计。如果使用高斯加权函数(Brown, Szeliski, and Winder 2005)，这个平均梯度就等同于一个一阶导向滤波器(3.2.3 节)，也就是说，可以通过使用水平和竖直方向上的高斯函数导数滤波器的图像卷积来求它(Freeman and Adelson 1991)。为了使这个估计更稳定，通常更倾向于选择一个比检测窗口大的聚合窗口(高斯核大小)(Brown, Szeliski, and Winder 2005)。图 4.10 中正方形框的方向就是使用这个方法计算得到的。

然而有时区域内的平均(带符号的)梯度可能很小，因而指示方向不稳定。一个更为稳定的方法是计算关键点周围的梯度方向直方图。Lowe(2004)计算一个 36 维的梯度直方图，该直方图同时由梯度大小和距离中心的高斯函数距离加权，寻找全局最大值 80%以内的所有峰值，然后使用一个三分抛物线拟合计算出一个更准确的方向估计(图 4.12)。

仿射不变

除了尺度和旋转不变非常有意义外，在很多应用中，比如"宽基线立体视觉匹配"(wide baseline stereo matching)(Pritchett and Zisserman 1998; Schaffalitzky and Zisserman 2002)或者位置识别(Chum, Philbin, Sivic et al. 2007)，全仿射不变同样很有意义。仿射不变检测器不仅在尺度和旋转变化后产生一致的检测位置，而且在仿射变换如(局部)透视收缩效应中仍产生一致的位置(图 4.13)。事实上，对于一个足够小的图像块，任何一个连续的图像变形都可以用仿射变形很好地近似。

图 4.13　用于匹配拍摄视角相差很大的两幅图像的仿射区域检测器(Mikolajczyk and Schmid 2004) © 2004 Springer

为了引入仿射不变性，一些学者提出先用一个椭圆拟合自相关矩阵或者Hessian 矩阵(使用特征值分析)，然后使用这个拟合的主要坐标轴和比例作为仿射坐标系(Lindeberg and Garding 1997; Baumberg 2000; Mikolajczyk and Schmid 2004; Mikolajczyk, Tuytelaars, Schmid et al. 2005; Tuytelaars and Mikolajczyk 2007)。图 4.14 表明了如何使用矩阵的平方根来将一个局部块变换到另一个在相差一个旋转意义下相似的框架。

图 4.14　使用二阶矩矩阵进行仿射归一化，该方法是 Mikolajczyk, Tuytelaars, Schmid et al. (2005)© 2005 Springer 描述的。使用矩阵 $A_0^{-1/2}$ 和 $A_1^{-1/2}$ 将图像坐标变换之后，它们通过一个纯旋转矩阵 R 关联起来，这个矩阵可以使用主方向方法估计出来

另一个重要的仿射不变区域检测器是 Matas, Chum, Urban et al.(2004)提出的"最稳定极值区域"(maximally stable extremal region，MSER)检测器。为了检测MSER，要在所有可能的灰度级别上(因此这个方法只适用于灰度图像)阈值化图像从而得到二值区域。这个操作可以按照下面的方式高效地完成。首先根据灰度值对所有像素排序，然后随着阈值的改变逐渐将像素添加到相应的连通分量中(Nister and Stewenius 2008)。随着阈值的改变，监控每一个分量(区域)的面积；那些相对于阈值的面积变化率最小的区域被定义为"最稳定的"(maximally stable)。因此这样的区域在仿射几何变换和光度(线性偏差-增益或光滑单调的)变换下不变(图 4.15)。如果需要，可以使用每一个区域的矩矩阵为其拟合出一个仿射坐标系。

图 4.15　从几幅图像中提取和匹配的最稳定极值区域(MSER)(Matas, Chum, Urban et al. 2004)©2004 Elsevier

特征点检测器这个研究方向依然非常活跃，每年在各个主要的计算机视觉会议中都有论文出现(Xiao and Shah 2003; Koethe 2003; Carneiro and Jepson 2005; Kenney, Zuliani, and Manjunath 2005; Bay, Tuytelaars, and Van Gool 2006; Platel, Balmachnova, Florack *et al*. 2006; Rosten and Drummond 2006)。Mikolajczyk, Tuytelaars, Schmid *et al*.(2005)回顾了许多常见的仿射区域检测器并且给出了它们针对常见图像变换比如尺度、旋转、噪声和模糊不变性的实验比较结果。在网址 *http://www.robots.ox.ac.uk/~vgg/research/affine/* 可以找到这些实验的结果、代码和指向那些综述论文的链接。

当然，关键点不是唯一可以用来进行图像配准的特征。Zoghlami, Faugeras, and Deriche(1997)使用直线片段和点状特征来估计图像对之间的同态映射，而 Bartoli, Coquerelle, and Sturm(2004)使用具有沿着边缘局部对应的直线片段来提取 3D 结构和运动。Tuytelaars and Van Gool(2004)使用仿射不变区域来检测宽基线立体视觉匹配中的对应，而 Kadir, Zisserman, and Brady(2004)则检测显著性区域，即其图像块的熵和其随尺度的变化率是局部最大的。Corso and Hager(2005)使用一个相关的方法给同态的区域拟合 2D 有向高斯核函数。在本节的后面，可以找到更多关于寻找和匹配曲线、直线和区域的更多方法的细节。

4.1.2 特征描述子

检测到特征(关键点)之后，我们必须匹配它们，也就是说，我们必须确定哪些特征来自于不同图像中的对应位置。在一些情况下，比如对于视频序列(Shi and Tomasi 1994)或者已经矫正过的立体对(Zhang, Deriche, Faugeras *et al*. 1995; Loop and Zhang 1999; Scharstein and Szeliski 2002)，每个特征局部附近的运动可能主要是平移性的。在这种情况下，简单的误差度量，比如 8.1 节描述的"差的平方和"(sum of squared differences)或者"规范化互相关"(normalized cross-correlation)可以用来直接比较特征点周围小图像块的亮度值。(下面讨论的 Mikolajczyk and Schmid(2005)所做的比较性研究使用的是互相关)。因为可能没有精确地定位特征点，可以通过使用 8.1.3 节中给出的增量运动求精方法来计算更精确的匹配值，但是这种方法非常耗时，而且有时甚至会降低性能(Brown, Szeliski, and Winder 2005)。

然而在很多情况下，特征点的局部表观可能会在方向和尺度上变化，有时甚至存在仿射变形。因此，提取出局部尺度、方向或仿射的框架估计，然后在形成特征描述子之前使用它来对图像块重新采样，这样的做法通常更可取(图 4.17)。

即便在对这些变化进行补偿之后，在不同图像之间，图像块的局部表观常常仍然会因图像不同而变化。我们怎样才能使图像描述子对这些变化具有更好的不变性但同时保持不同(非对应的)图像块之间的区分性(图 4.16)？Mikolajczyk and Schmid(2005)回顾了最近提出的一些视角不变的局部图像描述子，并且做实验比较了它们的性能。下面，我们详细描述几个这样的描述子。

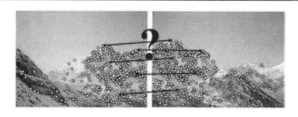

图 4.16 特征匹配:为了建立正确的对应,我们怎样提取对图像间的变化具有不变性而仍然具备足够的区分性的局部描述子?

偏差和增益规范化(MOPS)

对于那些未呈现出大量透视收缩的任务,比如图像拼接,仅仅规范化亮度图像块就可以得到相当好的结果且很容易实现(Brown, Szeliski, and Winder 2005)(图4.17)。为了弥补在特征点检测中小的错误(位置,方向和尺度),这些多尺度带方向的图像块(MOPS)是以在相对于检测尺度的五个像素间隔下进行采样的,使用图像金字塔的较粗层次来避免失真。为了弥补仿射照度不同(线性曝光变化或者偏差和增益 3.3),图像块的亮度被重新缩放以使其均值为 0,方差是 1。

 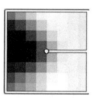

图 4.17 MOPS 描述子是使用一个 8×8 采样得到的,该采样使用了偏差和增益规范化了的亮度值,采样的间隔是相对于检测尺度的五个像素(Brown, Szeliski, and Winder 2005)© 2005 IEEE。这个低频采样使该特征对特征点位置错误具备了一定的鲁棒性,这是通过在比检测尺度较高的金字塔尺度上采样完成的

尺度不变特征变换(SIFT)

SIFT(scale invariant feature transform)特征通过计算在检测到的关键点周围 16×16 窗口内每一个像素的梯度得到,计算时使用检测到的关键点所在的高斯金字塔级别。因为远离中心的权重更容易受到位置不正的影响,为了减小这些影响,梯度值被通过一个高斯下降函数降低权重(如图 4.18a 中的蓝色圆圈所示)。

在每个 4×4 的四分之一象限,通过(在概念上)将加权梯度值加到直方图八个方向区间中的一个,计算出一个梯度方向直方图。为了减少位置和主方向估计偏差的影响,使用三线性插值将最初的 256 个加权梯度值平滑地添加到 2×2×2 直方图中。在需要计算直方图的应用中,将数值平滑分布于直方图相邻的区间通常是一个不错的想法,举个例子,如 Hough 变换(4.3.2 节)或者局部直方图均衡化(3.1.4 节)。

得到的 128 维的非负值形成了一个原始版本的 SIFT 描述子向量。为了减少对比度和增益的影响(加性变化已经通过梯度去掉了),将这个 128 维的向量归一化到单位长度。为了进一步使描述子对其他各种光度变化鲁棒,再将这些值以 0.2 截尾，然后将得到的向量再归一化到单位长度。

PCA-SIFT

受 SIFT 的启发，Ke and Sukthankar(2004)提出了一种更简单的计算描述子的方法。这种方法首先在一个 39×39 的图像块上计算 x 和 y 方向(梯度)导数，然后使用主成分分析(PCA)将得到的 3042 维向量降维到 36 维(14.2.1 节和附录 A.1.2)。另外一个常见的 SIFT 的变体是 SURF(Bay, Tuytelaars, and Van Gool 2006)，它使用方框型滤波器来近似 SIFT 中使用的导数和积分。

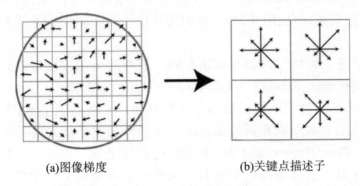

(a)图像梯度　　　　　　　　　(b)关键点描述子

图 4.18　Lowe(2004)的尺度不变特征变换(SIFT)的图解表示。(a)在每一个像素处计算梯度大小和方向，并使用一个高斯下降函数加权(蓝色圆圈)。(b)在每一个子区域内使用三线性插值来计算加权梯度方向直方图。这幅图中给出的是 8×8 像素块和一个 2×2 描述子数组，Lowe 的实际实现中使用的是 16×16 的像素块和 4×4 直方图数组，每个直方图分 8 个区间

梯度位置方向直方图(GLOH)

Mikolajczyk and Schmid(2005)提出了一种是 SIFT 变体的描述子，它使用对数极坐标分级结构来替代 Lowe(2004)使用的四象限(图 4.19)。空间上的区间半径分别为 6、11 和 15，角度上分八个区间(除了中间区域外)，然后使用在很大数据集上训练得到的 PCA 模型将这个 272 维直方图映射到一个 128 维的描述子。在他们的评测中，GLOH(gradient location-orientation histogram)比 SIFT 在性能上有小幅提高，整体上取得了最好的性能。

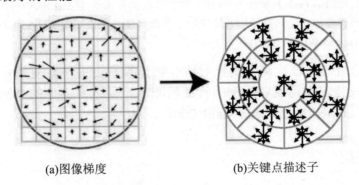

(a)图像梯度　　　　　　　　　(b)关键点描述子

图 4.19　梯度位置方向直方图(GLOH)描述子使用对数极坐标区间，而不是方形区间来计算方向直方图(Mikolajczyk and Schmid 2005)

导向滤波器

导向滤波器(steerable filter，3.2.3 节)是高斯导数滤波器的组合，这些滤波器能够很快地在所有可能的方向上计算奇和偶(对称和反对称)的边缘类特征和角点类特征(Freeman and Adelson 1991)。因为它们使用了相当宽的高斯函数，所以在一定程度上对位置和方向错误不敏感。

局部描述子的性能

在 Mikolajczyk and Schmid(2005)比较的局部描述子中，他们发现 GLOH 性能最好，SIFT 紧随其后(参见图 4.25)。他们还提供了很多本书没有提到的其他描述子的结果。

特征描述子这个领域目前仍在快速发展，最新的一些方法考虑到局部的颜色信息(van de Weijer and Schmid 2006; Abdel-Hakim and Farag 2006)，Winder and Brown(2007)提出了一种多阶段的描述子计算框架，它还涵盖了 SIFT 和 GLOH，而且允许学习最优的参数得到更新的描述子，这些新描述子比之前手工调参数得到描述子性能更好。Hua, Brown, and Winder(2007)扩展了这一工作，它是通过学习高维描述子的最具区分性的低维映射得到的。这些论文都使用了一个数据集，该数据集是由在如下位置处采样得到的真实世界的图像块(图 4.20b)组成的：对于互联网上的图像库，应用一种鲁棒的由运动到结构的算法，在这些位置上能够得到稳定的匹配。在同时期的一个工作中，Tola, Lepetit, and Fua(2010)提出了一种 DAISY 描述子来进行稠密立体视觉匹配，并基于标定好的立体数据来优化其参数。

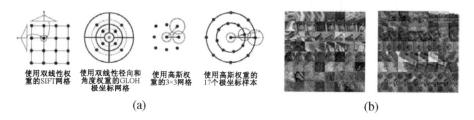

图 4.20　SIFT, GLOH 的空间求和块和一些最新发展的特征描述子(Winder and Brown 2007)© 2007 IEEE：(a)新特征的参数(比如高斯函数的权重)，是从(b)的训练数据库中学习得到的；(b)将鲁棒的由运动到结构方法应用于互联网图像集合后得到匹配的真实世界图像块(Hua, Brown, and Winder 2007)

这些方法构造的特征检测器是针对所有物体类别上的可重复性进行优化的，另一种可能是生成特定类别或者示例的特征检测器，使之与其他的类别之间的区分性最大(Ferencz, Learned-Miller, and Malik 2008)。

4.1.3　特征匹配

一旦我们从两幅或者多幅图像中提取特征及其描述子，下一步就是要在这些图像之间建立一些初始特征之间的匹配。在本节中，我们将这个问题分为两个独立的阶段，第一阶段是选择一个匹配策略，用来确定哪些匹配将被传到下一个阶段进行进一步处理。第二阶段是设计出有效的数据结构和算法来尽可能快的完成这个匹配。(请看 14.3.2 节相关方法的讨论。)

匹配策略和错误率

决定哪些特征间的匹配需要进一步处理取决于匹配操作所在的上下文。比如给我们的两幅图像有很大一部分重合(举个例子，在图 4.16 所示的图像拼接中，或者在视频中跟踪物体时)，我们知道尽管其中会有一些特征由于遮挡或者其表观发生了较大变化而没有匹配，但一幅图像中的大部分特征还是很有可能和另一幅图像匹配的。

另一方面，如果我们要在一个拥挤场景中识别有多少个已知的目标出现(图 4.21)，特征中的大多数可能不会匹配。进一步，需要搜索数目巨大的潜在匹配目标，这就需要下文描述的更有效的策略。

图 4.21 在拥挤场景中识别物体(Lowe 2004)© 2004 Springer。左边显示的是数据库中的两幅训练图像。使用 SIFT 特征将它们和中间的拥挤场景进行匹配，结果在右图中以小矩形显示。每一个识别出来的数据库图像在场景上的仿射变形在右图中用较大的平行四边形显示

开始之前，我们先假设已经设计好了特征描述子，所以特征空间的(向量幅值)欧式距离可以直接用来对潜在匹配进行排序。如果结果是一个描述子的某些参数(轴)比其他参数更稳定，通常倾向于提前对这些轴进行重缩放，比如，通过与其他已知好的匹配比较来确定它们有多大的变化(Hua, Brown, and Winder 2007)。在基于特征脸的人脸识别(14.2.1 节)中，我们将详细讨论一个更一般的过程，将一个特征向量转化到另一个新的缩放基上，这个过程称为"白化"(whitening)。

给定一个欧式距离度量，最简单的一个匹配策略就是先设定一个阈值(最大距离)，然后返回在这个阈值范围之内的其他图像中的所有匹配。这个阈值如果设的太高，就会产生很多"误报"(false positive)，也就是说，返回了不正确的匹配。这个阈值设的太低的话就会产生很多"漏报"(false negative)，也就是说，很多正确的匹配被丢失了(图 4.22)。

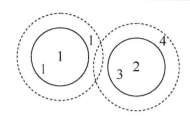

图 4.22 误报和漏报：黑色数字 1 和 2 是特征，将与其他图像中的一个特征的数据库进行匹配。在目前阈值设置下(实线圈)，绿色 1 是一个正确肯定(正确的匹配)，蓝色 1 是一个漏报(未能匹配)，红色 3 是一个误报(不正确的匹配)。如果我们将阈值设置设高一点(虚线圆圈)，蓝色 1 变成一个正确肯定但棕色 4 变成了一个额外的误报

我们可以在一个特定阈值上对一个匹配算法的性能进行量化，通过使用下面的定义(Fawcett 2006)来计算正确和错误匹配及匹配失败的数目。

TP：正确肯定，也就是正确匹配的数目；
FN：漏报，没有正确找到的匹配的数目；
FP：误报，给出的匹配是不正确的；
TN：正确否定，正确拒绝的非匹配对。

表 4.1 给出了一个包含这些数字的混淆矩阵(列联表)的样例。

表 4.1 使用一个特征匹配算法得到的正确和错误匹配数目，给出了正确肯定(TP)的数目、误报(FP)的数目、漏报(FN)的数目和正确否定(TN)的数目。每一列加起来对应于实际肯定(P)和否定(N)的数目，而每一行加起来对应于预测的肯定(P')和否定(N')的数目。在正文中给出了计算正确肯定率(TPR)，错误肯定率(FPR)，正确预测率(PPV)和精度(ACC)的公式

	正确的匹配数目	正确的未匹配数目		
预测的匹配数目	TP = 18	FP = 4	P' = 22	PPV = 0.82
预测的未匹配数目	FN = 2	TN = 76	N' = 78	
	P = 20	N = 80	Total = 100	
	TPR = 0.90	FPR = 0.05		ACC = 0.94

通过定义如下的量值(Fawcett 2006)，我们可以将这些数字转化为单位比率：

正确肯定率(TPR)

$$\text{TPR} = \frac{\text{TP}}{\text{TP+FN}} = \frac{\text{TP}}{\text{P}}; \quad (4.14)$$

错误肯定率(FPR)

$$\text{FPR} = \frac{\text{FP}}{\text{FP+TN}} = \frac{\text{FP}}{\text{N}}; \quad (4.15)$$

肯定预测值(PPV)，

$$\text{PPV} = \frac{\text{TP}}{\text{TP+FP}} = \frac{\text{TP}}{\text{P'}}; \quad (4.16)$$

精确度(ACC)，

$$\text{ACC} = \frac{\text{TP+TN}}{\text{P+N}}. \quad (4.17)$$

在信息检索(或者文本检索)研究中(Baeza-Yates and Ribeiro-Neto 1999; Manning, Raghavan, and Schütze 2008)，使用术语"精度"(返回的文本中有多少是相关的)来替代 PPV，使用术语"召回率"(找到了多少比例的相关文本)来替代 TPR。

任何一个匹配策略(在特定的阈值和参数设置下)都可以用 TPR 和 FPR 来进行评价；理想情况下，正确肯定率应该接近 1 而错误肯定率应该接近 0。随着匹配阈值的变化，我们获得了一系列这样的点，所有的这些点就是所谓的"接收器操作特性"(receiver operating characteristic，ROC 曲线)(Fawcett 2006)(图 4.23a)。这个曲线越接近于左上角，也就是曲线下的面积越大(AUC)，它的性能越好。图 4.23b 给出了我们怎样将匹配数目和非匹配数目作为特征间距离 d 的一个函数来绘制。然后这些曲线可以用来绘制一个 ROC 曲线(习题 4.3)。ROC 曲线也可以用来计算"均值平均精度"(mean average precision)，这是当你通过改变阈值来选择最好的结果(然后是最好的两个结果，等等)时所得到的平均精度(PPV)。

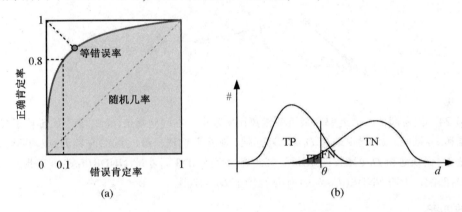

图 4.23 ROC 曲线和它相关的比率。(a)对于一特定的特征提取和匹配算法的组合 ROC 曲线绘制了正确肯定率相对于错误肯定率的曲线。理想情况下，正确肯定率应该接近 1，而错误肯定率应该接近 0。ROC 曲线下面的面积(AUC)通常用作算法性能的一个单一(标量)度量。有时也用等错误率作为另一个替换的度量。(b)肯定数目(匹配的数目)和否定数目(非匹配的数目)的分布作为特征间距离 d 的一个函数。随着阈值 θ 的增加，正确肯定(TP)和误报(FP)的数目也随着增加

使用一个固定阈值的问题是它很难设定。随着我们移动到特征空间的不同部分时，阈值的有效范围可能会变化很大(Lowe 2004; Mikolajczyk and Schmid 2005)。在这些情况下一个较好的策略是在特征空间中简单地做最近邻匹配。由于一些特征可能没有匹配(比如，它们可能是物体识别中的拥挤背景的一部分或者它们可能在其他图像中被遮挡了)，所以仍然要用一个阈值来减少误报的数目。

理想情况下，这个阈值本身应该自适应于特征空间的不通区域。如果有足够的训练数据(Hua, Brown, and Winder 2007)，有时为每个特征学习不同的阈值是可能的。然而，通常情况下给我们的都是一堆要匹配的数据，比如，在图像拼接时或者根据未排序的图片集构建 3D 模型时(Brown and Lowe 2007, 2003; Snavely, Seitz, and Szeliski 2006)。在这样的情况下，可以使用的一个启发式策略是比较最近邻距

离和次近邻距离，这个次近邻距离是从已知的和目标不匹配的一幅图像中取得的(比如，数据库中的另一个物体)(Brown and Lowe 2002; Lowe 2004)。我们可以定义这个最近邻距离比率(nearest neighbor distance ratio)(Mikolajczyk and Schmid 2005)为

$$\text{NNDR} = \frac{d_1}{d_2} = \frac{\|D_A - D_B\|}{\|D_A - D_C\|}, \tag{4.18}$$

其中 d_1 和 d_2 是最近邻的和次近邻距离，D_A 是目标描述子，D_B 和 D_C 是它的最近的两个邻居(图4.24)。

图 4.25 给出了 Mikolajczyk and Schmid(2005)评测的对特征描述子使用三个不同的匹配策略的效果。我们可以看出，最近邻和 NNDR 策略产生了性能改进的 ROC 曲线。

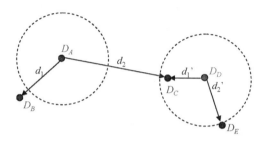

图 4.24 固定阈值，最近邻和最近邻距离比率匹配。在固定阈值(虚线圆)下，描述子 D_A 未能和 D_B 匹配，D_D 错误地和 D_C 和 D_E 匹配。如果我们选择最近邻，D_A 和 D_B 成功匹配，但是 D_D 错误地和 D_C 匹配。使用最近邻距离比率(NNDR)，小的 NNDR(d_1/d_2)正确地将 D_A 和 D_B 匹配，大的 NNDR(d'_1/d'_2)正确地拒绝了 D_D 的匹配

高效匹配

一旦确定匹配策略，我们仍需要在潜在的候选中高效地进行搜索。一种最简单的找到所有对应特征点的方式就是在每对潜在的匹配图像中将特征和其他所有特征进行匹配。不幸的是，这样的复杂度是提取特征数目的二次方，对大多数应用中都不现实。

一个较好的方法是导出一个索引结构(比如一个多维搜索树或者一个哈希表)，来快速寻找一个给定特征的邻近的特征。可以为每个图像单独建立这样的索引结构(如果我们只考虑某些特定的潜在匹配，比如搜索一特定物体，这样很有用)，也可以为一个给定数据库中的所有图像建立全局的索引结构，这样有可能加快速度，因为不需要在每个图像上迭代。对于非常大的数据库(几百万的图像或者更多)，可以使用基于源自于文本检索思想的更高效的结构(比如，词汇树，(Nistér and Stewénius 2006)(14.3.2 节)。

一个较简单的实现方法是多维散列，它基于施加在每一个描述子向量上的某些函数将描述子映射到一个固定大小的一些桶里。在匹配时，每一个特征被散列到一个桶中，然后在邻近的桶中进行搜索来返回潜在的候选者，然后将这些候选者排序或者打分以确定哪些是有效的匹配。

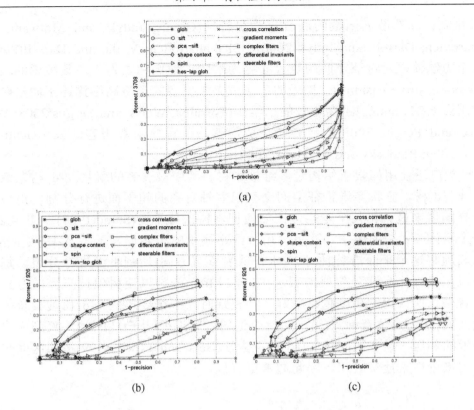

图 4.25 Mikolajczyk and Schmid(2005) © 2005 IEEE 评测的特征描述子的性能,给出了三种不同的匹配策略:(a)固定阈值;(b)最近邻;(c)最近邻距离比率(NNDR)。可以发现算法的排序变化不是很大,但是使用不同的匹配策略会使整体性能变化很大

散列的一个简单例子是 Brown, Szeliski, and Winder(2005)在其 MOPS 文章中使用的 Haar 小波。在匹配结果构造的过程中,每一个 8×8 的缩放过、有向的、归一化了的 MOPS 块被转换到一个通过在各个四分象限上求和得到的三元素索引上(图 4.26)。得到的三个值通过它们的期望标准偏差进行归一化,然后映射到两个(它的 b=10)最近的 1D 区间中。通过连接三个量化后的数值形成的三维索引可用来索引 2^3=8 个特征存储(累加)的区间。在查询时,只用那些最主要(最接近的)索引,因此只有一个三维区间需要检查。区间中的系数可以用于选择 k 个近似最近的邻居做进一步处理(比如计算 NNDR)。

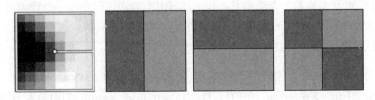

图 4.26 Brown, Szeliski, and Winder(2005)设计的散列 MOPS 描述子所使用的三种 Haar 小波系数,它们是通过对每个 8×8 规范化块的浅色和深色区域的和求差计算得到的

一个更复杂但更适用的散列版本叫"局部敏感散列"(locality sensitive hashing),

它使用独立计算的散列函数的联合来索引特征(Gionis, Indyk, and Motwani 1999; Shakhnarovich, Darrell, and Indyk 2006)。Shakhnarovich, Viola, and Darrell(2003)扩展了这个方法使之对特征空间的点分布更为敏感，他们称之为"参数敏感的散列"(parameter-sensitive hashing)。最近的工作成果中更多的是将高维描述子向量转化成可以使用海明(Hamming)距离比较的二维码(Torralba, Weiss, and Fergus 2008; Weiss, Torralba, and Fergus 2008)，或者转化成能够与任意核函数相容的表示(Kulis and Grauman 2009; Raginsky and Lazebnik 2009)。

另一类广泛使用的索引结构是多维搜索树。其中最著名的就是 k-d 树，或者经常也写作 kd-树，它将多维特征空间交替沿着轴对齐的超平面进行分割，沿着每一个轴选择阈值来最大化某个策略，比如使搜索树保持平衡(Samet1989)。图 4.27 给出了一个两维 k-d 树的例子。这里，用二维平面里的小菱形表示了 A-H 八个点。k-d 树递归地将这个平面沿着轴对齐(水平或者垂直方向)切割平面进行分裂。每一个分割可以用维度序号和分裂值来表示(图 4.27.b)。分裂以使树保持平衡的方式来安排，也就是，保持树的高度越小越好。在查询时，一个经典的 k-d 树搜索首先在它的合适区间(D)上定位查询点(+)，然后查询树上附近的叶子(C,B,...)，直到能够确保找到最近的邻居。最优区间优先(BBF)搜索法(Beis and Lowe 1999)按照区间与查询点的空间接近性来搜索区间，因此通常更有效。

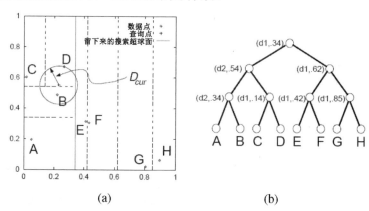

图 4.27 k-d 树和最优区间优先(BBF)搜索(Beis and Lowe 1999)© 1999 IEEE。(a)轴对齐切割平面的空间排列用虚线表示。每一个独立的数据点用小的菱形表示。(b)同一个细分可以用一个树表示，其中非叶子节点表示一个轴对齐切割平面(比如，根节点沿着 d1 维以数值 34 切割)，叶子节点是一个数据点。在一次 BBF 搜索过程中，一个查询点(用"+"表示)首先在包含它的区间(D)寻找然后再在它的最近邻区间(B)寻找，而不是在它在树上最近的邻居(C)中寻找

人们提出了许多其他的数据结构来解决最近邻问题(Arya, Mount, Netanyahu *et al*. 1998; Liang, Liu, Xu *et al*. 2001; Hjaltason and Samet 2003)。比如，Nene and Nayar(1997)提出了一种称为"切片"(slicing)的方法，该方法对在不同维上排序的点列表进行一系列的 1D 二分搜索来高效地剔除那些在查询点超立方体邻域内的候选点列表。Grauman and Darrell(2005)在一个索引树的不同级别上重新计算匹配的

权重,这使得他们的方法对于树构建过程中的离散化错误不是那么敏感。Nistér and Stewénius(2006)使用的是度量树(metric tree),它将特征描述子和一个层次结构中的每个级别上的少量的原型进行比较。得到的量化视觉词(visual word)可以结合经典的信息检索(文本相关)方法来从数百万幅图像组成的数据集中快速地剥离候选集合中"潜伏"的杂质(14.3.2 节)。Muja and Lowe(2009)比较了很多这类方法,并提出了他们的一种新方法(在层次性的 k-均值树中进行优先搜索),最后总结得出多个随机 k-d 树取得的性能往往最好。尽管这些工作取得了满意的结果,但图像特征对应的快速计算仍是一个极具挑战性的开放研究问题。

特征匹配验证和紧致化

一旦我们得到一些假设的(推断的)匹配,往往可以用几何配准来验证哪些匹配是"内点"(inlier),哪些是"外点"(outlier)。比如,如果我们预计出整个图像在匹配视角上平移或者旋转了,我们可以拟合出一个全局几何变换并仅保留那些与估计出来的变换足够接近的特征。这个选择一小部分种子匹配然后验证更大集合的过程叫"随机采样"(random sampling)或者 RANSAC(6.1.4 节)。一旦建立初始的对应集合,一些系统就寻找额外的对应,比如,通过在极线上(11.1 节)寻找额外的对应或者基于全局变换在估计位置的附近区域寻找对应。这些主题将在 6.1,11.2,和 14.3.1 节中进一步讨论。

4.1.4 特征跟踪

在所有候选图像中独立地寻找特征然后将它们进行匹配,另一种替代策略是,在第一幅图像中寻找可能的特征位置集合,然后在后续的图像中搜索它们的对应位置。这类"先检测后跟踪"(detect and track)的方法在视频跟踪应用中使用得非常广泛,这里,所期望的相邻帧之间的运动和表观的变形比较小。

选择好特征来跟踪的过程和选择好特征来进行更一般的识别应用紧密相关。在实际中,那些在两个方向上梯度值均大的区域,也就是,自相关矩阵拥有大的特征值的区域(4.8),提供了可用于寻找对应的稳定的位置(Shi and Tomasi 1994)。

在后续的帧中,搜索那些平方差小的对应图像块区域(4.1)通常很高效。但是,如果图像的光照发生了变化,明确地补偿这些变化(8.9)或者使用规范化互相关(normalized cross-correlation)(8.11)就可能很有用。如果搜索范围很大,使用一种分层搜索策略常常更有效,它使用低分辨率下的匹配来提供较好的初始猜测值,因此加速了搜索过程(8.1.1 节)。这种策略的替代策略包括了解被跟踪块的表观应该是什么,然后在预测的位置附近搜索它(Avidan 2001; Jurie and Dhome 2002; Williams, Blake, and Cipolla 2003)。这些主题都将在 8.1.3 节中详细描述。

如果特征需要在较长的图像序列中进行跟踪,它们的表观就会发生较大的变化。这样就需要决定是继续使用最初检测到的图像块(特征)进行匹配,还是在后续帧的匹配位置上重新采样特征。前一个策略很容易失败,因为最初的图像块会发生表观变化，比如透视收缩。后者则存在将特征从原始位置漂移到某个其他图像位置的风险(Shi and Tomasi 1994)。(从数学角度上讲，小的注册错误会组合产生一个马尔科夫随机游走(Markov Random Walk)，从而长时间下来会导致大的漂移。)

比较原始块和后续图像位置一个更可取的解决方案是使用一个仿射运动模型(8.2 节)。Shi and Tomasi(1994)首先使用一个平移模型来比较相邻帧间的图像块，然后将这个步骤产生的位置估计用于初始化当前帧的图像块和最初检测出特征的基准图像帧间的一个仿射注册模型(图 4.28)。在他们的系统中，特征仅在很少的情况下被检测，比如仅在跟踪失败的区域中。通常情况下，在特征的当前预测位置周围的一个区域内使用一种增量注册算法进行搜索(8.1.3)。得到的跟踪器通常叫"Kanade-Lucas-Tomasi(KLT)跟踪器"。

图 4.28 使用仿射运动模型进行特征跟踪(Shi and Tomasi 1994)© 1994 IEEE，上一行：跟踪到的特征位置附近的图像块。下一行：使用仿射模型往后朝着第一帧变形后的图像块。尽管速度指示符逐帧在变大，仿射变换在最初帧和后续跟踪帧之间维持了一个较好的相似性

自 Shi and Tomasi 在特征跟踪方面原创的工作之后，他们的方法已经产生了一系列有意义的后续论文和应用。Beardsley, Torr, and Zisserman(1996)使用扩展的特征跟踪与由运动到结构(第 7 章)相结合来从视频序列中递增地建立稀疏 3D 模型。Kang, Szeliski, and Shum(1997)将相邻(规范网格化的)块的角点配合在一起来提供附加的稳定跟踪能力，这种方法的代价是处理遮挡的能力变弱。Tommasini, Fusiello, Trucco et al.(1998)为基本的 Shi and Tomasi 算法提供了一个较好的剔除伪匹配的策略，Collins and Liu(2003)为特征选择和应对随时间推移而产生较大表观变化提供了一种改进的机制，Shafique and Shah(2005)为在包含大量运动目标或者点列的视频中进行特征匹配(数据关联)提出了新的算法。Yilmaz, Javed, and Shah(2006)和 Lepetit and Fua(2005)回顾了更广泛的物体跟踪领域，其中不仅包括基于特征点的跟踪，还包括基于轮廓和区域的可供选择的方法(5.1 节)。

特征跟踪最新的一个发展是使用学习算法来建立一个特殊目的识别器，以在图像中的任意位置处快速寻找匹配的特征(Lepetit, Pilet, and Fua 2006; Hinterstoisser, Benhimane, Navab et al. 2008; Rogez, Rihan, Ramalingam et al. 2008; Özuysal, Calonder, Lepetit et al. 2010)。[①]通过花时间训练样本块与其仿射变形间的分类器，

[①] 也请参考我关于基于学习的跟踪的早期工作的说明(Avidan 2001; Jurie and Dhome 2002; Williams, Blake, and Cipolla 2003)。

可以构建出非常快而稳定的特征检测器,这样就可以处理更快的运动变化(图 4.29)。将这些特征耦合到形变模型(Pilet, Lepetit, and Fua 2008)或者从运动到结构算法(Klein and Murray 2008)之中就可以产生更高的稳定性。

图 4.29　使用 Lepetit, Pilet, and Fua (2004)©2004 IEEE 训练的快速分类器进行实时的头部跟踪

4.1.5　应用:表演驱动的动画

快速特征跟踪的最有趣的一个应用就是表演驱动的动画(performance-driven animation),也就是,基于跟踪用户的运动,对一个 3D 图形模型进行交互式形变(Williams 1990; Litwinowicz and Williams 1994; Lepetit, Pilet, and Fua 2004)。

Buck, Finkelstein, Jacobs et al.(2000)展示了一个系统,该系统能够跟踪用户的脸部表情和头部运动,然后使用跟踪的结果来变形一系列手绘的草图。动画器首先提取每一幅草图中眼睛和嘴巴区域,然后在每一幅图像中绘制控制线(图 4.30a)。在运行时,一个脸部跟踪系统(Toyama 1998)确定这些特征的当前位置(图 4.30b)。这个动画系统基于特征表观的最近邻匹配和三角形质心插值来决定变形哪些输入图像。它同时还根据所跟踪的特征来计算全局的头部位置和方向。得到的变形了的眼睛和嘴巴区域随后被复合到整体头部模型来生成一帧手绘的动画(图 4.30d)。

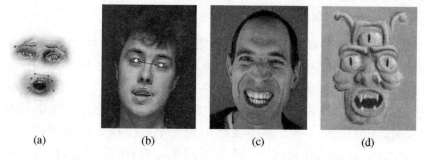

图 4.30　表演驱动的手绘动画(Buck, Finkelstein, Jacobs et al. 2000)© 2000 ACM。(a)手绘草图中的眼睛和嘴巴部分,上面叠加了控制线;(b)一幅输入视频帧,上面叠加了跟踪到的特征;(c)另一幅不同的输入视频帧;(d)对应手绘动画

在更多的近期工作中，Barnes, Jacobs, Sanders et al.(2008)观看用户在桌子表演纸张裁剪，然后将得到的运动和绘画转化成一个无缝的2D动画。

4.2 边　　缘

尽管特征点对于寻找 2D 中能够精确匹配的图像位置非常有用，但是边缘点更为丰富且常常包含重要的语义关联。比如，物体的边界，它与 3D 中的遮挡事件对应，通常可以用可见的轮廓描绘。其他一些类别的边缘对应着影子的边缘或者褶皱的边缘，这些边缘的表面方向变化都剧烈。孤立的边缘点可以被聚集成较长的曲线(curve)或者轮廓(contour)以及线段(straight line segment)(4.3 节)，有意思的是，即使是很小的孩子都可以从这些简单的线描图中轻松识别出熟悉的物体或者动物。

4.2.1　边缘检测

给定一幅图像，我们怎样找到显著的边缘呢？考虑图 4.31 中的彩色图像。如果有人让你指出最"显著"或者"强"的边缘或者物体边缘(Martin, Fowlkes, and Malik 2004; Arbeláez, Maire, Fowlkes et al. 2010)，你会追踪哪些边缘？你的感知和图 4.31 给出的边缘图像会有多接近？

图 4.31　人工边缘检测(Martin, Fowlkes, and Malik 2004)© 2004 IEEE。边缘的深色程度对应着有多少个人在那个位置标记了边缘

定性地说，边缘出现在那些颜色、亮度或者纹理不一样的区域之间。不幸的是，将图像分割成一致的区域是一个很困难的任务，详见在第 5 章的描述。通常，只使用局部信息来检测边缘更为合适。

在这样的情况下，较为合理的一种方式是将边缘定义为亮度剧烈变化①(rapid intensity variation)的位置。将一幅图像想象为一个高度域。在这样的一个表面，边缘出现在陡峭斜率(steep slopes)上，或者等同地，在(地形图上)有着紧密拥挤轮廓线的区域上。

---

① 我们将边缘检测推广在彩色图像上。

从数学角度定义一个表面的斜率和方向是通过它的梯度来实现的，

$$J(x) = \nabla I(x) = (\frac{\partial I}{\partial x}, \frac{\partial I}{\partial y})(x). \tag{4.19}$$

局部梯度向量 J 指向亮度函数的极速上升(steepest ascent)方向。它的幅值是其斜率或者变化强度的一个指示，它的方向指向了与其局部轮廓垂直的方向。

不幸的是，求取图像的导数强调了高频率的部分因而放大了噪声，因为噪声和信号的比例在高频部分较大。因此在计算梯度之前需要考虑将图像用一个低通滤波器进行平滑。因为我们希望边缘检测器的响应与方向无关，所以需要一个圆对称的平滑滤波器。如3.2节所述，高斯函数是唯一可分离的圆对称滤波器，所以大多数边缘检测算法中都使用了它。Canny(1986)讨论了其他可供选择的滤波器，其他许多研究者回顾了可供选择的边缘检测算法并比较了它们的性能(Davis 1975; Nalwa and Binford 1986; Nalwa 1987; Deriche 1987; Freeman and Adelson 1991; Nalwa 1993; Heath, Sarkar, Sanocki *et al.* 1998; Crane 1997; Ritter and Wilson 2000; Bowyer, Kranenburg, and Dougherty 2001; Arbeláez, Maire, Fowlkes *et al.* 2010)。

因为微分是一个线性操作，所以它和其他线性滤波操作可交换。因此一个平滑后的图像的梯度可以写作

$$J_\sigma(x) = \nabla[G_\sigma(x) * I(x)] = [\nabla G_\sigma](x) * I(x), \tag{4.20}$$

也就是说，我们可以用一个水平的和垂直的高斯核函数的导数来和图像进行卷积，

$$\nabla G_\sigma(x) = (\frac{\partial G_\sigma}{\partial x}, \frac{\partial G_\sigma}{\partial y})(x) = [-x \quad -y]\frac{1}{\sigma^3}\exp\left(-\frac{x^2+y^2}{2\sigma^2}\right) \tag{4.21}$$

(参数σ反映了高斯函数的带宽。)这和 Freeman and Adelsons(1991)使用的一阶导向滤波器是相同的计算，我们已经在3.2.3节中讲过。

然而，对于很多应用，我们希望将这个连续梯度图像稀疏化使之仅存在孤立的边缘，也就是说，仅以单个像素分布在沿着边缘轮廓的离散位置上。这个可以通过在与边缘方向(也就是梯度方向)垂直方向上寻找边缘响应(梯度幅值)的极大值来获得。

寻找最大值对应于将强度场沿着梯度方向求取方向导数，然后寻找过零点。这个要求的方向导数与第二个梯度算子和第一个梯度算子的结果的点积等价，

$$S_\sigma(x) = \nabla \cdot J_\sigma(x) = [\nabla^2 G_\sigma](x) * I(x)]. \tag{4.22}$$

这个梯度算子与梯度的点积称为"拉普拉斯"(Laplacian)，而卷积核

$$\nabla^2 G_\sigma(x) = \frac{1}{\sigma^3}\left(2 - \frac{x^2+y^2}{2\sigma^2}\right)\exp\left(-\frac{x^2+y^2}{2\sigma^2}\right) \tag{4.23}$$

称为"高斯二阶导数"(Laplacian of Gaussian)(LoG)核函数(Marr and Hildreth 1980)。这个核函数可以分成两个部分，

$$\nabla^2 G_\sigma(x) = \frac{1}{\sigma^3}\left(1 - \frac{x^2}{2\sigma^2}\right)G_\sigma(x)G_\sigma(y) + \frac{1}{\sigma^3}\left(1 - \frac{y^2}{2\sigma^2}\right)G_\sigma(y)G_\sigma(x) \tag{4.24}$$

(Wiejak, Buxton, and Buxton 1985)，因此可以使用分离的滤波来非常高效地实现(3.2.1节)。

在实践中，常用的做法将高斯二阶导数卷积换成高斯差分(DoG)来计算，因为

其核函数的形状在性质上相似(图 3.35)。如果已经计算了一个拉普拉斯金字塔(3.5节)，这就非常方便[①]。

事实上，在计算边缘场时，并没有严格要求在相邻高斯层间求取差值。考虑一下在一个"推广"的高斯差分图像中过零点表示什么。较精细(核小)的高斯函数是一个原始图像的降噪版本。较粗(核大)的高斯函数是一个较大区域平均亮度的估计。因此，每当 DoG 图像改变符号时，这对应着一幅(轻微模糊的)图像就其邻域中的平均亮度而言从相对较暗变到相对较亮。

一旦计算了符号函数 $S(x)$，我们就必须找到它的过零点(zero crossing)并把它们转化成边缘元素(边界基元)。一个检测和表示过零点的简单方式是寻找那些符号改变的相邻像素位置 x_i 和 x_j，也就是，$[S(x_i) > 0] \neq [S(x_j) > 0]$。

通过计算连接 $S(x_i)$ 和 $S(x_j)$ 的"直线"的"x-拦截点"，可以获得这个过零点的亚像素级位置，

$$x_z = \frac{x_i S(x_j) - x_j S(x_i)}{S(x_j) - S(x_i)}. \tag{4.25}$$

这样的边界元的方向和强度可以通过对在原始像素网格上计算的梯度值线性插值来获得。

另外一种可选的边界元表示可以通过在双重网格上连接相邻边界元来形成新的边界元，这些新边界元位于原始像素网格中四邻接像素形成的方形的内部[②]。这种表示的(潜在)优势是这些边界元现在位于的网格位移是原来网格的一半像素，因此更容易保存和读取。和前面一样，边界的方向和强度可以通过在梯度场上插值或者从高斯差分图像上估计来计算(参考习题 4.7)。

对于那些边界方向精度更为重要的应用来说，可以使用高阶导向滤波器(Freeman and Adelson 1991)(参见 3.2.3 节)。这些滤波器对于加长的边缘更具选择性且可能对曲线相交做更好的建模，因为它们可以在同一个像素处表示多个方向(图 3.16)。它们的缺点是计算更为复杂且边缘强度的方向导数没有一个简单的闭合解[③]。

尺度选择和模糊估计

正如我们前面所提到的，导数、拉普拉斯和高斯滤波器的差分(图 4.20～4.23)都需要选择空间尺度参数 σ。如果只对检测强烈的边缘感兴趣，滤波器的带宽可以从图像的噪声特性来确定(Canny 1986; Elder and Zucker 1998)。然而，如果我们想要检测出现在不同分辨率下的边缘(图 4.32b 和 c)，就需要一种尺度空间上的方法来寻找不同尺度上的边缘(Witkin 1983; Lindeberg 1994, 1998a; Nielsen, Florack, and Deriche 1997)。

Elder and Zucker(1998)给出了解决这个问题的一种理论方法。给定一个已知噪

[①] 回忆一下文献 Burt and Adelson(1983a)中的"拉普拉斯金字塔"，实际上计算的是不同层高斯滤波的差。
[②] 这个算法是 3D 移动立方体(marching cube)等高面提取算法(Lorensen and Cline 1987)的 2D 版本。
[③] 事实上，对于条状的边缘，其方向存在 180°的歧义性，这使得在导数中求取过零点很有技巧性。

声级别,他们的方法为每一个像素计算可以可靠地检测出边缘的最小尺度(图 4.32d)。通过根据梯度幅度值选择在不同尺度上计算出来的梯度估计,他们的方法是首先为整个图像计算稠密的梯度。然后它对二阶方向导数进行类似的最小尺度估计,并使用后者的过零点来鲁棒地选择边缘(图 4.32e 和 f)。作为可选的最后一步,每一个边缘的模糊宽度可以根据在二阶导数响应中的极值点间的距离减去高斯滤波器的宽度来计算。

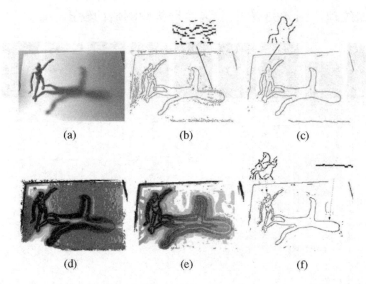

图 4.32 边缘检测中的尺度选择(Elder and Zucker 1998)© 1998 IEEE。(a)原始图像;(b 和 c)使用 Canny/Deriche 边缘检测器调整参数到精细的(模糊)和粗的(影子)尺度;(d)梯度估计的最小稳定尺度;(e)二阶导数估计的最小稳定尺度;(f)最终检测到的边缘

彩色边缘检测

尽管大多数边缘检测方法都是针对灰度图像开发的,彩色图像可以提供附加的信息。举个例子,等亮度线(iso-luminant)间的明显边缘是非常有用的信息,但是灰度边缘检测子就不能检测到。

一个简单方法是把灰度检测器在每个颜色空间独立输出的结果结合起来[①]。然而,需要注意一些问题。比如,如果我们仅简单地将每个颜色空间的梯度加起来,带符号的梯度实际上可能会互相抵消!(比如考虑一个纯的由红到绿的边缘。)我们也可以在每个颜色空间独立地检测边缘,然后将它们联合起来,但这样做可能导致边缘加粗或者出现双倍的边缘而难以连接。

一个较好的方法是在每个颜色空间上计算有向能量(oriented energy)(Morrone and Burr 1988; Perona and Malik 1990a),比如,使用一个二阶导向滤波器(3.2.3 节)(Freeman and Adelson 1991),然后将加权的有向能量加起来再寻找它们联合的

[①] 与其使用原始的 RGB 空间,不如换用一个在视觉上更均匀分布的颜色空间,比如 L*a*b*(参考 2.3.2 节)。当试图与人的表现进行比较时(Martin, Fowlkes, and Malik 2004),这是合理的。但是,就本质的图像形成和感知的物理属性而言,这可能是一个有问题的策略。

最佳方向。不幸的是，这个能量的方向导数可能没有一个闭合形式的解答(与带符号的一阶导向滤波器的情况一样)，因此不能使用简单的基于过零点的策略。但是，Elder and Zucker(1998)描述的方法可以用于这些过零点的数值计算。

一个可选的方法是在每个像素周围区域估计局部颜色统计信息(Ruzon and Tomasi 2001; Martin, Fowlkes, and Malik 2004)。这样做的优势是可以使用一些更复杂的方法(比如，3D 颜色直方图)来比较区域的统计信息，另外也可以考虑一些附加的度量，比如纹理。图 4.33 给出了这样一些检测器的输出。

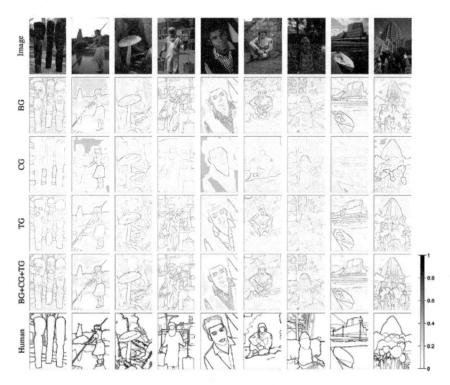

图 4.33 结合了亮度、颜色和纹理的边界检测器(Martin, Fowlkes, and Malik 2004) © 2004 IEEE。不同的行依次给出了亮度梯度(BG)，颜色梯度(CG)，纹理梯度(TG)以及融合后(BG+CG+TG)的检测器的输出。最后一行给出了从一个手工分割的图像中得到的人工标定的边界(Martin, Fowlkes, Tal *et al.* 2001)

当然，还有很多方法可以用来检测颜色边缘，这可以追溯到早期 Nevatia(1977)的工作。Ruzon and Tomasi(2001)和 Gevers, van de Weijer, and Stokman(2006)为这些工作做了较好的综述，其中包括各种想法，比如融合多通道输出，使用多维梯度和基于向量的方法。

结合边缘特征线索

如果边缘检测的目标是与人工边界检测(boundary detection)表现相匹配(Bowyer, Kranenburg, and Dougherty 2001;Martin, Fowlkes, and Malik 2004; Arbeláez,Maire, Fowlkes *et al.* 2010)，那么这不同于简单地寻找稳定的特征用于匹配，通过结合多

个低级视觉线索，比如亮度、颜色和纹理，我们可以构建出更好的检测器。

Martin, Fowlkes, and Malik(2004)描述了一个结合亮度、颜色和纹理边缘的系统，这个系统在一个手工分割的自然彩色图像数据库(Martin, Fowlkes, Tal *et al.* 2001)中取得了最佳的性能。首先，他们构建并训练[1]分离的有向半圆状检测器来衡量亮度(照度)、颜色(a*通道和 b*通道加起来的响应)和纹理(根据 Malik, Belongie, Leung *et al.*(2001)工作得到的未归一化滤波器组的响应)上的显著差别。然后使用一个软性非最大抑制方法来将这些响应中的一些锐化。最后，使用各种机器学习方法将这三个检测器的输出结合起来，发现"逻辑回归"(logistic regression)在速度、空间和准确度上的折中最理想。得到的系统(参见图 4.33 中给出的一些例子)表明优于之前发展的方法。Maire, Arbeláez, Fowlkes *et al.*(2008)通过融合基于局部表观的检测器以及一个谱(基于分割的)检测器(Belongie and Malik 1998)对这些结果进行了改进。在最近的工作中，Arbeláez, Maire, Fowlkes *et al.*(2010)在该边缘检测器的基础上使用分水岭算法的一个变种构建了一个层次性分割。

4.2.2 边缘连接

尽管孤立的边缘对于很多应用很有用，比如线检测(4.3 节)和稀疏立体视觉匹配(11.2 节)，但如果将它们连接成连续的轮廓会更有用。

如果边缘已经通过某个函数的过零点检测到，那么由于相邻的边界元拥有共同的端点，所以将它们连接起来也就非常直接。将边界元连接成链涉及寻找一个没有被连接的边界元并在两个方向上跟踪它的邻居。无论是用边界元的有序列表(比如，先根据 x 坐标排序然后再根据 y 坐标排序)，还是用 2D 数组，都可以加速这个邻居寻找过程。如果使用过零点没有检测到边缘，那么寻找一个边界元的接续就会很有技巧性。这种情况下，出现歧义时可以比较相邻边界元的方向(或者还可选择相位)。连通分量计算中使用的想法有时也可以用于使边缘连接的过程更加快速(参见习题 4.8)。

一旦将边界基元连成链状，我们就可以施加一个可选的带滞后作用的阈值化处理来去除低强度的轮廓片段(Canny 1986)。滞后作用的基本想法是设定两个不同的阈值，允许高于较高阈值的被跟踪曲线可以包含那些强度低至低阈值的边缘。

可以使用各种可选的表示将连接的边界基元列表进行更加紧凑的编码。对于位于八邻域($\mathcal{N}_8$)网格上的连通点列表，链码(chain code)方式使用一个三位码来对应该点和下一个点之间的八个基本方向(N，NE，E，SE，S，SW，W，NW)(图 4.34)。尽管相对于原始边界基元列表来说这种表示更为紧凑(特别是当使用预测变长编码时)，但它不便于进一步处理。

"弧长参数化"(arc length parameterization)$x(s)$是另一种更有用的表示，其中 s 表示一条曲线上的弧长度。请看图 4.35a 中给出的连在一起的边界基元集合。我们

---

[1] 训练使用了 200 张标定的图像，测试是在另外 100 张图像中进行的。

从一个点开始(图 4.35a 中位于(1.0, 0.5)的点)并将它画在 $s = 0$ 坐标处(图 4.35b)。下一个在(2.0, 0.5)点被画在 $s = 1$，下一个在(2.5, 1.0)处的点画在 1.7071，也就是说，我们根据每一个边缘片段的长度来增加 s。在进行进一步处理之前，得到的绘图可以重新采样到一个规则的(比如，整数的)网格上。

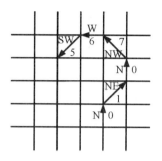

图 4.34 网格对齐的边缘连接链的链码表示。编码被表示成一系列的方向码，比如，0 1 1 7 6 5，它可以进一步用预测编码和行程编码压缩

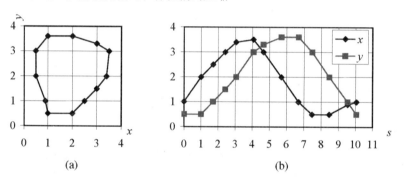

图 4.35 轮廓的弧长参数化。(a)沿着轮廓的离散点可以首先被转换为(b)中沿着弧长 s 的 (x,y) 对。然后可以对这条曲线规范地重采样或者将其转换为其他(比如，傅里叶)表示

"弧长参数化"表示的优势是它使得匹配和处理(比如，平滑)操作更加容易。请看图 4.36 显示的两个描述相似形状的曲线，为了比较这两条曲线，我们首先从每个描述子中减去平均值 $(x_0 = \int_s x(s))$，然后，我们对每个描述子进行缩放使得 s 在 0 到 1 之间变换，而不是 0 到 S，也就是说，我们要用 S 除 $x(s)$。最后，我们对每一个规范化后的描述子进行傅里叶变换，将每一个 $\boldsymbol{x} = (x,y)$ 看成一个复数。如果原始的两条曲线是相同的(只有未知的尺度和旋转变化)，得到的傅里叶变换应该只有反映在幅值上的尺度变化和一个恒定的复相位上的平移。弧长参数化可以用来平滑曲线从而去除数字化误差。然而，如果我们只是仅仅使用一个通用平滑滤波器，曲线本身就倾向于收缩(图 4.37a)。Lowe(1989) 和 Taubin(1995)给出了可以补偿这种收缩的方法，这是通过根据二阶导数估计加上一个偏移项或者使用一个较大的平滑核函数实现的(图 4.37b)。Finkelstein and Salesin(1994)基于选择性地修改一个小波分解中的不同频率，给出了另一种可选的方法。如图 4.38 所示，除了控制收缩而不影响它的"扫描"(sweep)，小波还允许对一条曲线的"特性"做交互式的修改。

曲线被平滑和简单化的演化过程和"烧草"(距离)变换和区域骨架(3.3.3 节)相关(Tek and Kimia 2003)，可用于基于物体的轮廓形状的物体识别(Sebastian and Kimia 2005)。更局部的曲线形状描述子——比如形状上下文(Belongie, Malik, and Puzicha 2002)可以用于识别，潜在地对于由于遮挡造成的部件丢失更为鲁棒。

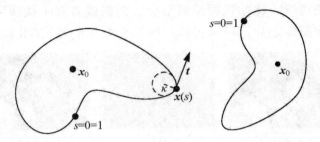

图 4.36 使用弧长参数化来匹配两个轮廓。如果两条曲线都被规范化到单位长度，$s \in [0,1]$，并且以它们的重心 x_0 为中心，它们就会在只差一个整体的"时序"偏移(由于 $s=0$ 的不同起始点造成的)和一个相位(x–y)偏移(由于旋转造成的)的意义下有相同的描述子

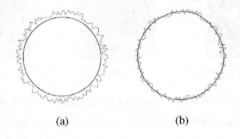

图 4.37 使用高斯核函数平滑曲线(Lowe 1988)© 1998 IEEE：(a)不使用收缩修正项；(b)使用收缩修正项

图 4.38 改变一条曲线的特性而不影响它的扫描(Finkelstein and Salesin 1994)© 1994 ACM：较高频率的小波可以用一个风格库中的样例替换，从而影响局部的表观

轮廓检测和连接研究方向仍然发展很快，目前发展的方法包括全局轮廓分组、边界完成和连接点检测(Maire, Arbeláez, Fowlkes *et al.* 2008)以及将轮廓聚集成相像的区域(Arbeláez, Maire, Fowlkes *et al.* 2010)和宽基线对应(Meltzer and Soatto 2008)。

4.2.3 应用：边缘编辑和增强

边缘不仅可以用作物体识别的一个组件或者是用作匹配的特征，它们还可以直接用于图像编辑。

事实上，如果每一个边缘保留了边缘幅值和模糊估计，使用这些信息就可以重建出一个类似的图像(Elder 1999)。基于这个原理，Elder and Goldberg(2001)提出了一个"在轮廓域中作图像编辑"的系统。他们的系统允许用户选择性地去除那些不想要的特征对应的边缘，比如镜面反射部分、阴影或者起干扰作用的视觉元素。根据剩下的边缘重建图像之后，可以去掉那些不合要求的特征(图 4.39)。

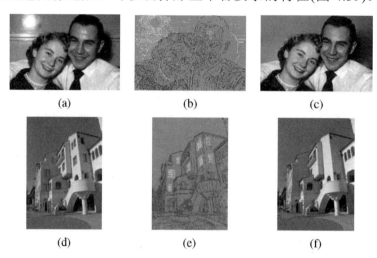

图 4.39 基于轮廓的图像编辑(Elder and Goldberg 2001)© 2001 IEEE：(a)和(d)是原始图像；(b)和(e)是提取的边缘(要删除的边缘用白色表示)；(e)和(f)是重建的编辑图像

另一个潜在的应用是在简化基本图像以产生类似于卡通或者"笔墨画"风格的图像时增强那些显著的边缘(DeCarlo and Santella 2002)。这个应用将在 10.5.2 节中进一步讨论。

4.3 线　　条

尽管边缘和一般曲线适合用来描述自然物体的轮廓，但人类世界中直线段到处都是。检测和匹配这些线条对于很多应用很有用，包括建筑学上的建模、城市环境下的姿态估计以及书面文档版式分析。

这本节中，我们给出了一些从前一节计算得到的曲线中提取分段线性(piecewise linear)描述的方法。我们开始先描述一些使用分段线性折线来近似曲线的算法，然后我们讲解一下哈夫变换(Hough transform)，即使在存在间隙和遮挡的时候，它都可以用于将边界元聚集成线片段，最后，我们描述怎样将拥有"共同消失点"(vanishing point)的 3D 直线聚集在一起。正如 6.3.2 节中所描述的那样，这些消失点可以用来标定摄像机，从而确定它相对于一个矩形状场景的方向。

4.3.1 逐次近似

我们根据 4.2.2 节可以知道，将一个曲线描述成一系列 2D 位置点 $x_i = x(s_i)$，给出了曲线的一个适合匹配和进一步处理的通用表示。然而，在很多应用中，我们优先选择用更简单的表示来近似这个曲线，举个例子，如图 4.40 所示，用一个分段线性的折线或者 B 样条曲线来近似(Farin 1996)。

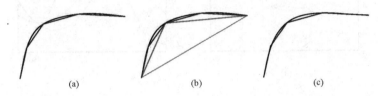

图 4.40 用折线或者 B 样条曲线来近似一个曲线(用黑色表示)：(a)原始曲线用一个红色的折线来近似；(b)通过迭代地寻找与目前的近似离得最远的点做逐次近似；(c)拟合折线顶点的光滑的内插样条用深蓝色显示

多年来已经发展了很多方法来完成这个近似，通常也称为"直线简化"(line simplification)。其中 Ramer(1972)和 Douglas and Peucker(1973)给出了最早同时也最简单的一个方法，如图 4.40 所示，它是通过递归地细分连接两个端点(或者当前粗略折线近似)的直线最远的曲线点来近似曲线的。Hershberger and Snoeyink(1992)给出了一个更高效的实现，它也引用了这个方向上其他的一些相关工作。

一旦完成了直线简化，就可以使用它来近似原始曲线。如图 4.40c 所示，如果需要一个更平滑的表示或者可视化，就可以使用近似线或者插值样条或者是曲线本身(3.5.1 节和 5.1.1 节)(Szeliski and Ito 1986; Bartels, Beatty, and Barsky 1987; Farin 1996)。

4.3.2 Hough 变换

使用折线进行曲线近似可以成功地得到直线提取，然而实际中的线段有时被断开为不连通的分量或者由多个共线的线段组成。在很多情况下，需要将这些共线的线段聚集成扩展的直线。在进一步处理阶段(在 4.3.3 节中描述)，我们可以将这些线段聚集成拥有公共消失点的集合。

以它的最初发明者命名(Hough 1962)的哈夫变换，是一种非常有名的根据边缘来对可能的直线位置进行"投票"的方法(Duda and Hart 1972; Ballard 1981; Illingworth and Kittler 1988)。在它最初的形式化中(图 4.41)，每一个边缘点为通过它的所有可能直线进行投票，然后检查那些对应着最高累加器(accumulator)或者区间(bin)的直线以寻找可能的线匹配[①]。除非这个直线上的点确实是点状的，否则(根

[①] 哈夫变换同样可以推广到寻找其他几何学特征，比如圆(Ballard 1981)，但本书中我们不讨论这些扩展。

据我的经验)更好的方法是下面所描述的使用每个边界基元的局部方向信息来对单个累加器单元进行投票(图 4.42)。一个混合的策略，也就是让每一个边界基元为一些可能的方向或者为以估计方向为中心的位置对进行投票，在一些情况下可能更可取。在我们对直线假设进行投票之前，首先必须选择一个合适的表示。

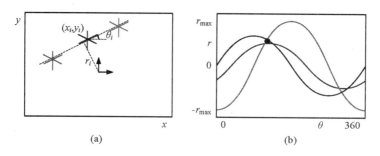

图 4.41 原始哈夫变换：(a)每一个点为所有可能的直线 $r_i\theta = x_i\cos\theta + y_i\sin\theta$ 进行投票进行；(b)每束线给出 (r, θ) 处的一个正弦函数；它们的交点确定了所要求的直线方程

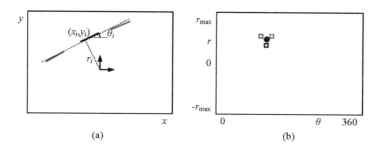

图 4.42 带方向的哈夫变换：(a)一个位于极坐标 (r, θ) 处的边界基元用 $\hat{n}_i = (\cos\theta_i, \sin\theta_i)$ 和 $r_i = \hat{n}_i \cdot x_i$ 重新参数化；(b) (r, θ) 显示了分别标记为红绿蓝三个边界基元投票的累加器阵列

$$\theta = \tan^{-1} n_y/n_x, \tag{4.26}$$

图 4.43(复制了图 2.2a)显示了一条直线的法向-距离 $(\hat{n}, d)$ 的参数化。由于直线是由边缘段组成的，我们根据惯例，直线的法向 $\hat{n}$ 指向和图像梯度 $J(x) = \nabla I(x)$ 一样的方向(也就是，有一样的符号)。为了获得直线的一个最小的两参数表示，我们将法向向量转化成一个角度

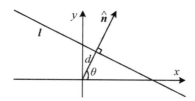

图 4.43 将 2D 直线表示成法向量 $\hat{n}$ 和到原点的距离 d 的形式

如图 4.43 所示。假设我们使用(2.61)中规范化后的在 $(-1,1)$ 内的像素坐标，(θ, d) 可能的范围是 $[-180°, 180°] \times [-\sqrt{2}, \sqrt{2}]$。每个轴使用的区间数目取决于每个边界基元可获得的位置和方向估计的精度以及期望的线密度，这个数目最好在样本图

像上测试运行几次,通过实验来设定。

给定了直线参数化,哈夫变换按照算法 4.2 所示那样进行。

需要注意的是,对于哈夫变换最初的形式化,由于假设中不知道边界基元方向的任何信息,所以需要在第 2 步中添加一个附加的循环,这个循环在 θ 所有的可能值上迭代并增加一个整个序列的累加器的值。

为了使哈夫变换能够很好地工作,需要考虑到很多细节问题,但最有效的方式就是写一个实现然后在样本数据上进行测试。习题 4.12 更详细地描述了其中的一些步骤,包括在投票过程中使用边缘片段长度或者强度,在累加器数组中维护一个组成边界基元的列表来使后处理更容易,以及选择性地将不同"极性"的边缘结合到一个线段中。

---

procedure Hough($\{(x, y, \theta)\}$):

1. 清空累加器数组。
2. 对于每一个检测到的位置为 (x, y) 和方向为 $\theta = \tan^{-1} \boldsymbol{n}_y / \boldsymbol{n}_x$ 的边界基元,计算如下值
$$d = x\boldsymbol{n}_x + y\boldsymbol{n}_y$$
增加对应于 (θ, d) 的累加器的值。
3. 寻找对应于直线的累加器峰值;
4. 对候选的边界基元重新进行最优拟合。

算法 4.2 基于有向边缘片段的哈夫变换算法流程

另一种可选的直线的 2D 极坐标 (θ, d) 表示是使用完整 3D 直线方程 $m = (\hat{n}, d)$ 并映射到单位球上。其中可以使用球面坐标系(2.8)来参数化球。

$$\hat{m} = (\cos\theta\cos\phi, \sin\theta\cos\phi, \sin\phi), \tag{4.27}$$

这种方式没有在球内均匀采样,并且需要使用三角函数。

通过使用一种立方图(cube map),也就是将 m 映射到单位立方体(图 4.44a),可以得到一种可选的表示。为了计算一个 3D 向量 m 的立方图坐标,需要先找到 m 的最大(绝对值)的分量,也就是 $m = \pm\max(|n_x|, |n_y|, |d|)$,然后使用这个来选择六个平面之一。用 m 来除剩下的两个坐标然后并用它们作为立方体表面的索引。这样就避免了使用三角函数,只需要一些决策逻辑。

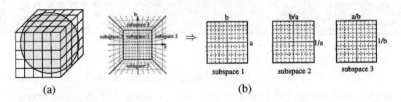

图 4.44 直线方程和消失点的立方图表示:(a)包围单位球的一个立方图;(b)将半立方体投影到三个子空间(Tuytelaars, Van Gool, and Proesmans 1997)© 1997 IEEE

Tuytelaars, Van Gool, and Proesmans(1997)最早指出使用立方图的一个好处，即通过一个点的所有直线都对应着立方体表面的线段，当使用原始(全投票)哈夫变换类算法时，这就非常有用。在他们的工作中，他们将直线方程表示为 $ax+b+y=0$，对 x 和 y 不是对称地对待。注意，如果我们忽略边界方向(梯度的方向)的极性，就可以限制 $d \geq 0$，如图 4.44b 所示，这样就可以使用一个由三个立方体表面表示的半立方体来替代(Tuytelaars, Van Gool, and Proesmans 1997)。

基于 RANSAC 的直线检测

除哈夫变换外，另一种可选的方法是 6.1.4 节详细描述的随机样本一致性(RANdom SAmple Consensus，RANSAC)算法。简而言之，RANSAC 随机选取一对边界基元来形成一个直线假设，然后测试其他的边界基元有多少落在这条直线上。(如果边缘方向足够精确，一个单独的边界基元就可以产生这个假设。)然后选择存在大量内点(匹配的边界基元)的线条作为想要的直线片段。

RANSAC 的一个优势是不需要使用累加器数组，因此这个算法在空间上更高效并且对区间大小的选择也不那么敏感。缺点是相对于直接在累加器数组中寻找峰值，需要产生和测试更多的假设。

一般来说，对于哪种直线估计方法效果最好没有一致的看法。因此一个好的想法是仔细考虑你要处理的问题并使用几种方法(逐次近似、哈夫和 RANSAC)来确定哪种方法最适合具体应用。

4.3.3 消失点

在很多场景中，一些直线由于在三维中平行，所以它们在结构上非常重要，所有这些直线拥有一个共同的消失点。这样的直线的例子包括平行和垂直的建筑物边缘、斑马线、火车轨道、诸如桌子碗柜等家具的边缘，当然还包括通用的标定模式(图 4.45)。找到这些直线集合的公共消失点可以有助于优化它们在图像中的位置，以及在特定场景下确定摄像机的内参数和外参数(6.3.2 节)。

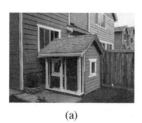
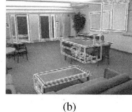
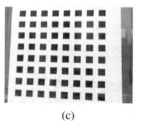

(a)　　　　　　　　(b)　　　　　　　　(c)

图 4.45 现实世界中的消失点：(a)建筑物(Sinha, Steedly, Szeliski *et al.* 2008)；(b)家具(Mičušik, Wildenauer, and Košecká 2008)© 2008 IEEE；(c)标定模式(Zhang 2000)

在过去的几十年里，产生了数量众多的寻找消失点的方法，其中包括(Quan and Mohr 1989; Collins and Weiss 1990; Brillaut-O'Mahoney 1991; McLean and Kotturi

1995; Becker and Bove 1995; Shufelt 1999; Tuytelaars, Van Gool, and Proesmans 1997; Schaffalitzky and Zisserman 2000; Antone and Teller 2002; Rother 2002; Košecká and Zhang 2005; Pflugfelder 2008; Tardif 2009)——请从其中较新的一些文献中找更多的参考文献。在本节中，我们给出一种简单的 Hough 方法，该方法基于已有的直线对来为潜在的消失点位置进行投票，然后再通过一个鲁棒的最小二乘拟合阶段。对于其他的一些方法，请参考上面所列文章中较新的那些文献。

我给出的消失点检测算法的第一个阶段是使用哈夫变换来投票累加那些可能的消失点候选。正如直线拟合那样，可能的方法是让每一条直线为所有可能的消失点方向进行投票，这可以使用立方图(Tuytelaars, Van Gool, and Proesmans 1997; Antone and Teller 2002)或者高斯球(Collins and Weiss 1990)，其中还可以选择是否根据消失点位置的不确定性来进行加权投票(Collins and Weiss 1990; Brillaut-O'Mahoney 1991; Shufelt 1999)。我更喜欢的方法是使用检测到的直线片段对来形成候选消失点位置。用 $\hat{m}_i$ 和 $\hat{m}_j$ 表示一对直线片段的(单位法向)直线方程，l_i 和 l_j 是它们对应的片段长度，对应的消失点假设位置可以按照如下的方式进行计算

$$v_{ij} = \hat{m}_i \times \hat{m}_j \tag{4.28}$$

对应的权重集合设为

$$w_{ij} = \|v_{ij}\| l_i l_j. \tag{4.29}$$

这样会有期望的效果：让(近似)共线的直线片段或者短的直线片段权重降低。哈夫空间本身可以使用球坐标(4.27)或者立方图(图 4.44a)来表示。

一旦哈夫累加器空间确定下来，就可以使用与之前讨论过的线检测方法相类似的方式来检测峰值。给定一个用于为消失点投票的线段候选集合，作为选项这个集合可以以一个列表的形式保存在每一个哈夫累加器单元中，然后我使用一个鲁棒的最小二乘拟合来为每一个消失点估计出一个更为精确的位置。

如图 4.46 所示，考虑到直线片段的两个端点 $\{p_{i0}, p_{i1}\}$ 和消失点 v 的关系，这三个点形成的三角形的面积 A，也就是它们三元积的幅值

$$A_i = |(p_{i0} \times p_{i1}) \cdot v|, \tag{4.30}$$

是与一个端点到消失点 v 和另一个端点形成的直线之间的垂直距离 d_1 成正比的，也与 p_{i0} 和 v 之间的距离成正比。假设一个拟合的直线片段的精度是和它们的端点的精度成正比的(习题 4.13)，因此这就形成了一个消失点与一个提取出的线段集合拟合程度好坏的最优度量(Leibowitz(2001, Section 3.6.1)和 Pflugfelder(2008, Section 2.1.1.3))。消失点的一个鲁棒化的最小二乘估计(附录 B.3)因此可以写作

$$\mathcal{E} = \sum_i \rho(A_i) = v^T \left(\sum_i w_i(A_i) m_i m_i^T \right) v = v^T M v, \tag{4.31}$$

其中 $m_i = p_{i0} \times p_{i1}$ 是根据它的长度 l_i 加权的线段方程，$\omega_i = \rho'(A_i)/A_i$ 是最终错误的一个鲁棒化的(重加权的)度量(附录 B.3)。注意，这个度量与成对加权哈夫变换累加器步骤的原始公式密切相关。v 的最终期望的值是由 M 的最小特征向量计算得到的。

前面描述的方法用两个独立的步骤完成，交替变换指定给消失点的直线和重新

拟合消失点的位置,可能会得到更好的结果(Antone and Teller 2002; Košecká and Zhang 2005; Pflugfelder 2008)。通过同时搜索两个或者三个相互正交的消失点可以使原本独立检测的消失点更鲁棒(Shufelt 1999; Antone and Teller 2002; Rother 2002; Sinha, Steedly, Szeliski et al. 2008)。在图 4.45 中,可以看到这样的消失点检测算法的一些结果。

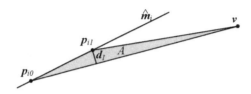

图 4.46 线段端点 p_{i0} 和 p_{i1} 以及消失点 v 的三重数积。面积 A 正比于垂直距离 d_1 及另一个点 p_{i0} 和消失点之间的距离

4.3.4 应用:矩形检测

一旦检测出相互正交的消失点的集合,就有可能在图像中搜索出 3D 的矩形结构(图 4.47)。在过去十年里,产生了各种方法来检测这样的矩形,这些方法主要关注的是在建筑物场景中检测这些矩形(Košecká and Zhang 2005; Han and Zhu 2005; Shaw and Barnes 2006; Mičušìk, Wildenauer, and Košecká 2008; Schindler, Krishnamurthy, Lublinerman et al. 2008)。

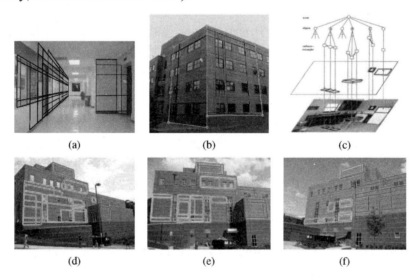

图 4.47 矩形检测:(a)室内走廊;(b)带有侧面分组的建筑物外部(Košecká and Zhang 2005)©1999 Elsevier;(c)基于语法的识别(Han and Zhu 2005) © 2005 IEEE;(d~f)使用平面扫描算法的矩形匹配(Mičušìk, Wildenauer, and Košecká 2008) © 2008 IEEE

检测到正交的消失点方向之后,Košecká and Zhang(2005)先优化拟合线的方程,搜索在直线交点附近的角点,然后通过调整对应的图像块和寻找占主导的水平

和垂直方向上的边缘来验证矩形假设(图 4.47a 和 b)。在后续的工作中,Mičušìk,Wildenauer, and Košecká(2008)使用一个马尔科夫随机场(MRF)来消除由于潜在重叠的矩形假设所造成的歧义。他们还使用一个平面扫描算法来匹配不同视角间的矩形(图 4.47d~f)。

Han and Zhu(2005)提出了一种不同的方法,这种方法使用潜在匹配的矩形形状和嵌套结构(在矩形和消失点之间)的语法来推导矩形框最有可能的对应线段(图 4.47c)。

4.4 扩 展 阅 读

Lowe(2004)是关于特征检测、描述和匹配影响最深远的文章之一。Schmid, Mohr, and Bauckhage(2000);Mikolajczyk and Schmid(2005);Mikolajczyk, Tuytelaars, Schmid et al.(2005);Tuytelaars and Mikolajczyk(2007)对相关方法做了全面的综述和评测。另外,Shi and Tomasi(1994)和 Triggs(2004)也提供了很好的回顾。

在特征检测领域(Mikolajczyk, Tuytelaars, Schmid et al. 2005),除了像 Förstner–Harris(Förstner 1986; Harris and Stephens 1988)和高斯差分(Lindeberg 1993, 1998b; Lowe 2004)这些传统方法外,最稳定极值区域(MSER)被广泛用于那些要求仿射不变的应用(Matas, Chum, Urban et al. 2004; Nistér and Stewénius 2008)。Xiao and Shah(2003);Koethe(2003) Carneiro and Jepson(2005);Kenney, Zuliani, and Manjunath(2005);Bay, Tuytelaars, and Van Gool(2006);Platel, Balmachnova, Florack et al.(2006);Rosten and Drummond(2006)讨论了最新的关于特征点检测器的方法,(Zoghlami, Faugeras, and Deriche 1997; Bartoli, Coquerelle, and Sturm 2004)给出了基于直线检测的方法,(Kadir, Zisserman, and Brady 2004; Matas, Chum, Urban et al. 2004; Tuytelaars and Van Gool 2004; Corso and Hager 2005)讨论了基于区域检测的方法。

Mikolajczyk and Schmid(2005)回顾和比较了数目众多的特征描述子(和匹配策略)。该领域最新发表的文章包括 van de Weijer and Schmid(2006);Abdel-Hakim and Farag(2006);Winder and Brown(2007);Hua, Brown, andWinder(2007)。高效特征匹配的方法包括 k-d 树(Beis and Lowe 1999; Lowe 2004; Muja and Lowe 2009)、金字塔状匹配核函数(Grauman and Darrell 2005)、度量(词汇)树(Nistér and Stewénius 2006)以及各种各样的多维哈希方法(Shakhnarovich, Viola, and Darrell 2003; Torralba, Weiss, and Fergus 2008; Weiss, Torralba, and Fergus 2008; Kulis and Grauman 2009; Raginsky and Lazebnik 2009)。

特征检测和跟踪的经典参考文献是(Shi and Tomasi 1994)。这个领域最新的工作集中于更好地学习特定特征的匹配函数(Avidan 2001; Jurie and Dhome 2002; Williams, Blake, and Cipolla 2003; Lepetit and Fua 2005; Lepetit, Pilet, and Fua 2006; Hinterstoisser, Benhimane, Navab et al. 2008; Rogez, Rihan, Ramalingam et al. 2008;

Özuysal, Calonder, Lepetit *et al.* 2010)。

　　Canny(1986)提出了一种被广泛引用和应用的边缘检测器。在下面的文献中可以找到可选的一些边缘检测器以及实验对比：Nalwa and Binford(1986)；Nalwa(1987)；Deriche(1987)；Freeman and Adelson(1991)；Nalwa(1993)；Heath, Sarkar, Sanocki *et al.*(1998)；Crane(1997)；Ritter and Wilson(2000)；Bowyer, Kranenburg, and Dougherty(2001)；Arbeláez, Maire, Fowlkes *et al.*(2010)。Elder and Zucker(1998)很好地处理了边缘检测中的尺度选择，而颜色和纹理边缘检测可以在(Ruzon and Tomasi 2001; Martin, Fowlkes, and Malik 2004; Gevers, van de Weijer, and Stokman 2006)中找到。最近，人们将边缘检测器和区域分割方法结合起来进一步改进了语义上显著的边界的检测(Maire, Arbeláez, Fowlkes *et al.* 2008; Arbeláez, Maire, Fowlkes *et al.* 2010)。连成轮廓的边缘可以经平滑后用于生成艺术效果(Lowe 1989; Finkelstein and Salesin 1994; Taubin 1995)和识别(Belongie, Malik, and Puzicha 2002; Tek and Kimia 2003; Sebastian and Kimia 2005)。

　　Burns, Hanson, and Riseman(1986)是早期关于直线提取的一篇广受关注的文章。最新的方法通常将直线检测和消失点检测结合起来(Quan and Mohr 1989; Collins and Weiss 1990; Brillaut-O'Mahoney 1991; McLean and Kotturi 1995; Becker and Bove 1995; Shufelt 1999; Tuytelaars, Van Gool, and Proesmans 1997; Schaffalitzky and Zisserman 2000; Antone and Teller 2002; Rother 2002; Košecká and Zhang 2005; Pflugfelder 2008; Sinha, Steedly, Szeliski *et al.* 2008; Tardif 2009)。

4.5　习　　题

习题 4.1：兴趣点检测器

　　实现一个或者多个关键点检测器，然后比较它们的性能(和自己的检测器进行比较或者和同学的检测器进行比较)。

　　可能有以下检测器：

- 拉普拉斯或者高斯差分；
- Förstner–Harris Hessian(尝试式(4.9～4.11)给出的不同的公式变种)；
- 有向/导向滤波器，如 4.1.1 节讨论的那样，寻找一个二阶的第二高的响应或者一个窗口内的两个边缘(Koethe 2003)。

其他检测器在文献 Mikolajczyk, Tuytelaars, Schmid *et al.*(2005)；Tuytelaars and Mikolajczyk(2007)中有叙述。另外可能还需要以下可选步骤。

1. 在次八度金字塔上进行检测并找 3D 最大值。
2. 使用导向滤波器响应或者一个梯度直方图方法来做局部方向估计。
3. 实现非最大抑制，如 Brown, Szeliski, and Winder(2005)给出的自适应方法。
4. 改变窗口的形状和大小(预滤波和聚集)。

为了测试可重复性，在网址 *http://www.robots.ox.ac.uk/~vgg/research/affine/* 下载代码(Mikolajczyk, Tuytelaars, Schmid et al. 2005; Tuytelaars and Mikolajczyk 2007)，或者简单的旋转和裁剪你自己的测试图像。(挑选一个你后来可能会使用的场景，比如，为了室外场景的拼接。)

记得衡量和报告尺度和方向估计的稳定性。

习题 4.2：特征点描述子

实现一个或者多个描述子(用来描述局部尺度和方向)并比较它们的性能(和你自己的描述子进行比较或者和同学的描述子进行比较)。

可能包括以下描述子：

- 对比度规范化了的块(Brown, Szeliski, and Winder 2005)；
- SIFT(Lowe 2004)；
- GLOH(Mikolajczyk and Schmid 2005)；
- DAISY(Winder and Brown 2007; Tola, Lepetit, and Fua 2010)。

文献 Mikolajczyk and Schmid(2005)描述了其他一些描述子。

习题 4.3：ROC 曲线计算

如图 4.23b 所示，将匹配和非匹配特征的数目作为欧式距离 d 的一个函数，画一对曲线(直方图)，在这个基础上，给出一个绘制 ROC 曲线的算法(图 4.23a)。特别地，令 $t(d)$ 和 $f(d)$ 分别表示(正确)匹配和(错误)未匹配的分布。写下 ROC 的公式，也就是，TPR(FPR)和 AUC。

(提示：绘制累积分布 $T(d) = \int t(d)$ 和 $F(d) = \int f(d)$，看看它们能否帮助你导出在一个给定阈值 θ 下的 TPR 和 FPR。)

习题 4.4：特征匹配器

从一些互相重合或者变形的图像集合中[1]提取特征之后，使用最近邻匹配或者其他更有效的匹配策略(比如 k-d 树)来根据它们的描述子匹配它们。

看看你能否使用最近邻距离比率方法来改进匹配的精度。

习题 4.5：特征跟踪器

与其在多幅图像中独立地寻找特征然后再匹配它们，不如在视频或者图像序列中的第一幅图像中寻找特征，然后使用搜索或者梯度下降(Shi and Tomasi 1994)或者学习到的特征检测器(Lepetit, Pilet, and Fua 2006; Fossati, Dimitrijevic, Lepetit et al. 2007)在下一帧中重新定位对应的特征点。跟踪的特征数目下降到一定数目或者图像中新的区域变得可见时，寻找更多的点来跟踪。

(可选)通过估计一个同态映射(6.19~6.23)或者基本矩阵(7.2.1 节)来排除不正确的匹配。

(可选)使用 8.2 节和习题 8.2 中描述的迭代注册算法来改进匹配精度。

---

[1] *http://www.robots.ox.ac.uk/~vgg/research/affine/*

习题 4.6：脸部特征跟踪器

应用你的特征跟踪器来跟踪一个人的脸部，初始化过程可以是手工指定兴趣位置比如眼角，或者自动使用兴趣点来初始化。

(可选)匹配两个人之间的特征，然后使用这些特征来进行图像变形(习题 3.25)。

习题 4.7：边缘检测器

实现一种你选择的边缘检测器。与同学的检测器或者从网上下载的代码比较它的性能。

一个简单但效果很好的亚像素边缘检测器可以按照如下方法创建。

1. 将输入图像模糊一下，
$$B_\sigma(\boldsymbol{x}) = G_\sigma(\boldsymbol{x}) * I(\boldsymbol{x}).$$

2. 构建一个高斯金字塔(习题 3.19)，
$$P = \text{Pyramid}\{B_\sigma(\boldsymbol{x})\}$$

3. 从原始分辨率的模糊图像中减去一个插值的粗层次的金字塔图像，
$$S(\boldsymbol{x}) = B_\sigma(\boldsymbol{x}) - P.\text{InterpolatedLevel}(L).$$

4. 对于每个像素四元组，$\{(i,j),(i+1,j),(i,j+1),(i+1,j+1)\}$，计数沿着四个边缘的过零点数目。

5. 恰好有两个过零点时，使用(4.25)计算它们的位置，然后将这些边界基元的端点和中间点存到边界基元结构中(图 4.48)。

6. 对于每一个边界基元，沿着过零点边缘，通过求取 S 的水平和垂直方向上的差异来计算梯度。

7. 将这个梯度的幅值存为边缘的强度，将它的方向或者连接边界基元端点的片段作为边缘的方向存储。

8. 将边界基元添加到边界基元列表中或者存到一个 2D 边界基元数组中(通过像素位置访问)。

   图 4.48 给出了每一个计算出的边界基元的可能表示。

```
struct SEdgel {
    float e(2) (2);            //边界基元端点(过零点)
    float x, y;                //亚像素边界位置(中间点)
    float n_x, n_y;            //方向，以法向量的形式
    floa theta;                //方向，以角度的形式(度)
    float length;              //边界基元的长度
    float strength;            //边界基元的强度(梯度幅值)
};

Struct SLine : public SEdge 1 {
    float line_length;         //直线的长度(从椭圆面估计得到的)
    float sigma;               //估计的边界基元噪声的标准方差
    float r;                   //直线方程: x * n_y - y * n_x = r
};
```

图 4.48 边界基元和直线元素的可能的 C++结构

习题 4.8：边缘连接和阈值化

将前面习题计算出来的边缘连接起来，然后可以选择进行滞后阈值。

可能包括以下步骤。

1. 用一个 2D 数组(比方说，一个指向边界基元列表的整数图像索引)或者一个预先排好序的边界基元列表，为了加速邻居搜索，该列表先按照(整数)x 坐标排序然后按照 y 坐标排序。
2. 从没有连接的边界基元列表中选取一个边界基元，然后从每个方向寻找它的邻居，直到找不到邻居或者得到一个闭合的轮廓为止。访问边界基元之后，就将它们标记一下，然后将它们放到已连接的边界基元列表中。
3. 另一种可选的方法是，扩展之前得到的连通域检测算法(习题 3.14)来通过两次图像扫描过程完成连接。
4. (可选)进行基于滞后作用的阈值化(Canny 1986)。使用两个衡量边缘强度的阈值 "hi" 和 "lo"。当一个候选边界基元的强度高于阈值 "hi" 或者高于阈值 "lo" 且(递归地)与之前检测的边缘相连时，就被认为是一个边缘。
5. (可选)将那些间隔很小且端点方向相近的轮廓连接在一起。
6. (可选)寻找相邻轮廓的联结点，比如，使用文献 Maire, Arbeláez, Fowlkes *et al.*(2008)中的想法或者它所引用的文献中的想法。

习题 4.9：轮廓匹配

将一个闭合轮廓(连接的边界基元列表)转化成它的弧长参数化表示，然后用这个表示来匹配物体的外形。

可能包括以下步骤。

1. 遍历整个轮廓，使用弧长公式建立一个(x_i,y_i,s_i)三元组列表
$$s_{i+1} = s_i + \|x_{i+1} - x_i\|. \tag{4.32}$$
2. 使用每个片段的线性插值来将这个列表重采样到一个规范化的(x_i,y_i,s_i)样本集合。
3. 计算 x 和 y 的平均值，即 $\bar{x}$ 和 $\bar{y}$，然后从采样的曲线点中减去它们。
4. 将原始(x_i,y_i,s_i)分段线性函数重采样到一个长度不变的样本集合，如 $j \in [0,1023]$。(使用的长度为 2 的幂会使后续的傅里叶变换更加方便。)
5. 将每一个 (x,y) 对看做一个复数，计算曲线的傅里叶变换。
6. 为了比较两条曲线，将两条曲线的相位差以一个线性函数拟合。(注意：相位差在 360° 重新循环。你也可以使用傅里叶的频谱幅值来加权样本——参见 8.1.2 节)
7. (可选)证明相位的常量部分对应着 s 在时序上的偏移，而线性部分对应着旋转。

当然，可以尝试计算机视觉文献中给出的任意其他曲线描述子和匹配方法(Tek and Kimia 2003; Sebastian and Kimia 2005)。

习题 4.10：拼图游戏求解器——有一定难度

写一个程序自动从扫描的拼图块集合中求解拼图问题。你的软件可能包括如下的组件。

1. 扫描平板上的拼图块(无论是正面朝上或者正面朝下)，平板具有可区分的背景颜色。
2. (可选)在盒子上方扫描得到一个低分辨率的参考图像。
3. 使用基于颜色的阈值化方法来孤立每个拼图块。
4. 使用边缘检测和连接来提取每一个块的轮廓。
5. (可选)以弧长或者其他再参数化的方法来重新表示每一个轮廓。将轮廓分成一些有意义的可匹配块。(这个困难吗？)
6. (可选)对每一个轮廓关联一种颜色以便于匹配。
7. (可选)使用一些旋转不变的特征描述子将这些块和参考图像进行匹配。
8. 求解一个全局优化或者(回溯)搜索问题将这些块结合在一起，并将它们放到相对于参考图像的合适位置。
9. 在一系列比较困难的拼图中测试你的算法，并将你的结果和其他人的进行比较。

习题 4.11：逐次近似直线检测器

实现一个直线简化算法(4.3.1 节)(Ramer 1972; Douglas and Peucker 1973)将一个手绘曲线(或者连接的边缘图像)转化为少量的折线。

(可选)使用一个近似算法或者插值样条或者贝塞尔曲线来重新绘制这个曲线(Szeliski and Ito 1986; Bartels, Beatty, and Barsky 1987; Farin 1996)。

习题 4.12：哈夫变换直线检测器

实现一个哈夫变换来检测图像中的直线。

1. 建立一个大小合适的累加器数组，并将它清空。用户能够指定方向区间之间的度数间隔以及距离区间之间的像素数目。这个数组可以分配为整数(做简单计数)、浮点数(做加权计数)或者一个向量数组来维护指向组成边缘的回溯指针。
2. 对于每一个在位置(x, y)处检测出来的方向为$\theta = \tan^{-1} n_y/n_x$的边界基元，计算下面的值

$$d = xn_x + yn_y \tag{4.33}$$

   然后增加(θ, d)对应的累加器值。

   (可选)根据每个边缘的长度(参考习题 4.7)或者其梯度的强度来加权每个边缘的投票。
3. (可选)通过加上它的直接邻居的值来平滑标量累加器数组。这样有助于抵消仅投票给单一区间所产生的离散化效应——参见习题 3.7。
4. 找到累加器中最大的峰值(局部最大值)，这些值就对应着直线。
5. (可选)对于每一个峰值，使用全部最小二乘(附录 A.2)重新拟合所有的构成

直线的边界基元得到直线。使用原始的边界基元长度或者强度权重来加权最小二乘拟合，同时还可以考虑假设的直线方向和边界基元方向之间的一致性。看看这些启发式策略能否增加拟合的精度。

6. 拟合每个峰值之后，将数组中这个峰值和它邻近的区间置为零或去除，继续处理下一个峰值。

在室内和室外拍摄的各种图像以及棋盘标定模式中测试你的哈夫变换。

对于棋盘模式，你可以修改你的哈夫变换，让角度正好相反的区间$(\theta \pm 180° - d)$和(θ, d)重合，从而不考虑找到的直线的极性。在现实中，可以找到类似的例子吗？

习题 4.13：直线拟合的不确定性

使用不确定性分析来估计直线拟合的不确定性(方差)。

1. 在确定了哪些边界基元属于一个线段(使用逐次近似或者哈夫变换)之后，使用全部最小二乘(Van Huffel and Vandewalle 1991; Van Huffel and Lemmerling 2002)重新拟合线段，也就是，找到边界基元的均值或者质心，然后使用特征值分析来找到主导方向。
2. 计算沿垂直方向到直线的误差(偏差)，然后使用一个估计器比如 MAD(附录 B.3)来鲁棒地估计拟合噪声的方差。
3. (可选)通过丢弃外点或者使用鲁棒范数或者影响函数来重新拟合直线的参数。
4. 估计直线片段在垂直位置上的误差和它的方向误差。

习题 4.14：消失点

使用 4.3.3 节中所描述的计算消失点的一种方法来计算消失点，可以选择性地优化与每一个消失点关联的原始直线方程。你的结果以后可以用于目标跟踪(习题 6.5)或者建筑物的重建(12.6.1 节)。

习题 4.15：消失点的不确定性

对你估计出来的消失点进行不确定性估计。可能需要决定怎样表示消失点，如使用球坐标系上的齐次坐标来处理无穷远处的消失点。

可以参考 Collins and Weiss(1990)对 Bingham 分布的讨论。

第 5 章
分　　割

5.1 活动轮廓
5.2 分裂与归并
5.3 均值移位和模态发现
5.4 规范图割
5.5 图割和基于能量的方法
5.6 补充阅读
5.7 习题

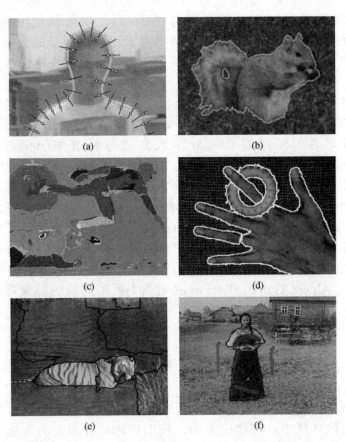

图 5.1 一些流行的图像分割技术：(a)活动轮廓(Isard and Blake 1998)© 1998 Springer；(b)水平集(Cremers, Rousson, and Deriche 2007)© 2007 Springer；(c)基于图的归并(Felzenszwalb and Huttenlocher 2004b)© 2004 Springer；(d)均值移位(Comaniciu and Meer 2002)© 2002 IEEE；(e)基于纹理和中间轮廓的规范图割(Malik, Belongie, Leung *et al.* 2001)© 2001 Springer；(f)利用图割求解的二值 MRF(Boykov and Funka-Lea 2006)© 2006 Springer

图像分割的任务是寻找"相互匹配"的像素组。在统计学中，该问题称为"聚类分析"(cluster analysis)，是一个拥有数以百计不同算法的被广泛研究的领域(Jain and Dubes 1988; Kaufman and Rousseeuw 1990; Jain, Duin, and Mao 2000; Jain, Topchy, Law et al. 2004)。

在计算机视觉中，图像分割是最古老的也是研究最广泛的问题之一(Brice and Fennema 1970; Pavlidis 1977; Riseman and Arbib 1977; Ohlander, Price, and Reddy 1978; Rosenfeld and Davis 1979; Haralick and Shapiro 1985)。早期的方法倾向于采用区域分裂与归并(Brice and Fennema 1970; Horowitz and Pavlidis 1976; Ohlander, Price, and Reddy 1978; Pavlidis and Liow 1990)，对应于聚类文献中的区分式(divisive)和聚集式(agglomerative)聚类(Jain, Topchy, Law et al. 2004)。更新的算法通常是优化某种全局准则，例如区域内部的一致性和区域间的边界长度或者不相似度(Leclerc 1989; Mumford and Shah 1989; Shi and Malik 2000; Comaniciu and Meer 2002; Felzenszwalb and Huttenlocher 2004b; Cremers, Rousson, and Deriche 2007)。

在 3.3.2 节与 3.7.2 节中，我们已经看到了图像分割的一些例子。在这一章中，我们回顾若干其他为图像分割而开发的方法。它们包括基于活动轮廓(5.1 节)和水平集(5.1.4 节)的算法，区域分裂与归并(5.2 节)，均值移位(mean shift)(模态发现)(5.3 节)，规范图割(normalized cut)(基于像素相似性度量的分裂)(5.4 节)以及用图割解决的二值马尔科夫随机场(5.5 节)。图 5.1 展示了一些将这些方法应用于不同图像的例子。

图像分割相关文献浩如烟海，因此，要掌握那些表现较好的算法，一个好方法就是看看它们在由人标注的数据集上的实验对比(Arbeláez, Maire, Fowlkes et al. 2010)。其中最著名的便是伯克利分割数据集与基准集(Berkeley Segmentation Dataset and Benchmark)[①](Martin, Fowlkes, Tal et al. 2001)，它由来自 Corel 图片集的 30 个人手工标注的 1000 张图片组成。许多最近的图像分割算法都在该数据集上报告比较结果。例如，Unnikrishnan，Pantofaru，and Hebert (2007)提出了新的比较这些算法的度量。Estrada and Jepson(2009)在伯克利数据集上比较了四种著名的分割算法，并得出结论：虽然他们自己的 SE-MinCut 算法(Estrada, Jepson, and Chennubhotla 2004)比其他算法稍胜一筹，但在自动与人工分割性能之间仍然存在着很大的差距[②]。Alpert, Galun, Basri et al. (2007)所用的新的前景背景分割集也现成可用[③]。

5.1 活动轮廓

线段、消失点和矩形在人造世界中司空见惯，但对应于物体边界的曲线更为普遍，尤其是在自然环境中。在这一节中，我们描述三种相关的方法以在图像中确定

[①] http://www.eecs.berkeley.edu/Research/Projects/CS/vision/grouping/segbench/
[②] 关于它们的 ROC 曲线有一个有趣的现象，即自动化的方法紧紧沿着相似的曲线聚集成簇，但人工性能却布满整幅图。
[③] http://www.wisdom.weizmann.ac.il/~vision/Seg_Evaluation_DB/index.html

边界曲线的位置。

第一，最早由其发明者称为"蛇行"(snake)(Kass, Witkin, and Terzopoulos 1988)(5.1.1 节)，它是一个能量最小化的二维样条曲线，向着诸如强边界这样的图像特征发展(移动)。第二，智能剪刀(intelligent scissor)(Mortensen and Barrett 1995)(5.1.3 节)，它允许用户实时地勾勒一条紧附在物体边界上的曲线。最后，水平集方法(5.1.4 节)逐步演化某个特征函数的零点集形成的曲线，这使其可以容易地改变拓扑结构和引入基于区域的统计特性。

所有这三种方法都是活动轮廓(active contour)的例子(Blake and Isard 1998; Mortensen 1999)，因为这些边界检测器都在图像和可选的用户指导约束下，迭代地移向其最终解。

5.1.1 蛇行

蛇行是在 3.7.1 节中首次提到的 1D 能量最小化样条的二维扩展情况。

$$\mathcal{E}_{\text{int}} = \int \alpha(s)\|\boldsymbol{f}_s(s)\|^2 + \beta(s)\|\boldsymbol{f}_{ss}(s)\|^2 \, ds, \tag{5.1}$$

其中，s 是沿曲线 $\boldsymbol{f}(s)=(x(s),y(s))$ 的弧长，而 $\alpha(s)$ 和 $\beta(s)$ 是类似于(3.100 和 3.101)中介绍的 $s(x,y)$ 和 $c(x,y)$ 的一阶和二阶连续性加权函数。我们可以通过沿曲线长度均匀采样起始曲线位置(图 4.35)来将此能量离散化，从而得到

$$\begin{aligned} E_{\text{int}} = & \sum_i \alpha(i)\|f(i+1)-f(i)\|^2/h^2 \\ & + \beta(i)\|f(i+1)-2f(i)+f(i-1)\|^2/h^4, \end{aligned} \tag{5.2}$$

其中 h 是步长大小，如果在每次迭代后，我们沿其弧长重采样曲线，则可以将其忽略。

除了这个内部样条能量，蛇行还同时最小化外部的基于图像的和基于约束的势函数。基于图像的势函数是若干项之和

$$\mathcal{E}_{\text{image}} = w_{\text{line}}\mathcal{E}_{\text{line}} + w_{\text{edge}}\mathcal{E}_{\text{edge}} + w_{\text{term}}\mathcal{E}_{\text{term}}, \tag{5.3}$$

其中，线条(line)项吸引蛇行到暗的脊，边缘(edge)项将其吸引向强梯度(边缘)，而端点(term)项将其吸引向线条端点。在实践中，大多数系统只采用边缘项，它可以直接正比于图像梯度

$$E_{\text{edge}} = \sum_i -\|\nabla I(\boldsymbol{f}(i))\|^2, \tag{5.4}$$

也可以正比于平滑后的图像拉普拉斯(二阶导数)

$$E_{\text{edge}} = \sum_i -|(G_\sigma * \nabla^2 I)(\boldsymbol{f}(i))|^2. \tag{5.5}$$

人们有时也抽取边缘，而后采用到这些边缘的距离图作为那两种最初提出的势函数的备选。

在交互式应用中，可以加入多种用户指定的约束，例如对锚点 $d(i)$ 的吸引(弹性)力

$$E_{\text{spring}} = k_i \|\boldsymbol{f}(i) - \boldsymbol{d}(i)\|^2, \tag{5.6}$$

和(火山)排斥力 $1/r$(图 5.2a)。因为蛇行通过最小化它们的能量来逐步演化,所以它们经常"扭动"和"滑动",这也解释了它们的得名。图 5.2b 展示了蛇行用于跟踪一个人的嘴唇。

因为规则的蛇行有收缩的趋势(习题 5.1),所以在要跟踪的兴趣物体外,最好能画出蛇行来进行初始化。或者在动力学(Cohen and Cohen 1993)中加入气球膨胀(ballooning)力,其本质上是将各个点沿其法向向外移动。

要想高效求解由蛇行能量最小化带来的稀疏线性系统,可以使用稀疏直接求解方法(附录 A.4 节),因为该线性系统本质上是五阶对角阵[①]。蛇行演化通常在该线性系统的解和非线性约束例如边缘能量的线性化之间交替进行。更直接的找出全局能量最小值的方法是采用动态规划(Amini, Weymouth, and Jain 1990; Williams and Shah 1992),但这种方法在实践中并不常采用,因为可以用智能剪刀(5.1.3 节)和GrabCut(5.5 节)之类的更有效的或者交互式的算法来取代。

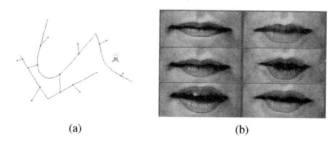

(a)　　　　　(b)

图 5.2　蛇行(Kass, Witkin, and Terzopoulos 1988)©1988 Springer:(a)用于交互控制形状的"蛇行凹陷";(b)嘴唇跟踪

弹性网和光滑弹簧

蛇行的一个有趣变种是旅行商问题(Traveling Salesman Problem,TSP)的弹性网形式化,它首先由 Durbin and Willshaw (1987)提出,而后由 Durbin, Szeliski, and Yuille (1989)在能量最小化框架中重新形式化。回想在 TSP 中,旅行者必须拜访每个城市一次,而同时要最小化他所经过的总旅程。将蛇行约束为要穿过每个城市,可以解决该问题(没有任何最优保证),但是不可能提前知道哪个蛇行控制点应该关联到每个城市。

城市被假设穿过沿旅程的某个点附近(图 5.3),而不是在蛇行节点与城市之间有确定的约束,就像在公式(5.6)中那样。在概率解释中,每个城市由以每个旅行点为中心的高斯混合生成,

$$p(\boldsymbol{d}(j)) = \sum_i p_{ij} \text{ with } p_{ij} = e^{-d_{ij}^2/(2\sigma^2)} \tag{5.7}$$

其中,σ 是高斯的标准差,而

[①] 闭合蛇行具有 Toeplitz 矩阵形式,可以在 $O(N)$ 时间内分解和求解。

$$d_{ij} = \|\boldsymbol{f}(i) - \boldsymbol{d}(j)\| \tag{5.8}$$

是旅行点 $\boldsymbol{f}(i)$ 和城市位置 $\boldsymbol{d}(j)$ 之间的欧式距离。对应的数据拟合能量(负对数似然度)是

$$E_{\text{slippery}} = -\sum_j \log p(\boldsymbol{d}(j)) = -\sum_j \log\left[\sum e^{-\|\boldsymbol{f}(i) - \boldsymbol{d}(j)\|^2/2\sigma^2}\right]. \tag{5.9}$$

该能量的命名源自于这样一个事实，不像规则弹簧将给定蛇行点与给定约束式(5.6)耦合在一起，这另外的能量定义了一种光滑弹簧(slippry spring)，它允许约束(城市)与曲线(旅程)点之间的关联随着时间逐渐演化(Szeliski 1989)。注意，这是流行的迭代最近点数据约束的软性变种，它常用于将表面拟合或配准到数据点或者彼此配准(12.2.1 节) (Besl and McKay 1992; Zhang 1994)。

为给旅行商问题求一个好的解，将光滑弹簧数据关联能量与规则的一阶内部平滑能量(5.3)结合起来定义旅程的费用。旅程 $\boldsymbol{f}(s)$ 初始化为围绕城市点均值的一个小圆，而且 σ 逐渐减小(图 5.3)。对于大的 σ 值，旅程尽力保持在这些点的质心附近，但是随着 σ 减小，每个城市越来越强烈地拉近它最近的旅行点(Durbin, Szeliski, and Yuille 1989)。在 $\sigma \to 0$ 的极限中，每个城市都保证至少抓住一个旅行点而且相邻城市之间的旅程为直线。

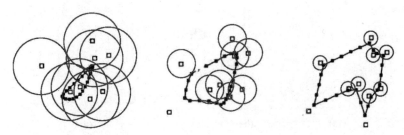

图 5.3 弹性网：空心正方形表示城市而通过直线线段连接起来的实心正方形是旅行点。蓝圈表示每个城市吸引力的近似范围，它随着时间递减。在该弹性网的贝叶斯解释下，蓝圈对应于圆形高斯的一个标准差，该高斯从某未知旅行点生成每座城市

样条和形状先验

虽然蛇行非常擅长于捕捉大量真实世界轮廓中的细小的不规则的细节，但有时候它们显示出太高的自由度，使其很可能在演化中陷入局部最小值。

这个问题的一个解决方法是通过使用 B-样条近似来控制蛇行使得自由度更少 (Menet, Saint-Marc, and Medioni 1990b,a; Cipolla and Blake 1990)。这样得到的 B-蛇行可以写为

$$\boldsymbol{f}(s) = \sum_k B_k(s)\boldsymbol{x}_k \tag{5.10}$$

或者用离散形式写为

$$\boldsymbol{F} = \boldsymbol{B}\boldsymbol{X} \tag{5.11}$$

其中

$$F = \begin{bmatrix} f^T(0) \\ \vdots \\ f^T(N) \end{bmatrix}, \quad B = \begin{bmatrix} B_0(s_0) & \ldots & B_K(s_0) \\ \vdots & \ddots & \vdots \\ B_0(s_N) & \ldots & B_K(s_N) \end{bmatrix}, \quad \text{and} \quad X = \begin{bmatrix} x^T(0) \\ \vdots \\ x^T(K) \end{bmatrix}. \quad (5.12)$$

如果被跟踪或识别的物体在位置、大小或者方向上有大的变化，它们可以模型化为控制点上的附加变换，例如 $x'_k = sRx_k + t$ (2.18)，它可以在估计控制点值的同时估计出来。另一种方法，可以采用分开的检测(detection)和配准(alignment)，先确定感兴趣物体的位置和朝向(Cootes, Cooper, Taylor *et al.* 1995)。

在 B-蛇行中，因为蛇行被更小的自由度所控制，所以对用在原始蛇行中的内部平滑力有更少的需求，虽然它们仍可以用有限元分析，即使用 B-样条基函数的微分和积分(Terzopoulos 1983; Bathe 2007)，得到和实现。

在实践中，更通用的做法是在控制点 $\{x_k\}$ 的典型分布上估计一组形状先验(shape prior)(Cootes, Cooper, Taylor *et al.* 1995)。考虑图 5.4a 所示的这组电阻形状。如果用图 5.4b 所示的这组控制点来描述每个轮廓，那么我们可以用一个散点图来描绘每个点的分布，如图 5.4c 所示。

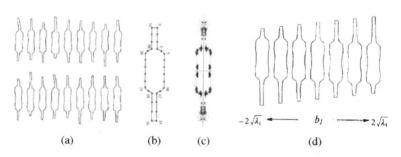

(a)　　(b)　　(c)　　(d)

图 5.4　一系列电阻的点分布模型(Cootes, Cooper, Taylor *et al.* 1995) © 1995 Elsevier：(a) 输入电阻形状集合；(b)控制点到边界的分配；(c)点位置的分布(散布图)；(d)整体形状中的第一个(最大的)变种模式

描述这种分布的可能的方法是采用每个点 x_k 的位置 $\bar{x}_k$ 和 2D 协方差 C_k。这可以进一步转化为在该点位置的二次惩罚(先验能量)，

$$E_{\text{loc}}(x_k) = \frac{1}{2}(x_k - \bar{x}_k)^T C_k^{-1}(x_k - \bar{x}_k). \quad (5.13)$$

然而在实践中，点位置的变化通常是高度相关的。

一个更可取的方法是同时估计所有点的联合协方差。首先将所有点位置 $\{x_k\}$ 连接成为一个单独的向量 x，例如，交错存储每个点的 x 和 y 位置。这些向量在所有训练数据上的分布(图 5.4a)可以用平均值 $\bar{x}$ 和协方差

$$C = \frac{1}{P} \sum_p (x_p - \bar{x})(x_p - \bar{x})^T, \quad (5.14)$$

描述，其中 x_p 是第 P 个训练样例。采用本征值分析(eigenvalue analysis)(附录 A.1.2)，也称"主分量分析"(Principle Component Analysis，PCA)(附录 B.1.1)，协

方差矩阵可以写为

$$C = \Phi \operatorname{diag}(\lambda_0 \ldots \lambda_{K-1}) \Phi^T. \tag{5.15}$$

在大部分情况下，这些点可能的表观可以只用几个具有最大本征值的本征向量模型化。这样产生的点分布模型(point distribution model)(Cootes, Taylor, Lanitis *et al.* 1993; Cootes, Cooper, Taylor *et al.* 1995)可以写为

$$x = \bar{x} + \hat{\Phi} b, \tag{5.16}$$

其中 b 是一个 $M \ll K$ 个元素的形状参数(shape parameter)向量，而 $\hat{\Phi}$ 是 Φ 的前 m 列。为了约束形状参数取合理的值，我们可以使用如下形式的二次惩罚

$$E_{\text{shape}} = \frac{1}{2} b^T \operatorname{diag}(\lambda_0 \ldots \lambda_{M-1}) b = \sum_m b_m^2 / 2\lambda_m. \tag{5.17}$$

另一种方法，将允许的 b_m 的取值范围限制在某特定区间，例如 $|b_m| \leq 3\sqrt{\lambda_m}$ (Cootes, Cooper, Taylor *et al.* 1995)。Isard and Blake(1998)回顾了产生一组形状向量的其他方法。

将每个形状参数 b_m 在范围 $-2\sqrt{\lambda_m} \leq 2\sqrt{\lambda_m}$ 中变化可以给出表观期望变化的良好表示，如图 5.4d 所示。另一个与脸部轮廓相关的例子如图 5.5a 所示。

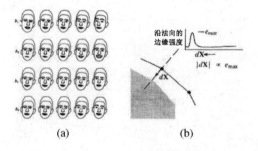

图 5.5 活动形状模型：(a)对一组脸变化前 4 个形状参数的影响(Cootes, Taylor, Lanitis *et al.* 1993)© 1993 IEEE; (b)沿每个控制点的法向搜索最强梯度(Cootes, Cooper, Taylor *et al.* 1995) © 1995 Elsevier

为了将一个点分布模型配准到一幅图像，每个控制点在轮廓法向方向上搜索最可能对应的图像边缘点(图 5.5b)。这些单个的度量可以与形状参数(如果需要，还有位置、尺度和方向参数)上的先验相结合来估计一组新的参数。如此产生的活动形状模型(Active Shape Model，ASM)可以被迭代最小化以将图像拟合到非刚体可变形物体上，例如医疗图像或者手这样的身体部件(Cootes, Cooper, Taylor *et al.* 1995)。ASM 还可以与基本灰度分布的 PCA 分析相结合从而创造出活动表观模型(Active Appearance Model，AAM)(Cootes, Edwards, and Taylor 2001)我们将在 14.2.2 节更详细地讨论它。

5.1.2 动态蛇行和 CONDENSATION

在很多活动轮廓应用中，感兴趣物体在逐帧跟踪中，变形演化。在这种情况下，根据之前帧得到的估计来预测和约束新的估计是合乎情理的。

做这件事的一个方法是采用卡尔曼滤波器，它产生了称为"卡尔曼蛇行"(Kalman snake)(Terzopoulos and Szeliski 1992; Blake, Curwen, and Zisserman 1993)的形式化表达。卡尔曼滤波器是基于形状参数演化的线性动态模型，

$$x_t = Ax_{t-1} + w_t, \tag{5.18}$$

其中 x_t 和 x_{t-1} 是当前和前一个状态变量，A 是线性传递矩阵(transition matrix)，而 w 是噪声(扰动)向量，它通常被模型化为高斯分布(Gelb 1974)。矩阵 A 和噪声协方差矩阵可以通过观察被跟踪物体的典型序列来提前学习(Blake and Isard 1998)。

卡尔曼滤波器的定性行为可以参见图 5.6a。该线性动态模型在之前的估计中引起确定性的改变(漂移)，而过程噪声(扰动)引起随机扩散增加了系统的熵(缺乏确定性)。来自当前帧的新的测量在更新估计中可以恢复一些确定性(尖峰)。

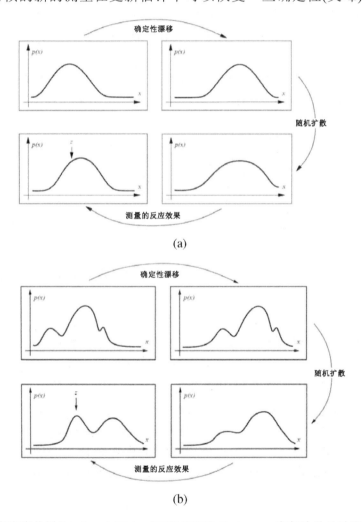

图 5.6 概率密度传播(Isard and Blake 1998)©1998 Springer。在每次估计阶段的开始，概率密度根据线性动态模型(确定性漂移)进行更新，而它的确定性因为过程噪声(随机扩散)而减小。新的测量引入额外的信息以帮助改善当前的估计。(a)卡尔曼滤波器将分布模型化为单一模式，即使用一个期望与协方差。(b)一些应用需要更一般的多模态分布

然而，在许多情况下，例如在拥挤环境中跟踪，如果我们去掉卡尔曼滤波器所需的分布是高斯分布的假设，可以得到更好的轮廓估计。在这种情况下，一般性的多模态分布被传播，如图 5.6b 所示。为了模型化这样的多模态分布，Isard and Blake (1998)向计算机视觉界引入了粒子滤波器(particle filter)的使用[①]。

粒子滤波器方法用一组带权样本点来表达一个概率分布(图 5.7a)(Andrieu, de Freitas, Doucet et al. 2003; Bishop 2006; Koller and Friedman 2009)。为了根据线性动态方程(确定性漂移)更新样本位置，样本中心根据(5.18)更新，并为每个点生成多个样本(图 5.7b)。它们进一步被扰动以解释随机扩散，即它们的位置被从分布 w 产生的随机向量移动。[②]最终，这些样本的权重乘以测量概率密度，即我们取每一个样本，并根据给定的当前(新的)测量估量其似然度。因为这些样本点表示并传播多模态密度的条件估计，Isard and Blake(1998)将他们的算法命名为 CONditional DENSity propagATION 或者 CONDENSATION。

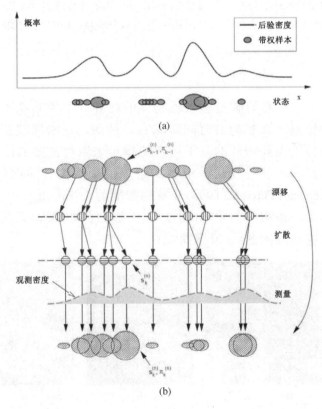

图 5.7 使用 CONDENSATION 算法中的粒子滤波器的分解采样(Isard and Blake 1998)© 1998 Springer：(a)每个密度分布都采用多个带权粒子的叠加来表示；(b)漂移—扩散—测量循环采用随机采样、扰动和重新设定权重的方法来实现

[①] 其他模型化多模态分布的选择包括高斯混合模型(mixtures of Gaussians)(Bishop 2006)和多假设跟踪(multiple hypothesis tracking)(Bar-Shalom and Fortmann 1988; Cham and Rehg 1999)。

[②] 注意，因为这些步骤的结构，可以使用非线性动态方程和非高斯噪声。

图 5.8a 通过对被跟踪的每个(每个子集的)粒子用红色的 B-样条画一条轮廓，表明一个头部跟踪器的分解采样是什么样的。图 5.8b 表示了为什么测量密度子集常常是多模态的：由于背景拥挤，垂直于样条曲线的边缘位置可以有多个局部最小值。最终，图 5.8c 展示了条件密度(头部和肩部跟踪器质心的 x 坐标)在它随时间跟踪若干个人时的时间演化。

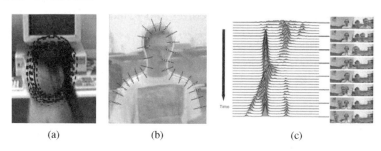

图 5.8 使用 CONDENSATION 的头部跟踪(Isard and Blake 1998)© 1998 Springer：(a)头部估计分布的采样集表示；(b)在每个控制顶点位置的多个测量；(c)随时间的多假设跟踪

5.1.3 剪刀

活动轮廓允许用户粗略地制定感兴趣物体的边界，并使系统不但将该轮廓演化到一个更加准确的位置，而且随着时间跟踪它。然而，这种曲线演化的结果可能是不可预测的，而且可能需要额外的基于用户的指示来取得需要的结果。

另一个可供选择的方法是使得系统随着用户描画而实时地优化轮廓(Mortensen 1999)。由 Mortensen and Barrett (1995)开发的智能剪刀(intelligent scissors)正是这样做的。随着用户画出一条粗略的轮廓(图 5.9a 中的白色曲线)，系统计算并画出一条更好的紧贴高对比度边缘的曲线(橙色曲线)。

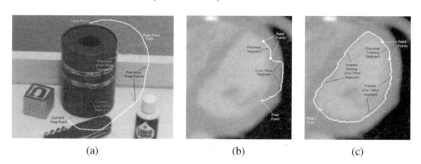

图 5.9 智能剪刀：(a)当鼠标沿白色路径移动，智能剪刀会跟随沿着物体边界的橙色路径移动(绿色曲线表明中间位置)(Mortensen and Barrett 1995)©1995 ACM；(b)规则的剪刀有时候会跳到强(错误)的边界；(c)在按照之前的分割训练后，倾向于产生相似的边缘特性(Mortensen and Barrett 1998) © 1995 Elsevier

为了计算最优的曲线路径(火线)，图像首先进行预处理给可能是边界元素的边缘关联上低的成本(邻近水平、垂直和对角的连接，即 $\mathcal{N}_8$ 邻域)。他们的系统使用过零点、梯度大小、梯度方向的组合来计算这些成本。

接下来，随着用户描绘一条粗略的曲线，系统采用 Dijkstra 算法连续不断地重新计算起始种子点(seed point)与当前鼠标位置之间的最低成本路径。Dijkstra 算法是一个在当前目标位置结束的宽度优先的动态规划算法。

为了保证系统不会不可预测地跳动，系统会在一段不活跃期之后，"冻结"当前的曲线(重置种子节点)。为了阻止火线跳转到邻近的高对比度轮廓上，系统也会"学习"当前优化的曲线下的亮度特性，并以此保持火线优先沿着相同(或者看起来相似)的边界移动(图 5.9b 和 c)。

针对该基本算法，已经提出若干扩展，它们即使在其原始形式下也依然工作得非常好。Mortensen and Barrett(1999)使用滑雪橇(tobogganing)——它是分水岭区域分割的简单形式——来预分割图像为区域，其边界被用作优化曲线路径的候选。如此产生的区域边界变成一个更小的图，其中节点位于任何三个或四个区域相接之处。而后 Dijkstra 算法在这个缩小的图上运行，因此有更快(也更加稳定)的性能。另一个对智能剪刀的扩展是采用概率框架将当前轨迹考虑进去，这就是 JetStream 的系统(Pérez, Blake, and Gangnet 2001)。

与其每个时刻都立即重新计算一条最优曲线，一个更简单的系统可以通过将当前鼠标位置"硬拽"到最近的可能边界点上来实现(Gleicher 1995)。10.4 节将展示这些边界提取方法在图像剪切粘贴中的应用。

5.1.4 水平集

基于形式为 $f(s)$ 的参数化曲线的活动轮廓，例如蛇行、B-蛇行和 CONDENSATION，它们的局限性在于在曲线的演化中改变其拓扑结构非常困难。(McInerney and Terzopoulos (1999, 2000)叙述了一种解决该问题的方法。)进一步，如果形状变化非常剧烈，可能需要曲线重新参数化。

针对这类闭合轮廓的另一种表示是使用水平集(Level Set)，其中一个特征(或者带符号的距离(3.3.3 节))函数的过零点(集)定义一条曲线。水平集通过调整其中嵌入函数(该 2D 函数的另一个名字)$\phi(x, y)$ 而不是曲线 $f(s)$ 来演化以拟合和跟踪感兴趣的物体(Malladi, Sethian, and Vemuri 1995; Sethian 1999; Sapiro 2001; Osher and Paragios 2003)。为了减少所需要的计算量，在每一步只有当前过零点位置附近很小的条带(前沿边界)需要被更新，这也就产生了所谓的快速行进法(fast marching method)(Sethian 1999)。

一个演化公式的例子是由 Caselles, Kimmel and Sapiro(1997) 和 Yezzi, Kichenassamy, Kumar et al. (1997)提出的测地活动轮廓(geodesic active contour)，

$$\begin{aligned} \frac{d\phi}{dt} &= |\nabla\phi|\mathrm{div}\left(g(I)\frac{\nabla\phi}{|\nabla\phi|}\right) \\ &= g(I)|\nabla\phi|\mathrm{div}\left(\frac{\nabla\phi}{|\nabla\phi|}\right) + \nabla g(I) \cdot \nabla\phi, \end{aligned} \tag{5.19}$$

其中 $g(I)$ 是蛇行边缘势函数(5.5)的推广版本。为得到该曲线行为的直观感受，假设嵌入函数 ϕ 是到曲线的带符号的距离函数(图 5.10)，其中 $|\phi|=1$。公式(5.19)的第一项将曲线沿其曲率方向移动，即在调整函数 $g(I)$ 的影响下拉直曲线。第二项沿 $g(I)$ 的梯度方向移动曲线，鼓励曲线移向 $g(I)$ 的极小值。

图 5.10 针对测地活动轮廓的水平集演化。嵌入函数 ϕ 基于下面由边缘/速度函数 $g(I)$ 以及梯度 $g(I)$ 调节的曲面的曲率进行更新，因此会吸引它到强边缘处

虽然该水平集的形式化可以容易地改变拓扑结构，但它仍然容易受到局部最小值的影响，因为它基于类似图像梯度这样的局部测量。一个可选的方法是在分割框架下重新塑造该问题，其中能量用于度量分割区域内外(Cremers, Rousson, and Deriche 2007; Rousson and Paragios 2008; Houhou, Thiran, and Bresson 2008)的图像统计量(例如颜色，纹理，运动)的一致性。这些方法建立在由 Leclerc(1989)，Mumford and Shah (1989)和 Chan and Vese(1992)引入的早期基于能量的分割框架上，在 5.5 节中会有更加详细的讨论。图 5.11 展示了这种水平集分割的例子，它展现了水平集从一系列散布的圆演化为最终二值分割的过程。

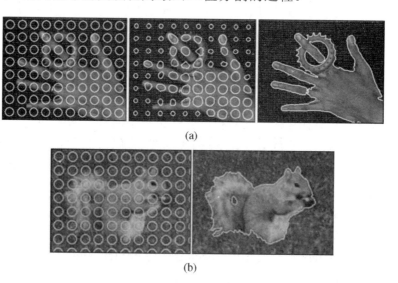

图 5.11 水平集分割(Cremers, Rousson, and Deriche 2007) ©2007 Springer：(a)灰度图像分割；(b)彩色图像分割。单变量和多变量高斯被用于模型化前景和背景像素分布。初始化的圆随着演化调整它们的拓扑结构形成前景和背景的精确分割

更多关于水平集及其应用的信息，请看由 Osher and Paragios(2003)编辑的论文集，还有计算机视觉中关于变分法和水平集法的系列研讨会(Paragios, Faugeras, Chan *et al.* 2005)以及计算机视觉中关于尺度空间和变分法的专题集(Paragios and Sgallari 2009)。

5.1.5 应用：轮廓跟踪和转描机

活动轮廓可以广泛用于多种物体跟踪应用中(Blake and Isard 1998; Yilmaz, Javed, and Shah 2006)。例如，它们可以用于跟踪脸部特征以实现表演驱动的动画处理(Terzopoulos andWaters 1990; Lee, Terzopoulos, and Waters 1995; Parke andWaters 1996; Bregler, Covell, and Slaney 1997) (图 5.2b)。它们也可以用于跟踪头部和人(如图 5.8 所示)以及运动车辆(Paragios and Deriche 2000)。其他的应用包括医学图像分割，其中轮廓在计算机断层显像(3D 医学图像)中被逐层跟踪(Cootes and Taylor 2001)或者随时间跟踪，正如在超声波扫描中那样。

一个与计算机卡通画和视觉特效更相关的有趣应用是转描机(rotoscoping)，它用跟踪到的轮廓变化一组手画的动画形象(或者调整或替换原始的视频帧)。[①] Agarwala, Hertzmann, Seitz *et al.* (2004)采用几何和表观相结合的准则，提出了一个基于在选择的关键帧上跟踪手画 B-样条轮廓的系统(图 5.12)。它们还提供了一个关于以前的转描机和基于图像的轮廓跟踪系统的优秀回顾。

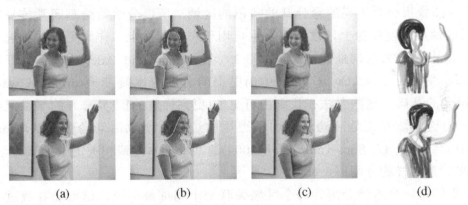

图 5.12 基于关键帧的转描机(Agarwala, Hertzmann, Seitz *et al.* 2004) ©2004 ACM：(a)原始图像；(b)转描的轮廓；(c)重新上色的上衣；(d)转描的手画动画

另外的转描应用(物体轮廓检测和分割)，例如从一张照片中剪切物体粘贴到另一张照片中，将会在 10.4 节中介绍。

---

① 这个术语来自于一个设备(一个转描机)，它将真人动作影片的帧投影在醋脂纤维上，这样画家可以直接在演员形状上画动画。

5.2 分裂与归并

正如在本章简介中提到的,对于灰度图像最简单的可能方法就是选择一个阈值,然后计算连通分量(3.3.2 节)。不幸的是,因为光照和物体内部统计特性的变化,对于整张图片而言单一阈值是不够的。

在本节中,我们介绍许多算法,它们要么递归地将整幅图像基于区域统计特性分裂成块,要么反过来将像素和区域以分层的方式归并起来。也可以通过一个中等粒度的分割(用四叉树表示)作为起始,然后同时允许归并和分裂操作(Horowitz and Pavlidis 1976; Pavlidis and Liow 1990),这样将分裂与归并结合起来。

5.2.1 分水岭

与阈值化相关的一个方法是分水岭(watershed)(Vincent and Soille 1991)计算,因为它是在灰度图像上进行操作的。这种方法将图像分割成若干集水盆地(catchment basin),它们是图像中雨水会流到相同湖泊的区域(解译为高度场或地形图)。计算这些区域的一个高效的方法是从所有的局部最小值处开始洪泛地形,并将演化出的不同部分交接的地方标记为脊。整个算法可以用像素点的优先级队列和宽度优先搜索来实现(Vincent and Soille 1991)。[①]

因为图像很少有被亮的脊分开的暗色的区域,所以分水岭分割通常施加在平滑处理后的梯度幅度图像上,这使得它也可用于彩色图像上。作为另外一种选择,导向滤波器中的最大有向能量(3.28 和 3.29) (Freeman and Adelson 1991)可以用作 Arbeláez, Maire, Fowlkes et al. (2010)提出的有向分水岭变换(oriented watershed transform)的基础。这些方法在找到由可见的(更高梯度的)边界分开的平滑区域时结束。因为这些边界正是活动轮廓通常遵循的边界,所以活动轮廓算法(Mortensen and Barrett 1999; Li, Sun, Tang et al. 2004)通常用分水岭或相关的滑雪橇方法(5.1.3 节)预先计算这样的分割。

不幸的是,分水岭分割为每个区域关联一个局部最小值,这可能导致过分割。因此,分水岭分割常用作交互系统的一部分,用户先标记一些种子位置(用点击或短线),它们对应于不同的期望分量的中心。图 5.13 展示了在共焦显微图像上使用一些手动放置的标记点的分水岭分割算法运行结果。它也展示了分水岭分割的一个优化版本,即通过使用局部形态学来平滑和优化分割区域的边界(Beare 2006)。

[①] 相关的算法可以用于高效的计算最稳定极值区域(maximally stable extremal regions,MSER) (4.1.1 节) (Nistér and Stewénius 2008)。

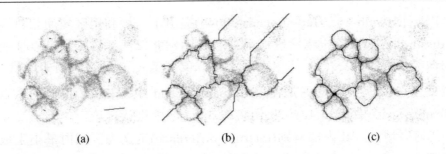

图 5.13 局部约束的分水岭分割(Beare 2006)©2006 IEEE：(a)带有标记种子(线段)的原始共焦显微图像；(b)标准分水岭分割；(c)局部约束的分水岭分割

5.2.2 区域分裂(区分式聚类)

将图像成功分裂成为更好的区域是计算机视觉中最古老的方法之一。Ohlander, Price, and Reddy(1978)提出了这样一种方法，它首先计算整幅图像的直方图，然后寻找最好的区分直方图中极大峰的阈值。该过程重复进行，直到所有区域要么相当一致，要么小于某个特定大小。

更多最新的分裂算法通常优化某种区域内部相似度和区域之间不相似度的衡量标准。这些将在 5.4 节和 5.5 节介绍。

5.2.3 区域归并(凝聚式聚类)

区域归并方法也可以追溯到计算机视觉的开端。Brice and Fennema(1970)使用双网格来表达像素之间的边界，并且根据它们相对边界长度和在这些边界上的可见边缘的强度来归并区域。

在数据聚类中，算法可以根据群簇最近点(单连接聚类)、最远点(全连接聚类)或者介于二者之间(Jain, Topchy, Law et al. 2004)的距离将群簇链接起来。Kamvar, Klein and Manning(2002)提出了这些算法的概率解释并且展示了如何将其他模型整合到这个框架中。

一个非常简单的像素级别的归并方法是将邻近区域中平均颜色差别小于某个阈值或者很小的那些区域归并起来。将图像分割成这样的没有语义含义的超像素(Superpixel)(Mori, Ren, Efros et al. 2004)，可以作为一种有用的预处理阶段，使得高级算法(如立体视觉匹配(Zitnick, Kang, Uyttendaele et al. 2004; Taguchi, Wilburn, and Zitnick 2008)、光流(Zitnick, Jojic, and Kang 2005; Brox, Bregler, and Malik 2009)和识别(Mori, Ren, Efros et al. 2004; Mori 2005; Gu, Lim, Arbelaez et al. 2009; Lim, Arbeláez, Gu et al. 2009))更快，也更鲁棒。

5.2.4 基于图的分割

当许多归并算法还只是简单地将一个固定的准则用于把像素和区域聚合到一起

的时候，Felzenszwalb and Huttenlocher(2004b)提出了一种利用区域间相对不相似性(relative dissimilarity)来确定哪些区域需要合并的归并算法；该算法可以证明优化了一个全局聚类度量标准。它们从像素间不相似度度量 $w(e)$ 开始，例如度量 $\mathcal{N}_8$ 邻域之间的亮度差别。(另一种选择是使用 Comaniciu and Meer(2002)提出的联合特征空间(joint feature space)距离(5.42)，我们将在 5.3.2 节中进行讨论)

对于任何区域 R，其内部差别(internal difference)定义为区域内最小生成树的最大边的权重，

$$Int(R) = \max_{e \in MST(R)} w(e). \tag{5.20}$$

对于任何两个至少有一条边连接它们的顶点的邻近区域，它们之间的差别定义为连接两个区域的最小权重边。

$$Dif(R_1, R_2) = \min_{e=(v_1,v_2)|v_1 \in R_1, v_2 \in R_2} w(e). \tag{5.21}$$

它们的算法将区域间差别小于这两个区域内部差别最小值的任何两个邻近区域归并，

$$MInt(R_1, R_2) = \min(Int(R_1) + \tau(R_1), Int(R_2) + \tau(R_2)), \tag{5.22}$$

其中 $\tau(R)$ 是启发式区域惩罚，Felzenszwalb and Huttenlocher(2004b)设置为 $k/|R|$，但它可以被设置为任何应用所特有的区域品质的量度。

通过按照分开区域的边缘的递减顺序归并区域(可以利用 Kruskal 的最小生成树算法的变种高效地估计)，可以证明它们产生的分割既不会太精细(存在本应该被归并的区域)也不会太粗糙(存在本应该被分开而不应该被归并的区域)。对于固定大小的像素邻域，该算法的执行时间为 $O(N \log N)$，其中 N 是像素数，这使得它成为最快的分割算法之一(Paris and Durand 2007)。图 5.14 展示了使用它们算法分割的图像例子。

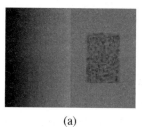
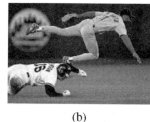
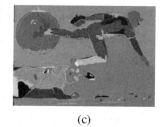

(a)   (b)   (c)

图 5.14 基于图的归并分割(Felzenszwalb and Huttenlocher 2004b)© 2004 Springer：(a)即使小矩形内部的变化大于中间边界之间的变化，输入亮度图像仍被成功分割成三个区域；(b)输入亮度图像；(c)使用 $\mathcal{N}_8$ 像素邻域产生的分割

5.2.5 概率聚集

Alpert, Galun, Basri *et al.* (2007)基于两条称为灰度相似度和纹理相似度的线索，提出了一种概率归并算法。两个区域 R_i 和 R_j 之间的灰度相似度基于与其他邻

居区域的最小外部差异(minimal external difference)，

$$\sigma_{local}^+ = \min(\Delta_i^+, \Delta_j^+), \tag{5.23}$$

其中 $\Delta_i^+ = \min_k |\Delta_{ik}|$，且 Δ_{ik} 为区域 R_i 和 R_k 之间的平均亮度差异。这被比作平均亮度差异(average intensity difference)

$$\sigma_{local}^- = \frac{\Delta_i^- + \Delta_j^-}{2}, \tag{5.24}$$

其中 $\Delta_i^- = \sum_k (\tau_{ik}\Delta_{ik})/\sum_k (\tau_{ik})$，且 τ_{ik} 是区域 R_i 与 R_k 之间的边界长度。纹理相似度采用简单有向 Sobel 滤波响应的直方图格子的相对差异来定义。成对统计量 σ_{local}^+ 和 σ_{local}^- 用于计算两个区域应该被归并的似然度 p_{ij}。(更多细节见 Alpert, Galun, Basri et al. (2007)的论文)。

归并采用分层的方式进行，这是受到代数多重网格法(Brandt 1986; Briggs, Henson, and McCormick 2000)的启发。先前由 Alpert, Galun, Basri et al. (2007)在他们使用加权聚类(SWA)算法(Sharon, Galun, Sharon et al. 2006)的分割中使用。我们将在 5.4 节中讨论 SWA 算法。节点中与所有原始节点(区域)(共同)强耦合(strongly coupled)的子集用于定义在更粗粒度尺度(图 5.15)的问题，其中强耦合定义为

$$\frac{\sum_{j \in C} p_{ij}}{\sum_{j \in V} p_{ij}} > \phi, \tag{5.25}$$

其中 ϕ 通常设为 0.2。更粗粒度节点的亮度和纹理相似度统计量采用加权平均的方式递归计算，其中粗粒度和细粒度层次节点间的相对强度(耦合)取决于它们的归并概率 p_{ij}。利用在基于金字塔滤波或预处理算法中使用的分层的凝聚操作，可以使算法本质上运行时间为 $O(N)$。一旦在更粗粒度上得到了分割结果，每个像素的准确隶属关系就可以通过将粗粒度层次的赋值传播给它们更精细粒度层次的"孩子"来计算得到(Sharon, Galun, Sharon et al. 2006; Alpert, Galun, Basri et al. 2007)。图 5.22 展示了由该算法得到的分割结果以及与其他流行算法的比较。

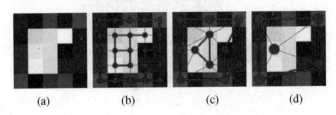

图 5.15 加权聚类分割(SWA)中的由粗到精节点聚类(Sharon, Galun, Sharon et al. 2006) © 2006 Macmillan Publishers 公司(Nature)：(a)原始亮度像素网格；(b)像素间耦合，其中粗连接表示强耦合；(c)一层粗化后，每个原始节点都强耦合到该粗层次的节点之一；(d)两层粗化后

5.3 均值移位和模态发现

均值移位和模态发现方法，例如 k-均值和高斯混合，将与每个像素关联的特征向量(例如颜色和位置)模型化为来自某个未知概率密度函数的样本，而后试图寻找

在此分布中的群簇(模态)。

考虑图 5.16a 中的彩色图像。你如何能只基于颜色分割该图像？图 5.16b 显示了在 L*u*v*空间中的像素分布，它等价于一个忽略空间位置的视觉算法所看到的情况。为了使得该可视化更加简单，我们只考虑 L*u*坐标，如图 5.16c 所示。你可以看到多少明显的(被拉长的)的群簇？你怎样才能找到这些群簇？

k-均值和高斯混合采用密度函数的参数化(parametric)模型来回答这个问题，即它们假设该密度是少量简单分布(例如高斯)的叠加，其位置(中心)和形状(协方差)可以被估计出来。另一方面，均值移位平滑该分布并寻找它的峰值以及对应于每个峰值的特征空间区域。因为模型化的是一个完整的密度，所以这种方法称为"非参数化的"(non-parametric)(Bishop 2006)。让我们更详细地看一下这些方法。

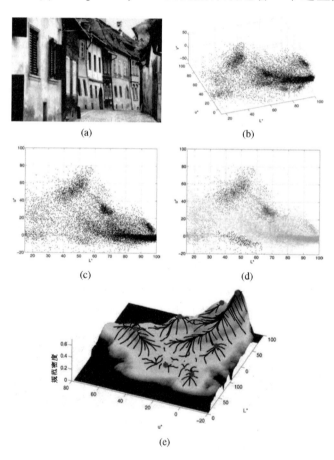

图 5.16 均值移位图像分割(Comaniciu and Meer 2002)©2002 IEEE：(a)输入彩色图像；(b)散布在 L*u*v*空间中的像素；(c)L*u*空间的分布；(d)159 次均值移位后的聚类结果；(e)对应的轨迹，其中峰值标记为红点

5.3.1 k-均值和高斯混合

当 k-均值默认将概率密度模型化为球形对称分布的叠加时，它并不需要任何概

率推理或模型化(Bishop 2006)。相反，指定该算法应该找到的群簇数 k，它会迭代地根据离每个聚类中心最近的样本来更新聚类中心位置。该算法可以通过从输入特征向量中随机采样 k 个中心来进行初始化。已开发了一些方法基于聚类中心的统计量进行分裂或归并以及加速寻找最近平均中心的过程(Bishop 2006)。

在高斯混合中，每个聚类中心都由一个协方差矩阵扩展，该矩阵的值通过对应的样本重新估计。Mahalanobis 距离(附录 B.1.1)代替最近邻被用于将输入样本点关联到聚类中心：

$$d(\boldsymbol{x}_i, \boldsymbol{\mu}_k; \boldsymbol{\Sigma}_k) = \|\boldsymbol{x}_i - \boldsymbol{\mu}_k\|_{\boldsymbol{\Sigma}_k^{-1}} = (\boldsymbol{x}_i - \boldsymbol{\mu}_k)^T \boldsymbol{\Sigma}_k^{-1}(\boldsymbol{x}_i - \boldsymbol{\mu}_k) \tag{5.26}$$

其中 $\boldsymbol{x}_i$ 为输入样本，$\boldsymbol{\mu}_k$ 为聚类中心，而 $\boldsymbol{\Sigma}_k$ 为它们的协方差估计。样本可以关联于最近的聚类中心(隶属关系的硬分配(hard assignment))或者软分配(softly assigned)给若干个邻近的聚类。

后者更常用，它相当于为高斯混和密度函数重新估计参数，

$$p(\boldsymbol{x}|\{\pi_k, \boldsymbol{\mu}_k, \boldsymbol{\Sigma}_k\}) = \sum_k \pi_k \mathcal{N}(\boldsymbol{x}|\boldsymbol{\mu}_k, \boldsymbol{\Sigma}_k), \tag{5.27}$$

其中 π_k 是混合参数，$\boldsymbol{\mu}_k$ 和 $\boldsymbol{\Sigma}_k$ 是高斯均值和协方差，且

$$\mathcal{N}(\boldsymbol{x}|\boldsymbol{\mu}_k, \boldsymbol{\Sigma}_k) = \frac{1}{|\boldsymbol{\Sigma}_k|} e^{-d(\boldsymbol{x}, \boldsymbol{\mu}_k; \boldsymbol{\Sigma}_k)} \tag{5.28}$$

是正态(高斯)分布(Bishop 2006)。

为了给未知混合参数 $\{\phi_k, \boldsymbol{\mu}_k, \boldsymbol{\Sigma}_k\}$ 迭代地计算(局部)最大可能估计，期望最大化算法(expectation maximization)(Dempster, Laird, and Rubin 1977)会在两个步骤中交替进行。

1. 段期望(expectation)阶段(E 步骤)估计责任(responsibilities)

$$z_{ik} = \frac{1}{Z_i} \pi_k \mathcal{N}(\boldsymbol{x}|\boldsymbol{\mu}_k, \boldsymbol{\Sigma}_k) \quad \text{with} \quad \sum_k z_{ik} = 1, \tag{5.29}$$

这是像素点 x_i 从第 k 个高斯聚类生成的可能性估计。

2. 段最大化(maximization)阶段(M 步骤)更新参数值

$$\boldsymbol{\mu}_k = \frac{1}{N_k} \sum_i z_{ik} \boldsymbol{x}_i, \tag{5.30}$$

$$\boldsymbol{\Sigma}_k = \frac{1}{N_k} \sum_i z_{ik} (\boldsymbol{x}_i - \boldsymbol{\mu}_k)(\boldsymbol{x}_i - \boldsymbol{\mu}_k)^T, \tag{5.31}$$

$$\pi_k = \frac{N_k}{N}, \tag{5.32}$$

其中

$$N_k = \sum_i z_{ik}. \tag{5.33}$$

是分配到每个群簇的样本点数估计。

Bishop(2006)对高斯混合估计和更加一般性的期望最大化主题有一个非常好的阐述。

在图像分割环境中，Ma, Derksen, Hong et al. (2007)提出了一个利用高斯混合进

行分割的出色的回顾,并发展了他们基于最小描述长度(MDL)编码的扩展,他们展示了在 Berkeley 分割数据集上得到的很好的结果。

5.3.2 均值移位

k-均值和高斯混合采用参数化形式为分割的概率密度函数建模,而均值移位默认用平滑的连续非参数化模型来模型化该分布。均值移位的关键在于在高维数据分布中高效地寻找峰值,而不用显式地计算完整的函数(Fukunaga and Hostetler 1975; Cheng 1995; Comaniciu and Meer 2002)。

再考虑一次图 5.16c 中的数据点。它们可以看作是从某个数据密度函数中获得的。如果我们可以计算该密度函数,如图 5.16e 所示,我们可以找到它的主要峰值(模态)并确定输入空间中攀登到同一个峰值而成为同一个区域部分的那些区域。这与 5.2.1 节中介绍的分水岭算法是相反的,它沿山而下来寻找吸引盆地(basins of attraction)。

那么,第一个问题就是给定稀疏样本集,如何估计密度函数。最简单的方法就是仅仅平滑该数据,例如将它用带宽为 h 的固定核进行卷积,

$$f(\boldsymbol{x}) = \sum_i K(\boldsymbol{x} - \boldsymbol{x}_i) = \sum_i k\left(\frac{\|\boldsymbol{x} - \boldsymbol{x}_i\|^2}{h^2}\right), \tag{5.34}$$

其中 x_i 是输入样本,而 $k(r)$ 是核函数(或者 Parzen 窗)。①该方法称为"核密度估计"(kernel density estimation)或者 Parzen 窗方法(Parzen window technique)(Duda, Hart, and Stork 2001, Section 4.3; Bishop 2006, Section 2.5.1)。一旦我们已经计算了 $f(\boldsymbol{x})$,如图 5.16e 和图 5.17 所示,我们就可以用梯度下降或者其他一些优化技巧来寻找它的局部极大值。

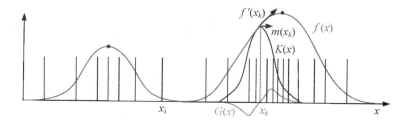

图 5.17 核密度估计,它的微分以及均值位移的一维可视化。核密度估计 $f(x)$ 通过将稀疏的输入样本 x_i 与和函数 $K(x)$ 卷积得到。该函数的微分 $f'(x)$ 可以通过将输入与微分核 $G(x)$ 卷积得到。在当前估计 x_k 附近估计局部替换向量会产生均值移位向量 $m(x_k)$,在多维设定中,它指向与函数梯度 $\nabla f(x_k)$ 相同的方向。红色点表示 $f(x)$ 的局部极大值,均值移位会收敛到该点

---

① 在这个简化的公式里,使用的是欧氏度量。我们稍后(5.42)将讨论如何将它推广到非均匀(缩放的或有向的)度量。还要注意,该分布可以不是正常分布,即积分为 1。因为我们只是在密度函数中寻找极大值,所以这没有关系。

"强力"方法的问题在于,对于高维度的情况,在完整搜索空间中估计 $f(x)$ 在计算上是不可行的。[①]相反,均值移位采用的是在优化领域中知名的多重启梯度下降(multiple restart gradient descent)的变种。从对一个局部最大值 y_k 的猜测开始,例如一个随机输入数据点 x_i,均值移位计算密度估计 $f(x)$ 在 y_k 处的梯度,并沿此方向上行进一步(图 5.17)。$f(x)$ 的梯度由下式计算得出

$$\nabla f(x) = \sum_i (x_i - x) G(x - x_i) = \sum_i (x_i - x) g\left(\frac{\|x - x_i\|^2}{h^2}\right), \tag{5.35}$$

其中

$$g(r) = -k'(r), \tag{5.36}$$

并且 $k'(r)$ 是 $k(r)$ 的一阶微分。我们可以把密度函数的梯度重写为

$$\nabla f(x) = \left[\sum_i G(x - x_i)\right] m(x), \tag{5.37}$$

其中向量

$$m(x) = \frac{\sum_i x_i G(x - x_i)}{\sum_i G(x - x_i)} - x \tag{5.38}$$

被称为"均值移位"(mean shift),因为它是 x 周围的邻居 x_i 的加权平均与 x 的当前值的差异。

在均值移位过程中,模态 y_k 的当前估计在第 k 次迭代中被其局部加权平均所取代,

$$y_{k+1} = y_k + m(y_k) = \frac{\sum_i x_i G(y_k - x_i)}{\sum_i G(y_k - x_i)}. \tag{5.39}$$

Comaniciu and Meer (2002) 证明在对核 $k(r)$ 有合理的弱约束,例如单调递减时,该算法会收敛到 $f(x)$ 的局部最大值。对于常规的梯度下降而言,不能保证收敛,除非采用合适的步长大小控制。

Comaniciu and Meer (2002)研究的两个核是 Epanechnikov 核,

$$k_E(r) = \max(0, 1 - r), \tag{5.40}$$

双线性核的径向推广,以及高斯(正态)核,

$$k_N(r) = \exp\left(-\frac{1}{2}r\right). \tag{5.41}$$

对应的微分核 $g(r)$ 分别是单位球和另一个高斯核。使用 Epanechnikov 核可以在有限步数内收敛;而高斯核有更平滑的轨迹(并且产生更好的结果),但在模态附近收敛非常慢(习题 5.5)。

使用均值移位的最简单的方式是,在每个输入点 x_i 上开始一个单独的均值移位模态估计 y 并且迭代固定步数,直到均值移位幅度小于一个阈值。更快的方法是随机二次采样一些输入点 x_i 并且保存每个点的临时演化轨迹。剩余的点可以根据最近

[①] 即使对于一维,如果空间极其稀疏,这样做的效率仍然很低。

演化路径分类(Comaniciu and Meer 2002)。Paris and Durand (2007)回顾了许多更高效的均值移位的实现,包括他们自己的方法。该方法基于利用$f(x)$的完整多维空间上有效的低分辨率估计和平稳点的方法。

图 5.16 所示的基于颜色的分割在确定最好的聚类时只看像素颜色。因此它可能将碰巧具有相同颜色的少量孤立像素点聚在一起,而可能并不对应于图像的具有语义含义的分割。

更好的结果通常通过在颜色与位置的联合域(joint domain)上进行聚类来得到。在这种方法中,图像 $x_s = (x, y)$ 的空间坐标,也称"空间域"(spatial domain),连接到颜色值 x_r,也称"值域"(range domain),上而均值移位就在这五维空间 x_j 上进行。因为位置和颜色可能有不同的尺度,核也相应的进行调整,即我们使用如下形式的核

$$K(\boldsymbol{x}_j) = k\left(\frac{\|\boldsymbol{x}_r\|^2}{h_r^2}\right) k\left(\frac{\|\boldsymbol{x}_s\|^2}{h_s^2}\right), \tag{5.42}$$

其中参数 h_s 和 h_r 分别用于控制滤波核的空间和值域带宽。图 5.18 展示了一个以参数 $(h_s, h_r, M) = (16, 19, 40)$ 在联合域中进行均值移位聚类的例子,其中包含像素少于 M 个的空间区域被消除掉。

图 5.18 以参数 $(h_s, h_r, M) = (16, 19, 40)$ 进行均值移位的彩色图像分割(Comaniciu and Meer 2002)© 2002 IEEE

联合域滤波核的形式让人联想到 3.3.1 节讨论的双边滤波器(3.34~3.37)。然而,均值移位与双边滤波器的区别在于,在均值移位中每个像素点的空间坐标是与其颜色值一道进行调整的,使得具有相似颜色的像素比其他像素迁移得更快,因此稍后可以用于聚类和分割。

虽然已经探索过很多方法,但确定均值移位中使用的最好的带宽参数 h 仍是比较困难的事情。这些方法包括优化偏差-方差的权衡,寻找聚类数变化较慢的参数区间,优化某种外部聚类标准,或者使用自顶向下(应用领域)的知识(Comaniciu and Meer 2003)。也可以改变在联合参数空间中核的方向以适应应用要求,比如空间-时间(视频)分割(Wang, Thiesson, Xu et al. 2004)。

均值移位已经被应用于计算机视觉中的很多不同问题,包括人脸跟踪、2D 形状提取和纹理分割(Comaniciu and Meer 2002)和更近一些用于立体视觉匹配(第 11 章)(Wei and Quan 2004)、非真实感绘制渲染(10.5.2 节) (DeCarlo and Santella 2002)

和视频编辑(10.4.5 节)(Wang, Bhat, Colburn et al. 2005)。Paris and Durand(2007) 提供了一个对这些应用的很好的回顾以及更多高效的求解均值移位方程与产生分层的分割结果的方法。

5.4 规范图割

自底向上的归并方法将区域聚合为已知的整体，均值移位方法试图利用模态发现找到相似像素的群簇，而由 Shi and Malik(2000)提出的规范图割方法则检测邻近像素点之间的亲和度(affinity，相似度)并试图分开那些被弱亲和度连接起来的像素集。

考虑图 5.19a 所示的简单图。在 A 组中的像素都以高吸引力链接起来，以粗红线表示，B 组中也是如此。而这两组之间的连接，以细蓝线表示，就要弱得多。这两组之间的规范图割(normalized cut)，以虚线表示，将它们分成两个群簇。

A 和 B 两组之间的图割定义为所有被切割的权重之和，

$$cut(A,B) = \sum_{i\in A, j\in B} w_{ij}, \tag{5.43}$$

其中两个像素(或者区域)i 和 j 之间的权重衡量它们的相似度。然而，采用最小割作为分割标准并不会产生合理的聚类结果，因为最小的割通常会包含孤立的像素。

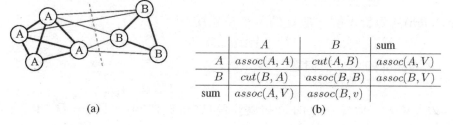

图 5.19 样本带权图和它的规范图割：(a)一个小的样本图和它最小的规范图割；(b)该图的关联和图割的表格。assoc 和 cut 的条目由关联权重矩阵 W (图 5.20)的区域和(area sum)计算而得。通过行或列之和来规范化表格条目，可以得到规范化的关联与图割 Nassoc 和 Ncut

更好的分割度量方法是规范图割，它定义为

$$Ncut(A,B) = \frac{cut(A,B)}{assoc(A,V)} + \frac{cut(A,B)}{assoc(B,V)}, \tag{5.44}$$

其中 $assoc(A,A) = \sum_{i\in A, j\in A} w_{ij}$ 是一个群簇中的关联度(所有权重之和)而 $assoc(A,V) = assoc(A,A) + cut(A,B)$ 是所有与 A 中节点相关的权重之和。图 5.19b 展示了割和关联如何可以被视为权重矩阵 $W = [w_{ij}]$ 的区域和，其中矩阵的项经安排使得 A 中的节点先出现而 B 中的节点第二个出现。图 5.20 展示了一个真正的这些区域和可以被计算的权重矩阵。将每一个这样的区域按照对应的行之和(图 5.19b 中最右边的列)分开就产生了规范图割和关联值。这些规范值更好地反映了一个特定分割的适合度，因为它们寻找一组边，使其与一个特定区域内部和从该区域出发的所有边都弱相关。

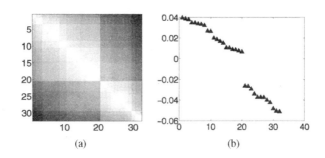

图 5.20 样本权重表及其第二小本征向量(Shi and Malik 2000)© 2000 IEEE：(a) 32×32 的样本权重矩阵 $\boldsymbol{W}$；(b)对应于推广的本征值问题 $(\boldsymbol{D}-\boldsymbol{W})\boldsymbol{y} = \lambda \boldsymbol{D}\boldsymbol{y}$ 的第二小本征值的本征向量

不幸的是，计算该最优规范图割是 NP-完全问题。作为替代，Shi and Malik(2000)建议计算节点到组的实值分配。令 $\boldsymbol{x}$ 为指标向量，其中 $x_i = +1$ 当且仅当 $i \in A$ 而 $x_i = -1$ 当且仅当 $i \in B$。令 $\boldsymbol{d} = \boldsymbol{W}\boldsymbol{1}$ 为对称矩阵 $\boldsymbol{W}$ 的行之和而 $\boldsymbol{D} = diag(\boldsymbol{d})$ 为对应的对角阵。Shi and Malik(2000)证明在所有指标向量 $\boldsymbol{x}$ 上最小化规范图割等价于最小化

$$\min_{\boldsymbol{y}} \frac{\boldsymbol{y}^T(\boldsymbol{D}-\boldsymbol{W})\boldsymbol{y}}{\boldsymbol{y}^T \boldsymbol{D} \boldsymbol{y}}, \tag{5.45}$$

其中 $\boldsymbol{y} = ((1+\boldsymbol{x}) - b(1-\boldsymbol{x}))/2$ 是有所有 1 和 $-b$ 组成的向量，使得 $\boldsymbol{y} \cdot \boldsymbol{d} = 0$。最小化该瑞利(Rayleigh)商等价于解一般化的本征值系统

$$(\boldsymbol{D}-\boldsymbol{W})\boldsymbol{y} = \lambda \boldsymbol{D}\boldsymbol{y}, \tag{5.46}$$

它可以转化为标准的本征值问题

$$(\boldsymbol{I}-\boldsymbol{N})\boldsymbol{z} = \lambda \boldsymbol{z}, \tag{5.47}$$

其中 $\boldsymbol{N} = \boldsymbol{D}^{-1/2} \boldsymbol{W} \boldsymbol{D}^{-1/2}$ 是规范化的亲和度矩阵(Weiss 1999)并且 $\boldsymbol{z} = \boldsymbol{D}^{1/2} \boldsymbol{y}$。因为这些本征向量可以被解读为弹簧-质量系统的大的振动模态，所以规范图割是图像分割中谱方法的一种。

扩展最初由 Scott and Longuet-Higgins(1990), Weiss(1999)提出的想法，建议规范化亲和度矩阵然后用其前 k 个本征向量来组成矩阵 Q。其他的论文通过用不同的方法来修改亲和度矩阵，寻求最小化问题更好的离散解或者应用多尺度方法(Meilă and Shi 2000, 2001; Ng, Jordan, and Weiss 2001; Yu and Shi 2003; Cour, Bénézit, and Shi 2005; Tolliver and Miller 2006)来扩展基本的规范图割框架。

图 5.20b 展示了对应于图 5.20a 中权重矩阵的第二小(实值)本征向量。(这里，行已经经过调换使得属于该本征向量不同部分的两组变量分开。)在这些实值向量计算之后，对应于正和负本征向量值的变量被关联到两个图割的部分。该过程可以进一步重复进行以分层细分图像，如图 5.21 所示。

Shi and Malik(2000)提出的原始算法使用空间位置和图像特征差异来计算半径 $\|\boldsymbol{x}_i - \boldsymbol{x}_j\| < r$ 内的像素间亲和度，

$$w_{ij} = \exp\left(-\frac{\|F_i - F_j\|^2}{\sigma_F^2} - \frac{\|x_i - x_j\|^2}{\sigma_s^2}\right), \tag{5.48}$$

其中 F 是由亮度、颜色或者有向滤波器直方图组成的特征向量。(注意(5.48)是如何成为联合特征空间距离(5.42)的负指数值的)。

图 5.21　规范图割分割结果(Shi and Malik 2000)© 2000 IEEE：输入图像和由规范图割算法返回的分量

在后续的工作中，Malik, Belongie, Leung $et\ al.$ (2001)寻找像素 i 和 j 之间的中间轮廓(intervening contours)并且定义中间轮廓权重

$$w_{ij}^{IC} = 1 - \max_{x \in l_{ij}} p_{con}(x), \tag{5.49}$$

其中，l_{ij} 是连接像素 i 和 j 的图像线，而 $p_{con}(x)$ 是一条中间轮廓垂直于这条图像线的概率，它可以定义为在此垂直方向上有向能量的负指数值。他们将这些权重乘以基于 texton 的纹理相似度度量，利用完全基于局部像素级的特征得到的初始过分割结果以基于区域的方式来重新估计中间轮廓和纹理统计量。图 5.22 展示了在大量测试图片上运行该改进算法得到的结果。

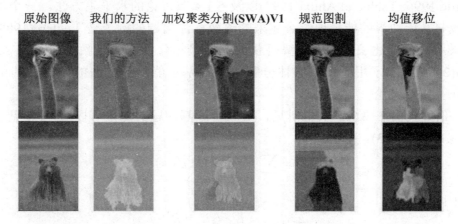

图 5.22　比较分割结果(Alpert, Galun, Basri $et\ al.$ 2007)© 2007 IEEE。"我们的方法"指的是由 Alpert 等提出的概率自底向上归并算法

因为规范图割需要求解巨大的稀疏本征值问题，所以会相当慢。Sharon, Galun, Sharon $et\ al.$ (2006)提出了一种利用受代数多重网格法(Brandt1986; Briggs, Henson, and

McCormick 2000)启发的方法来加速规范图割的计算。为了使得原问题的粒度更大，它们选择了更少数量的变量以使剩余细粒度层的变量会与至少一个粗粒度层的变量强耦合(strongly coupled)。图 5.15 用图展示了该过程，而(5.25)给出了强耦合的定义，只是在这种情况下使用原来规范图割中的原始权重 w_{ij} 而不是归并概率 p_{ij} 。

一旦选择了一组粗粒度的变量，一个元素类似于(5.25)左手边的中间层插值矩阵就可以用来定义该规范图割问题的简化版本。为了利用基于插值粒度粗化来计算权重矩阵，要用额外的区域统计量来调节权重。规范图割在更粗粒度层次的分析中计算出来之后，更细粒度层次节点的隶属值就可以通过插值父节点值来计算得到，并将以 0 和 1 的在 $\epsilon = 0.1$ 内的值映射到纯布尔值。

由加权聚类(SWA)得到的分割示例如图 5.22 所示，同时还有由 Alpert, Galun, Basri *et al.* (2007)提出的概率自底向上归并算法，参见 5.2 节的介绍。更新的工作成果中，Wang and Oliensis (2010)展示了如何直接从亲和度图中估计分割的统计量(例如平均区域大小)。它们以此得到比其他可能的分割更中心化(central)的分割。

5.5 图割和基于能量的方法

图像分割的公共主题是这样的需求，即将具有相似表观(统计量)的像素聚集起来，且使不同区域像素间的边界长度短且穿越可见的不连续处。如果我们约束边界度量是在直接邻居之间而且通过在像素级求和来计算区域隶属统计量，则可以用变分形式化(variational formulation)(正则化，3.7.1 节)或二值马尔科夫随机场(3.7.2 节)将其形式化为一个经典的基于像素的能量函数。

连续性方法的例子包括(Mumford and Shah 1989; Chan and Vese 1992; Zhu and Yuille 1996; Tabb and Ahuja 1997)以及 5.1.4 节讨论的水平集方法。一个早期将基于区域与基于边界的能量项结合起来的离散标注问题的例子是 Leclerc(1989)的工作，他采用最小描述长度(MDL)编码获取需要最小化的能量函数。Boykov and Funka-Lea(2006)给出了用于二值物体分割的多种基于能量的方法的调研，其中一些我们将在下文中讨论。

正如我们在 3.7.2 节中所见，对应于分割问题的能量可以写作(参见公式(3.100)与(3.108~3.113)，

$$E(f) = \sum_{i,j} E_r(i,j) + E_b(i,j), \tag{5.50}$$

其中区域项

$$E_r(i,j) = E_S(I(i,j); R(f(i,j))) \tag{5.51}$$

是像素亮度(或颜色) $I(i,j)$ 与区域 $R(f(i,j))$ 统计量一致的似然度的负对数值，而边界项

$$E_b(i,j) = s_x(i,j)\delta(f(i,j) - f(i+1,j)) + s_y(i,j)\delta(f(i,j) - f(i,j+1)) \tag{5.52}$$

度量的是 $\mathcal{N}_4$ 邻居之间由局部垂直水平平滑项 $s_x(i,j)$ 和 $s_y(i,j)$ 修正过的不一致性。

区域统计量可以是简单的平均灰度或颜色(Leclerc 1989)，这种情况下

$$E_S(I;\mu_k) = \|I - \mu_k\|^2. \tag{5.53}$$

作为另外的选择，它们可以更加复杂，例如区域亮度直方图(Boykov and Jolly 2001)或者颜色高斯混合模型(Rother, Kolmogorov, and Blake 2004)。对于平滑(边界)项，通常是使平滑强度 $s_x(i,j)$ 反比于局部边缘强度(Boykov, Veksler, and Zabih 2001)。

最初，基于能量的分割问题采用迭代梯度下降方法优化，该方法很慢而且容易陷于局部最小值。Boykov and Jolly(2001)率先将 Greig, Porteous, and Seheult(1989)提出的二值 MRF 优化算法用于二值物体分割。

在该方法中，用户首先用几下图像笔画描绘出图像背景与前景区域(图 3.61)。而后这些像素就成为 S-T 图中将节点连接到源与汇标签 S 和 T 的种子节点(图 5.23a)。种子像素被用于估计前景和背景区域统计量(亮度或者颜色直方图)。

在图中其他边的容量通过区域和边界能量项得到，即与前景或背景区域更兼容的像素与对应的源或汇有更强的连接关系；具有更大平滑性的邻近像素点也会有更强的连接。一旦采用多项式时间解决了最小割/最大流问题(Goldberg and Tarjan 1988; Boykov and Kolmogorov 2004)，在计算出来的图割的两侧像素根据它们仍然连接的源或汇进行标注(图 5.23b)虽然图割只是 MRF 能量最小化的若干著名方法之一(附录 B.5.4)，但它仍然是解决二值 MRF 问题最常用的方法。

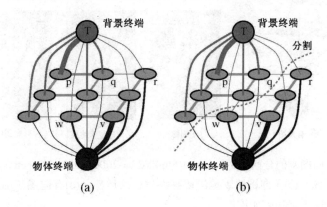

图 5.23 用于区域分割的图割(Boykov and Jolly 2001)© 2001 IEEE：(a)能量函数编码为最大流问题；(b)最小割确定区域边界

Boykov and Jolly(2001)的基本二值分割算法已经在多个方向上被扩展了。Rother, Kolmogorov, and Blake(2004)的 GrabCut 系统迭代地重新估计区域统计量，其模型是颜色空间中的高斯混合。这就使得他们的系统允许最少的用户输入，例如只是一个边界框(图 5.24a)——背景颜色模型通过围绕框轮廓的条带初始化(前景颜色模型根据内部像素进行初始化，但会很快收敛到更好的物体估计。)用户也可以在解决过程中加入额外的笔画来改善分割结果。在更近的工作中，Cui, Yang, Wen *et al.* (2008)利用从相似物体之前的分割结果中得到的颜色和边界模型来提升 GrabCut 中所用的局部模型。

(a)　　　　　　　　　(b)　　　　　　　　(c)

图 5.24　GrabCut 图像分割(Rother, Kolmogorov, and Blake 2004)© 2004 ACM：(a)用户画一个红色的边界框；(b)算法猜测物体与背景的颜色分布并执行二值分割；(c)该过程采用更好的区域统计量重复进行

另外对原始二值分割形式化的主要扩展是有向边的加入，它使得边界区域是有方向的，例如更喜欢由亮到暗的转换或反之(Kolmogorov and Boykov 2005)。图 5.25 展示了有向图割能正确地将亮灰色肝脏从其暗灰色环境中分割出来的例子。同样的方法可以用于度量离开一个区域的流量，即投影到区域边界法向的带符号梯度。将更大的邻域结合到有向图中能够近似估计连续问题，例如在全局最优图割框架中的那些用水平集解决的问题(Boykov and Kolmogorov 2003; Kolmogorov and Boykov 2005)。

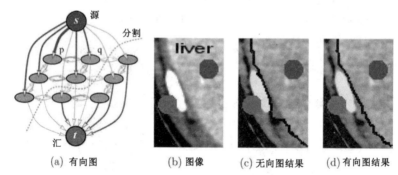

(a) 有向图　　(b) 图像　　(c) 无向图结果　　(d) 有向图结果

图 5.25　采用有向图割的分割(Boykov and Funka-Lea 2006)© 2006 Springer：(a)有向图；(b)带种子点的图像；(c)无向图错误地沿明亮物体延续边界；(d)有向图正确地将明亮灰色区域从它的更暗的背景中分割出来

再近一些的基于图割的分割方法的发展包括额外的连通性先验以强迫前景是单独的一块(Vicente, Kolmogorov, and Rother 2008)和在分割过程中利用关于物体形状知识的形状先验(Lempitsky and Boykov 2007; Lempitsky, Blake, and Rother 2008)。

优化二值 MRF 能量(5.50)需要利用组合优化方法，例如最大流，近似的解决方案可以通过将二值能量函数转化为定义在连续(0,1)区间上的二次能量随机场来获得，而后这就变成了经典的基于薄膜(membrane-based)的正则化问题(3.100～3.102)。这样产生的二次能量函数可以采用标准的线性系统解决方法(3.102～3.103)来解决，然而如果关心速度的话，你应该采用多重网格或者它的变种之一(附录

A.5)。一旦计算得出连续解，则将它按照 0.5 阈值化就可以得到二值分割结果。

该(0,1)连续优化问题也可以被解读为在每个像素点计算从该像素点出发最终停止在其中一个有标注的种子像素的随机游走(random walker)概率，也等价于计算在一个电阻值等于边权重的电阻网格中的势能(Grady 2006; Sinop and Grady 2007)。K 值分割也可以通过在种子标签中迭代来解决，即将一个标签设置为 1 而所有其他标签设置为 0 的二值问题来计算每个像素相对隶属关系的概率。在进一步的工作中，Grady and Ali (2008)利用预先计算的线性系统的本征向量来使得新的一组种子的求解更快，它与 10.4.3 节中提到的 Laplacian 抠图问题相关(Levin, Acha, and Lischinski 2008)。Couprie, Grady, Najman et al. (2009)将随机游走关联到分水岭和其他分割方法。Singaraju, Grady and Vidal(2008)添加有向能量约束来支持流量，这使能量分段二次化并且因此不能用单一的线性系统求解。随机游走算法也可以被用于解决 Mumford-Shah 分割问题(Grady and Alvino 2008)和快速多重网格求解(Grady 2008)。Singaraju, Grady, Sinop et al. (2010)给出了这些方法的一个很好的回顾。

计算连续(0,1)近似分割的更快的方法是计算 0 和 1 种子区域之间的加权测地距离(weighted geodesic distance)(Bai and Sapiro 2009)，它可以被用于估计软透明度遮罩(10.4.3 节)。由 Criminisi, Sharp and Blake(2008)提出的相关方法可以用于快速搜索一般二值马尔科夫随机场优化问题的近似解。

应用：医学图像分割

图像分割最有前途的应用之一就是在医学图像领域，它可以用于分割解剖切片以便于进一步的定量分析。图 5.25 展示了采用有向边的二值图割将肝脏切片(亮灰色)从围绕它的骨骼(白色)和肌肉(暗灰色)切片中分割出来。图 5.26 展示了在 $256\times256\times119$ 的 X 光断层扫描体的骨骼分割结果。没有今天图像分割算法强有力的优化方法，这样的处理在过去通常需要更多的人力劳动才能描绘每一幅 X 光图片。

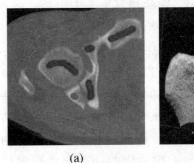
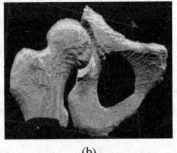

(a)　　　　　　　　　　(b)

图 5.26 采用图割的 3D 体医学图像分割(Boykov and Funka-Lea 2006)©2006 Springer：(a) 带有某些种子的计算机断层扫描(CT)切片；(b)恢复的 3D 体骨骼模型(在 $256\times256\times119$ 体素网格上)

医学图像分割领域(McInerney and Terzopoulos 1996)和医学图像注册(Kybic and

Unser 2003)(8.3.1 节)是具有丰富成果的研究领域，具有自己的专门会议——比如"医学影像计算与计算机辅助干预(MICCAI)"[①]和期刊——比如《医学影像分析》和《IEEE 医学成像》。对于在该领域的研究，这些会是参考文献和思路的理想来源。

5.6 补充阅读

图像分割主题与聚类算法息息相关，在大量专题著作和综述文章中都有提及(Jain and Dubes 1988; Kaufman and Rousseeuw 1990; Jain, Duin, and Mao 2000; Jain, Topchy, Law et al. 2004)。一些早期的分割方法包括由 Brice and Fennema(1970); Pavlidis(1977); Riseman and Arbib(1977); Ohlander, Price, and Reddy(1978); Rosenfeld and Davis(1979); Haralick and Shapiro(1985)引入，而更新的方法的例子则由 Leclerc(1989); Mumford and Shah(1989); Shi and Malik(2000); Felzenszwalb and Huttenlocher(2004b)研究发展。

Arbeláez, Maire, Fowlkes et al. (2010)给出了自动分割方法的优秀回顾，同时也比较了它们在 Berkeley Segmentation Dataset and Benchmark(Martin, Fowlkes, Tal et al. 2001)[②]上的性能。其他的比较论文和数据集包括 Unnikrishnan, Pantofaru, and Hebert (2007); Alpert, Galun, Basri et al. (2007); Estrada and Jepson (2009)。

活动轮廓主题已经有很长的历史，它起始于关于蛇行和其他能量最小化变分方法的很有潜力的一些工作(Kass, Witkin, and Terzopoulos 1988; Cootes, Cooper, Taylor et al. 1995; Blake and Isard 1998)，继而出现了诸如智能剪刀之类的方法(Mortensen and Barrett 1995, 1999; Pérez, Blake, and Gangnet 2001)，最终在水平集上达到巅峰(Malladi, Sethian, and Vemuri 1995; Caselles, Kimmel, and Sapiro 1997; Sethian 1999; Paragios and Deriche 2000; Sapiro 2001; Osher and Paragios 2003; Paragios, Faugeras, Chan et al. 2005; Cremers, Rousson, and Deriche 2007; Rousson and Paragios 2008; Paragios and Sgallari 2009)，这些都是当今应用最广的活动轮廓方法。

结合凝聚或分裂方法的基于局部像素相似度的图像分割方法包括分水岭(Vincent and Soille 1991; Beare 2006; Arbeláez, Maire, Fowlkes et al. 2010)、区域分裂(Ohlander, Price, and Reddy 1978)、区域归并(Brice and Fennema 1970; Pavlidis and Liow 1990; Jain, Topchy, Law et al. 2004)以及基于图和概率的多尺度方法(Felzenszwalb and Huttenlocher 2004b; Alpert, Galun, Basri et al. 2007)。

在像素的密度函数表示中寻找模态(峰值)的均值移位算法由 Comaniciu and Meer(2002); Paris and Durand(2007)提出。参数化的高斯混合也可以用来表达和分割这样的像素密度(Bishop 2006; Ma, Derksen, Hong et al. 2007)。

在用于图像分割的谱(本征值)方法中，大有潜力的工作是 Shi and Malik(2000)

[①] *http://www.miccai.org/*

[②] *http://www.eecs.berkeley.edu/Research/Projects/CS/vision/grouping/segbench/*

的规范图割算法。相关的工作包括 Weiss(1999); Meilă and Shi(2000, 2001); Malik, Belongie, Leung et al.(2001); Ng, Jordan, and Weiss(2001); Yu and Shi(2003); Cour, Bénézit, and Shi(2005); Sharon, Galun, Sharon et al.(2006); Tolliver and Miller(2006); Wang and Oliensis(2010)。

用于交互式分割的基于连续能量(变分法)的方法包括 Leclerc(1989); Mumford and Shah(1989); Chan and Vese(1992); Zhu and Yuille(1996); Tabb and Ahuja(1997)。这些问题的离散变种通常采用二值图割或其他组合能量最小化方法来优化 (Boykov and Jolly 2001; Boykov and Kolmogorov 2003; Rother, Kolmogorov, and Blake 2004; Kolmogorov and Boykov 2005; Cui, Yang, Wen et al. 2008; Vicente, Kolmogorov, and Rother 2008; Lempitsky and Boykov 2007; Lempitsky, Blake, and Rother 2008),但也可以使用连续优化方法再加阈值化(Grady 2006; Grady and Ali 2008; Singaraju, Grady, and Vidal 2008; Criminisi, Sharp, and Blake 2008; Grady 2008; Bai and Sapiro 2009; Couprie, Grady, Najman et al. 2009)。Boykov and Funka-Lea (2006) 给出了各种用于二值物体分割的基于能量方法的优秀回顾。

5.7 习 题

习题 5.1 蛇行演化

证明在没有外力作用下,不论使用一阶或二阶平滑(或者某种组合),蛇行总是会收缩到一个小圆而最终为一个点。

(提示:如果能证明 $x(s)$ 和 $y(s)$ 两部分的演化是独立的,那么你可以更容易地分析 1D 情况。)

习题 5.2 蛇行跟踪

实现一个基于蛇行的轮廓跟踪:

1. 决定使用大量的轮廓点还是少量的用 B-样条插值的点;
2. 定义你自己的平滑能量函数并确定使用哪种基于图像的吸引力;
3. 在每次迭代中,建立带状线性系统的方程(二次能量函数)并用带状 Cholesky 因子分解求解(附录 A.4)。

习题 5.3 智能剪刀

实现智能剪刀(火线)交互分割算法(Mortensen and Barrett 1995),并设计一个图形化用户交互界面用于在图像上描绘一些曲线并用它们进行分割。

习题 5.4 区域分割

实现本章中描述的一种区域分割算法,一些流行的分割算法包括:

- k-均值(5.3.1 节);
- 高斯混合(5.3.1 节);
- 均值移位(5.3.2 节)和习题 5.5;
- 规范图割(5.4 节);

- 基于相似度图的分割(5.2.4 节);
- 二值马尔科夫随机场,采用图割求解(5.5 节)。

将你的区域分割算法用于视频序列,并用它在帧与帧之间跟踪运动区域。

另外一个选择,在 Berkeley segmentation database(Martin, Fowlkes, Tal et al. 2001)上测试你的算法。

习题 5.5 均值移位

实现一个用于彩色图像的均值移位分割算法(Comaniciu and Meer 2002):将你的图像转化为 L*a*b*空间,或者保持原始的 RGB 空间,利用像素 (x,y) 位置提升它们。

1. 对于每个像素 (L,a,b,x,y),使用单位球(Epanechnikov 核)或有限半径高斯核或其他你选择的核计算它邻居的加权平均。对颜色和空间尺度作不同的加权,例如采用图 5.18 中所示的值 $(h_s, h_r, M) = (16, 19, 40)$。

2. 用该加权平均值替代当前值并迭代直到运动小于阈值或者达到预先设置的步数。

3. 将在阈值下的所有最终值(模态)聚类,即寻找连通分量。因为每个像素都关联到一个最终的均值移位(模态)值,这就产生了图像分割结果,即每个像素都标注为它的最终分量。

4. (可选)随机选取一组点作为起始点并找到每个未被标注的像素属于哪个分量,可以通过搜索最近邻或者迭代均值移位直到找到一个邻近的均值移位值的轨迹。描述你将使该过程高效的数据结构。

(可选)均值移位通过局部权重分开核密度函数估计来获取步长大小,这保证会收敛但可能很慢。使用优化相关文献中的其他步长估计算法,看看是否能使算法收敛更快。

第 6 章
基于特征的配准

6.1 基于 2D 和 3D 特征的配准
6.2 姿态估计
6.3 几何内参数标定
6.4 补充阅读
6.5 习题

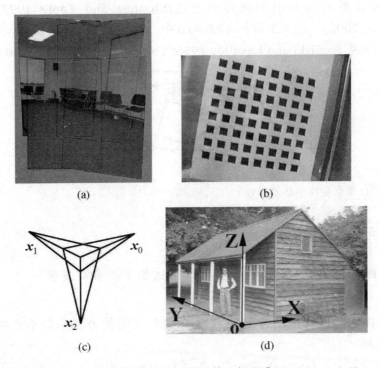

图 6.1 几何配准和标定：(a)用于拼接的 2D 图形的几何配准(Szeliski and Shum 1997)© 1997 ACM；(b)二维标定目标(Zhang 2000)© 2000 IEEE；(c)从消失点标定；(d)场景中存在容易找到的线和消失方向(Criminisi, Reid, and Zisserman 2000)© 2000 Springer

一旦从图像中提取出特征后,很多视觉算法的下一步是在不同图像中匹配这些特征(4.1.3 节)。匹配的一个重要成分是验证匹配的特征是否在几何上一致,即,特征位移可以用简单的 2D 或者 3D 几何变换描述。计算出的运动可以用于其他的应用,比如图像拼接(第 9 章)或者增强现实(6.2.3 节)。

在本章中,我们审视几何图像配准问题,即,2D 和 3D 的变换将一个图像中的特征映射到另一个图像中(6.1 节)。这个问题的一个特殊例子是姿态估计(pose estimation),确定摄像机相对于已知 3D 物体或者场景的位置(6.2 节)。另一种情况是摄像机内标定(intrinsic calibration)的计算,由内部参数组成,比如焦距和径向畸变(6.3 节)。在第 7 章中,我们考虑如下相关问题:如何从 2D 匹配估计 3D 点的结构(三角剖分);如何同时估计 3D 几何和摄像机运动(由运动到结构)。

6.1 基于 2D 和 3D 特征的配准

基于特征的配准是从两个或者多个匹配的 2D 或 3D 点的集合中估计运动的问题。在这一节,我们限定在整体的参数变换中,比如 2.1.2 节描述的,表 2.1 和图 6.2 显示的,或者更高阶的用于曲面的变换(Shashua and Toelg 1997; Can, Stewart, Roysam et al. 2002)。在 8.3 节和 12.6.4 节中,将论述在非刚性和弹性变换中的应用(Bookstein 1989; Szeliski and Lavallée 1996; Torresani, Hertzmann, and Bregler 2008)。

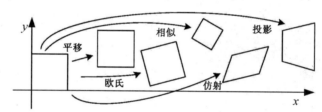

图 6.2 2D 平面变换的基本集合

6.1.1 使用最小二乘的 2D 配准

给定匹配的特征点集合 $\{x_i, x'_i\}$ 和以下形式的平面参数变换[①]
$$x' = f(x; p), \tag{6.1}$$

我们怎样才能最好地估计运动参数 p 呢?常用的方法是最小二乘,也就是最小化残差平方和
$$E_{\mathrm{LS}} = \sum_i \|r_i\|^2 = \sum_i \|f(x_i; p) - x'_i\|^2, \tag{6.2}$$

其中
$$r_i = x'_i - f(x_i; p) = \hat{x}'_i - \tilde{x}'_i \tag{6.3}$$

[①] 以非平面参数模型为例,比如二次型,参见 Shashua and Toelg (1997); Shashua and Wexler (2001)的工作。

是测量值 $\tilde{x}'_i$ 和其对应的当前预测值 $\tilde{x}'_i = f(x_i; p)$ 之间的残差(更多关于最小二乘的资料参见附录 4.2，其统计学依据参见附录 B.2)。

在 2.1.2 节和表 2.1 中描述的大多数运动模型，比如平移、相似和仿射变换，在运动 $\Delta x = x' - x$ 和未知参数 p 之间存在线性关系

$$\Delta x = x' - x = J(x)p, \tag{6.4}$$

其中 $J = \partial f / \partial p$ 是关于运动参数 p(见表 6.1)的变换矩阵 f 的雅克比矩阵。在这种情况下，简单的线性回归(线性最小二乘问题)可以形式化为

$$\begin{aligned} E_{\text{LLS}} &= \sum_i \|J(x_i)p - \Delta x_i\|^2 & (6.5) \\ &= p^T \left[\sum_i J^T(x_i)J(x_i)\right] p - 2p^T \left[\sum_i J^T(x_i)\Delta x_i\right] + \sum_i \|\Delta x_i\|^2 & (6.6) \\ &= p^T A p - 2p^T b + c. & (6.7) \end{aligned}$$

这个最小值可以通过求解正规方程(normal equations)[①]的对称正定(symmetric positive definite，SPD)系统得到

$$Ap = b, \tag{6.8}$$

其中

$$A = \sum_i J^T(x_i)J(x_i) \tag{6.9}$$

称为 Hessian 矩阵，$b = \sum_i J^T(x_i)\Delta x_i$。对于单纯的平移变换，最后的公式有一个更简单的形式，即平移是对应点之间平移的平均值，或者等价的质心点的平移。

表 6.1 2D 坐标变换的雅克比矩阵 $x' = f(x; p)$，如表 2.1 所示，其中我们重新参数化了运动，使其对于 $p = 0$ 是一致的

变换	矩阵	参数 p	雅克比矩阵 J
平移	$\begin{bmatrix} 1 & 0 & t_x \\ 0 & 1 & t_y \end{bmatrix}$	(t_x, t_y)	$\begin{bmatrix} 1 & 0 \\ 0 & 1 \end{bmatrix}$
欧几里得	$\begin{bmatrix} c_\theta & -s_\theta & t_x \\ s_\theta & c_\theta & t_y \end{bmatrix}$	(t_x, t_y, θ)	$\begin{bmatrix} 1 & 0 & -s_\theta x - c_\theta y \\ 0 & 1 & c_\theta x - s_\theta y \end{bmatrix}$
相似	$\begin{bmatrix} 1+a & -b & t_x \\ b & 1+a & t_y \end{bmatrix}$	(t_x, t_y, a, b)	$\begin{bmatrix} 1 & 0 & x & -y \\ 0 & 1 & y & x \end{bmatrix}$
仿射	$\begin{bmatrix} 1+a_{00} & a_{01} & t_x \\ a_{10} & 1+a_{11} & t_y \end{bmatrix}$	$(t_x, t_y, a_{00}, a_{01}, a_{10}, a_{11})$	$\begin{bmatrix} 1 & 0 & x & y & 0 & 0 \\ 0 & 1 & 0 & 0 & x & y \end{bmatrix}$
投影	$\begin{bmatrix} 1+h_{00} & h_{01} & h_{02} \\ h_{10} & 1+h_{11} & h_{12} \\ h_{20} & h_{21} & 1 \end{bmatrix}$	$(h_{00}, h_{01}, \ldots, h_{21})$	见 6.1.3 节

[①] 对于欠适定的问题，最好是在线性方程组 $J(x_i)p = \Delta x_i$ 上用 QR 分解代替正态方程 (Björck 1996; Golub and Van Loan 1996)。但是，在图像配准中，这样的情况很少发生。

不确定性加权　前面最小二乘形式化方法假设所有特征点都是在同样精度下匹配的。实际上通常都不是这样的，因为某些点可能会相较于其他点落在纹理比较多的区域中。如果我们给每个点关联一个标量方差估计(scalar variance estimate) σ_i^2，我们可以最小化加权最小二乘问题[①]

$$E_{\text{WLS}} = \sum_i \sigma_i^{-2} \|\boldsymbol{r}_i\|^2. \tag{6.10}$$

正如 8.1.3 节所示，基于块的匹配的方差估计，可以通过用块的 Hessian $\boldsymbol{A}_i$(8.55)的逆乘以每个像素的噪声方差 σ_n^2(8.44)得到。给每个残差平方添加其协方差的逆 $\Sigma_i^{-1} = \sigma_n^{-2}\boldsymbol{A}_i$ (也称为"信息矩阵")作为权重，我们得到

$$E_{\text{CWLS}} = \sum_i \|\boldsymbol{r}_i\|^2_{\Sigma_i^{-1}} = \sum_i \boldsymbol{r}_i^T \Sigma_i^{-1} \boldsymbol{r}_i = \sum_i \sigma_n^{-2} \boldsymbol{r}_i^T \boldsymbol{A}_i \boldsymbol{r}_i. \tag{6.11}$$

6.1.2　应用：全景图

图像配准一个最简单(也最有趣)的应用是一种特殊形式的图像拼接，称为"全景图"。在全景图里，在用简单的平均进行融合之前，图像会被平移，可能还有旋转和缩放(如图 6.3 所示)。这个过程模仿的是艺术家 David Hockney 设计的摄影拼贴(photographic collage)，虽然他的方法是使用不透明覆盖模型从一些常规照片中进行合成。

图 6.3　由三幅图像组成的用平移模型自动配准后平均在一起的全景图

网上的大部分例子都是手动配准的，以达到最佳的艺术效果[②]。但可以用特征匹配和配准方法进行自动的注册(registration)(Nomura, Zhang, and Nayar 2007; Zelnik-Manor and Perona 2007)。

考虑简单的平移模型。我们希望不同图像中的所有对应点尽可能排成一行。用 t_j 表示第 j 个图像坐标帧在全局合成帧中的位置，x_{ij} 是在第 j 个图像中第 i 个匹配的特征的位置。为了配准图像，我们希望最小化最小二乘误差

$$E_{\text{PLS}} = \sum_{ij} \|(\boldsymbol{t}_j + \boldsymbol{x}_{ij}) - \boldsymbol{x}_i\|^2, \tag{6.12}$$

---

① 每个测量值都可以有不同的方差或者确定性程度的问题称为"异方差模型"(heteroscedastic models)。
② http://www.flickr.com/groups/panography/

其中 x_i 是在全局坐标帧中特征 i 的一致(平均)位置(另外一个方法是分别注册每对重叠的图像，然后计算每帧的一致位置——参见习题6.2)。

前面的最小二乘问题是不确定的(你可以给所有帧和点的位置 t_j 和 x_i 添加一个偏移常量)。为了解决这个问题，可以选一帧作为起点或者添加一个约束使平均的帧偏移量是 0。

添加旋转和缩放变换的公式是简单的，留作习题(习题6.2)。看看你是否能创造一个你会乐意在网上共享的拼贴画。

6.1.3 迭代算法

线性最小二乘是最简单的参数估计方法，然而在计算机视觉中大多数问题在测量值和未知值之间不存在简单的线性关系。在这种情况下得到的问题称为"非线性最小二乘"(non-linear least squares)或者"非线性回归"(non-linear regression)。

考虑一个例子，比如在两组点之间的刚性 2D 欧式变换(平移加旋转)的估计问题。如果我们把这个变换参数化为平移量(t_x, t_y)和旋转角度θ，如表2.1所示，这个变换的雅克比矩阵在表 6.1 中给出，取决于当前值θ。在表 6.1 中，注意我们如何重新参数化运动矩阵，使其在原点 $p=0$ 总是单位矩阵，这样使得初始化运动参数更简单了。

为了最小化非线性最小二乘问题，我们对于当前估计的参数 p 迭代地找到更新 Δp，最小化

$$E_{\text{NLS}}(\Delta p) = \sum_i \|f(x_i; p + \Delta p) - x_i'\|^2 \tag{6.13}$$

$$\approx \sum_i \|J(x_i; p)\Delta p - r_i\|^2 \tag{6.14}$$

$$= \Delta p^T \left[\sum_i J^T J\right] \Delta p - 2\Delta p^T \left[\sum_i J^T r_i\right] + \sum_i \|r_i\|^2 \tag{6.15}$$

$$= \Delta p^T A \Delta p - 2\Delta p^T b + c, \tag{6.16}$$

其中"Hessian"[①]A 和公式(6.9)是一样的，右边的向量

$$b = \sum_i J^T(x_i) r_i \tag{6.17}$$

现在是以雅克比矩阵为权重的残差向量的和。这在直观上是容易理解的，因为参数以与雅克比矩阵成正比的强度被拉向预测误差方向。

一旦计算出 A 和 b，我们便使用下面的式子计算 Δp

$$(A + \lambda \text{diag}(A))\Delta p = b, \tag{6.18}$$

相应地更新参数向量 $p \leftarrow p + \Delta p$。参数$\lambda$是附加的阻尼参数，用于确保系统在能量上采取一个"下山"步骤，是 Levenberg–Marquardt 算法的必要构件(更多细节参见

[①] 这里的"Hessian"A 不是非线性最小二乘问题(6.13)真正的 Hessian(二阶导数)。相反，它是近似的 Hessian，在 $f(x_i; p + \Delta p)$ 中忽略二阶(和更高阶)的导数。

附录 A.3)。在很多应用中,如果系统能够成功地收敛,它则可以设为 0。

在 2D 平移加旋转的情况中,最终我们得到含有的未知参数 $(\delta t_x, \delta t_y, \delta\theta)$ 的 3×3 正规方程。(t_x, t_y, θ) 的初始值可以通过 (t_x, t_y, c, s) 的四参数相似变换拟合,然后令 $\theta = \tan^{-1}(s/c)$ 得到。另一个方法是用 2D 点的质心估计变换参数,然后用极坐标估计旋转角度(习题 6.3)。

对于其他的 2D 运动模型,表 6.1 中的导数是很直接的,除了 2D 投影运动(单应性)外,它出现在图像拼接中(第 9 章)。用新的参数形式,从(2.21)可以将这些公式重新写作

$$x' = \frac{(1+h_{00})x + h_{01}y + h_{02}}{h_{20}x + h_{21}y + 1} \quad \text{and} \quad y' = \frac{h_{10}x + (1+h_{11})y + h_{12}}{h_{20}x + h_{21}y + 1}. \tag{6.19}$$

因此雅克比矩阵是

$$J = \frac{\partial f}{\partial p} = \frac{1}{D}\begin{bmatrix} x & y & 1 & 0 & 0 & 0 & -x'x & -x'y \\ 0 & 0 & 0 & x & y & 1 & -y'x & -y'y \end{bmatrix}, \tag{6.20}$$

其中 $D = h_{20}x + h_{21}y + 1$ 是(6.19)的分母,取决于当前的参数设置(x' 和 y' 也是如此)。

8 个未知量 $\{h_{00}, h_{01}, \cdots, h_{21}\}$ 的初始推测可以按如下方式得到:给方程组(6.19)的两边同乘以这个分母,得到线性方程组

$$\begin{bmatrix} \hat{x}' - x \\ \hat{y}' - y \end{bmatrix} = \begin{bmatrix} x & y & 1 & 0 & 0 & 0 & -\hat{x}'x & -\hat{x}'y \\ 0 & 0 & 0 & x & y & 1 & -\hat{y}'x & -\hat{y}'y \end{bmatrix} \begin{bmatrix} h_{00} \\ \vdots \\ h_{21} \end{bmatrix}. \tag{6.21}$$

然而,从统计的角度看,这不是最合适的选择,因为给每个方程乘的分母 D,在点和点之间变化很大。[①]

对此,一个补偿方法是用当前估计的分母 D 的倒数给每个方程加权

$$\frac{1}{D}\begin{bmatrix} \hat{x}' - x \\ \hat{y}' - y \end{bmatrix} = \frac{1}{D}\begin{bmatrix} x & y & 1 & 0 & 0 & 0 & -\hat{x}'x & -\hat{x}'y \\ 0 & 0 & 0 & x & y & 1 & -\hat{y}'x & -\hat{y}'y \end{bmatrix} \begin{bmatrix} h_{00} \\ \vdots \\ h_{21} \end{bmatrix}. \tag{6.22}$$

初看起来这个和方程组(6.21)是完全一样的,因为用的是最小二乘来求解超定方程组,加权确实是不一样的,加权能够产生不同的正规方程组,在实际中表现更好。

做估计时,一个最基本的方法是用 Gauss–Newton 近似法直接最小化残差方程组(6.13),也就是对 p 做一阶泰勒展开,如公式(6.14)所示,得到方程组

$$\begin{bmatrix} \hat{x}' - \tilde{x}' \\ \hat{y}' - \tilde{y}' \end{bmatrix} = \frac{1}{D}\begin{bmatrix} x & y & 1 & 0 & 0 & 0 & -\tilde{x}'x & -\tilde{x}'y \\ 0 & 0 & 0 & x & y & 1 & -\tilde{y}'x & -\tilde{y}'y \end{bmatrix} \begin{bmatrix} \Delta h_{00} \\ \vdots \\ \Delta h_{21} \end{bmatrix}. \tag{6.23}$$

这与公式(6.22)看起来很像,但是在两个重要方面存在差异。首先,公式左边是由未加权的预测误差组成,而不是点的位移,解向量是参数向量 p 的扰动。其

[①] Hartley and Zisserman (2004) 把从有理方程组形成线性方程组的策略称为"直接线性变换"(direct linear transform),但这个名词在更多时候与姿态估计相关联(6.2 节)。也要注意,我们定义的 h_{ij} 参数与其书中的定义不同,因为我们定义 h_{ii} 是和单位矩阵的差,没有让 h_{22} 成为自由参数,这意味着我们不能处理某些极端的单应。

次，J 内部的数量涉及预测的特征位置 $(\tilde{x}', \tilde{y}')$ 而不是观察到的特征位置 $(\hat{x}', \hat{y}')$。虽然这两个差异都很细微，但是都导向一个算法，当与合适的下山步长核查相结合时（和 Levenberg–Marquardt 算法一样），将会收敛到一个局部最小值。注意，迭代方程组(6.22)不一定保证收敛，因为它不是在最小化一个定义明确的能量函数。

方程(6.23)类似于用于直接基于亮度注册的加性算法(8.2 节)，因为计算的是全部变换的变化。如果我们在当前单应的前面添加一个增量的单应，也就是说，用一个**合成**算法(在 8.2 节描述的)得到 $D = 1$ (由于 $p = 0$)，前面的公式简化为

$$\begin{bmatrix} \hat{x}' - x \\ \hat{y}' - y \end{bmatrix} = \begin{bmatrix} x & y & 1 & 0 & 0 & 0 & -x^2 & -xy \\ 0 & 0 & 0 & x & y & 1 & -xy & -y^2 \end{bmatrix} \begin{bmatrix} \Delta h_{00} \\ \vdots \\ \Delta h_{21} \end{bmatrix}, \tag{6.24}$$

其中为了简明，我们把 $(\tilde{x}', \tilde{y}')$ 代替为 (x,y)。(注意这是如何得到与(8.63)中雅克比矩阵一样结果的。)

6.1.4 鲁棒最小二乘和 RANSAC

常规的最小二乘对其中噪声符合正态(高斯)分布的量测来说是一个合适的选择，然而在对应点中有外点时(几乎总是有的)，需要更鲁棒的最小二乘。在这种情况下，最好用 M-估计(Huber 1981; Hampel, Ronchetti, Rousseeuw et al. 1986; Black and Rangarajan 1996; Stewart 1999)，它对残差施加一个鲁棒惩罚函数 $\rho(r)$

$$E_{\text{RLS}}(\Delta p) = \sum_i \rho(\|r_i\|) \tag{6.25}$$

来代替它们的平方。

相对于 p 对这个函数求偏导，并令其为 0，有

$$\sum_i \psi(\|r_i\|) \frac{\partial \|r_i\|}{\partial p} = \sum_i \frac{\psi(\|r_i\|)}{\|r_i\|} r_i^T \frac{\partial r_i}{\partial p} = 0, \tag{6.26}$$

其中 $\psi(r) = \rho'(r)$ 是 ρ 的倒数，称为影响函数(influence function)。如果引入加权函数(weight function) $w(r) = \psi(r)/r$，可以看出用(6.26)求解(6.25)的不动点等价于最小化迭代重加权最小二乘(iteratively reweighted least squares，IRLS))问题

$$E_{\text{IRLS}} = \sum_i w(\|r_i\|) \|r_i\|^2, \tag{6.27}$$

其中 $\omega(\|r_i\|)$ 和(6.10)中 σ_i^{-2} 都起着相同的局部加权作用。IRLS 算法在计算影响函数 $\omega(\|r_i\|)$ 和求解得到的加权最小二乘问题(固定的 ω 值)之间交替。其他增量的鲁棒最小二乘算法可以在如下工作中找到 Sawhney and Ayer(1996); Black and Anandan (1996); Black and Rangarajan(1996); Baker, Gross, Ishikawa et al. (2003)，也出现在鲁棒统计学方面的教材和辅导资料中(Huber 1981; Hampel, Ronchetti, Rousseeuw et al. 1986; Rousseeuw and Leroy 1987; Stewart 1999)。

M-估计一定能够减少外点的影响，但是在一些情况中，从太多外点起步的话会使得 IRLS(或者其他梯度下降算法)不能收敛到全局最优。一个时常更好的方法是

寻找一个起步的对应的内点集合，也就是与主运动估计一致的点[①]。

针对这个问题，两个常用的方法是 RANdom SAmple Consensus 或简称为 RANSAC(Fischler and Bolles 1981)和最小中位方差(least median of squares, LMS)(Rousseeuw 1984)。这两个方法都是先(随机的)选择 k 个对应点子集，然后计算初始的估计 p。再计算对应点全集的残差为

$$r_i = \tilde{x}'_i(x_i; p) - \hat{x}'_i, \tag{6.28}$$

其中 $\tilde{x}'_i$ 是估计的(映射的)位置，$\hat{x}'_i$ 是测量的(检测的)特征点的位置。

而后 RANSAC 方法计算内点(inlier)的数目，这里内点即为在预测位置距离 ε 以内的点，也就说 $\|r_i\| \leq \varepsilon$(ε 值和应用相关，通常为 1 至 3 个像素)。最小中位方差是找 $\|r_i\|^2$ 值的中值。随机选择过程重复 S 次，含有最多内点(或者具有最小中值残差)的样本集被作为最终的解。初始参数 p 或者计算出的内点的全部集合然后传到下一步的数据拟合步骤中。

当观测值很多时，最好是在初始阶段只评测观测值的一个子集，从这个子集中选择最可能的假设进行附加的评测和选择。这样改进的 RANSAC，能够大大提升其性能，称为抢先 RANSAC(Preemptive RANSAC) (Nistér 2003)。RANSAC 的另一个变种是 PROSAC(PROgressive SAmple Consensus)，在开始的时候从最"可信"的匹配中挑选样本，从而能加快寻找(统计上)可能好的内点集合的过程(Chum and Matas 2005)。

为了确保随机采样有较好的机会找到真正的内点集合，必须试验足够多的次数 S。令 p 为任意给定对应点合法的概率，P 是经过 S 次试验后成功的总体的概率。在某次试验中，k 个随机样本都是内点的可能性是 p^k。因此，S 次试验失败的可能性是

$$1 - P = (1 - p^k)^S \tag{6.29}$$

最少需要试验的次数是

$$S = \frac{\log(1-P)}{\log(1-p^k)}. \tag{6.30}$$

Stewart(1999)给出了达到 99%成功概率需要的试验次数 S 的例子。正如你可以从表 6.2 看出的那样，随着使用的样本点数的增加，试验次数增长得很快。这为在任意给定的试验中使用最少数目的样本点数 k 提供了很强的诱因，它也正是在实践中通常如何使用 RANSAC 的。

表 6.2 达到 99%成功概率需要的试验次数 S (Stewart 1999)

k	p	S
3	0.5	35
6	0.6	97
6	0.5	293

[①] 对于基于像素的配准方法(8.1.1 节)，分层(由粗到精)方法常用于锁定场景的主运动。

不确定性建模

除了鲁棒地计算好的配准之外，一些应用需要计算不确定性(见附录 B.6)。对于线性问题，可以通过求 Hessian 矩阵(6.9)的逆，然后用特征位置的噪声与它相乘得到(如果它们还没有被用于对各个测量加权)，如方程组(6.10)和(6.11)所示。在统计中，Hessian 是协方差矩阵的逆，有时也叫"Fisher 信息矩阵"(information matrix)(参见附录 B.1.1)。

当问题涉及非线性最小二乘时，Hessian 矩阵的逆提供了协方差矩阵的 Cramer–Rao 下界(Cramer–Rao lower bound)，即在给定解中它提供了最小协方差，如果能量从发现最优解的局部最小位置开始变平，那么它实际上可以有更广的延伸("长尾"(longer tail)。

6.1.5 3D 配准

不同于 2D 图像特征配准，很多计算机视觉应用需要配准 3D 点。在 3D 变换对运动参数是线性的情况下，比如对于平移变换、相似变换和仿射变换，可以使用常规的最小二乘(6.5)。

刚性(欧式)运动的情况，

$$E_{\text{R3D}} = \sum_i \|x'_i - Rx_i - t\|^2, \tag{6.31}$$

这是出现得比较频繁的问题，常称为"绝对方向"(absolute orientation)问题(Horn 1987)，需要稍微不同的方法。如果只用标量加权(与 3D 每个点各向异性协方差估计相对比)，两个点云 c 和 c' 的加权的质心可以用于估计平移 $t=c'-Rc$[①]。接下来的问题是从两个点的集合 $\{\hat{x}_i = x_i - c\}$ 和 $\{\hat{x}'_i = x'_i - c'\}$ 估计旋转，这个两个集合都是以原点为中心的。

一个常用的方法是正交 Procrustes 算法(Orthogonal Procrustes algorithm)(Golub and Van Loan 1996, p.601)，它计算 3×3 相关矩阵的 SVD 分解(singular value decomposition)

$$C = \sum_i \hat{x}' \hat{x}^T = U\Sigma V^T. \tag{6.32}$$

然后得到旋转矩阵 $R = UV^T$。(当 $\hat{x}' = R\hat{x}$ 时，你可以自己验证这一点)。

另外一个方法是绝对方向算法(absolute orientation algorithm)(Horn 1987)，估计对应与旋转矩阵 R 的单位四元组，即从 C 的项中形成 4×4 的矩阵，然后找到具有最大正特征值的特征向量。

Lorusso, Eggert, and Fisher (1995)用实验的方法将这两个方法与文献中的另外两种方法进行比较，但发现它们在精度方面的差异可以忽略不计(远低于测量噪声影响)。

[①] 使用完整协方差时，它们做的是旋转变换，因此闭合形式的平移解是不可能的。

在这些闭合形式算法不适用的情况下，比如使用完整 3D 协方差或者 3D 配准是一个更大优化问题的一部分时，可以使用 2.1.4 节(2.35 和 2.36)介绍的增量旋转更新，它是用即时旋转向量 ω 参数化的(在图像拼接中的应用可参见 9.1.3 节)。

在一些情况下，比如融合距离数据图(range data maps)时，数据点之间的对应是事先已知的。在这种情况下，使用的迭代算法可以先匹配邻近的点，然后更新最可能的对应点(Besl and McKay 1992; Zhang 1994; Szeliski and Lavallée 1996; Gold, Rangarajan, Lu et al. 1998; David, DeMenthon, Duraiswami et al. 2004; Li and Hartley 2007; Enqvist, Josephson, and Kahl 2009)。这些方法将在 12.2.1 节进一步讨论。

6.2 姿态估计

一个经常出现的特殊的基于特征的配准实例是从一组 2D 点的映射中估计物体的 3D 姿态。姿态估计问题也称为"外参数标定"(extrinsic calibration)，相对于 6.3 节讨论的摄像机内参数(比如焦距)的内标定过程。从三个对应点中恢复姿态，需要的信息是最少的，称为"透视三点问题"(perspective-3-point-problem，P3P)，扩展到更多的点，合起来称为"PnP"(Haralick, Lee, Ottenberg et al. 1994; Quan and Lan 1999; Moreno-Noguer, Lepetit, and Fua 2007)。

在这一节，我们看看解决这类问题的一些方法，首先是直接线性变换(direct linear transform，DLT)，它恢复 3×4 的摄像机矩阵，然后是其他的"线性"算法，最后我们看看统计上最优的迭代算法。

6.2.1 线性算法

类似于从透视投影的摄像机矩阵(2.55 和 2.56)来估计 2D 运动(6.19)的情形，恢复摄像机姿态最简单的方法是形成一组线性方程组。

$$x_i = \frac{p_{00}X_i + p_{01}Y_i + p_{02}Z_i + p_{03}}{p_{20}X_i + p_{21}Y_i + p_{22}Z_i + p_{23}} \tag{6.33}$$

$$y_i = \frac{p_{10}X_i + p_{11}Y_i + p_{12}Z_i + p_{13}}{p_{20}X_i + p_{21}Y_i + p_{22}Z_i + p_{23}}, \tag{6.34}$$

其中 (x_i, y_i) 是测量的 2D 特征位置，(X_i, Y_i, Z_i) 是已知的 3D 特征位置(图 6.4)。与(6.21)一样，这个方程组可以通过给公式两边同乘以这个分母，用一个线性方式求解出摄像机矩阵 P 中的未知量[①]。得到的算法称为"直接线性变换"(direct linear transform，DLT)，通常归功于 Sutherland(1974)。(更进一步的讨论参见 Hartley and Zisserman (2004))的工作。为了计算 P 中 12 个或者 11 个未知量，至少需要知道 3D 和 2D 位置间的 6 个对应点。

[①] 因为 P 是取决于尺度的未知量，所以我们固定一项，比如 $p_{23}=1$，或者找到线性方程组最小的单个向量。

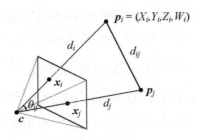

图 6.4 用直接线性变换和测量点对之间的可视角度和距离的方法进行姿态估计

在估计单应的情况中(6.21~6.23)，P 中各项更精确的结果可以用非线性最小二乘直接最小化(6.33 和 6.34)方程组经过少量迭代得到。

一旦恢复 P 的各项，观察方程(2.56)，就可能同时恢复内标定矩阵 K 和刚性变换$(R，t)$

$$P = K[R|t]. \tag{6.35}$$

由于 K 一般是上三角矩阵(参见 2.1.5 节的讨论)，K 和 R 可以用 P 前面的 3×3 子矩阵的 RQ 因子分解得到(Golub and Van Loan 1996)[①]。

然而在很多应用中，我们对于内参数标定矩阵 K 有一些先验信息，比如像素是正方形的，倾斜角度很小以及光心基本在图像的中心(2.57~2.59)。这些限制可以结合在 K 和$(R;t)$中参数的非线性最小化中，正如 6.2.2 节所描述的那样。

在摄像机已经标定的情况下，比如矩阵 K 是已知的(6.3 节)，我们可以用少到仅 3 个点来估计姿态(Fischler and Bolles 1981; Haralick, Lee, Ottenberg *et al.* 1994; Quan and Lan 1999)。基本观察结果是，这些线性 PnP(perspective n-point)算法的依据是，任意的 2D 点对 $\hat{x}_i$ 和 $\hat{x}_j$ 之间的可视角度必须和其对应的 3D 点对 p_i 和 p_j 之间的角度一样(图 6.4)。

给定一组 2D 和 3D 的对应点 $\{(\hat{x}_i, p_i)\}$，其中 $\hat{x}_i$ 是单位方向矢量，它是通过标定矩阵 K 的逆将 2D 像素测量值 x_i 变换为单位模长的 3D 方向矢量得到的 $\hat{x}_i$：

$$\hat{x}_i = \mathcal{N}(K^{-1}x_i) = K^{-1}x_i/\|K^{-1}x_i\|, \tag{6.36}$$

未知量是从摄像机中心 c 到 3D 点 p_i 的距离 d_i，其中

$$p_i = d_i\hat{x}_i + c \tag{6.37}$$

(图 6.4)。三角形 $\Delta(c, p_i, p_j)$ 的余弦定律告诉我们

$$f_{ij}(d_i, d_j) = d_i^2 + d_j^2 - 2d_id_jc_{ij} - d_{ij}^2 = 0, \tag{6.38}$$

其中

$$c_{ij} = \cos\theta_{ij} = \hat{x}_i \cdot \hat{x}_j \tag{6.39}$$

以及

$$d_{ij}^2 = \|p_i - p_j\|^2. \tag{6.40}$$

对于任意的三元组约束 (f_{ij}, f_{ik}, f_{jk})，用西尔维斯特法则(Sylvester resultant)(Cox,

---

[①] 不幸的是，有术语冲突：在矩阵代数教材中，R 代表上三角矩阵；在计算机视觉中，R 是正交旋转矩阵。

Little, and O'Shea 2007)消除 d_j 和 d_k，得到 d_i^2 的四次方程

$$g_{ijk}(d_i^2) = a_4 d_i^8 + a_3 d_i^6 + a_2 d_i^4 + a_1 d_i^2 + a_0 = 0. \tag{6.41}$$

给定五个或者更多的对应点，我们可以产生 $\frac{(n-1)(n-2)}{2}$ 个三元组来得到 $(d_i^8, d_i^6, d_i^4, d_i^2)$ 的值的线性估计(用 SVD)(Quan and Lan 1999)。d_i^2 可以用连续两个估计的比率 d_i^{2n+2}/d_i^{2n} 来估计，然后可以把这些值平均来获得最终的估计 d_i^2 (因此也就得到 d_i)。

一旦每个 d_i 距离的估计计算出来后，我们能够产生由缩放的点方向 $d_i \hat{x}_i$ 组成的 3D 结构，然后它可以用绝对方向(6.1.5 节)与 3D 点云 $\{p_i\}$ 配准，得到期望的姿态估计。Quan and Lan(1999)给出了这个和其他方法的精确结果，使用更少的点但是需要更复杂的代数运算。Moreno-Noguer, Lepetit, and Fua(2007)回顾了更多新近的其他方法，也给出了一个具有较低复杂度的一般能产生更精确结果的算法。

不幸的是，因为最小 PnP 解可能对噪声很敏感，而且还有浅浮雕歧义(即深度逆转)(7.4.3 节)，所以通常最好用线性的六点算法估计初始姿态，然后用 6.2.2 节介绍的迭代方法对该估计进行优化。

另一个姿态估计的算法涉及从缩放的正交投影模型起步，然后用更精确的透视投影模型迭代地改进这个初始估计(DeMenthon and Davis 1995)。这个模型吸引人的是，正如其论文题目所陈述的那样，可以"在 25 行(Mathematica)代码内"实现。

6.2.2 迭代算法

最精确(灵活)的估计姿态的方法是直接最小化 2D 点上的平方(或鲁棒)重投影误差，将其看作是在 $(\boldsymbol{R}, \boldsymbol{t})$ 和可选的 $\boldsymbol{K}$ 中的未知姿态参数的函数，用非线性最小二乘求解(Tsai 1987; Bogart 1991; Gleicher and Witkin 1992)。我们可以将投影方程写作

$$\boldsymbol{x}_i = \boldsymbol{f}(\boldsymbol{p}_i; \boldsymbol{R}, \boldsymbol{t}, \boldsymbol{K}) \tag{6.42}$$

并迭代地最小化鲁棒线性化(robustified linearized)的重投影误差

$$E_{\text{NLP}} = \sum_i \rho \left(\frac{\partial \boldsymbol{f}}{\partial \boldsymbol{R}} \Delta \boldsymbol{R} + \frac{\partial \boldsymbol{f}}{\partial \boldsymbol{t}} \Delta \boldsymbol{t} + \frac{\partial \boldsymbol{f}}{\partial \boldsymbol{K}} \Delta \boldsymbol{K} - \boldsymbol{r}_i \right), \tag{6.43}$$

其中 $r_i = \tilde{x}_i - \hat{x}_i$ 是当前的残差向量(预测位置中的 2D 误差)，偏导是相对于未知姿态参数(旋转、平移和可选的标定)的。注意，如果可以得到 2D 特征位置的完整 2D 协方差估计，那么前面模的平方可以用逆点协方差矩阵加权，就像公式(6.11)那样。

前面非线性回归问题的一个用重写更容易理解的版本是将投影方程组重写为一系列连续的更简单的步骤，每一步把一个 4D 齐次坐标 p_i 变为简单的变换，比如旋转，透视分解(图 6.5)。最后投影方程组可以写作

$$\boldsymbol{y}^{(1)} = \boldsymbol{f}_{\text{T}}(\boldsymbol{p}_i; \boldsymbol{c}_j) = \boldsymbol{p}_i - \boldsymbol{c}_j, \tag{6.44}$$

$$\boldsymbol{y}^{(2)} = \boldsymbol{f}_{\text{R}}(\boldsymbol{y}^{(1)}; \boldsymbol{q}_j) = \boldsymbol{R}(\boldsymbol{q}_j)\, \boldsymbol{y}^{(1)}, \tag{6.45}$$

$$\boldsymbol{y}^{(3)} = \boldsymbol{f}_{\text{P}}(\boldsymbol{y}^{(2)}) = \frac{\boldsymbol{y}^{(2)}}{z^{(2)}}, \tag{6.46}$$

$$x_i = f_C(y^{(3)}; k) = K(k)\, y^{(3)}. \tag{6.47}$$

注意，在这些方程中，我们用下标 j 索引摄像机中心 c_j 和摄像机旋转四元组 q_j，以应对要用到标定物体多个姿态的情况(参见 7.4 节)。我们也用摄像机中心 c_j 而不是世界坐标系平移 t_j，因为这是更自然的要估计的参数。

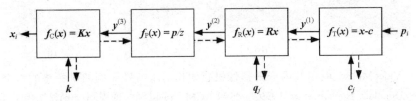

图 6.5 通过一系列的变换 $f^{(k)}$ 把 3D 点 p_i 映射到 2D 观测 x_i，其中每一个变换都是用自己的参数控制的。虚线表示在逆推计算偏导时的信息的流动

这种链式变换的优点是，每一个变换相对于自己的参数和输入都有简单的偏导。因此，一旦基于 3D 位置 p_i 和当前姿态参数 (c_j, q_j, k) 的预测值 $\tilde{x}_i$ 已经计算出来后，我们用链式法则可以得到所有需要的偏导

$$\frac{\partial r_i}{\partial p^{(k)}} = \frac{\partial r_i}{\partial y^{(k)}} \frac{\partial y^{(k)}}{\partial p^{(k)}}, \tag{6.48}$$

其中 $p^{(k)}$ 代表要优化的参数向量中的一个。(同样的"技巧"也用在神经网中，作为后向传播算法的一部分(Bishop 2006))。

在这个形式化下，一个可以明显简化的特殊例子是旋转更新的计算。与其把 3×3 旋转矩阵 $R(q)$ 的偏导直接当作单位四元组项的函数进行计算，你可以把公式 (2.35) 所给的增量旋转矩阵 $\Delta R(w)$ 添加到当前旋转矩阵之前，然后相对于这些参数计算变换的偏导，这样得到后向链式偏导(backward chaining partial derivative)和向外 3D 向量(outgoing 3D vector)(2.36)的简单叉积。

6.2.3 应用：增强现实

姿态估计的一个广泛应用是增强现实，其中虚拟的 3D 图像或者标注被重叠在一个实况视频订阅源之上，其实现不是通过使用透视眼镜(头戴式显示器)就是通过使用常规的电脑或者移动设备的屏幕(Azuma, Baillot, Behringer *et al.* 2001; Haller, Billinghurst, and Thomas 2007)。在一些应用中，跟踪印在卡片或者书上的特殊模式来实现增强现实(Kato, Billinghurst, Poupyrev *et al.* 2000; Billinghurst, Kato, and Poupyrev 2001)。在桌面应用中，印在鼠标垫上的网格点可以用嵌在增强鼠标中的摄像机来跟踪，以便为用户提供在 3D 空间中对于位置和角度的所有六个自由度的控制(Hinckley, Sinclair, Hanson *et al.* 1999)，如图 6.6 所示。

有时，场景本身会提供一个便于跟踪的对象，比如在通过镜头的摄像机控制(through-the-lens camera control)中所用的定义桌面的矩形(Gleicher and Witkin 1992)。在室外位置中，比如影片布景，更常用的方式是在场景中放一个特殊的标记(比如发光的彩球)，以便更容易发现并跟踪它们(Bogart 1991)。在更老的应用中，在拍摄之前使用测量方法来确定这些球的位置。现在，更常用的是将由运动到结构方法直接用于电影连续镜头本身(7.4.2 节)。

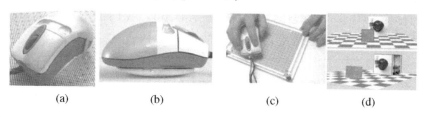

图 6.6 VideoMouse 能用其内嵌的摄像机感知相对于特殊印刷的鼠标垫的六个自由度(Hinckley, Sinclair, Hanson *et al.* 1999)© 1999 ACM：(a)鼠标的顶视图；(b)显示用于摇晃的曲面基底的鼠标视图；(c)用另一只手移动鼠标垫扩展交互能力；(d)在显示器上看到的移动结果

对任天堂 Wii 游戏机使用的手柄的位置和角度进行跟踪，最核心的任务便是快速的姿态估计。嵌入游戏手柄的高速摄像机，用于跟踪红外发光二极管(LED)的位置，该发光二极管条棒安装在电视机顶部或旁边。然后以很高的帧率使用姿态估计推断远程控制的位置和角度。正如 Johnny Lee[①]所描述的，可以把条棒安装在手持设备上，把 Wii 系统扩展到很多其他用户交互应用中。

习题 6.4 和 6.5 让你为增强现实应用实现两种不同的跟踪和姿态估计系统。第一个系统跟踪矩形物体的外轮廓，比如书的封面或者杂志的一页。第二个系统跟踪手持魔方的姿态。

6.3 几何内参数标定

正如前面公式(6.42 和 6.43)所描述的，相对于已知的标定对象，摄像机内标定参数的计算和摄像机的(外部的)姿态估计可以同时出现。这实际上是在摄影测量学(Slama 1980)和计算机视觉(Tsai 1987)两个领域中所使用的"经典的"摄像机标定方法。在这一节，我们考虑另外的形式(这可能不涉及非线性回归问题的完整解)、其他标定对象的使用和摄像机光学非线性部分的估计，比如径向畸变[②]。

6.3.1 标定模式

使用一个标定模式或者标记集合是更可靠的估计摄像机内参数的方法之一。在摄影测量学中，经常在很宽阔的田野架一个摄像机来观察远处的标定目标，其真实位置已经用观测工具事先计算了(Slama 1980; Atkinson 1996; Kraus 1997)。在这种情况下，姿态的平移部分变得不相关，只有摄像机的旋转参数和内参数需要恢复。

---

① *http://johnnylee.net/projects/wii/*
② 在一些应用中，你可以用和 JPEG 图像关联的 EXIF 标签得到摄像机焦距的粗略估计，但是得小心使用这个方法，因为其结果经常不精确。

如果需要使用较小的标定装备(calibration rig)，比如对于室内机器人应用或对于携带它们自己的标定对象的移动机器人来说，最好是标定物体可以尽可能地跨越其工作区域，因为平面状目标经常无法精确预测远离该平面的姿态的成分(图 6.8a)。判定是否成功进行标定的一个好方法是估计参数的协方差矩阵(6.1.4 节)，然后把工作区域内不同的 3D 点映射到图像中以估计其 2D 位置上的不确定性。

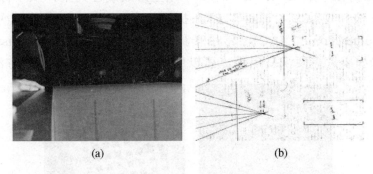

图 6.7 在厚纸板上画直线来标定透镜(Debevec, Wenger, Tchou *et al.* 2002) © 2002 ACM：(a)用视频摄像机拍摄的图像，显示一只握着金属尺子的手，尺子的右边界看起来在图像中是垂直的；(b)画在硬纸板上线的集合收敛到透镜的前节点(投影中心)，指明水平视野

另一个估计焦距和透镜投影中心的方法是把摄像机放在一张大型的平纸板上，用一个大的金属尺子在纸板上画一些看起来在图像中垂直的直线，如图 7.6a 所示(Debevec, Wenger, Tchou *et al.* 2002)。这些线在平面上平行于摄像机传感器的垂直轴方向，同时也通过透镜的前节点。节点的位置(垂直映射到纸板平面)和水平视野(由处于可视图像左右边缘的线决定)可以通过把这些线相交然后测量其角度范围得到(图 6.7b)。

如果没有可用的标定模式，也可以在结构和姿态恢复的同时进行标定(6.3.4 节和 7.4 节)，常称为"自标定"(Faugeras, Luong, and Maybank 1992; Hartley and Zisserman 2004; Moons, Van Gool, and Vergauwen 2010)。然而，这样的方法需要大量精确的图像。

平面标定模式

当使用的工作区有限且有精确的机械和运动控制平台时，进行标定的一个好的方法是在工作区空间域内以可控的方式移动平面标定目标。该方法有时称为"N-平面标定方法"(Gremban, Thorpe, and Kanade 1988; Champleboux, Lavallée, Szeliski *et al.* 1992; Grossberg and Nayar 2001)，其优点是摄像机的每个像素可以映射到空间中唯一的 3D 射线，这既考虑了由标定矩阵 K 建模的线性影响，也考虑了诸如径向畸变这样的非线性影响(6.3.5 节)。

一个不那么笨重但也不太精确的标定可以通过在摄像机前挥动平面标定模式得到(图 6.8b)。在这种情况下，模式的姿态(在原理上)必须和内参一道进行恢复。在这个方法里，每个输入的图像用于计算一个分开的单应(6.19~6.23) $\tilde{H}$，把平面

的标定点 $(X_i, Y_i, 0)$ 映射到图像点 (x_i, y_i)，

$$\boldsymbol{x}_i = \begin{bmatrix} x_i \\ y_i \\ 1 \end{bmatrix} \sim \boldsymbol{K} \begin{bmatrix} \boldsymbol{r}_0 & \boldsymbol{r}_1 & \boldsymbol{t} \end{bmatrix} \begin{bmatrix} X_i \\ Y_i \\ 1 \end{bmatrix} \sim \tilde{\boldsymbol{H}} \boldsymbol{p}_i, \tag{6.49}$$

其中 r_i 是 $\boldsymbol{R}$ 的前两列，~表明在相差一个尺度意义上是相等的。根据这些，Zhang (2000)给出了如何在矩阵 $\boldsymbol{B} = \boldsymbol{K}^{-T}\boldsymbol{K}^{-1}$ 的 9 个项上形成线性约束，从中标定矩阵 $\boldsymbol{K}$ 可以用矩阵平方根和逆恢复。(矩阵 $\boldsymbol{B}$ 在摄影几何中称为"绝对二次曲线的像"(image of absolute conic，IAC)，常用于摄像机标定(Hartley and Zisserman 2004, Section 7.5)。)如果只需要恢复焦距，可以用使用消失点的更简单的方法。

图 6.8 标定模式：(a)三维目标(Quan and Lan 1999)©1999 IEEE；(b)二维目标(Zhang 2000)©2000 IEEE。注意，在特征点用于标定之前需要去除这些图像中的径向畸变

6.3.2 消失点

在实际中经常发生的常见的标定情况是，摄像机观看人造的场景，其中有大幅扩展的矩形物体，比如盒子和房屋墙壁等。在这个情况下，我们可以把对应于 3D 平行线的 2D 线相交，计算它们的消失点，如 4.3.3 节所述，用它们来确定内外标定参数(Caprile and Torre 1990; Becker and Bove 1995; Liebowitz and Zisserman 1998; Cipolla, Drummond, and Robertson 1999; Antone and Teller 2002; Criminisi, Reid, and Zisserman 2000; Hartley and Zisserman 2004; Pflugfelder 2008)。

假设我们已经检测出两个或者多个正交的消失点，它们都是有限的，即它们不是从在图像平面中看起来平行的线得到的(图 6.9a)。我们还假设标定矩阵 $\boldsymbol{K}$ 的一个简化形式，其中只有焦距未知(2.59)。(对于粗略的 3D 建模来讲，假设光心在图像的中心、横纵比率为 1 且没有倾斜，这常常是安全的)。在这个情况下，消失点的投影方程可以写作

$$\hat{\boldsymbol{x}}_i = \begin{bmatrix} x_i - c_x \\ y_i - c_y \\ f \end{bmatrix} \sim \boldsymbol{R}\boldsymbol{p}_i = \boldsymbol{r}_i, \tag{6.50}$$

其中 p_i 对应其中一个基轴方向 $(1,0,0)$，$(0,1,0)$ 或者 $(0,0,1)$，r_i 是旋转矩阵 $\boldsymbol{R}$ 的第 i 列。

旋转矩阵列与列之间是正交的，我们由此得到

$$\boldsymbol{r}_i \cdot \boldsymbol{r}_j \sim (x_i - c_x)(x_j - c_y) + (y_i - c_y)(y_j - c_y) + f^2 = 0 \tag{6.51}$$

从中，我们可以得到 f^2 的估计。注意，随着消失点逐渐移近图像中心，估计的精

确度也在增加。

换句话说，最好是把标定模式以在 45°附近的合适角度倾斜，如图 6.9a 所示。一旦确定焦距 f，R 的每一列可以通过归一化公式(6.50)的左边然后用叉乘得到。另一个方法是用初始 R 估计的 SVD，它是正交 Procrustes (6.32)的一个变种。

如果在同一图像中，所有三个消失点都可见且有限，我们可以用三个消失点形成的三角形的垂心来估计光心(Caprile and Torre 1990; Hartley and Zisserman 2004, Section 7.6)(图 6.9b)。然而实际上，更精确的方法是用非线性最小二乘(6.42)重新估计任何未知的内标定参数。

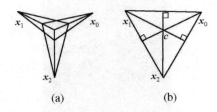

图 6.9 从消失点标定：(a)有限的消失点对 $(\hat{x}_i, \hat{x}_j)$ 可以用于估计焦距；(b)消失点三角形的垂心给出了图像光学中心 c

图 6.10 单视图测量学(Criminisi, Reid, and Zisserman 2000)©2000 Springer：(a)显示了用两个水平消失点计算的三个坐标轴(可以用小屋的侧边来确定)的图像；(b)3D 重建的新视角

6.3.3 应用：单视图测量学

消失点估计和摄像机标定的一个有趣的应用是 Criminisi, Reid, and Zisserman (2000)提出的单视图测量学系统。他们的系统允许人们交互测量高度或者其他维度，建立分片的平面 3D 模型，如图 6.10 所示。

他们的系统中，第一步是识别地面上两个正交的消失点和垂直方向上的消失点(可以在图像中画一些平行线得到)。(另一个方法使用自动的方法，比如 4.3.3 节讨论的或者使用 Schaffalitzky and Zisserman(2000)提出的)。然后用户在图像中标记几个维度，比如参照物体的高度，系统就能自动计算另一个物体的高度。墙和其他平面构件(planar impostor)(几何的)也可以粗略地描绘出来并重建。

在 Criminisi, Reid, and Zisserman(2000)最早的形式化中，系统产生了一个仿射

重建，即在相差一组独立的、沿每个轴的缩放因子的意义下确定的重建。假设摄像机在相差一个未知焦距意义下是标定好的，这样的标定可以从正交的(有限)消失方向恢复，正如 6.3.2 节所述，可以构建出一个潜在用处更大的系统。一旦这步完成，用户便可以在地面上指出一个原点和另一个已知到原点距离的点。由此，地面上的点可以直接映射到 3D，地面上方的点(与其地面上的投影形成一对的时候)也可以恢复。由此便能够实现全尺寸的重建。

图 6.11 手持摄像机拍摄的四张图像，用 3D 旋转运动模型注册，可以用于估计摄像机焦距。(Szeliski and Shum 1997)© 2000 ACM

习题 6.9 让你实现一个这样的系统，然后用它来建模一些简单的 3D 场景。12.6.1 节描述了其他对建筑物重建的可能是多视图的方法，包括使用消失点来建立 3D 线的方向和平面法向的交互式分片平面建模系统(Sinha, Steedly, Szeliski et al. 2008)。

6.3.4 旋转运动

当没有标定目标或者已知的结构可用但可以绕着前节点旋转摄像机(或差不多的情形，工作在一个大的空旷场景中，所有物体都距离较远)时，摄像机可以从一组假定在做纯旋转运动的重叠图像中标定，如图 6.11 所示(Stein 1995; Hartley 1997b; Hartley, Hayman, de Agapito et al. 2000; de Agapito, Hayman, and Reid 2001; Kang and Weiss 1999; Shum and Szeliski 2000; Frahm and Koch 2003)。如果在这个标定中使用了完整的 360° 运动，就可以获得焦距 f 的一个非常精确的估计，因为这个估计的精度和最终产生的柱面全景图的像素总数成正比(9.1.6 节) (Stein 1995; Shum and Szeliski 2000)。

为了使用这个方法，我们首先计算在所有重叠的图像对之间的单应 $\hat{H}_{ij}$，如公式(6.19～6.23)所描述。然后，我们使用如下的观察：每个单应通过(未知的)标定矩阵 K_i 和 K_j 与摄像机间的旋转矩阵 R_{ij} 相关联，这个观察首先出现在公式(2.72)中，然后在 9.1.3 节的公式(9.5)中给出了更多的细节，

$$\tilde{H}_{ij} = K_i R_i R_j^{-1} K_j^{-1} = K_i R_{ij} K_j^{-1}. \tag{6.52}$$

获得标定最简单的方法是使用标定矩阵(2.59)的简化形式，其中我们假设像素

是正方形的，光心在图像中心，即 $K_K = \text{diag}(f_k, f_k, 1)$。我们相应地计数像素的坐标，把像素 $(x, y) = (0, 0)$ 置于图像中心。然后我们可以重写公式(6.52)为

$$R_{10} \sim K_1^{-1} \tilde{H}_{10} K_0 \sim \begin{bmatrix} h_{00} & h_{01} & f_0^{-1} h_{02} \\ h_{10} & h_{11} & f_0^{-1} h_{12} \\ f_1 h_{20} & f_1 h_{21} & f_0^{-1} f_1 h_{22} \end{bmatrix}, \quad (6.53)$$

其中 h_{ij} 是 $\hat{H}_{10}$ 的元素。

利用旋转矩阵 R_{10} 的正交性和已知公式(6.53)右边只相差一个尺度量的事实，我们得到

$$h_{00}^2 + h_{01}^2 + f_0^{-2} h_{02}^2 = h_{10}^2 + h_{11}^2 + f_0^{-2} h_{12}^2 \quad (6.54)$$

和

$$h_{00} h_{10} + h_{01} h_{11} + f_0^{-2} h_{02} h_{12} = 0. \quad (6.55)$$

由此，我们可以算出

$$f_0^2 = \frac{h_{12}^2 - h_{02}^2}{h_{00}^2 + h_{01}^2 - h_{10}^2 - h_{11}^2} \quad \text{if } h_{00}^2 + h_{01}^2 \neq h_{10}^2 + h_{11}^2 \quad (6.56)$$

或者

$$f_0^2 = -\frac{h_{02} h_{12}}{h_{00} h_{10} + h_{01} h_{11}} \quad \text{if } h_{00} h_{10} \neq -h_{01} h_{11}. \quad (6.57)$$

的 f_0 的估计值(注意，Szeliski and Shum(1997)最早给出的公式有误，正确的公式由 Shum and Szeliski(2000)给出。)如果这两种情况都不满足，我们还可以计算第一行(或者第二行)和第三行的点积。通过分析 $\hat{H}_{10}$ 的列，对于 f_1 可以得到类似的结果。如果对于两张图像焦距都一样，我们可以用 f_0 和 f_1 的几何平均来估计焦距 $f = \sqrt{f_1 f_0}$。如果可以得到 f 的多个估计，比如从不同的单应，其均值可以作为最终的估计。

在固定参数的摄像机 $K_i = K$ 条件下，可以使用 Hartley(1997b)的方法获得更一般的(上三角形的)K 估计。从(6.52)观察到，$R_{ij} \sim K^{-1} \tilde{H}_{ij} K$ 和 $R_{ij}^{-T} \sim K^T \tilde{H}_{ij}^{-T} K^{-T}$。令 $R_{ij} = R_{ij}^{-T}$，可得 $K^{-1} \tilde{H}_{ij} K \sim K^T \tilde{H}_{ij}^{-T} K^{-T}$，从中我们得出

$$\tilde{H}_{ij} (KK^T) \sim (KK^T) \tilde{H}_{ij}^{-T}. \quad (6.58)$$

这给我们提供了对 $A = KK^T$ 中各项的一些同态线性限制，称为绝对二次曲线的像对偶(dual of the image of the absolute conic)(Hartley 1997b; Hartley and Zisserman 2004)。(回想一下，当我们估计单应的时候，只能在相差一个未知的尺度量的意义下恢复它。)给定足够数量的独立单应估计 $\tilde{H}_{ij}$，我们便可以用 SVD 或者特征值分析恢复 A(在相差一个尺度意义下)，然后通过 Cholesky 分解(附录 A.1.4 节)来恢复 K。推广到时变(temporally varying)标定参数和非静止摄像机的讨论参见 Hartley, Hayman, de Agapito et al. (2000)和 de Agapito, Hayman, and Reid(2001)。

构建整个 360° 全景图可以大大提高摄像机内参数的质量，因为序列中第一张图像拼接到自己时，焦距的误估计会导致一个缝隙(图 9.5)。正如 9.1.4 节所描述的，得到误配准可以用来改进焦距的估计和重新调整旋转的估计。把摄像机绕着其

光轴旋转 90°然后重新拍摄全景是检查横纵比和倾斜像素问题的好方法，正如有足够的纹理时形成完整的半球面全景图的情况一样。

但最终，标定参数(包括径向畸变)最精确的估计可以用内和外(旋转)参数的完全同时非线性最小化来获得，如 9.2 节所述。

6.3.5 径向畸变

当图像是通过广角镜头拍摄的时候，经常需要对镜头畸变进行建模，比如径向畸变。正如 2.1.6 节所讨论的，径向畸变模型是说，在观察到的图像中，坐标在显示的时候是远离(桶形畸变，barrel distortion)或者靠近(枕形畸变，pincushion distortion)图像中心一个量的，该量正比于它们的径向距离(图 2.13a 和 b)。简单的径向畸变用低阶多项式模型化(比照公式(2.78))

$$\begin{aligned} \hat{x} &= x(1 + \kappa_1 r^2 + \kappa_2 r^4) \\ \hat{y} &= y(1 + \kappa_1 r^2 + \kappa_2 r^4), \end{aligned} \quad (6.59)$$

其中 $r^2 = x^2 + y^2$，κ_1 和 κ_2 称为"径向畸变参数"(radial distortion parameters)(Brown 1971; Slama 1980)。[①]

针对给定镜片，有很多方法可以用于估计径向畸变参数。[②]其中一个最简单也最有用的方法是拍摄有许多直线，尤其是与图像边缘对齐或者贴近的线条的场景图像。然后可以调整径向畸变参数，直到图像中所有的线条都是直线，这通常称为"铅垂线方法"(plumb-line method)(Brown 1971; Kang 2001; El-Melegy and Farag 2003)。习题 6.10 给出了如何实现类似方法的一些细节。

另一个方法使用几个重叠的图像，把径向畸变参数的估计和图像配准过程结合起来，即对 9.2.1 节用于拼接的流水线进行扩展。Sawhney and Kumar(1999)把多层次的运动模型(平移、仿射、透视)连同二次径向畸变校正项用在一个从粗到精的策略中。他们用直接的(基于亮度的)最小化的方法计算配准。Stein(1997)用一个基于特征的方法，同时结合一般 3D 运动模型(和二次径向畸变)，后者需要的匹配更多(相对于视差无关旋转全景图)，但可能更具一般性。最近的方法有时同时计算未知的内参数和径向畸变系数(它们可能包括更高阶项或更复杂的有理式或者非参数形式)(Claus and Fitzgibbon 2005; Sturm 2005; Thirthala and Pollefeys 2005; Barreto and Daniilidis 2005; Hartley and Kang 2005; Steele and Jaynes 2006; Tardif, Sturm, Trudeau et al. 2009)。

使用一个已知的标定目标时(图 6.8)，径向畸变估计可以并入其他内外参数的估计中(Zhang 2000; Hartley and Kang 2007; Tardif, Sturm, Trudeau et al. 2009)。这可以看作是在图 6.5 所示的一般的非线性最小化流水线中，在内参数乘法方框 f_C 和透视

[①] 有时，x 和 $\hat{x}$ 之间的关系用另一种方式表示，即用右边的装填(最后)坐标 $x = \hat{x}(1 + \kappa_1 \hat{r}^2 + \kappa_2 \hat{r}^4)$。如果我们把图像像素映射到(卷绕的)射线，然后将射线去除畸变得到空间中的 3D 射线，即我们在用逆卷绕，这是方便的。

[②] 现在有一些数字摄像机开始使用内置软件去除径向畸变。

除法方框 f_p 之间，添加了另一个步骤。(更多关于平面标定目标情况的细节，可参见习题 6.11。)

当然，正如 2.1.6 节所讨论的，有时需要镜头畸变的更一般的模型，比如鱼眼镜头和非中心投影(non-central projection)。然而，这种镜头的参数化可能更复杂(2.1.6)，一般的方法不是使用已知 3D 位置的标定装置，就是使用场景的多个重叠图像进行自标定，这两种方法都可以用(Hartley and Kang 2007; Tardif, Sturm, and Roy 2007)。用于校正径向畸变的同样方法，也可以通过分别标定每个颜色通道，再将通道卷绕后放回去得到配准，用于减少色像差的量(习题 6.12)。

6.4 补充阅读

Hartley and Zisserman(2004)提供了基于特征的配准和最佳运动估计这些主题的非常好的介绍，同时也深入讨论了摄像机标定和姿态估计方法。

鲁棒估计方法的详细讨论参见附录 B.3 节，关于这个话题的专题论文以及综述文章可参见(Huber 1981; Hampel, Ronchetti, Rousseeuw *et al*. 1986; Rousseeuw and Leroy 1987; Black and Rangarajan 1996; Stewart 1999)。在计算机视觉中，最常用的鲁棒的初始化方法是 RANdom SAmple Consensus(RANSAC)(Fischler and Bolles 1981)，它产生了一系列更有效的变种(Nistér 2003; Chum and Matas 2005)。

注册 3D 点数据集合的主题称为"绝对定向"(Horn 1987)和"3D 姿态估计"(Lorusso, Eggert, and Fisher 1995)。人们提出了很多方法，用于同时计算 3D 点对应及其对应的刚性变换(Besl and McKay 1992; Zhang 1994; Szeliski and Lavallée 1996; Gold, Rangarajan, Lu *et al*. 1998; David, DeMenthon, Duraiswami *et al*. 2004; Li and Hartley 2007; Enqvist, Josephson, and Kahl 2009)。

摄像机标定最早是在摄像测量学中研究的(Brown 1971; Slama 1980; Atkinson 1996; Kraus 1997)，但它在计算机视觉中也有很多的研究(Tsai 1987; Gremban, Thorpe, and Kanade 1988; Champleboux, Lavallée, Szeliski *et al*. 1992; Zhang 2000; Grossberg and Nayar 2001)。消失点常用于进行基本的标定，而消失点的观察或者来自于矩形类标定物体或者来自于人造的建筑物(Caprile and Torre 1990; Becker and Bove 1995; Liebowitz and Zisserman 1998; Cipolla, Drummond, and Robertson 1999; Antone and Teller 2002; Criminisi, Reid, and Zisserman 2000; Hartley and Zisserman 2004; Pflugfelder 2008)。不使用已知目标就进行摄像机标定，称为自标定，在由运动到结构的相关教材和综述(Faugeras, Luong, and Maybank 1992; Hartley and Zisserman 2004; Moons, Van Gool, and Vergauwen 2010)中有讨论。这些方法中的一个流形的子集使用纯旋转运动(Stein 1995; Hartley 1997b; Hartley, Hayman, de Agapito *et al*. 2000; de Agapito, Hayman, and Reid 2001; Kang and Weiss 1999; Shum and Szeliski 2000; Frahm and Koch 2003)。

6.5 习 题

习题 6.1：基于特征的图像配准用于翻书动画

拍摄一个动作场景或者肖像的一组照片(最好是电机驱动的——连续拍摄——模式)并进行配准，用于形成一个合成的或者翻书动画。

1. 用 4.1.1 节到 4.1.2 节介绍的一些方法提取特征和特征描述子。
2. 用最近邻距离比测试(4.18)方式的最近邻匹配作特征匹配。
3. 使用最小二乘方法(6.1.1 节)计算在第一张图像和其他所有图像之间的最优 2D 平移变换和旋转变换，考虑到鲁棒性，可选择 RANSAC(6.1.4 节)。
4. 用双线性或者双三次重采样方法，将其他所有的图像重采样到第一张图像的坐标框架，可选择将它们裁剪到它们共同的区域上。
5. 把结果图像转化为动画 GIF(用网上的一些软件)或者可选地实现交叉淡入淡出(cross-dissolves)把它们变换为慢动作视频。
6. (选做)把这个方法和基于特征的变形(习题 3.25)结合起来。

习题 6.2：全景图 建立 6.1.2 讨论的那种全景图，通常可以在网上找到。

1. 收集一系列有趣的重叠的照片。
2. 用习题 4.1 到习题 4.4(或者已有的软件)中完成的特征检测器、描述子和匹配子在图像间匹配特征。
3. 把匹配特征的每一个连通分量变成一个轨迹(track)，即为每个轨迹分配一个唯一的索引 i，抛弃任何不一致的轨迹(在同一个图像中包含两个不同特征)。
4. 用公式(6.12)对于每张图像计算全局的平移变换。
5. 因为匹配可能有错，故把前面的最小二乘准则变为鲁棒准则(6.25)，用迭代重加权最小二乘重新求解你的系统。
6. 计算最后得到的合成画布的大小，把每张图像重采样到画布上的最终位置。(保持包围盒的轨迹能使其更高效。)
7. 计算所有图像的平均，或者选择某个顺序，然后实现半透明合成(3.8)。
8. (选做)把你的参数化的运动模型扩展到包含旋转和缩放，即表 6.1 给出的相似变换。讨论如何处理平移变换和旋转变换的情况(没有缩放)。
9. (选做)写一个简单的工具，使用户能调整顺序和不透明度，添加或者删除图像。
10. (选做)写下一个涉及逐对图像不同的匹配的最小二乘问题。讨论为什么这可能比(6.12)给出的全局匹配公式更好或者更差。

习题 6.3：2D 刚性/欧氏匹配

6.1.3 节给出了估计 2D 刚性(欧氏)配准的一些可选择的方法。

1. 在合成的数据上实现多种可选的方法并比较它们的精度。合成的数据就是带有噪声干扰的随机的 2D 点云。
2. 一个方法是从质心估计平移，然后在极坐标下估计旋转。需要以某种方式来给极坐标分解得到的角度加权以得到统计上正确的估计吗？
3. 考虑标量(6.10)或者二维点的协方差矩阵加权(6.11)，你如何修改你的方法？是所有前面完成的"捷径"仍然有效，还是完整的加权需要迭代的优化？

习题 6.4：2D 匹配移动/增强现实

用一个不同的图像或者视频替换杂志或者书上的一幅图像。
1. 用一个网络摄像头拍摄杂志或者书某页的一幅图像。
2. 给这页上的图或图像画一个矩形轮廓，即画出它们在图像中所出现的四条边。
3. 在这个区域中，与每个新的图像帧作特征匹配。
4. 用"广告"插入代替原图，用合适的单应变换新的图像。
5. 在一个运动视频(比如室内或者室外的足球比赛)的剪辑上试试你的方法，实现广告牌的替换。

习题 6.5：3D 操纵杆

跟踪魔方，实现一个 3D 操纵杆/鼠标控制器。
1. 拿出一个旧的魔方(或者从父母那里要一个)。
2. 写一个程序，检测每个彩色正方形的中心。
3. 把这些中心分组到线条上，然后找到每个面的消失点。
4. 从消失点中，估计旋转的角度和焦距。
5. 估计完整的 3D 姿态(包括平移)的方法是，找到一个或者更多个 3×3 的网格，用 Zhang(2000)提出的方法，从已知的单应中恢复平面的完整方程。
6. 另一个方法，由于你已经知道旋转角度，可以用线性或者非线性最小二乘，从立方体的已知 3D 角点和它们测量的 2D 位置，简单地估计未知的平移。
7. 用 3D 旋转和位置控制 VRML 或者 3D 游戏浏览器。

习题 6.6：基于旋转的标定

从一个旋转的摄像头中拍摄一段有很少视差的室外或者室内的视频序列，使用 6.3.4 节或者 9.1.3～9.2.1 节描述的方法标定摄像机的焦距。
1. 用习题 6.10 和 6.11 或者老师提供的给定摄像机的参数，去除图像中的径向畸变。
2. 在相邻帧之间检测并匹配特征点，然后把它们链接成一个特征轨迹。
3. 用公式(6.56 和 6.57)计算重叠帧之间的单应，估计焦距。
4. 计算完整的 360° 全景图，更新焦距的估计使缝隙关闭(9.1.4 节)。
5. (选做)对于旋转矩阵和焦距，进行完整的光束平差调整，以得到最高质量的估计(9.2.1 节)。

习题 6.7：基于目标的标定

用三维目标标定摄像机。

1. 构建一个已知 3D 位置的三维的标定模式。除非你使用一个机械车间，否则要得到高精度是不容易的，但是你可以用重型的胶合板和印刷图案来得到近似效果。
2. 找到角点，比如先找到线，然后找到线的交点。
3. 实现 Tsai(1987); Bogart(1991); Gleicher and Witkin(1992)中所描述的迭代标定和姿态估计算法中的一个或者 6.2.2 节所描述的系统。
4. 在相对于标定目标不同的距离和不同的角度拍摄多张照片，报告你的重投影误差和精度。(针对后者，你可能需要使用模拟的数据。)

习题 6.8：标定精度

比较三种标定方法(基于平面的、基于旋转的和基于 3D 目标的)。

一个方法是让不同的学生实现每个方法，然后比较结果。另一个方法是用合成的数据，可以再次使用习题 2.3 完成的软件。使用合成数据的优点是你知道标定和姿态参数的真值，因此能够很轻松地做很多实验，也可以用合成的方式改变测量中的噪声。

下面是构建测试集的一些可能的指导。

1. 假定中等范围的焦距(比如，50°视野)。
2. 对基于平面的方法，产生一个 2D 的网格状目标，在不同的倾斜度上对它进行投影。
3. 对于 3D 目标，产生一个立方体内角，把它放在一个位置使其占据大部分视野。
4. 对于旋转的方法，在一个球上均匀地撒点，直到得到和其他方法类似数目的点。

比较你的方法之前，预测哪一个将是最精确的(用点数目的平方根归一化你的结果)。

给测量中添加不同量的噪声，描述你的不同方法对噪声的敏感度。

习题 6.9：单视图度量

使用可见消失方向，从一幅人造场景图像，实现一个系统来测量维度并重建 3D 模型。(6.3.3 节) (Criminisi, Reid, and Zisserman 2000)。

1. 从平行线中找到三个正交的消失点，用它们建立三个坐标轴(相对于场景的摄像机的旋转矩阵 R)。如果消失点中的两个是有限的(没有在无穷远处)，则假设已知光心，用它们计算焦距。否则，用另外的方法标定摄像机，可以用 Schaffalitzky and Zisserman (2000)描述的某些方法。
2. 点击地平面上的一个点建立你的原点，然后点击已知距离的一个点建立场景的尺度。这样便可计算摄像机和场景之间的平移 t。另一个可选的方法是，点击一对点，一个在地平面上，一个在它上方，用已知的高度建立场景的尺度。
3. 写一个用户界面，允许你点击地平面上的点用以恢复它们的 3D 位置(提示：你已经知道摄像机矩阵，因此一个点 z 值的信息足以恢复其 3D 位置。)点击这一对点(一个点在地平面上，一个点在它上方)来测量垂直的高度。
4. 用 Sinha, Steedly, Szeliski *et al.* (2008)描述的一些方法扩展你的系统，允许在场景中画四边形(对应于与世界坐标系轴对齐的矩形)。将 3D 矩形导出到一个 VRML 或者 PLY[①]文件中。
5. (选做)通过使用正确的单应关系变换封闭四边形内的像素，对每个平面多边形产生一个纹理图。

习题 6.10：径向畸变铅垂线法

实现铅垂线算法确定径向畸变参数。
1. 拍摄一些包含很多直线的场景图，比如家里或者办公室的走廊，尝试尽可能使一些线贴近图像的边缘。
2. 提取边缘，并把它们连接为曲线，如 4.2.2 节和习题 4.8 所述。
3. 用 4.3.1 节和习题 4.11 描述的一般逐次线条近似算法，用二次或者椭圆曲线拟合连接的边缘，保留那些能更好拟合这个形式的曲线。
4. 对于每个曲线的片段，用直线拟合，在调整径向畸变参数期间最小化曲线和直线的垂直距离。
5. 轮流执行重新拟合直线和调整径向畸变参数，直到收敛。

习题 6.11：径向畸变标定目标法

用一个网格状标定目标确定径向畸变参数。
1. 打印出一个平面标定目标，把它装在一个硬板上，使其充满你的视野。
2. 检测标定目标的正方形、线和点。
3. 从没有任何径向畸变的图像的中央部分，估计把目标映射到摄像机的单应。
4. 预测剩余目标的位置，用观测位置和预测位置之间的差异估计径向畸变。
5. (选做)用一个一般样条模型拟合(对于严重失真)代替四次畸变模型。
6. (选做)把你的方法扩展到标定一个鱼眼镜头。

习题 6.12：色差

使用前面习题计算的每个颜色通道的径向畸变估计，通过卷绕所有通道得到配准，清理广角镜头图像。(选做)同时理顺这些图像。
你能想想为什么这样的变换策略并不总是见效吗？

[①] *http://meshlab.sf.net*

第 7 章
由运动到结构

7.1 三角测量
7.2 二视图由运动到结构
7.3 因子分解
7.4 光束平差法
7.5 限定的结构和运动
7.6 补充阅读
7.7 习题

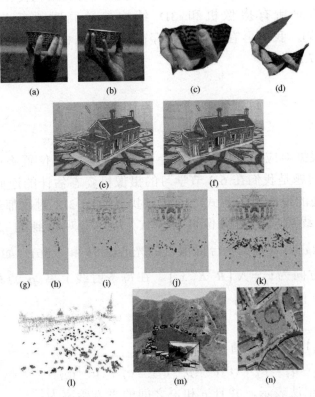

图 7.1 由运动到结构的系统：(a~d)正交因子分解(Tomasi and Kanade 1992)© 1992 Springer；(e 和 f)线条匹配(Schmid and Zisserman 1997)© 1997 IEEE；(g~k)增量由运动到结构(Snavely, Seitz, and Szeliski 2006)；(l)特拉法加广场的 3D 重建(Snavely, Seitz, and Szeliski 2006)；(m)中国长城的 3D 重建(Snavely, Seitz, and Szeliski 2006)；(n)布拉格旧城广场的 3D 重建(Snavely, Seitz, and Szeliski 2006)© 2006 ACM

从第 6 章中，我们了解到 2D 和 3D 点集是如何配准并使用配准结果来估计摄像机姿态及其内部标定参数的。在这一章中，我们处理这样一个逆问题，给出几幅图像及其图像特征的一个稀疏对应集合，如何估计 3D 点的位置。这个求解过程通常涉及 3D 几何(结构)和摄像机姿态(运动)的同时估计，这就是通常意义下的由运动到结构(Ullman 1979)。

射影几何和由运动到结构的内容非常丰富，也有一些非常好的教科书和综述报告(Faugeras and Luong 2001; Hartley and Zisserman 2004; Moons, Van Gool, and Vergauwen 2010)。这一章跳过这些书中的很多内容，例如三焦张量和完全自标定中的代数方法，将着重介绍那些我们在大规模基于图像的重建问题中发现的很有用的基本方法(Snavely, Seitz, and Szeliski 2006)。

我们首先简要讨论三角测量(7.1 节)，即估计从多台摄像机看到的一个点的 3D 位置。接下来，我们将介绍二视图(两帧)的由运动到结构问题(7.2 节)，它涉及确定两个摄像机之间的极线几何，并且它也可以使用自标定用于恢复摄像机内参数的某些信息(7.2.2 节)。7.3 节将介绍由大量点的轨迹同时估计结构和运动的因子分解方法，它使用投影模型的正交近似。然后我们对由运动到结构提出一个更通用的、更有用的方法，即对所有摄像机和 3D 结构参数同时进行光束平差法(bundle adjustment)(7.4 节)。我们再看看场景中存在较高层结构(比如线和面)的时候所出现的一些特殊情况(7.5 节)。

7.1 三角测量

三角测量是根据对应图像位置的集合和已知的摄像机位置确定一个点的 3D 位置的问题。这个问题是我们在 6.2 节学习的摄像机姿态估计的逆问题。

解决这一问题的最简单的方式之一是寻找离所有 3D 射线都最近的那个 3D 点 p，这些 3D 射线对应于由摄像机 $\{P_j = K_j[R_j|t_j]\}$ 观察到的 2D 匹配特征位置 $\{x_j\}$，其中 $t_j = -R_j c_j$，c_j 是第 j 台摄像机的光心(2.55 和 2.56)。如图 7.2 所示，这些射线原点在 c_j，方向为 $\hat{v}_j = \mathcal{N}(R_j^{-1} K_j^{-1} x_j)$。在每条射线上离 p 点最近的点，我们定义为 q_j，最小化下面这个距离

$$\|c_j + d_j \hat{v}_j - p\|^2, \qquad (7.1)$$

它在 $d_j = \hat{v}_j \cdot (p - c_j)$ 取得最小值。因此，

$$q_j = c_j + (\hat{v}_j \hat{v}_j^T)(p - c_j) = c_j + (p - c_j)_{\parallel}, \qquad (7.2)$$

用公式(2.29)的标记法表示，并且 p 和 q_j 之间的平方距离为：

$$r_j^2 = \|(I - \hat{v}_j \hat{v}_j^T)(p - c_j)\|^2 = \|(p - c_j)_{\perp}\|^2. \qquad (7.3)$$

那个离所有射线最近的最优点 p，可以当成常规最小二乘问题来计算，通过累加所有 r_j^2 的和找出 p 点的最优值，

$$\boldsymbol{p} = \left[\sum_j (\boldsymbol{I} - \hat{\boldsymbol{v}}_j \hat{\boldsymbol{v}}_j^T)\right]^{-1} \left[\sum_j (\boldsymbol{I} - \hat{\boldsymbol{v}}_j \hat{\boldsymbol{v}}_j^T) \boldsymbol{c}_j\right]. \tag{7.4}$$

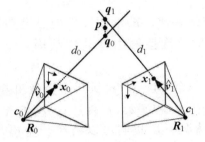

图 7.2 通过找到离所有光学射线 $c_j + d_j \hat{v}_j$ 最近的这点 p 而进行的 3D 点三角测量

另外一种形式化方法，在统计上更优，当有部分摄像机比其他摄像机更靠近那个 3D 点的时候，它能得到明显更好的估计结果，即最小化测量方程的残差

$$x_j = \frac{p_{00}^{(j)} X + p_{01}^{(j)} Y + p_{02}^{(j)} Z + p_{03}^{(j)} W}{p_{20}^{(j)} X + p_{21}^{(j)} Y + p_{22}^{(j)} Z + p_{23}^{(j)} W} \tag{7.5}$$

$$y_j = \frac{p_{10}^{(j)} X + p_{11}^{(j)} Y + p_{12}^{(j)} Z + p_{13}^{(j)} W}{p_{20}^{(j)} X + p_{21}^{(j)} Y + p_{22}^{(j)} Z + p_{23}^{(j)} W}, \tag{7.6}$$

其中 (x_j, y_j) 是那些 2D 特征点的坐标，$\{p_{00}^{(j)} \cdots p_{23}^{(j)}\}$ 是摄像机矩阵 P_j 中的已知项 (Sutherland 1974)。

如同公式(6.21，6.33 和 6.34)，通过在等号两边同时乘上分母，我们可以将这一非线性方程组转化成一个线性的最小二乘问题去求解。注意，如果我们采用齐次坐标 $\boldsymbol{p} = (X, Y, Z, W)$，所得到的方程组将是齐次的，可以用奇异值分解(SVD)或者本征值问题(找最小的奇异向量或者本征向量)很好地解决。假如令 $W = 1$，我们则可以用常规最小二乘法求解，不过结果有可能是奇异的或者病态的，即，假如所有的视线都是平行的，正如远离摄像机的点所呈现的那样。

由于这个原因，通常优先使用齐次坐标中参数化 3D 点，特别是当我们知道有些点离摄像机的距离有可能有很大差异的时候。当然，不管选择什么表示形式，用非线性最小二乘法最小化观测集(7.5 和 7.6)，如同(6.14 和 6.23)描述的那样，胜于用线性最小二乘法去求解。

对于两个观测的情况，准确最小化真实重投影误差(7.5 和 7.6)的点 p 的位置，可以用六阶方程组求解得到(Hartley and Sturm 1997)。三角测量要注意的另一个问题是手性(chirality)的影响，即，必须保证重建点在所有摄像机的前面(Hartley 1998)。然而这并不总是能够得到保证的，有一种有用的启发式方法，即，将那些在摄像机后面的点看作是因为它们的射线是发散的(想象一下图 7.2 中射线互相背离的情形)，并通过将它们的 W 值设为 0 来将这些点置于无穷远的平面上。

7.2 二视图由运动到结构

迄今为止我们学习的 3D 重建，都是假设事先已经知道 3D 点位置或者 3D 摄像机的姿态。在这节中，我们首次接触由运动到结构，它是从图像的对应同时恢复 3D 结构和姿态。

考虑图 7.3，它给出了一个 3D 点 p(从两个摄像机观察到的情况，这两个摄像机的相对位置可以通过旋转 $\boldsymbol{R}$ 和平移 $\boldsymbol{t}$ 表达)。因为我们对摄像机的位置一无所知，所以在不失一般性的情况下，我们可以将第一个摄像机设定在原点处 $c_0 = 0$，并且它的规范方向为 $\boldsymbol{R}_0 = \boldsymbol{I}$。

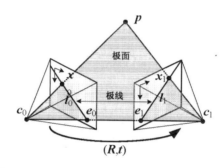

图 7.3 极线几何：向量 $\boldsymbol{t} = \boldsymbol{c}_1 - \boldsymbol{c}_0$，$\boldsymbol{p} - \boldsymbol{c}_0$ 和 $\boldsymbol{p} - \boldsymbol{c}_1$ 共面，并且用像素量测 x_0 和 x_1 表达了所定义了的基本极线约束

现在注意，点 p 在第一幅图像中的观测位置为 $p_0 = d_0\hat{x}_0$，可以通过下面的变换把它映射到第二幅图像中

$$d_1\hat{\boldsymbol{x}}_1 = \boldsymbol{p}_1 = \boldsymbol{R}\boldsymbol{p}_0 + \boldsymbol{t} = \boldsymbol{R}(d_0\hat{\boldsymbol{x}}_0) + \boldsymbol{t}, \tag{7.7}$$

其中 $\hat{x}_j = \boldsymbol{K}_j^{-1} x_j$ 是(局部的)射线方向向量。通过在公式两边同时取与 t 的叉积，以达到在公式右边去除 t 的效果①

$$d_1[\boldsymbol{t}]_\times \hat{\boldsymbol{x}}_1 = d_0[\boldsymbol{t}]_\times \boldsymbol{R}\hat{\boldsymbol{x}}_0. \tag{7.8}$$

在公式两边点积 $\hat{x}_1$ 可得

$$d_0\hat{\boldsymbol{x}}_1^T ([\boldsymbol{t}]_\times \boldsymbol{R})\hat{\boldsymbol{x}}_0 = d_1\hat{\boldsymbol{x}}_1^T [\boldsymbol{t}]_\times \hat{\boldsymbol{x}}_1 = 0, \tag{7.9}$$

上式为 0 是因为等号右边是有两个相等项的混合积(向量三重积)。(另外一种说法是，叉积矩阵 $[t]_\times$ 是反对称的，当在它左边和右边同乘上相同的向量时将返回 0。)

因此我们将得到基本的极线约束

$$\hat{\boldsymbol{x}}_1^T \boldsymbol{E} \hat{\boldsymbol{x}}_0 = 0, \tag{7.10}$$

其中

$$\boldsymbol{E} = [t]_\times \boldsymbol{R} \tag{7.11}$$

叫本质矩阵(essential matrix)(Longuet-Higgins 1981)。

---
① 叉积符号 $[\ ]_\times$ 在(2.32 节)中有介绍。

另一种导出极线约束的方法是，为了使摄像机朝向使射线 $\hat{x}_0$ 和 $\hat{x}_1$ 在 3D 空间相交于点 P 的方向，连接两个摄像机中心的向量 $c_1 - c_0 = -R_1^{-1}t$ 和与像素点 x_0 和 x_1 对应的射线(即 $R_j^{-1}\hat{x}_j$)，肯定是共面的。这就要求有一个混合积

$$(\hat{x}_0, R^{-1}\hat{x}_1, -R^{-1}t) = (R\hat{x}_0, \hat{x}_1, -t) = \hat{x}_1 \cdot (t \times R\hat{x}_0) = \hat{x}_1^T([t]_\times R)\hat{x}_0 = 0. \quad (7.12)$$

注意，因为 $\hat{x}_1^T l_1 = 0$，所以本质矩阵 E 将在图像 0 中的点 $\hat{x}_0$ 映射到图像 1 中的直线 $l_1 = E\hat{x}_0$(图 7.3)。所有这些直线都必须通过第二个极点 e_1，因此它被定义为具有 0 奇异值的 E 的左奇异向量，或者等价地，向量 t 到图像 1 的投影。这些关系的对偶(转置)给出图像 0 的极线为 $l_0 = E^T\hat{x}_1$，e_0 为 E 的 0 值右奇异向量。

给定这个基本关系(7.10)，怎样使用它来恢复编码在本质矩阵 E 中的摄像机运动呢？假如我们有 N 个对应测量 $\{(x_{i0}, x_{i1})\}$，我们可以形成 N 个关于 $E = \{e_{00}...e_{22}\}$ 的 9 个元素的齐次方程，

$$\begin{array}{cccc}
x_{i0}x_{i1}e_{00} + & y_{i0}x_{i1}e_{01} + & x_{i1}e_{02} + & \\
x_{i0}y_{i1}e_{00} + & y_{i0}y_{i1}e_{11} + & y_{i1}e_{12} + & \\
x_{i0}e_{20} + & y_{i0}e_{21} + & e_{22} & = 0
\end{array} \quad (7.13)$$

其中 $x_{ij} = (x_{ij}, y_{ij}, 1)$。可以更紧凑地写为

$$[x_{i1} x_{i0}^T] \otimes E = Z_i \otimes E = z_i \cdot f = 0, \quad (7.14)$$

其中 $\otimes$ 表示矩阵元素的逐元素乘积并求和，z_i 和 f 是 $Z_i = \hat{x}_{i1}\hat{x}_{i0}^T$ 和本质矩阵 E 的栅格(向量)化形式[①]。给出 $N \geqslant 8$ 这样的方程，我们可以用 SVD 计算本质矩阵每个项(在相差一个尺度量的意义下)的估计值。

在考虑到噪声的情况下，这个估计值与统计最优值有多接近？如果看(7.13)中的项，你会看到有些项是图像测量的乘积(即 $x_{i0}y_{i1}$)，而另外一些则是直接的图像测量(甚至是单位项)。假如这些测量具有大小相近的噪声，则这种测量相乘的项会通过乘上其他元素而放大其噪声，从而使缩放性变得很差，例如，有很大坐标值(距离图像中心很远)的点会产生非常大的影响。

为了消除这种趋势，Hartley(1997a)建议应该平移和缩放这些点的坐标值从而使其质心在原点处，而它们的方差为单位大小。即，

$$\tilde{x}_i = s(x_i - \mu_x) \quad (7.15)$$
$$\tilde{y}_i = s(y_i - \mu_y) \quad (7.16)$$

使得 $\sum_i \tilde{x}_i = \sum_i \tilde{y}_i = 0$ 和 $\sum_i \tilde{x}_i^2 + \sum_i \tilde{y}_i^2 = 2n$，其中 n 是点的个数[②]。

一旦从变换后的坐标 $\{(\tilde{x}_{i0}, \tilde{x}_{i1})\}$ 计算出本质矩阵 $\tilde{E}$，其中 $\tilde{x}_{ij} = T_j\hat{x}_{ij}$，原始的本质矩阵 E 便可以恢复为

$$E = T_1^T \tilde{E} T_0. \quad (7.17)$$

在 Hartley(1997a)论文中，比较了由于其重归一化(re-normalization)策略而得到的改

[①] 我们使用 f 而不是 e 来表示 E 的栅格化形式是为了避免和极点 e_j 产生混淆。
[②] 更准确地说，Hartley (1997a)建议缩放这些点"使其到原点的平均距离等于 $\sqrt{2}$"，但是使用单位方差在计算上更快(并不要求计算所有点的平方根)且在性能上有相应的提升。

进和其他人替换距离度量所得到的改进，例如 Zhang(1998a, b)，其结论是他的简单重归一化方法在大多数情况下与其他方法一样有效(或更好)。Torr and Fitzgibbon(2004)推荐了这个算法的一个变种，把本质矩阵 E 的左上方的 2×2 的子矩阵的模设为 1，这样对于 2D 坐标变换更稳定。

一旦恢复出本质矩阵 E 的一个估计值，就可以估计平移向量 t 的方向。注意，两个摄像机的绝对距离是不可能仅靠图像测量来恢复的，不论使用多少个摄像机或者多少个点。摄像机和点的绝对位置和距离，在摄影测绘学中通常被称作"地面控制点"，在确定最后的尺寸、位置和方向时总是需要中的。

要估计这个方向 $\hat{t}$，注意到在理想的无噪声条件下，本质矩阵 E 是奇异的，即 $\hat{t}^T E = 0$。这个奇异性表现在对本质矩阵 E 作 SVD 分解后有一个 0 奇异值，

$$E = [\hat{t}]_\times R = U\Sigma V^T = \begin{bmatrix} u_0 & u_1 & \hat{t} \end{bmatrix} \begin{bmatrix} 1 & & \\ & 1 & \\ & & 0 \end{bmatrix} \begin{bmatrix} v_0^T \\ v_1^T \\ v_2^T \end{bmatrix} \tag{7.18}$$

在有噪声的情况下对 E 进行分解时，最小奇异值对应的奇异向量给我们 $\hat{t}$。(另外两个奇异值会很接近，但是一般情况下不会等于 1，因为 E 只能在相差一个未知尺度的意义下计算出来。)

因为 E 是不满秩的，所以我们只需要公式(7.14)形式的七个对应而不是八个来估计这个矩阵(Hartley 1994a; Torr and Murray 1997; Hartley and Zisserman 2004)。(在 RANSAC 鲁棒拟合阶段使用更少对应的好处是只需要产生更少的随机样本。) 从这七个齐次方程(我们可以把它们组织成 7×9 的矩阵以作 SVD 分析)，我们可以得到两个独立的向量 f_0 和 f_1 使得 $Z_i \cdot f_j = 0$。这两个向量可以转换回 3×3 的矩阵 E_0 和 E_1，它们展开为解空间

$$E = \alpha E_0 + (1-\alpha)E_1. \tag{7.19}$$

为了找到 α 的正确数值，我们知道本质矩阵 E 不是满秩的，所以它的行列式为 0，因此

$$\det|\alpha E_0 + (1-\alpha)E_1| = 0. \tag{7.20}$$

根据上式，我们可以得到关于 α 的三次方程，它有一个或者三个解(根)。将所得的值代入公式(7.19)得到 E，我们可以用之前没用过的对应测试该本质矩阵，选择正确的一个。

一旦恢复了 $\hat{t}$，我们怎样估计对应的旋转矩阵 R？回忆一下叉积算子 $[\hat{t}]_\times$ (2.32)，它将一个向量投影到一组正交基向量上，其中包括 $\hat{t}$，让 $\hat{t}$ 这个方向的分量为零，另外两个向量旋转 90°，

$$[\hat{t}]_\times = SZR_{90°}S^T = \begin{bmatrix} s_0 & s_1 & \hat{t} \end{bmatrix} \begin{bmatrix} 1 & & \\ & 1 & \\ & & 0 \end{bmatrix} \begin{bmatrix} 0 & -1 & \\ 1 & 0 & \\ & & 1 \end{bmatrix} \begin{bmatrix} s_0^T \\ s_1^T \\ \hat{t}^T \end{bmatrix}, \tag{7.21}$$

其中 $\hat{t} = s_0 \times s_1$。从公式(7.18 和 7.21)，我们可以得到

$$E = [\hat{t}]_\times R = SZR_{90°}S^T R = U\Sigma V^T, \tag{7.22}$$

从上式我们可以得出结论 $S=U$。回忆一下没有噪声的本质矩阵的情况，$(\Sigma=Z)$，因此

$$R_{90°}U^TR = V^T \tag{7.23}$$

然后

$$R = UR_{90°}^TV^T. \tag{7.24}$$

不幸的是，我们仅在相差一个符号的意义下知道 E 和 $\hat{t}$。而且不能保证矩阵 U 和 V 是旋转矩阵(你可以翻转它们的符号而仍然能够得到有效的 SVD)。因此，我们必须产生四种可能的旋转矩阵

$$R = \pm UR_{\pm 90°}^TV^T \tag{7.25}$$

然后保留行列式 $|R|=1$ 的两个矩阵 R。为了消除剩余形成扭曲对(Hartley and Zisserman 2004, p.240)的两个潜在旋转矩阵之间的歧义，我们需要把它们与平移方向 $\pm\hat{t}$ 两种可能的组合都考虑到，选择两个摄像机①前面都能看到的点数最大的组合。

所有的观测点必须在摄像机前面，即，在从摄像机发出的视线上距离为正，这个性质就叫"手性"(chirality)(Hartley 1998)。除了确定旋转和平移的正负性之外，如前所述，重建过程中点的手性(距离的正负)可以用在 RANSAC 过程中(与重投影误差一起)以区分可能的和不可能的构型。②手性还可以用于把投影重建(7.2.1 节和 7.2.2 节)转化为拟仿射(quasi-affine)重建(Hartley 1998)。

前面描述的标准"八点算法"(Hartley 1997a)并不是通过对应关系来估计摄像机运动的唯一方法。变种方法有，在矩阵 E 中使用秩为二的约束(7.19~7.20)而使用七个点的方法，另外一种是五点算法，这个算法需要寻找一个十阶多项式的根(Nistér 2004)。因为这些算法使用更少的点来计算估计，所以作为随机采样(RANSAC)策略的一部分使用时它们对外点更不敏感。

单纯平移(已知旋转参数)

在已知旋转参数的情况下，我们可以预先旋转第二幅图像中的点使其与第一幅图像的视场方向匹配。得到的 3D 点集都移向(或远离)汇集点(延伸焦点，focus of expansion，FOE)，如图 7.4 所示。③所产生的本质矩阵 E(在没有噪声的情况下)是反对称的，所以可以通过在公式(7.13)中设置 $e_{ij}=-e_{ji}$ 和 $e_{ii}=0$，从而更直接地估计 E。现在，根据这两个具有非零视差的点足以估计 FOE。

一种更直接的推导出 FOE 估计的方法是最小化混合积

$$\sum_i (x_{i0}, x_{i1}, e)^2 = \sum_i ((x_{i0} \times x_{i1}) \cdot e)^2, \tag{7.26}$$

它等价于求解如下方程组的零空间

$$(y_{i0} - y_{i1})e_0 + (x_{i1} - x_{i0})e_1 + (x_{i0}y_{i1} - y_{i0}x_{i1})e_2 = 0. \tag{7.27}$$

---

① 在无噪声的情况下，一个点就足够了。但要测试全部的点或者足够的点的子集，更安全的做法是降低靠近无穷远处平面那些点的权重，因为这些点很容易导致深度反向。

② 注意，当点远离摄像机时，即，靠近无穷远处的平面时，手性错误更有可能发生。

③ 《星际迷航》和《星球大战》爱好者会认为这是"穿越时空"的视觉效果。

注意，正如八点算法中演示的那样，在计算估计之前，建议归一化 2D 点使其具有单位方差。

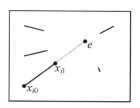

图 7.4 摄像机的纯平移运动将导致所有点都向汇集点 e (FOE)靠近(或者远离)的视觉运动。因此它们满足混合积条件 $(x_0, x_1, e) = e \cdot (x_0 \times x_1) = 0$

在有大量点位于无穷远处的情形下，例如，拍摄室外场景时或者摄像机的运动相对远距离的物体而言很小时，需要另外一种 RANSAC 策略来估计摄像机的运动。首先，选取一对点估计旋转矩阵，最好这两点都在无穷远处(距离摄像机非常远)。然后，计算 FOE 并检查残差是否很小(表明与旋转假设一致)且所有移向或者远离极点(FOE)的运动是否同向(忽略非常微小的运动，它们可能被噪声污染了)。

单纯旋转

纯旋转运动的情况将导致本质矩阵 E 和平移方向 $\hat{t}$ 估计退化。首先假设旋转矩阵已知的情况。FOE 的估计过程将退化，因为 $x_{i0} \approx x_{i1}$，从而导致公式(7.27)的退化。类似的推断表明本质矩阵的方程式(7.13)也是不满秩的。

这表明如下做法可能是明智的：在计算整个本质矩阵之前，先用公式(6.32)估计旋转矩阵 R，这可能只需要少量的点就能完成，然后计算旋转这些点之后的残差，再推进到计算整个本质矩阵 E 这一步。

7.2.1 投影(未标定的)重建

在很多情况下，比如要根据网上照片或老照片(由未知摄像机拍摄的、没有任何 EXIF(可交换照片文件资料)标签的)来建立 3D 模型时，我们事先并不知道与这些输入图像所关联的内标定参数。在类似情况下，尽管真正的度量结构可能无法获得(例如，在世界坐标系中正交的线或面在重建过程中最终也许不能恢复成正交的)，但我们还是可以估计出一个二视图重建。

考虑我们用来估计本质矩阵 E 的推导过程(7.10～1.12)。在未标定的情况下，我们不知道标定矩阵 K_j，所以无法使用归一化射线方向 $\hat{x}_j = K_j^{-1} x_j$。我们只能得到图像坐标 x_j，所以本质矩阵(7.10)变为

$$\hat{x}_1^T E \hat{x}_1 = x_1^T K_1^{-T} E K_0^{-1} x_0 = x_1^T F x_0 = 0, \tag{7.28}$$

其中

$$F = K_1^{-T} E K_0^{-1} = [e]_\times \tilde{H} \tag{7.29}$$

称为"基本矩阵"(fundamental matrix)(Faugeras 1992; Hartley, Gupta, and Chang 1992; Hartley and Zisserman 2004)。

与本质矩阵类似,基本矩阵(原理上)的秩为二,

$$F = [e]_\times \tilde{H} = U\Sigma V^T = \begin{bmatrix} u_0 & u_1 & e_1 \end{bmatrix} \begin{bmatrix} \sigma_0 & & \\ & \sigma_1 & \\ & & 0 \end{bmatrix} \begin{bmatrix} v_0^T \\ v_1^T \\ e_0^T \end{bmatrix}. \tag{7.30}$$

它左侧最小的奇异向量代表图像 1 中的极点 e_1 而右侧最小的奇异向量是 e_0 (如图 7.3 所示)。公式(7.29)中的单应矩阵 $\tilde{H}$,在原理上应该等于

$$\tilde{H} = K_1^{-T} R K_0^{-1}, \tag{7.31}$$

它并不能唯一地从 F 恢复出来,因为任何形式的 $\tilde{H}' = \tilde{H} + ev^T$ 单应矩阵最后都会得到同一个 F 矩阵。(注意,$[e]_\times$ 会使 e 的任何倍数都消失。)

这些有效的单应矩阵 $\tilde{H}$ 中的任何一个都将场景中的某个平面从一幅图像映射到另一幅图像。不可能事先知道究竟是哪一个,除非作为基本矩阵 F 估计过程的一部分(与估计本质矩阵 E 的旋转矩阵类似),我们选取四个或更多共面对应来计算单应矩阵 $\tilde{H}$,或者通过单应矩阵映射一幅图像上的所有点到另一幅图像,看哪幅图的点与其在另一幅图上的对应点排成线。①

为了进行场景的投影重建,我们可以选取满足公式(7.29)的任何单应矩阵 $\tilde{H}$。例如,通过类比公式(7.18~7.24)的方法,我们得到

$$F = [e]_\times \tilde{H} = SZR_{90°} S^T \tilde{H} = U\Sigma V^T \tag{7.32}$$

因此

$$\tilde{H} = U R_{90°}^T \hat{\Sigma} V^T, \tag{7.33}$$

将奇异值矩阵的最小值项替换成合理的数值(比如中间值)②就形成其中的 $\hat{\Sigma}$。这样,我们可以形成一对摄像机矩阵

$$P_0 = [I|0] \quad \text{and} \quad P_0 = [\tilde{H}|e], \tag{7.34}$$

根据这对摄像机矩阵,就可以使用三角测量(7.1 节)对场景进行投影重建。

尽管投影重建在实践中可能没什么用处,但是它通常可以升级为仿射或者度量重建,后文将给出详细内容。即使没有这个步骤,基本矩阵 F 在寻找其他对应时还是非常有用的,因为它们肯定都在对应的极线上,即,假设点的运动是由刚性变换引起的,那么图像 0 中任何特征点 x_0 的对应点都肯定位于图像 1 中的关联极线 $l_1 = Fx_0$ 上。

7.2.2 自标定

如果能够得到度量重建,即,平行线重建出来之后还是平行的,正交的墙面重建出来也是正交的,且重建的模型是真实情形的一个缩放,则由运动到结构的计算

---

① 这个过程有时被援引为平面加视差(2.1.5 节)(Kumar, Anandan, and Hanna 1994; Sawhney 1994)。
② Hartley and Zisserman (2004, p. 237)推荐使用 $\tilde{H} = [e]_\times F$ (Luong and Vieville 1996),它将摄像机放在无穷远处的平面上。

结果将非常有用(更易于理解)。多年来,为了把投影重建转换到度量重建,这等价于恢复关联于每幅图像的未知标定矩阵 K_j,人们提出了大量自标定(self-calibration)或者自动标定(auto-calibration)方法(Hartley and Zisserman 2004; Moons, Van Gool, and Vergauwen 2010)。

如果已知场景的其他特定信息,也许可以采用不同的方法。例如,假如场景中有平行线(通常,让若干条线汇聚到同一个消失点便是明证),三个或者更多的消失点(它们是无穷远处那些点的图像),可用于为无穷远处的平面建立单应矩阵,可以由它恢复出焦距和旋转矩阵。如果能够观测到两个或者更多有限正交消失点,就可以使用基于消失点的单幅图像标定方法(6.3.2 节)。

在缺乏这种外部信息的情况下,单纯依靠图像中的对应不可能恢复出每幅图像的整个参数化独立标定矩阵 K_j。要明白这一点,可考虑所有摄像机矩阵 $P_j = K_j [R_j | t_j]$ 的集合,将世界坐标系中的点 $p_i = (X_i, Y_i, Z_i, W_i)$ 投影到屏幕坐标 $x_{ij} \sim P_j p_i$。现在考虑通过任意的 4×4 投影变换矩阵 $\tilde{H}$ 变换 3D 场景中的点 $\{p_i\}$,能够得到新的点 $p_i' = \tilde{H} p_i$,从而构成新的模型。对每个摄像机矩阵 P_j 都后乘上 $\tilde{H}^{-1}$ 也会得到相同的屏幕坐标,而对新的摄像机矩阵 $P_j' = P_j \tilde{H}^{-1}$ 使用 RQ 分解,能够得到一组新的标定矩阵。

为此,所有自标定方法都假定标定矩阵的形式有限,不是设定一部分参数或使一部分参数相等,就是假设它们不随时间变化。尽管 Hartley and Zisserman (2004); Moons, Van Gool, and Vergauwen(2010)中讨论的大多数方法都要求三帧以上的图像,但在这一节中,我们提出一种简单的方法,它能够从二视图重建过程中的基本矩阵 F 中恢复出这两幅图像的焦距 (f_0, f_1) (Hartley and Zisserman 2004, p. 456)。

为了完成这个任务,我们假设摄像机的扭曲为零、已知的横纵比(通常设为 1)和一个已知的光心,如公式(2.59)所示。在实际中,这种假设的合理性如何?就像很多问题那样,答案视情况而定。

如果需要绝对的度量精度,就像在摄影测绘学应用中那样,则必须要预先标定摄像机,这可以用 6.3 节介绍的其中一种方法来进行,并使用地面控制点确定重建。如果我们只是为了可视化或者基于图像的绘制应用而希望重建世界,如同 Snavely, Seitz, and Szeliski(2006)的照片游览系统那样,这样的假设在实际中是很合理的。

现在大部分摄像机都有方形像素,而且光心就在图像中心附近,且很可能因为径向畸变(6.3.5 节)而与简单的摄像机模型有所不同,这意味着只要可能就应该进行补偿。最大的问题出现在图像的截取偏离中心的时候(即光心不再位于中间)或透视图片是从一个不同的图片拍摄得到的时候(这种情况下,需要有一般的摄像机矩阵)。[①]

鉴于以上警示,Hartley and Zisserman(2004, p. 456)提出了基于 Kruppa 公式的

[①] 在照片游览系统中,我们的系统用大教堂的真实图片注册巴黎圣母院外的信息标识照片。

二视图焦距估计算法，过程如下。取基本矩阵 $\boldsymbol{F}$ (7.30)的左右奇异向量 $\{\boldsymbol{u}_0, \boldsymbol{u}_1, \boldsymbol{v}_0, \boldsymbol{v}_1\}$ 和它们关联的奇异值 $\{\sigma_0, \sigma_1\}$，形成下面这组等式：

$$\frac{\boldsymbol{u}_1^T \boldsymbol{D}_0 \boldsymbol{u}_1}{\sigma_0^2 \boldsymbol{v}_0^T \boldsymbol{D}_1 \boldsymbol{v}_0} = -\frac{\boldsymbol{u}_0^T \boldsymbol{D}_0 \boldsymbol{u}_1}{\sigma_0 \sigma_1 \boldsymbol{v}_0^T \boldsymbol{D}_1 \boldsymbol{v}_1} = \frac{\boldsymbol{u}_0^T \boldsymbol{D}_0 \boldsymbol{u}_0}{\sigma_1^2 \boldsymbol{v}_1^T \boldsymbol{D}_1 \boldsymbol{v}_1}, \tag{7.35}$$

其中两个矩阵

$$\boldsymbol{D}_j = \boldsymbol{K}_j \boldsymbol{K}_j^T = \operatorname{diag}(f_j^2, f_j^2, 1) = \begin{bmatrix} f_j^2 & & \\ & f_j^2 & \\ & & 1 \end{bmatrix} \tag{7.36}$$

包含未知的焦距。为了简单起见，我们重写(7.35)的分子和分母得到

$$e_{ij0}(f_0^2) = \boldsymbol{u}_i^T \boldsymbol{D}_0 \boldsymbol{u}_j = a_{ij} + b_{ij} f_0^2, \tag{7.37}$$

$$e_{ij1}(f_1^2) = \sigma_i \sigma_j \boldsymbol{v}_i^T \boldsymbol{D}_1 \boldsymbol{v}_j = c_{ij} + d_{ij} f_1^2. \tag{7.38}$$

注意，两者中的任何一个都是 f_0^2 或者 f_1^2 的仿射变换(线性变化加常量)。因此，我们可以通过交叉相乘这些等式得到 f_j^2 的二次方程，后者很容易求解(还可以参考 Bougnoux(1998)的工作，他使用的是另一种形式化方法。)

另一种解决方案是，观察得到我们有一个关于未知标量 λ 的三公式方程组，即，

$$e_{ij0}(f_0^2) = \lambda e_{ij1}(f_1^2) \tag{7.39}$$

(Richard Hartley, personal communication, July 2009)。这很容易求解得到 $(f_0^2, \lambda f_1^2, \lambda)$ 进而得到 (f_0, f_1)。

在实际中这种方法的效果怎么样？有几种确定的退化构型，例如，没有旋转的时候，或者光轴交叉的时候，这种方法根本不起作用的时候。(在这种情况下，可以改变摄像机的焦距以获得更深的或者更浅的重建，这是浅浮雕歧义的一个例子(7.4.3 节)。)Hartley and Zisserman(2004)推荐使用基于至少三帧图像的重建方法。然而，如果能够找到从两幅图像可以很好地估计得到 $(f_0^2, \lambda f_1^2, \lambda)$，则可用它们来初始化对所有参数施加的更完全的光束平差法(7.4 节)。另一种方法，经常被用于诸如照片游览等系统，是用摄像机的 EXIF 标签信息或者一般默认值来初始化焦距的估计并作为光束平差法的一部分对它们求精。

7.2.3 应用：视图变形

视图变形是二视图由运动到结构的一个有趣的应用(也叫"视图插值"，参见 13.1 节)，它可以用于产生从 3D 场景的一个视角到其另一个视角的平滑的 3D 动画(Chen and Williams 1993; Seitz and Dyer 1996)。

为了实现这种转变，必须首先平滑地插值摄像机矩阵，即摄像机位置、角度和焦距。可以用简单的线性插值法(用四元组来表示旋转矩阵(2.1.4 节))，但是用渐入 (easing in)和渐出(easing out)的方法得到摄像机参数，效果更好，例如使用凸起的余弦或者沿比较圆的轨迹移动摄像机(Snavely, Seitz, and Szeliski 2006)。

为了生成中间帧，要么建立 3D 对应的完全集合，要么为每个参考视图创建 3D 模型(原型)。13.1 节介绍了几种解决这个问题的常用方法。其中最简单的方法是三

角剖分每幅图像上的匹配特征点集合，例如，使用 Delaunay 三角剖分。因为这些 3D 点被重投影到其中间视图，所以可以通过仿射变换或投影变换把像素从原始的源图像映射到它们的新视图(Szeliski and Shum 1997)。然后，最后的图像由两张参考图像的线性融合而成，就像使用通常的变形那样(3.6.3 节)。

7.3 因子分解

在处理视频序列的时候，我们通常会得到扩展的特征轨迹(4.1.4 节)，可从它使用一个称作"因子分解"的过程恢复出结构和运动。考虑由一个旋转乒乓球生成的轨迹，球上标注点以便形状和运动更易于分辨(图 7.5)。我们很容易从轨迹的形状判断出这个运动物体是球形的，但如何能从数学上推导出这个结论呢？

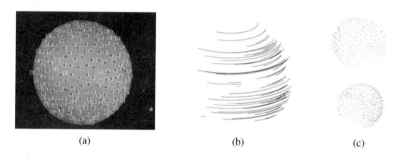

图 7.5 用因子分解的方法对旋转的乒乓球进行 3D 重建(Tomasi and Kanade 1992)© 1992 Springer：(a)覆盖所跟踪特征的样例图像；(b)子采样特征运动流；(c)重建后 3D 模型的两个视图

结果证明，在我们下面讨论的正交投影或者相关的模型下，使用奇异值分解方法可以同时恢复出形状和运动(Tomasi and Kanade 1992)。考虑公式(2.47～2.49)中介绍的正交和弱透视投影模型。因为最后一行通常是(0 0 0 1)，所以没有透视除法(perspective division)，我们可以写成

$$x_{ji} = \tilde{P}_j \bar{p}_i, \tag{7.40}$$

其中 x_{ji} 是第 j 帧的第 i 个点的位置，$\tilde{P}_j$ 是投影矩阵 P_j 的上面 2×4 的部分，$\bar{p}_i = (X_i, Y_i, Z_i, 1)$ 是增广(齐次)的 3D 点位置。①

我们假设(暂时)每一帧 j 的每个点 i 都是可见的。我们可以得到在 j 帧中的所有投影点 x_{ji} 的质心(平均值)，

$$\bar{x}_j = \frac{1}{N}\sum_i x_{ji} = \tilde{P}_j \frac{1}{N}\sum_i \bar{p}_i = \tilde{P}_j \bar{c}, \tag{7.41}$$

其中 $\bar{c} = (\bar{X}, \bar{Y}, \bar{Z}, 1)$ 是点云的 3D 质心的齐次坐标。

由于由运动到结构中世界坐标系总是任意的，即，我们不能在没有地面控制点(已知测量值)的情况下恢复出真正的 3D 位置，所以我们可以把世界坐标系的原点

---

① 在这一节中，我们用 x_{ji} 而不是 x_{ij} 表示 2D 坐标的位置，因为这是一般因子分解论文(Tomasi and Kanade 1992)所采用的惯例形式，并且这和(7.43)给出的因子分解形式是一致的。

移到点的质心位置，即，$\bar{X}=\bar{Y}=\bar{Z}=0$，所以 $\bar{c}=(0,0,0,1)$。因此，在每帧中 2D 点集的质心 $\bar{x}_j$ 是矩阵 $\tilde{P}_j$ 的最后的元素。

2D 点在减去其图像质心后的位置可以表示成 $\tilde{x}_{ji}=x_{ji}-\bar{x}_j$。现在我们可以写

$$\tilde{x}_{ji}=M_j p_i, \tag{7.42}$$

其中 M_j 是投影矩阵 P_j 的上面 2×3 的部分，$p_i=(X_i,Y_i,Z_i)$。联合所有这些测量公式可以得到一个的大矩阵

$$\hat{X}=\begin{bmatrix} \tilde{x}_{11} & \cdots & \tilde{x}_{1i} & \cdots & \tilde{x}_{1N} \\ \vdots & & \vdots & & \vdots \\ \tilde{x}_{j1} & \cdots & \tilde{x}_{ji} & \cdots & \tilde{x}_{jN} \\ \vdots & & \vdots & & \vdots \\ \tilde{x}_{M1} & \cdots & \tilde{x}_{Mi} & \cdots & \tilde{x}_{MN} \end{bmatrix}=\begin{bmatrix} M_1 \\ \vdots \\ M_j \\ \vdots \\ M_M \end{bmatrix}\begin{bmatrix} p_1 & \cdots & p_i & \cdots & p_N \end{bmatrix}=\hat{M}\hat{S}. \tag{7.43}$$

$\hat{X}$ 是"测量矩阵"，$\hat{M}$ 和 $\hat{S}$ 分别是运动矩阵和结构矩阵(Tomasi and Kanade 1992)。

因为运动矩阵 $\hat{M}$ 是 $2M\times 3$ 的，结构矩阵 $\hat{S}$ 是 $3\times N$ 的，所以对 $\hat{X}$ 进行 SVD 分解只能得到 3 个非零的奇异值。当 $\hat{X}$ 中的测量包含噪声时，SVD 返回在最小二乘意义上最接近 $\hat{X}$ 的秩为 3 的分解因子(Tomasi and Kanade 1992; Golub and Van Loan 1996 ; Hartley and Zisserman 2004)。

如果对 $\hat{X}=U\Sigma V^T$ 进行 SVD 分解能直接得到矩阵 $\hat{M}$ 和 $\hat{S}$ 就好了，但是得不到。我们只能写成如下关系

$$\hat{X}=U\Sigma V^T=[UQ][Q^{-1}\Sigma V^T] \tag{7.44}$$

且设 $\hat{M}=UQ$ 和 $\hat{S}=Q^{-1}\Sigma V^T$。①

我们怎样才能恢复出 3×3 的矩阵 Q？这取决于所采用的运动模型。在正交投影(2.47)中，M_j 中的项是旋转矩阵 R_j 的前两行，所以我们有

$$\begin{array}{rll} m_{j0}\cdot m_{j0}= & u_{2j}QQ^T u_{2j}^T & =1, \\ m_{j0}\cdot m_{j1}= & u_{2j}QQ^T u_{2j+1}^T & =0, \\ m_{j1}\cdot m_{j1}= & u_{2j+1}QQ^T u_{2j+1}^T & =1, \end{array} \tag{7.45}$$

其中 u_k 是矩阵 U 的行(3×1)。针对矩阵 QQ^T 中的所有项，我们可以得到大量的公式，并且应用一种叫矩阵平方根的方法(附录 A.1.4 节)可以恢复出矩阵 Q。如果处理归一的正交投影的情况，即 $M_j=s_j R_j$，第一和第三个公式等于 s_j，所以可以把它们设置成彼此相等。

注意，即使恢复出矩阵 Q，仍然存在着浅浮雕歧义，即，我们不可能确定物体是从左向右旋转还是在其深度反转后是向相反的方向运动。(这个问题在经典旋转 Necker Cube 视觉错觉中可以见到。)额外的线索，例如特征点或透视效果的出现和消失，这些都会在下文讨论，可以用来消除歧义。

对于不同于单纯正交投影的运动模型，例如，对于归一的正交投影和弱透视投

---

① Tomasi and Kanade(1992)首先求出 Σ 的平方根，再将它们分配给 U 和 V，但这样做并没有特别的根据。

影,前面这个方法必须以合适的方式加以扩展。从最初原理导出这类方法是相对直观的;(Poelman and Kanade 1997)扩展了基本的因子分解方法以适应更灵活的模型,更多的细节可以在其论文中找到。最初因子分解算法的其他扩展包括多体刚性运动(Costeira and Kanade 1995)、顺序更新分解因子(Morita and Kanade 1997)、增加线和面(Morris and Kanade 1998)以及重缩放测量以加入个体位置的不确定性(Anandan and Irani 2002)。

因子分解方法的一个缺点是它们要求有轨迹的完整集合,即为了让因子分解方法有效,每个特征点在每帧中必须都是可见的。Tomasi and Kanade(1992)提出了一种处理方法,首先在较小的稠密子集上应用因子分解,然后用已知的摄像机(运动)或者点(结构)的估计来幻生(hallucinate)出其余缺失的数值,从而以增量方式合并更多特征和摄像机。Huynh, Hartley, and Heyden(2003)扩展了这个方法,他们把缺失数据视为外点的特例。Buchanan and Fitzgibbon(2005)提出了一种快速迭代算法以处理含有缺失数据的大规模矩阵的因子分解。带有缺失数据的主分量分析(principal component analysis,PCA)主题也出现在其他计算机视觉问题中(Shum, Ikeuchi, and Reddy 1995; De la Torre and Black 2003; Gross, Matthews, and Baker 2006; Torresani, Hertzmann, and Bregler 2008; Vidal, Ma, and Sastry 2010)。

7.3.1 透视与投影因子分解

常规因子分解的另一个缺点是不能处理透视摄像机。绕过这个问题的一种途径是首先进行仿射(例如,正交)重建,然后以迭代方式校正这个结果以符合透视效果(Christy and Horaud 1996)。

注意,以物体为中心的投影模型(2.76)

$$x_{ji} = s_j \frac{\boldsymbol{r}_{xj} \cdot \boldsymbol{p}_i + t_{xj}}{1 + \eta_j \boldsymbol{r}_{zj} \cdot \boldsymbol{p}_i} \tag{7.46}$$

$$y_{ji} = s_j \frac{\boldsymbol{r}_{yj} \cdot \boldsymbol{p}_i + t_{yj}}{1 + \eta_j \boldsymbol{r}_{zj} \cdot \boldsymbol{p}_i} \tag{7.47}$$

与归一的正交投影模型(7.40)有所区别,它包含分母项 $(1+\eta_j r_{zj} \cdot p_i)$[①]。

如果已知 $\eta_j = t_{zj}^{-1}$、结构参数 p_i 和运动参数 (R_j, t_j) 的准确值,我们就可以在公式左边(可视点测量 x_{ji} 和 y_{ji})交叉乘上分母得到校正值,双线性投影模型(7.40)对它而言是精确的。在实践中,初始重建后,η_j 可以通过比较每帧的重建点和感知到的点的位置单独估计。(旋转矩阵 r_{zj} 的第三行作为前两行的叉积总是可以获得的。)注意,因为 η_j 是从图像测量确定的,所以摄像机不需要事先进行标定,即,它们的焦距可以通过 $f_j = s_j / \eta_j$ 恢复。

一旦估计出 η_j 的值,特征位置就能够得到更正以进行新一轮的因子分解。注意,由于初始深度反向二义性的缘故,在计算 η_j 时两个重建都需要尝试。(不正确

[①] 假设光心 (c_x, c_y) 在点 $(0,0)$ 位置,像素是方形的。

的重建会使 η_j 为负，这在物理上是没有意义的。)Christy and Horaud (1996)报告称他们的算法通常可以在三到五个迭代过程中收敛，大部分的时间都在进行 SVD 分解计算。

在没有假设已知部分摄像机参数的情况下(已知光心、正方形像素和零扭曲)，另一种方法是进行完全投影因子分解(Sturm and Triggs 1996; Triggs 1996)。在这种情况下，列入摄像机矩阵的第三行(7.40)等价于每个重建测量 $x_{ji} = M_j p_i$ 乘上该点(透视)深度的逆 $\eta_{ji} = d_{ji}^{-1} = 1/(P_{j2}p_i)$，或者等价于每个测量位置乘上其投影深度 d_{ji}，

$$\hat{X} = \begin{bmatrix} d_{11}\tilde{x}_{11} & \cdots & d_{1i}\tilde{x}_{1i} & \cdots & d_{1N}\tilde{x}_{1N} \\ \vdots & & \vdots & & \vdots \\ d_{j1}\tilde{x}_{j1} & \cdots & d_{ji}\tilde{x}_{ji} & \cdots & d_{jN}\tilde{x}_{jN} \\ \vdots & & \vdots & & \vdots \\ d_{M1}\tilde{x}_{M1} & \cdots & d_{Mi}\tilde{x}_{Mi} & \cdots & d_{MN}\tilde{x}_{MN} \end{bmatrix} = \hat{M}\hat{S}. \tag{7.48}$$

在 Sturm and Triggs (1996)最初的论文中，投影深度 d_{ji} 是通过二视图重建得到的，在近期的工作中(Triggs 1996; Oliensis and Hartley 2007)，他们把 d_{ji} 初始化为 1，然后在每一轮迭代中不断地更新。Oliensis and Hartley(2007)提出了一个更新的公式，能够保证收敛到一个稳定的值。所有作者都没有建议实际将矩阵 P_j 的第三行的估计作为投影深度计算的一部分。无论在哪种情况下，我们都不清楚什么时候完全投影重建胜于部分标定重建，特别是要用它们来初始化光束平差法全部参数的时候。

因子分解方法的魅力在于它提供一种"闭式"(有时叫"线性")方法来初始化迭代方法，例如光束平差法。另一种初始化方法是针对所有摄像机都能看到的某个公共平面估计其对应的单应(Rother and Carlsson 2002)。对于已标定的摄像机，这相当于为所有摄像机估计一致的旋转，例如，使用匹配的消失点(Antone and Teller 2002)。一旦恢复出这些参数，就可以通过求解一个线性系统来获得摄像机的位置(Antone and Teller 2002; Rother and Carlsson 2002; Rother 2003)。

7.3.2 应用：稀疏 3D 模型提取

一旦估计出场景的多视角 3D 重建，就可以创建物体的带有纹理映射的 3D 模型，并从新的角度观测它。

第一步是创建比由运动到结构产生的稀疏点云更稠密的 3D 模型。另外一种方法是使用稠密多视角立体视觉(11.3 节～11.6 节)。可替代的一种更简单的方法是使用 3D 三角剖分，如图 7.6 所示，对 207 个重建得到的 3D 点进行三角剖分，生成一个表面网格。

为了得到更真实的模型，可以为每个三角面片抽取出纹理图。把 3D 三角面片上的点映射到 2D 图像上的公式是直观易懂的：只需要通过 3×4 的投影矩阵得到三角面片的 2D 坐标即可获得一个 3×3 的单应矩阵(平面透视投影)。可获得多个视角的图像时，正如在多视角重建中的情况，或者使用最近的和最正面平行的(fronto-parallel)图像，或者融合多幅图像来处理视角相关的透视收缩(Wang, Kang, Szeliski et al. 2001)，或者获得超分辨(super-resolved)的结果(Goldluecke and Cremers 2009)。另一种方法是从每个参考摄像机创建一个分开的纹理图，然后在绘制阶段融合这些纹理图，这是视角相关的纹理映射(13.1.1 节)(Debevec, Taylor, and Malik 1996; Debevec, Yu, and Borshukov 1998)。

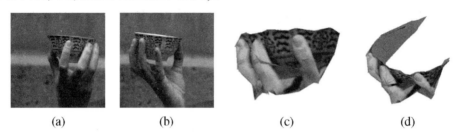

图 7.6 从一段 240 帧的视频序列中重建出茶杯的 3D 模型(Tomasi and Kanade 1992)© 1992 Springer：(a)视频的第一帧；(b)视频的最后一帧；(c)3D 模型的侧面图；(d)3D 模型的俯视图

7.4 光束平差法

我们多次在前面的内容中提到，最准确的估计结构和运动的方法是非线性最小化测量(重投影)误差，这是在摄影测绘学(和现在的计算机视觉)界中所说的光束平差法(bundle adjustment)[①]。Triggs, McLauchlan, Hartley et al. (1999) 给出了这个主题的概述，包括它的发展历程、指向摄影测绘学文献的索引(Slama 1980; Atkinson 1996; Kraus 1997)和有关测量歧义(gauge ambiguities)的微妙问题。这个主题在多视角几何学的相关教材和综述中有深入阐述(Faugeras and Luong 2001; Hartley and Zisserman 2004; Moons, Van Gool, and Vergauwen 2010)。

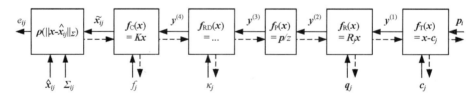

图 7.7 将 3D 点 p_i 通过一系列的变换 $f^{(k)}$ 变换到 2D 测量 x_{ij} 的变换链，每个变换由其自身的参数集控制。虚线表示在向后传递期间在计算偏导时的信息流。径向畸变函数的公式是 $f_{RD}(x) = (1+\kappa_1 r^2 + \kappa_2 r^4)x$

我们已经在前面的迭代姿态估计(6.2.2 节)的讨论中介绍了光束平差法的要素，即，公式(6.42～6.48)和图 6.5。这些公式和完全光束平差法的最大差别在于现在点

[①] "光束"(bundle)一词表示连接光心和 3D 点的射线束，"平差"(adjustment)一词表示通过迭代的方法最小化重投影误差。在视觉界中，另外的说法包括最优运动估计(Weng, Ahuja, and Huang 1993)和非线性最小二乘(附录 A.3 节)(Taylor, Kriegman, and Anandan 1991; Szeliski and Kang 1994)。

的测量 x_{ij} 不但和点(轨迹索引)i 有关, 还跟摄像机的姿态索引 j 有关,

$$x_{ij} = f(p_i, R_j, c_j, K_j), \tag{7.49}$$

且 3D 点位置 p_i 也同时得到更新。另外, 如果摄像机没有经过预先的标定, 如图 7.7 所示, 普遍的做法将是加入径向畸变参数估计(2.78)的环节,

$$f_{\mathrm{RD}}(x) = (1 + \kappa_1 r^2 + \kappa_2 r^4)x, \tag{7.50}$$

图 7.7 中的大多数盒子(变换)前面都已经介绍过, 就剩下最左边那个还没有解释。这个盒子是在用测量噪声方差 Σ_{ij} 对 2D 点位置的预测 $\hat{x}_{ij}$ 和实测 $\tilde{x}_{ij}$ 做过再缩放之后对它们进行鲁棒比较。更为详细一点儿, 这个操作可以写成

$$r_{ij} = \tilde{x}_{ij} - \hat{x}_{ij}, \tag{7.51}$$
$$s_{ij}^2 = r_{ij}^T \Sigma_{ij}^{-1} r_{ij}, \tag{7.52}$$
$$e_{ij} = \hat{\rho}(s_{ij}^2), \tag{7.53}$$

其中 $\hat{\rho}(r^2) = \rho(r)$。对应的雅可比矩阵(偏导数)可以写成

$$\frac{\partial e_{ij}}{\partial s_{ij}^2} = \hat{\rho}'(s_{ij}^2), \tag{7.54}$$
$$\frac{\partial s_{ij}^2}{\partial \tilde{x}_{ij}} = \Sigma_{ij}^{-1} r_{ij}. \tag{7.55}$$

前面提到, 链式表示法的优点是, 不仅使偏导数和雅可比矩阵的计算更简单, 而且它还可以被调整为适合于任何摄像机构型。作为例子, 我们来考虑安装在机器人上的一对摄像机, 该机器人在世界中移动, 如图 7.8a 所示。把图 7.7 最右边的两个变换用图 7.8b 中的变换替换, 除了恢复出世界的 3D 结构外, 我们还可以同时恢复出每一时刻机器人的位置和每个摄像机相对于平台的标定参数。

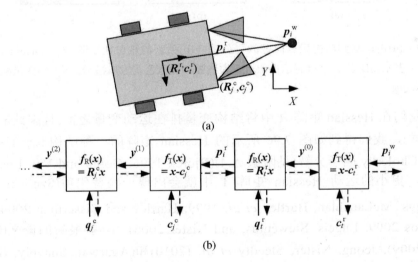

图 7.8 一个摄像机平台及其关联的变换链。(a)随着移动的平台(机器人)在世界中移动, 它在 t 时刻相对于世界坐标系的姿态可以表示为 (R_t^r, c_t^r)。每个摄像机相对于平台的姿态可以表示为 (R_j^c, c_j^c)。(b)一个世界坐标为 p_i^w 的 3D 点, 首先要变换成平台对应的坐标 p_i^r, 然后再通过该摄像机所对应的这个链的剩余部分, 如图 7.7 所示

7.4.1 挖掘稀疏性

大规模的光束平差法问题,例如要对因特网上成千上万的图像进行 3D 场景重建(Snavely, Seitz, and Szeliski 2008b; Agarwal, Snavely, Simon et al. 2009; Agarwal, Furukawa, Snavely et al. 2010; Snavely, Simon, Goesele et al. 2010),需要对上百万的测量(特征的匹配)和数以万计的未知参数(3D 点的位置和摄像机的姿态)这样的数据进行非线性最小二乘问题的求解。除非有特定的约束,否则这类问题很难求解,因为这种稠密最小二乘问题的(直接)求解的算法复杂度是未知量数目的立方级别的。

幸运的是,由运动到结构是一个二分问题。给定图像中的每个特征点 x_{ij} 依赖于一个 3D 点位置 p_i 和一个 3D 摄像机的姿态 (R_j, c_j)。正如图 7.9a 所示,每个圆(1~9)代表着一个 3D 点,每个正方形(A—D)代表一个摄像机,连线(边)表示哪些点在哪些摄像机(2D 特征)中是可见的。如果所有点的值都已知或者是固定的,则所有摄像机的公式都是独立的,反之亦然。

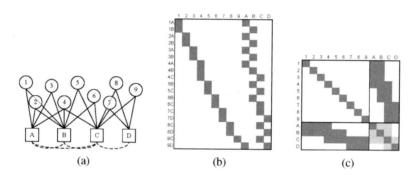

图 7.9 (a)由运动到结构玩具问题的二分图。(b)关联的雅可比矩阵 J。(c)Hessian 矩阵 A。数字表示 3D 点,字母表示摄像机。虚线弧和浅蓝色正方形指示结构(点)变量减少时出现的替代项

如果我们在 Hessian 矩阵 A 中将结构变量排在运动变量之前(且因此右手侧向量 b 也是一样),我们得到如图 7.9c 所示的 Hessian 矩阵的一种结构①。当这样的系统使用稀疏 Cholesky 因子分解(见附录 A.4)(Björck 1996; Golub and Van Loan 1996)进行求解后,更小的运动 Hessian 矩阵 A_{cc} 中就会出现一些替代项(Szeliski and Kang 1994; Triggs, McLauchlan, Hartley et al. 1999; Hartley and Zisserman 2004; Lourakis and Argyros 2009; Engels, Stewénius, and Nistér 2006)。一些最新的论文(Byrod and øAstrom 2009), Jeong, Nister, Steedly et al. (2010)和(Agarwal, Snavely, Seitz et al. 2010)考察了一些迭代(共轭梯度)方法对光束平差法进行求解。

更详细地说,缩减的运动 Hessian 矩阵通过 Schur 补集计算得到,

---

① 当摄像机的个数少于 3D 点数时(这种情况很常见),这种排序是可取的。例外的情况是,当我们在很多视频帧中跟踪少量的点时,应该反过来排序。

$$A'_{cc} = A_{cc} - A_{pc}^T A_{pp}^{-1} A_{pc}, \tag{7.56}$$

其中 A_{pp} 是点(结构)的 Hessian 矩阵(图 7.9c 的左上方区域)，A_{pc} 是点-摄像机 Hessian 矩阵(右上方区域)，而 A_{cc} 和 A'_{cc} 是点变量消减前和后的运动 Hessian 矩阵 (图 7.9c 中的右下方区域)。注意，如果两个摄像机能看到公共的 3D 点，则它们在 A'_{cc} 中对应的项就不为零。这在图 7.9a 中用虚弧线而在图 7.9c 中用浅蓝色方块标示。

不管什么时候，如果在重建算法中出现全局的参数，例如所有的摄像机拥有相同的内参数，或者是摄像机平台的标定参数，如图 7.8 所示，它们就会排在最后 (放在矩阵 A 的右下角)，达到减少填充的效果。

Engels, Stewénius, and Nistér(2006)介绍了一种很好的稀疏光束平差法，还介绍了迭代初始化所必须的所有步骤，还有系统实时设置下使用固定回溯帧数的典型计算时间。他们还推荐用齐次坐标来表示结构参数 p_i，这是一个好主意，因为它避免了接近无穷远处那些点的数值不稳定性。

光束平差法现在是许多由运动到结构问题的标准算法，普遍应用于含有几百个弱标定图像和几万个点的问题中，例如，在诸如 Photosynth 这样的系统中。(在摄影测绘学和航拍成像中，通常要解决更复杂的大问题，不过这些通常都经过仔细标定且使用的是测量过的地面控制点。)但是，随着问题的规模变大，在每次迭代中重新计算全部光束平差问题变得不太现实。

解决这个问题的一种方法是使用增量算法，其中随着时间的推移，加入新的摄像机。(如果处理的数据是从视频摄像机或移动车辆获得的，这种方法就特别有实际意义(Nistér, Naroditsky, and Bergen 2006; Pollefeys, Nistér, Frahm *et al.* 2008)。一旦采集到新的数据，卡尔曼滤波器便被用于增量地更新估计值。不幸的是，这种顺序的更新对线性最小二乘问题而言，只能说是统计上最优的。

对于非线性的问题，例如由运动到结构的问题，需要使用扩展的卡尔曼滤波器，它在当前估计附近对测量和更新方程进行线性化(Gelb 1974; Viéville and Faugeras 1990)。为了克服这个限制，可以对数据进行若干遍处理(Azarbayejani and Pentland 1995)。因为点会从视角中消失(且老的摄像机会变得不相关)，可变状态维数滤波器(VSDF)可以用于随时间而调整状态变量集合，例如，通过保留在最近 k 帧中可见的摄像机和点的轨迹(McLauchlan 2000)。使用数量固定的帧，一种更灵活的方法是通过点和摄像机向后传播更正，直至参数的变化低于阈值(Steedly and Essa 2001)。这些方法的一些变种，包括使用一个固定的时间窗口进行光束平差(Engels, Stewénius, and Nistér 2006)或者选取关键帧进行光束平差(Klein and Murray 2008)，现在被广泛用于实时跟踪和增强现实应用中，7.4.2 节将加以讨论。

在需要最大精确度的时候，最好的方法还是使用所有帧进行整体的光束平差。为了控制所产生的计算复杂度，一种方法是将这些帧分成具有局部刚性构型的子集，并优化这些子集的相对位置(Steedly, Essa, and Dellaert 2003)。另一种不同的方法是选取较少量的帧形成骨架集，使其仍然能够展开整个数据集并能产生相当精确度的重建(Snavely, Seitz, and Szeliski 2008b)。我们将在 7.4.4 节更详细地介绍这些新方法,其中将讨论由运动到结构应用于大规模图像集的情况。

尽管我们在解决大部分由运动到结构问题时选择了光束平差法和其他鲁棒的非线性最小二乘方法,但它们会受到初始化问题的影响,即,如果开始处不是充分靠近全局最优,它们可能会陷于局部能量最小值。许多系统为了缓解这种情况,在早期会保守地进行重建且谨慎地选取将合适的摄像机和点加入解(7.4.4)。然而还有另一种方法,即使用一种支持全局最优计算的范式(norm)重新定义这个问题。

Kahl and Hartley(2008)介绍了在几何重建问题中使用 L_∞ 范式的方法。这类范式的优点是能够用二次锥规划(second-order cone programming,SOCP)高效地求解全局最优值。缺点是 L_∞ 范式对外点特别敏感,所以在使用它们之前必须先用好的外点剔除方法进行处理。

7.4.2 应用:匹配运动和增强现实

由运动到结构一个有趣的应用是估计一个视频或者电影摄像机的 3D 运动,以及一个 3D 场景的几何信息,以便把 3D 图形或者计算机生成的图像(CGI)添加到场景中。在视觉特效行业,这称作"匹配运动"问题(Roble 1999),因为用来渲染图形的合成 3D 摄像机的运动必须与真实世界摄像机的运动匹配。对于轻微的运动,或者只涉及单纯摄像机旋转的运动,一个或者两个跟踪点就足以计算出必要的视觉运动。对于在 3D 场景中运动的平面,则需要用四个点来计算单应矩阵(然后可用于在场景中插入平面覆盖物),例如,在体育比赛中替换广告板的内容。

这个问题的一般化版本需要估计整个 3D 摄像机姿态及透镜的(变焦)焦距和可能的径向畸变参数(Roble 1999)。如果事先已知场景中的 3D 结构,则可以使用一些姿态估计方法,例如视图相关(Bogart 1991)或者通过镜头的摄像机控制(Gleicher and Witkin 1992),正如 6.2.3 节中介绍的那样。

对于更复杂的场景,通常更倾向于使用由运动到结构方法同时恢复出 3D 结构和摄像机运动。使用这种方法的技巧在于为了避免合成图像和真实场景之间的可视抖动,必须高精度地跟踪特征且在插入位置附近必须有足够多的特征跟踪轨迹。当今一些知名的匹配运动软件包都源于计算机视觉界由运动到结构的研究(Fitzgibbon and Zisserman 1998),例如 2d3[①]的 *boujou* 软件包,它赢得了 2002 年的艾美奖。

与匹配运动问题紧密相关的是机器人导航,其中机器人必须估计其相对于环境的位置,同时避开任何危险的障碍。这个问题通常称作"同时定位和制图"(SLAM)(Thrun, Burgard, and Fox 2005)或者"视觉测距法"(Levin and Szeliski 2004; Nistér, Naroditsky, and Bergen 2006; Maimone, Cheng, and Matthies 2007)。这些算法的早期版本使用距离传感(range-sensing)技术(例如超声波、激光测距器或者立体视觉匹配)来估计局部的 3D 几何信息,然后它们会被融入一个 3D 模型。一些更新的

---

① *http://www.2d3.com/*

方法能够单凭视觉特征跟踪完成相同的任务，有时甚至不需要立体视觉摄像机平台(Davison, Reid, Molton et al. 2007)。

另一种紧密相关的应用是增强现实，其中 3D 物体能够实时插入视频订阅源，通常用来标注或帮助用户了解一个场景(Azuma, Baillot, Behringer et al. 2001)。传统的系统需要知道场景或正在进行视觉跟踪的物体的相关先验知识(Rosten and Drummond 2005)，更新的系统能够同时建立 3D 环境的模型并且跟踪它，从而将图形叠加进去。

Klein and Murray(2007)介绍了一个并行跟踪和映射(PTAM)系统，这个系统能够同时对从视频序列中选取的关键帧进行完整的光束平差，并且在中间帧中实现鲁棒的实时姿态估计。图 7.10a 展示了他们这个系统的一个用法示例。一旦初始的 3D 场景被重建出来，就可以估计一个主平面(在这个例子中，桌面)，然后虚拟插入 3D 动画角色。Klein and Murray(2008)通过加入边缘特征扩展了以前的系统以处理更快的摄像机运动，这些边缘特征即便兴趣点变模糊依然能够被检测到。他们还使用一种直接(基于亮度)旋转估计算法来处理甚至更快的运动。

没有将整个场景建模为一个刚性参考帧，Gordon and Lowe(2006)首先使用特征匹配和由运动到结构方法为单个物体建立一个 3D 模型。一旦系统被初始化，他们就针对每一个新的帧，使用 3D 实例识别算法找出这个物体及其姿态，然后将图形物体叠加到那个模型上，如图 7.10b 所示。

图 7.10 3D 增强现实：(a)达斯·维达和一群伊沃克族人在桌面上战斗，这是用实时的、基于关键帧的由运动到结构恢复得到的(Klein and Murray 2007) © 2007 IEEE；(b)一个虚拟的茶壶放在一个真实世界的咖啡杯上，咖啡杯的姿态在每帧中做重新识别(Gordon and Lowe 2006)© 2007 Springer

尽管现在可靠地跟踪物体和环境是一个已经得到很好解决了的问题，但是确定哪些像素会被前景元素遮挡这个问题还是一个悬而未决的问题(Chuang, Agarwala, Curless et al. 2002; Wang and Cohen 2007a)。

7.4.3 不确定性和二义性

因为由运动到结构涉及对许多高度耦合的参数的估计，往往并不知道"真值"

分量，所以由运动到结构算法产生的估计经常呈现出大量的不确定性(Szeliski and Kang 1997)。一个例子是典型的浅浮雕歧义，它使我们很难同时估计出场景的 3D 深度和摄像机的运动量(Oliensis 2005)。[①]

正如之前所说的，单凭单目视觉测量本身，是不可能得到重建场景的唯一性的坐标框架和尺度的。(使用立体平台的时候，我们如果知道摄像机之间的距离(基线)，就能够恢复尺度。)这个七个自由度的测量二义性(gauge ambiguity)使计算与 3D 重建关联的协方差矩阵变得棘手(Triggs, McLauchlan, Hartley et al. 1999; Kanatani and Morris 2001)。一种忽略测量自由度(不确定性)的计算协方差矩阵的简单方法是，丢弃信息矩阵(协方差矩阵的逆)的七个最小的特征值，它们的数值在相差一个噪声缩放意义下等价于该问题的 Hessian 矩阵 A(参见 6.1.4 节和附录 B.6 节)。在这样做之后，对所得矩阵求逆即可得到参数协方差矩阵的一个估计值。

Szeliski and Kang(1997)使用这种方法来可视化由运动到结构典型问题中的最大变化方向。不足为奇的是，他们发现(忽略测量自由度)，对于从少量邻近视点观测一个物体的类似问题而言，最不确定的是相对于摄像机运动范围的3D结构的深度。[②]

我们还可以估计单个参数的局部和边迹不确定性简单对应于从整个协方差矩阵取出子矩阵块。在一些特定条件下，例如在摄像机的姿态相对 3D 点位置比较确定的时候，这种不确定性估计会很有意义。然而，在很多情况下，单独的不确定性测量可能罩住(mask)一个范围，在这个范围上重建误差是相互关联的，正因为此，考查最大联合变化的最初几个模态会很有帮助。

测量二义性对由运动到结构问题和特别对光束平差法的影响，另一种表现方式是它们使 Hessian 矩阵 A 变得不满秩，因而不能求逆。有许多方法可用来缓解这个问题(Triggs, McLauchlan, Hartley et al. 1999; Bartoli 2003)。但在实践中，把 Hessian 矩阵的对角元素乘上一个小常量 $\lambda\mathrm{diag}(A)$ 再加到 Hessian 矩阵 A，正如 Levenberg-Marquandt 的非线性最小二乘算法(附录 A.3 节)所做的那样，通常很有效。

7.4.4 应用：由因特网照片重建

由运动到结构最广泛的应用是使用视频序列和图像集进行物体和场景的 3D 重建(Pollefeys and Van Gool 2002)。最近十年，能够自动完成这个任务的方法呈现出爆炸性增长，这些方法不需要人工匹配或事先测量的地平面控制点。许多方法都假设使用相同的摄像机来取景，所以图像都有相同的内参数(Fitzgibbon and Zisserman 1998; Koch, Pollefeys, and Van Gool 2000; Schaffalitzky and Zisserman 2002; Tuytelaars and Van Gool 2004; Pollefeys, Nistér, Frahm et al. 2008; Moons, Van Gool, and Vergauwen 2010)。这些方法中，很多都会先进行稀疏特征匹配和由运动到结构

[①] "浅浮雕歧义"这个名字源于一些雕塑装饰品，通常出现在室内的横楣上，这些横楣雕刻得比其实际上所占据的浅。当阳光从上方照在横楣上时，相对深度和光照角度之间的二义性使其呈现出具有真实的3D深度的视觉效果(12.1.1 节)。
[②] 减少这种二义性的简单方法是使用广角摄像机(Antone and Teller 2002; Levin and Szeliski 2006)。

的计算,再使用多视角立体视觉方法对得到的结果计算得到稠密 3D 表面模型(11.6 节)(Koch, Pollefeys, and Van Gool 2000; Pollefeys and Van Gool 2002; Pollefeys, Nistér, Frahm *et al.* 2008; Moons, Van Gool, and Vergauwen 2010)。

在这个领域中,最新的进展是对互联网上成千上万的图像应用由运动到结构和多视角立体视觉方法来进行 3D 重建,在这样的情况下,对拍摄这些照片的摄像机几乎一无所知(Snavely, Seitz, and Szeliski 2008a)。在进行由运动到结构的计算之前,必须首先建立不同图像对的稀疏对应,然后将这些对应连接成特征轨迹(将单个的 2D 图像特征和全局 3D 点关联起来)。因为所有图像对的比较量级是 $O(N^2)$,速度很慢,所以在识别领域中提出了加速这一过程的一些方法(14.3.2 节)(Nistér and Stewénius 2006; Philbin, Chum, Sivic *et al.* 2008; Li, Wu, Zach *et al.* 2008; Chum, Philbin, and Zisserman 2008; Chum and Matas 2010)。

为了开始重建过程,选择一对好的图像是很重要的,既有大量一致的匹配(降低不正确对应的可能性,也有很明显的平面外视差①以确保可以获得稳定的重建(Snavely, Seitz, and Szeliski 2006)。照片所关联的 EXIF 标签可以用于得到一个很好的摄像机焦距的初始估计,但这并不总是必须的,因为这些参数会作为光束平差过程中的一部分被重新调整。

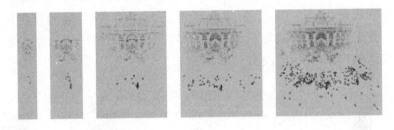

图 7.11 增量或由运动到结构(Snavely, Seitz, and Szeliski 2006)©2006 ACM:从最初的许愿泉两帧重建开始,使用姿态估计加入一组图像,使用光束平差法对它们的位置(和 3D 模型一起)进行优化

一旦重建出初始图像对,就可以估算出那些摄像机的姿态,只要它们看到的 3D 点足够多(6.2 节),而且整个摄像机集合和特征对应可以用于执行下一轮光束平差。图 7.11 展示了增量光束平差算法的进展过程,其中每次光束平差后都加入了更多的摄像机集合,图 7.12 展示了另外一些结果。另一种播种和生长的方法是,首先根据图像三元组重建,然后分层地将三元组并入更大的集合,正如 Fitzgibbon and Zisserman(1998)所描述的那样。

不幸的是,随着增量式由运动到结构算法不断加入新的摄像机和点,它会变得非常慢。为摄像机姿态更新直接求解一个包含 $O(N)$ 个方程的稠密系统,需要花 $O(N^3)$ 的时间;而由运动到结构问题很少是稠密的,像城市广场那样的场景,大部分摄像机看到的点都是一样的。每次加入少量摄像机后重新进行光束平差法,所产

---

① 计算这个的简单方法是为对应鲁棒拟合一个单应矩阵并测量重投影误差。

生的运行时间是数据集中图像数的二次方。解决这个问题的一种方法是选取更少量图像进行原始场景的重建而在最后阶段混入剩余图像。

图 7.12　Snavely, Seitz, and Szeliski(2006)©2006 ACM 提出的增量或由运动到结构算法产生的 3D 重建：(a)特拉法尔加广场的摄像机和点云；(b)在一幅中国长城图上叠加的摄像机与点；(c)注册到航拍照片的布拉格旧城广场重建的俯视图

Snavely, Seitz, and Szeliski(2008b)提出一种方法来计算这样的图像骨架集，保证产生的重建误差在最优重建精确度的限定范围内。他们的算法首先计算所有重叠图像间的逐对不确定性(位置协方差)，然后把它们链接起来为每对远距离图像的相对不确定性估计一个下界。构造这样的骨架集使其任意图像对之间最大不确定性的增长不超过一个常量因子。图 7.13 展示了罗马万神殿 784 幅图算得的骨架集的一个例子。可以看出，尽管骨架集只包含一小部分原始图像，但骨架集的形状和完整光束平差法重建的形状实际上没有什么区别。

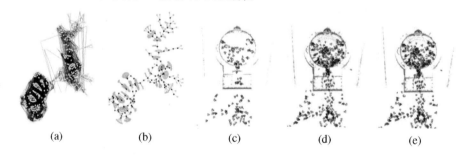

图 7.13　使用骨骼集的大规模从运动到结构(Snavely, Seitz, and Szeliski 2008b) © 2008 IEEE：(a)784 幅图的原始匹配图；(b)包含 101 幅图的骨骼集；(c)从骨骼集恢复出来的场景(万神殿)的从上到下视角；(d)通过姿态参数估计，加入除骨骼集以外余下各图后得的重建结果；(e)光束平差法最后一轮重建结果，和使用骨骼集得到的结果基本相同

能够使用大型非结构化图像集自动重建 3D 模型，引发了很多附加的应用，包括自动找到和标记感兴趣位置和区域(Simon, Snavely, and Seitz 2007; Simon and Seitz 2008; Gammeter, Bossard, Quack et al. 2009)和在大型图像集中聚类使其能被自动标注(Li, Wu, Zach et al. 2008; Quack, Leibe, and Van Gool 2008)。一些这样的应用将在 13.1.2 节中详细介绍。

7.5 限定结构和运动

最一般的由运动到结构算法对重建的物体或场景不做任何先验假设。然而，在很多情况下，场景中要包含较高层的几何基元，例如直线和平面。这些可以提供与兴趣点有关的补充信息，还可作为 3D 建模和可视化的有用构件。此外，这些基元还时常以特殊的关系排列，即，许多直线和平面不是互相平行就是彼此正交。建筑场景和模型尤其如此，详情将在 12.6.1 节介绍。

有时，我们不在场景结构中挖掘规整性，而利用限定运动模型的优势。例如，如果一个感兴趣物体在转盘上旋转(Szeliski 1991b)，即绕着固定但未知的轴旋转，我们可以运用特殊方法来恢复这种运动(Fitzgibbon, Cross, and Zisserman 1998)。在另外一些情况中，摄像机可能在绕着某个旋转中心的固定圆弧上运动(Shum and He 1999)。移动立体视觉摄像机平台或装备多个固定摄像机的运动车辆，这样的特殊获取装置可以充分利用单个摄像机相对获取平台(大多数)都是固定的这种知识，如图 7.8 所示。[①]

7.5.1 基于线条的方法

众所周知，成对图像的极线几何单靠线条匹配是不能恢复的，即便摄像机已经标定好。为了了解这种情况，可以想象一下将每幅图上的线条集投影到空间中的 3D 平面集。你可以移动两个摄像机到任意构型，仍然可以得到 3D 线条的有效重建。

当线条在三个或者更多视角可见时，可以用三视点(三焦点)张量(trifocal tensor)将线条从一对图转换到另一对图(Hartley and Zisserman 2004)。三视点张量还可以只根据线条匹配来计算。

Schmid and Zisserman(1997)介绍了一种广泛应用的 2D 线条匹配方法，这种方法基于 15×15 像素相关评分的平均值来进行匹配，平均值是根据其共有线段交集的所有像素计算得到的[②]。在他们的系统中，假设极线几何已知，例如，由点匹配计算得到。对于宽基线，与经过此 3D 线条的平面对应的所有可能的单应矩阵都用来卷绕像素，然后选取相关评分最高的。对于图像三元组，首先用三视点张量来验证线条在几何上的对应位置，然后再计算线段之间的相关评分。图 7.14 展示的是他们这个系统的结果。

Bartoli and Sturm(2003)介绍了这样一个完整的系统，该系统从手工线条对应计算出的三视角关系(三视点张量)扩展到对所有线条和摄像机参数进行完整的光束平差法。他们这个方法的关键在于使用普吕克坐标(2.12)来参数化线条并且直接最小

[①] 因为机械的咬合性和抖动，所以允许单个摄像机沿标称位置少量旋转可能是明智的。
[②] 由于线条时常出现在深度或者方向的不连续性处，所以计算相关评分或者匹配颜色直方图(Bay, Ferrari, and Van Gool 2005)，更好的方法可能是对线条的两边分别进行计算。

化重投影误差。还可以用线段的两个端点表示 3D 线段，衡量垂直于每幅图中检测到的 2D 线段的重投影误差，或者用一个与线段方向对齐的狭长不确定性椭圆求得 2D 误差(Szeliski and Kang 1994)。

图 7.14 两幅玩具房子的图像及其匹配的 3D 线段(Schmid and Zisserman 1997)© 1997 Springer

没有重建 3D 线条，Bay, Ferrari, and Van Gool(2005)使用 RANSAC 将线条组织为可能共面的子集。随机选取四条线来计算一个单应，再通过基于颜色直方图的相关度评分方法证实这些线段和其他合理线段的匹配关系来对单应进行验证。同属于一个平面的线条的 2D 交点将用作虚拟测量来估计极线几何，这比直接使用单应更精确。

不同于将线条分组为共面子集，另一种方法是基于平行性对线条进行分组。每当有三条或者更多 2D 线条交于一个共同的消失点时，它们很有可能在 3D 空间是平行的。通过找到一幅图像上的多个消失点(4.3.3 节)并建立不同图像中类似消失点之间的对应关系，可以直接估计出不同图像之间的相对旋转关系(常常还有摄像机的内参数)(6.3.2 节)。

Shum, Han, and Szeliski(1998)介绍了一个 3D 建模系统，首先用多幅图像构建已标定的全景图(7.4 节)，然后根据用户所画的垂直和水平线条给平面区域划界。这些线条最初用来为每一张全景图建立一个绝对旋转矩阵，然后(和推导所得的顶点和平面)用于推断一个 3D 结构，它可以从一幅或多幅图像在相差一个尺度因子的意义下恢复出来(12.15 节)。

Werner and Zisserman(2002)提出了一种基于线条的由运动到结构的全自动方法。在他们的系统中，他们首先找到线条，并根据每幅图像上的消失点将这些线条分组(4.3.3 节)。然后用消失点标定摄像机，即执行"度量升级"(6.3.2 节)。同时使用表观(Schmid and Zisserman 1997)和三视点张量来匹配有共同消失点的线条。图 12.16a 展示了根据共同消失点方向(3D 方向)来进行颜色编码的 3D 线条的结果集合。这些线条随后被用于推断平面和场景一个块结构模型，12.6.1 节将详细介绍。

7.5.2 基于平面的方法

在一些平面结构丰富的场景(例如，在建筑或家具之类的手工物体)中，使用基于特征或者基于亮度的方法直接估计不同平面之间的单应变换是有可能的。原则上，这个信息可用于同时推断摄像机的姿态和平面方程，即，计算基于平面的由运动到结构。

Luong and Faugeras(1996)介绍了怎样使用代数运算和最小二乘计算直接从两个或更多单应变换计算得到一个基本矩阵。不幸的是，这种方法的效果通常很差，因为代数误差与有意义的重投影误差并不对应(Szeliski and Torr 1998)。

一种更好的方法是在一定的区域内幻生虚拟的点对应，从中计算出每个单应，并将它们添加到一个标准的由运动到结构算法中(Szeliski and Torr 1998)。一种更好的方法是使用带有明确平面方程的完整光束平差法，并用额外的约束强制重建的共面特征准确位于其对应平面上。(一个合乎原则的做法是为每个平面建立一个坐标系，例如，取其中一个特征点作为原点，然后在这个平面内的其他点都用 2D 坐标表示。)Shum, Han, and Szeliski(1998)开发的系统是这种方法的实例，其中场景中线条的方向和平面的法向量都由用户预先设定。

7.6 补充阅读

由运动到结构这个主题在多视角几何的相关书籍和回顾论文中有广泛的论述(Faugeras and Luong 2001; Hartley and Zisserman 2004; Moons, Van Gool, and Vergauwen 2010)。对于二视图重建，Hartley(1997a)写了一篇引用率非常高的论文，内容是含有合理的点归一化的计算本质矩阵或基本矩阵的"八点算法"。当摄像机被标定后，Nistér(2004)的五点算法与 RANSAC 相结合能够获得由最少点恢复得到的初始的重建。当摄像机未被标定时，各种各样的自标定方法可以在如下的工作中找到：Hartley and Zisserman(2004)；Moons, Van Gool, and Vergauwen(2010)——我只简要地介绍了其中一种最简单的方法，即 Kruppa 方程(7.35)。

在某些应用中，需要逐帧跟踪点的轨迹，这时便可以使用基于正交的摄像机模型(Tomasi and Kanade 1992; Poelman and Kanade 1997; Costeira and Kanade 1995; Morita and Kanade 1997; Morris and Kanade 1998; Anandan and Irani 2002)或者投影扩展模型(Christy and Horaud 1996; Sturm and Triggs 1996; Triggs 1996; Oliensis and Hartley 2007)的因子分解方法。

Triggs, McLauchlan, Hartley et al. (1999)提供了很好的关于光束平差法的教程和综述，而 Lourakis and Argyros(2009)和 Engels, Stewénius, and Nistér(2006)提供了实现和高效实践所需的相关技巧。多视角几何的教材和综述也介绍了光束平差法(Faugeras and Luong 2001; Hartley and Zisserman 2004; Moons, Van Gool, and Vergauwen 2010)。处理更大规模问题的方法在如下论文中有论述：Snavely, Seitz, and Szeliski(2008b); Agarwal, Snavely, Simon et al. (2009); Jeong, Nistér, Steedly et al. (2010); Agarwal, Snavely, Seitz et al. (2010)。尽管光束平差法通常在增量重建算法的内循环中调用(Snavely, Seitz, and Szeliski 2006)，但分层的(Fitzgibbon and Zisserman 1998; Farenzena, Fusiello, and Gherardi 2009)和全局的(Rother and Carlsson 2002; Martinec and Pajdla 2007)初始化方法也是可能的且有时甚至是更好的选择。

由运动到结构最初应用于动态场景时，非刚性由运动到结构(Torresani, Hertzmann, and Bregler 2008)这个主题(本书未涉及)将变得更重要。

7.7 习　　题

习题 7.1：三角测量

用你在习题 6.7 中建立和测试的标定模式测试你的三角测量精度。或者，生成合成的 3D 点和摄像机并且在 2D 的点测量中加入噪声。

1. 假设你知道摄像机的姿态，即摄像机矩阵。使用到射线的 3D 距离或者方程(7.5 和 7.6)的线性化版本计算 3D 点位置的初始集合。将它们与你知道的真值位置进行对比。
2. 用迭代的非线性最小化方法改进初始估计并报告精度的改进情况。
3. (选做)用 Hartley and Sturm (1997)描述的方法进行二视图三角测量。
4. 看看 Hartley and Sturm (1997)或者 Hartley (1998)提到的失败情况在实践中是否出现。

习题 7.2：本质和基本矩阵

实现 7.2 节中论述的二视图本质矩阵 E 和基本矩阵 F 的估计算法，为了更好地克服噪声，应带有适当的缩放。

1. 用习题 7.1 的数据验证你的算法并且给出它们的精度。
2. (选做)实现其中一种改进的估计本质矩阵 E 或者基本矩阵 F 的算法，例如，使用重归一化方法(Zhang 1998b; Torr and Fitzgibbon 2004; Hartley and Zisserman 2004)，RANSAC(Torr and Murray 1997)，Nistér (2004)提出的最小均方差(LMS)或者五点算法。

习题 7.3：视角变形与插值

实现自动的视角变形，即计算二视图由运动到结构，然后使用这些结果针对从一幅图像到下一幅图像生成平滑的动画(7.2.3 节)。

1. 判断如何表示你的 3D 场景，例如，对已匹配点进行 Delaunay 三角剖分，并且判断如何处理靠近边界的三角形。(提示：为场景拟合一个平面，例如，在大多数点后面。)
2. 计算中间摄像机的位置和方向。
3. 将每个三角形卷绕到它的新位置，最好使用正确的透视投影(Szeliski and Shum 1997)。
4. (选做)假如你有一个稠密的 3D 模型(例如，从立体视觉得到)，决定如何处理"裂缝"。
5. (选做)对于一个非刚性场景，例如，人脸上带有不同表情的两张图片，不是所有的匹配点都遵循极线几何。决定如何处理这种情况以得到最好的效果。

习题 7.4：因子分解

用习题 4.5 计算得到的点迹实现 7.3 节中描述的因子分解算法。

1. (选做)实现不确定性缩放(Anandan and Irani 2002)，并说说是否改进了你的实验结果。

2. (选做)实现在 7.3.1 节中介绍的其中一种对因子分解的透视改进方法(Christy and Horaud 1996; Sturm and Triggs 1996; Triggs 1996)。这个能明显降低重投影误差吗？你能将这个重建升级为一个度量重建吗？

习题 7.5：光束平差器

实现一个完整的光束平差器。这个可能听起来困难，但实际上并不是这样。

1. 设计内部的数据结构和外部的文件表示方法，存放你的摄像机参数(位置、方向和焦距)、3D 点位置(欧氏或者齐次坐标)和 2D 点轨迹(帧和点的标识以及 2D 位置)。

2. 用一些其他的方法，例如因子分解，从 2D 点轨迹(例如，出现在所有帧中的点的子集)去初始化 3D 点和摄像机的位置。

3. 写出图 7.7 中的对应于向前变换的代码，即对于每个 2D 点测量，取对应的 3D 点，将其通过摄像机变换(包括透视投影和焦距缩放)映射到 2D 上，比较它和 2D 点测量以得到残余误差。

4. 取残余误差，并且用图 7.7 和公式(6.47)所示的反向链相对于所有未知的运动和结构参数计算它的导数。这给你提供了在公式(6.13~6.17)和公式(6.43)中所使用的稀疏的雅可比矩阵 J。

5. 用稀疏最小二乘或者线性系统的解决方法——例如 MATLAB，SparseSuite 或者 SPARSKIT(参见附录 A.4 节和 A.5 节)——求解对应的线性化系统，加上对角项乘上个小常量的和作为预处理，就像在 Levenberg–Marquardt 中那样。

6. 更新参数，确定旋转矩阵仍是标准正交的(例如，从你的四元组重新计算它们)，然后在监控残差的同时继续迭代。

7. (选做)应用"Schur 补集的技巧"(7.56)减小要解决系统的规模(Triggs, McLauchlan, Hartley *et al.* 1999; Hartley and Zisserman 2004; Lourakis and Argyros 2009; Engels, Stewénius, and Nistér 2006)。

8. (选做)实现你自己的迭代稀疏求解方法(例如共轭梯度)，和直接方法比较性能。

9. (选做)使你的光束平差器对外点有好的鲁棒性，或者尝试加入(Engels, Stewénius, and Nistér 2006)提到的其他改进方法。你能想到其他能使算法更快或更鲁棒的方法吗？

习题 7.6：匹配运动和增强现实

使用前面习题的结果把一个经过渲染的 3D 模型叠加在视频上方。更多细节和想法请查看 7.4.2 节。检查"锁定"物体的效果。

习题 7.7：基于线条的重建

增强前面完成的光束平差器以包含线条，还可能包含已知的 3D 方向。

或者，使用共面点线集形成面并强制共面(Schaffalitzky and Zisserman 2002; Robertson and Cipolla 2002)。

习题 7.8：灵活的光束平差法

设计一个光束平差法能够允许任意的变换链和关于未知量的先验知识，就像图 7.7 和 7.8 所展示的那样。

习题 7.9：无序图像匹配

从任意的照片汇集计算摄像机姿态和场景的 3D 结构(Brown and Lowe 2003; Snavely, Seitz, and Szeliski 2006)。

第 8 章
稠密运动估计

8.1 平移配准
8.2 参数化运动
8.3 基于样条的运动
8.4 光流
8.5 层次运动
8.6 补充阅读
8.7 习题

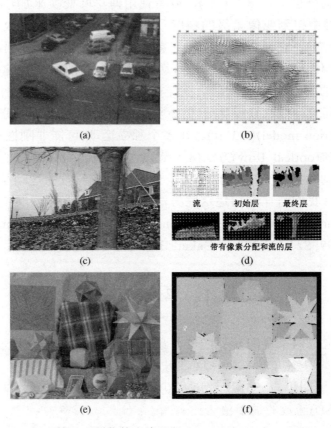

图 8.1 运动估计：(a 和 b)基于正则化的光流(Nagel and Enkelmann 1986)© 1986 IEEE；(c 和 d)分层运动估计(Wang and Adelson 1994)© 1994 IEEE；(e 和 f)测试数据集中的样例图片和真实光流(Baker, Black, Lewis *et al*. 2007) © 2007 IEEE

视频序列中的图像配准和运动估计算法广泛应用在计算机视觉中。例如，帧速率图像配准广泛应用于摄像机和数码摄像机中，从而实现图像稳定化(image stabilization, IS)。

一个早期广泛使用的图像注册算法(image registration)的例子是基于块的平移配准(光流)方法(Lucas and Kanade，1981)。该算法的大量变形应用在几乎所有的运动补偿视频压缩算法中，例如 MPEG 和 H.263(Le Gall 1991)。与此相同，参数化运动估计算法也有着大量的应用，包括视频摘要(Teodosio and Bender 1993；Irani and Anandan 1998)、视频稳定化(Hansen，Anandan，Dana et al.1994；Srinivasan, Chllappa, Veeraraghavan et al. 2005；Matsushita, Ofek, Ge et al. 2006)以及视频压缩(Irani, Hsu, and Anandan 1995；Lee, ge Chen, lung Bruce Lin et al. 1997)。医学图像处理和遥感中研究出更多复杂的图像注册算法。Brown(1992)、Zitov'aa and Flusser(2003)、Goshtasby(2005)和 Szeliski(2006a)对图像注册方法做了综述。

为了在两幅或更多的图像中做运动估计，必须挑选一个合适的误差度量来比较图像(8.1 节)。之后，需要有一个合适的搜索方法。最简单的方法是比较所有可能的配准，也称为全搜索(full search)。实际应用，这种方法太慢，一般使用基于金字塔的由粗到精的分层估计(8.1.1 节)，或者使用傅里叶变换来加速计算(8.1.2 节)。

为了在配准中得到亚像素级的精度，通常使用基于图像函数 Taylor 展开的逐次求精方法(8.1.3 节)。它同样可以应用在参数化运动模型(parametric motion model)中(8.2 节)，参数化运动模型对旋转、剪切之类的全局图像变换建模。通过学习被跟踪场景或物体的典型的动态或运动统计模型(如行人的自然步态)，可以使运动估计算法变得更为可靠(8.2.2 节)。对于更多复杂的运动，可以使用分段参数的样条运动模型(spline motion model)(8.3 节)。在多个独立运动(可能非刚性)的情况下，需要使用一般性的光流(optical flow)方法(8.4 节)。对于包含大量遮挡的更复杂的运动，可以使用层次运动模型(8.5 节)，它将场景分解为若干一致的运动层。

在本章中，我们将详细描述这些方法。更详细的内容可以参考运动估计的回顾和对比文章(Barron, Fleet, and Beauchemin 1994; Mitiche and Bouthemy 1996; Stiller and Konrad 1999; Szeliski 2006a; Baker, Black, Lewis et al. 2007)。

8.1 平移配准

在两幅图像或两个块间建立配准，最简单的方法是相对于一幅图平移另一幅。给定一个在离散像素位置上 $\{x_i = (x_i, y_i)\}$ 采样的模板图像 $I_0(x)$，我们希望能够找到其在图像 $I_1(x)$ 中的位置。该问题的最小二乘解是找到误差平方和(sum of squared differences，SSD)函数的最小值。

$$E_{\text{SSD}}(u) = \sum_i [I_1(x_i + u) - I_0(x_i)]^2 = \sum_i e_i^2, \tag{8.1}$$

其中 $u = (u,v)$ 是位移，$e_i = I_1(x_i + u) - I_0(x_i)$ 称为残差(residual error，在视频编码文献

中称为平移帧差分(displaced frame difference))[①]。(这里我们暂时忽略 I_0 可能落在 I_1 边界外而不可见的情况。)两幅图像中对应像素的值保持相同的假设称为"亮度恒常性约束"[②](brightness constancy constraint)。

一般来说，位移 u 是一个分数，所以 $I_1(x)$ 必须要使用一个合适的插值函数。实际中，常常会用到双线性插值，但双三次插值(bicubic interpolation)的结果更好一些(Szeliski and Scharstein 2004)。彩色图像可以用所有三色通道的差别和来处理，或者先将其转化到其他的颜色空间，抑或仅仅使用亮度(常用在视频编码中)。

鲁棒误差度量 将二次误差项替换为鲁棒函数 $\rho(e_i)$，可以使之前的误差度量相对于外点(异常值)更鲁棒(Huber 1981; Hampel, Ronchetti, Rousseeuw et al. 1986; Black and Anandan 1996; Stewart 1999):

$$E_{\mathrm{SRD}}(u) = \sum_i \rho(I_1(x_i+u) - I_0(x_i)) = \sum_i \rho(e_i). \tag{8.2}$$

鲁棒函数 $\rho(e)$ 比最小二乘所使用的二次惩罚函数的增长要慢。其中之一是绝对误差和(sum of absolute differences, SAD)[③]又叫 L_1 范数，由于其速度快，所以常被用于视频编码中的运动估计:

$$E_{\mathrm{SAD}}(u) = \sum_i |I_1(x_i+u) - I_0(x_i)| = \sum_i |e_i|. \tag{8.3}$$

但是，由于该函数在原点处不是可导的，它并不适合于梯度下降的方法，如 8.1.3 节中的算法。

相反，我们常用一种从原点处先二次增长之后减缓的平滑变化的函数。Black and Rangarajan(1996)讨论了很多此类函数，包括 Geman-McClure 函数:

$$\rho_{\mathrm{GM}}(x) = \frac{x^2}{1+x^2/a^2}, \tag{8.4}$$

其中 a 是用于外点阈值(outlier threshold)的常量。可以通过鲁棒统计来获得一个合适的该阈值(Huber 1981; Hampel, Ronchetti, Rousseeuw et al. 1986; Rousseeuw and Leroy 1987)。例如，计算平均绝对偏差(median absolute deviation, MAD)，MAD = $med_i|e_i|$，并乘以 1.4 得到一个内点噪声过程的标准差的鲁棒估计。

空间变化权重 在一个配准中，一些待比较的像素可能在原始图像边界之外，之前的误差度量忽略了这一点。此外，我们可能想部分地或者完全消除某些像素的影响。例如，拼接马赛克时，我们希望能选择性地"消除"图像的某些部分不做考虑，其中切除了不需要的前景物体。在背景稳定化的应用中，图像中部常常包含摄像机在跟踪的独立运动的物体，我们希望能降低其权重。

以上所有任务都可以通过对两幅要匹配的图像中的每个像素关联随空间变化的权重来解决。于是，误差度量变为带权重的(或带窗的)误差平方和(SSD)函数，

[①] 使用最小二乘的通常的理由是它是高斯噪声下最优的估计。参见后文关于鲁棒误差度量的讨论和附录 B.3。
[②] 亮度恒常性(Horn 1974)是物体在变化光照条件下保持其被感知的亮度的倾向。
[③] 在视频压缩中，如 H.264 标准(http://www.itu.int/rec/T-REC-H.264)，常用绝对变换误差和(sum of absolute transformed differences, SATD)，它在频率变换空间中计算误差，例如用 Hadamard 变换，因为它可以更准确地预测质量(Richardson 2003)。

$$E_{\text{WSSD}}(\boldsymbol{u}) = \sum_i w_0(\boldsymbol{x}_i)w_1(\boldsymbol{x}_i+\boldsymbol{u})[I_1(\boldsymbol{x}_i+\boldsymbol{u})-I_0(\boldsymbol{x}_i)]^2, \quad (8.5)$$

其中权函数 ω_0, ω_1 在图像边界外为 0。

如果允许大范围运动，上面的度量会有倾向于重叠范围较小的解的偏向。为此，将带窗的误差平方和除以重叠区域的面积

$$A = \sum_i w_0(\boldsymbol{x}_i)w_1(\boldsymbol{x}_i+\boldsymbol{u}) \quad (8.6)$$

以计算每个像素(或平均)的平方像素误差 E_{WSSD}/A。它的平方根称为"均方根"亮度误差

$$RMS = \sqrt{E_{\text{WSSD}}/A} \quad (8.7)$$

它常出现在对比研究中。

偏差与增益(曝光差) 通常，要对准的两幅图像有着不同的曝光值。偏差与增益(bias and gain)模型是两幅图像间的一种简单的线性(仿射)亮度变化模型，

$$I_1(\boldsymbol{x}+\boldsymbol{u}) = (1+\alpha)I_0(\boldsymbol{x})+\beta, \quad (8.8)$$

其中 β 是偏差，α 是增益(Lucas and Kanade 1981; Gennert 1988; Fuh and Maragos 1991; Baker, Gross, and Matthews 2003; Evangelidis and Psarakis 2008)。最小二乘公式变为

$$E_{\text{BG}}(\boldsymbol{u}) = \sum_i [I_1(\boldsymbol{x}_i+\boldsymbol{u})-(1+\alpha)I_0(\boldsymbol{x}_i)-\beta]^2 = \sum_i [\alpha I_0(\boldsymbol{x}_i)+\beta-e_i]^2. \quad (8.9)$$

和对应块的简单平方差相比，这里需要使用计算代价更大的线性回归(附录A.2)。注意，对于彩色图像，对每个通道分别计算偏差和增益来补偿一些数字摄像机的自动色彩修正(2.3.2 节)，可能是必要的。偏差与增益补偿在视频编码中也要用到，称为"加权预测"(Richardson 2003)。

更一般的(空间变化的，非参数的)亮度变化模型用在(Negahdaripour 1998; Jia and Tang 2003; Seitz and Baker 2009)，它是注册过程计算的一部分。它在处理由广角镜头、大光圈或者镜头箱等引起的诸如虚影的局部变化时很有用。它也可以用于比较图像值之前的预处理，例如使用带通滤波过的图像(Anandan 1989; Bergen, Anandan, Hanna et al. 1992)、梯度(Scharstein 1994; Papenberg, Bruhn, Brox et al. 2006)或其他如直方图或者排序的局部变换(Cox, Roy, and Hingorani 1995; Zabih and Woodfill 1994)或者最大化"互信息"(mutual information)(Viola and Wells III 1997; Kim, Kolmogorov, and Zabih 2003)。Hirschmuller and Scharstein(2009)对比了一些此类方法，并报告了它们在不同曝光场景中的相对性能。

相关性 相关性(correlation)是另一种获取亮度差的方法，即使两幅图像的内积(互相关)最大化，

$$E_{\text{CC}}(\boldsymbol{u}) = \sum_i I_0(\boldsymbol{x}_i)I_1(\boldsymbol{x}_i+\boldsymbol{u}). \quad (8.10)$$

这使图像倾向于排列成线而无论其相对大小和偏差，因而初看起来不再需要对偏差与增益建模了。但是，这实际上并不正确。如果在 $I_1(x)$ 中存在某个非常亮的块，最大的内积会出现在其上。

于是，我们往往更常使用规范化互相关(normalized cross-correlation，NCC)，

$$E_{\text{NCC}}(\boldsymbol{u}) = \frac{\sum_i [I_0(\boldsymbol{x}_i) - \overline{I_0}] [I_1(\boldsymbol{x}_i + \boldsymbol{u}) - \overline{I_1}]}{\sqrt{\sum_i [I_0(\boldsymbol{x}_i) - \overline{I_0}]^2} \sqrt{\sum_i [I_1(\boldsymbol{x}_i + \boldsymbol{u}) - \overline{I_1}]^2}}, \tag{8.11}$$

其中

$$\overline{I_0} = \frac{1}{N} \sum_i I_0(\boldsymbol{x}_i) \quad \text{and} \tag{8.12}$$

$$\overline{I_1} = \frac{1}{N} \sum_i I_1(\boldsymbol{x}_i + \boldsymbol{u}) \tag{8.13}$$

是对应块的图像均值，N 是块中的像素数量。规范化互相关的值在(−1, 1)区间中，这使得如决定哪个块真正匹配等某些上层应用更容易处理。规范互相关在匹配不同曝光下拍摄的图像时性能很好，例如，创建高动态范围图像时(10.2 节)。但是注意，当两块中有任一个块的方差为零时，NCC 值无定义(实际上，其性能在有噪声的低对比度区域上就会下降)。

规范化的 SSD 值(NSSD)是 NCC 的一个变形，它与在匹配值(8.9)中的隐式偏差-增益回归相关，

$$E_{\text{NSSD}}(\boldsymbol{u}) = \frac{1}{2} \frac{\sum_i \left[[I_0(\boldsymbol{x}_i) - \overline{I_0}] - [I_1(\boldsymbol{x}_i + \boldsymbol{u}) - \overline{I_1}] \right]^2}{\sqrt{\sum_i [I_0(\boldsymbol{x}_i) - \overline{I_0}]^2 + [I_1(\boldsymbol{x}_i + \boldsymbol{u}) - \overline{I_1}]^2}} \tag{8.14}$$

它在近期由 Criminisi, Shotton, Blake et al. (2007)提出。在实验中，他们发现其结果和 NCC 类似。但在使用移动平均方法施加在大量重叠的块上时，它更高效(3.2.2 节)。

8.1.1 分层运动估计

现在我们已经定义好要优化的配准损失函数，我们如何找到最小值呢？最简单的做法是在一定平移范围内做一个全搜索，整数或者亚像素级的步长都可以。这在运动补偿的视频压缩的块匹配中经常使用，其中要探查可能的运动范围(如±16 个像素)。[①]

为了加速搜索过程，我们常使用分层运动估计：建立一个图像金字塔(3.5 节)，首先在最粗糙的层上搜索数量较少的离散的像素(Quam 1984; Anandan 1989; Bergen, Anandan, Hanna et al. 1992)。之后，将该层的运动估计结果作为初始值来对下一层做更小范围局部的搜索。换言之，粗糙层的一些种子(好的结果)用来初始化下一层的搜索。尽管这不能保证和全搜索得到的结果相同，但通常其结果差不多且速度快很多。

更正式的情况，令

$$I_k^{(l)}(\boldsymbol{x}_j) \leftarrow \tilde{I}_k^{(l-1)}(2\boldsymbol{x}_j) \tag{8.15}$$

是 l 层降采样的图像，它由子采样(下采样)$l-1$ 层的平滑图像得到。3.5 节介绍了如

---

① 在立体视觉匹配中(11.1.2 节)，显示地搜索所有可能的视差(即平面扫描)几乎肯定会进行，这是因为搜索假设的数量非常少(可能是位移的 1D 本质使然)。

何进行所需要的下采样(构建金字塔)而避免产生大的走样。

在最粗糙的层,搜索最优的位移 $\boldsymbol{u}^{(l)}$,使得最小化 $I_0^{(l)}$ 和 $I_1^{(l)}$ 的差别。通常在 $\boldsymbol{u}^{(l)} \in 2^{-l}[-S,S]^2$ 的范围中做全搜索得到,其中 S 是在最高(原)分辨率下所期望的搜索范围,也可以紧接一个 8.1.3 节中所描述的逐次求精的步长。

估计好合适的运动向量时,它就可以用来预测下一层可能的位移[①]

$$\hat{\boldsymbol{u}}^{(l-1)} \leftarrow 2\boldsymbol{u}^{(l)} \tag{8.16}$$

之后在下一层重复位移搜索,这次的位移范围小得多,$\hat{\boldsymbol{u}}^{(l-1)} \pm 1$,同样也可以使用逐次求精步长(Anandan 1989)。换言之,图像可以用当前的运动估计来卷绕(重采样),这样只有精细层的小的增量运动需要重新计算。Bergen, Anandan, Hanna et al.(1992)在参数化运动估计中的扩展(8.2 节)更好地描述了整个过程。

8.1.2 基于傅里叶的配准

当搜索范围对应于更大图像的大部分区域时(如第 9 章中图像拼接的例子),分层的方法可能就没有那么好用了,因为常常无法将图像过于粗化,否则图像的重要特征会被模糊掉。在这种情况下,基于傅里叶的方法更好。

基于傅里叶的配准依靠傅里叶变换的一个特性,变换后的信号和原来的幅度相同,但相位发生线性变化(3.4 节),即

$$\mathcal{F}\{I_1(\boldsymbol{x}+\boldsymbol{u})\} = \mathcal{F}\{I_1(\boldsymbol{x})\}e^{-j\boldsymbol{u}\cdot\boldsymbol{\omega}} = \mathcal{I}_1(\boldsymbol{\omega})e^{-j\boldsymbol{u}\cdot\boldsymbol{\omega}}, \tag{8.17}$$

其中 ω 是傅里叶变换的角频率向量,我们用手写体 $\mathcal{I}_1(\omega) = F\{I_1(\boldsymbol{x})\}$ 来表示信号的傅里叶变换(3.4 节)。傅里叶变换另一个有用的性质是空间域卷积相应于傅里叶域的乘积(3.4 节)[②]。这样,傅里叶变换的互相关函数 E_{CC} 变为:

$$\mathcal{F}\{E_{CC}(\boldsymbol{u})\} = \mathcal{F}\left\{\sum_i I_0(\boldsymbol{x}_i)I_1(\boldsymbol{x}_i+\boldsymbol{u})\right\} = \mathcal{F}\{I_0(\boldsymbol{u})\bar{*}I_1(\boldsymbol{u})\} = \mathcal{I}_0(\boldsymbol{\omega})\mathcal{I}_1^*(\boldsymbol{\omega}), \tag{8.18}$$

其中

$$f(\boldsymbol{u})\bar{*}g(\boldsymbol{u}) = \sum_i f(\boldsymbol{x}_i)g(\boldsymbol{x}_i+\boldsymbol{u}) \tag{8.19}$$

是相关性函数,即,一个信号和另一个的翻转的卷积。$\mathcal{I}_1^*(\omega)$ 是 $\mathcal{I}_1(\omega)$ 的共轭复数。这是因为卷积的定义是一个信号和另一个信号翻转的卷积之和(3.4 节)。

于是,为了在所有 $\boldsymbol{u}$ 的有效值范围内有效的计算 E_{CC},我们对 $I_0(\boldsymbol{x})$ 和 $I_1(\boldsymbol{x})$ 同时计算傅里叶变换。再将两个变换相乘(第二个取共轭复数),之后对结果做逆变换。快速傅里叶变换算法(FFT)计算 $N \times M$ 的图像的复杂度是 $O(NM \log NM)$ (Bracewell 1986)。这比考虑完全范围图像重叠情况下全搜索的 $O(N^2M^2)$ 复杂度明显快很多。

[①] 这种翻倍的位移只有在整数像素坐标的定义下才是必须的。这是文献 Bergen, Anandan, Hanna et al. (1992)中的常用方法。如果使用规范设备坐标(2.1.5 节),位移(和搜索范围)并不需要每层都变化,但是步长还是需要调整的以保证基本一个像素的搜索步长。

[②] 事实上,观察卷积定理,傅里叶移位特性(8.17)和用位移后的脉冲函数 $\delta(\boldsymbol{x}-\boldsymbol{u})$ 卷积是等价的。

基于傅里叶的卷积常用来加速图像相关性计算,它也可以用于加速计算平方误差和(及其变形)。考虑 8.1 节中的 SSD 公式。它的傅里叶变换可以写为

$$\begin{aligned}\mathcal{F}\{E_{\mathrm{SSD}}(\boldsymbol{u})\} &= \mathcal{F}\left\{\sum_i [I_1(\boldsymbol{x}_i+\boldsymbol{u})-I_0(\boldsymbol{x}_i)]^2\right\} \\ &= \delta(\boldsymbol{\omega})\sum_i [I_0^2(\boldsymbol{x}_i)+I_1^2(\boldsymbol{x}_i)] - 2\mathcal{I}_0(\boldsymbol{\omega})\mathcal{I}_1^*(\boldsymbol{\omega}).\end{aligned} \quad (8.20)$$

于是,SSD 可以通过计算两幅图像的能量和再减去两次相关性函数得到。

窗口相关性 可惜,只有在 x_i 上的积分作用在两幅图像所有像素上时,才可以用傅里叶卷积定理,当有像素在边界外时,使用图像循环移位。尽管在小移位和图像大小相当时可以使用这种方法,但图像有部分重叠或者一个是另一个的子图像时,算法则没有什么意义。

该情况下,需要用窗口的(加权)互相关函数来代替互相关函数,

$$E_{\mathrm{WCC}}(\boldsymbol{u}) = \sum_i w_0(\boldsymbol{x}_i)I_0(\boldsymbol{x}_i)\,w_1(\boldsymbol{x}_i+\boldsymbol{u})I_1(\boldsymbol{x}_i+\boldsymbol{u}), \quad (8.21)$$

$$= [w_0(\boldsymbol{x})I_0(\boldsymbol{x})]\bar{*}[w_1(\boldsymbol{x})I_1(\boldsymbol{x})]. \quad (8.22)$$

其中权值函数 ω_0 和 ω_1 在图像的合法范围外都是 0,并且填充图像,使循环移位在原图外时返回 0。

计算式 8.5 中的加权 SSD 函数会显得更有趣,

$$E_{\mathrm{WSSD}}(\boldsymbol{u}) = \sum_i w_0(\boldsymbol{x}_i)w_1(\boldsymbol{x}_i+\boldsymbol{u})[I_1(\boldsymbol{x}_i+\boldsymbol{u})-I_0(\boldsymbol{x}_i)]^2. \quad (8.23)$$

将其展开为相关性的和并导出合适的傅里叶变换集留作习题 8.1。

该推导同样适用于偏差-增益修正的差平方和函数 E_{BG} (8.9)。还有,在偏差和增益参数的线性回归中所需要的所有相关性,可以用傅里叶变换高效地计算,从而为每个可能移位估计出曝光补偿差(习题 8.1)。

相位相关性 另一种有时用于运动估计的常规相关性(8.18)是相位相关性(Kuglin and Hines 1975; Brown 1992)。这里,将两个要匹配的信号的谱通过对每个频率下的乘积除以傅里叶变换的幅度来白化,

$$\mathcal{F}\{E_{\mathrm{PC}}(\boldsymbol{u})\} = \frac{\mathcal{I}_0(\boldsymbol{\omega})\mathcal{I}_1^*(\boldsymbol{\omega})}{\|\mathcal{I}_0(\boldsymbol{\omega})\|\|\mathcal{I}_1(\boldsymbol{\omega})\|} \quad (8.24)$$

在最后做逆傅里叶变换之前,如果是具有完美(循环)移位的无噪信号,我们有 $I_1(x+u)=I_0(x)$,所以从公式 8.17 我们得到

$$\begin{aligned}\mathcal{F}\{I_1(\boldsymbol{x}+\boldsymbol{u})\} &= \mathcal{I}_1(\boldsymbol{\omega})e^{-2\pi j\boldsymbol{u}\cdot\boldsymbol{\omega}}=\mathcal{I}_0(\boldsymbol{\omega}) \text{ 和}\\ \mathcal{F}\{E_{\mathrm{PC}}(\boldsymbol{u})\} &= e^{-2\pi j\boldsymbol{u}\cdot\boldsymbol{\omega}}.\end{aligned} \quad (8.25)$$

相位互相关的输出结果(在理想条件下)变为单峰(脉冲),位于 u 的正确值处。这使估计 u 的值变得简单。

相位互相关在某些情况下性能很好,但它的结果依赖于信号特性和噪声。如果图像受到某个很窄频段内的噪声污染(换言之,低频噪声或尖刺杂声),白化过程可以有效地消除这些区域中的噪声影响。但是,如果图像中某些区域信噪比很低(比

如说，带有很多高频噪声的两个模糊或者低纹理的图像)，白化过程反而会恶化结果(参见习题 8.1)。

近期，梯度互相关作为一种有望替代相位互相关的选项浮现出来(Argyriou and Vlachos 2003)，尽管可能还需要进一步的系统性研究论证。Fleet and Jepson (1990)也研究了相位互相关性，将其作为估计一般光流和立体视觉视差的方法。

旋转和尺度 尽管基于傅里叶变换的配准多用于估计图像的平移，但在特定的限制条件下，也可以用于估计平面内旋转和尺度。考虑两幅仅有旋转变换的图像，即

$$I_1(\hat{\boldsymbol{R}}\boldsymbol{x}) = I_0(\boldsymbol{x}). \tag{8.26}$$

如果将图像重采样为极坐标，

$$\tilde{I}_0(r,\theta) = I_0(r\cos\theta, r\sin\theta) \text{ and } \tilde{I}_1(r,\theta) = I_1(r\cos\theta, r\sin\theta), \tag{8.27}$$

就得到

$$\tilde{I}_1(r, \theta + \hat{\theta}) = \tilde{I}_0(r, \theta). \tag{8.28}$$

可以使用基于快速傅里叶变换(FFT)移位的方法来估计目标旋转。

如果两幅图像同时还有尺度关系，

$$I_1(e^{\hat{s}}\hat{\boldsymbol{R}}\boldsymbol{x}) = I_0(\boldsymbol{x}), \tag{8.29}$$

将其重采样为对数极坐标

$$\tilde{I}_0(s,\theta) = I_0(e^s\cos\theta, e^s\sin\theta) \text{ 和 } \tilde{I}_1(s,\theta) = I_1(e^s\cos\theta, e^s\sin\theta), \tag{8.30}$$

就得到

$$\tilde{I}_1(s+\hat{s}, \theta+\hat{\theta}) = \tilde{I}_0(s,\theta). \tag{8.31}$$

此时，我们必须小心选择合适的 s 值的范围来在原图中合理地采样。

对于还有微小平移的图像，

$$I_1(e^{\hat{s}}\hat{\boldsymbol{R}}\boldsymbol{x} + \boldsymbol{t}) = I_0(\boldsymbol{x}), \tag{8.32}$$

De Castro and Morandi(1987)独创性地提出了使用几个步骤的未知参数估计方法。首先，两幅图像都变换到傅里叶域，并且只记录变换图像的幅度。原理上，傅里叶幅度图像对图像平面内的平移并不敏感(尽管会有常见的边界效应警告)。接着，用极坐标或对数极坐标对两个幅度图像配准旋转和尺度。估计好旋转和尺度后，将其中一幅图像去除旋转并进行缩放，再运行常规的算法来估计平移移位。

不幸的是，这种巧妙的方法只能应用在有大量重叠的两幅图像中(小幅平移运动)。对于更一般的块和图像的运动，需要用 8.2 节介绍的参数化运动估计或 6.1 节描述的基于特征的方法。

8.1.3 逐次求精

迄今为止介绍的方法可以估计配准到最近的像素(如果使用更小的搜索步长可能到亚像素)。一般情况下，图像稳定化和拼接应用要想得到可以接受的结果，需要相当高的精度。

为了得到更好的亚像素估计，我们可以使用 Tian and Huhns(1986)介绍的方法

中的一种。一种可能是评估在当前最佳位置附近的若干离散 (u,v) 值，通过给匹配值插值来寻找解析的最小值。

更常见的方法是在 SSD 能量函数上应用梯度下降法(8.1)，它是 Lucas and Kanade(1981)提出的，对图像函数 Taylor 展开(图 8.2)，

$$E_{\text{LK-SSD}}(u+\Delta u) = \sum_i [I_1(x_i+u+\Delta u) - I_0(x_i)]^2 \tag{8.33}$$

$$\approx \sum_i [I_1(x_i+u) + J_1(x_i+u)\Delta u - I_0(x_i)]^2 \tag{8.34}$$

$$= \sum_i [J_1(x_i+u)\Delta u + e_i]^2, \tag{8.35}$$

其中

$$J_1(x_i+u) = \nabla I_1(x_i+u) = (\frac{\partial I_1}{\partial x}, \frac{\partial I_1}{\partial y})(x_i+u) \tag{8.36}$$

是图像的梯度或 (x_i+u) 处的雅克比，且最初在(8.1)中介绍的

$$e_i = I_1(x_i+u) - I_0(x_i), \tag{8.37}$$

是现有的亮度误差[①]。很多方法都可以计算亚像素 (x_i+u) 处的梯度，最简单的办法是在点 x、$x+(1,0)$、 $x+(0,1)$ 之间，计算水平和垂直方向的差。一些复杂的导数某些时候可以取得显著的性能改善。

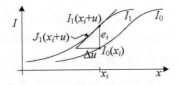

图 8.2 函数的 Taylor 近似和逐次计算光流修正量。$J_1(x_i+u)$ 是图像在 (x_i+u) 处的梯度，并且 e_i 是当前亮度差

SSD 误差的逐次更新的线性化形式(8.35)常被称为光流约束(optical flow constraint)或亮度恒常性约束(brightness constancy constraint)方程

$$I_x u + I_y v + I_t = 0, \tag{8.38}$$

其中 I_x 和 I_y 中的下标代表空间导数，I_t 称为时间导数(temporal derivative)，可以理解为视频序列中的瞬时速度。求其在一个区域上的平方并求和或者积分，可以用来计算光流(Horn and Schunck 1981)。

上述最小二乘问题(8.35)可以通过解关联的正规方程(附录 A.2)来解决。

$$A\Delta u = b \tag{8.39}$$

其中

$$A = \sum_i J_1^T(x_i+u) J_1(x_i+u) \tag{8.40}$$

并且

[①] 根据约定，机器人学中常用到，由 Baker and Matthews(2004)提出，列向量的导数为行向量，这样可以减少转置的使用。

$$b = -\sum_i e_i J_1^T(x_i + u) \tag{8.41}$$

分别称为(高斯-牛顿近似的)Hessian 和梯度加权残差向量(gradient-weighted residual vector)[①]。常记为

$$A = \begin{bmatrix} \sum I_x^2 & \sum I_x I_y \\ \sum I_x I_y & \sum I_y^2 \end{bmatrix} \text{ and } b = -\begin{bmatrix} \sum I_x I_t \\ \sum I_y I_t \end{bmatrix}. \tag{8.42}$$

计算 $J_1(x_i + u)$ 用的梯度，可以与在估计 $I_1(x_i + u)$ 时所需的图像卷绕同时计算得到(3.6.1 节，公式(3.89))。事实上，它是图像插值计算的副产品。如果考虑效率，这些梯度可以用模板图像的梯度来代替。

$$J_1(x_i + u) \approx J_0(x_i), \tag{8.43}$$

由于接近正确的配准，所以模板和位移后的图像应该近似。它的优点是，使我们可以预计算 Hessian 和雅克比图像，这将大大节省计算时间(Hager and Belhumeur 1998; Baker and Matthews 2004)。将公式(8.37)中计算 e_i 用的卷绕图像 $I_1(x_i + u)$，写作亚像素插值滤波和 I_1 中离散样本的卷积，可以进一步减少计算(Peleg and Rav-Acha 2006)。预计算梯度场和移位后的 I_1 的内积可以使得 e_i 的迭代重计算在常数时间完成(独立于像素数量)。

上述逐次修正的效率依赖于 Taylor 近似的质量。如果远离实际位移(比如说，一两个像素)，需要若干次迭代。但可以使用最小二乘拟合一系列较大的位移来估计 J_1 的值，从而加大收敛范围(Jurie and Dhome 2002)，或者如 4.1.4 节所述，"学习"一个给定块的特殊目的的识别器(Avidan 2001; Williams, Blake, and Cipolla 2003; Lepetit, Pilet, and Fua 2006; Hinterstoisser, Benhimane, Navab et al. 2008; Ozuysal, Calonder, Lepetit et al. 2010)。

一个常用的逐次求精的结束判断条件是监控位移修正 $\|u\|$ 的量级，当其小于某阈值(如一个像素的 1/10)则停止。对于较大的运动，常将逐次求精和分层的由粗到精搜索策略相结合，见 8.1.1 节。

适定和孔径问题 有时，由于配准的块缺少二维纹理，线性系统(8.39)的逆可能会很不适定。一个常见的例子是孔径问题，它先出现在一些早期光流文章中(Horn and Schunck 1981)，之后 Anandan (1989)做了更广泛的研究。考虑一个有斜边的向右移动的图像块(图 8.3)，此时只有速度(位移)的法向分量才能被可靠地计算出来。这表明在方程(8.39)中的矩阵 A 的秩不满，即它的较小的特征值接近于 0[②]。

解出方程(8.39)后，边界方向的位移数分量是很不适定的，在微小的扰动下会产生很大偏差。一种解决办法是加入先验(软约束)的运动的预期范围(Simoncelli, Adelson, and Heeger 1991; Baker, Gross, and Matthews 2004; Govindu 2006)。可以通过在 A 的对角元上加上很小的一个值来实现，它在本质上将会使解偏向于较小的但依然(很大程度地)最小化平方误差的 Δu 值。

[①] 真正的 Hessian 是误差函数 E 的完全二阶导数，可能并不正定，参见 6.1.3 节和附录 A.3。
[②] 由其构造定义矩阵 A 一定始终为对称半正定矩阵，即，特征值为非负实数。

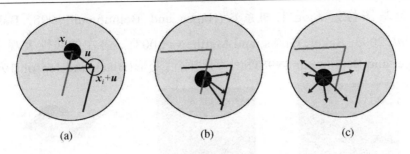

图 8.3 不同图像区域的孔径问题，由橙色和红色的 L 型结构指示，画在同一幅图像中以方便解释光流：(a)以 x_i 为中心的窗口 $w(x_i)$ (黑色)可以被唯一的匹配到第二幅图像(红色)中 x_i+u 为中心的对应结构上；(b)中心在边界上的窗口体现了传统的孔径问题，它可以匹配到一个 1D 可能位置的族上；(c)在完全无纹理的区域内，匹配变得毫无约束

但是，使用简单(固定)的二次先验时，其纯高斯模型的假设，如在(Simoncelli, Adelson, and Heeger 1991)中那样，实际中并不一定可行，比如由于沿强边缘的走样(Triggs 2004)。为此，最好在小特征值上加上大特征值的一个比率(如 5%)，再去矩阵求逆。

不确定性建模 使用不确定性模型可以更正式地获取特定的基于块的运动估计的可靠性。最简单的此类模型是协方差矩阵，它计算运动估计在所有可能方向上的预期方差。如 6.1.4 节和附录 B.6 所述，在少量的加性高斯噪声下，可以证明协方差矩阵 Σu 和 Hessian A 的逆成正比。

$$\Sigma u = \sigma_n^2 A^{-1}, \tag{8.44}$$

其中 σ_n^2 是加性高斯噪声的方差(Anandan 1989; Matthies, Kanade, and Szeliski 1989; Szeliski 1989)。

对于较大量的噪声，Lucas–Kanade 算法所做的线性化(8.35)只是一个近似，于是之前的数量变为实际协方差的 Cramer–Rao 下限。于是，Hessian 矩阵 A 的最大和最小特征值可以解释为在最确定和最不确定方向上的(缩放的)方差的逆。(Steele and Jaynes(2005)使用更实际的图像噪声模型给出了更详细的分析。图 8.4 展示了一幅图像中三个不同像素位置的局部 SSD 表面，可见在高纹理区域有清晰的最小值，而在强边缘处遭受孔径问题影响。

偏差和增益、加权和鲁棒误差度量。 Lucas–Kanade 修正法同样适用于偏差-增益方程(8.9)

$$E_{\text{LK-BG}}(u+\Delta u) = \sum_i [J_1(x_i+u)\Delta u + e_i - \alpha I_0(x_i) - \beta]^2 \tag{8.45}$$

(Lucas and Kanade 1981; Gennert 1988; Fuh and Maragos 1991; Baker, Gross, and Matthews 2003)。所产生的 4×4 的方程组可以同时估计出平移位移 Δu 以及偏差和增益参数 β 和 α。

对于有线性表观变化的图像(模板)可以得到类似的公式，

$$I_1(x+u) \approx I_0(x) + \sum_j \lambda_j B_j(x), \tag{8.46}$$

其中 $B_j(x)$ 是基图像，λ_j 是未知系数(Hager and Belhumeur 1998; Baker, Gross, Ishikawa *et al*. 2003; Baker, Gross, and Matthews 2003)。潜在的线性表观变化包括光照变化(Hager and Belhumeur 1998)和小的非刚性变形(Black and Jepson 1998)。

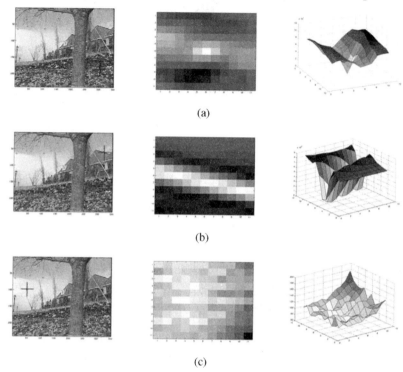

图 8.4　一幅图像中三处位置(红色十字)的 SSD 表面：(a)高纹理区域，清晰的最小值，不确定性低；(b)强边缘，孔径问题，某一个方向上不确定性高；(c)低纹理区域，无明显的最小值，不确定性大

Lucas–Kanade 算法的加权(窗口)的版本也是可能的：

$$E_{\text{LK-WSSD}}(u + \Delta u) = \sum_i w_0(x_i) w_1(x_i + u) [J_1(x_i + u) \Delta u + e_i]^2. \tag{8.47}$$

这里需要注意，在由原来的加权 SSD 函数(8.5)推导 Lucas–Kanade 修正中，我们忽略了 $\omega_1(x_i + u)$ 加权函数关于 u 的导数。在实践中这通常是可以接受的，特别是加权函数是略有转变的二值模板时。

Baker, Gross, Ishikawa *et al*. (2003)仅使用了 $\omega_0(x)$ 项，如果两幅图像范围相同：重叠部分没有(独立)缺失，这样是可以的。他们还讨论了将权重正比于 $\nabla I(x)$，这对噪声很大的图像很有用，此时的梯度本身噪声就很大。另一些研究光流的学者(Weber and Malik 1995; Bab-Hadiashar and Suter 1998b; Muhlich and Mester 1998)得到了类似的结果，并将其总结为全部最小二乘(total least square, Van Huffel and Vandewalle 1991; Van Huffel and Lemmerling 2002)。最后，Baker, Gross, Ishikawa *et al*. (2003)表明了对于足够大的图像，仅在最可靠(梯度最大)的像素上评估方程(8.47)，即使仅用 5%～10%的像素，也不会大幅降低性能。(该想法最早由 Dellaert and Collins(1999)提出，但他们用了更复杂的选择标准)。

Lucas–Kanade 逐次求精步骤同样可以用于 8.1 节介绍的鲁棒误差度量，
$$E_{\text{LK-SRD}}(\boldsymbol{u}+\Delta\boldsymbol{u}) = \sum_i \rho(\boldsymbol{J}_1(\boldsymbol{x}_i+\boldsymbol{u})\Delta\boldsymbol{u}+e_i), \tag{8.48}$$
可以用 6.1.4 节介绍的迭代重加权最小二乘方法(iteratively reweighted least squares)求解。

8.2 参数化运动

如 2.1.2 节所述，很多图像配准任务，如手持摄像机的图像拼接，需要更复杂的运动模型。由于这些模型(如仿射变形)，比单纯平移的参数更多，所以可能值范围内的全搜索是行不通的。而逐次求精的 Lucas-Kanade 算法可以扩展到参数化运动模型，并和分层搜索算法结合使用(Lucas and Kanade 1981; Rehg and Witkin 1991; Fuh and Maragos 1991; Bergen, Anandan, Hanna et al. 1992; Shashua and Toelg 1997; Shashua and Wexler 2001; Baker and Matthews 2004)。

对于参数化运动，不同于使用单个恒定平移向量 $\boldsymbol{u}$，我们使用空间变化运动场((motion field)或者对应图(correspondence map)，$\boldsymbol{x}'(\boldsymbol{x};\boldsymbol{p})$，用低维向量 $\boldsymbol{p}$ 来参数化，其中 $\boldsymbol{x}'$ 可以是 2.1.2 节中的任意运动模型。参数化逐步求精公式变为

$$\begin{aligned}E_{\text{LK-PM}}(\boldsymbol{p}+\Delta\boldsymbol{p}) &= \sum_i [I_1(\boldsymbol{x}'(\boldsymbol{x}_i;\boldsymbol{p}+\Delta\boldsymbol{p})) - I_0(\boldsymbol{x}_i)]^2 & (8.49)\\ &\approx \sum_i [I_1(\boldsymbol{x}'_i) + \boldsymbol{J}_1(\boldsymbol{x}'_i)\Delta\boldsymbol{p} - I_0(\boldsymbol{x}_i)]^2 & (8.50)\\ &= \sum_i [\boldsymbol{J}_1(\boldsymbol{x}'_i)\Delta\boldsymbol{p} + e_i]^2, & (8.51)\end{aligned}$$

其中 Jacobian 变为
$$\boldsymbol{J}_1(\boldsymbol{x}'_i) = \frac{\partial I_1}{\partial \boldsymbol{p}} = \nabla I_1(\boldsymbol{x}'_i)\frac{\partial \boldsymbol{x}'}{\partial \boldsymbol{p}}(\boldsymbol{x}_i), \tag{8.52}$$
即，图像梯度 ∇I_1 和对应场的 Jacobian 的内积，$J_{x'} = \partial \boldsymbol{x}'/\partial \boldsymbol{p}$。

表 6.1 是 2.1.2 节和表 2.1 中介绍的 2D 平面变形的运动 Jacobians $\boldsymbol{J}_{x'}$。注意我们如何重新参数化运动矩阵使其在原点 $\boldsymbol{p}=\boldsymbol{0}$ 时始终为单位矩阵。之后讨论复合和逆复合算法会用到。(它也使在运动中加入先验限制更为容易。)

对于参数化运动，(高斯-牛顿)Hessian 和梯度加权残差向量变为
$$\boldsymbol{A} = \sum_i \boldsymbol{J}_{\boldsymbol{x}'}^T(\boldsymbol{x}_i)[\nabla I_1^T(\boldsymbol{x}'_i)\nabla I_1(\boldsymbol{x}'_i)]\boldsymbol{J}_{\boldsymbol{x}'}(\boldsymbol{x}_i) \tag{8.53}$$
并且
$$\boldsymbol{b} = -\sum_i \boldsymbol{J}_{\boldsymbol{x}'}^T(\boldsymbol{x}_i)[e_i\nabla I_1^T(\boldsymbol{x}'_i)]. \tag{8.54}$$
注意，方括弧中的表达式和简单平移运动情况中(8.40 和 8.41)的表达是一样的。

基于块的近似 参数化运动的 Hessian 和残差向量的计算比平移情况要昂贵很多。对于 n 个参数，N 个像素的参数化运动，计算 $\boldsymbol{A}$ 和 $\boldsymbol{b}$ 需要 $\boldsymbol{O}(n^2N)$ 的操作(Baker and Matthews 2004)。一个显著降低计算量的办法是将图像划分为小的区域(块)P_j，并在像素层面，只计算方括弧中简单的 2×2 的量(Shum and Szeliski 2000)。

$$A_j = \sum_{i \in P_j} \nabla I_1^T(x_i') \nabla I_1(x_i') \tag{8.55}$$

$$b_j = \sum_{i \in P_j} e_i \nabla I_1^T(x_i'). \tag{8.56}$$

整个 Hessian 和残差可以近似为

$$A \approx \sum_j J_{x'}^T(\hat{x}_j)[\sum_{i \in P_j} \nabla I_1^T(x_i') \nabla I_1(x_i')] J_{x'}(\hat{x}_j) = \sum_j J_{x'}^T(\hat{x}_j) A_j J_{x'}(\hat{x}_j) \tag{8.57}$$

并且

$$b \approx -\sum_j J_{x'}^T(\hat{x}_j)[\sum_{i \in P_j} e_i \nabla I_1^T(x_i')] = -\sum_j J_{x'}^T(\hat{x}_j) b_j, \tag{8.58}$$

其中 $\hat{x}_j$ 是块 P_j 的中心(Shum and Szeliski 2000)。这和用分段常数近似真正的运动 Jacobian 相同。实际中，它的性能很好。9.2.4 节讨论了该近似和基于特征注册的关系。

复合方法 对于复杂的参数化运动，例如同构，运动 Jacobian 的计算变得复杂，而且可能包含逐像素分组(division)。Szeliski and Shum (1997)发现可以先将目标图像 I_2 根据当前运动估计 $x'(x;p)$ 卷绕，来简化计算

$$\tilde{I}_1(x) = I_1(x'(x;p)), \tag{8.59}$$

之后将卷绕后的图像和模板 $I_0(x)$ 对比，

$$E_{\text{LK-SS}}(\Delta p) = \sum_i [\tilde{I}_1(\tilde{x}(x_i; \Delta p)) - I_0(x_i)]^2 \tag{8.60}$$

$$\approx \sum_i [\tilde{J}_1(x_i) \Delta p + e_i]^2 \tag{8.61}$$

$$= \sum_i [\nabla \tilde{I}_1(x_i) J_{\tilde{x}}(x_i) \Delta p + e_i]^2. \tag{8.62}$$

注意，由于假设两幅图像相似，只需要逐次参数化运动，换言之，增量运动可以在 $p=0$ 附近计算，这可以大大简化计算过程。例如，平面投影变换的 Jacobian(6.19)变为

$$J_{\tilde{x}} = \frac{\partial \tilde{x}}{\partial p}\bigg|_{p=0} = \begin{bmatrix} x & y & 1 & 0 & 0 & 0 & -x^2 & -xy \\ 0 & 0 & 0 & x & y & 1 & -xy & -y^2 \end{bmatrix}. \tag{8.63}$$

一旦计算完增量运动 $\hat{x}$，可以前置到之前的运动估计，对于如表 2.1 和 6.1 中给出的由变换矩阵表示的运动这是容易计算的。Baker and Matthews (2004)称其为向前复合算法(forward compositional algorithm)，因为它重卷绕目标图像并构成最终运动估计。

如果卷绕图像和模板足够相似，如之前所述(8.43)，我们可以用 $I_0(x)$ 的梯度代替 $\tilde{I}_1(x)$ 的梯度。一个巨大的好处是我们可以预计算公式(8.53)中矩阵 A 的 Hession(和逆)。残差向量 b(8.54)也可以预计算，即，最速下降图像(steepest descent images) $\nabla I_0(x) J_{\tilde{x}}(x)$ 可以预计算，并保存下来为之后和 $e(x) = \tilde{I}_1(x) - I_0(x)$ 误差图像相乘(Baker and Matthews 2004)。该想法首先由 Hager and Belhumeur (1998)提出，Baker and Matthews(2004)称其为"逆加性"(inverse additive)方案。

Baker and Matthews(2004)介绍了另一个变型算法,他们称其为"逆复合"(inverse compositional)算法。他们并不(概念上)重卷绕已经卷绕后的目标图像 $\tilde{I}_1(\boldsymbol{x})$,而是卷绕模板图像 $I_0(\boldsymbol{x})$,并最小化

$$E_{\mathrm{LK-BM}}(\Delta \boldsymbol{p}) = \sum_i [\tilde{I}_1(\boldsymbol{x}_i) - I_0(\tilde{\boldsymbol{x}}(\boldsymbol{x}_i; \Delta \boldsymbol{p}))]^2 \tag{8.64}$$

$$\approx \sum_i [\nabla I_0(\boldsymbol{x}_i) \boldsymbol{J}_{\tilde{\boldsymbol{x}}}(\boldsymbol{x}_i) \Delta \boldsymbol{p} - e_i]^2. \tag{8.65}$$

这就是向前卷绕算法(8.62)中将梯度 $\nabla \tilde{I}_1(\boldsymbol{x})$ 换成 $\nabla I_0(\boldsymbol{x})$,除了 e_i 的符号外。得到的 $\Delta \boldsymbol{p}$ 是用变化了的公式(8.62)计算出来的结果求反,于是逐次变换的逆必须要前置到当前变换。因为逆复合方法可以预计算 Hessian 的逆和最速下降图像,这使其成为 Baker and Matthews(2004)所综述的算法中最应优先选择的算法。图 8.5(Baker, Gross, Ishikawa et al. 2003)优美地展示了实现逆复合算法所需的步骤。

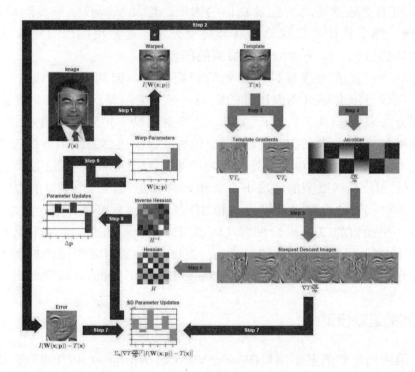

图 8.5 图解逆复合算法,Baker, Gross, Ishikawa et al. 2003 授权使用。3~6步(淡色箭头)作为预计算执行一次。主算法为简单迭代:图像卷绕(第1步),图像求差(第2步),图像点积(第7步),乘以 Hessian 的逆(第8步),更新卷绕图像(第9步)。以上步骤均可高效计算

Baker and Matthews(2004)还对比了使用高斯-牛顿迭代(即最小平方的一阶展开,first-order expansion of the least squares),如上)和其他方法,如最速下降、Levenberg–Marquardt。该系列的后续部分(Baker, Gross, Ishikawa et al. 2003; Baker, Gross, and Matthews 2003, 2004)讨论了更多高级方法,如逐像素加权、为提高效率的像素选取、鲁棒度量和算法的更深入的讨论、线性表观变化和形状先验。它对于有兴趣实现高效逐次图像注册的人都非常重要。Evangelidis and Psarakis(2008)提供

了这些算法及其他相关方法更详细的实验评测。

8.2.1 应用：视频稳定化

视频稳定化是参数化运动估计的广泛应用之一(Hansen, Anandan, Dana *et al.* 1994; Irani, Rousso, and Peleg 1997; Morimoto and Chellappa 1997; Srinivasan, Chellappa, Veeraraghavan *et al.* 2005)。稳定化算法既应用在摄像机、照像机等硬件里，也使用在改善抖动视频的视觉效果的软件包中。

Matsushita, Ofek, Ge *et al.*(2006)在全帧视频稳定化论文中提出了稳定化的三个主要步骤，即运动估计、运动平滑和图像卷绕。运动估计算法常用相似变换来处理摄像机平移、旋转和缩放。这里的棘手问题是将算法锁定在由摄像机运动产生的背景运动上，而不受独立运动的前景物体的影响。运动平滑算法恢复运动的低频(缓慢变化)部分，然后估计需要移除的高频抖动分量。最后用图像卷绕算法做高频修正，呈现仿佛摄像机仅仅平滑运动而拍摄的图像。

稳定化算法可以大大改善抖动视频的表观效果，但同时也会包含人工视觉现象。例如，图像卷绕会导致图像边界的消失，必须要用其他帧信息来剪裁、填充，或者用修图方法来处理(10.5.1 节)。另外，在快速运动下拍摄的视频帧往往很模糊。可以用去模糊方法(10.3 节)或者从其他运动较少或聚焦更好的帧中得到边界像素信息(Matsushita, Ofek, Ge *et al.* 2006)来处理。习题 8.3 会让你实现一些这方面的算法。

在摄像机 3D 平移很多的情况下，如摄影师在走动，一个更好的办法是通过完整的由运动到结构方法重建摄像机运动和 3D 场景。然后可以计算出一个平滑的 3D 摄像机路径，用插值的 3D 点云作为几何原型来做视图插值，重新渲染视频，并且同时保留突出特征(Liu, Gleicher, Jin *et al.* 2009)。如果有若干摄像机而不仅仅是一个，使用光场渲染方法(13.3 节) (Smith, Zhang, Jin *et al.* 2009)可以得到更好的结果。

8.2.2 学到的运动模型

学习一组应用定制的基本函数(Black, Yacoob, Jepson *et al.* 1997)是另一种用几何变形，如仿射变换，来参数化运动场的方法。首先，通过一组训练视频计算一组稠密运动估计场(8.4 节)。之后，对运动场栈 $u_t(x)$ 做奇异值分解(SVD)，得到开始的几个特征向量 $v_k(x)$。最后，对于新的测试序列，使用由粗到精算法，估计参数化光流场中的未知系数 a_k，以计算一个新的光流场

$$u(x) = \sum_k a_k v_k(x). \tag{8.66}$$

图 8.6a 展示了通过走路运动视频学习到的一组基场。图 8.6b 展示了基系数的时间演化和一些恢复的参数化运动场。注意，同样的想法也可以用于特征跟踪(Torresani, Hertzmann, and Bregler 2008)，4.1.4 节和 12.6.4 节更详细地讨论这部分内容。

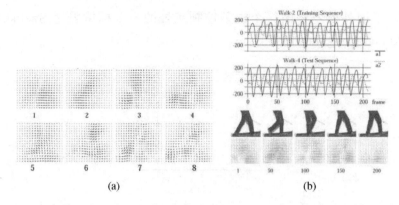

图 8.6 学习到的走路序列的参数化运动场(Black, Yacoob, Jepson *et al*. 1997)© 1997 IEEE：(a)学习到的基流场；(b)随时间的运动参数图，以及相应估计的运动

8.3 基于样条的运动

尽管参数化运动模型有着广泛应用(如视频稳定、2D 表面映射)，但大部分图像运动还是过于复杂，无法用这样低维的模型来捕获。

传统上，光流算法(8.4 节)计算每个像素的独立运动估计，换言之，计算的光流向量数量和输入像素的数量一样。类比于方程(8.1)一般的光流转换方程可以写为

$$E_{\text{SSD-OF}}(\{\boldsymbol{u}_i\}) = \sum_i [I_1(\boldsymbol{x}_i + \boldsymbol{u}_i) - I_0(\boldsymbol{x}_i)]^2. \tag{8.67}$$

注意，变量 $\{\boldsymbol{u}_i\}$ 的数量是量测数量的两倍，所以问题是欠约束的。

我们将在 8.4 节中介绍两种经典解决办法，在重叠区域(基于块的或者基于窗口的方法)求和，或者在 $\{\boldsymbol{u}_i\}$ 中通过正则化(regularization)或者马尔科夫随机场(3.7 节)加入平滑项。本节中，我们介绍另一种介于一般光流(像素独立光流)和参数化光流(少量全局参数)之间的方法。它用少量控制顶点 $\{\hat{\boldsymbol{u}}_j\}$ (图 8.7)控制的二维样条来代表运动场，

$$\boldsymbol{u}_i = \sum_j \hat{\boldsymbol{u}}_j B_j(\boldsymbol{x}_i) = \sum_j \hat{\boldsymbol{u}}_j w_{i,j}, \tag{8.68}$$

其中 $B_j(\boldsymbol{x}_i)$ 称为基函数(basis function)，且仅在一小段有限支持区间内非零(Szeliski and Coughlan 1997)。我们称 $w_{ij} = B_j(\boldsymbol{x}_i)$ 为权来强调 $\{\boldsymbol{u}_i\}$ 是 $\{\hat{\boldsymbol{u}}_j\}$ 已知的线性组合。图 8.8 是一些常用的样条基函数。

将各像素自己的光流向量 $\boldsymbol{u}_i$ (8.68)代入 SSD 误差度量(8.67)得到类似于方程(8.50)的参数化运动公式。最大差别在于 Jacobian $\boldsymbol{J}_1(\boldsymbol{x}_i')$ (8.52)现在由权重矩阵 $\boldsymbol{W} = [\omega_{ij}]$ 中的若干稀疏项构成。

在一些我们知道有关运动场更多信息的情况下，例如，运动是由摄像机在静态场景中移动产生的，我们可以用更特殊的运动模型。例如，平面加视差(plane plus parallax)模型(2.1.5 节)可以自然地融入基于样条的运动模型表达，其中单应(6.19)代

表平面运动,平面外视差 d 由每个样条控制点处的一个标量表达(Szeliski and Kang 1995; Szeliski and Coughlan 1997)。

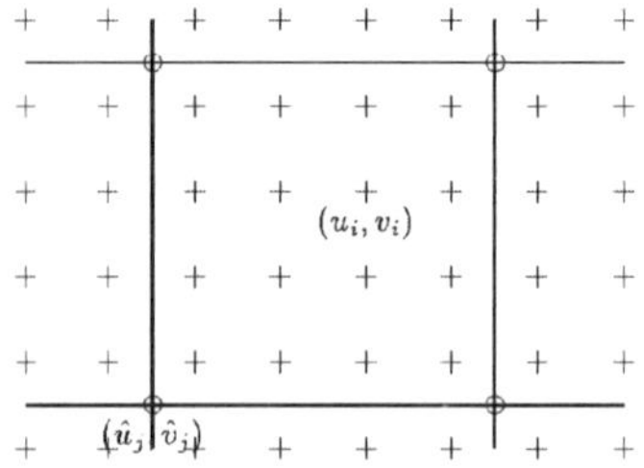

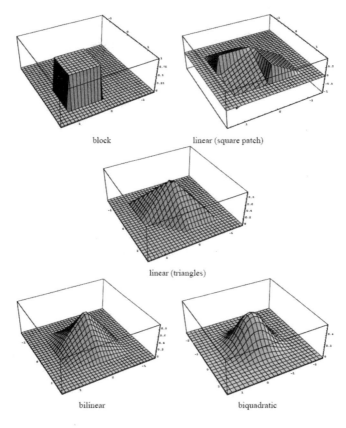

图 8.7 样条运动场:位移向量 $\boldsymbol{u}_i = (u_i, v_i)$ 如加号(+)所示,由少数由圆圈表示(○)的控制顶点 $\hat{\boldsymbol{u}}_j = (\hat{u}_j, \hat{v}_j)$ 控制

图 8.8 样条基函数样例(Szeliski and Coughlan 1997) © 1997 Springer。块(常数)插值/基函数对应于基于块的运动估计(Le Gall 1991)。3.5.1 节有更多关于样条函数的细节

很多情况下,少量样条点可以很好地处理运动估计问题。但是,如果大面积无纹理区域(或者遭受孔径问题影响的细长边界)在若干样条块中持续出现,就需要加入正则化项来使该问题适定化(3.7.1 节)。最简单的办法是直接在邻接样条控制网格点 $\{\hat{\boldsymbol{u}}_j\}$ 间加入误差平方惩罚,如(3.100)所示。如果使用多分辨率(由粗到精)策略,

那么在从一层到另一层时，需要注意重新缩放这些平滑项。

对应于基于样条运动估计的线性系统是稀疏且规则的。由于其一般是中等大小的，可以使用如 Cholesky 分解（附录 A.4)等方法直接解。另一种情况，如果问题变得太大，承受过多的填充，需要使用一些迭代方法（附录 A.5)，如分层预处理共轭梯度方法(Szeliski 1990b, 2006b)。

由于其鲁棒性，基于样条的运动估计有广泛应用，包括视觉特效(Roble 1999)和医学图像注册(8.3.1 节)(Szeliski and Lavallée 1996; Kybic and Unser 2003)。

但其基本方法的缺点之一是，在运动不连续处附近模型结果不好，除非使用更多节点。为了纠正这一问题，Szeliski and Shum (1996)提出了使用四叉树(quadtree)来计算样条控制网格(图 8.9a)。大单元(cell)用于表达平滑运动，在运动不连续区域加入小单元(图 8.9c)。

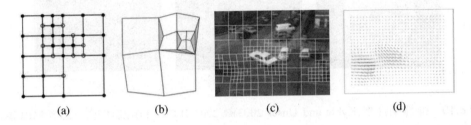

图 8.9 四叉基于树样条的运动估计(Szeliski and Shum 1996)© 1996 IEEE：(a)四叉树样条表达；(b)可能导致的缝隙，除非限制白色节点依赖于其父节点；(c)叠加在灰度图像上的变形四叉树样条网络；(d)显示成针状图的流场

为了估计运动，我们使用由粗到精的策略。开始在低分辨率图像上使用一个规则样条，得到一个初始运动估计。在运动不一致的样条块处，即平方残差(8.67)大于阈值，重新分割成更小的块。为了避免最后运动场有缝隙(图 8.9b)，在求精的网格中的某些节点的值需要进行限制，即靠近大块的点的值，需要依赖于其父节点的值。最简单的办法就是用分层的四叉树样条来实现(Szeliski 1990b)，并有选择地将一些分层基函数设为 0，具体可参见(Szeliski and Shum 1996)。

应用：医学图像注册

由于基于样条运动模型易于表示弹性形变场，它在医学图像注册中有广泛应用(Bajcsy and Kovacic 1989; Szeliski and Lavallée 1996; Christensen, Joshi, and Miller 1997)[①]。注册方法可以用于长期跟踪病人情况(纵向研究)，也可以用来匹配不同病人图像，以便找到共性及发现变种或病理(横向研究)。当要注册的是不同图像模态时，例如，计算机断层扫描(CT)和核磁共振成像(MRI)，相似性的互信息度量时常是必要的(Viola and Wells III 1997; Maes, Collignon, Vandermeulen et al. 1997)。

---

① 在计算机图形学中，这样的弹性形变称为自由形变(Sederberg and Parry 1986; Coquillart 1990; Celniker and Gossard 1991)。

Kybic and Unser(2003)写了篇很好的文献回顾,描述了一个完整的工作系统,他们将图像和形变场都表达成多分辨率样条。图 8.10 是这一系统的样例,它注册患者大脑 MRI 到标注的脑图谱图像。系统可以全自动模式运行,但人工定一些界标会更准确。近期的形变医学图像注册及其性能评测论文包括(Klein, Staring, and Pluim 2007; Glocker, Komodakis, Tziritas *et al*. 2008)。

和其他应用相比,规则体积样条可以用选择性求精来加强。在 3D 体积图像或者表面注册中,这称为"八叉树样条"(Szeliski and Lavallee 1996),并用来注册医学表面模型,如病人的脸、脊椎等(图 8.11)。

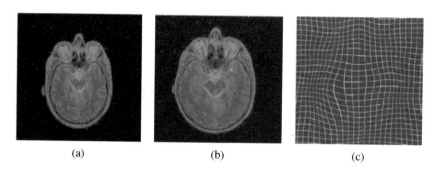

图 8.10 弹性脑注册(Kybic and Unser 2003)© 2003 IEEE:(a)原脑图谱,患者 MRI 图像为红色覆盖;(b)用户标注 8 个点(未显示)后弹性注册后的结果;(c)形变网格表示的三次 B 样条形变场

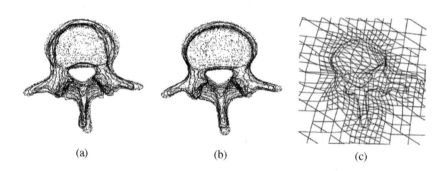

图 8.11 两个脊椎表面模型的八叉树样条图像注册(Szeliski and Lavallee 1996)©1996 Springer:(a)初始刚性配准后;(b)弹性配准后;(c)适定后的八叉树样条形变场的横横截面

8.4 光 流

最一般(最困难)的运动估计是独立估计每个像素运动,一般称作"光流"(optical 或 optic flow)。上节提到,它需要最小化整个图像中对应的像素亮度或色差之和,

$$E_{\text{SSD-OF}}(\{\boldsymbol{u}_i\}) = \sum_i [I_1(\boldsymbol{x}_i + \boldsymbol{u}_i) - I_0(\boldsymbol{x}_i)]^2. \tag{8.69}$$

由于变量 $\{u_i\}$ 的数量是量测的两倍，所以问题是欠约束的。两种经典办法是计算局部重叠区域的和(基于块或者基于窗口)，或使用正则化或马尔科夫随机场在 $\{u_i\}$ 中加入平滑项，并搜索全局最小值。

基于块的方法通常使用位移图像函数(8.35)的 Taylor 展开，由此得到亚像素估计(Lucas and Kanade 1981)。Anandan(1989)展示了如何在由粗到精金字塔中，将局部不连续搜索加入到 Lucas–Kanade 逐步求精算法中，以此来估计 8.1.1 节所述的大的运动。他同时分析了局部运动估计不确定性是如何与局部 Hessian 矩阵 A_i(8.44)的特征值有关的，如图 8.3 和图 8.4 所示。

Bergen, Anandan, Hanna et al.(1992)提出了一个统一的框架来同时描述参数化(8.2 节)和基于块的光流算法，并提供一个很棒的介绍。每次由粗到精金字塔光流估计迭代后，他们将一幅图像重新卷绕，这样只要计算逐次光流估计(8.1.1 节)。当使用重叠块时，首先计算每个像素处的梯度和亮度误差的外积(8.40 和 8.41)，再使用移动平均滤波器[①]计算重叠部分窗口的和，可以提高效率。

Horn and Schunck(1981)提出了一个基于正则化的框架，它使(8.69)同时最小化所有光流向量 $\{u_i\}$，而不是独立计算每个运动(更新运动)。为了约束该问题，在原来每个像素误差度量里加入了平滑约束项，即光流微分平方惩罚。由于该方法开始是为变分(连续函数)框架下的微小运动而设计的，所以线性化的亮度恒常性约束(brightness constancy constraint)对应于(8.35)，即(8.38)。更普遍的记为解析积分

$$E_{\mathrm{HS}} = \int (I_x u + I_y v + I_t)^2 \, dx \, dy, \tag{8.70}$$

其中 $(I_x, I_y) = \nabla I_1 = J_1$ 和 $I_t = e_i$ 是时间导数(temporal derivative)，即图像间的亮度变化。Horn and Schunck 模型可以看作基于样条的运动估计在样条变为1×1像素块时的情况。

使用局部累加的(和单个像素相对)Hessian 作为亮度恒常项，将局部和全局的光流估计思想合并为一个框架也是可能的(Bruhn, Weickert, and Schnorr 2005)。考虑全局解析能量(8.70)的离散模拟(8.35)，

$$E_{\mathrm{HSD}} = \sum_i u_i^T [J_i J_i^T] u_i + 2 e_i J_i^T u_i + e_i^2. \tag{8.71}$$

如果我们将每像素(一阶)Hessian $A_i = [J_i J_i^T]$ 和残差 $b_i = J_i e_i$ 替换为区域累加形式(8.40 和 8.41)，可以得到使用基于区域的亮度恒常约束的全局最小化算法。

基础光流模型的另一个扩展是使用全局(参数化)和局部运动模型的结合。例如，如果我们知道运动是由静态场景中的摄像机移动造成的(刚性运动)，我们可以把问题变为每像素深度和全局摄像机运动参数的估计(Adiv 1989; Hanna 1991; Bergen, Anandan, Hanna et al. 1992; Szeliski and Coughlan 1997; Nir, Bruckstein, and Kimmel 2008; Wedel, Cremers, Pock et al. 2009)。这样的方法与立体视觉匹配(第 11 章)紧密相关。另外，我们也可以与逐像素残差矫正相结合来估计每幅图像或者每

---

① 这个阶段也可以使用其他平滑、累加滤波器(Bruhn, Weickert, and Schnorr 2005)。

个片段的仿射运动模型(Black and Jepson 1996; Ju, Black, and Jepson 1996; Chang, Tekalp, and Sezan 1997; Memin and Perez 2002)。8.5 节将回顾这一问题。

当然，用图像亮度来度量表观一致性并不一定总是合适，比如在图像光照变化时。8.1 节讨论过，梯度匹配、滤波后的图像或者其他度量如图像 Hessian(二次导数度量)可能更合适。同样可以计算图像的局部导向滤波器的相位，它对偏差和增益变换不敏感(Fleet and Jepson 1990)。Papenberg, Bruhn, Brox et al. (2006)研究了这些限制，为逐次光流计算中的迭代重卷绕图像提出了更详细的分析和评判。

由于每个像素独立计算亮度恒常性，而不是在块上计算它们的和，每块上可能不符合常数光流假设，所以全局最优方法在运动不连续处可能会得到更好结果。在平滑约束上使用鲁棒度量时，该问题非常明显(Black and Anandan 1996; Bab-Hadiashar and Suter 1998a)[①]。一个常用的鲁棒度量方法是 L_1 范式，也叫"全变差"(total variation，TV)，它导出一个可求全局最小值的凸能量(Bruhn, Weickert, and Schnorr 2005; Papenberg, Bruhn, Brox et al. 2006)。先验各向异性平滑是另一种常用的方法，它对图像梯度的平行和垂直方向应用不同的平滑(Nagel and Enkelmann 1986; Sun, Roth, Lewis et al. 2008; M.Werlberger, Trobin, Pock et al. 2009)。从光流和亮度图像对中学习到一组更好的平滑约束(导数滤波器和鲁棒函数)也是可能的(Sun, Roth, Lewis et al. 2008)。Baker, Black, Lewis et al. (2007) 和 Baker, Scharstein, Lewis et al. (2009) 给出了这些方法的更多细节。

由于光流估计中有大的 2D 搜索空间，更多的算法使用梯度下降变型和由粗到精接续法来最小化全局能量函数。这和立体匹配("较简单"的一维视差估计问题)形成鲜明对比，组合优化方法已经成为近十年来立体匹配方法的选项。

幸运的是，基于马尔科夫随机场的组合优化方法在近期发布的光流数据集上显得且趋势是性能更好的方法(Baker, Black, Lewis et al. 2007)。[②]

这些方法的例子包括 Glocker, Paragios, Komodakis et al. (2008)，他们提出了一个由粗到精策略，使用逐像素 2D 不确定性估计，再用来指导优化下一层的搜索。不同于用梯度下降法来优化光流估计，组合搜索使用快速 PD 算法在离散位移标志上搜索(可以找到更好的能量极小值)(Komodakis, Tziritas, and Paragios2008)。

Lempitsky, Roth, and Rother(2008)在基本光流算法(Horn and Schunck 1981; Lucas and Kanade 1981)的结果上使用融合步骤(Lempitsky, Rother, and Blake 2007)来得到更好结果。融合步骤的基本思想是，用其他更基本方法(或者它们的变换版本)得到的假说来替换当前最优估计的部分，并使其与局部梯度下降交替从而更好地进行能量最小化。

精确运动估计领域仍在快速发展，性能每年都有显著提高。光流评估网站(*http://vision.middlebury.edu/flow/*)很好地介绍了近期的高效算法(图 8.12)。

[①] 鲁棒度量(8.1 节和 8.2 节)也可以提高基于窗口方法的性能(Black and Anandan 1996)。
[②] *http://vision.middlebury.edu/flow/*

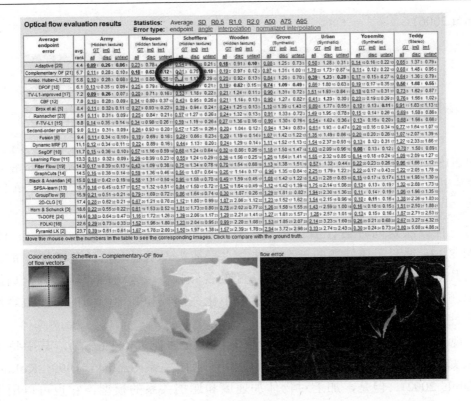

图 8.12 24 种光流算法评估结果，October 2009, *http://vision.middlebury.edu/flow/*, (Baker, Scharstein, Lewis *et al*. 2009)。移动鼠标到有下划线的分数之下，用户可以交互查看对应光流和误差图的关系。点击分数可以看计算结果和真实流。每个分数旁边，蓝色数字表示当前列对应的排名。粗体表示每列最小(优)值。表用平均排名排序(计算全部 24 列，八个序列，每个三个区域掩模)。平均排名近似在选定度量/统计方法下的性能量度

8.4.1 多帧运动估计

至今为止，我们将运动估计看作两帧问题，求一个将一幅图像中的像素配准到另一幅图像中去的运动场。实际中，运动估计多在视频中使用，整个图像序列都可以用来计算。

一个经典的多帧序列运动估计方法是用有向或导向滤波器(Heeger 1988)滤波时空体，类似于有向边缘检测(3.2.3 节)。图 8.13 展示了常用的花园序列中的两帧以及时空体上的水平切片，即所有视频帧叠合得到的 3D 体。因为像素运动多为水平，可以清楚地观察到单个(有纹理)像素轨迹的斜率，这对应于它们的水平速度。时空滤波通过使用每个像素周围的 3D 体来确定时空最优方向，直接对应于像素的速度。

不幸的是，为了得到图像中所有像素准确合理的速度估计，时空滤波器具有比较大的范围，这降低了运动不连续处的估计质量(基于窗口的 2D 运功估计器也有这样的问题)。不同于完全的时空滤波，另一种办法是估计更多的局部时空导数，并在全局最优框架中使用它们来填充无纹理区域(Bruhn, Weickert, and Schnorr 2005;

Govindu 2006)。

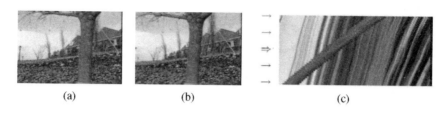

图 8.13 时空体中的切片(Szeliski 1999)© 1999 IEEE: (a 和 b)花园序列的两帧，(c)完全时空体中的一个水平切片，箭头表示估计流的潜在关键帧位置。注意花园序列的颜色不正确，正确的颜色(黄色花朵)参见图 8.15

另一种办法是同时估计多个运功估计，还可以选择处理遮挡(Szeliski 1999)问题。图 8.13c 展示了该问题的一个方法。水平箭头表示运动估计关键帧 s 的位置，其他切片表示和关键帧插值预测颜色匹配的视频帧 t。运动估计转化为全局能量最小化问题，在使用鲁棒平滑项之外同时最小化关键帧和其他帧间的亮度兼容性和光流兼容性项。

多视角框架更适合于刚性场景运动(多视角立体视觉)(11.6 节)，其中每个像素处的未知量是视差，而可以由像素深度得到遮挡关系(Szeliski 1999; Kolmogorov and Zabih 2002)。但是，加上对物体加速度和遮挡关系的模型，它也可以用于一般运动。

8.4.2 应用：视频去噪

视频去噪是指去除噪声及其他缺陷，如电影和视频划痕(Kokaram 2004)。不同于只有一幅图像信息的单个图像去噪，视频去噪可以取均值或使用邻近帧的信息。但是，为了不引入模糊或跳动(不正常运动)，需要准确的像素级的运动估计。

习题 8.7 有其需要的一些步骤，包括确定当前运动估计是否足够准确，可以和其他帧取均值。Gai and Kang(2009)描述了他们近期提出的还原过程，包括处理特殊性质旧电影的一系列步骤。

8.4.3 应用：去隔行扫描

像素运动估计另一种常见的应用是去隔行扫描，将一个奇偶行交换存储的视频转换为每帧包含所有无交错信号的视频(de Haan and Bellers 1998)。两个简单的去隔行扫描方法是摆动(bob)和波动(weave)，摆动将缺失行的上面或下面拷贝，波动将前一帧或后一帧的同一行拷贝。名称由这两种简单方法带来的视觉缺陷而来：摆动导致沿水平方向上下摆动运动；波动沿水平变化边缘产生"拉链"效应。将简单拷贝变为求均值会改善但不能完全解决这些缺陷。

为此，人们提出了很多方法，一般嵌入计算机视频数字化板卡的特制 DSP 芯片里(由于广播视频常常隔行扫描，而计算机显示器则不是)。此类方法的一大类是估计局部逐像素的运动并利用邻近时空场的信息来插值缺失数据。Dai, Baker, and Kang(2009)回顾了这方面的文献并提出了自己的算法，他们使用马尔科夫框架从七种不同插值函数中选择，对像素插值。

8.5 层次运动

很多情况下，视觉运动是由场景中少量不同深度的物体移动造成的。此时，如果将像素按合适的物体或层分组(Wang and Adelson 1994)，像素运动可以更简洁地描绘(并更可靠的估计)。

图 8.14 展示了这种方法。序列中的运动由背景的平移运动和前景手的转动产生。完整的运动序列可以通过前景和背景物体来重建，可以表达为透明度遮罩(alpha-matted)图像(子画面(sprite)或视频物体)和对应于每层的参数化运动。按从后到前顺序位移和合成这些层(3.1.3 节)重建了原始的视频序列。

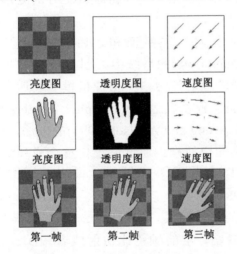

图 8.14 层次运动估计框架(Wang and Adelson 1994) © 1994 IEEE：前两行描述了两层，每层由亮度图、透明度图(黑色表示透明)和参数化运动场组成。两层用不同量的运动复合，重建运动序列

层次运动表达不仅仅使表达简洁(Wang and Adelson 1994; Lee, Chen, Lin *et al.* 1997)，也可以挖掘多个帧中的可用信息，还能在不连续运动处准确建模。这使其特别适合用于基于图像的渲染(13.2.1 节)(Shade, Gortler, He *et al.* 1998; Zitnick, Kang, Uyttendaele *et al.* 2004)和物体层的视频编辑。

为了计算视频序列的层次表达，Wang and Adelson(1994)首先在不重叠块上估计仿射运动模型，之后用 k 均值聚类结果。然后他们再从像素转换到层，使用 Darrell and Pentland(1991)的方法，根据像素结果重计算每层的运动估计。当计算完对于每个独立帧的参数化运动和按像素对齐的层后，卷绕并合并所有帧碎片，得到层结果。使用中值滤波器可以生成对微小亮度变化鲁棒的清晰的复合层，并能推断遮挡关系。图 8.15 展示了花园序列上的这一过程。你可以看到某帧的初始层和最终层的赋值，以及复合光流和带有对应光流向量覆盖的透明度遮罩层。

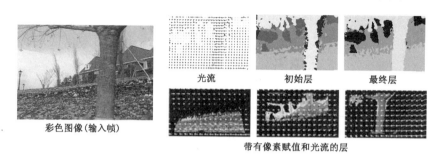

图 8.15 层次运动估计结果(Wang and Adelson 1994)© 1994 IEEE

之后的工作中，Weiss and Adelson(1996)使用形式化的概率混合模型来推断层的最优数量和每个像素属于哪个层。Weiss(1997)将每层仿射运动模型替换为像素规范化平滑运动估计把方法一般化，它使得系统更好的处理弯曲和波动的层，这在现实视频序列中很常见。

但是上述方法在估计运动、层赋值和之后的估计层颜色上仍旧有差别。在 Baker, Szeliski, and Anandan(1998)的系统中，图 8.14 的产生式模型被推广到真实世界的刚性运动场景。每帧的运动使用 3D 摄像机模型表达，并使用 3D 平面方程加上像素残余深度偏差(平面加视差表达，plane plus parallax)(2.1.5 节))来表达每层的运动。初始层估计类似于 Wang and Adelson(1994)的方法，但用刚性 2D 运动(同构)代替仿射运动模型。最终的模型修正通过最小化重合成和观测序列的差，来同时重优化层像素颜色、透明度 L_l 和 3D 深度、平面以及运动参数 z_l，n_l 和 P_t (Baker, Szeliski, and Anandan 1998)。

图 8.16 展示了算法的最终结果。如图所示，运动边界和层赋值比图 8.15 更清晰。由于使用像素深度偏差，层的颜色比仿射或者 2D 运动模型的结果更锐利。Baker, Szeliski, and Anandan(1998)最初的版本需要有大概的初始层赋值，而 Torr, Szeliski, and Anandan(2001)描述的自动贝叶斯方法，可以自动初始化系统并决定最优的层数。

层次运动估计一直都是研究热点。其中的代表论文有(Sawhney and Ayer 1996; Jojic and Frey 2001; Xiao and Shah 2005; Kumar, Torr, and Zisserman 2008; Thayananthan, Iwasaki, and Cipolla 2008; Schoenemann and Cremers 2008)。

当然，分层并不是把分割引入运动估计的唯一办法。有大量的算法把光流向量估计和将其分割为一致区域这两个问题交替进行(Black and Jepson 1996; Ju, Black, and Jepson 1996; Chang, Tekalp, and Sezan 1997; Memin and Perez 2002; Cremers and Soatto 2005)。近期的一些方法先分割输入彩色图像，再估计逐个分割部分的运动，产生一致的运动场，同时建立遮挡模型(Zitnick, Kang, Uyttendaele et al. 2004;

Zitnick, Jojic, and Kang 2005; Stein, Hoiem, and Hebert 2007; Thayananthan, Iwasaki, and Cipolla 2008)。

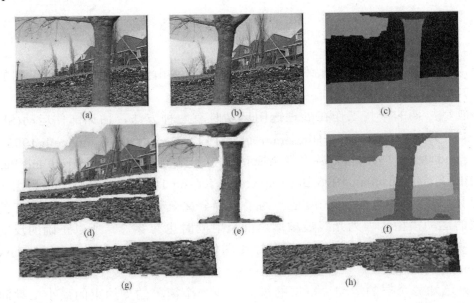

图 8.16 分层立体重建。(Baker, Szeliski, and Anandan 1998) © 1998 IEEE：(a)初始和(b)最终输入图像；(c)初始分为六层的分割；(d)和(e)六层的子画面；(f)平面形子画面的深度图(深色表示近处)；最前面一层在残差深度估计之前(g)和之后(h)。注意花园序列的色彩不正确；正确的颜色(黄花)见图 8.15

8.5.1 应用：帧插值

帧插值是另一种广泛使用的运动估计应用，通常和去隔行扫描在同一个电路里，用来匹配视频和显示器的实际刷新率。与去隔行扫描一样，新的帧间帧的信息需要从之前和之后的帧里插值。如果可以在每个未知像素的位置上计算出准确的运动估计，就可以得到最优结果。但除了计算运动之外，移动前景物体可能遮挡前后帧的个别像素造成污染，此时遮挡信息对于防止这种被污染的色彩十分重要。

考虑更多细节，在图 8.13c 中，假设箭头指示我们希望在其间插值图像的关键帧。图中条纹的方向代表像素的速度。如果在图像 I_0 的 x_0 处得到的运动估计 u_0 和在图像 I_1 的 $x_0 + u_0$ 处得到的运动估计相同，我们说光流向量一致。运动估计可以传导到图像 I_t 的 $x_0 + tu_0$ 处，其中 $t \in (0,1)$ 是时间插值。最终 $x_0 + tu_0$ 处颜色可以用线性混合得到，

$$I_t(\boldsymbol{x}_0 + t\boldsymbol{u}_0) = (1-t)I_0(\boldsymbol{x}_0) + tI_1(\boldsymbol{x}_0 + \boldsymbol{u}_0). \tag{8.72}$$

但是如果运动向量在对应位置不一致，必须用一些方法决定哪一个是正确的，哪个图的颜色是被遮挡的。真实的原因比这更微妙。此类插值算法的一个例子是基于早期工作的深度图插值(Shade, Gortler, He *et al.* 1998; Zitnick, Kang, Uyttendaele *et al.* 2004)。Baker, Black, Lewis *et al.* (2007); Baker, Scharstein, Lewis *et al.* (2009)的光

流评估论文中使用了它。Mahajan, Huang, Matusik *et al.* (2009)发表了一个使用基于梯度重建的更高质量的插值算法。

8.5.2 透明层和反射

分层运动一个常出现的特例是透明运动,通常由窗口和相框中所见到的反射(图 8.17 和图 8.18)造成。

该领域早期的一些工作在处理透明运动时有两种方法,估计部件运动(Shizawa and Mase 1991; Bergen, Burt, Hingorani *et al.* 1992; Darrell and Simoncelli 1993; Irani, Rousso, and Peleg 1994)和把各个像素赋值到相互竞争的运动层(Darrell and Pentland 1995; Black and Anandan 1996; Ju, Black, and Jepson 1996),后者对于场景由于遮挡部分可见(如叶子)比较合适。但是,为了准确分离真正的透明层,需要更好的与反射相关的运动模型。因为光线被玻璃表面反射同时也有穿透,所以正确的反射模型是一个加性的模型,每个运动层都对最终图像的某些亮度有贡献(Szeliski, Avidan, and Anandan 2000)。

如果已知每个层的运动,每层的恢复变成一个简单的受约束的最小二乘问题,每层的图像必须为正。但该问题受扩展低频不确定性影响很大,尤其是在层缺少深色(黑色)像素或者运动是单方向的时候。在他们的论文中,Szeliski, Avidan, and Anandan(2000)展示了通过交替进行鲁棒计算运动层和做层亮度的保守估计(上限或下限),来同时估计运动和层的值。在一个类似(Baker, Szeliski, and Anandan 1998)的联合约束最小二乘公式上,将其中覆盖合成操作替换为加法,并使用梯度下降法,可以优化最终的运动和层估计。

图 8.17 和图 8.18 展示了对两个带有反射的不同图像使用这一方法的结果。注意在第二个图序列中,反射光和透射光(女孩)相比很少,但算法依然能同时恢复两个层。

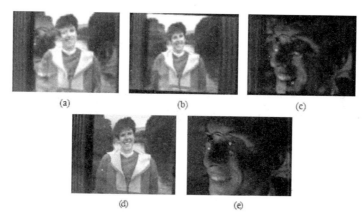

图 8.17 透明玻璃相框的光线反射:(a)输入序列的第一幅图; (b)最小合成的主运动层;(c)最大合成的次运动残余层; (d 和 e)最终估计的图片和反射层。原图像来自于 Black and Anandan (1996),分开的层来自于 Szeliski, Avidan, and Anandan (2000) © 2000 IEEE

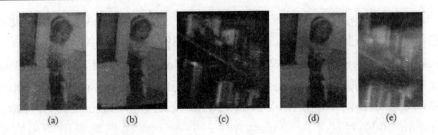

图 8.18 透明运动分离(Szeliski, Avidan, and Anandan 2000)© 2000 IEEE：(a)输入序列的第一幅图像；(b)最小合成的主运动层；(c)最大合成的次运动残余层；(d 和 e)最终估计的图片和反射层。注意，为了更好展示结构，(c)和(e)中的反射层的亮度被翻倍了

不幸的是，(Szeliski, Avidan, and Anandan 2000)中使用的简单参数化运动模型只对平面反射和深度较浅的场景有效。这些方法对曲面反射和非常深的场景的扩展的研究可见(Swaminathan, Kang, Szeliski et al. 2002; Criminisi, Kang, Swaminathan et al. 2005)，对具有更复杂 3D 深度的场景的扩展见(Tsin, Kang, and Szeliski 2006)。

8.6 补充阅读

一些早期的运动估计算法是为运动补偿视频编码提出来的(Netravali and Robbins 1979)，这些方法仍用在现代编码标准如 MPEG, H.263 和 H.264 中(Le Gall 1991; Richardson 2003)[①]。在计算机视觉中，该领域最初称为"图像序列分析"(Huang 1981)。一些早期开创性论文包括 Horn and Schunck(1981)和 Nagel and Enkelmann(1986)的变分方法以及 Lucas and Kanade(1981)的基于块的平移配准方法。Quam(1984), Anandan(1989)和 Bergen, Anandan, Hanna et al. (1992)发展了此类算法的分层(由粗到精)版本，它们同样也长期用于视频编码的运动估计。

Rehg and Witkin(1991), Fuh and Maragos(1991)，和 Bergen, Anandan, Hanna et al. (1992)将平移运动模型一般化为仿射运动，Shashua and Toelg(1997)和 Shashua and Wexler (2001)将其一般化到二次参考表面——Baker and Matthews(2004)对此进行了很好的综述。此类参数化运动估计算法广泛用于视频摘要(Teodosio and Bender 1993; Irani and Anandan 1998)、视频稳定化(Hansen, Anandan, Dana et al. 1994; Srinivasan, Chellappa, Veeraraghavan et al. 2005; Matsushita, Ofek, Ge et al. 2006)和视频压缩(Irani, Hsu, and Anandan 1995; Lee, ge Chen, lung Bruce Lin et al. 1997)。Brown(1992), Zitov'aa and Flusser(2003), Goshtasby(2005)和 Szeliski(2006a) 做了参数化图像注册的综述。

一般性综述和光流算法对比包括 Aggarwal and Nandhakumar(1988), Barron, Fleet, and Beauchemin(1994), Otte and Nagel(1994), Mitiche and Bouthemy(1996), Stiller and Konrad(1999), McCane, Novins, Crannitch et al.(2001), Szeliski(2006a)，以

---

① http://www.itu.int/rec/T-REC-H.264

及Baker, Black, Lewis et al. (2007)。匹配基元的主题，即在匹配前使用滤波或其他方法预变换图像，可参阅很多论文(Anandan 1989; Bergen, Anandan, Hanna et al. 1992; Scharstein 1994; Zabih andWoodfill 1994; Cox, Roy, and Hingorani 1995; Viola and Wells III 1997; Negahdaripour 1998; Kim, Kolmogorov, and Zabih 2003; Jia and Tang 2003; Papenberg, Bruhn, Brox et al. 2006; Seitz and Baker 2009)。Hirschmuller and Scharstein(2009)对比了一些该类方法，并在不同曝光场景中给出它们的相对性能。

评估光流算法的新基准的发表(Baker, Black, Lewis et al. 2007)使估计算法的质量得到快速发展，已经到达临界点使新的数据集变得必不可少。根据他们更新的技术报告(Baker, Scharstein, Lewis et al. 2009)，多数性能最好的算法使用鲁棒数据和平滑范式(一般为L_1或TV(全变差))以及连续变分优化方法，仍有些方法使用离散优化或分割(Papenberg, Bruhn, Brox et al. 2006; Trobin, Pock, Cremers et al. 2008; Xu, Chen, and Jia 2008; Lempitsky, Roth, and Rother. 2008; M.Werlberger, Trobin, Pock et al. 2009; Lei and Yang 2009; Wedel, Cremers, Pock et al. 2009)。

8.7 习 题

习题 8.1：相关性

实现并对比以下相关性算法的性能：
- 误差平方和(8.1)
- 鲁棒误差和(8.2)
- 绝对误差和(8.3)
- 偏差-增益补偿平方和(8.9)
- 规范化互相关(8.11)
- 以上的窗口版本(8.22 和 8.23)
- 以上度量的基于傅里叶的实现(8.18 和 8.20)
- 相位相关(8.24)
- 梯度互相关(Argyriou and Vlachos 2003)

在不同运动序列，不同的噪声、曝光变化、遮挡及频率变化(高频纹理，如沙或布，低频图像，如云或运动模糊视频)，对比你的一些算法。光照变化和对应真值(水平运动)的数据集见 *http://vision.middlebury.edu/stereo/data/*(2005 和 2006 数据集)。

一些附加想法、变形和问题如下所示。

1. 你认为什么时候相位相关性能优于一般相关或 SSD？你能做实验证明或者加以分析解释吗？
2. 对于基于傅里叶的掩模或窗口相关和误差平方和，结果应该与直接实现相同。注意，需要扩展(8.5)为成对相关的和，就像(8.22)那样(这是习题的一部分)。

3. 对于偏差-增益修正的平方误差变种(8.9)，你需要扩展其项为3×3(最小二乘)方程组。如果实现快速傅里叶变换版本，你需要指出为什么所有这些项都可以在傅里叶域评估。

4. (可选)实现 Hirschmuller and Scharstein(2009)研究的某些其他方法，看看你的结果和他们的是否一致。

习题 8.2：仿射注册

实现由粗到精方法来做仿射和投影图像注册。

1. 在金字塔较粗层使用较低阶(简单)的模型是否有帮助(Bergen, Anandan, Hanna *et al.* 1992)？

2. (可选)实现基于块的加速(Shumand Szeliski 2000; Baker andMatthews 2004)。

3. 阅读 Baker and Matthews (2004) 综述中的更多比较和想法。

习题 8.3：稳定化

写程序来稳定视频序列。你需要实现以下步骤，如 8.2.1 节所述。

1. 计算带有鲁棒外点剔除的连续帧间的平移(可选，旋转)。

2. 对运动参数做时域高通滤波，去除低频分量(使运动平滑)。

3. 高频运动补偿，缓慢缩放(用户指定值)避免丢失边缘像素。

4. (可选)不缩小，但从前帧或后续帧中取像素填充。

5. (可选)从邻近帧"偷取"较高频信息补偿由于快速运动模糊的图像。

习题 8.4：光流

计算两幅图像间的光流(基于样条或逐像素)，使用本章中一种或多种方法。

1. 使用 *http://vision.middlebury.edu/flow/* 或 *http://people.csail.mit.edu/celiu/motionAnnotation/* 的运动序列，测试你的算法，并与网上的结果(视觉上)比较你的性能。如果觉得你的算法和最好的结果有得一拼，可以考虑提交算法做正式评估。

2. 使用帧插值生成图像间的帧，可视化你的结果的质量(习题 8.5)。

3. 你觉得自己方法的相对效率(速度)如何？

习题 8.5：自动变形/帧插值

写程序自动在两幅图像间变形。完成下列步骤，如 8.5.1 节和 Baker, Scharstein, Lewis *et al.* (2009)所述。

1. 用两种方法计算光流(前一个习题)。考虑使用多帧($n>2$)方法更好的处理遮挡区域。

2. 对于每个中间(变形)图像，计算一组光流向量，确定哪个图像用来合成最终结果。

3. 混合(渐隐渐现)图像，并用序列查看器查看序列。

拿朋友和同事的图像尝试程序，看看能得到怎样的变形。还可以在视频序列中做高质量的慢动作效果。比较你的算法和简单的交互褪色。

习题 8.6：基于运动的用户交互

写程序计算低分辨率运动场来交互控制简单的应用(Cutler and Turk 1998)。示例如下。

1. 使用金字塔下采样每幅图像，计算从前一帧来的光流(基于样条或者像素)。
2. 将每个训练视频序列分割成不同的"动作"(如，手向内移动，向上移动，不动)，并"学习"与每个动作关联的速度场。(你可以简单计算每个运动场的均值和方差，或者使用更复杂的方法，如支持向量机(SVM))
3. 写一个识别器来找到近似正确的后续动作，并将其挂勾在交互应用上(如，发声器或者电脑游戏)。
4. 找你的朋友来测试。

习题 8.7：视频去噪

实现应用 8.4.2 中简述的算法。算法需要包括如下几步。

1. 计算准确的逐像素光流。
2. 决定参考图像中哪些点和其他帧有好的匹配。
3. 平均所有匹配点或选择最锐利的图像来做模糊补偿。作为解决方案的一部分，不要忘了使用常规的单帧去噪方法(见 3.4.4 节，3.7.3 节和习题 3.11)。
4. 对你认为光流估计不够准确的区域设计后退(fall-back)策略。

习题 8.8：运动分割

写程序把图像分割为分开的运动区域或寻找可靠的运动边界。

使用人工辅助的运动分割数据库 *http://people.csail.mit.edu/celiu/motionAnnotation/* 作为测试集的一部分。

习题 8.9：分层运动估计

将移动摄像机拍摄的视频序列分解成不同的层(8.5 节)。

1. 通过块计算或者鲁棒估计并迭代修正外点，找出主(仿射或平面透视)运动集。
2. 确定每个像素属于哪个运动。
3. 混合不同帧的像素来构建层。
4. (可选)加入像素光流残差或深度。
5. (可选)使用迭代全局最优方法来优化你的估计。
6. (可选)实现一个交互渲染器来生成帧间帧或从不同角度来看场景(Shade, Gortler, He et al. 1998)。
7. (可选)从更复杂的场景构造一个非卷绕马赛克，并用它来做视频编辑(Rav-Acha, Kohli, Fitzgibbon et al. 2008)。

习题 8.10：透明运动和反射估计

拍摄一个透过窗户(或相框)所看到的视频，你是否可以去除反射以便看清楚视频内容。

具体步骤见 8.5.2 节和 Szeliski, Avidan, and Anandan(2000)。其他方法见 Shizawa and Mase(1991), Bergen, Burt, Hingorani *et al.* (1992), Darrell and Simoncelli (1993), Darrell and Pentland(1995), Irani, Rousso, and Peleg(1994), Black and Anandan (1996)和 Ju, Black, and Jepson(1996)。

第 9 章

图 像 拼 接

9.1 运动模型
9.2 全局配准
9.3 合成
9.4 补充阅读
9.5 习题

图 9.1 图像拼接。(a)依据 54 张不同照片构造出来的局部圆柱形全景图；(b)球形全景图 (Szeliski and Shum 1997)；(c)从一个未排序的照片集中自动生成的多图像全景图；不带 运动物体去除方法的多图像拼接结果(d)和带有后的结果(e)(Uyttendaele, Eden, and Szeliski 2001)©2001 IEEE

配准图像并将其拼接成无缝拼图的算法是计算机视觉领域最古老且应用最广泛的算法之一(Milgram 1975; Peleg 1981)。图像拼接算法所生成高分辨率拼接图像可以用来产生数字地图和卫星照片。该算法还可以内嵌在大多数数码相机里面，用来生成特宽广角的全景图。

图像拼接方法起源于摄影测绘学领域，其中更多的是采用精细的手动处理方式，利用测量的地面控制点和连接点(tie point)注册方法，将多个航拍照片合成为大尺度拼接图像。该领域最关键的贡献之一就是发明了光束平差法算法(7.4 节)，同时求解所有摄像机位置的坐标，从而能够得到满足全局一致性的结果(Triggs, McLauchlan, Hartley et al. 1999)。另一个在生成拼接图像过程中反复遇到的问题就是如何消除可见的拼接缝隙，针对这个问题，多年以来人们已经提出了多种多样的解决方法(Milgram 1975, 1977; Peleg 1981; Davis 1998; Agarwala, Dontcheva, Agrawala et al. 2004)。

在胶片摄影领域，20 世纪 90 年代发明了可以拍摄特宽广角全景图的特殊摄像机。通常，可以一边让摄像机绕着固定轴旋转，一边同时让胶片在垂直的缝隙下曝光成像来实现(Meehan 1990)。20 世纪 90 年代中期，图像配准方法开始应用在普通手持摄像机的广角无缝全景图的构造上(Mann and Picard 1994; Chen 1995; Szeliski 1996)。该领域最近的研究工作关注的需求包括计算全局一致性的配准(Szeliski and Shum 1997; Sawhney and Kumar 1999; Shum and Szeliski 2000)，消除由于视差和物体运动导致的虚影(Davis 1998; Shum and Szeliski 2000; Uyttendaele, Eden, and Szeliski 2001; Agarwala, Dontcheva, Agrawala et al. 2004)，以及针对不同曝光条件下的处理(Mann and Picard 1994; Uyttendaele, Eden, and Szeliski 2001; Levin, Zomet, Peleg et al. 2004; Agarwala, Dontcheva, Agrawala et al. 2004; Eden, Uyttendaele, and Szeliski 2006; Kopf, Uyttendaele, Deussen et al. 2007)[①]。这些方法已经催生了大量的商业拼接产品(Chen 1995; Sawhney, Kumar, Gendel et al. 1998)，网上可以找到相关的评价和比较信息。[②]

不同于早期直接寻求最小化像素点到像素点之间差异的大多数方法，近期的算法通常需要提取一个稀疏的特征集，再对这些特征进行匹配，正如第 4 章中所描述的。这些基于特征的图像拼接方法对于场景运动具有较强的鲁棒性是其优势，而且如果实现方式得当，速度上也会更快。不过，这种方法最大的优势是能够"认出全景图"，即能够自动从未排序的图像集找出图像之间的邻接关系，从而使一般的使用者能够全自动地拼接这些图像(Brown and Lowe 2007)。

那么，对于图像拼接来说，有哪些最基本的问题呢？首先，为了进行图像的配准，我们必须确定一幅图像中的像素点坐标与另一图像中的像素点坐标之间关联的近似数学模型——本书的 9.1 节将回顾我们已经研究过的几种基本模型，并且阐述

---

① 这些论文中的一部分已经由 Benosman and Kang(2001)编著整理，相关的综述由 Szeliski(2006a)完成。
② Photosynth 网站，*http://photosynth.net*，允许用户免费生成和上传全景图。

与全景图拼接特别相关的一些新的运动模型。其次，我们必须估计相关联的各种图像对之间正确的配准参数。第 4 章讨论了如何从每一幅图像中找出具有区分性的特征，进而通过快速高效匹配找到图像对之间的对应关系。第 8 章讨论了如何通过直接比较像素点，并结合梯度下降法以及其他优化方法，同样可以估计这些参数。在一幅全景图中存在多个图像的情况下，我们可以利用光束平差法(见 7.4 节)来计算具有全局一致性的配准结果，并且能够有效地找出一幅图像和另一幅图像的重叠部分。在 9.2 节中，我们将分析如何在这些成熟方法的基础上进行修改，以利用常见的成像装置来构造全景图。

一旦完成图像的配准工作，我们就必须选择一个最终的合成表面来卷绕那些已经配准的图像(见 9.3.1 节)。我们还需要利用相关算法来无缝剪裁和融合重叠的图像，甚至处理存在视差、镜头畸变、场景运动和曝光差异情况下的图像拼接问题(参见 9.3.2 节～9.3.4 节)。

9.1 运动模型

为了能够完成图像的注册和配准，我们需要建立将一幅图像中的像素点坐标映射到另一幅图像中数学变换关系。可以用于描述上述变换关系的参数化运动模型有很多，比如简单 2D 变换、平面透视模型、3D 摄像机旋转、镜头畸变和非平面模型(比如圆柱形模型)。

在本书的 2.1 节和 6.1 节中，我们已经介绍了其中的几个模型。比如，我们在 2.1.5 节中介绍了如何用一种含有 8 个参数的同构(单应)关系(2.71)描述从不同位置观测一个特定平面的参数化运动(Mann and Picard 1994; Szeliski 1996)。我们还介绍了在单纯旋转运动下的摄像机服从另一种同构关系(2.72)。

在本节中，我们不仅会回顾这些运动模型，而且会介绍如何将这些模型应用于不同的拼接场景中。此外，我们还要介绍球形和圆柱形合成表面，并展示如何在有利的情况下将它们用于使用纯平移进行配准(9.1.6 节)。在给定场景或给定数据集的情况下，判断最佳配准模型的问题称为"模型选择问题"(Hastie, Tibshirani, and Friedman 2001; Torr 2002; Bishop 2006; Robert 2007)，这个重要问题在本书中并未涉及。

9.1.1 平面透视运动

可用于图像配准的最简单的运动模型就是在 2D 中进行单纯平移和旋转，如图 9.2(a)所示。这正是有一些有重叠的摄影照片时用到的运动模型。这也是 David Hockney 在制作拼贴画时用的方法，他将这些拼贴画称作"接合者"(joiner)(Zelnik-Manor and Perona 2007; Nomura, Zhang, and Nayar 2007)。在类似于 Flickr 这样的网站上，以"全景图法"(panography)(6.1.2 小节)之名制作这种带有可见缝隙和不一致性艺术效果的拼贴画是很流行的。平移和旋转通常还是对摄像机的微小运动进行补偿的合适的运动模型，可以应用于图像和视频的防抖动和合并(习题 6.1 和 8.2.1 节)。

(a)平移变换(2 自由度)　(b)仿射变换(6 自由度)　(c)透视变换(8 自由度)　(d)三维旋转(3+自由度)

图 9.2　二维运动模型及其在图像拼接中的应用

在 6.1.3 节中，我们介绍了将两个不同的摄像机对一个共同的平面成像其结果间的映射关系可以描述成一个 3×3 的同构矩阵(2.71)。考虑将一幅图像中的像素点映射到 3D 空间中，再反过来映射到第二幅图像中的过程，该过程对应一个矩阵 $\boldsymbol{M}_{10}$，

$$\tilde{x}_1 \sim \tilde{\boldsymbol{P}}_1 \tilde{\boldsymbol{P}}_0^{-1} \tilde{x}_0 = \boldsymbol{M}_{10} \tilde{x}_0. \tag{9.1}$$

当矩阵 $\boldsymbol{P}_0$ 的最后一行被替换为平面方程 $\hat{\boldsymbol{n}}_0 \cdot \boldsymbol{p} + c_0$ 且假设所有的点都位于同一个平面上，即它们的视差是 $d_0 = 0$ 时，我们就可以不用关心最后的深度缓冲(z-buffer)大小，从而可以忽略矩阵 $\boldsymbol{M}_{10}$ 的最后一列和最后一行。最终得到的同构矩阵 $\tilde{\boldsymbol{H}}_{10}$ ($\boldsymbol{M}_{10}$ 左上角的 3×3 子矩阵)描述了两幅图像中像素间的映射关系，

$$\tilde{x}_1 \sim \tilde{\boldsymbol{H}}_{10} \tilde{x}_0. \tag{9.2}$$

基于这种观察，形成了最早期的图像自动拼接算法的基础(Mann and Picard 1994; Szeliski 1994, 1996)。因为当时还没有产生可靠的特征匹配方法，这些算法都采用了直接像素值匹配方法，即直接的参数化运动估计方法，参见 8.2 节和公式(6.19 和 6.20)。

更近一些的拼接算法首先要抽取特征，然后用一些鲁棒性较强的方法，比如用 RANSAC 算法(6.1.4 节)来计算一个好的内点集合，对它们进行匹配。最后计算同构矩阵(9.2)的问题实际上就是在给定对应特征配对的条件下，求解最小二乘拟合的问题，

$$x_1 = \frac{(1+h_{00})x_0 + h_{01}y_0 + h_{02}}{h_{20}x_0 + h_{21}y_0 + 1} \quad \text{and} \quad y_1 = \frac{h_{10}x_0 + (1+h_{11})y_0 + h_{12}}{h_{20}x_0 + h_{21}y_0 + 1}, \tag{9.3}$$

可以用最小二乘迭代法进行求解，参见 6.1.3 节和公式(6.21～6.23)。

9.1.2　应用：白板和文档扫描

图像拼接算法最简单的一个应用就是将平板扫描仪得到的多个扫描图像拼接在一起。比如，你有一张大地图或者一张儿童绘画作品，但因为尺寸太大而不能直接放到扫描仪中进行扫描。你只要简单地对文档进行多次扫描，并且保证每次扫描的

内容之间有足够大的重叠部分,确保能够提供足够的共同特征。然后,只要依次取出一对已知有重叠的扫描图像,提取特征并进行特征匹配(必要时还可以用两点的 RANSAC 方法估计内点集合,找出特征的最佳匹配)进而估计出 2D 的刚性变换式(2.16)

$$x_{k+1} = R_k x_k + t_k, \tag{9.4}$$

最后,在最终的合成表面上(例如,与第一幅扫描图配准)对所有图像进行重采样(3.6.1 节)并对其做平均合在一起。你能看出这个方案存在哪些潜在的问题吗?

其中一个比较棘手的问题是,2D 刚性变换对旋转角度 θ 是非线性的,所以必须采用非线性最小二乘法或者将 R 约束在正交的情况下(参见 6.1.3 节)。

在像素点对的配准过程中,存在着一个更大的问题。随着配准的像素点对不断增多,解的数值可能会产生漂移,从而丧失全局的一致性。在这种情况下,需要 9.2 节中介绍的全局优化过程(global optimization procedure)。但在某些情况下,比如线性同构(9.1.3 节)和相似变换(6.1.2 节),这种全局优化用常规最小二乘法就可以求解。但是除此之外,这种全局优化通常都需要求解一个大型的非线性系统。

一个稍微复杂一些的情形是,有多个用手持摄像机拍摄的白板或其他大面积平坦物体的照片时(He and Zhang 2005; Zhang and He 2007)。尽管原则上可以用更复杂的模型来估计相对于平面的 3D 刚性运动(加上可能未知的焦距变化),但这里我们还是用同构关系作为自然的运动模型。

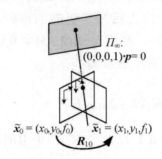

图 9.3 纯 3D 摄像机旋转。同构(映射)的形式非常简单,仅依赖于 3D 旋转矩阵和焦距

9.1.3 旋转全景图

在全景图拼接方法中,最典型的情况就是处理摄像机单纯的旋转运动。试想一下,你站在大峡谷的边上拍照,相对于场景中远处的物体来说,摄像机服从单纯的旋转运动。这相当于假设所有的点都距离摄像机很远,即位于无穷远平面(图 9.3)上。令 $t_0 = t_1 = 0$,我们可以得到简化的 3×3 单应(同构)

$$\tilde{H}_{10} = K_1 R_1 R_0^{-1} K_0^{-1} = K_1 R_{10} K_0^{-1}, \tag{9.5}$$

这里 $K_k = \mathrm{diag}(f_k, f_k, 1)$ 是在假设 $c_x = c_y = 0$ 的条件下,即假设我们从光学中心对像素进行索引(Szeliski 1996)的情况下,简化的摄像机内参数矩阵(2.59)。这样,我们还可以将上式重写为

$$\begin{bmatrix} x_1 \\ y_1 \\ 1 \end{bmatrix} \sim \begin{bmatrix} f_1 & & \\ & f_1 & \\ & & 1 \end{bmatrix} \boldsymbol{R}_{10} \begin{bmatrix} f_0^{-1} & & \\ & f_0^{-1} & \\ & & 1 \end{bmatrix} \begin{bmatrix} x_0 \\ y_0 \\ 1 \end{bmatrix} \tag{9.6}$$

或

$$\begin{bmatrix} x_1 \\ y_1 \\ f_1 \end{bmatrix} \sim \boldsymbol{R}_{10} \begin{bmatrix} x_0 \\ y_0 \\ f_0 \end{bmatrix}, \tag{9.7}$$

上述公式反映了映射方程的简单性,并显式地表达了所有的运动参数。这样,我们可以得到 3 个参数、4 个参数或 5 个参数的 3D 旋转运动模型,分别对应焦距 f 为已知焦距、固定焦距或可变焦距(Szeliski and Shum 1997)[①]的情况,用于取代两个图片之间一般的 8 参数同构关系。对每一幅图像自身的 3D 旋转矩阵(以及可选的焦距)估计,本质上来说比估计有 8 个自由度的同构关系更加稳定。因此,很多大尺度图像拼接算法(Szeliski and Shum 1997; Shum and Szeliski 2000; Brown and Lowe 2007)都采用了这种估计方法。

在这种表达的前提下,我们如何更新旋转矩阵来最好地配准两幅有重叠的图像呢?给定公式(9.5)中同构矩阵的一个当前估计 $\hat{\boldsymbol{H}}_{10}$,更新旋转矩阵 $\boldsymbol{R}_{10}$ 的最佳方式就是考虑相对于当前旋转矩阵 $\boldsymbol{R}_{10}$ 的增量旋转矩阵 $\boldsymbol{R}(w)$ (Szeliski and Shum 1997; Shum and Szeliski 2000),

$$\tilde{\boldsymbol{H}}(\omega) = \boldsymbol{K}_1 \boldsymbol{R}(\omega) \boldsymbol{R}_{10} \boldsymbol{K}_0^{-1} = [\boldsymbol{K}_1 \boldsymbol{R}(\omega) \boldsymbol{K}_1^{-1}][\boldsymbol{K}_1 \boldsymbol{R}_{10} \boldsymbol{K}_0^{-1}] = \boldsymbol{D}\tilde{\boldsymbol{H}}_{10}. \tag{9.8}$$

注意,这里我们已经用复合形式重写了更新的规则,其中增量更新矩阵 $\boldsymbol{D}$ 是相对于当前的同构矩阵 $\hat{\boldsymbol{H}}_{10}$ 而言的。利用公式(2.35)中介绍的小角度近似方法来近似描述 $\boldsymbol{R}_{(\omega)}$,我们可以将增量更新矩阵重写为

$$\boldsymbol{D} = \boldsymbol{K}_1 \boldsymbol{R}(\omega) \boldsymbol{K}_1^{-1} \approx \boldsymbol{K}_1 (\boldsymbol{I} + [\omega]_\times) \boldsymbol{K}_1^{-1} = \begin{bmatrix} 1 & -\omega_z & f_1\omega_y \\ \omega_z & 1 & -f_1\omega_x \\ -\omega_y/f_1 & \omega_x/f_1 & 1 \end{bmatrix}. \tag{9.9}$$

注意,矩阵 $\boldsymbol{D}$ 的元素和表 6.1 以及公式(6.19)中的参数序列 $h_{00},\cdots,h_{21}$ 之间存在着非常漂亮的一一对应关系,即

$$(h_{00}, h_{01}, h_{02}, h_{00}, h_{11}, h_{12}, h_{20}, h_{21}) = (0, -\omega_z, f_1\omega_y, \omega_z, 0, -f_1\omega_x, -\omega_y/f_1, \omega_x/f_1). \tag{9.10}$$

因此,我们将链式规则应用于公式(6.24 和 9.10),可以得到

$$\begin{bmatrix} \hat{x}' - x \\ \hat{y}' - y \end{bmatrix} = \begin{bmatrix} -xy/f_1 & f_1 + x^2/f_1 & -y \\ -(f_1 + y^2/f_1) & xy/f_1 & x \end{bmatrix} \begin{bmatrix} \omega_x \\ \omega_y \\ \omega_z \end{bmatrix}, \tag{9.11}$$

这样便得到一个用于估计 $\boldsymbol{\omega} = (\omega_x, \omega_y, \omega_z)$[②]的线性更新方程。注意,这个更新规则依赖于目标视角的焦距 f_1,而与模板视角的焦距 f_0 无关。因为这种复合算法实际上是对目标做了一些微小的扰动。一旦计算出增量旋转向量 $\boldsymbol{\omega}$,就可以利用公式 $\boldsymbol{R}_2 \leftarrow \boldsymbol{R}(\omega)\boldsymbol{R}_1$ 来更新旋转矩阵 $\boldsymbol{R}_1$。

更新焦距估计的公式实际上更加复杂一些,(Shum and Szeliski 2000)中有介

[①] 焦距的初始估计可以通过 6.3.4 节描述的内参数标定方法或 EXIF 图像格式的标记获得。
[②] 根据期望输出全景图的分辨率,对最终合成平面的缩放因子也可以设置一个略大或者略小的值,参见 9.3 节。

绍。由于我们在 9.2.1 节中会介绍另一种基于最小化 3D 光束向后投影误差的方法，也用于更新焦距的估计结果，因此这里我们就不再赘述了。图 9.4 显示了 3D 旋转运动模型下四幅图像的配准情况。

图 9.4　四幅利用手持摄像机拍摄的照片，通过 3D 旋转模型注册的图像(Szeliski and Shum 1997 ©1997ACM)。注意，这些同构关系(而不是任意的变换)是如何得到很合适的梯形图像，其宽度从原始图像扩展开

9.1.4　缝隙消除

本小节介绍的方法用来估计一系列的旋转矩阵和焦距，进而共同构造出大型的全景图。但是，由于累积误差的影响，这种方法很少能构造出一个闭合的 360° 全景图。实际上，结果中总是存在一些缝隙或者重叠(图 9.5)。

通过将序列中的第一幅图像与最后一幅图像进行匹配，我们可以解决这个问题。与重现的第一幅图相关的这两个旋转矩阵估计值之间的误差表明了注册错误量。如果将与这些旋转关联的两个四元数的商(即"误差四元数")除以序列中图像的数量，就可以将误差在整个序列上均匀地分布(假设相邻图像帧之间的旋转是相对固定的)。另外，还可以根据注册错误量来更新焦距的估计结果。实际操作过程中，我们首先将误差四元数转化为一个缝隙角(gap angle) θ_g，然后再利用公式 $f' = f(1 - \theta_g / 360°)$ 来对焦距进行更新。

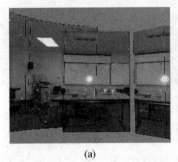　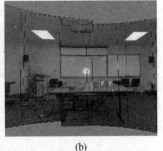

(a)　　　　　　　　(b)

图 9.5　缝隙消除方法(Szeliski and Shum 1997© 1997 ACM)：(a)当焦距错误($f = 510$)时，存在明显可见的缝隙；(b)当焦距正确($f = 468$)时，没有可见的缝隙

图 9.5a 给出了注册后的图像序列结果以及原始的第一幅图像。在最后一幅图像和第一幅图像之间存在着巨大的缝隙，而实际上它们应该是同一幅图像。由于错误的估计了焦距（$f=510$），所以产生了32°间隙缝角。图 9.5b 给出了利用正确的焦距（$f=468$）进行缝隙消除之后的注册结果。注意，两幅拼图都几乎没有可见的错误注册(除了缝隙)，但是图 9.5a 在计算过程中使用的焦距存在 9%的误差。为了解决焦距估计结果中存在的相关问题，可以采用单纯摇移运动和圆柱面图像等方法，相关方法是由 Hartley(1994b)，McMillan and Bishop(1995), Stein(1995)和 Kang and Weiss(1997)提出来的。

可惜，这种特殊的缝隙消除启发式方法，只适用于那种摄像机朝一个方向连续转动的"一维"全景图。在 9.2 节中，我们会介绍另一种在任意摄像机运动下消除缝隙和重叠的方法。

9.1.5 应用：视频摘要和压缩

图像拼接方法一种有趣的应用，就是能够对摇动的摄像机拍摄的视频进行摘要和压缩。这类应用最早是由 Teodosio and Bender(1993)提出来的，当时被他们称作"基于拼图的静态摘要照片"(Silient still)。这个原始想法，经过 Irani, Hsu, and Anandan(1995), Kumar, Anandan, Irani *et al.* (1995)和 Irani and Anandan(1998)等人的扩展，产生了很多附加的应用，比如视频压缩和视频索引。不过，这些早期的方法采用的都是仿射运动模型，所以只能应用于长焦距的情况。Lee, Chen, Lin *et al.* (1997)将其推广为 8 参数的完整同构运动模型，用于拼接背景图层，并且成为 MPEG-4 视频压缩标准的一部分，其中拼接而成的背景图层称为"视频子画面"(图 9.6)。

图 9.6 视频拼接背景图像的目的是构造一个子画面图像，从而能够用于传输和重建每一帧的背景(Lee, ge Chen, lung Bruce Lin*et al.* 1997)© 1997 IEEE

虽然在很多情况下，视频的拼接就是多个图像拼接的直接的推广(Steedly, Pal, and Szeliski 2005; Baudisch, Tan, Steedly *et al.* 2006)，但是却因可能存在大量的独立运动、镜头伸缩以及期望对动态事件进行可视化而面对诸多挑战。比如，运动的前景目标可以通过中值滤波(median filtering)来去除。另一方面，还可以提取前景目标(Sawhney and Ayer 1996)并将其放入一个独立的图层中，然后再融合进已经拼接好的全景图中。有时候可以同时放入多个实例，从而达到"连续照片(Chronophotograph)"的效果(Massey and Bender 1996)；有时候可以用于视频覆盖(Irani and Anandan 1998)。除此之外，视频还可以用来生成动态的全景视频纹理(panoramic video textures，参见13.5.2节)，其中全景场景中的不同部分都是随运动视频循环而相对独立动态变化的(Agarwala, Zheng, Pal et al. 2005; Rav-Acha, Pritch, Lischinski et al. 2005)，或者利用"视频闪光照片"合成拼接的场景(Sawhney, Arpa, Kumar et al. 2002)。

需要利用运动摄像机构造全景图的时候，视频也可以作为一种有趣的素材来源。尽管这些素材并不满足单个视点(光学中心)的假设，但是我们依然可以得到一些有趣的结果。比如，由Sawhney, Kumar, Gendel et al. (1998)开发的VideoBrush系统，就是将水平运动的摄像机获得的图像中央的细条抽出，用于构造全景图。利用自适应流形上的拼图方法(Peleg, Rousso, Rav-Acha et al. 2000)，这种思路可以推广应用于其他的摄像机运动模型和合成平面。此外，还可以利用这种思路来构造立体全景图(Peleg, Ben-Ezra, and Pritch 2001)。相关的思路还可以应用于为多平面单帧动画构造全景背景绘图(Wood, Finkelstein, Hughes et al. 1997)，或者为存在视差的场景构造拼接图像(Kumar, Anandan, Irani et al. 1995)，以及利用多投影中心图像(multiple-center-of-projection images)(Rademacher and Bishop 1998))和多透视全景图(multi-perspective panoramas)(Román, Garg, and Levoy 2004; Román and Lensch 2006; Agarwala, Agrawala, Cohen et al. 2006)方法为更复杂的场景构造3D表示。

另一个基于视频的全景图问题的变形是同心拼图问题(13.3.3节)(Shum and He 1999)。该问题的目标不是简单生成一个单独的全景图，而是根据完整的原始视频，利用光线重映射(光场渲染)方法，重新合成出完整的视图(从不同的摄像机源)，从而赋予全景图以三维景深的感觉。利用多基线立体视觉方法(Peleg, Ben-Ezra, and Pritch 2001; Li, Shum, Tang et al. 2004; Zheng, Kang, Cohen et al. 2007)，在相同的数据集上还可以明确地重建出深度图。

9.1.6 圆柱面和球面坐标

利用同构变换或者3D运动来进行图像配准时，可以首先将图像卷绕到柱面坐标系下，再使用单纯的平移模型来进行配准(Chen 1995; Szeliski 1996)。不过，只有所有图像都是用同一水平线或者同一已知倾斜角的摄像机拍摄时，这种方法才适用。

这里不妨假设摄像机处于规范(canonical)位置，即摄像机的旋转矩阵是单位矩阵，$R=I$，从而光轴平行于z轴而与y轴垂直。这样与像素点(x,y)对应的3D射线即为(x,y,f)。

我们希望将图像投影到一个单位半径的圆柱体表面(Szeliski 1996)。该表面上的

一个点是由角度 θ 和高度 h 参数化决定的，(θ, h) 与 3D 圆柱体坐标之间的对应关系为

$$(\sin\theta, h, \cos\theta) \propto (x, y, f), \tag{9.12}$$

如图 9.7a 所示。根据这种对应关系，我们计算出卷绕后或映射后的坐标形式 (Szeliski and Shum 1997)

$$x' = s\theta = s\tan^{-1}\frac{x}{f}, \tag{9.13}$$

$$y' = sh = s\frac{y}{\sqrt{x^2+f^2}}, \tag{9.14}$$

其中，s 是任意的缩放因子(有时也称为圆柱体的半径)，可以令 $s=f$ 以最小化图像中心的变形(缩放)程度[①]。逆映射公式为

$$x = f\tan\theta = f\tan\frac{x'}{s}, \tag{9.15}$$

$$y = h\sqrt{x^2+f^2} = \frac{y'}{s}f\sqrt{1+\tan^2 x'/s} = f\frac{y'}{s}\sec\frac{x'}{s}. \tag{9.16}$$

图像还可以投影到球面上(Szeliski and Shum 1997)。当最后的全景包括完整的球面或半球面视角，而不是条带状圆柱面时，这种做法是非常有用的。在这种情况下，球面是由两个角度 (θ,ϕ) 参数构成的，与 3D 球面坐标之间的对应关系为

$$(\sin\theta\cos\phi, \sin\phi, \cos\theta\cos\phi) \propto (x, y, f), \tag{9.17}$$

如图 9.7b[②]所示。坐标之间的映射关系由下式(Szeliski and Shum 1997)给出

$$x' = s\theta = s\tan^{-1}\frac{x}{f}, \tag{9.18}$$

$$y' = s\phi = s\tan^{-1}\frac{y}{\sqrt{x^2+f^2}}, \tag{9.19}$$

而逆映射公式为

$$x = f\tan\theta = f\tan\frac{x'}{s}, \tag{9.20}$$

$$y = \sqrt{x^2+f^2}\tan\phi = \tan\frac{y'}{s}f\sqrt{1+\tan^2 x'/s} = f\tan\frac{y'}{s}\sec\frac{x'}{s}. \tag{9.21}$$

注意，通过变换参数 z 和缩放因子 f，可以非常简单地从公式 9.17 产生缩放后的坐标 (x,y,f)。

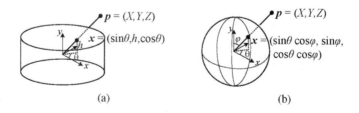

图 9.7 从 3D 到圆柱面(a)和球面的投影(b)

[①] 根据预期输出全景图的分辨率，还可以针对最终合成平面的缩放因子设置一个略大或者略小的值，详情参见 9.3 节。

[②] 注意，这里的坐标是与公式(2.8)中使用的普通球面坐标不同的。这里，由于我们习惯于看水平拍摄的照片，即 y 轴是指向重力向量的方向，所以指向北极点的是 y 轴而不是 z 轴。

当已知摄像机处于水平位置且只绕其垂直轴旋转时(Chen 1995)，圆柱面图像拼接算法的应用是最广泛的。在这种情况下，不同旋转角度的图像实际上是通过单纯的水平平移操作关联起来的[①]。由于这种方法可以避免透视变换配准算法(6.1 节，8.2 节和 9.1.3 节)中大量复杂计算，因而成为计算机视觉入门课程中非常具有吸引力的课程设计题目。图 9.8 中给出了水平旋转全景图中的两幅圆柱面卷绕后的图像是如何通过单纯的平移操作关联起来的(Szeliski and Shum 1997)。

专业的全景图摄影师经常使用摇动-俯仰平台(pan-tilt head)，使倾斜角的控制更容易，并且能够固定在特定的旋转角上。为了获得更大的全景图，有时候也会使用自动旋转镜头(Kopf, Uyttendaele, Deussen et al. 2007)[②]。这样不仅能够保证用必要数量的图像来对视野区域进行一致的覆盖，而且还能用圆柱面或者球面坐标和单纯的平移变换来拼接图像。在这种情况下，像素点坐标 (x,y,f) 必须根据已知的倾斜(tilt)角度和偏移(panning)角度进行旋转变换，然后才能投影到圆柱面或球面坐标系下(Chen 1995)。如果已知一个粗略的偏移角度，则输入图像之间的粗略相对位置关系事先就可以确定了，计算配准的过程会简单很多。图 9.9 给出了由整个 3D 旋转全景图反向卷绕到球面上的展示(Szeliski and Shum 1997)。

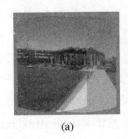 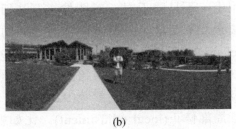

(a)　　　　　　　　　　　　(b)

图 9.8　圆柱形全景图(Szeliski and Shum 1997 © 1997 ACM)：(a)有水平平移关系的两幅圆柱面卷绕后的图像；(b)从图像序列中合成的圆柱形全景图的一部分

图 9.9　由 54 张相片构成的一张球面全景图(Szeliski and Shum 1997 © 1997 ACM)

最后一个值得一提的坐标映射方式是极坐标映射，其中北极点位于光轴上而不

[①] 微小的垂直倾斜有时可以通过垂直平移来进行补偿。
[②] 参考 http://gigapan.org

是位于垂直轴上，映射关系为

$$(\cos\theta\sin\phi, \sin\theta\sin\phi, \cos\phi) = s(x,y,z). \tag{9.22}$$

在这种情况下，映射方程的形式变为

$$x' = s\phi\cos\theta = s\frac{x}{r}\tan^{-1}\frac{r}{z}, \tag{9.23}$$

$$y' = s\phi\sin\theta = s\frac{y}{r}\tan^{-1}\frac{r}{z}, \tag{9.24}$$

其中 $r = \sqrt{x^2 + y^2}$ 是在 (x,y) 平面上的径向距离，而 $s\phi$ 在 (x',y') 平面上平面上起类似的作用。这种映射关系为某些广角全景图提供了一个极具吸引力的可视化表面，而且正如 2.1.6 节所述为鱼眼镜头(fisheye lense)引起的畸变(2.1.6 节)提供了一个良好的模型。注意，对于很小的 (x,y) 数值，该映射方程退化为 $x' \approx sx/z$，这表明 s 所扮演的角色和焦距 f 很类似

9.2 全 局 配 准

到目前为止，我们已经讨论了如何利用各种运动模型将一对图像进行注册。在大多数的应用中，不只限于对一对图像进行注册。真正的目标是找到一个全局一致的配准参数集合，从而最小化所有图像对之间的注册误差(Szeliski and Shum 1997; Shum and Szeliski 2000; Sawhney and Kumar 1999; Coorg and Teller 2000)。

在本节中，我们将两两配准的条件(6.2，8.1 和 8.50)扩展到一个全局的能量函数，其中涉及每一幅图像的姿态参数(9.2.1 节)。计算出全局配准的结果之后，我们经常需要进行局部修正(local adjustment)，比如视差消除(parallax removal)，即消除由于局部注册错误出现的重影和模糊现象(9.2.2 节)。最后，如果我们要对一个未排序的图像集合进行注册，就需要发现哪些图像放在一起构成一个或多个全景图。这个过程称为"认出全景图"(panorama recognition)，将在 9.2.3 节中进行介绍。

9.2.1 光束平差法

要将大量图像进行注册，一种方式就是每次向全景图中添加一幅新的图像，将最近添加的图像与集合中原有的图像进行配准(Szeliski and Shum 1997)，并且在必要时找出与之有重叠的图像(Sawhney and Kumar 1999)。在 360°全景图的情况下，累积误差可能会导致全景图的两端之间产生缝隙(或过度重叠)，这类问题可以通过拉伸所有图像的配准来实现缝隙消除(gap closing)(Szeliski and Shum 1997)。但是，解决这类问题的另一种更好的方式是，在最小均方误差框架下对图像进行同时配准，以正确地分配误注册偏差。

在摄影测绘学界，同时调整有重叠的大量图像的姿态参数的过程，称为"光束平差法"(bundle adjustment)(Triggs, McLauchlan, Hartley *et al.* 1999)。在计算机视觉领域，这种方法首先应用于一般的由运动到结构问题(Szeliski and Kang 1994)，

后来为全景图的拼接问题进行了特殊化(Shum and Szeliski 2000; Sawhney and Kumar 1999; Coorg and Teller 2000)。

在本节中，我们利用基于特征的方法来形式化全局配准的问题，因为这样做产生更简单的系统。无论是将图像分割为块状区域并为每一块区域建立虚拟的特征对应关系(如 9.2.4 节中讨论的情况，以及 Shum and Szeliski 2000)，或者利用每个像素的误差度量来替代每个特征的误差度量，都是等价的直接求解方法。

考虑公式(6.2)中给出的基于特征的配准问题，

$$E_{\text{pairwise-LS}} = \sum_i \|r_i\|^2 = \|\tilde{x}_i'(x_i; p) - \hat{x}_i'\|^2. \tag{9.25}$$

对于多幅图像的配准，与单独一对图像之间的特征对应关系 $\{(x_i, \hat{x}_i)\}$ 不同，我们用 n 个特征构成的集合，包括第 i 个特征点在第 j 幅图像中的位置 x_{ij}，以及自身的置信度(即方差的倒数) c_{ij}。每一幅图像还有一些相关联的姿态参数。

尽管依据同构矩阵来形式化也是可以的(Szeliski and Shum 1997; Sawhney and Kumar 1999)，但是在本小节中，我们假设图像的姿态是由旋转矩阵 R_j 和焦距 f_j 决定的。将 3D 空间中一个点 x_i 映射到第 j 帧图像中的一个点 x_{ij} 的等式，可以由公式(2.68)和公式(9.5)改写为

$$\tilde{x}_{ij} \sim K_j R_j x_i \text{ and } x_i \sim R_j^{-1} K_j^{-1} \tilde{x}_{ij}, \tag{9.26}$$

其中 $K_j = \text{diag}(f_j, f_j, 1)$ 是标定矩阵的简化形式。类似的，第 j 帧图像中的点 x_{ij} 和第 k 帧图像中的点 x_{ik} 的运动映射为

$$\tilde{x}_{ik} \sim \tilde{H}_{kj} \tilde{x}_{ij} = K_k R_k R_j^{-1} K_j^{-1} \tilde{x}_{ij}. \tag{9.27}$$

给定一个由链式成对配准估计得到的初始集合 $\{(R_j, f_j)\}$，我们如何精炼这些估计呢？

一种做法是直接将逐对的能量函数 $E_{\text{pairwise-LS}}$(公式 9.25)扩展为多视角的形式，

$$E_{\text{all-pairs-2D}} = \sum_i \sum_{jk} c_{ij} c_{ik} \|\tilde{x}_{ik}(\hat{x}_{ij}; R_j, f_j, R_k, f_k) - \hat{x}_{ik}\|^2, \tag{9.28}$$

其中 $\tilde{x}_{ik}$ 是第 k 帧图像中的特征 i 由公式(9.27)得到的预测位置，$\tilde{x}_{ij}$ 是观测位置，而下标中的"2D"表示在最小化图像平面的误差(Shum and Szeliski 2000)。注意，由于 $\tilde{x}_{ik}$ 是依赖于 $\tilde{x}_{ij}$ 的观测值，我们实际上面临着一个变量性误差(errors-in-variable)问题，理论上这需要比最小二乘法更复杂的方法来进行求解(Van Huffel and Lemmerling 2002; Matei and Meer 2006)。但是，在实践中，如果我们有足够多的特征，就可以直接用常规的非线性最小二乘法来最小化上述误差量，进而得到精确的多帧配准结果。

尽管这种方法在实践中效果很好，但是它有两个潜在的缺点。第一个缺点是，由于我们要对所有图像的配对特征求和，所以会导致观测次数较多的特征在最后的结果中所占的比重过大。(实际上，观测了 m 次的特征会被计数 $\binom{m}{2}$ 次，而不是 m 次。)第二个缺点是，$\tilde{x}_{ik}$ 相对于 $\{(R_j, f_j)\}$ 的导数是有点难以处理的，虽然这可以用 9.1.3 节中介绍的 R_j 增量修正方法来简化处理。

一种可能的方法就是用真正的光束平差法来形式化该最优化过程，即同时求解姿态参数 $\{(\boldsymbol{R}_j, f_j)\}$ 和 3D 坐标点 $\{x_i\}$，

$$E_{\text{BA-2D}} = \sum_i \sum_j c_{ij} \|\tilde{x}_{ij}(x_i; \boldsymbol{R}_j, f_j) - \hat{x}_{ij}\|^2, \tag{9.29}$$

其中 $\tilde{x}_{ij}(x_{wj}; R_j, f_j)$ 是由公式(9.26)给出的。完整光束平差法的缺点是，需要求解的变量更多，所以每一次迭代的收敛速度以及整体的收敛速度都会变慢。(想象每次更新旋转矩阵后 3D 空间中的点是如何"漂移"的。)但是，每次线性化高斯-牛顿迭代法的计算复杂度都可以用稀疏矩阵方法来简化(7.4.1 节)(Szeliski and Kang 1994; Triggs, McLauchlan, Hartley et al. 1999; Hartley and Zisserman 2004)。

另一种可能的方法就是在 3D 投影射线方向上最小化误差(Shum and Szeliski 2000)，即

$$E_{\text{BA-3D}} = \sum_i \sum_j c_{ij} \|\tilde{x}_i(\hat{x}_{ij}; \boldsymbol{R}_j, f_j) - x_i\|^2, \tag{9.30}$$

其中 $\tilde{x}_i(x_{wj}; R_j, f_j)$ 是由公式(9.26)的后半部分给出的。这与公式(9.29)相比，并没有特别的优势。实际上，由于最小化是在 3D 投影射线方向上的误差，且射线之间的夹角会随着焦距 f 的增大而减小，所以会倾向于估计出更长的焦距。

不过，如果消去 3D 射线 x_i，我们可以导出在 3D 射线空间中形式化的逐对的能量函数(Shum and Szeliski 2000)，

$$E_{\text{all-pairs-3D}} = \sum_i \sum_{jk} c_{ij} c_{ik} \|\tilde{x}_i(\hat{x}_{ij}; \boldsymbol{R}_j, f_j) - \tilde{x}_i(\hat{x}_{ik}; \boldsymbol{R}_k, f_k)\|^2. \tag{9.31}$$

由于焦距 f_k 可以包含在利用公式(9.7)构造齐次坐标向量的过程中，所以这个结果是更新方程的最简化集合(Shum and Szeliski 2000)。因此，尽管这种形式下发生特征权重过大的情况更加频繁，在 Shum and Szeliski (2000)和 Brown, Szeliski, and Winder(2005)都采用了这种方法。为了减少倾向更长焦距的偏向，我们可以将每次迭代的残差(3D 误差)乘以 $\sqrt{f_j f_k}$，这与将 3D 射线投影到具有中间焦距的"虚拟摄像机"中是类似的。

向上矢量的选择 正如前面所说，上述方法计算得到的 3D 摄像机姿态参数存在着全局性的歧义。尽管这个问题可能并不要紧，但是人们更喜欢最后的拼接图像是"直立"的，而不是扭曲或倾斜的。更具体一些，人们习惯于看到图像以竖直(重力)轴点直立向上的方式显示出来。考虑你通常是如何拍摄图像的：虽然你可能会以某种方式摇动或者倾斜镜头，但一般会保持摄像机的水平边界(x 轴)是平行于地表平面的(与实际的重力方向相垂直)。

数学上，关于旋转矩阵的约束可以用如下形式表示。回忆公式(9.26)中关于 3D 到 2D 映射的方程为

$$\tilde{x}_{ik} \sim \boldsymbol{K}_k \boldsymbol{R}_k x_i. \tag{9.32}$$

为了使全局 $\boldsymbol{Y}$ 轴的投影，$\hat{j} = (0,1,0)$ 是与图像的 $\boldsymbol{X}$ 轴，$\hat{i} = (0,1,0)$[①]相互垂直的，我们

[①] 注意，这里我们使用计算机图形学中的普通约定(即垂直的世界坐标是对应 y 轴的)。如果我们想让一个水平拍摄的"常规"图像，而不是一幅绕 x 轴旋转 90°的图像，所关联的旋转矩阵是单位矩阵，那么这是一种非常自然的选择。

希望对每个旋转矩阵 R_k 后乘以一个全局的旋转矩阵 R_g。

这种约束可以写成如下形式

$$\hat{i}^T R_k R_g \hat{j} = 0 \tag{9.33}$$

(注意，标定矩阵的缩放因素在这里是不相关的)。这相当于要求矩阵 R_k 的第一行 $r_{k0} = i^T R_k$ 与矩阵 R_g 的第二列 $r_{g1} = R_g \hat{j}$ 相互垂直。这些约束的集合(每一幅输入图像一个约束)可以写成一个最小二乘问题，

$$r_{g1} = \arg\min_r \sum_k (r^T r_{k0})^2 = \arg\min_r r^T \left[\sum_k r_{k0} r_{k0}^T\right] r. \tag{9.34}$$

因此，r_{g1} 就是由单个摄像机旋转 x-向量张成的散度矩阵即矩矩阵的最小特征向量，当摄像机直立时，其一般的形式应该是 $(c, 0, s)$。

为了完全确定全局的旋转矩阵 R_g，我们还需要指定一个附加的约束条件。这个条件与 9.3.1 节中讨论的视角选择问题相关。一个简单的启发式方法是，让单个旋转矩阵的 z 轴平均值 $\bar{k} = \sum_k \hat{k}^T R_k$ 尽量靠近世界坐标系的 z 轴 $r_{g2} = R_g \hat{k}$。这样，我们就可以用三个步骤来计算完整的旋转矩阵：

1. $r_{g1} = \min \text{ eigenvector} \left(\sum_k r_{k0} r_{k0}^T\right)$;
2. $r_{g0} = \mathcal{N}((\sum_k r_{k2}) \times r_{g1})$;
3. $r_{g2} = r_{g0} \times r_{g1}$,

其中 $\mathcal{N}(v) = v/\|v\|$ 是向量 v 的归一化表示。

9.2.2 视差消除

当我们已经最优化摄像机的全局方向和焦距之后，可能会发现图像依然没有完美对齐，即最后得到的拼接图在某些地方看起来仍然有些模糊或者重影。这种情况可能是由多种因素造成的，包括未建模的径向畸变、3D 视差(摄像机未能绕着其光学中心旋转)、微小的场景运动(比如摇晃的树枝)以及大尺度的场景运动(比如进出画面的行人)。

这些问题中的每一个都可以用不同的方法来解决。径向畸变可以用 2.1.6 节中介绍的几种方法之一来估计(也许能够预先估计)。例如，用铅垂线法(plumb-line method)(Brown 1971; Kang 2001; El-Melegy and Farag 2003)来调整径向畸变的参数，直到那些稍稍弯曲的线条都变成直线；而基于拼图的方法则是不断调整这些参数，直到图像重叠区域的误注册减到最少(Stein 1997; Sawhney and Kumar 1999)。

3D 视差可以用完整的 3D 光束平差法来处理，即将公式(9.29)中的投影公式(9.26)替换为公式(2.68)，从而可以对摄像机的平移进行建模。匹配特征点和摄像机的 3D 位置是可以同时恢复的，不过相比于无视差的图像注册来说，这种方法的代价明显很高。一旦已经恢复出 3D 结构，那么场景(理论上)就可以投影到一个不含有视差的单独(中心)视角上。但是，为了能够做到这一点，就需要进行稠密的立体

视觉对应(11.3 节)(Li, Shum, Tang et al. 2004; Zheng, Kang, Cohen et al. 2007),这在图像之间只有部分重叠的时候,也许是不可能做到的。在这种情况下,我们可能只需要修正重叠区域内的视差问题,这可以用多透视平面扫描算法(multi-perspective plane sweep,MPPS)来完成(Kang, Szeliski, and Uyttendaele 2004; Uyttendaele, Criminisi, Kang et al. 2004)。

场景的运动比较剧烈时,即有目标出现或完全消失的情况时,一个明智的解决方式是每次只从一幅图像挑选像素点,作为最后合成图像的来源(Milgram 1977; Davis 1998; Agarwala, Dontcheva, Agrawala et al. 2004),参见 9.3.2 节中的讨论。但是,当运动比较小(大概只有很少的一些像素点)的时候,在混合之前,可以用一般的 2D 运动估计方法(光流法)来进行适当的修正,然后再用局部配准(Shum and Szeliski 2000; Kang, Uyttendaele, Winder et al. 2003)进行处理。同样的过程还可以用于补偿径向畸变和 3D 视差,不过它使用比明确地对误差来源建模要弱一些的运动模型,所以可能会经常失败或者引入有害的畸变。

Shum and Szeliski(2000)提出的局部配准方法,第一步就是用全局光束平差法来最优化摄像机的姿态参数。在估计完这些参数之后,想得到的 3D 空间点 x_i 的位置就可以用 3D 位置向后投影的平均值来表示,

$$\bar{x}_i \sim \sum_j c_{ij}\tilde{x}_i(\hat{x}_{ij}; R_j, f_j) \bigg/ \sum_j c_{ij}, \tag{9.35}$$

并可以将其投影到每一帧图像 j 上,得到目标位置 $\bar{x}_{ij}$。目标位置 $\bar{x}_{ij}$ 和原始特征位置 x_{ij} 的差别构成一组局部运动估计

$$u_{ij} = \bar{x}_{ij} - x_{ij}, \tag{9.36}$$

可以以此为基础,进行内插,形成一个稠密的修正量场 $u_j(x_i)$。Shum and Szeliski (2000)的系统中,他们使用了一种逆卷绕(inverse warping)算法,即将稀疏的 u_{ij} 值置于新目标位置 $\bar{x}_{ij}$ 上,用双线性核函数(Nielson 1993)进行插值,然后在计算卷绕(修正)后的图像时,再将修正值叠加到原始图像坐标上。为了能够得到的用于插值的比较合适密度的特征集合,Shum and Szeliski(2000)选择在每一个图像块的中心位置放置一个特征点(图像块的尺寸控制了局部配准阶段的光滑程度),而不是依赖于某种特征点提取算子的结果(图 9.10)。

图 9.10 消除含有运动视差的拼接图像中的虚影 Shum and Szeliski(2000) © 2000 IEEE:(a) 含有视差的合成结果;(b)一次消除虚影步骤后的结果(图像块大小为 32 像素);(c)多次消除虚影步骤后的结果(图像块大小分别为 32,16 和 8 像素)

另一种基于运动估计的消除重影的方法是由 Kang, Uyttendaele,Winder *et al.*(2003)提出的，该方法是通过估计每幅输入图像和中心参考图像之间的稠密光流场来达到目标的。他们先用图像一致性(photo-consistency)度量来检查光流向量的精确度，再用一个给定的合法卷绕像素点，来计算高动态范围的辐射度估计，这是他们整个算法的目标。由于该方法需要一个参考图像，虽然可以通过扩展来缓解这一问题，这就使得该方法不太适用于一般的图像拼接问题。

9.2.3 认出全景图

为了进行全自动的图像拼接，最后一项必要的任务就是识别出哪些图像实际上是在一起的，Brown and Lowe(2007)称为"认出全景图"(recognizing panoramas)。如果用户是依次拍摄图像的，那么每一幅图像都和它的前一个有重叠的部分，而且第一幅图像和最后一幅图像应该拼接在一起。在这种情况下，结合"拓扑推理"(topology inference)过程的光束平差法可以用来自动生成全景图(Sawhney and Kumar 1999)。但是，用户在拍摄全景图的时候经常跳转，比如，他们可能在前一幅图像的上面新开始一行，或者跳回之前的某个地方重新拍摄一个镜头，或者需要发现360°全景图的端到端重叠部分。此外，如果能够发现用户在一段时间内拍摄的多个全景图，那么也会为用户提供很大的便利。

为了能够认出全景图，Brown and Lowe(2007)首先用基于特征的方法来找出所有存在重叠部分的图像对，然后再通过找出重叠图像之间的连接部分来"认出"单独的全景图(图 9.11)。在基于特征的匹配阶段中，首先要从所有输入图像中提取尺度不变特征(SIFT 特征)的位置和特征描述子(Lowe 2004)，并将其放入某种索引结构中，参见 4.1.3 节中的介绍。对于每一个需要考量的图像对，针对第一幅图像中的每一个特征找出最近匹配的邻居特征，即利用索引结构快速找出候选，然后通过比较特征描述子来找出最佳匹配。为了找出符合要求的内点匹配集合，常常使用 RANSAC 算法，即用几个匹配对来做相似运动模型假设，然后在用其计算内点的数目。最近，Brown, Hartley, and Nistér(2007)提出了一种特别针对旋转全景图的 RANSAC 算法。

在实践中，要让一个全自动的图像拼接算法能够正常工作，最困难的是决定哪些图像实际上对应场景中的同一个部分。场景中的重复结构，比如窗户(图 9.12)，采用基于特征的方法时可能会导致错误的匹配。一种缓解这个问题的方式是对注册后的图像直接进行像素级别的比较，判断它们是否实际上属于相同场景的不同视图。遗憾的是，当场景中有运动物体时(图 9.13)，这种启发式方法可能会失败。虽然在缺乏对完整场景理解的条件下，这个问题并没有完美的解决方法(magic bullet)，但是可以利用在特定领域上的启发式方法做出进一步的改进，比如关于典型摄像机运动的先验知识以及用于验证匹配结果的机器学习方法。

图 9.11 认出全景图(Brown, Szeliski, and Winder 2005)，图像来自 Matthew Brown：(a)成对匹配的输入图像；(b)根据连接部分(全景图)对图像进行分组；(c)单个全景图的注册和拼接融合结果

图 9.12 错误的匹配(Brown, Szeliski, and Winder 2004)：某些特征的偶然匹配可能会导致实际上并不重叠的图像对之间的匹配

图 9.13 当场景中存在运动物体时，直接通过像素级别的误差比较来进行图像匹配的验证可能会失败(Uyttendaele, Eden, and Szeliski 2001)©2001 IEEE

9.2.4 直接配准和基于特征的配准

考虑到现在有两种进行图像配准的方法，哪一种更好呢？

早期基于特征的匹配方法，可能会因为某些纹理太强或者太弱的区域而造成失败。在图像中，特征的分布常常是不均匀的，因此使那些原本应该对齐的图像对匹配失败。此外，一般是通过在特征点附近的图像块之间简单的互相关来找出对应关系，但是当图像发生旋转或者因为同构映射出现透视收缩时，这种方法并不能达到很好的效果。

现在的特征检测和匹配方案的鲁棒性已经非常高了，甚至可以用于存在很大间隔的不同视角下的已知物体的识别问题(Lowe 2004)。特征不仅仅在强"角点"区域有响应(Förstner 1986; Harris and Stephens 1988)，还在"团块状"(blob-like)区域(Lowe 2004)，以及均匀的区域有响应(Matas, Chum, Urban et al. 2004; Tuytelaars and Van Gool 2004)。此外，由于这些方案可以在尺度空间上进行操作，并提取主方向(或方向不变性描述子)，所以就可以对不同尺度、方向甚至透视收缩的图像对进行匹配。根据我们自己使用基于特征的匹配算法的经验，如果图像中的特征点分布很好，并且使用的特征描述子重复性很好，就可以找出充足的图像拼接所需要的对应(Brown, Szeliski, and Winder 2005)。

直接基于像素级别比较的配准方法最大的局限性在于，该方法的收敛范围有限。虽然理论上可以在分层(由粗到精)的框架下进行估计，但是在实践中很难使用两三层以上的图像金字塔，否则就会丢失很多重要的细节而产生模糊[①]。对于视频中图像序列的匹配问题，直接配准方法通常都是有效的。但是，对于图像只有部分重叠的照片全景图，或者对于其中对比度或内容变化太大的图像集合，它们常常无能为力，因此这时一般倾向于采用基于特征的方法。

---

① 基于傅里叶变换的相关性(Szeliski 1996; Szeliski and Shum 1997)可以扩展该范围，但前提是圆柱面图像或运动预测能有一定的效果。

9.3 合　　成

一旦将彼此相关的所有输入图像注册在一起之后，我们就需要考虑如何产生最后的拼接图像。这个过程包括选择一个最终的合成表面(平面、圆柱面、球面等)和视图(参考图像)。另外，它还包括挑选哪些像素点来合成最终的图像，如何以最优方式将这些像素融合在一起，以最小化可见的缝隙、模糊和重影现象。

在本节中，我们回顾一下用于处理合成表面的参数化、像素点和缝隙的选择、融合以及曝光补偿等问题的方法。我强调的是处理这些问题的全自动方法。创建一个高质量的全景图和合成贴图既是计算性任务，也是艺术性创作，人们开发出多种交互式的工具来协助这一过程(Agarwala, Dontcheva, Agrawala et al. 2004; Li, Sun, Tang et al. 2004; Rother, Kolmogorov, and Blake 2004)。部分工具的详细介绍参见10.4 节。

9.3.1 合成表面的选择

首先要做的一个决定是，如何表示最后的图像。如果只有少量图像要拼接在一起，最自然的方法就是选择一幅图像作为参考图像，然后将其他所有的图像都卷绕到参考坐标系中。由于到最终合成表面的投影仍然是透视投影，从而原来的直线变换后依然是直线(这通常是一种符合预期的属性)[①]，因此这种合成结果有时也称作"平坦"(flat)的全景图。

但是，对于视野更大的情况，如果我们不过分地将图像边界附近的像素拉伸的话，就不能保持用平坦的表示方法。(在实践中，一旦视角扩展到 90°附近，平坦全景图就开始会看起来变得严重扭曲。)对于合成大角度全景图来说，通常的选择是使用圆柱面投影(Chen 1995; Szeliski 1996)或者球面投影(Szeliski and Shum 1997)，参见 9.1.6 节中的介绍。实际上，可以使用任何计算机图形学中的环境映射(environment mapping)表面，比如立方贴图(cube map)就代表带有立方体六个正方形表面的完整球形视野(Greene 1986; Szeliski and Shum 1997)。另外，绘图师还开发出了很多其他用于表示整个球的方法(Bugayevskiy and Snyder 1995)。

在某种程度上，参数化的选择问题是依赖于具体应用的，它涉及权衡保持局部表观不失真(比如，保持直线依然是笔直的)和对环境进行适当的均匀采样之间存在的矛盾。基于全景图的范围，如何自动对此进行选择和如何在表达之间进行平滑的过渡，是当前研究的一个活跃领域(Kopf, Uyttendaele, Deussen et al. 2007)。

在全景摄影中最近出现了一个有趣的进展，即利用立体摄影投影方法从上方拍

---

① 最近，学者们提出了一些将圆柱形和球形全景图中的曲线变为直线的方法(Carroll, Agrawala, and Agarwala 2009; Kopf, Lischinski, Deussen et al. 2009)。

摄地面(在室外场景中)来创建"小行星"渲染效果。[①]

视角选择 我们在选择输出参数之后，还需要判断场景中的哪一部分要放在最后视野范围的中心位置。正如前面所提到的那样，对于平坦的合成表面，我们可以选择其中一幅图像作为参考图像。通常，选择一幅几乎位于几何中心的图像是一个比较合理的选择。比如，对于用一系列 3D 旋转矩阵表示的旋转全景图来说，我们可以选择那些 z 轴坐标最接近平均 z 轴坐标的图像(假设视野范围是合理的)。或者，我们可以用平均 z 轴坐标(或者用更巧妙的四元数法)来定义参考旋转矩阵。

对于范围更大的全景图，比如圆柱形全景图或球形全景图，如果球形视野的一个子集已经成像，我们就可以应用同样的启发式方法。对于 360° 全景图的情况，选择输入序列中间的图像可能效果更好，或者有时如果第一幅图像中含有最感兴趣的目标时，选择第一幅图像效果也比较好。在所有这些情况中，让用户控制最后的视角通常会得到更令人满意的结果。如果 9.2.1 节中介绍的"向上矢量"计算过程有效，这个问题甚至可以简化为摇动图像或者为最后的全景图设置一条"中垂线"。

坐标变换 在选择了参数和参考视角之后，我们还需要计算输入像素点坐标和输出像素点坐标之间的映射关系。

如果最后的合成表面是平坦的(比如，一个单独的平面或者立方体映射的一个面)，并且输入图像没有径向畸变，那么坐标变换就是公式(9.5)表述的那种简单同构。这种卷绕在图形硬件中可以通过合适地设置纹理映射坐标和渲染单个四边形来完成。

如果最后的合成表面有其他解析形式(比如，圆柱体表面或球体表面)，那么我们就需要将最后全景图中的每一个像素都转换为一视线(3D 点)，然后再根据投影方程(以及可能的径向畸变)将其映射回每一幅图像中。可以通过预先计算某些查找表的方式，来提高这一过程的效率，比如，预先计算圆柱体或球体坐标到 3D 坐标映射中所用到的部分三角函数值，或者每个像素点处的径向畸变场。另外，还可以通过在较粗的网格上计算精确的像素映射而后再对其值进行插值的方法，来对该过程进行加速。

最后的合成表面是一个纹理映射多面体时，必须使用一种稍微复杂一点的算法。我们不仅需要对 3D 坐标和纹理映射坐标适当的处理，而且为了保证 3D 渲染过程中被插值的纹理像素有有效的数值，还需要在纹理图中位于三角面片外部的少量扩展点。

采样问题 虽然上述计算过程可以用每一幅输入图像产生正确(部分)的像素点，但我们仍然需要注意采样的问题。例如，如果最后全景图的分辨率低于输入图像的分辨率，那么对输入图像进行预滤波便可以有效防止出现图像走样。针对这些

[①] 这是受 Antoine De Saint-Exupery 的《小王子》故事的启发。在 *http://www.flickr.com* 网站上搜索 "little planet projection" 可以看到很多结果。

问题，在图像处理和计算机图形学领域中都有广泛的研究。基本的问题就是，如何根据源图像中相邻采样点之间的距离(和排列)，计算合适的预滤波器。正如 3.5.2 节和 3.6.1 节介绍的那样，图形学领域已经提出了很多近似算法，比如 MIP 映射算法(Williams 1983)和椭圆加权高斯平滑算法(Greene and Heckbert 1986)。为了能够达到最高的视觉质量，可能需要高阶插值算子(比如三次插值)和空间适应的预滤波器。在某些情况下，可能也可以利用超分辨率(super-resolution)过程来产生比输入图像分辨率更高的图像。

9.3.2 像素选择和加权(去虚影)

将源像素点映射到最后的合成表面之后，我们还必须决定如何将它们融合在一起以产生一个漂亮的全景图。如果所有图像都完美注册在一起，并且像素点保持一致性，那么这个问题就非常简单了，即任意选择一个或综合多个像素点都可以。但是，对于真实的图像数据，可能会产生可见的缝隙(由曝光差异引起)、模糊(由错误注册引起)或虚影(由运动物体引起)。

要构造干净美观的全景图，涉及到决定使用哪些像素点，如何对它们进行加权或融合。这两个步骤之间的划分不是绝对的，因为确定每个像素点权重的过程可以看作选择过程和融合过程相结合的产物。在本节中，我们将讨论随空间变化的权重分配、像素点选择(缝隙连接方式)以及更复杂的融合问题。

羽化和中心加权

最简单的产生最终合成图像的方法，就是简单计算每个像素点所在位置的平均值，即

$$C(\boldsymbol{x}) = \sum_k w_k(\boldsymbol{x})\tilde{I}_k(\boldsymbol{x}) \bigg/ \sum_k w_k(\boldsymbol{x}), \tag{9.37}$$

其中 $\tilde{I}_k(x)$ 是卷绕(重采样)后的图像，而 $\omega_k(x)$ 在合法像素点处的值是 1，在其他像素点处的值是 0。在计算机图形硬件上，这种求和操作可以用累积缓冲(accumulation buffer)(用 A 通道作为权重)来实现。

由于曝光差异、错误注册和场景运动都非常显著(图 9.14a)，所以简单求平均值的方法常常得不到很好的结果。如果唯一的问题在于快速运动的物体，那么通常就可以利用中值滤波器(一种选择像素点的算子)来消除它们(图 9.14b)
(Irani and Anandan 1998)。相反地，中心加权法(参见下面的讨论)和最小似然度法(minimum likelihood)选择(Agarwala, Dontcheva, Agrawala et al. 2004)有时可以用来保留运动物体的多个复制(图 9.17)。

相对于直接计算均值而言，将图像中心附近的像素点权重增大，而将图像边缘附近的像素点权重降低，会得到更好的结果。图像包含某些截断区域时，应该将图像边缘和截断区域边缘附近的像素点权重都降低。这可以通过计算距离图(distance map)或烧草变换(grassfire transform)得到

$$w_k(\boldsymbol{x}) = \arg\min_{\boldsymbol{y}}\{\|\boldsymbol{y}\| \mid \tilde{I}_k(\boldsymbol{x}+\boldsymbol{y}) \text{ is invalid }\}, \tag{9.38}$$

其中每一个合法像素点都用其与最近的非法像素点的欧式距离来标记(3.3.3 节)。欧式距离图可以通过两遍光栅扫描算法(Danielsson 1980; Borgefors 1986)来快速计算得到。

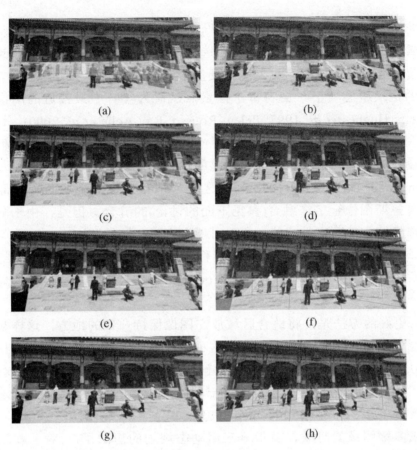

图 9.14 由不同算法计算得到的最后合成图像(Szeliski 2006a)：(a)均值法；(b)中值滤波法；(c)羽化均值法；(d)p 范数 $p = 10$；(e)Voronoi 多边形法；(f)带羽化的加权 ROD 顶点覆盖；(g)图割缝隙分割后使用泊松融合；(h)使用金字塔融合

利用距离图进行加权平均的做法，通常称为"羽化"(feathering)(Szeliski and Shum 1997; Chen and Klette 1999; Uyttendaele, Eden, and Szeliski 2001)。这种做法对于存在曝光差异的图像之间的融合效果不错。但是，依然还存在着模糊和虚影的问题(图 9.14c)。需要注意的是，即使将权重值(归一化之后)作为 alpha 通道(透明度通道)的值，加权平均的做法与单个图像合成里面经典的覆盖(over)操作(Porter and Duff 1984; Blinn 1994a)也是不同的。这是因为，覆盖操作削弱了更远表面上像素点的影响，从而与直接求和并不等价。

一种对羽化操作进行改进的做法是，将距离图像转化为更高幂次的数值，即使用公式(9.37)中的 $\omega_j^p(x)$。这种加权平均的方法会让贡献大的数值占据主导地位，即

更像 p 范数(p-norm)的结果。采用这种方法,最后的合成结果通常能在可见的曝光差异和模糊现象之间进行平衡(图 9.14d)。

在 $p \to \infty$ 的极限情况下,仅选择了最大权重的像素点,

$$C(\boldsymbol{x}) = \tilde{I}_{l(\boldsymbol{x})}(\boldsymbol{x}), \tag{9.39}$$

其中

$$l = \arg\max_k w_k(\boldsymbol{x}) \tag{9.40}$$

是标号赋值(label assignment)或像素点选择(pixel selection)函数,用于选择每一个像素点处使用哪一幅图像的信息。这种硬像素选择过程会产生所熟悉的 Voronoi 图的可见性的掩模敏感变形,会将每个像素点赋值为集合中最接近的图像中心(Wood, Finkelstein, Hughes et al. 1997; Peleg, Rousso, Rav-Acha et al. 2000)。这样得到的合成结果,虽然有一定的艺术指导意义,但对于有高度重叠的全景图(流形拼图)当曝光变化较大时,倾向于会产生非常明显的边缘和缝隙。

Xiong and Turkowski (1998)基于这种 Voronoi 思想(烧草变换的局部最大值)来选择拉普拉斯图像金字塔融合(参见下面的讨论)的缝合。但是,由于随着新图像的加入,选择缝合是顺序进行的,所以可能会产生一些缺陷。

选择最佳缝合。 针对最终合成图像重叠区域中不同图像的缝合问题,计算 Voronoi 图是一种解决方法。但是,Voronoi 图完全忽略了缝合中的局部图像结构问题。

一个更好的方法是,将缝合区域放在图像保持一致的地方,这样从一幅图像到另一幅图像的过渡就是不可见的。在这种情况下,该算法能够避免"切断"运动的物体,从而防止出现看起来不自然的缝合区域(Davis 1998)。对于一对图像来说,该过程可以形式化为一个从重叠区域一端到另一端的动态规划问题(Milgram 1975, 1977; Davis 1998; Efros and Freeman 2001)。

合成多幅图像的时候,就很难应用动态规划的想法了。(对于顺序合成方形纹理瓷砖的情况,Efros and Freeman(2001)采用了沿瓷砖四条边进行动态规划的算法。)

为了克服这个难题,Uyttendaele, Eden, and Szeliski(2001)观察到,对于那些很好地注册在一起的图像,绝大部分可见缺陷都是由运动物体引起的,也称为"半透明的虚影"。因此,他们的系统要决定哪些物体需要保留,哪些物体需要删除。首先,该算法对所有输入图像对,计算其不一致的重叠部分对应的差分区域(regions of difference,RODs)。其次,用顶点表示 RODs,用边表示 ROD 对在最后合成图像中的重叠关系,就可以构造出一个图(图 9.15)。由于存在一条边就代表存在一个不一致的区域,所以在最后的合成图像中,必须不断删除顶点(区域),直到剩下的顶点之间不存在边为止。满足这种要求的最小集合可以用顶点覆盖(vertex cover)算法来计算得到。由于这样的覆盖可能不唯一,所以可以用 ROD 中羽化权重的和来得到顶点的权重(Uyttendaele, Eden, and Szeliski 2001),再用加权顶点覆盖(weighted vertex cover)算法来代替。因此，该算法会倾向于删除图像边缘附近的区域，这样就会降低部分可见的物体在最后合成图像中出现的可能性。(通过比较跨越 ROD 边界的"边界强度"(edginess)(像素点差异性)，发现有物体出现时数值较大，从而就可以推断出差分区域内的哪些目标是属于前景的(Herley 2005)。)在删除了多余的差分区域之后，就可以通过羽化的方法来构造最后的合成图像(图 9.14f)。

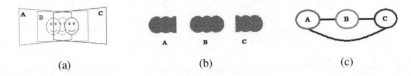

图 9.15 计算差分区域(RODs)(Uyttendaele, Eden, and Szeliski 2001)© 2001 IEEE：(a)包含一个运动人脸的三幅互相重叠的图像；(b)对应的差分区域；(c)一致的差分区域的图

另一种选择像素点和缝合位置的算法是由 Agarwala, Dontcheva, Agrawala *et al.* (2004)提出来的。在他们的系统中，是通过最优化两个目标函数的和来计算标号赋值的。目标函数的第一项是逐个像素的图像目标项(image objective)，它决定哪些像素点最可能得到好的合成图像，

$$\mathcal{C}_D = \sum_{x} D(x, l(x)), \tag{9.41}$$

其中 $D(x,l)$ 是在像素点 x 处挑选图像 l 时所关联的数据惩罚(data penalty)。在他们的系统中，用户可以通过在图像上"绘画"的方式来选择要使用的像素点，以得到期望的物体或外观。除了用户选择的标号之外，将其他所有标号的 $D(x,l)$ 设置成一个很大的值，就可以达到这种效果(图 9.16)。除此之外，还可以用一些自动化的挑选准则，比如最大似然度(maximum likelihood)，即更倾向于保留背景中重复出现的像素点(用于擦除物体)；或者最小似然度(minimum likelihood)，即更倾向于保留较少出现的运动物体，即用于保留物体运动遗迹。如果使用更传统的中心加权数据项，就会更倾向于选择输入图像中心附近的物体(图 9.17)。

图 9.16 照片蒙太奇(Agarwala, Dontcheva, Agrawala *et al.* 2004)© 2004 ACM。从一个包含五幅图像(其中四幅图像显示在左侧)的集合中，照片蒙太奇方法可以很快地构造出一幅合成的全家福，其中每个人都微笑着看着镜头(右侧)。用户可以只是浏览照片堆叠，粗略根据源图像中他们想加入最终合成图像中的人来描绘线条。这种用户施加的线条以及计算得到的区域(中间)是用左侧源图像的边界来标记颜色的

图 9.17 五张跟踪一个滑雪者跳跃的照片，拼接成了一个无缝的合成图像。由于该算法倾向于选择图像中心附近的像素点，所以最后保留了多幅图像的边界

目标函数的第二项是缝合目标项，即对于相邻图像间的标记差异的惩罚项，

$$C_S = \sum_{(\boldsymbol{x},\boldsymbol{y})\in\mathcal{N}} S(\boldsymbol{x},\boldsymbol{y},l(\boldsymbol{x}),l(\boldsymbol{y})), \quad (9.42)$$

其中 $S(x,y,l_x,l_y)$ 是图像依赖的在像素点 x 和像素点 y 之间放置缝合的交互惩罚 (interaction penalty)或缝合代价(seam cost)，而 $\mathcal{N}$ 是 $\mathcal{N}_4$ 邻域像素点的集合。例如，(Kwatra, Schödl, Essa et al. 2003; Agarwala, Dontcheva, Agrawala et al. 2004)中使用的最简单的基于颜色的缝合惩罚可以写成如下形式

$$S(\boldsymbol{x},\boldsymbol{y},l_x,l_y) = \|\tilde{I}_{l_x}(\boldsymbol{x}) - \tilde{I}_{l_y}(\boldsymbol{x})\| + \|\tilde{I}_{l_x}(\boldsymbol{y}) - \tilde{I}_{l_y}(\boldsymbol{y})\|. \quad (9.43)$$

更复杂的缝合惩罚可以考虑图像的梯度或者图像边缘所占的比例(Agarwala, Dontcheva, Agrawala et al. 2004)。缝合惩罚在其他计算机视觉应用中使用也很广泛，比如在立体视觉匹配问题(Boykov, Veksler, and Zabih 2001)中，可以用于描述标记函数的一致性或平滑性。另一种方法是，将缝合区域放在相互重叠图像之间具有强一致性的边界上，可以用 Soille (2006)描述的分水岭算法进行计算。

这两个目标函数的和构成一个马尔科夫随机场(Markov random field，MRF)，对它们有好的优化算法可用，比如 3.7.2 节、5.5 节以及附录 B.5 中介绍的算法。对于这类的标记计算(标注)问题，Boykov, Veksler, and Zabih (2001)提出的 α 扩展算法效果特别好(Szeliski, Zabih, Scharstein et al. 2008)。

对于图 9.14g 中显示的结果，Agarwala, Dontcheva, Agrawala et al. (2004)针对非法像素点使用很大的数据惩罚，而对于合法像素点，惩罚为 0。注意，缝合区域算法避开了差分区域，包括那些可能导致物体被切断的图像边界。图割(graph cut)算法(Agarwala, Dontcheva, Agrawala et al. 2004)和顶点覆盖算法(Uyttendaele, Eden, and Szeliski 2001)常常得到的结果看起来类似，不过前者由于要在所有像素点上进行优化，所以速度明显要慢一些，而后者对于判定差分区域阈值的变化更敏感。

9.3.3 应用：照片蒙太奇

虽然图像拼接方法通常用于合成有部分重叠的照片，但是它也可以用同一个场景的重复照片来合成图像,以获得图像中每一个元素最佳的合成效果。

图 9.16 演示了由 Agarwala, Dontcheva, Agrawala et al. (2004)开发的照片蒙太奇系统。在该系统中,用户可以在已经对齐的图像上,用画线的方式来表示他们想保留每一幅图像中的哪些部分。当系统求解出了这个多标签的图割问题(9.41 和 9.42)之后,就可以用一种的泊松图像融合方法(Poisson image blending)(9.44~9.46)的变形来将不同图像源的片段融合在一起。他们的系统还可以用来将一系列等焦距的图像融合成一幅全对焦的图像(Hasinoff, Kutulakos, Durand et al. 2009),或者从一个图像集合中去除电线或其他不想要的元素。习题 9.10 会让你实现这个系统,并尝试一些相关的变化。

9.3.4 融合

在确定图像之间的缝合方式并移除那些不想要的物体之后,我们还需要进行图像的融合,以补偿曝光差异和处理其他错误配准的问题。前面讨论的空间渐变权重(羽化)方法可以用来完成这个目标。但是,在实践中很难在平滑低频的曝光变化和保留锐化的过渡效果(虽然在羽化过程中使用高阶指数项有一定的帮助)之间取得令人满意的平衡。

拉普拉斯图像金字塔融合　解决该问题的一个吸引人的方法就是采用由 Burt and Adelson (1983b)提出的拉普拉斯图像金字塔融合方法,参见我们在 3.5.5 节中的讨论。与一般选用单个过渡宽度不同,我们用频域自适应的宽度来构造一个带通(拉普拉斯)金字塔,从而让每一层都有一个是层级的函数的过渡宽度,即在像素级别具有相同的宽度。在实践中,只要很少的金字塔层数,比如只要两层(Brown and Lowe 2007),就能够补偿曝光差异的影响。使用这种金字塔融合的结果如图 9.14h 所示。

梯度域融合　另一种进行多频段图像融合的方法是在梯度域(gradient domain)进行操作。在计算机视觉领域,从梯度场重建图像(Horn 1986)已经有很悠久的历史了,最早来源于亮度恒定性(Horn 1974),由明暗到形状(Horn and Brooks 1989)和光度测定学立体视觉(Woodham 1981)的相关工作。最近,相关的思路已经应用于从图像的边界重建图像(Elder and Goldberg 2001)、消除图像阴影(Weiss 2001)、从单幅图像分离反射(Levin, Zomet, and Weiss 2004; Levin and Weiss 2007)和通过减少图像边缘(梯度)的幅值来完成高动态范围图像的色调映射(Fattal, Lischinski, and Werman 2002)。

Perez, Gangnet, and Blake(2003)演示了如何将梯度域的重建方法应用于图像编辑应用中的无缝物体嵌入问题(图 9.18)。与直接复制像素点不同,他们将新图像片段的梯度部分进行复制。复制区域的实际像素点数值是通过求解一个泊松方程(Poisson equation)得到的,该方程需要在满足固定狄利克雷(Dirichlet)(严格匹配)条件下,在缝合区域附近对梯度进行匹配。Perez, Gangnet, and Blake(2003)证明了这是与计算源图像和目标图像沿着边界上的误差的加性薄膜插值(membrane interpolant)[①]等价的。在更早的相关工作中，Peleg(1981)也提出了通过附加一个平滑函数来强制沿着缝合曲线的一致性。

图9.18 泊松图像编辑(Perez, Gangnet, and Blake 2003)© 2003 ACM：(a)选择一只狗和两个儿童的图像作为源图像，然后粘贴到目标游泳池里面；(b)简单的粘贴算法不能在边界处匹配相应的颜色；(c)泊松图像融合可以掩盖这些差异

Agarwala, Dontcheva, Agrawala et al. (2004) 将这个思路扩展为多个图像源的形式，在这种情况下，再继续讨论目标图像中实际像素点必须在缝合边界上匹配这个问题就没有意义了。相反，每一个图像源都贡献了自己的梯度场，根据诺伊曼边界条件，即抛弃所有涉及到图像边界之外像素点的方程，就可以求解泊松方程。

Agarwala, Dontcheva, Agrawala et al. (2004)没有选择求解泊松偏微分方程，而是直接最小化一个变分问题(variational problem)

$$\min_{C(\boldsymbol{x})} \|\nabla C(\boldsymbol{x}) - \nabla \tilde{I}_{l(\boldsymbol{x})}(\boldsymbol{x})\|^2. \tag{9.44}$$

该方程的离散化形式是一组梯度约束方程

$$C(\boldsymbol{x}+\hat{\imath}) - C(\boldsymbol{x}) = \tilde{I}_{l(\boldsymbol{x})}(\boldsymbol{x}+\hat{\imath}) - \tilde{I}_{l(\boldsymbol{x})}(\boldsymbol{x}) \text{ and} \tag{9.45}$$

$$C(\boldsymbol{x}+\hat{\jmath}) - C(\boldsymbol{x}) = \tilde{I}_{l(\boldsymbol{x})}(\boldsymbol{x}+\hat{\jmath}) - \tilde{I}_{l(\boldsymbol{x})}(\boldsymbol{x}), \tag{9.46}$$

其中 $\hat{\imath}=(1,0)$ 和 $\hat{\jmath}=(0,1)$ 是 x 方向和 y 方向上的单位向量[②]。然后他们要求解一个关联的稀疏最小二乘问题。由于这个方程系统是在相差一个附加约束条件的意义下有定义的，所以 Agarwala, Dontcheva, Agrawala et al. (2004)要求用户选择一个像素点的值。在实践中，一个更好的选择是使结果略微倾向于复制出原始的颜色值。

为了加速求解这个稀疏线性系统的过程，Fattal, Lischinski, and Werman(2002)使用了多重栅格法(multigrid)，Agarwala, Dontcheva, Agrawala et al. (2004)使用的则是分层预处理共轭梯度下降法(Szeliski 1990b, 2006b)(附录A.5)。在随后的工作中，Agarwala(2007)演示了如何利用四叉树表示，在很小的精度损失下来进一步加速这个计算过程。Szeliski, Uyttendaele, and Steedly(2008)证明，用更粗糙的样条函数来表示每一幅图像的偏移场，求解的速度会更快。后来的工作还讨论了在对数域进行

---
① 薄膜插值因比频域插值具有更好的在任意形状约束下的插值特性而著名(Nielson 1993)。
② 在缝合位置，方程右边要用两幅源图像梯度的平均值来代替。

融合的情况，即使用乘法运算来代替加法的偏移量，从而能够更好地匹配缝合区域边界上的纹理变化。在实践中，缝合区域的融合结果非常好(图 9.14h)，但是在复制缝合区域附近大梯度值的时候必须非常小心，不要引入"双重边界"(double edge)。

在缝合区域，直接从源图像复制梯度场只是梯度域融合的一种方法。Levin, Zomet, Peleg et al. (2004) 的论文中调研了该方法的几种不同的变型，他们称之为"梯度域图像拼接"(Gradient-domain Image STitching，GIST)。他们所调研的方法包括源图像梯度的羽化(融合)以及在由梯度场重建图像的过程中使用 L1 范数代替公式(9.44)中的 L2 范数。他们所推荐的方法是，在原始图像的梯度场上，进行羽化(融合)代价函数的 L1 优化(他们称之为 GIST1-l_1)。由于使用线性规划来进行 L1 优化可能会很慢，他们在多重栅格算法的框架下，提出了一种基于中位数的快速迭代算法。他们所推荐的方法获得的结果与其称为"最优梯度缝合"(实际上与 Agarwala, Dontcheva, Agrawala et al. (2004)的方法等价)的结果看起来很相似，比金字塔融合和羽化算法都有明显的提高。

曝光补偿 金字塔和梯度域的融合方法，对于图像之间适度的曝光差异可以有较好的补偿效果。但是，当曝光差异变得比较大的时候，就需要采用其他的方法了。

Uyttendaele, Eden, and Szeliski(2001)对每一幅源图像和融合后的合成图像进行局部的迭代修正。首先，在每一幅源图像和初始羽化合成图像之间拟合一个基于块的转移二次方程。然后，对该转移方程进行邻域平均，从而就可以通过计算相邻图像块数值的样条(插值)，得到一个更平滑的逐像素点的转移函数。在每一幅图像都已经平滑调整之后，就可以计算新的羽化合成结果，然后再重复该过程(一般重复三次)。Uyttendaele, Eden, and Szeliski(2001)给出的结果表明了，这种方法在曝光补偿方面比简单的羽化方法有更好的效果，而且能够处理一些由比如镜头光晕效果引起的局部曝光变化。但是，最根本的问题在于，处理曝光差异问题最合理的方法是在辐射度域内对图像进行拼接，即根据图像自身的曝光度，将每一幅图像都转化为辐射度图像，然后再拼接得到一个高动态范围的图像，参见 10.2 节的讨论(Eden, Uyttendaele, and Szeliski 2006)。

9.4 补充阅读

图像拼接问题的历史可以追溯到 20 世纪 70 年代摄影界的相关工作(Milgram 1975, 1977; Slama 1980)。在计算机视觉领域，20 世纪 80 年代早期就开始出现一些相关的论文(Peleg 1981)，而大概十年之后才开始发展全自动的拼接方法(Mann and Picard 1994; Chen 1995; Szeliski 1996; Szeliski and Shum 1997; Sawhney and Kumar 1999; Shum and Szeliski 2000)。这些论文中使用的都是直接基于像素点的配准，而现在更普遍的是基于特征的方法(Zoghlami, Faugeras, and Deriche 1997; Capel and

Zisserman 1998; Cham and Cipolla 1998; Badra, Qumsieh, and Dudek 1998; McLauchlan and Jaenicke 2002; Brown and Lowe 2007)。这些论文中的一部分收录于 Benosman and Kang(2001)的书中。Szeliski (2006a)对于图像拼接问题做了全面的综述，本章的材料也主要基于这篇论文。

高质量的最佳缝合区域选择和融合方法是图像拼接系统中另一个重要的组成部分。该领域的重要进展包括 Milgram(1977), Burt and Adelson(1983b), Davis(1998), Uyttendaele, Eden, and Szeliski(2001), Pérez, Gangnet, and Blake(2003), Levin, Zomet, Peleg et al. (2004), Agarwala, Dontcheva, Agrawala et al. (2004), Eden, Uyttendaele, and Szeliski(2006)和 Kopf, Uyttendaele, Deussen et al. (2007)。

除了对创建航空或地球全景图所拍摄的多幅相互重叠的照片进行合并，拼接方法可以应用于自动的白板扫描(He and Zhang 2005; Zhang and He 2007)，鼠标扫描(Nakao, Kashitani, and Kaneyoshi 1998)和视网膜图像拼图(Can, Stewart, Roysam et al. 2002)。这些方法同样也可以应用于视频序列(Teodosio and Bender 1993; Irani, Hsu, and Anandan 1995; Kumar, Anandan, Irani et al. 1995; Sawhney and Ayer 1996; Massey and Bender 1996; Irani and Anandan 1998; Sawhney, Arpa, Kumar et al. 2002; Agarwala, Zheng, Pal et al. 2005; Rav-Acha, Pritch, Lischinski et al. 2005; Steedly, Pal, and Szeliski 2005; Baudisch, Tan, Steedly et al. 2006)，甚至可以用于视频的压缩(Lee, ge Chen, lung Bruce Lin et al. 1997)。

9.5 习　　题

习题 9.1：直接基于像素的配准

选取两幅图像，先计算它们之间由粗到精的仿射配准(习题 8.2)，再用平均值法(习题 6.2)或拉普拉斯图像金字塔法(习题 3.20)将它们融合在一起。另外，将你的运动模型由仿射变换扩展为透视变换(同构)，以更好地处理旋转拼图以及任意运动下的平面视图。

习题 9.2：基于特征的拼接

先用完整的透视变换模型，将习题 6.2 扩展为基于特征的配准方法，再用平均值法或更复杂的按距离羽化方法(习题 9.9)，将最后的拼图融合在一起。

习题 9.3：圆柱形带状全景图

为了能够从水平摇动(转动)的摄像机中获得圆柱形或球形的全景图，最好使用三脚架来固定摄像机。配置好摄像机，拍摄一系列相互之间有 50%重叠的图像，然后再根据下列步骤来构造全景图。

1. 拍摄一些边缘附近有很多长直线的照片，利用习题 6.10 的铅垂线法来估计径向畸变的程度。
2. 利用直尺和白纸，如图 6.7 所示(Debevec, Wenger, Tchou et al. 2002)计算

焦距；或者通过转动三脚架，使得相邻图像之间恰好相切且不重叠，计算构成 360° 全景图所需要的图像数量。

3. 利用公式(9.12～9.16)，将每一幅图像都转换到圆柱形坐标系下。
4. 利用直接基于像素点的方法，比如由粗到精的增量方法，或者快速傅里叶变换(FFT)，或者基于特征的方法，用平移运动模型来对齐这些图像。
5. (可选)如果要构造一个完整的 360 度全景图，就要将第一幅图像和最后一幅图像进行配准。计算垂直方向上的累积错误配准误差，然后在图像之间重新分配。
6. 利用羽化或者其他方法对结果图像进行融合。

习题 9.4：粗配准

利用快速傅里叶变换(FFT)或相位相关性(8.1.2 节)来估计连续图像之间的初始配准。注意观察这种方法的效果在何种重叠范围内？如果这种方法效果不好，在图像的一部分(比如四分之一)上进行配准是不是更好？

习题 9.5：自动拼图

利用基于特征的配准算法，和四个参考点的 RANSAC 同构算法(6.1.3 节，公式 (6.19～6.23))或者三个参考点的 RANSAC 旋转运动模型(Brown, Hartley, and Nistér 2007)来匹配所有相互重叠的图像对。

通过成对关系的生成树，将这些成对的估计结果合并在一起。将最后全局的配准结果进行可视化，即显示每一幅和其他所有与之重叠的图像融合之后的结果。

为了获得更好的鲁棒性，可以尝试不同的生成树(可能基于成对配准的置信度进行随机采样)，看看你是否可以从坏的匹配对造成的干扰中恢复(Zach, Klopschitz, and Pollefeys 2010)。可以统计多少匹配对的估计结果是和全局配准结果保持一致的，作为适合度的估计。

习题 9.6：全局优化

利用前一个算法的结果作为初始值，计算在所有摄像机旋转和焦距上的完整光束平差法，参见 7.4 节和 Shum and Szeliski (2000)的介绍。可选地，还可以估计径向畸变的参数或支持鱼眼广角镜头(2.1.6 节)。

类似于上一个习题，通过构造每一幅输入图像和其邻居的合成结果，将注册结果进行可视化。另外，还可以在原始图像和合成结果之间闪烁切换，从而更好地看见错误配准的缺陷。

习题 9.7：消除虚影

利用以上习题得到的光束平差法的结果来预测在一致几何约束下每一个特征点的位置。根据特征点的预测位置和实际位置之间的差异来修正小的错误注册误差，参见 9.2.2 节的介绍(Shum and Szeliski 2000)。

习题 9.8：合成表面

选择一个合成表面(9.3.1 节)，即一个由参考图像扩展而成的大平面、一个用圆柱形或球形坐标表示的球形、一个立体视觉成像的"小行星"投影或者一个立方体映射。

另一种方法是将你的所有图像都变换到这个表面上，再用相等的权重将它们融合在一起(看一看原始图像的缝合区域在哪里)。

习题 9.9：羽化和融合

对于每一个变换后的源图像，计算其羽化(距离)图像，然后用这些图像将变换后的图像融合在一起。另外，还可以使用拉普拉斯金字塔融合(习题 3.20)或者梯度域融合。

习题 9.10：照片蒙太奇和物体的移除

实现一个"照片蒙太奇"系统，其中用户可以通过画线或者其他几何基元(比如边界框)，在预先注册好的图像中，指定想要的或者不想要的区域。

(可选)给定配准图像中的某种一致性(比如，中值滤波)通过分析哪个不一致区域是或多或少典型的，设计一个自动移除(或"保留")运动物体的工具。图 9.17 演示了一个保留运动物体的例子。尝试将这个工作扩展到大量相互重叠的图像序列中，考虑利用图像平均值的方法来使运动物体的虚影看起来更虚。

第 10 章
计算摄影学

10.1 光度学标定
10.2 高动态范围成像
10.3 超分辨率和模糊去除
10.4 图像抠图和合成
10.5 纹理分析与合成
10.6 补充阅读
10.7 习题

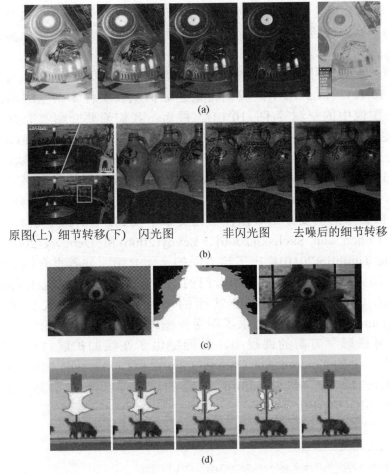

原图(上) 细节转移(下)　闪光图　　　非闪光图　　去噪后的细节转移

图 10.1 计算摄影学：(a)融合多个不同曝光程度的图像来生成高动态范围图像(Debevec and Malik 1997) ©1997 ACM；(b)融合闪光和非闪光图像(Petschnigg, Agrawala, Hoppe *et al.* 2004)©2004 ACM；(c)图像抠图和合成(Chuang, Curless, Salesin *et al.* 2001)©2001 IEEE；(d)用修图方法进行空洞填充(Criminsi, Pérez, and Toyama 2004)©2004 IEEE

在第 9 章中，我们介绍过将多幅图像拼接成宽视场全景图的方法，它允许我们生成用普通摄像机无法拍摄的图像。这仅仅是"计算摄影学"(computational photography)的一个实例，在计算摄影学中，常常会通过在一幅或多幅图像上应用图像分析与处理算法来获得超越传统成像系统能力的图像。现在，有些这类方法已经被直接嵌入到数码相机中，例如，有些新型数码相机有全景扫描模式，并可以在低光照的条件下，通过多次拍摄来降低图像噪声。

在本章中，我们将接触很多其他计算摄影学算法。首先，我们将回顾光度学图像标定(10.1 节)，即摄像机和镜头响应特性的测量，这是后续很多算法的先决条件。接着我们会讨论高动态范围成像(10.2 节)，这种方法通过使用多种曝光程度的图像来获取场景中的亮度幅度(图 10.1a)。我们还会讨论"色调映射算子"(tone mapping operator)，它将色彩丰富的图像映射回常规显示设备，如屏幕和打印机。另外，我们还会介绍通过合成闪光图像和正常图像来获得较好曝光效果的图像的算法(图 10.1b)。

接下来，我们会讨论如何通过融合多张照片或者巧妙利用先验知识来改善图像的分辨率(10.3 节)，包括从多数摄像机中存在的贝叶斯马赛克样式中提取全彩色图像的各种算法。

10.4 节中，我们将讨论从一幅图像中分离部分图像再将其粘合到其他图像中的各种算法(图 10.1c)。10.5 节中，我们将论述如何从真实样本中生成精细的纹理，使之用于修补图像中的空洞等(图 10.1d)。最后，我们简要介绍"非真实感绘制"(10.5.2 节)，这种方法可以将正常照片变为富有艺术效果的画作，类似传统绘画。

本书中没有广泛涉及的一个主题是新型计算(computational)传感器、光学器件及摄像机。Nayar(2006)、Raskar and Tumblin(2010)新近出版的一本书以及较新的研究论文(Levin, Fergus, Durand et al. 2007)。10.2 节和 13.3 节中也有部分相关介绍。

关于计算摄影学的简要介绍可以在下列文章中找到：Hayes(2008)写的文章，及 Nayar(2006)、Cohen and Szeliski(2006)、Levoy(2006) 和 Debevec(2006)[①]等综述论文。Raskar and Tumblin(2010)给出了这方面的全面介绍，并着重介绍了计算摄像机和传感器。高动态范围成像方面有本专门的书介绍它的研究情况(Reinhard, Ward, Pattanaik et al. 2005)，还有一本非常棒的书可供专业摄影人员参考(Freeman 2008)。[②]Wang and Cohen(2007a)提供了图像抠图方面很好的综述论文。

在一些计算摄影学方面的课程中，教师提供了在线的扩展材料，例如 Frédo Durand 在 MIT 开设的 Computation Photography 课程[③]、Alyosha Efros 在 CMU 开设的课程[④]、Marc Levoy 在 Standford 开设的课程[⑤]以及一系列 Computational Photography[⑥]的 SIGGRAPH 课程。

[①] 也可参考分别由 Bimber(2006)和 Durand and Szeliski(2007)编写的两期特刊。
[②] Gulbins and Gulbins(2009)讨论了相关的摄影方法。
[③] MIT 6.815/6.865, *http://stellar.mit.edu/S/course/6/sp08/6.815/materials.html*
[④] CMU 15-463, *http://graphics.cs.cmu.edu/courses/15-463/*
[⑤] Stanford CS 448A, *http://graphics.stanford.edu/courses/cs448a-10/*
[⑥] *http://web.media.mit.edu/~raskar/photo*

10.1 光度学标定

在可以将多幅图像成功融合前，我们需要先刻画从射入光线到像素值及图像噪声间的映射函数。在本节中，我们将深入了解成像流水线中影响这种映射关系的三个成分(图 10.2)。

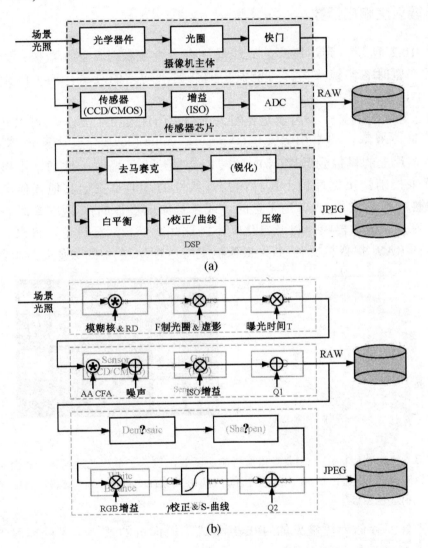

图 10.2 成像流水线。(a)展示了各种各样的噪声源及数字化后典型的处理步骤；(b)等价的信号变换，包括卷积、增益以及噪声引入。缩写词含义：RD=径向畸变(radial distortion)、AA=反走样滤波(anti-aliasing filter)、CFA=色彩滤波器矩阵(color filter array)、Q1 和 Q2=量化噪声(quantization noise)

第一个是辐射响应函数(Mitsunaga and Nayar 1999)，它将到达镜头的光子映射到存储在图像文件中的数值(10.1.1 节)。第二个是虚影(vignetting)，使靠近图像边缘的像素变暗，特别是对大孔径的光圈(10.1.3 节)。第三个是点扩散函数(point spread function)，它刻画了因镜头、反走样滤波和有限的传感区域导致的模糊问题(10.1.4 节)。[①]本节的内容基于 2.2.3 节和 2.3.3 节中描述的成像过程，所以如果需要，请回顾并复习它们。

10.1.1 辐射度响应函数

如图 10.2 所示，到达镜头的光线最终有多少被映射存储为数值，影响的因素很多。现在暂时忽略镜头内部可能发生的任何不均匀的衰减，这个因素我们会在 10.1.3 节中讨论。

影响这种映射的第一个因素是光圈大小和快门速度(2.3 节)，这可以建模为对入射光线的整体乘数，通常用曝光值(exposure value)($\log 2$ 亮度比率)来刻画。接下来，传感芯片上的模拟到数字(A/D)的转换会导致电子增益，这通常由摄像机中的 ISO 设置来控制。虽然理论上这种增益是线性的，但是正如其他任何电子方法一样，非线性也可能发生(或者是无意的，或者是如此设计的)。现在暂时忽略光子噪声、片上噪声、放大器噪声以及量化噪声，这些我们稍后会讨论，可以假定入射光线和存储在 RAW 摄像机文件(如果摄像机支持这项功能)中的数值是近似线性的。

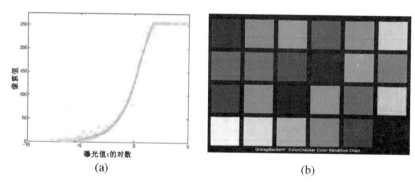

图 10.3 辐射响应标定：(a)典型摄像机响应函数，显示了单彩色通道下，入射辐照度的对数(曝光值)和输出的 8 位像素值间的映射关系(Devevec and Malik 1997)©1997 ACM；(b)选色图板

如果图像要存储为更常见的 JPEG 格式，摄像机的数字信号处理器(DSP)接下来就会进行贝叶斯样式的去马赛克(2.3.2 节和 10.3.1 节)，这主要是线性处理过程(但往往并不稳定)。有些锐化处理也常常在这一阶段进行。下一步，彩色值会被乘上不同的常量(或者有时是 3×3 的彩色扭曲矩阵)来完成色彩平衡，也就是说让白

---

① 其他由光度学摄像机和镜头产生的影响包括传感器眩光、模糊现象以及色像差，其中色像差也可以认为是几何畸变的光谱变化的一种形式(2.2.3 节)。

色点更接近纯白。最后，每个彩色通道的亮度会进行标准的伽马校正，并且彩色值会被转换为 YCbCr 格式后再进行 DCT 变换、量化以及压缩为 JPEG 格式(2.3.3节)。图 10.2 用图解的形式展示了所有这些步骤。

由于这些处理过程的复杂性，从基本原理上，很难对摄像机响应函数，即入射辐照度和数字 RGB 值间的映射关系，进行建模(图 10.3a)。更为实用的方法是通过测量入射光线和最终数值间的对应关系来标定摄像机。

最准确同时也最贵的方法是采用积分球(integrating sphere)，它是一个很大(直径往往为 1 米)的球体，在其内部仔细粉刷上白色油漆。在顶部有个精确标定的光源，它控制球体内部的光照量(由于球体的辐射，球体内部光照处为常量)，在球体边开一个大小正好的小口安装摄像机/镜头。通过慢慢改变进入球体的光线，就可以建立起准确的入射光照和测量所得像素值间的对应关系。摄像机的虚影和噪声等特性也可以同时确定。

另外一种更实用的方法是采用标定图板(图 10.3b)，例如 Macbeth 或者 Munsell 选色图板[①]。这种方法最大的难题是保证单一光源。一个解决方法是在一间大黑屋中设立一个远离(且垂直于)图板的高质量光源。另外一种方法是在室外将图板放置在远离任何阴影的地方。(这两种方法的结果不同，因为光源的色彩不同)。

最简单的方法大概是，将摄像机放置在三脚架上，对同一场景拍摄多个不同曝光度的图像，然后通过同时估计每个像素的入射光线和响应曲线来恢复响应函数。(Mann and Picard 1995；Debevec and Malik 1997；Mitsunaga and Nayar 1999)。这种方法在 10.2 节关于高动态范围成像的讨论中会有更详细的介绍。

如果这些方法都失效了，即只有一张或多张不相干的照片，你可以利用国际色彩联盟(ICC)为摄影机建立的文档(Fairchild 2005)。[②]或者甚至更简单的，针对 RAW 文件，假设响应是线性的；针对 JPEG 图像，假设对每个 RGB 彩色通道有 $\gamma = 2.2$ 的非线性(加上限幅)。

10.1.2 噪声水平估计

除了知道摄像机的响应函数外，知道在特定摄像机设定(例如，ISO/增益水平)下引入的噪声量也很重要。最简单的噪声特性描述是单个标准偏差，它通常以灰度阶为单位且与像素值无关。通过将噪声水平作为像素值的函数来估计，可以得到更精确的模型(图 10.4)，这称为噪声水平函数(Liu, Szeliski, Kang *et al.* 2008)。

对于摄像机响应函数，最简单的方法是在实验室中估计这些未知量，即采用积分球或标定图板。可以在单独考虑每个像素通过重复曝光并计算时间上测量值的变化来估计噪声，或者通过假设像素值在某个区域中(例如在一个彩色方格内)相同并计算在一些区域上空间的变化来估计。

[①] *http://www.xrite.com*
[②] 请参见 ICC 的相关文档信息，*http://www.color.org/info_profiles2.xalter*。

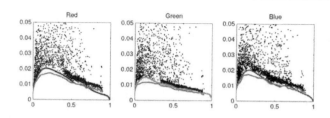

图 10.4 由单一彩色照片获得的噪声水平函数(NLF)的估计(Liu，Szeliski，Kang et al. 2008)©2008 IEEE。彩色曲线是 NLF 的估计，它被拟合为在带噪声的图像与分段光滑的图像之间测量得到的偏差的概率下包络。通过求取 29 幅图像均值而得到的 NLF 的真值在图中用灰色标出

这种方法可以推广到有亮度不变或者变化缓慢区域的照片(Liu，Szeliski，Kang et al. 2008)。首先，将图像分割成这样的一些区域，并在每个区域上拟合一个常数或者线性函数。接下来，针对带噪声的输入像素与远离较大梯度和区域边界的光滑拟合函数之间的差异，测量其(空间上的)标准偏差。对每个彩色通道，将这些测量值绘制成输出色阶(output level)的函数，如图 10.4 所示。最后，为了忽略那些是外点的像素值或偏差值，对这个分布拟合一个下包络(envelope)。(Liu，Szeliski，Kang et al. 2008)提出了解决这个问题的一种完全 Bayesian 方法，它对每个量的统计分布进行建模。另外一种更简单的方法是对这个下包络拟合一个低维的函数(例如正值 B-样条)，在大部分情况下，这种方法都能生成有用的结果(习题 10.2)。

在最近的工作中，Matsushita and Lin(2007)提出了一种同时估计摄像机响应和噪声水平的方法，这种方法基于水平依赖(level-dependent)的噪声分布中的倾斜(不对称)。他们的论文也包含对这些领域的前人工作成果的广泛引用。

10.1.3 虚影

使用广角和大光圈的镜头有个常见的问题就是图像的角落处会变暗(图 10.5a)。这个问题通常称为"虚影"(vignetting)，且有几种不同的形式，包括自然的、光学的、机械的虚影(2.2.3 节)(Ray 2002)。与辐射响应函数标定一样，最精确的标定虚影的方法是采用积分球或者一张同一颜色的、被照亮的空白墙的照片。

图 10.5 单幅图像虚影修正(Zheng，Yu，Kang et al. 2008)©2008 IEEE：(a)原图像有很明显的虚影；(b)Zheng，Zhou，Georgescu et al.(2006)中介绍的虚影补偿；(c 和 d)Zheng，Yu，Kang et al.(2008)介绍的虚影补偿

另外一种方法是拼接全景图并假设每个像素的真正光照来自每幅输入图像的中心部分。若辐射响应函数已知(例如，在 RAW 模式下摄像)且曝光值保持为常数，将更容易做到这一点。如果响应函数、图像曝光值和虚影函数未知，它们也可以通过最优化一个大的最小二乘拟合问题恢复出来(Litvinov and Schechner 2005；Goldman 2011)。图 10.6 展示了一个从一系列有重叠的照片中，同时估计虚影、曝光度、辐射响应的例子(Goldman 2011)。请注意除非虚影被建模并补偿，常规的梯度域图像混合(9.3.4 节)不会产生满意的结果。

如果只有单一的图像输入，虚影可以通过在径向寻找缓慢的一致的光照变化来估计。Zheng，Lin，and Kang(2006)提出的最初的算法先将图像预分割成变化平滑的区域再在每个区域里进行分析。而 Zheng，Yu，Kang *et al.*(2008)没有进行预分割而是算出所有像素的径向梯度并利用这种分布中的不对称性(因为远离中心的梯度平均上是偏负的)来估计虚影。图 10.5 展示了对一幅有大量虚影的图像分别应用这些算法的效果。习题 10.3 将让你实现上述的一些方法。

图 10.6 同时估计虚影、曝光度、辐射响应(Goldman 2011)©2011 IEEE：(a)输入图像的原均值；(b)虚影补偿后；(c)仅采用梯度域混合(注意它看起来依旧有杂色)；(d)采用虚影补偿和混合后

10.1.4 光学模糊(空间响应)估计

最后还有个需要标定的成像系统的特性是空间响应函数，它编码了光学模糊信息(用来和入射图像卷积以生成按点采样的图像)。卷积核，也称为"点扩散函数 PSF"(point spread function)或"光传递函数 OTF"(optical transfer function)，其形状取决于多个因素，包括镜头模糊和径向畸变(2.2.3 节)、传感器前端的反走样滤波以及每个活跃像素区域的形状和范围(2.3 节)，参见图 10.2。在一些应用——如多图像超分辨率和去模糊化(10.3 节)中，需要对这个函数有较好的估计。

理论上，只需要简单地观察图像上每个位置上无限小的点光源就可以估计出 PSF。然而，在实际中，通过在暗板上钻孔并在其背面用强光照射以产生所需的样本阵列是很难的。

更为实用的方法是通过观察由长直线或直条组成的图像，因为它们可以被拟合成任意精度。由于水平和垂直的边的位置在获取过程中会走样,稍微倾斜的边效果会更好。这种边的外形(profile)和位置的估计可以达到亚像素级,这样一来,在亚像素分辨率下估计 PSF 就成为可能(Reichenbach,Park,and Narayanswamy 1991; Burns and Williams 1999; Williams and Burns 2001; Goesele, Fuchs, and Seidel 2003)。Murphy(2005)的论文对摄像机标定的各个方面都做了很好的综述,包括空间频率响应(SFR)、空间均匀化、色调再现、彩色重建、噪声、动态范围、彩色通道注册以及深度场。它还介绍了被称为 sfrmat2 的一种倾斜边(slant-edge)标定算法。

倾斜边方法可以用来恢复 2D PSF 的 1D 投影,例如,偏垂直的直边可以用来恢复出水平的"线扩散函数(LSF)"(line spread function)(Williams 1999)。接着 LSF 通常会被转换到傅里叶域并且它的幅值会被描绘成一个一维的"调制传递函数(MTF)"(modulation transfer function),它可以反映出在获取过程中哪些图像频率丢失(模糊)或走样了(2.3.1 节)。对于大多数计算摄影学的应用,最好是直接完整估计 2D 的 PSF,因为它很难由它的投影恢复出来(Williams 1999)。

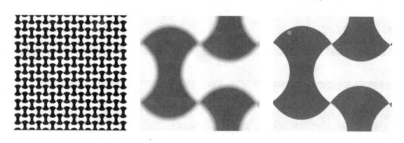

图 10.7 标定图案中的边缘在各个方向均匀分布,它可以用来估计 PSF 和径向畸变(Joshi,Szeliski,and Kriegman 2008)© 2008 IEEE。真实传感到的图像的一部分如上图中间所示,理想图像的特写在右边

图 10.7 显示了一个在各个方向上都有边缘的图案(pattern),它可以用来直接恢复出 2D PSF。第一步,图案中的角点可以通过在图像中提取边缘,连接它们并找到这些圆弧的交点来定位。接下来,理想图像的解析形式是已知的,将它卷绕(使用一个单应(同构映射、1-1 映射))拟合在输入图像的中心部分,并调整亮度使其适合所感知的图像。如果理想的话,图案可以以高于输入图像的分辨率描绘出,这样 PSF 的估计就可以达到亚像素分辨率(图 10.8a)。最后,通过解一个大的线性最小二乘系统可以求得未知的 PSF 核 K,

$$K = \arg\min_K \|B - D(I * K)\|^2, \tag{10.1}$$

其中,B 表示图像(模糊的),I 表示预测(清晰)的图像,D 是个可选的下采样算子,用来使理想的图像和感知的图像的分辨率相匹配(Joshi, Szeliski, and Kriegman 2008)。由 3.4.3 节中介绍的符号(3.75),上式也可写为 1

$$b = \arg\min_b \|o - D(s * b)\|^2, \tag{10.2}$$

其中,o 是观察到的图像,s 是清晰图像,b 是模糊核。

如果，估计 PSF 的过程是在局部的重叠图像块上完成的，它也可以用来估计由镜头引起的径向畸变和色像差(图 10.8b)。由于从理想目标到所感知的图像的单应映射是在图像的中心(无畸变的)区域估计的，所以任何由光学成像引进的(每个通道)偏移将体现为 PSF 中心的位移。[①]对这些偏移的补偿可以去除消色差径向畸变(achromatic radial distortion)和那些导致可见色差畸变(chromatic aberration)的通道间的偏移。由色差畸变引起的与色彩相关的模糊(图 2.21)也可以采用 10.3 节讨论的去模糊化方法消除。图 10.8b 显示了径向畸变和色差畸变如何表现在 PSF 的拉伸和移位上，以及在标定目标区域上消除这些影响后的效果。

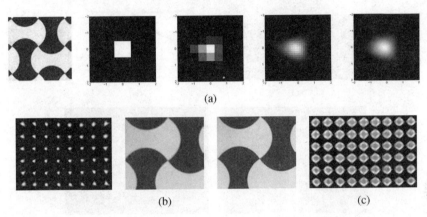

图 10.8　采用标定目标的点扩散函数估计(Joshi，Szelisik，and Kriegman 2008)©2008 IEEE：(a)分辨率逐渐升高下的亚像素 PSF(请注意方形感知区域和圆形镜头模糊间的相互作用)；(b)径向畸变和色差畸变也可以估计并消除；(c)对焦不良(模糊的)的镜头的 PSF 在角落上有些衍射和虚影效应

局部的 2D PSF 估计方法也可以用来估计虚影。图 10.8c 显示了机械虚影体现为在图像中角落处的 PSF 的限幅(clipping)。为了正确捕获与虚影关联的整体变暗，理想图案改变后的亮度需要根据图案的中心区域来外推，采用被一致照亮的目标效果最好。

处理 RAW Bayer pattern(样式)图像时，估计 PSF 的正确方法是在感知的像素值上只估计(10.1 式)中的最小二乘项，并将理想图像插值到所有值。对于 JPEG 图像，应当首先对亮度进行线性化，例如，去除 γ 和任何其他在估计得到的辐射响应函数中的非线性。

假如你有一幅由未标定摄像机拍摄的图像，还可以恢复出 PSF 并利用它来修正图像吗？实际上，只要稍加修改，上述的算法仍然是有效的。

如果没有已知的标定图像，你可以检测细长的边缘并在这些区域用理想的阶梯边缘拟合(图 10.9b)，就可获得清晰图像，如图 10.9d 所示。对每个完全被一组有效

[①] 这个处理过程并没有区分几何学和光度学标定。理论上，任何几何畸变都可以建模为 PSF 在空间上的位移。实践中，将所有大的偏移都折算入几何标定部分会更简单。

估计出来的邻居(图 10.9c 中的绿色像素)所包围的像素,应用最小二乘公式(10.1)来估计核 K。获得的局部估计的 PSF 可以被用来修正色差畸变(因为每个通道 PSF 间的相对位移可被算得),如 Joshi,Szeliski,and Kriegman(2008)所述。

习题 10.4 提供了一些更详细的实现和测试基于边缘的 PSF 估计算法的说明。另外有个方法不需要显式检测边缘,而是改用图像的统计信息(梯度分布),它由 Fergus,Singh,Hertzmann et al.(2006)提出。

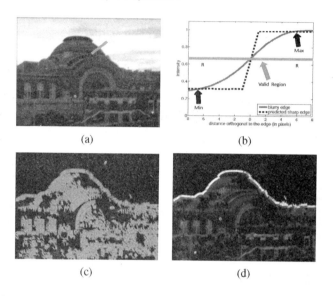

图 10.9 未采用标定图案的 PSF 估计(Joshi,Szeliski,and Kriegman 2008)©2008 IEEE:(a)输入图像和蓝色截面(轮廓)的位置;(b)拍摄的和预测的阶梯边缘的轮廓;(c 和 d)边缘附近所预测彩色的位置和值

10.2 高动态范围成像

正如我们在这一章早些时候所提到的那样,用不同曝光度拍摄的已注册图像可以用来标定摄像机的辐射响应函数。更重要的是,它们可以帮助你制作曝光良好的照片,甚至在一些富有挑战性的条件下,例如在明亮照明的场景下,任何单一曝光度都会使场景中既有饱和(过度曝光)区域也有阴暗(曝光不足)区域(图 10.10)。这个问题很常见,因为自然界包含的辐射值范围远远超过任何成像传感器或胶片可以拍摄的范围(图 10.11)。拍摄一组包围式曝光(bracketed exposures)图像(在摄像机的自动包围曝光(AEB)模式下来故意拍摄一些过度曝光或曝光不足的图像)可以提供一些可供创作合适曝光图像的素材,如图 10.12 所示(Reinhard,Ward,Pattanaik et al. 2005;Freeman 2008;Gulbins and Gulbins 2009;Hasinoff,Durand,and Freeman 2010)。

图 10.10　室内图像样例，其中窗外区域过度曝光而室内部分又太暗

图 10.11　不同场景的相对亮度，从在一间由监视器照亮的黑屋中的 1 到对着太阳时的 2 000 000。照片使用承蒙 Paul Debevec 的允许

图 10.12　一组包围式曝光的镜头(使用摄像机的自动包围曝光(AEB)模式)以及生成的高动态范围(HDR)合成图

虽然可以通过结合不同曝光度图像的像素直接生成最终的合成图(Burt and Kolczynski 1993；Mertens，Kautz，and Reeth 2007)，但是，这种方法存在导致对比度反转和光环的风险。更常用的方法是采用以下三个步骤进行处理。

1. 从配准的图像中估计出辐射响应函数。
2. 通过从不同曝光度的图像中选择或者混合像素来估计辐照度图(radiance map)。
3. 将生成的高动态(HDR)图像色调映射回可供显示的色域。

估计辐射响应函数背后的思想相对直观(Mann and Picard 1995；Debevec and Malik 1997；Mitsunaga and Nayar 1999；Reinhard，Ward，Pattanaik et al. 2005)。假设你拍摄三组不同曝光度的照片(快门速度)，例如在 ±2 曝光值间。[①]如果能够确定每个像素的辐照度(曝光值) E_i(2.101)，我们就可以绘制出它在每种曝光时间 t_j 下随测量得到的像素值 z_{ij} 的变化曲线，如图 10.13 所示。

不幸的是，我们不知道辐照度值 E_i，所以这些值不得不和辐射响应函数 f 同时估计，可以写成(Debevec and Malik 1997)

---

① 改变快门速度比改变光圈大小更好，因为后者会改变虚影和聚焦。采用 ±2 "F 制光圈"(技术上，曝光值或 EVs，由于 F 制光圈指的是光圈大小)通常是对拍摄效果良好的高动态范围图像和在任何地方像素都有良好曝光的一个正确的折中。

$$z_{ij} = f(E_i\, t_j), \tag{10.3}$$

其中，t_j 是第 j 幅图像的曝光时间。响应曲线的逆可表示成

$$f^{-1}(z_{ij}) = E_i\, t_j. \tag{10.4}$$

对两边同时取对数(以 2 为底较为方便，因为我们现在以 EVs 衡量数值)，可得

$$g(z_{ij}) = \log f^{-1}(z_{ij}) = \log E_i + \log t_j, \tag{10.5}$$

其中，$g = \log f^{-1}$(它将像素值 z_{ij} 映射成辐照度的对数)是我们要估计的曲线(图 10.13 将其放在侧面)。

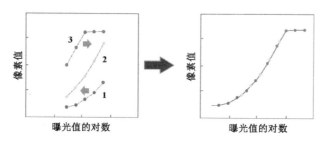

图 10.13　采用多种曝光度进行辐射度标定(Debevec and Malik 1997)。对应像素值被绘制为曝光度对数(辐照度)的函数。移动左边的曲线来调节每个像素未知的辐射亮度，直到它们连接成一条平滑的曲线

Debevec and Malik(1997)假设曝光时间 t_j 是已知的。(回顾一下，这些信息可以从摄像机的 EXIF 的标签上获得，但是它们实际上是按照 2 的乘方增长的 1 $\cdots$,1/128,1/64,1/32,1/16,1/8,$\cdots$，而不是像标号所示的那样 $\cdots$,1/125,1/60,1/30,1/15,1/8,$\cdots$ 数值见习题 2.5。)因此，未知量是每个像素的曝光度 E_i 和响应值 $g_k = g(k)$，其中 g 可以根据常见的 8-位图像的 256 个像素值进行离散化。(各个彩色通道的响应曲线分别进行标定。)

为使响应曲线平滑，Debevec and Malik(1997)增加了一个二阶光滑性约束：

$$\lambda \sum_k g''(k)^2 = \lambda \sum [g(k-1) - 2g(k) + g(k+1)]^2, \tag{10.6}$$

这个和蛇行(5.3)中采用的很相似。由于像素值在其取值范围中部的更可靠(并且 g 函数在饱和值附近变为单一的)，他们还增加了一个加权(帽子)函数 $\omega(k)$，它在像素值范围的两头衰减为 0，

$$w(z) = \begin{cases} z - z_{\min} & z \le (z_{\min} + z_{\max})/2 \\ z_{\max} - z & z > (z_{\min} + z_{\max})/2. \end{cases} \tag{10.7}$$

将这些项组合在一起，他们将问题转化为在未知量 $\{g_k\}$ 和 $\{E_i\}$ 上的最小二乘问题，

$$E = \sum_i \sum_j w(z_{i,j})[g(z_{i,j}) - \log E_i - \log t_j]^2 + \lambda \sum_k w(k)g''(k)^2. \tag{10.8}$$

(为了消除响应函数和辐照度值的全局偏移不确定性，响应曲线的中部被置为 0。)Debevec and Malik(1997)展示了这些工作如何用 21 行 MATLAB 代码实现，这也在一定程度上促进了其方法的流行。

虽然，Debevec and Malik(1997)假设曝光时间 t_j 是精确已知的，但是他们并没

有给出原因说明为什么这些附加的变量不能由最小二乘问题一并解决，这只需要再增加一项 $\eta\sum_j(t_j-\hat{t}_j)^2$ 来约束其最终估计值接近于标称值 $\hat{t}_j$。

图 10.14 显示了恢复得到的数码相机的辐射响应函数以及在整体辐照度图中选取的(相对)辐射值。图 10.15 显示了彩色胶片上拍摄的包围式曝光输入图像以及对应的辐照度图。

Debevec and Malik(1997)采用常规的二阶平滑曲线 g 来参数化他们的响应曲线，而 Mann and Picard(1995)采用了一个三参数的函数

$$f(E) = \alpha + \beta E^{\gamma}, \tag{10.9}$$

Mitsunaga and Nayar(1999)采用低阶($N\leqslant 10$)多项式刻画响应函数 g 的逆。Pal，Szeliski，Uyttendaele et al.(2004)推导出一个贝叶斯模型来对每幅图像估计一个独立的光滑响应函数，它可以更好地模拟当前数码相机中更复杂(也因此更难预测)的自动对比度和色调调节。

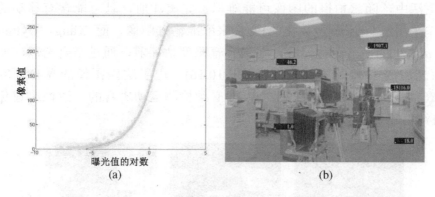

图 10.14 恢复得到的真实数码相机(DCS460)的响应函数和辐照度图像(Debevec and Malik 1997)© 1997 ACM

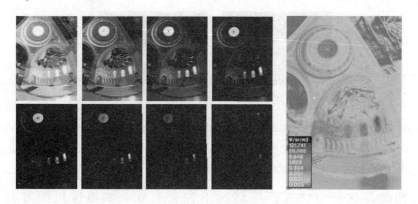

图 10.15 一组由胶片摄影机拍摄的包围曝光图像以及用伪彩色显示的结果辐照度图像(Debevec and Malik 1997)© 1997 ACM

一旦响应函数被估计出来，制作高动态图像的第二步就是将输入图像融合成一个合成的辐照度图(radiance map)。如果响应函数和图像都精确已知，也就是说假设它们都是没有噪声的，你就可以通过用逆响应曲线 $E = g(z)$ 来映射任何一个非饱和

的像素值从而估计出对应的辐照度。

不幸的是,像素都是含有噪声的,特别是在低光照条件下,只有很少的光子会到达传感器。为了对此进行补偿,Mann and Picard(1995)在确定最终的辐照度估计中采用响应函数的导数作为权重,因为曲线中更"平坦"的区域告诉我们有较少的入射辐照度。Debevec and Malik(1997)采用一个加权函数(10.7)来加强色调居中的像素并避免饱和像素值。Mitsunaga and Nayar(1999)指出要最大化信噪比(SNR),加权函数必须同时加强较高的像素值和传递函数中较大的梯度,即

$$w(z) = g(z)/g'(z), \tag{10.10}$$

其中权重 ω 用来形成最终的辐照度估计

$$\log E_i = \frac{\sum_j w(z_{ij})[g(z_{ij}) - \log t_j]}{\sum_j w(z_{ij})}. \tag{10.11}$$

习题 10.1 将要求你实现一种辐射响应函数标定方法并用它来制作辐照度图。

现实生活中,随意拍摄的图像可能难以被完美注册,且可能含有移动的物体。Ward(2003)采用一个全局(参数化的)变换来配准输入图像,而 Kang, Uyttendaele, Winder et al.(2003)提出一种算法,在混合辐照度估计前,通过结合全局注册和局部运动估计(光流)来精确配准输入图像(图 10.16)。由于这些图像的曝光程度差别很大,所以在估计运动时必须格外小心,即必须检测运动本身的一致性来避免造成重影和物体碎片。

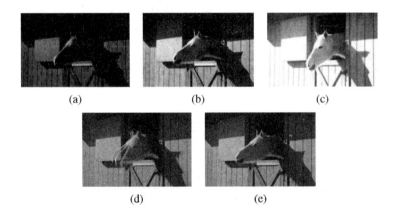

图 10.16 融合多种曝光图像来制作高动态范围合成图像(Kang,Uyttendaele,Winder *et al*. 2003):(a~c)三种曝光程度不同的图像;(d)采用传统算法进行合成(注意由于马头运动而导致的重影);(e)带有运动补偿的合成

然而,即使是这种方法在一些情况下也会失效,如摄像机同时有大的摇移(panning)和曝光改变时,这在随意拍摄的全景图中是很常见的。这种情况下,图像的不同部分会有不同的曝光度。设计出一个既能混合所有这些不同来源的图像,又可以避免突然转变并处理场景运动的算法是极具挑战的。一种可行的方法是,先找到一个一致的拼图(consensus mosaic),然后有选择地计算曝光不足和曝光过度区域的辐照度(Eden,Uyttendaele,and Szeliski 2006),如图 10.17 所示。

近来，一些摄像机，如 Sony α550 和 Pentax K-7，已经开始将多重曝光(multiple exposure)图像的融合和色调映射等功能直接集成到摄像机硬件中。将来，摄像机中传感方法的进步可能会淘汰这种从多重曝光的图像中计算高动态范围图像的需求(图 10.18)(Yang，E1 Gamal，Fowler et al. 1999；Nayar and Mitsunaga 2000；Nayar and Branzoi 2003；Kang，Uyttendaele，Winder et al. 2003；Narasimhan and Nayar 2005；Tumblin，Agrawal，and Raskar 2005)。不过，混合这种图像及将其色调映射到较低色域显示器的需求应该还会存在。

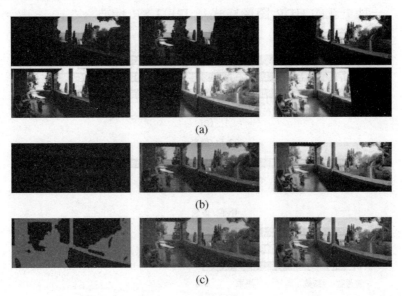

图 10.17 有大量运动的 HDR 融合(Eden，Uyttendaele，and Szeliski 2006)© 2006 IEEE。(a)已注册的包围曝光图像组；(b)第一轮图像选取后的结果：参考标签，图像和色调映射后的图像；(c)第二轮图像选取后的结果：最终标签，压缩后的 HDR 图像和色调映射后的图像

图 10.18 富士 SuperCCD 高动态范围成像传感器。成对的一大一小有效区域可造成两种不同的有效曝光

HDR 图像格式 在介绍将 HDR 图像映射回可供显示的色域的方法前，我们先来讨论常用的 HDR 图像存储格式。

如果不考虑存储空间的话，将图像的 R，G，B 各个值分别存储为 32 位 IEEE 标准浮点数是最好的解决方案。常用的 Portable PixMap 格式(.ppm)支持无压缩的

ASCII 码和未处理的二进制数值编码,改变它的文件头,就可以扩展为 Portable FloatMap 格式(.pfm)。TIFF 格式也支持全浮点数值。

另一种紧凑的图像格式是辐射亮度(Radiance)格式(.pic 和.hdr)(Ward 1994),它采用单一的共同指数和每个通道上各自的尾数(10.19b)。介于两者之间的编码方式是 ILM[①]公司的 OpenEXR,它每个通道用 16 位浮点数表示,这种格式在大部分现代 GPU 上都有天然的支持。Ward(2004)以及 Reinhard, Ward, Pattanaik *et al.* (2005)和 Freeman (2008)对以上数据格式和 Logluv(Larson 1998)之类的其他的一些格式有更详细的介绍。一种更新的 HDR 图像格式是 JPEG XR 标准[②]。

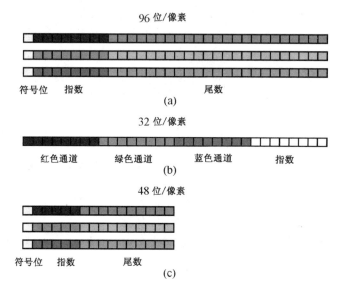

图 10.19 HDR 图像编码格式:(a)Portable PixMap 格式(.ppm);(b)辐射亮度格式(.pic,.hdr);(c)OpenEXR 格式(.exr)

10.2.1 色调映射

一旦计算出辐照度图,通常需要将它显示在较低的色域范围(例如,8 位)的屏幕或者显示器上。为了这个目标,已经开发了大量色调映射的方法,它们大都通过计算空间变化传递函数或者降低图像梯度来适合可用的动态范围(Reinhard, Ward, Pattanaik *et al.* 2005)。

将高动态范围辐照图像压缩到低动态范围色域,最简单的方法是采用全局变换曲线(Larson, Rushmeier, and Piatko 1997)。图 10.20 显示了这样一个例子,它用一条伽马曲线将 HDR 图像映射回到可供显示的色域。若伽马分别作用于各个通道(图 10.20b),色彩会变得柔和(欠饱和的),因为数值高的彩色通道对最终彩色的

[①] http://www.openexr.net/
[②] http://www.itu.int/rec/T-REC-T.832-200903-I/en

贡献(成比例地)较少。若先将图像拆分为它的亮度部分和色彩部分(即，$L*a*b*$)(2.3.2 节)，将全局映射应用在亮度通道上，然后再重构出彩色图，效果会更好(图 10.20c)。

图 10.20 全局色调映射：(a)输入的 HDR 图像，线性映射后的；(b)每个彩色通道分别独立作用 γ；(c)γ 只作用于亮度(色彩较少冲洗)。HDR 原图承蒙 Paul Debevec 许可使用，*http://ict.debevec.org/~debevec/Research/HDR/*。处理后的效果图承蒙 Frédo Durand 许可使用，MIT 6.815/6.865 计算摄影学课程

不幸的是，若图像的曝光变化范围实在很广，这种全局方法仍然无法保持在有剧烈曝光变化的区域内的细节。需要的是类似摄影师在暗房中的局部遮蔽(dodging)和高亮(burning)操作。数学上，这类似于把每个像素的值除以像素周边区域内的平均亮度值。

图 10.21 显示了具体过程。如上所述，图像被拆分为亮度通道和颜色通道。对亮度通道取对数

$$H(x,y) = \log L(x,y) \tag{10.12}$$

然后，通过低通滤波形成基底层(base layer)

$$H_L(x,y) = B(x,y) * H(x,y), \tag{10.13}$$

以及高通滤波的细节层(detail layer)

$$H_H(x,y) = H(x,y) - H_L(x,y). \tag{10.14}$$

接着，基底层通过缩放到所要求的亮度对数范围而减低对比度，

$$H'_H(x,y) = s\, H_H(x,y) \tag{10.15}$$

加上细节层形成新的亮度对数(log-luminance)图

$$I(x,y) = H'_H(x,y) + H_L(x,y), \tag{10.16}$$

这又可以通过指数化来形成色调映射后的(压缩的)亮度图。注意，这一过程等价于将每个像素的亮度值除以像素周围区域的亮度对数值的(单调映射的)平均值。

图 10.21 显示了低通和高通的亮度对数图和最终色调映射后的彩色效果图。注意，细节层在高对比度边界处的光晕(halos)非常明显，这在最后的色调映射结果中也很明显。这是因为线性滤波不保留边界，在细节层产生了光晕(图 10.23)。

解决方案是采用一个边界保持的滤波器来产生基底层。Durand and Dorsey(2002)研究了大量这种边界保持型滤波器，包括各向异性和鲁棒的各向异性(anisotropic and robust anisotropic)扩散，并选择了双边滤波器(3.3.1 节)作为他们的边界保持滤波器。(最近，Farbman, Fattal, Lischinski *et al.* (2008)的一篇论文认为应该用加权最

小二乘(WLS)滤波器取代双边滤波器，另外 Paris, Kornprobst, Tumblin et al. (2008)综述了双边滤波器及其在计算机视觉和计算摄影学中的应用。)图 10.22 显示了如何用双边滤波器取代线性低通滤波器来生成没有明显光晕的色调映射结果图。图 10.24 总结了这一过程中完整的信息流，从分解为亮度对数和色彩图、双边滤波、对比度下降以及重新合成到最终的输出图像。

图 10.21 采用线性滤波器的局部色调映射：(a)低通和高通滤波后的亮度对数图像和色彩(色度)图像；(b)色调映射后的效果图像(经衰减低通亮度对数图像后)在树的周边有明显的光晕。处理后的图像承蒙 Frédo Durand 许可使用，MIT 6.815/6.865 计算摄影学课程

图 10.22 采用双边滤波的局部色调映射(Durand and Dorsey 2002)：(a)低通和高通经双边滤波后的亮度对数图像和色彩(色度)图像；(b)色调映射后的效果图像(经衰减低通亮度对数图像后)没有光晕。处理后的图像承蒙 Frédo Durand 许可使用，MIT 6.815/6.865 计算摄影学课程

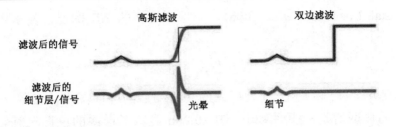

图 10.23 高斯滤波和双边滤波(Petschnigg, Agrawala, Hoppe *et al.* 2004)© 2004 ACM：高斯低通滤波模糊所有边缘，因此在细节层图像上有强烈的尖峰和低谷，从而造成光晕。双边滤波不平滑强烈的边缘，因而减少了光晕并仍能保留细节

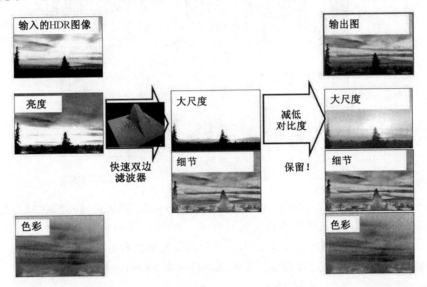

图 10.24 采用双边滤波器的色局部调映射(Durand and Dorsey 2002)：算法流程总结。图像承蒙 Frédo Durand 许可使用，MIT 6.815/6.865 计算摄影学课程

另一种压缩基底层的方法是压缩它的导数，即亮度对数图像的梯度(Fattal, Lischinski, and Werman 2002)。图 10.25 阐述了这一过程。对亮度对数图像求差分得到梯度图像

$$H'(x,y) = \nabla H(x,y). \tag{10.17}$$

再用空间变化衰减函数 $\Phi(x,y)$ 减弱得到的梯度图像，

$$G(x,y) = H'(x,y)\,\Phi(x,y). \tag{10.18}$$

衰减函数 $I(x,y)$ 被设计来减弱大范围的亮度变化(图 10.26a)并考虑了不同空间尺度下的梯度(Fattal, Lischinski, and Werman 2002)。

衰减后，最终的梯度场可通过求解一个一阶变分(最小二乘)问题重新积分得到，

$$\min \iint \|\nabla I(x,y) - G(x,y)\|^2 dx\,dy \tag{10.19}$$

来获得压缩后的亮度对数图像 $I(x,y)$。这个最小二乘问题和泊松混合中采用的是一样的(9.3.4 节)，它的第一次引入是在我们学习规范化时(3.7.1 节，3.100)。采用例如多重网格法和分层基础预处理(Fattal, Lischinski, and Werman 2002; Szeliski 2006b;

Farbman, Fattal, Lischinski et al. 2008)。一旦算出新的亮度图像，便采用下式和原彩色图像合并

$$C_{\text{out}} = \left(\frac{C_{\text{in}}}{L_{\text{in}}}\right)^s L_{\text{out}}, \tag{10.20}$$

其中 $C=(R,G,B)$，L_{in} 和 L_{out} 分别为原来的和压缩后的亮度图像。指数 s 调节颜色的饱和度，它的典型值是 $s\in[0.4,0.6]$。图 10.26b 显示了最终的色调映射彩色图像，尽管输入的辐照度值有极大的变化，但它并没有明显的光晕。

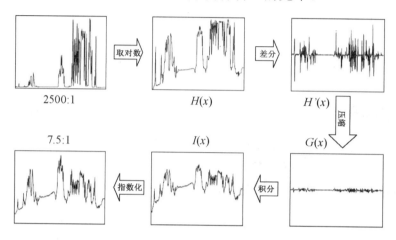

图 10.25　梯度域色调映射(Fattal, Lischinski, and Werman 2002)© 2002 ACM。原图的动态范围为 2415∶1，首先将其转化到对数域，$H(x)$，算出它的梯度，$H'(x)$。根据局部对比度，$G(x)$，将它们衰减(压缩)，并积分求得新的对数曝光度图像 $I(x)$，然后将其指数化以得到最终的亮度图像，它的动态范围是 7.5∶1

图 10.26　梯度域色调映射(Fattal, Lischinski, and Werman 2002)© 2002 ACM：(a)衰减图，彩色越深，表示衰减越多；(b)色调映射后的最终图像

除了这两种方法外，还有一种方法采用局部尺度选择性算子(Reinhard, Stark, Shirley et al. 2002)来进行局部遮蔽和高亮。图 10.27 显示了尺度选择算子是如何决定半径(尺度)以使内部圈只包含相近的颜色值，并避免外部圈中有过亮的像素。实际应用中，可在一系列尺度上评估由内部高斯响应规范化的高斯差分，并选取其度量在阈值下的最大尺度(Reinhard, Stark, Shirley et al. 2002)。

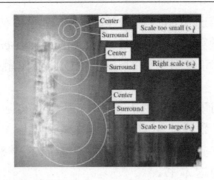

图 10.27　色调映射的尺度选择(Reinhard, Stark, Shirley *et al.* 2002) © 2002 ACM

这些方法的共同特点是它们都适应性地减弱或增强图像的不同区域，以使图像可以在不丧失对比度的情况下显示在有限的色域内。Lischinski, Farbman, Uyttendaele *et al.* (2006b)提出了一种基于交互的方法，它通过插值一系列稀疏的用户描画的调整信息(笔画和对应的曝光值修正)来得到分段连续的曝光度调整图(图 10.28)。插值可通过最小化局部加权最小二乘(WLS)变分问题实现，

$$\min \iint w_{\mathrm{d}}(x,y)\|f(x,y)-g(x,y)\|^2 dx\, dy + \lambda \iint w_{\mathrm{s}}(x,y)\|\nabla f(x,y)\|^2 dx\, dy, \quad (10.21)$$

其中，$g(x,y)$ 和 $f(x,y)$ 分别表示输入和输出对数曝光度(衰减)图(图 10.28)。数据加权项 $w_{\mathrm{d}}(x,y)$ 在笔画处为 1，其他地方为 0。光滑度加权项 $\omega_{\mathrm{s}}(x,y)$ 反比于对数亮度梯度，

$$w_{\mathrm{s}} = \frac{1}{\|\nabla H\|^\alpha + \epsilon} \quad (10.22)$$

因而导致 $f(x,y)$ 图像在低梯度区域较为光滑，而高梯度区域有些不连续。[①]这种方法也可以用于自动的色调映射，即在每个像素设定目标曝光值，并由加权最小二乘将它们转化为分段光滑的调整图像。

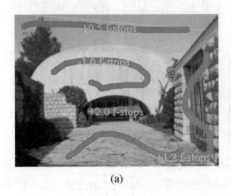
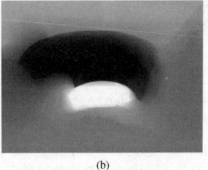

(a) (b)

图 10.28　交互式局部色调映射(Lischinski, Farbman, Uyttendaele *et al.* 2006b) © 2006 ACM：(a)用户描画的笔画及关联的曝光度值 $g(x,y)$；(b)对应的分段光滑曝光度调整图 $f(x,y)$

加权最小二乘算法最早是为图像彩色化应用开发的(Levin, Lischinski, and Weiss

---

① 实际应用中，对 x 和 y 的离散导数是分别加权的(Lischinski, Farbman, Uyttendaele *et al.* 2006b)。论文中默认的参数设置为 $\lambda = 0.2, \alpha = 1$ 及 $\varepsilon = 0.0001$。

2004)，近来已广泛应用于通用的边缘保持平滑处理中，例如对比度增强(Bae, Paris, and Durand 2006)和色调映射(Farbman, Fattal, Lischinski *et al.* 2008)，后者以前采用的是双边滤波方法。它还可以用来同时进行 HDR 图像融合和色调映射(Raman and Chaudhuri 2007, 2009)。

现有的局部适应性色调映射算法有很多，那么在实际应用中应该采用哪种呢？Freeman(2008)广泛讨论了商业应用算法及其效果和可调节参数。他还针对 HDR 摄影学和工作流程提出了很多小建议。我向所有想在这个领域深入研究(或有个人摄影爱好)的人强烈推荐他的这本书。

10.2.2 应用：闪影术

高动态范围成像能够结合同一场景的不同曝光度图像，那么也可能通过结合闪光和非闪光图像来达到更好的曝光效果和色彩平衡并减少噪声(Eisemann and Durand 2004; Petschnigg, Agrawala, Hoppe *et al.* 2004)。

闪光图像的问题是色彩通常不自然(它无法捕捉周围环境的光照)，可能会有很强的阴影或高光(镜面反射)，而且亮度上远离摄像机会有径向衰减(图 10.1b 和 10.29a)。光照不足条件下拍摄的非闪光照片又常常有过多的噪声(由于高 ISO 增益和低光子数)和模糊(由于较长的曝光时间)。有没有一种方法能结合在闪光前一瞬间拍摄的非闪光照片和接下来的闪光照片来生成既有良好的色彩值和清晰度又有较低噪声水平的照片呢？[1]

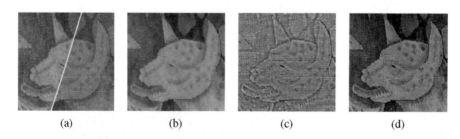

图 10.29 闪光/非闪光照片的细节转移(Detail transfer)(Petschnigg, Agrawala, Hoppe *et al.* 2004) © 2004 ACM：(a)输入环境光下的图像 A 和闪光图像 F 的细节；(b)经联合双边滤波后的非闪光图像 A^{NR}；(c)从闪光图像 F 中算出的细节层 F^{Detail}；(d)最终合成的图像 A^{Final}

Petschnigg, Agrawala, Hoppe *et al.* (2004)提出了处理这个问题的一种方法。第一步，他们采用双边滤波器的变形联合双边滤波器(joint bilateral filter)[2]对非曝光(环境光)图像 A 进行滤波，其中范围核(3.36)

$$r(i,j,k,l) = \exp\left(-\frac{\|f(i,j) - f(k,l)\|^2}{2\sigma_r^2}\right) \tag{10.23}$$

是在闪光图像 F 而不是非闪光图像 A 上进行评估，因为闪光图像带有较少噪声而

---

[1] 实际上，停产的富士摄像机 FinePix F40fd 能快速拍摄两张连续的闪光和非闪光照片，但它只让你决定保留其中之一。
[2] Eisemann and Durand (2004)称此为"交叉双边滤波器"(cross bilateral filter)。

有更可靠的边缘(图 10.29b)。因为在闪光图像中，阴影的边界和内部以及反光处都是不可靠的，所以这些都需要检测出来并用经过常规双边滤波的图像 A^{Base} 代替(图 10.30)。

下一步，他们的算法计算出一个闪光的细节图像

$$F^{Detail} = \frac{F + \epsilon}{F^{Base} + \epsilon}, \tag{10.24}$$

其中 F^{Base} 是闪光图像 F 经过双边滤波的结果，且 $\varepsilon = 0.02$。这个细节图像(图 10.29c)包含那些可能被滤去的不在去噪后的非闪光图像 A^{NR} 中的细节信息，以及闪光造成的通常带有清脆感(crispness)的附加细节。这个细节图像用来调节去噪后的非闪光图像 A^{NR} 以产生最终结果

$$A^{Final} = (1 - M) A^{NR} F^{Detail} + M A^{Base} \tag{10.25}$$

如图 10.1b 和图 10.29d 所示。

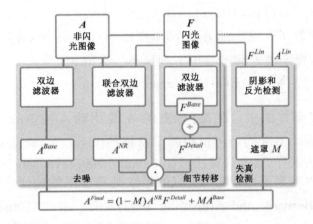

图 10.30 闪光/非闪光摄影算法(Petschnigg, Agrawala, Hoppe *et al.* 2004)© 2004 ACM。环境光(非闪光)图像 A 经常规双边滤波器滤波后得到 A^{Base}，这将被用在阴影和反光区域，经联合双边滤波器去噪后得到 A^{NR}。闪光图像 F 经双边滤波后得到基础图像 F^{Base} 和细节(比率)图像 F^{Detail}，这将用来调整去噪后的环境光图像。阴影/反光遮罩 M 是通过比较线性化后的闪光和非闪光图像获得

Eisemann and Durand(2004)提出了另一种方法，与以上方法有一些相似的基本概念。两篇文章非常值得阅读和比较(习题 10.6)。

闪光图像也可以用在很多其他的应用上，如用来提取更可靠的前景物体(Raskar, Tan, Feris *et al.* 2004; Sun, Li, Kang *et al.* 2006)。闪影术只是主动照明这一更大主题中的一个实例，它在 Raskar and Tumblin(2010)中有更详细的介绍。

10.3 超分辨率和模糊去除

高动态范围成像方法使我们可以获得比通常单幅图像更大的动态范围，而超分辨率可以让我们生成比普通照片空间分辨率更高、噪声更少的图像(Chaudhuri 2001;

Park, Park, and Kang 2003; Capel and Zisserman 2003; Capel 2004; van Ouwerkerk 2006)。通常，超分辨率指将几幅输入图像合并且配准以生成高分辨率的合成图像(Irani and Peleg 1991; Cheeseman, Kanefsky, Hanson *et al.* 1993; Pickup, Capel, Roberts *et al.* 2009)。然而，一些新方法可以对单一图像进行超分辨率处理(Freeman, Jones, and Pasztor 2002; Baker and Kanade 2002; Fattal 2007)，因此它们和模糊去除方法有紧密联系(3.4.3 节和 3.4.4 节)。

最自然的解决超分辨率问题的方法是写出随机的(stochastic)图像形成方程以及图像先验，然后利用贝叶斯推理来获得高分辨率的(原始)清晰图像。我们可以通过一般化图像去模糊(3.4.3 节)中用到的图像形成方程(3.75)来做到这点，这个在(10.2)中的模糊核(PSF)估计中也有用到(10.1.4 节)。这样，我们就有一些观察到的图像 $\{o_k(x)\}$，以及它们各自对应的图像卷绕函数 $\hat{h}_k(x)$(图 3.47)。结合这些要素，可得到(有噪声的)观测方程[1]

$$o_k(\boldsymbol{x}) = D\{b(\boldsymbol{x}) * s(\hat{\boldsymbol{h}}_k(\boldsymbol{x}))\} + n_k(\boldsymbol{x}), \tag{10.26}$$

其中，D 是下采样算子，它在超分辨率的(清晰的)卷绕图像 $s(\hat{\boldsymbol{h}}_k(x))$ 和模糊核 $b(x)$ 卷积后进行。上述图像形成方程可导出下列最小二乘问题，

$$\sum_k \|o_k(\boldsymbol{x}) - D\{b_k(\boldsymbol{x}) * s(\hat{\boldsymbol{h}}_k(\boldsymbol{x}))\}\|^2. \tag{10.27}$$

大多数超分辨率算法中，配准(卷绕)项 $\hat{\boldsymbol{h}}_k$ 是用输入帧中的某一帧作为参考帧(reference frame)估计出来的；可以用基于特征的(6.1.3 节)或直接(基于图像的)(8.2 节)参数化配准方法进行。(一些算法，如 Schultz and Stevenson(1996)和 Capel(2004)中的用的是稠密(逐像素流)估计。)更好的做法是算出原始估计 $s(x)$ 后直接进行最小化(10.27)(Hardie, Barnard, and Armstrong 1997)或者一并忽略(marginalize)运动参数(Pickup, Capel, Roberts *et al.* 2007)——相关的视频超分辨率工作另见 Protter and Elad (2009)。

点扩散函数(模糊核)b_k 可以从图像形成过程的信息推断出(例如，运动量的大小或者散焦造成的模糊以及摄像机传感器的光学特性)或者采用 10.1.4 节中的方法由测试图像或观测到的图像集 $\{o_k\}$ 标定出。同时推断模糊核和清晰图像的问题被称作"盲图像去卷积"(blind image deconvolution)(Kundur and Hatzinakos 1996; Levin 2006)。[2]

有了 $\hat{\boldsymbol{h}}_k$ 和 $b_k(x)$ 的估计，式 10.27 可重新用矩阵/向量符号写成一个对于未知的超分辨率的像素 s 的大型稀疏最小二乘问题，

$$\sum_k \|o_k - DB_k W_k s\|^2. \tag{10.28}$$

(回顾式(3.89)，一旦卷绕函数 $\hat{\boldsymbol{h}}_k$ 已知，$s(\hat{\boldsymbol{h}}_k(x))$ 和 $s(x)$ 便成线性关系)虽然一些早期

---

[1] 对每个观测增加一个未知的偏移-增益项(Capel 2004)也是可行的，这在运动估计的式(8.8)中用过。

[2] 请注意，如果模糊核和超分辨率图像都是未知的，那就有"先有鸡还是先有蛋"的问题。解放办法是采用对清晰图像的结构性假设，如有明显边缘(Joshi, Szeliski, and Kriegman 2008)，或者图像的先验模型，如边缘稀疏性(Fergus, Singh, Hertzmann *et al.* 2006)。

的算法,如 Irani and Peleg(1991)中使用的,采用常规梯度下降法来解决这个最小二乘问题(在计算机断层扫描文献中也被称作"迭代反向投影"(IBP)),但是,更有效的方法是用预处理的共轭梯度下降法(Capel 2004)。

上述方法假设卷绕可以表达为超分辨率清晰图像的简单(sinc 或者双三次函数)插值重采样,接着一个静态(空间不变)模糊(PSF)和区域积分过程。然而,如果表面是严重透视伸缩(foreshortened)的,我们在对由光学器件和摄像机传感器导致的 PSF 进行建模前,必须考虑图像卷绕过程中产生的随空间变化的滤波(3.6.1节)(Wang, Kang, Szeliski et al. 2001; Capel 2004)。

这种最小二乘(MLE)方法的最终效果如何呢?实际上,这在很大程度上取决于摄像机光学器件造成的模糊和走样程度以及运动和 PSF 估计的准确性(Baker and Kanade 2002; Jiang, Wong, and Bao 2003; Capel 2004)。更少的模糊和更多的走样意味着有更多的(走样的)高频信息有待修复。然而,由于最小二乘(最大似然)方法没有使用图像的先验,所以解决过程中可能会引入很多高频噪声(图 10.31c)。

因此,大部分超分辨率算法都假设某种形式的图像先验。最简单的方法是对图像导数(image derivative)设置一个类似于方程(3.105 和 3.113)的惩罚项,例如,

$$\sum_{(i,j)} \rho_p(s(i,j) - s(i+1,j)) + \rho_p(s(i,j) - s(i,j+1)). \tag{10.29}$$

如 3.7.2 节所述,当 ρ_p 是二次式时,这是一个 Tikhonov 正规化形式(3.7.1 节),且总体上还是个线性最小二乘问题。最后得到的先验图像模型是个高斯马尔科夫随机场(GMRF),它也可以扩展到其他的(如对角线)差异上,详见(Capel 2004)(图 10.31)。

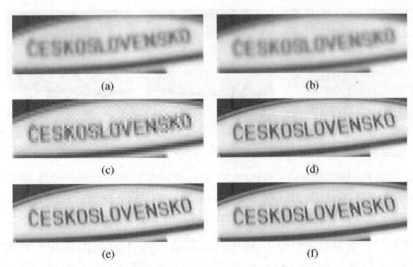

图 10.31 采用各种图像先验的生成的超分辨率效果图(Capel 2001):(a)低分辨率 ROI(双三次 3×zoom);(b)平均图像;(c)MLE@1.25×pixel-zoom;(d)简单 $\|x\|^2$ 先验 ($\lambda = 0.004$);(e)GMRF($\lambda = 0.003$);(f)HMRF($\lambda = 0.01, \alpha = 0.04$)。采用 10 幅图像作为输入,除(c)中的 MLE 外,每种生成一个 3 倍超分辨率图像

不幸的是,GMRF 容易产生明显波纹,这也可以看作是在中频段中的噪声敏感

度的增强(习题 3.17)。好的图像先验模型应该是能够导致分段连续结果的鲁棒的模型(Black and Rangarajan 1996),见附录 B.3。这样的模型如由高斯和长尾 Laplcian 的混合而成的 Huber 势(Schultz and Stevenson 1996; Capel and Zisserman 2003),和 Levin, Fergus, Durand *et al.* (2007) 及 Krishnan and Fergus (2009)中采用的甚至更稀疏的(更重拖尾的(heavier-tailed))Hyper-Laplacian。也可以通过交叉验证来学习得到这些先验模型的参数(Capel 2004; Pickup 2007)。

虽然稀疏(鲁棒的)导数先验可以减少波纹效应并增强边缘清晰度,它却不能幻生(hallucinate)出高频的纹理或者细节。为达到这个效果,可采用一个训练样本图像集来找出似乎可信的原有的低频和丢失的高频信息间的映射关系。受我们在 10.5 节中讨论的基于样例的纹理合成算法的启发,Freeman, Jones, and Pasztor(2002)开发了基于样例的超分辨率算法,它采用训练图像来学习局部块上的纹理和缺失的高频细节间的映射关系。为了保证重叠块外观相似,采用的是马尔科夫随机场,并且通过置信传播(Freeman, Pasztor, and Carmichael 2000)或者光栅扫描确定性变体(raster-scan deterministic variant)(Freeman, Jones, and Pasztor 2002)来进行优化。图 10.32 显示了采用这种方法幻生出丢失细节的效果并和更新的 Fattal(2007)算法进行了比较。后一种算法基于像素和检测到的最近的边缘间的相对位置关系和对应的边缘统计信息(大小和宽度)来学习预测更精细分辨率下的有向梯度的大小。组合使用稀疏(鲁棒的)导数先验和基于样例的超分辨率也是可行的,如 Tappen, Russell, and Freeman(2003)所述。

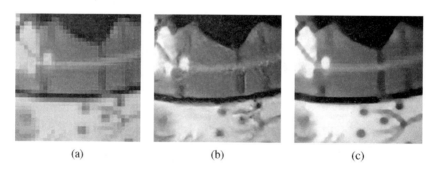

图 10.32 基于样例的超分辨:(a)32×32 的低分辨率原图;(b)256×256 的基于样例的超分辨率效果图(Freeman, Jones, and Pasztor 2002) © 2002 IEEE;(c)通过施加边缘统计信息进行上采样得到的效果图(Fattal 2007) © 2007 ACM

另外一种(但紧密相关的)幻生(hallucination)形式是识别出图像训练集中对应于低分辨率像素的部分。Baker and Kanade(2002)在他们的工作中采用局部高斯导数滤波响应作为特征,然后用与 De Bonet(1997)相似的方法匹配母结构(parent structure)向量。[①]在每个识别出的训练图像处的高频梯度将和通常的重建(预测)方程(10.27)一起作为超分辨率图像的约束。图 10.33 显示了从低分辨率的输入图像幻生

---

① 对于人脸超分辨率图像,所有图像都是预先配准的,只检查不同图像中的对应像素。

出高分辨率人脸的效果；Baker and Kanade(2002)还展示了一些超分辨率已知字体文本的例子。习题 10.7 给出了更具体的如何实现和测试一个或更多这些超分辨率方法的信息。

在适宜条件下，超分辨率和相应的上采样方法可以增加拍摄良好的一张或一组照片的分辨率。然而，如果输入图像一开始就是模糊的，那么最好的预期只能是减少模糊。这个问题和超分辨率密切相关，最大的区别是它的模糊核 b 通常要大得多而且下采样因子 D 是单一的。关于图像去模糊化，有很多文献；一些近期发表的文献做了很好的综述，如 Fergus, Singh, Hertzmann *et al.* (2006), Yuan, Sun, Quan *et al.* (2008)和 Joshi, Zitnick, Szeliski *et al.* (2009)。另外还有一些可行的减少模糊的方法，例如结合清晰(但有噪声)图像和模糊(但更纯净)图像(Yuan, Sun, Quan *et al.* 2007)，拍摄很多快速曝光的图像[①] (Hasinoff and Kutulakos 2008; Hasinoff, Kutulakos, Durand *et al.* 2009; Hasinoff, Durand, and Freeman 2010)，或者采用编码光圈(coded aperture)方法来同时估计深度和减少模糊(Levin, Fergus, Durand *et al.* 2007; Zhou, Lin, and Nayar 2009)。

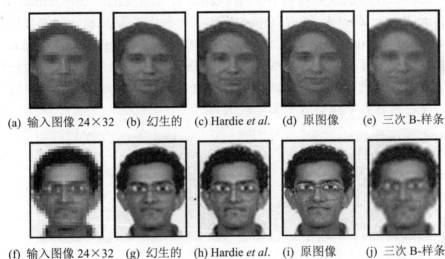

(a) 输入图像 24×32　(b) 幻生的　(c) Hardie *et al.*　(d) 原图像　(e) 三次 B-样条

(f) 输入图像 24×32　(g) 幻生的　(h) Hardie *et al.*　(i) 原图像　(j) 三次 B-样条

图 10.33　基于识别的超分辨率方法(Baker and Kanade 2002) © 2002 IEEE。幻生那一列展示了基于识别的算法并和基于正则化的方法(Hardie, Barnard and Armstrong(1997)进行了比较

10.3.1　彩色图像去马赛克

超分辨率有个特例是对色彩滤波器矩阵(CFA)中的样本进行去马赛克处理以转化为全彩 RGB 图像，这在日常生活中的大部分数码相机中都会用到。图 10.34 展示可最常用的 CFA——Bayer 样式，它的绿色传感器(G)的数量是红色和蓝色的两倍。

[①] SONY DSC-WX1 相机通过多次拍摄来在低光照条件下生成较好的照片。

G	R	G	R
B	G	B	G
G	R	G	R
B	G	B	G

(a)

rGb	Rgb	rGb	Rgb
rgB	rGb	rgB	rGb
rGb	Rgb	rGb	Rgb
rgB	rGb	rgB	rGb

(b)

图 10.34 Bayer RGB 样式：(a)色彩滤波器矩阵布局；(b)插值后的像素彩色值，其中未知(猜测)的值用小写字母表示

由已知的 CFA 像素值到全彩 RGB 图像的处理过程是相当有挑战性的。在常规的超分辨率处理中，对未知像素值猜测的小错误只是导致模糊或者走样，但这在去马赛克中却常常导致杂色或者高频的拉链(zippering)图案，这些在肉眼下都是非常明显的(图 10.35b)。

多年以来，已开发出了各种各样的解决图像去马赛克的方法(Kimmel 1999)。Bennett, Uyttendaele, Zitnick *et al.* (2006)提出了一种新的算法和一些很好的参考文献，而 Longere, Delahunt, Zhang *et al.* (2002)和 Tappen, Russell, and Freeman (2003)采用感知驱动的衡量标准对一些先前开发的方法进行了对比。为了减少拉链效应，大部分方法采用了绿色通道的边缘或者梯度信息来为红色和蓝色通道推断较为合理的值，因为绿色通道的采样更为密集，较为可信。

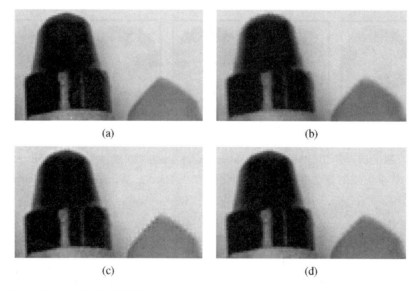

图 10.35 CFA 去马赛克效果(Bennett, Uyttendaele, Zitnick *et al.* 2006) © 2006 Springer：(a)完整分辨率原图像(经彩色重采样后的版本作为算法的输入)；(b)双线性插值的结果，在蓝色蜡笔的笔尖有彩色边缘现象并且左侧(垂直)边缘有拉链效应；(c)Malvar, He, and Cutler(2004)的高质量线性插值结果(注意黄色蜡笔上明显的光晕/格子状瑕疵)；(d)Bennett, Uyttendaele, Zitnick *et al.* (2006)中采用局部二彩色先验的效果

为减少彩色边缘现象，一些方法进行了彩色空间的分析，例如，在互补颜色通道(color opponent channel)上采用中值滤波(Longere, Delahunt, Zhang et al. 2002)。Bennett, Uyttendaele, Zitnick et al. (2006)中的方法用一个移动的 5×5 窗找出局部上主导的两种彩色，并由最初的去马赛克结果局部地建立一个二彩色模型(图 10.36)。①

一旦估计出每个像素上的局部彩色模型，就可以用贝叶斯方法来引导像素彩色值沿着彩色线分布并聚集在主导彩色周围，这样就减少了光晕(图 10.35d)。贝叶斯方法同时也可应用在去马赛克和超分辨率上，即用多 CFA 输入来融合成高质量的全彩色图像，随着附加处理功能逐渐嵌入到当今的摄像机中，这种方法变得越来越重要。

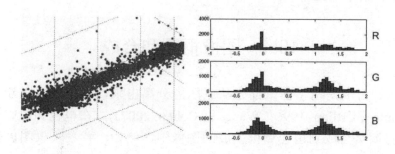

图 10.36 由 5×5 的局部邻域内计算得到的二彩色模型(Bennett, Uyttendaele, Zitnick et al. 2006)©2006 Springer。经过 2-均值聚类并重投影到连接两个主导彩色(图中红点)的直线上，可见大部分像素的彩色值落在设定的直线附近。沿这条直线的分布投影到 RGB 坐标轴上后可以看到峰值在 0 和 1 处，分别为两个主导彩色处

10.3.2 应用：彩色化

虽然彩色化，即手动给"黑白"(灰度)图像加入色彩，在严格意义上并不属于超分辨率，但它却是稀疏插值问题的另一个实例。在大部分彩色化应用中，用户通过手描一些笔画表明在某一区域上预想的彩色(图 10.37a)，然后系统将特定彩色值 (u, v) 插值到整幅图像并结合输入图像的亮度通道生成最终的彩色图像，如图 10.37b 所示。在 Levin, Lischinski, and Weiss (2004)开发的系统中，采用局部加权正则化来进行插值(3.100)，其局部光滑性权重反比于亮度梯度。这种实现局部加权正则化的方法激发了后来的高动态范围色调映射的算法(Lischinski, Farbman, Uyttendaele et al. 2006a)，参见 10.2.1 节，以及一些加权最小二乘法的其他应用(Farbman, Fattal, Lischinski et al. 2008)。另外一种实现稀疏彩色插值的方法是基于测地线(边缘识别，edge-aware))距离函数的，已由 Yatziv and Sapiro (2006)开发。

---

① 先前的关于局部线性颜色模型的工作(Klinker, Shafer, and Kanade 1990; Omer and Werman 2004)集中在单一材料的颜色和亮度变化上，然而，Bennett, Uyttendaele, Zitnick et al. (2006)使用二颜色模型来刻画横跨彩色(材料)边缘的变化。

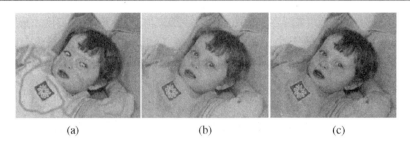

图 10.37 基于优化的彩色化方法(Levin, Lischinski, and Weiss 2004) © 2004 ACM：(a) 灰度图像及覆盖的彩色笔画；(b)彩色化后的效果图；(c)导出灰度图及笔画颜色值的原图。原图由 Rotem Weiss 提供

10.4 图像抠图和合成

图像抠图和合成是指将前景部分从原图中分离出来，再将其置于新背景之下的一种方法(Smith 和 Blinn 1996；Wang 和 Cohen 2007a)。在电视和电影制作中，它常常被用来将演员合成到由计算机生成的图像中，例如，背景气象图以及 3D 虚拟角色和场景(Wright 2006; Brinkmann 2008)。

在前面的章节中，我们已经介绍了很多交互式的分割物体的方法，包括蛇行(5.1.1 节)、剪刀(5.1.3 节)和 GrabCut 分割(5.5 节)等。虽然这些方法可以生成合理的像素精度的分割，但是对于边界处的混合像素(mixed pixel)，它们难以捕捉到背景色和前景色间微妙的相互作用(Szeliski and Golland 1999)(图 10.38a)。

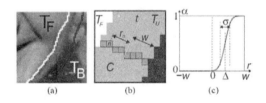

图 10.38 柔化硬分割边界(边界抠图)(Rother, Kolmogorov, and Blake 2004)©2004 ACM：(a)分割的边界区域，各像素的前景色和背景色混合明显；(b)边界上的像素颜色值被用来计算透明度遮罩；(c)估计曲线 t 上的每一点的偏转 Δ 和宽度 α

为了成功将前景物体从一幅图像复制到另一幅图像而不留有明显的人工痕迹，我们需要引入遮罩(matte)，即从输入的合成图像 C 中估计出柔和的不透明度通道 α 和纯正的前景颜色 F。回顾 3.1.3 节(图 3.4)，合成方程(3.8)可写成

$$C = (1 - \alpha)B + \alpha F. \tag{10.30}$$

公式中，背景颜色 B 的影响减弱为$(1-\alpha)$倍，并加上了(部分的)对应于前景颜色值的影响因素 F。

合成操作是容易实现的，但是与其互逆的抠图操作——从输入的图像 C 中估计 F、α和 B，却极具挑战性(图 10.39)。仔细观察，可以看到合成后的像素颜色 C 提供了

3 个测量值，而未知的 F、α、B 却有 7 个自由度。在问题高度缺乏约束的情况下，设计出估计这些未知量的方法正是图像抠图的难点及本质所在。

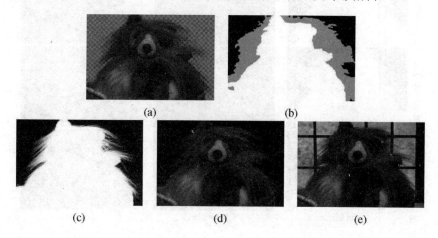

图 10.39 自然图像抠图(Chuang, Curless, Salesin et al. 2001)©2001 IEEE：(a)带有"自然"(非常量)背景的输入图像；(b)手绘的三值图——灰色表示未知区域；(c)提取得到的透明度遮罩图；(d)提取得到(未乘α(乘之前))的前景图；(e)新背景中的合成图

在本节中，我们将回顾大量图像抠图方法。首先是蓝屏抠图(blue screen matting)，这种方法假设背景为已知的单一颜色，还有它的变体——双屏抠图(可以使用多个背景)以及差分抠图(背景可以是任意的，已知即可)。然后，我们将讨论自然图像抠图(natural image matting)，这种情况下，前景和背景都是未知的。通常，在这种应用中会先给定一个三值图(trimap)，即标定了前景、背景和未知区域三个部分的图像(图 10.39b)。接下来，我们会介绍一些采用全局优化方法进行自然图像抠图的方法。最后，我们将讨论抠图问题的一些变体，包括阴影抠图、闪抠图和环境抠图。

10.4.1 蓝屏抠图

蓝屏抠图涉及在单一颜色的背景下对演员(或物体)进行拍摄。虽然，原来人们更喜欢用亮蓝色，但现在亮绿色更常用(Wright 2006; Brinkmann 2008)。Smith and Blinn(1996)论述了很多蓝屏抠图的方法，不过，它们大部分是在专利中表述而不是发表在公开的科研文献上。早期的方法采用物体的各个彩色通道间的线性组合，由用户调整参数来估计不透明度α。

Chuang，Curless，Salesin et al.(2001)描述了一种更新的方法——Mishima算法，这种方法涉及拟合两个多面体表面(多面体以背景颜色在颜色空间中的均值为中心)，再区分前景和背景的颜色分布，最后通过度量特定颜色到这些表面的相对距离来估计α(图 10.41e)。虽然这种方法在很多摄影棚环境设定下工作的很好，但它仍然存在蓝溢出(blue spill)的问题，即物体边界的半透明像素会受背景影响而有些发蓝(图 10.40)。

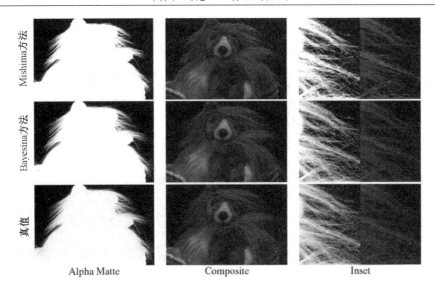

图 10.40 蓝屏抠图效果(Chuang, Curless, Salesin et al. 2001)©2001 IEEE。Mishima 算法的结果有明显的蓝溢出(毛发处有色彩扩散)，而 Chuang 等提出的 Bayesian 抠图方法能生成更为精确的结果

双屏抠图 Smith and Blinn(1996)在他们的论文中介绍了三角剖分抠图(triangulation matting)，这种方法采用多种已知背景颜色以达到对求取不透明度α和前景颜色 F 的方程施加更强的约束。

例如，假设在合成方程(10.30)中，设背景颜色为黑色，即 $B=0$。那么合成的结果图像 C 就应等于αF。现在替换背景颜色，使 B 为已知的非零值，则有

$$C - \alpha F = (1-\alpha)B, \tag{10.31}$$

这就为估计α提供了充分的约束。在实践中，B 的选取应当保证 C 不饱和，而且为了达到更好的精度，应采用多个不同的 B 值。在处理前，使色彩线性化也很重要，这一点对所有的图像抠图算法都适用。在论文中，为了评价效果而生成的真值(ground truth)一般都采用这些方法来获得精确的遮罩(Chuang, Curless, Salesin et al. 2001; Wang and Cohen 2007b; Levin，Acha, and Lischinski 2008; Rhemann, Rother, Rav-Acha et al. 2008; Rhemann, Rother, Wang et al. 2009)[①]。习题 10.8 要求你做这些工作。

差分抠图 差分抠图(Difference Matting)适用于背景已知但并不规则的情况(Wright 2006; Brinkmann 2008)。这种方法通常用于在静态背景下拍摄演员或者物体，例如，视频会议、人物跟踪(Toyama, Krumm, Brumitt et al. 1999)以及在体积的三维重构方法中用于生成轮廓(11.6.2 节)(Szeliski 1993; Seitz and Dyer 1997; Seitz, Curless, Diebel et al. 2006)。它还可以配合遥移摄像机使用，这种情况下，一些帧的前景通过垃圾遮罩(garbage matte)(10.4.5 节)去除，并由这些帧融合(Chuang, Agarwala, Curless et al. 2002)得到背景。另外一个新兴的应用是用来检测电影制作

[①] 请参见抠图评测网站，网址为 *http://alphamatting.com/*。

中的视觉连续性错误，即用来找出重拍镜头中不合理的背景差异(Pickup and Zisserman 2009)。

当前景和背景的运动都可以由参数变换来刻画时，三角剖分抠图方法可以生成高质量的遮罩(Wexler, Fitzgibbon and Zisserman 2002)。但是，当帧与帧之间需要独立处理时，结果将会很差(图 10.42)。这种情况下，采用一对立体摄像机作为输入设备可以显著改善效果(Criminisi, Cross, Blake *et al.* 2006; Yin, Criminisi, Winn *et al.* 2007)。

10.4.2 自然图像抠图

图像抠图中最一般的情况是背景完全未知，除了可能有个三值图(trimap) (图 10.39b)，它将图像粗略的分割为前景、背景和未知区域。不过，在一些最新的方法中，对这种硬分割的要求放宽了，只需要用户在图像上做一些简单的笔画，如图 10.45 和图 10.46 所示(Wang and Cohen 2005; Wang, Agrawala, and Cohen 2007; Levin, Lischinski, and Weiss 2008; Rhemann, Rother, Rav-Acha *et al.* 2008; Rhemann, Rother, and Gelautz 2008)。另外，仅根据一幅图像进行全自动抠图的方法也已被提出(Levin, Acha, and Lischinski 2008; Singaraju, Rother, and Rhemann 2009)。在 Wang and Cohen(2007a)的综述论文中，对这些方法有详细的描述和对比，现节选如下。

Knockout 方法是一种相对简单的自然图像抠图算法，它由 Chuang, Curless, Salesin *et al.* (2001)提出，如图 10.41f 所示。对于未知区域的每个像素，算法首先确定它邻近的前景和背景像素集合(在图像空间上)，并根据这些邻近的前景和背景的颜色均值来分别估计该像素的前景色 F 和背景色 B。然后，修正背景色，使得观测到的像素颜色 C 落在前景色 F 和背景色 B 的连线上。最后，对每个通道分别估计不透明度 α，并根据每个通道的颜色差对 α 进行加权平均(这是对 α 的最小平方根的估计)。图 10.42 表明，当背景在局部上有多于一种的主导颜色时，Knockout 算法会出问题。

如果我们在某些局部区域考虑颜色分布，我们可以获得更精确的抠图效果(图 10.41g 和 h)。Ruzon and Tomasi(2000)将局部颜色分布建模为一些(非相关的)高斯混合模型，这些模型按条带计算。接着他们找到最适合描述观测到的颜色 C 的局部前景模型 F 和局部背景模型 B，然后，通过计算这些模型均值间的相对距离获得 α，并修正 F 和 B 的估计，使得它们与 C 共线。

Chuang, Curless, Salesin *et al.* (2001)和 Hillman, Hannah, and Renshaw(2001)采用 3×3 的颜色协方差矩阵来建立一系列相关的高斯混合模型，并对每个像素分别估值。遮罩的估计由已知区域按条带向未知区域推进，这样最新算出的 F 和 B 值就可以用于后续阶段。

为了获得最准确的未知像素点的不透明度及(未混合的)前、背景颜色估计，Chuang 等人采用完全贝叶斯方程来最大化

$$P(F,B,\alpha|C) = P(C|F,B,\alpha)P(F)P(B)P(\alpha)/P(C). \tag{10.32}$$

这等价于最小化概率的负对数

$$L(F,B,\alpha|C) = L(C|F,B,\alpha) + L(F) + L(B) + L(\alpha) \tag{10.33}$$

(舍去常数项 $L(C)$)。

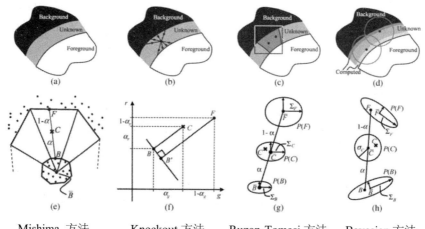

图 10.41　图像抠图算法(Chuang, Curless, Salesin *et al.* 2001)©2001 IEEE。Mishima 算法将前景和背景颜色分布建模为中心在背景(蓝色)颜色分布中心的多面体表面。Knockout 算法对每个未知区域的像素采用局部颜色信息来估计其前景和背景颜色，并对每个通道分别计算 α。Ruzon-Tomasi 算法对前景和背景颜色及其方差进行局部建模。Chuang 等人提出的 Bayesian 抠图方法根据前景和背景颜色的局部分布，通过 MAP 估计来计算前景色和不透明度

让我们来依次分析这些项。首先，$L(C|F,B,\alpha)$ 表示给定未知量 (F,B,α)，观测到像素颜色为 C 的可能性。若假设我们观测中的高斯噪声的方差为 σ_C^2，概率的负对数(数据项)为

$$L(C) = 1/2\|C - [\alpha F + (1-\alpha)B]\|^2/\sigma_C^2, \tag{10.34}$$

如图 10.41h 所示。

第二项 $L(F)$，对应于在高斯混合模型中出现特定前景颜色 F 的概率。先对前景颜色样本进行划分集群(cluster)，然后，算出加权均值和协方差，其中权重正比于给定前景像素的不透明度和它与未知像素间的距离。这样每个集群概率的负对数可以写成

$$L(F) = (F - \overline{F})^T \Sigma_F^{-1}(F - \overline{F}). \tag{10.35}$$

同理，可以估计未知区域的背景颜色分布。如果背景为已知，例如，在蓝屏或者差分抠图中，则直接采用测定的颜色值和协方差。

除了将前景和背景颜色分布建模为高斯混合模型，还可以在原颜色周边样本中计算出最适合解释观测到的颜色 C 的样本对(Wang and Cohen 2005, 2007b)。这些方法在(Wang and Cohen 2007a)中有更详细的介绍。

在他们的 Bayesian 抠图论文中，Chuang, Curless, Salesin *et al.*(2001)假设 $L(\alpha)$ 是一常数(无信息的)分布。更新的论文中假设这一分布在 0 和 1 周围有更高的尖峰，或者有时也采用马尔科夫随机场(MRF)来确定一个全局相关的 $P(\alpha)$ 的先验分布(Wang and Cohen 2007a)。

为了计算得到 (F,B,α) 的最可能估计值，Bayesian 抠图算法转而求取 (F,B) 和 α，因为这两个问题都是二次的，可以当作小型的线性系统来求解。算出一些颜色集群后，就可以选用最有可能的前景和背景颜色集群对。

Bayesian 图像抠图方法改善了 Ruzon and Tomasi(2000)原有的自然图像抠图算法的效果，如图 10.42 所示。但是，相较于一些更新的方法(Wang and Cohen 2007a)，对复杂的背景或不精确的三值图，它的效果并不理想(图 10.44)。

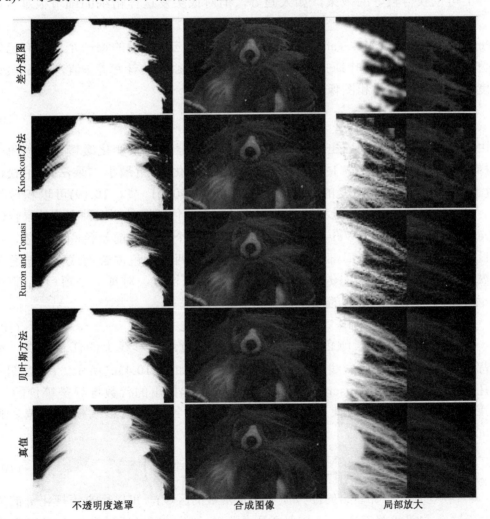

图 10.42　自然图像抠图结果(Chuang, Curless, Salesin *et al.* 2001)©2001 IEEE。差分抠图和 Knockout 方法在这种背景下效果都很差，而更新近的自然图像抠图方法表现良好。Chuang 等人的结果稍微光滑些，和真值更接近

10.4.3 基于优化的抠图

另外，通过考虑 α 值的局部关联，求取全局最优的遮罩，也可以分别估计出每个像素的不透明度及其前景颜色的。这种方法的两个典型代表是 GrabCut 交互分割系统中的边界抠图(Rother, Knlmogorov, and Blake 2004)以及 Poisson 抠图(Sun, Jia, Tang et al. 2004)。

边界抠图首先膨胀由 GrabCut(5.5 节)生成的硬分割边界区域，接着求出一个亚像素的边界位置 Δ，并对边界上的每个点求模糊宽度 σ(图 10.38)。这些参数在边界上的光滑性由正规化保证，并由动态规划算法求取最优解。虽然，这种方法在光滑的边界上可以获得较好的效果，如人脸等，但对像头发这样精细的细节很难有好的表现。

Poisson 抠图(Sun, Jia, Tang et al. 2004)假定对三值图中的每一个像素有已知的前景色和背景色(类似贝叶斯抠图)。然而，它没有独立估计每个 α 值，而是假设不透明度遮罩的梯度变化和图像梯度有如下关系：

$$\nabla \alpha = \frac{F-B}{\|F-B\|^2} \cdot \nabla C, \tag{10.36}$$

这可由(10.30)两边求梯度导出，而且，它还假设背景和前景变化缓慢。求得每个像素的梯度后，先后采用 3.7.1 节(3.100)介绍的正规化方法(最小二乘法)、泊松混合(9.3.4 节，9.44)和基于梯度的动态范围压缩映射(10.2.1 节，10.19)可以积分得到 $\alpha(x)$。在可以较好估计出前景和背景颜色并且它们变化缓慢时，这种方法效果较好。

Levin, Lischinskiand Weiss(2008)没有去算出每个像素的前景色和背景色，而只是假设前景色和背景色在局部上分别可以近似是由两种颜色混合而成，这被称为色彩线模型(color line model)(图 10.43a~c)。在这种假设下，对每个小窗口 W_k(例如，3×3)内的每个像素 i 的 α 值的闭环估计可表示为

$$\alpha_i = \boldsymbol{a}_k \cdot (\boldsymbol{C}_i - \boldsymbol{B}_0) = \boldsymbol{a}_k \cdot \boldsymbol{C} + b_k, \tag{10.37}$$

其中，C_i 为表示每个像素颜色的三维向量，B_0 是背景色彩线上的任意一点，$\boldsymbol{a}_k$ 是连接背景色彩线和前景色彩线间最近两点的向量，如图 10.43c 所示。(注意图中所示的几何推导和 Levin, Lischinski and Weiss(2008)提出的代数推导是等价的。)在所有重叠的窗口 W_k 上，欲使不透明度值 α_i 和它们各自的色彩线模型偏差之和最小，可以构造代价函数

$$E_\alpha = \sum_k \left(\sum_{i \in W_k} (\alpha_i - \boldsymbol{a}_k \cdot \boldsymbol{C}_i - b_k)^2 + \epsilon \|\boldsymbol{a}_k\| \right), \tag{10.38}$$

其中，ε 在两个颜色分布重合时(例如，在 α 是常数的区域)，可以用来正规化 a_k 的值。

由于这个方程式关于未知量 $\{a_k, b_k\}$ 是二次的，所以可以通过在每个窗口 W_k 内取极值而消去它们，这样能量函数变为

$$E_\alpha = \boldsymbol{\alpha}^T \boldsymbol{L} \boldsymbol{\alpha}, \tag{10.39}$$

其中，L 矩阵中的各项可表示为

$$L_{ij} = \sum_{k:i\in W_k \wedge j\in W_k} \left(\delta_{ij} - \frac{1}{M}\left(1 + (C_i - \mu_k)^T \hat{\Sigma}_k^{-1} (C_j - \mu_k)\right) \right), \tag{10.40}$$

其中，$M = |W_k|$ 表示每个(重叠的)窗口中像素的个数，μ_k 是窗口 W_k 内各像素颜色的均值，$\hat{\Sigma}_k$ 是 3×3 的像素颜色的协方差矩阵再加上 $\varepsilon/_M I$。

图 10.43d 解释了这个亲和性矩阵(affinity matrix)中各项的直观含义，它被称为抠图拉普拉斯(matting Laplacian)。请注意，在一个窗口 W_k 中，从均值 μ_k 分别指向两个像素颜色 C_i 和 C_j 的向量方向相反时，它们间的加权点积接近-1，因此它们的亲和度(affinity)接近 0。当像素在颜色空间上相互靠近时(因此有相似的 α 值)，亲和度将接近-2/M。

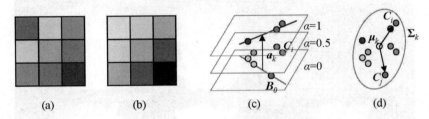

图 10.43 色彩线抠图(Levin, Lischinski, Weiss 2008)：(a)3×3 的局部小色块；(b)可能的 α 值；(c)前景和背景色彩线、连接它们间最近两点的向量 a_k 以及一组 α 为常量的平行平面，$\alpha_i = a_k \cdot (C_i - B_0)$。(d)颜色样本的散布图和两个颜色 C_i 和 C_j 到平均值 μ_k 的偏移

考虑笔画区域已知的约束 $\alpha = \{0,1\}$，最小化二次能量函数(10.39)只需要解一组稀疏线性方程，这就是为什么作者称他们的方法为自然图像抠图的一种闭环的解决方案。一旦 α 值被算出，考虑空间上的一阶光滑性，对合成方程(10.30)可以通过最小二乘法求解前景色和背景色，

$$E_{B,F} = \sum_i \|C_i - [\alpha + F_i + (1-\alpha_i)B_i]\|^2 + \lambda|\nabla\alpha_i|(\|\nabla F_i\|^2 + \|\nabla B_i\|^2), \tag{10.41}$$

其中，权重 $|\nabla\alpha_i|$ 分别应用于 F 和 B 导数的 x 和 y 分量(Levin, Lischinski, and Weiss 2008)。

Wang and Cohen(2007a)总结并比较了大量基于优化的方法，拉普拉斯(闭环)抠图只是其中一种。它们中有些采用不同的亲和度表示方法，有些对保证 α 遮罩的光滑性用不同的方法，或者采用别的估计方法(如置信传播)等，或者用别的表示方法(例如局部颜色直方图等)来刻画局部前景色和背景色的分布(Wang and Cohen 2005, 2007b, c)。这些方法中有些也能够实现实时的抠图，用户只需要描画出轮廓线，或者提供一些稀疏的笔画(Wang, Agrawala, and Cohen 2007; Rhemann, Rother, Rav-Acha et al. 2008)，甚至也可以先将图像划分成小块，用户只需要简单的点击选择(Levin, Acha, and Lischinski 2008)。

图 10.44 展示了一些总结过的算法在给定相同三值图下，对一毛绒玩具进行局部抠图的效果。图 10.45 展示了一些仅由稀疏笔画作为输入生成的抠图效果。图 10.46 是一个更新的算法的抠图效果(Rhemann, Rother, Rav-Acha et al. 2008)，这种方法声称超过了 Wang and Cohen(2007a)综述中的所有算法。

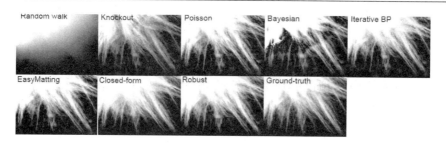

图 10.44 由中等精度的三值图生成的抠图效果对比。Wang and Cohen(2007a)对这些方法逐个进行了比较

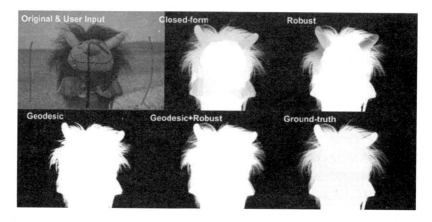

图 10.45 以笔画为输入的抠图效果比较。Wang and Cohen(2007a)逐个介绍了这些方法

图 10.46 基于笔画的分割结果(Themann, Rother, Rav-Acha et al. 2008)©2008 IEEE

粘合(Pasting) 一旦从图像中提取出遮罩,通常就可以将它直接合成到一个新的背景中了,除非是需要隐藏提取出的部分和背景间的接缝处,这种情况下可以采用 Poisson 融合(Poisson blending)(Pérez, Gangnet, and Blake 2003)(9.3.4 节)。

在后一种情况下,如果遮罩边界所经过的区域中几乎没有纹理或者新、旧背景相似,将对粘合很有帮助。Jian, Sun, Tang et al.(2006)和 Wang and Cohen(2007c)在论文中解释了具体做法。

10.4.4 烟、阴影和闪抠图

除了抠出只有少量边界的实心图形,抠出像烟雾这样的半透明的介质也可以实现(Chuang, Agarwala, Curless et al.2002)。首先有一段视频序列,将每个像素建模为(未知的)背景色和恒定的前景(烟)色间的线性组合,注意每个像素都有前景色。采用颜色空间上投票的方式来估计前景色,沿每个色彩线的距离被用来估计每个像素不透明度随时间的变化(图10.47)。

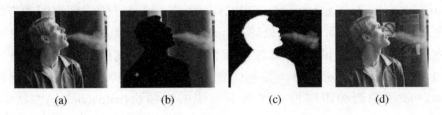

图 10.47 烟抠图(Chuang, Agarwala, Curless *et al*. 2002)©2002 ACM：(a)输入的视频帧；(b)移去前景物体后；(c)估计的透明度遮罩；(d)将新物体插入到背景中

提取和重新插入阴影也可使用相关的方法实现(Chuang, Goldman, Curless *et al*. 2003)。这里没有假定有恒定的前景色，而是假设每个像素都在它完全处于光亮下的颜色和完全处于阴影下的颜色间变化，这个可以通过选取(鲁棒的)阴影掠过场景时每个像素颜色可能的最大值和最小值来获得(习题 10.9)。获得的部分(fractional)阴影遮罩(shadow matte)可以用来重新投射阴影到新的场景中。如果目标场景中有非平面的几何体，可以通过在场景中扫过一个直棍形的阴影来检测。然后，新的阴影遮罩就可以由算得的变形场卷绕得到，从而正确地披放在新场景上(图10.48)。

(a) 前景图　　(b) 背景图　　(c) 蓝屏合成图　　(d) 我们的方法　　(e) 参考图

图 10.48 阴影抠图(Chuang, Goldman, Curless *et al*. 2003)©2003 ACM。阴影抠图能够正确使用阴影减弱场景中的光亮，并在 3D 几何体上披上阴影(d)，而不是简单地使新场景中的阴影部分变暗(c)

抠图算法的质量和可靠性也可以通过采用更为高级的拍摄系统来加强。例如，拍摄一对闪光和非闪光的图像，前景可由两幅图像中亮度变化巨大的区域体现出，这样就保证了可靠的前景遮罩提取(Sun, Li, Kang *et al*. 2006)。由聚焦不同的两台摄影机同时拍摄(McGuire, Matusik, Pfiste *et al*. 2005)或者采用多个摄像机阵列(Joshi, Matusik, and Avidan 2006)等也是获取高质量遮罩的好方法。这些方法的详细论述可参见(Wang and Cohen 2007a)。

最近，在一系列模式化的背景前拍摄有折射能力的物体，使得将这种物体放入复杂三维环境中成为可能。这种环境抠图方法将在 13.4 节讨论。

10.4.5 视频抠图

虽然通常的单帧抠图方法，如蓝屏或者绿屏抠图(Smith and Blinn 1996; Wright 2006; Brinkmann 2008)可以应用于视频序列中，但是，移动的物体有时可以让抠图变得更简单,因为背景的一些信息可以透过前后帧获得。

Chuang, Agarwala, Curless et al.(2002)论述了一种很好的视频抠图的方法,这种方法首先采用一个相对保守的垃圾遮罩(garbage matte)除去前景物体,然后对得到的背景薄板(background plate)通过配准整合以产生一个高质量的背景估计。他们还论述了如何采用双向光流法将关键帧上的三值图插入到帧间。其他的视频抠图方法,例如,涉及描画和跟踪视频序列中曲线的转描机(rotoscoping)方法(Agarwala, Hertzmann, Seitz et al. 2004)等,在 Wang and Cohen(2007a)的综述论文中都有介绍。

10.5 纹理分析与合成

虽然,纹理分析与合成方法最初看起来不像是属于计算摄影学的范畴,但实际上,它们被广泛应用于修复小洞等类似的图像瑕疵,用来由常规照片生成非真实感的画家风格绘制效果。

纹理合成问题可描述为:给定一小块"纹理"样本(图 10.49a),生成大片的外观相似的图像(图 10.49b)。可以想象对于特定的纹理样本,这个问题是相当有挑战性的。

图 10.49 纹理分析:(a)给定一小块纹理,目标是合成大片纹理;(b)外观相似的大片纹理;(c)一些难以合成的半结构化纹理(图像承蒙 Alyosha Efros 许可使用)

传统的纹理分析与合成方法试图去匹配原图的频谱并产生一些有形的噪声。匹配频率特性,也就等价于匹配空间上的相关性,以其本身而言是不够的。不同频率上的响应分布也必须匹配。Heeger and Bergen(1995)开发出的一种算法交替匹配多尺度(导向金字塔)响应的直方图和最终图像的直方图。Portilla and Simoncelli (2000)又增加了对成对的尺度和方向上的统计信息的匹配,改进了这一算法。De Bonet (1997)采用由粗到精的策略在原纹理中找出一些位置,它们具有相似的母结构(parent structure),即有相似的多尺度有向滤波响应。然后随机选取一个匹配的位置作为当前样本的值。

更新的纹理合成算法通过在原纹理中寻找与已合成图像有相近邻域的像素来一步步地生成纹理(Efros and Leung 1999)。考虑图 10.50 中左侧的部分合成的纹理中(暂时)未知的像素 p。由于该像素邻域内的一些像素已经合成完毕，我们可以在右图中寻找有相近邻域的像素，并随机选取一个作为 p 的值。这个过程可以这么一点一点重复下去，可以用光栅扫描的方式，或者在补洞时用由外围向内推进的方式("剥洋葱")，相关讨论参见 10.5.1 节。在实际实现中，Efros and Leung(1999)先找出最相近的邻域，然后将所有相似度距离在其 $d = (1+\varepsilon)$ 倍内的都包括进来，其中 $\varepsilon = 0.1$。另外，它们还可选地根据相似度 d 对像素的随机选取进行加权。

图 10.50　采用非参数化采样的纹理合成(Efros and Leung 1999)。最新的像素 p 的值从原纹理(输入图像)中有相似局部(部分)块的像素中随机选取(图像承蒙 Alyosha Efros 许可使用)

为了加速这一处理过程并改善视觉效果，Wei and Levoy (2000)采用了由粗到精的生成过程来改进这一方法，已合成的较粗层次在接下来的匹配中也会被纳入考虑(De Bonet 1997)。并且为了加速最相近邻域的寻找，算法采用了树形的分层向量结构。

Efros and Freeman(2001)提出了另一种加速及改善效果的方法。这种方法不是一次只合成一个像素，而是通过采用和先前合成的区域间的相似性信息选出相互重叠的正方形小图像块(图 10.51)。一旦选定合适的小图像块，可以通过动态规划确定重叠小图像块间的接缝处。(由于垂直边界上每行只需要一个接缝位置，故不需要采用完全图割接缝选择方法。)由于这种处理过程涉及了小图像块的选取和它们间的拼接，Efros and Freeman(2001)称他们的处理系统为图像绗缝(image quilting)。Komodakis and Tziritas(2007b)提出了这种图像块合成算法的基于 MRF 的版本，它采用一种被称为"优先级(Priority)-BP"的新的、高效的带环的置信传播方法。

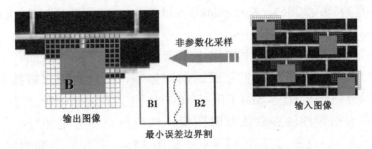

图 10.51　用图像绗缝算法进行纹理合成(Efros and Freeman 2001)。算法从原纹理中一次复制一个图像块，而不是每次只生成单一像素。选定图像块间的重叠区域的过渡将由动态规划进行优化(图像承蒙 Alyosha Efros 许可使用)

10.5.1 应用：空洞填充与修图

从照片中去除物体或者瑕疵后，需要做的就是空洞填充，这称为"修图"(inpainting)，是纹理合成最常见的应用之一。这种方法不但可以用来从照片中去除不想要的人或者闯入者(King 1997)，也可以用来修复一些老照片或者电影中的小瑕疵(划痕去除，scratch removal)或者在电影制作中去除支撑支柱或演员的绳索(线去除，wire removal)。Bertalmio, Sapiro, Caselles et al. (2000)通过沿等照度线(恒定值)传播像素值并夹杂一些各向异性扩散步骤来解决这个问题(图 10.52a 和 b)。Telea (2004)从水平集(level set)(5.1.4 节)采用快速行进法，开发出了一种更快的方法。然而，这些方法都不能在缺失区域幻生出纹理。Bertalmio, Vese, Sapiro et al. (2003)通过在需填充区域加入人工纹理增强了他们的早期方法。

图 10.52　图像修复(空洞填充)：(a 和 b)沿等照度线方向的传播(Bertalmio, Sapiro, Caselles et al. 2000)© 2000 ACM；(c 和 d)基于样例的修图，填充顺序以置信度为基础 (Criminisi, Pérez, and Toyama 2004)

前一小节中讨论的基于例子的(非参数化的)纹理生成方法也可以用于由外而内("剥洋葱"法或定序法)填充空洞。然而，这种方法难以传播有很强方向性的结构。Criminisi, Pérez, and Toyama(2004)采用了一种基于样例的纹理合成方法，它根据区域边界上的梯度强度来确定合成的顺序(图 10.1 和 10.52c 和 d)。Sun, Yuan, Jia et al. (2004)提出了一种相关的方法，用户可以交互的画上线条来表明哪些部位需要优先传播。和这些方法相关的方法还包括：Drori, Cohen-Or, and Yeshurun (2003)，Kwatra, Schödl, Essa et al. (2003)，Kwatra, Essa, Bobick et al. (2005)，Wilczkowiak, Brostow, Tordoff et al. (2005)，Komodakis and Tziritas(2007b)和 Wexler, Shechtman, and Irani(2007)。

大多数空洞填充算法从原图中借用小图像块来填充空洞。然而，有大型图像数据库时，如来自图像分享网站或者互联网的图像等，有时也可以直接复制单一连续的图像区域来填充空洞。Hays and Efros(2007)就提出了这样一种方法，它采用图像的上下文和边界的相容性信息来选取源图像，然后再用图分割和 Poisson 混合将其与原图(多洞的)混合。这种方法在 14.4.4 节和图 14.46 中有更详细的讨论。

10.5.2 应用：非真实感绘制

基于样例的纹理合成思想还有两个应用分别是纹理转移(Efros and Freeman 2001)和图像模拟(Hertzmann, Jacobs, Oliver *et al.* 2001)，它们都属于非真实感绘制(Gooch and Gooch 2001)。

除了采用源纹理图像，纹理转移还使用参考(或目标)图像，并试图将目标图像的某些特性和新合成的图像进行匹配。例如，图 10.53c 中新绘制的图像不仅试图满足和图 10.53b 中源纹理的相似性约束，还试图匹配参考图像的亮度特点。Efros and Freeman(2001)提到模糊图像的亮度或者局部图像的方向角也可以作为被匹配的量。

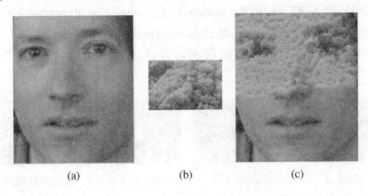

图 10.53 纹理转移(Efros and Freeman 2001)© 2001 ACM：(a)参考(目标)图像；(b)纹理源；(c)用纹理(部分)绘制后的图像

Hertzmann, Jacobs, Oliver *et al.* (2001) 形式化地表述了下列问题：

给定一对图像 A 和 A' (分别为未处理和处理后的源图像)，以及某个别的未处理目标图像 B，合成新的处理后的图像 B'，使得

$$A : A' :: B : B'.$$

这种方法并不需要用户设定某种非真实感绘制效果。只需给系统提供一对处理前后的图像样例，系统就会采用基于样例的合成方法自动生成新图像，如图 10.54 所示。

图 10.54 图像模拟(Hertzmann, Jacobs, Oliver *et al.* 2001) © 2001 ACM。给定一对样例，源图像 A 及其绘制后的(处理后的)版本 A'，由另外一幅未处理的源图像 B 生成绘制后的版本 B'

解决图像模拟的算法流程类似于(Efros and Leung 1999; Wei and Levoy 2000)中的纹理合成算法。一旦算出源图像和参考图像的高斯金字塔，算法就在 A' 生成的金字塔中寻找邻域满足其与已部分生成的 B' 中的邻域相似并在 A 和 B 中的对应位

置有相似的多分辨率外观。对于纹理转移,外观特性不仅包括(模糊的)颜色和亮度值,还有方向等。

这种通用的框架为图像模拟在各种不同绘制任务上的应用提供了便利。除了应用于基于样例的非真实感绘制,图像模拟还可用在传统的纹理合成、超分辨率以及纹理转移(对 A 和 A' 采用相同的纹理图像)。如果只有处理后的(绘制后的)图像 A' 已知,如画作,缺失的参考图像 A 可以采用巧妙的(边缘保持的)模糊算子幻生出来。最后,根据自然图像上各个像素对应的语义,如水、树木和草地等,用伪彩色进行手工着色(图 10.55a 和 b),有可能训练出能够完成由数字指定纹理(texture-by-numbers)的系统。获得的系统可以把新描绘的草图转化成绘制良好的合成照片(图 10.55c 和 d)。最近的工作中,Cheng, Vishwanathan, and Zhang(2008)借鉴了图像绗缝(Efros and Freeman 2001)和 MRF 推理(Komodakis, Tziritas, and Paragios 2008)的思想,将其融入到基础的图像模拟算法中,而 Ramanarayanan and Bala(2007)将这种处理重新考虑为能量最小化问题,这意味着它也可以被视为条件随机场(3.7.2 节),另外他们还设计了一个高效的寻找最小值的算法。

原始图像 A'　　绘制图像 A　　新的绘制图像 B　　新的带纹理图像 B'

图 10.55　由数字指定纹理(Hertzmann, Jacobs, Oliver et al. 2001) © 2001 ACM. 给定纹理图像 A' 和手工标注的(描绘的)版本 A,只给定描绘版本 B,合成新的图像 B'

还有一些传统的滤波和特征检测方法也可以用在非真实感绘制中。[①] 例如,钢笔画绘制(Winkenbach and Salesin 1994)和画家风格绘画方法(Litwinowicz 1997)采用局部颜色、亮度和方向估计作为他们绘制程序算法的输入。风格化和简单化照片及视频的方法(DeCarlo and Santella 2002; Winnemöller, Olsen, and Gooch 2006; Farbman, Fattal, Lischinski et al. 2008),如图 10.56 所示,结合了边缘保持模糊方法(3.3.1 节)和边缘检测与增强方法(4.2.3 节)。

(a)　　(b)

图 10.56　照片的非真实感抽象:(a)DeCarlo and Santella(2002)© 2002 ACM 和 (b)Farbman, Fattal, Lischinski et al. (2008) © 2008 ACM

---

① 论文精选请参见网站 *http://www.npar.org/*的 Symposia on Non-Photorealistic Animation and Rendering (NPAR)。

10.6 补充阅读

在以下文献中都可以找到很好的计算摄影学概述：Raskar and Tumblin(2010)的书，Nayar(2006), Cohen and Szeliski(2006), Levoy(2006)和 Hayes(2008)的综述文章，以及分别由 Bimber(2006)和 Durand and Szeliski(2007)编辑的两期特刊。本章开头提到的计算摄影学课程上的教案也是非常好的材料和参考文献。[①]

高动态范围成像这一子领域有它自己的讨论该领域研究的书籍(Reinhard, Ward, Pattanaik et al. 2005)，以及一些介绍相关摄影方法的书(Freeman 2008; Gulbins and Gulbins 2009)。标定摄像机辐射响应函数的相关算法介绍可以在 Mann and Picard (1995), Debevec and Malik(1997)和 Mitsunaga and Nayar(1999)等文章中找到。

色调映射这一课题在(Reinhard, Ward, Pattanaik et al. 2005)中有广泛的讨论。在大量的关于这一课题的文献中，具有代表性的论文有 Tumblin and Rushmeier(1993), Larson, Rushmeier, and Piatko(1997), Pattanaik, Ferwerda, Fairchild et al.(1998), Tumblin and Turk(1999), Durand and Dorsey(2002), Fattal, Lischinski, and Werman (2002), Reinhard, Stark, Shirley et al.(2002), Lischinski, Farbman, Uyttendaele et al. (2006b), 以及 Farbman, Fattal, Lischinski et al.(2008)。

超分辨率的相关文献非常多(Chaudhuri 2001; Park, Park, and Kang 2003; Capel and Zisserman 2003; Capel 2004; van Ouwerkerk 2006)。超分辨率这一术语通常用于描述配准和融合多幅图像以生成高分辨率合成图像的方法(Keren, Peleg, and Brada 1988; Irani and Peleg 1991; Cheeseman, Kanefsky, Hanson et al. 1993; Mann and Picard 1994; Chiang and Boult 1996; Bascle, Blake, and Zisserman 1996; Capel and Zisserman 1998; Smelyanskiy, Cheeseman, Maluf et al. 2000; Capel and Zisserman 2000; Pickup, Capel, Roberts et al. 2009; Gulbins and Gulbins 2009)。不过，单一图像的超分辨率方法也已经被开发出来(Freeman, Jones, and Pasztor 2002; Baker and Kanade 2002; Fattal 2007)。

Wang and Cohen(2007a)给出了图像抠图领域的很好的综述。代表性的论文(它们都包含与先前工作的广泛比较)包括 Chuang, Curless, Salesin et al. (2001),Wang and Cohen(2007b), Levin, Acha, and Lischinski(2008), Rhemann, Rother, Rav-Acha et al. (2008), and Rhemann, Rother, Wang et al. (2009)。

纹理合成和空洞填充方面的文献包括传统的纹理合成方法，即尝试匹配源图像和目标图像间的统计信息(Heeger and Bergen 1995; De Bonet 1997; Portilla and

---

[①]
- MIT 6.815/6.865, *http://stellar.mit.edu/S/course/6/sp08/6.815/materials.html*
- CMU 15-463, *http://graphics.cs.cmu.edu/courses/15-463/2008 fall/*
- Stanford CS 448A, *http://graphics.stanford.edu/courses/cs448a-08-spring/*
- SIGGRAPH courses, *http://web.media.mit.edu/_raskar/photo/*

Simoncelli 2000)，以及一些新的方法，即在源样本中搜索匹配的邻域或小图像块(Efros and Leung 1999; Wei and Levoy 2000; Efros and Freeman 2001)。类似地，在空洞填充中，传统的方法涉及局部变分(光滑的连续)问题的解决(Bertalmio, Sapiro, Caselles et al. 2000; Bertalmio, Vese, Sapiro et al. 2003; Telea 2004)，而更新近的方法采用数据驱动的纹理合成方法(Drori, Cohen-Or, and Yeshurun 2003; Kwatra, Schödl, Essa et al. 2003; Criminisi, Pérez, and Toyama 2004; Sun, Yuan, Jia et al. 2004; Kwatra, Essa, Bobick et al. 2005; Wilczkowiak, Brostow, Tordoff et al. 2005; Komodakis and Tziritas 2007b; Wexler, Shechtman, and Irani 2007)。

10.7 习　　题

习题 10.1：辐射测量标定

实现 10.2 节中介绍的多重曝光辐射测量标定算法中的一种(Debevec and Malik 1997; Mitsunaga and Nayar 1999; Reinhard, Ward, Pattanaik et al. 2005)。这种标定在很多不同的应用中都很有用，例如图像拼接、带有不同曝光的图像间的立体视觉匹配以及由阴影到形状。

1. 用带三脚架的摄像机拍摄一系列包围图像。如果你的摄像机有自动包围曝光(AEB)模式，拍摄三幅图像应该就足够标定大多数摄像机的动态范围，特别是场景中有很多明暗分明的区域时。(在户外拍摄或者在晴天透过窗户拍摄效果最佳。)
2. 如果你拍照时没有用三脚架，那么先进行全局配准(相似变换)。
3. 用上面提到的一种方法估计辐射响应函数。
4. 通过选取或混合不同曝光度下的像素估计高动态范围辐照度图(Debevec and Malik 1997; Mitsunaga and Nayar 1999; Eden, Uyttendaele, and Szeliski 2006)。
5. 在不同条件下，如在白炽灯照明的室内，重复你的标定实验，从而获知所使用的摄像机的色彩平衡效果。
6. 如果你的摄像机支持 RAW 和 JPEG 模式，同时标定两组图像并相互标定(每个像素处的辐射亮度将会对应起来)。尝试找出你所用的摄像机的工作方式，例如，它将色彩平衡作为一个对角矩阵还是 3×3 全矩阵相乘、除了伽马校正它是否还采用其他非线性方法、生成 JPEG 图像时它是否锐化图像等。
7. 开发一个交互式的查看器，根据鼠标周围的平均曝光度来改变图像的曝光程度。(可以在鼠标旁边弹出一个窗口来显示调整后的图像，也可以根据鼠标位置直接调整整幅图像。)
8. 实现一个色调映射算子(习题 10.5)，并用它把你的辐照度图像映射到可供显示的色域。

习题 10.2：噪声水平函数

通过拍摄多个镜头或者分析光滑的区域来确定所用摄像机的噪声水平函数。

1. 在三角架上放好你的摄像机，对准一个标定目标或者有合适输入色阶和彩色变化的静态场景。(检查摄像机的直方图确保所有数值上都有采样。)
2. 重复拍摄相同场景的多幅图像(最好有远程快门遥控器)，并进行平均化以算得每个像素上的方差。丢弃高梯度(因摄像机晃动造成的)区域的像素，根据算得的值，在每个彩色通道上绘制出各像素对应的标准偏差。
3. 对这些测量值拟合一个下包络，并用它作为你的噪声水平函数。观察噪声随输入色阶的变化有多大？其中有多少是显著的(即偏离了摄像机响应函数中的平坦区域，在其上你并不想进行采样)？
4. (可选)使用相同的图像，开发出一种方法将图像分割成一块块近似恒定的区域(Liu, Szeliski, Kang et al. 2008)。(如果拍摄的是标定图案，这会比较容易。)对单一图像算出每个区域上的偏差并用它们来估计 NLF。这和用多幅图像的方法比怎样？对不同的图像，你的估计有多稳定？

习题 10.3：虚影

采用下列三种方法中的一种(或自己另选)，估计所用镜头的虚影量。

1. 拍摄大块亮度均匀的区域(如照明良好的墙或者蓝天——但请留心亮度梯度)的图像，并拟合一个径向多项式曲线来估计虚影。
2. 构造一幅中心加权(center-weighted)的全景图并把这些像素值和输入图像进行比较，以估计虚影函数。由于高梯度区域上很小的配准偏差就会导致大的误差，故对变化缓慢区域的像素赋予更高的权重。还可以选做对辐射响应函数的估计(Litvinov and Schechner 2005; Goldman 2011)。
3. 分析径向梯度(特别是在低梯度区域)并将这些梯度的鲁棒的均值拟合到虚影函数的导数上，如 Zheng, Yu, Kang et al. (2008)所述。

对于虚影函数的参数形式，你可以采用简单的径向函数，如

$$f(r) = 1 + \alpha_1 r + \alpha_2 r^2 + \cdots \tag{10.42}$$

或者由 Kang and Weiss(2000)和 Zheng, Lin, and Kang(2006)提出的特殊公式。

不论采用哪种方法，请确保使用的是线性化亮度测量值，或者至少保证 γ 曲线已去除的图像，前者可以通过使用 RAW 图像或者经辐射响应函数线性化后的图像来达到。

(可选)如果你忘记取消伽马校正就开始拟合(乘积的)虚影函数，会发生什么情况？

习题 10.4：光学模糊(PSF)估计

采用已知的目标图(图 10.7)或者通过检测和拟合阶梯边缘(10.1.4 节)(Joshi, Szeliski, and Kriegman 2008)来计算光学 PSF。

1. 在亚像素精度下检测明显边缘。
2. 对每个有向边缘拟合一个局部轮廓，并将这些像素填充到理想目标图像

中，可以在图像自身分辨率上也可以在更高的分辨率上进行(图 10.9c 和 d)。
3. 在有效像素上用最小二乘(10.1)估计 PSF 的核 K，可以对全局进行，也可以在图像局部的重叠子区域上进行。
4. 画出恢复的 PSF 并用它们去除色差畸变或对图像进行去模糊处理。

习题 10.5：色调映射

实现 10.2.1 节中讨论的一种色调映射算法(Durand and Dorsey 2002; Fattal, Lischinski, and Werman 2002; Reinhard, Stark, Shirley et al. 2002; Lischinski, Farbman, Uyttendaele et al. 2006b)，或者 Reinhard, Ward, Pattanaik et al. (2005) 和 *http://stellar.mit.edu/S/course/6/sp08/6.815/materials.html* 讨论的任何一种算法。

(可选)将你实现的算法与局部直方图均衡化进行比较(3.1.4 节)。

习题 10.6：闪光增强

开发一种算法结合闪光和非闪光图像以达到最好效果。你可以借鉴 Eisemann and Durand (2004) 和 Petschnigg, Agrawala, Hoppe et al. (2004)中的思想，或者采用其他任何你觉得更有效的方法。

习题 10.7：超分辨率

实现一种或多种超分辨率算法并比较它们的性能。
1. 用手持摄像机对相同场景拍摄一组图像(确保图像间有些许晃动)。
2. 用习题 10.4 中的一种方法确定需要进行超分辨处理的图像的 PSF。
3. 或者，通过拍摄高质量图像(避免压缩失真)并应用你自己的预滤波核及下采样模拟出一组低分辨率的图像。
4. 用参数化平移和旋转运动估计算法(6.1.3 节或 8.2 节)估计图像间的相对运动。
5. 通过最小化观察到的和下采样后的图像间的区别，实现一个基本的最小二乘超分辨率算法。
6. 加入梯度图像先验作为另一个最小二乘项，或者作为一个鲁棒项用迭代重加权最小二乘算法进行最小化(附录 A.3)。
7. (可选)实现一种基于样例的超分辨率方法。这种方法对一组样例图像进行匹配，用于推断出重建过程中需要的高频信息(Freeman, Jones, and Pasztor 2002)，或者超分辨率图像中需要匹配的高频梯度(Baker and Kanade 2002)。
8. (可选)采用局部边缘统计信息来改善超分辨率图像的质量(Fattal 2007)。

习题 10.8：图像抠图

开发一种算法用于从自然图像中提取前景遮罩，如 10.4 节所述。
1. 请确保你拍摄的图像是线性化后的(习题 10.1 和 10.1 节)，并且摄像机的曝光度是固定的(全手动模式)，至少在对同样场景拍摄多个镜头时保证这点。
2. 为了获得真值数据，将你的物体放在电脑屏幕前，并显示各种纯色背景以及一些自然图像。

3. 拿去你的物体，然后重新显示相同的图像以获得已知的背景颜色。
4. 使用三角剖分抠图(Smith and Blinn 1996)来估计物体的真值的不透明度 α 以及乘上 α 后的前景颜色 αF。
5. 实现一种或者多种 10.4 节中介绍的自然图像抠图算法，并将它们的结果和你算得的真值进行比较。另外，也可以使用网站 *http://alphamatting.com/* 上的抠图测试图像。
6. (可选)在用相同的标定过的摄像机拍摄的其他图像(或者其他任何你觉得有趣的图像)上运行你的算法。

习题 10.9：烟和阴影抠图

从一个场景中将烟和阴影的遮罩抠出，再将其嵌入别的场景(Chuang, Agarwala, Curless *et al.* 2002; Chuang, Goldman, Curless *et al.* 2003)。

1. 找一些静态图像或者视频帧，它们中有些有断续的烟和阴影，有些没有。(记住在做任何计算处理前先线性化你的图像)。
2. 对每个像素，对观察到的颜色值拟合一条直线。
3. 如果是做烟抠图，那么鲁棒地算出这些线的交点来获得对烟颜色的估计。然后，将背景颜色作为另外一个极值来估计(除非你已经拍摄了一幅无烟的背景图像)。

    如果是做阴影抠图，为每个像素分别算出鲁棒阴影值(最小值)和照亮值(最大值)。
4. 从每个帧中提取烟或阴影遮罩，作为这两个值(背景和烟或者阴影的和照亮的)的分数混合(比例折算)。
5. 寻找新的(目标)场景或者用图像编辑器修改原背景。
6. 将烟和阴影遮罩以及任何你抠出的前景物体嵌入新场景。
7. (可选)使用一系列木棍投影来估计目标场景的变形场，这样就可以在新的几何体上正确卷绕(褶绕)阴影。(这和 Bouguet and Perona(1999)开发的以及习题 12.2 中实现的阴影扫描算法相关。)
8. (可选)Chuang, Goldman, Curless *et al.* (2003)只是将他们的方法用在源几何体平坦的情况中。你是否可以将他们的方法扩展到可以捕获不规则源几何体上的阴影？
9. (可选)你是否可以改变阴影的方向，即模拟出改变光源方向的效果？

习题 10.10：纹理合成

实现 10.5 节中介绍的一种纹理合成或空洞填充算法。可行的步骤如下。

1. 实现 Efros and Leung(1999)的基本算法，即从外围开始(对于空洞填充)或者按光栅扫描顺序(对于纹理合成)，在源纹理图像中搜索有相似邻域的像素，并将其复制过来。
2. 加入 Wei and Levoy(2000)中按由粗到精方式生成像素的扩展，即先生成低分辨率的合成纹理(或填充后的图像)，再用这个指导更精细的分辨率下区域的匹配。

3. 加入 Criminisi, Pérez, and Toyama(2004)中的用一些局部结构的某种函数(梯度或者方向性强度)对像素填充顺序进行优先权指定的思想。
4. 选择上述任何一种算法进行扩展，在源纹理中选取子图像块并采用优化的方法来确定新图像块和已有图像间重叠区域的接缝(Efros and Freeman 2001)。
5. (可选)实现一种等照度线(光滑连续)修图算法(Bertalmio, Sapiro, Caselles *et al*. 2000; Telea 2004)。
6. (可选)增加支持提供目标(参考)图像(Efros and Freeman 2001)，或提供滤波后或未滤波的(参考和绘制后的)样本图像的能力(Hertzmann, Jacobs, Oliver *et al*. 2001)，参见 10.5.2 节。

习题 10.11：彩色化

实现 10.3.2 节和图 10.37 中概述的 Levin, Lischinski, and Weiss(2004)彩色化算法。找一些早期的黑白照片以及现代的彩色照片。写一个交互工具来从彩色照片中"挑选"颜色并绘制在旧照片上。调整算法的参数以达到良好效果。你是否对结果感到满意？你能想出方法让它们看起来更"古旧"吗，例如带有更柔和的(欠饱和而又有些轮廓过于鲜明)颜色？

第 11 章
立体视觉对应

11.1　极线几何学
11.2　稀疏对应
11.3　稠密对应
11.4　局部方法
11.5　全局优化
11.6　多视图立体视觉
11.7　补充阅读
11.8　习题

图 11.1　立体视觉重建方法可以把(a 和 b)中的一对图像转化成(c)中的一幅深度图 (*http://vision.middlebury.edu/stereo/data/scenes*2003/)或者把(d 和 e)中的一个图像序列转化到(f)中的一个 3D 模型(*http://vision.middlebury.edu/mview/data/*)。(g)中的一个立体视觉分析绘图仪，可以生成(h)中的等高线图(Kenney Aerial Mapping 公司授权使用)

立体视觉匹配是这样的一个过程，通过寻找两幅或者多幅图像间的匹配像素点，然后将它们的 2D 位置转化到 3D 深度，从而估计出场景的一个 3D 模型。在第 6 章和第 7 章，我们描述了恢复摄像机位置和建立场景或目标的稀疏 3D 模型的方法。在本章中，我们将讨论如何建立一个更完整的 3D 模型，比如为输入图像中的像素赋予一个相对深度的深度图(depth map)，它可以是稀疏或稠密的。我们同时还将讨论关于多视图立体视觉(multi-view stereo)算法的主题，这些算法可以生成完整的基于 3D 体积或表面的物体模型。

为什么研究人员对立体视觉匹配感兴趣呢？从最早的关于视觉感知的研究中、研究人员发现我们是根据基于左眼和右眼之前看到的表观差异来感知深度的[①]。作为一个简单的实验，在你的眼睛前竖直举起你的手指，然后交替地闭上你的眼睛，你会注意到你的手指相对于场景的背景向左和向右跳动，同样的现象在图 11.1a 和 b 中也可以看到，在这两幅图像中，前景目标相对于背景向左和向右移动。

我们很快会知道，在简单的成像设置下(两个眼睛或者摄像机都直视前方)，水平上的运动或者视差(disparity)是与观测者的距离成反比的。尽管人们已经对场景结构产生视差所相关的基本物理和几何原理很清楚(11.1 节)，但是自动地通过建立稠密而精确的图像间对应(correspondence)来测量视差仍然是一个很有挑战性的工作。

最早的立体视觉匹配算法是从摄影测量学(photogrammetry)这个领域发展起来的，它的目标是从重叠航空影像中自动地构建出地形学上的高度图。在此之前，操作员要使用一种摄影测量学上的立体视觉绘图仪来绘制，这种绘图仪给每只眼睛显示这类图像的移动过的版本，并允许操作员在等高的轮廓周围浮动一个点状的光标(图 11.1g)。这个领域一个主要的进步是开发出了完全自动的立体视觉匹配算法，能使处理航空影像的速度更快、成本更低(Hannah 1974；Hsieh, McKeown, and Perlant 1992)。

在计算机视觉领域，关于立体视觉匹配的主题是研究最广和最基础的问题之一(Marr and Poggio 1976；Barnard and Fischler 1982；Dhond and Aggarwal 1989；Scharstein and Szeliski 2002；Brown, Burschka, and Hager 2003；Seitz, Curless, Diebel et al. 2006)，而且仍将是最活跃的研究方向之一。摄影测地学上的匹配主要关注航空影像，计算机视觉中的应用包括对人的视觉系统建模(Marr 1982)，机器人导航和操作(Moravec 1983；Konolige 1997；Thrun, Montemerlo, Dahlkamp et al. 2006)，以及视图插值和基于图像的渲染(图 11.2.a~d)，3D 模型建立(图 11.2e 和 f 和 h~j)以及在计算机中生成的影像中混合实景动作(图 11.2g)。

在本章中，我们首先按照 Scharstein and Szeliski(2002)提出的一般分类方法描述立体视觉背后的一些基本原理。我们先在 11.1 节中给出立体视觉图像匹配中的几何学(geometry)回顾，即给定一幅图像中的一个像素，怎样计算这个像素在另一

[①] 立体视觉(stereo)这个词最早来自于希腊语中"固体"(solid)这个词汇，立体视觉探讨的是我们如何感知固体形状的问题(Koenderink 1990)。

幅图像中可能的位置范围，即它的极线(epipolar line)。我们描述如何预卷绕图像使它们对应的极线是一致的(矫正，rectification)。我们同时给出一个称为"平面扫描"(plane sweep)的通用重采样算法，它可以用来对具有任意摄像机构型的多图像立体视觉进行匹配。

其次，我们要对兴趣点和边缘类特征的稀疏(sparse)立体视觉匹配方法做一个简要的回顾。然后我们开始本章的主要主题，也就是在视差图形式下来估计出像素间对应的一个稠密(dense)集合(图 11.1c)。具体内容涉及：首先选择一个像素匹配准则(11.3 节)，然后使用局部的基于区域的聚集(11.4 节)或者全局优化(11.5 节)来帮助消除潜在匹配中的歧义。在 11.6 节，我们讨论多视图立体视觉(multi-view stereo)方法，这类方法的目的是建立一个完整的 3D 模型，而不只是建立一个单独的视差图像(图 11.1d~f)。

图 11.2　立体视觉的应用：(a)输入图像；(b)计算出来的深度图；(c)根据多视图立体视觉生成的新的视图(Matthies, Kanade, and Szeliski 1989)© 1989 Springer；(d)在两幅图像之间进行视图变形(Seitz and Dyer 1996)© 1996 ACM；(e 和 f)3D 人脸建模(图像经由 Frédéric Devernay 授权)；(g)z-键控(z-keying)实景和计算机生成的影像(Kanade, Yoshida, Oda et al. 1996)© 1996 IEEE；(h~j)在虚拟化的现实中从多个视频流中建立 3D 表面模型(Kanade, Rander, and Narayanan 1997)

11.1 极线几何学

给定一幅图像中的一个像素,我们怎样计算它在另一幅图像中的对应呢?在第 8 章中,我们可以找到基于像素的局部表观以及基于相邻像素的运动的各种搜索方法,这些搜索方法可以用来进行像素匹配。然而,对于立体视觉匹配这个问题,我们有一些附加的可用信息,也就是在同一个静态场景中拍摄图像所用的摄像机的位置和标定数据(7.2 节)。

我们怎样利用这些信息来减少潜在匹配数目,从而同时加速匹配和提高稳定性呢?图 11.3a 显示了一幅图像中的一个像素 x_0 怎样映射到另一幅图像中的一个极线片段(epipolar line segment)的。这个片段的一个端点是以原始观察射线上的无穷远处的投影为界,另一个端点是以原始摄像机中心在第二个摄像机中的投影为界,这个投影也就是所谓的极点(epipole) e_1。如果我们将第二幅图像中的极线反向投影到第一幅图像中,我们可以得到另一个直线(片段),这一次是以另一个对应的极点 e_0 为界。将两个直线片段都扩展到无限远,我们就可以得到对应的极线对(图 11.3b),这个极线对是两个图像平面与穿过两个摄像机中心 c_0 和 c_1 以及兴趣点 p 的极平面的交线(Faugeras and Luong 2001;Hartley and Zisserman 2004)。

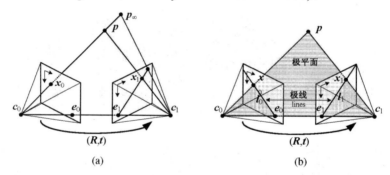

图 11.3 极线几何:(a)对应于一条光线的极线片段;(b)极线的对应集合及其极平面

11.1.1 矫正

正如我们在 7.2 节中所看到的那样,一对摄像机相对的姿态和标定隐含这对摄像机的极线几何,它可以通过七个或者更多对应点使用基本矩阵(或者通过五个或者更多点使用标定后的本质矩阵)轻松计算得到(Zhang 1998a,b;Faugeras and Luong 2001;Hartley and Zisserman 2004)。一旦计算得到这个几何关系,我们可以使用一幅图像中一个像素点对应的极线来限制在另一幅图像中对应像素的搜索。完成这个过程的一种途径是使用一个通用的对应算法,比如光流(8.4 节),但是只考虑在极线上的位置(或者将所有可能落到直线上的光流向量进行投影)。

更高效的一个算法可以这样得到,首先矫正(rectify,也就是卷绕)输入图像,使得对应的水平扫描线是极线(Loop and Zhang 1999；Faugeras and Luong 2001；Hartley and Zisserman 2004)，[1]然后，就能够独立地匹配水平的扫描线或者在计算匹配分数时水平移动图像(图 11.4)。

矫正两幅图像的一种简单方式是首先将两个摄像机旋转使它们看起来垂直于连接它们的摄像机中心 c_0 和 c_1 的直线，由于倾斜(tilt)有一个自由度，所以应该使用能够达到这个要求的最小旋转尺度。其次，为了确定所期望的光轴周围的扭曲，使向上向量(摄像机 y 坐标)和摄像机的中心线垂直。这样就确保了对应的极线对是水平的并且在无穷远处点的偏离是 0。最后，如果有必要，重新缩放图像以考虑不同的焦距长度，放大较小的图像以避免走样。(这个过程的完整细节可以参见文献 Fusiello, Trucco and Verri(2000)和习题 11.1。)注意在一般情况下，虽然旋转摄像机使它们都朝向相同的方向可以减少因缩放和平移所造成的摄像机间的像素移动，但是对于任意的图像集合，除非它们的光心共线，否则不可能对它们同时进行矫正的。

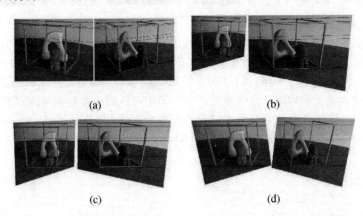

图 11.4 Loop and Zhang(1999)中的多阶段立体视觉矫正算法©1999 IEEE：(a)叠加了几个极线的初始图像对；(b)转化到使极线平行的图像；(c)矫正后使得极线平行并且在垂直上对应的图像；(d)使水平畸变最小的最终的矫正结果

前面得到的标准矫正几何(standard rectified geometry)用于大量立体视觉摄像机设置和立体视觉算法中，并且可以得到一个非常简单的 3D 深度 Z 和视差 d 之间的反向关系，

$$d = f\frac{B}{Z}, \tag{11.1}$$

其中 f 是焦距长度(用像素测量)，B 是基线，而

$$x' = x + d(x,y),\ y' = y \tag{11.2}$$

描述了左右图像对应像素间坐标的关系(Bolles, Baker, and Marimont 1987；Okutomi and Kanade 1993；Scharstein and Szeliski 2002)。[2]从一个图像集合中提取深度图这

---

[1] 尽管通过旋转摄像机可以对任意图像对进行矫正——只要这对图像不是过于处于边缘或者存在太大的尺度变化——但只有在摄像机彼此相邻时，这个做法才最有意义。对于存在大尺度变化的情况，使用平面扫描(下面将要讲到)或者对 3D 中小的平面块位置做假设(Goesele, Snavely, Curless et al. 2007)可能更可取。

[2] 术语"视差"(disparity)最早是在人类视觉文献中被引入的，它用来描述左右眼看到的对应特征的位置差异(Marr 1982)。水平视差是研究得最多的现象，但是当两个眼睛处于边缘时，竖直视差也有可能出现。

个任务就变成了一个估计视差图 $d(x,y)$ 的问题。

在矫正之后，我们可以非常容易地比较位置 (x,y) 和 $(x',y')=(x+d,y)$ 的对应像素的相似度并将它们保存在一个视差空间图像(disparity space image，DSI) $C(x,y,d)$ 供进一步处理(图 11.5)。"视差空间"这个概念可以追溯到立体视觉匹配的早期工作(Marr and Poggio 1976)，而视差空间图像(体)通常和文献 Yang, Yuille and Lu(1993) 和 Intille and Bobick(1994) 联系在一起。

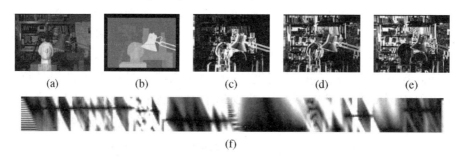

图 11.5 一个典型视差空间图像(DSI)的切片(Scharstein and Szeliski 2002)© 2002 Springer：(a)原始彩色图像；(b)视差图真值；(c～e)对于 $d=10,16,21$ 的三个 (x,y) 切片；(f)对于 $y=151$ 的一个 (x,d) 切片((b)中的虚线)。在(c～e)中各种深色(匹配)区域都可见，比如书架、桌子和罐以及头像雕塑，在图(f)中可以明显看到表现为水平直线的三个不同视差水平。这些 DIS 中深色的条纹表示在这个视差上匹配的区域。(小的深色区域通常是由于无明显纹理结构区域所造成的)Bobick and Intille(1999)讨论了更多 DSI 的例子

11.1.2 平面扫描

在图像匹配之前进行预矫正的另一种方式是扫描场景中的一组平面并测量不同图像重新映射到这些平面上的影像一致性(photoconsistency)，如图 11.6 所示。这个过程一般称为"平面扫描"(plane sweep)算法(Collins 1996；Szeliski and Golland 1999；Saito and Kanade 1999)。

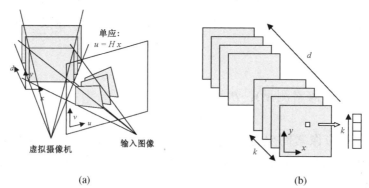

图 11.6 在场景中扫描一组平面(Szeliski and Golland 1999)© 1999 Springer：(a)从一个虚拟摄像机观察的一组平面在其他任一个源(输入)摄像机图像中产生了一组单应；(b)从其他所有摄像机卷绕的图像可以存储到一个推广的视差空间体中 $\tilde{I}(x,y,d,k)$，这个视差空间体是以像素位置 (x,y)，视差 d 和摄像机 k 来索引的

与我们在 2.1.5 节所介绍的投影深度(也称"平面加上视差"，plane plus parallx)(Kumar, Anandan, and Hanna 1994；Sawhney 1994；Szeliski and Coughlan，1997)一样，一个满秩的 4×4 投影矩阵 $\tilde{\boldsymbol{P}}$ 的最后一行可以被设置到任何一个平面方程 $\boldsymbol{p}_3 = s_3[\hat{n}_0 | c_0]$ 中，得到的四维投影变换(直射映射)(2.68)将 3D 世界中的点 $\boldsymbol{p} = (X,Y,Z,1)$ 映射到屏幕坐标系 $x_s = (x_s, y_s, 1, d)$，其中投影深度(projective depth)，或者视差(parallax) d (2.66)在参考平面处值为 0(图 2.11)。

通过一系列视差假设扫描 d，如图 11.6a 所示，对应着将每一个输入图像映射到虚拟摄像机 $\tilde{\boldsymbol{P}}$，它通过一系列单应定义了视差空间(2.68~2.71)，

$$\tilde{\boldsymbol{x}}_k \sim \tilde{\boldsymbol{P}}_k \tilde{\boldsymbol{P}}^{-1} \boldsymbol{x}_s = \tilde{\boldsymbol{H}}_k \tilde{\boldsymbol{x}} + \boldsymbol{t}_k d = (\tilde{\boldsymbol{H}}_k + \boldsymbol{t}_k [0\ 0\ d])\tilde{\boldsymbol{x}}, \tag{11.3}$$

如图 2.12b 所示，其中 $\hat{\boldsymbol{x}}_k$ 和 $\hat{\boldsymbol{x}}$ 分别是源图像和虚拟(参考)图像中的单应像素坐标(Szeliski and Golland 1999)。单应集合 $\tilde{\boldsymbol{H}}_k(d) = \tilde{\boldsymbol{H}}_k + \boldsymbol{t}_k [0\ 0\ d]$ 的成员是通过相加一个秩为 1 的矩阵来参数化的，并且每一个通过一个平面同源映射(planar homology)来相互关联(Hartley and Zisserman 2004, A5.2)。

虚拟摄像机和参数化的选择是依赖于具体的应用，从而使这个框架有很大的灵活性。在很多应用之中，因为会使用其中一个输入摄像机(作为参考摄像机)，因此计算出的深度图是注册在其中一个输入图像上的，这在后面可以用于基于图像的渲染(13.1 节和 13.2 节)。在其他的一些应用中，比如用于视频会议中视点修正的视图插值(11.4.2 节)(Ott, Lewis, and Cox 1993；Criminisi, Shotton, Blake et al. 2003)，将一个摄像机置于两个输入摄像机之间的中心通常更合适，因为它为产生虚拟的中间图像提供了所需要的逐像素(per-pixel)视差。

如何选择视差采样，即零视差平面的设置和整数视差的缩放，也依赖于具体应用，通常被设置为包括兴趣范围，即工作容积(working volume)，且同时缩放视差来对图像在像素(亚像素)移动(shift)上采样。例如，将立体视觉用于机器人导航中的障碍规避时，将视差设置成在地面上逐像素测量的高度是最方便的(Ivanchenko, Shen, and Coughlan 2009)。

当每幅图像被卷绕到以视差 d 进行参数化的当前平面时，它就可以被堆积到一个广义视差空间图像(generalized disparity space image) $\tilde{I}(x,y,d,k)$ 供进一步处理(图 11.6b)(Szeliski and Golland 1999)。在大多数立体视觉算法中，都要计算相对于参考图像 I_r 的影像一致性(比如，平方或者鲁棒差值的和)，并将它保存在 DSI 中

$$C(x,y,d) = \sum_k \rho(\tilde{I}(x,y,d,k) - I_r(x,y)). \tag{11.4}$$

然而，也可以计算一些其他可选的统计值，比如鲁棒方差、焦距或者熵(11.3.1 节)(Vaish, Szeliski, Zitnick et al. 2006)，或者使用这个表示来推断遮挡的原因(Szeliski and Golland 1999；Kang and Szeliski 2004)。到 11.6 节中我们讲多视图立体视觉这个主题时，广义 DSI 就会变得非常容易理解了。

当然，平面并不是能够用来在兴趣空间定义一个 3D 扫描的唯一表面。圆柱形表面，特别是与全景摄影结合在一起时(第 9 章)，也很常用(Ishiguro, Yamamoto, and Tsuji 1992；Kang and Szeliski 1997；Shum and Szeliski 1999；Li, Shum, Tang et al. 2004；Zheng, Kang, Cohen et al. 2007)。也可以定义其他流形拓扑，例如，摄像机绕着一个固定轴旋转的情况(Seitz 2001)。

一旦计算出 DSI，在大多数立体视觉匹配算中的下一步便是在视差空间 $d(x, y)$ 中生成一个单值函数来最恰当地描述场景中表面的形状。这可以看作是在视差空间图像中寻找一个具备某项最优特性的嵌入平面，比如最小代价或者最佳(分段)光滑性。图 11.5 显示了一个典型 DSI 的切片示例。在 Bobick and Intille(1999)论文中可以找到更多的这类图像。

11.2 稀疏对应

早期的立体视觉匹配算法是基于特征的(feature-based)，即它们首先使用兴趣算子或者边缘检测器来提取潜在能够匹配的图像位置集合，然后使用基于块的度量在其他图像中搜索对应的位置(Hannah 1974；Marr and Poggio 1979；Mayhew and Frisby 1980；Baker and Binford 1981；Arnold 1983；Grimson 1985；Ohta and Kanade 1985；Bolles, Baker, and Marimont 1987；Matthies, Kanade, and Szeliski 1989；Hsieh, McKeown, and Perlant 1992；Bolles, Baker, and Hannah 1993)。局限于稀疏对应的部分原因是由于计算资源的限制，但也是由于希望立体视觉算法产生的匹配响应具有高确定性。在一些应用中还希望在可能存在很大光照差异的情况下进行匹配，其中可能只有边缘才是唯一稳定的特征(Collins 1996)。这样得到的稀疏 3D 重建可以使用 3.7.1 节和 2.3.1 节讨论的表面拟合算法来进行进一步处理。

这个方向最新的工作主要关注于如何提取出高稳定性的特征并使用它们作为种子以生长出更多的匹配(Zhang and Shan 2000；Lhuillier and Quan 2002)。类似的方法也已经被扩展到宽基线多视图立体视觉问题中，并与 3D 表面重建(Lhuillier and Quan 2005；Strecha, Tuytelaars, and Van Gool 2003；Goesele, Snavely, Curless et al. 2007)或者自由空间推理(Taylor 2003)结合在一起，这些方法在 11.6 节中有更详细的讨论。

3D 曲线与轮廓

稀疏对应的另一个例子是匹配轮廓曲线(profile curve)——或者遮挡轮廓(occluding contour)——它主要发生在物体的边界处(图 11.7)和内部的自遮挡处，在这些位置处表面向远离摄像机的视角方向弯曲。

匹配轮廓曲线的困难在于轮廓曲线的位置一般来说是关于摄像机视角变化的一个函数。因此，在两幅图像中直接匹配曲线然后再对这些匹配做三角剖分会导致错误的形状测量。幸运的是，如果可以获取三帧或者更多的在空间上邻近的图像，就能够拟合出对应边界基元位置的一个局部圆弧(图 11.7a)，因此就可以从这些匹配中得到半稠密的曲面网格(图 11.7c 和图 11.7g)。匹配这些曲线的另一个好处是,只要背景和前景的颜色存在明显的差异,它们就可以用来重建无纹理表面的表面形状。

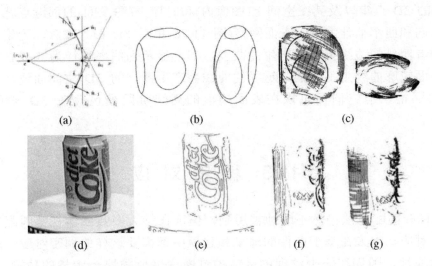

图 11.7 从遮挡轮廓中重建表面(Szeliski and Weiss 1998)©1998 Springer:(a)极平面中拟合的圆弧;(b)带有截断侧面和椭圆形表面标记的椭球体的合成样例;(c)部分重建的表面网格的倾斜视图和俯视图;(d)一个旋转台上的饮料罐的真实世界图像序列;(e)提取的边缘;(f)部分重建的轮廓曲线;(g)部分重建的表面网格(这里只显示了部分重建以免图像显得杂乱)

多年以来,产生了很多根据轮廓线曲线来重建表面形状的方法(Giblin and Weiss 1987;Cipolla and Blake 1992;Vaillant and Faugeras 1992;Zheng 1994;Boyer and Berger 1997;Szeliski and Weiss 1998)。Cipolla and Giblin(2000)描述了很多这样的方法,以及一些相关的主题(比如根据轮廓曲线序列推断摄像机运动)。在下面,我们总结了 Szeliski and Weiss(1998)提出的方法,它假设一个离散的图像集合,而不是将问题形式化在一个连续微分框架中。

让我们假设摄像机的运动足够平滑从而局部的极线几何变化很慢,即连续的摄像机中心所产生的极平面和所考察的边界基元几乎是共面的。整个处理流程的第一步是在每一幅输入图像中提取和连接边缘(图 11.7b 和图 11.7e),然后,使用逐对极线几何、邻近性和(可选)表观来匹配图像序列中的边界基元。这提供了一个在时空体中相连的边缘集合,它有时也称为"交织墙"(weaving wall)(Baker 1989)。

为了重建出每一个独立边界基元的 3D 位置以及它的局部平面内法向和曲率,我们将对应于其相邻边界基元的观察射线投影到由摄像机中心、观察射线和摄像机速度所定义的即时极平面上,如图 11.7a 所示。我们然后根据投影曲线拟合出一个密切圆(osculating circle),根据圆的中心点 $c = (x_c, y_c)$ 和半径 r 对它进行参数化,

$$c_i x_c + s_i y_c + r = d_i, \tag{11.5}$$

其中 $c_i = \hat{\boldsymbol{t}}_i \cdot \hat{\boldsymbol{t}}_0$ 和 $s_i = -\hat{\boldsymbol{t}}_i \cdot \hat{\boldsymbol{n}}_0$ 分别是观察射线 i 和中心观察射线 0 之间的角度的余弦和正弦,$d_i = (\boldsymbol{q}_i - \boldsymbol{q}_0) \cdot \hat{\boldsymbol{n}}_0$ 是观察射线 i 和局部原点 $\boldsymbol{q}_0$ 间的垂直距离,这个局部原点是

选择在中心观察射线与直线的交点接近的点(Szeliski and Weiss 1998)。得到的线性方程组可以使用最小二乘来求解,解的好坏(残差)可以用来检查错误的对应。

得到的 3D 点集以及其在空间上(图像内)和时序上(图像间)的邻近点就形成了包含局部法向和曲率估计的 3D 表面网格(图 11.7c 和图 11.7g)。注意,无论是由于表面标记或者明显的边缘折痕所产生的曲线,都会显示为曲率半径为 0 或者很小。这与连续表面轮廓曲线是不同的。这样的曲线会产生孤立的 3D 空间曲线,而不是光滑的表面网格,但它们仍可以在表面插值的后续阶段被包含在 3D 表面模型中(12.3.1 节)。

11.3 稠密对应

尽管稀疏匹配算法仍然不时地被用到,但现在绝大多数立体视觉匹配算法都集中于稠密对应,因为在基于图像的渲染和建模中都需要这样的稠密对应。这个问题更具有挑战性,因为在缺少纹理的区域中推断深度值需要一定量的猜测。(试想从一个栅栏中看一个有颜色的固体背景,它的深度是什么呢?)

在本节中,我们回顾了由 Scharstein and Szeliski(2002)首先提出来的关于稠密立体视觉匹配的分门别类方法。这个分类中包含了一个算法的"构建模块"的集合,根据这些模块,可以构建出非常多的算法。它是基于这么一个事实,立体视觉匹配算法通常都要进行如下四个步骤中的部分操作:

1. 匹配代价(cost)计算;
2. 代价(支持)聚集(aggregation);
3. 视差计算和优化;
4. 视差求精(refinement)。

例如,对于局部(基于窗口的)算法(11.4 节),它在计算一个给定点的视差时仅仅依赖于局部有限窗口内的亮度值,通常都隐含假设聚集支持具备平滑性。这些算法中的一些可以非常清楚地分成步骤 1、步骤 2、步骤 3。例如,传统的误差平方和(SSD)算法可以如下描述:

1. 匹配代价是在给定视差上的亮度值的误差平方;
2. 通过将视差为常量的方形窗口内所有匹配代价求和来进行聚集;
3. 在每一个像素选择最小的(获胜的)聚集值来计算视差。

一些局部算法,将步骤 1 和 2 结合起来并使用基于支持区域的匹配代价,比如规范化的交叉验证(Hannah 1974;Bolles, Baker, and Hannah 1993)和秩变换(Zabih and Woodfill 1994)以及其他的顺序度量(Bhat and Nayar 1998)。(这也可以被看成是一个预处理步骤;参考 11.3.1 节)

另一方面,全局算法通常明确地做出平滑性假设,然后求解一个全局优化问题(11.5 节)。这些算法通常都不进行聚集操作这一步,而是求取一个视差赋值(步骤 3)

来最小化一个包含数据(步骤 1)项和平滑项的全局代价函数。这些算法间的主要区别在于所使用的最小化过程，例如，模拟退火(Marroquin, Mitter, and Poggio 1987; Barnard 1989)、概率(均值域)传播(Scharstein and Szeliski 1998)、期望最大化(EM)(Birchfield, Natarajan, and Tomasi 2007)、图割(Boykov, Veksler, and Zabih 2001)或者带环置信度传播算法(Sun, Zheng, and Shum 2003)。

在这两大类方法之间有一些迭代算法并没有明确指定一个全局函数来最小化，但实际上它们的运行情况非常类似地模拟了一些迭代优化算法(Marr and Poggio 1976; Zitnick and Kanade 2000)。分层的(由粗到精的)算法和这些迭代算法类似，但是它们通常在图像金字塔上进行操作，粗层次的结果用来限制精细层在更局部的位置进行搜索(Witkin, Terzopoulos, and Kass 1987; Quam 1984; Bergen, Anandan, Hanna et al. 1992)。

相似性度量

任何一个稠密立体视觉匹配算法的第一个组件是一个相似性度量，用来比较像素值从而确定它们在对应中有多像，在本节中，我们简要回顾一下 8.1 节中介绍的相似性度量，然后介绍其他的一些专门为立体视觉匹配而设计的相似性度量(Scharstein and Szeliski 2002; Hirschmüller and Scharstein 2009)。

最常见的基于像素的匹配代价包括亮度平方差(squared intensity differences，SSD)和(Hannah 1974)以及亮度绝对值差(absolute intensity difference)(Kanade 1994)。在视频处理领域，这些匹配准则称为"均方误差"(mean-squared error, MSE)和平均绝对差(mean absolute difference, MAD)度量；术语"平移""帧差分"(Displaced frame difference, DFD)也很常用(Tekalp 1995)。

最近，研究人员提出了鲁棒度量(8.2)，它包括了截断二次多项式和受污染的(contaminated)高斯函数。这些度量由于限制了错误匹配在聚集过程中的影响，因而非常有用。Vaish, Szeliski, Zitnick et al.(2006)比较了很多种这类鲁棒度量，其中包括一种新的度量，它基于在特定视差假设处的像素值的熵，对多视图立体视觉尤其有用。

其他传统匹配代价包括归一化交叉验证(8.11)(Hannah 1974; Bolles, Baker, and Hannah 1993; Evangelidis and Psarakis 2008)，它与误差平方和(SSD)效果类似，还包括二值匹配代价(即匹配或者未匹配)(Marr and Poggio 1976)，它基于二值特征，例如边缘(Baker and Binford 1981; Grimson 1985)或者拉普拉斯函数的符号(Nishihara 1984)，由于它们的区分性很弱，所以简单的二值匹配代价现在已经不再用于稠密立体视觉匹配。

一些匹配代价对于摄像机的增益和偏差差异不敏感，比如基于梯度的度量(Seitz 1989; Scharstein 1994)，相位和滤波器组响应(Marr and Poggio 1979; Kass 1988; Jenkin, Jepson, and Tsotsos 1991; Jones and Malik 1992)，常规或鲁棒(双边滤

波)均值移除滤波(Ansar, Castano, and Matthies 2004；Hirschmüller and Scharstein 2009)，稠密特征描述子(Tola, Lepetit, and Fua 2010)，以及非参数度量如秩和统计(census)变换(Zabih and Woodfill 1994)、顺序度量(Bhat and Nayar 1998)或者熵(Zitnick, Kang, Uyttendaele et al. 2004；Zitnick and Kang 2007)。其中统计度量是将一个移动窗口内的每一个像素转化成一位向量，这个向量表示那些邻居高于或低于中心像素，Hirschmüller and Scharstein(2009)发现它对于大尺度、非静态曝光和光照变化非常鲁棒。

同样，通过执行一个预处理或迭代求精过程，也可以纠正有差异的全局摄像机属性，这个过程使用全局回归(Gennert 1988)、直方图均衡化(Cox, Roy, and Hingorani 1995)或者互信息(Kim, Kolmogorov, and Zabih 2003；Hirschmüller 2008)来估计图像间偏差-增益方差。另外，一种局部平滑变化补偿场方法也被提出来(Strecha, Tuytelaars, and Van Gool 2003；Zhang, McMillan, and Yu 2006)。

为了弥补采样中的一些问题，即在高频区域中像素值的差异非常明显，Birchfield and Tomasi(1998)提出了一种对于图像采样中的移动不是那么敏感的匹配代价。他们将一幅参考图像中的每一个像素与其他图像的一个线性插值函数进行比较，这与仅仅通过积分比较移动的像素值不同(因为可能会漏掉合理的匹配)。(Szeliski and Scharstein 2004；Hirschmüller and Scharstein 2009)讨论了更多的细节和其他的一些匹配代价。特别地，如果你觉得你所匹配的图像中存在明显的曝光和表观变化，就应该使用在 Hirschmüller and Scharstein(2009)评测的鲁棒度量中的一些效果很好的度量，比如可以使用统计变换(Zabih and Woodfill 1994)、顺序度量(Bhat and Nayar 1998)、双边差分(Ansar, Castano, and Matthies 2004)或者分层互信息(Hirschmüller 2008)。

11.4 局 部 方 法

局部和基于窗口的方法是通过在 DSI $C(x,y,d)$ 上的一个支持区域(support region)求取代价和或者平均值来聚集代价的。[①]一个支持区域可以是一个固定视差上的二维区域(这对正向平行表面很有用)，也可以是 $x-y-d$ 空间上的三维区域(支持倾斜的表面)。两维的证据聚集已经通过使用如下方法被实现，包括平方窗口或高斯卷积(传统的)、固定在不同点的多窗口，即可移动的窗口(Arnold 1983；Fusiello, Roberto, and Trucco 1997；Bobick and Intille 1999)、自适应大小窗口(Okutomi and Kanade 1992；Kanade and Okutomi 1994；Kang, Szeliski, and Chai 2001；Veksler 2001, 2003)，基于常量视差的连通域的窗口(Boykov, Veksler, and Zabih 1998)或者是基于彩色的分割结果(Yoon and Kweon 2006；Tombari, Mattoccia, Di Stefano et al.

---

① 关于这些方法最新的两个综述和比较，请参考 Gong, Yang, Wang et al.(2007)和 Tombari, Mattoccia, Di Stefano et al.(2008)的研究成果。

2008)。对于三维支持函数已经提出了下面的一些方法,包括受限视差差异(Grimson 1985)、受限视差梯度(Pollard, Mayhew, and Frisby 1985)、Prazdny 的一致性准则(Prazdny 1985)以及 Zitnick and Kanade(2000)的最新工作(其中包括可见性和遮挡推理)。

可以使用 2D 或者 3D 卷积来进行固定支持区域上的聚集:

$$C(x,y,d) = w(x,y,d) * C_0(x,y,d),\qquad(11.6)$$

或者,针对矩形窗口的情况,使用高效的移动平均箱式滤波(3.2.2 节)(Kanade, Yoshida, Oda *et al.* 1996;Kimura, Shinbo, Yamaguchi *et al.* 1999)。可移动窗口也可以使用独立的滑动最小滤波来高效地实现(图 11.8)(Scharstein and Szeliski 2002, 4.2 节)。通过首先计算一个求和区域表,可以更高效地来完成不同形状和大小的窗口选择问题(3.2.3 节,3.30–3.32)(Veksler 2003)。选择合适的窗口很重要,因为窗口必须大足以包含充分的纹理,同时又要足够小从而使其中区域的深度连贯(图 11.9)。另外一种可选的聚集方法是迭代扩散(iterative diffusion),即不断地向每一个像素的代价中加入其邻居的代价(Szeliski and Hinton 1985;Shah 1993;Scharstein and Szeliski 1998)。

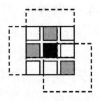

图 11.8 可移动窗口(Scharstein and Szeliski 2002)© 2002 Springer:尝试黑色像素周围所有的 3×3 移动窗口的效果与求取相同邻居的所有居中(未移动的)窗口的最小值的效果是一样的(为清晰起见,这里只显示了相邻移动窗口中的三个)

Gong, Yang, Wang *et al.*(2007)和 Tombari, Mattoccia, Di Stefano *et al.*(2008)比较的所有局部聚集方法中,Veksler(2003)的快速可变窗口方法和 Yoon and Kweon(2006)提出的局部加权方法在平衡性能和速度两方面具有突出的效果。[①] 其中需要特别指出的是局部加权方法,因为它没有使用均匀权重的矩形窗口,一个聚集窗口内的每一个像素根据它的颜色相似度和空间距离来影响最终的匹配代价,这与双线性滤波器(图 11.9c)是一样的。(事实上,除了其聚集过程是同时在参考图像和目标图像上对称地进行操作外,它们的聚集过程很类似于一个颜色/视差图上的联合双线性滤波。)Tombari, Mattoccia, and Di Stefano(2007)的基于分割的聚集方法效果更好,但目前还没有它的快速实现。

在局部方法中,问题的关键是在匹配代价计算和代价聚集这两步。而计算最终视差则不是很难:在每一处简单地选择代价最小的视差即可。因此,这些方法都在每一个像素使用一个局部的 "赢者通吃"(winner-take-all,WTA)优化策略。这种

[①] *http://www.vision.deis.unibo.it/spe/SPEHome.aspx* 有 Tombari, Mattoccia, DiStefano *et al.* (2008)论文更新和更多的结果。

方法(以及其他的很多对应方法)的一个限制是只在一个图像(参考图像)上施加了匹配唯一性，而在其他图像上的匹配点可能有多个，除非使用交叉检验和后续空洞填充(Fua 1993；Hirschmüller and Scharstein 2009)。

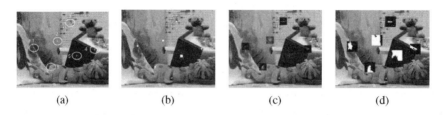

图 11.9 适应图像内容的聚集窗口大小和权重(Tombari, Mattoccia, Di Stefano et al. 2008)© 2008 IEEE：(a)选择评测点的原始图像；(b)可变窗口(Veksler 2003)；(c)自适应权重(Yoon and Kweon 2006)；(d)基于分割(Tombari, Mattoccia, and Di Stefano 2007)。注意自适应权重和基于分割的方法如何使其支持适应类似的彩色像素

11.4.1 亚像素估计与不确定性

大多数立体视觉匹配算法都在某个离散化空间上计算一个视差估计集合，例如整数视差(例外情况包括光流(Bergen, Anandan, Hanna et al. 1992)或样条(Szeliski and Coughlan 1997)之类的连续优化方法)。对于如机器人导航或者人体跟踪之类的应用，这些算法就足够了。然而对于基于图像的渲染，这样量化的视差图会导致非常不好的视图合成结果，即得到的场景好像是由许多剪切层组成的。为了消除这种情况，很多算法都在最初的离散对应阶段之后使用一个亚像素求精阶段。(另外一种备选方法是简单地在一开始就使用更多的离散视差级别(Szeliski and Scharstein 2004)。)

亚像素视差估计可以使用多种方式来计算，包括迭代梯度下降和在离散视差级别上对匹配代价进行曲线拟合(Ryan, Gray, and Hunt 1980；Lucas and Kanade 1981；Tian and Huhns 1986；Matthies, Kanade, and Szeliski 1989；Kanade and Okutomi 1994)。它提供了一种提高立体视觉算法分辨率的简单方法，同时却只用了很少的计算量，然而，为了达到好的效果，待匹配的亮度必须变化平缓，并且这些估计计算所在的区域必须在同一个(正确的)表面上。

最近，研究人员提出了关于从整数采样的匹配代价拟合相关曲线是否可行的一些问题(Shimizu and Okutomi 2001)。使用采样不敏感的相异性度量时，情况会变得更糟。Szeliski and Scharstein(2004)深入地探讨了这些问题。

除了亚像素计算以外，还有其他的一些方法可以用来对计算得到的视差进行后处理。使用交叉检验可以检测出遮挡的区域，即比较从左到右和从右到左的视差图(Fua 1993)。对于可疑的误匹配，可以使用一个中值滤波器来清除，而对于由于遮挡产生的空洞，可以使用表面拟合或者分配邻居视差估计来进行填充(Birchfield and Tomasi 1999；Scharstein 1999；Hirschmüller and Scharstein 2009)。

另外一种后处理方式是将置信度(confidence)和逐像素的深度估计(图 11.10)关联起来，这种方法对于后续的处理过程非常有用。它可以通过考察相关表面的曲率来完成，即查看 DSI 图像中在获胜视差处的最小值有多强。Matthies, Kanade and Szeliski(1989)指出在小噪声、图像进行了光度学标定和视差采样稠密的假设下，局部深度估计的方差可以按照如下方式估计

$$Var(d) = \frac{\sigma_I^2}{a}, \tag{11.7}$$

其中 a 是 DSI 的曲率，它是 d 的一个函数，可以使用局部抛物线拟合或者计算窗口内所有水平梯度的平方来度量，σ_I^2 是图像噪声的方差，可以通过最小 SSD 值来估计。(也请参考 6.1.4 节，(8.44)和附录 B.6。)

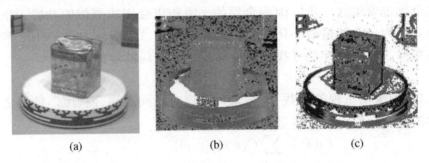

图 11.10 立体视觉深度估计中的不确定性(Szeliski 1991b)：(a)输入图像；(b)估计的深度图像(蓝色表示深度更近)；(c)估计的置信度(红色表示置信度更高)。从中可以看出，纹理更明显的区域拥有更高的置信度

11.4.2 应用：基于立体视觉的头部跟踪

实时立体视觉算法的一个常见应用是可以用来跟踪一个与计算机或者游戏系统交互的用户。与仅仅尝试使用单目颜色和亮度信息相比，使用立体视觉可以显著改进这类系统的稳定性(Darrell, Gordon, Harville et al. 2000)。一旦获得这个信息，就可以将它用到各种各样的应用之中，包括控制一个虚拟环境或者游戏，在视频会议中修正明显的注视方向以及背景替换。我们在下面讨论前两个应用，将背景替换的讨论放到 11.5.3 节中。

当一个用户注视计算机显示器上的一个 3D 物体或者场景时，使用头部跟踪来控制他的虚拟视角有时也叫鱼缸虚拟现实(fish tank virtual reality)，因为用户观察的 3D 世界就像是包含在一个鱼缸中一样(Ware, Arthur, and Booth 1993)。早期的这类视觉系统使用的是机械式头部跟踪和立体视觉眼镜。现在，这样的系统可以使用基于立体视觉的头部跟踪来控制，而立体视觉眼镜可以用自由立体显示器来替换。头部跟踪也可以用来构建一个"虚拟镜子"，其中用户的头部可以使用各种视觉效果来实时地修改(Darrell, Baker, Crow et al. 1997)。

立体视觉头部跟踪和 3D 重建的另一个应用是视角修正(Ott, Lewis, and Cox 1993)。当一个用户参加一个桌面视频会议或者视频聊天时，摄像机通常位于显示

器的顶部。由于人注视着屏幕的某个位置，这样看起来好像他们像是看着下面并远离了其他参与者，而不是直视着他们。将这个单摄像头换成两个或者多个使得能够构建一个虚拟的视图，这个虚拟视图正是他们看的位置，从而产生了虚拟的眼睛交流。实时立体视觉匹配可以用来构建一个精确的 3D 头部模型，视图插值(13.1 节)可以用来合成新的中间视图(Criminisi, Shotton, Blake et al. 2003)。

11.5 全局优化

全局立体视觉匹配方法在视差计算阶段之后会进行一些优化或迭代步骤，而且往往会完全跳过聚集这一步，因为全局的平滑约束会完成类似的功能。很多全局方法都被形式化为一个能量最小化框架，正如我们在 3.7 节(3.100～3.102)和 8.4 节中所看到的那样，它的目标是寻找一个能够最小化全局能量的解 d，

$$E(d) = E_d(d) + \lambda E_s(d). \tag{11.8}$$

数据项 $E_d(d)$ 用来衡量视差函数 d 与输入图像对之间的吻合程度。使用之前定义的视差空间图像，我们将这个能量函数定义为

$$E_d(d) = \sum_{(x,y)} C(x, y, d(x, y)), \tag{11.9}$$

其中 C 是(最初的或者聚集的)匹配代价 DSI。

平滑项 $E_s(d)$ 用来编码算法定义的平滑假设。为了使优化计算容易处理，平滑项通常被限定为只用来衡量相邻像素之间视差的差别，

$$E_s(d) = \sum_{(x,y)} \rho(d(x,y) - d(x+1,y)) + \rho(d(x,y) - d(x,y+1)), \tag{11.10}$$

其中 ρ 是某个单调递增的视差差别的函数。平滑项可以使用范围更大一些的邻居，比如八邻域 $\mathcal{N}_8$，这样可以得到更好的边界(Boykov and Kolmogorov 2003)，或者使用二阶平滑项(Woodford, Reid, Torr et al. 2008)，但是这样的项需要更复杂的优化方法。平滑性泛函的一个替代方法是使用如样条函数这样的一个较低维的表示(Szeliski and Coughlan 1997)。

在标准正规化中(3.7.1 节)，ρ 是一个二次函数，这使 d 对每一处都进行平滑，这会导致在物体边界处的结果很差。不会产生这种问题的能量函数称为"不连续性保持"(discontinuity-preserving)，它们基于鲁棒的 ρ 函数(Terzopoulos 1986b；Black and Rangarajan 1996)。在 Geman and Geman(1984)这篇重要的论文中给出了这种能量函数的一个贝叶斯解释，并且提出了一个基于马尔科夫随机场(MRF)和一个附加的线过程(line process)的不连续性保持能量函数，这个线过程是附加的二值变量，用来控制是否施加平滑性惩罚。Black and Rangarajan(1996)给出了如何使用鲁棒逐对视差项来替换独立的线过程变量。

同样，也可以使项 E_s 依赖于亮度差别，例如，

$$\rho_d(d(x,y) - d(x+1,y)) \cdot \rho_I(\|I(x,y) - I(x+1,y)\|), \tag{11.11}$$

其中 ρ_l 是某个关于亮度差别的单调递减函数，它在高亮度梯度处具有小的平滑性代价。这个想法(Gamble and Poggio 1987；Fua 1993；Bobick and Intille 1999；Boykov, Veksler, and Zabih 2001)使得视差不连续性与亮度或颜色边缘保持一致，而且看起来是使一些全局优化方法性能不错的主要原因。大多数研究人员采用基于探索的方式来设置这些参数，Scharstein and Pal(2007)给出了如何从真实的视差图中学习得到这样的条件随机场(3.7.2 节，(3.118))中的自由参数。

一旦定义了全局能量，就可以使用各种各样的算法来寻找(局部)最小值。这些方法将正规化和马尔科夫随机场与传统方法相结合，包括延续(Blake and Zisserman 1987)、模拟退火(Geman and Geman 1984；Marroquin, Mitter, and Poggio 1987；Barnard 1989)、最高置信度优先(Chou and Brown 1990)和均值域退火(Geiger and Girosi 1991)。

最近，研究人员提出了最大流(max-flow)和图割(graph cut)算法来解决一类特殊的全局优化问题，这些方法比模拟退火更有效，并且产生了很好的结果，研究人员还提出了一些基于带环置信度传播的方法(Sun, Zheng, and Shum2003；Tappen and Freeman 2003)。附录 B.5 及最近一篇关于 MRF 推断的综述论文(Szeliski, Zabih, Scharstein *et al*. 2008)更详细地讨论和比较了这些方法。

尽管全局优化方法目前得到了最好的立体视觉匹配结果，但还有其他一些可选的方法值得研究。

协作算法 受人类立体视觉的计算模型启发的协作算法，是最早提出的计算视差的方法之一(Dev 1974；Marr and Poggio 1976；Marroquin 1983；Szeliski and Hinton 1985；Zitnick and Kanade 2000)。这类算法使用非线性操作来迭代地更新视差估计，从而产生类似于全局优化算法的整体效果。事实上，对于这类算法中的一些，可以明确地声明一个被最小化的全局函数(Scharstein and Szeliski 1998)。

由粗到精和增量卷绕 目前最好的算法中，大多数都先在所有可能的视差上枚举所有可能的匹配，然后使用某种方式来选择最好的匹配集合。有时可以使用受经典(极小)光流计算启发的方法来获得更快的办法。在这里，图像被连续地卷绕，视差估计被增量地更新直到得到一个满意的注册。这些方法中，大多数通常使用一个由粗到精的分层求精框架来实现(Quam 1984；Bergen, Anandan, Hanna *et al*. 1992；Barron, Fleet, and Beauchemin 1994；Szeliski and Coughlan 1997)。

11.5.1 动态规划

另一类全局优化算法基于动态规划(dynamic programming)。尽管对于普通类别的平滑函数来说，可以证明公式(11.8)的 2D 优化是 NP-难题(Veksler 1999)，动态规划可以在多项式时间内找到独立扫描线的全局最小值。动态规划最早用于立体视觉是在稀疏的基于边缘的方法中(Baker and Binford 1981；Ohta and Kanade 1985)。最新的一些方法集中在稠密(基于亮度的)扫描线匹配问题(Belhumeur 1996；Geiger,

Ladendorf, and Yuille 1992；Cox, Hingorani, Rao et al. 1996；Bobick and Intille 1999；Birchfield and Tomasi 1999)。这些方法的原理是计算通过一个矩阵的最小代价路径，这个矩阵是由所有的成对匹配代价构成的，而匹配代价是指在两条对应的扫描线(也就是通过 DSI 的一个水平切片)之间的。通过分配一幅图像中的一组像素到另一幅图像中的一个像素，就可以明确地处理部分遮挡问题。图 11.11 用示意图显示了 DP 是如何工作的，而图 11.5f 则显示了应用 DP 得到的一个实际的 DSI 切片。

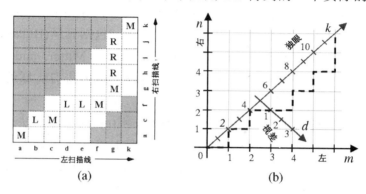

图 11.11 (a)Scharstein and Szeliski(2002)© 2002 Springer 和(b)Kolmogorov, Criminisi, Blake et al.(2006)© 2006 IEEE 中描述的使用动态规划的立体视觉匹配。对于每一对对应的扫描线，选择所有逐对匹配代价(DSI)构成的矩阵的最小代价路径。小写字母(a～k)用来符号化沿着每一个扫描线的强度。大写字母表示选择的通过矩阵的路径。用 M 表示匹配成功，而用 L 和 R 分别表示仅在左边或者右边图像可见的部分遮挡点(这些点有固定的代价)。通常只考虑一定的视差范围(图中的 0～4，用非实心矩形表示)。(a)中的表示允许对角移动而(b)中的表示不行。注意，这些图表使用独眼(cyclopean)表示深度，即相对于在两个输入摄像机之间的一个摄像机的深度，它们显示了一个"非倾斜"的通过 DSI 的 $x-d$ 切片

为了实现对于扫描线 y 的动态规划，2D 代价矩阵 $D(m,n)$ 中每一项的计算结合了它的 DSI 值

$$C'(m,n) = C(m+n, m-n, y) \tag{11.12}$$

和它的一个前驱代价值。使用图 11.11a 中给出的允许"对角"移动的表示，聚集的代价匹配可以按照下面的方式迭代地计算

$$\begin{aligned}
D(m,n,M) &= \min(D(m-1,n-1,M), D(m-1,n,L), D(m-1,n-1,R)) + C'(m,n)\\
D(m,n,L) &= \min(D(m-1,n-1,M), D(m-1,n,L)) + O\\
D(m,n,R) &= \min(D(m,n-1,M), D(m,n-1,R)) + O,
\end{aligned} \tag{11.13}$$

其中 O 是逐像素的遮挡代价。图 11.11b 对应的聚集规则是由 Kolmogorov, Criminisi, Blake et al.(2006)给出的，它也使用一个两阶段的背景-前景模型来进行双层分割。

动态规划存在的主要问题包括如何为遮挡像素选择合适的代价函数，另外如何在扫描线之间施加一致性也比较困难，尽管一些方法已经提出了处理后一种问题的方式(Ohta and Kanade 1985；Belhumeur 1996；Cox, Hingorani, Rao et al. 1996；

Bobick and Intille 1999；Birchfield and Tomasi 1999；Kolmogorov, Criminisi, Blake *et al*. 2006)。动态规划方法的另一个问题是它要求施加单调性(monotonicity)或有序性限制(ordering constraint)(Yuille and Poggio 1984)。这个限制要求像素在一条扫描线上的相对顺序在两个不同的视角之间保持一致，对于包含狭窄前景目标的实际场景来说，情况可能不是这样的。

Scharstein and Szeliski(2002)介绍了传统动态规划方法的一种替代算法，它忽略了(11.10)中的垂直平滑限制并使用全局能量方程(11.8)来简单地优化独立的扫描线，这可以使用一个递归算法，

$$D(x,y,d) = C(x,y,d) + \min_{d'}\{D(x-1,y,d') + \rho_d(d-d')\}. \tag{11.14}$$

这种扫描线优化算法(scanline optimization)的优点是它的计算表示与完整2D能量函数(11.8)相同且最小化的是同一个能量函数的简约版本。不幸的是，它也像动态规划那样，存在条纹现象。

一个更好的方法是从多个方向来衡量累计代价函数，比如，从八个顺序方向：N、E、W、S、NE、SE、SW 和 NW(Hirschmüller 2008)。得到的半-全局(semi-global)优化结果非常不错且能够相当高效地实现。

尽管动态规划和扫描线优化算法一般不能产生最精确的立体视觉重建，但与一些复杂的聚集策略结合使用，它们可以产生高效和高质量的结果。

11.5.2 基于分割的方法

尽管大多数立体视觉匹配算法是在逐像素基础上进行计算的，但是最新的一些方法首先将图像分割成区域，然后尝试为每一个区域标记一个视差。

例如，Tao, Sawhney, and Kumar(2001)先分割参考图像，再使用一种局部方法来估计逐像素的视差，然后再在每个片段(segment)内进行局部平面拟合，最后再对相邻的片段施加平滑性约束。Zitnick, Kang, Uyttendaele *et al*.(2004)和Zitnick and Kang(2007)使用了过分割来减轻初始化分割中的错误影响。在把每个片段的初始代价值集合存到视差空间分布(disparity space distribution，DSD)中之后，使用松弛迭代(或者使用 Zitnick and Kang(2007)最新工作中的带环置信度传播)来调整每个片段的视差估计，如图 11.12 所示。Taguchi, Wilburn, and Zitnick(2008)将片段形状求精作为优化过程的一部分，使得到的结果提高很多，如图 11.13 所示。

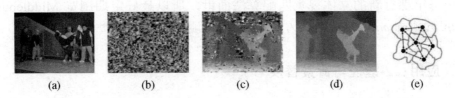

图 11.12　基于分割的立体视觉匹配(Zitnick, Kang, Uyttendaele *et al*. 2004)© 2004 ACM：(a)输入彩色图像；(b)基于颜色的分割；(c)初始视差估计；(d)最终分段平滑视差；(e)在视差空间分布中的片段上的 MRF 邻居定义(Zitnick and Kang 2007)© 2007 Springer

Klaus, Sormann, and Karner(2006)得到了更精确的结果,他们首先使用均值移位来分割参考图像,再使用一个小的(3×3)SAD 加上梯度 SAD(用交叉检验加权)来得到初始的视差估计,然后拟合局部平面,再重拟合全局平面,最后使用带环置信度传播的 MRF 来得到平面赋值。该算法于 2006 年被首次提出,当时是评测网址 *http://vision.middlebury.edu/stereo* 中排名最靠前的算法,在 2010 年初,它在新的测量数据集上的排名仍然最靠前。

目前排名第一的算法,是由 Wang and Zheng(2008)给出的,在分割上采用了类似的方法,然后做局部平面拟合,然后再对相邻平面的拟合参数做协作优化。Yang, Wang, Yang *et al.*(2009)是另外一个排名非常靠前的算法,使用 Yoon and Kweon(2006)的颜色相关方法和分层置信度传播来得到初始的视差估计集合,用左一右一致性检查来检测遮挡像素,然后使用基于分割的平面拟合以及一轮或多轮分层置信度传播来重新计算低置信度的数据项和遮挡的像素,从而得到最终的视差估计。

基于分割的立体视觉算法的另一个重要功能是能够在深度不连续区域提取像素的分数的透明度遮罩(Bleyer, Gelautz, Rother *et al.* 2009),这与使用明确层次(Baker, Szeliski, and Anandan 1998;Szeliski and Golland 1999)或者边界提取(Hasinoff, Kang, and Szeliski 2006)的算法相同。如图 11.13b 所示,对于试图创建没有紧贴边缘或者撕裂现象的虚拟视图插值(Zitnick, Kang, Uyttendaele *et al.* 2004)和无缝插入虚拟物体(Taguchi, Wilburn, and Zitnick 2008)的任务,这个能力非常关键。

图 11.13 使用自适应过分割和抠图的立体视觉匹配(Taguchi, Wilburn, and Zitnick 2008)© 2008 IEEE:(a)分割边界在优化过程中被求精,这样能够得到更精确的结果(例如,底部行中的薄的绿叶);(b)在片段边界抽取的透明度遮罩,得到视觉效果更好的合成结果(中间列)

由于每年都有新的立体视觉匹配算法面世,所以最好定期浏览 Middlebury 评测网址 *http://vision.middlebury.edu/*,得到最新评测算法列表。

11.5.3 应用:z-键控与背景替换

实时立体视觉的另一个应用是 z-键控,即使用深度信息将一个前景行动者(actor)从背景中分割出来,通常的目的是将背景换成计算机生成的图像,如图 11.2g 所示。

最初,z-键控系统需要昂贵的定制设备来实时产生所需要的深度图,因此其应用就被限定于广播演播室应用(Kanade, Yoshida, Oda et al. 1996; Iddan and Yahav 2001)。人们也开发出了从视频流中估计 3D 多视角几何(13.5.4 节)的离线系统(Kanade, Rander, and Narayanan 1997; Carranza, Theobalt, Magnor et al. 2003; Zitnick, Kang, Uyttendaele et al. 2004; Vedula, Baker, and Kanade 2005)。高精度实时立体视觉匹配的最新发展使 Z-键控有可能在普通台式机上进行，从而使图 11.14 所示的桌面视频会议应用成为可能(Kolmogorov, Criminisi, Blake et al. 2006)。

图 11.14　使用 z-键控和双层分割算法的背景替换(Kolmogorov, Criminisi, Blake et al. 2006)© 2006 IEEE

11.6　多视图立体视觉

尽管匹配一对图像是获得深度信息的一种有用途径，但是匹配多幅图像可以得到更好的结果。在本节中，我们不仅要回顾建立完整 3D 物体模型的方法，还将讨论一些使用多幅源图像来改进深度图质量的简单方法。

正如我们在平面扫描(11.1.2 节)中所讨论的那样，可以在每一个视差假设 d 处将所有的 k 个相邻图像采样到一个广义视差空间体 $\tilde{I}(x,y,d,k)$ 上。利用这些附加图像的一种最简单的方式是像公式(11.4)那样将它们相对于参考图像 I_r 的差值加起来，

$$C(x,y,d) = \sum_k \rho(\tilde{I}(x,y,d,k) - I_r(x,y)). \tag{11.15}$$

这就是众所周知的误差平方和的和(SSSD)和 SSAD 的基础(Okutomi and Kanade 1993; Kang, Webb, Zitnick et al. 1995)，它可以被扩展用来推导遮挡的可能模式(Nakamura, Matsuura, Satoh et al. 1996)。Gallup, Frahm, Mordohai et al.(2008)的最新工作成果给出了如何使基线适应期望深度，从而得到几何精度(宽基线)和遮挡鲁棒性(窄基线)之间最好的折中。另一些可替代的多视图代价度量包括合成焦点清晰度和像素颜色分布的熵等(Vaish, Szeliski, Zitnick et al. 2006)。

检查将所有图像中对应扫描线堆叠起来所得到的极平面图像(epipolar plane image，EPI)，是可视化多帧立体视觉估计问题的一个有用方式，如图 8.13c 和图 11.15 所示(Bolles, Baker, and Marimont 1987; Baker and Bolles 1989; Baker 1989)。从图 11.15 中可以看出，随着摄像机水平平移(以一种标准的水平矫正后的几何形式)，位于不

同深度的物体以与其深度(11.6)成反比的速度横向运动。① 前景目标遮住背景目标可以看作是一个 EPI-条带(EPI-strip)遮住了 EPI 上的另一个条带。如果给我们足够稠密的图像集合,我们便可以找到这样的条带并且推导它们之间的关系从而同时重建 3D 场景和推断半透明物体(Tsin, Kang, and Szeliski 2006)和镜面反射(Swaminathan, Kang, Szeliski et al. 2002;Criminisi, Kang, Swaminathan et al. 2005)。作为另一种选择,我们还可以将这些图像序列看作是顺序的观测并使用卡尔曼滤波器(Matthies, Kanade, and Szeliski 1989)或者最大似然推断(Cox 1994)来合并它们。

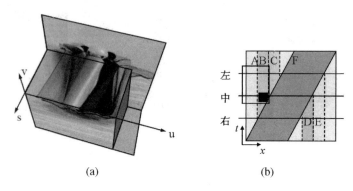

图 11.15 极面图像(EPI)(Gortler, Grzeszczuk, Szeliski et al. 1996)©1996 ACM 和一个 EPI 原理图(Kang, Szeliski, and Chai 2001)© 2001 IEEE。(a)照度图(光场)(13.3 节)是由通过一个体空间的所有光线构成的 4D 空间。从它取一个 2D 切片即可得到包含在一个平面上的所有光线,这与从一个堆积的 EPI 体上得到的扫描线是等同的。在不同深度的物体以与其深度的逆成正比的速度横向运动。遮挡(和半透明)效果可以从这种表示中轻松观察到。(b)对应于图 11.16 的 EPI,它显示了作为通过 EPI 体切片的 3 幅图像(中、左、右)。黑色像素周围的在空间上和时间上移动的窗口用矩形框表示,显示表明右边的图像没有用来进行匹配

当只有少量图像可用时,就很有必要借助于聚集方法(如滑动窗口或者全局优化)。而有附加可用的图像时,遮挡的可能性又会同时增大。因此,同时使用如图 11.16a 所示的可移动窗口方式来调整最佳的窗口位置以及选择性地挑选相邻图像子集来降低图像中的兴趣区域遮挡的程度(如图 11.16b 所示(Kang, Szeliski, and Chai 2001)),就变得很重要。图 11.15b 显示了这样的窗口时空选择或移动是如何对应于在极平面图像体中选择最有可能的无遮挡体区域的。

图 11.17 显示了对多帧花园图像序列应用这些方法得到的结果,并与使用常规(非移动)的窗口 SSSD 方法、空间上移动的窗口和完全时空窗口选择方法进行了比较。(将立体视觉应用于用移动摄像机拍摄的刚性场景也叫"运动立体视觉"(motion stereo))。(Kang and Szeliski 2004)报告了类似的使用时空选择所获得的改进

---

① EPI 的四维推广是我们在 13.3 节中研究的光场(light field)。原理上,虽然光场中有足够的信息可以用来同时恢复物体的形状和 BRDF(Soatto, Yezzi, and Jin 2003),但目前这个课题还没有什么进展。

结果，结合使用局部测量和全局优化时，改进效果更明显。

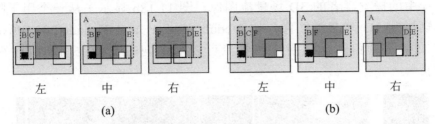

图 11.16 时空可移动窗口(Kang, Szeliski, and Chai 2001)© 2001 IEEE。一个简单的三帧图像序列(中间的图像是参考图像)，它有一个移动的正面灰色方块(用 F 标记)和一个固定的背景。区域 B、C、D 和 E 被部分遮挡：(a)常规 SSD 算法在匹配这些区域的像素(比如以区域 B 中的黑色像素为中心的窗口)和跨越不同深度区域的窗口(区域 F 中的白色像素为中心的窗口)时会产生错误；(b)在部分遮挡的区域和深度不连续的区域附近，可移动窗口有助于缓解这个问题。以区域 F 中的白色像素为中心的可移动窗口在所有图像中都匹配正确。以区域 B 中的黑色像素为中心的可移动窗口在左图中匹配成功，但是需要时序选择来使右图匹配失效。图 11.15b 显示了对应于这个序列的一个 EPI，它更详细地描述了时序选择是如何工作的

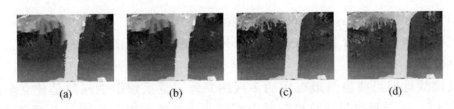

图 11.17 局部(基于 5×5 窗口的)匹配结果(Kang, Szeliski, and Chai 2001)© 2001 IEEE：(a)没有在空间上扰动(居中)的窗口；(b)空间上扰动的窗口；(c)使用 10 个相邻帧中最好的 5 个；(d)使用较好的一半序列。注意时序选择是如何改进树干附近的结果的

尽管从多幅输入图像计算得到深度图在性能上胜于从一对图像中计算得到深度图，但是同时估计多个深度图会得到更显著的性能改进(Szeliski 1999；Kang and Szeliski 2004)。多个深度图的存在使遮挡推断更为精确，因为一幅图像中被遮挡的区域在其他图像中有可能是可见的(可匹配的)。多视图重建问题可以被形式化为在多个关键帧中同时估计深度图(图 8.13c)，估计过程中同时最大化影像一致性和不同帧估计出来的视差之间的一致性。Szeliski(1999)和 Kang and Szeliski(2004)使用软(基于惩罚的)约束来使多个视差图保持一致，而 Kolmogorov and Zabih(2002)给出了如何将一致性编码成硬性约束，它能够保证多个深度图不仅相似而且在重叠区域完全一样。同时估计多个视差图的较新算法包括 Maitre, Shinagawa, and Do(2008)和 Zhang, Jia, Wong et al.(2008)所写的论文。

与多帧立体视觉估计紧密相关的一个问题是场景流(scene flow)，它使用多个摄像机来拍摄动态场景，其目标是同时恢复出每一个时间点的物体 3D 形状和相邻帧之间每一个表面点的完整 3D 运动。这个领域的代表性论文包括 Vedula, Baker, Rander et al.(2005), Zhang and Kambhamettu(2003), Pons, Keriven, and Faugeras(2007),

Huguet and Devernay(2007)和 Wedel, Rabe, Vaudrey *et al.*(2008)。图 11.18a 显示了图 11.2h～j 中的探戈舞者的 3D 场景流图像，图 11.18b 显示了从一个用于障碍规避的移动车辆拍摄 3D 场景流图像。如 Vedula, Baker, and Kanade(2005)所演示的那样，为了支持测量和安全应用，场景流可以用来同时支持空间和时序视图插值(13.5.4 节)。

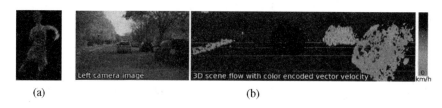

图 11.18 三维场景流：(a)从悬于图 11.2h～j 中跳舞者屋顶周围的多个摄像机计算出来的场景流(Vedula, Baker, Rander *et al.* 2005)©2005 IEEE；(b)从固定在一个移动车辆上的立体视觉摄像机计算出来的场景流(Wedel, Rabe, Vaudrey *et al.* 2008)© 2008 Springer

11.6.1　体积与 3D 表面重建

根据文献 Seitz, Curless, Diebel *et al.*(2006)：

多视图立体视觉的目标是由已知视角的摄像机拍摄的图像集合中重建出完整的 3D 物体模型。

最具挑战性但可能也是最有用的多视图立体视觉重建衍生问题是建立全局一致的 3D 模型。这个问题在计算机视觉上有着非常长的历史，开始于表面网格重建方法，如 Fua and Leclerc(1995)给出的方法(图 11.19a)。研究人员使用各种各样的方法和表示来解决这个问题，其中包括 3D 体素(voxel)表示(Seitz and Dyer 1999；Szeliski and Golland 1999；De Bonet and Viola 1999；Kutulakos and Seitz 2000；Eisert, Steinbach, and Girod 2000；Slabaugh, Culbertson, Slabaugh *et al.* 2004；Sinha and Pollefeys 2005；Vogiatzis, Hernandez, Torr *et al.* 2007；Hiep, Keriven, Pons *et al.* 2009)、水平集(Faugeras and Keriven 1998；Pons, Keriven, and Faugeras 2007)，多边形网格(Fua and Leclerc 1995；Narayanan, Rander, and Kanade 1998；Hernandez and Schmitt 2004；Furukawa and Ponce 2009)和多个深度图(Kolmogorov and Zabih 2002)。图 11.19 给出了使用这些方法重建出来的 3D 物体模型的一些代表样例。

为了组织和比较所有这些方法，Seitz, Curless, Diebel *et al.*(2006)提出了一个六点分类法，对这些算法根据场景表示(scene representation)、影像一致性测量(photoconsistency measure)、可见性模型(visibility model)、形状先验(shape priors)、重建算法(reconstruction algorithm)及其使用的初始化要求(initialization requirement)进行分类。下面，我们对这些选择中的一些进行总结，并列举其中的一些典型论文。更多的细节，请参考(Seitz, Curless, Diebel *et al.* 2006)这篇完整的综述论文以及评测网址 http://vision.middlebury.edu/mview/，其中包含指向更多最新论文和结果的链接。

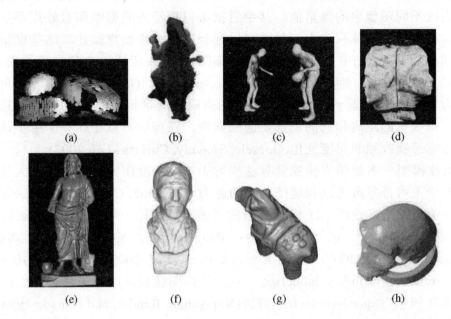

图 11.19 多视图立体视觉算法：(a)基于表面的立体视觉(Fua and Leclerc 1995)；(b)体素着色(Seitz and Dyer 1999)© 1999 Springer；(c)深度图合并(Narayanan, Rander, and Kanade 1998)；(d)水平集演进(Faugeras and Keriven 1998)© 1998 IEEE；(e)轮廓和立体视觉融合(Hernandez and Schmitt 2004)© 2004 Elsevier；(f)多视图图像匹配(Pons, Keriven, and Faugeras 2005)© 2005 IEEE；(g)体积的图割(Vogiatzis, Torr, and Cipolla 2005)© 2005 IEEE；(h)雕刻的视觉外形(Furukawa and Ponce 2009)© 2009 Springer

场景表达 3D 体素的均匀网格是最流行的 3D 表达方法之一[①]，它可以使用各种雕刻(Seitz and Dyer 1999；Kutulakos and Seitz 2000)或者优化(Sinha and Pollefeys 2005；Vogiatzis, Hernandez, Torr *et al.* 2007；Hiep, Keriven, Pons *et al.* 2009)方法来进行重建。水平集方法(5.1.4 节)也是在一个均匀网格上操作，但是它们并没有使用二值的占位图来表示，而是使用到表面的一个带符号的距离来表示(Faugeras and Keriven 1998；Pons, Keriven, and Faugeras 2007)，这种表示可以编码更精细的细节。多边形网格是另一种常见的表示(Fua and Leclerc 1995；Narayanan, Rander, and Kanade 1998；Isidoro and Sclaroff 2003；Hernandez and Schmitt 2004；Furukawa and Ponce 2009；Hiep, Keriven, Pons *et al.* 2009)。网格是计算机图形学中的标准表示方法，它也可轻松支持可见性和遮挡计算。最后，正如我们在前一节中所讨论的那样，可以使用多个深度图(Szeliski 1999；Kolmogorov and Zabih 2002；Kang and Szeliski 2004)。很多算法同时使用了多种单独的表示，例如，它们开始时计算多个深度图，然后将它们合并为一个 3D 物体模型(Narayanan, Rander, and Kanade 1998；Furukawa and Ponce 2009；Goesele, Curless, and Seitz 2006；Goesele, Snavely, Curless *et al.* 2007；Furukawa, Curless, Seitz *et al.* 2010)。

[①] 对于包含无穷远处的室外场景，空间的非均匀网格可能更合适(Slabaugh, Culbertson, Slabaugh *et al.* 2004)。

影像一致性 正如我们在 11.3.1 节中讨论的那样，可以使用各种各样的相似性度量来比较不同图像中的像素值，其中包括可以抵消光照影响和对外点不那么敏感的度量。在多视图立体视觉中，算法可以选择在模型表面直接计算这些度量，即在场景空间(scene space)中，将像素值从一幅图像中(或者从一个带纹理的模型)反向映射到另一幅图像上，即在图像空间(image space)中。(后者更类似于一个贝叶斯方法，因为输入的图像是一个彩色 3D 模型的带噪声的量测。)物体的几何信息，即它与每一个摄像机的距离和它的局部表面法向量，获得后可以在计算中用于调整匹配窗口以考虑透视收缩和尺度变化(Goesele, Snavely, Curless et al. 2007)。

可见性模型 多视图立体视觉算法相对于单一深度图方法的一个巨大优势是它们具备理论上可推导可见性和遮挡关系的能力。在 Seitz, Curless, Diebel et al.(2006)给出的分类方法中，使用 3D 模型的当前状态来预测每一幅图像中哪些表面像素可见的方法(Kutulakos and Seitz 2000；Faugeras and Keriven 1998；Vogiatzis, Hernandez, Torr et al. 2007；Hiep, Keriven, Pons et al. 2009)被归为使用几何学可见性模型(geometric visibility model)这一类。选择图像邻居的子集进行匹配的方法被称为"准几何学"(quasi-geometric)方法(Narayanan, Rander, and Kanade 1998；Kang and Szeliski 2004；Hernandez and Schmitt 2004)，而使用传统的鲁棒相似性度量的方法则被称为"基于外点"的(outlier-based)方法。尽管完整的几何学推导是原理上最精确的方法，但它计算起来会很慢并且依赖于为预测可见性而进行的当前表面估计的演进质量，而这可能有点像是先有鸡还是先有蛋的问题，除非使用收敛假设，参见 Kutulakos and Seitz(2000)。

形状先验 因为立体视觉匹配往往约束不够，特别是在无纹理的区域，所以大多数匹配算法都为期望形状使用了(无论是明确地还是隐含地)某种形式的先验模型。这些方法中的很多都依赖于使用 3D 平滑性或者基于区域影像一致性限制的优化，由于平滑表面向内收缩的自然趋势，这通常会产生一个极小曲面(minimal surface)先验(Faugeras and Keriven 1998；Sinha and Pollefeys 2005；Vogiatzis, Hernandez, Torr et al. 2007)。一些方法割开空间体直到找到一个影像一致性的解(Seitz and Dyer 1999；Kutulakos and Seitz 2000)，这对应于最大表面(maximal surface)偏置，即这些方法倾向于夸大估计真实形状。最后，多深度图方法通常可以采用传统的基于图像的平滑性(正规化的)限制。

重建算法 根据重建算法如何实施的实际细节，可以判断各种多视图立体视觉算法的最大变化点(最突出的创新)。

一些方法使用定义在三维影像一致性体上的全局优化来恢复完整的表面。基于图割的方法使用多项式复杂度的二值分割算法来恢复定义在体素网格上的物体模型(Sinha and Pollefeys 2005；Vogiatzis, Hernandez, Torr et al. 2007；Hiep, Keriven, Pons et al. 2009)。水平集方法使用一个连续表面演进来找到潜在表面构型空间上的一个好的最小值，因此它需要相当合理的初始化(Faugeras and Keriven 1998；Pons, Keriven, and Faugeras 2007)。为了使影像一致性体有意义，匹配代价需要以某种鲁棒的方式进行计算，比如，使用多个有限的视图或者通过聚集多个深度图。

全局优化的一个替代方法是在 3D 体中扫描并且同时计算出影像一致性及可见性。Seitz and Dyer(1999)中的体基元着色(voxel coloring)算法执行的是一次从头到尾的平面扫描操作。在每一个平面内，所有影像足够一致性的体素被标为物体的一部分。然后，源图像中对应的像素就可以"擦除"，因为它们已经被计算在内，所以不会再对影像一致性计算有贡献。(Szeliski and Golland(1999)使用了一个类似的方法，尽管没有从头到尾的扫描顺序。)所得到的 3D 体，在没有噪声和重采样的情况下，可以保证产生影像一致的 3D 模型并且围住产生这些图像的任意真实的 3D 物体模型。

不幸的是，体素着色算法只有在所有摄像机都位于扫描平面同一侧时才能保证有效，通常的环状摄像机布置是不可能做到这样的。Kutulakos and Seitz(2000)将体素着色推广到空间雕刻(space carving)，它选择满足体素着色限制的摄像机的子集来迭代地雕刻掉沿着不同轴的 3D 网格。

另一种常见的多视图立体视觉方法首先单独计算多个深度图，然后将这些部分图合并成一个完整的 3D 模型。深度图合并的方法在 12.2.1 节中有详细的讨论，其中包括 Goesele, Curless, and Seitz(2006)使用的带符号距离函数(Curless and Levoy 1996)和 Goesele, Snavely, Curless et al.(2007)使用的泊松表面重建(Kazhdan, Bolitho, and Hoppe 2006)。同样，也可以只重建稀疏的表示，比如 3D 点和线，然后将它们插值成完整的 3D 表面(12.3.1 节)(Taylor 2003)。

初始化要求 Seitz, Curless, Diebel et al.(2006)讨论的最后一点是不同算法所需要的不同程度的初始化。因为一些算法要求精和演进一个粗的 3D 模型，所以它们需要有一个适当精确(或者过完备)的初始模型，这个初始模型通常可以通过 11.6.2 节讨论的从物体轮廓重建的立体得到。然而，如果算法要进行全局优化(Kolev, Klodt, Brox et al. 2009; Kolev and Cremers 2009)，这个初始化依赖就不是问题。

实验评测 为了对多视图立体视觉中大量的设计方案进行评测，Seitz, Curless, Diebel et al.(2006)收集了一个使用球形台架拍摄的标定好的图像数据集。图 11.20 显示了六个数据集各自的代表图像，但目前只有前两个数据集已经完全处理并用于评测。图 11.21 显示了在 Temple 数据集上运行七个不同算法得到的结果。正如你所看到的那样，大多数方法在捕捉圆柱体细节方面都做得不错，但同时也可以非常清楚地看到这些方法使用不同程度的平滑来获得这些结果。

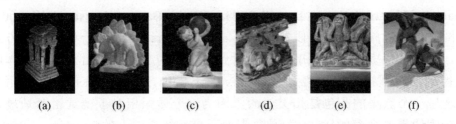

图 11.20 Seitz, Curless, Diebel et al.(2006)拍摄的多视图立体视觉数据集© 2009 Springer。目前只有(a)和(b)用于评测

图 11.21　(Hernandez and Schmitt 2004; Furukawa and Ponce 2009; Pons, Keriven, and Faugeras 2005; Goesele, Curless, and Seitz 2006; Vogiatzis, Torr, and Cipolla 2005; Tran and Davis 2002; Kolmogorov and Zabih 2002)这七种算法由 Seitz, Curless, Diebel *et al.*(2006)在包含 47 幅图像的 Temple Ring 数据集上进行评测，得到重建结果(细节)。每一个细节图下面的数字是每种方法以毫米度量的精确度

自从 Seitz, Curless, Diebel *et al.*(2006)这篇综述论文发表之后，多视图立体视觉匹配一直在快速发展(Strecha, Fransens, and Van Gool 2006；Hernandez, Vogiatzis, and Cipolla 2007；Habbecke and Kobbelt 2007；Furukawa and Ponce 2007；Vogiatzis, Hernandez, Torr *et al.* 2007；Goesele, Snavely, Curless *et al.* 2007；Sinha, Mordohai, and Pollefeys 2007；Gargallo, Prados, and Sturm 2007；Merrell, Akbarzadeh, Wang *et al.* 2007；Zach, Pock, and Bischof 2007b；Furukawa and Ponce 2008；Hornung, Zeng, and Kobbelt 2008；Bradley, Boubekeur, and Heidrich 2008；Zach 2008；Campbell, Vogiatzis, Hern´andez *et al.* 2008；Kolev, Klodt, Brox *et al.* 2009；Hiep, Keriven, Pons *et al.* 2009；Furukawa, Curless, Seitz *et al.* 2010)。多视图立体视觉评测网址 *http://vision.middlebury.edu/mview/* 提供了这些算法的定量结果，并给出了这些论文的链接。

11.6.2　由轮廓到形状

在很多情况下，对兴趣目标进行前景-背景分割是初始化或拟合一个 3D 模型(Grauman, Shakhnarovich, and Darrell 2003；Vlasic, Baran, Matusik *et al.* 2008)或者向多视图立体视觉施加凸集合限制(Kolev and Cremers 2008)的一种很好的方式。多年以来，产生了很多从二值轮廓图像映射到 3D 后的交集重建出一个 3D 体模型的方法。得到的模型被称为"视觉外形"(visual hull)，有时也称作"线外形"(line hull)，这与一个点集的凸包是相类似的，因为这个体积相对于视觉轮廓是最大的并且沿着轮廓边界的观察射线(直线)是在表面元素切线方向上的(Laurentini 1994)。使用多视图立体视觉(Sinha and Pollefeys 2005)或者通过分析投射的影子(Sinha and

Pollefeys 2005)来雕刻出一个更精确的重建也是可能的。

一些方法首先使用一个多边形表示来近似每一个轮廓，然后将所产生的面片状锥形区域在三维空间中相交来产生一个多面体模型(Baumgart 1974；Martin and Aggarwal 1983；Matusik, Buehler, and McMillan 2001)，这个模型在后面可以使用三角形样条来进一步求精(Sullivan and Ponce 1998)。其他一些方法使用基于体素的表示，通常编码为八叉树(Samet 1989)，这是由于其结果在空间-时间方面的高效性。图 11.22a～b 给出了一个 3D 八叉树模型和它关联的颜色树的一个例子，其中黑色节点在模型的内部，白色节点在模型的外部，灰色节点拥有混合的内外关系。基于八叉树的重建方法的例子包括 Potmesil(1987), Noborio, Fukada, and Arimoto(1988), Srivasan, Liang, and Hackwood(1990)和 Szeliski(1993)。

Szeliski(1993)中的方法首先将每一个二值轮廓转化成距离图的一个单边变形，其中图中的每一个像素指明完全位于轮廓内部(或外部)的最大正方形。这样，能够很快将一个八叉树单元投影到轮廓上，从而确定它是完全在内部还是外部，然后就可以被着成黑色、白色或者留下来当作灰色(混合的)以便在更小的网格上进一步求精。八叉树构建算法以一种由粗到精的方式执行，首先在一个相对粗糙的分辨率上建立八叉树，然后通过重新访问并细分所有输入图像中灰色的(混合的)尚未确定内外关系(占位)的单元对其求精。图 11.22d 显示了从转盘上的一个咖啡杯计算得到的八叉树模型。

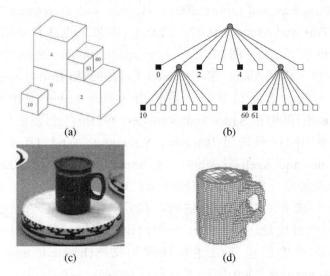

图 11.22　根据二值轮廓来进行物体的八叉树重建(Szeliski 1993)©1993 Elsevier：(a)八叉树表示及其对应的(b)树结构；(c)转盘上的物体输入图像；(d)计算后所得到的 3D 物体八叉树模型

视觉外形计算最新的一些工作从基于图像的渲染那里借鉴了一些想法，因此被称为"基于图像的视觉外形"(image-based visual hull)(Matusik, Buehler, Raskar *et al.* 2000)，它并没有事先计算一个全局 3D 模型，而是针对每一个新的视点通过连续将观察射线片段和每一个图像中的二值轮廓相交来重新计算一个基于图像的视觉外形。这样一来，不仅可以得到一个快速计算算法，而且同时能够用来自输入图像的颜色对恢复的模型快速添加纹理。这个方法还可以与高质量可变形模板相结合以捕捉和重新用动画绘制整个身体的运动(Vlasic, Baran, Matusik et al. 2008)。

11.7 补充阅读

立体视觉匹配和深度估计这个领域是计算机视觉中最古老、研究最广泛的课题之一。多年来，研究人员已经写了很多很好的综述论文(Marr and Poggio 1976；Barnard and Fischler 1982；Dhond and Aggarwal 1989；Scharstein and Szeliski 2002；Brown, Burschka, and Hager 2003；Seitz, Curless, Diebel et al. 2006)，这些论文可以作为此类文献很好的指引。

由于计算能力的限制以及期望找到与表观无关的对应，早期的算法通常关注的是找到稀疏对应(sparse correspondence)(Hannah 1974；Marr and Poggio 1979；Mayhew and Frisby 1980；Baker and Binford 1981；Arnold 1983；Grimson 1985；Ohta and Kanade 1985；Bolles, Baker, and Marimont 1987；Matthies, Kanade, and Szeliski 1989；Hsieh, McKeown, and Perlant 1992；Bolles, Baker, and Hannah 1993)。

极线几何计算和图像预矫正的主题分别在 7.2 节和 11.1 节进行了讨论，多视角几何的相关教材(Faugeras and Luong 2001；Hartley and Zisserman 2004)和专门讨论这个主题的论文(Torr and Murray 1997；Zhang 1998a,b)也都有详细介绍。视差空间(disparity space)和视差空间图像(disparity space image)这两个概念通常也与Marr(1982)的开创性工作和 Yang, Yuille, and Lu(1993)和 Intille and Bobick(1994)这两篇论文联系在一起。平面扫描算法最初由 Collins(1996)提出而流行，然后由 Szeliski and Golland(1999)和 Saito and Kanade(1999)推广到完全任意的投影设置。平面扫描还可以使用圆柱形表面(Ishiguro, Yamamoto, and Tsuji 1992；Kang and Szeliski 1997；Shum and Szeliski 1999；Li, Shum, Tang et al. 2004；Zheng, Kang, Cohen et al. 2007)或者更一般的拓扑结构(Seitz 2001)来形式化。

一旦建立代价体或者 DSI 的拓扑结构，我们就需要为每一个像素和潜在的深度计算实际的影像一致性度量。如 11.3.1 节所讨论的那样，已经提出了各种各样此类的度量。关于匹配代价的最新综述和评测论文(Scharstein and Szeliski 2002；Hirschmüller and Scharstein 2009)比较了这些度量中的一些。

为了从这些代价中计算出实际的深度图，需要使用某种形式的优化或者选择准则。其中最简单的是各种各样的滑动窗口方法，这些方法的讨论参见 11.4 节，Gong, Yang, Wang et al.(2007)和 Tombari, Mattoccia, Di Stefano et al.(2008)给出了它们的综述。更常见的是，如 11.5 节所述，全局优化框架可以用来计算最佳视差场，这些方法包括动态规划和真正的全局优化算法，如图割和带环置信度传播。由于这个主题的文献非常多，所以 11.5 节中进行了详细的描述。Middlebury 立体视觉页面(*http://vision.middlebury.edu/stereo*)是找到指向这个领域最新结果链接的好地方。

多视图立体视觉算法通常可以分为两类。第一类算法使用多幅图像计算影像一致性度量从而计算出传统的深度图(Okutomi and Kanade 1993；Kang, Webb, Zitnick et al. 1995；Nakamura, Matsuura, Satoh et al. 1996；Szeliski and Golland 1999；Kang, Szeliski, and Chai 2001；Vaish, Szeliski, Zitnick et al. 2006；Gallup, Frahm, Mordohai et al. 2008)。可选择地，其中的一些方法计算多个深度图然后使用附加的约束来鼓励不同的深度图保持一致(Szeliski 1999；Kolmogorov and Zabih 2002；Kang and Szeliski 2004；Maitre, Shinagawa, and Do 2008；Zhang, Jia, Wong et al. 2008)。

第二类算法包括计算真实 3D 体积的或者基于表面的物体模型的这些论文。由于针对这个主题已经发表了大量的论文，所以与其在这里引用它们，我们推荐你参考 11.6.1 节中的材料和 Seitz, Curless, Diebel et al.(2006)给出的综述论文以及在线评测网址(*http://vision.middlebury.edu/mview/*)。

11.8 习　　题

习题 11.1：立体视觉对矫正
实现下面的简单算法(11.1.1 节)。
1. 同时旋转两个摄像机使其看起来与连接其中心 c_0 和 c_1 的直线垂直。最小旋转角度可以通过最初光轴和期望光轴之间的向量叉积计算得到。
2. 扭动光轴使每一个摄像机的水平光轴看起来在另一个摄像机的方向上。(同样，第一次旋转后的当前 x 轴与连接两个摄像机的直线之间的向量叉积给出了旋转角度。)
3. 如果需要，按比例放大较小(不够清晰)的图像使其与另一幅图像分辨率相同(从而有线到线的对应)。

现在将你的结果和 Loop and Zhang(1999)中提出的算法进行比较。你能不能想出使他们的方法更可取的情况？

习题 11.2：刚性直接配准
使用极线几何来修改习题 8.4 中实现的基于样条或者光流的运动估计，即只估计视差。

(可选)扩展你的算法使其同时估计极线几何(不需要先用点对应)和残余视差(运动)的极点。极线几何估计，请通过估计对应于参考平面的主运动的基本单应来完成。

习题 11.3：从轮廓(剖面)到形状
根据一系列边缘图像来重建一个表面模型(11.2.1 节)。
1. 提取边缘并将它们连接起来(习题 4.7 和 4.8)。
2. 基于之前计算的极线几何，在三幅(或者更多的)图像间匹配边缘。
3. 使用密切圆来重建曲线的 3D 位置(11.5)。

4. 将得到的 3D 表面当作一个稀疏网格进行渲染，即绘制重建出来的轮廓曲线然后将相邻图像间密切圆相似的 3D 点连接起来。

习题 11.4：平面扫描

实现一个平面扫描算法(11.1.2 节)。

如果图像已经预先矫正，这就只是一个简单将图像相对移动并比较像素的过程。如果图像没有预先矫正，则为每个平面计算将目标图像映射到参考图像坐标系上的重采样单应。

评测下面一组相似性度量(11.3.1 节)并通过可视化视差空间图像(DSI)来比较它们的性能，DSI 上正确深度处的像素应为深色：

- 平方差(SD)；
- 绝对值差(AD)；
- 截断的或鲁棒的度量；
- 梯度差；
- 秩或统计变换(后者通常性能更好)；
- 根据预计算出来的联合密度函数的互信息。

考虑使用 Birchfield and Tomasi(1998)中比较相邻像素间(不同移动或卷绕下的图像)的范围的方法。也尝试使用 11.3.1 节讨论的一种或多种方法针对偏差或者增益变化对图像进行预补偿。

习题 11.5：聚集和基于窗口的立体视觉

实现 11.4 节描述的一种或多种匹配代价聚集策略：

- 使用一个方框型(箱式)或者高斯核函数来卷积；
- 通过应用一个最小值滤波器来移动窗口位置(Scharstein and Szeliski 2002)；
- 挑选一个使某种匹配稳定性度量最大化的窗口(Veksler 2001, 2003)；
- 根据像素与中心像素的相似度来加权像素(Yoon and Kweon 2006)；

一旦你在 DSI 中聚集了代价，就在每一个像素处选出胜利者(赢者通吃)，然后选择性地进行如下的一个或多个后处理阶段：

1. 双向计算匹配，然后只选择鲁棒的匹配(用另一种颜色绘制其他匹配)；
2. 标记不确定的匹配(那些置信度过低的匹配)；
3. 使用邻近处的值填充不确定的匹配；
4. 通过对胜利者周围的 DSI 值拟合一个抛物线或者使用 Lukas-Kanade 的一次迭代来将匹配求精到亚像素视差。

习题 11.6：基于优化的立体视觉

使用习题 11.4 中实现的方法之一来计算视差空间图像(DSI)体，然后实现 11.5 节中的一种(或多种)全局优化方法来计算深度图。有下面四种可能的选择包括：

- 动态规划或扫描线优化算法(相对比较容易)；
- 半全局优化(Hirschmüller 2008)，它是扫描线优化的一个简单推广并且效果很好；

- 使用 alfa 扩展的图割(Boykov, Veksler, and Zabih 2001)，为此你需要找一个最大流或者最小割算法(*http://vision.middlebury.edu/stereo*);
- 带环置信度传播(附录 B.5.3)。

在 Middlebury 立体视觉数据集上评测你的算法。

相对于局部聚集(Yoon and Kweon 2006)，你的算法有多好？能不能想出一些扩展或者修正使其更好？

习题 11.7：再议视角插值

使用前面实现的方法来计算一个稠密深度图，并且用它(或者，更好的是，每一个源图像一个深度图)从一个立体视觉数据集生成平滑的中间视图。

使用真实深度数据(如果可以获得的话)来比较你的结果。

你看到了哪些种类的缺陷？你可以想出一些方法来消除它们吗？

实现这个算法的更多细节可以在 13.1 节和习题 13.1～13.4 中找到。

习题 11.8：多帧立体视觉

为使用多个输入帧而扩展之前的一种方法(11.6 节)并尝试改进只用两个视角所得到的结果。

如果有必要，尝试使用时序选择(Kang and Szeliski 2004)来处理在多帧数据集合中增加的遮挡量。

你也可以尝试同时估计多个深度图并使其一致(Kolmogorov and Zabih 2002; Kang and Szeliski 2004)。

在一些标准多视图数据集合上测试你的算法。

习题 11.9：体积立体视觉

作为习题 11.4 实现的平面扫描算法的一个简单扩展，实现体素着色(Seitz and Dyer 1999)。

1. 不要一次全部计算出整个 DSI，从头到尾每次只计算一个平面。
2. 标记每一个影像一致性低于特定阈值的体素，将它们作为物体的一部分并且记住它们的平均(或鲁棒)颜色(Seitz and Dyer 1999；Eisert, Steinbach, and Girod 2000；Kutulakos 2000；Slabaugh, Culbertson, Slabaugh *et al*. 2004)。
3. 在输入图像中擦除已标记体素对应的输入像素，例如，通过设置其 alpha 值为 0(或者根据占位图设为某个减少的数)。
4. 评价下一个平面时，使用源图像的 alpha 值修改所得到的影像一致性得分，例如，只考虑拥有完全 alpha 值的像素或者根据像素的 alpha 值来加权像素。
5. 如果摄像机没有全部都在扫描平面的同一侧，就使用空间雕刻(Kutulako and Seitz，2000)在源图像的不同子集中循环雕刻掉来自不同方向的体积(volume)。

习题 11.10：深度图合并

使用你在习题 11.8 中实现的多视图立体视觉方法或者其他不同的方法，比如

Goesele, Snavely, Curless *et al.*(2007)中描述的方法，为每一幅输入图像计算深度图。

将这些深度图合并成一个连贯的 3D 模型，比如，使用泊松表面重建(Kazhdan, Bolitho, and Hoppe 2006)。

习题 11.11：从轮廓到形状

使用八叉树或者你自己选择的其他表示来建立一个基于轮廓的立体重建算法(11.6.2 节)。

第 12 章
3D 重建

12.1 由 X 到形状
12.2 主动距离获取
12.3 表面表达
12.4 基于点的表达
12.5 体积表达
12.6 基于模型的重建
12.7 恢复纹理映射与反照率
12.8 补充阅读
12.9 习题

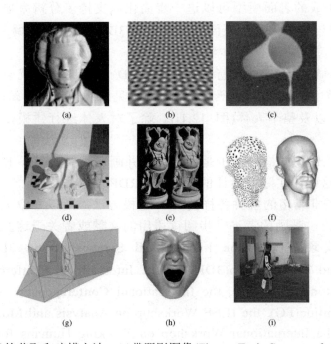

图 12.1 3D 形状的获取和建模方法：(a)带阴影图像(Zhang, Tsai, Cryer *et al*. 1999)© 1999 IEEE；(b)纹理梯度(Garding 1992) © 1992 Springer；(c)实时由聚焦到深度(Nayar,Watanabe, and Noguchi 1996) © 1996 IEEE；(d)利用一个棍子的阴影扫描场景(Bouguet and Perona 1999) ©1999 Springer；(e)将距离图合并为一个 3D 模型(Curless and Levoy 1996) © 1996 ACM；(f)基于点的表面建模(Pauly, Keiser, Kobbelt *et al*. 2003)©2003 ACM；(g)利用直线和平面对 3D 建筑自动建模(Werner and Zisserman 2002)©2002 Springer；(h)从时空立体视觉中得到的 3D 人脸模型(Zhang, Snavely, Curless *et al*. 2004)©2004 ACM；(i)人物跟踪(Sigal, Bhatia, Roth *et al*. 2004) © 2004 IEEE

如第 11 章所述，各种立体视觉匹配方法已被用于从两幅或者更多幅图像中重建高质量的 3D 模型。然而，立体视觉只是可以用于从图像中推测形状的许多线索之一。在本章中，我们研究一定数量的方法，它们不仅包括如阴影和聚焦等视觉线索，而且可以将多个距离图像或深度图像融合为 3D 模型，还可以重建指定的模型，例如头部、身体或者建筑。

在可以用于推测形状的视觉线索中，表面上的阴影(明暗度)(图 12.1a)可以提供很多关于局部表面方向甚至整体表面形状的信息(12.1.1 节)。当不同方向照射的光源可以分别打开或者熄灭时，这种方法会变得更强大(光度测量立体视觉)。纹理梯度(图 12.1b)，即规则模式随着表面倾斜或者弯曲远离摄像机时产生的透视收缩，可以提供有关局部表面方向的类似的线索(12.1.2 节)。聚焦则是场景深度的另一个强大的线索，特别是同时使用两幅或多幅不同焦距的图像时(12.1.3 节)。

3D 形状也可以使用主动的照明方法来估计，例如光带(图 12.1d)或者飞行时间测距仪(time of flight range finders)(12.2 节)。如 12.2.1 节所述，使用这样的方法(或者被动的基于图像的立体视觉)获得的部分表面模型可以合并为更一致的模型(图 12.1e)。这种方法已经被用于建立文化遗产(例如古迹)的详细及精确的模型(12.2.2 节)。生成的表面模型可以进一步简化以支持多分辨率展现以及网络流媒体(12.3.2 节)。另一种处理连续表面的方式是将 3D 表面表示为稠密的带方向的 3D 点的汇集(12.4 节)或者一系列基本体元(12.5 节)。

如果我们对试图重建的目标有所了解，3D 建模的工作就会变得更加快速而有效。在 12.6 节中，我们着眼于三个特殊但常见的例子，即建筑(图 12.1g)、头部和人脸(图 12.1h)以及整个人体(图 12.1i)。除了对人体进行建模，我们还将讨论人体跟踪方法。

形状和表观建模的最后一步是提取纹理并画在我们的 3D 模型上(12.7 节)。有些方法超越了这些，实际上估计的是整个 BRDF(12.7.1 节)。

由于存在着非常多的各种各样的 3D 建模方法，所以本章无法对其中任何一个进行深入的讨论。鼓励读者进一步阅读引用的文献或者关于这些主题的更专业的出版物以及会议录，例如 the International Symposium on 3D Data Processing, Visualization,and Transmission(3DPVT), the International Conference on 3D Digital Imaging and Modeling(3DIM), the International Conference on Automatic Face and Gesture Recognition(FG), the IEEE Workshop on Analysis and Modeling of Faces and Gestures 以及 the International Workshop on Tracking Humans for the Evaluation of their Motion in Image Sequences(THEMIS)。

12.1 由 X 到形状

除了双眼视差外，阴影、纹理和聚焦都参与了我们对形状的感知。从这些线索来推断形状的研究有时称为"由 X 到形状"，因为具体说法有由阴影到形状、由纹理到形状和由聚焦到形状。[①]本节中，我们将研究这三种线索以及如何使用它们重建 3D 几何。Wolff，Shafer and Healey(1992b)编辑的基于物理学的形状推断论文集中有对这些主题的一个好的综述。

12.1.1 由阴影到形状与光度测量立体视觉

我们观察一幅含有光滑阴影物体的图像时，例如图 12.2 中的图像，我们可以清楚地从阴影的变化中看到物体的形状。这怎么可能呢？答案在于当表面法向沿着物体改变时，表现出来的亮度同样发生变化，并且是局部表面方向和入射光夹角的函数，如图 2.15 所示(2.2.2 节)。

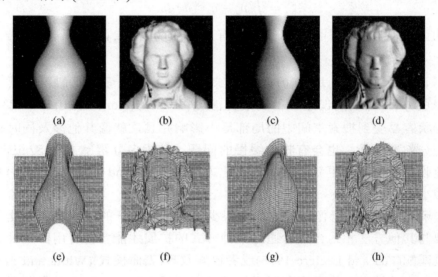

图 12.2 从阴影合成形状(Zhang, Tsai, Cryer *et al.* 1999)© 1999 IEEE：带阴影图像：(a 和 b)光源在正面(0, 0, 1)；(c 和 d)光源在右前方(1; 0; 1)；(e 和 f)利用 Tsai and Shah (1994)方法从阴影重建的对应形状

从图像亮度变化推断出表面形状的问题称为"由阴影到形状"，它是计算机视觉的一个经典问题(Horn 1975)。Horn and Brooks(1989)的论文汇编，特别是其中关于变分法的章节，是这个问题的一个很好的信息来源。Zhang，Tsai，Cryer，*et al.*(1999)的综述不仅回顾了近期的一些方法，还提供了一些比较结果。

大部分由阴影到形状的算法都假设所考虑的表面具有均匀的反照率和反射率，且光源的方向要么已知，要么可以使用参考目标标定得到。在假设光源和观察者都在远处的前提下，亮度的变化(辐照度方程)变成单纯的局部表面方向的函数，

$$I(x,y) = R(p(x,y), q(x,y)), \tag{12.1}$$

其中$(p, q)=(z_x, z_y)$是深度图的导数而 $R(p,q)$ 被称为反射图。例如，一个漫反射(郎伯，Lambertian)表面的反射图是表面法向 $\hat{n}=(p,q,1)/\sqrt{1+p^2+q^2}$ 和光源方向

[①] 在第 11 章中我们已经见到了由立体视觉到形状、由剖面到形状和由轮廓到形状的例子。

$\boldsymbol{v} = (v_x, v_y, v_z)$ 的(非负)点积(2.88)，

$$R(p,q) = \max\left(0, \rho\frac{pv_x + qv_y + v_z}{\sqrt{1+p^2+q^2}}\right), \tag{12.2}$$

其中 ρ 为表面反射率因子(反照率)。

原则上，公式(12.1 和 12.2)可以用于使用非线性最小二乘或者其他方法估计 (p,q)。但不幸的是，除非加入额外的约束，否则每个像素上的未知数 (p,q) 会多于测量值 (I)。一个常用的约束是光滑性约束，

$$\mathcal{E}_s = \int p_x^2 + p_y^2 + q_x^2 + q_y^2 \, dx \, dy = \int \|\nabla p\|^2 + \|\nabla q\|^2 \, dx \, dy, \tag{12.3}$$

我们在 3.7.1 节中(3.94)中已经见过。另外一个是可积分性约束：

$$\mathcal{E}_i = \int (p_y - q_x)^2 \, dx \, dy, \tag{12.4}$$

这个约束是很自然的，由于对于一个有效的深度图 $z(x,y)$ 且 $(p,q) = (z_x, z_y)$，所以我们有 $p_y = z_{xy} = z_{yx} = q_x$。

除了首先恢复方向场 (p,q) 并将其综合得到一个表面，我们还可以直接最小化图像形成方程(12.1)的误差以得到最优深度图 $z(x,y)$ (Horn 1990)。不幸的是，由阴影到形状容易受到搜索空间中的局部最小影响并且，就像其他涉及同时估计多个变量的变分问题一样，也会有收敛缓慢的问题。使用多分辨率方法(Szeliski 1991a)可以加速收敛，而使用更复杂的随机优化方法(Dupuis and Oliensis 1994)有助于避免局部最小。

实际上，除了石膏像以外的表面很少具有一个统一的反照率。因此由阴影到形状需要与其他方法相结合或者通过某种方式的扩展才能变得有用。一种方式是与立体视觉匹配(Fua and Leclerc 1995)或者已知纹理(表面模式)(White and Forsyth 2006)相结合。立体视觉和纹理成分可以提供纹理区域的信息，而由阴影到形状则填补了具有均匀颜色的区域的信息而且也可以提供关于表面形状的更精确的信息。

光度测量立体视觉 使由阴影到形状方法更加可靠的另一种方式是使用多个可以选择性开关的光源。由于光源在其中起到了类似于传统立体视觉(Woodham 1981)[①]中不同位置的摄像机的作用，这种方法称为"光度测量立体视觉"。对每个光源，我们有一个不同的反射图，$R_1(p,q), R_2(p,q)$，等等。给定一个像素点上相应的亮度 I_1, I_2，等等，我们可以大体上同时恢复出一个未知的反照率 ρ 和表面方向估计 (p,q)。

对于漫反射表面(12.2)，如果用 $\hat{\boldsymbol{n}}$ 参数化局部方向，我们可以(对非阴影像素)得到一组具有如下形式的线性方程

$$I_k = \rho\hat{\boldsymbol{n}} \cdot \boldsymbol{v}_k, \tag{12.5}$$

从中我们可以使用线性最小二乘恢复出 $\rho\hat{\boldsymbol{n}}$。只要(三个或者以上)向量 $\boldsymbol{v}_k$ 是线性无关的，即它们不在同一个方位角(远离观察者的方向)上，这些方程是适定的(well

---

[①] 利用三色光源可以替代开关光源(Woodham 1994; Hernandez, Vogiatzis, Brostow et al. 2007; Hernandez and Vogiatzis 2010)。

conditioned)。

一旦每个像素上的表面法向或者梯度被恢复出来，就可以使用正规化表面拟合的一个变种(3.100)将它们综合得到一个深度图。(Nehab, Rusinkiewicz, Davis et al.(2005)及 Harker and O'Leary(2008)在这方面做出了一些更新的工作成果)。

当表面是镜像反射时，我们需要大于 3 个方向的光源。事实上，(12.1)给出的辐照度方程不仅仅要求光源和摄像机远离表面，而且忽略了物体间的相互反射，这是我们观察到的物体表面阴影的一个主要来源，例如在凹面结构如凹槽和裂缝内部看到的变黑的部分(Nayar, Ikeuchi, and Kanade 1991)。

12.1.2 由纹理到形状

我们观察到的规则纹理的透视收缩变化还可以对局部表面方向提供有用的信息。图 12.3 显示了这样一个模式及估计得到的局部表面方向。由纹理到形状的算法需要几个处理步骤(包括抽取重复模式或者局部频率的测量以计算局部仿射变形)和接下来推测局部表面方向的步骤。关于这些不同步骤的细节可以参见研究文献(Witkin 1981; Ikeuchi 1981; Blostein and Ahuja 1987; Garding 1992; Malik and Rosenholtz 1997; Lobay and Forsyth 2006)。

当原始模式有规则时，可以将一个规则但轻微变形的网格拟合到图像上，并通过这个网格来进行各种图像替换或者分析任务(Liu, Collins, and Tsin 2004; Liu, Lin, and Hays 2004; Hays, Leordeanu, Efros et al. 2006; Lin, Hays, Wu et al. 2006; Park, Brocklehurst, Collins et al. 2009)。如果使用具有特殊印刷纹理的布料模式，这个过程将更加容易。

在曲面镜的反射中观察规则模式的形变，如图 12.3c～d 所示，可用于恢复表面的形状(Savarese, Chen, and Perona 2005; Rozenfeld, Shimshoni, and Lindenbaum 2007)。我们也可以从高光流(specular flow)，即从移动的摄像机观察到的高光运动推测局部形状(Oren and Nayar 1997; Zisserman, Giblin, and Blake 1989; Swaminathan, Kang, Szeliski et al. 2002)。

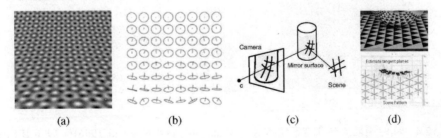

图 12.3 由纹理到形状的合成示例(Garding 1992)© 1992 Springer：(a)卷绕在曲面上的规则纹理和(b)对应的表面法向估计。由镜面反射到形状(Savarese, Chen, and Perona 2005) © 2005 Springer：(c)曲面镜反射出的规则模式产生(d)中的曲线，从中可以推断出 3D 点位置和法向

12.1.3　由聚焦到形状

物体深度的一个强线索是模糊的程度，它随着物体表面远离摄像机焦距而增加。如图 2.19 所示，将物体表面移动远离焦平面时散光圈依据一个公式增大，该公式很容易用相似三角形建立(习题 2.4)。

有很多根据散焦程度(散焦深度)估计深度的工作(Pentland 1987; Nayar and Nakagawa 1994; Nayar, Watanabe, and Noguchi 1996; Watanabe and Nayar 1998; Chaudhuri and Rajagopalan 1999; Favaro and Soatto 2006)。为了使这样的方法成为现实，需要研究下面这些问题。

- 在向远离焦平面方向移动时，两个方向上的模糊程度都会增加。因此，我们需要使用两幅或更多幅使用不同焦距设置捕捉的图像(Pentland 1987; Nayar, Watanabe, and Noguchi 1996)或者在深度方向上平移目标并寻找具有最大锐利度(sharpness)的点(Nayar and Nakagawa 1994)。
- 物体的放大率可以随着焦距的变化或物体的移动而变化。这种变化可以显式地描述(使得对应更加困难)或者使用远心光学描述，它近似一个正交投影摄像机并需要在镜头前有一个光圈(Nayar, Watanabe, and Noguchi 1996)。
- 散焦的程度必须可靠地估计出来。一个简单的方式是计算一个区域内梯度平方的均值，但这种方法有几个问题，包括上面提到的图像放大问题。更好的解决方法是使用精心设计的有理多项式滤波器(Watanabe and Nayar 1998)。

图 12.4 给出一个实时的由散焦到深度的传感器，它采用共享同一个光学路径但深度略有差别的两个成像芯片，并使用一个主动照明系统在同一个方向上投影一个棋盘图像。如图 12.4b~g 所示，这个系统可以在静态或动态的场景中实时产生高精度深度图。

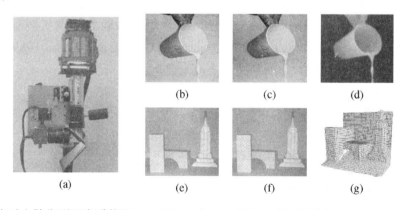

图 12.4　实时由散焦到深度系统(Nayar, Watanabe, and Noguchi 1996)© 1996 IEEE：(a)实时聚焦距离传感器，包括一个位于两个远心镜头之间的半涂银镜(右下)、一个将图像分裂到两个 CCD 传感器上的棱镜(左下)以及一个利用氙灯照明的有边棋盘模式(上部); (b 和 c)两个摄像机中的输入视频帧及其对应深度图(d); (e 和 f)其中两帧(放大后可以看到纹理)以及对应的 3D 网格模型(g)

12.2 主动距离获取

如 12.1 节所述，主动场景照明，无论是为了使用光度测量立体视觉估计法还是为由散焦到形状提供人工纹理，都可以大大提高视觉系统的性能。这种主动照明从机器视觉的早期年代就开始使用了，用于构建估计 3D 深度图像的高可靠性传感器，其中使用了各种距离获取(或距离感知)方法(Besl 1989; Curless 1999; Hebert 2000)。

激光或光带传感器是最流行的主动照明传感器之一，它在一个偏离的视点观察物体时使用一个光平面扫描场景或物体，如图 12.5b 所示(Rioux and Bird 1993; Curless and Levoy 1995)。光带落在物体上时，会根据它所照射的表面的形状发生形变。这样一来，使用光学三角测量法估计特定光带上观察到的所有点的 3D 位置就很容易了。具体而言，有关光带的 3D 平面方程的信息使我们可以推断出每个被照明的像素的 3D 位置，如公式(2.70 和 2.71)中所述。光带方法的精确度可以通过为每个像素的照明寻找精确的瞬时峰值来提高(Curless and Levoy 1995)。扫描仪的最终精确度决定于斜棱(slant edge)调制方法，即在标定物体上投射尖锐的折线(Goesele, Fuchs, and Seidel 2003)。

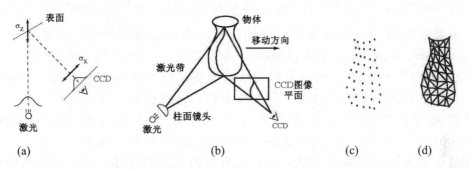

图 12.5 距离数据扫描(Curless and Levoy 1996)© 1996 ACM：(a)表面上的激光点在 CCD 传感器上成像；(b)一个激光带(光片)在传感器上成像(光带的形变刻画了到物体的距离)；(c)产生的 3D 点集被转换为(d)中的三角网格

Bouguet and Perona(1999)提出了光带距离获取的一个有趣的变种。他们简单地挥动一个根子在一个由点光源如灯或太阳照射下的场景中或物体上留下阴影(图 12.6a)，而不是投影光带。当阴影落在两个相对于摄像机而言其方向已知(或通过预标定推断出)的背景平面上时，可以根据这两个 3D 方程已知的投影直线推断出每个光带的平面方程(图 12.6b)。阴影沿着扫描物体的形变就揭示了其 3D 形状，类似于通常的光带主动距离获取(习题 12.2)。这种方法也可用于估计背景场景的 3D 几何及其进入阴影时外观的变化，从而在场景中加入新的阴影(Chuang, Goldman, Curless et al. 2003)(10.4.3 节)。

使用光带方法扫描一个物体所需的时间正比于所使用的深度平面的个数，通常可以与一幅图像上的像素点个数比较。通过使用一个结构化的方式，例如，使用二进制码或格雷码，开关不同的投影器像素可以建立一个更快的扫描仪(Besl 1989)。例如，假设我们使用的液晶投影仪有 1024 列像素。使用 10 位二进制码对应每一列的地址 (0...1023)，我们先投影第一位，再投影第二位，依此类推。经过 10 次投影 (30Hz 的摄像机-投影仪系统只需三分之一秒)，摄像机中的每个像素都知道它看见的是投影光源的 1024 列中的哪一列。我们也可以使用类似的方法通过在物体后面放置一个监视器以估计物体的折射性质(Zongker, Werner, Curless et al. 1999; Chuang, Zongker, Hindorff et al. 2000)(13.4 节)。使用单束激光也可以构造一个很快的扫描仪，即实时飞点(flying spot)光学三角扫描仪(Rioux, Bechthold, Taylor et al. 1987)。

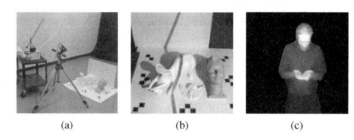

图 12.6 使用投射阴影扫描形状(Bouguet and Perona 1999)© 1999 Springer：(a)摄像机装置与一个点光源(不带反射罩的台灯)和一个用于投射阴影的手持棍子以及(b)两个平面背景前的被扫描的物体；(c)使用脉冲照明系统的实时深度图(Iddan and Yahav 2001)© 2001 SPIE

如果需要更快的，也就是帧速率的扫描，我们可以在场景上投影单个纹理模式。Proesmans, Van Gool, and Defoort(1998) 描述了一个系统，该系统将一个棋盘格投影到物体(如人脸)上，而这个棋盘格的形变则用以推测 3D 形状。不幸的是，这样的方法仅在表面足够连续以至所有棋盘格点能够相互连接时有效。

使用高速自定义照明和感知硬件可以构建一个更好的系统。Iddan and Yahav(2001)介绍了他们的 3DV Zcam 视频速率深度感知摄像机，该系统将一个平面的光线脉冲投影到场景中，然后累积短时间内返回的光线，以获得场景中每个像素的距离的飞行时间度量。关于早期飞行时间系统，包括 LIDAR 的幅度和频率调制方式，可以参见(Besl 1989)。

除了使用单个摄像机，我们也可以使用立体视觉成像装置构建一个主动照明距离传感器。最简单的方法是在场景中投影随机的带状模式以产生人工纹理，这对无纹理表面的匹配有帮助(Kang, Webb, Zitnick et al. 1995)。而投影一系列已知的光带，就像在编码模式单摄像机距离获取中一样，可使像素间的对应没有歧义且支持恢复仅在单个摄像机中可见的像素的深度(Scharstein and Szeliski 2003)。这种方法已被用于产生大量高精度注册的多图像立体视觉对及其深度图以评测立体视觉对应算法(Scharstein and Szeliski 2002; Hirschmüller and Scharstein 2009)及学习深度图的先验和参数(Scharstein and Pal 2007)。

虽然投影多个模式通常需要场景或物体保持静止，但使用附加的处理可以产生动态场景的实时深度图。基本的想法(Davis, Ramamoorthi, and Rusinkiewicz 2003;

Zhang, Curless, and Seitz 2003)是假设在每个像素周围的 3D 时空窗内深度近似恒定,并使用 3D 时空窗进行匹配及重建。这个假设在某些表面形状和运动下是易于出错的,如(Davis, Nahab, Ramamoorthi *et al.* 2005)中所示。为了更精确地对形状进行建模,Zhang, Curless, and Seitz(2003)对时空窗内的线性视差变化建模,发现通过在视频体上对视差和视差梯度进行全局优化可以得到更好的结果(Zhang, Snavely, Curless *et al.* 2004)。图 12.7 显示了该系统在人脸上应用的结果。这样,帧速率 3D 表面建模可以进一步用于基于模型的拟合及计算机图形学操作(12.6.2 节)。

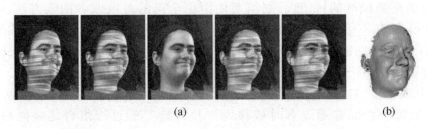

图 12.7 利用时空立体视觉方法捕捉的实时稠密 3D 人脸(Zhang, Snavely, Curless *et al.* 2004)© 2004 ACM:(a)两个立体视觉摄像机中的连续 5 帧视频(每 5 帧中有一帧没有条状模式,用于提取纹理);(b)生成的高质量 3D 表面模型(作为阴影表面渲染的深度图)

12.2.1 距离数据归并

单个距离图像可用于各种应用,如实时 z 键控或人脸运动捕捉,而它们也可以作为更加完整的 3D 物体建模的构件。在这样的应用中,接下来的两个处理步骤是局部 3D 表面模型的注册(配准(对齐))以及将它们整合为一致的 3D 表面(Curless 1999)。如果需要,可以接着一个模型拟合的步骤,使用参数化表示(例如广义柱(Agin and Binford 1976; Nevatia and Binford 1977; Marr and Nishihara 1978; Brooks 1981)和超二次型(Pentland 1986; Solina and Bajcsy 1990; Terzopoulos and Metaxas 1991),或者非参数化模型(如三角网格 (Boissonat 1984)),或者基于物理的模型(Terzopoulos, Witkin, and Kass 1988; Delingette, Hebert, and Ikeuichi 1992; Terzopoulos and Metaxas 1991; McInerney and Terzopoulos 1993; Terzopoulos 1999))进行模型拟合。还有一些方法将距离图像分割为更简单的表面构件(Hoover, Jean-Baptiste, Jiang *et al.* 1996)。

3D 注册方法中应用最广泛的是迭代最近点(iterated closet point,ICP)算法,它交替地在待注册的两个表面上寻找最近点,然后解一个 3D 绝对定向问题(6.1.5 节,(6.31 和 6.32)(Besl and McKay 1992; Chen and Medioni 1992; Zhang 1994; Szeliski and Lavallée 1996; Gold, Rangarajan, Lu *et al.* 1998; David, DeMenthon, Duraiswami *et al.* 2004; Li and Hartley 2007; Enqvist, Josephson, and Kahl 2009)。[①]由于待注册的两个表面通常仅有部分重叠并也许有很多外点,所以我们通常使用鲁棒

---

① 某些方法,例如 Chen and Medioni (1992)提出的方法,利用局部表面的切平面来使计算更加精确并加速收敛。

的匹配准则(6.1.4 节及附录 B.3)。为了加速最近点的确定及使到表面距离的计算更加精确,两个点集中的一个(例如当前的归并了的模型)可以转化为一个带符号的距离函数,可以选择使用八叉树样条来紧凑地表示(Lavallée and Szeliski 1995)。基本 ICP 算法的变种可用于注册具有非刚性形变的 3D 点集,例如医学应用(Feldmar and Ayache 1996; Szeliski and Lavallée 1996)。点上的颜色值或者距离度量也可作为注册过程的一部分使用以提高鲁棒性(Johnson and Kang 1997; Pulli 1999)。

不幸的是,ICP 算法及其变种只能找到 3D 表面的局部最优配准。如果没有已知的(关于全局配准的)先验,则需要使用更全局的对应或者搜索方法,基于对 3D 刚体变换具有不变性的局部描述子。此类描述子的一个例子是旋转图像(spin image),它是在局部法向轴周围的 3D 表面块的圆投影(Johnson and Hebert 1999)。另一个(早期的)例子是 Stein and Medioni(1992)引入的飞溅(splash)表示。

两个或更多 3D 表面配准后,它们可以归并为单个模型。一种方法是使用三角网格表示每个表面并通过有时被称为"拉拉链"的过程组合这些网格(Soucy and Laurendeau 1992; Turk and Levoy 1994)。另一个应用更广泛的方法是计算一个带符号的距离函数以拟合所有的 3D 数据点(Hoppe, DeRose, Duchamp et al. 1992; Curless and Levoy 1996; Hilton, Stoddart, Illingworth et al. 1996; Wheeler, Sato, and Ikeuchi 1998)。

图 12.8 显示了一个这样的方法,Curless and Levoy(1996)提出的体积的距离图像处理(volumetric range image processing,VRIP),它首先根据每个距离图像计算一个带权有符号的距离函数,然后使用加权平均过程归并它们。为使表达更加简洁,可使用行程编码来编码那些空的、可见的和变化(带符号距离)的体素(voxel),而仅存储每个表面附近的带符号距离值①。一旦计算得到归并后的带符号距离函数,便可用一个过零点表面提取算法,如移动立方体法(marching cube)(Lorensen and Cline 1987),来恢复一个网格表面模型。图 12.9 显示一个完整的距离数据归并和等值面提取流程。

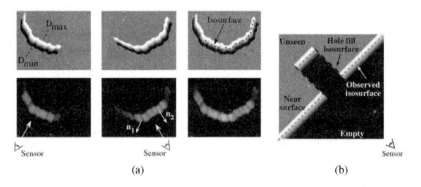

图 12.8　距离图像归并(Curless and Levoy 1996) © 1996 ACM:(a)两个带符号距离函数(左上)利用它们的权重(左下)合并产生一个函数的合并集(右列),从中可以提取等值面(绿色虚线);(b)带符号距离函数与空及不可见空间标签相结合以在等值面中填补空洞

---

① 另一个更加紧凑的表示方式是八叉树(Lavallée and Szeliski 1995)。

基于带符号距离或特征(内-外)函数的体积的距离数据归并方法广泛应用于根据有向或无向点集提取光滑的良性的表面(Hoppe, DeRose, Duchamp *et al.* 1992; Ohtake, Belyaev, Alexa *et al.* 2003; Kazhdan, Bolitho, and Hoppe 2006; Lempitsky and Boykov 2007; Zach, Pock, and Bischof 2007b; Zach 2008)，如 12.5.1 节所述。

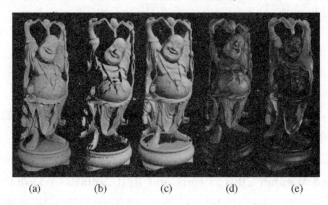

图 12.9　弥勒佛雕塑的重建和复制(Curless and Levoy 1996)© 1996 ACM：(a)原始雕塑喷涂灰色漆后的照片；(b)部分距离扫描数据；(c)合并的距离扫描数据；(d)重建模型的彩色渲染；(e)使用立体光刻得到的重建模型的复制品

12.2.2　应用：数字遗产

主动距离获取方法，与表面及表观建模方法结合(12.7 节)，广泛应用于考古学及文物保护领域，这些应用也被称为"数字遗产"(MacDonald 2006)。在这种应用中，我们获得文化遗产的详细 3D 模型以便以后的应用，例如分析、保护、复原及生产艺术品的复制品(Rioux and Bird 1993)。

这项工作最近的一个例子是数字米开朗基罗计划，它使用安装在高架上的 Cyberware 激光带扫描仪及高清数字单反相机获取米开朗基罗的大卫像及佛罗伦萨的其他雕塑的扫描数据。此计划还扫描了古罗马城图志——一个破碎的古代石材质的罗马地图，并使用数字方法获得这些碎片的新的匹配。整个过程，从初始计划到软件开发、获取及后处理，消耗数年时间(且有众多志愿者参与)，作为结果最终产生了极为丰富的 3D 形状及表观建模方法。

图 12.10　大吴哥巴戎寺的激光距离建模(Banno, Masuda, Oishi *et al.* 2008)© 2008 Springer：(a)现场的照片样例；(b)从地面扫描的详细头部模型；(c)利用气球上装载的一个激光距离传感器扫描并最终融合的 3D 寺庙模型

现在有更大规模的计划正在进行中，例如，扫描大吴哥这样的整座寺庙(Ikeuchi and Sato 2001; Ikeuchi and Miyazaki 2007; Banno, Masuda, Oishi *et al.* 2008)。图 12.10 给出了这个计划的细节，包括一个照片示例，一个从地平面扫描得到的详细的 3D(雕刻的)头部模型，及最终归并的 3D 遗址模型的空中概览——这是使用一个气球得到的。

12.3 表面表达

在前面的小节中，我们已经看到了用于综合 3D 距离扫描的不同表达。我们现在详细考虑其中一些表达方式。显式表面表达，如三角网格、样条(Farin 1992, 1996)及细分表面(Stollnitz, DeRose, and Salesin 1996; Zorin, Schröder, and Sweldens 1996; Warren and Weimer 2001; Peters and Reif 2008)，不仅可以创建高度详细的模型，还可以处理操作，如插值(12.3.1 节)、整形或平滑，以及抽取(decimation)和简化(12.3.2 节)。我们还调查了基于点的离散表达(12.4 节)及体积表达(12.5 节)。

12.3.1 表面插值

针对表面，一个最普遍的操作是从一系列稀疏数据约束中重建表面，即散列数据插值。为了表示这样的问题，可以将表面表示为高度场 $f(x)$，或 3D 参数化表面 $f(x)$，或非参数化模型(如一系列三角面片)。

在图像处理的章节中，我们介绍了二维函数插值及近似问题 $\{d_i\} \rightarrow f(x)$ 可使用正规化转化为能量最小化问题(3.7.1 节(3.94~3.98)。[①])这种问题也可同时指定曲面上不连续处与局部方向约束(Terzopoulos 1986b; Zhang, Dugas-Phocion, Samson *et al.* 2002)。

解决这种问题的一个方法是将表面和能量函数都使用有限元分析(3.100~3.102)在一个离散栅格或网格(grid or mesh)上离散化(Terzopoulos 1986b)。然后可以稀疏系统解法，如多重网格(Briggs, Henson, and McCormick 2000)或分层前置条件共轭梯度法(Szeliski 2006b)，来解决这类问题。表面也可使用一个多级 B 样条的分层组合(Lee, Wolberg, and Shin 1996)表示。

另外一种方法是使用径向基函数(或者核函数)(Boult and Kender 1986; Nielson 1993)。径向基函数用以下公式插值一个场 $f(x)$ 使之经过(或靠近)位于 x_i 的数据值 d_i：

$$f(\boldsymbol{x}) = \frac{\sum_i w_i(\boldsymbol{x})\boldsymbol{d}_i}{\sum_i w_i(\boldsymbol{x})}, \tag{12.6}$$

其中权重，

$$w_i(\boldsymbol{x}) = K(\|\boldsymbol{x} - \boldsymbol{x}_i\|), \tag{12.7}$$

[①] 插值与近似的区别在于，前者要求表面或者函数经过这些数据，而后者允许函数经过数据附近，因此也可以用于表面平滑。

使用径向基(球对称)函数 $K(r)$ 计算。

如果要求函数 $f(x)$ 精确地插值数据点，那么核函数在原点必须是奇异的，$\lim_{r\to 0} K(r) \to \infty$，否则必须解一个稠密线性系统以确定每个基函数的幅度(Boult and Kender 1986)。其实，对于特定的正规化问题，例如(3.94～3.96)，存在径向基函数(核函数)，能给出纯解析解法那样的结果(Boult and Kender 1986)。不幸的是，由于稠密系统的解法的复杂度是数据点数的立方，所以基函数方法只适用于诸如基于特征的图像变形等小规模问题(Beier and Neely 1992)。

当要对三维参数化表面进行建模时，(12.6)或(3.94～3.102)中的向量值函数 f 编码了表面上的 3D 坐标 (x,y,z)，且域 $x=(s,t)$ 编码了表面参数。对称性搜索(symmetry-seeking)模型就是这种表面的一个例子，它是广义柱[①]的弹性形变版本(Terzopoulos, Witkin, and Kass 1987)。在这些模型中，s 是沿着可变形管状体脊方向的参数而 t 是环绕着管状体的参数。当使用模型拟合基于图像的轮廓曲线时，需要各种光滑及径向对称力的约束。

我们也可以定义非参数表面模型(如广义三角网格)并在这种网格中(使用有限元分析)同时加入内部光滑性度量和外部数据拟合度量(Sander and Zucker 1990; Fua and Sander 1992; Delingette, Hebert, and Ikeuichi 1992; McInerney and Terzopoulos 1993)。这些方法大部分都假定一个弹性的形变模型，这个模型使用一个二次内部光滑项，也可以使用一个次线性(sub-linear)能量模型以更好地保持表面的褶皱(Diebel, Thrun, and Brünig 2006)。三角网格也可以使用样条元素(Sullivan and Ponce 1998)或细分表面(Stollnitz, DeRose, and Salesin 1996; Zorin, Schröder, and Sweldens 1996; Warren and Weimer 2001; Peters and Reif 2008)进行增强以对表面提供更好的光滑控制。参数化和非参数化表面模型都假定表面的拓扑结构事先已知而且固定。为了表示更灵活的表面，我们可以使用有向点的汇集(12.4 节)或 3D 隐式函数(12.5.1 节)表示表面，还可以结合弹性 3D 表面模型(McInerney and Terzopoulos 1993)。

12.3.2 表面简化

从 3D 数据建立三角网格后，我们通常希望建立一个层次的网格模型，例如，用以在计算机图形学应用中控制细节水平(level of detail，LOD)。(本质上，这是图像金字塔(3.5 节)的 3D 模拟)。一种方法是用一个具有细分连接(subdivision connectivity)的网格近似给定网格，在这个近似网格上我们可以计算一组三角小波系数(Eck, DeRose, Duchamp et al. 1995)。一个更连续的方法是使用顺序的边折叠(edge collapse)操作将原始的高分辨率网格变为一个粗糙的基准级网格(Hoppe 1996)。得到的渐进网格(progressive mesh，PM)表示可用于在任意细节层次上渲染

[①] 广义柱(Brooks 1981)是一个旋转体，即一个(通常是光滑的)曲线绕一个轴旋转得到的结果。它也可以通过一个变化的圆截面沿着一个轴缓慢扫过得到(这两种解释是等价的)。

3D 模型，如图 12.11 所示。

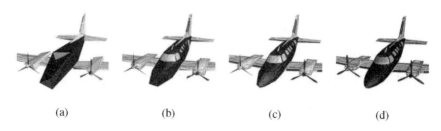

图 12.11　一个飞机模型的渐进网格表示(Hoppe 1996)© 1996 ACM：(a)基础网格 M^0(150 面)；(b)网格 M^{175} (500 面)；(c)网格 M^{425} (1000 面)；(d)原始网格 $M = M^n$ (13 546 面)

12.3.3　几何图像

虽然多分辨率表面表达，比如(Eck, DeRose, Duchamp et al. 1995; Hoppe 1996)，支持细节等级操作，但它们仍然由不规则的一组三角形组成，这使它们难以对缓存高效的方式压缩及存储[①]。

为使三角剖分完全正规(均匀的及栅格状的)，Gu, Gortler, and Hoppe(2002)描述了如何通过沿着精选的线切分表面网格及将结果表达"展平"为一个正方形来创建几何图像。图 12.12a 显示了表面网格映射在一个单位正方形上的 (x,y,z) 的值，而图 12.12b 显示了关联的 (n_x,n_y,n_z) 法向图，即每个网格顶点关联的表面法向，如果原始几何图像经过严重的压缩，它们就可以用于补偿视觉真实度的损失。

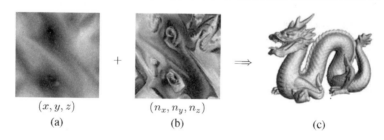

图 12.12　几何图像(Gu, Gortler, and Hoppe 2002)© 2002 ACM：(a)257×257 的几何图像定义一个表面上的网格；(b)512×512 的法向图定义顶点法向；(c)最终光照 3D 模型

12.4　基于点的表达

如前所述，基于三角形的表面模型假设 3D 模型的拓扑结构(通常以及粗略的形状)事先已知。尽管模型在形变或被拟合时可以对模型重新网格化，但一个更简单的解决方法是整个免除显式的三角网格，而是使用有向点(或称粒子)，或表面元

[①] 细分三角剖分，如(Eck, DeRose, Duchamp et al. 1995)中的那些，是半正规的，即在每个细分基本三角形中是正规的(有序及内嵌的)。

(surfel)(Szeliski and Tonnesen 1992)代替三角形的顶点。

为给得到的粒子系统赋予内部光滑性约束，我们可以定义成对的交互作用势函数以近似等价弹性弯曲能量，这可以使用局部有限元分析得到。①我们使用一个软性影响函数来耦合附近的粒子，而不是事先定义每个粒子(顶点)的有限元近邻。得到的 3D 模型可以在演化时同时改变拓扑结构和粒子密度，因此可以用于插值带孔的部分 3D 数据(Szeliski, Tonnesen, and Terzopoulos 1993b)。还可以对表面方向和褶皱曲线的不连续性进行建模(Szeliski, Tonnesen, and Terzopoulos 1993a)。

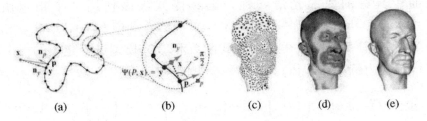

图 12.13 利用移动最小二乘(MLS)建立的基于点的表面模型(Pauly, Keiser, Kobbelt et al. 2003)© 2003 ACM：(a)一组点(黑色点)被转换为一个隐式内-外部函数(黑色曲线)；(b)到最近有向点的带符号距离可用于近似内部外部函数；(c)具有可变采样密度的一组有向点表示一个 3D 表面(头部模型)；(d)采样密度的局部估计，用于移动最小二乘；(e)重建的连续 3D 表面

为将此粒子系统渲染为一个连续的表面，我们可以使用局部动态三角化试探法(Szeliski and Tonnesen 1992)或者直接表面元素的体空间扫描(splatting)算法(Pfister, Zwicker, van Baar et al. 2000)。另一种方法是首先通过到有向点的最小带符号距离(Hoppe, DeRose, Duchamp et al. 1992)或以径向基函数插值一个特征(内部-外部)函数(Turk and O'Brien 2002; Dinh, Turk, and Slabaugh 2002)将点云转换为一个隐式带符号距离或者内部-外部函数。更精确的隐式函数拟合，包括处理非正规点密度的能力，可以通过计算带符号距离函数的一个移动最小二乘估计得到(Alexa, Behr, Cohen-Or et al. 2003; Pauly, Keiser, Kobbelt et al. 2003)，如图 12.13 所示。进一步的改进措施有使用局部球拟合(Guennebaud and Gross 2007)，更快、更精确的重采样(Guennebaud, Germann, and Gross 2008)以及使用核回归以更好地容忍外点(Oztireli, Guennebaud, and Gross 2008)来得到。

12.5 体积表达

3D 表面建模的第三个选项是建立 3D 体积内部-外部函数。在 11.6.1 节中，我们已经看到过这种例子，当时我们着眼于立体视觉匹配中的体元着色(Seitz and Dyer 1999)、空间雕刻(Kutulakos and Seitz 2000)以及水平集(Faugeras and Keriven

---

① 如前所述，一个替代的方法是使用次线性交互作用势函数，它支持保存表面褶皱(Diebel, Thrun, and Brünig 2006)。

1998; Pons, Keriven, and Faugeras 2007)方法，在 11.6.2 节中，我们也讨论了使用二进制轮廓图像来重建体积。在本节中，我们着眼于连续隐式(内部-外部)函数以表示 3D 形状。

隐式表面和水平集

虽然基于多面体和体元的表达可以以任意精度表达 3D 形状，但它们缺少连续隐式表面中具有的内部光滑属性，连续隐式表面使用一个指示函数(特征函数) $F(x,y,z)$ 指示 3D 点在物体内部 $F(x,y,z)<0$ 或者外部 $F(x,y,z)>0$。

超二次型是计算机视觉中使用隐式函数对 3D 模型进行建模的一个早期例子，它是一个二次型(例如椭圆体)参数体积模型的推广，

$$F(x,y,z) = \left(\left(\frac{x}{a_1}\right)^{2/\epsilon_2} + \left(\frac{y}{a_2}\right)^{2/\epsilon_2}\right)^{\epsilon_2/\epsilon_1} + \left(\frac{x}{a_1}\right)^{2/\epsilon_1} - 1 = 0 \quad (12.8)$$

(Pentland 1986; Solina and Bajcsy 1990; Waithe and Ferrie 1991; Leonardis, Jaklič, and Solina 1997)。(a_1,a_2,a_3) 的值控制模型沿着每个 (x,y,z) 轴伸展，而 $(\varepsilon_1,\varepsilon_2)$ 控制模型有多"方"。为了对变化范围更大的形状进行建模，超二次型通常与刚性或非刚性变形结合使用(Terzopoulos and Metaxas 1991; Metaxas and Terzopoulos 2002)。超二次型模型可以拟合距离数据或直接用于立体视觉匹配。

另一种隐式形状模型可以通过在一个正规 3D 网格上定义一个带符号的距离函数，还可以选择使用八叉树样条在远离表面(零集)的地方更粗略地表示这个函数(Lavallée and Szeliski 1995; Szeliski and Lavallée 1996; Frisken, Perry, Rockwood et al. 2000; Ohtake, Belyaev, Alexa et al. 2003)。我们已经见过一些例子：带符号距离函数用于表示距离变换(3.3.3 节)；水平集用于二维轮廓拟合及跟踪(5.1.4)；体积的立体视觉(11.6.1 节)；距离数据归并(12.2.1 节)及基于点的建模(12.4 节)。在网格上直接表示这种函数的优势是能够快速轻易地查找任意 (x,y,z) 位置的距离函数值，同时也可以很容易使用移动立方体算法提取等值面(Lorensen and Cline 1987)。Ohtake, Belyaev, Alexa et al.(2003)的工作最值得一提，因为它允许同时使用几个距离函数并在局部结合以产生尖锐的特征(如褶皱)。

泊松表面重建(Kazhdan, Bolitho, and Hoppe 2006)使用一个紧密相关的体积函数，即一个经过平滑的 0/1 内部-外部(特征)函数，它可以被看作一个经过裁剪的带符号距离函数。这个函数的梯度在已知表面点附近设为沿着有向表面法向，在其他地方设为 0。这个函数本身使用一个八叉树上的二次张量乘积 B 样条表示，它在远离表面或点密度较低的区域使用更大的单元从而提供一个紧致的表示，同时允许相关泊松公式的高效解法(3.100～3.102)，参见 9.3.4 节(Pérez, Gangnet, and Blake 2003)。

我们也可以使用 L_1(全变差)约束代替泊松方程中使用的二次惩罚项并得到一个凸优化问题，这个问题可以用连续(Zach, Pock, and Bischof 2007b; Zach 2008)或离

散图割(Zach, Pock, and Bischof 2007b; Zach 2008)方法解决。

带符号的距离函数对水平集演化方程(5.1.4 节和 11.6.1.节)是不可或缺的,其中距离函数在网格上的值随着表面的演化更新以拟合多视角立体视觉的光度一致性(photoconsistency)度量(Faugeras and Keriven 1998)。

12.6 基于模型的重建

如果我们事先对待建模的物体有所了解,我们可以使用专门的方法及表达建立更详细而可靠的 3D 模型。例如,建筑物通常由大块平面区域及其他参数化形式(例如旋转曲面)构成,通常方向垂直于重力方向并且相互垂直(12.6.1 节)。由于人脸形状和表观的变化虽然很大,但仍是有界的,头部和人脸可以使用低维非刚性形状模型表示(12.6.2 节)。人体或者部件,例如手部,是高度关节式的结构,可以使用分段刚性骨架元素通过关节链接形成的运动链表示(12.6.4 节)。

本节中,我们强调某些主要用于这三种情况的某些想法、表示和建模算法。额外的细节及引用可以参考专注于这些主题的会议及研讨会,例如 the International Symposium on 3D Data Processing, Visualization, and Transmission(3DPVT), the International Conference on 3D Digital Imaging and Modeling(3DIM), the International Conference on Automatic Face and Gesture Recognition(FG), the IEEE Workshop on Analysis and Modeling of Faces and Gestures 和 the InternationalWorkshop on Tracking Humans for the Evaluation of their Motion in Image Sequences(THEMIS)。

12.6.1 建筑结构

建筑建模,特别是从航拍图像,是摄影制图学及计算机视觉中研究时间最长的问题之一(Walker and Herman 1988)。最近,可靠的基于图像建模方法的发展,以及数码摄像机和 3D 电脑游戏的普及,刺激了这个领域的复兴。

Debevec, Taylor, and Malik(1996)的工作是最早的混合式基于几何和基于图像的建模及渲染的系统之一,他们的 Façade 系统结合了一个交互的以图像为指导的几何建模工具和一个基于模型(局部平面加上视差)的立体视觉匹配以及视角相关纹理匹配。在交互式摄影制图建模过程中,用户选择块元素并将它们的边缘与输入图像上的可见边缘对齐(图 12.14a)。这个系统接着使用约束优化自动地计算这些块的尺寸和位置以及摄像机位置(图 12.14b 和 c)。这个方法本质上比一般基于特征的由运动到结构算法更可靠,因为它使用了块基元中可用的强几何信息。Becker and Bove(1995), Horry, Anjyo, and Arai(1997)和 Criminisi, Reid, and Zisserman(2000)的相关工作使用了从消失点中可获得的类似信息。在 Sinha, Steedly, Szeliski et al.(2008)的基于图像的交互建模系统中,消失点方向被用于指导用户画出多边形,这些多边形接着自动拟合到使用由运动到结构得到的稀疏 3D 点上。

图 12.14 利用 Façade 系统进行交互式建筑建模(Debevec, Taylor, and Malik 1996)© 1996 ACM：(a)一幅带有用户描绘的边缘(用绿色表示)的输入图像；(b)带阴影 3D 实体模型；(c)几何基元覆盖在输入图像上；(d)最终视角相关的纹理映射的 3D 模型

估计出粗糙的几何结构后，可以使用一个局部平面扫描算法计算每个平面的更详细的偏移图(offset map)，这在 Debevec, Taylor, and Malik(1996)中被称为"基于模型的立体视觉"。最后，在渲染时，使用一个叫"视角相关的纹理映射"过程(与光场以及光图法渲染相关，见 13.3 节)，将摄像机在场景中移动时所获得的不同视角的图像卷绕后混合在一起(图 12.14d)。

对于内部建模，相对于使用单幅图片，使用全景图更加有效，因为可以看到墙或其他结构的更大的范围。Shum, Han, and Szeliski(1998)开发的 3D 建模系统首先从多幅图像中建立标定的全景图(7.4 节)，然后让用户在图像上画出垂直和水平线以标定平面区域的边界。这些直线起初用于对每个全景图建立一个绝对旋转量并稍后用于(与推测出的顶点和平面一起)优化 3D 结构，这些 3D 结构可以从一幅或多幅图像中恢复出来(图 12.15)。360°高动态范围全景图也可以用于户外建模，由于它们提供了有关摄像机相对方向以及消失点方向的高度可靠的估计(Antone and Teller 2002; Teller, Antone, Bodnar *et al.* 2003)。

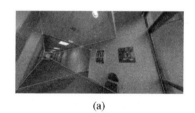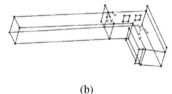

(a)　　　　　　　　　　　　(b)

图 12.15 全景图中的交互式 3D 建模(Shum, Han, and Szeliski 1998)© 1998 IEEE：(a)全景图的广角视角以及用户描绘的垂直和水平(对齐到坐标轴的)线；(b)走廊的单视角重建

虽然早期基于图像的建模系统需要一些用户监督，Werner and Zisserman(2002)展现了一个全自动基于直线的重建系统。如 7.5.1 节所述，他们首先检测直线和消失点并用它们标定摄像机，然后他们使用表观匹配和三视张量建立直线对应，这使他们能够重建 3D 线段簇，如图 12.16a 所示。他们接着同时使用共面 3D 直线和一个基于在兴趣点处求出的互相关分数的平面扫描算法生成平面假设。平面的交线被用于确定每个平面的范围，即一个初始的几何结构，然后加上矩形或楔形形状的缺

口和突出部分以改进这个结构(图 12.16c)。注意，如果有正在建模建筑的自顶向下的图，可以用于进一步约束 3D 建模过程(Robertson and Cipolla 2002, 2009)。使用匹配的 3D 直线以估计消失点方向和主平面，这一想法仍然继续应用在一些近期全自动基于图像的建筑建模系统中(Zebedin, Bauer, Karner et al. 2008; Mičušík and Košeckà 2009; Furukawa, Curless, Seitz et al. 2009b; Sinha, Steedly, and Szeliski 2009)。

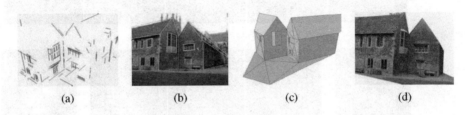

图 12.16 利用 3D 直线和平面的自动建筑重建(Werner and Zisserman 2002)© 2002 Springer：(a)重建的 3D 直线，根据其消失方向着色；(b)叠加在输入图像上的线框模型；(c)带窗户的三角面片模型；(d)最终带纹理映射的模型

建筑结构的另一个共性是基元(如窗户、门和柱廊)的重复使用。可以设计一个建筑建模系统来搜索这样的重复元素并将它们作为结构推断过程的一部分(Dick, Torr, and Cipolla 2004; Mueller, Zeng, Wonka et al. 2007; Schindler, Krishnamurthy, Lublinerman et al. 2008; Sinha, Steedly, Szeliski et al. 2008)。

这些方法的组合现在使大型 3D 场景的结构重建成为可能(Zhu and Kanade 2008)。例如，Pollefeys, Nistér, Frahm et al.(2008)的 Urbanscan 系统从装有 GPS 的车辆所拍视频重建城市街道纹理映射 3D 模型。为了得到实时的性能，他们既使用了最优在线由运动到结构算法以及 GPU 实现的对齐于主平面的平面扫描立体视觉算法和深度图融合。Cornelis, Leibe, Cornelis et al.(2008)提出一个相关的系统，也使用平面扫描立体视觉(对齐到建筑的外立面)，同时结合针对车辆的物体识别和分割方法。Mičušík and Košeckà(2009)在这些结果的基础上使用沿主平面方向的全方向图像和基于超像素立体视觉匹配算法。我们也可以直接从主动距离扫描数据同时结合经过曝光和光线变化补偿的彩色图像重建(Chen and Chen 2008; Stamos, Liu, Chen et al. 2008; Troccoli and Allen 2008)。

12.6.2 头部和人脸

特定形状和表观模型非常有用的另一个领域是头部和人脸建模。即使人们的表观初看是千差万别的，但一个人的头部和人脸的实际形状可以使用几十个参数很好地描述(Pighin, Hecker, Lischinski et al. 1998; Guenter, Grimm, Wood et al. 1998; DeCarlo, Metaxas, and Stone 1998; Blanz and Vetter 1999; Shan, Liu, and Zhang 2001)。

图 12.17 是一个基于图像的建模系统的示例，其中几幅图像中用户指定的关键

点被用于将一个通用头部模型拟合到人脸上。如图 12.17c 所示，指定刚刚超过 100 个关键点后，人脸的形状已经非常合适并可以辨认。从原始图像中提取纹理图然后应用于头部模型，便可以得到一个具有高视觉保真度的可动画的模型(图 12.18a)。

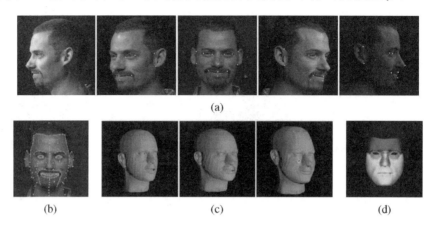

图 12.17 拟合在一组图像上的 3D 模型(Pighin, Hecker, Lischinski et al. 1998)© 1998 ACM：(a)带有用户选定关键点的五幅输入图像集；(b)关键点和曲线的完整集合；(c)三个网格——原始网格、利用 13 个关键点调整的网格以及加上另外 99 个关键点后的网格；(d)将图像划分为可分开动画的区域

通过在一组 3D 扫描人脸上应用主分量分析(prinapal component analysis，PCA)我们可以建立一个更强大的系统，我们将在 12.6.3 节中讨论这个问题。如图 12.19 所示，我们可以将可变形 3D 模型拟合到单幅图像上将这个模型用于各种动画及视觉特效(Blanz and Vetter 1999)。我们也可以设计立体视觉匹配算法直接优化头部模型参数(Shan, Liu, and Zhang 2001; Kang and Jones 2002)或使用带主动照明的实时立体视觉的输出(Zhang, Snavely, Curless et al. 2004)(如图 12.7 和图 12.18b 所示)。

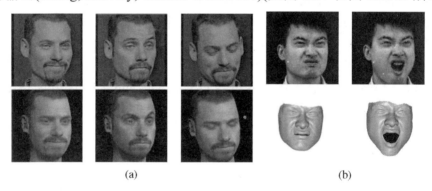

图 12.18 利用 3D 形变模型的头部和表情跟踪与动画再现：(a)在五段输入视频流中直接拟合的模型(Pighin, Szeliski, and Salesin 2002) © 2002 Springer，下面一行显示动画再现的合成的纹理映射的 3D 模型，其姿态和表情参数与上面一行的输入图像相符；(b)帧速率时空立体视觉表面模型拟合的模型(Zhang, Snavely, Curless et al. 2004)© 2004 ACM，上面一行显示覆盖了合成绿色标记点的输入图像，下一行显示拟合的 3D 表面模型

随着 3D 脸部捕捉系统的发展，重建模型的细致和真实程度也在不断进化。更新的系统不仅可以(实时)捕捉表面细节如皱纹和褶皱，还可以建立准确的皮肤反射、半透明以及次表面漫反射模型(Weyrich, Matusik, Pfister et al. 2006; Golovinskiy, Matusik, ster et al. 2006; Bickel, Botsch, Angst et al. 2007; Igarashi, Nishino, and Nayar 2007)。

3D 头部模型建立后，可以用于各种应用，如头部跟踪(Toyama 1998; Lepetit, Pilet, and Fua 2004; Matthews, Xiao, and Baker 2007)，如图 4.29 和图 14.24 所示，人脸转移，即在视频中将一个人的脸部替换为另一个人的(Bregler, Covell, and Slaney 1997; Vlasic, Brand, Pfister et al. 2005)。另外的应用包括通过将人脸图像卷绕到一个更加迷人的"标准"上进行脸部美化(Leyvand, Cohen-Or, Dror et al. 2008)、用于隐私保护的人脸去识别(de-indentification)(Gross, Sweeney, De la Torre et al. 2008)以及人脸交换(Bitouk, Kumar, Dhillon et al. 2008)。

12.6.3 应用：脸部动画

3D 头部建模最广泛的应用大概是脸部动画。一旦我们建立了形状和表观(表面纹理)的参数化 3D 模型，就可以直接使用它来跟踪人的脸部运动(图 12.18a)并驱动一个不同的角色做同样的运动和表情(Pighin, Szeliski, and Salesin 2002)。

可以通过首先应用主分量分析(PCA)对可能的头部形状和脸部表情进行建模来建立一个这种系统的改进版。Blanz and Vetter(1999)描述了一个系统，其中他们首先捕捉了一个有 200 张人脸的着色距离扫描图的集合，这个集合可表示为一大批 (X,Y,Z,R,G,B) 样本(顶点)。[①]为使 3D 形变有意义，首先必须找出不同人的扫描数据之间的对应(Pighin, Hecker, Lischinski et al. 1998)。这个过程完成之后，可以应用主分量分析来更自然的参数化 3D 可变形模型。这个模型的灵活性可以通过在不同的子区域，如眼睛、鼻子和嘴，进行不同的分析来提高，就像在模块化本征空间中一样(Moghaddam and Pentland 1997)。

子空间表示计算完成后，这个空间中的不同的方向可以关联到不同的特征上，比如性别、脸部表情或者脸部特征(图 12.19a)。如在 Rowland and Perrett(1995)的工作中，可以通过放大人脸与平均图像之间的位移以生成漫画。

在手动将模型初始化调整到大概正确的姿态、大小和位置之后，3D 可变形模型可对输入图像和重新合成的模型图像之间的误差应用梯度下降法拟合到一幅图像上(图 12.19b 和 c)。这个拟合过程的效率可以使用反向合成图像配准算法提高，具体描述参见 Romdhani and Vetter(2003)。

我们可以修改得到的纹理映射 3D 模型以生成各种视觉效果，包括改变一个人的体重或者表情或者三维效果，比如重新设置照明或 3D 基于视频的动画(13.5.1 节)。这种模型也可以用于视频压缩，例如通过仅传送少量脸部表情及姿态参数来

[①] 圆柱坐标系提供这个集合的自然二维嵌入，但这样的嵌入对于主分量分析来说是不必要的。

驱动一个合成的头像(Eisert, Wiegand, and Girod 2000; Gao, Chen, Wang et al. 2003)。

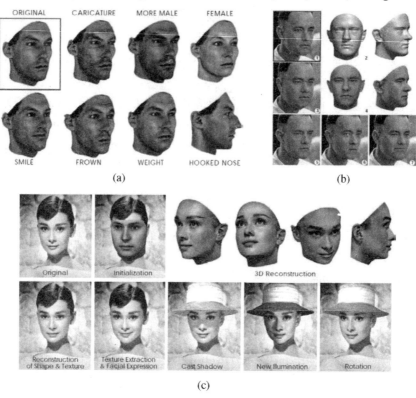

图 12.19　3D 可变形人脸模型(Blanz and Vetter 1999)© 1999 ACM：(a)附加了指定属性上形状和纹理变化的原始 3D 模型：性别、表情、体重和鼻子形状到均值的差异；(b)与单幅图像匹配的 3D 可变形模型，拟合后可以控制其重量和表情；(c)带有另一组 3D 操作比如照明和姿态变化的 3D 重建的另一个例子

3D 脸部动画通常与一个演员的表演相匹配，被称为"表演驱动的动画"(4.1.5 节)(Williams 1990)。传统的表演驱动的动画系统使用基于标记的动作捕捉(Ma, Jones, Chiang et al. 2008)，而一些较新的系统使用视频录像(video footage)来控制动画(Buck, Finkelstein, Jacobs et al. 2000; Pighin, Szeliski, and Salesin 2002; Zhang, Snavely, Curless et al. 2004; Vlasic, Brand, Pfister et al. 2005)。

后一种方法的一个例子是为电影《本杰明·巴顿》(又译"巴顿奇事"和"返老还童")开发的系统，其中 Digital Domain(数字领域)公司使用 Mova[①]的 CONTOUR(轮廓)系统来捕捉演员 Brad Pitt(布拉德·皮特)的脸部运动和表情(Roble and Zafar 2009)。CONTOUR 系统使用结合了磷光漆和多个高清视频摄像机来捕捉演员的实时 3D 深度扫描数据。这些 3D 模型紧接着被转换为脸部动作编码系统(Facial Action Coding System，FACS))中的形状和表情参数(Ekman and Friesen 1978)以驱动一个不同的(较旧的)合成动画的计算机生成的形象(CGI)角色。

---

① *http://www.mova.com*

12.6.4 完整人体建模与跟踪

关于跟踪人体，对它们的形状和表观进行建模，并识别人体活动的主题是计算机视觉中最活跃的研究领域之一。年度会议[①]和杂志的特刊(Hilton, Fua, and Ronfard 2006)专注于这个问题，以及近期两个综述(Forsyth, Arikan, Ikemoto *et al.* 2006; Moeslund, Hilton, and Krüger 2006)各自列举了400多篇关于这些主题的论文[②]。人体运动的 HumanEva 数据集[③]包含关节型(articulated)人体活动的多视角视频序列以及运动捕捉数据、评估代码和一个供参考的基于粒子滤波的 3D 跟踪算法。随数据库提供的 Sigal, Balan, and Black(2010)的论文不仅描述了这个数据集以及评估方法，还包含这个领域的重要工作的一个很好的综述。

由于这个领域过于宽广，我们很难对这些研究进行分类，尤其是由于不同的方法经常建立在彼此的基础上。Moeslund, Hilton, and Krüger(2006)把他们的综述划分为初始化、跟踪(包括背景建模和分割)、姿态估计和活动(行为)识别。Forsyth, Arikan, Ikemoto *et al.*(2006)的综述划分为关于跟踪(背景差分、可变形模板、流和概率模型)、从 2D 观测中恢复 3D 姿态、数据关联和身体部件。其中还包含一个关于运动合成的章节，运动合成在计算机图形学中有更加广泛的研究(Arikan and Forsyth 2002; Kovar, Gleicher, and Pighin 2002; Lee, Chai, Reitsma *et al.* 2002; Li, Wang, and Shum 2002; Pullen and Bregler 2002)，详情参见 13.5.2 节。这个领域工作的另一个潜在的分类方法是按照是否使用 2D 还是 3D(或多视角)图像作为输入和是否使用 2D 或 3D 运动学模型。

本节中，我们简要回顾背景差分、初始化和检测、使用流跟踪、3D 运动学模型、概率模型、自适应形状建模以及动作识别领域中的一些原创及被广泛引用的论文。对于其他问题和更多的细节，我们推荐读者参考前面提到的综述。

背景差分 很多(但肯定不是全部)人体跟踪系统的第一步是建立背景模型以提取对应于人的移动前景物体(轮廓)。Toyama, Krumm, Brumitt *et al.*(1999)回顾了一些差分抠图和背景维护(建模)方法并提供了对这个主题的一个好的介绍。Stauffer and Grimson(1999)描述了一些基于混合模型的方法，Sidenbladh and Black(2003)则开发了一个更综合的方法，它不仅对背景图像统计信息建模，还对前景物体的表观进行建模，例如它们的边缘和运动(帧间差异)的统计信息。

一旦轮廓一个或多个摄像机中被提取出来，就可以用可变形模板或其他轮廓模型进行建模(Baumberg and Hogg 1996; Wren, Azarbayejani, Darrell *et al.* 1997)。在时

[①] International Conference on Automatic Face and Gesture Recognition (FG), IEEE Workshop on Analysis and Modeling of Faces and Gestures 和 International Workshop on Tracking Humans for the Evaluation of their Motion in Image Sequences(THEMIS)。
[②] 更早的综述包括 Gavrila(1999) 和 Moeslund and Granum(2001)。一些关于姿势识别的综述，本书中并没有覆盖，包括 Pavlović, Sharma, and Huang(1997) and Yang, Ahuja, and Tabb(2002)。
[③] http://vision.cs.brown.edu/humaneva/

间上跟踪这样的轮廓可以支持对在场景内移动的多个人的分析,包括建立形状和表观模型以及检测他们是否携带物品(Haritaoglu, Harwood, and Davis 2000; Mittal and Davis 2003; Dimitrijevic, Lepetit, and Fua 2006)。

初始化和检测 为了以全自动的方式跟踪人,有必要首先在各个视频帧中检测(或者重新获得)他们的存在。这个问题与行人检测紧密相关,行人检测通常被当作一种物体识别(Mori, Ren, Efros *et al.* 2004; Felzenszwalb and Huttenlocher 2005; Felzenszwalb, McAllester, and Ramanan 2008),因此在14.1.2节中将深入讨论。其他基于2D图像初始化3D跟踪器的方法包括Howe, Leventon, and Freeman(2000), Rosales and Sclaroff(2000), Shakhnarovich, Viola, and Darrell(2003), Sminchisescu, Kanaujia, Li *et al.*(2005), Agarwal and Triggs(2006), Lee and Cohen(2006), Sigal and Black(2006)以及Stenger, Thayananthan, Torr *et al.*(2006)。

单帧人体检测和姿态估计算法有时可以使用自己进行跟踪(Ramanan, Forsyth, and Zisserman 2005; Rogez, Rihan, Ramalingam *et al.* 2008; Bourdev and Malik 2009),如4.1.4节所述。然而更多的时候,它们和逐帧跟踪方法结合使用以提供更强的可靠性(Fossati, Dimitrijevic, Lepetit *et al.* 2007; Andriluka, Roth, and Schiele 2008; Ferrari, Marin-Jimenez, and Zisserman 2008; Andriluka, Roth, and Schiele 2009)。

使用流跟踪 人体及其姿态的逐帧跟踪可以通过计算光流或逐帧匹配人体肢体的表观得以增强。例如,Ju, Black, and Yacoob(1996)的纸板小人模型将每个腿的部件(上部和下部)的表观建模为一个移动的矩形,并使用光流来估计它们在后续每一帧中的位置。Cham and Rehg(1999)和Sidenbladh, Black, and Fleet(2000)使用光流和模板跟踪肢体,并提供了处理多个假设和不确定性的方法。Bregler, Malik, and Pullen(2004)使用一个肢体及人体运动的全3D模型,如下文所述。我们也可以将估计的运动场本身和某些准则进行匹配以识别跑步动作中的特定阶段或匹配两个低分辨率视频片段以进行视频替换(Efros, Berg, Mori *et al.* 2003)。

3D运动学模型 使用人体形状和运动的一个更精确的3D模型,人体建模和跟踪的效果可以大大提高。这种表示(在游戏中的3D计算机动画以及特效中是普遍存在的)的本质是一个运动学模型或运动链,它指定了骨架中每个肢体的长度以及肢体或片段之间的2D或3D旋转角(图12.20a–b)。从可见表面点的位置推测关节角度值的过程被称为"逆运动学"(inverse kinematics,IK),它在计算机图形学中得到了广泛的研究。

图12.20a显示了Rehg, Morris, and Kanade(2003)用以在视频中跟踪手部运动的手部的运动学模型。如图所示,手指之间的结合点有两个自由度,而手指关节本身仅有一个自由度。使用这类模型可以大大增强基于边缘的跟踪器处理快速运动、3D姿态歧义以及部分遮挡的能力。

运动学链式模型在全身建模及跟踪中的应用甚至更广泛(O'Rourke and Badler 1980; Hogg 1983; Rohr 1994)。一个流行的方法是在运动学模型中将每个刚性肢体与一个椭圆体或超二次型结合,如图12.20b所示。然后,这个模型可以通过匹配从已知背景中抽取出来的轮廓或者通过匹配和跟踪遮挡边缘的位置，被拟合到一个或多个视频流的每一帧上(Gavrila and Davis 1996; Kakadiaris and Metaxas 2000; Bregler, Malik, and Pullen 2004; Kehl and Van Gool 2006)。注意，有的方法使用与2D测量的2D模型，有的使用 3D 测量(距离数据或多视角视频)和 3D 模型，有的使用单目视频直接推测及跟踪 3D 模型。

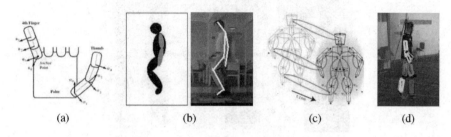

图 12.20 跟踪 3D 人体运动：(a)人类手部的运动学链式模型(Rehg, Morris, and Kanade 2003)© 2003, reprinted by permission of SAGE；(b)在一个视频序列中跟踪一个运动学链链滴(kinematic chain blob)模型(Bregler, Malik, and Pullen 2004)© 2004 Springer；(c 和 d)人体部件的概率松散肢体汇集(Sigal, Bhatia, Roth et al. 2004)

我们也可以使用时序模型，通过分析关节角度对时间的函数，以改进周期性运动，如步行的跟踪(Polana and Nelson 1997; Seitz and Dyer 1997; Cutler and Davis 2000)。这类方法的一般性和适应性可以通过使用主分量分析学习典型运动模式来提高(Sidenbladh, Black, and Fleet 2000; Urtasun, Fleet, and Fua 2006)。

概率模型　由于跟踪可以是一项非常困难的任务，所以复杂的概率推断方法经常被用于估计被跟踪人的可能状态。一个流行的方法，称为"粒子滤波"(Isard and Blake 1998)，最初用于跟踪人和手的轮廓，如 5.1.2 节(图 5.6～5.8)所述。它随后被用于全身跟踪(Deutscher, Blake, and Reid 2000; Sidenbladh, Black, and Fleet 2000; Deutscher and Reid 2005)并继续用在现代跟踪器中(Ong, Micilotta, Bowden et al. 2006)。处理跟踪问题中固有的不确定性的其他方法包括多假设跟踪(Cham and Rehg 1999)和膨胀协方差(Sminchisescu and Triggs 2001)。

图 12.20c 和 d 显示了一个被称为"松散肢体人"(loose-limbed people)的复杂时空概率图模型，它对不同肢体之间的几何约束和它们可能的时间动力学都进行了建模(Sigal, Bhatia, Roth et al. 2004)。他们从训练数据中学习出不同肢体在不同时间的条件概率，并使用粒子滤波进行最终的姿态推断。

自适应形状建模　全身建模和跟踪的另一个重要部分是参数模型到视觉数据的拟合。如 12.6.3 节所述(图 12.19)，注册的大量 3D 距离扫描数据可用于建立形状和表观的可变形模型(Allen, Curless, and Popović 2003)。Anguelov, Srinivasan, Koller et al.(2005)在这个工作的基础上提出一个被称为"SCAPE"(Shape Completion and Animation for PEople，人的形状补全和动画)的复杂系统，它首先获取不同人和一个人不同姿态的距离扫描数据，然后使用半自动标记布置将这些数据注册。注册的数据集被用于建立作为人物特征和骨骼姿态(例如某个关节上的肌肉是收缩的(图 12.21,

顶行))的函数的形状变化模型。这个系统可以用于形状补全，即通过同时在形状和姿态空间中寻找最符合观测数据的模型参数以便能从捕捉到的少量标记点恢复出整个儿 3D 网格模型。

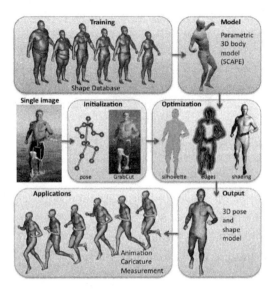

图 12.21 使用参数化 3D 模型从单幅图像中估计人体形状和姿态(Guan, Weiss, Bălan *et al.* 2009)© 2009 IEEE

由于该系统完全由身穿紧身衣的人体扫描数据建立并且使用一个参数化形状模型，所以它不能处理身穿宽松服饰的人。Bălan and Black(2008)通过估计一个适合于从多个视角观测同一个人得到的视觉外型(visual hull)的人体形状突破了这个局限。而 Vlasic, Baran, Matusik *et al.*(2008)则调整一个配备了一个参数化形状模型的初始表面网格以更好的匹配视觉外型。

前述的人体拟合和姿态估计系统使用多视角图像来估计人体形状，Guan, Weiss, Bălan *et al.*(2009)更近期的工作可以将一个人体形状和姿态模型拟合到人在自然背景下的单幅图像上。它使用手动初始化来估计一个粗略姿态(骨架)和高度模型，然后使用它们通过 Grab Cut 分割算法(5.5 节)分割出人的轮廓。然后结合轮廓边缘信息和阴影信息来改善形状和姿态估计的结果(图 12.21)。结果的 3D 模型可以用于创作新的动画。

活动识别 人体建模中最后一个广泛研究的主题是运动(motion)、活动(activity)和动作(action)识别(Bobick 1997; Hu, Tan, Wang *et al.* 2004; Hilton, Fua, and Ronfard 2006)。经常识别的动作包括行走和奔跑、跳跃、跳舞、拾取物品、坐下和站起以及挥手。关于这些问题近期的代表性的论文有 Robertson and Reid(2006), Sminchisescu, Kanaujia, and Metaxas(2006), Weinland, Ronfard, and Boyer(2006), Yilmaz and Shah(2006), and Gorelick, Blank, Shechtman *et al.*(2007)。

12.7 恢复纹理映射与反照率

得到物体或人体的 3D 模型后，建模的最后一步通常是恢复纹理映射以描述物体的表面表观。这首先需要建立 (u,v) 纹理坐标的参数化，将其表示为 3D 表面位置的函数。一个简单的方法是对每个三角形(或每对三角形)关联一个独立的纹理映射。更省空间的方法包括将表面反卷绕到一个或多个纹理映射上，例如，使用细分网格(12.3.2 节)(Eck, DeRose, Duchamp *et al.* 1995)或者几何图像(12.3.3 节)(Gu, Gortler, and Hoppe 2002)。

一旦确定每个三角形的 (u,v) 坐标后，从纹理 (u,v) 到图像 j 的像素 (u_j,v_j) 的透视投影方程可以通过连接仿射映射 $(u,v) \rightarrow (X,Y,Z)$ 和透视单应(同态) $(X,Y,Z) \rightarrow (u_j,v_j)$ 得到(Szeliski and Shum 1997)。然后可以重采样并存储 (u,v) 纹理映射的颜色值，或原始图像本身使用投影纹理映射也可当作纹理源(OpenGL-ARB 1997)。

当有多个源图像可用于表观恢复时(这是通常情况)，情况更加复杂。一种方法是使用视角相关的纹理映射(13.1.1 节)，其中根据虚拟摄像机与表面法向以及源图像的角度，每个多边形面使用不同的源图像(或源图像的组合)。另一个方法是为每个表面点估计完整的表面光场(Wood, Azuma, Aldinger *et al.* 2000)，如 13.3.2 节所述。在某些情况下，例如，在传统 3D 游戏中使用模型时，在预处理阶段将所有的源图像合并为一个一致的纹理映射则更为适合。理想情况下，每个表面三角应选择它在其中最直接可见(垂直于它的法向)并最能匹配纹理映射的分辨率的源图像。[①] 这个问题可以看作一个图割优化问题，其中光滑项鼓励邻接三角形使用相似的源图像，接着是混合以补偿曝光差异(Lempitsky and Ivanov 2007; Sinha, Steedly, Szeliski *et al.* 2008)。通过对几何和源图像间的光度测量学误配准(misalignment)进行建模，可以得到更好的结果(Shum and Szeliski 2000; Gal, Wexler, Ofek *et al.* 2010)。

这类方法在光源相对于物体保持固定时，即摄像机在物体或空间周围移动时，可以得到好结果。然而，当照明具有很强的方向性时以及物体相对光源移动时，可能会出现强阴影效果或高光，这会干扰纹理(反照率)映射的可靠恢复。在这种情况下，更可取的做法是恢复纹理映射时通过对光源方向建模并估计表面反射属性来显式去除阴影效果(12.1 节)(Sato and Ikeuchi 1996; Sato, Wheeler, and Ikeuchi 1997; Yu and Malik 1998; Yu, Debevec, Malik *et al.* 1999)。图 12.22 显示了这类方法的一个结果，其中，在估计满反射(matte reflectance)成分(反照率)时，高光首先被去除，并稍后通过估计 Torrance–Sparrow 反射模型(2.91)中的镜面反射成分 k_s 高光被重新引入。

[①] 当从倾斜的角度观察表面时，也许需要混合不同的图像以得到最好的分辨率(Wang, Kang, Szeliski *et al.* 2001)。

图 12.22 估计一个扫描 3D 模型的漫反射反照率和反射参数(Sato, Wheeler, and Ikeuchi 1997)© 1997 ACM：(a)投影在模型上的输入图像集合；(b)完整漫反射(反照率)模型；(c)包含镜面反射成分的反射模型的渲染

12.7.1 估计 BRDF

解决视角相关表观建模的一个更进取的方法是为物体表面上每个点估计一个一般双向反射分布函数(BRDF)。Dana, van Ginneken, Nayar et al.(1999), Jensen, Marschner, Levoy et al.(2001)，以及 Lensch, Kautz, Goesele et al.(2003)提出了不同的估计这种函数的方法，Dorsey, Rushmeier, and Sillion(2007)和 Weyrich, Lawrence, Lensch et al.(2008)给出了关于 BRDF 建模、恢复和渲染这个主题的最新综述。

如 2.2.2 节(2.81)所述，BRDF 可表示为

$$f_r(\theta_i, \phi_i, \theta_r, \phi_r; \lambda), \tag{12.9}$$

其中 (θ_i, ϕ_i) 和 (θ_r, ϕ_r) 是入射光线 $\hat{v}_i$ 和反射光线 $\hat{v}_r$ 在局部表面坐标系 $(\hat{d}_x, \hat{d}_y, \hat{n})$ 下的角度，如图 2.15 所示。对物体的表观进行建模时，相对于一块材料的表观，我们需要在物体表面上每个点 (x,y) 上估计这个函数，这给出了空间变化的 BRDF，即 SVBRDF(Weyrich, Lawrence, Lensch et al. 2008)

$$f_v(x, y, \theta_i, \phi_i, \theta_r, \phi_r; \lambda). \tag{12.10}$$

如果要对次表面(sub-surface)散射效果建模，如光通过如雪花石膏等材料长距离传播时，需要使用 8 维双向散射表面反射分布函数(BSSRDF)：

$$f_e(x_i, y_i, \theta_i, \phi_i, x_e, y_e, \theta_e, \phi_e; \lambda), \tag{12.11}$$

其中下标 e 在这里表示发射光而不是反射光方向。

Weyrich, Lawrence, Lensch et al.(2008)针对这些问题以及相关问题提供了一个很好的综述，包括基本光度学、BRDF 模型、使用反射测向(视觉角度和反射量的精确测量)的传统 BRDF 获取、皮肤建模(Debevec, Hawkins, Tchou et al. 2000; Weyrich, Matusik, Pfister et al. 2006)以及基于图像的获取方法，它从多张照片中同时恢复物体的 3D 形状和反射率。

后一种方法的一个很好的例子是由 Lensch, Kautz, Goesele et al.(2003)开发的系统，他们估计局部变化的 BRDF 并使用对表面法向的局部估计来改进他们的形状模型。为了建立模型，他们首先为每个表面点关联一个亮度纹理(lumitexel)，lumitexel 包括一个 3D 位置、一个表面法向和一个稀疏辐射率样本集。接着，他们使用 Lafortune 反射模型(Lafortune, Foo, Torrance et al. 1997)和分裂(divisive)聚类

法将这些 lumitexel 聚类为具有共同属性的材质(图 12.23a)。最后，为了对细致空间可变表观进行建模，每个 lumitexel(表面点)被投影到聚类表观模型的基上(图 12.23b)。

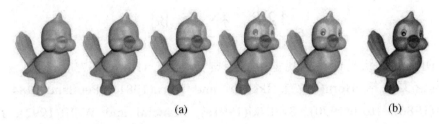

图 12.23 基于图像的表观及细致几何重建(Lensch, Kautz, Goesele et al. 2003)© 2003 ACM：(a)利用分裂(divisive)聚类重新估计表观模型(BRDFs)；(b)为了刻画细致的空间变化表观，每个亮度纹理(lumitexel)被投影到由聚类材质形成的基上

本节中讨论的大多数方法需要大量视角以估计表面属性，未来的一个具有挑战性的方向是将这些方法搬出实验室放在真实世界中，并将它们与常规的和因特网图像之基于图像的建模方法相结合。

12.7.2 应用：3D 摄影学

本章描述的从多幅图像中建立完整 3D 模型并接着恢复它们的表面表观的方法开创了一个全新的应用领域，该领域通常被称为"3D 摄像学"。Pollefeys and Van Gool(2002)提供了关于这个领域的很好的介绍，包括特征匹配的处理步骤、由运动到结构的恢复[①]、稠密深度图估计、3D 建模和纹理映射恢复。Vergauwen and Van Gool(2006) 和 Moons, Van Gool, and Vergauwen(2010)描述了一个完整的基于网络的系统来自动完成这些任务，称为 ARC3D。后一篇论文不仅提供了整个领域的深度综述，而且提供了完整的端到端系统的详细描述。

除了这样的全自动系统，我们还可以将用户加入循环，这有时被称为"交互式计算机视觉"。van den Hengel, Dick, Thormhlen et al.(2007)描述了他们的 VideoTrace 系统，该系统进行视频中的自动的点跟踪和 3D 结构恢复，然后让用户在得到的点云上画三角形和表面，并以交互方式调整模型点的位置。Sinha, Steedly,Szeliski et al.(2008)描述了一个系统，该系统在多幅图像(图 4.45)中使用匹配消失点推测 3D 直线方向和平面法向，然后用于指导用户画出轴对齐的平面，这些平面被自动拟合到 3D 点云上。Zebedin, Bauer, Karner et al.(2008), Furukawa, Curless, Seitz et al.(2009a), Furukawa, Curless, Seitz et al.(2009b), Mičušík and Košeckà(2009)以及 Sinha, Steedly, and Szeliski(2009)则提出了基于这些想法的全自动系统。

[①] 这些早期步骤在 7.4.4 节中也有讨论。

随着这些方法的复杂度和可靠性不断改进，我们有望看到更加用户友好的从图像作真实感 3D 建模的应用(习题 12.8)。

12.8 补充阅读

由阴影到形状是计算机视觉中的经典问题之一(Horn 1975)。该领域中的一些代表性的论文包括 Horn(1977), Ikeuchi and Horn(1981), Pentland(1984), Horn and Brooks(1986), Horn(1990), Szeliski(1991a), Mancini and Wolff(1992), Dupuis and Oliensis(1994), 以及 Fua and Leclerc(1995)。Horn and Brooks(1989)编辑的论文集，尤其是关于变分法的那一章，是关于这个主题的一个很好的资料。Zhang, Tsai, Cryer et al.(1999)的综述不仅回顾了更近期的方法，还提供了一些比较结果。Woodham(1981)则写了关于光度测量立体视觉的开创性论文。由纹理到形状方法包括 Witkin(1981), Ikeuchi(1981), Blostein and Ahuja(1987), Garding(1992), Malik and Rosenholtz(1997), Liu, Collins, and Tsin(2004), Liu, Lin, and Hays(2004), Hays, Leordeanu, Efros et al.(2006), Lin, Hays, Wu et al.(2006), Lobay and Forsyth(2006), White and Forsyth(2006), White, Crane, and Forsyth(2007)以及 Park, Brocklehurst, Collins et al.(2009)。关于由散焦到深度，好的论文和图书包括 Pentland(1987), Nayar and Nakagawa(1994), Nayar, Watanabe, and Noguchi(1996), Watanabe and Nayar(1998), Chaudhuri and Rajagopalan(1999)以及 Favaro and Soatto(2006)。从各种照明效果中恢复形状的其他方法，包括互反射(Nayar, Ikeuchi, and Kanade 1991)，Wolff, Shafer, and Healey(1992b)编辑的形状恢复相关图书中有讨论。

主动距离获取系统(使用投影在场景中的激光或自然光照明)的相关描述可参见 Besl(1989), Rioux and Bird(1993), Kang, Webb, Zitnick et al.(1995), Curless and Levoy(1995), Curless and Levoy(1996), Proesmans, Van Gool, and Defoort(1998), Bouguet and Perona(1999), Curless(1999), Hebert(2000), Iddan and Yahav(2001), Goesele, Fuchs, and Seidel(2003), Scharstein and Szeliski(2003), Davis, Ramamoorthi, and Rusinkiewicz(2003), Zhang, Curless, and Seitz(2003), Zhang, Snavely, Curless et al.(2004), 以及 Moons, Van Gool, and Vergauwen(2010)。独立的距离获取数据可以使用 3D 对应和距离优化方法(如迭代最近点及其变种)进行对齐(Besl and McKay 1992; Zhang 1994; Szeliski and Lavallée 1996; Johnson and Kang 1997; Gold, Rangarajan, Lu et al. 1998; Johnson and Hebert 1999; Pulli 1999; David, DeMenthon, Duraiswami et al. 2004; Li and Hartley 2007; Enqvist, Josephson, and Kahl 2009)。距离扫描数据对齐后，可以使用对体积样本点到表面带符号距离函数建模的方法进行归并(Hoppe, DeRose, Duchamp et al. 1992; Curless and Levoy 1996; Hilton, Stoddart, Illingworth et al. 1996; Wheeler, Sato, and Ikeuchi 1998; Kazhdan, Bolitho, and Hoppe 2006; Lempitsky and Boykov 2007; Zach, Pock, and Bischof 2007b; Zach 2008)。

一旦 3D 表面建立后，可以使用各种三维表达进行建模和操作，这些三维表达包括三角网格(Eck, DeRose, Duchamp et al.1995; Hoppe 1996)、样条(Farin 1992, 1996; Lee, Wolberg, and Shin 1996)、细分表面(Stollnitz, DeRose, and Salesin 1996; Zorin, Schröder, and Sweldens 1996; Warren and Weimer 2001; Peters and Reif 2008)和几何图像(Gu, Gortler, and Hoppe 2002)。作为选择，它们还可以表示为一系列带局部方向估计的点样本(Hoppe, DeRose, Duchamp et al. 1992; Szeliski and Tonnesen 1992; Turk and O'Brien 2002; Pfister, Zwicker, van Baar et al. 2000; Alexa, Behr, Cohen-Or et al. 2003; Pauly, Keiser, Kobbelt et al. 2003; Diebel, Thrun, and Brünig 2006; Guennebaud and Gross 2007; Guennebaud, Germann, and Gross 2008; Oztireli, Guennebaud, and Gross 2008)。表面还可以使用在正规或非正规(八叉树)体积网格上采样的隐式内部-外部特征函数或带符号距离函数建模(Lavallée and Szeliski 1995; Szeliski and Lavallée 1996; Frisken, Perry, Rockwood et al. 2000; Dinh, Turk, and Slabaugh 2002; Kazhdan, Bolitho, and Hoppe 2006; Lempitsky and Boykov 2007; Zach, Pock, and Bischof 2007b; Zach 2008)。

基于模型的 3D 重建，相关文献有很多。对于建筑及城市场景建模，人们开发出了交互式的或全自动的系统。专注于大型 3D 场景重建的特刊(Zhu and Kanade 2008)是一个很好的参考资料，而 Robertson and Cipolla(2009)给出了一个完整系统的很好的描述。12.6.1 节中可以找到很多额外的参考文献。人脸和全身建模及跟踪是计算机视觉中的一个很活跃的子领域，它有自己的会议和研讨会，例如 the International Conference on Automatic Face and Gesture Recognition(FG), the IEEE Workshop on Analysis and Modeling of Faces and Gestures, and the InternationalWorkshop on Tracking Humans for the Evaluation of their Motion in Image Sequences(THEMIS)。近期关于全身建模和跟踪主题的综述论文包括 Forsyth, Arikan, Ikemoto et al.(2006), Moeslund, Hilton, and Krüger(2006), and Sigal, Balan, and Black(2010)。

12.9 习 题

习题 12.1：由聚焦到形状

使用数码单反相机的手动对焦档(或使用允许程序对焦的相机)拍摄一系列图像并恢复一个物体的形状。

1. 拍摄一些标定物体，例如一个棋盘，以便可以计算散焦程度和焦距设置间的映射。
2. 尝试一个正面平行的平面目标和一个倾斜的、能够覆盖传感器工作范围的目标。哪一个更好一些？
3. 现在在场景中放入一个真实物体并进行一个类似的聚焦扫描。
4. 对每个像素，计算局部锐利度并拟合一个关于焦距的抛物曲线以寻找最清晰的焦距设置。
5. 将这些焦距设置映射为深度并比较你的结果和真实值。如果使用的是一个已知简单物体，如球或圆柱(一个球或者苏打罐)，很容易测量它的真实形状。
6. (可选)看看你是否能仅从两个或三个焦距设置恢复出深度图。
7. (可选)使用 LCD 投影仪在场景上投影人工纹理。使用一对摄像机来比较你的由聚焦到形状和由立体视觉到形状方法。
8. (可选)使用 Agarwala, Dontcheva, Agrawala et al.(2004)的方法建立一个全在焦图像。

习题 12.2：阴影条纹

实现 Bouguet and Perona(1999)的手持阴影条纹系统。基本步骤如下。

1. 在目标物体后设置两个背景平面并计算它们相对于观察者的方向，例如，使用基准标记。
2. 使用一个棍子在场景中投射阴影，使用摄像机记录视频或捕捉数据。
3. 根据阴影在两个背景上的投影估计每个光平面方程。
4. 在每条曲线上对剩下的点进行三角剖分以得到 3D 条纹并使用 3D 图像引擎显示这些条纹。
5. (可选)去除对已知第二个(垂直)平面的要求并使用 Bouguet and Perona(1999)所描述的方法推测它的位置(或者光源的位置)。习题 10.9 中的方法在这里同样有所帮助。

习题 12.3：距离数据注册

使用迭代最近点(ICP)算法(Besl and McKay 1992; Zhang 1994; Gold, Rangarajan, Lu et al. 1998)或八叉树带符号距离场(Szeliski and Lavallée 1996)来注册两个或多个 3D 数据集(12.2.1 节)。

将你的系统应用于窄基线立体视觉对，例如，通过在物体周围移动摄像机而获得的数据，使用由运动到结构恢复摄像机姿态，并使用一个标准的立体视觉匹配算法。

习题 12.4：距离图像归并

使用带符号距离场(Curless and Levoy 1996; Hilton, Stoddart, Illingworth et al. 1996)对你在上一题中注册的数据集进行归并。如果在上一个注册阶段已经实现八叉树，你可以选择使用它来压缩表示这个场。

使用移动立方体算法从带符号距离场中抽取网格表面模型并显示结果模型。

习题 12.5：表面简化

使用渐进网格(Hoppe 1996)或 12.3.2 节中的一些其他方法建立表面模型的分层简化模型。

习题 12.6：建筑建模

根据一系列手持广角照片建立某个建筑结构(如你家)的一个 3D 内部或外部模型。

1. 抽取直线和消失点(习题 4.11~4.15)以估计每个图像的主方向。
2. 使用由运动到形状恢复所有的摄像机姿态并与消失点相配。
3. 让用户通过在直线上描绘对应于墙壁底部、顶部和水平范围的直线来确定墙壁的位置(Sinha, Steedly, Szeliski *et al.* 2008),参见习题 6.9。对开口处(门和窗)和简单的家具(桌子和柜台面)做类似的工作。
4. 将得到的多边形网格转换为 3D 模型(如 VRML)并可选作对这些表面用图像进行纹理映射。

习题 12.7:人体跟踪

从 HumanEva 网站下载视频序列。[①]从头开始实现一个人体运动跟踪系统或以某种有趣的方式扩展网站上的代码(Sigal, Balan, and Black 2010)。

习题 12.8:3D 摄影学

结合你前面所开发的所有方法,生成一个系统,拍摄一系列照片或视频并建立具有真实感纹理映射的 3D 模型。

---

① *http://vision.cs.brown.edu/humaneva/*

第 13 章

基于图像的绘制

13.1 视图插值
13.2 层次深度图像
13.3 光场与发光图
13.4 环境影像形板
13.5 基于视频的绘制
13.6 补充阅读
13.7 习题

图 13.1 基于图像的绘制和基于视频的绘制：(a)照片游览重建系统的 3D 视图(Snavely, Seitz, and Szeliski 2006)© 2006 ACM；(b)4D 光场切片(Gortler, Grzeszczuk, Szeliski et al. 1996)© 1996 ACM；(c)带深度的子画面(Shade, Gortler, He et al. 1998)© 1998 ACM；(d)表面光场(Wood, Azuma, Aldinger et al. 2000) © 2000 ACM；(e)新背景前的环境影像形板(Zongker, Werner, Curless et al. 1999)© 1999 ACM；(f)实时视频环境影像形板(Chuang, Zongker, Hindorff et al. 2000)© 2000 ACM；(g)用于对原始视频重新进行动画绘制的视频重写(Bregler, Covell, and Slaney 1997)© 1997 ACM；(h)蜡烛火焰的视频纹理(Schodl, Szeliski, Salesin et al. 2000)© 2000 ACM；(i)视频中的视图插值(Zitnick, Kang, Uyttendaele et al. 2004)© ACM

在过去的二十年中，基于图像的绘制方法(Image-based Rendering)逐渐成为计算机视觉领域中最让人兴奋的应用之一(Kang, Li, Tong et al. 2006; Shum, Chan, and Kang 2007)。基于图像的绘制方法将计算机视觉领域中的三维重建(3D reconstruction)方法与计算机图形学中的绘制方法结合起来，使得计算机能够使用一个场景之多个视角的图像生成交互式的真实感体验，如图 13.1a 所示的照片游览系统。一些相关的商用系统已经出现，如在线地图系统中的沉浸式街景浏览[①]以及由大量随意拍摄的照片组合而成的 3D Photosynths[②]。

在本章中，我们将探讨各种基于图像的绘制方法，如图 13.1 所示。首先我们将介绍视图插值(view interpolation)方法(13.1 节)，这种方法使用单张或多张预先计算好的深度图将两幅参考图像无缝地拼接起来。与这个想法密切相关的一个工具是视图相关的纹理映射(13.1.1 节)，它是多个纹理映射在一个 3D 模型的表面上的混合。在视图插值中，有很多灵巧的方式可以用来描述彩色影像和 3D 几何信息，例如层次深度图像(13.2 节)和带深度的子画面(13.2.1 节)。

接着，我们将介绍另一类基于图像的绘制方法——使用光场(light field)和场景外观的四维发光图(Lumigraph)表示(13.3 节)，这类方法可以用于绘制任意视点的场景图像。这种表示的一些变体包括非结构化的发光图(13.3.1 节)、表面光场(13.3.2 节)、同心拼图(13.3.3 节)和环境影像形板(13.4 节)。

本章的最后一部分将讨论基于视频的绘制，即使用一个或多个视频来生成全新的视频体验(13.5 节)。我们将讨论两个具体的问题，一是基于视频的面部动画(13.5.1 节)，二是视频纹理(13.5.2 节)，即将一些较短的视频剪辑无缝地连接成一段动态的、实时的、基于视频的场景绘制。在本章末尾，我们将讨论由多个视频流生成 3D 视频(13.5.4 节)的方法以及基于视频的游览(13.5.5 节)，其中基于视频的游览在沉浸式户外地图系统和驾驶导向系统中已经得到了广泛的应用。

13.1 视图插值

尽管基于图像的绘制这个术语首次出现在 Chen(1995) 和 McMillan and Bishop(1995)的论文中，但 Chen and Williams(1993)关于视图插值的工作被认为是这个领域的奠基之作。视图插值方法是将两幅原始彩色图像和它们对应的预先计算好的深度图结合起来，通过插值方法生成这两张图之间视角的图像，即模拟一个虚拟摄像机在这两幅参考图像之间所拍摄到的画面。

视图插值方法融合了计算机视觉和计算机图形学领域已有研究中的两个基本想法：一是在计算过程中将参考图像和计算得到的深度图结合起来，然后使用纹理映射得到的 3D 模型生成新视角的图像(图 11.1)；二是变形(3.6.3 节)(图 3.53)，即使用

---

① http://maps.bing.com and http://maps.google.com
② http://photosynth.net

两幅图像之间的对应关系,分别将每幅参考图像卷绕(warp)到一个中间位置,同时将这两个卷绕后的图像交溶(cross-dissolve)(淡入淡出)到一起。

图 13.2 更加详细地展现了这个过程。首先,两幅源图像被分别卷绕到新的视角,这个过程中用到了参考摄像机和虚拟 3D 摄像机的姿态信息以及每幅图像的深度图(2.68~2.70)。在论文 Chen and Williams(1993)中使用了一种叫"前向卷绕"的算法(算法 3.1 和图 3.46)。考虑到空间效率和绘制的时间效率,深度图是用四叉树表示的(Samet 1989)。

在前向卷绕过程中,多个像素(相互间有遮挡)可能会落在同一个目标像素上。为了解决这个问题,可以为每个目标像素维护一个关联的 z-缓冲深度值,或者根据极线几何的知识计算参考图像的深度顺序,然后将这些图像按照深度从后到前的顺序进行卷绕(Chen and Williams 1993; Laveau and Faugeras 1994; McMillan and Bishop 1995)。

当两幅参考图像被卷绕到新的视角(图 13.2a 和 b)后,我们就可以把它们合并到一起形成一个合成图像(图 13.2c)。合并时,在卷绕后的两幅图像中,若其中一幅的某个像素为空洞(如图中青色像素所示),则我们使用另一幅图像中该像素点的颜色值;若两幅图像对应像素均不是空洞时,则可以按照通常做变形的方法进行混合,即根据虚拟摄像机与源摄像机间的相对距离进行混合。需要注意的一点是,在对实际图像做视图插值时,两幅图像的曝光量可能相差很大,由此,按照上面的做法,在新合成的视角图像中,被填充过的空洞像素点与混合得到的像素点曝光量就会不同,导致一些细小的缺陷。

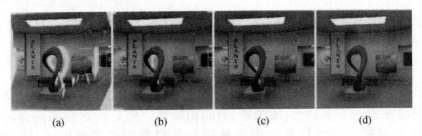

图 13.2 视图插值(Chen and Williams 1993)© 1993 ACM:(a)一张原始图像中的空洞(蓝色部分);(b)将两张相隔较远的图像进行组合后结果中的空洞;(c)将两张相隔较近的图像进行组合后结果中的空洞;(d)插值(空洞填补)后的结果

视图插值的最后一步(图 13.2d)是填充剩下的所有空洞和裂纹,这些空洞和裂纹通常是由前向卷绕过程或源数据的缺失(场景可见性问题)产生的。为了填补这些空洞和裂纹,我们可以从与这个空洞相邻的更远的(further)像素复制像素颜色值(否则,前景物体将产生"肥胖效果")。

上面的视图插值算法对于刚性场景效果已经很好,不过其视觉效果(缺乏频谱混叠)还可以通过使用两遍的前向-后向(forward-backward)算法(13.2.1 节)(Shade, Gortler, He et al. 1998)或者全 3D 绘制(Zitnick, Kang, Uyttendaele et al. 2004)进一步提高。当两幅参考图像是一个非刚性场景的两个视图时,例如一幅图像中一个人面带笑容，而另一幅图像中这个人却皱着眉，对于这样的情况我们可以使用视图变形(view morphing)算法(Seitz and Dyer 1996)，该算法融合了视图插值算法和一般变形算法的思想。

尽管最初的视图插值论文描述了如何基于类似的预计算(线性透视)图像生成新视图的方法，但全光函数建模的论文 McMillan and Bishop (1995)认为应该使用圆柱形图像存储预计算绘制或真实世界图像。(Chen 1995)也提出了使用(圆柱形的、立方形的或球形的)环境贴图作为视图插值的源图像。

13.1.1 视图相关的纹理映射

视图相关的纹理映射(view-dependent texture map)(Debevec, Taylor, and Malik 1996)与视图插值算法密切相关。不同于为同一场景每个视图的输入图像分别关联一个深度图，在视图相关的纹理映射中，我们只为场景生成一个唯一的 3D 模型，但是根据对应的虚拟摄像机的位置使用不同的图像作为纹理映射源(图 13.3a)。①

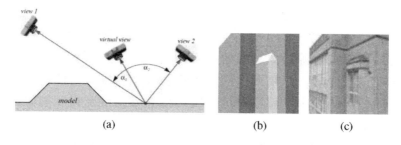

图 13.3 视图相关纹理映射(Debevec, Taylor, and Malik 1996)© 1996 ACM：(a)每个源视图的权重和源视图与新(虚拟)视图的相对角度有关；(b)简化的 3D 几何模型；(c)使用视图相关纹理映射后，几何模型看起来更加细致(如内嵌的窗户)

具体来说，给定一个虚拟摄像机的位置，该算法将比较这个虚拟摄像机视图和可能的源图像在每一个多边形(或像素)上的相似性。然后，这些源图像将按照一定的权重被混合起来，这些权重与虚拟视图和源视图所成的角度 α_i (图 13.3a)成反比。尽管原始的几何模型可能相当粗糙(图 13.3b)，但是混合不同视角图像产生了强烈的具有更细致几何模型的感觉，这是因为对应像素间的视差(视觉运动)的原因。在提出视图相关纹理映射的原始论文中，加权混合的计算是在每个像素或粗糙多边形表面分别进行的，而作为后续的相关研究，Debevec, Yu, and Borshukov(1998)提出了一个更加高效的实现，在他们的实现中，他们为视图空间的各个部分预先计算了贡献，然后使用投影纹理映射(OpenGL-ARB 1997)得到结果。

视图相关的纹理映射被提出后，这个思想就被广泛应用于后来的基于图像的绘

---

① 目前，人们普遍使用"基于图像的建模"(image-based modeling)这个术语来描述由多个图像生成一个带有纹理映射的 3D 模型的过程，这个术语首次出现在 Debevec, Taylor and Malik(1996)的论文中。作者同时也使用了"摄影测量建模"(photogrammetric modeling)这个词来描述这个过程。

制系统中，包括人脸建模和动画(Pighin, Hecker, Lischinski *et al.* 1998)以及 3D 扫描和可视化(Pulli, Abi-Rached, Duchamp *et al.* 1998)。后面将要介绍的发光图和非结构化发光图系统(13.3 节)(Gortler, Grzeszczuk, Szeliski *et al.* 1996; Buehler, Bosse, McMillan *et al.* 2001)的基础——4D 空间中的光线混合与视图相关纹理映射有着密切的联系。

Debevec, Taylor and Malik(1996)为了使他们的外墙绘制系统显得更真实，他们使用了基于模型的立体视觉(model-based stereo)组件，该组件选择性地为 3D 模型中的每一个粗糙平面计算一个偏移量(视差)图。他们将使用这种方法得到的分析和绘制系统称为"基于几何-图像混合"的绘制系统，因为该系统既使用了传统的 3D 几何建模方法来生成一个全局的 3D 模型，也使用了局部视差信息和视图插值方法来增加视觉上的真实度。

13.1.2 应用：照片游览

视图插值方法最早被提出时，其初衷是为了在没有图形加速的低能耗处理器和系统上对 3D 场景绘制进行加速，但是该方法被提出后，人们发现视图插值能够直接应用在大量随意拍摄的照片集上进行场景绘制。Snavely, Seitz, and Szeliski(2006)在此基础上开发了一个照片游览系统，系统使用由运动到结构信息计算出所有拍摄照片的摄像机在三维空间中的位置和姿态，同时也计算出场景的稀疏 3D 点云模型(sparse 3D point-cloud model)(7.4.4 节，图 7.11)。

为了实现在海量图片(Aliaga, Funkhouser, Yanovsky *et al.* 2003)中进行基于图像的游览，照片游览系统首先为每幅图像关联一个 3D 代理。虽然由点云得到的三角网格有时可以作为合适的代理，例如用于户外地形模型，但是使用一个拟合在图像中可见的 3D 点云上的简单主导平面作为代理往往更好，因为这样的平面中不会出现导致视觉缺陷的错误线段或连接。但是，随着 3D 自动建模方法的不断发展，钟摆可能朝向使用更详细的 3D 几何模型回摆(Goesele, Snavely, Curless *et al*; Sinha, Steedly, and Szeliski 2009)。

在这种基于图像的游览系统中，用户可以通过以下几种方式在不同的照片之间漫游：在场景的俯视视角中选择虚拟摄像机位置(图 13.4a)，或者选择图像中感兴趣的区域以切换到一个附近的视图，或选择相关的缩略图(图 13.4b)。在绘制中，为了给一个特定 3D 场景生成相应背景，例如为俯视视图的场景生成背景，系统可以使用诸如半透明刷色或者高亮 3D 线段等的非真实感绘制方法(10.5.2 节)(图 13.4a)。基于图像的游览系统还可用于标注图像区域进而自动将标记传播给其他相关照片。

在照片游览系统以及微软的 Photosynth 系统中使用 3D 平面代理(替代，proxy)会造成不同照片的过渡中存在类似"翻页"效果的非真实感，为所有代理平面选择一个稳定的 3D 轴可以减少这种晃动感，从而增强游览的 3D 立体感(图 13.4c)(Snavely, Garg, Seitz *et al.* 2008)。另一种可能的解决方法是自动检测出多视角场景中出现的物体并围绕这些物体生成一条"视点轨道"。在 3D 位置和视角方向上都相近的图像还可以被联系在一起形成"虚拟路径",进而被用于在任意图像对之间进行游览,这些图像可能来自于你在某个著名旅游景点游览时自己拍摄的若干照片(Snavely, Garg, Seitz, et al. 2008)。

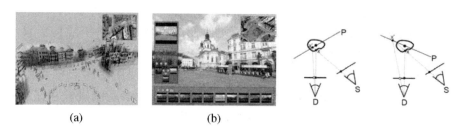

图 13.4 照片游览系统(Snavely, Seitz, and Szeliski 2006)© 2006 ACM:(a)场景的 3D 概览图,半透明刷色和线段被绘制在平面代理上;(b)用户选择感兴趣区域时,系统底部会显示一些对应的缩略图;(c)最佳稳定的平面代理(替代)选择(Snavely, Garg, Seitz et al. 2008)© 2008 ACM

照片游览系统中进行的图像特征与区域的空间匹配还可用于从大量的图像集里挖掘更多的信息。例如,Simon, Snavely and Seitz(2007)展示了如何使用旅游景点图像间的匹配图来寻找图像集中最具代表性(被拍摄得较多)的物体,并且找出与它们相关的标注。Simon and Seitz(2008)在后续的工作中进一步展示了如何通过分析出现在照片中心部分的 3D 点,将这些标注信息传播到每幅图像的子区域中。还有许多研究将这些方法扩展应用到真实世界的图像中,包括使用可以获得的 GPS 标注信息来提高照片游览系统的体验(Li, Wu, Zach et al. 2008; Quack, Leibe, and Van Gool 2008; Crandall, Backstrom, Huttenlocher et al. 2009; Li, Crandall, and Huttenlocher 2009; Zheng, Zhao, Song et al. 2009)。

13.2 层次深度图像

传统的视图插值方法需要为每一幅源图像或参考图像都关联一幅深度图像。不幸的是,当我们把这些深度图像卷绕到新的视角时,前景物体后将不可避免地出现一些空洞和裂缝。缓解该问题的一个办法是在参考图像中为每一个像素(或者至少在前景-背景过渡区域的一些像素)关联多个深度和颜色值(深度像素)(图 13.5)。这种数据结构叫"层次深度图像"(layered depth image, LDI)。使用层次深度图像和从后向前的前向卷绕算法(抛雪球算法)(Shade, Gortler, He et al. 1998)就可以绘制出新的视图。

替代、子画面、层次

除了像层次深度图像(LDI)那样给每一个像素维护一个颜色-深度值列表,另一个可以选择的办法是将物体组织成不同的层次(图层,即 layer)或子画面(sprite)。"子画面"这个术语来源于计算机游戏领域,用来指定游戏中不同平面的动画角色,例如吃豆人和马里奥兄弟。在放进 3D 环境设置中时,一般把这些物体称为"代理"(impostor),因为人们使用了一种平面的、透明遮罩(alpha-matted)的几何模型来表示那些远离摄像机的 3D 物体的简化版本(Shade, Lischinski, Salesin et al. 1996; Lengyel and Snyder 1997; Torborg and Kajiya 1996)。在计算机视觉领域中,一般把这种表示叫"层次"(Wang and Adelson 1994; Baker, Szeliski and Anandan 1998; Torr, Szeliski and Anandan 1999; Birchfield, Natarajan and Tomasi 2007)。在 8.5.2 节已经讨论过一些关于可能出现在镜面和透明表面(如玻璃)上的透明图层和反射的问题。

虽然平面图层已经足以表示远处物体的几何和外观,但是使用每个像素相对一个基准平面的偏移量进行建模,我们可以获得更好的几何真实感,如图 13.5 和图 13.6a—b 所示。正如 8.5 节所讨论的(图 8.16),在计算机视觉文献中,这种表示被称作"平面加视差"(plane plus parallax)(Kumar, Anandan, and Hanna 1994; Sawhney 1994; Szeliski and Coughlan 1997; Baker, Szeliski, and Anandan 1998)。除了全自动化的立体视觉方法外,一些可能的方法还包括绘制深度图层(Kang 1998; Oh, Chen, Dorsey et al. 2001; Shum, Sun, Yamazaki et al. 2004)以及使用单目图像线索推测场景的三维结构(参见 14.4.4 节)(Hoiem, Efros, and Hebert 2005b; Saxena, Sun, and Ng 2009)。

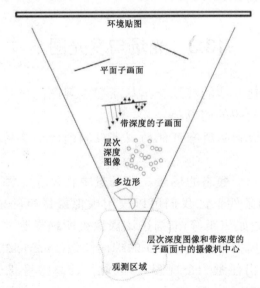

图 13.5 各种基于图像绘制的基元,可以根据摄像机和感兴趣物体的距离远近使用相应的基元(Shade, Gortler, He et al. 1998)© 1998 ACM。距离摄像机较近的物体需要更加精细的多边形表示,中距离的物体可以使用层次深度图像(LDI),而远距离的物体可以使用(可带深度)子画面和环境贴图

那么,我们应该如何绘制一个新视点的带深度信息的子画面呢?一个方法是像使用一般的深度图那样,将每个像素都前向卷绕到对应的新位置,只是这样做可能

会产生空洞而失真；另一个更好的方法已经在 3.6.2 节提到过，我们可以首先将深度(或者(u,v)位移)图卷绕到新的视图并且填补空洞，然后使用更高质量的反向卷绕在彩色图像中重采样(Shade, Gortler, He et al. 1998)。图 13.6d 演示了应用这种两遍绘制算法得到的结果。从这个静态图像的例子你可以欣赏到，使用带深度的子画面使得前景子画面看起来更圆滑；但是如果要充分地感受其在实际中的改进效果，你需要看实际的动画序列。

(a)   (b)   (c)   (d)

图 13.6 带深度的子画面(Shade, Gortler, He et al. 1988)©1998 ACM：(a)Alpha 遮罩的彩色子画面；(b)对应的相对深度或视差；(c)不使用相对深度的绘制结果；(d)使用相对深度的绘制结果(注意物体边界的弧线部分)

带深度的子画面同样可以用传统的图形硬件渲染得到(Zitnick, Kang, Uyttendaele et al. 2004)。Rogmans, Lu, Bekaert et al. (2009)给出了实时立体视觉匹配(real-time stereo matching)和实时前向和后向绘制算法的 GPU 实现。

13.3 光场与发光图

既然基于图像的绘制方法能够综合出从新视点观测的场景绘制结果，那么这些方法引出了下面这个更一般的问题：

是否有可能捕捉和绘制从所有可能的视点观测到的场景的外观，如果能，得到的结构复杂度如何？

假设我们观察的是一个静态的场景，即场景中物体和光源都是固定的，只有观察者在运动。在这样的条件下，我们可以仅用虚拟摄像机的位置、方向(6 个自由度)以及摄像机的内参数(如焦距等)来描述该摄像机所拍摄到的图像。而如果我们在每个可能的摄像机位置周围拍摄一幅 2D 的球面图像(spherical image)，我们就能够使用这些信息重新绘制出任意一个视图[①]。因此，计算摄像机位置在 3D 空间中的坐标与球面图像上 2D 坐标的叉积(cross-product)，我们就得到 Adelson, Bergen(1991) 提出的 5D 全光函数(plenoptic function)，这正是 McMillan, Bishop(1995)设计的基于图像的绘制系统的基础。

然而需要注意的是，当场景中没有光散射时，即没有烟或雾时，所有在自由空间(在固体或折射体之间)某部分重合的光线颜色应该相同。在该条件的约束下，我

---

① 由于我们在计数维度，所以在此暂时忽略采样和分辨率的问题。

们可以将 5D 全光函数降维成为光线的 4D 光场(Gortler, Grzeszczuk, Szeliski et al. 1996; Levoy and Hanrahan 1996; Levoy 2006)。①

为了让这个 4D 函数的参数化更加简单，我们可以在 3D 场景中放置两个平面把感兴趣的区域大致包围起来，如图 13.7a 所示。如果一条光线在一个位于 st 平面前方(假设这个空间是空的)的摄像机处终止(即被该摄像机捕捉到)，那么这条光线一定穿过了这两个平面，设光线与平面相交的坐标分别为 (x,t) 和 (u,v)，于是这条光线就可以用一个 4D 坐标 (s,t,u,v) 来表示。这个模型(以及参数化过程)可以被看作是空间中有一系列位于 st 平面上的摄像机，而这些摄像机的像平面是 uv 平面。uv 平面可以位于无穷远处，相当于所有的虚拟摄像机都朝着同一个方向拍摄。

在实际应用中，如果平面是有限范围的，那么我们就可以使用有限光片(light slab) $L(s,t,u,v)$ 来生成任意的视图，就像一个摄像机从 st 平面上一个有限的观察孔观察到的与 uv 平面相交的一个视截锥(view frustum)一样。为了使摄像机可以在物体的整个周围移动，这个物体周围的 3D 空间可以被分割成多个区域，其中每个区域都有其自身的光片参数化方式。相反地，如果摄像机在一个向外看的有界的自由空间体内部运动，围绕摄像机的多个立方体面可用作(s,t)平面。

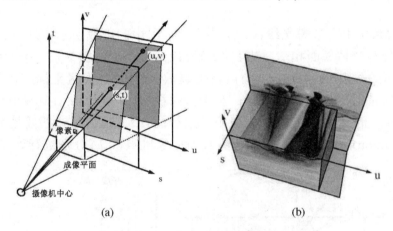

图 13.7 发光图(Gortler, Grzeszczuk, Szeliski et al. 1996)© 1996 ACM：(a)光线可以用其四维双平面参数 (s,t) 和 (u,v) 表示；(b)3D 光场子集 (u,v,s) 的一个切片

由于在 4D 空间中考虑问题比较困难，因此我们在图示中减少一个维度使得示意图更加直观。如果我们固定坐标 t 的值，即限定摄像机沿 s 轴方向移动并且朝着 uv 平面拍摄，那么我们就可以把摄像机拍摄到的所有静态图像"堆"起来从而得到我们在 11.6 节中讨论过的 (u,v,s) 极线容积(epipolar volume)。"极线容积"中的一个水平截面即是大家熟知的极面图像(epipolar plane image)(Bolles, Baker, and Marimont 1987)，如图 13.7b 中的 us 切面所示。

从这个截面中我们可以看到，每一个彩色像素都落在一条直线轨迹上，且这条

---

① Levoy and Hanrahan(1996)从 Gershun(1939)的论文中借用了"光场"(light field)这个术语。另一个表达这个概念的术语是"感光场"(photic field)(Moon and Spencer 1981)。

直线的斜率与这个像素距离 uv 平面的深度(视差)有关。(那些正好位于 uv 平面上的像素在 us 截面图上的轨迹看起来是垂直的，即当摄像机沿着 s 轴移动时，它们的 u 坐标是不变的。)此外，像素的轨迹之间可能会互相挡住，因为与这些像素对应的 3D 表面元素之间存在遮挡。半透明的像素则是与背景像素混合在一起(3.1.3 节，(3.8))，而不是直接挡住背景像素。因此，我们可以认为相邻的像素间具有相似的平面几何，如 EPI 条或 EPI 管(EPI 即极面图像，Epipolar Plane Image)(Criminisi, Kang, Swaminathan et al. 2005)。

从虚拟摄像机像素点 (x,y) 到对应 (s,t,u,v) 坐标的映射方程的推导比较简单，你会在习题 13.7 中完成其推导。我们还可以证明，对应于常规的正交投影或透视投影(即 3D 点与像素点(x,y)之间具有线性映射关系)，摄像机的像素点在 (s,t,u,v) 光场中都落在同一个 2D 超平面上(习题 13.7)。

虽然使用光场可以从任意一个新视点绘制一个复杂的 3D 场景，但是如果我们知道场景的某些 3D 几何信息，我们就可以得到更好(重影更少)的绘制结果。例如 Gortler, Grzeszczuk, Szeliski et al. (1996)提出的发光图系统对基本的光场绘制方法做了一定的扩展，考虑了每个 3D 光线对应表面点的 3D 位置，从而得到了更好的绘制效果。

在他们的系统中，考虑光线 (s,u)，如图 13.8 中的虚线所示，这条光线在距 uv 平面距离 z 处与物体表面相交。当我们想要找到这个点在摄像机 s_i 中的像素颜色值时(假设光场是在规则的 4D (s,t,u,v) 网格上离散采样的)，像素坐标应该是 u'，而不是光线 (s,u) 给定的坐标 u。类似的，对于摄像机 $s_{i+1}(s_i \leq s \leq s_{i+1})$，我们应该使用 u'' 作为像素坐标。因此，(u,v) 值并非由相邻采样的 (s,t,u,v) 值通过四线性插值(quadri-linear interpolation)得到，而是对每个离散的摄像机点 (s_i,t_i) 分别进行修正得到。

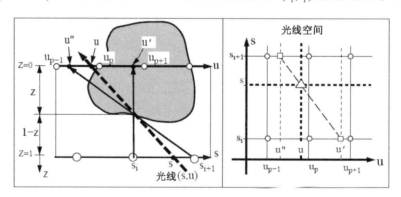

图 13.8　发光图中的深度补偿(Gortler, Grzeszczuk, Szeliski et al. 1996)© 1996 ACM。为了重采样虚线所示的光线 (s,u)，需要对每个离散摄像机位置 s_i 对应的 u 参数根据平面外深度 z 进行修正，得到新的坐标 u' 和 u''；在 (u,s) 光线空间中，原始采样($\triangle$)会从 (s_i,u') 和 (s_{i+1},u'') 进行重采样，而这两个新的坐标则是它们相邻原始采样点(○○)的线性组合

图 13.8 还给出了在光线空间(ray space)中同样的推导。其中，原始的离散取值的光线 (s,u) 在图中用一个三角形表示，在其附近离散采样的值则如小圆圈所示。

这时，绘制系统并没有直接将四个邻近的离散采样点组合起来，而是首先计算得到修正光线坐标 (s_i, u') 和 (s_{i+1}, u'')，再将这两个坐标进行混合。

相比使用无代理(proxy-free)光场得到的绘制系统，如上方法得到的绘制系统能够得到质量高得多的图像，因此只要能够推测场景的 3D 几何信息，都应该选择使用这种有代理的绘制系统。在后续的相关工作中，Isaksen, McMillan and Gortler(2000)提出，使用场景的平面代理这样一种更简单的 3D 模型能够简化重采样方程。他们还描述了如何通过混合更邻近的采样点(Levoy and Hanrahan 1996)来生成可以模拟大孔径镜头拍摄效果的合成孔径照片。类似的方法还可用于对全光(plenoptic)摄像机(微镜头阵列)或光场显微镜(light field microscope)拍摄的照片重调焦距(Ng, Levoy, Bréedif et al. 2005; Ng 2005)。这种方法还可用于看穿障碍物：使用一些超大的合成孔径摄像机，将它们对焦到能够虚化前景物体的背景上，从而使前景物体看起来像是透明的(Wilburn, Joshi, Vaish et al. 2005; Vaish, Szeliski, Zitnick et al. 2006)。

既然我们已经知道了如何使用光场绘制新的图像，那么我们应该如何拍摄得到需要的数据集呢？一个方法是将一个标定好的摄像机放置在运动控制装置或者运动支架(gantry)[①]上进行移动。另一个办法是用手持摄像机拍摄照片然后使用标定阶段或由运动到结构来确定照片对应的摄像机姿态和摄像机内标定参数。对于第二种方法，在进行绘制前，需要将拍摄到的照片重分片(rebinned)到一个正规的 4D (s,t,u,v) 空间中(Gortler, Grzeszczuk, Szeliski et al. 1996)。或者，也可以直接使用原始图像，这需要使用下面即将讨论的非结构化发光图(unstructured Lumigraph)。

因为涉及的图像非常多，所以光场和发光图会变得非常巨大而难以存储和传输。幸运的是，从图 13.7b 可以看到，光场中存在大量的冗余(内在相关性)。更直观地来讲，我们可以首先计算出一个 3D 模型，就像在发光图中一样。人们已经提出了许多方法来压缩和渐进式地传输这些大量的数据(Gortler, Grzeszczuk, Szeliski et al. 1996; Levoy and Hanrahan 1996; Rademacher and Bishop 1998; Magnor and Girod 2000; Wood, Azuma, Aldinger et al. 2000; Shum, Kang, and Chan 2003; Magnor, Ramanathan, and Girod 2003; Shum, Chan, and Kang 2007)。

13.3.1 非结构化发光图

当发光图中的源图像是以非结构化(非正规)的方式采得时，我们就无法将光线重采样得到一个正规分片的 (s,t,u,v) 数据结构。这一方面是因为重采样总会导致一定量的走样(aliasing)，另一方面是因为得到的网格化光场可能会非常稀疏或者不规则。

在这种情况下，一个办法是直接使用源图像进行绘制，即为虚拟摄像机中的每

[①] 参见 http://lightfield.stanford.edu/acq.html，您可以查看一些斯坦福大学计算机图形学实验室制造的运动支架和摄像机阵列。该网站还提供了很多光场数据集，是有关研究和项目资料的很好的来源。

一条光线在源图像中找到最近的像素。Buehler, Bosse, McMillan et al. (2001)提出的非结构化发光图绘制(Unstructured Lumigraph Rendering, ULR)系统描述了如何结合多个保真度准则来选择此类最近的像素，这些准则包括极点一致性(epipole consistency)(光线到源摄像机中心的距离)、角度偏差(物体表面光线入射方向的相似性)、分辨率(表面采样密度的相似性)、连续性(和邻近像素)以及一致性(沿着光线)。将这些准则结合起来就能得到一个加权函数，这个函数的输入是提供绘制信息的若干个候选输入摄像机，输出是每个虚拟摄像机像素。为了使算法更加高效，计算过程中需要通过在物体的多面体网格模型和输入摄像机的投影中心上覆盖一个规则网格将虚拟摄像机的成像平面离散化，并且对顶点间的加权函数进行插值。

非结构化发光图是之前基于图像的绘制和光场绘制两者的推广。如果输入摄像机构成一个网格，那么非结构化发光图方法就等价于一般的发光图绘制方法；但是当输入摄像机个数较少但能够获得精确的场景几何信息时，非结构化发光图方法就类似于视图相关的纹理映射(13.1.1节)。

13.3.2 表面光场

当然，基于两个平面进行光场的参数化并非唯一的选择。(但是通常我们会首先介绍这种方法，因为这种方法的映射方程最容易推导且示意图最容易理解。)正如我们在光场压缩中所提到的，如果我们能够知道需要进行光场建模的物体或场景的3D形状，我们就能对光场进行有效的压缩，因为相邻的表面元素发出的光线应该具有相似的颜色值。

事实上，如果物体是完全漫反射的，忽略遮挡，这可以用3D图形学算法或z缓冲处理，所有通过一个给定表面点的所有光线颜色值应该是相同的。因此，在这种情况下，光场退化为通常意义下定义在物体表面的二维纹理映射。相反的，如果物体表面是完全镜面反射的(例如镜子)，那么每一个表面点将反射该点周围环境的一个缩影。如果物体的存在不会导致相互反射(例如一个非常大的开放空间中的一个凸物体)，则每个表面点就只反射仍然是2D的远场环境贴图(environment map)(2.2.1节)。因此，它看起来对4D光场在物体表面上进行重参数化非常有用。

这些观察构成了Wood, Azuma, Aldinger et al.(2000)提出的表面光场(surface light field)表示的基础。在他们的系统中，他们为物体构造了一个精确的3D模型，然后通过估计或者直接拍摄得到了每个表面点发出的所有光线的光球(Lumisphere)(图13.9)。相邻的光球具有高度的相关性，因此可被压缩和修改。

为了估计每个光球的漫反射成分，该方法对每个颜色通道在所有可见方向上都进行了中值滤波，滤波结果即是估计漫反射成分。从光球中减去估计得到的漫反射成分，剩下的几乎就是围绕局部表面法向的镜面反射成分了，这样光球就成为这个表面点周围局部环境的副本。然后邻近的光球就可以使用预测编码、向量量化或主成分分析等方法进行压缩了。

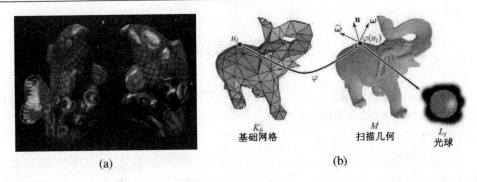

图13.9 表面光场(Wood Azuma, Aldinger *et al.* 2000)© 2000 ACM：(a)一个带有强相互反射的高度镜面物体的例子；(b)表面光场储存了从每个表面点发出的所有可见方向的光线，即"光球"

将光球分解为漫反射成分和镜面反射成分的想法还可用于编辑和处理操作，如重绘表面、修改反射的镜面分量(例如模糊或锐化光球的镜面反射成分)，甚至在保留物体表面外观细节的同时对物体进行几何变形。

13.3.3 应用：同心拼图

基于光场的绘制，一个简单且实际的应用是带视差的全景图，即用一个绕固定旋转点旋转的摄像机拍摄的一段视频或一系列照片。这些视频和照片可以通过将摄像机固定于架在三脚架上的云台上拍摄得到，甚至更简单的，还可以通过手持摄像机伸直手臂然后绕一个固定的竖直轴旋转自己的身体拍摄得到。

得到的图片集可以看作是一个同心拼图(concentric mosaic)(Shum and He 1999; Shum, Wang, Chai *et al.* 2002)或叫"层次深度全景图"(layered depth panorama) (Zheng, Kang, Cohen *et al.* 2007)。"同心拼图"这个术语来源于一种可用于将所有样本光线重新分片的特殊结构，这种结构本质上是将每一列像素和与之相切的同心圆半径联系起来(Shum and He 1999; Peleg, Ben-Ezra, and Pritch 2001)。

用这样的数据结构进行绘制是快速且直观的。我们假设场景距虚拟摄像机位置足够远，我们就能够将虚拟摄像机中的每一列像素和输入图片集中最邻近的一列像素联系起来。(对于一组规则地拍摄到的图像，该运算可以通过解析方法实现。)如果我们能够得到关于这些像素深度的一些粗略的信息，就可以将一列像素在垂直方向上进行拉伸，从而补偿两个摄像机深度的不同。如果我们能够获得更详细的深度图(Peleg, Ben-Ezra, and Pritch 2001; Li, Shum, Tang *et al.* 2004; Zheng, Kang Cohen *et al.* 2007)，我们就可以进行更加精细的逐像素深度修正了。

虽然虚拟摄像机的移动被限制在源摄像机平面上且在源摄像机的拍摄半径内，但是拼接结果可以表现出各种复杂的绘制效果，如倒影和半透明，这是纹理映射3D模型无法得到的。在习题13.10中，让你用一系列手持摄像机拍摄的照片或视频构建一个同心拼图绘制系统。

13.4 环境影像形板

到目前为止，我们已经在本章中讨论了视图插值和光场法这两种对复杂静态场景从不同视点进行建模和绘制的方法。

下面我们考虑这样一个问题：如果我们不是移动虚拟摄像机，而是将一个复杂的、折射性的物体(例如高脚水杯，如图 13.10 所示)放置在一个新的背景前面。这时我们不是对场景发出的所有光线构成的四维空间进行建模，而是要对我们视图中这个物体的每个像素如何折射从其周围环境入射的光线进行建模。

这种表示内在的维度是多少呢？我们又如何刻画它呢？如果我们跟踪一条由摄像机的像素点 (x,y) 射向物体的光线，那么这条光线将被物体以角度 (ϕ,θ) 重新反射或折射回到环境中。如果我们假设场景中其他物体和光源都离得足够远(我们在13.3.2 节中讨论表面光场时已做过同样的假设)，那么四维映射 $(x,y) \to (\phi,\theta)$ 就可以刻画出一个折射性物体与其环境之间的所有信息。Zongker, Werner, Curless *et al.* (1999)将这种表示方法称为"环境影像形板"(environment matte)，因为这种表示可以看作是对物体抠图(object matting)的推广：它不仅将物体从一幅图像中剪切粘贴到另一幅图像，还考虑了物体和环境之间折射、反射等精细的相互影响。

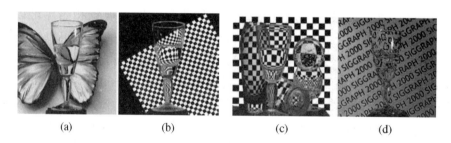

图 13.10 环境影像形板：(a 和 b)将一个折射物体放在不同的背景前，这些背景的光照模式会被正确地折射(Zongker, Werner, Curless *et al.* 1999)；(c)多次折射可以用混合高斯模型处理；(d)实时影像形板可以利用单个分级彩色背景抽取得到(Chuang, Zongker, Hindorff *et al.* 2000)© 2000 ACM

回忆公式(3.8)和(10.30)，一个前景物体可以用它的预乘(premultiplied)颜色和不透明度来表示($\alpha F,\alpha$)。然后这个影像形板(抠图)可以被合成到一个新的背景上：

$$C_i = \alpha_i F_i + (1-\alpha_i)B_i, \qquad (13.1)$$

其中 i 是像素索引。对环境影像形板，我们在这个公式中增加一个反射或折射项来模型化环境和摄像机之间的间接光路。在 Zongker, Werner, Curless *et al.*(1999)的工作中，这个间接成分 I_i 表示为如下形式：

$$I_i = R_i \int A_i(\boldsymbol{x})B(\boldsymbol{x})d\boldsymbol{x}, \qquad (13.2)$$

其中 A_i 是这个像素的矩形支撑面积(area of support)，R_i 是折射率或透射率(对彩色光泽表面或玻璃)，$B(x)$ 是背景(环境)图像，I_i 即是背景图像 $B(x)$ 在面积 $A_i(x)$ 上的积分。在后续工作中，Chuang, Zongker, Hindorff *et al.* (2000)使用了一种高斯叠加模型：

$$I_i = \sum_j R_{ij} \int G_{ij}(\boldsymbol{x}) B(\boldsymbol{x}) d\boldsymbol{x}, \tag{13.3}$$

其中，每个二维高斯分布

$$G_{ij}(\boldsymbol{x}) = G_{2\mathrm{D}}(\boldsymbol{x}; \boldsymbol{c}_{ij}, \boldsymbol{\sigma}_{ij}, \theta_{ij}) \tag{13.4}$$

是通过其中心 c_{ij}、非旋转宽度 $\sigma_{ij} = (\sigma_{ij}^x, \sigma_{ij}^y)$ 和方向 θ_{ij} 建模得到的。

按照这种环境影像形板的表示方式，我们如何对一个特定的物体估计其环境影像形板呢？我们可以将物体放置在一个监视器前(或者在物体周围放置多个监视器)，然后我们调节光照模式 $B(x)$，观察每一个复合像素 c_i 颜色值的变化。[①]

和传统的双屏幕抠图(two-screen matting)(10.4.1 节)一样，我们可以使用各种各样彩色背景来估计每个像素的前景色彩 $\alpha_i F_i$ 和不透明度 α_i。为了估计公式(13.2)中的支撑面积 A_i，Zongker, Werner, Curless *et al.* (1999)使用了一系列频率和相位不同的周期性水平和垂直条纹，这让人联想到主动距离测定(12.2 节)中所使用的结构光模式。关于公式(13.3)中给出的更复杂的高斯混合模型，Chuang, Zongker, Hindorff *et al.* (2000)使用了四个方向(水平、竖直以及两条对角线)上的窄高斯条带(Gaussian stripe)进行扫描，从而可以估计每个像素处多个方向上的高斯响应。

只要抽取出环境影像形板，我们就很容易将背景用另一幅图像 $B(x)$ 进行替换，从而得到将物体放到不同环境中的新合成图像(图 13.10a~c)。但是，在抠图过程中使用多个背景使得这种方法无法用于动态场景，例如水倒入玻璃杯(图 13.10d)。在这种情况下，可以使用一个分级的彩色背景来估计每一个像素上的二维单色替换(Chuang, Zongker, Hindorff *et al.* 2000)。

13.4.1 更高维光场

如前面的讨论，从原理上来讲，环境影像形板就是将每个像素 (x,y) 映射到一个光线的 4D 分布，因此，这实际上是一个 6D 表示。(而在实践中，每个 2D 像素的响应需要十多个参数，例如 $\{F, \alpha, B, R, A\}$，而不是全映射。)如果我们想对物体在每个可能的视点表现出的折射性质进行建模，我们应该怎么做呢？这时，我们需要建立从每条入射 4D 光线到每条可能的 4D 出射光线的映射，这是一个 8D 的表示。如果我们使用在讨论表面光场时使用过的方法，每一个表面点可用其 4D BRDF 表示进行参数化，从而将这个映射降维到 6D，但是这样会牺牲处理多折射路径的能力。

如果我们想处理动态光场的情况，我们还需要增加一个时间维。(Wenger,

[①] 如果我们放宽"环境是远距离的"这个假设，那么我们可以将监视器放在多个不同的深度处，从而估计出一个深度相关的映射函数(Zongker, Werner, Curless *et al.* 1999)。

Gardner, Tchou et al. (2005)给出了很好的动态景象和光照采集系统。)类似地，如果我们还希望光的波长符合一个连续的分布，则波长成为另一个维度。

这些例子显示了通过采样对一个视觉场景进行全面的建模代价有多么高。幸运的是，我们可以通过使用光传播的物理规律和真实世界物体的自然一致性构造专门的模型，使得这些问题更容易处理。

13.4.2 从建模到绘制

到此为止，关于基于图像的绘制这个主题，我们在本章中讨论过的各种表示和算法涵盖了从传统 3D 纹理映射模型到纯粹基于光线采样的表示的一系列方法(图 13.11)。视图相关纹理映射和发光图这样的表示仍然使用的是单个全局几何模型，但是映射到物体表面的颜色值是从邻近的图像中选出的。例如在视图相关的几何中，例如多层深度图的采用，使得我们不再需要一个明确清晰的 3D 几何模型，有时还能对局部非刚性效果(如高光移动)(Swaminathan, Kang, Szeliski et al. 2002; Criminisi, Kang, Swaminathan et al. 2005)。带深度的子图像和层次深度图像同时使用了基于图像的色彩和几何表示，可以使用卷绕操作来代替 3D 几何光栅化进行高效的绘制。

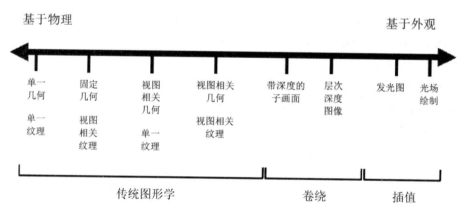

图 13.11 基于图像的绘制方法中从几何到图像的过渡(Kang, Szeliski, and Anandan 2000)© 2000 IEEE。靠左的表示使用更详细的几何模型和更简单的图像表示，而靠右的表示算法使用更多图像信息和更少几何信息

如何选择最合适的表示和绘制算法依赖于输入图像的数量和质量以及实际应用的目的。当绘制邻近的视图时，基于图像的表示能够更好的保留真实世界的视觉真实度，因为基于图像的表示是直接对场景景观进行采样得到的。而另一方面，如果输入图像非常少，或者基于图像的模型需要处理(如调整形状或外观)，那么几何模型和局部反射模型这样的更加抽象的 3D 表示将会更加合适。随着我们拍摄和处理的视觉数据(图像)越来越多，基于图像的建模和绘制中包括以上问题在内的各方面的研究也定将进一步发展。

13.5 基于视频的绘制

既然使用一个图像集可以绘制新的图像或提供交互式体验，那么使用视频能否进行类似的绘制呢？实际上，在基于视频的绘制(video-based rendering)和基于视频的动画(video-based animation)领域，研究人员已经做了许多工作。这两个术语由Schodl, Szeliski, Salesin et al. (2000)第一次提出，用以表示从拍摄得到的视频片段生成新视频的过程。该领域一个较早期的工作是视频改写(Video Rewrite)(Bregler, Covell, and Slaney 1997)，即将存档中的视频片段重新进行动画绘制以改变演员的口型，以配合不同的台词(图 13.12)。更近些时候，基于视频的绘制这个术语被研究人员用来表示使用工作室中的一组同步摄像机生成虚拟摄像机视频的过程(Magnor 2005)。(人们有时也使用自由视点视频(free-viewpoint video)和 3D 视频来表示这一过程，见 13.5.4 节。)

本小节中，我们将给出一些基于视频的绘制系统和应用。首先我们介绍基于视频的动画(13.5.1 节)，它对原有的视频片段进行重新排列或修改，例如人脸表情的拍摄和重绘。视频纹理(video texture)(13.5.2 节)是基于视频的动画的一个特例，在视频纹理中，源视频被自动地切割为若干片段，然后这些片段被重新连接起来形成一段无限长的视频动画。我们还可以使用静态图片生成这样的动画，这需要将静态图片分割成独立的运动区域，然后使用随机运动场生成动画。

在接下来的小节中，我们将注意力转向 3D 视频(13.5.4 节)，3D 视频是指使用多个同步的摄像机从不同的角度拍摄同一个场景。然后我们可对拍摄得到的源视频帧使用视图插值等基于图像的绘制方法进行重新组合，从而生成源摄像机位置之间的虚拟摄像机路径，提供更多的实时视觉体验。最后，我们将讨论通过手持全景摄像机穿过一个场景或者将全景摄像机放置在车中穿过场景来拍摄环境，从而生成基于视频的交互式游览系统(13.5.5 节)。

13.5.1 基于视频的动画

正如前面所提到的，基于视频的动画领域的一个早期工作是视频改写，它可以将源视频片段中的帧进行重新排列从而使视频可以与新的台词对应，例如电影配音(图 13.12)。从想法来说，这和连接式语音合成(concatenative speech synthesis)系统的工作原理是相似的(Taylor 2009)。

在 Bregler, Covell and Slaney(1997)的工作中，他们首先使用了语音识别方法来从源视频片段和新的音频流中抽取音素(phoneme)。这些音素被分组为三音子(triphone)(三个音素的组合)，因为三音子能够更好地描述人说话时协同发音的影响。然后系统在源视频片段和音频流中寻找匹配的三音子，将对应于选出的视频帧的口形图像剪切并粘贴到被重绘或配音的视频片段中，在此过程中，系统还会根据头部运动进行必要的几何变换。在分析阶段中，对应于嘴唇、下巴和头部的特征是使用计算机视觉的方法跟踪得到的。而在合成阶段，系统使用了图像变形(image morphing)方法来将相邻的口形混合和拼接成为一个更加一致的整体。在 Ezzat, Geiger, and Poggio(2002)更新的工作中，他们描述了如何将多维形变模型(multidimensional morphable model)(12.6.2 节)和规则化的轨迹合成结合起来以获得更好的结果。

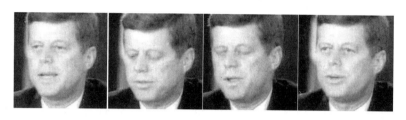

图 13.12　视频重写(Bregler, Covell, and Slaney 1997)© 1997 ACM：新的视频帧是由与新音频匹配的旧视频片段合成的

这个系统还有一个更加复杂而精细的版本叫"人脸转移"(face transfer)，该系统使用了一段新的源视频代替上面系统中的音频流来指导一段先前拍摄的视频的动画重绘，例如使用合适的视觉语音(visual speech)特征、表情和头部姿态元素(Vlasic, Brand, Pfister et al. 2005)来重新绘制一段说话人头部的视频。该系统是一种表演驱动的动画绘制(performance-driven animation)系统，这种系统经常被用于对 3D 人脸模型进行动画绘制(图 12.18 和图 12.19)。传统的表演驱动动画绘制系统使用的是基于标记点的运动捕捉(Williams 1990; Litwinowicz and Williams 1994; Ma, Jones, Chiang et al. 2008)，而在现在系统中，视频片段常常可直接控制动画绘制过程(Buck, Finkelstein, Jacobs et al. 2000; Pighin, Szeliski, and Salesin 2000; Zhang, Snavely, Curless et al. 2004; Vlasic, Brand, Pfister et al. 2005; Roble and Zafar 2009)。

除了最常见的人脸动画绘制，基于视频的动画绘制还可用于全身运动(12.6.4 节)，例如在两段不同的源视频之间寻找匹配光流场，从而可以使用其中一段来指导另一段的绘制(Efros, Berg, Mori et al. 2003)。基于视频的绘制中另一个方法是使用光流或 3D 建模来对表面纹理进行去卷绕(unwarp)得到稳定图像，然后进行进一步的修改并重绘制到源视频中(Pighin, Szeliski, and Salesin 2002; Rav-Acha, Kohli, Fitzgibbon et al. 2008)。

13.5.2　视频纹理

通过将已有的视频片段改写成另一个期望的活动或剧本，基于视频的绘制可以作为生成真实感视频(photo-realistic video)的有力工具。如果我们并不是要创建一段特定的动画或情景，只是简单地希望视频能够以一种合理的方式连续地播放下去，我们应该怎么做呢？很多网站使用图像或者视频来突出网站的目的和内容，例如使用冲浪和风中飘动的棕榈树来描绘迷人的海滩。我们能否将视频片段转换成一段无限长、一直播放的连续动画，以此来代替静态图像或在循环播放过程中会出现间断的视频呢？

这就是视频纹理(video texture)的基本概念，即可以通过重排视频中的帧来将较短的视频剪辑任意地进行延展，同时保留视觉连续性(Schodl, Szeliski, Salesin et al. 2000)。生成视频纹理的基本问题是如何进行重排使得结果中不会出现人工修改痕迹。如果让你来解决这个问题，你会怎么做呢？

最简单的方法是通过视觉相似性(例如L_2距离)来进行帧的匹配，然后在相似的帧之间进行切换。不幸的是，如果这两帧中的运动情况是不同的，结果中将会出现戏剧性的缺陷(例如视频将会看起来断断续续)。例如，如果我们无法对图 13.13a 中的钟摆运动做出很好的匹配，视频中的钟摆可能在摆动到中间时突然改变方向。

于是我们需要考虑，如何将基本的帧匹配方法进行拓展使其能进行运动匹配呢？原理上我们可以计算并匹配每一帧的光流。但是，光流估计常常不太可靠(尤其是在无明显纹理的区域)，而且衡量视觉和运动彼此的相对的相似性并不容易。于是，Schodl, Szeliski, Salesin et al. (2000)提出对三元组(triplet)或更多相邻视频帧进行匹配，正如视频改写中匹配三音子一样。只要我们构造出了 $n \times n$ 的帧间相似性矩阵(其中 n 是帧的个数)，那么我们就可以对每个待匹配的视频序列应用有限脉冲响应(FIR)滤波器以突出匹配度较高的子序列。

这种匹配运算的结果是一个跳转表(jump table)，这等价于源视频任意两帧之间的一个转移概率(transition probability)。图 13.13b 中的红色弧线示意性地显示了这个关系，其中红色的条带是当前正在播放的帧，当某个前向或后向转换发生时红色弧线将会变亮。我们可以把这个转移概率看作是对一个隐马尔可夫模型(HMM)的编码，如果把视频的生成看做一个随机过程，那么隐马尔可夫模型就是这个过程的数学表示。

有时候从源视频中寻找精确匹配的子序列非常困难。这时候，我们可以使用变形方法，即在转换中对多个帧进行卷绕和混合(3.6.3 节)来隐藏视觉上的差别(图 13.13c)。当运动足够混乱时，如篝火和瀑布(图 13.13d 和 e)，使用简单的混合(扩展交融(extended dissolves))就足够了。效果更好的转换可以通过对随机场中该转换周围的时空体进行 3D 图割得到(Kwatra, Schödl, Essa et al. 2003)。

不过，视频纹理并不只被限制用于火、风和水这样一些混乱随机的现象。我们也能生成以人为中心的视频纹理，例如一张笑脸(图 13.13f)或者一个骑车的人(Schodl, Szeliski, Salesin et al. 2000)。当视频中有多个人或物体独立运动时，如图 13.13g 和 h 所示，我们必须先将视频分割成独立的运动区域，然后对每个区域分别进行动画绘制。我们还可以用慢速摇动摄像机拍摄的视频来生成较大的全景视频纹理(panoramic video texture)(Agarwala, Zheng, Pal et al. 2005)。

除了仅仅将源视频中的帧以某种随机方式进行回放外，视频纹理还可用于生成按照剧本编排的动画或交互动画。如果我们将视频中的独立元素(如鱼缸中的鱼，

图 13.13i)抽取出来形成独立的视频子画面(video sprite)，我们就可以按照预先给定的路径对其进行动画绘制(匹配路径方向与原始子画面的运动)，使得视频中的元素按照希望的方式运动(Schodl and Essa 2002)。实际上，视频纹理的相关工作还给关于如何从运动捕捉得到的数据重新合成新运动序列(一些人将其称为"动态捕捉汤(mocap soup)")的研究带来了很多启发(Arikan and Forsyth 2002; Kovar, Gleicher, and Pighin 2002; Lee, Chai, Reitsma *et al.* 2002; Li, Wang, and Shum 2002; Pullen and Bregler 2002)。

图 13.13 视频纹理(Schodl, Szeliski, Salesin *et al.*2000)© 2000 ACM：(a)一个运动方向正确匹配的钟摆；(b)蜡烛火焰，其中弧线显示了时间转移；(c)画面中旗子是通过在跳转处进行变形操作得到的；(d)在生成篝火的视频纹理时，使用了更长的淡入淡出；(e)生成瀑布的视频纹理时，多个序列被淡入淡出在一起；(f)动画笑脸；(g)对两个荡秋千的小孩分别进行动画绘制；(h)多个气球被自动分割成独立的运动区域；(i)一个合成的包括水泡、植物和鱼的鱼缸。与这些图片对应的视频可以在 *http://www.cc.gatech.edu/gvu/perception/projects/videotexture/* 找到

视频纹理本质上是将视频看作可以在时间上进行重新排列的帧(区域)序列来进行分析，而时域纹理(temporal textures)(Szummer and Picard 1996; Bar-Joseph, El-Yaniv, Lischinski *et al.* 2001)和动态纹理(dynamic textures)(Doretto, Chiuso, Wu *et al.* 2003; Yuan, Wen, Liu *et al.* 2004; Doretto and Soatto 2006)则是把视频看作具有纹理属性的 3D 时空体进行分析，而 3D 时空体可以用自回归时域模型(auto-regressive temporal model)来描述。

13.5.3 应用：图片动画

前面已经介绍过，视频纹理可以将一段较短的视频片段重新绘制成一段无限长的视频，那么我们能否将单张静态图片也重新绘制成无限长的视频呢？答案是肯定的，你可以通过首先将图片分解成多个图层，然后对每个图层分别进行动画绘制来做到这一点。

Chuang, Goldman, Zheng *et al.* (2005)描述了如何使用交互抠图(interactive matting)方法将图片分解成多个独立的图层，然后使用特定类别的运动合成对每个图层进行动画绘制。如图 13.14 所示，我们可以使用有形的噪声位移图(shaped noise displacement map)使得船来回摇晃，树在风中摇摆，云朵水平地运动，而水面泛着波纹。所有这些效果可以用一些全局控制参数进行约束，如虚拟风的速度和风向。当所有独立图层的动画绘制完成后，这些图层就可以被重新组合起来生成最后的动态绘制结果。

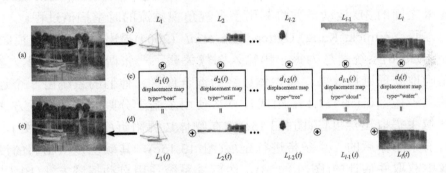

图 13.14 静态图片的动画绘制(Chuang, Goldman, Zheng *et al.* 2005)© 2005 ACM。(a)输入静态图像被人工分割成；(b)多个图层；(c)使用不同的随机运动纹理对每个图层进行动画绘制；(d)绘制好的各个图层被组合在一起生成；(e)最终绘制结果

13.5.4 3D 视频

近年来，3D 电影引人注目地流行起来了，从最近发行的《汉娜·蒙塔娜》，到 U2 的 3D 演唱会视频，再到詹姆斯·卡梅隆(James Cameron)的《阿凡达》，以及将于 2012 年上映的 3D 版《泰坦尼克号》。目前，这些影片和视频都是通过立体视觉摄像设备拍摄的，在影院(或家中)放映室的观看者需要佩戴偏振眼镜。[①]然而在未来，观众可能希望在家里能通过多区段自动立体视觉显示器(multi-zone autostereoscopic display)来观看此类影片，其中不同位置的人可以看到与其所在视角对应的立体视频流，因此人们可以在绕着场景移动来从不同视角观看影片场景。[②]

---

① *http://www.3d-summit.com/*
② *http://www.siggraph.org/s2008/attendees/caf/3d/*

这种 3D 视频有时也被称为"自由视点视频"(free-viewpoint video)(Carranza, Theobalt, Magnor et al. 2003)或"虚拟视点视频"(virtual viewpoint video)(Zitnick, Kang, Uyttendaele et al. 2004)，其中计算机视觉领域中的立体视觉匹配方法和图形学领域中的基于图像的绘制(视图插值)方法都是不可或缺的。除了需要解决一系列逐帧重建和视图插值的问题，分析阶段得到的深度图或代理还必须在时间上一致以避免出现闪烁现象。

Shum, Chan, ang Kang(2007)和 Magnor(2005)给出了关于各种视图插值方法和系统的很好的综述。这些系统中包括 Kanade, Rander, and Narayanan(1997)和 Vedula, Baker, and Kanade(2005)提出的虚拟现实(Virtualized Reality)系统、沉浸式视频(Immersive Video)(Moezzi, Katkere, Kuramura et al. 1996)，基于图像的视觉外形(Image-based Visual Hulls)(Matusik, Buehler, Raskar et al. 2000; Matusik, Buehler, and McMillan 2001)和自由视点视频(Carranza, Theobalt, Magnor et al. 2003)，这些系统都使用了全局 3D 几何模型(基于表面的(12.3 节)或基于体积的(12.5 节))作为绘制时的代理。Vedula, Baker, and Kanade(2005)的工作还计算了场景流(scene flow)，即对应表面元素之间的 3D 运动，并将其用于多视角视频流的时空插值过程。

另一方面，Zitnick, Kang, Uyttendaele et al. (2004)提出的虚拟视点视频(Virtual Viewpoint Video)系统则是为每一幅输入图像关联了一张两层的深度图，这使得他们能够精确地处理遮挡效果，如物体边缘的混合像素。他们的系统由 8 个连接到磁盘阵列的同步摄像机(图 13.15a)组成，首次使用了基于分割的立体视觉方法来从每幅输入图像中提取出深度图(图 13.15e)。在物体边界附近(深度不连续处)，背景图层被沿着前景物体后的一个条带进行延展(图 13.15c)，其颜色是从相邻的没有被遮挡的图像中获取并估计的(图 13.15d)。然后，系统使用自动抠图方法(10.4 节)来估计前景图层中边界像素的不透明度和颜色(图 13.15f)。

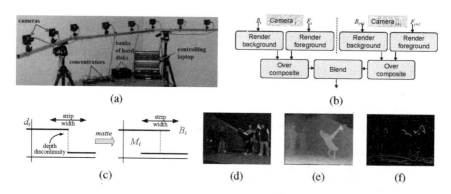

图 13.15 视频视图插值(Zitnick, Kang, Uyttendaele et al. 2004)© 2004 ACM：(a)捕捉系统的硬件由 8 个同步摄像机组成；(b)不同视图混合前，每个摄像机拍摄到的前景图像和背景图像首先被绘制和组合；(c)边缘遮罩前后的两层表示；(d)背景颜色估计；(e)背景深度估计；(f)前景颜色估计

在绘制时，给定位于两个源摄像机之间的一个新的虚拟摄像机，相邻这两个源

摄像机中的图层作为纹理映射三角形首先被绘制，即前景图层(可能存在部分不透明)和背景图层被合成到一起(图 13.15b)。然后所得到的两幅图像通过比较 z 缓冲值被混合到一起。(当两个 z 缓冲值非常接近时，可以使用两个颜色值的线性组合。)在通常的显卡上，这种交互式绘制系统可以实时地运行，因此它可被用来在播放视频的同时改变观看者的视角或者将场景定格而进行 3D 浏览。最近，Rogmans, Lu, Bekaert et al. (2009)给出了实时立体视觉匹配和实时绘制算法的 GPU 实现，使得在实时环境下观察各种算法的效果成为可能。

目前，使用离线(off-line)立体视觉匹配算法从 8 个立体视觉摄像机计算得到的深度图是所有与活动视频相关的深度图中质量最好的。[1]因此这种设备和方法常常被用在热门的 3D 视频压缩的研究中(Smolic and Kauff 2005; Gotchev and Rosenbahn 2009)。另外，主动视频速率(active video-rate)深度传感摄像机，如 12.2.1 节讨论过的 3DV Zcam(Iddan and Yahav 2001)，也是一种可以采集类似数据的设备。

如果可以使用大量彼此离得很近的摄像机，如斯坦福光场摄像机(Stanford Light Field Camera)(Wilburn, Joshi, Vaish et al. 2005)，那么在生成基于视频的绘制效果时，计算出显式的深度图并非总是必须的，当然，如果使用显式深度图绘制，效果通常会更好(Vaish, Szeliski, Zitnick et al. 2006)。

13.5.5 应用：基于视频的游览

摄像机阵列设备的出现使得同时从多个视点拍摄一个 3D 动态场景成为可能，这使得观看者可以在原始拍摄位置附近改变视点来游览场景。如果我们希望拍摄更大的区域，如住宅、影片集或甚至整个城市呢？

这时，在目标环境中移动摄像机进行拍摄，然后回放视频进行交互式视频游览就显得更有意义。为了使观看者能够朝各个方向观察，最好使用一个全景摄像机进行拍摄(Uyttendaele, Criminisi, Kang et al. 2004)。[2]

组织获取过程结构的一种方式是在一个二维水平平面上拍摄这些图像，例如将摄像机安放在室内的一个网格上。然后，我们可以使用得到的海量图像(Aliaga, Funkhouser, Yanovsky et al. 2003)使得在拍摄位置之间的连续运动成为可能。[3]然而，将这个想法扩展到更大的环境中(比如比一个屋子大得多的环境)并不容易，因为这样工作将变得非常繁琐，且数据量巨大。

取而代之的是，一种自然的在空间中游览的方式是沿着某些预先指定的路径进行游览，就像在博物馆或者家里，讲解员或主人通常带着参观者沿着一条特定的线路进行参观，如沿着每个屋子的中间走过。[4]类似的，城市级别的游览可以限制游

---

[1] http://research.microsoft.com/en-us/um/redmond/groups/ivm/vvv/
[2] 从 http://www.cis.upenn.edu/~kostas/omni.html 可以找到全景(全方位)视觉系统的介绍和相关的专题会议论文。
[3] Snavely, Seitz, and Szeliski (2006)设计的照片游览(Photo Tourism)系统就是将这个想法应用于一些较少结构化的图片集上。
[4] 在计算机游戏中，将玩家限制在预先设定的一些路径上前后移动的游戏被称为"基于轨道的游戏"(rail-based gaming)。

览者开车沿着每条街道的中心行驶而只允许游览者在每个路口转弯。这样的想法可以追溯到 Aspen MovieMap 项目(Lippman 1980)，该系统从移动的车中拍摄模拟视频，并记录影碟中用于后续的交互式回放。

近来视频方法的发展使我们可以使用小型同地协作的摄像机阵列，如 Uyttendael, Criminisi, Kang *et al.* (2004)为他们交互式基于视频的游览项目设计的灰点瓢虫(Point Grey Ladybug)摄像机[①](图 13.16b)来拍摄全景(球状)视频了。在他们的系统中，从 6 个摄像机拍摄到的同步视频流通过一系列为该项目专门设计的方法被拼接成 360°的全景图。

由于不同摄像机的投影中心并不相同，所以摄像机之间的视差可能导致视图中的重叠区域出现重影(图 13.16c)。为了消除该影响，系统使用了一个多视角平面扫描(multi-perspective plane sweep)立体视觉算法来估计重叠区域的每一列中每个像素的深度。为了标定每个摄像机相对其他摄像机的参数，系统会在空间中旋转每个摄像机，然后使用一个受限的由运动到结构算法(图 7.8)来估计摄像机的相对姿态和内参数。然后，系统在游览视频中使用特征跟踪(feature tracking)方法来稳定视频帧序列——Liu, Gleicher, Jin *et al.* (2009)在这方面做了一些最新的研究工作。

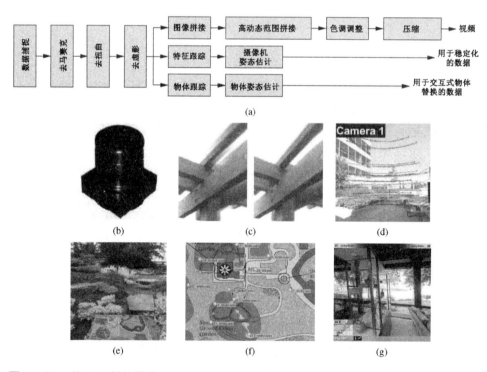

图 13.16 基于视频的游览(Uyttendaele, Criminisi, Kang *et al.* 2004)© 2004 IEEE：(a)视频预处理的系统框图；(b)灰点瓢虫摄像机；(c)利用多视角平面扫描算法进行重影消除；(d)点跟踪，用于标定和稳定化；(e)交互式花园游览，下方是概览地图；(f)俯视地图的生成和声音的嵌入；(g)交互式住宅游览，系统上方有导航栏，下方有兴趣图标

---

① http://www.ptgrey.com/

有窗户的室内场景以及有显著阴影的晴朗室外环境往往具有较高的动态光照范围，远远超过了一般的视频传感器。因此，Ladybug 摄像机具有可编程曝光控制功能，使得系统可以对后续的视频帧进行包围式曝光(bracketing of exposure)。在将拍摄得到的帧合并成高动态范围(HDR)视频前，相邻帧中的像素需要先进行运动补偿(Kang, Uyttendaele, Winder et al. 2003)。

如果系统能够为用户提供一幅概览地图，交互式视频游览系统的使用将更加方便。在图 13.16f 中，概览地图中有很多会在游览过程中显示出来的标注和当用户靠近时会播放出来(音量会随着距离变化)的声音源。将视频序列与概览地图进行匹配的工作可以通过一个叫"地图相关"(map correlation)的过程自动完成(Levin and Szeliski 2004)。

将以上所有这些方法组合在一起就能够为用户提供一个丰富的、交互式的、身临其境的体验。图 13.16e 显示了一个贝尔维尤植物园(Bellevue Botanical Gardens)中的游览，其中视频窗口下部的一个视图中显示了植物园的概览地图。地面上的箭头指示了一些游览的路径。用户只需要简单地将其目光投向一个箭头(该体验可以通过一个游戏控制设备提供)然后沿着箭头所指的路径进行"游览"即可。

图 13.16g 显示了一个住宅室内游览体验。除了在左下角提供一个概览图和在顶部的导航栏提供相邻房间的名字，当一些可能让参观者感兴趣的物体，如主人的艺术品等在窗口中可见时，系统还会在底部显示一个该物体的图标。用户点击该图标便能获得该物体更详细的信息和 3D 视图。

交互式视频游览的发展刺激了研究者们开始重新关注 360°基于视频的虚拟游览和地图体验，Google 街景(Google Street View)和微软的 Bing 地图(Bing Maps)这些商业应用的出现证明了这一点。同样的视频还可以用于生成路线规划驾驶导航(turn-by-turn driving directions)，并使用扩展视野和基于图像的绘制方法来提升用户体验(Chen, Neubert, Ofek et al. 2009)。

随着我们不断拍摄更多真实世界中的景象，我们就会得到大量高质量的图片和视频，这时，本章中介绍的交互式建模、游览和绘制等方法将会在把世界上遥远地方的景象带到用户眼前的努力中发挥更大的作用。

13.6 补充阅读

近年来还出现了两篇很好的关于基于图像的绘制方法的综述文章，分别由 Kang, Li, Tong et al. (2006)和 Shum, Chan, and Kang(2007)发表。另外还有几篇较早的由 Kang(1999)，McMillan and Gortler(1999)和 Debevec(1999)发表的综述。"基于图像的绘制"这个术语是由 McMillan and Bishop(1995)引入的，虽然该领域开创性的研究是 Chen and Williams(1993)发表的论文中提出的视图插值方法。Debevec, Taylor and Malik(1996)描述了他们的外墙绘制系统，他们不仅创建了很多基于图像建模的工具，还提出了在该领域被广泛使用的视图相关的纹理映射方法。

平面替代和图层的早期研究包括 Shade, Lischinski, Salesin et al.(1996)、Lengyel and Snyder(1997)和 Torborg and Kajiya(1996)的工作,而这方面较新的工作是由 Shade, Gortler, He et al.(1998)提出的带深度的子画面。

基于图像的绘制领域中两篇基础性的论文分别是 Levoy and Hanrahan(1996)提出的光场绘制和 Gortler, Grzeszczuk, Szeliski et al.(1996)提出的发光图。后来,Buehler, Bosse, McMillan et al. (2001)将发光图方法推广到了不规则放置的摄像机拍摄的图片集,而 Levoy (2006)则给出了关于光场和基于图像的绘制的更正式的综述和介绍。

表面光场(Wood, Azuma, Aldinger et al. 2000)提供了另外一种使用精确已知的表面几何模型对光场进行参数化的方法,并且支持更好的压缩以及对表面性质进行编辑。同心拼图(Shum and He 1999; Shum, Wang, Chai et al. 2002)与带深度的全景图(Peleg, Ben-Ezra, and Pritch 2001)为通过摇动摄像机拍摄得到的光场提供了一种有用的参数化方法。多视角图像(Rademacher and Bishop 1998)和流形投影(Peleg and Herman 1997)这两种表示虽然不是严格意义上的光场,但是也和上面这些想法和做法密切相关。

在众多将光场扩展到更高维结构的工作中,最有用的便是环境影像形板(Zongker, Werner, Curless et al. 1999; Chuang, Zongker, Hindorff et al. 2000),尤其是我们需要将拍摄到的物体放入一个新的环境时。

基于视频的绘制是指对已有的视频片段进行重新使用,来生成新的动画或虚拟体验。这方面的早期工作包括 Szummer and Picard(1996)、Bregler, Covell, and Slaney(1997),和 Schödl, Szeliski, Salesin et al.(2000)的工作。一些重要的后续工作有 Schödl and Essa(2002)、Kwatra, Schödl, Essa et al.(2003)、Doretto, Chiuso, Wu et al. (2003)、Wang and Zhu (2003)、Zhong and Sclaroff (2003)、Yuan, Wen, Liu et al. (2004)、Doretto and Soatto(2006)、Zhao and Pietikainen(2007)和 Chan and Vasconcelos(2009)的工作。

基于多个同步视频流为用户提供切换 3D 视点体验的系统包括 Moezzi, Katkere, Kuramura et al.(1996)、Kanade, Rander, and Narayanan(1997)、Matusik, Buehler, Raskar et al.(2000)、Matusik Buehler, and McMillan(2001)、Carranza, Theobalt, Magnor et al.(2003)、Zitnick, Kang, Uyttendaele et al.(2004)、Magnor(2005)以及 Vedula, Baker, and Kanade(2005)的工作。3D(多视角)视频的编码和压缩算法也是研究中的热门领域(Smolic and Kauff 2005; Gotchev and Rosenhahn 2009),使用对 H.264/MPEG-4 AVC 扩展的多视角视频编码(MVC)的 3D 蓝光硬碟可在 2010 年年底研究完成。

13.7 习　　题

习题 13.1：深度图绘制

实现"视图外插"算法并使用该算法对一个之前已经计算好的立体视觉深度图和对应的彩色参考图像进行重新绘制。

1. 使用图形学中的 3D 网格绘制系统，如 OpenGL 或 Direct3D，其中每四个像素使用两个三角形，并且使用透视(投影)纹理映射(Debevec, Yu, and Borshukov 1998)。
2. 或者，你可以使用在习题 3.24 中实现的一遍或两遍前向卷绕算法，并使用公式(2.68～2.70)将视差或深度转换为位移。
3. (可选)在视图插值和外插中引入的直线缺陷在视觉上非常显著，这正是一般图像变形系统都会让你指定直线对应关系的原因(Beier and Neely 1992)。修改你的深度估计算法使其能够匹配和估计直线的几何特征，并将其运用到你的基于图像的绘制系统中。

习题 13.2：视图插值

扩展你在上题中实现的系统使其能够绘制两幅参考图像，并结合 z 缓冲、空洞填充和混合(变形)方法将两幅图像进行融合生成最终的结果图像(13.1 节)。

1. (可选)如果两幅源图像的曝光量差别非常大，这时进行了空洞填充的区域和进行了混合的区域也会有不同的曝光量。你能否改进你的算法减轻该影响呢？
2. (可选)扩展你的算法，对相邻的视图进行三路(三线性)插值。你可以考虑将参考摄像机姿态三角网格化并且对虚拟摄像机使用重心坐标系来确定混合权重。

习题 13.3：视图变形

修改你的视图插值算法，使其能在非刚体(如表情变化的人)的视图间进行变形操作。

1. 使用一般的光流算法代替纯立体视觉算法来计算位移，其中位移需要被分解为由摄像机移动造成的刚性位移和变形造成的非刚性位移。
2. 在绘制过程中，首先使用刚性几何来确定新像素的位置，然后再加上非刚性位移部分。
3. 或者，你可以计算得到一个立体视觉深度图，而让用户来指定额外的对应关系或者你直接使用基于特征的匹配算法自动提供这些对应关系。
4. (可选)选择单幅图像，如蒙娜丽莎或你朋友的照片，生成 3D 视图变形动画。
   (a) 找到图像的竖直对称轴，沿着这条对称轴翻转图像，得到一幅虚拟的配对图像(假设这个人的头发基本上是对称的)。
(b) 使用由运动到结构确定这两幅图像的相对摄像机姿态。
(c) 使用稠密立体视觉匹配算法估计 3D 形状。
(d) 使用视图变形算法生成 3D 动画。

习题 13.4：视图相关的纹理映射

使用原始图像和你自己创建的 3D 模型实现一个视图相关的纹理映射系统。

1. 使用你在习题 7.3、11.9、11.10 或 12.8 中实现的任意一个 3D 重建方法，由多幅照片构造一个基于图像的三角面片 3D 模型。
2. 对照片中的每一个模型面进行纹理抽取，你可以进行合适的重采样或者使用纹理映射软件直接操作源图像。
3. 在算法的执行阶段，对每个新的摄像机视图，为每个可见的模型面选择最佳的源图像。
4. 将该算法扩展到能够对最好的两三个纹理进行混合。这个任务更加复杂，因为它涉及纹理融合或像素着色(Debevec, Taylor, and Malik 1996; Debevec, Yu, and Borshukov 1998; Pighin, Hecker, Lischinski et al. 1998)。

习题 13.5：层次深度图像

将你实现的视图插值算法(习题 13.2)进行扩展，使其能为每个像素存储多个(多于一个)深度或颜色值(Shade, Gortler, He et al. 1998)，即层次深度图像(LDI)。相应的，你需要对你的绘制算法进行修改。对于数据，你可以使用光线跟踪、层次重建模型或体重建。

习题 13.6：使用子画面和图层的绘制

扩展你的视图插值算法使其能够处理多个平面或子画面(13.2.1 节)(Shade, Gortler, He et al. 1998)。

1. 使用你在习题 8.9 实现的算法抽取图层。
2. 或者，你可以使用交互绘画或 3D 布局系统来抽取图层(Kang 1998; Oh, Chen, Dorsey et al. 2001; Shum, Sun, Yamazake et al. 2004)。
3. 基于期望可见性确定层次从后到前的顺序，或者将 z 缓冲引入你的绘制算法以处理遮挡。
4. 绘制每个图层，然后将它们组合起来，你可能会用到 alpha 抠图方法来处理图层或子画面的边缘。

习题 13.7：光场变换

推导正规图像和 4D 光场坐标之间的变换公式。

1. 确定远处平面 (u,v) 坐标到虚拟摄像机 (x,y) 坐标的映射公式。
   (a) 按照 uv 平面的 (u,v) 坐标对该平面上的 3D 点进行参数化。
   (b) 使用 3×4 的摄像机矩阵 P(2.63)将得到的 3D 点投影到摄像机像素 $(x,y,1)$。
   (c) 推导联系 (u,v) 坐标和 (x,y) 坐标的 2D 单应矩阵(homography)。
2. 写出从 (s,t) 坐标到 (x,y) 坐标的类似变换。

3. 证明如果虚拟摄像机恰好位于 (s,t) 平面上，则 (s,t) 坐标只依赖于摄像机的光心，而与 (x,y) 坐标无关。
4. 证明用正规的正交投影摄像机或透视摄像机拍摄的图像，即在 3D 点和 (x,y) 坐标间具有线性投影关系(2.63)的图像，就是 (s,t,u,v) 光场沿着一个二维超平面的采样。

习题 13.8：光场和发光图绘制

实现一个光场或发光图绘制系统。

1. 从 *http://lightfield.stanford.edu* 下载一个光场数据集。
2. 设计一个算法，以该光场为基础合成一个新的视图。对 (s,t,u,v) 光线样本使用四线性插值。
3. 尝试改变对应于你期望的视图的焦平面(Isaken, McMillan, and Gortler 2000)，观察绘制结果是否看起来更加清晰。
4. 为场景中的物体确定一个 3D 代理。为此，你可以对光场使用多视角立体视觉算法来计算每幅图像的深度图。
5. 实现发光图绘制算法，该算法根据每个表面元素的 3D 元素对光线采样进行修改。
6. 自己收集一组图像，使用由运动到结构算法来确定它们的姿态。
7. 实现 Buehler, Bosse, McMillan *et al.* (2001)提出的非结构化发光图绘制算法。

习题 13.9：表面光场

构造一个表面光场(Wood, Azuma, Aldinger *et al.* 2000)，并且看看你能如何对其进行尽可能多的压缩。

1. 采集一个高光（镜面）场景或物体的光场，或者从 *http://lightfield.stanford.edu/* 下载一个。
2. 使用对由高光造成的外点具有较强鲁棒性的多视角立体视觉算法为物体构造 3D 模型。
3. 估计物体每个表面点的光球。
4. 估计漫反射成分。中值滤波是最好的方法吗？为什么不使用最小的颜色值呢？如果漫反射成分中存在朗伯阴影(Lambertian shading)又如何呢？
5. 使用 Wood, Azuma, Aldinger *et al.* (2000)提出的算法对光球中除去漫反射成分后剩余的部分进行建模和压缩。当然，你也可以自己发明一个算法。
6. 研究你的压缩算法的效果及其引入的缺陷。
7. (可选)开发一个能够编辑和修改表面光场的系统。

习题 13.10：手持拍摄同心拼图

实现一个系统，对手持摄像机拍摄的同心拼图进行浏览。

1. 手持一个便携式摄像机站在一间屋子的中央，将手臂向身体前方伸直，然后原地旋转一圈。

2. 使用一个由运动到结构系统确定每一输入帧的摄像机姿态和稀疏 3D 结构。
3. (可选)将你拍摄的图像像素重新分片到一个更加规则的同心拼图结构中。
4. 在绘制新视图(该视点应该在原始拍摄平面的附近)时，确定应该显示哪个源像素。你也可以简化你的算法，只对每一个输出列确定对应的原始像素列(和放缩)。
5. (可选)对你之前得到的稀疏 3D 结构进行插值得到一个稠密深度图，提高绘制质量(Zheng, kang, Cohen *et al*. 2007)。

习题 13.11：视频纹理

拍摄一些自然现象的视频，例如喷泉、火焰或笑脸，然后将视频无缝拼接成一段无限长的视频(Schodl, Szeliski, Salesin *et al*. 2000)。

1. 对视频片段中的每帧之间计算(平方差的和)相似度。(假设视频是用三脚架拍摄的，或者已经做过稳定化处理的。)
2. 在时域上对相似度表进行过滤，找出匹配较好的子序列。
3. 使用指数分布将相似度表转换为跳转概率表。请一定对视频末尾附近的转换进行修改，以免出现视频在最后帧"卡住"的现象。
4. 从第一帧开始，使用转移概率表决定向前跳转、向后跳转还是继续播放下一帧。
5. (可选)对原始的视频纹理算法进行任一扩展，例如处理多个运动区域、交互式控制或使用图割在时空域上进行纹理拼接。

第 14 章

识 别

14.1 物体检测
14.2 人脸识别
14.3 示例识别
14.4 类别识别
14.5 上下文与场景理解
14.6 识别数据库和测试集
14.7 补充阅读
14.8 习题

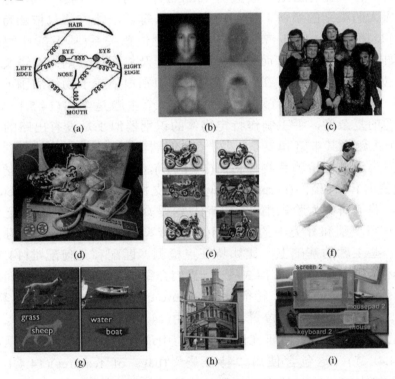

图 14.1 识别：(a)基于图案结构(pictorial structures)的人脸识别(Fischler and Elschlager 1973)© 1973 IEEE；(b)基于特征脸的人脸识别(Turk and Pentland 1991b)；(c)实时人脸检测(Viola and Jones 2004)；(d)(已知物体)示例的识别(Lowe 1999)© 1999 IEEE；(e)基于特征的识别(Fergus, Perona, and Zisserman 2007)；(f)基于区域的识别(Mori, Ren, Efros *et al.* 2004) © 2004 IEEE；(g)同时识别和分割(Shotton, Winn, Rother *et al.* 2009) © 2009 Springer；(h)位置识别(Philbin, Chum, Isard *et al.* 2007)© 2007 IEEE；(i)使用上下文信息(Russell, Torralba, Liu *et al.* 2007)

在我们也许会让计算机完成的所有视觉任务中，分析场景并识别所有构成场景的物体仍然是最有挑战性的任务。虽然计算机擅长从多个视角拍摄的图像中精确地重建出场景的 3D 形状，但是它们仍然无法叫出出现在图像中的物体和动物的名称，即便是以两岁小孩的水平来衡量也是如此。研究人员甚至对于何时能够达到这个水平的性能还没有达成共识。

识别为什么这么难呢？现实世界是由混杂的物体组成的，所有物体会遮挡其他物体，还经常呈现出不同的姿态。另外，由于复杂的非刚性关节连接关系以及形状和表观的极端变化(比如不同品种之间)，一类物体(例如狗)内固有的变化使得我们不可能简单地用样例数据库进行彻底的匹配。[①]

识别问题可以沿几个不同的轴拆解。比如，如果我们知道正在找什么，这个问题就是物体检测(object detection)(14.1 节)，它涉及快速扫描图像以确定可能出现匹配的位置(如图 14.1.c)。如果有将要识别的一个特定的刚性物体(示例识别，14.3 节)，我们可以寻找一些特有的特征点，然后验证它们在几何上可行的对齐方式(14.3.1 节)(图 14.1.d)。

最有挑战性的识别问题是一般类别(category)识别问题(14.4 节)，这个问题可能涉及到差异极大的类别的示例，比如动物、家具等。一些方法只依赖特征出现的频率(称为"词袋模型"(14.4.1 节))、它们的相对位置关系(基于部件的模型，参见 14.4.2 节，如图 14.1e 所示)，而另外一些方法涉及将图像分割为有语义含义的区域(14.4.3 节)，如图 14.1f 所示。在很多情况下，识别很依赖于物体周围的上下文和场景元素(14.5 节)。交织在所有这些方法中的一个主题是学习(14.5.1 节)，这是因为，给定问题的复杂度，手工制作特定物体的识别器似乎是没有出路的方法。

由于这个主题极其丰富和复杂，这一章的结构安排是从简单的概念到复杂的概念。我们首先讨论人脸检测和物体检测(14.1 节)，其中介绍一些机器学习方法，比如 boosting(提升)，神经网(nerural network)和支持向量机(support vector machine)等。接下来，我们研究人脸识别(14.2 节)，这是众所周知的识别应用之一。我们用这个主题来介绍识别和分类中的子空间(PCA)模型和贝叶斯(Bayesia)方法。在本书已经介绍的一些主题的基础上，比如特征点检测、匹配和几何配准(14.3.1 节)，我们接下来描述示例识别的方法(14.3 节)。我们介绍信息和文档检索方面的主题，比如频率向量(frequency vector)、特征量化(feature quantization)以及反向索引(inverted index)(14.3.2 节)。我们也介绍位置识别的应用(14.3.3 节)。

在本章的第二部分，我们研究识别的变型中最有挑战性的一个主题，称为"类别识别"(14.4 节)。这包含使用"特征袋"(bags of features)(14.4.1 节)、部件(14.4.2 节)和分割(14.4.3 节)的方法。我们将展示如何把这些方法用到自动图像编辑任务中，比如 3D 建模、场景修复(scene completion)和创建拼贴画(creating collages)等(14.4.4 节)。接下来，我们讨论上下文在单个物体识别和全局场景理解中的作用

---

① 但是，最近一些研究指出，对于足够大的数据库，直接的图像匹配也许是可行的(Russell, Torralba, Liu et al. 2007; Malisiewicz and Efros 2008; Torralba, Freeman, and Fergus 2008)。

(14.5 节)。在本章的结尾，我们讨论构建和评测识别系统的数据库和测试集。

尽管物体识别没有全面的参考资料，一些比较出色的解释说明有 ICCV2009 的短期课程(Fei-Fei, Fergus, and Torralba 2009)、Antonio Torralba 的 MIT 综合课程(Torralba 2008)、两个近期的论文汇编(Ponce, Hebert, Schmid *et al*. 2006; Dickinson, Leonardis, Schiele *et al*. 2007)和物体分类的综述(Pinz 2005)。一些性能最好的识别算法的评测可以在 PASCAL 视觉物体类(Visual Object Classes，VOC)网站(*http://pascallin.ecs.soton.ac.uk/challenges/VOC/*)找到。

14.1 物体检测

假如给定一幅图像让我们分析，比如图 14.2 中的合影照片，我们可以在这个图像中对每个可能的子窗口应用某个识别算法。这类算法可能既慢又易于出错。相反，构建特殊目的的检测器(detector)是更有效的方法，它们的任务是快速找到可能出现物体的区域。

在这一节，我们先介绍人脸检测，人脸检测是识别中比较成功的一个例子。这些算法已经嵌入到当今大部分数码相机的增强自动对焦以及视频会议系统中的前端摇移-倾斜云台控制里面。然后，作为更一般的物体检测方法的例子我们介绍行人检测。这类检测器能够用在汽车的安全性应用中，比如从移动车辆中检测行人和其他车辆(Leibe, Cornelis *et al*. 2007)。

14.1.1 人脸检测

在任意一幅图像中进行人脸识别之前，需要定位人脸的位置和大小(图 14.1c 和图 14.2)。从理论上讲，我们可以在每个像素每个尺度上进行人脸识别(Moghaddam and Pentland 1997)，但是在实际中，这样的过程太慢了。

图 14.2 Rowley, Baluja, and Kanade(1998a)© 1998 IEEE 的人脸检测结果。你能在 57 个正确结果中间找到一个误报(矩形框内不是人脸)吗？

多年来，产生了很多快速人脸检测的算法。Yang, Kriegman, and Ahuja(2002)提供了这个领域前期工作的全面综述；Yang 的 ICPR 2004 教程[①]和 Torralba(2007)短期课程提供了最新的回顾[②]。

根据 Yang, Kriegman, and Ahuja(2002)的分类，人脸检测的方法分为基于特征的、基于模板的和基于表观的方法。基于特征的方法，尝试寻找有区分性的图像特征的位置，比如眼睛、鼻子和嘴，然后在合理的几何布局上验证这些特征是否存在。这类方法包括一些早期的人脸识别方法(Fischler and Elschlager 1973; Kanade 1977; Yuille 1991)以及最新的一些基于模块特征空间(modular eigenspace)的方法(Moghaddam and Pentland 1997)、局部滤波器束方法(local filter jet) (Leung, Burl, and Perona 1995; Penev and Atick 1996; Wiskott, Fellous, Krüger et al. 1997)、支持向量机方法(Heisele, Ho, Wu et al. 2003; Heisele, Serre, and Poggio 2007)和 boosting 方法(Schneiderman and Kanade 2004)。

基于模板的方法，比如活动表观模型(active appearance model，AAM)(14.2.2 节)，能够处理姿态和表情较大范围的变化。它们通常需要接近真正人脸的好的初始化，因此对快速人脸检测器不合适。

基于表观的方法扫描图像的小的有重叠的矩形区域寻找似人脸的候选区域，然后用一组更昂贵但具选择性的检测算法的层叠(cascade)求精(Sung and Poggio 1998; Rowley, Baluja, and Kanade 1998a; Romdhani, Torr, Schölkopf et al. 2001; Fleuret and Geman 2001; Viola and Jones 2004)。为了处理尺度缩放，图像通常转化为一个子八度金字塔(sub-octave pyramid)，然后在每层上分别扫描。现在大多数基于表观的方法很依赖于在标定的人脸和非人脸的集合上训练检测器。

Sung and Poggio(1998)和 Rowley, Baluja, and Kanade(1998a)提出了最早的两个基于表观的人脸检测器，并提出了很多在其他人的后续工作中广为使用的新方法。

首先，这两个系统都收集了标定的人脸区域集合(图 14.2)和不包含人脸的图像区域集合，比如航空影像(aerial image)或者植被(图 14.3b)。收集的人脸图像通过镜像、旋转、尺度缩放和少量的平移进行扩大，使得人脸检测器对这些影响不敏感(图 14.3a)。

收集了初始训练样本之后，可以做一些可选的预处理，比如从图像中减去平均梯度(线性函数)来补偿整体光照影响；用直方图均衡化补偿摄像机对比度的变化(图 14.3c)。

聚类和 PCA 人脸和非人脸模式经过预处理后，Sung and Poggio(1998)先用 k 均值算法把这些数据集的每个集合聚为 6 个不同的聚类，然后对得到的 12 个聚类中的每一个用 PCA 子空间拟合(图 14.4)。在检测的时候，使用由 Moghaddam and Pentland(1997)(见图 14.14)首先提出的 DIFS 和 DFFS 准则，产生 24 个马氏距离

[①] http://vision.ai.uiuc.edu/mhyang/face-detection-survey.html
[②] 另一个人脸检测的方法是在图像中寻找肤色的区域(Forsyth and Fleck 1999; Jones and Rehg 2001)。更多的讨论和参考文献参见习题 2.8。

(Mahalanobis distance)度量(每个聚类两个)。得到的 24 个度量是多层感知器(multi-layer perceptron，MLP)的输入，这个多层感知器包含一个有加权和的交替层和用后向传播(back propagation)算法学习的 S 型非线性的神经网(Rumelhart, Hinton, and Williams 1986)。

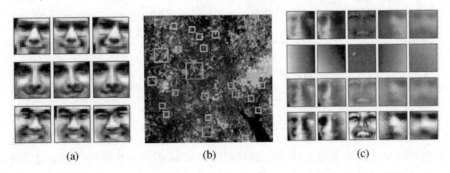

图 14.3 人脸检测器训练的预处理(Rowley, Baluja, and Kanade 1998a)© 1998 IEEE：(a)人工镜像、旋转、尺度缩放以及平移训练图像得到更大变化的训练样本；(b)用没有人脸的图像(比如向上看一棵树)产生非人脸样本；(c)预处理这些区域的方法是减去最佳拟合的线性函数(等梯度)以及直方图均衡化

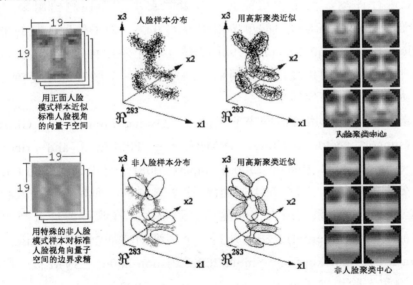

图 14.4 为人脸检测学习的高斯混合模型(Sung and Poggio 1998)© 1998 IEEE。先用 k 均值算法，把人脸和非人脸图像(19^2-长向量)聚类成 6 个不同的类别，然后用 PCA 进行分析。聚类中心显示在右面一列中

神经网 与先对数据进行聚类然后计算到聚类中心的马氏距离不同，Rowley, Baluja and Kanade(1998a)直接把神经网(MLP)用到 20×20 像素的灰度强度的区域中，用多种不同大小的手工设定的"感受野"(receptive field)刻画大尺度和小尺度的结构(如图 14.5 所示)。得到的神经网能够直接输出在多分辨率金字塔中每个重叠的区域中心存在人脸的似然率。由于几个重叠的区域(在空间和分辨率两者上)可能会在人脸附近激活，因此使用附加的融合神经网融合重叠的检测结果。论文作者也试验

了用多个神经网训练和融合检测的输出。图 14.2 显示了他们的人脸检测器的样例结果。

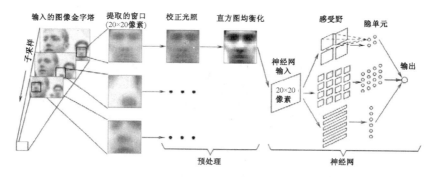

图 14.5 用于人脸检测的神经网(Rowley, Baluja, and Kanade 1998a)© 1998 IEEE。在金字塔的不同层上提取重叠的区域，然后用图 14.3b 所示的方法进行预处理。最后，用三层神经网检测可能出现人脸的位置

为了使检测器更快，训练了一个在 30×30 区域上的单独的神经网来共同检测人脸和平移±5 个像素的人脸。这个神经网在图像的每 10 个像素(垂直方向和水平方向)间隔上进行评测，这个"粗糙的"或者"草率的"检测器用来选择一些区域，在这些选择的区域上运行慢的单像素重叠的方法。为了处理人脸的平面内旋转问题，Rowley, Baluja, and Kanade(1998b)训练了一个路由神经网(router network)，从输入区域中估计旋转的角度，然后用估计的角度对每个区域进行旋转，最后运行正面向上的人脸检测器得到结果。

支持向量机 不同于用神经网来分类区域，Osuna, Freund, and Girosi(1997)用支持向量机(support vector machine，SVM)(Hastie, Tibshirani, and Friedman 2001; Schölkopf and Smola 2002; Bishop 2006; Lampert 2008)分类和 Sung and Poggio(1998)经过同样预处理的区域。SVM 在不同类(在这个问题中，即为人脸区域和非人脸区域)的特征空间中寻找一系列最大化间隔的分隔平面。在那些线性分类边界失效的情况下，特征空间用核(kernel)方法上升到一个高维的特征空间中(Hastie, Tibshirani, and Friedman 2001; Schölkopf and Smola 2002; Bishop 2006)。SVM 被一些研究人员用于人脸检测和人脸识别(Heisele, Ho, Wu *et al.* 2003;Heisele, Serre, and Poggio 2007)，SVM 也是广泛应用的一般物体识别的工具。

boosting 在现在使用的所有人脸检测器中，Viola and Jones(2004)提出的很可能是最有名而且应用最广的。他们的方法第一次在计算机视觉领域中引入了 boosting 的概念，它涉及训练一系列区分性渐增的简单分类器，并融合它们的结果 (Hastie, Tibshirani, and Friedman 2001; Bishop 2006)。

从细节上讲，boosting 涉及建立一个由简单的弱学习器(weak learner)的和组成的分类器 $h(x)$，

$$h(\boldsymbol{x}) = \text{sign}\left[\sum_{j=0}^{m-1} \alpha_j h_j(\boldsymbol{x})\right], \tag{14.1}$$

其中每个弱学习器 $h_j(\boldsymbol{x})$ 是输入的一个极为简单的函数，因此并不期望它(单独)对于分类性能有很大的贡献。

在 boosting 的多数变形中，弱学习器是个阈值型的函数，

$$h_j(\boldsymbol{x}) = a_j[f_j < \theta_j] + b_j[f_j \geq \theta_j] = \begin{cases} a_j & \text{if } f_j < \theta_j \\ b_j & \text{otherwise,} \end{cases} \tag{14.2}$$

它们也称为"决策树桩"(decision stump)(从本质上讲，是决策树最简单的版本)。在大多数情况中，习惯性地(简单的)把 a_j 和 b_j 设为 ±1，即 $a_j=-s_j$，$b_j=s_j$，以便只需要选择特征 f_j、阈值 θ_j 和阈值极性 $s_j \in \pm 1$。①

在 boosting 的很多应用中，特征是简单的坐标轴 x_k，即 boosting 算法选择输入向量的一个分量作为阈值化的最佳目标。在 Viola 和 Jones 的人脸检测器中，特征是输入区域的矩形块的差，如图 14.6 所示。使用这些特征的优点是，它们不仅比单个像素更有区分性，而且一旦经过预计算阶段得到求和的区域表(summed area table)(见 3.2.3 节(3.31 和 3.32))后，计算极快。本质上，在预计算阶段需要的代价为 $O(N)$(其中 N 是图像中像素的个数)，随后矩形的差可以用 $4r$ 次加法或者减法得到，其中 $r \in \{2,3,4\}$，是特征中矩形的个数。

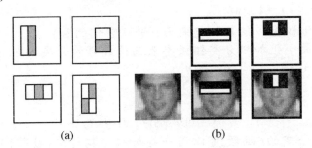

图 14.6 基于 boosting 的人脸检测器所使用简单特征(Viola and Jones 2004)© 2004 Springer：(a)由 2 到 4 个不同的矩形组成的矩形特征的差(用白色矩形中的像素值减去在灰色矩形中的像素值)；(b)AdaBoost 选择的第一和第二个特征。第一个特征度量眼睛和脸颊的颜色差异，第二个特征度量眼睛和鼻梁的颜色差异

boosting 成功的关键是不断选择弱分类器并在每个阶段后调整训练样本的权重(图 14.7)。AdaBoost (Adaptive Boosting)算法(Hastie, Tibshirani, and Friedman 2001; Bishop 2006)的做法是：通过将每个样本的权重调整作为其在每个阶段是否被正确分类的函数，然后用每个阶段的平均分类错误率在弱分类器中间决定最终权重 α_j，如算法 14.1 所描述。在实践中，尽管最后的分类器极快，但是训练时间可能会相当慢(以星期为单位)，这是因为每个阶段需要测试大量的特征(矩形的差)假设。

---

① 有些变形，比如 Viola and Jones (2004)，使用$(a_j,b_j) \in [0,1]$来相应地调整学习算法。

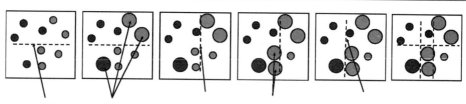

弱分类器 1　权重增加　　　弱分类器 2　权重增加　　　弱分类器 3　最终的分类器

图 14.7　boosting 的图解说明，经 Svetlana Lazebnik 授权使用，源自 Paul Viola 和 David Lowe 的解释。在选择每个弱分类器(决策树桩或者超平面)后，提高了被分错的数据点的权重。最终的检测器是简单弱分类器的线性组合

为了进一步提高检测器的速度，可以建立一个瀑布型的检测器，其中每个弱分类器用测试的一小部分(比如说两项的 AdaBoost 分类器)抛弃很大比例的非人脸且同时让所有潜在的人脸候选都通过(Fleuret and Geman 2001; Viola and Jones 2004)。Brubaker, Wu, Sun *et al.*(2008)最近提出了一个更快的进行瀑布型学习的算法。

1. 输入训练的正样本和负样本以及它们的标签$\{(x_i,y_i)\}$，其中$y_i=1$表示正样本(人脸)，$y_i=-1$是负样本。
2. 初始化样本权重$\omega_{i,1} \leftarrow \frac{1}{N}$，其中$N$是训练样本的数目。(Viola and Jones (2004)用N_1和N_2分别表示正负样本的数目。
3. 对训练的每一步 $j = 1\ldots M$
   (a) 重新归一化权重，使其加起来为1(除以它们的和)。
   (b) 通过找到最小化加权分类错误率来选择最好的分类器 $h_j(x;f_j,\theta_j,s_j)$
   $$e_j = \sum_{i=0}^{N-1} w_{i,j} e_{i,j}, \qquad (14.3)$$
   $$e_{i,j} = 1 - \delta(y_i, h_j(\boldsymbol{x}_i; f_j, \theta_j, s_j)). \qquad (14.4)$$
   对于任何给定的f_j函数,可以用一些加权平均计算(习题14.2)在线性时间内找到最优值(θ_j, s_j)。
   (c) 计算修正误差率β_j和分类器的权重α_j
   $$\beta_j = \frac{e_j}{1-e_j} \quad \text{and} \quad \alpha_j = -\log \beta_j. \qquad (14.5)$$
   (d) 根据分类错误率$e_{i,j}$更新权重
   $$w_{i,j+1} \leftarrow w_{i,j} \beta_j^{1-e_{i,j}}, \qquad (14.6)$$
   即将正确分类的训练样本与整体分类错误率成比例地降低权重。
4. 最终的分类器设置为
$$h(\boldsymbol{x}) = \text{sign}\left[\sum_{j=0}^{m-1} \alpha_j h_j(\boldsymbol{x})\right]. \qquad (14.7)$$

算法 14.1　AdaBoost 学习算法，从 Hastie, Tibshirani, and Friedman (2001), Viola and Jones (2004)和 Bishop (2006)修改而来

14.1.2 行人检测

尽管很多物体检测的研究关注于人脸检测，还有一些其他物体的检测，比如行人和汽车，也得到了广泛的关注(Gavrila and Philomin 1999; Gavrila 1999; Papageorgiou and Poggio 2000; Mohan, Papageorgiou, and Poggio 2001; Schneiderman and Kanade 2004)。一些方法和人脸检测一样强调速度和效率。但是，还有一些强调精度，把检测问题看作更有挑战性的一般类物体识别的一个变形(14.4 节)，在这些方法中，物体的位置以及其范围要尽可能的精确(比如，参见 PASCAL VOC 检测挑战，http://pascallin.ecs.soton.ac.uk/ challenges/VOC/)。

一个非常有名的行人检测的例子是 Dalal and Triggs(2005)，他们把一些重叠的梯度方向直方图(HOG)描述子送入 SVM(14.8)。和我们在 4.1.2 节和图 4.18 讨论的 Lowe(2004)提出的尺度不变特征变换(SIFT)一样，每个 HOG 包含多个单元，每个单元累加特定方向的以幅度加权的梯度值投票。但是和 SIFT 只在兴趣点位置计算不同，HOG 在规则的重叠的网格上计算，这些描述子的幅值用更粗糙的网格进行归一化，而且只在单尺度固定方向上计算。为了得到沿着人轮廓的方向上的精细变化，可以把方向划分为更多的箱子(bin)，在计算中心梯度差时没有进行平滑等——更多实现细节参见 Dalal and Triggs(2005)的工作。图 14.8d 显示了一个输入的样本图像，图 14.8e 显示了关联的 HOG 描述子。

一旦计算出描述子，在得到高维连续描述子向量上训练 SVM。图 14.8b 和 c 示意性地显示了每个块中(大部分)正负 SVM 权重，图 14.8f 和 g 显示了输入图像中心相应的加权 HOG 响应。正如你所看到的，在人的头、躯干和脚附近有很多正的响应，相对来说，在这些位置几乎没有负的响应(主要在运动衫的中间和衣领附近)。

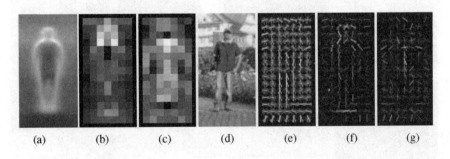

图 14.8 使用梯度方向直方图的行人检测(Dalal and Triggs 2005)© 2005 IEEE：(a)训练样本的平均梯度图像；(b)每个像素显示了以它为中心的块的最大的正 SVM 权重；(c)同样，在负 SVM 权重上的情况；(d)测试图像；(e)计算的 R-HOG (矩形的梯度直方图) 描述子；(f)用正的 SVM 权重加权了的 R-HOG 描述子；(g)用负的 SVM 权重加权了的 R-HOG 描述子

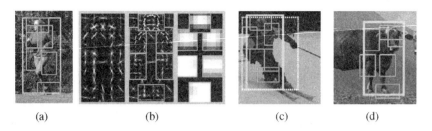

图 14.9 基于部件的物体检测(Felzenszwalb, McAllester, and Ramanan 2008)© 2008 IEEE：(a)输入照片及其关联的人(蓝色)和部件(黄色)检测结果；(b)检测模型用一个粗糙的模板、几个较高分辨率的部件模板及关于每个部件位置的空间模型定义；(c)滑雪者正确的检测结果；(d)作为错误检测结果的奶牛(当作人标注出来)

在过去十年，行人检测和一般物体检测领域还在持续地快速发展(Belongie, Malik, and Puzicha 2002; Mikolajczyk, Schmid, and Zisserman 2004; Leibe, Seemann, and Schiele 2005; Opelt, Pinz, and Zisserman 2006; Torralba 2007; Andriluka, Roth, and Schiele 2008, 2009; Dollàr, Belongie, and Perona 2010)。Munder and Gavrila(2006)比较了很多行人检测器，得出结论：基于局部感受野和 SVM 的方法得到的效果最好，其次是基于 boosting 的方法。Maji, Berg, and Malik(2008)用不重叠的多分辨率的 HOG 描述子和基于 Lazebnik, Schmid, and Ponce(2006)提出的空间金字塔匹配核的直方图相交核 SVM，对这些最好的结果进行了改进。

同时构建多类物体的检测器时，Torralba, Murphy, and Freeman(2007)表明在检测器之间共享特征和弱分类器可以得到更好的性能，不仅能够加快计算，还可以使用少量的训练样本。为了找到在共享方面做得最好的特征和决策树桩，他们提出了一个新颖的联合 boosting 算法，用 Friedman, Hastie, and Tibshirani(2000)提出的"gentleboost"算法，在每一轮优化的是期望指数损失函数的和。

图 14.10 针对人、自行车和马的基于部件的物体检测结果(Felzenszwalb, McAllester, and Ramanan 2008)© 2008 IEEE。前三列给出的是正确的检测结果，而最右列给出的是错误的检测结果

在最近的工作中，Felzenszwalb, McAllester, and Ramanan (2008)把梯度方向直方图的行人检测器扩展到包含柔性部件模型(14.4.2 节)。每个部件在整体物体模型下的两层金字塔上用 HOG 训练和检测，而且相对于父节点(整体的包围盒)的部件的位置也由学习得到并在识别中使用(图 14.9b)。为了弥补训练样本的包围盒(图 14.9c 中的白虚线)的不精确和不一致，父节点(蓝色包围盒)的真正位置被认为是一个隐变量，在训练和识别时推断得到。由于部件的位置也是隐变量，这个系统可以用半监督的方法训练，不需要在训练样本中给部件添加标签。这个系统的扩展(Felzenszwalb, Girshick, McAllester *et al.* 2010)，在它其改进中包括简单的上下文模型，在 2008 视觉物体类检测挑战(Visual Object Classes Detection Challenge)中是两个最好的物体检测的系统之一。其他最近的基于部件的人的检测和姿态估计的改进工作有 Andriluka, Roth, and Schiele(2009) 和 Kumar, Zisserman, and H.S.Torr (2009)。

Rogez, Rihan, Ramalingam *et al.*(2008)提出了更精确的估计人的姿态和位置的方法，他们在 HOG 基础上用随机森林计算在一个步行周期内的人和各关节的位置(图 14.11)。因为他们的系统产生了完整的 3D 姿态信息，在其应用领域更接近于 3D 的人体跟踪(Sidenbladh, Black, and Fleet 2000; Andriluka, Roth, and Schiele 2008)，这个我们在 12.6.4 节讨论过。

图 14.11 用随机森林做姿态检测(Rogez, Rihan, Ramalingam *et al.* 2008)© 2008 IEEE。估计的姿态(运动学模型的状态)画在每个输入帧上

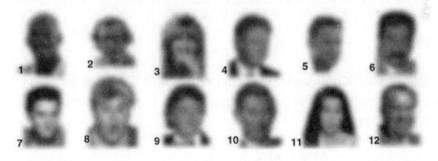

图 14.12 人类可以识别熟悉的人的低分辨率人脸(Sinha, Balas, Ostrovsky *et al.* 2006)© 2006 IEEE

最后一个是关于人和物体检测的说明。当可以得到视频序列时，光流和运动不连续性等额外信息对检测问题会有很大的帮助，正如 Efros, Berg, Mori *et al.*(2003), Viola, Jones, and Snow(2003), and Dalal, Triggs, and Schmid(2006)中讨论的。

14.2 人脸识别

在可能要求计算机处理的各种识别任务中,人脸识别是可商榷的最成功的一个。[1]虽然计算机不能从成千上万的从视频摄像机前走过的人中找出嫌疑人(即便是人也不能很容易地区分他们所不熟悉的相像的人(O'Toole Jiang, Roark et al. 2006; O'Toole, Phillips, Jiang et al. 2009)),但是它们在少数家庭成员和朋友中的区分能力为其在消费层照片的应用中找到了用武之地,例如 Picasa 和 iPhoto。人脸识别也可以用于各种其他应用,包括人机交互(HCI)、身份验证(Kirovski, Jojic, and Jancke 2004)、桌面登陆、家长控制和病患监控(Zhao, Chellappa, Phillips et al. 2003)。

虽然人们已经收集包括大量姿态和光照变化的数据集(Phillips, Moon, Rizvi et al. 2000; Sim, Baker, and Bsat 2003; Gross, Shi, and Cohn 2005; Huang, Ramesh, Berg et al. 2007; Phillips, Scruggs, O'Toole et al. 2010)(更多细节请参见 14.6 节的表 14.1),但是如今的人脸识别算法还是在给定人脸处于在相对均匀的光照条件下的全正面图像时表现最好。

一些最早期的人脸识别方法包括寻找区分性图像特征的位置,如眼睛、鼻子和嘴,并测量这些特征位置之间的距离(Fischler and Elschlager 1973; Kanade 1977; Yuille 1991),更近期的方法依赖于比较灰度图像在被称为"特征脸"(14.2.1 节)的低维空间的投影并利用活动表观模型(14.2.2 节)并对形状和表观变化(而忽略姿态变化)进行联合建模。

关于其他人脸识别方法的描述可参见一些关于这个主题的综述和书籍(Chellappa,Wilson, and Sirohey 1995; Zhao, Chellappa, Phillips et al. 2003; Li and Jain 2005)以及人脸识别网站。[2]Sinha, Balas, Ostrovsky et al(2006)的关于人类如何识别人脸的综述也值得一读;它包括一些令人惊讶的结果,比如人类识别低分辨率相似人脸的能力(图 14.12)和识别中眉毛的重要性。

14.2.1 特征脸

特征脸依赖于由 Kirby and Sirovich(1990)首先观察到的现象,任意人脸图像 x 可以通过从一个平均图像 m(图 14.1b)开始并加上少数有定比带符号的(scaled signed)图像 u_i 来压缩和重建,[3]

$$\tilde{x} = m + \sum_{i=0}^{M-1} a_i u_i, \qquad (14.8)$$

---

[1] 实例识别,即已知物体如地理位置和平面物体的再识别,是一般图像识别的另一个最成功的应用。在生物信息测定学的一般领域,即身份识别,特殊的图像如虹膜和指纹表现更好(Jain, Bolle, and Pankanti 1999; Pankanti, Bolle, and Jain 2000; Daugman 2004)。

[2] http://www.face-rec.org/。

[3] 前文中我们使用 I 表示图像,在本章中,我们使用更加抽象的量 x 和 u 来表示一幅转换为向量的图像中像素的汇集。

其中带符号的基图像(图 14.13b)可以利用主分量分析(principle component analysis)(也称"特征值分析"或 Karhunen–Loêaaave 变换)从全体训练图像中导出。Turk and Pentland(1991a)发现特征脸展开中的系数 a_i 本身可以用于构建一个快速图像匹配算法。

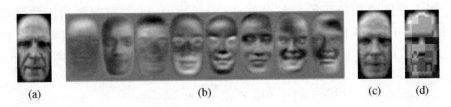

图 14.13 利用特征脸的人脸建模及压缩(Moghaddam and Pentland 1997)© 1997 IEEE: (a)输入图像；(b)前 8 个特征脸；(c)将图像投影到基上并用 85 字节压缩后重建的图像；(d)利用 JPEG(530 字节)重建的图像

详细来说，让我们首先从一系列训练图像 $\{x_j\}$ 开始，我们可以从它们中计算平均图像 m 和一个散布矩阵或协方差矩阵

$$C = \frac{1}{N} \sum_{j=0}^{N-1} (x_j - m)(x_j - m)^T. \tag{14.9}$$

我们可以应用特征值分解(A.6)将这个矩阵表示为

$$C = U\Lambda U^T = \sum_{i=0}^{N-1} \lambda_i u_i u_i^T, \tag{14.10}$$

其中 λ_i 是 C 的特征值而 u_i 是特征向量。对于一般图像，Kirby and Sirovich (1990)称这些向量为特征图像；对于人脸，Turk and Pentland (1991a)称之为"特征脸"(eigenface)(图 14.13b)。[①]

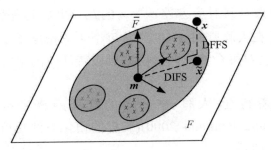

图 14.14 向由特征脸图像张成的线性子空间投影(Moghaddam and Pentland 1997)© 1997 IEEE. 人脸空间外距离(DFFS) 是到平面的垂直距离，而人脸空间内距离(DIFS) 是沿着平面到平均图像的距离。这两种距离可以变为马氏距离并给出概率解释

特征值分解的两个重要属性是任意新图像 x 的最优(最佳近似)系数 a_i 可以如下计算

[①] 在实际应用中，根本不需要计算 $P \times P$ 的散布矩阵公式(14.9)，而是累加一个由所有带符号偏差 $(x_i - m)$ 之间的内积组成的较小的 $N \times N$ 的矩阵。详见附录 A.1.2(A.13 和 A.14)。

$$a_i = (\boldsymbol{x} - \boldsymbol{m}) \cdot \boldsymbol{u}_i, \tag{14.11}$$

以及，假设特征值 $\{\lambda_i\}$ 按降序排序，在任意点 M 处对公式(14.8)中的近似做截断给出的是 $\tilde{x}$ 和 x 间的最佳近似(最小误差)。图 14.13c 显示了对图 14.13a 的近似结果并显示了这种方法在压缩人脸图像时远远好于 JPEG。

将一幅人脸图像的特征脸分解公式(14.8)截断为 M 个分量等价于将图像投影到一个线性子空间 F，我们可以将这个子空间称为"人脸空间"(图 14.14)。由于特征向量(特征脸)彼此正交且模为 1，投影人脸 $\tilde{x}$ 与平均脸 $\boldsymbol{m}$ 之间的距离可如下计算：

$$\text{DIFS} = \|\tilde{\boldsymbol{x}} - \boldsymbol{m}\| = \sqrt{\sum_{i=0}^{M-1} a_i^2}, \tag{14.12}$$

其中 DIFS 代表人脸空间内距离(distance in face space)(Moghaddam and Pentland 1997)。剩下的原始图像 x 与其投影 $\tilde{x}$ 之间距离，即人脸空间外距离(distance from face space)，可以在像素空间内计算并代表特定图像"像人脸"的程度。[①]在人脸空间中，也可以如下度量两个不同人脸之间的距离

$$\text{DIFS}(\boldsymbol{x}, \boldsymbol{y}) = \|\tilde{\boldsymbol{x}} - \tilde{\boldsymbol{y}}\| = \sqrt{\sum_{i=0}^{M-1} (a_i - b_i)^2}, \tag{14.13}$$

其中 $b_i = (\boldsymbol{y} - \boldsymbol{m}) \cdot \boldsymbol{u}_i$ 是 y 的特征脸系数。

然而在欧氏向量空间中计算这种距离并不能利用我们的协方差矩阵的特征值分解提供的额外信息。如果我们将协方差矩阵看作一个多维高斯分布的协方差(附录 B.1.1)，[②]我们可以通过计算马氏距离将 DIFS 变为一个对数似然度

$$\text{DIFS}' = \|\tilde{\boldsymbol{x}} - \boldsymbol{m}\|_{C^{-1}} = \sqrt{\sum_{i=0}^{M-1} a_i^2 / \lambda_i^2}. \tag{14.14}$$

马氏距离计算平方距离及其对应方差 $\sigma_i^2 = \lambda_i$ 的比率并将这些比率(每个分量的对数似然度)相加，而不是在人脸空间 F 中沿着每个主分量度量平方距离。另一个实现这个计算过程的方法是对每个特征向量根据对应特征值的平方根倒数进行放缩。

$$\hat{\boldsymbol{U}} = \boldsymbol{U}\boldsymbol{\Lambda}^{-1/2}. \tag{14.15}$$

这种白化变换意味着特征(人脸)空间中的欧氏距离直接对应了对数似然度(Moghaddam, Jebara, and Pentland 2000)(同样的白化方法也可以应用于基于特征的匹配算法，如 4.1.3 节所述。)

如果特征脸空间的分布十分狭长，马氏距离很好的对分量进行放缩以得到到平均值的一个可感知的(概率)距离。计算一个可感知的人脸空间外距离(DFFS)时也可以应用类似的分析方法(Moghaddam and Pentland 1997)，并且可以结合这两个距离来估计是真实人脸的似然度，这在做人脸检测时可以用到(14.1.1 节)。关于概率和贝叶斯主分量分析(Bayesian PCA)的细节可参见统计学习的教材(Hastie, Tibshirani,

---

[①] 这可以用于建立一个简单的人脸检测器，如 14.1.1 节所述。
[②] 图 14.14 中的椭圆表示这个多维高斯分布的等概率轮廓。

and Friedman 2001; Bishop 2006)，其中还讨论了在对一个分布进行建模时选择主分量个数 M 的技巧。

使用特征脸最大的优势之一是它可以将一幅新的人脸图像 x 和一个原型(训练)人脸图像 x_k(图 14.14 中的彩色 x 之一)之间的比较从像素空间中的 P 维偏差简化为人脸空间中的一个 M 维偏差

$$\|x - x_k\| = \|a - a_k\|, \tag{14.16}$$

其中 $a = U^T(x-m)$ 由带符号的与平均图像的差 $(x-m)$ 和特征脸 u_i 的点积组成。然而，这个欧式距离再一次忽略了我们在训练图像的分布中有更多关于人脸似然度信息可用的事实。

考虑图 14.15 中一个人在各种不同的照明条件下的图像集合。如图所示，这些图像中类内变化远大于典型的任意两个人在相同照明条件下拍摄的两张照片之间的类间变化。普通的主分量分析不能区分这两个变化的来源并事实上有可能用它大部分的主分量来刻画类内变化。

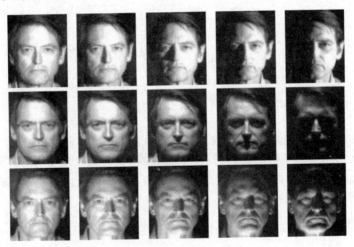

图 14.15　Belhumeur, Hespanha, and Kriegman(1997)© 1997 IEEE 使用的哈佛数据库(Harvard database)中的图像。注意其中的大范围光照变化，比不同人之间的变化更剧烈

如果我们利用线性子空间近似人脸，那么使用一个能够区分不同类别(人)并且对类别内变化不那么敏感的空间更加合适(Belhumeur, Hespanha, and Kriegman 1997)。考虑图 14.16 中用不同颜色表示的三个类别。如图所示，类内分布(用倾斜的带颜色的坐标轴表示)相对于主人脸空间(对齐到黑色 x,y 轴)是狭长而倾斜的。我们可以计算全部类内散布矩阵：

$$S_W = \sum_{k=0}^{K-1} S_k = \sum_{k=0}^{K-1} \sum_{i \in C_k} (x_i - m_k)(x_i - m_k)^T, \tag{14.17}$$

其中 m_k 是类 k 的均值而 S_k 是其类内散布矩阵。①类似的，我们可以计算类间散布

---

① 为与 Belhumeur, Hespanha, and Kriegman(1997)一致，我们用 S_W 和 S_B 表示散布矩阵，即使我们在其他地方(公式(14.9))使用 C。

矩阵

$$S_B = \sum_{k=0}^{K-1} N_k(m_k - m)(m_k - m)^T, \tag{14.18}$$

其中 N_k 是每个类别中的样本数量而 m 是整体的均值。对于图 14.16 中的三个分布，我们有

$$S_W = 3N \begin{bmatrix} 0.246 & 0.183 \\ 0.183 & 0.457 \end{bmatrix} \quad \text{and} \quad S_B = N \begin{bmatrix} 6.125 & 0 \\ 0 & 0.375 \end{bmatrix}, \tag{14.19}$$

其中 $N = N_k = 13$ 是每个类中的样本数量。

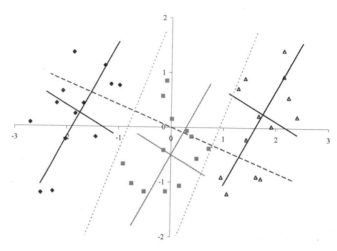

图 14.16 Fisher 线性鉴别分析的简单示例。样本来自三个不同的类别，在图中沿着其放缩到 $2\sigma_i$ 的主轴用不同的颜色表示。(倾斜坐标轴的交点是类别均值 m_k)短划线是(主要)Fisher 线性鉴别方向而虚线表示类间的线性区分。注意区分性方向如何融合类间和类内散布矩阵的主方向

为计算最具区分性的方向，Fisher 线性鉴别分析(Fisher's linear discriminant, FLD)(Belhumeur, Hespanha, and Kriegman 1997; Hastie, Tibshirani, and Friedman 2001; Bishop 2006)，也被称为"线性鉴别分析"(discriminant analysis，LDA)，选择使投影后的类间距离和类内距离之间的比值最大的方向

$$u^* = \arg\max_u \frac{u^T S_B u}{u^T S_W u}, \tag{14.20}$$

这等价于寻找下列广义特征值问题的最大特征值对应的特征向量

$$S_B u = \lambda S_W u \quad \text{或} \quad \lambda u = S_W^{-1} S_B u. \tag{14.21}$$

对于图 14.16 中的问题

$$S_W^{-1} S_B = \begin{bmatrix} 11.796 & -0.289 \\ -4.715 & 0.3889 \end{bmatrix} \quad \text{and} \quad u = \begin{bmatrix} 0.926 \\ -0.379 \end{bmatrix} \tag{14.22}$$

可见，利用这个方向可以得到比利用主要 PCA 方向(水平轴)更好的类间区分。Belhumeur, Hespanha, and Kriegman (1997)在他们的论文中说明 Fisherface 显著地胜过原始特征脸算法，尤其是当人脸具有大量光照变化时，如图 14.15 所示。

另一个刻画类内(相同人之间)和类间(不同人之间)变化的方法是对每个分布分

别建模，然后利用贝叶斯方法来寻找最近的样例(Moghaddam, Jebara, and Pentland 2000)。不需要计算每类的均值和类内类间分布，而是考虑评价全部每对训练图像 (x_i, x_j) 之间的差图像

$$\Delta_{ij} = x_i - x_j \tag{14.23}$$

具有相同类别(同一个人)的图像对之间的差用于估计类内协方差Σ_I，而不同人之间的差则用于估计类间协方差Σ_E。① 图 14.17 显示了这两类对应的主分量(特征脸)。

在识别时，我们计算一个新的人脸 x 和一个储存的训练图像 x_i 之间的距离并估计其类内似然度

$$p_I(\Delta_i) = p_\mathcal{N}(\Delta_i; \Sigma_I) = \frac{1}{|2\pi\Sigma_I|^{1/2}} \exp -\|\Delta_i\|^2_{\Sigma_I^{-1}}, \tag{14.24}$$

其中 $p_\mathcal{N}$ 是一个协方差为 Σ_I 的正态(高斯)分布，而

$$|2\pi\Sigma_I|^{1/2} = (2\pi)^{M/2} \prod_{j=1}^{M} \lambda_j^{1/2} \tag{14.25}$$

为其体积。马氏距离

$$\|\Delta_i\|^2_{\Sigma_I^{-1}} = \Delta_i^T \Sigma_I^{-1} \Delta_i = \|a^I - a_i^I\|^2 \tag{14.26}$$

可以通过首先将新图像 x 投影到白化类内人脸空间公式(14.15)

$$a^I = \hat{U}^I x \tag{14.27}$$

然后计算到训练图像向量 a_i^I 的欧式距离来更加高效的得到，a_i^I 可以离线预先计算。类间似然度 $p_E(\Delta_i)$ 可以用类似的方法计算。

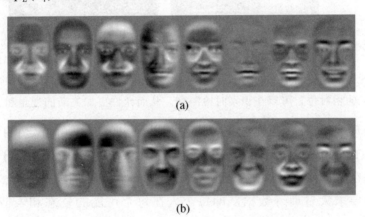

(a)

(b)

图 14.17 "双"特征脸(Moghaddam, Jebara, and Pentland 2000)© 2000 Elsevier：(a)相同人之间的；(b)不同人之间的

一旦得到类内和类间似然度后，我们可以计算一幅新图像 x 与一个训练样本 x_i 匹配的贝叶斯似然度

$$p(\Delta_i) = \frac{p_I(\Delta_i) l_I}{p_I(\Delta_i) l_I + p_E(\Delta_i) l_E}, \tag{14.28}$$

---

① 注意差图像的分布均值为 0，因为每个 Δ_{ij} 都对应一个相反的 Δ_{ji}。

其中 l_I 和 l_E 是这两幅图像是同类或异类的先验概率(Moghaddam, Jebara, and Pentland 2000)。一个不需要衡量类间概率的更简单的方法是简单的选取具有最高似然度 $p_I(\Delta_i)$ 的训练图像。这种情况下，在由预先计算的 $\{a_i^I\}$ 向量构成的空间中的最近邻搜索算法可以用于加速最佳匹配的搜索。①

另一个改进基于特征脸方法性能的途经是将图像划分为不同的区域，比如眼睛、鼻子和嘴(图 14.18)，然后独立匹配每个模块化特征空间(Moghaddam and Pentland 1997; Heisele, Ho, Wu et al. 2003; Heisele, Serre, and Poggio 2007)。这种模块化方法的优势在于它可以处理更广泛的视角，由于每个部分可以相对其他移动。它还支持大量不同的组合，例如，我们可以刻画一个细鼻浓眉的人，并不需要特征脸来扩展出鼻子、嘴和眉毛的所有可能组合。(如果你记得儿童图书中被称为"马铃薯先生的头"(Mr. Potato Head)的纸板游戏，其中你可以选择不同的上半脸和下半脸，你会明白。)

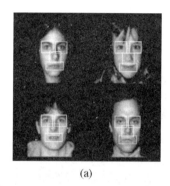 

图 14.18 用于人脸识别的模块化特征脸(Moghaddam and Pentland 1997)© 1997 IEEE：(a)通过分别检测不同的脸部特征(眼睛、鼻子和嘴)，可以估计各自的特征空间；(b)每个特征的相对位置可以在识别时检测到，从而允许更加灵活的视角和表情变化

另一个处理表观剧烈变化的方法是建立基于视角(视角特定)的特征空间，如图 14.19 所示(Moghaddam and Pentland 1997)。我们可以把这些基于视角的特征空间看作根据你在人脸空间中的位置选择不同坐标轴的局部描述子。然而要注意，这样的方法可能要求大量训练数据，即每个人在每个可能的姿态和表情下的照片。这与我们在 14.2.2 节研究的形状和表观模型形成鲜明对比，其中我们可以学习所有个体的形变。

我们也可以将 PCA 和 SVD 方法中隐含的双线性因子分解推广到可以同时刻画一些相互作用的因子的多重线性(张量)形式(Vasilescu and Terzopoulos 2007)。这些想法与当今机器学习中一些活跃的主题相关，如子空间学习(Cai, He, Hu et al. 2007)、局部距离函数(Frome, Singer, Sha et al. 2007)以及度量学习(Ramanan and

---

① 注意尽管协方差矩阵 $\sum_I$ 和 $\sum_E$ 是根据所有图像对之间的差异计算的，而运行时评价选择最近的图像以确定人脸身份。这是否在统计正确将在习题 14.4 进行探索。

Baker 2009)。学习方法在人脸识别中扮演着越来越重要的角色,如在 Sivic, Everingham, and Zisserman(2009)及 Guillaumin, Verbeek, and Schmid (2009)的工作中。

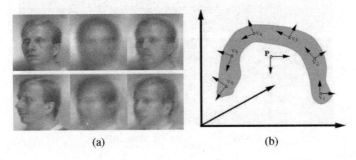

图 14.19 基于视角的特征空间(Moghaddam and Pentland 1997)© 1997 IEEE:(a)输入图像(左列)分别利用一个正规(参数化)的特征空间重建(中间一列)和一个基于视角的特征空间的重建(右列)的比较。上面一行是训练图像而下面一行是测试集;(b)两种方法的图解,显示每个视角如何计算其局部基表示

14.2.2 活动表观与 3D 形状模型

人脸识别中使用模块化或基于视角的特征值空间的需要是一个更一般的现象的征兆,即人脸表观和可识别性不仅依赖于颜色和纹理(这是特征脸所刻画的),同样依赖于形状。进一步,当处理 3D 头部旋转时,人头部的姿态在识别时是要忽略的。

事实上,最早的人脸识别系统,如 Fischler and Elschlager (1973), Kanade (1977) 和 Yuille(1991)所述,在人脸图像中寻找具有区分性的特征点并且在其相对位置或距离的基础上进行识别。较新的方法如局部特征分析(Penev and Atick 1996)和弹性图匹配(Wiskott, Fellous, Krüger *et al.* 1997)结合了区分性特征点位置上的局部滤波器响应(束(jet))和形状模型来进行识别。

Rowland and Perrett (1995)的工作是形状和纹理为何都如此重要的一个引人注目的例子,他们跟踪脸部特征的轮廓然后使用这些轮廓来将每幅图像归一化(卷绕)到规范形状上去。通过分析形状和颜色图像与均值的差异,可以将一些形状和颜色变化与个人特性如年龄和性别联系起来(图 14.20)。他们的工作表明形状和颜色对这类特性的理解都有着重要的影响。

差不多同时,计算机视觉的研究人员正在开始同时使用形变和纹理插值来刻画脸部由于身份或表情导致的表观变化(Beymer 1996; Vetter and Poggio 1997),并提出诸如活动形状模型(Active Shape Model)(Lanitis, Taylor, and Cootes 1997)、3D 变形模型(3D Morphable Model)(Blanz and Vetter 1999)以及弹性图匹配(Wiskott, Fellous, Krüger *et al.* 1997)之类的方法。①

---

① 在 12.6.3 节中,我们已经见过 PCA 在 3D 头部和人脸建模以及动画中的应用。

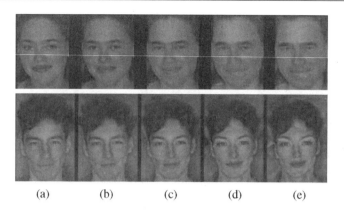

图 14.20 控制脸部表观的形状和颜色(Rowland and Perrett 1995)© 1995 IEEE。通过在输入图像(b)中加入或减去性别特异的形状和颜色特性，可以引入不同程度的性别变化。(在均值基础上)增加的数量为(a)+50%(性别增强)、(c)-50%(接近双性)，(d)-100%(性别切换)和-150%(相反的性别属性增强)

在所有这些方法中，Cootes, Edwards, and Taylor (2001)的活动表观模型(active appearance model，AAM)是在人脸识别和跟踪中使用最广泛的方法。如其他形状和纹理模型，AAM 模型同时刻画一幅图像中的形状 s 的变化(通常用图像中关键特征点的位置表示)和纹理 t 的变化(在分析前被归一化到规范形状上)(图 14.21c)。[①]

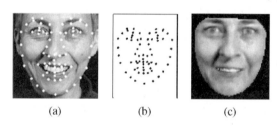

图 14.21 活动表观模型(Cootes, Edwards, and Taylor 2001)© 2001 IEEE：(a)输入图像与注册的特征点；(b)特征点(形状向量 s)；(c)形状无关表观图像(纹理向量 t)

形状和纹理都被表示为到一个平均形状 $\bar{s}$ 和平均纹理 $\bar{t}$ 的偏差

$$s = \bar{s} + U_s a \qquad (14.29)$$
$$t = \bar{t} + U_t a, \qquad (14.30)$$

其中 U_s 和 U_t 中的特征向量已经经过预先放缩(白化)使得单位向量 a 表示训练数据中观测得到的一个标准差。除了这些主要变形量之外，形状参数还通过一个全局相似性变换以匹配给定人脸的位置、大小和方向。类似，纹理图像包括一个尺度和一个偏移以最佳匹配新的光照条件。

如你所见，公式(14.29 和 14.30)中的表观参数 a 同时控制形状和纹理相对均值的变化，如果我们相信它们是相关的，这是合理的。图 14.22 显示了沿着前四个主分量方向移动三个标准差使得一个人表观的几个相关因素(包括表情、性别、年龄

---

① 当仅刻画形状变化时，这样的模型称为"活动形状模"(active shape model，ASM) (Cootes,Cooper, Taylor *et al*. 1995; Davies, Twining, and Taylor 2008)。在 5.1.1 节(公式(5.13~5.17))中已经讨论过。

和身份)发生变化的情况。

为了在一幅新图像上拟合活动表观模型，Cootes, Edwards, and Taylor (2001)使用一个与其他用于增量跟踪的快速算法相关的方法预先计算了一组"差分解"图像。这些方法包括我们在 4.1.4 节、8.1.3 节以及 8.2 节中讨论的算法(Gleicher 1997; Hager and Belhumeur 1998)，它们通常学习从匹配误差到增量位移的区分性映射(Avidan 2001; Jurie and Dhome 2002; Liu, Chen, and Kumar 2003; Sclaroff and Isidoro 2003; Romdhani and Vetter 2003; Williams, Blake, and Cipolla 2003)。

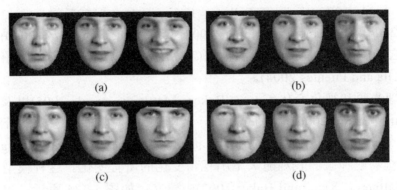

图 14.22 活动表观模型的主要变化模态(Cootes, Edwards, and Taylor 2001)© 2001 IEEE。这四幅图像显示了同时对纹理和形状在均值上对前 4 个模态做 $\pm\sigma_i$ 的变化的效果。可以明显看出人脸的形状和明暗如何同时受到影响

具体而言，Cootes, Edwards, and Taylor (2001)使用有限差分计算训练图像集相对于 a 中每个参数的导数，然后计算一组位移权重图像

$$W = \left[\frac{\partial x^T}{\partial a}\frac{\partial x}{\partial a}\right]^{-1}\frac{\partial x^T}{\partial a}, \tag{14.31}$$

它乘以当前误差残余可以产生参数的更新量，$\delta a = -Wr$。Matthews and Baker (2004)使用了他们的反向合成算法进一步加速活动表观模型的拟合和跟踪(这个算法最初他们是为参数化光流公式(8.64 和 8.65)提出来的)。图 14.23 显示了 AAM 在两个输入图像上拟合的例子。

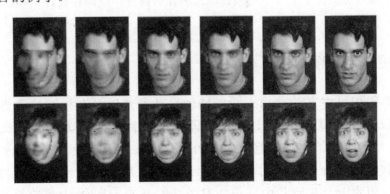

图 14.23 活动表观模型中的多分辨率模型拟合(搜索)(Cootes, Edwards, and Taylor 2001)© 2001 IEEE。每一列分别代表原始模型和经过 3，8，11 次迭代后的结果以及最终收敛的结果。最右边一列是输入图像

虽然活动表观模型起初被设计用于精确刻画关于人脸特点的形状和表观变化，但是它可以通过计算一个区分身份的变化和其他的变化来源(例如光照、姿态和表情)的身份子空间以适用于人脸识别。这种基本想法与特征脸中的类似工作(Belhumeur, Hespanha, and Kriegman 1997; Moghaddam, Jebara, and Pentland 2000)一样，是计算相同人和不同人之间的变化信息然后找到在这些子空间中的区分性方向。虽然有时直接用于识别(Blanz and Vetter 2003)，AAM 在识别中的主要应用是将人脸对齐到标准姿态上(Liang, Xiao, Wen et al. 2008)使得传统的人脸识别方法(Penev and Atick 1996; Wiskott, Fellous, Krüger et al. 1997; Ahonen, Hadid, and Pietikäinen 2006; Zhao and Pietikäinen 2007; Cao, Yin, Sun et al. 2010)能够适用。AAM(或者它们的简化版，活动形状模型(ASM))也可以用于对齐人脸图像以进行自动变形(Zanella and Fuentes 2004)。

随着处理光照和视角变化(Gross, Baker, Matthews et al. 2005)以及遮挡(Gross, Matthews, and Baker 2006)能力的改进，活动表观模型仍然是一个活跃的研究领域。其中一个最显著的扩展是建立 3D 形状模型(Matthews, Xiao, and Baker 2007)，它更能够刻画和解释脸部表观随大范围视角变化的所有变化。这种模型可以从单目视频序列(Matthews, Xiao, and Baker 2007)中建立，如图 14.24 所示。也可以从多视角视频序列(Ramnath, Koterba, Xiao et al. 2008)中建立，这提供了更可靠和精确的重建和跟踪。(关于头部姿态估计的最近的回顾，请参考 Murphy-Chutorian and Trivedi (2009)的综述。)

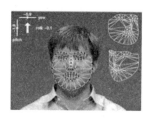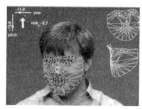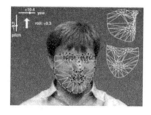

图 14.24　利用 3D 活动表观模型跟踪头部(Matthews, Xiao, and Baker 2007)© 2007 Springer。每幅图像显示了一个视频帧以及估计的摇摆、俯仰和旋转参数以及相应的 3D 可变形网格

14.2.3　应用：个人照片收藏

除了在数码相机中自动寻找人脸以辅助自动聚焦以及在视频摄像机中寻找人脸以在视频会议中将演讲者自动居中(机械居中或数字居中)，人脸检测在多数消费层照片管理包中找到了用武之地，如 iPhoto，Picasa 以及 Windows Live Photo Gallery。寻找人脸并允许用户对其进行标记使得以后寻找选定人物或与朋友自动分享更为简单。事实上，在照片中标记朋友的功能是 Facebook 中最流行的特性之一。

然而有时人脸很难寻找和识别，尤其是如果他们非常小、转离镜头或者被遮挡时。在这种情况下,人脸识别与人物跟踪以及服饰识别相结合将会非常有效,如图 14.25 所示(Sivic, Zitnick, and Szeliski 2006)。人物识别与其他上下文(如位置识别(14.3.3 节)或活动或事件识别)的结合也可以帮助提高性能(Lin, Kapoor, Hua et al. 2010)。

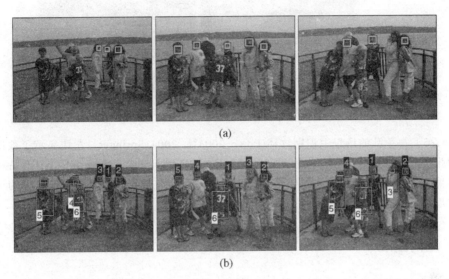

图 14.25 结合脸部、发型和躯干模型的人物检测和重识别(Sivic, Zitnick, and Szeliski 2006)© 2006 Springer:(a)仅使用人脸检测,有些头部被漏检;(b)人脸和服饰模型的结合成功地重新找出所有人

14.3 实 例 识 别

一般物体识别分为两大类,称为实例识别(instance recognition)和类别识别(class recognition)。前一个问题涉及到重新识别已知的 2D 或者 3D 刚性物体,这个物体可能是被从一个新的视点在杂乱的背景中有部分遮挡的情况下看到。后一个问题,常称为"类别级别"(category-level)的或者一般物体的识别,是识别特定的一般类别的示例(比如猫、汽车或者自行车)中,是一个更有挑战性的问题。

多年来,对于示例识别提出了很多不同的算法。Mundy (2006)综述了以前的方法,这些方法关注于从图像中提取线、轮廓或者 3D 表面,然后把它们和已知的 3D 物体模型匹配。另一个流行的方法是需要各个角度各种光照的图像,用本征空间分解的方法表达图像(Murase and Nayar 1995)。最近的方法(Lowe 2004; Rothganger, Lazebnik, Schmid et al. 2006; Ferrari, Tuytelaars, and Van Gool 2006b; Gordon and Lowe 2006; Obdržálek and Matas 2006; Sivic and Zisserman 2009)倾向于用视角无关的 2D 特征,比如我们在 4.1.2 节看到的那些。从新的图像和数据库中的图像提取了富有信息的稀疏 2D 特征后,用 4.1.3 节描述的一种稀疏特征匹配策略,在物体数据库中匹配图像特征。每当找到足够多的匹配时,通过寻找一个几何变换对齐两组特征来验证它们(图 14.26)。

接下来，我们描述一些刻画这些特征的几何关系的方法(14.3.1 节)。我们也介绍如何用文本和信息检索的想法来使特征匹配的过程更有效(14.3.2 节)。

图 14.26　在杂乱的场景中识别物体(Lowe 2004)© 2004 Springer。左边显示了数据库中两张训练图像。它们和在中间的杂乱场景用 SIFT 特征进行匹配，其中 SIFT 特征用右边图像中的小矩形显示。每个识别的数据库图像到场景中的仿射卷绕用更大的平行四边形显示在右边的图像中

14.3.1　几何配准

为了识别某类已知物体的一个或者多个示例，比如图 14.26 左列所示的那样，识别系统先在每个数据库图像中提取一组兴趣点，用索引树结构比如搜索树(4.1.3 节)保存关联的描述子(和原始的位置)。在识别的时候，先从新图像中提取特征，然后和存储的物体特征相比较。当找到和给定物体足够多的匹配特征时(比如，3 个或者更多)，然后系统调用匹配验证步骤，其目标是确定匹配的特征的空间排列和数据库图像中的是否一致。

因为图像可能会非常杂乱以及类似的特征可能属于几个物体，所以特征匹配的原始集合可能有很多外点。由于这个原因，Lowe (2004)建议用 Hough 变换(4.3.2 节)对类似的几何变换累计投票。在他的系统中，他在数据库物体和场景特征的集合之间用一个仿射变换，对几乎平坦或者至少几个对应特征共享一个准平面几何的物体有很好的效果。①

因为 SIFT 特征携带其自己的位置、尺度和角度，Lowe 用一个四维的相似变换作为原始 Hough 变换的分箱结构(original Hough binning structure)，即每个箱代表了物体中心、尺度和平面内旋转中的一个特定位置。每个匹配的特征投影到最近的 2^4 个箱中，然后选择变换的峰值用于拟合更细致的仿射运动。图 14.26(右图)显示了用这个系统识别出来的左图中两个物体的三个示例。Obdržálek and Matas (2006)用带有完整局部仿射框架的特征描述子把 Lowe 的方法一般化，在一些物体识别数据库中评测了他们的方法。

---

① 如果可以得到大量的特征，可以使用完整的基本矩阵(Brown and Lowe 2002; Gordon and Lowe 2006)。在拼接图像(Brown and Lowe 2007)时，改为使用 9.1 节讨论的运动模型。

另一个使用局部仿射框架的系统是 Rothganger, Lazebnik, Schmid et al. (2006)提出的。在他们的系统中，用 Mikolajczyk and Schmid(2004)提出的仿射区域检测器纠正局部图像块(图 14.27d)，从中可以计算一个 SIFT 描述子和一个 10×10 UV 颜色直方图，用于匹配和识别。对应于同一物体不同视角的块以及它们的局部仿射形变，用 7.3 节的因子分解算法的扩展，计算物体的 3D 仿射模型，然后可以提升到欧式重建(Tomasi and Kanade 1992)。

识别的时候，使用局部欧式邻域限制过滤可能的匹配，类似于 Lowe(2004)和 Obdrzalek and Matas(2006)使用的仿射几何限制方法。图 14.27 显示了用这个方法在杂乱环境中识别五个物体的结果。

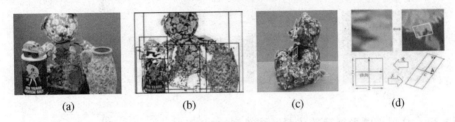

图 14.27 基于仿射区域的 3D 物体识别(Rothganger, Lazebnik, Schmid et al. 2006)© 2006 Springer：(a)样本输入图像；(b)五个识别的(重投影的)物体和它们的包围盒；(c)几个局部仿射区域；(d)把局部仿射区域(块)重投影到一个标准的(正方形的)框中，及其几何仿射变换

尽管基于特征的方法常用于检测和定位场景中的已知物体，但是基于这些匹配也可能得到场景像素级别的分割。Ferrari, Tuytelaars, and Van Gool(2006b)描述了这样一个系统，同时识别物体和分割场景。Kannala, Rahtu, Brandt et al. (2008)把这个方法扩展到非刚性形变中。在一般类别识别的背景下，14.4.3 节将再次讨论联合识别和分割的主题。

14.3.2 大型数据库

随着数据库中物体数量增加到很大(比如搜索百万个对象或者视频帧)，对每个数据库中图像匹配新的图像所花的时间会变得非常多。不同于每次比较一个图像，需要一些方法将搜索快速收窄到一些可能的图像，然后用更详细和更保守的验证阶段比较它们。

在文档中快速找到部分匹配的问题，是信息检索(IR)的核心问题之一(Baeza-Yates and Ribeiro-Neto 1999; Manning, Raghavan, and Schütze 2008)。快速文档索引算法中的基本方法是预计算每个词和它们出现的文档(或者网页或者新闻故事)的倒排索引。更准确地讲，用文档中某个特定词出现的频率快速找到与特定查询匹配的文档。

Sivic and Zisserman(2009)是第一个把 IR 方法用到视觉搜索中的。在他们的 Video Google 系统中，在所有视频帧中首先检测了仿射不变特征，它们被用 Harris

特征点附近的形状适应的区域(Schaffalitzky and Zisserman 2002; Mikolajczyk and Schmid 2004)和最大稳定极值区域(maximally stable extremal regions)(Matas, Chum, Urban *et al.* 2004) (4.1.1 节)进行索引，如图 14.28 所示。接下来，从每个归一化的区域(即图 14.28b 所示的块)计算 128 维的 SIFT 描述子。然后，通过对逐帧跟踪的特征累计统计信息估计这些描述子的平均协方差矩阵。然后用特征描述子协方差 Σ 定义特征描述子之间的马氏距离

$$d(\boldsymbol{x}_0, \boldsymbol{x}_1) = \|\boldsymbol{x}_0 - \boldsymbol{x}_1\|_{\Sigma^{-1}} = \sqrt{(\boldsymbol{x}_0 - \boldsymbol{x}_1)^T \Sigma^{-1} (\boldsymbol{x}_0 - \boldsymbol{x}_1)}. \tag{14.32}$$

在实际中，用 $\Sigma^{-1/2}$ 预乘以特征描述子进行白化(whitened)，使得可以使用欧式距离。①

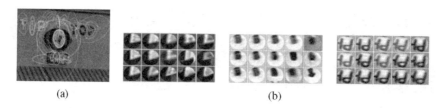

图 14.28　从用椭圆形规范的仿射区域得到的视觉词(Sivic and Zisserman 2009)© 2009 IEEE：(a)从每帧中提取仿射协变区域，然后在 SIFT 描述子上以学习得到的马氏距离用 k 均值聚类算法把它们聚类成一些视觉词；(b)每个网格中心的块显示的是查询，附近的块显示的是最近的邻居

图 14.29　基于视觉词的匹配(Sivic and Zisserman 2009)© 2009 IEEE：(a)左边查询区域的特征被匹配到排序很靠前的视频帧的对应特征上；(b)去掉停止词和用空间一致性过滤结果后的结果

为了在图像中应用快速信息检索方法，出现在每个图像中的高维特征描述子首先要映射到离散的视觉词中。Sivic and Zisserman(2003)用 k 均值算法做这一步映射，然而下面要讨论的一些更新的方法(Nistér and Stewénius 2006; Philbin, Chum, Isard *et al.* 2007)用的是其他的方法，比如词汇树(vocabulary tree)或者随机森林(randomized forest)。为了使得聚类时间是可控的，只用几百个视频帧学习聚类中心，即使是这样仍然需要从大约 300 000 个描述子中估计几千个中心。在视觉查询时，新的查询区域的每个特征(例如，图 14.28a 是从一个更大的视频帧中裁剪出来的区域)映射到它对应的视觉词上。为了不使很常见的模式污染所得到的结果，建立最常见视觉词的一个停止表(stop list)，在进一步的考虑中舍弃这些词。

---

① 注意，从匹配的特征点中计算特征协方差比简单地在描述空间上进行 PCA 要明智得多(Winder and Brown 2007)。这大致与我们在 14.2.1 节学习的类内散度矩阵(14.17)相对应。

一旦查询的图像或者区域映射到它组成的视觉词之后，必须从数据库中检索出可能的匹配图像或者视频帧来。信息检索系统通过在查询和目标文档之间匹配词的分布(词频)n_{id}/n_d来做到这一点，其中n_{id}是词i在文档d中出现的次数，n_d是文档d中词的总数。为了降低频繁出现的词的权重，并使得搜索时候关注出现较少(从而更有信息量)的词，应用逆文本频率(inverse document frequency)权重$\log N/N_i$，其中N_i是包含词i的文档的总数，N是数据库中文档的总数。这两个因素结合可以得到词频率-逆文本频率(the term frequency-inverse document frequency (*tf-idf*))衡量标准

$$t_i = \frac{n_{id}}{n_d} \log \frac{N}{N_i}. \tag{14.33}$$

匹配的时候，每个文档(或查询区域)表示为它的 *tf-idf* 向量，

$$\boldsymbol{t} = (t_1, \ldots, t_i, \ldots t_m). \tag{14.34}$$

两个文档的相似度用它们对应的归一化向量 $\hat{\boldsymbol{t}} = \boldsymbol{t}/\|\boldsymbol{t}\|$ 衡量，这意味着其不相似性和其欧式距离成正比。在 Sivic and Zisserman(2009)的期刊论文中，作者比较了这个简单的标准和很多其他的标准，结论是它得到的结果和其他复杂度量的结果类似。因为和视觉词数目($V \approx 20,000$)相比，一般查询或者文档中的非零词项t_i很小($M \approx 200$)，因此一对(稀疏的)*tf-idf* 向量间的距离可以非常快的计算。

基于词频检索到最好的$N_s = 500$个文档后，Sivic and Zisserman(2009)用空间一致性对结果重新排序。这一步涉及考虑每一个匹配的特征，计数在$k = 15$个最近邻特征中有多少在这两个文档间也是匹配的。(后一个过程可以用倒排文件加速，我们在下面会给出更多的细节。)正如图 14.29 所示，这一步能够帮助去掉不正确的匹配，并产生对视频中的帧和区域的真实匹配的更好的估计。算法 14.2 总结了使用视觉词进行图像检索的处理步骤。

虽然这个方法在数万视觉词和数千关键帧上可以做得很好，但是随着数据库大小的持续增大，量化每个特征和找到可能的匹配帧或者图像都会变得很难。Nister and Stewenius(2006)构建层次词汇树(vocabulary tree)来解决这个问题，其中特征向量层次聚类成k路原型树(k-way tree of prototypes)。(这个方法又叫"树结构的向量量化"(tree-structured vector quantization)(Gersho and Gray 1991)。)在数据库建立时和查询时，每个描述向量与词汇树中给定层的几个原型比较，选择最近的原型进一步细化(图 14.30)。这个方法可以支持百万(10^6)量级的词，能够使得每个词更加有区分性，而且对于每个描述子的量化只需要 10·6 次比较。

在查询的时候，词汇树中的每个节点保持它自己的倒排序文件的索引，因此匹配树中特定节点的特征可以很快映射到可能的匹配图像中。(内部的叶子节点只用它们对应的叶子节点的后代的倒排索引。)对所有的文档向量$\{t_j\}$用L_p标准给一个特定的查询 *tf-idf* 向量t_q评分[①]，用t_q中的非零t_{iq}项得到对应的非零t_{ij}项，L_p形式

[①] 在他们实际实现的时候，Nistér and Stewénius(2006)用的是L_1准则。

可以很有效地计算为

$$\|\boldsymbol{t}_q - \boldsymbol{t}_j\|_p^p = 2 + \sum_{i | t_{iq}>0 \wedge t_{ij}>0} (|t_{iq} - t_{ij}|^p - |t_{iq}|^p - |t_{ij}|^p). \tag{14.35}$$

为了减少描述子向量中噪声引起的量化误差，Nistér and Stewénius(2006)不仅给词汇树的叶子节点(对应视觉词)评分，也给树的内部节点评分，内部节点对应于相似视觉词的聚类中心。

---

1. 构造词汇 (离线)
(a) 从每个数据库图像中提取仿射协变区域。
(b) 计算描述子，可选择性地对它们进行白化，使得计算欧氏距离有意义 (Sivic and Zisserman 2009)。
(c) 用 k 均值算法(Sivic and Zisserman 2009)，或者层次聚类算法(Nistér and Stewénius 2006)或者随机的 k-d 树(Philbin, Chum, Isard et al. 2007)，把这些描述子聚类成一些视觉词。
(d) 确定哪些词是过于普通的，把它们放入停止表中。

2. 构造数据库(离线)
(a) 计算每个图像中的视觉词的词频，每个词的文档频率和每个文档的归一化的 *tf-idf* 向量。
(b) 计算从视觉词到图像的倒排索引(带有词的计数)。

3. 图像检索 (在线)
(a) 对每个查询的图像或者区域，提取区域、描述子和视觉词，并计算 *tf-idf* 向量。
(b) 通过详尽的比较稀疏 *tf-idf* 向量(Sivic and Zisserman 2009)或者使用倒排索引只检查图像的一个子集(Nistér and Stewénius 2006)，检索最相似的图像候选。
(c) 用空间一致性(Sivic and Zisserman 2009)或者仿射变换(或更简单的)模型(Philbin, Chum, Isard et al. 2007)，可选择性地再排序或者验证所有候选匹配。
(d) 可选择性地将排序很靠前的匹配再次作为新的查询来扩展答案集 (Chum, Philbin, Sivic et al. 2007)。

算法 14.2 使用视觉词的图像检索(Sivic and Zisserman 2009; Nistér and Stewénius 2006; Philbin, Chum, Isard et al. 2007; Chum, Philbin, Sivic et al. 2007; Philbin, Chum, Sivic et al. 2008)

因为量化和评价特征都很高效，所以他们的基于词汇树的识别系统可以在包含 4 万张 CD 封面的数据库上进行实时处理，并能以 1Hz 的速度匹配从六个完整长度的电影中取出的一百万帧的数据库。图 14.30b 显示了用于训练和测试词汇树识别系统的数据库中的变化视角和变化光照的物体的典型图像。

示例识别的最新发展水平在持续快速提高。Philbin, Chum, Isard *et al.* (2007)表明，在大型位置识别任务中，*k*-d 树的随机森林比词汇树的性能好(图 14.31)。他们同样比较了在验证阶段用不同的 2D 运动模型(2.1.2 节)的效果。在后续工作中，Chum, Philbin, Sivic *et al.* (2007)应用了信息检索中的另一个想法，称为"查询扩展"(query expansion)，把初始查询排名很靠前的图像重新提交，作为附加的查询，然后产生附加的候选结果，进一步提高难(遮挡或者倾斜的)的样本的识别率。Philbin, Chum, Sivic *et al.* (2008)说明了如何用软赋值的方法缓和视觉词选择的量化问题，其中，每个特征描述子映射到基于它到聚类原型的距离的一些视觉词中。为了后续验证检索附加的图像，反过来，用来源于这些距离的软权重，对 *tf-idf* 向量中用到的计数进行加权。综合起来，这些最新的发展具有把当前示例识别算法扩展到网络规模检索和匹配任务中的潜力(Agarwal, Snavely, Simon *et al.* 2009; Agarwal, Furukawa, Snavely *et al.* 2010; Snavely, Simon, Goesele *et al.* 2010)。

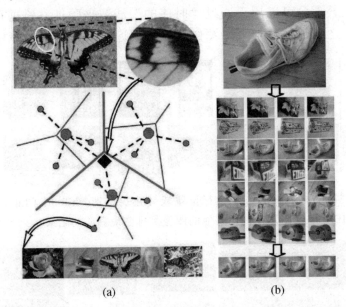

图 14.30 使用词汇树的可扩展的识别(Nistér and Stewénius 2006)© 2006 IEEE。(a)每个 MSER 椭圆区域转换为 SIFT 描述子，然后通过在词汇树中分层地与一些原型描述子比较来量化。每个叶子节点把它自己的倒排索引(非零 *tf-idf* 计数的稀疏表)存储到包含该特征的图像中。(b)识别的结果，显示了查询图像(最上面一行)被索引到 6000 个测试图像的数据库，正确地找到了对应的四个图像

14.3.3 应用：位置识别

当前，示例识别最令人兴奋的应用之一是在位置识别领域中，它可以用于桌面应用(我在什么地方拍摄的假日快照？)和移动(手机)应用。后一个例子不仅包括找到基于手机图像的你当前的位置，还包括为你提供导航的方向或者用有用的信息对你的图像进行标注，比如建筑名称和餐厅评论(即增强现实)的一个可移动的形式)。

一些位置定位的方法假设照片是由建筑场景组成的，因而可以用消失方向预先校正图像使得匹配更简单(Robertson and Cipolla 2004)。另外的方法，用一般的仿射协变兴趣点进行宽基线匹配(Schaffalitzky and Zisserman 2002)。Snavely, Seitz, and Szeliski (2006)的照片游览系统(The Photo Tourism system)(13.1.2 节)率先把这些想法应用于大规模图像匹配和(隐含的)位置识别，这些图像都是从网上收集的包含很大的视角变化的照片。

位置识别最大的困难是处理网站(比如 Flickr)上非常大的社区(用户生成的)照片集合(Philbin, Chum, Isard *et al.* 2007; Chum, Philbin, Sivic *et al.* 2007; Philbin, Chum, Sivic *et al.* 2008; Turcot and Lowe 2009)或者处理商业拍摄的数据库(Schindler, Brown, and Szeliski 2007)。通常广泛出现的元素，比如植被、标识以及常见的建筑元素，使这个问题更加复杂。图 14.31 显示了社区图像集上的位置识别的一些结果，图 14.32 显示了商业获取的稠密数据集上的样例输出。在后一个情况下，数据库相邻图像的重叠部分可以用于通过"时序的"滤波来验证并修剪可能的匹配，即在接受一个匹配之前，需要把查询图像匹配到相近的重叠的数据库图像上。

图 14.31 使用随机森林的方法进行位置或者建筑识别(Philbin, Chum, Isard *et al.* 2007)© 2007 IEEE。左图是查询，其他的图像是排名最靠前的结果

图 14.32 基于特征的位置识别(Schindler, Brown, and Szeliski 2007) © 2007 IEEE：(a) 三个典型的重叠的街景照片序列；(b)手持摄像机拍摄的照片和(c)它们对应的数据库中的照片

另一个位置识别的变型是自动发现地标(landmark)，即经常拍摄的物体和位置。Simon, Snavely, and Seitz (2007)显示了如何通过简单地分析匹配图发现这些类

别的物体，这里的匹配图是照片游览(Photo Tourism)中的 3D 建模过程的一部分。最新的一些工作，用高效的聚类方法把这个方法扩展到更大的数据集合上(Philbin and Zisserman 2008; Li, Wu, Zach *et al.* 2008; Chum, Philbin, and Zisserman 2008; Chum and Matas 2010)，以及把元数据(比如 GPS 和纹理标签)和视觉搜索结合起来(Quack, Leibe, and Van Gool 2008; Crandall, Backstrom, Huttenlocher *et al.* 2009)，如图 14.33 所示。基于物体标签在多个松散带有标签的图像中同时出现的关系(cooccurrence)，现在还可以自动给图像关联物体标签(Simon and Seitz 2008; Gammeter, Bossard, Quack *et al.* 2009)。

图 14.33 社区照片集合的自动挖掘、标注和定位(Quack, Leibe, and Van Gool 2008)© 2008 ACM。这个图没有显示文本标注或者对应的维基词条，这些也被发现了

最近，用位置组织世界的照片集的想法被扩展到组织宇宙中的(天文学的)照片中，这样的应用称为"天体测定学"(*http://astrometry.net/*)。如图 14.34 所示，匹配任何两个星场(star field)的方法是将相邻星体(包括两对星体，其中第二对星体在第一对星体的直径内)的四边形(quadruplet)形成 30 位(比特)的几何哈希，具体做法是以内接正方形为参照系，编码第二对点的相对位置。然后使用传统的信息检索方法(给天空地图集的不同部分建立不同的 k-d 树)，找到匹配的四边形作为可能的星场位置预测，然后用相似变换验证。

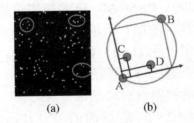

图 14.34 使用天体测量学定位星场，*http://astrometry.net/*：(a)输入的星场和一些选择的星体四边形；(b)恒星 C 和 D 的 2D 坐标相对于 A 和 B 定义的单位正方形进行编码

14.4 类别识别

虽然示例识别的方法相对成熟，也用到了很多商业应用中，比如照片合成(photosynth) (13.1.2 节)，但是一般类别(分类)识别问题仍然是一个悬而未决的问

题。考虑一个例子，图 14.35 的图像集显示了来自 10 个不同的视觉类的物体。(你可以自己发现这 10 个类别。)你如何写程序将这些图像分类成合适的类别，尤其是如果给你"不属于以上任何一个类别"的选择时？

图 14.35 Xerox 10 类数据集的样例图像(Csurka, Dance, Perronnin et al. 2006)© 2007 Springer。想象一下尝试写一个程序，从其他的图像中区分这样的图像

从这个例子可以看出，视觉分类识别问题是一个非常有挑战性的问题。还没有人能建立一个系统其性能能够接近两岁小孩的水平。然而，如果将现在的算法和十年前的算法相比，这个领域的进步还是很显著的。

图 14.54 显示了用于 2008 PASCAL 视觉物体类挑战(Visual Object Classes Challenge)中的 20 个类别的样例图像。黄颜色的矩形显示了在给定的图像中找到每类物体的范围。在这样的封闭世界(closed world)集合中，其任务是决定属于 20 个类别中的哪一个，现在的分类算法可以做的非常好。

在这一节，我们审视解决类别识别问题的多种方法。尽管从历史角度讲，基于部件的表达和识别算法(14.4.2 节)是首选的方法(Fischler and Elschlager 1973; Felzenszwalb and Huttenlocher 2005; Fergus, Perona, and Zisserman 2007)，但是我们先表述更简单的词袋(bag-of-words)方法(14.4.1 节)，它用没有顺序的特征描述子的集合来表示物体和图像。然后审视同时自动分割图像和识别物体的问题(14.4.3 节)，也介绍一些将这类方法用于照片处理的应用(14.4.4 节)。在 14.5 节，我们看一看如何用上下文和场景理解以及机器学习，提高整体的识别结果。更多关于这一节方法的细节，请参见(Pinz 2005; Ponce, Hebert, Schmid et al. 2006; Dickinson, Leonardis, Schiele et al. 2007; Fei-Fei, Fergus, and Torralba 2009)。

14.4.1 词袋

类别识别的最简单的算法之一是词袋(也称为"特征袋",即 bag of features 或者"关键点袋" bag of keypoints)方法(Csurka, Dance, Fan *et al.* 2004; Lazebnik, Schmid, and Ponce 2006; Csurka, Dance, Perronnin *et al.* 2006; Zhang, Marszalek, Lazebnik *et al.* 2007)。正如图 14.36 所示,这个算法简单的计算在查询图像中找到的视觉词的分布(直方图),比较这个分布和训练图像中的分布。在 14.3.2 节,公式 (14.33~14.35)和算法 14.2 中,我们已经见过这个方法的要素了。类别识别问题和示例识别问题最大的不同是没有几何验证步骤(14.3.1 节),因为相对来讲,如图 14.35 所示的一般视觉类的不同示例,它们的特征几乎没有空间关系(但是可以参见工作 Lazebnik, Schmid, and Ponce (2006))。

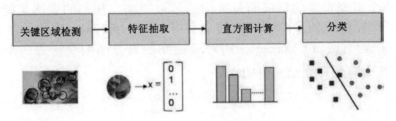

图 14.36 词袋类别识别系统的典型处理流水线(Csurka, Dance, Perronnin *et al.* 2006)© 2007 Springer。首先在每个关键点抽取特征,然后量化得到在学习来的视觉词(特征聚类中心)上的分布(直方图)。用分类算法,比如支持向量机,在特征分布直方图上学习决策面

Csurka, Dance, Fan *et al.* (2004)率先使用术语"关键点袋"描述这类方法,也是最先利用频率统计方法解决类别识别问题。他们的原始系统用仿射协变区域和 SIFT 描述子,k 均值视觉词汇构建,并结合朴素贝叶斯分类器和支持向量机来分类。(发现后者性能更好。)他们更新的系统(Csurka, Dance, Perronnin *et al.* 2006)用正规(非仿射)SIFT 区域,用 boosting 方法代替 SVM,并结合少量几何一致性信息。

Zhang, Marszalek, Lazebnik *et al.*(2007)对这类特征袋(bag of features)系统进行了更详细的研究。他们比较了很多特征检测器(Harris–Laplace (Mikolajczyk and Schmid 2004)和 Laplacian(Lindeberg 1998b))、描述子(SIFT,RIFT 和 SPIN (Lazebnik, Schmid, and Ponce 2005))和 SVM 核函数。为了估计核函数的距离,他们形成了图像签名(image signature)

$$S = ((t_1, \boldsymbol{m}_1), \ldots, (t_m, \boldsymbol{m}_m)), \tag{14.36}$$

类似于(14.34)中的 *tf-idf* 向量 t,其中聚类中心 m_i 是明确得到的。然后,他们研究了两种不同的核来比较这些图像签名。第一种是推土机距离(the earth mover's distance,EMD)(Rubner, Tomasi, and Guibas 2000),

$$EMD(S, S') = \frac{\sum_i \sum_j f_{ij} d(\boldsymbol{m}_i, \boldsymbol{m}'_j)}{\sum_i \sum_j f_{ij}}, \tag{14.37}$$

其中 f_{ij} 是可以用线性规划计算的一个流(flow)值，$d(m_i, m'_j)$ 是 m_i 和 m'_j 的测地(ground distance)距离(欧氏距离)。注意，EMD 可以用于比较两个不同长度的签名，其中，输入的项不需要对应。第二种是 χ^2 距离

$$\chi^2(S, S') = \frac{1}{2} \sum_i \frac{(t_i - t'_i)^2}{t_i + t'_i}, \tag{14.38}$$

度量的是从一致随机过程中产生的两个签名的似然性。然后，这些距离度量使用广义高斯核转化到 SVM 核中

$$K(S, S') = \exp\left(-\frac{1}{A} D(S, S')\right), \tag{14.39}$$

其中 A 是赋给训练图像间的均值距离的缩放参数集合。在他们的实验中，他们发现对于视觉类别识别，EMD 效果最好；对于纹理识别，χ^2 效果最好。

与把特征向量量化为视觉词不同，Grauman and Darrell(2007b)提出了一个方法，只计算两个可变长度的特征向量之间的近似距离。他们的方法是把特征向量划分到定义在特征空间上的多个分辨率的金字塔中(图 14.37a)，再统计落在对应箱 B_{il} 和 B'_{il} 中的特征数目(图 14.38a~c)。用对应箱间的直方图的交计算两个特征向量集合的距离(可以认为是高维空间中的点)

$$C_l = \sum_i \min(B_{il}, B'_{il}) \tag{14.40}$$

(图 14.38d)。然后把每一层的计数用带权的方式累加起来

$$D_\Delta = \sum_l w_l N_l \quad \text{with} \quad N_l = C_l - C_{l-1} \quad \text{and} \quad w_l = \frac{1}{d2^l} \tag{14.41}$$

(图 14.38e)，在使更细致层中的匹配权重更大的同时不计数已经在更细致层中找到的匹配。(d 是嵌入空间的维数，即特征向量的长度。)在后续的工作中，Grauman and Darrell(2007a)显示了如何用哈希的方法来避免明确地构建金字塔。

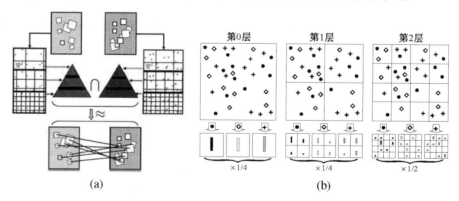

图 14.37 用金字塔匹配方法比较特征向量的集合：(a)特征空间金字塔匹配核(Grauman and Darrell 2007b)构建高维特征空间的一个金字塔，并用它计算特征向量集合之间的距离(和隐含的对应)；(b)空间金字塔匹配(Lazebnik, Schmid, and Ponce 2006)© 2006 IEEE 把图像分成区域池的金字塔，然后计算每个空间箱内单独的视觉词的直方图(分布)

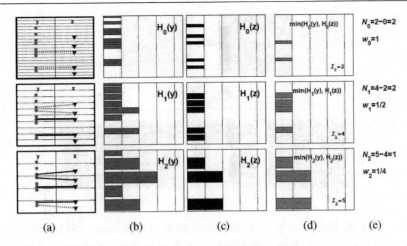

图 14.38 使用金字塔匹配核比较特征向量的汇集的一维图解(Grauman and Darrell 2007b)：(a)特征向量(点集)在金字塔箱中的分布；(b 和 c)对于两幅图像，在 B_{ii} 和 B_{ii}' 箱中点数的直方图；(d)直方图的交(最小值)；(e)每一层相似性的分值，加权相加得到最终的距离/相似性度量

受这个工作的启发，Lazebnik, Schmid, and Ponce(2006)显示了如何用类似的想法，将 2D 空间位置概念用于扩展关键点袋，类似于 SIFT(Lowe 2004)和"gist" (Torralba, Murphy, Freeman et al. 2003)汇集的方式。在他们的工作中，他们提取仿射区域描述子(Lazebnik, Schmid, and Ponce 2005)，把它们量化为视觉词。(基于以前 Fei-Fei and Perona(2005)的结果，在图像上稠密提取(在规则网格上)特征描述子，这样的方法对于描述无明显结构的区域(比如天空)很有用。然后，他们形成了包含词计数(直方图)的箱(bin)的空间金字塔，如 14.37b 所示，用类似的金字塔匹配核，以分层方式把直方图交的计数结合起来。

现在仍然在争论，是用量化的特征描述子还是连续的特征描述子，是用稀疏的特征还是稠密的特征。Boiman, Shechtman, and Irani(2008)显示了如果将一个查询图像和一个表达给定类别的所有特征相比，而不只是和每个类别图像单独比较，用最近邻匹配和朴素贝叶斯分类方法比量化的视觉词效果好(图 14.39)。不同于使用一般的特征检测子和描述子，一些作者已经研究了学习特定类的特征(Ferencz, Learned-Miller, and Malik 2008)，通常用随机森林(randomized forest)方法(Philbin, Chum, Isard et al. 2007; Moosmann, Nowak, and Jurie 2008; Shotton, Johnson, and Cipolla 2008)或者把特征产生和图像分类步骤结合起来(Yang, Jin, Sukthankar et al. 2008)。另外一些工作，比如 Serre, Wolf, and Poggio(2005)和 Mutch and Lowe (2008)，受生物学的(视觉皮层)加工启发，提出了稠密特征变换的层次方法，结合 SVM 得到最终的结果。

图 14.39 "图像-图像"对比"图像-类别"距离比较(Boiman, Shechtman, and Irani 2008)© 2008 IEEE。在左上角的查询图像，可能不与下面一行中任何一个数据库图像里的的特征分布相匹配。但是，如果将查询中的每个特征匹配到所有类别图像中最类似的那一个，就可以找到一个好的匹配

14.4.2 基于部件的模型

找到物体的组成部件并度量它们的几何关系的识别方法是物体识别最古老的方法(Fischler and Elschlager 1973; Kanade 1977; Yuille 1991)。我们已经看到很多基于部件的方法的例子，比如用于人脸识别的(图 14.18) (Moghaddam and Pentland 1997; Heisele, Ho, Wu et al. 2003; Heisele, Serre, and Poggio 2007)和行人检测的(图 14.9) (Felzenszwalb, McAllester, and Ramanan 2008)。

在这一节中，我们更仔细地审视基于部件的识别中的一些关键问题，也就是几何关系的表达、不同部件的表达、学习这类描述的算法以及在运行时识别它们。更多关于基于部件的模型的识别方法的细节可以在课程笔记 Fergus (2007b, 2009)中找到。

表示几何关系最早的方法称为"图案结构"(pictorial structure)Fischler and Elschlager(1973)，由不同特征位置间的弹性连接组成(图 14.1a)。为了使图案结构拟合一个图像，在所有可能的部件位置或者姿态$\{l_i\}$和部件对(i, j)上，最小化如下形式的能量函数

$$E = \sum_i V_i(l_i) + \sum_{ij \in E} V_{ij}(l_i, l_j) \tag{14.42}$$

这里的一对部件是在 E 中存在的一条边(几何关系)。注意这个能量和用于 MRF (3.108 和 3.109)的能量的紧密关系，使用 MRF 可以将图案结构嵌入到一个概率框架中，从而使得参数学习更容易(Felzenszwalb and Huttenlocher 2005)。

对于部件之间的几何联系，基于部件的模型可以有不同的拓扑(图 14.41)。比如，Felzenszwalb and Huttenlocher (2005)把这些联系限制为一棵树(图 14.41d)，这样使得学习和推断更方便。一个树形拓扑可以用递归 Viterbi(动态规划)算法(Pearl 1988; Bishop 2006)，其中叶子节点先作为它们父节点的函数进行优化，然后得到的值再插入到能量函数中并从能量函数中去掉——见附录 B.5.2。Viterbi 计算一个最

优的匹配需要 $O(N^2|E|+NP)$ 时间，其中 N 是每个部件可能的位置或者姿态的数目，$|E|$是边(成对约束)的数目，$P=|V|$是部件(图模型中的顶点，对于一个棵树，等于$|E|+1$)的数目。为了进一步提高推断的效率，Felzenszwalb and Huttenlocher (2005) 限制成对能量函数 $V_{ij}(l_i,l_j)$ 为位置变量上的马氏距离，然后用快速距离变换算法，在接近 N 的线性时间内最小化逐对的相互作用。

图 14.40 显示了使用图案结构算法把关节连接(articulated)部件模型拟合到一个背景差分得到的二值图上的结果。在这个图案结构的应用中，部件的参数有位置、大小和它们的近似矩形的方向。一元匹配的势函数 $V_i(l_i)$ 是由表示每个部件的倾斜矩形的内部和外部的前景和背景像素的百分比决定的。

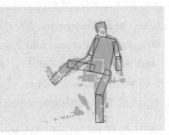

图 14.40 用图案结构定位并跟踪一个人(Felzenszwalb and Huttenlocher 2005)© 2005 Springer。结构由关节连接的矩形人体部件(躯干、头和四肢)组成，这些部件用一个编码了相对的部件位置和方向的树形拓扑连接起来。为了拟合图案结构，先用背景差分计算出来一个二值轮廓图

如图 14.41 所示，在过去的十年中，对于基于部件的识别，提出了很多不同的图模型。Carneiro and Lowe(2006)讨论了很多这样的模型，并提出了他们自己的模型，他们称为稀疏灵活模型(sparse flexible model)，这个模型将部件排序，而且每个部件的位置决定于最多 k 个其祖先的位置。

在 14.4.1 节，我们看到了最简单的模型，即词袋模型，其中不同的部件或者特征之间没有几何关系。虽然这样的模型可能很有效，但是它们表达部件的空间排列的能力很有限。树形或者星形(一种特殊的树形，其中所有的叶子节点直接和一个公共的根节点相连)在推断还有学习方面是最有效的(Felzenszwalb and Huttenlocher 2005; Fergus, Perona, and Zisserman 2005; Felzenszwalb, McAllester, and Ramanan 2008)。在复杂性方面其次的是有向无环图(Directed acyclic graph)(图 14.41f 和 g)，虽然以在部件模型上强加了因果结构，它仍支持有效的推断(Bouchard and Triggs 2005; Carneiro and Lowe 2006)。k-扇形(k-fan)，是由大小为 k 的团簇形成的星型模型的根(图 14.41c)，其推断复杂度为 $O(N^{k+1})$，尽管考虑距离变换和高斯先验这个复杂度可以降为 $O(N^k)$(Crandall, Felzenszwalb, and Huttenlocher 2005; Crandall and Huttenlocher 2006)。最后，全连接的星座模型(fully connected constellation model)(图 14.41a)是最一般的，但是由于将特征分配给部件的复杂度是 $O(N^P)$，其中 P 是适度的部件数目，这个分配过程会变得很棘手(Fergus, Perona, and Zisserman 2007)。

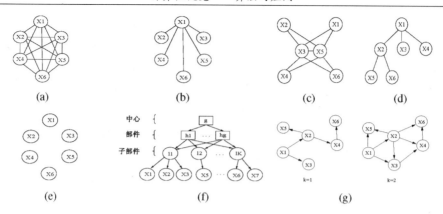

图 14.41 几何空间先验的图模型(Carneiro and Lowe 2006)© 2006 Springer：(a)星座模型(Fergus, Perona, and Zisserman 2007)；(b)星型模型(Crandall, Felzenszwalb, and Huttenlocher 2005; Fergus, Perona, and Zisserman 2005)；(c)k扇形模型(k = 2) (Crandall, Felzenszwalb, and Huttenlocher 2005)；(d)树形模型(Felzenszwalb and Huttenlocher 2005)；(e)特征袋模型(Csurka, Dance, Fan *et al.* 2004)；(f)层次模型(Bouchard and Triggs 2005)；(g)稀疏灵活模型(sparse flexible model)(Carneiro and Lowe 2006)

最原始的星座模型是由 Burl, Weber, and Perona(1998)提出的，它包含很多部件，其部件的相对位置由它们的平均位置和一个完整的协方差矩阵进行编码，其中协方差矩阵不仅用于表示位置的不确定性，也用于表示不同部件之间的潜在的相关性(协方差)(图 14.42a)。Weber, Welling, and Perona(2000)把这个方法扩展到一个弱监督的设置中，只给定整体的图像标签，能够自动学习每个部件的表观和位置。Fergus, Perona, and Zisserman(2007)进一步把这个方法扩展到从尺度不变的关键点检测中同时学习表观模型和形状模型。

图 14.42a 显示了对于摩托车类别学习到的形状模型。上面的图像显示了每个部件的平均相对位置和它们的位置协方差(没有显示部件之间的协方差)，以及出现的似然率。下面的曲线显示了在对数尺度上，相对于"地标"特征，每个部件相对的高斯概率密度函数(PDF)。图 14.42b 显示了为每个部件学习得到的表观模型，可视化为训练数据库中环绕检测得到的特征的块，这些块和表观模型最匹配。图 14.42c 显示了测试集中检测得到的特征(粉红色的点)以及它们被赋值的对应的部件(彩色的圆)。可以看出，这个系统成功地学习了并且使用了一个相当复杂的摩托车表观的模型。

建立在为其他多类提出的识别部件的基础上，基于部件的识别方法也已经扩展到从少量例子中学习新的类别(Fei-Fei, Fergus, and Perona 2006)。用语法的概念可以提出更复杂的层次化的基于部件的模型(Bouchard and Triggs 2005; Zhu and Mumford 2006)。正如图 14.43 最上面一行所示，一个更简单的使用部件的方法是把要识别的关键点作为对估计部件位置的类投票的一部分(Leibe, Leonardis, and Schiele 2008)。(这个方法隐含着对应于使用一个星型几何模型。)

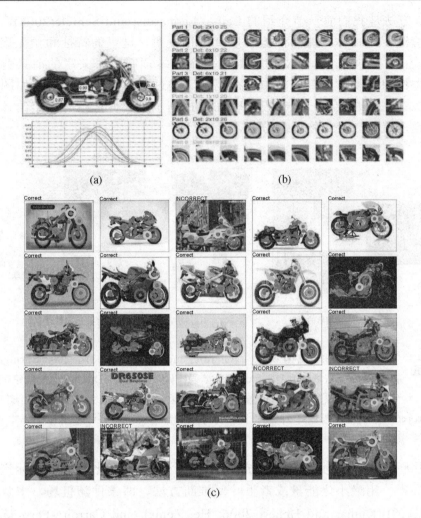

图 14.42 基于部件的识别(Fergus, Perona, and Zisserman 2007)© 2007 Springer：(a)每个部件的位置和协方差椭圆，以及它们出现的概率(上)和相对的对数空间内的概率密度(下)；(b)训练图像中抽取和平均表观最匹配的部件示例；(c)摩托车类别的识别结果，显示了检测的特征(粉红色的点)和部件(彩色的圆)

14.4.3 基于分割的识别

最有挑战性的一般物体识别是同时进行识别和精确的边界分割(Fergus 2007a)。以示例识别(14.3.1 节)为例，有时可以通过把物体模型逆向投影到场景中得到(如图 14.1d)，或者通过把新场景的部分和事先学习(分割的)的物体模型匹配得到 (Ferrari, Tuytelaars, and Van Gool 2006b; Kannala, Rahtu, Brandt *et al.* 2008)。

对于更复杂的(灵活的)物体模型，比如图 14.1f 所示的人的模型，一个不同的方法是把图像预分割为更大或者更小的块(第 5 章)，然后将每个块和模型的部分匹配(Mori, Ren, Efros *et al.* 2004; Mori 2005; He, Zemel, and Ray 2006; Gu, Lim, Arbelaez *et al.* 2009)。

另一个方法是我们前一节介绍的 Leibe, Leonardis, and Schiele(2008), 该方法是基于特征检测, 对物体可能的位置和尺度进行投票, 这里的特征和预先聚类的视觉码本项对应(图 14.43)。为了支持分割, 每个码本项有一个关联的前景背景掩模, 这是从预先标记的物体分割掩模中, 在码本聚类过程中学习得到的。如图 14.43 左侧所示, 在识别的时候, 一旦在投票空间中找到了最大值, 结合投票给该示例的相关联的掩模, 就可以得到物体的分割。

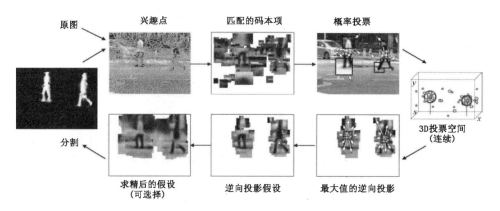

图 14.43 交错的识别和分割(Leibe, Leonardis, and Schiele 2008)© 2008 Springer。这个过程首先是在新图像(场景)中重新识别视觉词(码本项), 在 3D(x, y, s)投票空间(上一行)上, 将每个部件投票到可能的位置和大小。一旦找到一个最大值, 可以用逆向投影(backprojecting)所属的投票, 确定这个示例对应的部件(特征)。可以用每个码本项关联的逆向投影掩模得到每个物体的前景背景分割。然后重复整个识别和分割过程

一个更整体性的识别和分割的方法是把这个问题形式化为用类别标签标记图像中的每个像素, 用最小化能量或者贝叶斯推断方法, 即条件随机场, 求解这个问题(3.7.2 节(3.118))(Kumar and Hebert 2006; He, Zemel, and Carreira-Perpiñán 2004)。Shotton, Winn, Rother *et al.* (2009)的 TextonBoost 系统使用一元(逐像素)势, 基于特定图像的颜色分布(5.5 节)(Boykov and Jolly 2001; Rother, Kolmogorov, and Blake 2004)、位置信息(即前景物体更可能出现在图像的中间、天空更可能高一些、路更可能低一些)和用共享 boosting 学习的新的纹理布局(texture-layout)分类器。它也使用了传统的考虑图像颜色梯度的二元势(Veksler 2001; Boykov and Jolly 2001; Rother, Kolmogorov, and Blake 2004)。纹理基元布局(Texton-layout)特征先用一系列的 17 个方向滤波器组对图像进行滤波, 然后聚类这些响应, 把每个像素分为 30 个不同的纹理基元类别(Malik, Belongie, Leung *et al.* 2001)。然后用联合 boosting 算法(Viola and Jones 2004)训练得到的偏移矩形区域对这些响应进行滤波, 得到的纹理基元布局看作是一元势。

图 14.44a 显示了用 TextonBoost 成功标记图像的一些例子, 图 14.44b 显示了它不太成功的例子。正如你所看到的, 这种语义标注是极有挑战性的。

图 14.44 使用 TextonBoost 的同时识别和分割(Shotton, Winn, Rother et al. 2009)© 2009 Springer：(a)成功的识别结果；(b)较不成功的结果

Winn and Shotton(2006)将 TextonBoost 条件随机场框架扩展到 LayoutCRFs，他们将额外的限制合并到识别多个物体示例和处理遮挡中(图 14.45)。最近 Hoiem, Rother, and Winn(2007)更进一步结合了完整的 3D 模型。

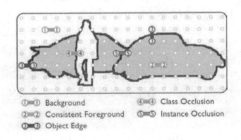

图 14.45 布局一致性随机场(Winn and Shotton 2006)© 2006 IEEE。数字指示了赋值为相同或者不同类别的像素之间的相邻关系类型。每一对关系有自己的似然度(能量惩罚)

条件随机场的使用仍然非常广泛并被扩展到同时识别和分割的应用中(Kumar

and Hebert 2006; He, Zemel, and Ray 2006; Levin and Weiss 2006; Verbeek and Triggs 2007; Yang, Meer, and Foran 2007; Rabinovich, Vedaldi, Galleguillos et al. 2007; Batra, Sukthankar, and Chen 2008; Larlus and Jurie 2008; He and Zemel 2008; Kumar, Torr, and Zisserman 2010)，在困难的 PASCAL VOC 分割挑战数据集上取得了一些最好的结果(Shotton, Johnson, and Cipolla 2008; Kohli, Ladický, and Torr 2009)。先将图像分割为单一或者多个分割部分的方法(Borenstein and Ullman 2008; He, Zemel, and Ray 2006; Russell, Efros, Sivic et al. 2006) (潜在地和 CRF 模型结合起来)也取得了很好的效果；Csurka and Perronnin(2008)在 VOC 分割挑战数据集上提出了性能最好的方法之一。有时也使用层次(多尺度)和语法(解析)模型(Tu, Chen, Yuille et al. 2005; Zhu, Chen, Lin et al. 2008)。

14.4.4 应用：智能照片编辑

最近，在物体识别和场景理解上的进展极大地加强了智能(半自动)图像编辑应用的能力。一个例子是 Lalonde, Hoiem, Efros et al. (2007)的图像剪贴画系统(the Photo Clip Art system)，该系统在网络图像集合上识别并分割感兴趣的对象，比如行人，然后允许用户把它们粘贴到他们自己的图像中。另一个例子是 Hays and Efros (2007)的场景补全系统(the scene completion system)，该系统解决的问题和我们在 10.5 节研究的修复(inpainting)问题是一样的。给定一幅图像，我们希望擦除并填充一个大片段(图 14.46a 和 b)，在哪里得到在编辑的图像中要填充空白的像素？传统的方法或者使用光滑一致性(smooth continuation)(Bertalmio, Sapiro, Caselles et al. 2000)或者是从图像的其他部分取像素(Efros and Leung 1999; Criminisi, Pérez, and Toyama 2004; Efros and Freeman 2001)。随着网络上大量的图像数据库的出现(回到我们在 14.5.1 节的主题)，寻找不同的图像用作缺失像素的来源通常更有意义。

在他们的系统中，Hays and Efros(2007)计算了每个图像的 gist(Oliva and Torralba 2001; Torralba, Murphy, Freeman et al. 2003)来寻找相似颜色和组合的图像。然后，他们运行一个图割算法，最小化图像梯度的差，并用 Poisson 图像融合把新的替换块合成到原始的图像中(9.3.4)(Pérez, Gangnet, and Blake 2003)。图 14.46d 显示了用帆船替换擦除的前景屋顶区域的结果。

图 14.46 用几百万幅照片进行场景补全(Hays and Efros 2007)© 2007 ACM：(a)原图；(b)消除不需要的前景后；(c)比较真实的场景匹配，用户选择的匹配用红色高亮显示出来了；(d)经过替换和融合的输出图像

图像识别和分割的另一个不同的应用是从单个照片中通过识别特定场景结构的方法来推断 3D 结构。比如为了推断 3D 几何结构,Criminisi, Reid, and Zisserman (2000)检测消失点,让用户画出基本结构,比如墙(6.3.3 节)。另一方面,Hoiem, Efros, and Hebert (2005a)的工作处理如图 14.47 所示的更"有组织的"场景。他们的系统从标记的图像中学习了各种各样的分类器和统计情况,把每个像素分类为地面、直立物(vertical)或者天空(图 14.47d)。为了做到这一点,他们首先计算超像素(图 14.47b),然后把超像素分成可能共享相似几何标签的合理区域 (图 14.47c)。如图 14.47e 所示,在所有像素被标记后,直立物和地面像素之间的边界可以用于推断 3D 直线,沿着这些直线,图像可以折叠为一个"弹出"(在移除天空像素后)。在相关的工作中,Saxena, Sun, and Ng(2009)提出了一个系统,直接推断每个像素的深度和方向,而不只是三个几何类别标签。

图 14.47 自动照片弹出(Hoiem, Efros, and Hebert 2005a)© 2005 ACM:(a)输入图像;(b)超像素(superpixels)分为;(c)多个区域;(d)标签代表的地面(绿色)、直立物(红色)和天空(蓝色);(e)产生的分段线性 3D 模型的新视图

人脸检测和定位也可以用于很多图像编辑应用中(除了用于相机中提供更好的对焦、曝光和闪光灯设计外)。Zanella and Fuentes(2004)用活动形状模型(active shape model)(14.2.2 节)对面部特征进行注册,创建了自动变形。为了决定图像数据库中哪些块可以拼接成一个拼贴画(collage),Rother, Bordeaux, Hamadi et al.(2006)用人脸检测和天空检测来决定感兴趣的区域。Bitouk, Kumar, Dhillon et al.(2008)描述了一个把给定的人脸图像匹配到网上大型人脸图像集合的系统,然后在原始图像中(结合仔细的重光照(relighting)算法)用于替换人脸。他们描述的应用包括去身份化(de-identification)和得到在"连拍模式"的合影快照中每个人带有最好可能笑脸的照片。Leyvand, Cohen-Or, Dror et al.(2008)显示了如何精确地用活动形状模型(Cootes, Edwards, and Taylor 2001; Zhou, Gu, and Zhang 2003)定位面部特征,依照面部吸引力很高的图像的类似配置情况,把这些特征(既而该图像)进行卷绕,从而"美化"图像而并没有完全丢失一个人的身份特性。

绝大多数这些方法或者依赖于标记的训练图像集(训练图像集对于所有学习的方法都是很关键的部分),或者依赖于最近从互联网上可以获得的成爆炸性增长的图像。一些这类工作(以及基于这类超大规模数据库的识别系统(14.5.1 节)),假设随着可以获得的(以及可能的部分标记的)图像集合的增大,找到接近的匹配变得更容易了。正如 Hays and Efros(2007)在他们的摘要中的陈述:"我们的主要洞察是,虽然图像空间是无限的,但是实际上,语义可区分的场景的空间没那么大。"

在 Levoy(2008)论文的有趣的评论中，作者对这个断言提出质疑，并声称："自然场景中的特征形成了重尾分布(heavy-tailed distribution)，这意味着虽然图像中的一些特征比其他特征更常见，但是不常见特征相对出现的概率下降得缓慢。换句话说，在世界上存在很多不同寻常的图像。"然而他也同意，和很多其他应用(比如语音识别、合成和翻译)相同，在计算摄影学(computational photography)上，"如果在足够大的数据库上训练，简单的机器学习的算法常常比更复杂的算法更有效。"接着，他还指出了这些系统的潜在优势，比如更好的自动色彩平衡，以及这些新方法带来的诸如图像伪造这类问题和陷阱。

更多因互联网计算机视觉而产生的照片编辑和计算摄影学的应用例子，请参见最近关于这个主题的研讨会[①]和特刊(Avidan, Baker, and Shan 2010)以及 Tamara Berg (2008)关于互联网视觉的课程。

14.5　上下文与场景理解

到目前为止，我们主要考虑的是单独地识别和定位物体的任务，这是从理解物体出现的场景(上下文)任务下进行的。这是很严格的限制，因为上下文在人类物体识别中起着很重要的作用(Oliva and Torralba 2007)。在这一节中我们将会看到，上下文能够很大程度地提高物体识别算法的性能(Divvala, Hoiem, Hays et al. 2009)，也对一般场景理解提供了有用的语义线索(Torralba 2008)。

考虑图 14.48a 和 b 中所示的两个照片。你能叫出所有物体的名字，尤其是图像(c 和 d)中用圆标出的物体吗？现在，更仔细地看看用圆标出的物体。是否看出它们的形状的一些相似性？实际上，如果你把它们旋转 90°，它们都和图 14.48e 中所示的"斑点(blob)"相同。我们用形状识别物体的能力就这么大！上下文识别的另一个(也许是更人工的)例子如图 14.49 所示。尝试叫出所有的字母和数字，然后看看你猜的是否是对的。

图 14.48　上下文的重要性(图像来源于 Antonio Torralba)。你能叫出图像(a 和 b)中物体的名字，尤其是(c 和 d)中用圆形标出的物体吗？仔细看用圆形标出的物体。正如列(e)所示，你发现它们有相同的形状(经过旋转后)了吗？

---

① http://www.internetvisioner.org/

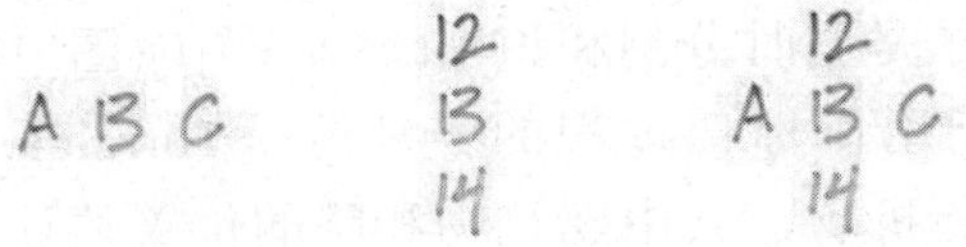

图 14.49 上下文的更多例子：在第一组中读出字母，在第二组中读出数字，在第三组中读出字母和数字(图像来源于 Antonio Torralba)

尽管我们没有在本章更早的时候明确地讨论上下文，我们已经看到了这个一般想法应用的一些示例。一个简单的方法是把空间信息结合到识别算法中，在不同的区域计算特征统计量，比如 Lazebnik, Schmid, and Ponce (2006)的空间金字塔系统。基于部件的模型(14.4.2 节，图 14.40~图 14.43)使用了一种局部上下文，其中不同的部件需要按照合适的几何关系排列组成一个物体。

基于部件的模型和上下文模型最大的区别是后者把物体和场景结合起来，每个类别中构成物体(构件)的数目事先不知道。实际上，可以把基于部件的模型和上下文模型结合在同一个识别架构内(Murphy, Torralba, and Freeman 2003; Sudderth, Torralba, Freeman *et al.* 2008; Crandall and Huttenlocher 2007)。

考虑图 14.50a 和 b 所示的街景和办公场景。如果我们有足够的标记区域的训练图像，比如建筑、汽车和路，或者显示器、键盘和鼠标，我们可以提出描述它们相对位置的几何模型。Sudderth, Torralba, Freeman *et al.* (2008)提出了这样的模型，这可以认为是一个两层的星座模型。在第一层，物体相对其他物体的分布以高斯函数为模型(图 14.50c，右上角)。在第二层，相对于物体中心部件的分布(仿射协变特征)以高斯混合模型为模型(图 14.50c，下面两行)。然而，由于场景中物体的数目以及每个物体的部件数目都是未知的，在一般框架下，物体和部件的创建模型是隐 Dirichlet 过程(Latent Dirichlet Process，LDP)。所有物体和部件的分布是从大量标记的数据库中学习到的，随后在推断(识别)中用于标记场景的元素。

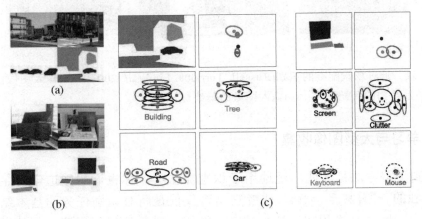

图 14.50 物体识别的上下文场景模型(Sudderth, Torralba, Freeman *et al.* 2008)© 2008 Springer：(a)一些街景和它们对应的标签(紫 = 建筑，红 = 汽车，绿 = 树，蓝 = 路)；(b)一些办公场景(红 = 计算机屏幕，绿 = 键盘，蓝 = 鼠标)；(c)从这些标记的场景中学习到的上下文模型。上面一行显示了样例标记图像和其他物体相对于中心红色(汽车或者屏幕)物体的分布情况。下面一行显示了组成每个物体的部件分布情况

上下文的另一个例子是同时分割和识别(14.4.3 节)(图 14.44 和图 14.45)，其中场景中不同物体的排列用作标记过程的一部分。Torralba, Murphy, and Freeman (2004)描述了一个条件随机场，其中建筑物和路的位置估计影响了汽车检测，其中将 boosting 用于学习 CRF 的结构。Rabinovich, Vedaldi, Galleguillos *et al.* (2007) 注意到某些邻接(关系)比其他的更易出现，比如人在马上比人在狗上更容易出现，用上下文改进了 CRF 的结果。

上下文同样在单幅图像的 3D 推断中起着重要作用(图 14.47)，将计算机视觉方法用于把每个像素标记为地面、直立物或者天空(Hoiem, Efros, and Hebert 2005a, b)。这项工作已经扩展到了更整体性的方法，同时推断物体标识、定位、表面方向、遮挡以及摄像机视图参数(Hoiem, Efros, and Hebert 2008a, b)。

一些方法用场景的 gist(Torralba 2003; Torralba, Murphy, Freeman *et al.* 2003)来决定特定物体示例更容易出现的位置。比如，Murphy, Torralba, and Freeman (2003)，基于图像的 gist，训练了一个回归器预测直立物的位置，比如行人、汽车和建筑物 (或者室内办公场景中的屏幕和键盘)。然后，这些位置分布和经典的物体检测器一起使用，提高检测的性能。正如我们在 Hays and Efros(2007)的场景补全工作中看到的，Gist 还可以用于直接匹配完整的图像。

最后，最近的一些场景理解的工作，利用大量标注的(或者甚至未标注的)图像，直接对整体图像进行匹配，其中图像自身隐式地编码了所期望的物体间的关系(图 14.51) (Russell, Torralba, Liu *et al.* 2007; Malisiewicz and Efros 2008)。在下一节我们讨论这样的方法，其中我们审视大型图像数据库对物体识别和场景理解的影响。

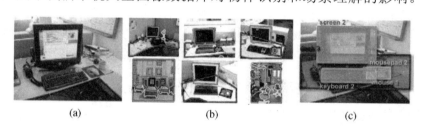

图 14.51 通过场景配准的识别(Russell, Torralba, Liu *et al.* 2007)：(a)输入图像；(b)有相似场景结构的匹配图像；(c)输入图像的最终标记

14.5.1 学习与大型图像收集

考虑到学习方法是如此广泛地用于识别算法中，你也许想知道学习主题是否值得作为单独的一节(甚至一章)，或者它是否仅仅是所有识别任务的基本结构的一部分。实际上，除了基本模式，比如 UPC 编码，对其他任何事情，不使用大量的训练数据，尝试建立识别系统已经被证明是一个令人沮丧的失败。

在这一章中，我们已经看到了来自机器学习、统计学和模式识别领域的很多方法。这些方法包括主分量分析(principal component analysis)、子空间分析(subspace analysis)和鉴别分析(discriminant analysis)(14.2.1 节)，以及更复杂的判别性分类算法，比如神经网、支持向量机和 boosting(14.1.1 节)。在具有挑战性的识别标准集上的一些性能最好的方法(Varma and Ray 2007; Felzenszwalb, McAllester, and Ramanan 2008; Fritz and Schiele 2008; Vedaldi, Gulshan, Varma et al. 2009)很依赖于最新的机器学习方法，这些机器学习方法的发展常常是受具有挑战性的视觉问题所驱动的(Freeman, Perona, and Schölkopf 2008)。

在识别领域中，有时将下面两个问题区别开，一个是大部分感兴趣的变量(比如部件)已经(部分)标注的问题，另一个是在少量监督条件下学习更多问题结构的系统(Fergus, Perona, and Zisserman 2007; Fei-Fei, Fergus, and Perona 2006)。实际上，Sivic, Russell, Zisserman et al. (2008)最新的工作展示了在一个完全无监督的框架内，学习视觉层次(物体部件的层次以及相关的视觉表观)和场景分割的能力。

识别领域最戏剧性的变化也许是大型训练图像的数据库的出现[①]。早期基于学习的算法，比如用于人脸和行人检测的算法(14.1 节)，使用相对来说少量的(几百个)标记的样本来训练识别算法的参数(比如 boosting 中的阈值)。现在，一些识别算法用(Russell, Torralba, Murphy et al. 2008)LabelMe 这样的数据库，在这些数据库中，有数万个标记的样本。

这种大型数据库的存在开辟了对训练图像直接进行匹配的可能性，而不仅仅是使用它们来训练识别算法的参数。Russell, Torralba, Liu et al. (2007)描述了一个系统，其中一个新的图像和训练图像中的每一个进行匹配，从而对场景中的未知物体可以推断得到具有一致性的标记，如图 14.51 所示。Malisiewicz and Efros (2008)首先对每个图像进行过分割，然后为了得到每个像素的类别标签，他们用 LabelMe 数据库查询相似的图像和构型。也可以把基于特征的对应算法和大型标注数据库结合起来，同时进行识别和分割(Liu, Yuen, and Torralba 2009)。

当图像数据库变得足够大的时候，甚至可以直接匹配完整的图像以期望找到好的匹配。Torralba, Freeman, and Fergus(2008)收集了有 8 千万幅小(32×32)图像的数据库，弥补了他们的图像标签的准确度差的缺点，这些数据是从互联网上自动收集的，他们的方法是用语义分类(Wordnet)对一个新的图像推断得到最可能的标签。在这 8 千万幅图像中的某些地方，存在足够的样例把某些图像的集合和他们系统中使用的 Wordnet 中的 75 000 个非抽象名词中的每一个关联起来。图 14.52 显示了一些样例识别结果。

另一个大型标记图像数据库的例子是 ImageNet (Deng, Dong, Socher et al. 2009)，它给 WordNet 中的 80 000 个名词(同义词集 synonym set)收集了图像(Fellbaum 1998)。截止 2010 年 4 月，对于 14 841 个同义词集(synsets)的每一个，已经收集了大约 500 到 1000 个经过仔细核实的样例(图 14.53)。Deng, Dong, Socher et al. (2009)的论文也很好地回顾了相关的数据库。

如我们在 14.4.3 节提到的，部分标注的互联网图像的大型数据库的存在已经产

[①] 在 14.4.4 节，我们已经看到了这样的数据库在计算摄像学上的一些应用。

生了互联网计算机视觉的一个新的子领域，有它自己的研讨会[①]和特刊(Avidan, Baker, and Shan 2010)。

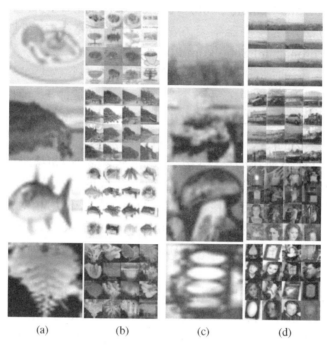

图 14.52 使用小图像的识别(Torralba, Freeman, and Fergus 2008)© 2008 IEEE：(a)和(c)列显示了输入图像样例,(b)和(d)列显示了在有 8 千万个小图像的数据库中其对应的 16 个最近邻

图 14.53 ImageNet(Deng, Dong, Socher et al. 2009)© 2009 IEEE。这个数据库包含对每个名词有 500 个仔细核实的图像，其中每个名词是 WordNet 层次结构的 14 841 个名词(截止 2010 年 4 月)中的一个

14.5.2 应用：图像搜索

尽管从一些衡量标准来看视觉识别算法还处于萌芽状态，但是它们已经对图像搜索产生了一些影响，即用关键词和视觉相似性结合的方法从互联网上进行图像检

---
① http://www.internetvisioner.org/

索。现在，大多数图像搜索引擎主要依赖于标题、附近文本和文件名称的文本关键字的发现，通过用户点击数据来加强文本关键字的发现(Craswell and Szummer 2007)。然而，随着识别算法持续改进，视觉特征和视觉相似性将开始用于关键词缺失或者错误的图像的识别。

用视觉相似性进行搜索的主题有很长的历史，也被冠上了各种各样的名称，包括基于内容的图像检索(content-based image retrieval，CBIR)(Smeulders, Worring, Santini et al. 2000; Lew, Sebe, Djeraba et al. 2006; Vasconcelos 2007; Datta, Joshi, Li et al. 2008)和基于图像内容的查询(query by image content，QBIC)(Flickner, Sawhney, Niblack et al. 1995)。这一领域早期发表的论文主要基于简单的整体图像的相似性度量，比如颜色和纹理(Swain and Ballard 1991; Jacobs, Finkelstein, and Salesin 1995; Manjunathi and Ma 1996)。

在最近的工作中，Fergus, Perona, and Zisserman(2004)用一个基于特征的学习和识别算法把来自传统基于关键词图像搜索引擎的结果重新排序。在后续的工作中，Fergus, Fei-Fei, Perona et al. (2005)用概率最近语义分析(probabilistic latest semantic analysis，PLSA)(Hofmann 1999)的扩展把图像搜索返回的结果聚类，然后把和排序最靠前的结果相关联的聚类中心作为那个类别有代表性的图像。

最近更新的工作依赖于仔细标注的图像数据库，比如 LabelMe(Russell, Torralba, Murphy et al. 2008)。比如 Malisiewicz and Efros(2008)描述了一个系统，给定一幅查询图像，该系统能够找到相似的 LabelMe 图像，而 Liu, Yuen, and Torralba (2009)把基于特征的对应算法和标注的数据库结合起来，同时进行识别和分割。

14.6 识别数据库和测试集

除了机器学习和统计建模方法的快速发展外，识别算法的持续发展的一个关键因素是图像识别数据库的可获得性和质量都增加了。

表 14.1 和表 14.2 是基于 Fei-Fei, Fergus, and Torralba(2009)中的类似表格，用更多最新的项和 URL(网址)进行了更新，显示了一些广泛使用的数据库。这些数据库中的一些，比如用于人脸识别和定位的，追溯到十几年前。最近的数据库，比如 PASCAL 数据库，每年都用更有挑战性的问题进行更新。表 14.1 显示了主要用于(整体图像)识别的数据库的例子，而表 14.2 显示了一些数据库，其中可以获得和期望具有更高精确的定位或者分割的信息。

表 14.1 用于识别的图像数据库，扩展并修改了 Fei-Fei, Fergus, and Torralba (2009)

名称 / URL	范围	内容/参考文献
物体检测/定位		
Yale 人脸数据库 http://www1.cs.columbia.edu/_belhumeur/	中心化的人脸图像	正面人脸 Belhumeur, Hespanha, and Kriegman(1997)

名称 / URL	范围	内容/参考文献
人脸检测的资源	各种数据库	各种姿态的人脸
http://vision.ai.uiuc.edu/mhyang/face-detection-survey.html		Yang, Kriegman, and Ahuja (2002)
FERET	中心化的人脸图像	正面人脸
http://www.frvt.org/FERET		Phillips, Moon, Rizvi *et al.* (2000)
FRVT	中心化的人脸图像	各种姿态的人脸
http://www.frvt.org/		Phillips, Scruggs, O'Toole *et al.* (2010)
CMU PIE 数据库	中心化的人脸图像	各种姿态的人脸
http://www.ri.cmu.edu/projects/project 418.html		Sim, Baker, and Bsat (2003)
CMU Multi-PIE 数据库	中心化的人脸图像	各种姿态的人脸
http://multipie.org		Gross, Matthews, Cohn *et al.* (2010)
Faces in the Wild	因特网图像	各种姿态的人脸
http://vis-www.cs.umass.edu/lfw/		Huang, Ramesh, Berg *et al.* (2007)
消费者图像人数据库	完整图像	人
http://chenlab.ece.cornell.edu/people/Andy/GallagherDataset.html		Gallagher and Chen (2008)
物体识别		
Caltech 101	分割掩模	101 类别
http://www.vision.caltech.edu/Image Datasets/Caltech101/		Fei-Fei, Fergus, and Perona (2006)
Caltech 256	中心化的物体	256 个类别和杂项
http://www.vision.caltech.edu/Image Datasets/Caltech256/		Griffin, Holub, and Perona (2007)
COIL-100	中心化的物体	100 示例
http://www1.cs.columbia.edu/CAVE/software/softlib/coil100.php		Nene, Nayar, and Murase (1996)
ETH-80	中心化的物体	8 个示例, 10 个视角
http://www.mis.tu-darmstadt.de/datasets		Leibe and Schiele (2003)
示例识别比较基准	各种姿态的物体	2550 个物体
http://vis.uky.edu/_stewe/ukbench/		Nistér and Stewénius (2006)
Oxford 建筑物数据集	建筑物图片	5062 个图像
http://www.robots.ox.ac.uk/_vgg/data/oxbuildings/		Philbin, Chum, Isard *et al.* (2007)
NORB	包围盒	50 玩具
http://www.cs.nyu.edu/_ylclab/data/norb-v1.0/		LeCun, Huang, and Bottou (2004)
小图像	完整图像	75 000 (Wordnet) 个事物
http://people.csail.mit.edu/torralba/tinyimages/		Torralba, Freeman, and Fergus (2008)
ImageNet	完整图像	14 000 (Wordnet) 个事物
http://www.image-net.org/		Deng, Dong, Socher *et al.* (2009)

表 14.2 用于检测和定位的图像数据库，扩展并修改了 Fei-Fei, Fergus, and Torralba (2009)

名称 / URL	范围	内容/参考文献
物体检测/定位		
CMU 正面人脸	图像块	正面人脸
http://vasc.ri.cmu.edu/idb/html/face/frontal_images		Rowley, Baluja, and Kanade (1998a)
MIT 正面人脸	图像块	正面人脸
http://cbcl.mit.edu/software-datasets/FaceData2.html		Sung and Poggio (1998)
CMU 人脸检测数据库	多人脸	各种姿态的人脸
http://www.ri.cmu.edu/research_project_detail.html?project_id=419		Schneiderman and Kanade (2004)
UIUC Image DB	包围盒	汽车
http://l2r.cs.uiuc.edu/_cogcomp/Data/Car/		Agarwal and Roth (2002)
Caltech Pedestrian Dataset(行人数据集)	包围盒	行人
http://www.vision.caltech.edu/Image_Datasets/CaltechPedestrians/		Dollàr, Wojek, Schiele et al. (2009)
Graz-02 Database	分割掩模	Bikes, cars, people
http://www.emt.tugraz.at/_pinz/data/GRAZ_02/		Opelt, Pinz, Fussenegger et al. (2006)
ETHZ Toys	杂乱图像	玩具、盒子、杂志
http://www.vision.ee.ethz.ch/_calvin/datasets.html		Ferrari, Tuytelaars, and Van Gool (2006b)
TU Darmstadt DB	分割掩模	摩托车、汽车、奶牛
http://www.vision.ee.ethz.ch/_bleibe/data/datasets.html		Leibe, Leonardis, and Schiele (2008)
MSR Cambridge	分割掩模	23 个类
http://research.microsoft.com/enus/projects/objectclassrecognition/		Shotton, Winn, Rother et al. (2009)
LabelMe dataset	多边形边界	>500 个类别
http://labelme.csail.mit.edu/		Russell, Torralba, Murphy et al. (2008)
Lotus Hill	分割掩模	场景和分层
http://www.imageparsing.com/		Yao, Yang, Lin et al. (2010)
在线标注工具		
ESP game	图像描述	Web 图像
http://www.gwap.com/gwap/		von Ahn and Dabbish(2004)
Peekaboom	标注区域	Web 图像
http://www.gwap.com/gwap/		von Ahn, Liu, and Blum(2006)
LabelMe	多边形边界	高分辨率图像
http://labelme.csail.mit.edu/		Russell, Torralba, Murphy et al. (2008)
挑战集		
PASCAL	分割, 盒	各种
http://pascallin.ecs.soton.ac.uk/challenges/VOC/		Everingham, Van Gool, Williams et al. (2010)

Ponce, Berg, Everingham *et al.* (2006)在更早的数据集上讨论了一些问题,并描述了最新的 PASCAL 视觉物体类挑战(Visual Object Classes Challenge)如何致力于解决这些问题。2008 挑战中的 20 个视觉类的一些例子显示在图 14.54 中。VOC 研讨会的幻灯片[①]是目前可获得的指向最佳识别方法的最好来源。

识别数据库中的两个最新的趋势是基于互联网的标注和数据收集工具的出现,以及用搜索和识别算法来建立数据库(Ponce, Berg, Everingham *et al.* 2006)。一些最有趣的工作在图像上进行人的标注,这些图像来源于一系列交互的多个人的游戏,比如 ESP (von Ahn and Dabbish 2004)和 Peekaboom(von Ahn, Liu, and Blum 2006)。在这些游戏中,通过给出和其内容相关的纹理线索,它隐式地标记了整个图像或者只是一些区域,人们互相帮助,猜测隐藏图像的特性。一个更"认真"的志愿工作是 LabelMe 数据库,其中为了获得对数据库的访问,视觉研究者手动标注多边形区域(Russell, Torralba, Murphy *et al.* 2008)。

图 14.54　PASCAL Visual Object Classes Challenge 2008 (VOC2008)数据库中的样例图像(Everingham, Van Gool, Williams *et al.* 2008)。原始图像来自 flickr (*http://www.flickr.com/*)。数据库的权利在 *http://pascallin.ecs.soton.ac.uk/challenges/VOC/voc2008/* 中解释了说明

用计算机视觉算法来收集识别数据库的工作可以追溯到 Fergus, Fei-Fei, Perona *et al.* (2005),他们用 PLSA 的扩展将 Google 图像搜索返回的结果聚类,然后选择和排序最靠前的结果关联的聚类中心。相关方法的最新例子包括 Berg and Forsyth (2006)和 Li and Fei-Fei(2010)的工作。

不管用什么样的方法来收集和验证识别数据库,每年它们都将在大小、可用性和难度上持续增加。它们也还将继续是研究识别和场景理解问题的重要组成部分,

---

[①] *http://pascallin.ecs.soton.ac.uk/challenges/VOC/*

识别和场景理解问题仍然一如既往地是计算机视觉的巨大挑战。

14.7 补充阅读

虽然现在关于图像识别和场景理解还没有专门的教材,但是可以找到一些描述最新方法的综述(Pinz 2005)和论文集(Ponce, Hebert, Schmid et al. 2006; Dickinson, Leonardis, Schiele et al. 2007)。关于这个主题,最近研究的其他好的资源,比如ICCV2009 短期课程(Fei-Fei, Fergus, and Torralba 2009)和 Antonio Torralba 的更全面的 MIT 课程(Torralba 2008)。PASCAL VOC Challenge 网站包含了总结当前最佳性能算法研讨会的幻灯片。

关于人脸、行人、汽车和其他物体检测的文献相当多。人脸检测中影响深远的论文包括 Osuna, Freund, and Girosi(1997), Sung and Poggio(1998), Rowley, Baluja, and Kanade(1998a), 和 Viola and Jones(2004),Yang, Kriegman, and Ahuja(2002)在这方面对早期工作提供了全面的综述研究。最近的例子包括(Heisele, Ho, Wu et al. 2003; Heisele, Serre, and Poggio 2007)。

早期的行人和汽车检测的工作有 Gavrila and Philomin(1999), Gavrila(1999), Papageorgiou and Poggio(2000), Mohan, Papageorgiou, and Poggio(2001), 和 Schneiderman and Kanade(2004)。最近的工作包括 Belongie, Malik, and Puzicha (2002), Mikolajczyk, Schmid, and Zisserman(2004), Dalal and Triggs(2005), Leibe, Seemann, and Schiele(2005), Dalal, Triggs, and Schmid(2006), Opelt, Pinz, and Zisserman(2006), Torralba(2007), Andriluka, Roth, and Schiele(2008), Felzenszwalb, McAllester, and Ramanan(2008), Rogez, Rihan, Ramalingam et al. (2008), Andriluka, Roth, and Schiele(2009), Kumar, Zisserman, and H.S.Torr(2009), Dollàr, Belongie, and Perona(2010)和 Felzenszwalb, Girshick, McAllester et al. (2010)。

一些人脸识别早期的方法涉及寻找区分性的图像特征和度量它们之间的距离(Fischler and Elschlager 1973; Kanade 1977; Yuille 1991),然而最近的方法依赖于比较灰度图像,通常将其映射到低维子空间中(Turk and Pentland 1991a; Belhumeur, Hespanha, and Kriegman 1997; Moghaddam and Pentland 1997; Moghaddam, Jebara, and Pentland 2000; Heisele, Ho, Wu et al. 2003; Heisele, Serre, and Poggio 2007)。关于主分量分析(principal component analysis,PCA)和其对应的贝叶斯方法的更多细节,可以在附录 B.1.1 和关于这个主题的书以及论文中找到(Hastie, Tibshirani, and Friedman 2001; Bishop 2006; Roweis 1998; Tipping and Bishop 1999; Leonardis and Bischof 2000; Vidal, Ma, and Sastry 2010)。Cai, He, Hu et al. (2007), Frome, Singer, Sha et al. (2007), Guillaumin, Verbeek, and Schmid(2009), Ramanan and Baker(2009), and Sivic, Everingham, and Zisserman(2009)涵盖了子空间学习、局部距离函数以及度量学习等主题。另一种直接匹配灰度级别图像或者块的方法是使用非线性局部变

换，比如局部二值模式(local binary patterns)(Ahonen, Hadid, and Pietikäinen 2006; Zhao and Pietik¨ainen 2007; Cao, Yin, Tang et al. 2010)。

为了提升那些本质上是基于 2D 表观的模型的性能，提出了各种形状和姿态变形模型(Beymer 1996; Vetter and Poggio 1997)，包括活动形状模型(Lanitis, Taylor, and Cootes 1997; Cootes, Cooper, Taylor et al. 1995; Davies, Twining, and Taylor 2008)，弹性束图匹配(Elastic Bunch Graph Matching)(Wiskott, Fellous, Krüger et al. 1997)，3D 变形模型(3D Morphable Model)(Blanz and Vetter 1999)和活动表观模型(Costen, Cootes, Edwards et al. 1999; Cootes, Edwards, and Taylor 2001; Gross, Baker, Matthews et al. 2005; Gross, Matthews, and Baker 2006; Matthews, Xiao, and Baker 2007; Liang, Xiao, Wen et al. 2008; Ramnath, Koterba, Xiao et al. 2008)。最近的 Murphy-Chutorian and Trivedi (2009)的综述专门覆盖了头部姿态估计的主题。

人脸识别的其他信息可以在关于这个主题的一些综述和书中找到(Chellappa, Wilson, and Sirohey 1995; Zhao, Chellappa, Phillips et al. 2003; Li and Jain 2005)，同样在人脸识别网站[①]也可以找到。Phillips, Moon, Rizvi et al. (2000), Sim, Baker, and Bsat(2003), Gross, Shi, and Cohn(2005), Huang, Ramesh, Berg et al. (2007), 和 Phillips, Scruggs, O'Toole et al. (2010)讨论了人脸识别的数据库。

示例识别，即检测静止的人造物体，这些物体在表观上变化稍有不同，但是在 3D 姿态上可能有变化，其算法通常是基于检测 2D 兴趣点，并描述为视角无关的描述子(Lowe 2004; Rothganger, Lazebnik, Schmid et al. 2006; Ferrari, Tuytelaars, and Van Gool 2006b; Gordon and Lowe 2006; Obdržálek and Matas 2006; Kannala, Rahtu, Brandt et al. 2008; Sivic and Zisserman 2009)。

随着被匹配的数据库的大小的增加，把视觉描述子量化为词变得更重要了(Sivic and Zisserman 2003; Schindler, Brown, and Szeliski 2007; Sivic and Zisserman 2009; Turcot and Lowe 2009)，然后用信息检索方法，比如倒排表(inverted index)(Nistér and Stewénius 2006; Philbin, Chum, Isard et al. 2007; Philbin, Chum, Sivic et al. 2008)，查询扩展(query expansion)(Chum, Philbin, Sivic et al. 2007; Agarwal, Snavely, Simon et al. 2009)和最小哈希(min hashing)(Philbin and Zisserman 2008; Li, Wu, Zach et al. 2008; Chum, Philbin, and Zisserman 2008; Chum and Matas 2010)执行有效的检索和聚类。

有很多为类别识别这个主题而写的综述、论文集和课程笔记(Pinz 2005; Ponce, Hebert, Schmid et al. 2006; Dickinson, Leonardis, Schiele et al. 2007; Fei-Fei, Fergus, and Torralba 2009)。对整个图像的类别识别，在词袋(关键点袋)的方法上，一些影响深远的论文是 Csurka, Dance, Fan et al. (2004), Lazebnik, Schmid, and Ponce(2006), Csurka, Dance, Perronnin et al. (2006), Grauman and Darrell(2007b)和 Zhang, Marszalek, Lazebnik et al. (2007)。这个领域更多最新的论文包括 Sivic, Russell,

---

[①] http://www.face-rec.org/

Efros et al. (2005), Serre, Wolf, and Poggio(2005), Opelt, Pinz, Fussenegger et al. (2006), Grauman and Darrell(2007a), Torralba, Murphy, and Freeman(2007), Boiman, Shechtman, and Irani(2008), Ferencz, Learned-Miller, and Malik(2008)和 Mutch and Lowe(2008)。基于物体轮廓的识别也是有可能的，比如用形状上下文(shape contexts)的方法(Belongie, Malik, and Puzicha 2002)或者其他方法(Jurie and Schmid 2004; Shotton, Blake, and Cipolla 2005; Opelt, Pinz, and Zisserman 2006; Ferrari, Tuytelaars, and Van Gool 2006a)。

很多物体识别算法用基于部件的分解办法，对关节连接和姿态提供更大的不变性。早期的算法着重于部件的相对位置(Fischler and Elschlager 1973; Kanade 1977; Yuille 1991)，然而更新的算法用更复杂的表观模型(Felzenszwalb and Huttenlocher 2005; Fergus, Perona, and Zisserman 2007; Felzenszwalb, McAllester, and Ramanan 2008)。关于基于部件模型的识别的好的综述可以在课程笔记 Fergus 2007b; 2009 中找到。

Carneiro and Lowe(2006)讨论了一些用于基于部件识别的图模型方法，包括树结构和星型结构(Felzenszwalb and Huttenlocher 2005; Fergus, Perona, and Zisserman 2005; Felzenszwalb, McAllester, and Ramanan 2008)，k 扇形结构(Crandall, Felzenszwalb, and Huttenlocher 2005; Crandall and Huttenlocher 2006)和星座结构(Burl, Weber, and Perona 1998; Weber, Welling, and Perona 2000; Fergus, Perona, and Zisserman 2007)。其他基于部件模型的识别方法包括 Dorkó and Schmid(2003)和 Bar-Hillel, Hertz, and Weinshall(2005)提出的方法。

把物体识别和场景分割结合起来有很大的好处。一个方法是把图像预分割为多块，然后把块匹配到部分模型上(Mori, Ren, Efros et al. 2004; Mori 2005; He, Zemel, and Ray 2006; Russell, Efros, Sivic et al. 2006; Borenstein and Ullman 2008; Csurka and Perronnin 2008; Gu, Lim, Arbelaez et al. 2009)。另一个方法是基于物体检测，投票得到物体可能的位置和尺度(Leibe, Leonardis, and Schiele 2008)。当前最流行的一个方法是用条件随机场(Kumar and Hebert 2006; He, Zemel, and Carreira-Perpiñán 2004; He, Zemel, and Ray 2006; Levin andWeiss 2006;Winn and Shotton 2006; Hoiem, Rother, andWinn 2007; Rabinovich, Vedaldi, Galleguillos et al. 2007; Verbeek and Triggs 2007; Yang, Meer, and Foran 2007; Batra, Sukthankar, and Chen 2008; Larlus and Jurie 2008; He and Zemel 2008; Shotton, Winn, Rother et al. 2009; Kumar, Torr, and Zisserman 2010)，这个方法在困难的 PASCAL VOC 分割挑战数据集上得到了一些最好的结果(Shotton, Johnson, and Cipolla 2008; Kohli, Ladický, and Torr 2009)。

更多的识别算法开始将场景上下文作为它们识别策略的一部分。这个方面代表性的论文有 Torralba(2003), Torralba, Murphy, Freeman et al. (2003), Murphy, Torralba, and Freeman(2003), Torralba, Murphy, and Freeman(2004), Crandall and Huttenlocher(2007), Rabinovich, Vedaldi, Galleguillos et al. (2007), Russell, Torralba, Liu et al. (2007), Hoiem, Efros, and Hebert(2008a), Hoiem, Efros, and Hebert(2008b),

Sudderth, Torralba, Freeman *et al.* (2008)和 Divvala, Hoiem, Hays *et al.* (2009)。

正如利用大型人工标注的数据库一样(Russell, Torralba, Liu *et al.* 2007; Malisiewicz and Efros 2008; Torralba, Freeman, and Fergus 2008; Liu, Yuen, and Torralba 2009)，复杂的机器学习方法也变为成功的物体检测和识别算法的关键因素(Varma and Ray 2007; Felzenszwalb, McAllester, and Ramanan 2008; Fritz and Schiele 2008; Sivic, Russell, Zisserman *et al.* 2008; Vedaldi, Gulshan, Varma *et al.* 2009)。粗糙的3D模型(Rough three-dimensional models)也重新被用于识别，体现在一些最近的论文上(Savarese and Fei-Fei 2007, 2008; Sun, Su, Savarese *et al.* 2009; Su, Sun, Fei-Fei *et al.* 2009)。一如既往，关于这个快速发展领域的最新算法，最好参考最近召开的计算机视觉会议。

14.8 习　　题

习题 14.1：人脸检测

建立并测试一个 14.1.1 节描述的人脸检测器。

1. 下载表 14.2 所示的一个或者多个标注的人脸检测数据库。
2. 用不包含任何人的照片，产生你自己的负样本。
3. 实现下面人脸检测器中的一个(或者你自己设计一个)。

- boosting (算法 14.1)基于简单的区域特征和可选的瀑布型检测器(Viola and Jones 2004)。
- PCA 人脸子空间(Moghaddam and Pentland 1997)。
- 计算到聚类的人脸原型和非人脸原型的距离，随后训练神经网分类器(Sung and Poggio 1998)或者 SVM(Osuna, Freund, and Girosi 1997)分类器。
- 在归一化的灰度级别的块上，直接训练多分辨率的神经网(Rowley, Baluja, and Kanade 1998a)。

4. 通过在子八度金字塔的每个位置上使用你的检测器，测试你的检测器在数据库上的性能。可选择性地做，在非人脸图像中得到的误报上重新训练你的检测器。

习题 14.2：确定 AdaBoost 的阈值

如算法 14.1 描述的，给定在训练样本 x_i 上的评价函数集合 $f_i = f(x_i) \in \pm 1$，训练标签 $y_i \in \pm 1$ 和权重 $w_i \in (0,1)$，设计一个有效的算法找到值 θ 和 $s = \pm 1$，最大化

$$\sum_i w_i y_i h(sf_i, \theta), \tag{14.43}$$

其中 $h(x, \theta) = \mathrm{sign}(x - \theta)$。

习题 14.3：用特征脸进行人脸识别

收集一个面部照片的集合，然后建立一个识别系统，重新识别同一个人。

1. 给你的每个同学拍几张照片，并保存。
2. 自动或者手动配准图像，方法是检测眼角点，然后用相似变换把每个图像

拉伸和旋转到标准的位置。
3. 计算人脸图像的平均图像和 PCA 子空间。
4. 在一周之后拍摄一个新的照片集合，作为你的测试集。
5. 将新的图像和数据库中每个图像进行比较，选择最近的一个作为识别的身份。验证在 PCA 空间中的距离和用完整 SSD(误差平方和，sum of squared difference)度量计算的距离是很接近的。
6. (可选)对不同的人和表情计算不同的主分量，用它们来改进你的识别结果。

习题 14.4：贝叶斯人脸识别

Moghaddam, Jebara, and Pentland(2000)计算各自的协方差矩阵 $\sum_I$ 和 $\sum_E$ 来审查所有图像对之间的差异。在运行的时候，他们选择最近的图像来确定面部身份。估计所有图像对之间的统计信息，并用它们测试到最近样例(exemplar)的距离的方法是合理的吗？讨论这在统计上是否正确。

所有同一人的图像对间的协方差矩阵 $\sum_I$ 是如何和类内散布矩阵 S_W 相联系的？在 $\sum_E$ 和 S_B 之间也存在相似的关系吗？

习题 14.5：模块化的特征脸

把你的人脸识别系统扩展到各自地匹配眼睛、鼻子和嘴的区域，如图 14.18 所示。
1. 把人脸图像归一化到一个标准的尺度和位置，手工地分割出眼睛、鼻子和人脸的某些区域。
2. 对这三个(或四个)类别的区域，使用子空间(PCA)方法或 14.1.1 节描述的一种方法建立各自的检测器。
3. 对每个要识别的新图像，先检测面部特征的位置。
4. 然后，分别在数据库中匹配各个特征，并注意这些特征的位置。
5. 使用各个特征匹配标识(ID)以及(可选)特征位置训练并测试分类器进行人脸识别。

习题 14.6：基于识别的色彩平衡

建立一个系统，使其能识别一般图像中大多数重要的颜色区域(天空、草地和皮肤)，然后相应地对图像进行色彩平衡。习题 2.8 以及 Forsyth and Fleck(1999), Jones and Rehg(2001), Vezhnevets, Sazonov, and Andreeva(2003) 和 Kakumanu, Makrogiannis, and Bourbakis(2007)给出了肤色检测的一些参照和想法。这些方法也许会给你提供如何检测其他区域的想法，或者你可以尝试更复杂的基于 MRF 的方法(Shotton, Winn, Rother *et al.* 2009)。

习题 14.7：行人检测

建立并测试 14.1.2 节描述的一种行人检测器。

习题 14.8：简单的示例识别

用你在习题 4.1~4.4 和 9.2 中开发的特征检测、匹配和配准算法，找到给定查询图像或者区域的匹配图像(图 14.26)。

你自己评价几种特征检测子、描述子和鲁棒验证策略，或者把你的结果和同学们的结果相比。

习题 14.9：大型数据库和位置识别

如算法 14.2 描述的，用量化视觉词和信息检索的方法，把前一个习题扩展到更大规模的数据库中。

在大型数据库上测试你的算法，比如 Nister and Stewenius(2006)或 Philbin, Chum, Sivic et al. (2008)用到的那个数据库，这些数据库在表 14.1 中列出。另一种选择，在互联网上或照片共享网站中(比如，对一座城市)用关键词查询，从而建立你自己的数据库。

习题 14.10：词袋

用 14.4.1 节描述的方法，把习题 14.8 开发的特征提取和匹配的流水线，改写到类别识别中。

1. 从表 14.1 和表 14.2 列出的一个或者多个数据库，下载训练和测试的图像，比如 Caltech 101, Caltech 256 或 PASCAL VOC。
2. 从每个训练图像中提取特征，量化它们，计算 *tf-idf* 向量(词袋直方图)。
3. 另一个选项是，考虑不量化特征并使用金字塔匹配(14.40 和 14.41)(Grauman and Darrell 2007b)或者使用空间金字塔以达到更大的选择性(Lazebnik, Schmid, and Ponce 2006)。
4. 选择一个分类算法(比如，最近邻分类或者支持向量机分类)，"训练"你的识别器，即建立合适的数据结构(比如 *k*-d 树)或者设置合适的分类器参数。
5. 用你在第 2 到 4 步开发的同样的流水线，在测试数据集上测试你的算法，把你的结果和报告的最好结果比较。
6. 解释为什么你的结果和先前报告的结果不同，给出如何改进你的系统的一些想法。

可以在 PASCAL 视觉物体类挑战(Visual Object Classes Challenge)的报告中找到性能最好的分类算法和他们的方法的概要，其网址是 *http://pascallin.ecs.soton.ac.uk/challenges/VOC/*。

习题 14.11：物体检测和定位

扩展上一个习题中开发的在图像中定位物体的分类算法，要求是报告每个检测出的物体的包围盒。做到这一点最简单的办法是用滑动窗口的方法。在和 PASCAL VOC 2008 Challenge 关联的研讨会中，可以找到一些指向这个领域的最新方法的指针。

习题 14.12：基于部件的识别

选择 14.4.2 节描述的一个或者多个方法，实现基于部件的识别系统。由于这些方法相当复杂，你将需要阅读这方面一些研究论文，选择一个你想遵循的一般方法，然后实现你的算法。一个好的出发点可以是论文 Felzenszwalb, McAllester, and Ramanan(2008)，因为它在 PASCAL VOC 2008 检测挑战上性能很好。

习题 14.13：识别和分割

选择 14.4.3 节描述的一个或者多个方法，实现同时的识别和分割系统。由于这些方法相当复杂，你将需要阅读这方面一些研究论文，选择一个你想遵循的一般方法，然后实现你的算法。在表 14.2 中的一个或者多个分割数据集上测试你的算法。

习题 14.14：上下文

实现 14.5 节描述的一个或者多个上下文和场景理解系统，并报告你的经验。上下文或者整个场景理解在识别物体上比独立的系统有更好的性能吗？

习题 14.15：小图像

从 *http://people.csail.mit.edu/torralba/tinyimages/* 下载小图像数据库，基于直接将你的测试图像与所有标注的训练图像做比较建立一个分类器。这个方法看起来有前途吗？

第 15 章
结　　语

在本书中，我们涵盖了计算机视觉主题的一个宽广范围。从图像形成开始，我们看到了为了去除噪声或模糊如何对图像进行预处理，如何将其分割为区域或将其转换为特征描述子。我们可以匹配或注册多幅图像，将其结果用于运动估计、跟踪人物、重建 3D 模型或将图像融合成更吸引人而有趣的合成和绘制。我们也可以分析图像以产生其内容的语义描述。但是，在该领域计算机和人之间的性能差距还十分巨大，且可能在很多年里仍然会如此。

我们的研究也涉及了数学方法的一个宽广范围。它们包括连续数学，比如信号处理、变分方法、三维和摄影几何、线性代数和最小二乘。我们也研究了离散数学和计算机科学中的主题，比如图算法、组合优化以及甚至于信息检索的数据库方法。由于计算视觉中的很多问题是涉及从带噪声的输入数据中估计未知量的逆问题，我们也检视了贝叶斯统计推断方法以及从大量训练数据学习概率模型的机器学习方法。由于在因特网上可获得的部分标注了的视觉影像不断地成指数增长，这后一种方法将对我们的领域持续产生主要的影响。

你可能要问：我们的领域为什么这么宽广？是否有某种统一的原理可用于简化我们的研究？问题的部分回答在于计算机视觉的广阔定义，它是图像和视频的分析，还在于视觉影像形成中所固有的难以置信的复杂度。以某种方式，我们的领域是与汽车工程的研究一样复杂，它要求理解内燃烧、机械学、空气动力学、人机工程学、电子线路和控制系统以及其他的诸多主题。类似地计算机视觉涉及多种多样的分学科，使其作为一个学期的课变得具有挑战性。相反地，也正是计算机视觉问题的极为宽广和技术复杂性吸引着众多研究人员投入该研究领域。

因为这一领域是如此的丰富，再加上取得并评价进展所存在的困难性，我试图在我的学生和本书的读者中灌输这一学科，它是建立在来自于工程、科学和统计学的原理基础上的。

解决问题的工程方法是首先仔细地定义要解决的整个问题，并质疑这一过程中的基本假设和固有目标。一旦完成这一过程，我们要实现几种不同的解决方案或方法并仔细地进行测试，关注于包括可靠性和计算代价在内的事项。最后，采纳一种或多种解决方案并在实际世界设定中评测。因为这个原因，本书含有多种不同的解决视觉问题的可选方案，其中的很多描绘在习题中供学生自己来实现并测试。

科学方法建立在对物理原理的基本理解基础之上。对于计算机视觉而言，这包括关于人造结构和自然结构的物理学，关于图像形成的物理学，包括照明和大气效应、光学和带噪声的传感器。我们的任务是这个形成过程的逆，即使用稳定而有效率的算法来获得对场景和其他感兴趣的量的可靠描述。科学方法也鼓励我们形成假设并对其进行测试，这与工程学科中固有的广泛测试和评估相似。

最后，因为太多的图像形成过程是内在地不确定性的和含糊不清的，对世界中的不确定性(例如，图片中动物的数量和类型)和图像形成过程中的噪声都进行建模的统计方法时常是必要的。贝叶斯推断方法可用于将先验和测量模型结合起来估计未知量并对其不确定性建模。机器学习方法可用于创建最初的概率模型。在这一过程中，诸如动态规划、图割和信度传播等高效的学习和推断算法往往起着关键作用。

考虑到在本书中我们涵盖了很广的材料，将来我们可能会看到哪些新的发展呢？正如我们以前曾提到的，计算机视觉领域的近期趋势之一是，使用因特网上的大规模部分标注了的视觉数据作为学习场景和物体的视觉模型的来源。我们已经看到数据驱动的方法在相关领域所取得的成功，包括语音识别、机器翻译、语音和音乐合成以及甚至计算机图形学(在基于图像的绘制和从运动捕捉来的动画制作两个方面)。类似的过程已经出现在计算机视觉中，其中最令人兴奋的新工作出现在物体识别和机器学习领域的交叉处。

在计算机视觉中更为传统的定量化方法，比如运动估计、立体视觉对应和图像增强，都受益于更好的关于图像、运动和视差的先验模型，也受益于高效的统计推断方法，比如用于不均匀和更高阶马尔科夫随机场的那些推断方法。有些方法，比如特征匹配和由运动到结构，已经成熟到可用于几乎任意的静态场景图像的汇集。这使得从因特网数据库来做 3D 建模的工作成爆炸性增长，这同样也是与从大规模数据来做视觉识别的工作相关的。

尽管这些都是令人鼓舞的进展，但是，在计算机和人之间的对语义场景理解的性能差距还十分巨大。在计算机对照片中的所有物体进行命名并给出其轮廓能够达到两岁大的孩子的技能水平之前，恐怕还有很多年。但是，我们必须记住人的技能常常是多年的训练和熟悉的结果，且通常在特殊的在生态上具重要价值的情况下工作得最出色。例如，尽管人在识别人脸方面看起来好像是专家，但当给我们看并不十分认识的人时我们实际的识别性能并不那么好。将视觉算法与对真实世界进行推理的一般的推断方法相结合，可能会带来更多的突破，尽管某些问题就人的经验和智能的完全仿真可能是必要的意义而言，可能会变为"人工智能-完全问题(AI-complete)"，

无论这些研究努力的结果如何，计算机视觉已经在很多领域产生了巨大影响，包括数字摄影学、视觉特效、医学成像、保安和监控以及基于网页的搜索。这个领域所固有的问题和方法的广度，与数学的丰富度和产生的算法的效用一道，确保其在未来的很长一段时间里仍将继续成为一个令人兴奋的研究领域。

附录 A
线性代数与数值方法

A.1 矩阵分解
A.2 线性最小二乘
A.3 非线性最小二乘
A.4 直接稀疏矩阵方法
A.5 迭代方法

在附录 A 中，我们将介绍线性代数和数值方法中的一些基本内容，本书的其他部分将会使用这些知识。首先，我们从矩阵代数中的一些基本分解开始，包括奇异值分解(SVD)、特征值分解及其他的矩阵(因子)分解。然后，我们查看线性最小二乘问题，它可以通过 QR 分解或正则方程(normal equation)解决。紧随其后的便是非线性最小二乘问题，如果测量方程对未知量是非线性的，或者使用了鲁棒误差函数，这种问题就会产生。这种问题需要用迭代法求解。接下来，我们会介绍稀疏问题的直接求解(因子分解)方法，其中变量的顺序将对计算和内存需求产生很大的影响。最后我们将讨论求解大规模线性(或线性化的)最小二乘问题的迭代方法。涵盖这些问题中绝大部分的好的一般性参考资料包括 Björck(1996)，Golub and Van Loan(1996)，Trefethen and Bau(1997)，Meyer(2000)，Nocedal and Wright(2006)以及 Björck and Dahlquist(2010)等人的工作。

向量和矩阵索引注释 为了与本书的其他部分一致，也为了与计算机科学及计算机视觉领域内的通常用法一致，对于向量和矩阵元素的索引，我采用了从 0 开始的索引模式。请注意，几乎所有数学参考书和论文使用的都是从 1 开始的索引模式，因此，当你阅读本书时，需要注意这些区别。

软件实现 本附录描述的所有算法的代码库是可以获得的，这些代码库都经过高度优化和详尽测试，附录 C.2 给出了它们的列表。

A.1 矩阵分解

为了更好地理解矩阵结构，更稳定地进行诸如求逆、系统求解等运算，可以使用许多矩阵(因子)分解方法。在这一节中，我们将回顾奇异值分解(SVD)、特征值分解，QR 因子分解和乔里斯基(Cholesky)因子分解。

A.1.1 奇异值分解

矩阵代数中一个最常用的分解就是奇异值分解(SVD)，即任意一个 $M \times N$ 的实数值矩阵 A 均可写成

$$A_{m \times n} = U_{m \times p} \Sigma_{p \times p} V_{p \times n}^T$$
$$= \begin{bmatrix} u_0 & \cdots & u_{p-1} \end{bmatrix} \begin{bmatrix} \sigma_0 & & \\ & \ddots & \\ & & \sigma_{p-1} \end{bmatrix} \begin{bmatrix} v_0^T \\ \cdots \\ v_{p-1}^T \end{bmatrix}, \quad (A.1)$$

其中，$p = \min(m,n)$，矩阵 U 和 V 是正交矩阵，即有 $U^T U = I$ 和 $V^T V = I$，同样，它们的列向量满足

$$u_i \cdot u_j = v_i \cdot v_j = \delta_{ij}. \quad (A.2)$$

所有的奇异值均是非负的，并可以按降序排列

$$\sigma_0 \geq \sigma_1 \geq \cdots \geq \sigma_{p-1} \geq 0. \tag{A.3}$$

关于矩阵 A 的 SVD 的一个几何直观解释可通过对(A.2)中的公式 $A = U\Sigma V^T$ 进行重写获得:

$$AV = U\Sigma \quad \text{or} \quad Av_j = \sigma_j u_j. \tag{A.4}$$

这个公式表明,当矩阵 A 作用于任何基向量 v_j 时,会把 v_j 变换到 u_j 的方向,同时将 v_j 的长度变为 σ_j,如图 A.1 所示。

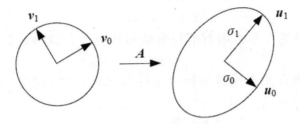

图 A.1 矩阵 A 的作用可以按如下方式进行可视化表示,考虑由一组正交向量 v_j 转换成的域,在每一个正交向量被变换成一个长度为 σ_j 的新的正交向量 u_j 后的形状变化。如果 A 表示的是一个协方差矩阵,并进行了特征值分解,那么每一个由 u_j 代表的轴均指向了一个主方向(分量),同时 σ_j 表示了沿着这个方向的标准差

如果仅有前 r 个奇异值是正的,那么矩阵 A 的秩即为 r,同时 SVD 分解中的下标 p 也可被 r 取代。(也就是说,我们可以丢弃 U 和 V 的最后 $p-r$ 列。)

矩阵奇异值分解的一个重要性质(这个性质同样适用于实对称非负定矩阵的特征值分解)是,如果我们对下面的展开式进行截尾计算,

$$A = \sum_{j=0}^{t} \sigma_j u_j v_j^T, \tag{A.5}$$

我们将获得原始矩阵 A 在最小二乘意义下的最佳逼近。这一性质被用于基于特征脸的人脸识别系统(14.2.1 节)和卷积核的可分性估计(3.21)。

A.1.2 特征值分解

如果矩阵 C 是对称矩阵 $(m = n)$[①],那么它可以写成特征值分解的形式:

$$
\begin{aligned}
C &= U\Lambda U^T = \begin{bmatrix} u_0 & | & \cdots & | & u_{n-1} \end{bmatrix} \begin{bmatrix} \lambda_0 & & \\ & \ddots & \\ & & \lambda_{n-1} \end{bmatrix} \begin{bmatrix} u_0^T \\ \cdots \\ u_{n-1}^T \end{bmatrix} \\
&= \sum_{i=0}^{n-1} \lambda_i u_i u_i^T.
\end{aligned} \tag{A.6}
$$

(特征矩阵 U 有时会记为 Φ,而特征向量 u 则记做 ϕ。)在这里,特征值

$$\lambda_0 \geq \lambda_1 \geq \cdots \geq \lambda_{n-1} \tag{A.7}$$

[①] 在这一附录中,我们用 C 表示对称矩阵,用 A 表示一般的矩阵。

可以为正，也可以为负[①]。

对称矩阵 C 的一个特殊情况是，它可以通过一系列外积之和来进行构造
$$C = \sum_i a_i a_i^T = AA^T, \tag{A.8}$$

求解最小二乘问题时，这种情况经常发生(附录 A.2)，这里的 A 是由所有的 a_i 列向量一列一列堆叠而成的。在这种情况下，我们可以保证所有的特征值 λ_i 都是非负的。由此产生的矩阵 C 也是半正定的
$$x^T C x \geq 0, \ \forall x. \tag{A.9}$$
如果矩阵 C 是满秩的，那么所有的特征值将均为正，这种矩阵被称作对称正定矩阵(SPD)。

对称半正定矩阵在数据统计分析中也经常出现，它们表示一组数据点 $\{x_i\}$ 围绕其中心 $\bar{x}$ 产生的协方差
$$C = \frac{1}{n} \sum_i (x_i - \bar{x})(x_i - \bar{x})^T. \tag{A.10}$$

此时，进行特征值分解就是通常所说的主分量分析(PCA)，因为它完成了对数据点分布在其中心周围变化的主方向(和幅度)的建模，正如在 5.1.1 节(5.13～5.15)，14.2.1 节(14.9)以及附录 B.1(B.10)中所展示的那样。图 A.1 表明了协方差矩阵 C 的主分量是如何指出表示数据点分布不确定性的椭圆域的主轴的，同时也说明了 $\sigma_j = \sqrt{\lambda_i}$ 是如何表示沿每一主轴的标准差的。

矩阵 C 的特征值和特征向量与 A 的奇异值和奇异向量有着紧密的联系。若有
$$A = U \Sigma V^T, \tag{A.11}$$
则有
$$C = AA^T = U \Sigma V^T V \Sigma U^T = U \Lambda U^T. \tag{A.12}$$
由此我们可以得到 $\lambda_i = \sigma_i^2$。同时，我们也可知道矩阵 A 的左奇异向量就是矩阵 C 的特征向量。

这一关系给出了一种计算大规模非满秩矩阵特征值分解的有效方法，在计算特征脸(14.2.1 节)时得到的散度矩阵就是这样一种矩阵。注意，(14.9)中的协方差矩阵 C 实际上就是(A.8)中的矩阵 C。同时也应注意到每个相对于均值的差分图像 $a_i = x_i - \bar{x}$ 就是长度为 P(即图像的像素数)的长向量，而样本的总数 N(训练数据库中的人脸数)则相对小得多。因此，为了避免计算 $C = AA^T$ (规模为 $P \times P$)，我们换而计算
$$\hat{C} = A^T A, \tag{A.13}$$
此时矩阵的规模为 $N \times N$。(这实际上是计算每对差分图像 a_i 和 a_j 的点积。)而矩阵 $\hat{C}$ 的特征值正是 A 的奇异值的平方，即 Σ^2，因此也是矩阵 C 的特征值。矩阵 $\hat{C}$ 的特征向量正是矩阵 A 的右奇异向量 V，需要的特征脸向量 U，同时也是矩阵 A 的左特征向量，可以根据 V 按如下公式算出：

[①] 非对称矩阵也可进行特征值分解，但此时特征值和特征向量中可能会出现复数项。

$$U = A V \Sigma^{-1}. \tag{A.14}$$

最后这一步实际上是将特征脸作为差分图像的线性组合进行计算。如果你拥有高质量的线性代数包，如 LAPACK，计算诸如 A 的矩形阵的一小部分左奇异向量和奇异值的例程通常是已提供的(附录 C.2)。然而如果将所有图像都储存在内存中的代价过于高昂，我们可以用构造(A.13)中的 $\hat{C}$ 来代替这种做法。

那么，特征值分解和奇异值分解是如何计算的呢？注意，特征向量的定义由下面的方程给出：

$$\lambda_i \boldsymbol{u}_i = \boldsymbol{C} \boldsymbol{u}_i \quad \text{or} \quad (\lambda_i \boldsymbol{I} - \boldsymbol{C}) \boldsymbol{u}_i = 0. \tag{A.15}$$

(此方程可以通过在式(A.6)的左右两边同时右乘 $\boldsymbol{u}_i$ 得到。)由于(A.15)中后面的方程是齐次的，即方程的右端为零，所以只有系统的秩不满，即

$$|(\lambda \boldsymbol{I} - \boldsymbol{C})| = 0. \tag{A.16}$$

这个方程才会有非零(非平凡)解 $\boldsymbol{u}_i$。解这个判定条件时会产生关于 λ 的特征多项式方程。如果问题的规模很小，如 2×2 或 3×3 的矩阵，这个方程是可以有固定解法的。

如果矩阵的规模较大，我们则经常使用一种迭代算法。这种算法首先使用正交变换将矩阵 $\boldsymbol{C}$ 简化为实对称三角阵，然后再进行 QR 迭代求解(Golub and Van Loan 1996; Trefethen and Bau 1997; Björck and Dahlquist 2010)。由于这些方法都相当复杂，所以最好使用一些线性代数工具包如 LAPACK(Anderson，Bai，Bischof et al. 1999)求解这些问题(参见附录 C.2)。

对有缺失数据项的矩阵进行因子分解会用到各种不同的迭代算法，这经常涉及对缺失项进行假设或最小化一些加权的重构度量，因此相较于一般的因式分解，它也更具有挑战性。这一领域在计算机视觉中得到广泛研究(Shum，Ikeuchi，and Reddy 1995; De la Torre and Black 2003; Huynh，Hartley，and Heyden 2003; Buchanan and Fitzgibbon 2005; Gross，Matthews，and Baker 2006; Torresani，Hertzmann，and Bregler 2008)，它有时称作"广义"PCA。但需要注意的是，有时这一术语也用来描述代数子空间中的聚类方法，这也是 Vidal, Ma, and Sastry (2010) 最新专著中的主题之一。

A.1.3 QR 因子分解

QR 因子分解是一项广泛应用于稳定求解病态最小二乘问题的方法(Björck 1996)，同时它也是一些更复杂的算法的基础，比如计算 SVD 及特征值分解：

$$\boldsymbol{A} = \boldsymbol{Q} \boldsymbol{R}, \tag{A.17}$$

其中 $\boldsymbol{Q}$ 是正交矩阵(或酉矩阵)，$\boldsymbol{Q} \boldsymbol{Q}^T = \boldsymbol{I}$，$\boldsymbol{R}$ 是上三角矩阵[①]。在计算机视觉中，QR 因子分解可以用于将摄像机矩阵转换为旋转矩阵和一个上三角的标定矩阵(6.35)，也可以用于各种自标定算法(7.2.2 节)。一些计算 QR 分解的常用算法，如改进的 Gram-Schmidt 算法、Householder 变换算法，已经被 Golub and Van Loan

---

[①] 记号 "R" 源自于 LU(上三角和下三角，lower-upper)分解的德文翻译，即 LR，全称为 "links" 和 "rechts" (左三角和右三角)。

(1996)、Trefethen and Bau(1997)以及 Björck and Dahlquist(2010)很好地描述过了，这些算法当然也可以在 LAPACK 中找到。与 SVD 和特征值分解不同，QR 因子分解并不需要迭代计算，它可以严格地在 $O(MN^2+N^3)$ 次运算内计算出来，其中 M 是矩阵的行数，N 是矩阵的列数(适用于行数大于列数的情况)。

A.1.4 乔里斯基分解

乔里斯基(Cholesky)因子分解可以用于将任意的对称正定矩阵 C 转换成上三角矩阵或下三角矩阵的乘积，其中的上三角矩阵和下三角矩阵是对称的

$$C = LL^T = R^T R, \tag{A.18}$$

其中的 L 是下三角矩阵，R 是上三角矩阵。不同于高斯消元法需要进行主元素的选取(行与列的重排序)，及有时可能变得不稳定(对舍入误差或行与列的顺序过于敏感)，乔里斯基因子分解对正定矩阵——例如那些出现在最小二乘问题的正则方程中的矩阵(附录 A.2)——都是稳定的。鉴于公式(A.18)的形式，矩阵 L 和 R 有时也被称作"矩阵的平方根"[①]。

计算矩阵的上三角乔里斯基分解的算法是对高斯消元法的一个直观的推广，这种推广具有对称性。它基于如下分解(Björck 1996; Golub and Van Loan 1996)

$$C = \begin{bmatrix} \gamma & c^T \\ c & C_{11} \end{bmatrix} \tag{A.19}$$

$$= \begin{bmatrix} \gamma^{1/2} & 0^T \\ c\gamma^{-1/2} & I \end{bmatrix} \begin{bmatrix} 1 & 0^T \\ 0 & C_{11} - c\gamma^{-1}c^T \end{bmatrix} \begin{bmatrix} \gamma^{1/2} & \gamma^{-1/2}c^T \\ 0 & I \end{bmatrix} \tag{A.20}$$

$$= R_0^T C_1 R_0, \tag{A.21}$$

经过迭代，上式可变为

$$C = R_0^T \ldots R_{n-1}^T R_{n-1} \ldots R_0 = R^T R. \tag{A.22}$$

算法 A.1 给出了一个更加过程化的定义，如果需要，我们可以将上三角矩阵 R 存储在矩阵 C 所在的内存空间中。对稠密矩阵而言，乔里斯基因子分解的总操作数为 $O(N^2)$，但对元素较少的稀疏矩阵来说(附录 A.4)，总操作数会大大降低。

---

procedure *Cholesky*(C, R):

$R = C$

for $i = 0 \ldots n-1$

    for $j = i+1 \ldots n-1$

        $R_{j,j:n-1} = R_{j,j:n-1} - r_{ij} r_{ii}^{-1} R_{i,j:n-1}$

    $R_{i,i:n-1} = r_{ii}^{-1/2} R_{i,i:n-1}$

算法 A.1 应用乔里斯基分解将矩阵 C 分解为它的上三角形式 R

[①] 实际上，一个矩阵平方根的族是存在的。任何具有 LQ 或 QR 形式的矩阵，其中 Q 为酉矩阵，都是 C 的平方根。

注意，乔里斯基分解也可应用于块状矩阵，此时(A.19)中的γ项是一个正方形的块状子矩阵，而 c 则是一个长方形的矩阵(Golub and Van Loan 1996)。如果要避免计算平方根，我们可以将γ留在(A.20)中位于中间的矩阵的对角线上，这就是 $C = LDL^T$ 因子分解，其中 D 为对角矩阵。但由于现在的计算机计算平方根相对较快，所以没有必要进行这种不必要的变形。因此，在多数情况下，我们任然使用乔里斯基分解。

A.2 线性最小二乘

线性最小二乘问题在计算机视觉中是无处不在的。例如，在基于特征点匹配的图像对齐中，我们就会最小化一个平方距离目标函数(6.2)，

$$E_{\text{LS}} = \sum_i \|r_i\|^2 = \sum_i \|f(x_i; p) - x_i'\|^2, \quad (\text{A.23})$$

其中

$$r_i = x_i' - f(x_i; p)\, r_i = f(x_i; p) - x_i' = \hat{x}_i' - \tilde{x}_i' \quad (\text{A.24})$$

是测量得到的位置 $\hat{x}_i'$ 和它对应的当前预测位置 $\tilde{x}_i' = f(x_i; p)$ 之间的残差。另外还有一些更加复杂的最小二乘问题，例如大规模的由运动到结构问题(7.4 节)，则可能会涉及最小化有上千个变量的函数。甚至像图像滤波(3.4.3 节)、正则化(3.7.1 节)之类的问题都可能牵扯到最小化平方误差的和。

图 A.2a 显示了一个简单的基于最小二乘的直线拟合问题的例子，其中要估计的参数是直线方程的参数 (m,b)。如果我们认为采样得到的点的纵坐标值 y_i 是直线 $y = mx + b$ 上点的纵坐标值加上一个随机噪声的结果，那么 (m,b) 的最佳估计就可以通过最小化纵坐标的平方误差得到，

$$E_{\text{VLS}} = \sum_i |y_i - (mx_i + b)|^2. \quad (\text{A.25})$$

需要注意的是，使用线性最小二乘进行拟合的函数本身并不需要是线性的。只要函数相对未知参数而言是线性的就足够了。例如，多项式拟合可以写成

$$E_{\text{PLS}} = \sum_i |y_i - (\sum_{j=0}^{p} a_j x_i^j)|^2, \quad (\text{A.26})$$

而关于未知幅值 A 和相位 ϕ(频率 f 已知)的正弦拟合则可表示为

$$E_{\text{SLS}} = \sum_i |y_i - A\sin(2\pi f x_i + \phi)|^2 = \sum_i |y_i - (B\sin 2\pi f x_i + C\cos 2\pi f x_i)|^2, \quad (\text{A.27})$$

它对 (B,C) 而言是线性的。

更一般的情况是，我们通常用 x 表示未知参数，进而线性最小二乘问题的一般形式可写成[1]

---

[1] 这里变量名的解释需要格外注意。在 2D 直线拟合问题中，x 一般用来表示横轴，但在一般的最小二乘问题中，$x = (m,b)$ 表示未知参数向量。

$$E_{\text{LLS}} = \sum_i |a_i x - b_i|^2 = \|Ax - b\|^2. \tag{A.28}$$

将上面的方程展开，我们可以得到

$$E_{\text{LLS}} = x^T(A^T A)x - 2x^T(A^T b) + \|b\|^2, \tag{A.29}$$

我们可以通过求解相应的**正则方程**来求得 x 的最小值(Björck 1996; Golub and Van Loan 1996)：

$$(A^T A)x = A^T b. \tag{A.30}$$

正则方程常见的解法是使用乔里斯基分解。令

$$C = A^T A = R^T R, \tag{A.31}$$

其中 R 是 Hessian 矩阵 C 的上三角乔里斯基因子。又有

$$d = A^T b. \tag{A.32}$$

因此，经过因子分解后，x 可以通过解下面的方程获得

$$R^T z = d, \quad Rx = z, \tag{A.33}$$

这只需求解两个三角系统，即一系列的向前和向后代换计算(Björck 1996)。

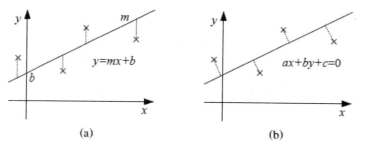

图 A.2 最小二乘回归。(a)直线 $y = mx + b$ 是四个有噪声的数据点 $\{(x_i, y_i)\}$（在图上以×标出）的拟合结果，它通过最小化数据点和直线上点的纵坐标的平方误差 $\sum_i \|y_i - (mx_i + b)\|^2$ 得到。(b)如果我们认为观测到的点 $\{(x_i, y_i)\}$ 在各个方向上都有随机噪声，那么在应用最小二乘时，我们应该最小化数据点到直线的垂直距离的平方和 $\sum_i \|ax_i + by_i + c\|^2$

如果一个最小二乘问题在数值上是病态的(一般来讲，这种情况可以通过引入足够的正则化条件或关于参数的先验知识来避免(附录 A.3))，那么我们可以直接对矩阵 A 进行 QR 因子分解或 SVD (Björck 1996; Golub and Van Loan 1996; Trefethen and Bau 1997; Nocedal and Wright 2006; Björck and Dahlquist 2010)，即

$$Ax = QRx = b \quad \longrightarrow \quad Rx = Q^T b. \tag{A.34}$$

需要注意的是，对矩阵 $C = A^T A$ 进行乔里斯基分解时产生的上三角矩阵 R 和这里对 A 进行 QR 因子分解时产生的上三角矩阵 R 是一样的，但求解方程(A.34)通常会更加稳定(对舍入误差更加不敏感)，同时速度也更慢(会慢上一个常数倍)。

全部最小二乘

在一些问题(如在 2D 图像中进行几何曲线拟合或在三维空间中用平面拟合一组

数据点云)中，测量误差并不是只在某一个特定的方向上存在，测量得到的点在各个方向上都具有一定的不确定度，这种情况又被称为"变量含误差模型"(errors-in-variables model) (Van Huffel and Lemmerling 2002; Matei and Meer 2006)。此时，最小化一组具有如下形式的齐次平方误差就会更有意义，

$$E_{\text{TLS}} = \sum_i (a_i x)^2 = \|Ax\|^2, \tag{A.35}$$

它被称为"全部最小二乘"(TLS) (Van Huffel and Vandewalle 1991; Björck 1996; Golub and Van Loan 1996; Van Huffel and Lemmerling 2002)。

上面的误差度量函数有一个平凡的最小值解 $x = 0$，同时它对 x 而言也的确是齐次的。正因为此，在这一最小化问题中，我们要求 x 必须满足 $\|x\|^2 = 1$。于是我们得到了下面的特征值问题：

$$x = \arg\min_x x^T(A^T A)x \quad \text{such that} \quad \|x\|^2 = 1. \tag{A.36}$$

可以最小化这个约束问题的 x 就是与 $A^T A$ 最小的特征值对应的特征向量。它和 A 的最末一个右特征向量是相同的。这是因为

$$A = U\Sigma V^T, \tag{A.37}$$

$$A^T A = V\Sigma^2 V^T, \tag{A.38}$$

$$A^T A v_k = \sigma_k^2 \cdot v_k, \tag{A.39}$$

通过选择最小的 σ_k 值，我们就可以达到最小化这个约束问题的目的。

图 A.2b 展示了一个直线拟合问题，其中测量误差被认为在 (x, y) 上是有各向同性的。此时最佳的直线方程 $ax + by + c = 0$ 可通过最小化下面的函数得到：

$$E_{\text{TLS-2D}} = \sum_i (ax_i + by_i + c)^2, \tag{A.40}$$

即找出与下面矩阵最小特征值对应的特征向量[①]

$$C = A^T A = \sum_i \begin{bmatrix} x_i \\ y_i \\ 1 \end{bmatrix} \begin{bmatrix} x_i & y_i & 1 \end{bmatrix}. \tag{A.41}$$

但要注意，只有 a_i ——类似于上例中的测量结果 $(x_i, y_i, 1)$ ——中所有测量项都具有相同的噪声，最小化(A.35)中的 $\sum_i (a_i x)^2$ 才得到统计最优的结果(附录 B.1.1)。显然，这个条件在图 A.2b 中直线拟合的例子(A.40)里是不成立的，因为 1 这个值没有噪声。为了满足这个条件，我们首先从所有测量得到的点中减去 x 的平均值和 y 的平均值，

$$\hat{x}_i = x_i - \bar{x} \tag{A.42}$$

$$\hat{y}_i = y_i - \bar{y} \tag{A.43}$$

然后通过最小化

$$E_{\text{TLS-2Dm}} = \sum_i (a\hat{x}_i + b\hat{y}_i)^2. \tag{A.44}$$

来拟合 2D 直线方程 $a(x - \bar{x}) + b(y - \bar{y}) = 0$。

[①] 同样，我们需要注意这里的变量名。测量方程是 $a_i = (x_i, y_i, 1)$，而未知参数为 $x = (a, b, c)$。

更一般的情况是每个独立的测量分量中含有噪声的情况各不相同，在估计本质矩阵和基本矩阵(7.2 节)时就如此。这种情况被称为"异方差的变量含噪声模型"(heteroscedastic errors-in-variable，HEIV)。Matei and Meer (2006)对它有详细的讨论。

A.3 非线性最小二乘

在很多视觉问题中，如由运动到结构问题，形如(A.23)的最小二乘问题中的函数 $f(x_i; p)$ 对未知参数 p 而言并不是线性的。这种问题称为"非线性最小二乘问题或非线性回归"(Björck 1996; Madsen, Nielsen, and Tingleff 2004; Nocedal and Wright 2006)。求解这种问题通常需要使用梯度导数(雅克比矩阵) $J = \partial f / \partial p$ 在当前估计的参数 p 附近对(A.23)迭代地进行线性化，并计算出一个增量改进 Δp。

如方程(6.13~6.17)，此时有

$$E_{\text{NLS}}(\Delta p) = \sum_i \|f(x_i; p + \Delta p) - x'_i\|^2 \tag{A.45}$$

$$\approx \sum_i \|J(x_i; p)\Delta p - r_i\|^2, \tag{A.46}$$

在构造正则方程时，这里的雅克比矩阵 $J(x_i; p)$ 和残差向量 r_i 与(A.28)中的 a_i 和 b_i 的作用相同。

由于上面的近似仅在局部极小值附近或在 Δp 很小时成立，因此 $p \leftarrow p + \Delta p$ 这种更新并不是总能减少(A.45)中累加来的平方残差。一种缓解这个问题的方法是使用更小的步长，

$$p \leftarrow p + \alpha \Delta p, \quad 0 < \alpha \leq 1. \tag{A.47}$$

一种确定 α 的合理值的方法是从 1 开始连续的将值减半，这其实是直线搜索的一个简单形式(Al-Baali and Fletcher. 1986; Björck 1996; Nocedal and Wright 2006)。

另一种确保误差逐步下降的方法是将对角阻尼项加到近似 Hessian 矩阵

$$C = \sum_i J^T(x_i) J(x_i), \tag{A.48}$$

上，即求解方程

$$[C + \lambda \operatorname{diag}(C)]\Delta p = d, \tag{A.49}$$

其中

$$d = \sum_i J^T(x_i) r_i, \tag{A.50}$$

这种方法叫"带阻尼的高斯-牛顿法"(damped Gauss-Newton method)。如果平方残差下降得不如预期的(例如可以通过(A.46)进行预测)快，那么阻尼参数 λ 就要增大。而如果达到预期的下降效果，那么阻尼系数就要减小(Madsen, Nielsen, and Tingleff 2004)。这种将牛顿逼近(A.46)(一阶泰勒系数)与自适应阻尼参数 λ 结合的方法通常称为"Levenberg–Marquardt 算法"(Levenberg 1944; Marquardt 1963)，它也是更加通用的信任区域方法(trust region method)的一个例子。有关信任区域方法

的更详细的讨论可参见(Björck 1996; Conn，Gould，and Toint 2000; Madsen，Nielsen，and Tingleff 2004; Nocedal and Wright 2006)。

如果最初的解距离局部最小值周围的二次收敛域较远，大残差方法——如 Newton-type 方法——可能会收敛得更快。这种方法在(A.46)的泰勒系数展开中加入了二阶项。如果内存空间有问题，Quasi-Newton 方法——如 BFGS——会比较有用。这种方法只需要估计梯度。这些方法在有关数值优化的教科书和论文中都有一定的论述(Toint 1987; Björck 1996; Conn，Gould，and Toint 2000; Nocedal and Wright 2006)。

A.4　直接稀疏矩阵方法

与计算机视觉有关的很多优化问题——如光速平差法(Szeliski and Kang 1994; Triggs，McLauchlan，Hartley et al. 1999; Hartley and Zisserman 2004; Snavely，Seitz，and Szeliski 2008b; Agarwal，Snavely，Simon et al. 2009)——都会涉及极度稀疏的雅克比矩阵和(近似的)Hessian 矩阵(7.4.1 节)。例如，图 7.9a 中显示的由运动到结构问题中十分典型的二分模型中，大多数点都仅仅被所有相机的一个子集观察到，这就使得图 7.9b 和 c 中的雅克比矩阵和 Hession 矩阵具有比较稀疏的模式。

只要 Hessian 矩阵足够稀疏，使用稀疏的乔里斯基因子分解就会比使用一般的乔里斯基因子分解效率更高。在有关稀疏矩阵直接解法的一些方法中，Hessian 矩阵 C 和它对应的乔里斯基因子 R 都以一种压缩的形式存储，此时存储的总量与(潜在的)非零项的数目成正比(Björck 1996; Davis 2006)[①]。从稀疏的雅克比矩阵 J 计算 C 和 R 的非零项的算法由 Björck (1996，Secton 6.4)给出，而计算乔里斯基和 QR 分解的数值算法也由 Björck (1996，Secton 6.5)进行了一些讨论。

变量重排

高效地使用直接(非迭代)解法求解稀疏矩阵问题的关键是为变量确定一个合理的顺序，这可以减少填充的数量。所谓填充，即指 R 中的一些非零项，这些非零项在原始矩阵 C 中的对应项为零。我们已经在 7.4.1 节中看到，将数量更多的 3D 点的参数存储在相机参数之前，加上舒尔补(Schur complement，7.56)运算是如何构成一个高效的算法的。类似，在基于视频的重建问题中，将参数按照时间排序通常会产生更少的填充。此外，对任何一个问题而言，如果它的邻接图(与稀疏模式对应的图)是一棵树，那么通过对变量进行适当的重排序(将所有的子节点置于其父节

[①] 例如，你可以存储一个 (i, j, c_{ij}) 三元组的列表。这种模式的一个例子就是压缩稀疏行(compressed spares row, CPR)存储。另外一种存储方法称为 skyline，它存储的是相邻非零元素的竖直区间(Bathe, 2007)，有时它会被用于有限元分析。诸如蛇行(5.3)一类的带状系统可以仅存储非零的条带元素(Björck 1996, Section 6.2)，并可以在 $O(nb^2)$ 的复杂度内进行求解，其中 n 为变量的个数，b 为带宽。

点之前)，这个问题就肯定可以在线性时间内解决。这些都是很好的重排序方法的例子。

对一般的无结构数据而言，仍有很多启发信息可以帮助我们找到一个很好的重排序(Björck 1996; Davis 2006)[①]。对一般的邻接(稀疏)图而言，最小维数排序通常会产生不错的效果。对经常在图像或样条网格(8.3 节)中出现平面图而言，嵌套分解——即沿小尺寸前沿(或边界)将图递归地分成两个相等的部分——通常工作得很好。这种域分解(或多前沿)方法也使并行处理变得可能，因为这些独立的子图可以并行地由不同的处理器进行处理(Davis 2008)。

算法 A.2 总结了用直接解法求解稀疏最小二乘问题的全部步骤，它其实是由 Björck(1996，6.6 节)提出的算法 6.6.1 的一个修改版本。如果正在求解的是一系列相关的最小二乘问题——迭代的非线性最小二乘问题(附录 A.3)就是这种情况——步骤 1~3 可以提前执行，并在每次用不同的 C 和 d 调用该算法时直接进行重用。如果求解的问题具有块状结构(在由运动到结构问题中，每个点(结构)变量在 C 中都对应一个 3×3 子块，每个相机在 C 中都对应一个 6×6(或更大)的子块)，那么相对于实际的数值因子分解计算，计算重排序的代价是很小的。这得益于块状矩阵的操作(Golub and Van Loan 1996)。当然，我们也可以对 QR 因子分解施用稀疏重排和多前沿方法(Davis 2008)，当最小二乘问题是病态的时候，这种方法可能会受到欢迎。

---

procedure: SparseCholeskySolve(C,d)

1. 符号化地给出 C 的结构，如邻接图。
2. (可选)计算变量的重排，此时需要考虑该问题可能含有的全部块状结构。
3. 确定 R 的填充模式，并为 R 分配压缩存储所需的空间，同时也要为经过重排序的右操作数 $\hat{d}$ 分配存储空间。
4. 将 C 和 d 中的元素根据计算出的重排结果进行排序，并将结果复制到 R 和 $\hat{d}$ 中。
5. 使用算法 A.1 对 R 进行数值因子分解。
6. 求解经过因子分解的系统(A.33)，即，
$$R^T z = \hat{d}, Rx = z.$$
7. 撤销重排后，将结果 x 返回。

算法 A.2 通过对矩阵 C 进行稀疏乔里斯基分解求解稀疏最小二乘问题

A.5 迭 代 方 法

当问题的规模很大时，无论是存储 Hessian 矩阵 C 及其因子 R 所需的存储代价，还是计算因子分解时所需的时间代价，都是十分高昂的。特别是有很多填充

[①] 寻找拥有最小填充的最佳重排是一个可被证明的 NP-难问题。

时，这种情况将变得更加明显。但是在像素网格基础上的图像处理问题却时常要面对这些问题，因为，即使我们对问题使用了最佳的重排序(嵌套分解)，填充的数量仍然很大。

求解这种线性系统一种更好的方法就是使用迭代方法，它通过某些方法——如在类似(A.29)的能量函数中选择一系列的下山步长——计算出一个估计值的序列，进而逼近最终结果。

近年来，人们已经发展了很多迭代方法，其中包括一些著名的算法，如连续超松弛算法及多重网格算法。这些算法在专门介绍迭代方法的教科书(Axelsson 1996; Saad 2003)中都有介绍，一些关于数值线性代数和最小二乘方法的更通用的教科书(Björck 1996;Golub and Van Loan 1996; Trefethen and Bau 1997; Nocedal and Wright 2006; Björck and Dahlquist 2010)中也可以见到它们的身影。

A.5.1 共轭梯度

通常，在迭代方法中，性能最好的就是共轭梯度下降法。它选择一系列的相互 C-共轭的下降步长，即 u 和 v，它们都指向下降方向并满足 $u^T C v = 0$。在实际应用中，共轭梯度下降法通常优于其他梯度下降法，因为它的收敛率与 C 的条件数的平方根成正比，而不是与 C 的条件数成正比[①]。在这个问题上，Shewchuk (1994)给出了一个很好的介绍，其中包括对共轭梯度算法背后的原理的一个清晰而直观的解释，共轭梯度法的效率也是其中一个主题。

算法 A.3 描述了共轭梯度算法和与之对应的最小二乘形式，如果原始的最小二乘线性方程组有类似于 $Ax = b$(A.28)的形式，那么我们就可以使用这个算法。虽然你很容易确定这两种形式在数学上是等价的，但如果由于矩阵的病态性质，舍入误差已经开始对结果造成影响，我们会更倾向于使用共轭梯度算法的最小二乘形式。另外，如果由于 A 的稀疏结构，相较于和矩阵 C 进行乘法操作，和原始矩阵 A 进行乘法操作在时间和空间上都更加高效，那么共轭梯度算法的最小二乘形式也是一种更好的选择。

共轭梯度算法首先计算当前的残差 $r_0 = d - Cx_0$，这也是能量函数(A.28)最陡峭的下降方向。我们将其作为初始下降方向 $p_0 = r_0$。接下来，我们将这个下降方向与具有二次型的(Hessian)矩阵 C 相乘，并将结果与当前残差进行组合来估计最佳步长 α_k。然后我们用这个步长去更新解向量 x_k 和残差向量 r_k(在步骤 4~8 中，我们需要注意共轭梯度算法的最小二乘版本是如何利用 $C = A^T A$ 来分解矩阵乘法的)。最后，我们需要计算新的搜索方向。首先我们通过计算当前残差和之前残差的幅度比得到因子 β，然后设置新的搜索方向 p_{k+1}，被设置为当前残差加上原来搜索方向 p_k 的 β 倍。此时新的搜索方向和原来的搜索方向保持 C-共轭。

[①] 矩阵的条件数 $\kappa(C)$ 是 C 最大特征值和最小特征值的比值。实际的收敛率依赖于特征值的聚合性，本节引用的参考文献中有详细的讨论。

```
ConjugateGradient(C, d, x₀)              ConjugateGradientLS(A, b, x₀)
  1. r₀ = d - Cx₀                          1. q₀ = b - Ax₀,  r₀ = Aᵀq₀
  2. p₀ = r₀                               2. p₀ = r₀
  3. for k = 0...                          3. for k = 0...
  4.   wₖ = Cpₖ                            4.   vₖ = Apₖ
  5.   αₖ = ‖rₖ‖²/(pₖ · wₖ)                5.   αₖ = ‖rₖ‖²/‖vₖ‖²
  6.   xₖ₊₁ = xₖ + αₖpₖ                    6.   xₖ₊₁ = xₖ + αₖpₖ
  7.   rₖ₊₁ = rₖ - αₖwₖ                    7.   qₖ₊₁ = qₖ - αₖvₖ
  8.                                       8.   rₖ₊₁ = Aᵀqₖ₊₁
  9.   βₖ₊₁ = ‖rₖ₊₁‖²/‖rₖ‖²                9.   βₖ₊₁ = ‖rₖ₊₁‖²/‖rₖ‖²
 10.   pₖ₊₁ = rₖ₊₁ + βₖ₊₁pₖ               10.   pₖ₊₁ = rₖ₊₁ + βₖ₊₁pₖ
```

算法 A.3：共轭梯度和共轭梯度最小二乘算法。算法在正文中有比较详细的描述，但大致说来，它首先计算出一个 β 因子，然后通过将原搜索方向的重要性降低 β 倍来得到新的搜索方向 p_k，所有这些搜索方向之间都是 C-共轭的，最后算法找到最优步长 α_k，并进行一个 $\alpha_k p_k$ 的下山操作

事实证明，共轭梯度下降法可以直接应用于非二次型的能量函数，例如那些在非线性最小二乘(附录 A.3)中出现的能量函数。我们并不会显示地构建一个局部的二次型 C 用于近似并计算残差 r_k，在非线性共轭梯度下降中，我们会在每次迭代过程中直接计算能量函数 E(A.45)的梯度，并用其设置搜索方向(Nocedal and Wright 2006)。由于能量函数的二次型近似可能并不存在或者并不精确，所以在决定步长 α_k 时，我们经常使用直线搜索。另外，为了补偿在寻找函数真正的最小值过程中产生的错误，我们也会经常使用一些其他的公式来计算 β_{k+1}，如 Polak-Ribière 公式 (Nocedal and Wright 2006)

$$\beta_{k+1} = \frac{\nabla E(x_{k+1})[\nabla E(x_{k+1}) - \nabla E(x_k)]}{\|\nabla E(x_k)\|^2} \tag{A.51}$$

A.5.2 预处理

如前所述，共轭梯度算法的收敛速度在很大程度上受限于条件数 $\kappa(x)$。因此，如果能够减小条件数，例如，通过改变 x 中元素的数目进而来改变矩阵 C 的行数和列数，那么算法的效率会得到极大的提升。

一般来讲，预处理通常被认为是改变坐标基，进而使向量 x 变成新的向量，

$$\hat{x} = Sx. \tag{A.52}$$

与之对应的需要求解的线性系统变为
$$AS^{-1}\hat{x} = S^{-1}b \quad \text{or} \quad \hat{A}\hat{x} = \hat{b}, \tag{A.53}$$
同样，与之对应的最小二乘能量函数(A.29)将具有如下形式，
$$E_{\text{PLS}} = \hat{x}^T(S^{-T}CS^{-1})\hat{x} - 2\hat{x}^T(S^{-T}d) + \|\hat{b}\|^2. \tag{A.54}$$
通常并不显示计算真正的预处理矩阵 $\hat{C} = S^{-T}CS^{-1}$，而是通过扩展算法 A.3 以在适当的地方引入 S^T 和 S^{-T} 操作(Björck 1996; Golub and Van Loan 1996; Trefethen and Bau 1997; Saad 2003; Nocedal and Wright 2006)。

好的预处理矩阵 S 应该是比较简单且容易计算的，但它同时也应该是对 C 的平方根的一个不错的逼近，使得 $\kappa(S^{-T}CS^{-1})$ 接近于 1。符合这些条件的最简单的选择就是对角矩阵的平方根 $S = D^{1/2}$，其中 $D = \text{diag}(C)$。这样做的一个好处是变量的任何尺度变化(如在角的度量中用弧度来代替角度)都不会影响迭代方法的收敛域。而至于那些本身就具有块状结构的问题——如由运动到结构问题，在其中我们要估计 3D 的点位置或 6D 的相机姿态——一个块状对角的预处理矩阵通常是一个不错的选择。

近年来，人们陆续研究出各种各样更加复杂的预处理矩阵(Björck 1996; Golub and Van Loan 1996; Trefethen and Bau 1997; Saad 2003; Nocedal and Wright 2006)。在这些矩阵中，有很多都可以直接应用于计算机视觉问题(Byröd and øAström 2009; Jeong，Nistér，Steedly et al. 2010; Agarwal，Snavely，Seitz et al. 2010)。它们当中有一些是基于矩阵 C 的不完全乔里斯基分解的，即分解时对 R 中的填充数量有严格的限制。这种限制的一个例子是 R 中的填充只能恰好对应 C 中的非零元素[①]。其他预处理矩阵是基于矩阵 C 的稀疏近似——如基于树形结构或基于聚类的近似(Koutis 2007; Koutis and Miller 2008; Grady 2008; Koutis，Miller，and Tolliver 2009)——因为这些形式有着广为人知的高效的求逆性质。

对基于网格的图像处理应用而言，并行化或层次化的预处理矩阵经常表现得十分出色(Yserentant 1986; Szeliski 1990b; Pentland 1994; Saad 2003; Szeliski 2006b)。这些方法对基变换矩阵 S 做了一些改变，使之更像 3.5 节讨论的图像金字塔或小波的表示形式，因此，这些方法也更加适用于并行和 GPU 实现。在这种新的表示中，粗粒度的元素可以很快收敛到解的低频分量，而更加精细的元素则对高频分量进行编码。至于层次化预处理矩阵、不完全的乔里斯基分解和多重网格方法之间的关系，则有 Saad (2003)和 Szeliski(2006b)进行深入的探索。

A.5.3 多重网格

在计算机视觉中广泛应用的另一种迭代方法称为"多重网格方法"(Briggs，

[①] 如果我们使用完整的乔里斯基因子分解 $C = R^T R$，那么我们将会得到 $\hat{C} = R^{-T}CR^{-1} = I$，于是所有的迭代方法仅用一步就达到了收敛。这明显背离了使用它们的初衷，更何况完整分解的代价通常也十分昂贵。注意，不完整的因子分解也可以从变量重排中获得好处。

Henson, and McCormick 2000; Trottenberg, Oosterlee, and Schuller 2000), 这种方法已经被应用于表面插值(Terzopoulos 1986a)、光流(Terzopoulos 1986a; Bruhn, Weickert, Kohlberger *et al.* 2006)、高动态范围色调映射(Fattal, Lischinski, and Werman 2002)、彩色化(Levin, Lischinski, and Weiss 2004)、自然图像抠图(Levin, Lischinski, and Weiss 2008)及分割(Grady 2008)等一系列问题。

多重网格背后的主要思想是构造原始问题的一个粗粒度(低分辨率)版本,并用它去求解问题的低频分量。然而,不同于用粗糙解初始化精细解的简单的由粗到精的方法,多重网格方法仅仅修正当前解的低频分量,然后通过多次求粗和求精的循环(一般称为"图像金字塔上的 V 型或 W 型运动")来获取快速的收敛。

在一些简单的齐次问题(如求解泊松方程)中,多重网格方法可以获得最优的性能,即计算时间与变量数成线性关系。但对更多的非齐次问题或在不规则网格上的问题而言,多重网格方法的一些变种,如通过观察矩阵 C 的结构来构建粗粒度问题的代数多重网格(AMG)方法,则可能更受欢迎。Saad(2003)对多重网格和并行预处理矩阵的联系以及使用多重网格和共轭梯度方法所具备的好处做了很好的讨论。

附录 B

贝叶斯建模与推断

B.1 估计理论
B.2 最大似然估计与最小二乘
B.3 鲁棒统计学
B.4 先验模型与贝叶斯推断
B.5 马尔科夫随机场
B.6 不确定性估计(误差分析)

下面这种问题在本书中不断地重现：已知一些测量量(图像、特征位置等)，估计出一些未知的结构或参数(摄像机位置、物体形状等)的值。这种问题通常称为"逆问题"，因为它们是在估计未知的模型参数，而不是在模拟正向的构造方程。[①] 计算机图形学就是一个典型的正向建模问题(已知一些物体、摄像机和光照，模拟出在这些条件下产生的图像)，而计算机视觉则通常是它的逆问题(已知一幅或一些图像，重现产生这些图像的场景)。

对一个具体的逆问题而言，在通常情况下，会有多种方法可以对它进行求解。比如通过巧妙的(有时也是很直截了当的)代数运算，我们有时就可以得到未知量的解析解。作为一个例子，我们考虑如下的摄像机矩阵标定问题(6.2.1 节)：任给一幅某一标定模式对应的图像，该标定模式由一些已知的三维点的坐标组成，计算出 3×4 的摄像机矩阵 P，使得该矩阵可以将这些点映射到图像平面上。

进一步，我们可以将这个问题写成(6.33 和 6.34)的形式

$$x_i = \frac{p_{00}X_i + p_{01}Y_i + p_{02}Z_i + p_{03}}{p_{20}X_i + p_{21}Y_i + p_{22}Z_i + p_{23}} \tag{B.1}$$

$$y_i = \frac{p_{10}X_i + p_{11}Y_i + p_{12}Z_i + p_{13}}{p_{20}X_i + p_{21}Y_i + p_{22}Z_i + p_{23}}, \tag{B.2}$$

这里 (x_i, y_i) 是在图像平面内测量到的第 i 个点的特征坐标， (X_i, Y_i, Z_i) 则是它对应的三维点坐标， p_{ij} 是摄像机矩阵 P 中的未知项。将分母移到等式的左边，我们就可以得到一组联立的线性方程组，

$$x_i(p_{20}X_i + p_{21}Y_i + p_{22}Z_i + p_{23}) = p_{00}X_i + p_{01}Y_i + p_{02}Z_i + p_{03}, \tag{B.3}$$

$$y_i(p_{20}X_i + p_{21}Y_i + p_{22}Z_i + p_{23}) = p_{10}X_i + p_{11}Y_i + p_{12}Z_i + p_{13}, \tag{B.4}$$

我们可以使用线性最小二乘(附录 A.2)对其进行求解，从而得到 P 的一个估计。

问题随之而来了：这个方程组真的是我们需要求解的那个方程组吗？如果观测是完全无噪声的，或者我们根本不在乎能否得到最优解，那么答案是肯定的。但是，一般而言，除非我们对可能的误差来源建立了模型，并设计了一个在这些潜在误差影响下也能工作的足够好的算法，否则我们无法确定我们得到的是一个合理的算法。

B.1 估 计 理 论

这种针对有噪声数据的推断问题的研究通常称为"估计理论"(Gelb 1974)，它的一个延伸称为"统计决策理论"(Berger 1993; Hastie, Tibshirani, and Friedman 2001; Bishop 2006; Robert 2007)，在统计决策理论中我们会显式地选择一个损失函数。首先，我们会写出一个正向过程，这个过程可以从我们的未知量(和已知量)演

[①] 在机器学习领域中，这些问题称为"回归问题"，因为我们试图从有噪声的输入中估计一个连续量，与之相对的是离散的分类问题(Bishop 2006)。

绎到一组有噪声污染的观测量。接下来，我们会设计一个算法来给出一个估计(或一组估计)，这种估计既对噪声不敏感(这是最好情况)，同时也可以量化这些估计的可靠性。

前面给出的方程组(B.1)仅仅是更一般的测量方程组的一个特殊实例：

$$y_i = f_i(x) + n_i. \tag{B.5}$$

其中，y_i是含有噪声污染的测量量，也就是方程(B.1)中的(x_i, y_i)，x则是未知的状态向量[①]。

每个测量量都有其对应的测量模型$f_i(x)$，它将未知量映射到该测量量上。还有另外一种公式，它使用了更一般的函数形式$f_i(x, p_i)$，并使用与测量量相关的参数向量p_i——如公式(B.1)中的(X_i, Y_i, Z_i)——来对不同的测量量进行区分。注意，使用$f_i(x)$这种形式可以让我们更直接地使用不同维度的测量量，这一性质在我们开始增加先验信息(附录B.4)时会变得十分有用。

每个测量量同时也被一些噪声n_i所污染。在公式(B.5)中，我们已经指出，n_i是一个零均值的正态(高斯)随机变量，其协方差矩阵为Σ_i。虽然在一般情况下，噪声并不一定服从高斯分布，并且在实际应用中假定一些测量值可能出现错误也是明智之举，但我们仍会将这方面的讨论推迟到附录B.3中进行。我们首先需要更详尽地探讨比较简单的关于高斯噪声的情况。我们同时假定噪声向量n_i是相互独立的。而在一些不符合假定的情况中(例如一些常量增益或偏移污染了某一图像中所有的像素)，我们可以将这种影响作为多余的参数加到我们的状态向量x中，稍后再估计它们的值(并将它们舍弃，如果需要的话)。

多变量高斯噪声的似然度

给定所有有噪声的测量结果$y = \{y_i\}$，我们希望能够推断出未知向量x的概率分布。在给定x特定值的情况下观测到$\{y_i\}$的似然度可以写为

$$L = p(y|x) = \prod_i p(y_i|x) = \prod_i p(y_i|f_i(x)) = \prod_i p(n_i). \tag{B.6}$$

每个噪声向量n_i都服从协方差为Σ_i的多变量高斯分布，

$$n_i \sim \mathcal{N}(0, \Sigma_i), \tag{B.7}$$

我们可以将该似然度写为

$$\begin{aligned} L &= \prod_i |2\pi\Sigma_i|^{-1/2} \exp\left(-\frac{1}{2}(y_i - f_i(x))^T \Sigma_i^{-1} (y_i - f_i(x))\right) \\ &= \prod_i |2\pi\Sigma_i|^{-1/2} \exp\left(-\frac{1}{2}\|y_i - f_i(x)\|^2_{\Sigma_i^{-1}}\right), \end{aligned} \tag{B.8}$$

其中矩阵范数$\|x\|^2_A$是$x^T A x$的简短表示。

[①] 在关于卡曼(Kalman)滤波的著作(Gelb 1974)中，更常用z而不是用y来表示测量量。

范数 $\|y_i - \bar{y}_i\|_{\Sigma_i^{-1}}$ 通常称为"Mahalanobis 距离"(5.26 和 14.14),它用于度量测量量与多变量高斯分布的中心的距离。在 Mahalanobis 距离意义下的等距离线就是等概率线。注意,如果测量值的协方差具有各向同性(在各个方向上均相同),即 $\Sigma_i = \sigma_i^2 I$,那么这个似然度可以写成

$$L = \prod_i (2\pi\sigma_i^2)^{-N_i/2} \exp\left(-\frac{1}{2\sigma_i^2}\|y_i - f_i(x)\|^2\right), \quad (B.9)$$

其中,N_i 是第 i 个测量向量 y_i 的长度。

如果我们先对协方差矩阵进行一次特征值或主分量分析(PCA)(A.6),

$$\Sigma = \Phi \operatorname{diag}(\lambda_0 \ldots \lambda_{N-1}) \Phi^T. \quad (B.10)$$

就可以更容易地对协方差矩阵的结构和相应的 Mahalanobis 距离进行形象化的描述。某一多变量高斯分布的等概率线——也就是 Mahalanobis 距离下的等距离线(图 14.14)——是一些多维椭圆,这些椭圆的轴的方向由 Φ 中的列(特征向量)决定,而轴的长度则由 $\sigma_j = \sqrt{\lambda_j}$ 给出(图 A.1)。

通常,使用似然度的负对数会更方便一些,我们可以将它看成是一种代价或一种能量

$$E = -\log L = \frac{1}{2}\sum_i (y_i - f_i(x))^T \Sigma_i^{-1}(y_i - f_i(x)) + k \quad (B.11)$$

$$= \frac{1}{2}\sum_i \|y_i - f_i(x)\|^2_{\Sigma_i^{-1}} + k, \quad (B.12)$$

这里的 $k = \sum_i \log|2\pi\Sigma_i|$ 是一个常量,它只依赖于测量的方差,而与 x 无关。

需要注意的是,协方差矩阵的逆矩阵 $C_i = \Sigma_i^{-1}$ 起着对每个测量误差余项(residual)加权的作用。所谓的测量误差余项,就是被污染的测量值 y_i 与未被污染的(预测得到的)值 $f_i(x)$ 之间的差。实际上,协方差矩阵的逆矩阵通常被称为"(Fisher)信息矩阵"(Bishop 2006),因为它可以让我们知道在某一给定的测量值中到底蕴含有多少信息。我们也可以认为这一矩阵给出了与每一个测量值相关的置信度(因此用 C 作为这个矩阵的记号)。

在这一公式中,一些信息矩阵奇异(秩退化),甚至全零(如果测量值都缺失,这种现象就会发生)都是很合理的。例如,用线条特征或边缘去测量三维空间里的类边特征时,亏秩的测量就会经常出现,因为边上的准确位置是未知的(有着无限或者特别大的方差,8.1.3 节)。

为了更显著地区分受到噪声污染的测量值,以及在某一特定的 x 值下该测量值对应的期望值,我们使用记号 $\tilde{y}$ 来表示前者(认为有波浪符的是近似或者有噪声的值),$\hat{y} = f_i(x)$ 代表后者(认为有小尖帽的是预测或期望值)。这样一来,我们可以将似然度的负对数写为

$$E = -\log L = \sum_i \|\tilde{y}_i - \hat{y}_i\|_{\Sigma_i^{-1}} + k. \quad (B.13)$$

B.2 最大似然估计与最小二乘

既然我们已经给出了似然度和对数似然度函数，那么我们怎样才能找到最优的状态估计值 x 呢？一个貌似合理的选择可能是去选择最大化 $L = p(x|y)$ 的 x 值。实际上，在没有任何关于 x 的先验模型(附录 B.4)时，我们有

$$L = p(\boldsymbol{y}|\boldsymbol{x}) = p(\boldsymbol{y},\boldsymbol{x}) = p(\boldsymbol{x}|\boldsymbol{y}).$$

因此，选择最大化似然度的 x 值也就等效于选择 x 的概率密度估计值取得最大值时对应的 x 值。

这种方法在什么时候会是一个不错的主意呢？如果数据(测量值)对 x 的可能值有一个约束，使所有这些可能值都紧密地聚集在某一值周围(例如，如果分布 $p(x|y)$ 是一个单高斯分布)，那么最大似然度估计就是在所有既没有偏置，又具有最小的可能的方差的估计值中最优的一个。另外，在很多其他情况中，例如我们所需要的仅仅只是一个单独的估计，通常最大似然度估计也还是最佳的估计[①]。但是，如果概率分布是多峰值的，即对数似然度函数有多个局部极小值点(图 5.7)，那我们就需要更加小心一点。特别的，我们可能需要延迟某些决策(如正在跟踪的物体的最终位置)，直到做出更多的观测。5.1.2 节给出的 CONDENSATION 算法是一种对这种多峰值分布进行建模和更新的可能的方法，但它也仅仅是更一般的粒子滤波器和马尔科夫链蒙特卡洛方法(Markov Chain Monte Carlo，MCMC)的一个例子而已(Andrieu, de Freitas, Doucet *et al.* 2003; Bishop 2006; Koller and Friedman 2009)。

另一种选择最佳估计的可能方法是最大化与获得正确估计相关的期望效用(或者反过来说，最小化期望风险或损失)，即最小化

$$E_{\text{loss}}(\boldsymbol{x},\boldsymbol{y}) = \int l(\boldsymbol{x} - \boldsymbol{z}) p(\boldsymbol{z}|\boldsymbol{y}) d\boldsymbol{z}. \tag{B.14}$$

例如，如果一个机器人希望不惜任何代价也要避免与墙相撞，那么只要某一估计低估机器人与墙的真实距离，损失函数的值就会变得很大。当 $l(x-y) = \delta(x-y)$ 时，我们就得到了最大似然度估计，而当 $l(x-y) = \|x-y\|^2$ 时，我们会得到均方误差(mean square error, MSE)或期望值估计。对效用或损失函数进行显式建模正是统计决策理论的一大特征(Berger 1993; Hastie, Tibshirani, and Friedman 2001; Bishop 2006; Robert 2007)。

我们该如何获得最大似然度估计呢？如果测量噪声服从高斯分布，那么我们可以最小化二次目标函数(B.13)。它甚至还可以进一步简化，如果测量方程是线性的，即

$$\boldsymbol{f}_i(\boldsymbol{x}) = \boldsymbol{H}_i \boldsymbol{x}, \tag{B.15}$$

[①] 根据高斯-马尔科夫理论，在忽略实际的噪声分布并假设噪声满足零均值且不相关的条件下，对线性测量模型而言，最小二乘可以产生最佳的线性无偏估计器(best linear unbiased estimator, BLUE)。

其中，H 是测量矩阵，它将未知的状态变量 x 与测量值 $\tilde{y}$ 联系了起来。在这种情况下，(B.13)变为

$$E = \sum_i \|\tilde{y}_i - H_i x\|_{\Sigma_i^{-1}} = \sum_i (\tilde{y}_i - H_i x)^T C_i (\tilde{y}_i - H_i x), \tag{B.16}$$

这是一个关于 x 的简单的二次型，可以使用线性最小二乘(附录 A.2)进行求解。当测量是非线性的时候，整个系统必须通过使用非线性最小二乘(附录 A.3)迭代地进行求解。

B.3 鲁棒统计学

在附录 B.1.1 中，我们假定作用于每一个观测上的噪声(B.5)都服从多变量的高斯分布(B.7)。如果噪声是很多微小的错误——如来自硅片成像器的热噪声——累加起来的结果，那么这个模型还是很合适的。但在大部分情况下，测量都要受到更大的外点(异常值)——即测量过程中出现的严重错误——的污染。这种外点的例子包括糟糕的特征匹配(6.1.4 节)、在立体视觉匹配中的遮挡(第 11 章)以及一些存在于原本很光滑的图像、深度图和标记图像中的不连续区域(3.7.1 节和 3.7.2 节)等。

在这些情况下，用长尾部的污染误差模型——如拉普拉斯模型——对测量误差进行建模会更加合理一些。在这种情况下，为了反映出现严重错误的似然度的增大，负对数似然度——不再是关于测量余项的二次型(B.12~B.16)——在其惩罚函数中有一个缓慢的增加。

在鲁棒统计学的著作中，这种形式的推断问题被称为"M-估计器"(Huber 1981; Hampel, Ronchetti, Rousseeuw et al. 1986; Black and Rangarajan1996; Stewart 1999)，它需要对余项施用一个鲁棒损失函数 $\rho(r)$

$$E_{\text{RLS}}(\Delta p) = \sum_i \rho(\|r_i\|) \tag{B.17}$$

而不是对它们进行平方操作。

正如我们在 6.1.4 节中提到的，我们可以对这个函数求 ρ 的导数并令其为 0，

$$\sum_i \psi(\|r_i\|) \frac{\partial \|r_i\|}{\partial p} = \sum_i \frac{\psi(\|r_i\|)}{\|r_i\|} r_i^T \frac{\partial r_i}{\partial p} = 0, \tag{B.18}$$

其中 $\psi(r) = \rho'(r)$ 是 ρ 的导数，叫影响函数。如果我们引入加权函数 $\omega(r) = \psi(r)/r$，我们就会发现用(B.18)寻找(B.17)的驻点与最小化下面迭代重加权最小二乘问题(IRLS)是等效的，

$$E_{\text{IRLS}} = \sum_i w(\|r_i\|) \|r_i\|^2, \tag{B.19}$$

其中， $\omega(\|r_i\|)$ 与(B.12)中的 $C_i = \Sigma_i^{-1}$ 一样，都起着局部加权的作用。Black and Anandan(1996)描述了多种鲁棒惩罚函数及其对应的影响函数和加权函数。

IRLS 算法交替进行计算加权函数 $\omega(\|r_i\|)$ 和求解加权最小二乘问题(w 值固定)这两种操作。其他关于鲁棒最小二乘的增量算法可以在 Sawhney and Ayer(1996);

Black and Anandan (1996); Black and Rangarajan (1996); Baker, Gross, Ishikawa *et al*. (2003)等人的工作中找到，关于鲁棒统计学的教材和教程(Huber 1981; Hampel, Ronchetti,Rousseeuw *et al*. 1986; Rousseeuw and Leroy 1987;Stewart 1999)也有相应的论述。当然，直接对方程(B.19)中给出的非线性代价函数使用一般的优化方法也是可行的，并且这样做有时可能会得到更好的收敛性质。

　　大部分鲁棒损失函数都含有一个尺度参数，它通常被设置为无污染(内点)噪声的方差(或标准差，这要根据公式的形式确定)。然而直接从测量值或它们的余项中估计这种噪声的噪声级可能是有问题的，因为这种估计本身也会被外点污染。鲁棒统计学文献中有很多估计这种参数的方法。一种最简单也最有效的方法是中位数绝对偏差法(median absolute deviation，MAD)，

$$MAD = \mathrm{med}_i \|r_i\|, \qquad (B.20)$$

乘以 1.4 以后，这个值便提供了对内点噪声过程的标准差的一个鲁棒估计。

　　正如 6.1.4 节中所说，通常在一个优质解的附近启动迭代非线性最小化方法，如 IRLS，会更好一些。这可以通过首先不断地随机选取测量值的小的子集，直到找到一个较好的内点集合来实现。在这些方法中，最著名的要数随机样本一致性方法(RANdom SAmple Consensus，RANSAC))(Fischler and Bolles 1981)，现在已经发展出了它的一些更好的变体，如优先选择权 RANSAC(Nistér 2003)和渐进样本一致性方法(PROgressive Sample Consensus，PROSAC)(Chum and Matas 2005)。

B.4　先验模型与贝叶斯推断

　　虽然最大似然估计通常都可以产生很好的结果，但在某些情况下，与测量值一致的可能解的范围过大而导致这种可能解没有什么用处。例如，考虑图像去噪问题(3.4.4 节与 3.7.3 节)。如果只根据每个像素上的噪声类型独立地估计每个像素，我们将无法取得任何进展，因为对任意一个有噪声的测量值而言，都有大量的真实值可以导致这一结果[①]。取而代之的做法是，我们需要依赖于图像的一些典型性质，如图像是趋向于分块光滑的(3.7.1 节)。

　　图像分块光滑的性质可以编码成一个先验分布 $p(x)$，它度量了一幅图像是自然的图像的似然度。例如，为了编码分段光滑性，我们可以使用马尔科夫随机场模型(3.109 和 B.24)，它的负对数似然度是与图像光滑性(梯度大小)的鲁棒化的度量值成比例的。

　　先验模型不仅仅局限于有关图像处理的应用。例如，我们可能拥有一些关于扫描物体的大致尺寸、正在被标定的镜头的焦距的外部知识或者是知道某一特定物体在图像中可能出现的似然度。所有的这些都是先验分布或概率的例子，它们都可以用于产生更加可靠的估计。

---

① 事实上，最大似然估计的结果恰恰就是噪声图像本身。

正如我们在(3.68)和(3.106)中已经看到的，贝叶斯规则表明给定测量值 y，关于未知量 x 的后验概率 $p(x|y)$ 可以用先验分布 $p(x)$ 乘以测量似然度 $p(y|x)$ 这种方法得到

$$p(\boldsymbol{x}|\boldsymbol{y}) = \frac{p(\boldsymbol{y}|\boldsymbol{x})p(\boldsymbol{x})}{p(\boldsymbol{y})}, \tag{B.21}$$

其中 $p(y) = \int_x p(y|x)p(x)$ 是一个用于使 $p(x|y)$ 分布变得严格(积分结果为 1)的归一化常数。对等式(B.21)的两边取负对数，我们得到

$$-\log p(\boldsymbol{x}|\boldsymbol{y}) = -\log p(\boldsymbol{y}|\boldsymbol{x}) - \log p(\boldsymbol{x}) + \log p(\boldsymbol{y}), \tag{B.22}$$

这是负后验对数似然度。通常去掉常数 $\log p(y)$，因为在能量最小化过程中，它的值不起任何作用。但是，如果先验分布 $p(x)$ 依赖于一些未知的参数，我们则可能会希望保留 $\log p(y)$ 这一项，从而利用奥卡姆剃刀(Occam's razor)原则，即最大化观测值的似然度，计算出这些参数的最可能的值，或通过模型选择(Hastie, Tibshirani, and Friedman 2001; Torr 2002; Bishop 2006; Robert 2007)来选出这些自由参数的正确数量。

为了在给定一些测量值 y 的条件下找出 x 的最可能(最大化后验概率，MAP)解，我们可以简单地最小化负对数似然度，也可以把这种负对数似然度看成一种能量，

$$E(\boldsymbol{x}, \boldsymbol{y}) = E_d(\boldsymbol{x}, \boldsymbol{y}) + E_p(\boldsymbol{x}). \tag{B.23}$$

第一项 $E_d(x,y)$ 是数据能量或数据惩罚，它度量了给定未知状态 x 时，观测到的测量值 y 的似然度。第二项 $E_p(x)$ 是先验能量，它与正则化过程中的平滑能量的作用类似。需要注意的是，MAP 估计可能并不总是那么令人满意，因为它选择的"峰值点"是关于后验分布的，而不是一些更加稳定的统计量，如 MSE——参见附录 B.2 中关于损失函数和决策理论的讨论——的"峰值点"。

B.5 马尔科夫随机场

马尔科夫随机场(Blake, Kohli, and Rother 2010)是最流行的适用于网格状、类似于图像的数据的先验模型①。这种数据不仅包括通常意义下的自然图像(3.7.2 节)，也包括类似于光流的二维场(第 8 章)、深度图(第 11 章)或二值场，如图像分割结果(5.5 节)。

正如我们在 3.7.2 节中讨论的，马尔科夫随机场的先验概率 $p(x)$ 服从吉布斯(Gibbs)或玻尔兹曼(Boltzmann)分布，其负对数似然度(根据 Hammer-sley–Clifford 定理)可以写作成对的交互作用势的和，

$$E_p(\boldsymbol{x}) = \sum_{\{(i,j),(k,l)\} \in \mathcal{N}} V_{i,j,k,l}(f(i,j), f(k,l)), \tag{B.24}$$

其中 $\mathcal{N}(i,j)$ 表示像素 (i,j) 的邻居像素。在更一般的情况下，MRF 还可以含有一元

---

① 其他的形式化包括能量谱(3.4.3 节)和非局部均值(Buades, Coll, and Morel 2008)。

势,以及定义在更大的基团上的更高阶的势(Kindermann and Snell 1980; Geman and Geman 1984; Bishop 2006; Potetz and Lee2008; Kohli, Kumar, and Torr 2009; Kohli, Ladický, and Torr 2009; Rother, Kohli, Feng *et al.* 2009; Alahari, Kohli, and Torr 2011)。它也可能包含线过程,即额外增加的用于缓和相邻元素间不连续性的二值变量(Geman and Geman1984)。Black and Rangarajan (1996)展示了如何消去独立的线过程变量,并将其合并到使用鲁棒成对惩罚函数的常规 MRF 中。

在马尔科夫随机场模型中,最常用的邻域是 $\mathcal{N}_4$ 邻域。在这种邻域模型中,随机场中的每个像素 $f(x,y)$ 只和它的直接邻居相互作用——图 B.1 展示了一个这样的 $\mathcal{N}_4$ MRF。黑色方块 $s_x(i,j)$ 和 $s_y(i,j)$ 表示随机场中相邻节点间任意大小的交互作用势,$\omega(i,j)$ 表示(B.23)中的数据惩罚项 E_d。这些方块形节点也可以被解释成是无向图模型版本的因子图中的因子(Bishop 2006; Wainwright and Jordan 2008; Koller and Friedman 2009),后者是相互作用势的另一个名字。(严格来讲,因子是不严格的概率函数,因子的乘积是未经过归一化的后验概率)。

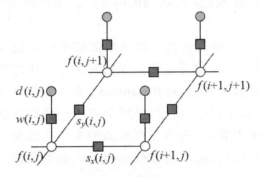

图 B.1 $\mathcal{N}_4$ 邻域的马尔科夫随机场的图模型。白色圆圈表示未知量 $f(i,j)$,黑色圆圈表示输入数据 $d(i,j)$,黑色方块 $s_x(i,j)$ 和 $s_y(i,j)$ 表示随机场中相邻节点间任意大小的交互作用势, $\omega(i,j)$ 表示数据惩罚函数。它们都是在方程(B.24)中使用的通用势能函数 $V_{i,j,k,l}(f(i,j),f(k,l))$ 的特例

更复杂、更高维的相互作用模型和邻域都是可能出现的。例如,我们可以用额外增加的对角连接($\mathcal{N}_8$ 邻域),甚至更多成对出现的项来对 2D 网格进行强化(Boykov and Kolmogorov 2003; Rother, Kolmogorov, Lempitsky *et al.* 2007)。而 3D 网格也可以用于计算 3D 容积医学图像的全局优化分割(Boykov and Funka-Lea 2006)(5.5.1 节)。另外,我们同样可以使用高阶团簇(Higher-order clique)来构建更复杂的模型(Potetz and Lee 2008; Kohli, Ladický, and Torr 2009; Kohli, Kumar, and Torr 2009)。

使用马尔科夫随机场模型时,我们需要面对的一个最大的挑战就是开发出高效的可以找到低能量解的推断算法(Veksler 1999; Boykov, Veksler, and Zabih 2001; Kohli 2007; Kumar 2008)。近年来,人们已经找到了很多种类的这类算法,包括模拟退火、图分割及带环的置信传播。推断方法的选择极大地影响着一个视觉系统的整体性能。例如,在 Middlebury Stereo Evaluation(Middlebury 立体视觉评测)页面

上，大多数性能极佳的算法都选择使用置信传播或图割算法。

在接下来的几个小节中，我们将回顾一些应用比较广泛的马尔科夫随机场推断方法。这些算法中绝大部分更加深入的介绍可以在最近出版的一本关于高级马尔科夫随机场方法的书(Blake, Kohli, and Rother 2010)中找到。而一些实验比较，包括测试数据集和参考软件，则由 Szeliski, Zabih, Scharstein et al. (2008)提供。[①]

B.5.1 梯度下降与模拟退火

最简单的优化方法就是梯度下降法，它通过改变一些相互独立的节点子集中的节点状态使这些节点具有更低的能量，来最小化能量函数。这种方法有各种各样不同的名字，包括上下文有关分类(Kittler and Föglein 1984)和迭代条件模型(ICM)(Besag 1986)。[②]变量既可以顺序地进行更新，如按照光栅扫描的方式，也可以并行地进行更新，如使用棋盘红黑着色方法。Chou and Brown(1990)则建议使用高置信度优先(highest confidence first, HCF)方法，即依据它们在减少能量方面的贡献挑选变量。

梯度下降法的一个问题就是它易于陷入局部极小值，这种问题在马尔科夫随机场问题中几乎随处可见。一种绕开这个问题的方法是使用随机梯度下降法或马尔科夫链蒙特卡洛法(MCMC)(Metropolis, Rosenbluth, Rosenbluth et al. 1953)，即随机地采用一些临时的上山步长来摆脱这种局部极小值。一种流行的更新规则是吉布斯采样器(Geman and Geman 1984)；在为正在更新的变量先选择一个变量的状态时，我们不是选择具有最低能量的状态，而是按照如下的概率选择一个状态

$$p(\boldsymbol{x}) \propto e^{-E(\boldsymbol{x})/T}, \tag{B.25}$$

这里的 T 叫温度，它控制着系统选择一个更随机的更新的可能性。随机梯度下降通常与模拟退火联合使用(Kirkpatrick, Gelatt, and Vecchi 1983)，它从一个相对较高的温度开始，从而随机地探索很大一部分状态空间，然后逐渐地降低温度(退火)去寻找一个好的局部极小值。在 20 世纪 80 年代后期，模拟退火是解决马尔科夫随机场推断问题的不二选择(Szeliski 1986; Marroquin, Mitter, and Poggio 1985; Barnard 1989)。

模拟退火的另一个变种是 Swendsen–Wang 算法(Swendsen and Wang 1987; Barbu and Zhu 2003, 2005)。在这种算法中，为了取代"翻转"(改变)单一变量这种做法，一组经过选择的相互连接的变量子集被选为基本的更新单元。我们通过基于马尔科夫随机场连接强度的随机徘徊来挑选这些变量子集。有些时候，这种方法可以有助于做出更大的状态改变，因此可以在更短的时间内找到更高质量的解。

虽然模拟退火在很大程度上已经被新近的图割(graph cut)法和带环的置信度传

[①] http://vision.middlebury.edu/MRF/
[②] 之所以有这个名字，是因为这种方法需要迭代地根据当前固定邻居变量的状态，将变量设置到模式(最可能，如最低能量)状态。

播方法取代，但偶尔我们还是可以找到使用它的地方的，特别是在那些高度联通和高度无子结构的图中(Rother, Kolmogorov, Lempitsky et al. 2007)。

B.5.2 动态规划

动态规划(DP)是一种高效推断算法，适用于任何具有树状结构的图模型，即没有任何环的图模型。任给这样的一棵树，我们挑选任意一个节点作为根节点 r，然后想象通过根节点将这棵树提起来的情形。这时所有其他节点到根节点的深度或距离就给出了节点间的一个偏序关系，而从这个偏序关系出发，以任意方式打破偏序关系中的结，我们就可以得到一个节点间的全序关系。我们现在将图按照树的方式进行布局，树的根在右边，节点下标从左到右依次增加，如图 B.2a 所示。

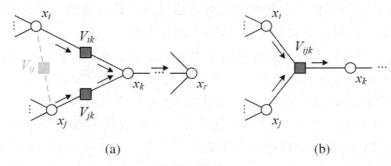

图 B.2 在以因子图形式表示的树上进行动态规划。(a)为了根据节点 x_k 左边所有节点的最优解计算 x_k 的最低能量解 $\hat{E}_k(x_k)$，我们枚举 $\hat{E}_k(x_k)+V_{ik}(x_i,x_k)$ 的所有可能值并选择最小的一个(对 j 的做法类似)。(b)对高阶的团簇而言，我们需要尝试所有的 (x_i,x_j) 组合来选择最优解。箭头显示了计算的基本流向。(a)中阴影较浅的因子 V_{ij} 显示了一个额外的连接，它将树变成有环图，而在有环图中，精确的推断是不能被高效计算出来的

在描述 DP 算法之前，我们先将等式(B.24)中的势能函数重写成一种更通用同时也更简洁的形式，

$$E(\boldsymbol{x}) = \sum_{(i,j)\in\mathcal{N}} V_{i,j}(x_i,x_j) + \sum_i V_i(x_i), \quad (B.26)$$

这里，我们不再使用像素下标 (i,j) 和 (k,λ)，而是使用标量下标变量 i 和 j。我们也使用更加简洁的记号 x_i 去代替函数值 $f(i,j)$，此时变量集合 $\{x_i\}$ 便构成了状态向量 $\boldsymbol{x}$。我们还可以进一步对这个函数进行简化，我们可以为每个拥有非零 $V_i(x_i)$ 的节点增加一个虚拟节点 i^-，并令 $V_{i,i^-}(x_i,x_{i^-})=V_i(x_i)$，这样我们就可以从(B.26)中去掉 V_i 这一项了。

动态规划通过按照从左到右的顺序，即按照下标变量递增的顺序，计算部分和的方式进行工作。令 C_k 表示 k 的所有子节点的下标集合，即 $i<k, (i,k)\in N$。然后定义

$$\tilde{E}_k(\boldsymbol{x}) = \sum_{i<k,\,j\le k} V_{i,j}(x_i,x_j) = \sum_{i\in\mathcal{C}_k}\left[V_{i,k}(x_i,x_k)+\tilde{E}_i(\boldsymbol{x})\right], \quad (B.27)$$

为(B.26)在所有下标的小于等于 k 的变量上，即在图 B.2a 中 x_k 左侧的所有部分上的

部分和。这个部分和依赖于 x 中所有下标 i 满足 $i<k$ 的未知变量的状态。

现在假设我们希望找到所有下标满足 $i<k$ 的变量的状态，以最小化这个部分和。事实证明，我们可以使用一个简单的递归公式

$$\hat{E}_k(x_k) = \min_{\{x_i, i<k\}} \tilde{E}_k(x) = \sum_{i \in C_k} \min_{x_i} \left[V_{i,k}(x_i, x_k) + \hat{E}_i(x_i) \right] \quad \text{(B.28)}$$

来找到这个最小值。直观上，这很容易理解。在图 B.2a 中，我们根据节点左边所有变量的最优状态来为每个节点 k 的每个可能状态 x_k 关联一个能量值 $\hat{E}_k(x_k)$。在这幅图中，我们很容易让自己确信，为了计算出这个值，只需要知道 $\hat{E}_i(x_i)$ 和 $\hat{E}_j(x_j)$。

一旦树中的信息流从左边传到右边，那么根节点对应的最小能量值 $\hat{E}_r(x_r)$ 就给出了 $E(x)$ 的一个 MAP(最低能量)解。同时，根节点的状态也被设定为能够最小化这个函数的状态 x_r，而其他节点的状态则可以通过反向推理传递，即选择原始递归函数(B.28)达到最小值时子节点 $i \in C_k$ 的状态，进行设置。

动态规划并不是只适用于那些拥有成对势能的树。图 B.2b 给出了一个包含在树中的三路势能 $V_{ijk}(x_i, x_j, x_k)$ 的例子。在这种情况下，为了计算 $\hat{E}_k(x_k)$ 的最优值，(B.28)中的递归公式必须在因子节点(灰色方块)所有前驱节点的所有状态的所有可能组合上估计最小值。因此，动态规划的计算复杂度通常随团簇(clique)大小的增加而呈指数性增长，即一个具有 n 个节点：每个节点有 l 种可能标记的团簇需要评价 l^{n-1} 种可能的状态(Potetz and Lee 2008; Kohli, Kumar, and Torr 2009)。然而，针对某些种类的势能函数 $V_{i,k}(i,k)$，包括 Potts 模型(delta 函数)、绝对值函数(总变差)、二次函数(高斯 MRF)，Felzenszwalb and Huttenlocher(2006)给出了将最小值搜索步骤(B.28)的复杂度从 $O(l^2)$ 降到 $O(l)$ 的方法。在附录 B.5.3 中，我们还会讨论 Potetz and Lee(2008)是如何降低特定种类的高阶团簇(即线性和接续非线性和)的复杂度的。

图 B.2a 同样给出了我们在节点 i 和 j 之间添加一个额外的因子时发生的事情。在这种情况下，图就不再是一棵树了，即它包含回路。我们也就不能再使用递归公式(B.28)了，因为 $\hat{E}_i(x_i)$ 会出现在部分和两个不同的项中，即同时作为节点 j 和节点 k 的子节点，同样的 x_i 的状态可能不会同时对二者进行最小化。换言之，当存在回路时，没有一种对变量的排序能使(B.28)中的递归(消去)正确无误。

但是，如图 B.2b 所示，还是有可能将小的回路转化为高阶因子进而对其进行求解的。然而，拥有很长的回路或网格的图会使团簇的规模变得极其巨大，从而可能使计算复杂度与图的规模呈指数关系。

B.5.3 置信传播

置信传播是一种推断方法，起初它是针对树状图的(Pearl 1988)，但最近它也被扩展到了"带环"(环状)图，如 MRF(Frey and MacKay 1997; Freeman, Pasztor, and

Carmichael 2000; Yedidia, Freeman, and Weiss 2001; Weiss and Freeman2001a, b; Yuille 2002; Sun, Zheng, and Shum 2003; Felzenszwalb and Huttenlocher 2006)。它与动态规划紧密相关，这表现为两种方法都会在树或图中对消息进行前向和后向传递。事实上，置信传播的两个变体之一，乘积最大化规则，与动态规划进行的计算(推断)是完全一样的，只不过它用概率代替了能量。

回忆一下，我们在 MAP 估计(B.26)中最小化的能量函数是经过因子分解的吉布斯后验分布

$$p(\boldsymbol{x}) = \prod_{(i,j)\in\mathcal{N}} \phi_{i,j}(x_i, x_j), \tag{B.29}$$

的负对数似然度(B.12，B.13 和 B.22)，其中

$$\phi_{i,j}(x_i, x_j) = e^{-V_{i,j}(x_i, x_j)} \tag{B.30}$$

是成对的交互作用势。我们可以将(B.27)重写为

$$\tilde{p}_k(\boldsymbol{x}) = \prod_{i<k,\, j\leq k} \phi_{i,j}(x_i, x_j) = \prod_{i\in\mathcal{C}_k} \tilde{p}_{i,k}(\boldsymbol{x}), \tag{B.31}$$

其中

$$\tilde{p}_{i,k}(\boldsymbol{x}) = \phi_{i,k}(x_i, x_k)\tilde{p}_i(\boldsymbol{x}). \tag{B.32}$$

据此，我们可将(B.28)重写为

$$\hat{p}_k(x_k) = \max_{\{x_i,\, i<k\}} \tilde{p}_k(\boldsymbol{x}) = \prod_{i\in\mathcal{C}_k} \hat{p}_{i,k}(\boldsymbol{x}), \tag{B.33}$$

其中

$$\hat{p}_{i,k}(\boldsymbol{x}) = \max_{x_i} \phi_{i,k}(x_i, x_k)\hat{p}_i(\boldsymbol{x}). \tag{B.34}$$

等式(B.34)为最大化更新规则，我们需要在图 B.2a 中所有用方块表示的因子上对其进行求值，而(B.33)则是乘积规则，我们需要在节点上对其进行求值。概率分布 $\hat{p}_{i,k}(x)$ 通常被看作是子节点 i 到父节点 k 的消息传递信息，因此通常把它写作 $m_{i,k}(x_k)$ (Yedidia, Freeman, and Weiss 2001)或 $\mu_{i->k}(x_k)$ (Bishop 2006)。

我们可以使用与动态规划(有时也称"最大化求和算法"(Bishop 2006))中的前向和后向扫描一样的扫描算法，利用最大化乘积规则计算树形图的 MAP 估计。另外还有一条规则称为"求和-乘积规则"。它将(B.34)中所有的可能值相加，而不是取它们的最大值。这实际上是在计算期望分布，而不是在计算最大似然分布。但这也就给出了一组可以用于计算边缘分布 $b_i = \sum_{x\setminus x_i} p(x)$ 的概率估计(Pearl 1988; Yedidia, Freeman, and Weiss 2001; Bishop 2006)。

置信传播可能不会产生关于含有回路的图的最佳估计，原因与动态规划针对含有回路的图无效的原因相同，即一个具有很多父节点的节点对每个父节点而言都会有一个最优值，而这些最优值不一定相等，即没有一个唯一的消去顺序。早期将置信传播推广到含有回路的图上的算法，被称为"带环的置信传播算法"，在图上以并行的方式执行更新，即它们使用的是同步更新策略(Frey and MacKay 1997; Freeman, Pasztor, and Carmichael 2000; Yedidia, Freeman, and Weiss 2001; Weiss and Freeman 2001a,b; Yuille 2002; Sun, Zheng, and Shum 2003; Felzenszwalb and

Huttenlocher 2006)。

例如，Felzenszwalb and Huttenlocher(2006)将 $\mathcal{N}_4$ 图分解为红色和黑色(棋盘)分量，然后交替地进行将消息从红色节点传递到黑色节点及反方向的操作。他们还使用了多重网格(较粗层次)更新来加快收敛的速度。正如前面讨论的，为了把基本的最大化乘积更新规则(B.28)的复杂度从 $O(l^2)$ 降到 $O(l)$，他们针对一些代价函数 $V_{i,k}(x_i,x_k)$，包括 Potts 模型(delta 函数)、绝对值函数(总变差)、二次函数(高斯 MRF)，提出了特殊的更新算法。一种与这种算法相关的算法，平均场扩散(mean field diffusion)(Scharstein and Szeliski 1998)，也使用节点间的同步更新策略计算边缘分布。Yuille(2010)讨论了平均场理论和带环的置信传播之间的关系。

最近的多环置信传播算法及其变体都使用了对图进行顺序扫描的方法(Szeliski, Zabih, Scharstein et al. 2008)。例如，Tappen and Freeman(2003)沿每一行将消息从左边传递到右边，当到达行的末端时再将消息传递的方向反转。这类似于将每一行作为一棵独立的树(链)进行处理，只是需要包含来自该行上面和下面节点的信息。然后他们再沿每一列进行类似的计算。相对于同步更新而言，顺序更新可以使信息在图像中传播得更快。

经过 Szeliski, Zabih, Scharstein et al. (2008)等人测试的置信传播的另一个变体，他们称之为"BP-S"或"TRW-S"，是基于 Kolmogorov(2006)对 Wainwright, Jaakkola, and Willsky(2005)等人的树形重加权消息传递的顺序扩展的。TRW 首先从邻接图中选择一组树，并在每棵树上计算一组概率分布。然后，用这些概率分布对带环置信传播中传递的消息进行重新加权。TRW 的顺序版本，叫 TRW-S，则以扫描线的方式，使用前向和后向传递对节点进行处理。在前向传递中，每个节点向它的右邻居和下邻居发送消息。而在后向传递中，消息则被发送到左邻居和上邻居。TRW-S 同样还计算了能量值的下界，Szeliski, Zabih, Scharstein et al. (2008)用它来估计这种算法得到的结果与所有评测 MRF 推断算法可能得到的最好结果。

与动态规划一样，随着每个因子团簇的阶增大，置信传播方法的效率也会变低。Potetz and Lee(2008)展示了在满足其因子涉及线性和接续非线性和条件的连续值问题中，如何将置信传播的复杂度降回到与团簇的阶呈线性关系，而这种条件在更复杂的 MRF 模型——如专家场(fields of experts)(Roth and Black 2009)和可操控随机场(steerable random fields)(Roth and Black 2007b)——中是很典型的。Kohli, Kumar, and Torr(2009)和 Alahari, Kohli, and Torr(2011)则在图割的背景下发展出了另一种处理高阶团簇的方法。

B.5.4 图割

计算机视觉领域将"图割"(graph cut)作为描述一大类基于求解一个或多个最小割或最大流问题的 MRF 推断算法的非正式名称(Boykov, Veksler, and Zabih 2001; Boykov and Kolmogorov 2010; Boykov,Veksler, and Zabih 2010; Ishikawa and Veksler

2010)。

MRF 图割最简单的例子是一种用于对二值 MRF 进行精确最小化的多项式时间复杂度的算法。该算法最初由 Greig, Porteous, and Seheult(1989)等人提出，Boykov, Veksler, and Zabih (2001)和 Boykov and Jolly (2001)则让计算机视觉领域开始关注这个算法。从 MRF 能量函数到最小割图的基本构造方法如图 B.3 所示，详情参见 3.7.2 节和 5.5 节。简单说来，首先将 MRF 中的节点连接到特殊的源节点和漏(sink)节点，然后使用多项式时间复杂度的最大流算法计算这两个节点间的最小割，最小割的总代价恰好就是在某一种二值赋值下 MRF 的能量值(Goldberg and Tarjan 1988; Boykov and Kolmogorov 2004)。

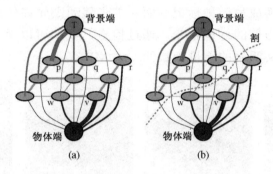

图 B.3 用于最小化二值子模块 MRF 能量的图割(Boykov and Jolly 2001)© 2001 IEEE：(a) 按照最大流问题方式编码的能量函数；(b)决定区域边界的最小割

正如在 5.5 节中讨论的那样，人们已经对这个基本算法做出了很多重要的扩展，这些扩展涉及有向边(Kolmogorov and Boykov 2005)、更大规模的邻域(Boykov and Kolmogorov 2003; Kolmogorov and Boykov 2005)、关于连通性的先验知识(Vicente, Kolmogorov and Rother 2008)、关于形状的先验知识(Lempitsky and Boykov 2007; Lempitsky, Blake, and Rother 2008)。Kolmogorov and Zabih(2004)形式化地描绘了一类二值能量势函数(正则化条件)的特征，在这一类能量势函数的约束下，这些算法可以找到全局最小值。Komodakis, Tziritas and Paragios(2008)，以及 Rother, Kolmogorov, Lempitsky et al. (2007)则给出了在这些算法未能找到全局最小值的情况下可以使用的一些很好的算法。

二值 MRF 问题也可以通过如下方法得到近似的解决。首先将其转化为连续的(0,1)问题，然后或者按照线性系统的方法对其进行求解(Grady 2006; Sinop and Grady 2007; Grady and Alvino 2008; Grady 2008; Grady and Ali 2008; Singaraju, Grady, and Vidal 2008; Couprie, Grady, Najman et al. 2009)(随机游走模型)，或者通过计算测地线距离对其进行求解(Bai and Sapiro 2009; Criminisi, Sharp, and Blake 2008)，最后对结果进行二值化。5.5 节提供了更多关于这些方法的细节描述，而一份不错的关于这些方法的回顾则可以在 Singaraju, Grady, Sinop et al. (2010)的工作中找到。与连续分割方法的一个不同的联系，由 Boykov, Kolmogorov, Cremers et al. (2006)提出，是与水平集文献的联系(5.1.4 节)，它提出了一种基于组合图割算法求解表面传播 PDE 的算法——Boykov and Funka-Lea(2006)讨论了这个方法及其他相关的方法。

多值 MRF 推断问题通常需要求解一系列相关的二值 MRF 问题(Boykov, Veksler, and Zabih 2001),虽然在某些特殊情况下,如势能函数是某些凸函数,单独一次图分割已经足够了(Ishikawa 2003; Schlesinger and Flach 2006)。这一领域内的开创性工作是由 Boykov, Veksler and Zabih(2001)做出的,他们引入了两个算法,分别称为"交换移动"和"扩张移动"。图 B.4 对这两个算法进行了简要的描述。$\alpha-\beta$ 交换移动首先选择两个标记(通常通过在所有可能的标记对中循环得到),然后形式化出一个二值 MRF 问题,在这个问题中,当前被标记为 α 或 β 的像素可以任意将自己的标记转换成另一种标记。而 α 扩张移动则允许任意位于 MRF 中的像素采用 α 标记或保持当前的标记不变。通过检查,我们可以很容易地发现这两种移动都可以产生带有定义明确的能量函数的二值 MRF。

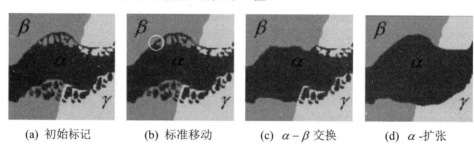

(a) 初始标记　　(b) 标准移动　　(c) $\alpha-\beta$ 交换　　(d) α-扩张

图 B.4　来自(Boykov, Veksler, and Zabih 2001)的多层图优化© 2001 IEEE:(a)初始的问题结构;(b)标准移动只改变一个像素;(c) $\alpha-\beta$ 交换对所有标记为 α 和标记为 β 的像素的标签进行了最佳交换;(d) α 扩张移动从当前像素的标记和 α 标记中作出最优选择

由于这些算法在它们的内循环中使用了二值 MRF 的最优化过程,因此它们也需要满足出现在二值标记问题中的关于能量函数的约束(Kolmogorov and Zabih 2004)。不过,最近的一些算法,如由 Komodakis, Tziritas, and Paragios (2008)及 Rother, Kolmogorov, Lempitsky et al. (2007)提出的算法,也可以用于为更一般的能量函数提供近似解。一些适用于在线应用的用于重用之前求解结果(流或割的再利用)的高效算法也已经被开发出来,如动态 MRF 和由粗到精的分带图割(Agarwala, Zheng, Pal et al. 2005; Lombaert, Sun, Grady et al. 2005; Juan and Boykov 2006)。同时,现在还可以对在 α 扩张移动过程中使用的标记的数量进行最小化求解(Delong, Osokin, Isack et al. 2010)。

在实验对比中,α 扩张移动通常比 $\alpha-\beta$ 交换移动更快地收敛到一个不错的解(Szeliski, Zabih, Scharstein et al. 2008),对那些包含大量具有相同标记的区域的问题,如在图像拼接中对源图像的标记(图 3.60)问题,这种优势更为明显。另外,针对在顺序值上定义的截断的凸能量函数,人们也已经提出了更加精确的算法。这种算法在每次最小割过程中都会考虑在取值范围内的全部标签,并且通常会产生更低的能量值(Veksler 2007; Kumar and Torr 2008; Kumar,Veksler, and Torr 2010)。整个高效 MRF 推断算法领域正在快速发展中，最近的特刊(Kohli and Torr 2008; Komodakis,Tziritas, and Paragios 2008; Olsson, Eriksson, and Kahl 2008; Potetz and Lee 2008)、文章(Alahari, Kohli, and Torr 2011)和一本即将出版的著作(Blake, Kohli, and Rother 2010)便是明证。

B.5.5 线性规划

这一节由 Vladimir Kolmogorov 提供，非常感谢！

许多成功的 MRF 最优化算法都基于对能量函数进行线性规划(LP)松弛(Weiss, Yanover, and Meltzer 2010)。虽然一般的 MRF 推断问题隶属为 NP-难题，但对某些实际的 MRF 问题，基于 LP 的方法可以产生全局最小解(Meltzer, Yanover, and Weiss 2005)。为了描述这种松弛，我们首先按如下方式重写(B.26)中的能量函数

$$E(\boldsymbol{x}) = \sum_{(i,j)\in\mathcal{N}} V_{i,j}(x_i,x_j) + \sum_i V_i(x_i) \tag{B.35}$$

$$= \sum_{i,j,\alpha,\beta} V_{i,j}(\alpha,\beta) x_{i,j;\alpha,\beta} + \sum_{i,\alpha} V_i(\alpha) x_{i;\alpha} \tag{B.36}$$

$$\text{subject to} \quad x_{i;\alpha} = \sum_{\beta} x_{i,j;\alpha,\beta} \quad \forall (i,j)\in\mathcal{N},\alpha, \tag{B.37}$$

$$x_{j;\beta} = \sum_{\alpha} x_{i,j;\alpha,\beta} \quad \forall (i,j)\in\mathcal{N},\beta, \quad \text{and} \tag{B.38}$$

$$x_{i,\alpha}, \ x_{i,j;\alpha,\beta} \ \in \ \{0,1\}. \tag{B.39}$$

其中，α 和 β 在整个标记的值域上取值，另外 $x_{i;\alpha} = \delta(x_i - \alpha)$ 和 $x_{ij;\alpha\beta} = \delta(x_i - \alpha)\delta(x_j - \beta)$ 分别是赋值 $x_i = \alpha$ 和 $(x_i, x_j) = (\alpha, \beta)$ 的指示器变量。将(B.39)中的离散约束换成线性约束 $x_{ij;\alpha\beta} \in [0,1]$，我们就可以得到 LP 松弛。可以很容易地看出，(B.36)的最优值也就是(B.36)的一个下界。

从 Schlesinger(1976)的工作开始，松弛问题已经在各种著作中被广泛研究。一个重要的问题是如何有效地求解这个 LP。不幸的是，通用的 LP 求解方法不能处理视觉中的大规模问题(Yanover, Meltzer, and Weiss 2006)。于是人们便提出了很多定制的迭代方法。这些迭代方法中的大部分都是在求解对偶问题，即它们首先形式化出(B.36)的一个下界，然后尝试去最大化这个下界。通常会使用对树进行凸组合这种方法对下界进行形式化，该方法是由 Wainwright, Jaakkola, and Willsky 2005 提出的。

LP 的下界可以通过很多种方法进行最大化，如最大和扩散(Werner 2007)、基于树的重加权的消息传递(TRW)(Wainwright, Jaakkola, and Willsky 2005; Kolmogorov 2006)、次梯度方法(Schlesinger and Giginyak 2007a,b; Komodakis, Paragios, and Tziritas 2007)和 Bregman 投影(Ravikumar, Agarwal, and Wainwright 2008)。注意，最大和传播和 TRW 算法并不能保证收敛到 LP 的全局最大值——它们可能会陷入一个次优点(Kolmogorov 2006; Werner 2007)。但在实际应用中，这似乎并不算是一个问题(Kolmogorov 2006)。

对一些视觉应用而言，基于松弛法(B.36)的算法可以产生极好的结果。但并不

能保证在所有的情况下都会有这种结果——毕竟，这个问题是 NP-难题。最近，研究人员又发明了另外一些线性规划松弛方法(Sontagand Jaakkola 2007; Sontag, Meltzer, Globerson et al. 2008; Komodakis and Paragios 2008; Schraudolph 2010)。相较于(B.36)，这些算法产生更紧的界，但也需要花费额外的计算代价。

LP 松弛与 α 扩张 求解一个线性规划问题会产生原始解和对偶解，并且这两个解满足互补松弛条件。一般而言，(B.36)的原始解并不是一个整数解，因此在实际应用中，我们可能要通过对其进行舍入来获得一个有效的标记 x。还有一种由 Komodakis and Tziritas (2007a); Komodakis, Tziritas, and Paragios (2007)提出的可以选用的方法是搜索原始解和对偶解，使得这两个解满足近似互补松弛条件且原始解是一个整数值。为了实现这个方法，(Komodakis andTziritas 2007a; Komodakis, Tziritas, and Paragios 2007)提出了一些基于最大流的算法，并且在这些算法中，Fast-PD 方法(Komodakis, Tziritas, and Paragios 2007)表现得最为出色。在度量相互作用(metric interactions)的情况下，Fast-PD 的默认版本与 α 扩张算法(Boykov, Veksler, and Zabih 2001)产生的原始解相同。这也给出了 α 扩张算法的一种很有趣的解释，即 α 扩张算法试图近似地求解(B.36)中的松弛问题。

与标准的 α 扩张算法不同，Fast-FD 还会维护一个对偶解，因此在实际中它运行得更快。Fast-FD 算法可以扩展到半度量相互作用(semi-metric interactions)的情形下(Komodakis, Tziritas, and Paragios 2007)。这种扩展的原始版本也是由 Rother, Kumar, Kolmogorov et al. (2005)给出的。

B.6 不确定性估计(误差分析)

除了计算最可能的估计值，很多应用也需要一个对这种估计的不确定程度的评估[①]。进行这种评估最一般的方法是计算出所有未知量上的一个完全的概率分布，但通常这是不可行的。不过也存在一种特例，可以轻松得到这种分布的简单描述，其中之一便是对含有高斯噪声的数据进行线性估计的问题，这种问题的联合能量函数(后验估计的负对数似然度)是一个二次函数。在这种条件下，后验分布是一个多变量的高斯分布，并且协方差矩阵可以直接通过对问题的 Hessian 矩阵求逆计算出来。(协方差矩阵的逆矩阵——在这种简单的情况下，它与 Hessian 矩阵相等——的另一个名字是信息矩阵)。

但即使在这种情况下，整个协方差矩阵的规模还是可能十分庞大，以致不能够计算和存储。例如，在大规模的由运动到结构的问题中，一个大规模的稀疏的 Hessian 矩阵通常会产生一个满秩且稠密的协方差矩阵。在这种情况下，只报告估计量的方差或简单的关于一些独立参数(如 3D 点的位置或摄像机的姿态估计)的协方差估计，往往是可以接受的(Szeliski 1990a)。对这个问题更加深刻的理解，如不

---

① 在经典的摄影测量学应用中尤其如此，在这种应用中，对精度的报告几乎总是必不可少的(Förstner 2005)。

确定性的主要模态，可以通过使用特征值分析的方法获得(Szeliski and Kang 1997)。

对那些后验能量函数不是二次函数的问题，如非线性的或鲁棒的最小二乘问题，通常还是可以在最优解的附近得到 Hessian 矩阵的估计的。在这种情况下，不确定性(协方差)的 Cramer–Rao 下界在计算时可以被认为是 Hessian 矩阵的逆矩阵。换句话说，虽然局部的 Hessian 矩阵可能会对能量函数的"宽度"估计不足，但协方差不可能小于基于这种局部二次拟合得到的估计。在一般的 MRF 问题中，如果 MAP 推断是使用图割进行的，那么我们还是可以估计出一种不同种类的不确定性(最小边际能量)的(Kohli and Torr 2008)。

虽然很多计算机视觉应用都会忽略不确定性模型，但计算出这些不确定性估计，然后通过它们来直观地体会当前估计值的可靠性，通常还是十分有用的。而某些应用，如卡曼滤波，为了以一种最优方式把新的测量值与以前计算出来的估计值整合在一起，则需要这种不确定性(或显式形式的后验协方差，或隐式形式的协方差矩阵的逆)计算。

附录 C
补 充 材 料

C.1 数据集
C.2 软件
C.3 幻灯片与讲座
C.4 参考文献

在这最后一个附录中，我整理了一些补充材料，这些材料可能会对学生、教师及研究人员有所帮助。本书网站 *http://szeliski.org/Book* 列有一份数据集和软件最新列表，所以也请同时查看一下。

C.1 数 据 集

在开发可靠的视觉算法过程中，关键的一点是在具有挑战性和代表性的数据集上测试你的程序。如果可以获得已知的正确结果或者其他人的结果，那么这种测试就可以提供更多的信息(同时也更为量化)。

近年来，为了测试和评价计算机视觉算法，人们已建立了大量的数据集。这些数据集(和软件)中的一些在 Computer Vision 主页[①]中列出。而一些较新的网站，例如 CVonline (*http://homepages.inf.ed.ac.uk/rbf/CVonline/*)，VisionBib.Com (*http://datasets.visionbib.com/*) 和 Computer Vision online(*http://computervisiononline.com/*)，则有指向较新资源的链接。

下面我列出一些较为流行的数据集，并按照本书的章节对它们进行分组，分组结果表明它们与书中的哪一章联系最为紧密。

第 2 章：图像形成

- CUReT：Columbia-Utrecht Reflectance and Texture Database，*http://www1.cs.columbia.edu/CAVE/software/curet/*，(Dana, van Ginneken, Nayar *et al.* 1999)。
- Middlebury Color Datasets：包含不同相机拍摄的彩色图片，用于研究相机如何对色域和颜色进行变换，*http://vision.middlebury.edu/color/data/*，(Chakrabarti, Scharstein, and Zickler 2009)。

第 3 章：图像处理

- Middlebury 测试数据集：用于评价 MRF 最小化/MRF 推断算法，*http://vision.middlebury.edu/MRF/results/*，(Szeliski, Zabih, Scharstein *et al.* 2008)。

第 4 章：特征检测与匹配

- Affine Covariant Features 数据库：用于评测特征检测器与特征描述子的质量和可重复性，*http://www.robots.ox.ac.uk/_vgg/research/affine/*，(Mikolajczyk and Schmid 2005; Mikolajczyk, Tuytelaars, Schmid *et al.* 2005)。
- 一个用于学习和特征描述子评测的已匹配的图像块数据库，

[①] *http://www.cs.cmu.edu/_cil/vision.html*，但 2004 年后没有继续维护。

http://cvlab.epfl.ch/_brown/patchdata/patchdata.html，(Winder and Brown 2007; Hua, Brown, and Winder 2007)。

第 5 章：分割

- Berkeley Segmentation Dataset and Benchmark：拥有 1000 幅标注好的图像，这些图像由 30 个人标注，同时也提供了测试程序，*http://www.eecs.berkeley.edu/Research/Projects/CS/vision/grouping/segbench/*，(Martin, Fowlkes, Tal *et al.* 2001)。
- Weizmann 分割测试数据库，包括 100 幅灰度图像，同时附有标准分割结果，*http://www.wisdom.weizmann.ac.il/_vision/Seg Evaluation DB/index.html*，(Alpert, Galun, Basri *et al.* 2007)。

第 8 章：稠密运动估计

- The Middlebury 光流测试网站，*http://vision.middlebury.edu/flow/data*，(Baker, Scharstein, Lewis *et al.* 2009)。
- The Human-Assisted Motion Annotation 数据库，*http://people.csail.mit.edu/celiu/motionAnnotation/*，(Liu, Freeman, Adelson *et al.* 2008)。

第 10 章：计算摄影学

- High Dynamic Range 辐照图集，*http://www.debevec.org/Research/HDR/*，(Debevec and Malik 1997)。
- Alpha 抠图评测网站，*http://alphamatting.com/*，(Rhemann, Rother, Wang *et al.* 2009)。

第 11 章：立体视觉对应

- Middlebury Stereo Datasets and Evaluation，*http://vision.middlebury.edu/stereo/*，(Scharstein and Szeliski 2002)。
- Stereo Classification and Performance Evaluation，用于评测立体视觉对应中不同的聚类代价，*http://www.vision.deis.unibo.it/spe/SPEHome.aspx*，(Tombari, Mattoccia, Di Stefano *et al.* 2008)。
- Middlebury Multi-View Stereo Datasets，*http://vision.middlebury.edu/mview/data/*，(Seitz, Curless, Diebel *et al.* 2006)。
- Multi-view and Oxford Colleges 建筑物重建，*http://www.robots.ox.ac.uk/~vgg/data/data-mview.html*。
- Multi-View Stereo Datasets，*http://cvlab.epfl.ch/data/strechamvs/*，(Strecha, Fransens, and Van Gool 2006)。
- Multi-View Evaluation，*http://cvlab.epfl.ch/_strecha/multiview/*，(Strecha, von Hansen, Van Gool *et al.* 2008)。

第 12 章：3D 重建

- HumanEva：人工合成的视频及运动捕捉数据库，用于评测关节型的人体活动，*http://vision.cs.brown.edu/humaneva/*，(Sigal, Balan, and Black 2010)。

第 13 章：基于图像的绘制

- The (New) Stanford Light Field Archive，*http://lightfield.stanford.edu/*，(Wilburn, Joshi, Vaish *et al.* 2005)。
- Virtual Viewpoint Video：多视点视频，同时附有每帧的深度图，*http://research.microsoft.com/en-us/um/redmond/groups/ivm/vvv/*，(Zitnick, Kang, Uyttendaele *et al.* 2004)。

第 14 章：识别

如果需要有关视觉识别数据库的列表，请参见表 14.1–14.2。除了这些以外，还有：

- Buffy 姿态类别集，*http://www.robots.ox.ac.uk/~vgg/data/buffy_pose_classes/* 和 Buffy 棒状人 V2.1 *http://www.robots.ox.ac.uk/~vgg/data/stickmen/index.html*，(Ferrari, Marin-Jimenez, and Zisserman 2009; Eichner and Ferrari 2009)。
- H3D：带有标注的人体姿态/关节照片集 *http://www.eecs.berkeley.edu/_lbourdev/h3d/*，(Bourdev and Malik 2009)。
- Action Recognition Datasets，*http://www.cs.berkeley.edu/projects/vision/action*，拥有指向其他一些动作和活动识别数据库的链接，同时也包含指向一些论文的链接。*http://www.nada.kth.se/cvap/actions/* 的人体运动数据库则拥有更多的动作序列。

C.2 软　　件

计算机视觉算法最好的一个资源为 Open Source Computer Vision(OpenCV)[①]库(*http://opencv.willowgarage.com/wiki/*)，它由 Gary Bradski 与其 Intel 的同事一同开发，现在 Willow Garage(Bradsky and Kaehler 2008)负责对其进行维护和扩展。它的可用函数的部分列表(通过 *http://opencv.willowgarage.com/documentation/cpp/* 得到)如下：

- 图像处理及变换(滤波，形态学，金字塔)；
- 图像的几何变换(旋转，缩放)；
- 其他五花八门的图像变换(傅里叶变换，距离变换)；

[①] 编注：有关 Open CV，可以参见 Gary Bradski 的著作《学习 Open CV(中文版)》。

- 直方图；
- 分割(分水岭，均值移位)；
- 特征检测(Canny, Harris, Hough, MSER, SURF)；
- 运动分析及物体跟踪(Lucas–Kanade，均值移位)；
- 相机标定和 3D 重建；
- 机器学习(k 近邻、支持向量机、决策树、boosting、随机森林、期望-最大化以及神经网络)；

The Intel Performance Primitives(IPP)库(http://software.intel.com/en-us/intel-ipp/)包含进行多种图像处理任务的高度优化的代码。一旦完成安装，许多 OpenCV 的例程就会凭借这个库运行得更快。就功能而言，它包含很多与 OpenCV 一致的运算符，除此之外，它还包括图像和视频压缩、信号和语音处理以及矩阵代数方面的库。

The MATLAB Image Processing Toolbox(http://www.mathworks.com/products/image/)包含的例程可用于空域变换(旋转、缩放)，计算归一化互相关，图像分析和统计(边缘检测，Hough 变换)，图像增强(自适应直方图均衡化，中值滤波)以及图像恢复(去模糊)，线性滤波(卷积)，图像变换(Fourier 和 DCT)，形态学操作(连通分量和距离变换)等方面。

另外两个较早的库是 VXL(C++ Libraries for Computer Vision Research and Implementation，http://vxl.sourceforge.net/)和 LTI-Lib 2(http://www.ie.itcr.ac.cr/palvarado/ltilib-2/homepage/)。这两个库现在已经不再进行开发了，但它们依旧含有很多有用的例程。

照片编辑和查看包，例如 Windows Live Photo Gallery，iPhoto，Picasa，GIMP 以及 IrfanView，在进行常用处理任务、格式转换以及查看结果等方面还是很有用的。另外，如果你尝试着从头实现一些图像处理算法(如颜色校正或去除噪声)，那么它们同样也是有趣的参考实现。

也有一些软件包和基本结构对构建实时视频处理样例有很大帮助。Vision on Tap(http://www.visionontap.com/)提供了一种 Web 服务可以实时处理你通过网络摄像头拍摄的视频(Chiu and Raskar 2009)。Video-Man(VideoManager，http://videomanlib.sourceforge.net/)在运行基于视频的样例和应用时十分有用。在 MATLAB 中，还可以使用 imread 直接通过任何 URL 读取图片，例如你可以直接读取网络摄像机的结果。

下面我列出了其他一些可以在网上找到的软件，并按照本书的章节对它们进行了分组，分组结果表明了它们与书中的哪一章联系最为紧密。

第 3 章：图像处理

- matlabPyrTools—Laplacian 金字塔、QMF/小波变换和导向金字塔的 MATLAB 代码，http://www.cns.nyu.edu/~lcv/software.php，(Simoncelli and Adelson 1990a; Simoncelli, Freeman, Adelson et al. 1992)。

- BLS-GSM 图像去噪，*http://decsai.ugr.es/~javier/denoise/*，(Portilla, Strela, Wainwright *et al.* 2003)。
- 快速双边滤波代码，*http://people.csail.mit.edu/jiawen/#code*，(Chen, Paris, and Durand 2007)。
- 快速距离变换算法的 C++实现，*http://people.cs.uchicago.edu/_pff/dt/*，(Felzenszwalb and Huttenlocher 2004a)。
- GREYC's Magic Image Converter，包括使用正则化和各向异性扩散实现图像恢复的软件，*http://gmic.sourceforge.net/gimp.shtml*，(Tschumperlé and Deriche 2005)。

第 4 章：特征检测与匹配

- VLFeat，一个开源且可移植的计算机视觉算法库，*http://vlfeat.org/*，(Vedaldi and Fulkerson 2008)。
- SiftGPU：Scale Invariant Feature Transform (SIFT)的一种 GPU 实现，*http://www.cs.unc.edu/_ccwu/siftgpu/*，(Wu 2010)。
- SURF：Speeded Up Robust Features，*http://www.vision.ee.ethz.ch/~surf/*，(Bay, Tuytelaars, and Van Gool 2006)。
- FAST 角点检测，*http://mi.eng.cam.ac.uk/~er258/work/fast.html*，(Rosten and Drummond 2005, 2006)。
- 适用于 Linux 的仿射区域检测器与描述子的二进制文件，另外还有用于计算可重复性和匹配分数的 MATLAB 文件，*http://www.robots.ox.ac.uk/_vgg/research/affine/*。
- Kanade–Lucas–Tomasi 特征跟踪器：KLT，*http://www.ces.clemson.edu/_stb/klt/*，(Shi and Tomasi 1994)；GPU-KLT，*http://cs.unc.edu/~cmzach/opensource.html*. (Zach, Gallup, and Frahm 2008)；Lucas–Kanade 20 Years On，*http://www.ri.cmu.edu/projects/project_515.html/*，(Baker and Matthews 2004)。

第 5 章：分割

- 基于图的高效图像分割，*http://people.cs.uchicago.edu/_pff/segment/*，(Felzenszwalb and Huttenlocher 2004b)。
- EDISON，边缘检测和图像分割，*http://coewww.rutgers.edu/riul/research/code/EDISON/*，(Meer and Georgescu 2001; Comaniciu and Meer 2002)。
- 规范割分割，包括中间轮廓，*http://www.cis.upenn.edu/_jshi/software/*，(Shi and Malik 2000; Malik, Belongie, Leung *et al.* 2001)。
- 基于加权聚类的分割(SWA)，*http://www.cs.weizmann.ac.il/_vision/SWA/*，(Alpert, Galun, Basri *et al.* 2007)。

第 6 章：基于特征的配准

- 非递归的 PnP 算法，*http://cvlab.epfl.ch/software/EPnP/*，(Moreno-Noguer, Lepetit, and Fua 2007)。
- Tsai Camera Calibration Software，*http://www-2.cs.cmu.edu/~rgw/TsaiCode.html*，(Tsai 1987)。
- Easy Camera Calibration Toolkit，*http://research.microsoft.com/en-us/um/people/zhang/Calib/*，(Zhang 2000)。
- Camera Calibration Toolbox for MATLAB，*http://www.vision.caltech.edu/bouguetj/calib_doc/*；OpenCV 中包含一个 C 版本的实现。
- 用于多视角几何学的 MATLAB 函数，*http://www.robots.ox.ac.uk/_vgg/hzbook/code/*，(Hartley and Zisserman 2004)。

第 7 章：由运动到结构

- SBA：一个基于 Levenberg–Marquardt 算法的通用稀疏光束平差 C/C++包，*http://www.ics.forth.gr/lourakissba/*，(Lourakis and Argyros 2009)。
- 简单稀疏光束平差(SSBA)，*http://cs.unc.edu/cmzach/opensource.html*。
- Bundler，无序图像集的由运动到结构系统，*http://phototour.cs.washington.edu/bundler/*，(Snavely, Seitz, and Szeliski 2006)。

第 8 章：稠密运动分析

- 光流软件，*http://www.cs.brown.edu/~black/code.html*，(Black and Anandan 1996)。
- 基于全变分和共轭梯度下降的光流，*http://people.csail.mit.edu/celiu/OpticalFlow/*，(Liu 2009)。
- TV-L1 光流的 GPU 实现，*http://cs.unc.edu/~cmzach/opensource.html*，(Zach, Pock, and Bischof 2007a)。
- Elastix：一个用于图像刚性及非刚性注册的工具箱，*http://elastix.isi.uu.nl/*，(Klein, Staring, and Pluim 2007)。
- 基于离散最优化的图像可变形性注册，*http://www.mrf-registration.net/deformable/index.html*，(Glocker, Komodakis, Tziritas et al. 2008)。

第 9 章：图像拼接

- 用于拼接图像的 Microsoft Research Image Compositing Editor，*http://research.microsoft.com/en-us/um/redmond/groups/ivm/ice/*。

第 10 章：计算摄影学

- HDRShop：将包围曝光加入到高动态范围辐照图中的软件，

http://projects.ict.usc.edu/graphics/HDRShop/。

- 超分辨率代码，*http://www.robots.ox.ac.uk/~vgg/software/SR/*，(Pickup 2007; Pickup, Capel, Roberts *et al.* 2007, 2009)。

第 11 章：立体视觉对应

- StereoMatcher，独立的 C++ 立体视觉对应代码，*http://vision.middlebury.edu/stereo/code/*，(Scharstein and Szeliski 2002)。
- 基于块的多视角立体视觉软件 (PMVS Version 2)，*http://grail.cs.washington.edu/software/pmvs*，(Furukawa and Ponce 2011)。

第 12 章：3D 重建

- Scanalyze：一个用于对齐及合并距离数据的系统，*http://graphics.stanford.edu/software/scanalyze/*，(Curless and Levoy 1996)。
- MeshLab：用于处理、编辑和可视化无结构 3D 三角网格的软件，*http://meshlab.sourceforge.net/*。
- 使用 VRML 查看器(种类不胜枚举)也是一种较好的可视化 3D 纹理映射模型的方法。

12.6.4 节：完整人体建模与跟踪

- 贝叶斯 3D 人体跟踪，*http://www.cs.brown.edu/~black/code.html*，(Sidenbladh, Black, and Fleet 2000; Sidenbladh and Black 2003)。
- HumanEva：跟踪关节型人体运动的基本代码，*http://vision.cs.brown.edu/humaneva/*，(Sigal, Balan, and Black 2010)。

14.1.1 节：人脸检测

- 人脸检测的样例代码和评测工具，*http://vision.ai.uiuc.edu/mhyang/face-detection-survey.html*。

14.1.2 节：行人检测

- 一个使用 boosting 算法的简单物体检测器，*http://people.csail.mit.edu/torralba/shortCourseRLOC/boosting/boosting.html*，(Hastie, Tibshirani, and Friedman 2001; Torralba, Murphy, and Freeman 2007)。
- 区分式训练的可变形部件模型，*http://people.cs.uchicago.edu/~pff/latent/*，(Felzenszwalb, Girshick, McAllester *et al.* 2010)。
- 上身检测器，*http://www.robots.ox.ac.uk/~vgg/software/UpperBody/*，(Ferrari, Marin-Jimenez, and Zisserman 2008)。
- 2D 关节型人体姿态估计软件，*http://www.vision.ee.ethz.ch/~calvin/articulated_human_pose_estimation_code/*，(Eichner and Ferrari 2009)。

14.2.2 节：活动表观与 3D 形状模型

- AAMtools：一个活动表现建模工具箱，*http://cvsp.cs.ntua.gr/software/AAMtools/*，(Papandreou and Maragos 2008)。

14.3 节：实例识别

- FASTANN 和 FASTCLUSTER，用于近似 k 均值(AKM)，*http://www.robots.ox.ac.uk/~vgg/software/*，(Philbin, Chum, Isard et al. 2007)。
- 使用快速近似最近邻算法的特征匹配，*http://people.cs.ubc.ca/~mariusm/index.php/FLANN/FLANN*，(Muja and Lowe 2009)。

14.4.1 节：词袋

- 两个词袋分类器，*http://people.csail.mit.edu/fergus/iccv2005/bagwords.html*，(Fei-Fei and Perona 2005; Sivic, Russell, Efros et al. 2005)。
- 特征袋和层次 k 均值，*http://www.vlfeat.org/*，(Nistér and Stewénius 2006; Nowak, Jurie, and Triggs 2006)。

14.4.2 节：基于部件的模型

- 一个简单的部件和结构化物体检测器，*http://people.csail.mit.edu/fergus/iccv2005/partsstructure.html*，(Fischler and Elschlager 1973; Felzenszwalb and Huttenlocher 2005)。

14.5.1 节：机器学习软件

- 支持向量机(SVM)软件 (*http://www.support-vector-machines.org/SVMsoft.html*) 拥有指向很多 SVM 库的链接，包括 SVM^{light}，*http://svmlight.joachims.org/*；LIBSVM，*http://www.csie.ntu.edu.tw/~cjlin/libsvm/*，(Fan, Chen and Lin 2005)；LIBLINEAR，*http://www.csie.ntu.edu.tw/~cjlin/liblinear/*，(Fan, Chang, Hsieh et al. 2008)。
- Kernel Machines：连接有 SVM，高斯过程，boosting 及其他机器学习算法，*http://www.kernel-machines.org/software*。
- 关于图像分类的诸多核函数，*http://www.robots.ox.ac.uk/~vgg/software/MKL/*，(Varma and Ray 2007; Vedaldi, Gulshan, Varma et al. 2009)。

附录 A.1 和 A.2：矩阵分解和线性最小二乘[①]

- BLAS(Basic Linear Algebra Subprograms)，*http://www.netlib.org/blas/*，(Blackford, Demmel, Dongarra et al. 2002)。
- LAPACK(Linear Algebra PACKage)，*http://www.netlib.org/lapack/*，(Anderson,

---

[①] 感谢 Sameer Agarwal 建议和描述了其中大部分网站。

Bai, Bischof *et al*. 1999)。

- GotoBLAS，*http://www.tacc.utexas.edu/tacc-projects/*。
- ATLAS(Automatically Tuned Linear Algebra Software)，*http://math-atlas.sourceforge.net/*，(Demmel, Dongarra, Eijkhout *et al*. 2005)。
- Intel Math Kernel Library (MKL)，*http://software.intel.com/en-us/intel-mkl/*。
- AMDCore Math Library (ACML)，*http://developer.amd.com/cpu/Libraries/acml/Pages/default.aspx*。
- 鲁棒 PCA 代码，*http://www.salle.url.edu/_ftorre/papers/rpca2.html*，(De la Torre and Black 2003)。

附录 A.3：非线性最小二乘

- MINPACK，*http://www.netlib.org/minpack/*。
- Levmar：Levenberg–Marquardt 非线性最小二乘算法，*http://www.ics.forth.gr/~lourakis/levmar/*，(Madsen, Nielsen, and Tingleff 2004)。

附录 A.4 和 A.5：稀疏矩阵的直接和迭代解法

- SuiteSparse(多种重排算法，CHOLMOD)和 SuiteSparse QR，*http://www.cise.ufl.edu/research/sparse/SuiteSparse/*，(Davis 2006, 2008)。
- PARDISO(迭代法和稀疏直接解法)，*http://www.pardiso-project.org/*。
- TAUCS(稀疏直接解法，迭代法，基于外存，具有预处理器)，*http://www.tau.ac.il/~stoledo/taucs/*。
- HSL Mathematical Software Library，*http://www.hsl.rl.ac.uk/index.html*。
- 一些解线性系统的模板，*http://www.netlib.org/linalg/html templates/Templates.html*，(Barrett, Berry, Chan *et al*. 1994)。可以下载关于如何使用这些软件的 PDF 指导。
- ITSOL，MIQR 和其他解稀疏矩阵的程序，*http://www.users.cs.umn.edu/~saad/software/*(Saad 2003)。
- ILUPACK，*http://www-public.tu-bs.de/~bolle/ilupack/*。

附录 B：贝叶斯建模与推断

- Middlebury 的 MRF 最小化源代码，*http://vision.middlebury.edu/MRF/code/*，(Szeliski, Zabih, Scharstein *et al*. 2008)。
- 高效置信传播算法的 C++ 代码的一个早期版本，*http://people.cs.uchicago.edu/~pff/bp/*，(Felzenszwalb and Huttenlocher 2006)。
- FastPD MRF 最优化代码，*http://www.csd.uoc.gr/~komod/FastPD*，(Komodakis and Tziritas 2007a; Komodakis, Tziritas, and Paragios 2008)。

高斯噪声生成 大部分基本软件包中都包含均匀随机噪声生成器(如 Unix 中的 rand()例程)，但并非所有软件包都包含高斯随机噪声生成器。如果需要计算模拟一个正态分布的随机变量,可以使用 Box-Muller 变换(Box and Muller 1958),它对应的 C 代码已经在算法 C.1 中给出——请注意这段例程的返回值是随机变量对。另外一种可以生成高斯随机数的方法由 Thomas, Luk, Leong et al. (2007)给出。

```
double urand()
{
  return ((double) rand()) / ((double) RAND_MAX);
}
void grand(double& g1, double& g2)
{
#ifndef M_PI
#define M_PI 3.14159265358979323846
#endif // M_PI

  double n1 = urand();
  double n2 = urand();
  double x1 = n1 + (n1 == 0); /* guard against log(0) */
  double sqlogn1 = sqrt(-2.0 * log (x1));
  double angl = (2.0 * M_PI) * n2;
  g1 = sqlogn1 * cos(angl);
  g2 = sqlogn1 * sin(angl);
}
```

算法 C.1:高斯随机噪声产生算法,其中使用了 Box-Muller 变换

伪彩色生成 在很多应用中,对赋给图像(或图像特征,如各种线条)的标记做可视化是十分方便的。一种早期的方法是为每个整数标签赋予一个唯一的颜色。在我的工作过程中,我发现用下述方法将这些标签以一种准均匀的方式分布在 RGB 颜色立方体中是非常方便的。

对每个(非负)标签值,首先将它的各个比特位拆分到三个颜色通道上。例如,对一个 9 比特位的值来说,所有的位应被标记为 RGBRGBRGB。然后收集三个颜色通道对应的比特位,并将所得的比特位进行反转,使低位的比特位变化最快。在实际应用中,对 8 位颜色通道而言,这种位反转的结果可以存储在一个表格中,或者事先计算出一个完整的标记-伪彩色映射表(假设有 4092 项)来。图 8.16 显示了一个这种伪彩色映射的例子。

GPU 实现 借助于拥有像素着色器、计算着色器等功能的可编程 GPU,一些用于实时应用的快速计算机视觉算法,如分割、跟踪、立体视觉及运动估计等(Pock, Unger, Cremers *et al.* 2008; Vineet and Narayanan 2008; Zach, Gallup, and Frahm 2008),正在蓬勃发展。如果要学习这方面的算法,一个很好的资源是 CVPR 2008 关于 Visual Computer Vision on GPUs (CVGPU)的研讨会,*http://www.cs.unc.edu/jmf/Workshop on Computer Vision on GPU.html/*,其中的论文可以在 CVPR 2008 会议录 DVD 中找到。其他一些 GPU 算法资源包括 GPGPU 网站和研讨会,*http://gpgpu.org/*,以及 OpenVIDIA 网站,*http://openvidia.sourceforge.net/index.php/OpenVIDIA*。

C.3 幻灯片与讲座

正如我在前言中所说，我希望上传一些与这本书内容相关的幻灯片。在它们准备好之前，你最好查看我在华盛顿大学共同教学时所用的幻灯片，或者其他拥有相似大纲的相关课程的幻灯片。这里有一个这种课程的部分列表：

- UW 455: Undergraduate Computer Vision, *http://www.cs.washington.edu/education/courses/455/*。
- UW576: Graduate Computer Vision, *http://www.cs.washington.edu/education/courses/576/*。
- Stanford CS233B: Introduction to Computer Vision, *http://vision.stanford.edu/teaching/cs223b/*。
- MIT 6.869: Advances in Computer Vision, *http://people.csail.mit.edu/torralba/courses/6.869/6.869.computervision.htm*。
- Berkeley CS 280: Computer Vision, *http://www.eecs.berkeley.edu/~trevor/CS280.html*。
- UNC COMP 776: Computer Vision, *http://www.cs.unc.edu/~lazebnik/spring10/*。
- Middlebury CS 453: Computer Vision, *http://www.cs.middlebury.edu/_schar/courses/cs453-s10/*。

相关课程也在计算摄影学这个主题下讲授，例如：

- CMU 15-463: Computational Photography, *http://graphics.cs.cmu.edu/courses/15-463/*。
- MIT 6.815/6.865: Advanced Computational Photography, *http://stellar.mit.edu/S/course/6/sp09/6.815/*。
- Stanford CS 448A: Computational photography on cell phones, *http://graphics.stanford.edu/courses/cs448a-10/*。
- SIGGRAPH courses on Computational Photography, *http://web.media.mit.edu/raskar/photo/*。

这里也有一份可以获得的极好的在线讲座集，它覆盖计算机视觉中的很多领域，如置信传播、图分割等。UW-MSR Course of Vision Algorithms 的网址是 *http://www.cs.washington.edu/education/courses/577/04sp/*。

C.4 参考文献

本书引用的所有参考文献的目录(BibTex.bib 文件)可在本书网站找到，而一份内容更全面的、关于几乎所有计算机视觉出版物的部分带注释的文献目录由 Keith

Price 在 *http://iris.usc.edu/Vision-Notes/bibliography/contents.html* 维护。另外，一份可搜索的计算机图形学的参考文献目录存在于 *http://www.siggraph.org/publications/bibliography/*。其他一些不错的科技论文来源包括 Google Scholar 和 CiteSeer[x]。

词 汇 表

3D Rotations see Rotations 3D 旋转，见 Rotations
3D alignment 3D 配准
    absolute orientation   绝对方向
    orthogonal Procrustes   正交 Procrustes
3D photography 3D 摄影学
3D video 3D 视频

A

Absolute orientation 绝对方向
Active appearance model (AAM) 活动表观模型(AAM)
Active contours 活动轮廓
Active illumination 主动照明
Active rangefinding 主动距离测定
Active shape model (ASM) 活动形状模型(ASM)
Activity recognition 活动识别
Adaptive smoothing 自适应平滑
Affine transforms 仿射变换
Affinities (segmentation) 相似性(分割)
    normalizing   规范化
Algebraic multigrid 代数多重网格法
Algorithms 算法
    testing   测试
Aliasing 走样(混叠、混淆)
Alignment see Image alignment 配准，见 Image alignment
Alpha 透明度问题
    opacity   不透明度
    pre-multiplied   乘之前
Alpha matte 透明度遮罩
Ambient illumination 环境照明
Analog to digital conversion (ADC) 模数转换(ADC)
Anisotropic diffusion 各向异性扩散
Anisotropic filtering 各向异性滤波
Anti-aliasing filter 反走样滤波

Aperture		孔径光阑
Aperture problem		孔径问题
Applications		应用
	3D model reconstruction	3D 模型重建
	3D photography	3D 摄影学
	augmented reality	增强现实
	automotive safety	汽车安全性
	background replacement	背景替换
	biometrics	生物测定学
	colorization	彩色化
	de-interlacing	去隔行扫描
	digital heritage	数字遗产
	document scanning	文本扫描
	edge editing	边缘编辑
	facial animation	面部动画
	flash photography	闪影术
	frame interpolation	帧插值
	gaze correction	注视矫正
	head tracking	头部跟踪
	hole filling	空洞填充
	image restoration	图像复原
	image search	图像搜索
	industrial	工业
	intelligent photo editing	智能照片编辑
	Internet photos	因特网照片
	location recognition	位置识别
	machine inspection	机器检验
	match move	匹配移动
	medical imaging	医学成像
	morphing	变形
	mosaic-based video compression	基于全景拼图的视频压缩
	non-photorealistic rendering	非真实感绘制
	Optical character recognition (OCR)	光学字符识别(OCR)
	panography	全景图法
	performance-driven animation	基于表演的动画
	photo pop-up	照片弹出
	Photo Tourism	照片游览

Photomontage	照片蒙太奇
planar pattern tracking	平面模式跟踪
rotoscoping	转描机
scene completion	场景补全
scratch removal	划痕去除
single view reconstruction	单视图重建
tonal adjustment	色调调整
video denoising	视频去噪
video stabilization	视频稳定化
video summarization	视频摘要
video-based walkthroughs	基于视频的排演
VideoMouse	视频鼠标
view morphing	视图变形
visual effects	视觉效果
whiteboard scanning	白板扫描
z-keying	Z-键控
Arc length parameterization of a curve	曲线的弧长参数化
Architectural reconstruction	结构重建
Area statistics	区域统计
mean (centroid)	均值(质心)
perimeter	周长
second moment (inertia)	二阶矩(惯量)
Aspect ratio	横纵比
Augmented reality	增强现实
Auto-calibration	自标定
Automatic gain control (AGC)	自动增益控制(AGC)
Axis/angle representation of rotations	旋转的轴/角表达

B

B-snake,	B-蛇形
B-spline	B-样条
cubic	三次
multilevel	多层
octree	八叉树
Background plate	背景底片
Background subtraction (maintenance)	背景消减(维护)
Bag of words (keypoints)	词袋(关键点)
distance metrics	距离度量

Band-pass filter	带通滤波器
Bartlett filter, see Bilinear kernel	巴特利滤波器，见 Bilinear Kernel
Bayer pattern (RGB sensor mosaic)	贝叶斯样式(RGB 滤光片排列样式)
demosaicing	去马赛克
Bayes' rule	贝叶斯法则
MAP (maximum a posteriori) estimate	MAP(最大后验)估计
posterior distribution	后验分布
Bayesian modeling	贝叶斯建模
MAP estimate	MAP 估计
matting	抠图
posterior distribution	后验分布
prior distribution	先验分布
uncertainty	不确定性
Belief propagation (BP)	置信传播(BP)
update rule,	更新法则
Bias	偏差
Bidirectional Reflectance Distribution Function, see BRDF	双向反射分布函数，见 BRDF
Bilateral filter	双边滤波器
joint	联合
Range kernel	范围核
tone mapping	色调映射
Bilinear blending	双线性混合
Bilinear kernel	双线性核
Biometrics	生物测定学
Bipartite problem	二分问题
Blind image deconvolution	盲图像去卷积
Block-based motion estimation，(block matching)	基于块的运动估计，(块匹配)
Blocks world	积木世界
Blue screen matting	蓝屏抠图
Blur kernel	模糊核
estimation	估计
Blur removal	模糊去除
Body color	体色
Boltzmann distribution	玻尔兹曼分布
Boosting	提升
AdaBoost algorithm	自适应提升算法
decision stump	决策树桩

weak learner　　　　　　　　　　　　　　弱学习器
Border (boundary) effects　　　　　　　　边界效应
Boundary detection　　　　　　　　　　　边界检测
Box filter　　　　　　　　　　　　　　　方框型滤波器
Boxlet　　　　　　　　　　　　　　　　方框型小波
BRDF　　　　　　　　　　　　　　　　BRDF
　　anisotropic　　　　　　　　　　　　各向异性
　　isotropic　　　　　　　　　　　　　各向同性
　　recovery　　　　　　　　　　　　　复原
　　spatially varying (SVBRDF)　　　　　空间变化(SVBRDF)
Brightness　　　　　　　　　　　　　　亮的
Brightness constancy　　　　　　　　　　亮度恒常性
Brightness constancy constraint　　　　　亮度恒常性约束
Bundle adjustment　　　　　　　　　　　光束平差法，光束法平差

C

Calibration, see Camera calibration　　　　标定，见 Camera calibration
Calibration matrix　　　　　　　　　　　标定矩阵
Camera calibration　　　　　　　　　　　摄像机标定
　　accuracy　　　　　　　　　　　　　精度
　　aliasing　　　　　　　　　　　　　走样(混叠、混淆)
　　extrinsic　　　　　　　　　　　　　外部的
　　intrinsic　　　　　　　　　　　　　内在的，本征的
　　optical blur　　　　　　　　　　　　光学模糊
　　patterns　　　　　　　　　　　　　模式，样式
　　photometric　　　　　　　　　　　光度测定的
　　plumb-line method　　　　　　　　　垂直线方法
　　point spread function　　　　　　　　点扩散函数
　　radial distortion　　　　　　　　　　径向畸变
　　radiometric　　　　　　　　　　　　辐射测量
　　rotational motion　　　　　　　　　　旋转运动
　　slant edge　　　　　　　　　　　　斜边
　　vanishing points　　　　　　　　　　消失点
　　vignetting　　　　　　　　　　　　虚影
Camera matrix　　　　　　　　　　　　摄像机矩阵
Catadioptric optics　　　　　　　　　　　兼反射折射光学
Category-level recognition　　　　　　　　类别层次识别
　　bag of words　　　　　　　　　　　词袋

  data sets   数据集
  part-based   基于部件的
  segmentation   分割
  surveys   综述
CCD   CCD
  blooming   模糊现象
Central difference   中心差分
Chained transformations   连锁变换
Chamfer matching   斜切匹配
Characteristic function   特征函数
Characteristic polynomial   特征多项式
Chirality   手性
Cholesky factorization   Cholesky 因子分解
  algorithm   算法
  incomplete   不完全的
  sparse   稀疏
Chromatic aberration   色像差
Chromaticity coordinates   色度坐标
CIE L*a*b*, see Color   CIE L*a*b*, 见 Color
CIE L*u*v*, see Color   CIE L*u*v*, 见 Color
CIE XYZ, see Color   CIE XYZ, 见 Color
Circle of confusion   模糊圈
CLAHE, see Histogram equalization   CLAHE, 见 Histogram equalization
Clustering   聚类
  agglomerative   聚集的
  cluster analysis   聚类分析
  divisive   区分的
CMOS   CMOS
Co-vector   辅助矢量
Coefficient matrix   系数矩阵
Collineation   共线，直射映射
Color   色彩
  balance   平衡
  camera   摄像机
  demosaicing   去马赛克
  fringing   边缘现象
  hue, saturation, value (HSV)   色调，饱和度，强度(HSV)

L*a*b* L*a*b*
　　L*u*v* L*u*v*
　　primaries 基色
　　profile 表述
　　ratios 比值
　　RGB RGB
　　transform 变换
　　twist 扭曲
　　XYZ XYZ
　　YIQ YIQ
　　YUV YUV
Color filter array (CFA) 色彩滤波器矩阵(CFA)
Color line model 色彩线模型
ColorChecker chart 选色板图
Colorization 彩色化
Compositing 合成
　　image stitching 图像拼接
　　opacity 不透明度
　　over operator 覆盖算子
　　surface 表面
　　transparency 透明
Compression 压缩
Computational photography 计算摄影学
　　active illumination 主动照明
　　flash and non-flash 闪光和无闪光
　　high dynamic range 高动态范围
　　references 参照
　　tone mapping 色调映射
Concentric mosaic 同心拼图
CONDENSATION CONDENSATION
Condition number 条件数
Conditional random field (CRF) 条件随机场(CRF)
Confusion matrix (table) 混淆矩阵(表)
Conic section 圆锥截面
Conjugate gradient descent (CG) 共轭梯度下降(CG)
　　algorithm 算法
　　non-linear 非线性

preconditioned	预处理
Connected components	连通分量
Constellation model	星群模型
Content based image retrieval (CBIR)	基于内容的图像检索(CBIR)
Continuation method	连续法
Contour	轮廓
arc length parameterization	弧长参数化
chain code	链码
matching	匹配
smoothing	平滑
Contrast	对比度
Controlled-continuity spline	控制连续性样条
Convolution	卷积
kernel	核
mask	掩模
superposition	叠加
Coring	核化降噪
Correlation	相关性
windowed	窗口的
Correspondence map	对应图
Cramer–Rao lower bound	Cramer–Rao 下限
Cube map	立方体贴图
Hough transform	Hough 变换，哈夫变换
image stitching	图像拼接
Curve	曲线
arc length parameterization	弧长参数化
evolution	演化
matching	匹配
smoothing	平滑
Cylindrical coordinates	柱面坐标

D

Data energy (term)	数据能量(项)
Data sets and test databases	数据集和测试数据库
recognition	识别
De-interlacing	去隔行扫描
Decimation	降采样
Decimation kernels	降采样核

  bicubic   双三次

  binomial   二项式

  QMF   QMF

  windowed sinc   窗 sinc

Demosaicing (Bayer)   去马赛克(Bayer)

Depth from defocus   由散焦到深度

Depth map, see Disparity map   深度图，见 视差图

Depth of field   景深

Di-chromatic reflection model   色散反射模型

Difference matting (keying)   差分抠图(键控特效)

Difference of Gaussians (DoG)   高斯差分(DoG)

Difference of low-pass (DOLP)   低通差分(DOLP)

Diffuse reflection   漫反射

Diffusion   扩散

  anisotropic   各向异性的

Digital camera   数字相机

  color   彩色

  color filter array (CFA)   色彩滤波器矩阵(CFA)

  compression   压缩

Direct current (DC)   直流(DC)

Direct linear transform (DLT)   直接线性变换(DLT)

Direct sparse matrix techniques   直接稀疏矩阵方法

Directional derivative   方向导数

  selectivity   选择性

Discrete cosine transform (DCT)   离散余弦变换(DCT)

Discrete Fourier transform (DFT)   离散*傅里叶*变换(DFT)

Discriminative random field (DRF)   区分性随机场(DRF)

Disparity   视差

Disparity map   视差图

  multiple   多重

Disparity space image (DSI)   视差空间图像(DSI)

  generalized   广义的

Displaced frame difference (DFD)   平移帧差分(DFD)

Displacement field   位移场

Distance from face space (DFFS)   到人脸空间的距离(DFFS)

Distance in face space (DIFS)   在人脸空间内的距离(DIFS)

Distance map, see Distance transform   距离图，见 Distance transform

Distance transform 距离变换
    Euclidean 欧氏
    image stitching 图像拼接
    Manhattan (city block) 曼哈顿(城市街区)
    signed 有符号的
Domain (of a function) (函数)定义域
Domain scaling law 定义域伸缩法则
Downsampling, see Decimation 下采样，见 Decimation
Dynamic programming (DP) 动态规划(DP)
    monotonicity 单调性
    ordering constraint 序约束
    scanline optimization 扫描线优化
Dynamic snake 动态蛇行
Dynamic texture 动态纹理

E

Earth mover's distance (EMD) 推土机距离(EMD)
Edge detection 边缘检测
    boundary detection 边界检测
    Canny Canny
    chain code 链码
    color 色彩
    Difference of Gaussian 高斯差分
    edgel (edge element) 边界元
    hysteresis 滞后处理
    Laplacian of Gaussian 高斯二阶导数
    linking 链接
    marching cubes 移动立方体法
    scale selection 尺度选择
    steerable filter 导向滤波器
    zero crossing 过零点
Eigenface 特征脸
Eigenvalue decomposition 本征值分解
Eigenvector 特征向量
Elastic deformations 弹性变形
    image registration 图像注册
Elastic nets 弹性网
Elliptical weighted average (EWA) 椭圆加权平均(EWA)

Environment map	环境贴图
Environment matte	环境影像形板
Epanechnikov kernel	Epanechnikov 核
Epipolar constraint	极线约束
Epipolar geometry	极线几何
pure rotation	纯旋转
pure translation	纯平移
Epipolar line	极线
Epipolar plane	极面
image (EPI)	图像(EPI)
Epipolar volume	极线容积
Epipole	极点
Error rates	误差率
accuracy (ACC)	精度(ACC)
false negative (FN)	错误否定,漏失(FN)
false positive (FP)	错误肯定,误报(FP)
positive predictive value (PPV)	阳性预测值(PPV)
precision	精度
recall	召回率
ROC curve	ROC 曲线
true negative (TN)	正确否定(TN)
true positive (TP)	正确肯定(TP)
Errors-in-variable model	变量误差模型
heteroscedastic	不等值变异
Essential matrix	本质矩阵
-point algorithm	点算法
eight-point algorithm	点算法
re-normalization	再规范化
seven-point algorithm	点算法
twisted pair	扭曲对
Estimation theory	估计理论
Euclidean transformation	欧氏变换
Euler angles	欧拉角
Expectation maximization (EM)	期望最大化方法(EM)
Exponential twist	指数扭曲
Exposure bracketing	曝光包围
Exposure value (EV)	曝光值

F

F-number (stop)	光圈号数，有效孔径
Face detection	人脸检测
boosting	提升
cascade of classifiers	瀑布型分类器
clustering and PCA	聚类和 PCA
data sets	数据集
neural networks	神经网
support vector machines	支持向量机
Face modeling	人脸建模
Face recognition	人脸识别
active appearance model	活动表观模型
data sets	数据集
eigenface	特征脸
elastic bunch graph matching	弹性束图匹配
local binary patterns (LBP)	局部二值模式(LBP)
local feature analysis	局部特征分析
Face transfer	人脸迁移
Facial motion capture	脸部运动捕捉
Factor graph	因子图
Factorization	因子分解
missing data	缺失数据
projective	投影的
Fast Fourier transform (FFT)	快速傅里叶变换(FFT)
Fast marching method (FMM)	快速行进法(FMM)
Feature descriptor	特征描述子
bias and gain normalization	偏差和增益规范化
GLOH	GLOH
patch	块
PCA-SIFT	PCA-SIFT
performance (evaluation)	性能(评测)
quantization	定量化
SIFT	SIFT
steerable filter	导向滤波器
Feature detection	特征检测
Adaptive non-maximal suppression	自适应非最大抑制
affine invariance	仿射不变性

auto-correlation	自相关
Forstner	Forstner
Harris	Harris
Laplacian of Gaussian	高斯二阶导数
MSER	MSER
region	区域
repeatability	重复性
rotation invariance	旋转不变性
scale invariance	尺度不变性
Feature matching	特征匹配
densification	致密化
efficiency	效率
error rates	错误率
hashing	散列法
indexing structure	索引结构
k-d trees	k-d 树
locality sensitive hashing	局部敏感散列
nearest neighbor	最近邻
strategy	策略
verification	验证
Feature tracking	特征跟踪
affine	仿射
learning	学习
Feature tracks	特征迹
Feature-based alignment	基于特征的配准
2D	2D
3D	3D
iterative	迭代
Jacobian	雅克比
least squares	最小二乘
match verification	匹配验证
RANSAC	RANSAC
robust	鲁棒
Field of Experts (FoE)	专家域(FoE)
Fill factor	填充系数
Fill-in	填充
Filter	滤波器

  adaptive 自适应

  band-pass 带通

  bilateral 双边

  directional derivative 方向导数

  edge-preserving 边缘保持

  Laplacian of Gaussian 高斯二阶导数

  median 中值

  moving average 移动平均

  non-linear 非线性

  separable 可分离的

  steerable 导向的

Filter coefficients 滤波器系数

Filter kernel, see Kernel 滤波器核，见 Kernel

Finding faces, see Face detection 发现人脸，见 人脸检测

Finite element analysis 有限元分析

  stiffness matrix 刚度矩阵

Finite impulse response (FIR) filter 有限脉冲响应(FIR)滤波器

Fisher information matrix Fisher 信息矩阵

Fisher's linear discriminant (FLD) Fisher 线性鉴别分析(FLD)

Fisheye lens Fisheye 镜头

Flash and non-flash merging 闪光和非闪光融合

Flash matting 闪抠图

Flip-book animation 活页动画

Flying spot scanner 飞点扫描器

Focal length 焦距

Focus 焦点

  shape-from 由……到形状

Focus of expansion (FOE) 延伸焦点(汇集点)

Form factor 形状系数

Forward mapping, see Forward warping 前向映射，见 Forward warping

Forward warping 前向卷绕

Fourier transform *傅里叶*变换

  discrete 离散

  examples 例子

  magnitude (gain) 幅度(增益)

  pairs 对

  Parseval's Theorem 帕萨瓦尔定理(Parseval 定理)

  phase (shift) 相位(偏移)
  power spectrum 功率谱
  properties 特性
  two-dimensional 二维
Fourier-based motion estimation 基于*傅里叶*变换的运动估计
  rotations and scale 旋转和尺度
Frame interpolation 帧插值
Free-viewpoint video 自由视点视频
Fundamental matrix 基本矩阵
estimation, see Essential matrix 估计，见 Essential matrix
Fundamental radiometric relation 基本辐射学关系

G

Gain 增益
Gamma γ, 伽马
Gamma correction γ校正
Gap closing (image stitching) 缝隙缝合(图像拼接)
Garbage matte 垃圾遮罩
Gaussian kernel 高斯核
Gaussian Markov random field (GMRF) 高斯马尔科夫随机场(GMRF)
Gaussian mixtures, see Mixture of Gaussians 高斯混合，见 Mixture of Gaussians
Gaussian pyramid 高斯金字塔
Gaussian scale mixtures (GSM) 高斯尺度混合(GSM)
Gaze correction 注视矫正
Geman–McClure function Geman–McClure 函数
Generalized cylinders 广义圆柱
Geodesic active contour 测地活动轮廓
Geodesic distance (segmentation) 测地距离(分割)
Geometric image formation 几何图像形成
Geometric lens aberrations 镜头几何畸变
Geometric primitives 几何基元
  homogeneous coordinates 齐次坐标
  lines 线条
  normal vector 法向量
  planes 平面
  points 点
Geometric transformations 几何变换
  2D 2D

3D	3D
2D perspective	2D 透视
3D rotations	3D 旋转
affine	仿射
bilinear	双线性
calibration matrix	标定矩阵
collineation	共线
Euclidean	欧氏
forward warping	前向卷绕
hierarchy	分层
homography	同构，单应，1-1 映射
inverse warping	逆卷绕
perspective	透视
projections	投影
projective	投影的
rigid body	刚体
scaled rotation	归一的旋转
similarity	相似性
translation	平移
Geometry image	几何图像
Gesture recognition	手势识别
Gibbs distribution	吉布斯分布
Gibbs sampler	吉布斯采样器
Gimbal lock	常平架锁定
Gist (of a scene)	(场景的)Gist
Global illumination	全局光照
Global optimization	全局优化
GPU algorithms	GPU 算法
Gradient location-orientation histogram(GLOH)	梯度位置-方向直方图(GLOH)
Graduated non-convexity (GNC)	分度非凸性(GNC)
Graph cuts	图割
MRF inference	MRF 推断
normalized cuts	规范割
Graph-based segmentation	基于图的分割
Grassfire transform	Grassfire 变换
Ground control points	地平面控制点

H

Hammersley–Clifford theorem	Hammersley–Clifford 定理
Hann window	Hann 窗
Harris corner detector, see Feature detection	Harris 角点检测，见 Feature detection
Head tracking	头部跟踪
active appearance model (AAM)	活动表观模型(AAM)
Helmholtz reciprocity	Helmholtz 互易性
Hessian	Hessian
eigenvalues	本征值
image	图像
inverse	逆
local	局部
patch-based	基于块的
rank-deficient	秩亏
reduced motion	弱化的运动
sparse	稀疏
Heteroscedastic	异方差
Hidden Markov model (HMM)	隐马尔科夫模型(HMM)
Hierarchical motion estimation	分层的运动估计
High dynamic range (HDR) imaging	高动态范围(HDR)成像
formats	格式
tone mapping	色调映射
Highest confidence first (HCF)	首先最高信度(HCF)
Hilbert transform pair	希尔伯特变换对
Histogram equalization	直方图均衡化
locally adaptive	局部自适应
Histogram intersection	直方图交
Histogram of oriented gradients (HOG)	有向梯度直方图(HOG)
History of computer vision	计算机视觉的历史
Hole filling	空洞填充
Homogeneous coordinates	齐次坐标
Homography	同构，单应，1-1 映射
Hough transform	Hough 变换
cascaded	瀑布式，级联
cube map	立方图
generalized	广义
Human body shape modeling	人体形状建模

Human motion tracking	人体运动跟踪
activity recognition	活动识别
adaptive shape modeling	自适应形状建模
background subtraction	背景消减
flow-based	基于流的
initialization	初始化
kinematic models	运动学模型
particle filtering	粒子滤波
probabilistic models	概率模型
Hyper-Laplacian	Hyper-Laplacian
Ideal points	理想点
Ill-posed (ill-conditioned) problems	不适定(病态的)问题
Illusions	幻觉
Image alignment	图像配准
feature-based	基于特征的
intensity-based	基于亮度的
intensity-based vs. feature-based	基于亮度的相对于基于特征的
Image analogies	图像模拟
Image blending	图像混合
feathering	羽化
GIST	GIST
gradient domain	梯度域
image stitching	图像拼接
Poisson	Poisson(泊松)
pyramid	金字塔
Image compositing, see Compositing	图像合成，见 合成
Image compression	图像压缩
Image decimation	图像降采样
Image deconvolution, see Blur removal	图像去卷积，见 Blur removal
Image filtering	图像滤波
Image formation	图像形成
geometric	几何的
photometric	光度学的
Image gradient	图像梯度
constraint	约束
Image interpolation	图像插值

Image matting	图像抠图
Image processing	图像处理
textbooks	教材
Image pyramid	图像金字塔
Image resampling	图像重采样
test images	测试图像
Image restoration	图像复原
blur removal	模糊去除
deblocking	去块效应
inpainting	修图
noise removal	去噪
using MRFs	使用 MRFs
Image search	图像搜索
Image segmentation, see Segmentation	图像分割，见 Segmentation
Image sensing, see Sensing	图像感知，见 Sensing
Image statistics	图像统计学
Image stitching	图像拼接
automatic	自动的
bundle adjustment	光束平差法，光束法平差
compositing	合成
coordinate transformations	坐标变换
cube map	立方图
cylindrical	柱面的
de-ghosting	去虚影
direct vs. feature-based	直接相对于基于特征的
exposure compensation	曝光补偿
feathering	羽化
gap closing	缝隙缝合
global alignment	全局配准
homography	同构，单应，1-1 映射
motion models	运动模型
panography	全景图法
parallax removal	视差消除
photogrammetry	摄影测量学
pixel selection	像素选择
planar perspective motion	平面透视运动
recognizing panoramas	识别全景画

rotational motion	选转运动
seam selection	接缝选择
spherical	球面的
up vector selection	向上矢量选择
Image warping	图像卷绕
Image-based modeling	基于图像的建模
Image-based rendering	基于图像的绘制
concentric mosaic	同心拼图
environment matte	环境影像形板
impostors	替换，冒充
layered depth image	分层的深度图像
layers	层次
light field	光场
Lumigraph	照度图
modeling vs. rendering continuum	建模相对于绘制连续体
sprites	子画面
surface light field	表面光场
unstructured Lumigraph	非结构化的照度图
view interpolation	视图插值
view-dependent texture maps	独立于视图的纹理映射
Image-based visual hull	基于图像的视觉外型
Implicit surface	隐含表面
Impostors, see Sprites	替换，冒充，见 Sprites
Impulse response	脉冲响应
Incremental refinement	逐次求精
motion estimation	运动估计
Incremental rotation	逐次旋转
Indexing structure	索引结构
Indicator function	指标函数
Industrial applications	工业应用
Infinite impulse response (IIR) filter	无限脉冲响应(IIR)滤波器
Influence function	影响函数
Information matrix	信息矩阵
Inpainting	修图
Instance recognition	示例识别
algorithm	算法
data sets	数据集

geometric alignment	几何配准
inverted index	逆索引
large scale	大尺度
match verification	匹配验证
query expansion	查询扩展
stop list	停止表
visual words	视觉词
vocabulary tree	词汇树
Integrability constraint	完整性约束
Integral image	积分图像
Integrating sphere	积分球
Intelligent scissors	智能剪刀
Interaction potential	交互作用势
Interactive computer vision	交互式计算机视觉
International Color Consortium (ICC)	国际色彩联盟(ICC)
Internet photos	因特网照片
Interpolation	插值
Interpolation kernels	插值核
bicubic	双三次
bilinear	双线性
binomial	二项的
sinc	sinc
spline	样条
Intrinsic camera calibration	摄像机内标定
Intrinsic images	本征图像
Inverse kinematics (IK)	逆运动学(IK)
Inverse mapping, see Inverse warping	逆映射，见 Inverse warping
Inverse problems	逆问题
Inverse warping	逆卷绕
ISO setting	ISO 设置
Iterated closest point (ICP)	迭代最近点(ICP)
Iterated conditional modes (ICM)	迭代条件模态(ICM)
Iterative back projection (IBP)	迭代反向投影(IBP)
Iterative feature-based alignment	迭代的基于特征的配准
Iterative sparse matrix techniques	迭代的稀疏矩阵方法
conjugate gradient	共轭梯度
Iteratively reweighted least squares(IRLS)	迭代重加权最小二乘(IRLS)

J

Jacobian	雅克比
image	图像
motion	运动
sparse	稀疏
Joint bilateral filter	联合双边滤波器
Joint domain (feature space)	联合域(特征空间)

K

K-d trees	K-d 树
K-means	K-均值
Kalman snakes	卡尔曼蛇行
Kanade–Lucas–Tomasi (KLT) tracker	Kanade–Lucas–Tomasi (KLT)跟踪器
Karhunen-Loeve transform (KLT)	Karhunen-Loeve 变换 (KLT)
Kernel	核
Bilinear	双线性
Gaussian	高斯
low-pass	低通
Sobel operator	Sobel 算子
unsharp mask	不锐化掩模
Kernel basis function	核基函数
Kernel density estimation	核密度估计
Keypoint detection, see Feature detection	关键点检查，见 Feature detection
Kinematic model (chain)	运动学模型(链)
Kruppa equations	Kruppa 公式

L

L*a*b*, see Color	L*a*b*，见 Color
L*u*v*, see Color	L*u*v*，见 Color
L norm	L 范数
Lambertian reflection	朗伯反射
Laplacian matting	Laplacian 抠图
Laplacian of Gaussian (LoG) filter	高斯二阶导数(LoG)滤波器
Laplacian pyramid	Laplacian 金字塔
Blending	混合
perfect reconstruction	完美重建
Latent Dirichlet process (LDP)	隐 Dirichlet 过程(LDP)
Layered depth image (LDI)	图像
Layered depth panorama	分层深度全景图

Layered motion estimation   分层运动估计
    transparent   透明的
Layers   分层
    image-based rendering   基于图像的绘制
Layout consistent random field   布局一致随机场
Learning in computer vision   计算机视觉中的学习
Least median of squares (LMS)   最小均方差(LMS)
Least squares   最小二乘
    iterative solvers   迭代解法
    Linear   线性
    non-linear   非线性
    robust, see Robust least squares   鲁棒，见 Robust least squares
    sparse   稀疏
    total   全部
    weighted   加权的
Lens   镜头
    Compound   复合
    nodal point   节点
    Thin   薄
Lens distortions   镜头畸变
    calibration   标定
    decentering   偏心
    radial   径向
    spline-based   基于样条的
    tangential   切向的
Lens law   镜头法则
Level of detail (LOD)   细节水平(LOD)
Level sets   水平集
    fast marching method   快速行进方法
    geodesic active contour   测地活动轮廓
Lens law   Lens law
Lifting, see Wavelets   提升，见 Wavelets
Light field   光场
    higher dimensional   高维
    light slab   光片
    ray space   射线空间
    rendering   绘制

surface	表面
Lightness	亮度
Line at infinity	无穷远处的线
Line detection	线条检测
Hough transform	Hough 变换
RANSAC	RANSAC
simplification	简化
successive approximation	逐次近似
Line equation	线方程
Line fitting	线拟合
uncertainty	不确定性
Line hull, see Visual hull	线外型，见 Visual hull
Line labeling	线条标注
Line process	线过程
Line spread function (LSF)	线扩散函数(LSF)
Line-based structure from motion	基于线条的由运动到结构
Linear algebra	线性代数
least squares	最小二乘
matrix decompositions	矩阵分解
references	参考
Linear blend	线性混合
Linear discriminant analysis (LDA)	线性鉴别分析(LDA)
Linear filtering	线性滤波
Linear operator	线性算子
superposition	叠加
Linear shift invariant (LSI) filter	线性平移不变(LSI)滤波器
Live-wire	火线
Local distance functions	局部距离函数
Local operator	局部算子
Locality sensitive hashing (LSH)	局部敏感散列(LSH)
Locally adaptive histogram equalization	局部自适应直方图均衡化
Location recognition	位置识别
Loopy belief propagation (LBP)	带环的置信传播
Low-pass filter	低通滤波器
sinc	sinc
Lumigraph	照度图
unstructured	非结构化的

Luminance		亮度
Lumisphere		亮度球

M

M-estimator		M-估计器
Mahalanobis distance		Mahalanobis 距离
Manifold mosaic		流形拼图
Markov chain Monte Carlo (MCMC)		马尔科夫链蒙特卡洛(MCMC)
Markov random field		马尔科夫随机场
cliques		团簇
directed edges		有向边
dynamic		动态
flux		连续的改变；不稳定的状态
inference, see MRF inference		推断，见 MRF inference
layout consistent		布局一致
learning parameters		参数学习
line process		线过程
neighborhood		邻域
order		序
random walker		随机游走
stereo matching		立体匹配
Marr's framework		Marr 框架
computational theory		计算理论
hardware implementation		硬件实现
representations and algorithms		表达与算法
Match move		匹配移动
Matrix decompositions		矩阵分解
Cholesky		Cholesky 分解
eigenvalue (ED)		本征值分解 (ED)
QR		QR 分解
singular value (SVD)		奇异值分解 (SVD)
square root		平方根分解
reflection		漫反射
Matting		抠图
alpha matte		透明度遮罩
Bayesian		贝叶斯
blue screen		蓝屏
difference		差分

flash	闪
GrabCut	GrabCut
Laplacian	Laplacian
natural	自然
optimization-based	基于优化的
Poisson	Poisson(泊松)
shadow	阴影
smoke	烟
triangulation	三角剖分
trimap	三值图
two screen	双屏
video	视频
Maximally stable extremal region (MSER)	最稳定极值区域(MSER)
Maximum a posteriori (MAP) estimate	最大后验(MAP)估计
Mean absolute difference (MAD)	平均绝对差
Mean average precision	平均精度
Mean shift	均值移位
bandwidth selection	带宽选择
Mean square error (MSE)	均方误差(MSE)
Measurement equation (model)	测量方程(模型)
Measurement matrix	测量矩阵
Measurement model, see Bayesian model	测量模型，见 Bayesian model
Medial axis transform (MAT)	中轴变换(MAT)
Median absolute deviation (MAD)	平均绝对偏差(MAD)
Median filter	中值滤波器
weighted	加权的
Medical image registration	医学图像注册
Medical image segmentation	医学图像分割
Membrane	薄膜
Mesh-based warping	基于网格的卷绕
Metamer	异性体
Metric learning	度量学习
Metric tree	度量树
MIP-mapping	MIP-映射
trilinear	三线性
Mixture of Gaussians	高斯混合
color model	彩色模型

expectation maximization (EM)	期望最大化(EM)
mixing coefficient	混合系数
soft assignment	软赋值
Model selection	模型选择
Model-based reconstruction	基于模型的重建
architecture	结构
heads and faces	头部和人脸
human body	人体
Model-based stereo	基于模型的立体视觉
Models	模型
Bayesian	贝叶斯
forward	前向
physically based	基于物理量的
physics-based	基于物理的
probabilistic	概率的
Modular eigenspace	模块本征空间
Modulation transfer function (MTF)	模块传递函数(MTF)
Morphable model	变形模型
body	人体
face	人脸
multidimensional	多维
Morphing	变形
3D body	3D 人体
3D face	3D 人脸
automated	自动的
facial feature	面部特征
feature-based	基于特征的
flow-based	基于流的
video textures	视频纹理
view morphing	视图变形
Morphological operator	形态学算子
closing	闭
dilation	膨胀
erosion	腐蚀
opening	开
Morphology	形态学
Mosaic, see Image stitching	拼图，马赛克，见 Image stitching

Mosaics	拼图
motion models	运动模型
video compression	视频压缩
whiteboard and document scanning	白板与文本扫描
Motion compensated video compression	运动补偿的视频压缩
Motion compensation	运动补偿
Motion estimation	运动估计
affine	仿射
aperture problem	孔径问题
compositional	复合的，合成的
Fourier-based	基于*傅里叶*变换的
frame interpolation	帧插值
hierarchical	分层的
incremental refinement	逐次求精
layered	层次的
learning	学习
linear appearance variation	线性表观变化
optical flow	光流
parametric	参数的
patch-based	基于块的
phase correlation	相位相关性
quadtree spline-based	基于四叉树样条
reflection	反射
spline-based	基于样条的
translational	平移的
transparent	透明的
uncertainty modeling	不确定性建模
Motion field	运动场
Motion models	运动模型
learned	学习到的
Motion segmentation	运动分割
Motion stereo	运动立体视觉
Motion-based user interaction	基于运动的用户交互
Moving least squares (MLS)	移动最小二乘(MLS)
MRF inference	MRF 推断
alpha expansion	A 扩张
belief propagation	置信传播

dynamic programming	动态规划
expansion move	扩张移动
gradient descent	梯度下降
graph cuts	图割
highest confidence first (HCF)	首先最高信度(HCF)
iterated conditional modes	迭代条件模态
linear programming (LP)	线性规划(LP)
loopy belief propagation	带环的置信传播
Markov chain Monte Carlo	马尔科夫链蒙特卡洛
simulated annealing	模拟退火
stochastic gradient descent	随机梯度下降
swap move (alpha-beta)	交换移动(α-β)
Swendsen-Wang	Swendsen-Wang
Multi-frame motion estimation	多帧运动估计
Multi-pass transforms	多通道变换
Multi-perspective panoramas	多透视全景图
Multi-perspective plane sweep (MPPS)	多透视面扫描
Multi-view stereo	多视角立体视觉
epipolar plane image	极面图像
evaluation	评价
initialization requirements	初始化要求
reconstruction algorithm	重建算法
scene representation	场景表达
shape priors	形状先验
silhouettes	轮廓
space carving	空间雕刻
spatio-temporally shiftable window	时空可变窗口
taxonomy	分类学
visibility	可见性
volumetric	体积的
voxel coloring	体素着色
Multigrid	多重网格
algebraic (AMG)	代数的(AMG)
Multiple hypothesis tracking	多假设跟踪
Multiple-center-of-projection images	多投影中心图像
Multiresolution representation	多分辨率表达
Mutual information	互信息

N

Natural image matting	自然图像抠图
Nearest neighbor	最近邻
distance ratio (NNDR)	距离比率(NNDR)
matching, see Feature matching	匹配，见 Feature matching
Negative posterior log likelihood	负对数后验似然度
Neighborhood operator	邻域算子
Neural networks	神经网络
Nintendo Wii	任天堂游戏机 Wii
Nodal point	节点
Noise	噪声
sensor	传感
Noise level function (NLF)	噪声水平函数(NLF)
Noise removal	去噪
Non-linear filter	非线性滤波器
Non-linear least squares, see Least squares	非线性最小二乘，见 Least squares
Non-maximal suppression, see Feature detection	非最大抑制，见 Feature detection
Non-parametric density modeling	非参数密度建模
Non-photorealistic rendering (NPR)	非真实感绘制(NPR)
Non-rigid motion	非刚性运动
Normal equations	正态方程
Normal map (geometry image)	正态图(几何图像)
Normal vector	法向量
Normalized cross-correlation (NCC)	规范化互相关(NCC)
Normalized cuts	规范割
intervening contour	中间轮廓
Normalized device coordinates (NDC)	规范设备坐标(NDC)
Normalized sum of squared differences	规范化的误差平方和(NSSD)
Norms	范数
L_1	L_1
L_∞	L_∞
Nyquist rate / frequency	奈奎斯特频率

O

Object detection	物体检测
car	汽车
face	人脸
part-based	基于部件的

pedestrian	行人
Object-centered projection	以物体为中心的投影
Occluding contours	遮挡轮廓
Octree reconstruction	八叉树重建
Octree spline	八叉树样条
Omnidirectional vision systems	全向视觉系统
Opacity	不透明度
Operator	算子
linearity	线性
Optic flow, see Optical flow	光流，见 Optical flow
Optical center	光心
Optical flow	光流
anisotropic smoothness	各向异性平滑
evaluation	评测
fusion move	融合步骤
global and local	全局和局部
Markov random field	马尔科夫随机场
multi-frame	多帧
normal flow	法向流
patch-based	基于块的
region-based	基于区域的
regularization	正则化
robust regularization	鲁棒正则化
smoothness	平滑
total variation	总变差
Optical flow constraint equation	光流约束方程
Optical illusions	视觉幻象
Optical transfer function (OTF)	光传递函数(OTF)
Optical triangulation	光学三角测量
Optics	光学
chromatic aberration	色像差
Seidel aberrations	塞德耳像差
vignetting	虚影
Optimal motion estimation	最优运动估计
Oriented particles (points)	有向粒子(点)
Orthogonal Procrustes	正交 Procrustes
Orthographic projection	正交投影

Osculating circle	密切圆
Over operator	覆盖算子
Overview	概述

P

Padding	填塞，统调，使……平直
Panography	全景图法
Panorama, see Image stitching	全景图，见 Image stitching
Panorama with depth	带深度的全景图
Para-perspective projection	类透视投影
Parallel tracking and mapping (PTAM)	并行跟踪和映射(PTAM)
Parameter sensitive hashing	参数敏感的散列
Parametric motion estimation	参数化运动估计
Parametric surface	参数曲面
Parametric transformation	参数化变换
Parseval's Theorem, see Fourier transform	帕萨瓦尔定理(Parseval 定理)，见 Fourier transform
Part-based recognition	基于部件的识别
constellation model	星群模型
Particle filtering	粒子滤波
Parzen window	Parzen 窗
PASCAL Visual Object Classes Challenge (VOC)	PASCAL 视觉物体类挑战(VOC)
Patch-based motion estimation	基于块的运动估计
Peak signal-to-noise Ratio (PSNR)	峰值信噪比(PSNR)
Pedestrian detection	行人检测
Penumbra	半影
Performance-driven animation	基于表演的动画
Perspective n-point problem (PnP)	n 点透视问题(PnP)
Perspective projection	透视投影
Perspective transform (D)	透视变换(D)
Phase correlation	相位相关
Phong shading	补色渲染
Photo pop-up	照片弹出
Photo Tourism	照片游览
Photo-mosaic	照片拼图
Photoconsistency	影像一致性
Photometric image formation	光度测定学图像形成
calibration	标定

global illumination	全局照明
lighting	照明
optics	光学
radiosity	辐射度
reflectance	反射
shading	阴影
Photometric stereo	光度测定学立体视觉
Photometry	光度学
Photomontage	照片蒙太奇
Physically based models	基于物理量的模型
Physics-based vision	基于物理的视觉
Pictorial structures	图案结构
Pixel transform	像素变换
Plucker coordinates	普吕克(Plucker)坐标
Planar pattern tracking	平面模式跟踪
Plane at infinity	无穷远平面
Plane equation	平面方程
Plane plus parallax	平面加上视差
Plane sweep	平面扫描
Plane-based structure from motion	基于平面的由运动到结构
Plenoptic function	全光函数
Plenoptic modeling	全光建模
Plumb-line calibration method	垂直线标定方法
Point distribution model	点分布函数
Point operator	点算子
Point process	点过程
Point spread function (PSF)	点扩散函数(PSF)
estimation	估计
Point-based representations	基于点的表达
Points at infinity	无穷远处的点
Poisson	Poisson(泊松)
blending	混合
equations	方程
matting	抠图
noise	噪声
surface reconstruction	表面重建
Polar coordinates	极坐标

Polar projection	极投影
Polyphase filter	多相滤波器
Pop-out effect	弹出效应
Pose estimation	姿态估计
iterative	迭代的
Power spectrum	功率谱
Precision, see Error rates	精度，见 Error rates
mean average	平均
Preconditioning	预处理
Principal component analysis (PCA)	主分量分析(PCA)
face modeling	人脸建模
generalized	广义的
missing data	缺失数据
Prior energy (term)	先验能量(项)
Prior model, see Bayesian model	先验模型，见 Bayesian model
Profile curves	轮廓曲线
Progressive mesh (PM)	累进网格(PM)
Projections	投影
object-centered	以物体为中心的
orthographic	正交的
para-perspective	类透视
Perspective	透视
Projective (uncalibrated) reconstruction	投影(未标定)重建
Projective depth	投影深度
Projective disparity	投影视差
Projective space	投影空间
PROSAC (PROgressive SAmple Consensus),	PROSAC(累进采样一致性)
PSNR, see Peak signal-to-noise ratio	PSNR，见 Peak signal-to-noise ratio
Pyramid	金字塔
blending	混合
Gaussian	高斯
half-octave	半-倍频程
Laplacian	Laplacian
motion estimation	运动估计
octave	倍频程
radial frequency implementation	径向频率实现
steerable	导向

Pyramid match kernel	金字塔匹配核

Q

QR factorization	QR 因子分解
Quadratic form	二次型
Quadrature mirror filter (QMF)	正交镜象滤波器(QMR)
Quadric equation	二次方程
Quadtree spline	二次树样条
motion estimation	运动估计
restricted	限制的
Quaternions	四元数
antipodal	正相反
multiplication	乘
Query by image content (QBIC)	图像内容查询(QBIC)
Query expansion	查询扩展
Quincunx sampling	五角采样

R

Radial basis function	径向基函数
Radial distortion	径向畸变
barrel	桶
calibration	标定
parameters	参数
pincushion	枕形失真
Radiance map	辐照度图
Radiometric image formation	放射(辐射)图像形成
Radiometric response function	放射(辐射)响应函数
Radiometry	辐射测量学
Radiosity	辐射度
Random walker	随机游走
Range (of a function)	(函数的)值域
Range data, see Range scan	距离数据，见 Range scan
Range image, see Range scan	距离图像，见 Range scan
Range scan	距离扫描
alignment	配准
large scenes	大场景
merging	归并
registration	注册
segmentation	分割

volumetric	体积的
Range sensing (rangefinding)	距离感知(距离发现)
coded pattern	编码的样式
light stripe	光带
shadow stripe	阴影带
space–time stereo	空间-时间立体视觉
stereo	立体视觉
texture pattern (checkerboard)	纹理模式(棋盘格板)
time of flight	飞行时间
RANSAC (RAndom SAmple Consensus)	RANSAC(随机采样一致性)
inliers	内点
preemptive	抢先，先占
progressive (PROSAC)	逐次(PROSAC)
RAW image format	原始图像数据存储格式
Ray space (light field)	射线空间(光场)
Ray tracing	光线追踪
Rayleigh quotient	瑞利(Rayleigh)商
Recall, see Error rates	召回率，见 Error rates
Receiver Operating Characteristic	接收器操作特性
area under the curve (AUC)	曲线下的面积(AUC)
mean average precision	平均精度
ROC curve	ROC 曲线
Recognition	识别
D models	D 模型
category (class)	物体类
color similarity	色彩相似性
context	上下文
contour-based	基于轮廓的
data sets	数据集
face	人脸
instance	示例
large scale	大尺度，大规模
learning	学习
part-based	基于部件的
scene understanding	场景理解
segmentation	分割
shape context	形状上下文

Rectangle detection	矩形检测
Rectification	矫正
standard rectified geometry	标准的矫正了的几何
Recursive filter	递归滤波器
Reference plane	参考平面
Reflectance	反射
Reflectance map	反射图
Reflectance modeling	反射建模
Reflection	反射
di-chromatic	色散
diffuse	扩散
specular	镜面反射
Region	区域
merging	归并
splitting	分裂
Region segmentation, see Segmentation	区域分割，见 Segmentation
Registration, see Image Alignment	注册，见 Image Alignment
feature-based	基于特征的
intensity-based	基于亮度的
medical image	医学图像
Regularization	正则化
robust	鲁棒
Regularization parameter	正则化参数
Residual error,	残差
RGB (red green blue), see Color	RGB(红绿蓝)，见 Color
Rigid body transformation	刚体变换
Robust error metric, see Robust penalty function	鲁棒误差度量，见 Robust penalty function
Robust least squares	鲁棒最小二乘
iteratively reweighted	迭代重加权的
Robust penalty function,	鲁棒惩罚函数
Robust regularization	鲁棒正则化
Robust statistics	鲁棒统计学
inliers	内点
M-estimator	M-估计器
Rodriguez's formula	Rodriguez 公式
Root mean square error (RMS)	均方根误差(RMS)
Rotations	旋转

Euler angles	欧拉角
axis/angle	轴/角
exponential twist	指数扭曲
incremental	逐次
interpolation	插值
quaternions	四元数

S

Sampling	采样
Scale invariant feature transform (SIFT)	尺度不变特征变换(SIFT)
Scale-space	尺度空间
Scatter matrix	散度矩阵
between-class	类间
within-class	类内
Scattered data interpolation	分散数据插值
Scene completion	场景补全
Scene flow	场景流
Scene understanding	场景理解
gist	gist
scene alignment	场景配准
Schur complement	Schur 补集
Scratch removal	划痕去除
Seam selection	接缝选择
image stitching	图像拼接
Second-order cone programming (SOCP)	二次锥规划(SOCP)
Seed and grow	播种和生长
stereo	立体视觉
structure from motion	由运动到结构
Segmentation	分割
active contours	活动轮廓
affinities	相似性
binary MRF	二值 MRF
CONDENSATION	CONDENSATION
connected components	连通分量
energy-based	基于能量的
for recognition	用于识别的
geodesic active contour	测地活动轮廓
geodesic distance	测地距离

GrabCut	GrabCut
graph cuts	图割
graph-based	基于图的
hierarchical	分层的
intelligent scissors	智能剪刀
joint feature space	联合特征空间
k-means	k-均值
level sets	水平集
mean shift	均值移位
medical image	医学图像
merging	归并
minimum description length (MDL)	最小描述长度(MDL)
mixture of Gaussians	高斯混合
Mumford-Shah	Mumford-Shah
non-parametric	非参数
normalized cuts	规范割
probabilistic aggregation	概率聚类
random walker	随机游走
snakes	蛇行
splitting	分裂
stereo matching	立体视觉匹配
thresholding	阈值化
tobogganing	平底雪橇滑雪
watershed	分水岭
weighted aggregation (SWA)	加权聚类(SWA)
Seidel aberrations	塞德耳像差
Self-calibration	自标定
bundle adjustment	光束平差法,光束法平差
Kruppa equations	Kruppa 方程
Sensing	传感
aliasing	走样(混叠、混淆)
color	彩色
color balance	色彩平衡
gamma	γ
pipeline	流水线
sampling	采样
sampling pitch	采样间距(步进式扫描采样)

Sensor noise	传感噪声
amplifier	放大器
dark current	暗电流
fixed pattern	固定样式
shot noise	散粒噪声
Separable filtering	可分离滤波
Shading	阴影
equation	公式
shape-from	由-到形状
Shadow matting	阴影抠图
Shape context	形状上下文
Shape from	由……到形状
focus	聚焦
photometric stereo	光度测定学立体视觉
profiles	轮廓
shading	阴影
silhouettes	轮廓
specularities	镜面反射性
stereo	立体视觉
texture	纹理
Shape parameters	形状参数
Shape-from-X	由 X 到形状
focus	聚焦
photometric stereo	光度测定学立体视觉
shading	阴影
texture	纹埋
Shift invariance	移位不变性
Shiftable multi-scale transform	可变多尺度变换
Shutter speed	快门速度
Signed distance function	带符号的距离函数
Silhouette-based reconstruction	基于轮廓的重建
octree	八叉树
visual hull	视觉外型
Similarity transform	相似变换
Simulated annealing	模拟退火
Simultaneous localization and mapping(SLAM)	同时定位和制图(SLAM)
Sinc filter	Sinc 滤波器

interpolation	插值
low-pass	低通
windowed	开窗的
Single view metrology	单视图度量
Singular value decomposition (SVD)	奇异值分解(SVD)
Skeletal set	骨骼集
Skeleton	骨架
Skew	倾斜
Skin color detection	肤色检测
Slant edge calibration	倾斜边界标定
Slippery spring	光滑弹簧
Smoke matting	烟抠图
Smoothness constraint	平滑约束
Smoothness penalty	平滑惩罚
Snakes	蛇行
ballooning	鼓胀
dynamic	动态
internal energy	内部能量
Kalman	卡尔曼
shape priors	形状先验
slippery spring	光滑弹簧
Soft assignment	软赋值
Software	软件
Space carving	空间雕刻
multi-view stereo	多视角立体视觉
Spacetime stereo	空间时间立体视觉
Sparse flexible model	稀疏柔性模型
Sparse matrices	稀疏矩阵
compressed sparse row (CSR)	压缩稀疏行(CSR)
skyline storage	天际线存储法
Sparse methods	稀疏方法
direct	直接
iterative	迭代
Spatial pyramid matching	空间金字塔匹配
Spectral response function	谱响应函数
Spectral sensitivity	谱敏感度
Specular flow	谱流

Specular reflection	镜面反射
Spherical coordinates	球面坐标
Spherical linear interpolation	球面线性插值
Spin image	旋转图像
Splatting see Forward warping	体空间扫描，见 Forward warping
volumetric	体积的
Spline	样条
controlled continuity	控制连续性
octree	八叉树
quadtree	四叉树
thin plate	薄板
Spline-based motion estimation	基于样条的运动估计
Splining images see Laplacian pyramid blending	拼合图像，见 Laplacian pyramid blending
Sprites	子画面
image-based rendering	基于图像的绘制
motion estimation	运动估计
video	视频
video compression	视频压缩
with depth	带深度的
Statistical decision theory	统计推断理论
Steerable filter	导向滤波器
Steerable pyramid	导向金字塔
Steerable random field	导向随机场
Stereo	立体视觉
aggregation methods	聚类方法
coarse-to-fine	由粗到精
cooperative algorithms	协同方法
correspondence	对应
curve-based	基于曲线的
dense correspondence	稠密对应
depth map	深度图
dynamic programming	动态规划
edge-based	基于边缘的
epipolar geometry	极线几何
feature-based	基于特征的
global optimization	全局优化
graph cut	图割

layers	层次
local methods	局部方法
model-based	基于模型的
multi-view	多视角，多视图
non-parametric similarity measures	非参数相似性
photoconsistency	影像一致性
plane sweep	平面扫描
rectification	矫正
region-based	基于区域的
scanline optimization	扫描线优化
seed and grow	播种和生长
segmentation-based	基于分割的
semi-global optimization	部分全局优化
shiftable window	可变窗口
similarity measure	相似性度量
space–time	空间-时间
sparse correspondence	稀疏对应
sub-pixel refinement	亚像素求精
support region	支持区域
taxonomy	分类学
uncertainty	不确定性
window-based	基于窗口的
winner-take-all (WTA)	赢者通吃(WTA)
Stereo-based head tracking	基于立体视觉的头部跟踪
Stiffness matrix	刚度矩阵
Stitching see Image stitching	拼图，见 Image stitching
Stochastic gradient descent	随机梯度下降
Structural Similarity (SSIM) index	结构相似性(SSIM)索引
Structure from motion	由运动到结构
affine	仿射
bas-relief ambiguity	浅浮雕歧义
bundle adjustment	光束平差法，光束法平差
constrained	约束的
factorization	因子分解
feature tracks	特征迹
iterative factorization	迭代因子分解
line-based	基于线条的

multi-frame	多帧
non-rigid	非刚性
orthographic	正交
plane-based	基于平面的
projective factorization	投影因子分解
seed and grow	播种和生长
self-calibration	自标定
skeletal set	骨骼集
two-frame	两帧
uncertainty	不确定性
Subdivision surface	细分曲面
subdivision connectivity	细分连通性
Subspace learning	子空间学习
Sum of absolute differences (SAD),	绝对误差和(SAD)
Sum of squared differences (SSD),	误差平方和(SSD)
bias and gain	偏差与增益
Fourier-based computation	基于*傅里叶*变换的计算
normalized	规范化的
surface	表面
weighted	加权的
windowed	开窗的
Sum of sum of squared differences (SSSD)	误差平方和的和(SSSD)
Summed area table	求和的区域表
Super-resolution	超分辨率
example-based	基于例子的
faces	人脸
hallucination	幻生
prior	先验
Superposition principle	叠加原理
Superquadric	超二次的
Support vector machine (SVM)	支持向量机(SVM)
Surface element (surfel)	面元(surfel)
Surface interpolation	表面插值
Surface light field	表面光场
Surface representations	表面表达
non-parametric	非参数的
parametric	参数的

point-based	基于点的
simplification	简化
splines	样条
subdivision surface	细分曲面
symmetry-seeking	对称性追寻
triangle mesh	三角网格
Surface simplification	表面简化
Swendsen-Wang algorithm	Swendsen-Wang 算法

T

Telecentric lens	远心镜头
Temporal derivative	时间导数
Temporal texture	时间纹理
Term frequency-inverse document frequency	项频率-逆文本频率(TF-IDF)
Testing algorithmsviii	测试算法
TextonBoost	TextonBoost
Texture	纹理
shape-from	由……到形状
Texture addressing mode	纹理寻址模态
Texture map	纹理图(映射)
recovery	恢复
view-dependent	视角相关的
Texture mapping	纹理映射
anisotropic filtering	各向异性滤波
MIP-mapping	MIP-映射
multi-pass	多通道
trilinear interpolation	三线性插值
Texture synthesis	纹理合成
by numbers	按数
hole filling	空洞填充
image quilting	图像绗缝
non-parametric	非参数的
transfer	转移
Thin lens	薄透镜
Thin-plate spline	薄板样条
Thresholding	阈值化
Through-the-lens camera control	经由镜头的摄像机控制
Tobogganing	Tobogganing

Tonal adjustment	色调调整
Tone mapping	色调映射
adaptive	自适应
bilateral filter	双边滤波器
global	全局
gradient domain	梯度域
halos	光晕
interactive	交互式
local	局部
scale selection	尺度选择
Total least squares (TLS)	全部最小二乘(TLS)
Total variation	全部变化
Tracking	跟踪
feature	特征
head	头部
human motion	人体运动
multiple hypothesis	多假设
planar pattern	平面模式
Translational motion estimation	平移运动估计
bias and gain	偏差与增益
Transparency	透明性
Travelling salesman problem (TSP)	旅行商问题(TSP)
Tri-chromatic sensing	三基色传感
Tri-stimulus values	三激励值
Triangulation	三角剖分
Trilinear interpolationsee MIP-mapping	三线性插值，见 MIP-mapping
Trimap (matting)	三值图(抠图)
Trust region method	信任区域方法
Two-dimensional Fourier transform	二维*傅里叶*变换

U

Uncanny valley	非 Canny 谷
Uncertainty	不确定性
correspondence	对应
modeling	建模
weighting	加权
Unsharp mask	不锐化掩模
Upsamplingsee Interpolation	上采样，见 Interpolation

V

Vanishing point	消失点
detection	检测
Hough	Hough
least squares	最小二乘
modeling	建模
uncertainty	不确定性
Variable reordering	变量重排
minimum degree	最小维数
multi-frontal	多前沿
nested dissection	嵌套分解
Variable state dimension filter (VSDF)	可变状态维数滤波(VSDF)
Variational method	变分法
Video compression	视频压缩
motion compensated	运动补偿
Video compression (coding)	视频压缩(编码)
Video denoising	视频去噪
Video matting	视频抠图
Video objects (coding)	视频物体(编码)
Video sprites	视频子画面
Video stabilization	视频稳定化
Video texture	视频纹理
Video-based animation	基于视频的动画
Video-based rendering	基于视频的绘制
D video	D 视频
animating pictures	动画画面
sprites	子画面
video texture	视频纹理
virtual viewpoint video	虚拟视点视频
walkthroughs	排演
VideoMouse	视频鼠标
View correlation	视图相关
View interpolation	视图插值
View morphing	视图变形
View-based eigenspace	基于视图的本证空间
View-dependent texture maps	视角相关的纹理映射

Vignetting 虚影
    mechanical 机械的
    natural 自然的
Virtual viewpoint video 虚拟视点视频
Visual hull 视觉外型
    image-based 基于图像的
Visual illusions 视觉幻象
Visual odometry 视觉测距法
Visual words 视觉词
Vocabulary tree 词汇树
Volumetric D reconstruction 体积 D 重建
Volumetric range image processing (VRIP) 体积距离图像处理(VRIP)
Volumetric representations 体积表达
Voronoi diagram Voronoi 图
Voxel coloring 体素着色
    multi-view stereo 多视角(视图)立体视觉

W

Watershed 分水岭
    basins 盆地
    oriented 有向的
Wavelets 小波
    compression 压缩
    lifting 提升
    overcomplete 过完备
    second generation 二代
    self-inverting 自反向
    tight frame 紧致框架
weighted 加权
Weaving wall 交织墙
Weighted least squares (WLS) 加权最小二乘(WLS)
Weighted prediction (bias and gain) 加权预测(偏差与增益)
White balance 白平衡
Whitening transform 白化变换
Wiener filter 维纳滤波
Wire removal 线去除
Wrapping mode 贴图模式

X

XYZsee Color XYZ，见 Color

Z

Zippering 拉链作用